# Handbook of Cyclization Reactions

*Edited by*
*Shengming Ma*

## Related Titles

Carreira, Erick M. / Kvaerno, Lisbet

**Classics in Stereoselective Synthesis**

2008

ISBN: 978-3-527-32452-1

Crabtree, R. H. (ed.)

**Handbook of Green Chemistry Set 1: Green Catalysis**

Series editor: Anastas, P. T. (ed.)

2009

ISBN: 978-3-527-31577-2

Dubois, P. / Coulembier, O. / Raquez, J.-M. (eds.)

**Handbook of Ring-Opening Polymerization**

2009

ISBN: 978-3-527-31953-4

Dupont, J. / Pfeffer, M. (eds.)

**Palladacycles**

Synthesis, Characterization and Applications

2008

ISBN: 978-3-527-31781-3

# Handbook of Cyclization Reactions

Volume 2

*Edited by*
*Shengming Ma*

WILEY-
VCH

WILEY-VCH Verlag GmbH & Co. KGaA

**The Editor**

*Prof. Dr. Shengming Ma*
State Key Labor. of Organomet.
Shanghai Institute of Organo.
354 Fenglin Lu
Shanghai 200032
Peoples Republic of China

All books published by Wiley-VCH are carefully produced. Nevertheless, authors, editors, and publisher do not warrant the information contained in these books, including this book, to be free of errors. Readers are advised to keep in mind that statements, data, illustrations, procedural details or other items may inadvertently be inaccurate.

**Library of Congress Card No.:** applied for

**British Library Cataloguing-in-Publication Data**
A catalogue record for this book is available from the British Library.

**Bibliographic information published by the Deutsche Nationalbibliothek**
The Deutsche Nationalbibliothek lists this publication in the Deutsche Nationalbibliografie; detailed bibliographic data are available on the Internet at http://dnb.d-nb.de.

© 2010 WILEY-VCH Verlag GmbH & Co. KGaA, Weinheim

All rights reserved (including those of translation into other languages). No part of this book may be reproduced in any form – by photoprinting, microfilm, or any other means – nor transmitted or translated into a machine language without written permission from the publishers. Registered names, trademarks, etc. used in this book, even when not specifically marked as such, are not to be considered unprotected by law.

Printed in the Federal Republic of Germany
Printed on acid-free paper

**Typesetting** Thomson Digital, Noida, India
**Printing** betz-Druck GmbH, Darmstadt
**Binding** Litges & Dopf Buchbinderei GmbH, Heppenheim
**Cover Design** Formgeber, Eppelheim

**ISBN:** 978-3-527-32088-2

# Contents

Contents to Volume 1  *XIX*
List of Contributors to Volume 2  *XXI*

**13** **Transition-Metal-Catalyzed Cycloisomerizations and Nucleophilic Cyclization of Enynes**  625
*Elena Herrero-Gómez and Antonio M. Echavarren*
13.1  Introduction  625
13.2  Cycloisomerizations of 1,*n*-Enynes  626
13.2.1  Metal-Catalyzed Alder-Ene Cycloisomerization and Related Processes  626
13.2.2  Cycloisomerizations of Enynes Via Propargylic 1,2- and 1,3-Acyl Migration  635
13.2.3  Cycloisomerizations of Hydroxy- and Alkoxy-Substituted Enynes  644
13.3  Addition of Nucleophiles to 1,*n*-Enynes  644
13.3.1  Hydroxy- and Alkoxycyclizations of 1,6- and 1,7-Enynes  644
13.3.2  Hydroxy- and Alkoxycyclizations of 1,5-Enynes  647
13.3.3  Amination of Enynes  649
13.3.4  Inter- and Intramolecular Additions of Carbon Nucleophiles to 1,6-Enynes  650
13.3.5  Other Inter- and Intramolecular Nucleophile Additions to 1,6-Enynes  654
13.3.6  Inter- and Intramolecular Additions of Carbon Nucleophiles to 1,5-Enynes  656
13.4  Skeletal Rearrangement of 1,*n*-Enynes  657
13.4.1  Single- and Double-Cleavage Skeletal Rearrangement of 1,6-Enynes  657
13.4.2  Skeletal Rearrangement of 1,7-Enynes  661
13.4.3  Formation of Cyclobutenes  663
13.4.4  Endocyclic Rearrangement of 1,6-Enynes  665
13.4.5  Skeletal Rearrangement of 1,5-Enynes  667
13.4.6  Inter- and Intramolecular Cyclopropanation of 1,6-Enynes  667

| 13.4.7 | Reactions of Furans with Alkynes  *671* |
| --- | --- |
| 13.5 | Summary and Outlook  *672* |
| 13.6 | Experimental: Selected Procedures  *672* |
| 13.6.1 | Synthesis of Selected Au(I) Complexes  *672* |
| 13.6.1.1 | General Procedure for the Synthesis of Neutral Au(I) Complexes 17a–d and 18a  *672* |
| 13.6.1.2 | Synthesis of Cationic Gold(I) Complexes 17e,f  *673* |
| 13.6.1.3 | Synthesis of Cationic Gold(I) Complex 18b  *673* |
| 13.6.1.4 | Synthesis of Neutral Gold(I) Complexes 19a–d  *673* |
| 13.6.1.5 | Synthesis of Cationic Gold(I) Complexes 19e  *673* |
| 13.6.2 | Selected Procedures for Metal-Catalyzed Cycloisomerizations  *673* |
| 13.6.2.1 | Synthesis of Compounds 11 and 13 Using $PtCl_2$ as Catalyst  *673* |
| 13.6.2.2 | Synthesis of 27 and 28 Using Catalyst 17e  *673* |
| 13.6.2.3 | Synthesis of 32a–c Using $AuCl(PPh_3)/AgSbF_6$ as Catalyst  *674* |
| 13.6.2.4 | Synthesis of 42a Using $AuCl(PPh_3)/AgOTf$ as Catalyst  *674* |
| 13.6.2.5 | Synthesis of Cyclopentenes from Alkynes and Enamines Formed In Situ Using $Cu(OTf)_2$ as Catalyst  *674* |
| 13.6.2.6 | Synthesis of 52 Using $AuCl(PPh_3)/AgOTf$ as Catalyst  *674* |
| 13.6.2.7 | General Procedure for Au(I)-Catalyzed Skeletal Rearrangement of Enynes (5a–l, 6a and 213a–e)  *674* |
| 13.6.2.8 | Synthesis of Product 198 Using $GaCl_3$  *675* |
| 13.6.2.9 | Synthesis of Product 203 with Ru(II) Catalyst  *675* |
| 13.6.3 | Selected Procedure for Metal-Catalyzed Cyclopropanation Using Propargylic Carboxylates  *675* |
| 13.6.3.1 | Synthesis of 58a,b Using $[RuCl_2(CO)_3]_2$ as Catalyst  *675* |
| 13.6.4 | Selected Procedure for Metal-Catalyzed Rautenstrauch Rearrangement  *676* |
| 13.6.4.1 | Synthesis of 65 Using $AuCl(PPh_3)/AgSbF_6$ as Catalyst  *676* |
| 13.6.5 | Selected Procedures for the Metal-Catalyzed Inter- and Intramolecular Alcoxycyclization  *676* |
| 13.6.5.1 | General Procedure for the Au(I)-Catalyzed Intermolecular Alkoxycyclization to Yield Products 125a–c  *676* |
| 13.6.5.2 | General Procedure for Au(I)-Catalyzed Intramolecular Alkoxycyclization (125d)  *676* |
| 13.6.6 | Experimental: Selected Procedure  *676* |
| 13.6.6.1 | Procedure for the for Au(I)-Catalyzed Addition of Carbon Nucleophiles to 1,6-Enynes  *676* |
| | References  *678* |
| | |
| **14** | **Cyclopropanation, Epoxidation and Aziridination Reactions**  *687* |
| | *Song Ye and Yong Tang* |
| 14.1 | Introduction  *687* |
| 14.2 | Cyclopropanation  *688* |
| 14.2.1 | Cyclopropanation with Halomethylmetal Compounds  *688* |

| | | |
|---|---|---|
| 14.2.2 | Cyclopropanation with Transition Metal-Catalyzed Decomposition of Diazo Compounds *690* | |
| 14.2.2.1 | Diazoalkanes *690* | |
| 14.2.2.2 | Diazocarbonyl Compounds *691* | |
| 14.2.3 | Cyclopropanation with Ylides, α-Halocarbanions and Related Species *694* | |
| 14.2.4 | Cyclopropanation with Metalocarbenoids *698* | |
| 14.3 | Epoxidations *701* | |
| 14.3.1 | Metal-Catalyzed Epoxidation of Alkenes *701* | |
| 14.3.1.1 | Titanium-Catalyzed Epoxidations *701* | |
| 14.3.1.2 | Vanadium-Catalyzed Epoxidations *702* | |
| 14.3.1.3 | Manganese-Catalyzed Epoxidations *703* | |
| 14.3.1.4 | Iron-Catalyzed Epoxidations *703* | |
| 14.3.1.5 | Ruthenium and Platinum-Catalyzed Epoxidations *704* | |
| 14.3.1.6 | Lanthanide-Catalyzed Epoxidations *705* | |
| 14.3.2 | Organocatalytic Epoxidation of Alkenes *706* | |
| 14.3.2.1 | Epoxidation Catalyzed by Ketones *706* | |
| 14.3.2.2 | Epoxidation Catalyzed by Iminium Salts *707* | |
| 14.3.2.3 | Epoxidation Catalyzed by Amines *708* | |
| 14.3.2.4 | Phase-Transfer Catalysis *709* | |
| 14.3.2.5 | Miscellaneous *710* | |
| 14.3.3 | Epoxidaton of Carbonyl Compounds *710* | |
| 14.3.3.1 | Ylide Epoxidation *710* | |
| 14.3.3.2 | Darzens Reaction *712* | |
| 14.4 | Aziridinations *713* | |
| 14.4.1 | Aziridination of Alkenes *713* | |
| 14.4.1.1 | Metal-Catalyzed/Mediated Aziridination *713* | |
| 14.4.1.2 | Metal-Free Aziridination *716* | |
| 14.4.2 | Aziridination of Imines *717* | |
| 14.4.2.1 | With Carbenes *718* | |
| 14.4.2.2 | With Ylides *719* | |
| 14.5 | Experimental: Selected Procedures *721* | |
| 14.5.1 | Synthesis of Compound 14 with Catalyst 12 *721* | |
| 14.5.2 | Synthesis of Compound 27 with Catalyst 24 *721* | |
| 14.5.3 | Synthesis of Compound 56 with Sulfonium salt 54 *722* | |
| 14.5.4 | Synthesis of Compound 79 with Catalyst 76 *722* | |
| 14.5.5 | Synthesis of Compound 90 with Catalyst 88 *722* | |
| 14.5.6 | Synthesis of Compound 107 with (+)-104 *722* | |
| 14.5.7 | Synthesis of Compound 120 with catalyst 118 *723* | |
| 14.5.8 | Synthesis of Compound 141 with Catalyst 139 (General Procedure) *723* | |
| 14.5.9 | Synthesis of Compound 157 with Catalyst 155 *724* | |
| 14.5.10 | Synthesis of Compound 162 with Catalyst 161 *724* |  |
| 14.5.11 | Synthesis of Compound 174 with Sulfide 172 *724* | |

| 14.5.12 | Synthesis of Compound 190 with Trichloroethyl N-tosyloxycarbamates 188  724 |
| 14.5.13 | Synthesis of Compound 210 with Catalyst 207  725 |
| 14.5.14 | Synthesis of Compound 219 with (S)-VAPOL Catalyst 217  725 |
| | References  726 |

**15  Cyclization of Cyclopropane- or Cyclopropene-Containing Compounds**  733
*Junliang Zhang and Yuanjing Xiao*

| 15.1 | Introduction  733 |
| 15.2 | Cycloisomerization Reactions  733 |
| 15.2.1 | Cycloisomerization of Vinylcyclopropanes and Their Hetero-Analogues  733 |
| 15.2.1.1 | Cycloisomerization of Vinylcyclopropanes to Cyclopentenes  733 |
| 15.2.1.2 | Cycloisomerization of Hetero-Analogues of VCPs to Five-Membered Heterocycles  736 |
| 15.2.2 | Cycloisomerization of Vinyl-, Allenyl-, and Alkynylcyclopropanes  738 |
| 15.2.3 | Cycloisomerization of Other Cyclopropanes  740 |
| 15.2.4 | Cycloisomerization of Methylene- and Alkylidenecyclopropanes  743 |
| 15.2.4.1 | Cycloisomerization of Acyl Methylene- and Alkylidenecyclopropanes  744 |
| 15.2.4.2 | Cycloisomerization via a Cascade Process  746 |
| 15.2.5 | Cycloisomerization of Vinylidenecyclopropanes  749 |
| 15.2.6 | Cycloisomerization of Cyclopropenes  749 |
| 15.2.6.1 | Cycloisomerization of 3-alkoylcarbonylcyclopropenes to Furans  749 |
| 15.2.6.2 | Cycloisomerization of 3-acylcyclopropenes to Furans  749 |
| 15.2.6.3 | Cycloisomerization of Cyclopropenyl Imines to Pyrroloheterocycles  753 |
| 15.3 | Cycloaddition Reactions  754 |
| 15.3.1 | [1 + 2] Cycloaddition Reaction  754 |
| 15.3.2 | [2 + m + n] Cycloaddition  754 |
| 15.3.3 | [3 + 1] cycloaddition of MCPs  755 |
| 15.3.4 | [3 + 2] Cycloaddition Reaction  756 |
| 15.3.4.1 | [3 + 2] Cycloaddition Reaction of Methylene- and Alkylidene-Cyclopropanes  756 |
| 15.3.4.2 | [3 + 2] Cycloaddition Reaction of Vinylidenecyclopropanes  761 |
| 15.3.4.3 | [3 + 2] Cycloaddition Reaction of Activated Cyclopropanes  763 |
| 15.3.4.4 | [3 + 2] Cycloaddition of VCPs  772 |
| 15.3.5 | [3 + 3] Cycloaddition of Activated Cyclopropanes  774 |
| 15.3.6 | [3 + 4] Cycloaddition  777 |
| 15.3.7 | [4 + 1] Cycloaddition  778 |
| 15.3.8 | [4 + 2] Cycloaddition of 1-(1-Alkynyl)cyclopropyl Ketones  780 |
| 15.3.9 | [5 + 1] Cycloaddition  780 |
| 15.3.10 | [5 + 2] Cycloaddition of VCPs  784 |

| | | |
|---|---|---|
| 15.3.10.1 | Intramolecular [5 + 2] Cycloaddition of VCPs  *784* | |
| 15.3.10.2 | Intermolecular [5 + 2] Cycloaddition of VCPs with Alkynes and Allenes  *791* | |
| 15.4 | Multicomponent Cyclization Reactions  *794* | |
| 15.4.1 | [5 + 2 + 1] Cycloaddition of Alkene-VCPs  *794* | |
| 15.4.2 | [5 + 1 + 2 + 1] Cycloaddition of Alkene-VCPs  *794* | |
| 15.4.3 | [3 + 2 + 2] Cycloaddition of MCPs  *794* | |
| 15.4.4 | [(3 + 3) + 1] Cycloaddition  *798* | |
| 15.5 | Cyclization Involving Miscellaneous and Cascade Reactions  *799* | |
| 15.5.1 | Miscellaneous Reactions Involving [4 + 2] Cycloaddition  *799* | |
| 15.5.2 | Cyclization Involving a Cascade Reaction  *800* | |
| 15.6 | Experimental: Selected Procedures  *804* | |
| 15.6.1 | Synthesis of Compound 16 via Vinylcyclopropane–Cyclopentene Rearrangement  *804* | |
| 15.6.2 | Regioselective Cycloisomerization of Cyclopropenyl Ketone 171  *804* | |
| 15.6.2.1 | Isomerization under Catalysis of CuI  *804* | |
| 15.6.2.2 | Isomerization under Catalysis of $PdCl_2(CH_3CN)_2$  *804* | |
| 15.6.3 | Synthesis of Compound 194 via [3 + 2] Cycloaddition Reaction of Alkylidenecyclopropanes with Aldehyde  *804* | |
| 15.6.4 | Stereoselective Synthesis of 2, 5-*cis* or 2, 5-*trans*-pyrrolo-isoxazolidines 272  *805* | |
| 15.6.4.1 | Synthesis of 2, 5-*cis*-pyrrolo-isoxazolidines *cis*-272  *805* | |
| 15.6.4.2 | Synthesis of 2, 5-*trans*-pyrrolo-isoxazolidines *trans*-272  *805* | |
| 15.6.5 | Synthesis of Optically Active Tetrahydrooxazines 322  *805* | |
| 15.6.6 | Synthesis of 5, 7-fused Bicycles 326 via Intramolecular [3 + 4] Cycloaddition  *805* | |
| 15.6.7 | [4 + 2] Cycloaddition of 1-(1-Alkynyl)cyclopropyl Ketones 346 with Aldehyde  *806* | |
| 15.6.8 | Asymmetric Rh(I)-Catalyzed Intramolecular [5 + 2] Cycloaddition of Alkene-VCPs 392  *806* | |
| 15.6.9 | [(5 + 2) + 1] Cycloaddition of Alkene-VCPs 435  *806* | |
| | References  *807* | |
| | | |
| **16** | **Transition Metal-Catalyzed Ring Expansion Cyclization Reactions**  *813* | |
| | *Masahiro Yoshida and Yoshimitsu Nagao* | |
| 16.1 | Introduction  *813* | |
| 16.2 | Palladium-Catalyzed Ring Expansions  *813* | |
| 16.2.1 | Reactivities of Alkenyl Substrates  *813* | |
| 16.2.2 | Reactivities of Dienyl Substrates  *820* | |
| 16.2.3 | Reactivities of Alkynyl Substrates  *827* | |
| 16.3 | Ring Expansions with Other Transition Metals  *832* | |
| 16.3.1 | Ruthenium-Catalyzed Reactions  *832* | |
| 16.3.2 | Gold-Catalyzed Reactions  *834* | |
| 16.3.3 | Reactions with Fischer-Carbene Complexes  *835* |

| | | |
|---|---|---|
| 16.4 | Conclusion  *837* | |
| 16.5 | Experimental: Selected Procedures  *837* | |
| 16.5.1 | General  *837* | |
| 16.5.2 | Synthesis of Compound 10  *837* | |
| 16.5.3 | Synthesis of Compounds 24 and 25  *837* | |
| 16.5.4 | Synthesis of Compound 32  *838* | |
| 16.5.5 | Synthesis of Compound 71  *838* | |
| 16.5.6 | Synthesis of Compound 91  *838* | |
| 16.5.7 | Synthesis of Compound 105  *838* | |
| 16.5.8 | Synthesis of Compound 118  *838* | |
| 16.5.9 | Synthesis of Compound 141  *839* | |
| 16.5.10 | Synthesis of Compound 156  *839* | |
| 16.5.11 | Synthesis of Compound 162  *839* | |
| | References  *840* | |
| | | |
| **17** | **Hydrometallation-Initiated Cyclization Reactions**  *843* | |
| | *Isamu Matsuda* | |
| 17.1 | Introduction  *843* | |
| 17.2 | Self-Feeding Cyclization  *843* | |
| 17.2.1 | Intramolecular Hydrosilylation to a Double Bond  *843* | |
| 17.2.2 | Intramolecular Hydrosilylation to a C–C Triple Bond  *849* | |
| 17.2.2.1 | syn-Addition  *849* | |
| 17.2.2.2 | Anti-Addition  *850* | |
| 17.2.3 | Intramolecular Version of Silylformylation of Alkynes and Alkenes  *853* | |
| 17.3 | Cyclization Tied by Outsourcing Reagents  *858* | |
| 17.3.1 | Hydrosilane  *858* | |
| 17.3.1.1 | Diynes  *858* | |
| 17.3.1.2 | Enynes  *863* | |
| 17.3.1.3 | Dienes  *867* | |
| 17.3.1.4 | Compounds Involving Three C–C Unsaturated Bonds  *877* | |
| 17.3.1.5 | Carbonyl Compounds Tethered by a C–C Unsaturated Bond  *881* | |
| 17.3.1.6 | Aldehydes Tethered by an α,β-Unsaturated Carbonyl Group  *884* | |
| 17.4 | Cascade Reactions Incorporating CO  *890* | |
| 17.4.1 | Annulative Silylcarbonylation  *890* | |
| 17.4.1.1 | 1,6-Diynes  *890* | |
| 17.4.1.2 | 1,6-Enynes  *896* | |
| 17.4.2 | Silylative Lactonization of Alkynols  *899* | |
| 17.4.3 | Silylative Carbamoylation of Alkynes  *902* | |
| 17.5 | Conclusion  *904* | |
| 17.6 | Experimental: Selected Procedures  *905* | |
| 17.6.1 | Intramolecular Hydrosilylation of 9 Catalyzed by 2  *905* | |
| 17.6.2 | Intramolecular Hydrosilylation of 9 Catalyzed by [RhCl(CH$_2$=CH$_2$)$_2$]$_2$  *905* | |
| 17.6.3 | Intramolecular Hydrosilylation of 34 to Form 35  *905* | |
| 17.6.4 | Synthesis of Spectaline 55 from 52  *906* | |
| 17.6.5 | Rh$_4$(CO)$_{12}$-Catalyzed Intramolecular Silylformylation of 56  *906* | |

| | | |
|---|---|---|
| 17.6.6 | Reaction of 101 with $Et_3SiH$ to form (Z)-111 | 907 |
| 17.6.7 | $Rh_4(CO)_{12}$-Catalyzed Silylative Annulation of 143 with $Me_2PhSiH$ | 907 |
| 17.6.8 | [Pd]-Catalyzed Silylative Annulation of 188 to 189 | 907 |
| 17.6.9 | Silylative Cascade Cyclization of 217 with the Catalysis of $Cp^*_2YCH_3 \cdot THF$ | 907 |
| 17.6.10 | Silylative Cyclization of 245 to Form 246 | 908 |
| 17.6.11 | Rh-Catalyzed Aldol Type Cyclization of 259 | 908 |
| 17.6.12 | $Rh_4(CO)_{12}$-Catalyzed Bicyclocarbonylation of 130 | 908 |
| 17.6.13 | $Rh_4(CO)_{12}$-Catalyzed Silylcarbonylation of 157 $\{X = (MeO_2C)_2C, R^1 = H\}$ | 909 |
| 17.6.14 | $Rh_4(CO)_{12}$-Catalyzed Silyl Carbonylation of 314 to Form 316 | 909 |
| 17.6.15 | Rh-Catalyzed Silylcarbamoylation of 346 to Form β-Lactam 347 | 909 |
| | References 911 | |
| **18** | **Catalytic Dipolar Cycloadditions of Alkynes with Azides and Nitrile Oxides** | **917** |
| | *Valery V. Fokin* | |
| 18.1 | Introduction | 917 |
| 18.2 | Azide–Alkyne Cycloaddition: Basics | 919 |
| 18.3 | Copper-Catalyzed Cycloadditions | 920 |
| 18.3.1 | Catalysts and Ligands | 920 |
| 18.3.2 | CuAAC with *In Situ* Generated Azides | 924 |
| 18.3.3 | Mechanistic Aspects of the CuAAC | 925 |
| 18.3.4 | Reactions of Sulfonyl Azides | 929 |
| 18.3.5 | Examples of Application of the CuAAC Reaction | 930 |
| 18.3.5.1 | Synthesis of Compound Libraries for Biological Screening | 930 |
| 18.3.5.2 | Synthesis of Glycoconjugates | 931 |
| 18.3.5.3 | Modification and Biological Profiling of Natural Products | 932 |
| 18.3.6 | Copper-Catalyzed Synthesis of Isoxazoles | 934 |
| 18.4 | Ruthenium-Catalyzed Cycloadditions | 935 |
| 18.4.1 | Ruthenium-Catalyzed Azide-Alkyne Cycloaddition | 935 |
| 18.4.2 | Ruthenium-Catalyzed Synthesis of Isoxazoles | 937 |
| 18.4.3 | Mechanistic Studies of the Ruthenium-Catalyzed Cycloadditions | 938 |
| 18.5 | Experimental: Selected Procedures | 944 |
| 18.5.1 | Copper-Catalyzed Cycloadditions | 944 |
| 18.5.2 | Ruthenium-Catalyzed Cycloadditions | 945 |
| | References 946 | |
| **19** | **Electrophilic Cyclizations** | **951** |
| | *Félix Rodríguez and Francisco J. Fañanás* | |
| 19.1 | Introduction | 951 |
| 19.2 | Mechanism, Regio- and Stereoselectivity | 952 |
| 19.3 | Halocyclizations | 953 |
| 19.3.1 | Oxygen-Centered Nucleophiles | 954 |
| 19.3.1.1 | Cyclization of Alcohols and Ethers | 954 |

| 19.3.1.2 | Cyclization of Carboxylic Acids and Carboxylic Acid Derivatives  958 |
| 19.3.1.3 | Cyclization of Carbonates and Carbamates  962 |
| 19.3.1.4 | Cyclization of Aldehydes and Ketones  963 |
| 19.3.1.5 | Cyclizations with Other Oxygen-Centered Nucleophiles  964 |
| 19.3.2 | Nitrogen-Centered Nucleophiles  965 |
| 19.3.2.1 | Cyclization of Amines  965 |
| 19.3.2.2 | Cyclization of Amides  967 |
| 19.3.2.3 | Cyclization of Carbamates, Amidines, Ureas and Isoureas  967 |
| 19.3.2.4 | Cyclization of Imines  968 |
| 19.3.2.5 | Cyclization of Azides  969 |
| 19.3.3 | Sulfur- and Selenium-Centered Nucleophiles  969 |
| 19.3.4 | Carbon-Centered Nucleophiles  970 |
| 19.3.4.1 | Cyclization of Compounds Containing Active Methine Groups  970 |
| 19.3.4.2 | Alkynes as Nucleophiles  971 |
| 19.3.4.3 | Aryl Groups as Nucleophiles  972 |
| 19.4 | Selenocyclizations  972 |
| 19.4.1 | Oxygen-Centered Nucleophiles  973 |
| 19.4.1.1 | Cyclization of Alcohols  973 |
| 19.4.1.2 | Cyclization of Aldehydes and Ketones  976 |
| 19.4.1.3 | Cyclization of Carboxylic Acids and Carboxylic Acid Derivatives  977 |
| 19.4.2 | Nitrogen-Centered Nucleophiles  979 |
| 19.4.2.1 | Cyclization of Amines  980 |
| 19.4.2.2 | Cyclization of Amides and Imidates  980 |
| 19.4.2.3 | Cyclization of Imines and Oximes  981 |
| 19.4.3 | Sulfur- and Selenium-Centered Nucleophiles  982 |
| 19.4.4 | Carbon-Centered Nucleophiles  982 |
| 19.5 | Conclusion and Perspectives  983 |
| 19.6 | Experimental: Selected Procedures  984 |
| 19.6.1 | Halocyclization Reactions  984 |
| 19.6.1.1 | $IPy_2BF_4$-Promoted Intramolecular Addition of Anilines to Alkynes  984 |
| 19.6.2 | Selenocyclization Reactions  984 |
| 19.6.2.1 | PhSeCl-Promoted Intramolecular Addition of Ethers to Alkynes  984 |
| | References  984 |

**20 Cyclizations Based on C–H Activation  991**
*David A. Capretto, Zigang Li, and Chuan He*

| 20.1 | Introduction  991 |
| 20.2 | Aryl C–H Bond Activation  994 |
| 20.2.1 | Rhodium  994 |
| 20.2.2 | Palladium  997 |
| 20.2.3 | Gold  1002 |
| 20.2.4 | Other Metals  1004 |

| | | |
|---|---|---|
| 20.3 | Alkyl C–H Bond Activation | *1005* |
| 20.3.1 | C–C Bond Formation | *1005* |
| 20.3.1.1 | Rhodium | *1005* |
| 20.3.1.2 | Palladium | *1008* |
| 20.3.1.3 | Other Metals | *1009* |
| 20.3.2 | C–N Bond Formation | *1010* |
| 20.3.2.1 | Rhodium | *1011* |
| 20.3.2.2 | Palladium | *1013* |
| 20.3.2.3 | Other Metals | *1015* |
| 20.3.3 | C–O Bond Formation | *1016* |
| 20.4 | Experimental: Selected Procedures | *1017* |
| 20.4.1 | Experimental Protocol for Reaction Shown in Scheme 20.4 | *1017* |
| 20.4.2 | Experimental Procedure for Reaction Shown in Scheme 20.13 | *1017* |
| 20.4.3 | Experimental Protocol for Reaction Shown in Scheme 20.19 | *1017* |
| 20.4.4 | Experimental Protocol for Reaction Shown in Scheme 20.23 | *1018* |
| 20.4.5 | Experimental Protocol for Reaction Shown in Scheme 20.24 | *1018* |
| 20.4.6 | Experimental Protocol for Reaction Shown in Scheme 20.28 | *1018* |
| 20.4.7 | Experimental Protocol for Reaction Shown in Scheme 20.33 | *1018* |
| | References | *1019* |
| | | |
| **21** | **Friedel–Crafts Type Cyclizations** | ***1025*** |
| | *Ryuji Hayashi and Gregory R. Cook* | |
| 21.1 | Introduction | *1025* |
| 21.2 | Classic Intramolecular Friedel–Crafts Cyclizations | *1026* |
| 21.2.1 | RCFC Reactions via Activations of Alkyl Halides | *1026* |
| 21.2.2 | RCFC Reactions of Carboxylic Acid Derivatives | *1026* |
| 21.3 | RCFC Reactions via Activation of Carbonyl Derivatives and Alcohols | *1029* |
| 21.3.1 | RCFC Reactions of Aldehydes and Ketones (1,2-Addition) | *1029* |
| 21.3.2 | RCFC Reactions of Imines (1,2-Addition) | *1030* |
| 21.3.3 | RCFC Reactions of Acetals | *1031* |
| 21.3.4 | RCFC Reactions via Conjugate Additions | *1033* |
| 21.3.5 | $S_N 1$ Type Reactions and Direct Substitution of Alcohols | *1033* |
| 21.3.6 | RCFC Reactions of Epoxide Derivatives | *1035* |
| 21.4 | RCFC Reactions of C–C π Bonds | *1036* |
| 21.4.1 | RCFC Reactions of Alkenes | *1036* |
| 21.4.2 | RCFC Reactions of Alkynes | *1039* |
| 21.5 | Miscellaneous Reactions | *1042* |
| 21.6 | Catalytic Enantioselective RCFC Reactions | *1044* |
| 21.7 | Conclusion | *1046* |
| 21.8 | Experimental: Selected Procedures | *1047* |
| 21.8.1 | In(III)-Catalyzed RCFC Reaction for Synthesis of 5 | *1047* |
| 21.8.2 | RCFC Reaction of Arene-Ynamides for Synthesis of 35 | *1047* |
| 21.8.3 | Ru(III)-Catalyzed Hydroarylation | *1047* |
| 21.8.4 | Acid-Catalyzed RCFC Reaction via Ketenium Ions | *1047* |

21.8.5 Enantioselective Organocatalytic RCFC of Indole  *1048*
21.8.6 Thiourea-Catalyzed Pictet–Spengler Reaction  *1048*
21.8.7 Asymmetric Pd-Catalyzed Pictet–Spengler RCFC  *1048*
21.8.8 Au-Catalyzed RCFC of Indole with Allenes  *1049*
References  *1050*

## 22 Macrolactones and Macrolactams  *1055*
*Chang-Liang Sun, Bi-Jie Li, and Zhang-Jie Shi*

22.1 Macrolactones  *1055*
22.1.1 Introduction  *1055*
22.1.2 Macrolactonization via Formation of an Acyl–Oxygen Bond  *1056*
22.1.2.1 Corey–Nicolaou Macrolactonization  *1056*
22.1.2.2 Masamune Macrolactonization  *1058*
22.1.2.3 Mukaiyama Macrolactonization  *1059*
22.1.2.4 Yamaguchi Macrolactonization  *1062*
22.1.2.5 Keck Macrolactonization  *1063*
22.1.2.6 Shiina Macrolactonization  *1065*
22.1.3 Macrolactonization Via Formation of an Alkyl C–O Bond  *1066*
22.1.3.1 Mitsunobu Reaction  *1066*
22.1.3.2 Other Methods  *1068*
22.1.4 Macrolactonization in Biosystems  *1070*
22.2 Macrolactams  *1072*
22.2.1 Introduction  *1072*
22.2.2 Formation of Macrolactams Via Direct Amidation  *1073*
22.2.2.1 Acyl Halides  *1073*
22.2.2.2 Acylimidazoles  *1074*
22.2.2.3 Anhydrides  *1074*
22.2.2.4 Esters  *1077*
22.2.3 Formation of Macrolactam via Aryl C–N Bond Formation  *1086*
22.3 Conclusion and Perspectives  *1086*
22.4 Experimental: Selected Procedures  *1088*
22.4.1 Synthesis of Compound 7 through Corey-Nicolaou Macrolactonization  *1088*
22.4.2 Synthesis of Brefeldin A 16  *1089*
22.4.3 Synthesis of Avermectin $A_1$ 19  *1089*
22.4.4 Synthesis of (−)-Macrolactin A 25  *1089*
22.4.5 Synthesis of Oleandolide 28  *1090*
22.4.6 Synthesis of Colletodiol 31  *1090*
22.4.7 Synthesis of Compound Aleuritic Acid Lactone 35  *1090*
22.4.7.1 Method A  *1090*
22.4.7.2 Method B  *1091*
22.4.8 Synthesis of Maduropeptin Chromophore Aglycon 60  *1091*
22.4.9 Preparation of Macrocyclic Lactam 70  *1091*

| 22.4.10 | Preparation of Macrocyclic Lactam 85   *1092* |
|---|---|
| 22.4.11 | Preparation of Macrocyclic Lactam 87   *1092* |
| | References   *1093* |

**23   Free Radical Cyclization Reactions**   *1099*
*Jake Zimmerman, Amanda Halloway, and Mukund P. Sibi*

| 23.1 | Radical Cyclizations: General Definition   *1099* |
|---|---|
| 23.1.1 | Formation of a Single Ring   *1100* |
| 23.1.2 | Tandem Cyclizations   *1101* |
| 23.2 | Different Methods for Radical Cyclizations   *1105* |
| 23.2.1 | Reductive Cyclizations   *1105* |
| 23.2.1.1 | Tin Hydride   *1106* |
| 23.2.1.2 | Silicon   *1108* |
| 23.2.1.3 | Germanium   *1109* |
| 23.2.1.4 | Indium   *1110* |
| 23.2.1.5 | Phosphorous Acid   *1111* |
| 23.2.1.6 | Thiols   *1112* |
| 23.2.2 | Atom-Transfer Cyclizations   *1113* |
| 23.2.2.1 | Copper   *1113* |
| 23.2.2.2 | Ruthenium   *1115* |
| 23.2.2.3 | Organotin   *1116* |
| 23.2.2.4 | Peroxides   *1116* |
| 23.2.3 | Manganese Triacetate-Mediated Cyclization   *1117* |
| 23.2.4 | Samarium Diiodide-Mediated Cyclization   *1119* |
| 23.2.5 | Titanium (III)-Mediated Cyclization   *1120* |
| 23.2.6 | Photochemical Cyclizations   *1122* |
| 23.3 | Annulation Reactions   *1124* |
| 23.4 | Diastereoselective Radical Cyclizations   *1127* |
| 23.5 | Enantioselective Radical Cyclizations   *1129* |
| 23.6 | Applications of Radical Cyclizations in Natural Product Synthesis   *1135* |
| 23.6.1 | Single Ring Formation – Amphidinolide E   *1135* |
| 23.6.2 | Group-Transfer Radical Cyclizations – Aphanorphine   *1135* |
| 23.6.3 | Tandem Radical Cyclizations – Luotonin A, Azadirachtin, Aspidospermidine, and Fortucine   *1136* |
| 23.6.4 | Manganese Triacetate-Mediated Synthesis – Mersicarpine   *1138* |
| 23.6.5 | Titanium(III)-Mediated Cyclization – Smenospondiol   *1139* |
| 23.6.6 | Samarium Iodide in Total Synthesis – Paeonilactone B and Platensimycin   *1139* |
| 23.7 | Experimental: Selected Procedures   *1140* |
| 23.7.1 | Transformation of 55 to 56, Reductive Radical Cyclization Using Tributyltin Hydride   *1140* |
| 23.7.2 | Transformation of 121 to 122, Manganese Triacetate-Mediated Cyclization   *1141* |
| 23.7.3 | Transformation of 186 to 187, Radical Annulation   *1141* |
| 23.7.4 | Transformation of 202 to 203, Atom-Transfer Radical Cyclization   *1141* |

| 23.7.5 | Transformation of 258 to 259 and 260, Titanium(III)-Mediated Radial Cyclization  *1142* |
| 23.7.6 | Transformation of 261 to 262 and 263, Samarium Diiodide-Mediated Radical Cyclization  *1142* |
| 23.8 | Conclusions  *1142* |
| | References  *1143* |

## 24 Photocyclization Reactions  *1149*
*Axel G. Griesbeck*

| 24.1 | Introduction  *1149* |
| 24.2 | Several Ways to Initiate Photocyclization  *1151* |
| 24.2.1 | Direct Excitation and Direct or Sequential Addition  *1152* |
| 24.2.2 | Direct Excitation and Group Transfer  *1152* |
| 24.2.3 | Sensitization: Energy Transfer  *1152* |
| 24.2.4 | Sensitization: Electron Transfer (PET)  *1153* |
| 24.2.5 | Substrate Tuning  *1154* |
| 24.3 | Norrish Type II Initiated Reactions: H Transfer  *1154* |
| 24.3.1 | Cyclization (Yang Process) Versus Cleavage  *1154* |
| 24.3.2 | Formation of Four-Membered Rings  *1156* |
| 24.3.3 | Formation of Three-Membered Rings: Direct vs. Indirect  *1157* |
| 24.3.4 | Formation of Larger Ring Products  *1161* |
| 24.4 | Photoinduced Electron Transfer Cyclizations  *1163* |
| 24.4.1 | Formation of Three- to Six-membered Rings  *1163* |
| 24.4.2 | Formation of Larger Ring Products  *1165* |
| 24.4.2.1 | Oxidative PET  *1169* |
| 24.4.2.2 | Reductive PET  *1173* |
| 24.5 | $\pi$–$\pi$-Interactions  *1177* |
| 24.5.1 | $\pi^n$–$\pi^m$-Cycloadditions  *1177* |
| 24.5.2 | $\pi^2$-$\pi^2$-Rearrangements: Di-$\pi$-Methane Rearrangement  *1179* |
| 24.5.3 | $\pi^2$–$\pi^2$-Rearrangements: Oxa-di-$\pi$-Methane Rearrangement  *1180* |
| 24.6 | $\pi^6$-Photocyclization  *1189* |
| 24.6.1 | Classical Examples  *1189* |
| 24.6.2 | Configurationally Restrained 1,2-Diarylethenes  *1190* |
| 24.7 | Experimental: Selected Procedures  *1191* |
| 24.7.1 | Oxa-di-$\pi$-Methane-Rearrangement  *1191* |
| 24.7.1.1 | Synthesis of Enantiomerically Pure Tricyclo[3.3.0.0.0]octane-3-ones  *1191* |
| 24.7.1.2 | Synthesis of 4,4,9,9-Tetramethyltetracyclo[6.4.0.0.0]dodecane-7,12-dione  *1191* |
| 24.7.2 | Paternò–Büchi Photocycloaddition  *1192* |
| 24.7.2.1 | Photocycloaddition of Benzaldehyde to 2,4-Dimethyl-5-methoxyoxazole with Subsequent Ring Opening  *1192* |
| 24.7.2.2 | Photocycloaddition of Benzaldehyde to 2,3-Dihydrofuran [308 nmXeCl Excimer Photolysis on the 1mol Scale]  *1192* |
| | References  *1193* |

| | | |
|---|---|---|
| **25** | **Asymmetric Organocatalyzed Cyclization Reactions** *1199* | |

*Liu-Zhu Gong, Jun Jiang, Meng-Xia Xue, and Shi-Wei Luo*

- 25.1 Introduction  *1199*
- 25.2 Enamine-Catalyzed Asymmetric Cyclization Reactions  *1200*
- 25.2.1 Intramolecular Aldol Reactions  *1200*
- 25.2.2 Intramolecular Michael Additions  *1202*
- 25.2.3 Asymmetric Cycloaddition and Domino Reactions  *1203*
- 25.3 Iminium-Catalyzed Asymmetric Cyclization Reactions  *1207*
- 25.3.1 Asymmetric Cycloadditions  *1207*
- 25.3.2 Asymmetric Intramolecular 1,4-Additions  *1210*
- 25.3.3 Asymmetric Domino Reactions  *1211*
- 25.4 Sequential Catalysis with Chiral Amines for Domino Cyclization Reactions  *1213*
- 25.4.1 Enamine/Iminium Catalytic Sequences  *1213*
- 25.4.2 Enamine/Enamine Catalytic Sequences  *1214*
- 25.4.3 Iminium/Enamine Catalytic Sequences  *1215*
- 25.5 Brønsted Acid-Catalyzed Asymmetric Cyclization Reactions  *1218*
- 25.5.1 Diels–Alder and Hetero-Diels–Alder Reactions  *1218*
- 25.5.2 Pictet–Spengler Reactions  *1222*
- 25.5.3 Nazarov Reactions  *1224*
- 25.5.4 Multicomponent and Domino Reactions  *1224*
- 25.6 Lewis Base-Catalyzed Asymmetric Cycloadditions  *1225*
- 25.6.1 Cycloaddition Reactions Catalyzed by Chiral Phosphines  *1225*
- 25.6.2 Cycloaddition Reactions Catalyzed by Cinchona Alkaloids  *1227*
- 25.7 Asymmetric Cyclization Reactions by Bifunctional Catalysts  *1228*
- 25.8 Asymmetric Cyclizations Catalyzed by Chiral Nucleophilic Carbenes  *1229*
- 25.8.1 Intramolecular Stetter Reactions  *1229*
- 25.8.2 Benzoin Reactions  *1231*
- 25.8.3 Cycloaddition Reactions  *1231*
- 25.9 Conclusion  *1233*
- 25.10 Experimental: Selected Procedures  *1233*
- 25.10.1 Synthesis of Compounds 8 via Intramolecular Aldol Reaction  *1233*
- 25.10.2 Synthesis of Compounds 19 via Intramolecular Michael Additions  *1234*
- 25.10.3 Typical Procedure for the Preparation of 62  *1234*
- 25.10.4 Synthesis of 66 via Organocatalytic Diels–Alder Reaction by 18a  *1234*
- 25.10.5 Typical Procedure for the Synthesis of 104  *1234*
- 25.10.6 Synthesis of 110 via Three-Component Domino Reaction  *1235*
- 25.10.7 Synthesis of 2,3-Dihydropyranones 148 Through Hetero-Diels–Alder Reaction by TADDOL  *1235*
- 25.10.8 Synthesis of 161 via Phosphoric Acid-Catalyzed Hetero-Diels–Alder Reaction  *1235*
- 25.10.9 Synthesis of 172 via an Asymmetric Pictet–Spengler Reaction  *1235*
- 25.10.10 Synthesis of DHPMs via an Asymmetric Biginelli Reaction  *1236*

25.10.11 Typical Procedure for the Synthesis of 192  *1236*
25.10.12 Typical Procedure for the Synthesis of 212  *1236*
25.10.13 Typical Procedure for Diels–Alder Reaction Catalyzed by 226 for the Synthesis of 227  *1237*
25.10.14 Intramolecular Stetter Reaction for the Synthesis of 232  *1237*
25.10.15 Synthesis of α-Hydroxy-α-alkyl Tetralones 243 via the Benzoin Reaction  *1237*
25.10.16 Synthesis of Chiral Dihydropyridinones 247 via an Aza-Diels–Alder Reaction Catalyzed by Carbene  *1237*
References  *1238*

**Index**  *1243*

## Contents to Volume 1

1 **Asymmetric Catalysis of Diels–Alder Reaction** *1*
   *Haifeng Du and Kuiling Ding*

2 **Catalytic Asymmetric Aza Diels–Alder Reactions** *59*
   *Yasuhiro Yamashita and Shū Kobayashi*

3 **1,3-Dipolar Cycloaddition** *87*
   *Takuya Hashimoto and Keiji Maruoka*

4 **Intramolecular 1,2-Addition and 1,4-Addition Reactions** *169*
   *Xiyan Lu and Xiuling Han*

5 **Cyclic Carbometallation of Alkenes, Arenes, Alkynes and Allenes** *227*
   *Ron Grigg and Martyn Inman*

6 **Transition Metal-Catalyzed Intramolecular Allylation Reactions** *271*
   *Zhan Lu and Shengming Ma*

7 **Cyclic Coupling Reactions** *315*
   *Xuefeng Jiang and Shengming Ma*

8 **Transition Metal-Mediated [2 + 2 + 2] Cycloadditions** *367*
   *David Lebœuf, Vincent Gandon, and Max Malacria*

9 **Cyclizations Based on Cyclometallation** *407*
   *Hao Guo and Shengming Ma*

10 **Transition Metal Catalyzed Cyclization Reactions of Functionalized Alkenes, Alkynes, and Allenes** *457*
   *Nitin T. Patil and Yoshinori Yamamoto*

| 11 | **Ring-Closing Metathesis of Dienes and Enynes**  527
*Miwako Mori* |

| 12 | **Ring-Closing Metathesis of Alkynes**  599
*Paul W. Davies* |

# List of Contributors to Volume 2

**David A. Capretto**
University of Chicago
Department of Chemistry
5735 S. Ellis Avenue
Chicago, IL 60637
USA

**Gregory R. Cook**
North Dakota State University
Department of Chemistry and
Molecular Biology
Fargo, ND 58105
USA

**Antonio M. Echavarren**
Institute of Chemical Research
of Catalonia (ICIQ)
Avinguda Països Catalans 16
43007 Tarragona
Spain

**Francisco J. Fañanás**
Universidad de Oviedo
Instituto Universitario de Química
Organometálica "Enrique Moles"
Julián Clavería 8
33006 Oviedo
Spain

**Valery V. Fokin**
The Scripps Research Institute
Department of Chemistry
10550 North Torrey Pines Road
La Jolla, CA
USA

**Liu-Zhu Gong**
University of Science and
Technology of China
Hefei National Laboratory for Physical
Sciences at the Microscale
Department of Chemistry
Hefei 230026
China

**Axel G. Griesbeck**
University of Cologne
Department of Chemistry
Organic Chemistry
Greinstr. 4
50939 Köln
Germany

**Amanda Halloway**
Ohio Northern University
Department of Chemistry
252 South Main Street
Ada, OH
USA

*Handbook of Cyclization Reactions. Volume 2.*
Edited by Shengming Ma
Copyright © 2010 WILEY-VCH Verlag GmbH & Co. KGaA, Weinheim
ISBN: 978-3-527-32088-2

**Ryuji Hayashi**
North Dakota State University
Department of Chemistry and
Molecular Biology
Fargo, ND 58105
USA

**Chuan He**
University of Chicago
Department of Chemistry
5735 S. Ellis Avenue
Chicago, IL 60637
USA

**Elena Herrero-Gómez**
Institute of Chemical Research
of Catalonia (ICIQ)
Avinguda Països Catalans 16
43007 Tarragona
Spain

**Jun Jiang**
Chinese Academy of Sciences
Chengdu Institute of Organic
Chemistry
Chengdu 610041
China

**Bi-Jie Li**
Peking University
College of Chemistry and
Molecular Engineering
Beijing National Laboratory of
Molecular Sciences (BNLMS)
PKU Green Chemistry Centre and Key
Laboratory of Bioorganic Chemistry and
Molecular Engineering of Ministry of
Education
100871 Beijing
China

**Zigang Li**
University of Chicago
Department of Chemistry
5735 S. Ellis Avenue
Chicago, IL 60637
USA

**Shi-Wei Luo**
University of Science and Technology
of China
Hefei National Laboratory for Physical
Sciences at the Microscale
Department of Chemistry
Hefei 230026
China

**Isamu Matsuda**
Nagoya University
Graduate School of Engineering
Department of Applied Chemistry
Chikusa
Nagoya 464-8603
Japan

**Yoshimitsu Nagao**
The University of Tokushima
Graduate School of Pharmaceutical
Sciences
1-78-1 Sho-machi
Tokushima 770-8505
Japan

**Félix Rodríguez**
Universidad de Oviedo
Instituto Universitario de Química
Organometálica "Enrique Moles"
Julián Clavería 8
33006 Oviedo
Spain

**Zhang-Jie Shi**
Peking University
College of Chemistry and
Molecular Engineering
Beijing National Laboratory of
Molecular Sciences (BNLMS)
PKU Green Chemistry Centre and Key
Laboratory of Bioorganic Chemistry and
Molecular Engineering of Ministry of
Education
100871 Beijing
China

**Mukund P. Sibi**
North Dakota State University
Department of Chemistry
Fargo, ND
USA

**Chang-Liang Sun**
Peking University
College of Chemistry and
Molecular Engineering
Beijing National Laboratory of
Molecular Sciences (BNLMS)
PKU Green Chemistry Centre and Key
Laboratory of Bioorganic Chemistry and
Molecular Engineering of Ministry of
Education
100871 Beijing
China

**Yong Tang**
Chinese Academy of Sciences
Shanghai Institute of
Organic Chemistry
State Key Laboratory of
Organometallic Chemistry
No. 354 Feng Lin Road
Shanghai 200032
China

**Yuanjing Xiao**
East China Normal University
Department of Chemistry
Shanghai Key Laboratory of Green
Chemistry and Chemical Processes
3663 North Zhongshan Road
Shanghai 200062
China

**Meng-Xia Xue**
Chinese Academy of Sciences
Chengdu Institute of
Organic Chemistry
Chengdu 610041
China

**Song Ye**
Chinese Academy of Sciences
Institute of Chemistry
Beijing National Laboratory for
Molecular Sciences
No. 2, 1st North Street,
Zhongguancun Beijing 100080
China

**Masahiro Yoshida**
The University of Tokushima
Graduate School of Pharmaceutical
Sciences
1-78-1 Shō-machi
Tokushima 770-8505
Japan

**Junliang Zhang**
East China Normal University
Department of Chemistry
Shanghai Key Laboratory of Green
Chemistry and Chemical Processes
3663 North Zhongshan Road
Shanghai 200062
China

**Jake Zimmerman**
Ohio Northern University
Department of Chemistry
525 South Main Street
Ada, OH
USA

# 13
# Transition-Metal-Catalyzed Cycloisomerizations and Nucleophilic Cyclization of Enynes

*Elena Herrero-Gómez and Antonio M. Echavarren*

## 13.1
### Introduction

Metal-catalyzed reactions of enynes proceed by two major pathways via coordination complexes **1** or **2** (Scheme 13.1) [1–7] using a variety of electrophilic salts or complexes $MX_n$ [8–15]. In the first pathway, coordination of the alkyne to the metal leads to complexes of type **1** that evolve via intermediates of **3** by an exo-dig cyclization or via **4** by an endo-dig process [16]. In the absence of nucleophiles, intermediates **3** usually evolve by skeletal rearrangement to afford dienes **5** and/or **6**, whereas intermediates **4** lead to bicyclic compounds **7**, which correspond to an intramolecular cyclopropanation of the alkene by the alkyne [17–19]. Alternatively, an Alder-ene cycloisomerization process can take place by coordination of the metal fragment to the alkyne and the alkene in complexes **2** which is followed by an oxidative cyclometallation to give **8** [3] that undergoes β-hydrogen elimination and reductive elimination to give dienes **9** [1–3, 5].

Gold(I) complexes usually surpass the reactivity shown by Pt(II) and other electrophilic metal salts and complexes for the activation of enynes. This has been attributed to relativistic effects, which are maximum for gold [6]. However, on occasion, the higher Lewis acidity of gold complexes can be detrimental in terms of selectivity and because of their low tolerance to certain functional groups. In these instances, the less Lewis-acidic Pt(II) complexes could be the catalysts of choice [20–22].

Cyclizations of enynes have been extensively reviewed in the last few years [1–7]. In this chapter we center on the more recent aspects of reactions of 1,n-enynes catalyzed by electrophilic transition metals (mainly gold and platinum) following the reaction pathways shown in Scheme 13.1 with a focus on those taken place in the presence of nucleophiles, leading to formation of the corresponding adducts. This review does not cover the enyne metathesis with Grubbs-type metal carbenes as catalysts [1c, 23, 24]. Reactions of arenes with alkynes are also not discussed [25]. However, reactions of furans with alkynes are included because of their close mechanistic similarity to those of enynes.

*Handbook of Cyclization Reactions. Volume 2.*
Edited by Shengming Ma
Copyright © 2010 WILEY-VCH Verlag GmbH & Co. KGaA, Weinheim
ISBN: 978-3-527-32088-2

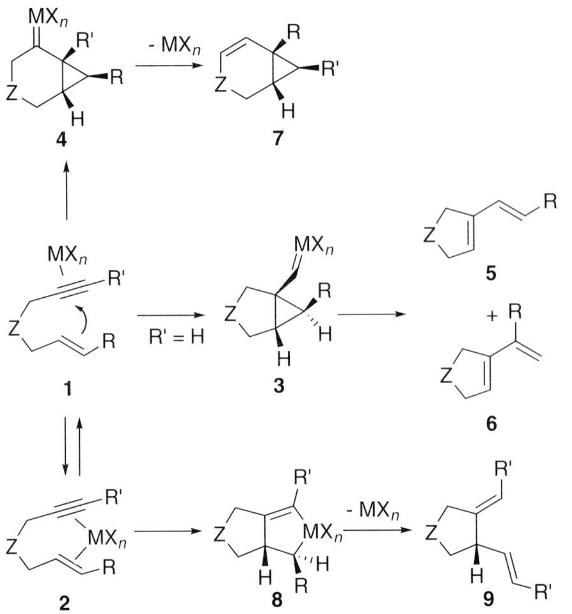

**Scheme 13.1** Major pathways for the metal-catalyzed cyclization of enynes.

## 13.2
## Cycloisomerizations of 1,n-Enynes

### 13.2.1
### Metal-Catalyzed Alder-Ene Cycloisomerization and Related Processes

The metal-catalyzed Alder-ene cycloisomerization of enynes was initially developed using Pd(II) salts or complexes as catalysts [8, 9, 26, 27]. This reaction is very general and can be performed with many metal catalysts. Thus, similar results were obtained using catalysts such as [CpRu(cod)Cl] [28], [CpRu(CH$_3$CN$_3$)$_3$]$^+$PF$_6^-$ [29, 30], Rh(I) complexes [31], [CpCo(CO)$_2$] [32, 33] and [Cp$_2$Ti(CO)$_2$] [34]. A general reaction also takes place using Pt(II) in non-nucleophilic solvents [35]. Certain Fe(0)-ate complexes are also good catalysts for a variety of Alder-ene, [4 + 2], [5 + 2], and [2 + 2 + 2] cycloaddition and cycloisomerization reactions of polyunsaturated substrates [36]. Enantioselective metal-catalyzed Alder-ene cycloisomerization has been achieved using Pd(OCOCF$_3$)$_2$ and chiral bidentate phosphines such as (R)-BINAP as ligands [37] as well as with Rh(I)/Me-DuPhos [38].

Deuteration studies demonstrate that the Alder-ene cycloisomerization is an intramolecular process [35]. According to DFT calculations, the reaction requires coordination of the metal to both the alkyne and the alkene, which is followed by an oxidative cyclometallation to form metallacycle **8** (Scheme 13.1). Mechanistically related processes take place in other important organometallic transformations.

Thus, oxidative cyclometallation is one of the key steps in the synthetically useful Pauson–Khand synthesis of cyclopentenones with $Co_2(CO)_8$ [39] or other transition metals [40].

For Pt(II), metallacycles of type **8** were shown to evolve by β-hydrogen elimination from one of the alkyl chains, which was followed by reductive elimination to give the product of cycloisomerization **9** (Scheme 13.1) [35]. Thus, dimethyl 2-geranyl-2-propargylmalonate (**10**) led to triene **11** as a 3:1 mixture of E and Z isomers, by the selective abstraction of a hydrogen from the allylic methylene (Scheme 13.2), whereas the neryl derivative **12** afforded product **13** by hydrogen abstraction from the methyl group. Interestingly, the Ru(II)-catalyzed cycloisomerization of enynes takes place via related ruthenacycles that can suffer elimination from either *trans-* or *cis-*allylic positions [35b].

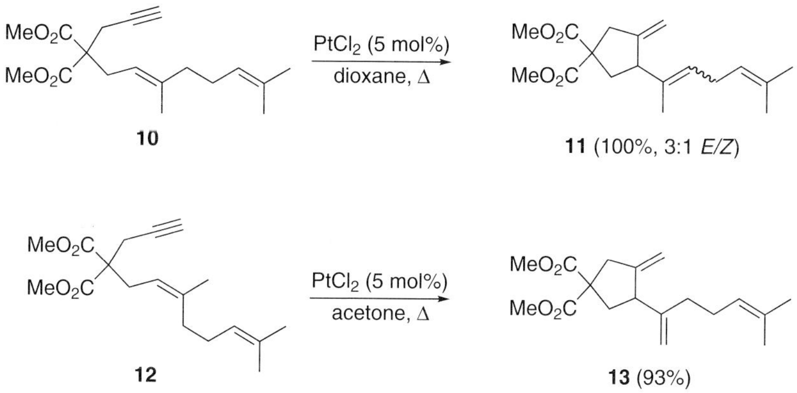

**Scheme 13.2** Pt(II)-catalyzed cycloisomerization of 2-geranyl-2-propargylmalonate **10** and 2-neryl-2-propargylmalonate **12**.

Other pathways may be followed with different metal catalysts. Thus, for example, the cobalt-catalyzed cycloisomerization of 1,6-enyne **14** leads to product **15** by an allylic activation at the alkenyl chain (Scheme 13.3) [41].

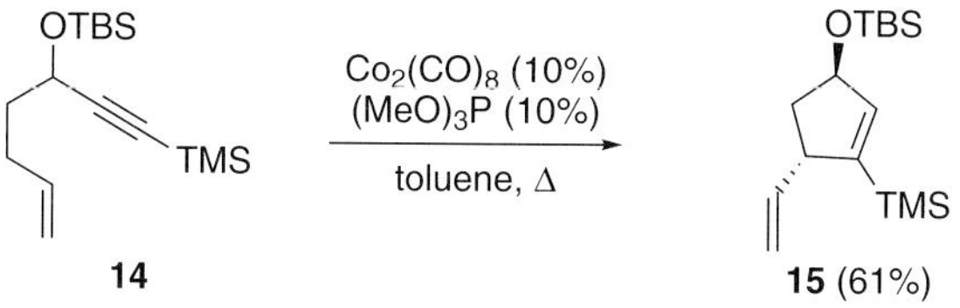

**Scheme 13.3** Cyclization by allylic activation using Co(II) catalyst.

The proposal of involvement of cyclopropyl metal carbenes of type **3** in the electrophilic activation of enynes by transition metals was first substantiated in reactions catalyzed by Pd(II), in which the initially formed cyclopropyl palladium carbenes undergo a [4 + 2] cycloaddition with the double bond of the conjugate enyne [42]. While other transition metals can react in both manners, the Alder-Ene cycloisomerization does not occur for gold(I) [43, 44] since oxidative addition processes are not facile for this metal [6]. Thus, cycloisomerizations of enynes catalyzed by gold(I) proceed exclusively by initial coordination of the metal to the alkyne, as shown in complex **1** (Scheme 13.1).

Simple AuCl or $AuCl_3$ are alkynophilic enough to catalyze many reactions of the more reactive enynes. However, for many gold-catalyzed transformations, the most convenient catalysts are cationic complexes generated by chloride abstraction from [AuCl(L)] (L = phosphine) using a silver salt with a non-coordinating anion to generate *in situ* the corresponding cationic derivative [Au(L)(S)]X (S = solvent or substrate molecule) [43, 44, 68]. Cationic complexes can be obtained *in situ* by cleavage of the Au−Me bond in [AuMe(PPh$_3$)] with a protic acid [43–46]. The cationic complex [Au(PPh$_3$)(MeCN)]SbF$_6$**16** has been prepared as a stable white solid [44]. The gold-oxo complex [(Ph$_3$PAu)$_3$O]BF$_4$ [47] has also been used as a catalyst in certain reactions of enynes [48]. Gold(I) complexes **17a–d** bearing bulky, biphenyl-based phosphines, lead to very active catalysts upon being mixed with Ag(I) salts [49]. More convenient are cationic complexes **17e–f** [44], stable crystalline solids, which are very reactive as catalysts in a variety of transformations (Figure 13.1) [50–52]. The structures of **16** and **17a–f** have been confirmed by X-ray crystallography [53, 54]. Related complexes **17g–h** with weakly coordinated bis(trifluoromethanesulfonyl)amide NTf$_2$ (Tf = CF$_3$SO$_2$) have also been prepared [55]. The gold(I) complex **18a** bearing tris(2,6-di-*tert*-butylphenyl)phosphite as a bulky ligand leads to a highly electrophilic cationic Au(I) catalyst *in situ* by chloride abstraction with AgSbF$_6$ [56, 57]. Recently, stable cationic **18b** has also been prepared [58]. Gold complexes with highly donating *N*-heterocyclic ligands (NHC) such as **19a–d** are also good precatalysts [49, 59–61]. Stable, easy to handle, cationic complexes bearing NHC ligands such as **19e–f** [58], and those with the NTf$_2$ ligand (**19g–h**) have also been reported [62, 63].

Fundamental differences exist between homogeneous and heterogeneous gold catalysis [64]. The selective alkyne activation in complex mixtures and/or enyne compounds is controlled by distinct mechanisms in homo- and heterogeneous gold catalysts. For heterogeneous catalysis, differential reactant adsorption is the key, leading to bonded, active species that give rise to fast hydrogenation or hydration reactions. In contrast, for homogeneous catalysis, alkenes are preferentially coordinated to gold, but electro- and nucleophilic attacks are more thermodynamically favored for alkynes. Enyne cyclizations in homogeneous systems are mainly driven by the exceptionally low LUMO of the coordinated C≡C. Therefore, the alkynophilicity in homogeneous systems is not related to the relative strength of the bonding of the active alkyne species but to its activation for a desired reaction. Experimentally, heterogeneous catalysts that are effective in CO oxidations and hydrogenation reactions have been shown to be inert in the activation of enynes [64].

Importantly, the fragment [Au(PR$_3$)]$^+$ cannot coordinate simultaneously to the alkene and the alkyne and, in consequence, the Alder-ene cycloisomerization does not compete and the cyclizations proceed exclusively through complexes of type **3** or **4** (Scheme 13.1). However, when the reaction of enyne **20a** was performed in DMSO at 50 °C with catalyst **17e**, a mixture of dienes **21** and **22** was obtained (Scheme 13.4) [44].

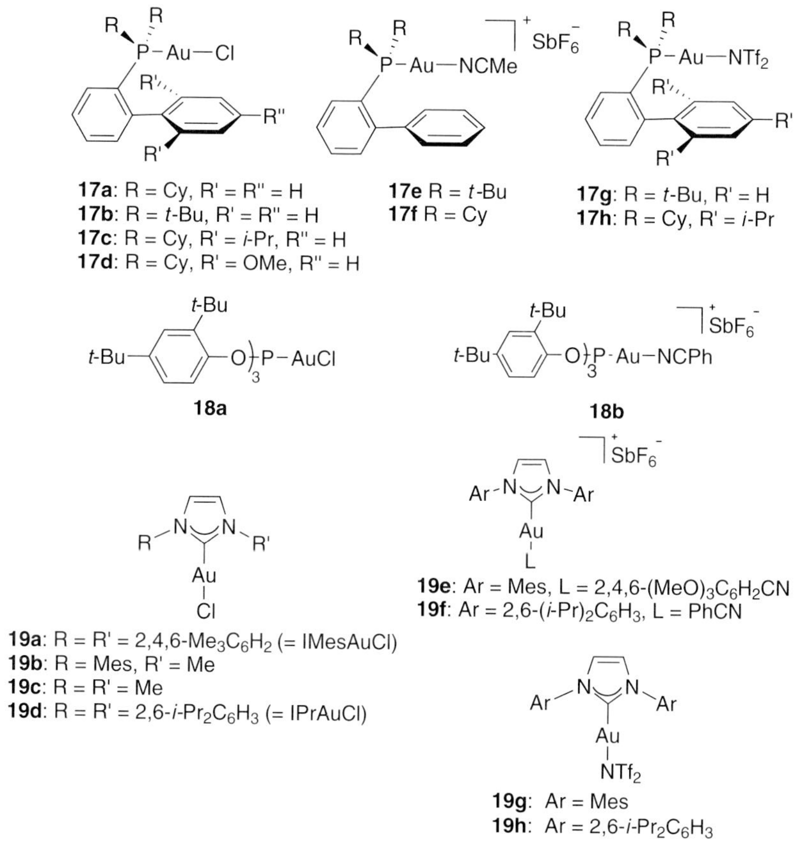

**Figure 13.1** Representative examples of Au(I)-complexes used in homogeneous catalysis.

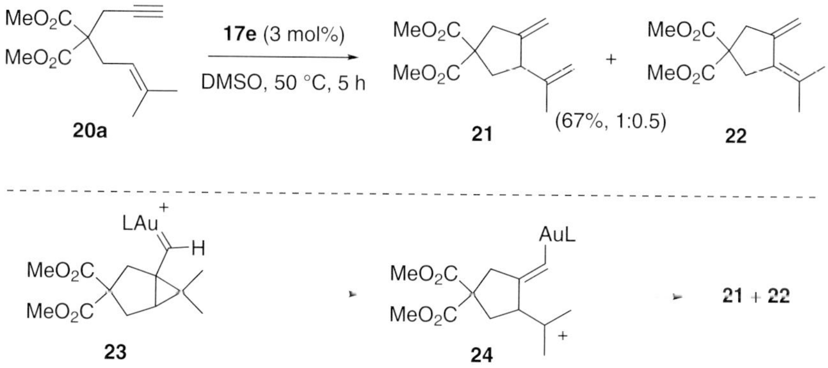

**Scheme 13.4** Alder-ene type dienes obtained in Au(I)-catalyzed reactions performed in DMSO.

Although diene **21** appears to be the product of a formal Alder-ene type cycloisomerization of **20a**, deuteration experiments excluded the participation of Alder-ene type cycloisomerization in this transformation. Instead, this transformation was proposed to proceed by opening of intermediate **23** to give cation **24**, followed by proton loss to form a mixture of **21** and **22**. Other enynes related to **20a**, with two methyl substituents at the terminal alkene carbon, also suffer proton loss to form dienes such as **28** by using a catalyst generated from [AuCl(PPh$_3$)] (5 mol%) and AgOTf (7 mol%) [65]. Products of type **21** have also been obtained in intramolecular reactions of allylstannanes and allylsilanes with alkynes using different metals as catalysts [66].

In the absence of nucleophiles, 1,6-enynes can give rise to bicyclo[4.1.0]hept-4-ene derivatives **7** (Scheme 13.1) by cyclopropanation of the alkene by the alkyne. These products are formed by endo-cyclization via intermediates **4** by proton loss and protodemetallation (Scheme 13.5) [11a,b, 17, 19f, 67]. Addition of heteronucleophiles at the terminal alkene carbon, by cleavage of the bond labeled **a** in intermediate **4** has been observed using PtCl$_2$ or AuCl$_3$ as catalysts [19c]. The alternative cleavage of bond **b** in intermediate **4** to form derivatives **25** has also been observed in a few cases using electrophilic Au(I) or Pt(II) catalysts [68]. Thus, for example, enyne **26** reacts with gold catalyst **17e** to give the product of intramolecular cyclopropanation **27** along with 3,4,7,9-tetrahydro-2H-pyrano[2,3-c]oxepine **28**, a derivative of type **25**.

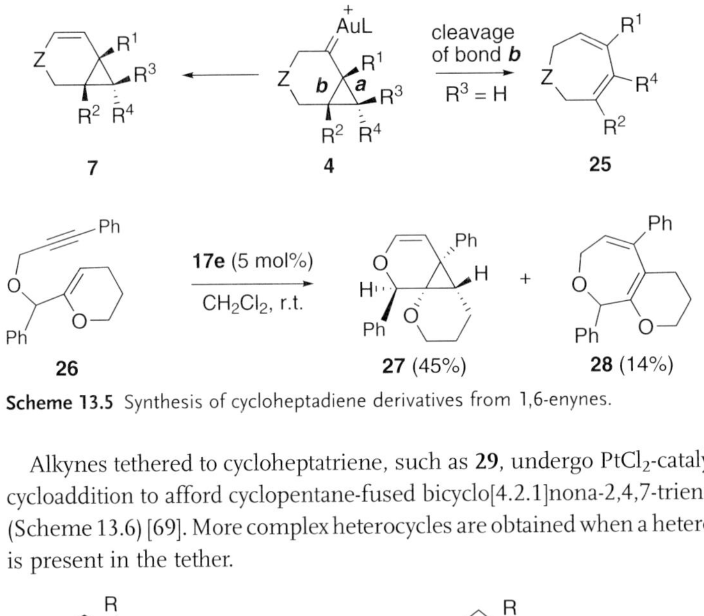

**Scheme 13.5** Synthesis of cycloheptadiene derivatives from 1,6-enynes.

Alkynes tethered to cycloheptatriene, such as **29**, undergo PtCl$_2$-catalyzed [6 + 2] cycloaddition to afford cyclopentane-fused bicyclo[4.2.1]nona-2,4,7-trienes of type **30** (Scheme 13.6) [69]. More complex heterocycles are obtained when a heteroatom (O, N) is present in the tether.

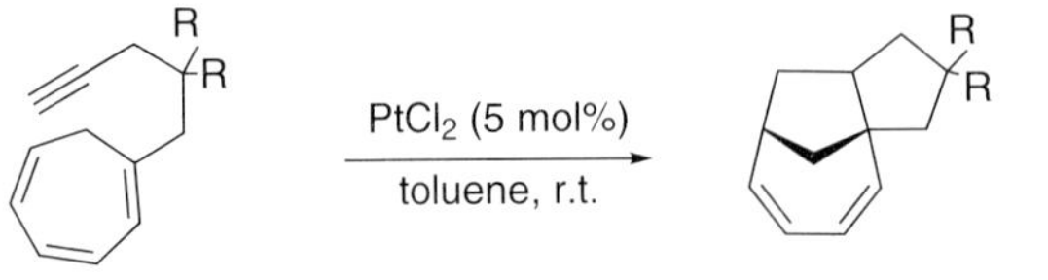

**29**: R = CH$_2$OAc       **30** (92%)

**Scheme 13.6** Formation of cyclopentane-fused bicyclo[4.2.1]nona-2,4,7-trienes using Pt(II).

Although the first examples of cyclization of 1,5-enynes with gold were actually reported in the context of a new synthesis of pyridines by cyclization of propargyl enanimes catalyzed by NaAuCl$_4$·2H$_2$O [70], the significant breakthrough in this area was made simultaneously by three groups which independently reported in 2004 that the gold- or platinum-catalyzed cyclization of 1,5-enynes gives bicyclo[3.1.0]hexanes [71, 72, 197b]. Thus, for example, simple enynes **31** afford derivatives **32** under mild conditions (Scheme 13.7) [72]. Interestingly, substrates **31b,c** react with concomitant ring expansion to form tricyclic derivatives **32b,c**. This reaction of 1,5-enynes presumably proceeds by formation of cyclopropyl gold carbene **33**, which evolves by proton loss to form the alkenyl-gold complex **34**, followed by protonolysis to form bicyclo[3.1.0]hexanes **35**. In the case of **31b,c**, intermediates of type **33** undergo ring expansion.

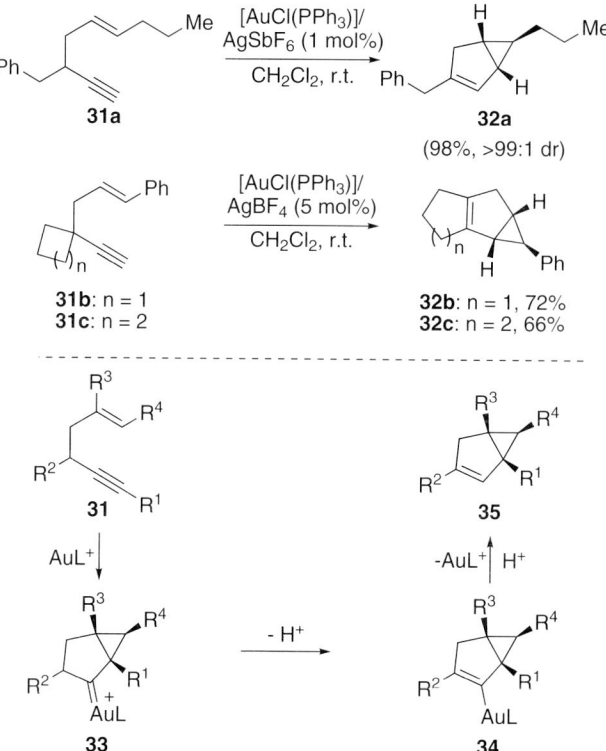

**Scheme 13.7** Au(I)-catalyzed cyclizations of 1,5-enynes.

Similarly, enynes **36** gave bicyclo[3.1.0]hexanes **37** (Scheme 13.8) [55, 73, 74]. In this case, products of skeletal rearrangement such as **38a,b** were also obtained (see Section 13.4.5). Although the presence of an additional stereogenic center in **36a,b** leads to mixtures of diastereomers, formation of **37c′** as a minor product in the cycloisomerization of substrates **36c** is noteworthy since this product corresponds to that expected from the Z-isomer of **36c**. This lack of stereoespecificity was explained by the equilibrium between two rapidly interconverting six-membered ring cations [73]. This result can also be accounted for by the opening of the initially formed cyclopropyl gold carbene **39a** to give the six-membered ring tertiary cation **40**, which can cyclize to form diastereomeric cyclopropyl gold carbene **39b**, the precursor of **37c'**. Intermediate six-membered ring cations of type **40** can be intercepted by an O-Boc group in reactions of Boc-protected hex-1-en-5-yn-3-ols, which leads to cyclohex-4-ene-1,2-diol derivatives [75].

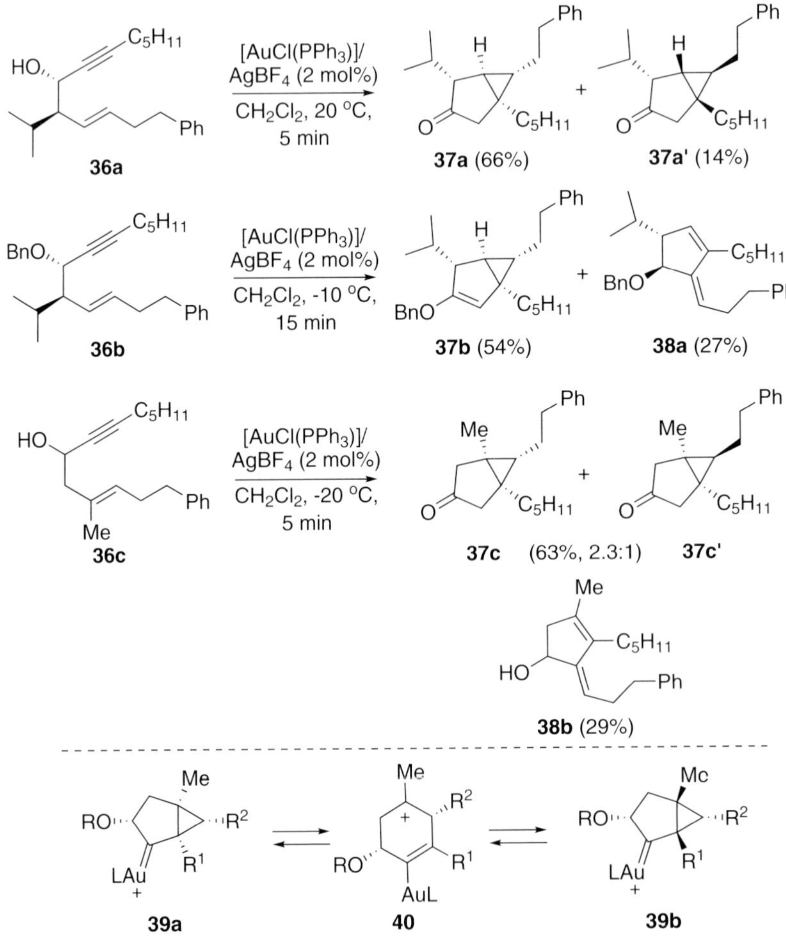

**Scheme 13.8** Au(I)-catalyzed cycloisomerizations of 1-en-5-yn-4-ols (**36**).

In these reactions, 1,5-enynes react exclusively by endocyclic pathways. This was rationalized for a Pt(II)-catalyzed reaction as a result of the formation of a bicyclo[3.1.0]hexane system in the endo-cyclization, which is more favorable than the strained bicyclo[2.1.0]pentane system that would be formed in the exo-cyclization [14c]. Endo-cyclization was also observed in the gold(I)-catalyzed cyclization of

1-alkynyl-2-alkenylbenzenes to form naphthalenes [76, 77]. However, in this case, the alternative exo-cyclization was also observed as a minor pathway from substrates bearing terminally substituted alkynes [77]. The exo derivatives were the major products only with substrates with terminal alkynes or iodoalkynes. 3-Hydroxy-1,5-enynes also react with a catalyst generated from [AuCl(PPh$_3$)] and AgOTf by an endocyclic process, which was applied for the synthesis of tetrahydronaphthalenes [78]. A totally different result was observed in the cyclization of 1,6-enynes with Grubbs catalysts, which leads to vinyl cyclobutenes [79].

The addition of β-ketoesters to alkynes (Conia-ene cyclization) is also catalyzed by gold(I) [80–83]. Thus, substrates **41a** react in an exo fashion to give derivatives **42a**, whereas **41b,c** react by endocyclic pathways to give **42b,c** (Scheme 13.9). These transformations are metal-catalyzed intramolecular reactions of enols with alkynes proceeding exo- or endocyclically depending on the length of the tether. An enantioselective version of this reaction has been developed using chiral palladium complexes [84–86]. Silylketene amides or carbamates react similarly with alkynes to form cyclopentanes or dehydro-δ-lactams [87].

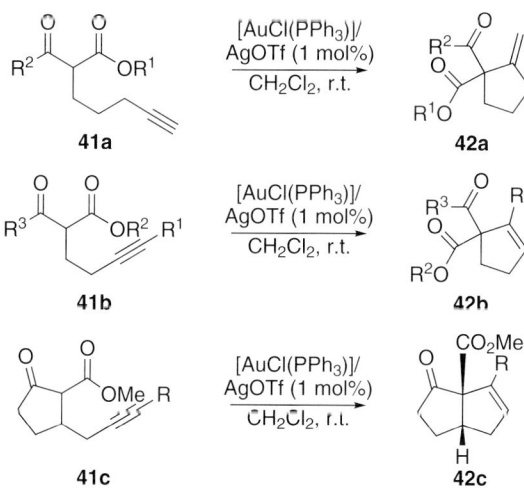

**Scheme 13.9** Au(I)-catalyzed Conia-ene cyclization.

Silyl enol ethers also react intermolecularly with alkynes, which has been applied for the synthesis of the alkaloids (+)-licopladine A [88] and (+)-fawcettimine **45** (Scheme 13.10) [89]. Thus, enantiomerically pure dienyne **43** was efficiently cyclized to give **44**, a key bicyclic intermediate in the synthesis of (+)-fawcettimine **45**. This reaction has also been applied in a recent synthesis of the antibiotic platencin [90]. Reactions of similar substrates bearing a 1,5- or 1,6-relation between the silyl enol ether and the alkyne take place in an exo fashion to form five- or six-membered rings [91]. The cyclization of alkynes with enamines formed *in situ* from aldehydes of secondary amines also takes place in the presence of gold [92] or other metal catalysts [93].

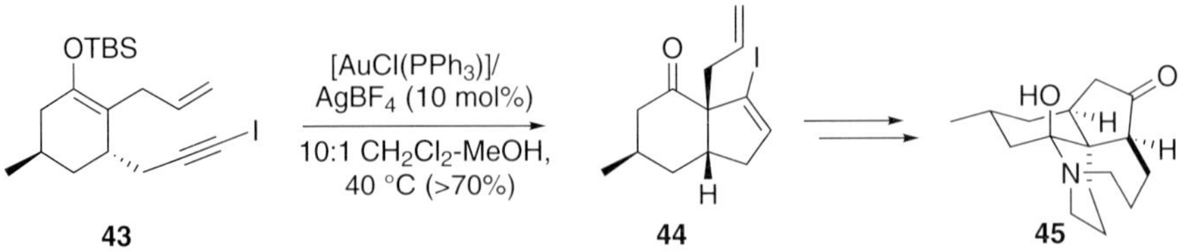

**Scheme 13.10** Synthesis of (+)-fawcettimine by a Au(I)-catalyzed reaction.

Propargyl vinyl ethers **46** undergo Claisen rearrangement with Au(I) catalysts to give allenes **47**, which were isolated as the corresponding alcohols (Scheme 13.11) [48, 94, 95]. A similar gold-catalyzed transformation of propargyl allyl ethers, which proceeds by initial isomerization to the propargyl vinyl ethers, has also been observed [96]. In the presence of water or alcohols, dihydropyrans **48** were obtained from **46** in good yields [97]. These results can be interpreted by a ring expansion of the initial cyclopropyl gold carbene **49** to form the oxonium cation **50**, which can undergo C–O fragmentation to afford the allenes **47** or trapping with heteronucleophiles to give **48**.

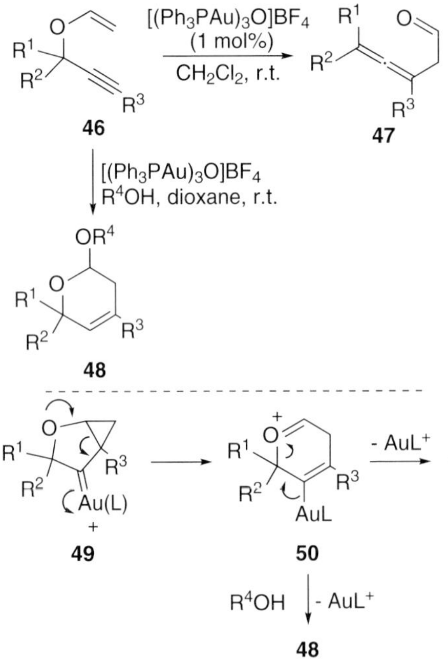

**Scheme 13.11** Au(I)-catalyzed reactions of propargyl vinyl ethers.

In the above transformations either 1,5- or 1,6-enynes were cycloisomerized with gold catalysts. Although relatively less studied, 1,7-enynes also react uneventfully with gold catalysts in hydroxy and alkoxycyclizations (see Section 13.3.1) as well as in skeletal rearrangements (see Section 13.4.2). However, no example of cyclization of a simple 1,8-enyne has yet been reported with gold catalysts (see however, Section 13.2.2, Scheme 13.21). Recently a remarkable cyclization of the 1,9-enyne **51** to form the 10-membered ring **52** has been reported using 50 mol% gold(I) complex (Scheme 13.12) [98]. This reaction presumably occurs via intermediates **53** and **54** and demonstrates that formation of large rings from 1,n-enynes ($n \geq 7$) by using gold chemistry could be possible.

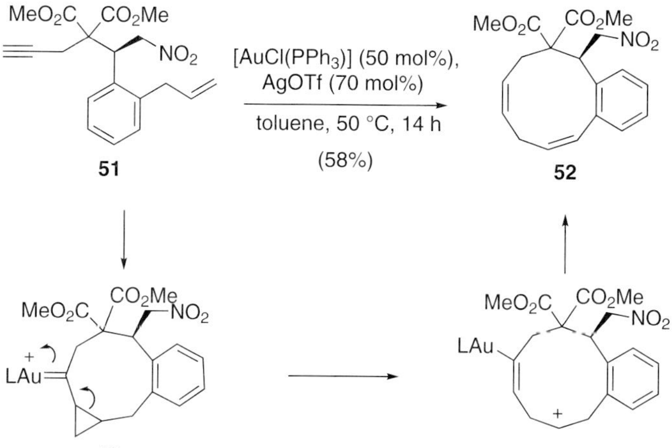

**Scheme 13.12** Au(I)-catalyzed cyclization of a 1,9-enyne.

## 13.2.2
### Cycloisomerizations of Enynes Via Propargylic 1,2- and 1,3-Acyl Migration

Propargylic esters coordinate to gold to form complexes **55** that can undergo 1,2- or 1,3-acyl migrations to form α acyloxy α,β unsaturated carbenes **56** or allene-gold complexes **57** (Scheme 13.13) [99–101]. According to DFT calculations, formation

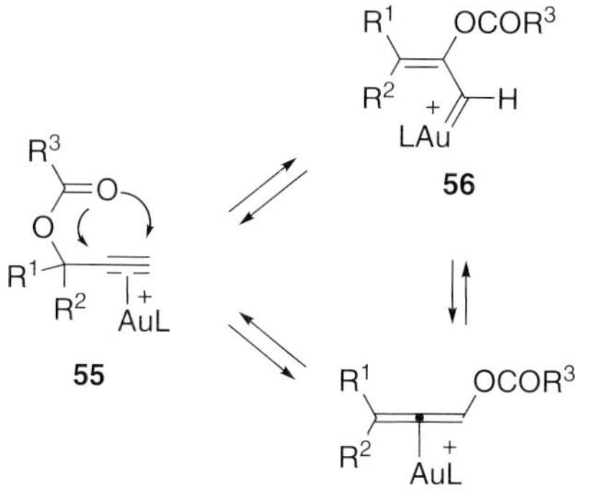

**Scheme 13.13** 1,2- and 1,3-acyl migration in Au(I)-catalyzed reactions.

of **56** and **57** is not concerted, but proceeds in two steps through five- and six-membered ring intermediates, respectively [100, 101]. Similar pathways have been described with other metal catalysts [100–104, 197a,b].

Evidence for the involvement of metal carbenes **56** has been obtained in intermolecular cyclopropanation reactions with alkenes [56, 104, 105, 203], trapping with carbon nucleophiles [106], sulfides [107], imines [108], and oxidation with $Ph_2SO$ to form α,β-unsaturated carbonyl compounds [178]. Cyclopropanation reactions have also been reported with other metals. Thus, for example, the ruthenium-catalyzed reaction between propargylic carboxylates and alkenes leads to vinyl cyclopropanes **58a,b** (Scheme 13.14) [104]. This type of reaction has been extended to the cyclopropanation of heteroaromatic compounds using propargylic carboxylates with Ru and Pt as catalysts [104c].

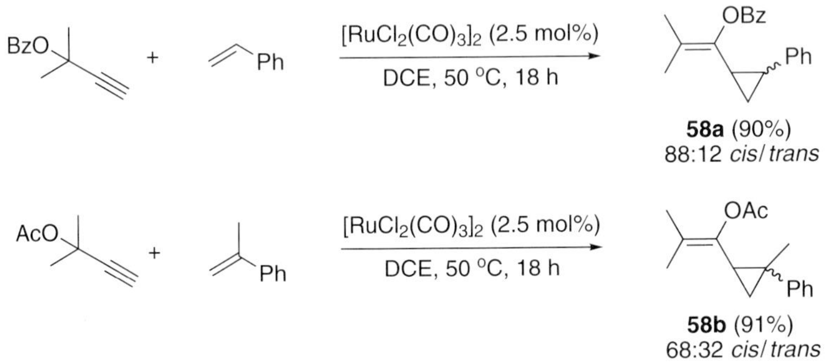

**Scheme 13.14** Ru(II)-catalyzed cyclopropanation reactions.

For the gold-catalyzed cyclization of enynes bearing α-acyloxy substituents at the propargylic position, in principle, two mechanistically distinct possibilities probably exist, depending on the order of attack of the acyloxy group and the alkene to the alkyne in complexes of type **59** (Scheme 13.15). If the alkene reacts first, the cyclopropyl metal carbene **60** would be formed, which could then suffer an intramolecular attack of the acyl to the carbene, followed by an elimination (formal 1,2-migration of the acyl) to give **61** after metal loss. Alternatively, the carboxylate group might undergo first the 1,2-migration to form carbene **62**, which would then form **61** by intramolecular cyclopropanation. The overall process is known as the Ohloff–Rautenstrauch rearrangement, or more simply as the Rautenstrauch rearrangement, following the original discovery in the context of zinc- or palladium-catalyzed cyclizations of similar systems [109, 110]. A theoretical study of these cyclization reactions catalyzed by $PtCl_2$ supports a mechanism taking place through intermediates of type **62** [111, 112]. The gold-[71] and platinum-catalyzed [197b] Rautenstrauch rearrangement allows one to access bicyclo[3.1.0]hexan-2-ones by cleavage of the intermediate enol carboxylates. Thus, for example, the gold-catalyzed cyclization of propargylic acetate **63** gives the product of cyclization and 1,2-acetate migration **64**, which reacts with methanol under basic conditions to yield ketone **65** [71, 113].

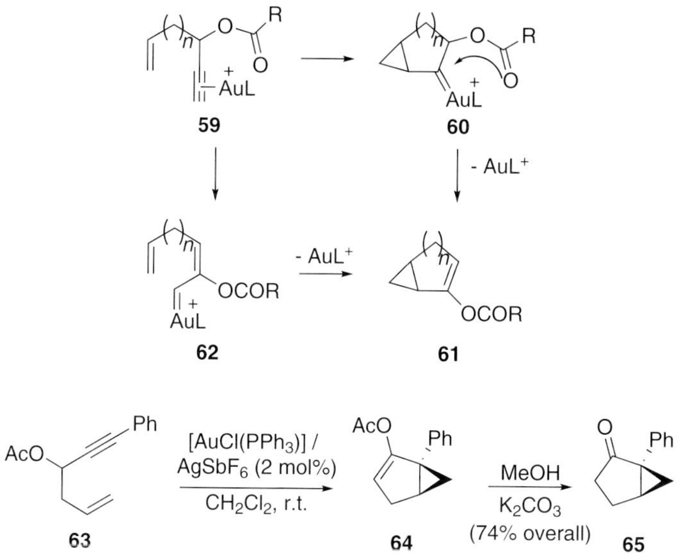

**Scheme 13.15** Rautenstrauch rearrangement in α-acyloxy enynes.

A pathway via intermediate **60** was suggested when $n = 2$ in a key Pt(II)-catalyzed cyclization for the synthesis of (−)-α-cubebene and (−)-cubebol [114]. Methanolysis of **67** affords ketone **68**, which was later converted into (−)-α-cubebene and (−)-cubebol. (−)-Cubebol was also synthesized using a similar cyclization catalyzed by Pt(II), Au(I), or Cu(I) [115]. A cyclization of this type catalyzed by AuCl$_3$ was used for the synthesis of 2-sesquicarene **71** and related compounds [114, 116]. The key step in the synthesis of **71** is the stereoselective cyclization of propargyl acetate **69** to give bicyclic enol acetate, which was methanolized to give ketone **70** (Scheme 13.16).

1,4-Enynes **72** substituted at the propargylic position with a carboxylate undergo the Rautenstrauch rearrangement to afford cyclopentenones **73** in a general way (Scheme 13.17) [117]. The reaction proceeds with remarkable transfer of chirality allowing the efficient enantioselective synthesis of cyclopentenones **73a–c** from the corresponding enantiomerically enriched propargylic pivalates **72a–c**.

This reaction was proposed to proceed by 1,2-acyl migration via intermediates **74** and **75** to give pentadienyl cation **76** (Scheme 13.18). This intermediate can also be viewed as a conjugated gold carbene **76′**. Cation **76** then undergoes an electrocyclization to form **77**, which suffers metal loss to give cyclopentadiene **78**. The cyclopentenones are obtained by hydrolysis of the enol acetates. The reaction can be envisioned as the intramolecular attack of the enol acetate on the allyl cation, in which the chiral information is transmitted from the stereocenter to the helical conformation of the pentadienyl cation. As the cyclization is faster than the rates of helix interconversion and carboxylate rotation, the products are obtained enantioselectively [118]

A synthesis of cyclopentenones somewhat related to that shown in Scheme 13.18 was found from conjugated 1,3-enynes **79** (Scheme 13.19) [119]. Cyclopentenones **80** were finally obtained by the hydrolysis of the intermediate enol acetates. This

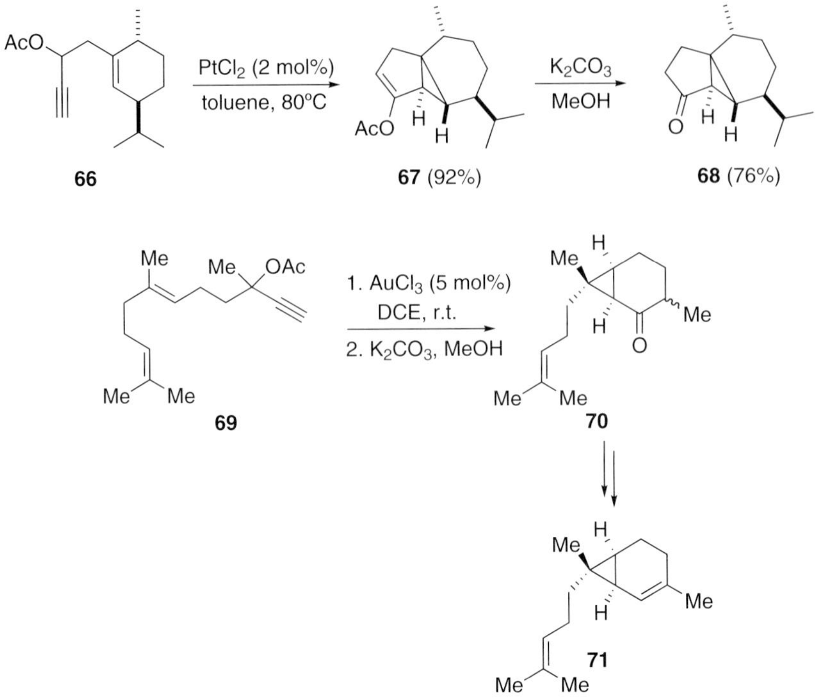

Scheme 13.16 Metal-catalyzed cyclizations in the total synthesis of (−)-α-cubebene, (−)-cubebol (**68**) and 2-sesquicarene (**71**).

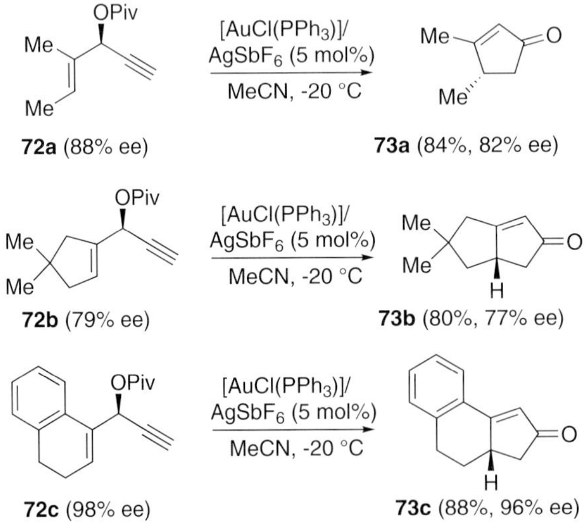

Scheme 13.17 Enantioselective synthesis of cyclopentenones via Au(I)-catalyzed Rautenstrauch rearrangement.

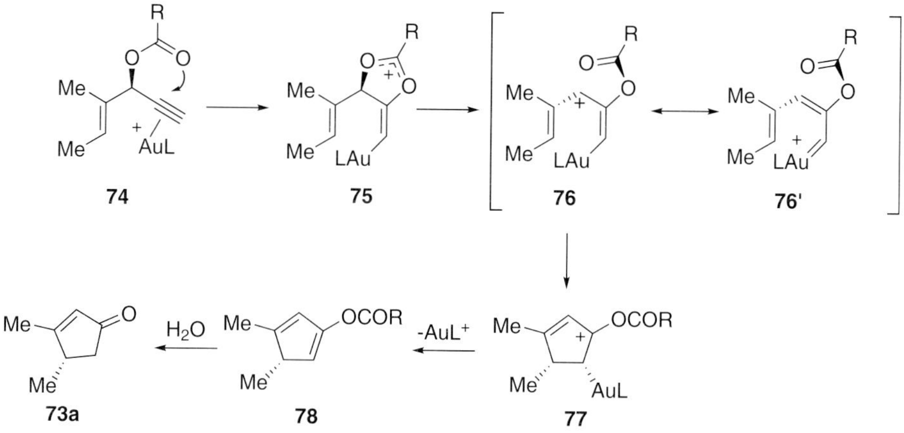

**Scheme 13.18** Proposed mechanism for the enantioselective Au(I)-catalyzed Rautenstrauch rearrangement.

transformation involves a 1,3-migration of the acetate to form the pentadienyl cation **81**, which undergoes a Nazarov-type cyclization to form the gold carbene **82**. A formal 1,2-hydrogen migration then forms **83**, which is followed by metal loss and hydrolysis to give cyclopentenones **80**. DFT calculations support this mechanism and provide interesting insight into the mechanism of the final stages of the process [100]. Thus, this theoretical study shows that although the 1,2-hydrogen migration to form **83** is possible, in the presence of water, a more facile proton loss assisted by $H_2O$, followed by a protodemetallation is preferred.

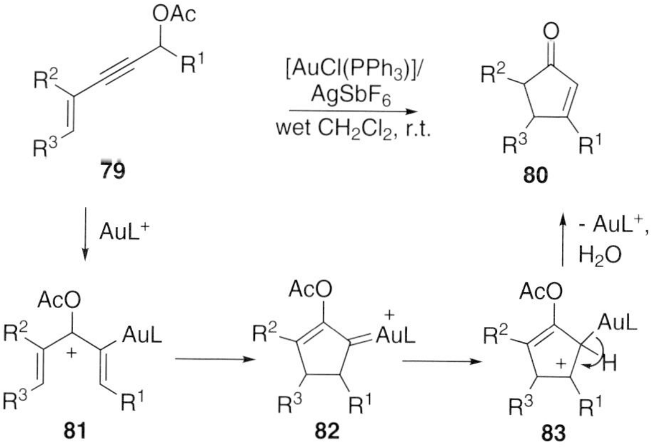

**Scheme 13.19** Synthesis of cyclopentenones from 1,3-enynes.

Propargylic acetates **84** react differently in a process that also takes place via a 1,3-acyl migration (Scheme 13.20) [120]. Thus, reaction of **84a–c** gives bicyclo[3.1.0] hexenes **85a–c**, which can be transformed into cyclohexenones by treatment with $K_2CO_3$ in MeOH. As shown in the transformation of **84c** into **85c**, the reaction takes place with nearly complete transfer of the stereochemical information. The reaction

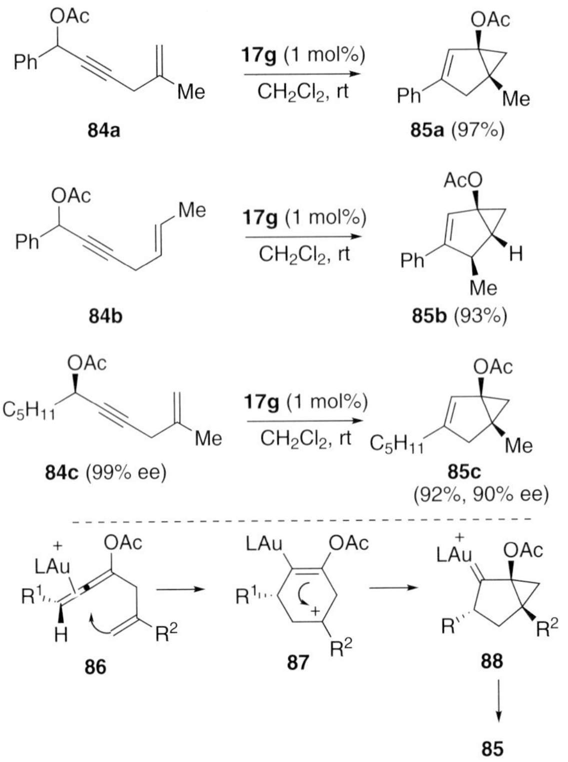

**Scheme 13.20** Au(I)-catalyzed reaction of propargylic acetates via 1,3-acyl migration.

was proposed to take place via the allene-gold complex **86**, which reacts with the alkene to form cyclopropyl gold carbene **88** via the cyclohexyl carbocation **87**.

The cyclization of substrates such as **89a,b** gives bicyclic derivatives **90a,b** with high stereoselectivity by an endocyclic pathway (Scheme 13.21) [121]. In the cyclization of

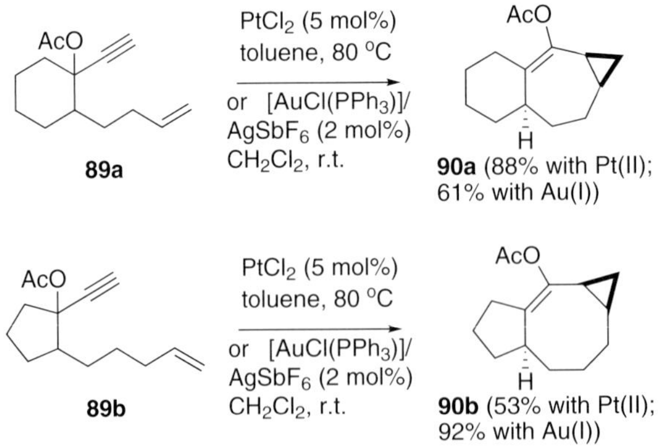

**Scheme 13.21** Pt(II)- and Au(I)-catalyzed cyclizations of 1,8-enynes.

this type of substrate, better yields were usually obtained with PtCl$_2$ or AuCl$_3$ as catalysts. Remarkably, the cyclization of **89b** is an example of a cyclization of a 1,8-enyne, although mechanistically this transformation could be interpreted as intramolecular cyclopropanation of the terminal alkene by the gold carbene formed in a 1,2-acyl migration. This type of cyclization has been applied for the synthesis of an allocolchicinoid [122].

Two remarkable transformations involving the indole nucleus were found from propargylic carboxylates **91** to give tetracyclic compounds **92** or **96** (Scheme 13.22)

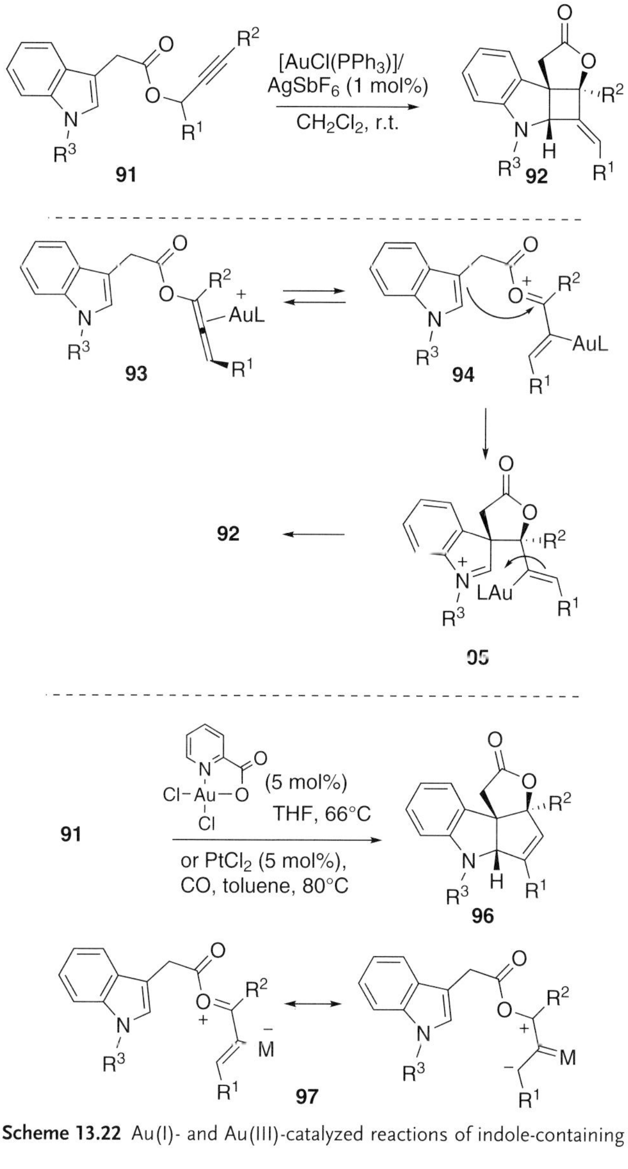

**Scheme 13.22** Au(I)- and Au(III)-catalyzed reactions of indole-containing substrates.

[123, 124]. The first reaction proceeds by an allene-gold complex **93** in equilibrium with **94**, which reacts intramolecularly with the indole to form **95** and then **92**. An allene of type **93** was isolated when the reaction was performed with AuCl$_3$ as catalyst [123]. On the other hand, when the reaction of **91** was performed with dichloro(pyridine-2-carboxylato)gold(III) or Pt(II) as catalysts, products **96** were instead obtained by a formal [3 + 2] cycloaddition of 1,3-dipoles **97** with the indole nucleus.

Simple propargylic esters **98** also undergo a 1,3-acyl migration catalyzed by Au (III), leading to 1,3-dicarbonyl compounds **99** as mixtures of E and Z isomers (Scheme 13.23) [125]. The reaction presumably proceeds by initial formation of complex **100**, followed by opening to give **101** [126]. The allene-gold complex **101** then generates **102**, which undergoes intramolecular acylation via cyclic intermediate **103** to form *E*-**99**. 1,3-Dicarbonyl compounds *E*-**99** could equilibrate readily with their Z isomers. Interestingly, propargylic esters **98** react with [Cu(MeCN)$_4$]BF$_4$ as catalyst to form *E*-**99** products, whereas Z isomers were preferentially formed with PtCl$_2$ [127]. Other 1,3-acyl migrations of propargylic carboxylates catalyzed by gold give intermediate allenes that can afford heterocyclic compounds by intramolecular attack of the appropriate nucleophiles [128]. The intermediate allenes formed by 1,3-acyl migrations can also react intramolecularly with alkynes to form naphthalenes, although this reaction proceeds more efficiently with Ag(I) catalysts [129].

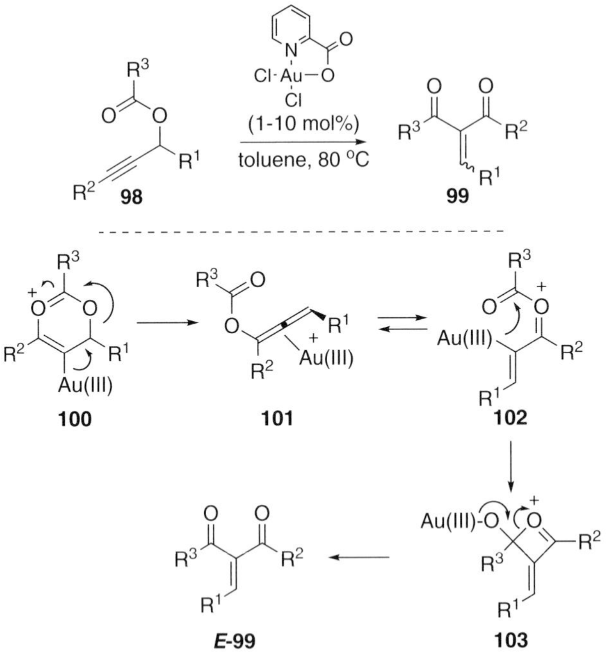

**Scheme 13.23** Au(III)-catalyzed 1,3-acyl migration.

Following the related synthesis of indenes using Ru(II) [130] or Pt(II) [103] catalysts, the acyl migration of substrates **104** was applied for the synthesis of **105** (Scheme 13.24) [131]. Isomeric indenes **106** were also obtained as minor products. Formation of **105** proceeds via allenes **107**, which could be independently prepared by Ag(I)-catalyzed 1,3-acetate migration from **104**. Propargylic sulfides and dithioacetals undergo similar transformations to give indene derivatives with Au(I) or Au(III) catalysts [132].

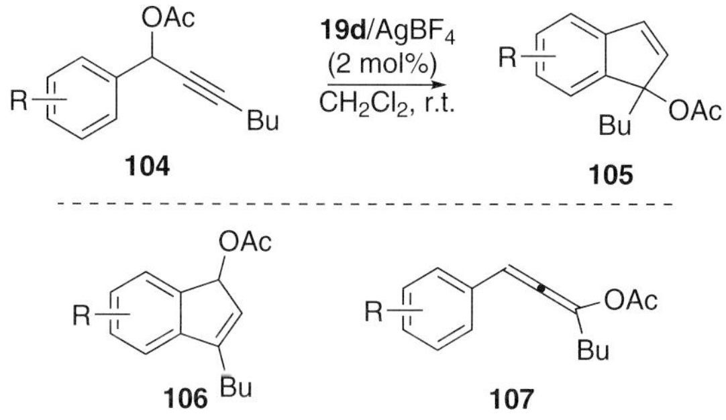

**Scheme 13.24** Au(I)-catalyzed 1,3-acyl migration applied to the synthesis of indene derivatives.

A different type of 1,3-acyl migration occurs from homopropargyl acetates **108** (Scheme 13.25) [133]. In this case, upon coordination of the alkyne with Au(III) in **110** and migration of the acetate via **111**, the zwitterionic species **112** is probably formed, which then collapses to form **109**. A somewhat related transformation of an alkenyl-gold moiety with a benzylic carbocation is involved in the gold(I)-catalyzed intramolecular reaction of benzylic ethers with alkynes [134].

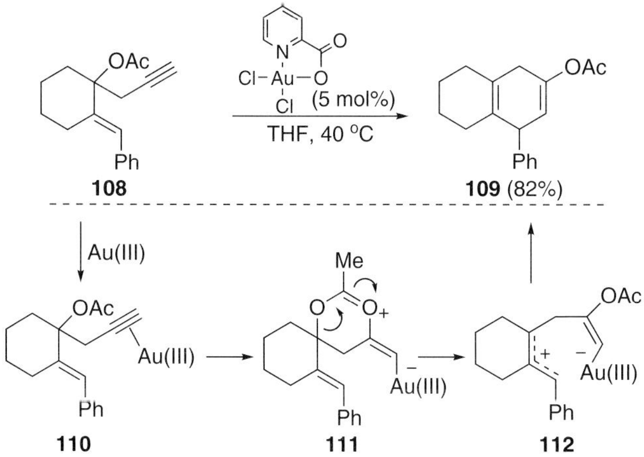

**Scheme 13.25** Au(III)-catalyzed 1,3-acyl migration in homopropargyl acetates.

## 13.2.3
### Cycloisomerizations of Hydroxy- and Alkoxy-Substituted Enynes

An interesting method has been developed for the synthesis of carbonyl compounds based on the cyclization of 3-silyloxy-1,5-enynes **113** under mild conditions with gold (I) catalysts in a process that involves a pinacol rearrangement (Scheme 13.26) [135, 136]. Carbonyl compounds such as **114a–d** are probably formed by a 5-endo-dig cyclization of **115** to form **116**. These intermediates may open to form six-membered ring carbocations **117**, which would undergo a pinacol rearrangement to form intermediates of type **118**. As shown in other cases, a final protodemetallation gives **114**. A direct transformation of **116** to **118** could also be conceived. When the reaction is carried out in the presence of 1 equiv of N-iodosuccinimide, the corresponding iodoalkenes can be obtained in moderate yields by a final iododemetallation, which corresponds to a an overall iodocyclization catalyzed by Au(I) [135]. Pinacol rearrangements have been also observed in reactions of cis-4,6-dien-1-yn-ols catalyzed by Au(III) [194b].

A similar cyclization/pinacol rearrangement takes place in the cyclization of substrates of type **119** to **120** (Scheme 13.27) [137]. Interestingly, a different bicyclic product **121**, resulting from a heterocyclization followed by a [3,3]-sigmatropic rearrangement via **122**, is obtained as the major compound when the reaction is performed with a complex bearing an electron-deficient phosphine. Pinacol-type reactions are also involved in the Cu(I)-catalyzed reactions of propargylic alcohols [138].

The Meyer–Schuster rearrangement (isomerization of propargylic alcohols to α, β-unsaturated carbonyl compounds) can be efficiently promoted by $AuCl_3$ as the catalyst in the presence of EtOH [139, 140]. Related transformations take place from propargylic carboxylates by using Au(I) catalysts [141, 142]. On the other hand, cyclopropanols and cyclobutanols **123** undergo ring expansion to form cyclobutanones and cyclopentanones **124** with Au(I) catalysts (Scheme 13.28) [143, 144]. A 1,2-alkyl shift is also involved in an interesting synthesis of 3(2H)-furanones with $AuCl_3$ or $PtCl_2$ [145, 146].

## 13.3
### Addition of Nucleophiles to 1,n-Enynes

## 13.3.1
### Hydroxy- and Alkoxycyclizations of 1,6- and 1,7-Enynes

The Pt(II)-catalyzed addition of water or alcohols to 1,6-enynes is a very general reaction [35, 147–149]. Mechanistically, this reaction is identical to that catalyzed by Pd(II), which shows a more limited scope [35e, 147]. Ru(II) complexes have also been used for the hydroxycyclization of 1,6-enynes [150].

1,6-Enynes also react stereospecifically with alcohols or water in the presence of Au(I) catalysts under milder conditions than with other metal catalysts

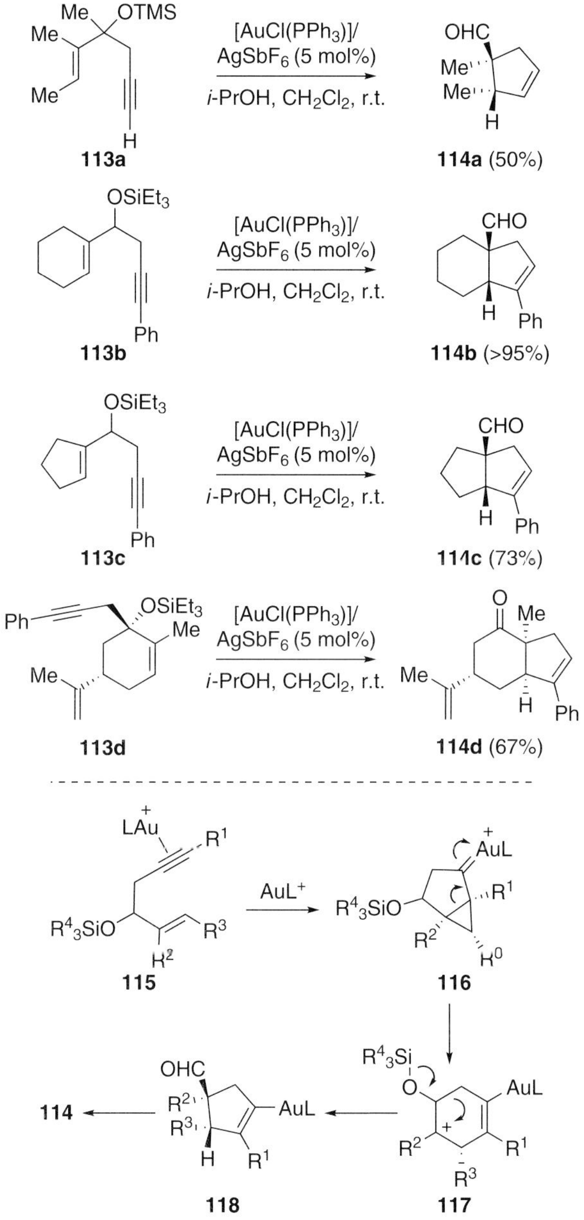

**Scheme 13.26** Tandem Au(I)-catalyzed cyclization/pinacol rearrangement.

(Scheme 13.29) [43, 44]. This reaction can be performed inter- or intramolecularly. In the latter case, enynes bearing hydroxy groups, such as **20d**, react with Au(I) to give cyclic ethers of type **125d**. Methoxycyclization of substrate **20b** with a chiral gold catalyst gives (−)-**125b** with good enantioselectivity, although the reaction was rather

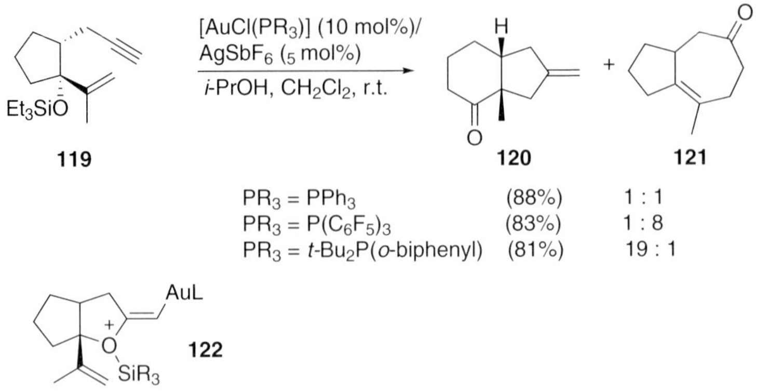

**Scheme 13.27** Ligands effects on the selectivity in the cyclization of substrate **119**.

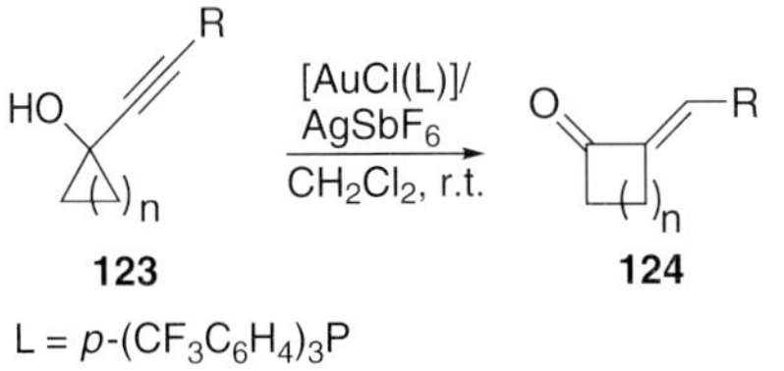

| PR₃ | | |
|---|---|---|
| PR₃ = PPh₃ | (88%) | 1 : 1 |
| PR₃ = P(C₆F₅)₃ | (83%) | 1 : 8 |
| PR₃ = t-Bu₂P(o-biphenyl) | (81%) | 19 : 1 |

**Scheme 13.28** Au(I)-catalyzed ring expansion access to cyclobutanones and cyclopentanones.

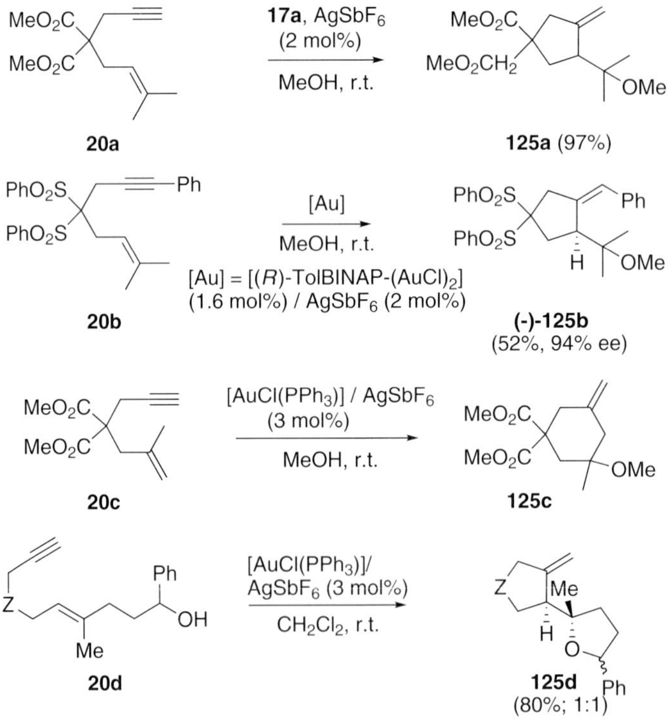

L = p-(CF₃C₆H₄)₃P

**Scheme 13.29** Au(I)-catalyzed hydroxy- and alkoxycyclizations of 1,6-enynes.

slow (reaction time = 7 days) [151]. Terminally unsubstituted enynes led to lower enantioselectivities (30–49% ee) with the same catalyst.

Although similar results were obtained from catalysts generated *in situ* from [AuMe(PPh$_3$)] and a protic acid or [AuCl(PPh$_3$)] and AgSbF$_6$, the catalysts of choice for the hydroxy- and alkoxycyclizations of 1,6-enynes are those bearing bulky biphenyl phosphines **17a–f** [16, 55]. Similar results can be obtained with NHC-Au(I) [63] or Au (III) complexes as catalysts [14c, 152]. The hydroxy- and alkoxycyclizations of 1,7-enynes take place similarly 184].

The intramolecular reaction of carboxylic acids to enynes gives rise to lactones in a process that is similar to that shown in the formation of **125d** from **20d** (Scheme 13.29) [153]. These cyclizations exhibit the characteristics associated with cationic polyene cyclization reactions, which strongly suggests that cationic species are involved in these processes (see Section 13.4.1).

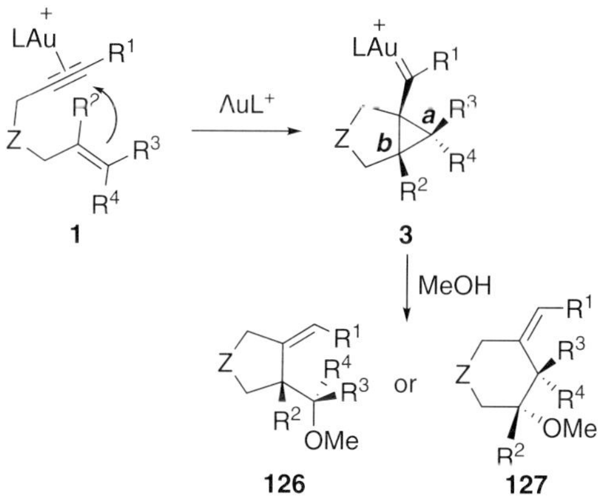

**Scheme 13.30** Exo-trig and endo-trig cyclization products formed by the attack of an external nucleophile in Au(I)-catalyzed cyclizations of 1,6-enynes.

Formation of products of exo-trig (**125a**, **125b**, and **125d**) and endo-trig cyclization (**125c**) shows that intermediates **3** can react at carbons **a** or **b** with ROH nucleophiles to form products **126** or **127** (Scheme 13.30), which is similar to that found for Pt (II) [14, 154].

### 13.3.2
### Hydroxy- and Alkoxycyclizations of 1,5-Enynes

1,5-Enynes such as **128a–c** also react in a similar manner with MeOH or H$_2$O in the presence of Au(I) to give adducts **129a–c** (Scheme 13.31) [155]. Whereas substrates **129a–c** are formed by cleavage of bond *a* in intermediates **130**, formation of products of type **132** by cleavage of bond *b* in **130** is also possible, as shown in the formation in the intramolecular hydroxycyclizations of 1,5-enynes **128d,e** to give six-membered ring derivatives **129d–e** [156].

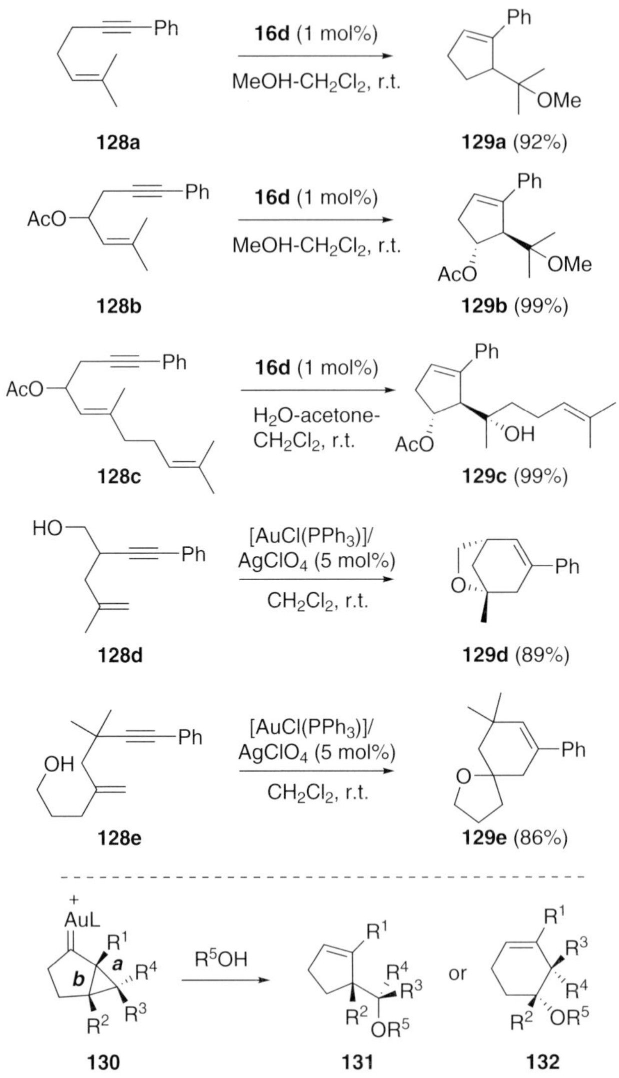

**Scheme 13.31** Au(I)-catalyzed hydroxy- and alkoxycyclizations of 1,5-enynes.

Allyl propargyl ethers **6** (oxygen-tethered 1,5-enynes) also react intermolecularly with alcohols or water to give six-membered ring acetals or hemiacetals [97]. Allyl silyl alkynes **133** react similarly with Au(I) catalysts in the presence of alcohols to give alkenylsilanes **134** and/or products **135** (Scheme 13.32) [157, 158]. These products are formed by an endocyclic cleavage of the initial intermediate **136** by attack of the nucleophile at the silicon atom or at the cyclopropane carbon. Alternatively, ring expansion of **136** to **137**, followed by ring opening or nucleophilic attack would form **134** and **135**.

**Scheme 13.32** Au(I)-catalyzed reaction of allyl silyl alkynes.

### 13.3.3
### Amination of Enynes

The intramolecular amination of 1,5-enynes **138a–c** proceeds similarly to the intramolecular hydroxycyclizations of related substrates to give products **139a–c** (Scheme 13.33) [156]. The intermolecular reaction of 1,6-enynes with a series of carbamates RO$_2$CNH$_2$ or anilines ArNH$_2$ substituted at the para or ortho positions

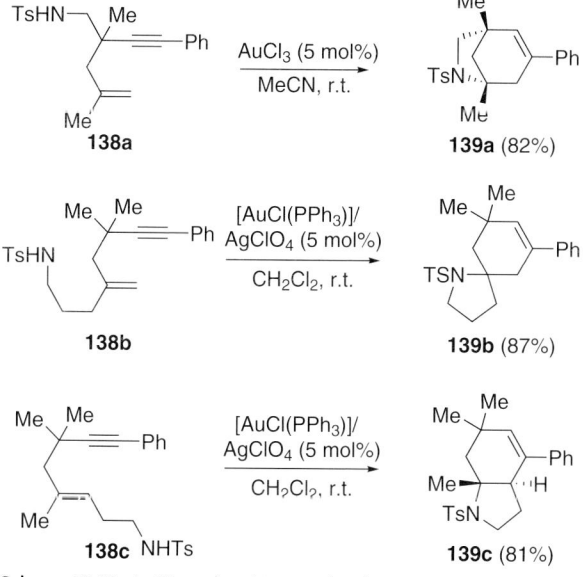

**Scheme 13.33** Au(I)-catalyzed intramolecular amination of 1,5-enynes.

with strong electron-withdrawing groups has also been reported by a process that is mechanistically similar to the gold-catalyzed hydroxy- or alkoxycyclization of 1,6-enynes [159].

### 13.3.4
### Inter- and Intramolecular Additions of Carbon Nucleophiles to 1,6-Enynes

Electron-rich aromatic and heteroaromatic compounds add to 1,6-enynes **20** in the presence of Au(I) catalysts (Scheme 13.34) [58, 160, 161]. This reaction leads stereospecifically to adducts of type **140** by nucleophilic addition on intermediates of type **3** by a process that is mechanistically related to that of the hydroxy- and alkoxycyclizations of 1,5-enynes. Interestingly, addition of the nucleophile can also take place in a different fashion to give adducts of type **141** [161]. Formation of products in which the cyclopropane is retained can be explained by attack of the carbon nucleophile at the metal carbene carbon to form intermediates such as **142**. Attack at the gold carbene carbon has also been observed in gold-catalyzed reactions of propargyl acetates with indole [106].

Additions of allylsilanes and 1,3-dicarbonyl compounds to 1,6-enyns is also possible [58]. Thus, for example, dibenzoylmethane reacts with enyne **20g** to give adducts **143** and **144** (Scheme 13.35). Interestingly, site selectivity was strongly dependent on the ligand on the gold. Using cationic gold complex **18b**, with a bulky phosphite as ligand, the major product was **143**, whereas **144** was formed almost exclusively with the NHC-gold complex **19e**.

Enynes **145** substituted at the alkyne with an aryl group react with a variety of gold catalysts to provide products **146** resulting from a formal intramolecular [4 + 2] cycloaddition occurring at an unusually low temperature (Scheme 13.36) [49, 57]. This reaction tolerates both electron-releasing and electron-withdrawing substituents at several positions of the arene. According to the experimental results and DFT calculations, the mechanism of the [4 + 2] cycloaddition of arylenynes proceeds by opening of **147** to form **148**, which reacts by a Friedel–Crafts-type reaction to form **149**, which aromatizes to form the alkenyl gold intermediate **150**. It is important to note that **147** and open carbocation **148** were found to be two different minima and not resonance forms and displayed very different bond lengths and angles. A final protodemetallation then forms products **146**. The rate-determining step is the attack of the alkene on the alkyne coordinated to Au(I) (**147** to **148**) and not the electrophilic aromatic substitution, which explains the relative insensitivity to the presence of electron-withdrawing substituents at the arene. Related cyclizations of allenes with alkynes [162, 163] and diynes [164, 165] have been described.

Products of type **152** and cyclobutenes **153** are also formed in a few cases in the cyclizations of arylenynes (Scheme 13.37) [49, 57]. Formation of cycloadducts **152** shows, that in competition with the 5-exo-dig cyclization via **148**, 6-endo-dig cyclization via **157** can also take place. Cyclobutenes **153** can also arise from intermediates **157** via benzylic carbocations **158**. Alternatively, a rotation of *anti*-

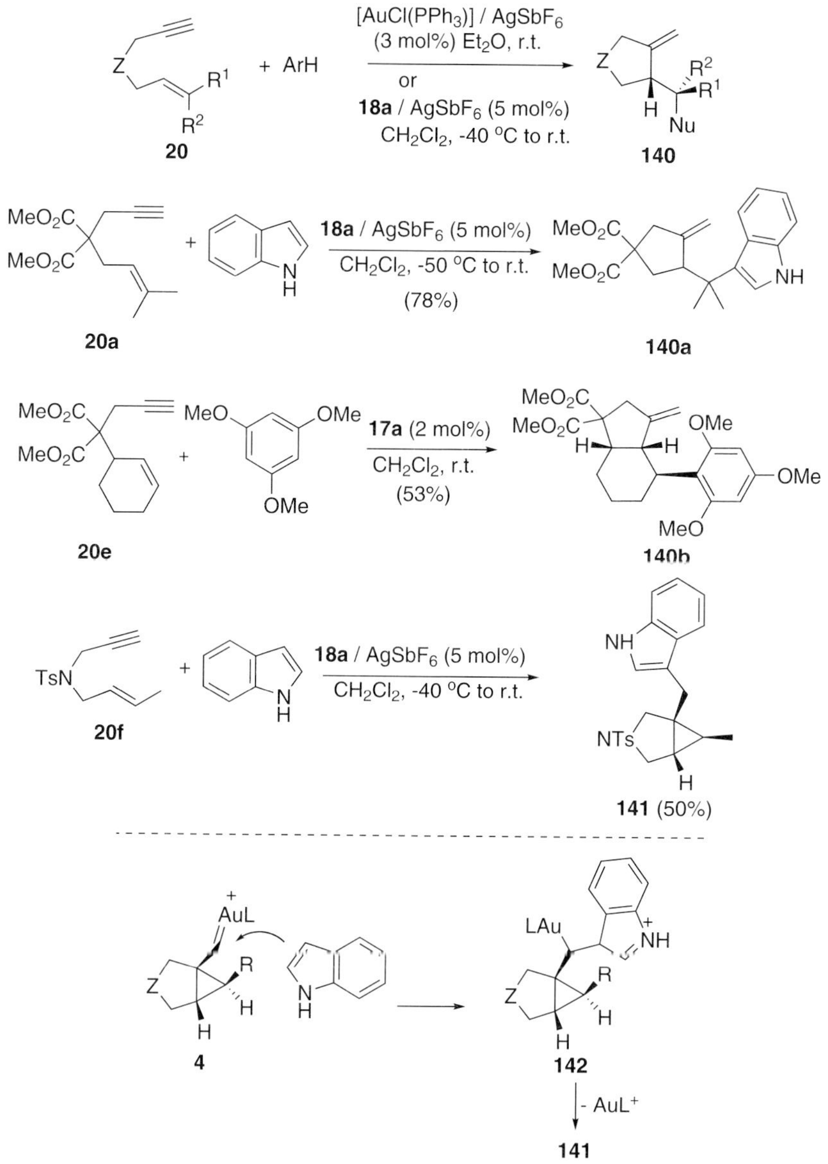

**Scheme 13.34** Addition of aromatic nucleophiles to 1,6-enynes catalyzed by Au(I) complexes.

cyclopropyl gold carbene **155** can provide *syn*-cyclopropyl gold carbene **156**, which then undergoes ring expansion to form **158**. In contrast with that found for the corresponding cyclopropyl gold(I) carbene formed from (*E*)-oct-6-en-1-yne, for which the anti to syn rotation requires 24.7 kcal mol$^{-1}$ [173], the phenyl-substituted system requires only 8.6 kcal mol$^{-1}$ for the anti to syn isomerization.

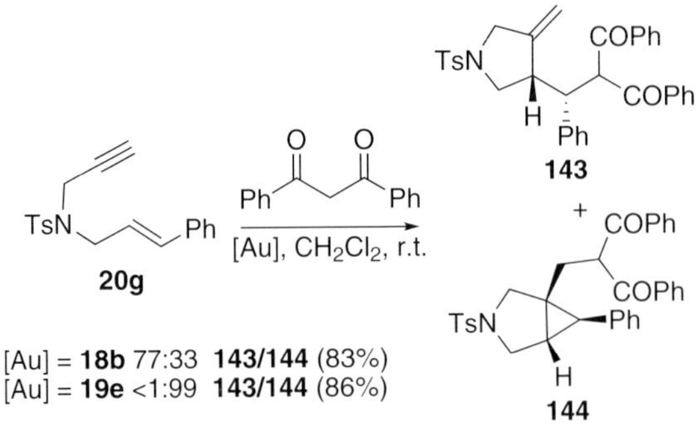

**Scheme 13.35** Au(I)-catalyzed addition of 1,3-dicarbonyl nucleophiles to 1,6-enynes.

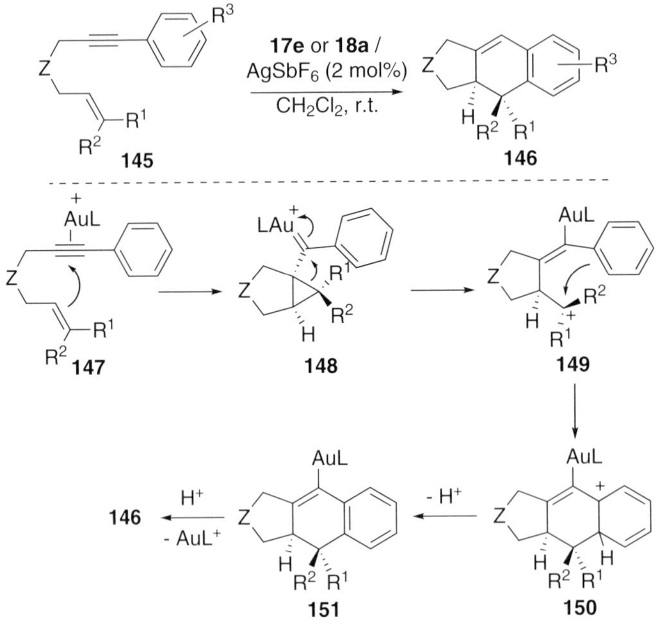

**Scheme 13.36** Au(I)-catalyzed [4 + 2] cycloaddition of aryl substituted enynes.

Cyclobutenes of type **153** are obtained in a more general way in the cyclization of enynes **159** using Pt(II) catalysts (Scheme 13.38) [20, 166]. Interestingly, formation of products of 6-endo-dig cyclization is the major pathway in the Pt(II)-catalyzed cycloaddition of arylalkynes with enesulfonamides or enamines (**161** to **162**) [167, 168].

Dienynes **163** also cyclize by a similar mechanism to give hydrindanes **164** as the major products (Scheme 13.39) [49, 57]. This reaction presumably proceeds through intermediates such as **165**, which undergo ring expansion in a process that is reminiscent of the Nazarov cyclization to form the allyl cation **166**, which is followed by loss of a proton followed by proto-demetallation to give **164**.

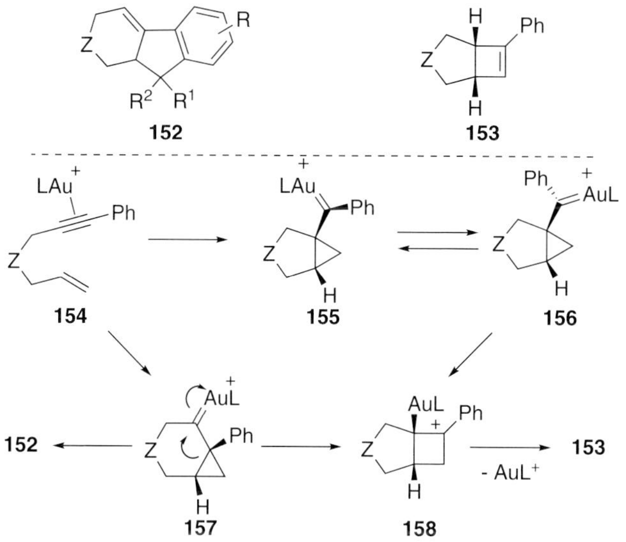

**Scheme 13.37** 5-exo-dig and 6-endo-dig pathways for the Au(I)-catalyzed [4 + 2] cycloaddition.

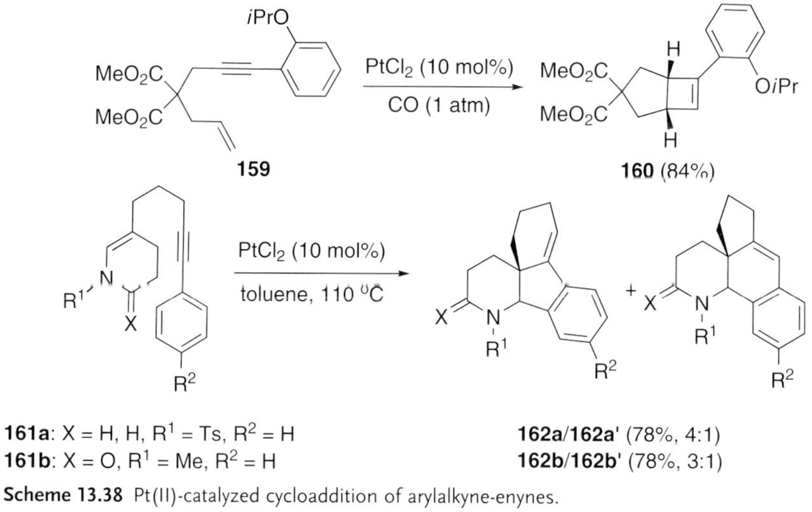

**161a**: X = H, H, R$^1$ = Ts, R$^2$ = H
**161b**: X = O, R$^1$ = Me, R$^2$ = H

**162a/162a'** (78%, 4:1)
**162b/162b'** (78%, 3:1)

**Scheme 13.38** Pt(II)-catalyzed cycloaddition of arylalkyne-enynes.

1,3-Dien-8-ynes **167** undergo formal intramolecular Diels–Alder reactions catalyzed by Au(I) to give adducts **168** (Scheme 13.40) [169]. This reaction was proposed to take place by evolution of the intermediate vinylcyclopropyl gold carbenes by a metalla-Cope rearrangement or by formation of a six-membered ring cation. Indeed, vinylcyclopropyl gold carbenes could be trapped intramolecularly by a hydroxy group at the diene chain or by a phenyl at the alkyne in a [4 + 2] cycloaddition process as in

**Scheme 13.39** Au(I)-catalyzed cyclization of dienynes.

**Scheme 13.40** Au(I) and Cu(I)-catalyzed formal Diels–Alder reaction of dienynes.

Scheme 13.36. The Diels–Alder reaction of substrates **167** (R = H) could also be efficiently catalyzed by Cu(I) by a different mechanism involving a copper acetylide [169].

### 13.3.5
### Other Inter- and Intramolecular Nucleophile Additions to 1,6-Enynes

Cyclizations of enynes **169** bearing carbonyl groups with Au(I) catalysts provide tricyclic compounds **170** along with ketones **171**, as minor products (Scheme 13.41) [51]. In these cyclizations, the carbonyl group acts as an internal nucleophile, as shown in **172**, to form oxonium **173**, which undergoes an intramolecular Prins reaction to give **174**. Elimination of the metal fragment forms tricyclic compounds **170**. Formation of rearranged ketones can be explained by the alternative elimination of the metal with fragmentation of the seven-membered ring via **175**.

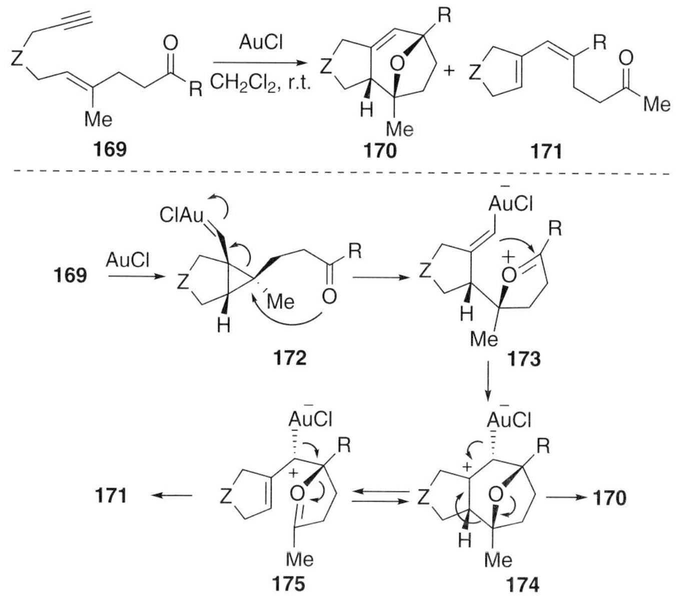

**Scheme 13.41** Au(I)-catalyzed intramolecular addition of carbonyl groups.

Carbonyl compounds can also act as the nucleophiles in an intermolecular process with 1,6-enynes [170]. Thus, the gold(I)-catalyzed reaction between enynes **20h** and **20i** with aldehydes or acetone gives, stereoselectively, tricyclic compounds **176a–c** (Scheme 13.42). The transformation is mechanistically intriguing as it proceeds by a rearrangement of the initially formed cyclopropyl gold carbene **3** to **177** (the intermediate in the double cleavage mechanism, see Scheme 13.48, Section 13.4.1), which is then trapped by the carbonyl compound to form **178**. This transformation is similar to that observed in the trapping of the same rearranged gold carbene **177** with styrene to form cyclopropanes **236** (Scheme 13.63, Section 13.4.6) [56]. It was suggested that a six-membered ring cation **179** could then be formed by attack of the alkene on the oxonium cation, which would be followed by trapping of the cation by the alkyl-gold. Formation of the 1,3-dipole **180** from **179**, followed by an intramolecular 1,3-dipolar cycloaddition, was also considered as an alternative.

A Prins cyclization is also involved in the reaction of cyclopropylenynes **181** with AuCl or cationic Au(I) complexes to give tricyclic derivatives **182/182′** with a octahydrocyclobuta[*a*]pentalene skeleton (Scheme 13.43) [51]. Formation of stereosiomers **182′** with more reactive cationic gold(I) catalysts was unexpected and suggests that two different pathways are competing in this process. Accordingly, complex **183** forms the cyclopropyl metal carbene **184**, which undergoes ring expansion to form **185**. The alkenyl-gold of **185** could undergo a Prins reaction with the oxonium to form the carbocation **186**, which upon demetallation forms tricycles **182**. The concerted pathway (**184** to **185**) is favored for AuCl as catalyst, whereas

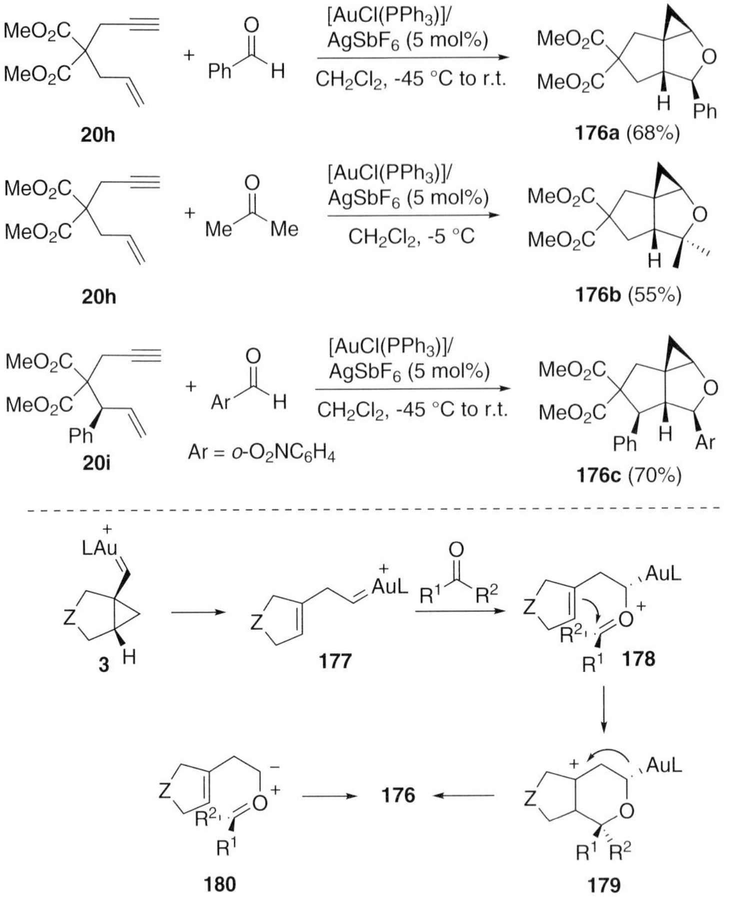

**Scheme 13.42** Au(I)-catalyzed intermolecular reaction of enynes with aldehydes.

cationic Au(I) complexes apparently favor a non-concerted reaction via the cyclopropyl-stabilized cation **187**, which undergoes a non-stereospecific ring expansion to give mixtures of **185/185′**. Cyclobutanones were also formed as minor side products in this reaction [51].

### 13.3.6
### Inter- and Intramolecular Additions of Carbon Nucleophiles to 1,5-Enynes

Trapping by aryl groups has also been reported in the Hg(II)-catalyzed cyclization of 1,5-enyne **188** that leads to **189** in excellent yield by an endo-cyclization (Scheme 13.44) in a transformation that mechanistically is probably related to that of Scheme 13.36 [171].

The intermolecular reaction of electron-rich arenes with 1,5-enynes proceeds similarly to that of the hydroxy- and alkoxycyclizations of 1,5-enynes (Section 13.3.2)

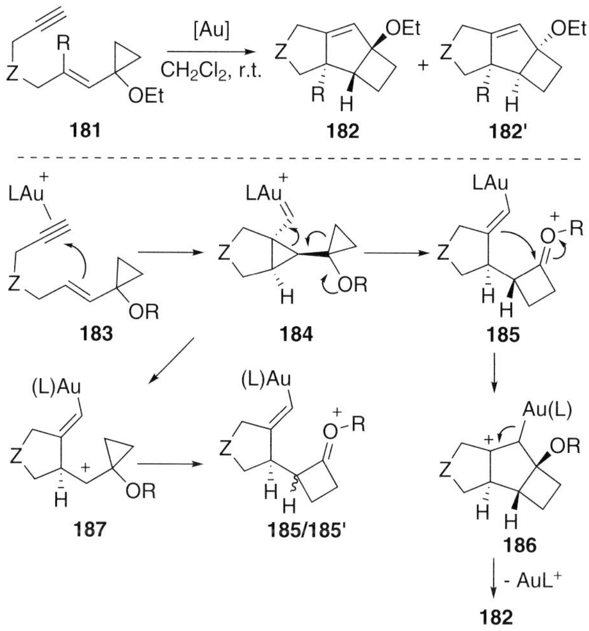

**Scheme 13.43** Au(I)-catalyzed cyclization of cyclopropylenynes.

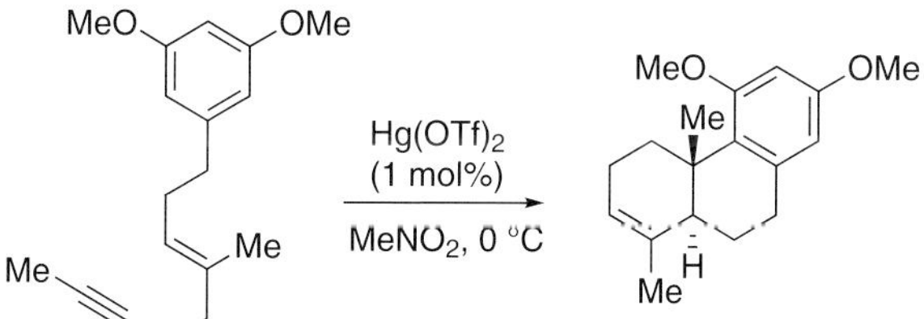

**Scheme 13.44** Hg(II)-catalyzed cyclization of 1,5-arylenynes.

to give adducts of type **191a–c** by using the cationic gold complex **17e** as catalyst (Scheme 13.45) [58].

## 13.4
## Skeletal Rearrangement of 1,*n*-Enynes

### 13.4.1
### Single- and Double-Cleavage Skeletal Rearrangement of 1,6-Enynes

Skeletal rearrangements are probably the most emblematic transformations of enynes catalyzed by electrophilic metal salts and complexes [3, 4]. For 1,6-enynes,

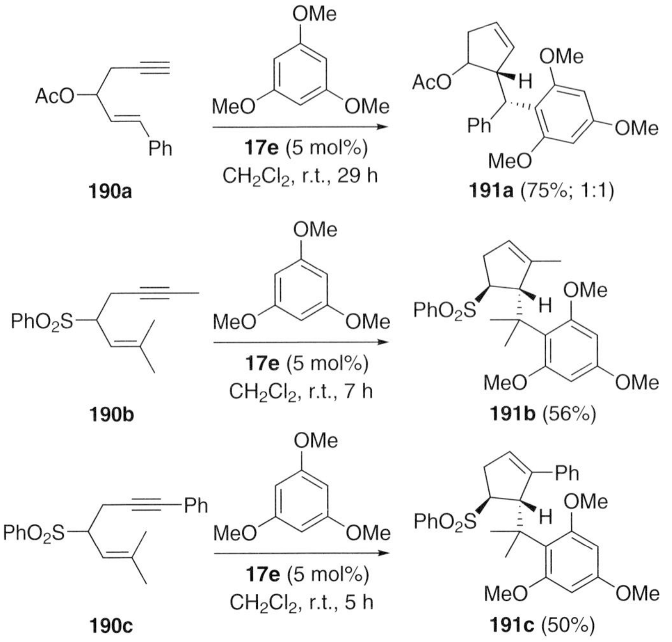

**Scheme 13.45** Au(I)-catalyzed trapping of 1,5-enynes with aromatic nucleophiles.

two main types of products **5** (single exo-cleavage) and **6** (double exo-cleavage) were initially identified (Scheme 13.46, see also Scheme 13.1) [9–16]. A third type of product **192** (single endo-cleavage) was found using Au(I) [43, 44, 172], InCl$_3$ [10], Fe (III) [44], or Ru(II) [150] as the catalysts.

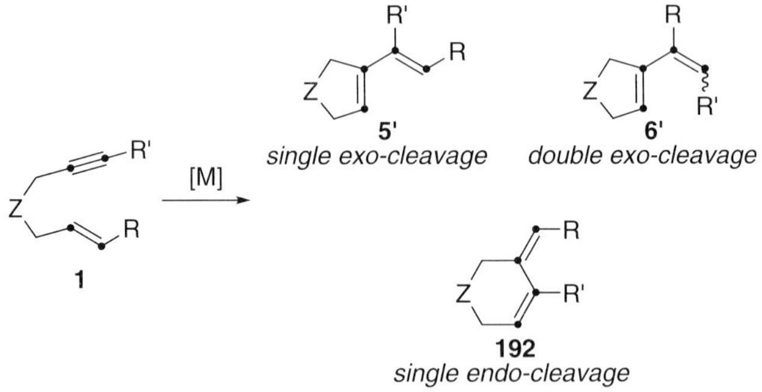

**Scheme 13.46** Dienes obtained by skeletal rearrangement of 1,6-enynes.

The single exo-cleavage rearrangement leading to dienes **5** is superficially similar to the metathesis of enynes [1c, 23, 24], although these reactions are very different mechanistically [16]. Gold(I) catalysts allow one to perform the skeletal rearrangement under the mildest conditions (Scheme 13.47) [3c, 43, 44]. Formation of **5a**

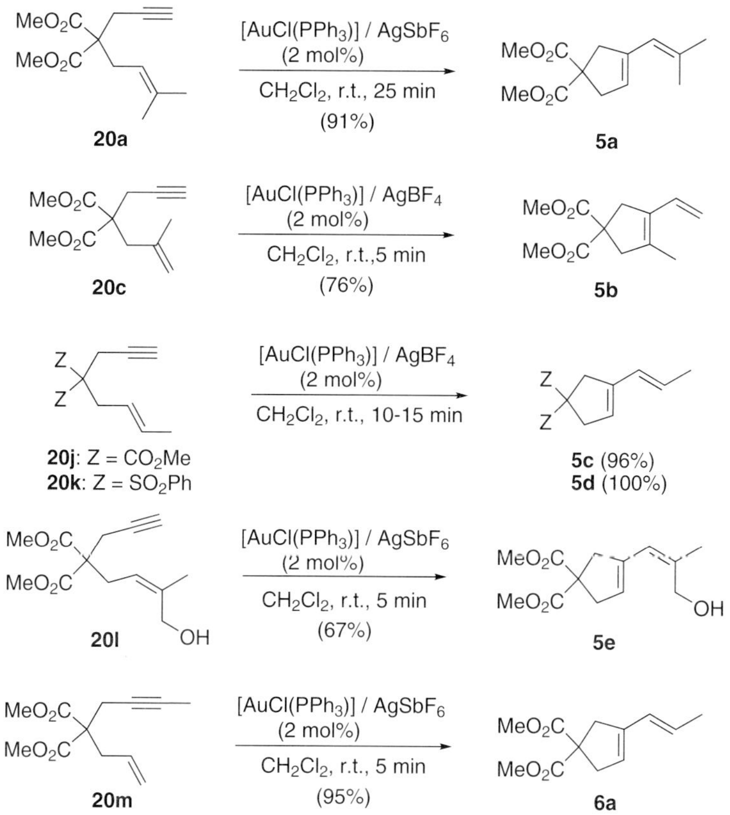

**Scheme 13.47** Au(I)-catalyzed skeletal rearrangement of 1,6-enynes.

from **20a** can be carried out even at −40 to −60 °C using catalysts **17e** or **17f** [173]. Dienes **5a–d** are products of single-cleavage rearrangements, in which only the alkene is cleaved. On the other hand, enyne **20m**, with a methyl substituent at the alkyne, reacts by a double cleavage rearrangement to afford exclusively diene **6a**, in which both the alkene and the alkyne have suffered cleavage [173]. Similar transformations have been carried out with other gold(I) catalysts [55, 63, 174].

Formation of products of single cleavage in the metal-catalyzed reactions of enynes had been proposed to take place by conrotatory ring opening of intermediate cyclobutenes. However, experimental and theoretical calculations do not support that proposal [173]. Thus, a detailed study of the rearrangement of enyne **20a** to give diene **5a** allowed determination of the activation parameters. This reaction proceeds with low enthalpies of activation ($\Delta H^{\ddagger} = 6.2$ and $3.7$ kcal mol$^{-1}$ for **17e** and **17f**, respectively) and negative entropies of activation ($\Delta S^{\ddagger} = -50.3$ and $-60.6$ cal mol$^{-1}$ K$^{-1}$ for **17e** and **17f**, respectively). These results establish a very low activation energy for the hypothetical conrotatory opening of a cyclobutene, which should be a fast process at temperatures as low as −63 °C. This is not consistent with theoretical data for the ring opening of cyclobutenes related to bicyclo[3.2.0]hept-5-ene.

Overall, the mechanism for the skeletal rearrangement is consistent with the evolution of the initial *anti*-cyclopropyl gold carbenes **3**, formed in the 5-exo-dig cyclization, to give carbocations **193**, which then undergo metal-elimination to give dienes **5** (single cleavage) (Scheme 13.48). For the double cleavage rearrangement, intermediates **3′** can suffer a diotropic rearrangement [175, 176] to give new carbenes **194**, which lose an α-hydrogen and then undergo protodemetallation to form dienes **6** (double cleavage). Intermediates **194** can also be formed by a carbocationic 1,2-shift of the cyclic alkenyl group in **193** [173]. The double exo-cleavage skeletal rearrangement often leads to dienes **6** with predominant [172] Z configuration. Interestingly, by using Rh(I) catalysts, Z-configured enynes have been exclusively obtained by the double exo-rearrangement [177].

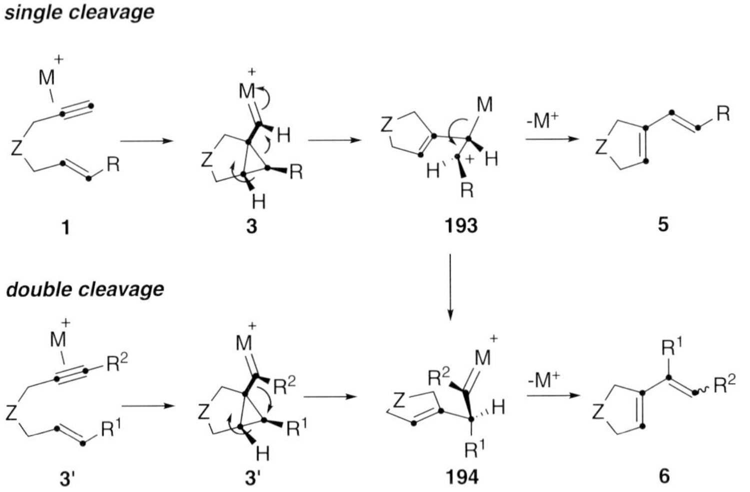

**Scheme 13.48** Single and double cleavage pathways proceed through cyclopropyl carbene intermediates.

Evidence for the involvement of cyclopropyl gold carbenes **3** and **194** has been obtained in the intra- and intermolecular cyclopropanations (see Section 13.4.6). Further evidence for the involvement of these species was provided in the formation of the corresponding aldehydes by performing the cyclizations with Au(I) catalysts in the presence of $Ph_2SO$ as a mild oxidant [178].

Many of the intermediates in gold-catalyzed reactions of enynes are conveniently drawn as gold carbenes, since backbonding in Au(I) has been shown not to be insignificant [5, 6, 179]. However, it is important to recall that no intermediate has been spectroscopically characterized in the reaction of enynes [180]. Interestingly, a gold carbene has been generated recently in the gas phase showing the reactivity expected for a metal carbene [181].

Although intermediates **3** are conventionally drawn as cyclopropyl gold carbenes, DFT calculations show highly distorted structures with a rather short C—C bond connecting the carbene and the cyclopropane consistent with a substantial double bond character [172, 173]. The structures of **3a,b** are actually consistent with a

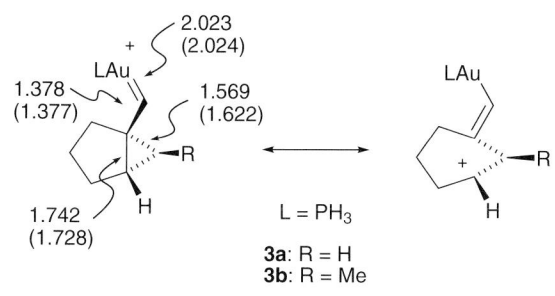

**Figure 13.2** Calculated bond distances (Å) for **3a,b** (values in parentheses are for **3b**). DFT calculations at the B3LYP/6-31G(d) (C,H,P), LANL2DZ (Au) level.

delocalized cyclopropylmethyl/cyclobutyl/homoallyl carbocation [172, 182] stabilized by gold (Figure 13.2).

The proposal that delocalized cyclopropylmethyl/cyclobutyl/homoallyl carbocations are the actual intermediates in these reactions [153], is supported by the recent finding that *trans*-1,6-enynes bearing electron-donating R groups at the alkene give rise to *cis*-dienes **5** in a reaction that is not stereospecific [183]. Thus, enynes **20n–q** with trans double bonds led to dienes **Z-5f–i** (Scheme 13.49). Enynes with cis double bonds were shown to give also dienes **Z-5**. This is in contrast with that previously found with other substrates for this single cleavage mechanism [3, 4].

These results with enynes bearing electron-donating groups at the alkene are consistent with the formation of open carbocations **3a′** that undergo facile bond rotation to **3b′** prior to the rearrangement (Scheme 13.50). The origin of the high cis-selectivity in the rearrangement of these enynes is still not clear. Nevertheless, the involvement of an equilibrium between species **3a′** and **3b′** was strongly suggested by the finding that this type of enynes undergoes non-stereospecific methoxycyclization and [4 + 2] cycloaddition reactions [183].

## 13.4.2
### Skeletal Rearrangement of 1,7-Enynes

Only a few examples of skeletal rearrangement of 1,7-enynes have been reported by using [RuCl(CO)$_2$]$_2$ [10a], PtCl$_2$ [10b,f], [IrCl(CO)$_3$]$_n$ [10d], GaCl$_3$ [10e, 185, 186], and InCl$_3$ [10g]. With the exception of rearrangements catalyzed by GaCl$_3$, which can proceed with 10–20 mol catalyst at 23–40 °C, higher temperatures were required with all other catalysts. Gold(I) complexes are again the best catalysts to effect this skeletal rearrangement [82, 184]. Thus, a variety of enynes **195** differently substituted at the alkene react at room temperature with the cationic catalyst [Au(PPh$_3$)(MeCN)]PF$_6$ (**16**) to provide dienes **196** in good yields at room temperature (Scheme 13.51) [184].

Although, in general, less reactive than Au(I) cationic catalysts, GaCl$_3$ is also an active catalyst in skeletal rearrangements [8e]. This catalyst was used in the rearrangement of enyne **197** to give **198**, a key intermediate in the total synthesis of synthesis of salviasperanol (**199**) and related compounds (Scheme 13.52) [185, 186].

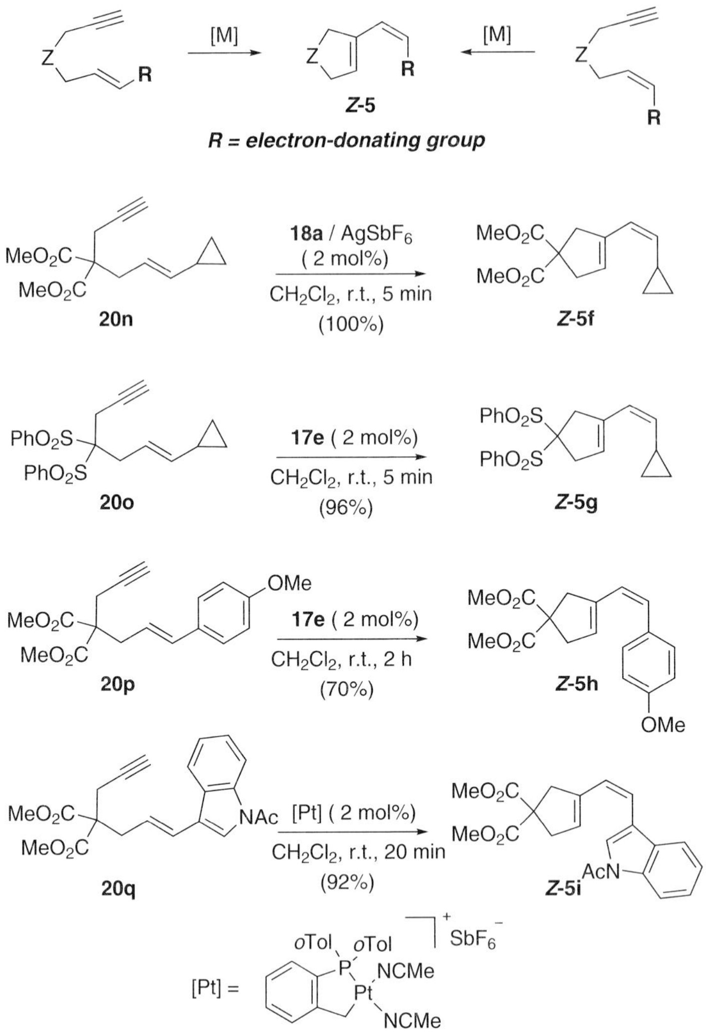

**Scheme 13.49** *Cis*-dienes formed in Au(I)- and Pt(II)-catalyzed cyclization of *trans*-enynes bearing electron-donating groups.

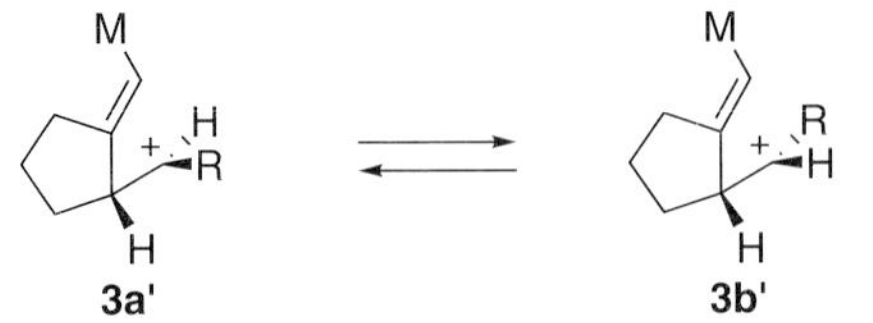

**Scheme 13.50** Interconversion between open carbocations.

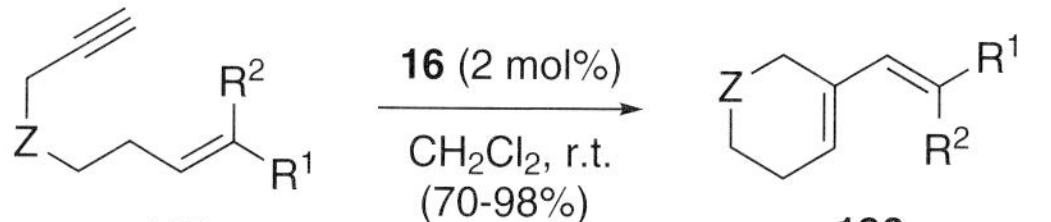

## 13.4 Skeletal Rearrangement of 1,n-Enynes

**Scheme 13.51** Au(I)-catalyzed skeletal rearrangement of 1,7-enynes.

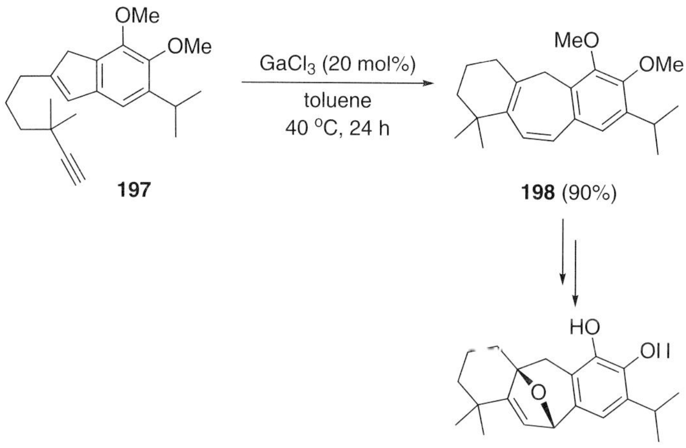

**Scheme 13.52** Ga(III)-catalyzed cyclization of 1,7-enynes applied to the total synthesis of salviasperanol (**199**).

### 13.4.3
### Formation of Cyclobutenes

Cyclobutenes resulting from a formal [2 + 2] cycloaddition have been obtained in certain reactions of 1,7-enynes catalyzed by electrophilic transition metals [9, 10e, 11b, 187]. These bicyclo[4.2.0]oct-6-ene are quite stable compounds. Thus, for example, the gold-catalyzed reaction of 1,7-enynes **200a,b** gives stereoselectively tricyclic compounds **210a,b** (Scheme 13.53), which do not undergo ring opening at 120–150 °C to form 1,3-dienes [173].

Cyclobutenes of type **203** are also obtained in the Ru(II)-catalyzed reaction of enynes such as **202** (Scheme 13.54) [187].

The PtBr$_2$-catalyzed consecutive enyne metathesis–aromatization of 1-(1-methoxybut-3-enyl)-2-(1-alkynyl)benzenes **204** leads to naphthalenes **205** (Scheme 13.55). In this reaction, PtBr$_2$ catalyzes the enyne metathesis and, as a Lewis acid, assists in the final elimination of MeOH [14]).

Other types of cyclobutenes have also been obtained from 1,6-enynes (see Section 13.3.4). However, only one example of a stable [3.3.0]hept-5-ene has been reported in a metal-catalyzed reaction from a 1,6-enyne [67, 188, 189]. A 3-oxabicyclo[3.2.0]hept-5-ene was proposed to be an intermediate in a Pt(IV)-catalyzed cyclization, but it could not be isolated [17]. Interestingly, following earlier work done with PtCl$_2$

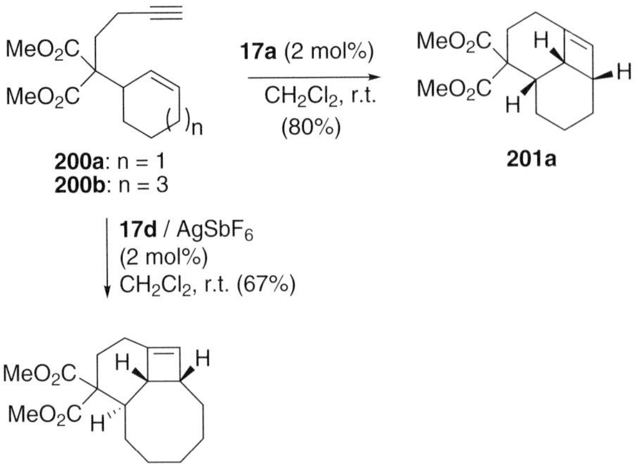


**201b**

**Scheme 13.53** Synthesis of cyclobutenes from 1,7-enynes using Au(I)-catalysts.

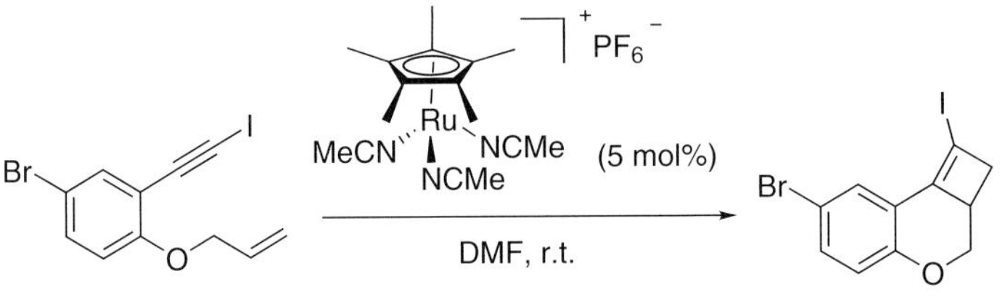


**202** **203** (90%)

**Scheme 13.54** Ru(II)-catalyzed cyclization of 1,7-enynes.

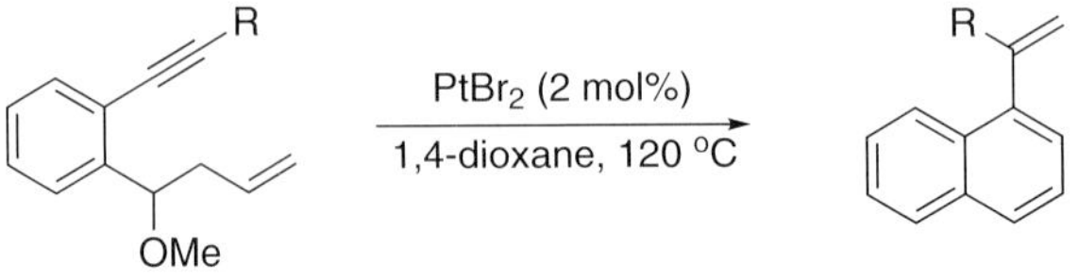


**204a**: R = *n*-Hex
**204b**: R = *Ph*

**205a** (60%)
**205b** (76%)

**Scheme 13.55** Pt(II)-catalyzed enyne metathesis followed by aromatization.

as catalyst [190], it was found that the gold-catalyzed reaction of 1,6-ene-ynamindes **206** affords cyclobutanones **207** (Scheme 13.56) [191], in a process that probably takes place through unstable bicyclic enamines **209**. In the reaction of **206a**, the skeletal rearrangement diene **208** was obtained as a minor product, whereas in the reaction of the same substrate catalyzed by $PtCl_2$, **208** was obtained in 98% yield [190].

No direct pathway for the formation of cyclobutenes from *anti*-cyclopropyl gold or platinum carbenes such as **3** was found by a low energy process [16, 173]. Similar

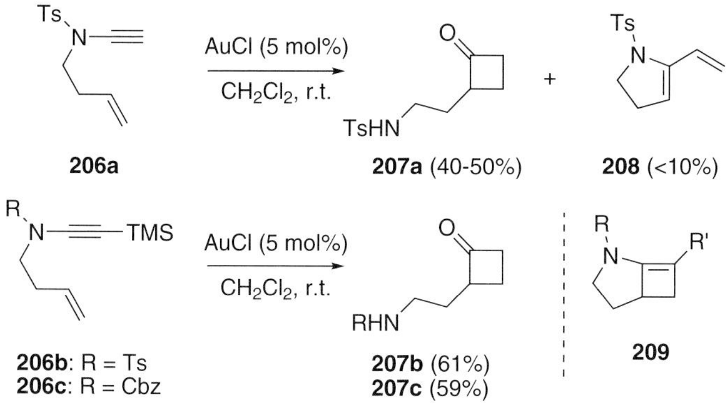

**Scheme 13.56** Au(I)-catalyzed cyclization of 1,6-ene-ynamindes.

results were obtained by the analogous platinum carbenes. DFT calculations carried out on a different system found a relatively high barrier for the formation of a cyclobutene from an *anti*-cyclopropyl platinum carbene (25.43 kcal mol$^{-1}$, for an endothermic process with $\Delta H = 11.48$ kcal mol$^{-1}$) [192, 193]. In contrast, a syn attack of the alkene as shown in **210** can form *syn*-cyclopropyl gold carbene **211** ($E_a = 9.4$ kcal mol$^{-1}$), which then undergoes smooth ring expansion to form cyclobutenes **212** ($E_a = 7.8$ kcal mol$^{-1}$) (Scheme 13.57). The anti to syn isomerization from **3** to **211** requires higher activation energy (*ca.* 25 kcal mol$^{-1}$) [16, 173].

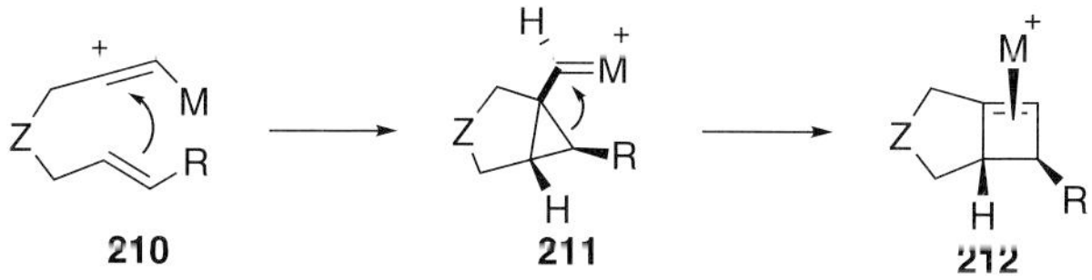

**Scheme 13.57** Cyclobutenes formed from *syn*-cyclopropyl gold carbene intermediates. DFT calculations at the B3LYP/6-31G(d) (C,H,P), LANL2DZ (Au) level.

## 13.4.4
### Endocyclic Rearrangement of 1,6-Enynes

Endo-cyclizations to give products of type **213** were first observed using gold(I) as catalysts (Scheme 13.58) [43, 44, 172]. As shown in the reactions of **27r** and **27v**, this endocyclic rearrangement is stereospecific. Only a few additional examples of this type of rearrangement have been reported from simple 1,6-enynes using FeCl$_3$ [44], InCl$_3$ [10g], or Ru(II) [150] as the catalyst. Endocyclic rearrangements have also been observed in the reaction of *cis*-4,6-dien-1-yl-3-ol derivatives with gold or platinum catalysts [194]. Products of type **213** are obtained in the cyclization of 1,6-enynes

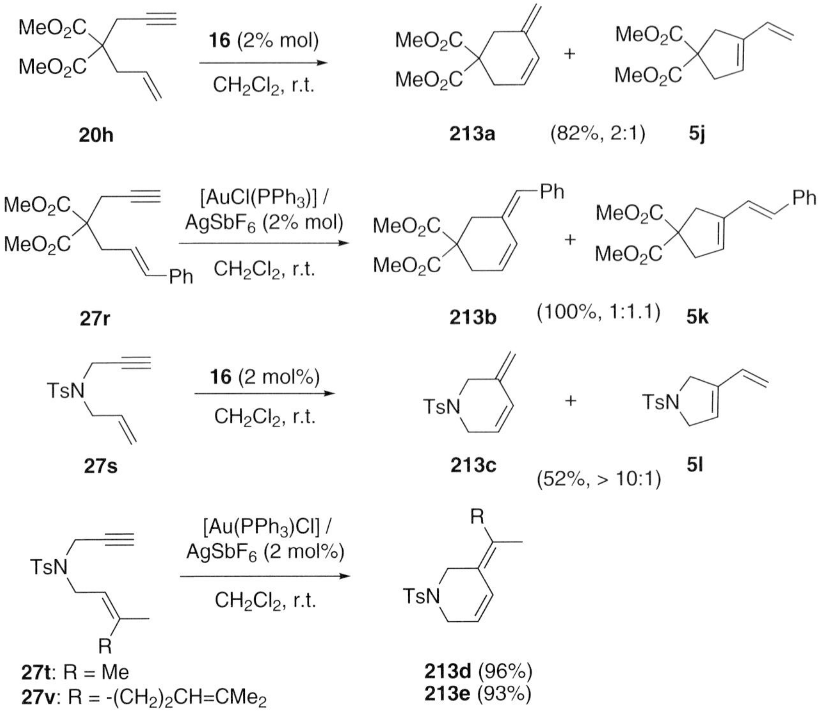

**Scheme 13.58** Examples of Au(I)-catalyzed endocyclic rearrangement of 1,6-enynes.

catalyzed by rhodium(I), although labeling experiments show that in that case the mechanism proceeds through vinylidene intermediates [195].

Labeling experiments are consistent with an intramolecular process in which the terminal carbon of the alkene is attached to C-2 of the alkyne [172]. DFT calculations for the Au(I)-catalyzed process support a mechanism in which cyclopropyl gold(I) carbene **3** rearranges with ring opening to give cation **214** in a moderately endothermic process with an activation energy of only 9.6 kcal mol$^{-1}$ (Scheme 13.59). Metal loss from **214** gives products **213**. Interestingly, the mechanism of this endocyclic rearrangement is just a variation of the single cleavage rearrangement of **3** to give **5** via intermediate **193** (see Scheme 13.48).

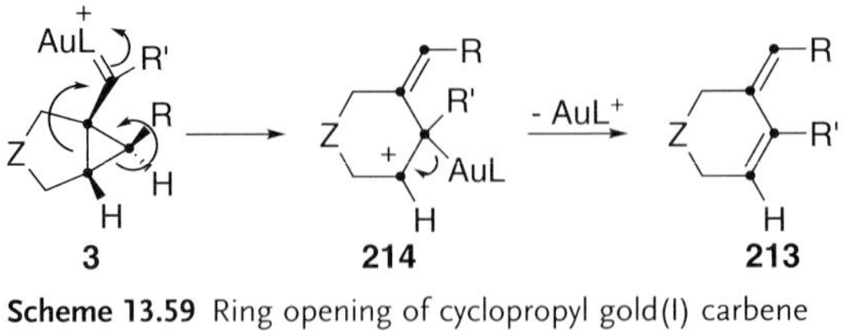

**Scheme 13.59** Ring opening of cyclopropyl gold(I) carbene provides six-membered ring intermediates.

## 13.4.5
### Skeletal Rearrangement of 1,5-Enynes

Products of type **218** have been obtained from 1,5-enynes [73], which can be rationalized as shown in Scheme 13.60 by a skeletal rearrangement. Thus, upon complexation of Au(I) to the alkyne in **215**, an endo-cyclization could occur to form cyclopropyl gold carbene **216**, which might suffer a skeletal rearrangement via **217** to give **218**. This is an example of a single cleavage rearrangement occurring in 1,5-enynes.

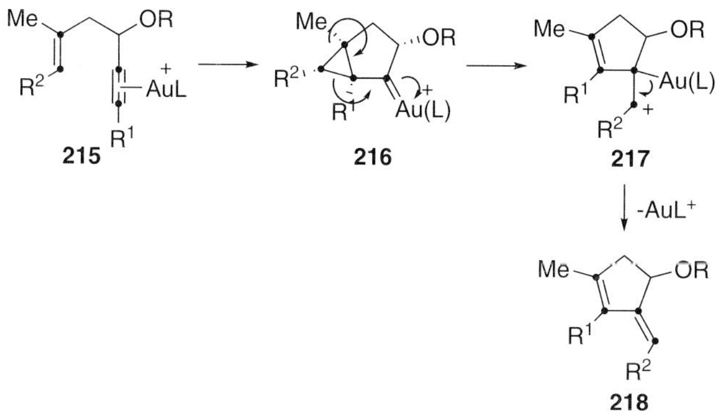

**Scheme 13.60** Proposed mechanism for the Au(I)-catalyzed single cleavage rearrangement of 1,5-enynes.

Siloxy 1,5-enynes like **219a–c** react with AuCl as catalyst to give cyclohexadienes **221a–c** (Scheme 13.61) [196]. The analogous reaction of 1,5-enynes in which the OTIPS group has been replaced by an alkyl or aryl group is better performed with the less electrophilic PtCl$_2$ [196b]. This mechanistically intriguing transformation probably involves a double cleavage rearrangement of 1,5-enynes. Thus, complex **221** (X = OTIPS) could form gold carbene **222**, which then undergoes ring expansion in a diotropic rearrangement to give **223** by a process reminiscent of that found for **3′** in the double cleavage rearrangement of 1,6-enynes (Scheme 13.48). Proton loss and demetallation then affords either 1,4-cyclohexadienes **220a** or **220b,c**. Similar transformations have been found with other 1,5-enynes [73].

## 13.4.6
### Inter- and Intramolecular Cyclopropanation of 1,6-Enynes

Additional evidence for the existence of cyclopropyl metal carbenes as intermediates was also obtained in the reaction of enynes bearing additional double bonds at the alkenyl chain with Ru(II) [10c], Pt(II) [197, 198], and Au(I) [50, 199] catalysts (Scheme 13.62). In these reactions, the intermediate cyclopropyl metal carbenes, such as **233** and **234**, are trapped intramolecularly by the terminal alkene to give tetracycles containing two cyclopropanes. The biscyclopropanation of **229a** also takes place with

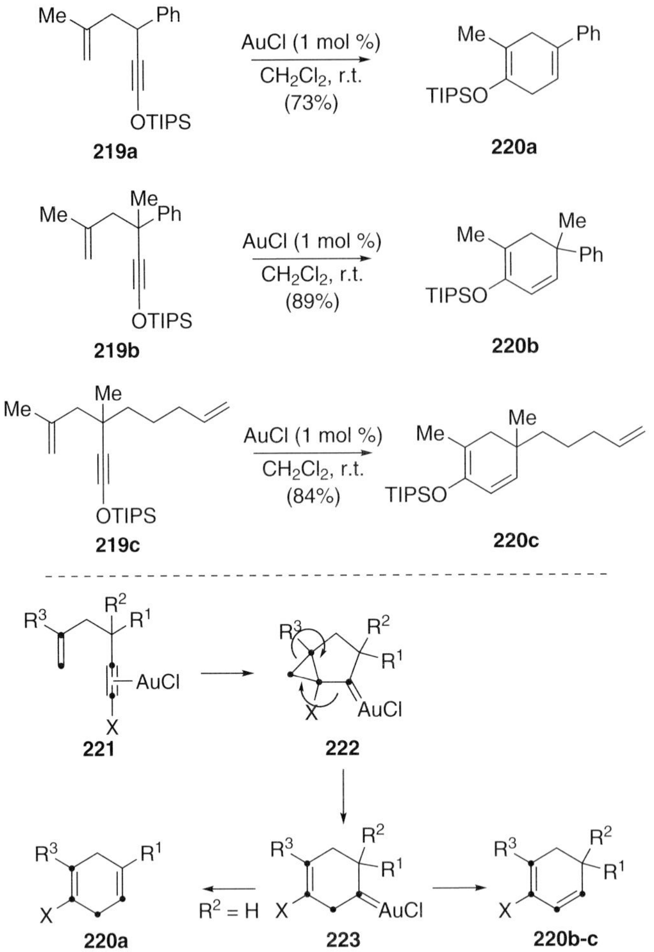

**Scheme 13.61** Au(I)-catalyzed reaction of siloxy 1,5-enynes.

Ag(I), albeit less efficiently [200]. Intramolecular cyclopropanations of gold carbenes formed in other reactions have also been observed [201]. Products of intramolecular cyclopropanation have also been obtained in the reaction of dienynes catalyzed by [Cp*RuCl(cod)], although a different mechanism has been proposed [202].

The intermolecular reaction between 1,6-enynes and alkenes is also possible with catalysts formed *in situ* from complexes **18a** or **19a** (Scheme 13.63) [56]. Interestingly, whereas enyne **27r** reacts with norbornene to give **235** via an intermediate of type **3** (see Scheme 13.48), **20h** and **20c** react with styrene and norbornene to give **236** and **237**, respectively, via rearranged gold carbenes of type **177**. Gold carbene intermediates formed by 1,2-acyl migration (see Section 13.2.2) also undergo cyclopropanation reactions with olefins [203]. In this case, by using DTBM-SEGPHOS as the chiral ligand for gold(I) 60–85% ee were obtained.

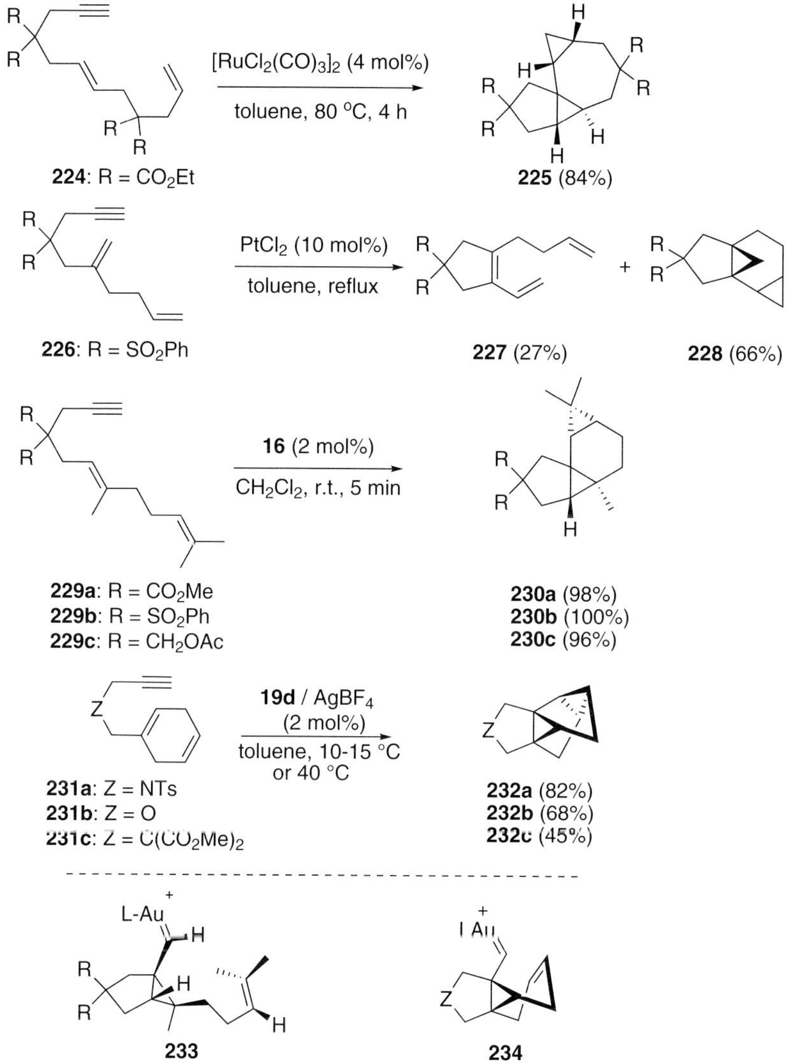

**Scheme 13.62** Intramolecular cyclopropanation in metal-catalyzed cyclization of dienynes.

Additional evidence for the involvement of metal carbenes in these reactions was obtained in the reaction of dimeric substrates **238** with cationic Au(I) catalyst **17e** to give **239/239′** (Scheme 13.64) [56]. These reactions can be explained by isomerization of the initially formed cyclopropyl gold carbene **240** to form **241** by a [1,3] metallotropic shift, followed by intramolecular trapping of the gold carbene by the alkene. This type of [1,3] metallotropic shift [204] has been observed before for Rh [205], Cr, Mo, W [206], and Ru [207–209] carbene complexes. Additional examples of [1,3] metallotropic shift in gold chemistry have also been observed [210–212].

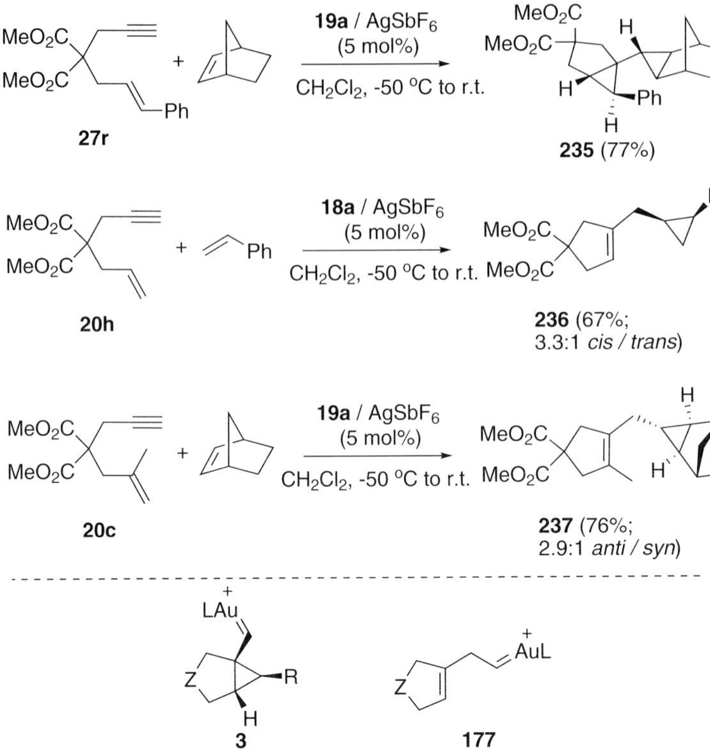

**670** | *13 Transition-Metal-Catalyzed Cycloisomerizations and Nucleophilic Cyclization of Enynes*

**Scheme 13.63** Au(I)-catalyzed intramolecular biscyclopropanation of 1,6-enynes with alkenes.

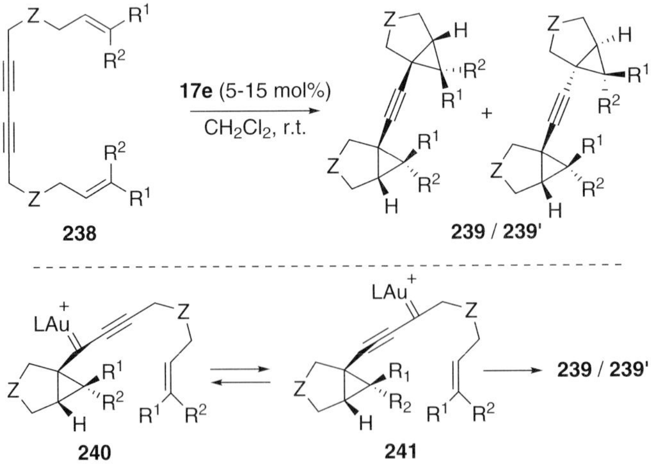

**Scheme 13.64** Au(I)-catalyzed [1,3]-metallatropic shift in dimeric enynes.

## 13.4.7
### Reactions of Furans with Alkynes

In contrast to the usual Friedel–Crafts-like cyclizations of arenes with alkynes [25], alkynyl furans **242** afford phenols **243** in good to excellent yields by using $AuCl_3$ as catalyst (Scheme 13.65) [213]. Au(I) [214], heterogeneous gold [215], and Pt(II) [216] can also be used as catalysts. Phenols bearing bulky groups at the ortho position can also be prepared by this method [217]. In addition, the gold-catalyzed Michael addition of furans to ethynyl vinyl ketones gives substrates [218–220] that can undergo *in situ* cyclization leading to hydroxyindanones in a domino process [219].

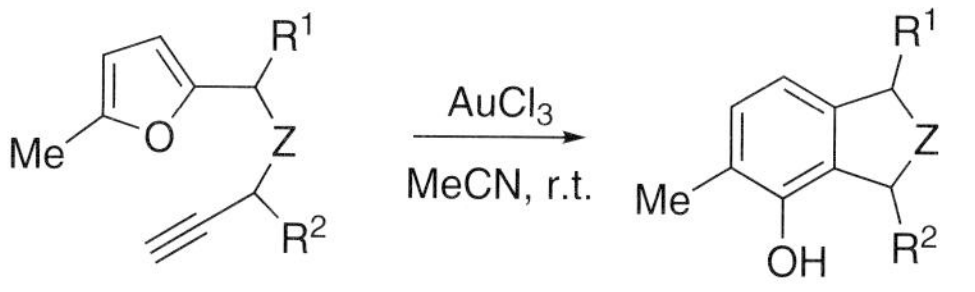

**Scheme 13.65** Synthesis of phenols using a Au(I)-catalyzed reaction of alkynyl furans.

According to experimental and theoretical studies on gold- and platinum-catalyzed reactions [213b,e,f, 216], the phenol synthesis proceeds by nucleophilic attack of the furan on the ($\eta^2$-alkyne)-metal complex **244** to form carbene **245**, similar to the intermediates formed in reactions of enynes with Au(I) or other metal complexes (Scheme 13.66). After cleavage of a C–C and C–O bond of the tricyclic intermediate, a new carbene **246** is formed, which cyclizes to form **247**. Elimination of the metal

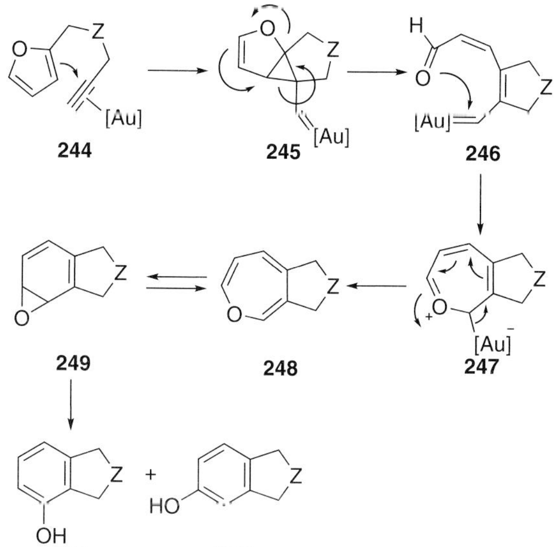

**Scheme 13.66** Proposed mechanism for the Au(I)-catalyzed synthesis of phenols.

forms oxepine **248**, which is in equilibrium with the arene oxide **249**, whose opening leads to the formation of phenols **250a**, the major compounds of this phenol synthesis, and their regioisomers **250b**. Oxepines **248** and arene oxides **249** have been observed in the reaction catalyzed by Au(III) [213f, 221, 222].

Although proceeding in moderate yield and with a long reaction time, the formation of phenol **251** in the intermolecular reaction of 2,5-dimethylfuran with phenylacetylene is remarkable (Scheme 13.67) [223]. This reaction required the use as Au(I) catalyst of the Schmidbaur–Bayler salt [(Mes$_3$PAu)$_2$Cl]BF$_4$ [224], and also provided furan **252**, the product of a Friedel–Crafts-type process.

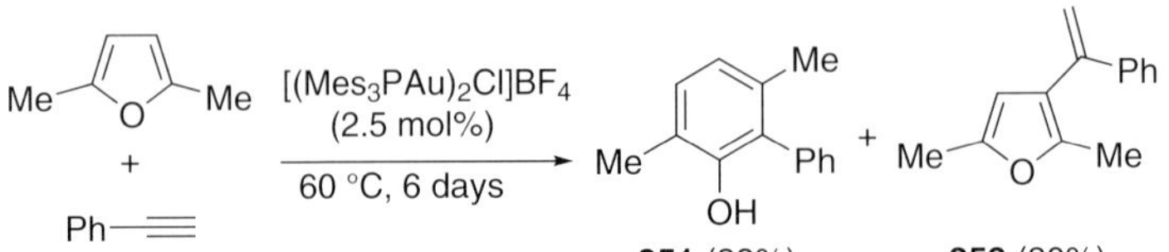

**Scheme 13.67** Intermolecular Au(I)-catalyzed reaction of a furan with phenylacetylene.

## 13.5
## Summary and Outlook

The similarities between reactions of enynes catalyzed by electrophilic transition metals and the carbocationic rearrangements of the cyclopropylmethyl–cyclobutyl manifold are striking. However, stabilization by the different metals confers special reactivity to the intermediate cationic species, leading to highly selective transformations. Gold complexes, in particular, cationic complexes bearing bulky phosphines, phosphites, or NHC ligands, usually surpass the most active platinum complexes reported for the activation of enynes. Although the basic pathways in the cycloisomerization of enynes are now better understood mechanistically, there are still many unresolved questions in this area. Thus, for example, factors that control the selectivity in skeletal rearrangements of enynes are not clear. Ligand control of closely related pathways is still very limited. Application of metal-catalyzed cycloisomerization of enynes to the formation of medium or large rings, and the development of general methods for the intermolecular reactions of alkynes with alkenes are still important challenges.

## 13.6
## Experimental: Selected Procedures

### 13.6.1
### Synthesis of Selected Au(I) Complexes

#### 13.6.1.1 General Procedure for the Synthesis of Neutral Au(I) Complexes 17a–d and 18a [44]

Sodium tetrachloroaurate(III) dihydrate or tetrachloroauric acid (1 mmol) was dissolved in water, and the orange solution was cooled in ice. 2,2′-Thiodiethanol

(3 mmol) was slowly added (about 45 min) to this solution with stirring. A solution of the phosphine ligand (1 mmol) in EtOH was added dropwise to give a white solid. The solid was filtered off, washed with MeOH, and dried *in vacuo*.

### 13.6.1.2 Synthesis of Cationic Gold(I) Complexes 17e,f [44]

A mixture of [Au(L)Cl] complex (0.5 mmol) and $AgSbF_6$ (0.5 mmol) was suspended in MeCN (4 mL). The reaction mixture was stirred at 23 °C for 12 h, the solvent was evaporated, and the crude product was dissolved in MeCN (1 mL). The mixture was then filtered through a pad of Celite, and the solvent was evaporated under reduced pressure to give the cationic complexes as white solids.

### 13.6.1.3 Synthesis of Cationic Gold(I) Complex 18b [58]

A solution of gold(I) chloro(tris(2,4-di*tert*-butylphenyl)phosphite) (**18a**) (0.88 g, 1.00 mmol) and PhCN (0.11 mL, 1.1 mmol) in $CH_2Cl_2$ (11 mL) was added to a solution of $AgSbF_6$ (0.344 g, 1.00 mmol) in $CH_2Cl_2$ (6 mL). A white precipitate appeared immediately. After stirring for 5 min, the mixture was filtered (2 Teflon filters), evaporated and vacuum-dried (50 °C overnight). The cationic complex was obtained as a white, foamy solid (1.05 g, 88%).

### 13.6.1.4 Synthesis of Neutral Gold(I) Complexes 19a–d [49]

$Ag_2O$ (0.5 mmol) was added to a solution of the imidazolium ligand (1 mmol) in $CH_2Cl_2$. The suspension became clear after stirring for 3 h at room temperature. Then a solution of $Me_2SAuCl$ (1 mmol) in $CH_2Cl_2$ was added drop-wise. The reaction mixture was stirred for another 4 h. The solution was then filtered through Celite and the solvent was partially evaporated. Addition of hexane led to the precipitation of the methiodides.

### 13.6.1.5 Synthesis of Cationic Gold(I) Complex 19e [58]

A solution of IMesAuCl (54.0 mg, 0.101 mmol) and 2,4,6-trimethoxybenzonitrile (19.4 mg, 0.101 mmol) in $CH_2Cl_2$ (1 mL) was added over solid $AgSbF_6$ (34.6 mg, 0.101 mmol) and stirred for 5 min. The mixture was filtered (HPLC Teflon filter) and the solid residue washed with $CH_2Cl_2$ (2 × 0.2 mL). The gold complex precipitated from the filtrate upon addition of $Et_2O$ (5–8 mL). Filtration and air-drying furnished a bright white solid (67.2 mg, 72%).

## 13.6.2
### Selected Procedures for Metal-Catalyzed Cycloisomerizations

### 13.6.2.1 Synthesis of Compounds 11 and 13 Using $PtCl_2$ as Catalyst [35b]

Enynes (**10**, **12**) and the catalyst (5 mol%) were stirred at 70 °C in acetone or 1,4-dioxane for 17–20 h. The solvent was evaporated and the residue was chromatographed (flash grade silica gel, EtOAc–hexane mixtures) to give the dienes **11** (100%) and **13** (93%). 4 Å molecular sieves were added in the reactions carried out in acetone.

### 13.6.2.2 Synthesis of 27 and 28 Using Catalyst 17e [68]

A mixture of enol ether **26** (0.5 mmol, 1 equiv) and the catalyst **17e** (0.05 equiv) was dissolved in $CH_2Cl_2$ (2.5 mL). The solution was stirred at room temperature for 0.3 h and then filtered through a short path of Celite. The solvent was removed under reduced pressure and the residue was purified by silica gel chromatography (100:1 hexane/EtOAc containing 5% Et$_3$N) to yield products **27** (45%) and **28** (14%).

### 13.6.2.3 Synthesis of 32a–c Using AuCl(PPh$_3$)/AgSbF$_6$ as Catalyst [72]

To a 1-dram vial with a threaded cap containing a magnetic stir bar and 1,5-enyne (100 mg, 1 equiv.) in CH$_2$Cl$_2$ (0.5 M) was added AgSbF$_6$ (1–5 mol%) and (Ph$_3$P)AuCl (1–5 mol%) sequentially. A cloudy white mixture formed during the course of the reaction. The mixture was stirred at room temperature and monitored by TLC analysis. Upon completion, the mixture was filtered through a short silica plug and eluted with CH$_2$Cl$_2$. Evaporation of the solvent, followed by column chromatography afforded the desired bicyclo[3.1.0]hexene.

### 13.6.2.4 Synthesis of 42a Using AuCl(PPh$_3$)/AgOTf as Catalyst [80]

To a small screw-cap scintillation vial equipped with a magnetic stir bar and charged with a solution of alkynyl β-ketoester (~150 mg, 1 equiv.) in CH$_2$Cl$_2$ (0.4 M) was added AuCl(PPh$_3$) (1 mol%) followed by AgOTf (1 mol%). The cloudy white reaction mixture was then stirred at room temperature and monitored periodically by TLC. Upon completion of reaction, the reaction mixture was loaded directly onto a silica gel column and chromatographed with the appropriate mixture of hexanes and EtOAc to give the Conia-ene products (79–99% yields).

### 13.6.2.5 Synthesis of Cyclopentenes from Alkynes and Enamines Formed In Situ Using Cu(OTf)$_2$ as Catalyst [93]

To a solution of the nucleophile (0.62 mmol), Cu(OTf)$_2$ (7.4 mg, 0.020 mmol) and PPh$_3$ (22 mg, 0.082 mmol) in anhydrous MeOH (1 mL) ps-BEMP (19 mg, 0.041 mmol) was suspended, then the conjugated ketone (0.41 mmol) and pyrrolidine (7 µL, 0.082 mmol) were added; the mixture was stirred at rt till complete conversion of the ketone by TLC, then diluted with CH$_2$Cl$_2$ and filtered to eliminate ps-BEMP. Solvents were evaporated and the obtained crude was purified by chromatography.

### 13.6.2.6 Synthesis of 52 Using AuCl(PPh$_3$)/AgOTf as Catalyst [98]

To a solution of nitro malonate **51** (40 mg, 0.11 mmol) in toluene (1.1 mL) was added AuCl(PPh$_3$) (30 mg, 0.4 mmol) at room temperature. To the resulting mixture was added AgOTf (20 mg, 0.5 mmol) and the reaction was heated at 50 °C for 14 h. The reaction mixture was concentrated *in vacuo*. Chromatography over SiO$_2$ (10% EtOAc in pet. ether) provided **52** (23 mg, 58%) as a white film.

### 13.6.2.7 General Procedure for Au(I)-Catalyzed Skeletal Rearrangement of Enynes (5a–l, 6a and 213a–e) [44]

The enyne (0.10–0.50 mmol) in CH$_2$Cl$_2$ (1 mL) was added to a mixture of [Au(PPh$_3$)Cl] (2 mol%) and silver(I) salt (2 mol%) (or the corresponding cationic Au(I) complex) in CH$_2$Cl$_2$ (2 mL) and the mixture was stirred at room temperature until TLC showed full conversion. The resulting mixture was filtered through SiO$_2$ and the solvent was evaporated to give the corresponding product.

### 13.6.2.8 Synthesis of Product 198 Using GaCl₃ [185]

In an $N_2$ drybox, a Schlenk flask was charged with $GaCl_3$ (66.7 mg, 0.38 mmol) and powdered 4 Å molecular sieves (133 mg). The flask was removed from the drybox and a solution of 2-(4′,4′-dimethyl-5′-hexynyl)-6-isopropyl-4,5- dimethoxyindene (**197**, 618 mg, 1.89 mmol) in freshly distilled benzene (47 mL) was added via syringe under $N_2$ to give a dark orange–red solution. The flask was sealed and placed in an oil bath that was preheated to 40 °C. After stirring at 40 °C for 40 h, the mixture was allowed to cool to rt. The solution was added to sat. aq. $NaHCO_3$ (50 mL) and extracted with $Et_2O$ (3 × 40 mL). The combined organic layers were washed (50 mL brine), dried ($MgSO_4$) and concentrated to give 640 mg of a yellow oil, which was purified by flash chromatography (30 : 1 hexanes/EtOAc) to give 554 mg (1.70 mmol, 90%) of **198** as a viscous pale yellow oil. This crystallized to a white solid on standing overnight in a freezer. (In subsequent runs, it was found that the reaction was complete after 24 h. Alternatively, quantitative conversion could be achieved with only 10 mol% catalyst, though this required a reaction time of 72 h. Thus, in subsequent reactions 20 mol% catalyst was used and the mixture was heated for 24 h.).

### 13.6.2.9 Synthesis of Product 203 with Ru(II) Catalyst [187]

[Cp*Ru(MeCN)₃ PF₆] (29.8 mg, 0.06 mmol) was added to a solution of enyne **202** (1.2 mmol) in DMF (6 mL) and the resulting mixture was stirred for 5 min. For work-up, the reaction was quenched with brine (10 mL) and extracted with EtOAc (3 × 10 mL), the combined organic layers were dried over $Na_2SO_4$ and evaporated, and the residue was purified by flash chromatography (neutral Alox, pentane/Et₂O, 4 : 1) to give iodocyclobutene **203** as a colorless oil (90%).

## 13.6.3
### Selected Procedure for Metal-Catalyzed Cyclopropanation Using Propargylic Carboxylates

### 13.6.3.1 Synthesis of 58a,b Using [RuCl₂(CO)₃]₂ as Catalyst [104b]

The complex [RuCl₂(CO)₃]₂ (2.6 mg, 0.005 mmol) was placed in a flame-dried Schlenk flask under $N_2$. A solution of substrate 1 (0.20 mmol) and alkene (1.0–4.0 mmol) in solvent (1.0 mL) was added to the flask at room temperature. After the mixture was stirred at 50 °C for 18 h, it was cooled to room temperature and the amount of products was determined by gas liquid chromatography (GLC) analysis using 2,6-dimethylnaphthalene as an internal standard. All trans-cyclopropanes could not be easily separated by column chromatography on $SiO_2$ (hexane/AcOEt) 15/1). Pure cis-isomers were partially separated as a first fraction of column chromatography, whereas trans-isomers as minor products were eluted together with their cis-isomers. Because of this contamination, GPC on $CHCl_3$ was required to obtain pure trans-isomers in each case. The configuration of the cyclopropane ring could be determined by 1H NMR coupling constants between protons in a cyclopropane ring. Generally, coupling constants of $J = 7.0–9.0$ Hz between two protons in a cyclopropane ring indicate that the configuration is cis, while those of $J = 4.0–6.0$ Hz correspond to trans.

## 13.6.4
## Selected Procedure for Metal-Catalyzed Rautenstrauch Rearrangement

### 13.6.4.1 Synthesis of 65 Using AuCl(PPh$_3$)/AgSbF$_6$ as Catalyst [71]

A solution of acetate **63** (0.200 g, 0.933 mmol) in CH$_2$Cl$_2$ (2.4 mL) was added to a suspension of (Ph$_3$P)AuCl (9.25 mg, 18.7 μmol) and AgSbF$_6$ (6.43 mg, 18.7 μmol) in CH$_2$Cl$_2$ (21 mL). After stirring at ambient temperature for 15 min, the solvent was evaporated and the crude product was dissolved in methanol (8 mL). K$_2$CO$_3$ (12.9 mg) was added and the suspension was stirred for 1 h before the reaction was quenched with water and the aqueous phase was extracted with *tert*-butyl methyl ether. The combined organic layers were dried (MgSO$_4$), filtered and evaporated, and the residue was purified by flash chromatography (hexane/ethyl acetate, 4 : 1 + 1% triethylamine, $v/v$) to give ketone **22** as a yellow liquid (0.119 g, 74%).

## 13.6.5
## Selected Procedures for the Metal-Catalyzed Inter- and Intramolecular Alcoxycyclization

### 13.6.5.1 General Procedure for the Au(I)-Catalyzed Intermolecular Alkoxycyclization to Yield Products 125a–c [44]

The protic acid (6 mol%) was added to a mixture of the enyne (0.2–0.5 mmol) and gold (I) catalyst [Au(PPh$_3$)Me] (3 mol%) in the stated solvent (3–5 mL of MeOH, EtOH, or aqueous acetone) and the mixtures were stirred at roon temperature until TLC showed complete conversion. The resulting mixture was filtered through Celite, the solvent was evaporated, and the residue was chromatographed (hexane/EtOAc) to give the corresponding carbo- or heterocycle. Alternatively, complex [Au(L)Cl] and a Ag(I) salt or the corresponding cationic Au(I) complex were used as the catalyst. Routinely, synthetic gradeMeOH or EtOH (>99.8% purity) were used as solvents, without further drying.

### 13.6.5.2 General Procedure for Au(I)-Catalyzed Intramolecular Alkoxycyclization (125d) [44]

The enyne (0.10–0.50 mmol) in CH$_2$Cl$_2$ (1 mL) was added to a mixture of [Au(PPh$_3$)Cl] (2 mol%) and silver(I) salt (2 mol%) (or the corresponding cationic Au(I) complex) in CH$_2$Cl$_2$ (2 mL) and the mixture was stirred at room temperature until TLC showed full conversion. The resulting mixture was filtered through SiO$_2$ and the solvent was evaporated to give the corresponding product.

## 13.6.6
## Experimental: Selected Procedure

### 13.6.6.1 Procedure for the for Au(I)-Catalyzed Addition of Carbon Nucleophiles to 1,6-Enynes

**General Procedure for the Synthesis of Products (140a,b,141) [161]** A solution of enyne **20** and the nucleophile in CH$_2$Cl$_2$ (3 mL) was added to a previously prepared mixture of the gold catalyst (5 mol% rel. to **20**) and AgSbF$_6$ (5 mol% rel. to **20**) in CH$_2$Cl$_2$ (1 mL). The reaction mixture was stirred at room temperature (unless stated otherwise). The mixture was filtered through silica gel with $CH_2Cl_2$ and the solvents evaporated. The residue was chromatographed to give the desired product.

**Abbreviations**

| | |
|---|---|
| Ac | acetyl |
| BINAP | 2,2′-bis(diphenylphosphino)-1,1′-binaphthyl |
| Boc | *tert*-butyloxycarbonyl |
| Bn | benzyl |
| Bz | benzoyl |
| Cod | cyclooctadiene |
| Cp | cyclopentadienyl |
| Cp* | pentamethylcyclopentadienyl |
| Cy | cyclohexyl |
| DCE | 1,2-dichloroethane |
| DFT | density functional theory |
| *dig* | digonal |
| DMF | N,N-Dimethyl formamide |
| DMSO | Dimethyl sulfoxide |
| DTBM-SEGPHOS | (R)-(−)-5,5′-Bis[di(3,5-di-*tert*-butyl-4-methoxyphenyl) phosphino]-4,4′-bi-1,3-benzodioxole |
| endo | endocyclic |
| exo | exocyclic |
| GLC | gas liquid chromatography |
| LUMO | lower unoccupied molecular orbital |
| Me-DuPhos | (−)-1,2-bis((2R,5R)-2,5-dimethylphospholano)benzene |
| Mes | mesityl |
| NHC | N-heterocyclic carbene |
| ps-BEMP | 2-*tert*-butylimino-2-diethylamino-1,3-dimethylperhydro-1,3,2-diazaphosphorine |
| TBS | *tert*-butyldimethylsilyl |
| *t*-Bu | *tert*-butyl |
| TES | triethylsilyl |
| Tf | trifluoromethanesulfonate |
| TIPS | triisopropylsilyl |
| TLC | thin layer chromatography |
| Tol-BINAP | 2,2′-bis(di-p-tolylphosphino)-1,1′-binaphthyl |
| *trig* | trigonal |

**Acknowledgment**

This work was supported by the MEC (projects CTQ2007-60745/BQU, Consolider Ingenio 2010 Grant CSD2006-003 and predoctoral fellowship to E.H-G.) the AGAUR (2005 SGR 00993), and the ICIQ Foundation.

## References

1. (a) Aubert, C., Buisine, O. and Malacria, M. (2002) *Chem. Rev.*, **102**, 813–834; (b) Lloyd-Jones, G.C. (2003) *Org. Biomol. Chem.*, 215–236; (c) Diver, S.T. and Giessert, A.J. (2004) *Chem. Rev.*, **104**, 1317–1382; (d) Hoffmann-Röder, A. and Krause, N. (2005) *Org. Biomol. Chem.*, **3**, 387–391; (e) Ma, S., Yu, S. and Gu, Z. (2006) *Angew. Chem. Int. Ed.*, **45**, 200–203; (f) Bruneau, C. (2005) *Angew. Chem. Int. Ed.*, **44**, 2328–2334.
2. (a) Hashmi, A.S.K. and Hutchings, G.J. (2006) *Angew. Chem. Int. Ed.*, **45**, 7896–7936; (b) Hashmi, A.S.K. (2007) *Chem. Rev.*, **107**, 3180–3211.
3. (a) Echavarren, A.M. and Nevado, C. (2004) *Chem. Soc. Rev.*, **33**, 431–436; (b) Echavarren, A.M., Méndez, M., Muñoz, M.P., Nevado, C., Martín-Matute, B., Nieto-Oberhuber, C. and Cárdenas, D.J. (2004) *Pure Appl. Chem.*, **76**, 453–463; (c) Jiménez-Núñez, E. and Echavarren, A.M. (2007) *Chem. Commun.*, 333–346; (d) Jiménez-Núñez, E. and Echavarren, A.M. (2008) *Chem. Rev.*, **108**, 3326–3350.
4. Zhang, L., Sun, J. and Kozmin, S.A. (2006) *Adv. Synth. Catal.*, **348**, 2271–2296.
5. Fürstner, A. and Davies, P.W. (2007) *Angew. Chem. Int. Ed.*, **46**, 3410–3449.
6. Gorin, D.J. and Toste, F.D. (2007) *Nature*, **446**, 395–403.
7. Michelet, V., Toullec, P.Y. and Genêt, J.P. (2008) *Angew. Chem. Int. Ed.*, **47**, 4268–4315.
8. (a) Trost, B.M. and Lautens, M. (1985) *J. Am. Chem. Soc.*, **107**, 1781–1783; (b) Trost, B.M., Lautens, M., Chan, C., Jebaratnam, D.J. and Mueller, T. (1991) *J. Am. Chem. Soc.*, **113**, 636–644; (c) Trost, B.M., Hipskind, P.A., Chung, J.Y.L. and Chan, C. (1989) *Angew. Chem., Int. Ed. Engl.*, **28**, 1502–1504; (d) Trost, B.M. and Gelling, O.J. (1993) *Tetrahedron Lett.*, **34**, 8233–8236.
9. (a) Trost, B.M. and Tanoury, G.J. (1988) *J. Am. Chem. Soc.*, **110**, 1636–1638; (b) Trost, B.M. and Trost, M.K. (1991) *Tetrahedron Lett.*, **32**, 3647–3650; (c) Trost, B.M. and Doherty, G.A. (2000) *J. Am. Chem. Soc.*, **122**, 3801–3810; (d) Trost, B.M., Yanai, M. and Hoogsteed, K. (1993) *J. Am. Chem. Soc.*, **115**, 5294–5295.
10. (a) Chatani, N., Morimoto, T., Muto, T. and Murai, S. (1994) *J. Am. Chem. Soc.*, **116**, 6049–6050; (b) Chatani, N., Furukawa, N., Sakurai, H. and Murai, S. (1996) *Organometallics*, **15**, 901–903; (c) Chatani, N., Kataoka, K., Murai, S., Furukawa, N. and Seki, Y. (1998) *J. Am. Chem. Soc.*, **120**, 9104–9105; (d) Chatani, N., Inoue, H., Morimoto, T., Muto, T. and Murai, S. (2001) *J. Org. Chem.*, **66**, 4433–4436; (e) Chatani, N., Inoue, H., Kotsuma, T. and Murai, S. (2002) *J. Am. Chem. Soc.*, **124**, 10294–10295; (f) Miyanohana, Y., Inoue, H. and Chatani, N. (2004) *J. Org. Chem.*, **69**, 8541–8543; (g) Miyanohana, Y. and Chatani, N. (2006) *Org. Lett.*, **8**, 2155–2158; (h) Nakai, H. and Chatani, N. (2007) *Chem. Lett.*, 1494–1495.
11. (a) Fürstner, A., Szillat, H. and Stelzer, F. (2000) *J. Am. Chem. Soc.*, **122**, 6785–6786; (b) Fürstner, A., Stelzer, F. and Szillat, H. (2001) *J. Am. Chem. Soc.*, **123**, 11863–11869; (c) Fürstner, A., Szillat, H.F., Gabor, B. and Mynott, R. (1998) *J. Am. Chem. Soc.*, **120**, 8305–8314.
12. Oi, S., Tsukamoto, I., Miyano, S. and Inoue, Y. (2001) *Organometallics*, **20**, 3704–3709.
13. Oh, C.H., Bang, S.Y. and Rhim, C.Y. (2003) *Bull. Korean Chem. Soc.*, **24**, 887–888.
14. Bajracharya, G.B., Nakamura, I. and Yamamoto, Y. (2005) *J. Org. Chem.*, **70**, 892–897.
15. Kim, S.M., Lee, S.I. and Chung, Y.K. (2006) *Org. Lett.*, **8**, 5425–5427.
16. Nieto-Oberhuber, C., López, S., Jiménez-Núñez, E. and Echavarren, A.M. (2006) *Chem. Eur. J.*, **12**, 5916–5923.
17. Blum, J., Beer-Kraft, H. and Badrieh, Y. (1995) *J. Org. Chem.*, **60**, 5567–5569.

18. Borodkin, V.S., Shpiro, N.A., Azov, V.A. and Kochetkov, N.K. (1996) *Tetrahedon Lett.*, **37**, 1489–1492.
19. Nevado, C., Ferrer, C. and Echavarren, A.M. (2004) *Org. Lett.*, **6**, 3191–3194.
20. Fürstner, A., Davies, P.W. and Gress, T. (2005) *J. Am. Chem. Soc.*, **127**, 8244–8245.
21. Fürstner, A. and Aïssa, C. (2006) *J. Am. Chem. Soc.*, **128**, 6306–6307.
22. Fürstner, A. and Davies, P.W. (2005) *J. Am. Chem. Soc.*, **127**, 15024–15025.
23. Mori, M. (2007) *Adv. Synth. Catal.*, **349**, 121–135.
24. Grela, K. (2008) *Angew. Chem. Int. Ed.*, **47**, 5504–5507.
25. Nevado, C. and Echavarren, A.M. (2005) *Synthesis*, 167–182.
26. Trost, B.M., Dumas, J. and Villa, M. (1992) *J. Am. Chem. Soc.*, **114**, 9836–9845.
27. Harada, K., Tonoi, Y., Kato, H. and Fukuyama, Y. (2002) *Tetrahedron Lett.*, **43**, 3829–3832.
28. Trost, B.M., Indolese, A., Müller, T.J.J. and Treptow, B. (1995) *Angew. Chem. Int. Ed.*, **117**, 615–623.
29. Trost, B.M. and Toste, F.D. (2000) *J. Am. Chem. Soc.*, **122**, 714–715.
30. See also: Le Paih, J., Cuervo-Rodríguez, D., Dérien, S. and Dixneuf, P.H. (2000) *Synlett*, **1**, 95–97.
31. Cao, P., Wang, B. and Zhang, X. (2000) *J. Am. Chem. Soc.*, **122**, 6490–6491.
32. Llerena, D., Aubert, C. and Malacria, M. (1996) *Tetrahedron Lett.*, **37**, 3517–3525.
33. Buisine, O., Aubert, C. and Malacria, M. (2001) *Chem. Eur. J.*, **7**, 7353–7356.
34. Sturla, S.J., Kablaoui, N.M. and Buchwald, S.L. (1999) *J. Am. Chem. Soc.*, **121**, 1976–1977.
35. (a) Méndez, M., Muñoz, M.P. and Echavarren, A.M. (2000) *J. Am. Chem. Soc.*, **122**, 11549–11550; (b) Méndez, M., Muñoz, M.P., Nevado, C., Cárdenas, D.J. and Echavarren, A.M. (2001) *J. Am. Chem. Soc.*, **123**, 10511–10520; (c) Nevado, C., Cárdenas, D.J. and Echavarren, A.M. (2003) *Chem. Eur. J.*, **9**, 2627–2635; (d) Muñoz, M.P., Méndez, M., Nevado, C., Cárdenas, D.J. and Echavarren, A.M. (2003) *Synthesis*, 2898–2902; (e) Nevado, C., Charruault, L., Michelet, V., Nieto-Oberhuber, C., Muñoz, M.P., Méndez, M., Rager, M.N., Genêt, J.-P. and Echavarren, A.M. (2003) *Eur. J. Org. Chem.*, 706–713.
36. Fürstner, A., Majima, K., Martín, R., Krause, H., Katting, E., Goddard, R. and Lehmann, C.W. (2008) *J. Am. Chem. Soc.*, **130**, 1992–2004.
37. Hatano, M., Terada, M. and Mikami, K. (2001) *Angew. Chem. Int. Ed.*, **40**, 249–253.
38. Cao, P. and Zhang, X. (2000) *J. Angew. Chem., Int. Ed.*, **39**, 4104–4106.
39. Yamanaka, M. and Nakamura, E. (2001) *J. Am. Chem. Soc.*, **123**, 1703–1708.
40. Shibata, T., Toshida, N. and Takagi, K. (2002) *J. Org. Chem.*, **67**, 7446–7450, and references therein.
41. Ajamian, A. and Gleason, J.L. (2003) *Org. Lett.*, **5**, 2409–2411.
42. (a) Trost, B.M. and Hashmi, A.S.K. (1993) *Angew. Chem. Int. Ed. Engl.*, **32**, 1085–1087; (b) Trost, B.M. and Hashmi, A.S.K. (1994) *J. Am. Chem. Soc.*, **116**, 2183–2184; (c) Trost, B.M., Hashmi, A.S.K. and Ball, R.G. (2001) *Adv. Synth. Catal.*, **343**, 490–494.
43. Nieto-Oberhuber, C., Muñoz, M.P., Buñuel, E., Nevado, C., Cárdenas, D.J. and Echavarren, A.M. (2004) *Angew. Chem. Int. Ed.*, **43**, 2402–2406.
44. Nieto-Oberhuber, C., Muñoz, M.P., López, S., Jiménez-Núñez, E., Nevado, C., Herrero-Gómez, E., Raducan, M. and Echavarren, A.M. (2006) *Chem. Eur. J.*, **12**, 1677–1693.
45. Teles, J.H., Brode, S. and Chabanas, M. (1998) *Angew. Chem. Int. Ed.*, **37**, 1415–1418.
46. (a) Mizushima, E., Sato, K., Hayashi, T. and Tanaka, M. (2002) *Angew. Chem. Int. Ed.*, **41**, 4563–4565; (b) Mizushima, E., Hayashi, T. and Tanaka, M. (2003) *Org. Lett.*, **5**, 3349–3352.
47. (a) Yang, Y., Ramamoorthy, V. and Sharp, P.R. (1993) *Inorg. Chem.*, **32**, 1946–1950; (b) Nesmeyanov, A.N., Perevalova, E.G.,

Struchkov, Y.T., Antipin, M.Y., Grandberg, K.I. and Dyadchenko, V.P. (1980) *J. Organomet. Chem.*, **201**, 343–349.

48 Sherry, B.D. and Toste, F.D. (2004) *J. Am. Chem. Soc.*, **126**, 15978–15979.

49 Nieto-Oberhuber, C., López, S. and Echavarren, A.M. (2005) *J. Am. Chem. Soc.*, **127**, 6178–6179.

50 Nieto-Oberhuber, C., López, S., Muñoz, M.P., Jiménez-Núñez, E., Buñuel, E., Cárdenas, D.J. and Echavarren, A.M. (2006) *Chem. Eur. J.*, **11**, 1694–1702.

51 Jiménez-Núñez, E., Claverie, C.K., Nieto-Oberhuber, C. and Echavarren, A.M. (2006) *Angew. Chem. Int. Ed.*, **45**, 5452–5455.

52 (a) Ferrer, C. and Echavarren, A.M. (2006) *Angew. Chem. Int. Ed.*, **45**, 1105–1109; (b) Ferrer, C., Amijs, C.H.M. and Echavarren, A.M. (2007) *Chem. Eur. J.*, **13**, 1358–1373.

53 Herrero-Gómez, E., Nieto-Oberhuber, C., López, S., Benet-Buchholz, J. and Echavarren, A.M. (2006) *Angew. Chem. Int. Ed.*, **45**, 5455–5459.

54 Partyka, D.V., Robilotto, T.J., Zeller, M., Hunter, A.D. and Gray, T.G. (2008) *Organometallics*, **27**, 28–32.

55 Mézailles, N., Ricard, L. and Gagosz, F. (2005) *Org. Lett.*, **7**, 4133–4136.

56 López, S., Herrero-Gómez, E., Pérez-Galán, P., Nieto-Oberhuber, C. and Echavarren, A.M. (2006) *Angew. Chem. Int. Ed.*, **45**, 6029–6032.

57 Nieto-Oberhuber, C., Pérez-Galán, P., Herrero-Gómez, E., Lauterbach, T., Rodríguez, C., López, C., Bour, C., Rosellón, A., Cárdenas, D.J. and Echavarren, A.M. (2008) *J. Am. Chem. Soc.*, **130**, 269–279.

58 Amijs, C.H.M., López-Carrillo, V., Raducan, M., Pérez-Galán, P., Ferrer, C. and Echavarren, A.M. (2008) *J. Org. Chem.*, **73**, 7721–7730.

59 de Fremont, P., Scott, N.M., Stevens, E.D. and Nolan, S.P. (2005) *Organometallics*, **24**, 2411–2418.

60 de Frémont, P., Stevens, E.D., Fructos, M.R., Diaz-Requejo, M.M., Perez, P.J. and Nolan, S.P. (2006) *Chem. Commun*, 2045–2047.

61 de Frémont, P., Singh, R., Stevens, E.D., Petersen, J.L. and Nolan, S.P. (2007) *Organometallics*, **26**, 1376–1385.

62 Li, G. and Zhang, L. (2007) *Angew. Chem. Int. Ed.*, **46**, 5156–5159.

63 Ricard, L. and Gagosz, F. (2007) *Organometallics*, **26**, 4704–4707.

64 García-Mota, M., Cabello, N., Maseras, F., Echavarren, A.M., Pérez-Ramírez, J. and López, N. (2008) *ChemPhysChem.*, **11**, 1624–1629.

65 Lee, S.I., Kim, S.M., Kim, S.Y. and Chung, Y.K. (2006) *Synlett*, 2256–2260.

66 (a) Fernández-Rivas, C., Méndez, M. and Echavarren, A.M. (2000) *J. Am. Chem. Soc.*, **122**, 1221–1222; (b) Fernández-Rivas, C., Méndez, M., Nieto-Oberhuber, C. and Echavarren, A.M. (2002) *J. Org. Chem.*, **67**, 5197–5201.

67 Lee, S.I., Kim, S.M., Choi, M.R., Kim, S.Y., Chung, Y.K., Han, W.-S. and Kang, S.O. (2006) *J. Org. Chem.*, **71**, 9366–9372.

68 Ferrer, C., Raducan, M., Nevado, C., Claverie, C.K. and Echavarren, A.M. (2007) *Tetrahedron*, **63**, 6306–6316.

69 Tenaglia, A. and Gaillard, S. (2008) *Angew. Chem. Int. Ed.*, **47**, 2454–2457.

70 Abbiati, G., Arcadi, A., Bianchi, G., Di Giuseppe, S., Marinelli, F. and Rossi, E. (2003) *J. Org. Chem.*, **68**, 6959–6966.

71 Mamane, V., Gress, T., Krause, H. and Fürstner, A. (2004) *J. Am. Chem. Soc.*, **126**, 8654–8655.

72 Luzung, M.R., Markham, J.P. and Toste, F.D. (2004) *J. Am. Chem. Soc.*, **126**, 10858–10859.

73 Gagosz, F. (2005) *Org. Lett.*, **7**, 4129–4132.

74 Zriba, R., Gandon, V., Aubert, C., Fensterbank, L. and Malacria, M. (2008) *Chem. Eur. J.*, **14**, 1482–1491.

75 Lim, C., Kang, J.-E., Lee, J.-E. and Shin, S. (2007) *Org. Lett.*, **9**, 3539–3542.

76 Dankwardt, J.W. (2001) *Tetrahedron Lett.*, **42**, 5809–5812.

77 Shibata, T., Ueno, Y. and Kanda, K. (2006) *Synlett*, 411–414.
78 Grisé, C.M. and Barriault, L. (2006) *Org. Lett.*, **8**, 5905–5908.
79 Debleds, O. and Campagne, J.M. (2008) *J. Am. Chem. Soc.*, **130**, 1562–1563.
80 Kennedy-Smith, J.J., Staben, S.T. and Toste, F.D. (2004) *J. Am. Chem. Soc.*, **126**, 4526–4527.
81 Staben, S.T., Kennedy-Smith, J.J. and Toste, F.D. (2004) *Angew. Chem. Int. Ed.*, **43**, 5350–5352.
82 Ochida, A., Ito, H. and Sawamura, M. (2006) *J. Am. Chem. Soc.*, **128**, 16486–16487.
83 Pan, J.-H., Yang, M., Gao, Q., Zhu, N.-Y. and Yang, D. (2007) *Synthesis*, 2539–2544.
84 Corkey, B.K. and Toste, F.D. (2005) *J. Am. Chem. Soc.*, **127**, 17168–17169.
85 Copper(i) and silver(i) co-catalyzed Conia-ene reactions of linear alkynyl β-ketoesters can proceed by exo- or endocyclic pathways: Deng, C.-L., Song, R.-J., Guo, S.-M., Wang, Z.-Q. and Li, J.-H. (2007) *Org. Lett.*, **9**, 5111–5114.
86 For other metal catalyzed Conia-ene reactions of lalkynyl β-ketoesters: (a) $TiCl_4$-caytalyzed: Kitagawa, O., Suzuki, T., Inoue, T., Watanabe, Y. and Taguchi, T. (1998) *J. Org. Chem.*, **63**, 9470–9475; (b) Pd-catalyzed: Lomberget, T., Bouyssi, D. and Balme, G. (2005) *Synthesis*, 311–329; (c) In(III)-catalyzed: Tsuji, H., Yamagata, K., Itoh, Y., Endo, K., Nakamura, M. and Nakamura, E. (2007) *Angew. Chem. Int. Ed.*, **46**, 8060–8062.
87 Minnihan, E.C., Colletti, S.L., Toste, F.D. and Shen, H.C. (2007) *J. Org. Chem.*, **72**, 6287–6289.
88 Staben, S.T., Kennedy-Smith, J.J., Huang, D., Corkey, B.K., LaLonde, R.L. and Toste, F.D. (2006) *Angew. Chem. Int. Ed.*, **45**, 5991–5994.
89 Linghu, X., Kennedy-Smith, J.J. and Toste, F.D. (2007) *Angew. Chem. Int. Ed.*, **46**, 7671–7673.
90 Nicolaou, K.C., Tria, G.S. and Edmonds, D.J. (2008) *Angew. Chem. Int. Ed.*, **120**, 1804–1807.

91 Lee, K. and Lee, P.H. (2007) *Adv. Synth. Catal.*, **349**, 2092–2096.
92 Binder, J.T., Crone, B., Haug, T.T., Menz, H. and Kirsch, S.F. (2008) *Org. Lett.*, **10**, 1025–1028.
93 Yang, T., Ferrali, A., Campbell, L. and Dixon, D.J. (2008) *Chem. Commun*, 2923–2925.
94 Suhre, M.H., Reif, M. and Kirsch, S.F. (2005) *Org. Lett.*, **7**, 3925–3927.
95 See also Binder, J.T. and Kirsch, S.F. (2006) *Org. Lett.*, **8**, 2151–2153.
96 For a related palladium-catalyzed process: Nevado, C. and Echavarren, A.M. (2004) *Tetrahedron*, **60**, 9735–9744.
97 Sherry, B.D., Maus, L., Laforteza, B.N. and Toste, F.D. (2006) *J. Am. Chem. Soc.*, **128**, 8132–8133.
98 Comer, E., Rohan, E., Deng, L. and Porco, J.A. (2007) *Org. Lett.*, **9**, 2123–2126.
99 Brief review: Marion, N. and Nolan, S.P. (2007) *Angew. Chem. Int. Ed.*, **46**, 2750–2752.
100 Shi, F.-Q., Li, X., Xia, Y., Zhang, L. and Yu, Z.-X. (2007) *J. Am. Chem. Soc.*, **129**, 15503–15512.
101 Correa, A., Marion, N., Fensterbank, L., Malacria, M., Nolan, S.P. and Cavallo, L. (2008) *Angew. Chem. Int. Ed.*, **47**, 718–721.
102 (a) Cariou, K., Mainetti, E., Fensterbank, L. and Malacria, M. (2004) *Tetrahedron*, **60**, 9745–9755; (b) Harrak, Y., Blasykowski, C., Bernard, M., Cariou, K., Mainetti, E., Mouriès, V., Dhimane, A.-L., Fensterbank, L. and Malacria, M. (2004) *J. Am. Chem. Soc.*, **126**, 8656–8657.
103 Bhanu Prasad, B.A., Yoshimoto, F.K. and Sarpong, R. (2005) *J. Am. Chem. Soc.*, **127**, 12468–12469.
104 (a) Miki, K., Ohe, K. and Uemura, S. (2003) *Tetrahedron Lett.*, **44**, 2019–2022; (b) Miki, K., Ohe, K. and Uemura, S. (2003) *J. Org. Chem.*, **68**, 8505–8513; (c) Miki, K., Fujita, M., Uemura, S. and Ohe, K. (2006) *Org. Lett.*, **8**, 1741–1743.
105 See also: Fructos, M.R., Belderrain, T.R., de Frémont, P., Scott, N.M., Nolan, S.P., Díaz-Requejo, M.M. and Pérez, P.J.

(2005) *Angew. Chem. Int. Ed.*, **44**, 5284–5288.

106 Amijs, C.H.M., López-Carrillo, V. and Echavarren, A.M. (2007) *Org. Lett.*, **9**, 4021–4024.

107 Davies, P.W. and Albrecht, S.J.-C. (2008) *Chem. Commun.*, 238–240.

108 Shapiro, N.D. and Toste, F.D. (2008) *J. Am. Chem. Soc.*, **130**, 9244–9245.

109 Strickler, H., Davis, J.B. and Ohloff, G. (1976) *Helv. Chim. Acta*, **59**, 1328–1332.

110 Rautenstrauch, V. (1984) *J. Org. Chem.*, **49**, 950–952.

111 Soriano, E. and Marco-Contelles, J. (2007) *J. Org. Chem.*, **72**, 2651–2654.

112 See also: (a) Soriano, E. and Marco-Contelles, J. (2007) *J. Org. Chem.*, **72**, 1443–1448; (b) Soriano, E. and Marco-Contelles, J. (2008) *Chem. Eur. J.*, **14**, 6771–6779.

113 For a more complex transformation: Marion, N., de Frémont, P., Lemière, G., Stevens, E.D., Fensterbank, L., Malacria, M. and Nolan, S.P. (2006) *Chem. Commun.*, 2048–2050.

114 Fürstner, A. and Hannen, P. (2006) *Chem. Eur. J.*, **12**, 3006–3019.

115 Fehr, C. and Galindo, J. (2006) *Angew. Chem. Int. Ed.*, **45**, 2901–2904.

116 Fürstner, A. and Hannen, P. (2004) *Chem. Commun.*, 2546–2547.

117 Shi, X., Gorin, D.J. and Toste, F.D. (2005) *J. Am. Chem. Soc.*, **127**, 5802–5803.

118 Nieto Faza, O., Silva López, C., Álvarez, R. and de Lera, A.R. (2006) *J. Am. Chem. Soc.*, **128**, 2434–2437.

119 Zhang, L. and Wang, S. (2006) *J. Am. Chem. Soc.*, **128**, 1442–1443.

120 Buzas, A. and Gagosz, F. (2006) *J. Am. Chem. Soc.*, **128**, 12614–12615.

121 Moreau, X., Goddard, J.-P., Bernard, M., Lemière, G., López-Romero, J.M., Mainetti, E., Marion, N., Mouriès, V., Thorimbert, S., Fensterbank, L. and Malacria, M. (2008) *Adv. Synth. Catal.*, **350**, 43–48.

122 Boyer, F.D., Le Goff, X. and Hanna, I. (2008) *J. Org. Chem.*, **73**, 5163–5166.

123 Zhang, L. (2005) *J. Am. Chem. Soc.*, **127**, 16804–16805.

124 Zhang, G., Catalano, V.J. and Zhang, L. (2007) *J. Am. Chem. Soc.*, **129**, 11358–11359.

125 Wang, S. and Zhang, L. (2006) *J. Am. Chem. Soc.*, **128**, 8414–8415; (2006) *J. Am. Chem. Soc.* [addition/correction], **128**, 9979.

126 (a) Wang, S. and Zhang, L. (2006) *Org. Lett.*, **8**, 4585–4587; (b) Yu, M., Zhang, G. and Zhang, L. (2007) *Org. Lett.*, **9**, 2146–2150.

127 Barluenga, J., Riesgo, L., Vicente, R., López, L.A. and Tomás, M. (2007) *J. Am. Chem. Soc.*, **129**, 7772–7773.

128 Buzas, A., Istrate, F. and Gagosz, F. (2006) *Org. Lett.*, **8**, 1957–1959.

129 Zhao, J., Hughes, C.O. and Toste, F.D. (2006) *J. Am. Chem. Soc.*, **128**, 7436–7437.

130 Miki, K., Ohe, K. and Uemura, S. (2003) *J. Org. Chem.*, **68**, 8505–8513.

131 Marion, N., Díez-González, S., de Frémont, P., Noble, A.R. and Nolan, S.P. (2006) *Angew. Chem. Int. Ed.*, **45**, 3647–3650.

132 Peng, L., Zhang, X., Zhang, S. and Wang, J. (2007) *J. Org. Chem.*, **72**, 1192–1197.

133 Wang, S. and Zhang, L. (2006) *J. Am. Chem. Soc.*, **128**, 14274–14275.

134 Dubé, P. and Toste, F.D. (2006) *J. Am. Chem. Soc.*, **128**, 12062–12063.

135 Kirsch, S.F., Binder, J.T., Crone, B., Duschek, A., Haug, T.T., Liébert, C. and Menz, H. (2007) *Angew. Chem. Int. Ed.*, **46**, 2310–2313.

136 Crone, B. and Kirsch, S.F. (2008) *Chem. Eur. J.*, **14**, 3514–3522.

137 Baskar, B., Bae, H.J., An, S.E., Cheong, J.Y., Rhee, Y.H., Duschek, A. and Kirsch, S.F. (2008) *Org. Lett.*, **10**, 2605–2607.

138 Fehr, C., Farris, I. and Sommer, H. (2006) *Org. Lett.*, **8**, 1839–1841.

139 Georgy, M., Boucard, V. and Campagne, J.-M. (2005) *J. Am. Chem. Soc.*, **127**, 14180–14181.

140 Engel, D.A. and Dudley, G.B. (2006) *Org. Lett.*, **8**, 4027–4029.

141 Yu, M., Li, G., Wang, S. and Zhang, L. (2007) *Adv. Synth. Catal.*, **349**, 871–875.

142 Marion, N., Carlqvist, P., Gealageas, R., de Frémont, P., Maseras, F. and Nolan, S.P. (2007) *Chem. Eur. J.*, **13**, 6437–6451.

143 Markham, J.P., Staben, S.T. and Toste, F.D. (2005) *J. Am. Chem. Soc.*, **127**, 9708–9709.

144 Sordo, T.L. and Ardura, D. (2008) *Eur. J. Org. Chem.*, 3004–3013.

145 Kirsch, S.F., Binder, J.T., Liébert, C. and Menz, H. (2006) *Angew. Chem. Int. Ed.*, **45**, 5878–5880.

146 Zhang, J. and Schmalz, H.-G. (2006) *Angew. Chem. Int. Ed.*, **45**, 6704–6707.

147 Michelet, V., Charruault, L., Gladiali, S. and Genêt, J.P. (2006) *Pure Appl. Chem.*, **78**, 397–407.

148 Charruault, L., Michelet, V., Taras, R., Gladiali, S. and Genêt, J.P. (2004) *Chem. Commun.*, 850–851.

149 (a) Galland, J.-C., Savignac, M. and Genêt, J.P. (1997) *Tetrahedron Lett.*, **38**, 8695–8698; (b) Galland, J.-C., Dias, S., Savignac, M. and Genêt, J.P. (2001) *Tetrahedron*, **57**, 5137–5148; (c) Charruault, L., Michelet, V. and Genêt, J.P. (2002) *Tetrahedron Lett.*, **43**, 4757–4760.

150 Faller, J.W. and Fontaine, P.P. (2006) *J. Organomet. Chem.*, **691**, 1912–1918.

151 Muñoz, M.P., Adrio, J., Carretero, J.C. and Echavarren, A.M. (2005) *Organometallics*, **24**, 1293–1300.

152 Genin, E., Leseurre, L., Toullec, P.Y., Genêt, J.P. and Michelet, V. (2007) *Synlett*, 1780–1784.

153 Fürstner, A. and Morency, L. (2008) *Angew. Chem. Int. Ed.*, **47**, 5030–5033.

154 An example of gold-catalyzed methoxycyclization of an allenene: Luzung, M.R., Mauleón, P. and Toste, F.D. (2007) *J. Am. Chem. Soc.*, **129**, 12402–12403.

155 Buzas, A.K., Istrate, F.M. and Gagosz, F. (2006) *Angew. Chem. Int. Ed.*, **46**, 1141–1144.

156 Zhang, L. and Kozmin, S.A. (2005) *J. Am. Chem. Soc.*, **127**, 6962–6963.

157 Park, S. and Lee, D. (2006) *J. Am. Chem. Soc.*, **128**, 10664–10665.

158 Horino, Y., Luzung, M.R. and Toste, F.D. (2006) *J. Am. Chem. Soc.*, **128**, 11364–11365.

159 Leseurre, L., Toullec, P.Y., Genêt, J.-P. and Michelet, V. (2007) *Org. Lett.*, **9**, 4049–4052.

160 Toullec, P.Y., Genin, E., Leseurre, L., Genêt, J.-P. and Michelet, V. (2006) *Angew. Chem. Int. Ed.*, **45**, 7427–7430.

161 Amijs, C.H.M., Ferrer, C. and Echavarren, A.M. (2007) *Chem. Commun.*, 698–700.

162 Lemière, G., Gandon, V., Agenet, N., Goddard, J.-P., de Kozak, A., Aubert, C., Fensterbank, L. and Malacria, M. (2006) *Angew. Chem. Int. Ed.*, **45**, 7596–7599.

163 Lin, G.-Y., Yang, C.-Y. and Liu, R.-S. (2007) *J. Org. Chem.*, **72**, 6753–6757.

164 Lian, J.-J., Chen, P.-C., Lin, Y.-P., Ting, H.-C. and Liu, R.S. (2006) *J. Am. Chem. Soc.*, **128**, 11372–11373.

165 Shibata, T., Fujiwara, R. and Takano, D. (2005) *Synlett*, 2062–2066. Erratum: (2007) *Synlett*, 2766.

166 Similar cyclobutenes are obtained in the Pt[II]-catalyzed reaction of 1,2-dien-7-ynes: Matsuda, T., Kadowaki, S., Goya, T. and Murakami, M. (2006) *Synlett*, 575–578.

167 Harrison, T.J., Patrick, B.O. and Dake, G.R. (2007) *Org. Lett.*, **9**, 367–370.

168 Deng, H., Yang, X., Tong, Z., Li, Z. and Zhai, H. (2008) *Org. Lett.*, **10**, 1791–1793.

169 Fürstner, A. and Stimson, C.C. (2007) *Angew. Chem. Int. Ed.*, **46**, 8845–8849.

170 Schelwies, M., Dempwolff, A.L., Rominger, F. and Helmchen, G. (2007) *Angew. Chem. Int. Ed.*, **46**, 5598–5601.

171 Imagawa, H., Iyenaga, T. and Nishizawa, M. (2005) *Org. Lett.*, **7**, 451–453.

172 Cabello, N., Jiménez-Núñez, E., Buñuel, E., Cárdenas, D.J. and Echavarren, A.M. (2007) *Eur. J. Org. Chem.*, 4217–4223.

173 Nieto-Oberhuber, C., López, S., Muñoz, M.P., Cárdenas, D.J., Buñuel, E., Nevado,

C. and Echavarren, A.M. (2005) *Angew. Chem. Int. Ed.*, **44**, 6146–6148.

174 Freytag, M., Ito, S. and Yoshifuji, M. (2006) *Chem. Asian J.*, **1**, 693–700.

175 (a) Reetz, M.T. (1972) *Angew. Chem. Int. Ed.*, **11**, 129–130; (b) Reetz, M.T. (1972) *Angew. Chem. Int. Ed.*, **11**, 130–131.

176 Nouri, D.H. and Tantillo, D.J. (2006) *J. Org. Chem.*, **71**, 3686–3695.

177 Ota, K. and Chatani, N. (2008) *Chem. Commun.*, 2906–2907.

178 Witham, C.A., Mauleón, P., Shapiro, N.D., Sherry, B.D. and Toste, F.D. (2007) *J. Am. Chem. Soc.*, **129**, 5838–5839.

179 (a) Irikura, K.K. and Goddard, W.A. (1994) *J. Am. Chem. Soc.*, **116**, 8733–8740; (b) Heinemann, C., Hertwig, R.H., Wesendrup, R., Knoch, W. and Schwarz, H. (1995) *J. Am. Chem. Soc.*, **117**, 495–500.

180 Anti-cyclopropyl iron and ruthenium carbenes have been characterized spectroscopically at low temperatures: Brookhart, M., Studabaker, W.B. and Husk, G.R. (1987) *Organometallics*, **6**, 1141–1145.

181 Fedorov, A., Moret, M.-E. and Chen, P. (2008) *J. Am. Chem. Soc.*, **130**, 8880–8881.

182 Casanova, J., Kent, D.R., Goddard, W.A. and Roberts, J.D. (2003) *Proc. Nat. Acad. Sci. USA*, **100**, 15–19.

183 Jiménez-Núñez, E., Claverie, C.K., Bour, C., Cárdenas, D.J. and Echavarren, A.M. (2008) *Angew. Chem. Int. Ed.*, **47**, anie.200803269.

184 Cabello, N., Rodríguez, C. and Echavarren, A.M. (2007) *Synlett*, 1753–1758.

185 Simmons, E.M. and Sarpong, R. (2006) *Org. Lett.*, **8**, 2883–2886.

186 Simmons, E.M., Yen, J.R. and Sarpong, R. (2007) *Org. Lett.*, **9**, 2705–2708.

187 Fürstner, A., Schlecker, A. and Lehmann, C.W. (2007) *Chem. Commun.*, 4277–4279.

188 Formation of a bicyclo[3.2.0]hept-5-ene by the thermal reaction of a 1,6-fullerenyne was shown to occur through a stepwise diradical reaction mechanism: (a) Martín, N., Altable, M., Filippone, S., Martín-Domenech, Á., Guell, M. and Solà, M. (2006) *Angew. Chem. Int. Ed.*, **45**, 1439–1442; (b) Guell, M., Martín, N., Altable, M., Filippone, S., Martín-Domenech, Á. and Solà, M. (2007) *J. Phys. Chem. A*, **111**, 5253–5258.

189 A silabicyclo[3.2.0]hept-5-ene has been isolated as a stable compound: Oba, G., Moreira, G., Manuel, G. and M. Koenig, M. (2002) *J. Organomet. Chem.*, **643-644**, 324–330.

190 Marion, F., Coulomb, J., Courillon, C., Fernsternbank, L. and Malacria, M. (2004) *Org. Lett.*, **6**, 1509–1511.

191 Couty, S., Meyer, C. and Cossy, J. (2006) *Angew. Chem. Int. Ed.*, **45**, 6726–6730.

192 Soriano, E., Ballesteros, P. and Marco-Contelles, J. (2005) *Organometallics*, **24**, 3182–3191.

193 See also: (a) Soriano, E., Ballesteros, P. and Marco-Contelles, J. (2005) *Organometallics*, **24**, 3172–3181; (b) Soriano, E., Ballesteros, P. and Marco-Contelles, J. (2004) *J. Org. Chem.*, **69**, 8018–8023; (c) Soriano, E. and Marco-Contelles, J. (2005) *J. Org. Chem.*, **70**, 9345–9353.

194 (a) Lin, M.-Y., Das, A. and Liu, R.-S. (2006) *J. Am. Chem. Soc.*, **128**, 9340–9341; (b) Tang, J.-M., Bhunia, S., Sohel, S.M.A., Lin, M.-Y., Liao, H.Y., Datta, S., Das, A. and Liu, R.-S. (2007) *J. Am. Chem. Soc.*, **129**, 15677–15683.

195 Kim, H. and Lee, C. (2005) *J. Am. Chem. Soc.*, **127**, 10180–10181.

196 (a) Zhang, L. and Kozmin, S. (2004) *J. Am. Chem. Soc.*, **126**, 11806–11807; (b) Sun, J., Conley, M.P., Zhang, L. and Kozmin, S.A. (2006) *J. Am. Chem. Soc.*, **128**, 9705–9710.

197 (a) Mainetti, E., Mouriès, V., Fenstebank, L., Malacria, M. and Marco-Contelles, J. (2002) *Angew. Chem. Int. Ed.*, **41**, 2132–2135; (b) Harrak, Y., Blaszykowski, C., Bernard, M., Cariou, K., Mainetti, E., Mouriès, V., Dhimane, A.-L., Fenstebank, L. and Malacria, M. (2004) *J. Am. Chem. Soc.*, **126**, 8656–8657; (c) Blaszykowski, C., Harrak, Y., Gonçalves, M.-H., Cloarec, J.-M.,

Dhimane, A.-L., Fensterbank, L. and Malacria, M. (2004) *Org. Lett.*, **6**, 3771–3774; (d) Marco-Contelles, J., Arroyo, N., Anjum, S., Mainetti, E., Marion, N., Cariou, K., Lemière, G., Mouriés, V., Fensterbank, L. and Malacria, M. (2006) *Eur. J. Org. Chem.*, 4618–4633; (e) Blaszykowski, C., Harrak, Y., Brancour, C., Nakama, K., Dhimane, A.-L., Fensterbank, L. and Malacria, M. (2007) *Synthesis*, 2037–2049.
198 Peppers, B.P. and Divers, S.T. (2004) *J. Am. Chem. Soc.*, **126**, 9524–9525.
199 Kim, S.M., Park, J.H., Choi, S.Y. and Chung, Y.K. (2007) *Angew. Chem. Int. Ed.*, **46**, 6172–6175.
200 Porcel, S. and Echavarren, A.M. (2007) *Angew. Chem. Int. Ed.*, **46**, 2672–2676.
201 Lemière, G., Gandon, V., Cariou, K., Fukuyama, T., Dhimana, A.-L., Fensterbank, L. and Malacria, M. (2007) *Org. Lett.*, **9**, 2207–2209.
202 Tanaka, D., Sato, Y. and Mori, M. (2007) *J. Am. Chem. Soc.*, **129**, 7730–7731.
203 Johansson, M.J., Gorin, D.J., Staben, S.T. and Toste, F.D. (2005) *J. Am. Chem. Soc.*, **127**, 18002–18003.
204 For a review see: Lee, D. and Kim, M. (2007) *Org. Biomol. Chem.*, **5**, 3418–3427.
205 Padwa, A., Austin, D.J., Gareau, Y., Kassir, J.M. and Xu, S.L. (1993) *J. Am. Chem. Soc.*, **115**, 2637–2647.
206 Barluenga, J., de la Rúa, R.D., de Saá, D., Ballesteros, A. and Tomás, M. (2005) *Angew. Chem. Int. Ed.*, **44**, 4981–4983.
207 (a) Kim, M., Miller, R.L. and Lee, D. (2005) *J. Am. Chem. Soc.*, **127**, 12818–12819; (b) Kim, M. and Lee, D. (2005) *J. Am. Chem. Soc.*, **127**, 18024–18025.
208 van Otterlo, W.A.L., Lindani Ngidi, E., de Koning, C.B. and Fernandes, M.A. (2004) *Tetrahedron Lett.*, **45**, 659–662.
209 Related metallotropic rearrangements of alkynyl Re and Mn carbenes: (a) Casey, C.P., Kraft, S. and Powell, D.R. (2000) *J. Am. Chem. Soc.*, **122**, 3771–3772; (b) Casey, C.P., Kraft, S. and Powell, D.R. (2001) *Organometallics*, **20**, 2651–2653; (c) Casey, C.P., Kraft, S. and Kavana, M. (2001) *Organometallics*, **20**, 3795–3799; (d) Casey, C.P., Kraft, S. and Powell, D.R. (2002) *J. Am. Chem. Soc.*, **124**, 2584–2594; (e) Casey, C.P., Dzwiniel, T.L., Kraft, S. and Guzei, I.A. (2003) *Organometallics*, **22**, 3915–3929; (f) Casey, C.P. and Dzwiniel, T.L. (2003) *Organometallics*, **22**, 5285–5290; (g) Ortin, Y., Sournia-Saquet, A., Lugan, N. and Mathieu, R. (2003) *Chem. Commun*, 1060–1061.
210 Gorin, D.J., Dubé, P. and Toste, F.D. (2006) *J. Am. Chem. Soc.*, **128**, 14480–14481.
211 (a) Cho, E.J., Kim, M. and Lee, D. (2006) *Eur. J. Org. Chem.*, 3074–3078; (b) For similar reactions catalyzed by Pt[II]: Cho, E.J., Kim, M. and Lee, D. (2006) *Org. Lett.*, **8**, 5413–5416.
212 Ohe, K., Fujita, M., Matsumoto, H., Tai, Y. and Miki, K. (2006) *J. Am. Chem. Soc.*, **128**, 9270–9271.
213 (a) Hashmi, A.S.K., Frost, T.M. and Bats, J.W. (2000) *J. Am. Chem. Soc.*, **122**, 11553–11554; (b) Hashmi, A.S.K., Frost, T.M. and Bats, J.W. (2001) *Org. Lett.*, **3**, 3769–3771; (c) Hashmi, A.S.K., Frost, T.M. and Bats, J.W. (2001) *Catal. Today*, **72**, 19–27; (d) Hashmi, A.S.K., Ding, L., Bats, J.W., Fischer, P. and Frey, W. (2003) *Chem. Eur. J.*, **9**, 4339–4345; (e) Hashmi, A.S.K., Weyrauch, J.P., Rudolph, M. and Kurpejovic, E. (2004) *Angew. Chem. Int. Ed.*, **43**, 6545–6547; (f) Hashmi, A.S.K., Rudolph, M., Weyrauch, J.P., Wölfle, M., Frey, W. and Bats, J.W. (2005) *Angew. Chem. Int. Ed.*, **44**, 2798–2801; (g) Hashmi, A.S.K., Weyrauch, J.P., Kurpejovic, E., Frost, T.M., Miehlich, B., Frey, W. and Bats, J.W. (2006) *Chem. Eur. J.*, **12**, 5806–5814; (h) Hashmi, A.S.K., Kurpejovic, E., Frey, W. and Bats, J.W. (2007) *Tetrahedron*, **63**, 5879–5885.
214 Hashmi, A.S.K., Haufe, P., Schmid, C., Rivas Nass, A. and Frey, W. (2006) *Chem. Eur. J.*, **12**, 5376–5382.
215 Carrettin, S., Blanco, M.C., Corma, A. and Hashmi, A.S.K. (2006) *Adv. Synth. Catal.*, **348**, 1283–1288.
216 (a) Martín-Matute, B., Cárdenas, D.J. and Echavarren, A.M. (2001) *Angew. Chem.*

*Int. Ed.*, **40**, 4754–4757; (b) Martín-Matute, B., Nevado, C., Cárdenas, D.J. and Echavarren, A.M. (2003) *J. Am. Chem. Soc.*, **125**, 5757–5766.
217 Hashmi, A.S.K., Salathé, R. and Frey, W. (2006) *Chem. Eur. J.*, **12**, 6991–6996.
218 Dyker, G., Muth, E., Hashmi, A.S.K. and Ding, L. (2003) *Adv. Synth. Catal.*, **345**, 1247–1252.
219 Hashmi, A.S.K. and Grundl, L. (2005) *Tetrahedron*, **61**, 6231–6236.
220 Aguilar, D., Contel, M., Navarro, R. and Urriolabeitia, E.P. (2007) *Organometallics*, **26**, 4604–4611.
221 Hashmi, A.S.K., Kurpejovic, E., Wölfle, M., Frey, W. and Bats, J.W. (2007) *Adv. Synth. Catal.*, **349**, 1743–1750.
222 Hashmi, A.S.K., Rudolph, M., Siehl, H.-U., Tanaka, M., Bats, J.W. and Frey, W. (2008) *Chem. Eur. J.*, **14**, 3703–3708.
223 (a) Hashmi, A.S.K., Blanco, M.C., Kurpejovic, E., Frey, W. and Bats, J.W. (2006) *Adv. Synth. Catal.*, **348**, 709–713; (b) Hashmi, A.S.K. and Blanco, M.C. (2006) *Eur. J. Org. Chem.*, 4340–4342.
224 Bayler, A., Bauer, A. and Schmidbaur, H. (1997) *Chem. Ber/Recueil*, **130**, 115–118.

# 14
# Cyclopropanation, Epoxidation and Aziridination Reactions
*Song Ye and Yong Tang*

## 14.1
## Introduction

Cyclopropanes, expoxides and aziridines occur frequently in a wide range of natural and synthetic compounds with important biological and pharmacological applications. Furthermore, due to the strain associated with the three-membered ring, they are also widely applied as important synthetic key intermediates in different processes. Thus the synthesis of cyclopropanes, epoxides, and aziridines has been heavily pursued, and many kinds of synthetic methodology have been developed.

Generally, the methodologies to synthesize cyclopropanes, epoxides, and aziridines can be summarized as in Scheme 14.1. Alkenes are the common starting material for all these three kinds of compounds, while epoxides and aziridines can also be prepared from aldehydes and imines, respectively.

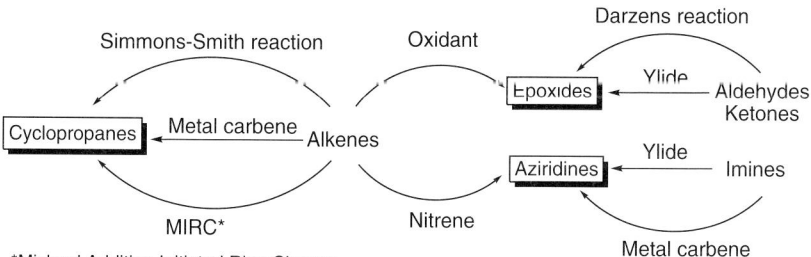

**Scheme 14.1** Synthesis of cyclopropanes, epoxides and aziridines.

There are hundreds of papers published every year on the synthesis and application of these three-membered ring compounds. Only an overview and selected recent examples will be given in this chapter. More detailed information can be found in recent reviews and book chapters dealing with cyclopropanation [1], epoxidation [2], and aziridination reactions [3].

## 14.2
## Cyclopropanation

Compounds containing cyclopropane fragments have received considerable attention because of their frequent occurrence in biologically active compounds such as terpenes, pheromones, fatty acid metabolites and unusual amino acids, as well as their utility as synthetic intermediates. In this chapter, four approaches to cyclcopropanes will be discussed: (i) cyclopropanation with halomethylmetal compounds; (ii) cyclopropanation with transition metal-catalyzed decomposition of diazo compounds; (iii) cyclopropanation with ylides, α-halocarbanions, and related species; (iv) cyclopropanation with metalocarbenoids.

### 14.2.1
### Cyclopropanation with Halomethylmetal Compounds

Simmons and Smith in 1958 reported that iodomethylzinc species, generated *in situ* from a zinc-copper couple and diiodomethane, reacted with alkenes to give cyclopropanes (Eq. 14.1) [4]. This transformation has become a powerful method for cyclopropane formation. The advantages of the Simmons–Smith reaction can be summarized as follows: (i) stereospecific transfer of a methylene group to a variety of alkenes including electron-rich and electron-deficient ones under mild conditions; (ii) strong directing effects of functional groups containing oxygen or nitrogen atoms, resulting in regio- and stereo-selective cyclopropanations; (iii) tolerant of most functional groups; (iv) no serious side reactions such as olefin isomerization and insertion of the carbenoid into C–H bonds.

$$\underset{1}{R^1\text{—CH=CH—}R^2} \xrightarrow[\text{Ether}]{\text{Zn/Cu, CH}_2\text{I}_2} \underset{2}{\left[ \begin{array}{c} \text{I—Zn—I} \\ \text{CH}_2 \\ R^1\text{, H} \cdots R^2\text{, H} \end{array} \right]^{\ddagger}} \longrightarrow \underset{3}{R^1\text{—△—}R^2} \quad (14.1)$$

Following the ground-breaking work of Simmons and Smith, many efforts were made to modify the synthesis of halomethylzinc reagents for cyclopropantion of olefins [5]. Other halomethylmetal reagents, such as samarium carbenoid [6], aluminum carbenoid [7], lithium carbenoid [8], chromium carbenoid [9] and titanium carbenoid [10] have also been developed.

Recent advancements in the Simmons–Smith reaction focus on its asymmetric version. Allylic alcohols proved to be optimal substrates for the stereoselective Simmons–Smith reaction. For example, the enantioselective cyclopropanation of allylic alcohols was achieved in the presence of stoichiometric dioxaborolane **4**, prepared from tetramethyltartaric acid diamide and butylboronic acid (Eq. 14.2) [11]. A catalytic asymmetric version has been realized in the presence of substoichiometric TADDOL-Ti compound **7** [12], or substoichiometric bis(sulfonamide) **8** [13] or **9** [14] (Eqs. 14.3 and 14.4).

$$\text{(14.2)}$$

Scheme for Eq. 14.2: Allylic alcohol **5** (with R¹, R², R³ substituents) is treated with:
1. Dioxaborolane **4** (1.1 equiv), Bu, CH₂Cl₂
2. Zn(CH₂I)₂ (2.2 equiv), CH₂Cl₂, 0 to 25 °C

giving cyclopropane **6**, > 80%, up to 94% ee.

$$\text{(14.3)}$$

Allylic alcohol (R¹, R², R³) with Ti–TADDOLate catalyst **7** (0.25 equiv), Zn(CH₂I)₂ (1 equiv), 4Å MS, CH₂Cl₂, 0 °C → cyclopropane product, 56–90%, up to 94% ee.

$$\text{(14.4)}$$

Allylic alcohol **10** with chiral disulfonamide catalyst (0.12 equiv):
**8** R = p-NO₂C₆H₄
**9** R = CH₃
ZnEt₂, CH₂I₂ → cyclopropane **11**, 36–100%, up to 86% ee (Cat **8**), 89% ee (Cat **9**).

Allylic amines [15], α, β-unsaturated carboxylic acids [16], α, β-unsaturated amides [17], enamides [18] and silyl enol ethers [19] are also good substrates for the Simmons–Smith reaction. By employing dipeptide **12** as a chiral ligand, for example, the asymmetric cyclopropanation reaction of silyl enol ethers proceeded smoothly with up to 96% ee (Eq. 14.5). In addition, the dipeptide can be recovered in good yield and reused without obvious loss of reactivity or enantioselectivity.

$$\text{(14.5)}$$

Silyl enol ether **13** (R₃SiO, R¹, R², R³) + ZnEt₂, CH₂I₂ with dipeptide ligand **12** (Boc-Val-Pro-derived) → cyclopropane **14**, 77–99%, up to 96% ee.

The asymmetric Simmons–Smith reaction of unfunctionalized olefins is a challenge. Shi et al. recently reported that stoichiometric dipeptide N-Boc-L-Val-L-Pro-OMe **15** could promote the asymmetric cyclopropanation of olefins without functional groups and up to 91% ee was attained (Eq. 14.6). In the presence of achiral carbonyl compounds, the amount of dipeptide **15** could be reduced to 0.25 equiv [20].

$$(14.6)$$

### 14.2.2
### Cyclopropanation with Transition Metal-Catalyzed Decomposition of Diazo Compounds

The cyclopropanation of olefins using the transition metal-catalyzed decomposition of diazoalkanes has been investigated extensively (Eq. 14.7). Copper, cobalt, ruthenium, rhodium and palladium compounds have been used as highly efficient catalysts. This methodology has been applied in laboratory-scale synthesis as well as in the fine chemical industry.

$$(14.7)$$

#### 14.2.2.1 Diazoalkanes
Diazoalkanes can efficiently transfer an alkylene unit to olefinic double bonds to form cyclopropane derivatives via metal catalyzed processes. Various alkenes can be employed as the substrates for this reaction. Copper(II) triflate, copper(II) acetylacetonate, and copper(II) hexafluoroacetylacetonate proved to be good catalysts. It is suggested that the real active species is copper(I) which is readily generated by the reduction of copper(II) with the diazo reagent. In the presence of 5 mol% of chiral bis(oxazoline), copper(I) triflate is also effective in catalysis of the enantioselective cyclopropanation reaction of trans-cinnamate esters with moderate to good ees (Eq. 14.8) [21].

$$\text{(14.8)}$$

Aggarwal and coworkers documented a cyclopropanation reaction of alkenes with aryl and alkenyl diazomethanes generated *in situ* from stable hydrazone derivatives [22]. This method was successfully employed for the preparation of cyclopropanylamino acid [23].

### 14.2.2.2 Diazocarbonyl Compounds

The reactions of alkenes with diazocarbonyl compounds in the presence of a metal complex provide a facile method for the construction of functionalized cyclopropanes. The commonly-used diazo compounds are diazoesters, diazoacetamides, diazoketones and dicarbonyl diazomethanes. Metal complexes derived from Cu, Rh, Ru, Co, Fe, Os, Pd, Pt, Cr, and so on have been reported to catalyze the decomposition of diazo compounds for cyclopropanation [24]. Generally, Rh, Ru, Co and Cu metal carbenes can react smoothly with electron-rich alkenes, whereas Pd metal carbenes are optimal for electron-deficient alkenes.

Since the first example of an enantioselective copper-based cyclopropanation reaction was reported [25], the metal-catalyzed cyclopropanation between alkenes and alkyl diazoesters has been extensively explored. The copper(I) complex of bis(oxazoline) 21 [26] is one of the optimal catalysts for cyclopropanations (Eq. 14.9). To achieve both high diastereoselectivities and excellent enantioselectivities, bulky diazoacetate is usually required. Recently, CuOTf/trisoxazoline ligand 24 (TOX) was found to catalyze the enantioselective cyclopropanation of alkenes with aryldiazoacetate with high ees (Eq. 14.10) [27]. In this reaction, trisoxazoline gave much higher ees and yields than the parent bisoxazolines.

$$\text{(14.9)}$$

23a R = Me, 70%, > 99% ee
23b R = Ph, 91%, > 99% ee

**692** | *14 Cyclopropanation, Epoxidation and Aziridination Reactions*

$$\text{R}^1\!\!\overset{}{\diagup}\!\!\text{R}^2 + \text{Ph}\underset{\underset{\text{O}}{\|}}{\overset{\text{N}_2}{\text{C}}}\text{OEt} \xrightarrow[\text{CH}_3\text{COOEt, 3Å MS}]{\textbf{24} \ (5\ \text{mol\%}) \atop \text{CuPF}_6} \overset{\text{R}^2}{\underset{\text{R}^1}{\triangle}}\!\!\overset{\text{COOEt}}{\text{Ph}}$$

**25**  **26**  **27**

51–99%
up to 95% ee

(14.10)

In general, ruthenium carbenes bearing chiral ligands are less reactive than those derived from copper and rhodium. PYBOX-Ru is an efficient catalyst for cyclopropanation reactions. For example, the reaction of alkenes with diazoacetate, catalyzed by PYBOX-Ru **28**, provided the desired cyclopropanes with enantiomeric excesses up to 98% (Eq. 14.11) [28]. With Cu(I)/BOX **21**, the trans/cis ratio of the cyclopropanes produced was improved with bulky diazoacetates (Me < Et < *t*-Bu < Ment; Ment = menthyl). Very recently, various ruthenium complexes derived from porphyrin [29], chiral Schiff base [30], and salen [31] have been synthesized and applied in asymmetric cyclopropanation reactions with excellent enantio- and diastereo-control.

$$\text{R}^1\!\!\diagup \xrightarrow[\substack{\text{N}_2\text{CHCO}_2\text{R} \\ \text{R = Me, Et, }^t\text{Bu, }l\text{-menthyl}}]{\textbf{28}} \text{R}^1\!\!\overset{}{\diagup}\!\!\overset{\text{CO}_2\text{R}}{} + \text{R}^1\!\!\overset{}{\triangle}\!\!\overset{}{\text{CO}_2\text{R}}$$

**29**  **30**

R¹ = Ph, **29**:**30** = >90:10; 86–96% ee (**29**);
R¹ = $^n$C$_5$H$_{11}$, R = *l*-menthyl, **29**:**30** = 94:6; 99% ee (**29**)

(14.11)

Yamada, Katsuki, and Zhang demonstrated that cobalt catalysts are efficient for trans- or cis-selective cyclopropanation reactions. Most catalytic systems worked exceptionally well with electron-rich olefins due to the electrophilic nature of the metal-carbene intermediates in the catalytic cycles. However, Zhang *et al.* disclosed that the cobalt-porphyrin complex **31** was highly efficient for the asymmetric cyclopropanation of electron-deficient olefins such as α, β-unsaturated esters, amides, ketones, and nitriles (Eq. 14.12) [32].

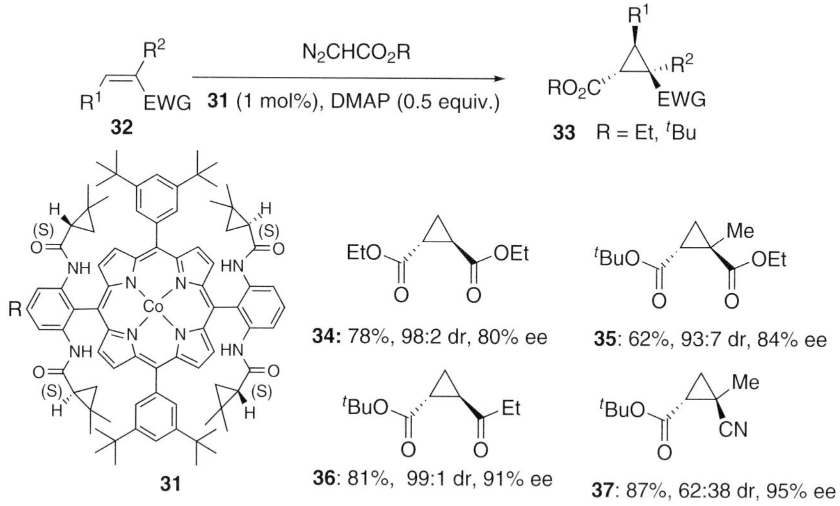

## 14.2 Cyclopropanation

$$\text{(14.12)}$$

**32**: R¹, R² alkene with EWG
**31**: (1 mol%), DMAP (0.5 equiv.), N₂CHCO₂R
**33**: R = Et, $^t$Bu

**34**: 78%, 98:2 dr, 80% ee (EtO₂C–cyclopropane–CO₂Et)
**35**: 62%, 93:7 dr, 84% ee ($^t$BuO₂C–cyclopropane(Me)–CO₂Et)
**36**: 81%, 99:1 dr, 91% ee ($^t$BuO₂C–cyclopropane–Et)
**37**: 87%, 62:38 dr, 95% ee ($^t$BuO₂C–cyclopropane(Me)–CN)

**31**: Co porphyrin catalyst

A rhodium-based chiral complex has been intensively investigated in both inter- and intramolecular cyclopropanations. The *N*-confused porphyrin rhodium complexes are highly efficient catalysts of the reaction of styrene with diazoacetate, affording the trans-disubstituted cyclopropanes in excellent yields with up to 98/2 dr (Eq. 14.13) [33]. Doyle and coworkers reported that the *l*-menthyl ester-derived azetidine-ligated dirhodium catalyst led to cis-diastereo-control and excellent enantio-control in the cyclopropanation of substituted styrenes with *tert*-butyl diazoacetate (Eq. 14.14) [34].

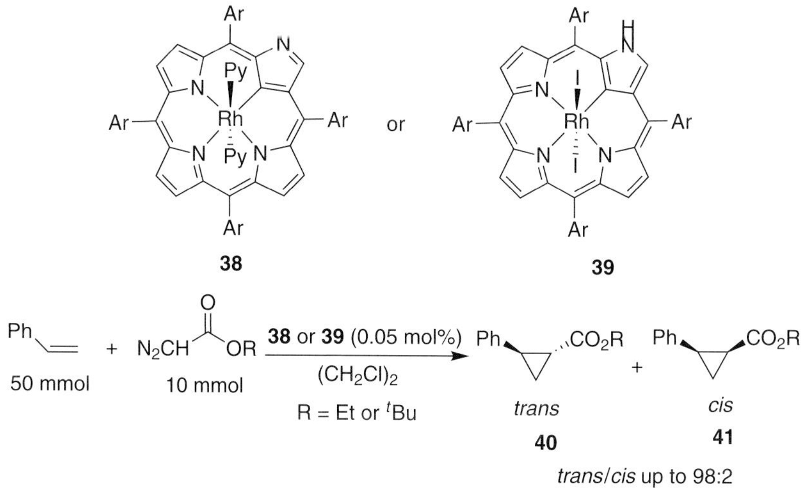

**38** and **39**: Rh N-confused porphyrin complexes

Ph–CH=CH₂ (50 mmol) + N₂CH–C(O)OR (10 mmol) → **38** or **39** (0.05 mol%), (CH₂Cl₂)₂, R = Et or $^t$Bu → **40** trans (Ph, CO₂R) + **41** cis (Ph, CO₂R)

trans/cis up to 98:2

$$\text{(14.13)}$$

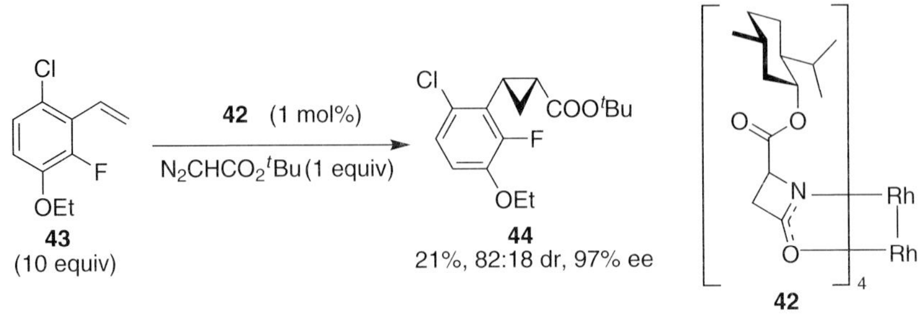

$$\text{43} \xrightarrow[\text{N}_2\text{CHCO}_2{}^t\text{Bu (1 equiv)}]{\textbf{42} \ (1 \ \text{mol\%})} \textbf{44}$$

**43** (10 equiv) — aryl vinyl with Cl, F, OEt substituents

**44**: 21%, 82:18 dr, 97% ee

**42**: Rh catalyst (azetidinone-based, ×4)

(14.14)

Aryliridinium-salen complexes (**45–47**) can catalyze the cyclopropanation of styrenes, substituted styrenes, indene and benzofuran, and both high cis-selectivity and excellent enantioselectivity could be achieved when **45** or **47** was used (Eq. 14.15) [35].

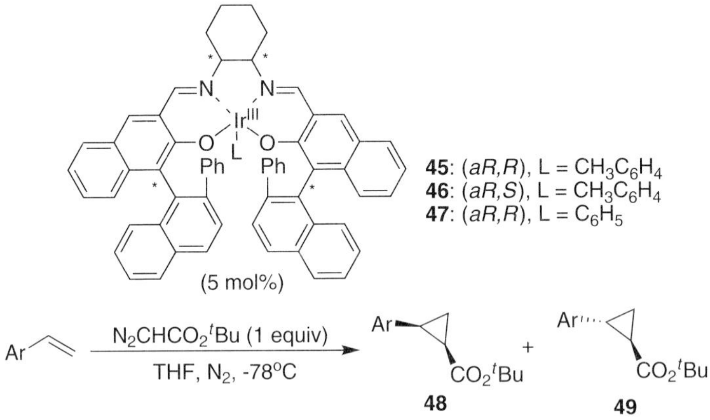

**45**: (aR,R), L = CH$_3$C$_6$H$_4$
**46**: (aR,S), L = CH$_3$C$_6$H$_4$
**47**: (aR,R), L = C$_6$H$_5$

$$\text{Ar}\diagup\!\!= \xrightarrow[\text{THF, N}_2,\ -78°\text{C}]{\text{N}_2\text{CHCO}_2{}^t\text{Bu (1 equiv)}} \textbf{48} + \textbf{49}$$

(5 mol%)

**48**: Ar—cyclopropane—CO$_2{}^t$Bu (cis)
**49**: Ar····cyclopropane—CO$_2{}^t$Bu (trans)

up to 99% yield, 99:1 cis/trans, 99% ee

(14.15)

Very recently, AgSbF$_6$ has also been demonstrated to be an efficient catalyst for the synthesis of donor/acceptor cyclopropanes and shows a different reactivity profile from the traditional rhodium-catalyzed reactions [36].

## 14.2.3
### Cyclopropanation with Ylides, α-Halocarbanions and Related Species

Cyclopropanation reactions that involve a conjugate addition to electron-deficient alkenes, followed by an intramolecular ring closure, are defined as Michael initiated ring closure reactions (MIRC). To date, two types of MIRC reactions have been reported. The first type involves the formation of cyclopropanes by nucleophilic addition to substrates containing a leaving group (Eq. 14.16). The nucleophiles reported include alkoxides, thiolates, cyanides, enolates, Grignard reagents, hydrides, phosphites, and phosphonites. The second is the addition of the nucleophile with a leaving group such as α-halo carbanions and ylides to electron-deficient alkenes (Eq. 14.17). Amongst these, ylide-initiated ring closure reactions represent an effective method for the synthesis of functionalized cyclopropanes and the studies in the past decade focus mainly on the development of its asymmetric and catalytic version.

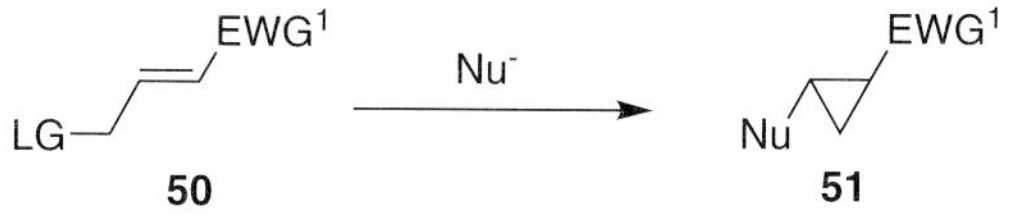

$$\text{LG}\diagup\!\!=\!\!\diagdown\text{EWG}^1 \xrightarrow{\text{Nu}^-} \underset{\text{Nu}}{\triangle}\text{EWG}^1 \quad (14.16)$$

**50**        **51**

$$=\!\!\diagdown\text{EWG}^1 \xrightarrow{\text{LG}\bar{\text{C}}\text{HR}} \underset{\text{R}}{\triangle}\text{EWG}^1 \quad (14.17)$$

**52**        **53**

Tang *et al.* reported that the chiral sulfonium salts, derived from readily available D-camphor, were novel reagents to react with a variety of Michael acceptors to give vinylcyclopropane derivatives. For example, the exo-type sulfonium salts **54** reacted with α, β-unsaturated esters, amides, ketones, and nitriles to give 1, 3-*trans*-2-vinylcyclopropanes with high diastereoselectivities and enantioselectivities. When endo-isomers **55** were employed, the corresponding cyclopropanes were obtained with opposite enantioselection (Eqs. 14.18 and 14.19) [37].

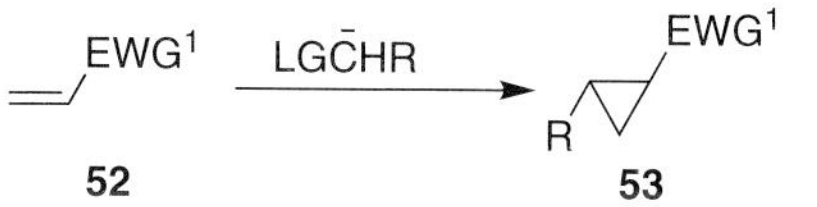

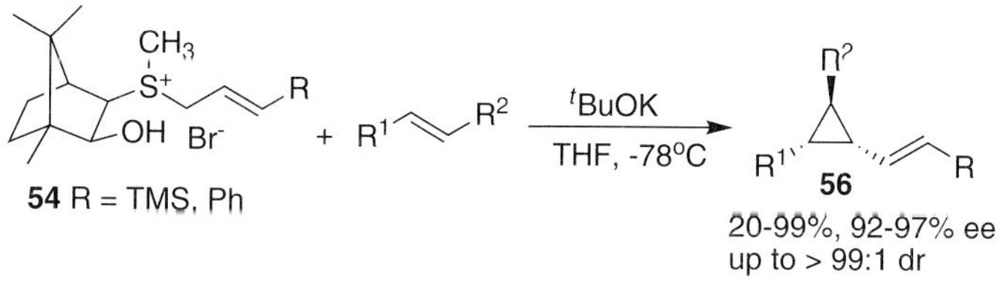

**54** R = TMS, Ph + R¹⟶R² →(tBuOK, THF, -78°C)→ **56**

20–99%, 92–97% ee
up to > 99:1 dr

(14.18)

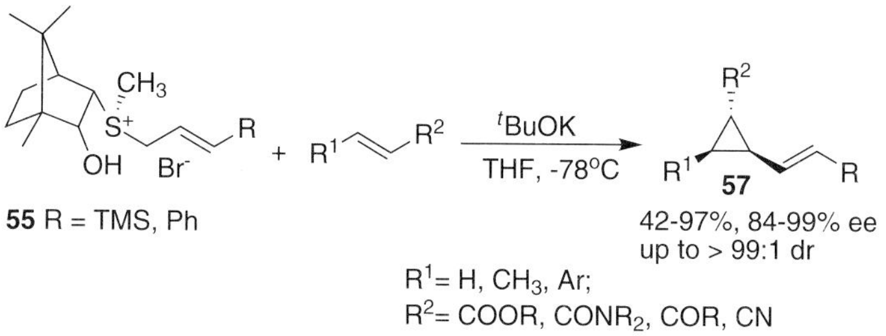

**55** R = TMS, Ph + R¹⟶R² →(tBuOK, THF, -78°C)→ **57**

42–97%, 84–99% ee
up to > 99:1 dr

R¹= H, CH₃, Ar;
R²= COOR, CONR₂, COR, CN

(14.19)

Aggarwal *et al.* found that sulfur ylides could be generated by the reaction of a sulfide with a diazo compound in the presence of a metal catalyst. This process has been successfully applied in asymmetric catalytic cyclopropanation of electron-deficient alkenes with diazo compounds (Eq. 14.20) [38].

$$\underset{59}{\underset{Ph}{\overset{H}{>}}\!\!=\!\!\underset{H}{\overset{COPh}{<}}} + \underset{\underset{1.5\ equiv}{60}}{Ph\!\!-\!\!CH\!\!=\!\!N\!\!-\!\!N^-\!\!-\!\!Ts\ Na^+} \xrightarrow[\substack{Rh_2(OAc)_4\ (1\ mol\%) \\ BnEt_3N^+Cl^-\ (20\ mol\%) \\ 1,4\text{-dioxane},\ 40\ ^\circ C}]{\mathbf{58}\ (20\ mol\%)} \underset{\underset{73\%,\ 4:1\ dr,\ 91\%\ ee}{61}}{\text{cyclopropane product}} \quad (14.20)$$

Madalengoitia disclosed the first examples of chiral Lewis acid-mediated asymmetric cyclopropanations of a Michael acceptor with sulfur ylides. The enantioselectivity depends on the loading of Lewis acid and up to 95% ee was obtained in the presence of stoichiometric $Zn(OTf)_2$/BOX (Eq. 14.21) [39]. Very recently, Shibasaki reported a catalytic asymmetric cyclopropanation of enones with dimethyloxosulfonium methylide using 1–10 mol% of La-Li$_3$-(biphenyldiolate)$_3$ and NaI complex as a catalyst, affording cyclopropanes in good yields with high ees (Eq. 14.22) [40].

$$\underset{63}{\text{oxazolidinone-CO-CH=CH-Me}} \xrightarrow[\substack{Ph_2S=CMe_2 \\ Zn(OTf)_2/THF,\ -78\ ^\circ C}]{\mathbf{62}} \underset{\underset{63\%,\ 95\%\ ee}{64}}{\text{cyclopropane}} \quad (14.21)$$

**65** Ligand

$$\underset{66}{R^2\!\!-\!\!CH\!\!=\!\!CH\!\!-\!\!CO\!\!-\!\!R^1} + \underset{(1.2\ equiv)}{(H_3C)_2\overset{+}{S}\!\!-\!\!CH_2^-} \xrightarrow[\substack{(S)\text{-La-Li}_3\text{-L}_3\ (5\ mol\%) \\ \text{NaI}\ (5\ mol\%) \\ \text{THF/toluene}\ (4:5) \\ \text{MS 4Å, -55 }^\circ C}]{} \underset{\underset{\substack{73\text{-}96\% \\ \text{up to 99\% ee}}}{67}}{R^2\text{-cyclopropyl-CO-}R^1} \quad (14.22)$$

Organocatalytic enantioselective cyclopropanation catalyzed by chiral secondary amine has been developed. The cyclopropanation reaction of α, β-unsaturated aldehydes with sulfur yildes proceeded smoothly in high yields with high diastereoselecitivities and enantioselecitivities in the presence of catalytic (2S)-2-carboxylic acid dihydroindole **68** (Eq. 14.23) [41]. Tetrazolic acid functionalized dihydroindole, and aryl sulfonamides derived from dihydroindole were also demonstrated to be efficient catalysts in the same reactions [42].

$$R^1\diagup\!\!\!\diagdown\!\!\!O + Me\overset{Me}{\underset{-}{S^+}}\!R^2 \xrightarrow[\text{CHCl}_3,\ -10\ °C]{\textbf{68}\ 20\ mol\%} R^1\triangledown R^2$$

**69**  **70**  up to 96% ee  **71** 64–85%, up to 96% ee

(14.23)

Telluronium allylides were developed as efficient reagents for diastereoselective and/or enantioselective cyclopropanation of a variety of Michael acceptors [43]. A recent study showed that telluronium salts **72**, in the presence of base, reacted with α,β-unsaturated esters, ketones, and amides with high diastereoselectivities and excellent enantioselectivities in good to high yields. This reaction provided an easy access to cis-2-silylvinyl-trans-3-substituted or trans-2-silylvinyl-trans-3-substituted cyclopropane derivatives by the choice of reaction conditions (Eqs. 14.24 and 14.25) [44].

**72** + **73** → (LiTMP, HMPA) **74** 42–99%, up to 99% ee, 99% de (14.24)

→ (LDA, LiBr) **75** 58–98%, up to 92% ee, 99% de (14.25)

The chiral ammonium ylides derived from cinchona alkaloid worked very well in the enantioselective intermolecular and intramolecular cyclopropanations (Eqs. 14.26 and 14.27) [45]. By selecting 2′-methyl-9-O-methylquinine (Me-MQ) and 2′-methyl-9-O-methylquinidine (Me-MQD), both enantiomers are obtainable in high yields with excellent enantioselectivities.

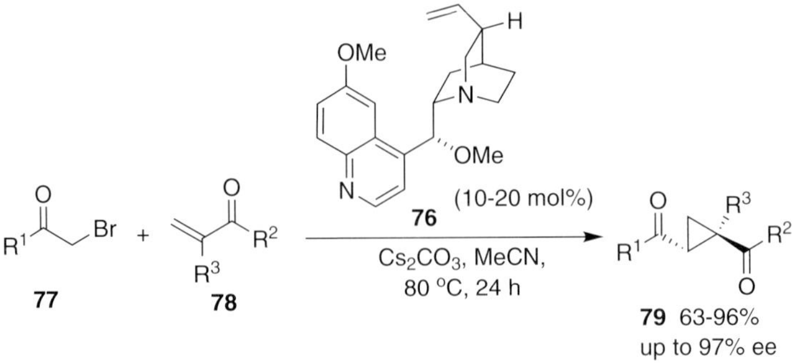

(14.26)

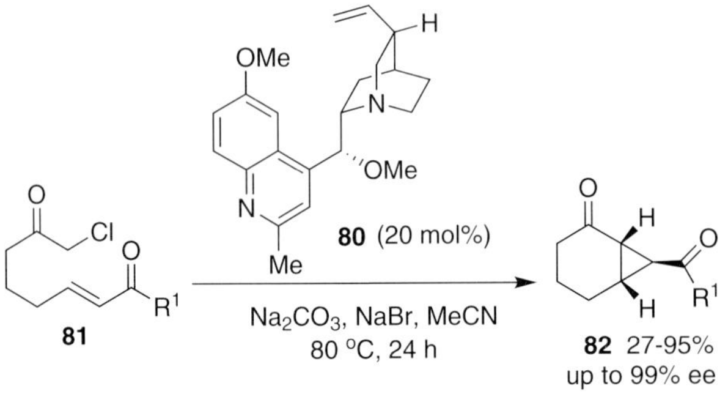

(14.27)

### 14.2.4
### Cyclopropanation with Metalocarbenoids

Fischer *et al.* first described the reactions of chromium and tungsten arylalkoxycarbene complexes with alkenes to produce substituted cyclopropanes (Eq. 14.28) [46]. Both electron-rich and electron-deficient alkenes are suitable substrates. Thus, a broad variety of substituted cyclopropanes are accessible by this approach [47]. The asymmetric cyclopropanation with Fischer carbenes has been explored recently [48].

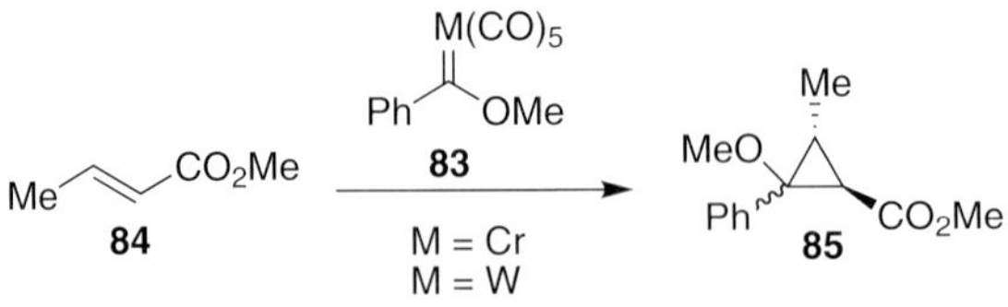

(14.28)

The metalocarbenoids can be generated from metal-catalyzed cycloisomerization of enynes and a catalytic version of this process has been developed for the synthesis of cyclopropanes very recently. Barluenga reported a tandem tungsten-catalyzed cycloisomerization–cyclopropanation reaction, which has been used as the key step for the synthesis of eight-membered carbocycles (Eq. 14.29) [49].

$$\mathbf{86} \xrightarrow{\text{[(THF)W(CO)}_5\text{] (25 mol\%)}}_{\text{THF, RT}} \mathbf{87}\ 60\text{–}91\% \qquad (14.29)$$

Gold(I)-catalyzed intramolecular cyclopropanation of dienynes and intermolecular cyclopropanation of enynes with alkenes has also been reported (Eqs. 14.30 and 14.31) [50]. Ohe and Uemura developed a rhodium-catalyzed cyclopropanation using ene-yne-imino ether compounds as precursors of (2-pyrrolyl)carbenoids (Eq. 14.32) [51]. The key intermediate (2-pyrrolyl)carbenoid in this cyclopropanation is generated by the nucleophilic attack of the nitrogen atom of an imine at a rhodium complex activated internal alkyne carbon.

$$\mathbf{89} \xrightarrow{\text{[Ph}_3\text{P-Au-NCMe]}^+ \text{SbF}_6^-\ (2\ \text{mol\%})\ (\mathbf{88})}_{\text{CH}_2\text{Cl}_2,\ 23\ °\text{C},\ 5\ \text{min}} \mathbf{90}\ 100\% \qquad (14.30)$$

$$\mathbf{92} + \text{cyclohexene} \xrightarrow[\text{CH}_2\text{Cl}_2,\ -50\ \text{to}\ 23\ °\text{C}]{\text{AuCl/AgSbF}_6\ \mathbf{91}\ (5\ \text{mol\%})} \mathbf{93}\ 77\% \qquad (14.31)$$

$$\mathbf{94} + \text{styrene} \xrightarrow[\text{CH}_2\text{Cl}_2,\ \text{RT},\ 2\text{h}]{\text{[Rh(OAc)}_2\text{]}_2\ (2.5\ \text{mol\%})} \mathbf{95}\ 99\%,\ cis/trans = 55{:}45 \qquad (14.32)$$

## 14 Cyclopropanation, Epoxidation and Aziridination Reactions

Metal-catalyzed cyclopropanation of metallocarbenoids from propargylic esters was first reported by Rautenstrauch in 1984 (Eq. 14.33) [52]. This reaction was recently reinvestigated by Marco-Contelles, Malacira, Nolan, Fürstner, and other groups. Pd(II), Pt(II), Au(I), Au(III) complexes proved efficient catalysts for the intramolecular cycloisomerization of propargylic carboxylates, providing functionalized bicyclo[n.1.0]enol esters [53]. Very recently, Ole and Uemura developed an intermolecular Rautenstrauch cyclopropanation reaction, allowing the synthesis of vinyl cyclopropanes from propargylic carboxylates and alkenes (Eq. 14.34) [54] [RuCp-(MeCN)$_3$]PF$_6$ and [RuClCp-(PPh$_3$)$_2$] complexes can catalyze the cyclopropanation of strained norbornenes, norbornadienes, and benzonorbornadienes [55].

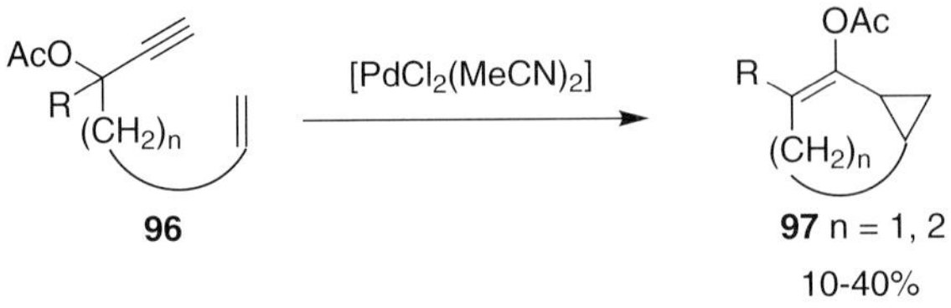

$$\text{96} \xrightarrow{[PdCl_2(MeCN)_2]} \text{97} \quad n = 1, 2 \quad 10\text{-}40\% \tag{14.33}$$

$$\text{98} + \text{99} \xrightarrow[\text{toluene, 60 °C, 18 h}]{[RuCl_2(CO)_3]_2 \text{ (5 mol\%)}} \text{100}$$

R = Me, Ph; R$^1$, R$^2$ = Ph, alkyl, cycloalkyl

60-91%, 50-84% de

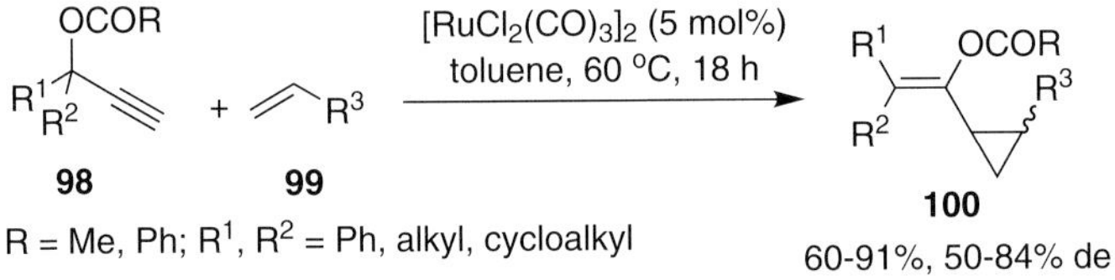

(14.34)

Using chiral ligands such as (R)-DTBM-SEGPHOS(AuCl)$_2$ (**101**), Toste *et al.* have developed an enantioselective gold (I)-catalyzed cyclopropanation of alkenes from proparylic esters with up to 81% ee (Eq. 14.35) [56].

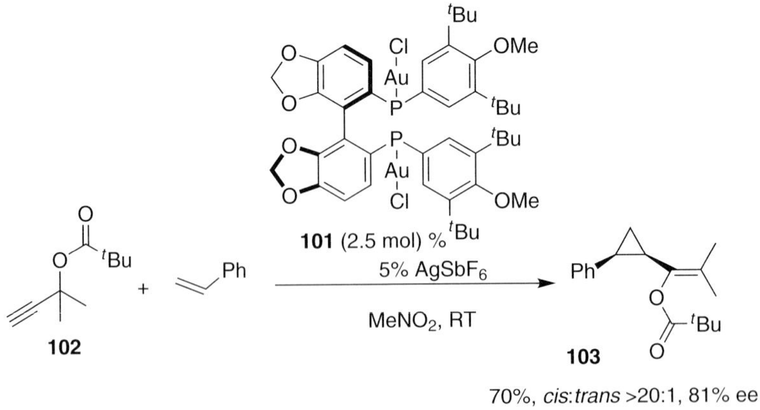

$$\text{102} + \text{Ph} \xrightarrow[\text{MeNO}_2, \text{RT}]{\text{101 (2.5 mol) \%}, 5\% \text{AgSbF}_6} \text{103}$$

70%, *cis*:*trans* >20:1, 81% ee

(14.35)

## 14.3
## Epoxidations

Epoxides have been proven useful as building blocks for the preparation of pharmaceutical targets and natural products, and they are highly versatile intermediates in the formation of bifunctional compounds. Various nucleophiles react with epoxides stereospecifically to give 1,2-difunctionalized products, establishing the stereochemistry of two vicinal carbons. Thus, optically active epoxides serve as useful synthetic intermediates for the construction of a wide range of enantiopure compounds.

### 14.3.1
### Metal-Catalyzed Epoxidation of Alkenes

#### 14.3.1.1 Titanium-Catalyzed Epoxidations

The Sharpless–Katsuki asymmetric epoxidation of allylic alcohols is a milestone for the enantioselective formation of epoxides [57]. This asymmetric transformation has extraordinary features of being easily accessible and performable, highly predictable in its enantioselection, and applicable to a wide range of substrates. The titanium tartrate catalyst is readily accessible from dialkyl tartrate and titanium alkoxide (most commonly diethyl tartrate (DET) and titanium tetraisopropoxide), and the alkyl hydroperoxide used as an oxidant is, in general, commercially available. As shown in Scheme 14.2, 'oxygen' transfers from "above" when $(S,S)$-(−)-DET is employed as a chiral ligand and from "below" when $(R,R)$-(+)-DET is used to generate the chiral titanium complex.

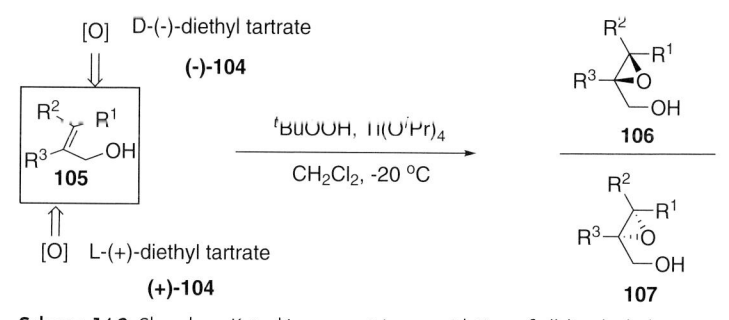

**Scheme 14.2** Sharpless–Katsuki asymmetric epoxidation of allylic alcohols.

Both di-μ-oxo titanium(salen) complex **108** and di-μ-oxo titanium(salan) complex **109** serve as efficient catalysts for asymmetric epoxidation of olefins using aqueous hydrogen peroxide as an oxidant [58]. Catalyst **108** is a chiral di-μ-oxo titanium half-reduced salen complex, which was prepared via *in situ* Meerwein–Ponndorf–Verley reduction and can be applied to a wide range of non-activated olefins with up to 97% ee (Eq. 14.36). Catalyst **109** is easily available but less robust and selective than catalyst **108**.

[(μ-O)₂Ti(salen)] **108**   [(μ-O)₂Ti(salan)] **109**

$$\underset{\textbf{110}}{\text{naphthalene-dihydro}} \xrightarrow[\text{CH}_2\text{Cl}_2,\ \text{RT}]{\textbf{108}\ (1\ \text{mol\%}),\ 30\%\ \text{H}_2\text{O}_2} \underset{\substack{\textbf{111} \\ > 99\%,\ > 99\%\ ee}}{\text{epoxide}} \quad (14.36)$$

## 14.3.1.2 Vanadium-Catalyzed Epoxidations

The phenomenon that early transition metals in combination with alkyl hydroperoxides could participate in olefin epoxidation was discovered in the early 1970s [59]. Vanadium-catalyzed asymmetric epoxidation was reported as early as 1977 [60]. Chiral hydroxamic acid derivatives are the optimal ligands for vanadium-catalyzed epoxidation of various allylic alcohols. For example, Yamamoto *et al.* reported recently that the vanadium complex **112** could catalyze the epoxidation of allylic alcohols with up to 97% ee (Eq. 14.37) [61]. Catalytic system **112** has the following features: (i) high enantioselectivity for a wide range of allylic alcohols; (ii) less than 1 mol% catalyst loading; (iii) mild reaction conditions, and (iv) use of aqueous *tert*-butyl hydroperoxide (v) as an oxidant instead of anhydrous TBHP. Complex **112** was also successfully applied to the kinetic resolution of a secondary allylic alcohol. Ligand **115** was efficient for the epoxidation of homoallylic alcohols with excellent enantioselectivities (Eq. 14.38) [62].

**112a**   **112b**

$$\underset{\textbf{113}}{\overset{R^2}{\underset{R^3}{\bigvee}}\!\!\!\!\text{—OH}} \xrightarrow[\text{TBHP (70\% aq), CH}_2\text{Cl}_2]{\text{catalyst } \textbf{112}(1\ \text{mol\%})} \underset{\substack{\textbf{114} \\ 24\text{-}91\%,\ \text{up to 97\% ee}}}{\overset{R^2}{\underset{R^3}{\bigvee}}\!\!\!\!\overset{\text{OH}}{\underset{\text{O}}{}}} \quad (14.37)$$

## 14.3 Epoxidations

[Structure 115: bis-hydroxamic acid ligand with cyclohexane backbone, CR₃ groups, R = 3,3'',5,5''-tetraethylbiphenyl group]

$$\text{Et}\diagup\!\!\!\diagup\!\!\diagdown\text{OH} \xrightarrow[\text{CHP, toluene} \atop \text{rt, 24h}]{\text{VO(O}^i\text{Pr)}_3 \text{ (1 mol\%)} \atop \text{115 (2 mol\%)}} \text{Et}\diagup\!\!\!\diagup\!\!\diagdown\text{OH}$$

116 → 117 (epoxide)

CHP = cumene hydroperoxide     85%, 92% ee

(14.38)

### 14.3.1.3 Manganese-Catalyzed Epoxidations

Historically, the interest in using manganese complexes as catalysts for the epoxidation of olefins comes from biologically relevant oxidative manganese porphyrins. Simple achiral Mn-Salen complexes proved to resemble closely those of the porphyrin systems as catalysts for olefin epoxidation [63]. A breakthrough in the asymmetric epoxidation came at the beginning of the 1990s, when the groups of Jacobsen and Katsuki more or less simultaneously discovered that chiral Mn-Salen complexes catalyzed the enantioselective formation of epoxides [64]. Unlike the titanium-catalyzed asymmetric epoxidation discovered by Sharpless, the Mn-salen system does not require precoordination of the olefin substrate to the catalyst. The enantioselectivity was shown to be highly sensitive to the substitution pattern of the olefin substrate. Excellent selectivities (>90% ee) were obtained for aryl- or alkynyl-substituted terminal-, cis-disubstituted-, and trisubstituted olefins (Eq. 14.39) [64a].

[Structure 118: chiral Mn-Salen complex with Ph, Ph on diamine backbone, X and $^t$Bu substituents, Mn-Cl-O]

119 (dihydronaphthalene) $\xrightarrow[\text{NaOCl, pyridine N-oxide additive} \atop \text{chlorobenzene}]{X = \text{OMe}, {}^t\text{Bu (0.5-1 mol\%)}}$ 120 (epoxide)

52%, 98% ee

(14.39)

### 14.3.1.4 Iron-Catalyzed Epoxidations

Iron porphyrins can be used as epoxidation catalysts but the conversion and the selectivity are often inferior to those of the manganese analogues. A number of biomimetic non-heme iron complexes have also proven useful as catalysts for olefin epoxidation [65]. For instance, Jacobsen and coworkers reported that a tetradentate ligand (BPMEN = N,N′-dimethyl-N,N′-bis(2-pyridylmethyl)ethylene-1,2-diamine) was combined with an iron(II) precursor and acetic acid to yield a self-assembled μ-oxo, carboxylate-bridged diiron(III) complex **121** [66]. This dimeric iron complex could catalytically epoxidize olefins efficiently in the presence of 50% aqueous hydrogen peroxide (Eq. 14.40).

$$\text{n-}C_8H_{17}\text{-CH=CH}_2 \quad \xrightarrow[\text{Acetic acid (30 mol\%), CH}_3\text{CN}]{\textbf{121} \text{ (3 mol\%)}, 1.5 \text{ equiv H}_2\text{O}_2 \text{ (aq)}} \quad \text{n-}C_8H_{17}\text{-epoxide}$$

**122** → **123** 85%

(14.40)

Very recently, Beller et al. developed a highly enantioselective iron-catalyzed epoxidation of aromatic alkenes using hydrogen peroxide (Eq. 14.41). This oxidation catalysis was realized by combining $FeCl_3$ with chiral diamine **124** in the presence of catalytic amount of pyridine-2,6-dicarboxylic acid ($H_2$pydic) [67].

$$Ar\text{-CH=CH-}Ar \quad \xrightarrow[\text{H}_2\text{O}_2 \text{ (2 equiv), 2-methylbutan-2-ol, RT, 1 h}]{(S,S)\text{-}\textbf{124} \text{ (12 mol\%)}, FeCl_3 \cdot 6H_2O \text{ (5 mol\%)}, H_2\text{pydic (5 mol\%)}} \quad Ar\text{-epoxide-}Ar$$

**125** → **126** Ar = 4-$^t$BuC$_6$H$_4$

82%, 81% ee

(14.41)

### 14.3.1.5 Ruthenium and Platinum-Catalyzed Epoxidations

High-valent ruthenium oxides are powerful reagents for oxidization of olefins, mostly resulting in cleavage of the double bond. The use of less reactive, low-valent ruthenium complexes in combination with various oxidants for the preparation of epoxides from simple olefins has been described [68]. Complex [Ru(tpy)(pydic)] **127** is an active catalyst for epoxidation of alkenes by aqueous 30% hydrogen peroxide in tertiary alcohols (Eq. 14.42) [69]. Asymmetric epoxidation of olefins with ruthenium catalysts based on chiral porphyrins, pyridine-2,6-bisoxazoline, and pyridinebisimidazolines has been reported, and moderate to good enantioselectivities were achieved in most cases [70].

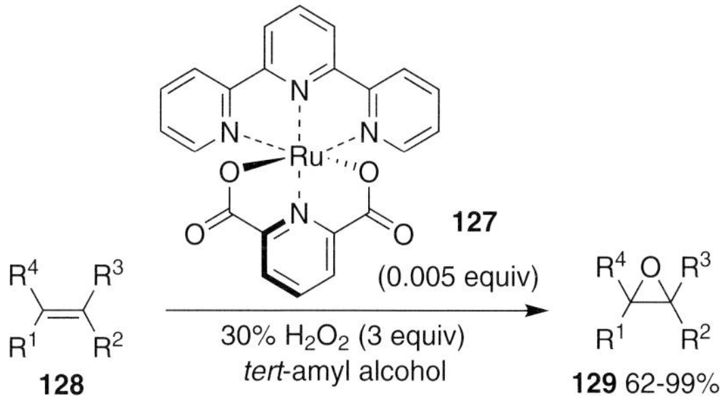

$$\begin{array}{c} \text{127 (0.005 equiv)} \\ \underset{128}{\text{R}^1\text{R}^2\text{C}=\text{CR}^3\text{R}^4} \xrightarrow{\text{30\% H}_2\text{O}_2 \text{ (3 equiv)}, \text{ tert-amyl alcohol}} \underset{129 \ 62\text{-}99\%}{\text{epoxide}} \end{array} \qquad (14.42)$$

Strukul *et al.* reported an asymmetric epoxidation of terminal alkenes with hydrogen peroxide catalyzed by pentafluorophenyl Pt(II) complex **130**, affording the epoxides in moderate to good yields with good enantioselectivities (Eq. 14.43). It is noteworthy that terminal double bonds were epoxidized with complete regioselectivity and ees up to 98% when dienes were employed [71].

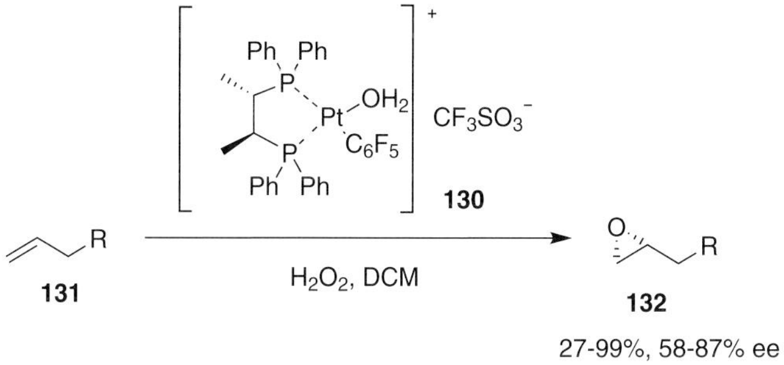

$$\underset{131}{\text{CH}_2=\text{CHCH}_2\text{R}} \xrightarrow[\text{H}_2\text{O}_2, \text{ DCM}]{\text{130}} \underset{132}{\text{epoxide-R}} \qquad (14.43)$$

27-99%, 58-87% ee

### 14.3.1.6 Lanthanide-Catalyzed Epoxidations

Alkali-metal free lanthanide-BINOL-Ph$_3$As=O complexes, generated from Ln(O-*i*-Pr)$_3$, BINOL, and Ph$_3$As=O in a ratio of 1:1:1, are useful catalysts for the asymmetric epoxidation of enones, α,β-unsaturated amides, and ester surrogates [72]. Recently, a catalytic asymmetric epoxidation of α,β-unsaturated esters using an yttrium-biphenyldiol complex has also been developed (Eq. 14.44) [73]. Both β-aromatic and β-aliphatic α,β-unsaturated esters gave the corresponding epoxides in high yields with excellent enantioselectivities.

$$\underset{134}{\text{R-CH=CH-CO}_2\text{Et}} \xrightarrow[\text{TBHP (1.2 equiv), THF, MS 4Å, RT}]{\text{Y-131-Ph}_3\text{As=O (1:1:1) (2-10 mol \%)}} \underset{135 \ R = \text{aryl, alkyl}}{\text{epoxide-CO}_2\text{Et}}$$

62-97%, up to 99% ee

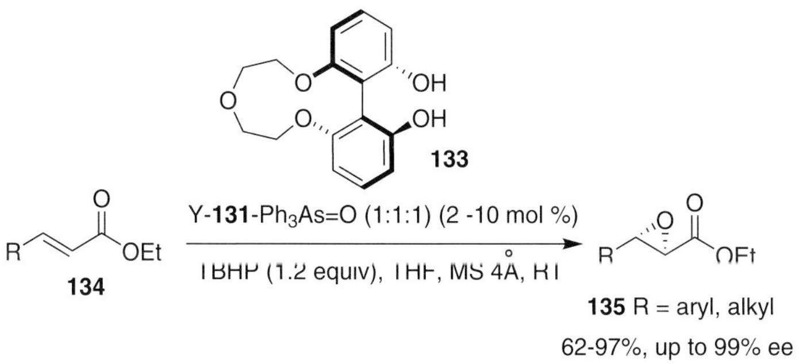

$$(14.44)$$

## 14.3.2
## Organocatalytic Epoxidation of Alkenes

### 14.3.2.1 Epoxidation Catalyzed by Ketones

The epoxidation of olefins with a ketone in the presence of potassium peroxomonosulfate (Oxone), has emerged as an important synthetic method. The potential for chiral ketone-mediated asymmetric epoxidation of olefins was first demonstrated by Curci in 1984 [74]. An important advance was made in 1996 by Yang, who observed that ketones bearing electron-withdrawing groups such as halogen atoms and OAc adjacent to the carbonyl showed higher catalytic activity in epoxidations of trans-olefins with Oxone [75]. On the basis of this finding, she designed the $C_2$-symmetric chiral ketone **136** and found that it could catalyze the epoxidation of unfunctionalized olefins with good enantioselectivities (Eq. 14.45) [76].

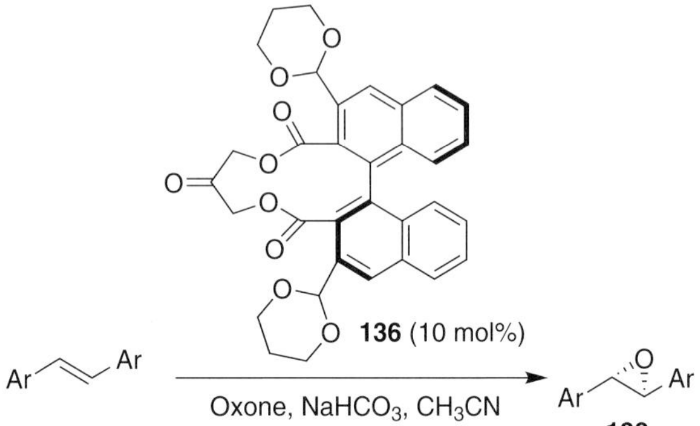

(14.45)

Later, Shi reported that fructose-derived ketone **139** promoted the asymmetric epoxidation of trans-olefins, trisubstituted olefins, conjugated dienes and enynes, and enol esters (Eq. 14.46) [77]. The features of this reaction are: (i) both enantiomers of the catalyst can be prepared easily from D- or L-fructose in two steps; (ii) a strong pH effect on catalytic efficiency is observed and the epoxidation reaction proceeds with high

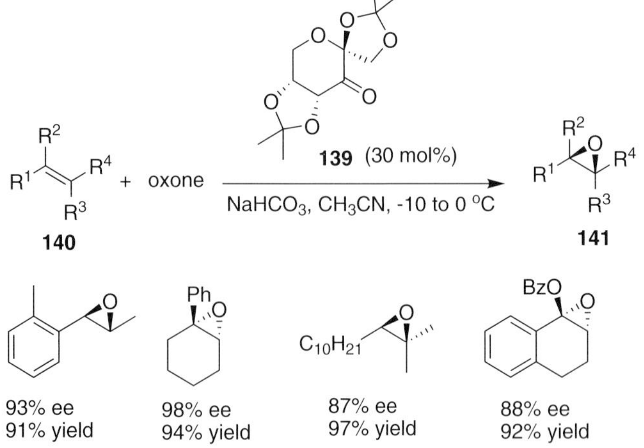

(14.46)

93% ee
91% yield

98% ee
94% yield

87% ee
97% yield

88% ee
92% yield

enantiomeric excess at relatively low catalyst loadings (20–30 mol%) without the need to use a large excess of Oxone, by keeping the pH at an optimum (~10.5).

A series of chiral ketones derived from fructose has been intensively investigated for epoxidation of different alkenes. Ketone **142** was effective for the epoxidation of α,β-unsaturated esters [78]. Oxazolindinone-containing ketone **143** proved effective for *cis* olefins [79], certain terminal olefins, and chromenes [80]. Ketone **143** was also efficient for the epoxidation of dienes to give cis-epoxides with high regio- and enantio-selectivities [81].

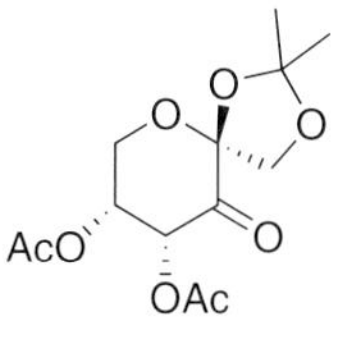

**142**

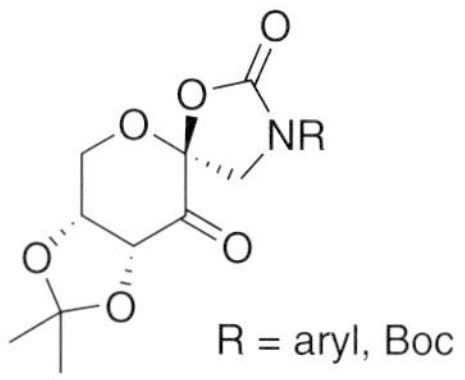

**143** R = aryl, Boc

### 14.3.2.2 Epoxidation Catalyzed by Iminium Salts

Oxaziridinium salts, generated *in situ* from iminium salts with Oxone, were shown to be effective electrophilic oxidants for epoxidation of alkenes [82]. Aggarwal *et al.* reported that 5 mol% of binaphthyl-based iminium salt **144** catalyzed the oxidation of alkenes to epoxides with moderate to good enantioselectivities [83]. By employing salt **145**, up to 84% ee was achieved [84]. Page *et al.* reported that the iminium salt **146** excelled in the epoxidation of cyclic cis-alkenes, affording the epoxides with up to 97% ee (Eq. 14.47) [85]. Catalyst **149**, bearing a binaphthalene backbone and exocyclic N-aminoacetal group, proved to be effective for the asymmetric epoxidation of unfunctionalized alkenes with up to 95% ee (Eq. 14.48) [86].

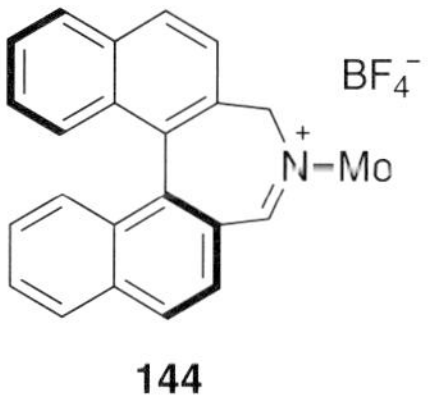

**144**

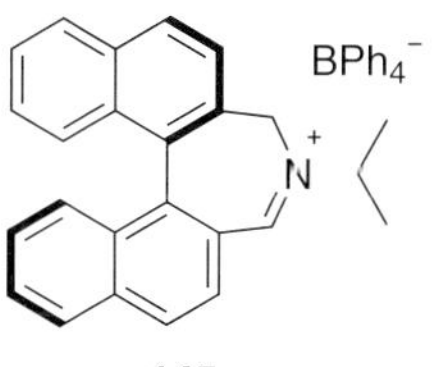

**145**

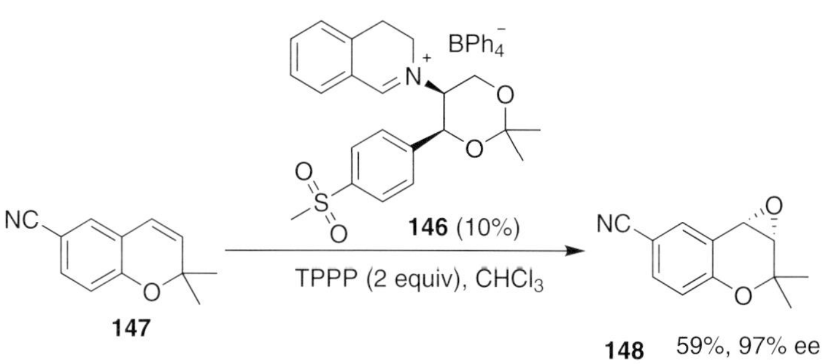

**147** → **148** 59%, 97% ee

**146** (10%), TPPP (2 equiv), CHCl$_3$

TPPP = tetraphenylphosphonium monoperoxysulfate

(14.47)

**708** | *14 Cyclopropanation, Epoxidation and Aziridination Reactions*

$$\underset{\mathbf{150}}{\text{Ph-cyclohexene}} \xrightarrow[\text{MeCN/H}_2\text{O (1:1), 0 °C}]{\mathbf{149}\ (5\ \text{mol\%})} \underset{\mathbf{151}}{\text{Ph-epoxide}} \quad (14.48)$$

69%, 91% ee

### 14.3.2.3 Epoxidation Catalyzed by Amines

Chiral amines are efficient catalysts for asymmetric epoxidation of alkenes. Using oxone as an oxidant, Aggarwal reported that (S)-2-(diphenylmethyl)pyrrolidine could promote the epoxidation of 1-phenylcyclohexene and up to 66% ee was achieved [87]. Using α,α-diphenyl-L-prolinol **152** as bifunctional organocatalyst and *tert*-butyl hydroperoxide (TBHP) instead of oxone, Lattnazi developed a catalytic enantioselective epoxidation of α,β-unsaturated ketones in good yields with up to 80% ee (Eq. 14.49). Zhao *et al.* further modified this reaction and improved the ee values to higher than 89% for chalcones by employing 4-cis-substituted pyrrolinol derivatives as a catalyst [88].

$$\underset{\mathbf{153}}{R^1\text{-CH=CH-}R^2\text{(C=O)}} \xrightarrow[\text{hexane, RT, TBHP}]{\mathbf{152}\ (30\ \text{mol\%})} \underset{\mathbf{154}}{R^1\text{-epoxyketone-}R^2} \quad (14.49)$$

up to 87% yield, 80% ee

Jørgensen and coworkers documented a prolinol **155**-catalyzed addition of hydrogen peroxide to enals **156**, affording 2,3-epoxyaldehydes in moderate yields with excellent stereoselectivities (Eq. 14.50) [89]. Using organic peroxides instead of hydrogen peroxide, the conversions were clearly reduced although high enantiomeric excesses were achieved. Pihko *et al.* developed an ortho-substituted aniline catalyst that readily transformed α-substituted aliphatic enals to the corresponding epoxides [90].

$$\underset{\mathbf{156}}{R\text{-CH=CH-CHO}} \xrightarrow[\text{H}_2\text{O}_2,\ \text{CH}_2\text{Cl}_2,\ \text{RT}]{\mathbf{155}\ (10\ \text{mol\%})\ \text{Ar}=3,5\text{-(CF}_3)_2\text{-C}_6\text{H}_3} \underset{\mathbf{157}}{R\text{-epoxyaldehyde}} \quad (14.50)$$

63-90%, 94-98% ee

Very recently, List *et al.* reported a chiral counteranion-directed catalysis for the asymmetric epoxidation of 2-enals. In the case of 2-aryl-2-enals, high enantiomeric excesses are always achieved (Eq. 14.51) [91].

$$\text{OHC}\diagup\!\!\diagdown\text{Ar} \xrightarrow[\text{\textit{t}BuOOH (1.1 equiv), Dioxane, 35 °C, 72 h}]{\textbf{158} \text{ (10 mol\%)}} \text{OHC}\overset{O}{\diagup\!\!\diagdown}\text{Ar}$$

**159** → **160**

R = 2,4,6-($^i$Pr)$_3$C$_6$H$_2$; Ar = 3,5-(CF$_3$)$_2$C$_6$H$_2$

60–84%
dr > 94:6, > 85% ee

(14.51)

### 14.3.2.4 Phase-Transfer Catalysis

Phase-transfer catalysis (PTC) features simple experimental operations, mild reaction conditions, inexpensive and environmentally benign reagents and solvents, and the possibility to conduct large-scale preparations [92]. The first example of asymmetric epoxidation by PTC was reported by Wynberg using alkylated cinchona alkaloids [93]. The quaternary cinchonidinium salt attaching a 9-anthracenylmethyl group was an efficient catalyst in the epoxidation of substituted chalcones with hypochlorite [94]. The chiral dimeric cinchona-derived quaternary ammonium salt has also been used for the asymmetric epoxidation of 2,4-diarylenones, providing the desired epoxides in excellent yields with high selectivities [95].

Recently, a new chiral quaternary ammonium bromide **161**, possessing a diarylmethanol functionality as a substrate recognition site, has been designed by Maruoka *et al.* The ammonium salt **161** has proved to be a promising, dual-functioning catalyst for the highly enantioselective epoxidation of α,β-unsaturated ketones under mild phase-transfer conditions. Using 3 mol% ammonium bromide **161** as a catalyst and 13% NaOCl as an oxidant, for instance, chalcone gave epoxychalcone quantitatively with 96% ee (Eq. 14.52) [96].

**161** Ar = R = 3,5-Ph$_2$-C$_6$H$_3$ (3 mol%)

13% NaOCl, toluene, 0 °C

R$^1$–CO–CH=CH–R$^2$ → R$^1$–CO–CH(O)CH–R$^2$ **162**

80–99%, 91–99% ee

(14.52)

### 14.3.2.5 Miscellaneous

In 2007, Miller et al. reported an enantioselective epoxidation reaction catalyzed by aspartic acid-containing peptides. The mechanism involves the transient conversion of the aspartate carboxylic acid in the peptide **163** to the corresponding peracid. Highly enantioselectivities were observed for a class of urethane-substituted olefins (Eq. 14.53) [97].

$$(14.53)$$

### 14.3.3
### Epoxidaton of Carbonyl Compounds

There are two general approaches to the direct epoxidation of carbonyl compounds: ylide epoxidation for the construction of aryl and vinyl epoxides, and α-halo enolate-mediated epoxidation (Darzens reaction) for the construction of epoxy esters, acids, amides and sulfones. The topic of chalcogenides-catalyzed epoxidation has been extensively reviewed by Aggarwal recently [98]. Herein, only selected examples for typical methodology are summarized.

### 14.3.3.1 Ylide Epoxidation

Stoichiometric sulfur ylide-mediated epoxidations were first reported in 1958 by Johnson and LaCount [99]. Since then, many improvements have been made to these reactions. The first catalytic sulfur ylide epoxidation was reported by Furukawa et al. in 1987 [100]. Two years later, an enantioselective version of this reaction was described by the same group (Eq. 14.54) [101]. To date, a number of research groups have modified this reaction, providing an easy access to epoxides from aldehydes.

$$(14.54)$$

Hou et al. reported that aldedydes reacted smoothly with a silylated dimethylsulfonium allylide. By a simple choice of sulfonium salts and the reaction conditions, both *trans*-vinylepoxides and *cis*-vinylepoxides were obtained selectively [102]. Later, Dai and Huang developed a catalytic enantioselective epoxidation of aldehydes with benzyl bromide using camphor-derived chiral sulfide **168** as a catalyst (Eq. 14.55) [103].

$$4\text{-ClC}_6\text{H}_4\text{CHO} + \text{PhCH}_2\text{Br} \xrightarrow[\text{KOH(s), CH}_3\text{CN, rt}]{\textbf{168 (20 mol\%)}} \underset{\textbf{169}}{4\text{-ClC}_6\text{H}_4\overset{\text{H}}{\underset{\text{O}}{\triangle}}\text{Ph}}$$

93%, *trans*/*cis* = 86:14, 60% ee

(14.55)

Metzner and coworkers found that $C_2$-symmetric chiral 2,5-dimethylthiolanes **170** with a locked conformation could promote the reaction of benzyl bromide with aldehydes, affording *trans*-stilbene oxide derivatives with enantiomeric ratios ranging from 95:5 to 98:2 (Eq. 14.56) [104].

$$\text{BnBr} + \text{ArCHO} \xrightarrow[\text{NaOH, rt}]{\textbf{170 (10-20 mol\%)}} \underset{\textbf{171}}{\text{Ar}\overset{\text{O}}{\triangle}\text{Ph}}$$

(14.56)

53-92% yields
76-90% de, 90-96% ee

Sulfonium ylides could be generated via a metal-catalyzed decomposition of diazo compounds with sulfide [105]. In a modified procedure, tosylhydrazone salts were used as the source of diazo compounds [106]. For example, Aggarwal et al. reported that sulfide **172** could catalyze epoxidation of aldehydes with high enantio- and diastereo-selectivies (Eq. 14.57) [107]. In this catalytic cycle, the sulfur ylide is generated through the reaction between chiral sulfide **172** and a metallocarbene that is formed by the *in situ* decomposition of the tosylhydrazone salt in the presence of a phase-transfer catalyst (Scheme 14.3).

$$\text{RCHO} + \underset{\substack{(2 \text{ equiv}) \\ \textbf{173}}}{\text{Ph}\overset{\text{Na}^+}{\underset{}{\diagup\!\!\!\diagdown}}\text{N}\overset{\text{N}^-}{\diagdown}\text{Ts}} \xrightarrow[\substack{\text{Rh}_2\text{(OAc)}_4 \text{ (1 mol\%)}, \\ \text{BnEt}_3\text{N}^+\text{Cl}^- \text{ (5 mol\%)} \\ \text{CH}_3\text{CN, 40 °C}}]{\textbf{172 (5 mol\%)}} \underset{\textbf{174}}{\text{Ph}\overset{\text{O}}{\triangle}\text{R}}$$

58-84% yields
76-96% de, 90-94% ee

(14.57)

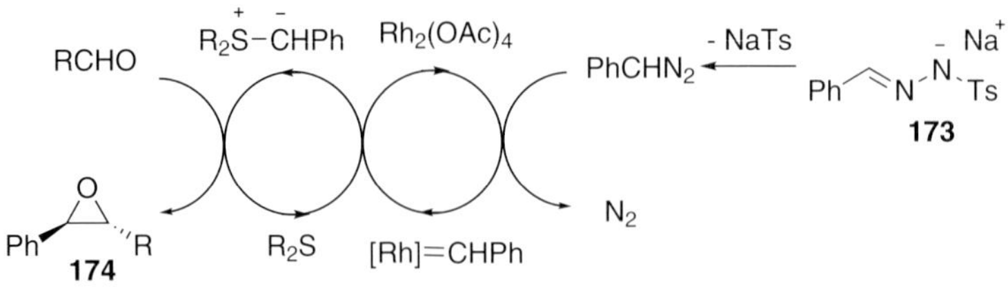

**Scheme 14.3** One-pot epoxidation of aldehydes with a tosylhydrazone.

The first example of telluride-catalyzed ylide epoxidation appeared in 1990 [108]. By employing chiral tellurides, optically active epoxides could be obtained with good diasteroselectivities and enantioselectivities, but in low yields [109].

### 14.3.3.2 Darzens Reaction

Epoxides bearing electron-withdrawing groups can be commonly synthesized by the Darzens reaction. The Darzens reaction involves the initial addition of α-halo enolate to the carbonyl compound, followed by ring-closure of alkoxides to give epoxides (Eq. 14.58).

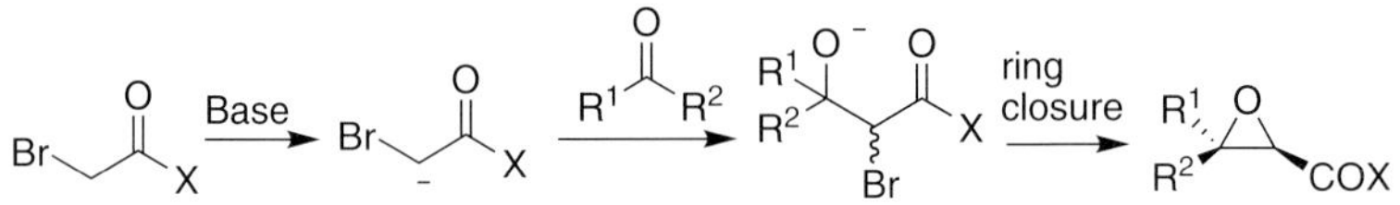

(14.58)

The asymmetric Darzens reaction has been less explored [110]. Chiral bromoborane **175** (1.1 equiv) was found to promote the addition of *tert*-butyl bromoacetate to aromatic, aliphatic, and α,β-unsaturated aldehydes to give α-bromohydrins **176** with high enantio- and diastereo-selectivities, which were cyclized to give the epoxy esters in the presence of potassium *tert*-butoxide (Eq. 14.59) [111].

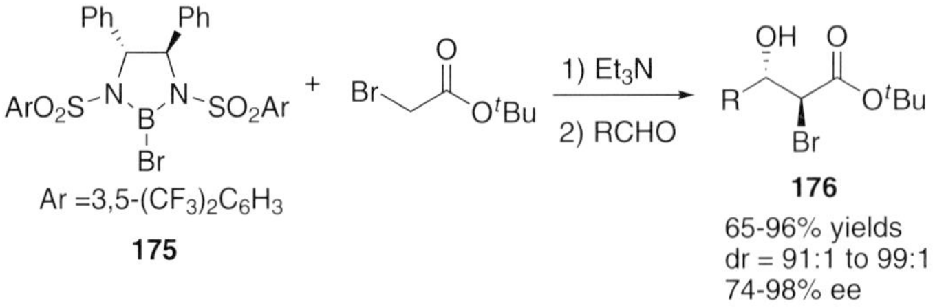

65-96% yields
dr = 91:1 to 99:1
74-98% ee

(14.59)

The asymmetric Darzens reaction catalyzed by chiral phase-transfer catalysts was a notable approach to optically active epoxides [112]. For instance, Arai et al. reported that the chiral ammonium salt **177**, derived from cinchona alkaloids, catalyzed the reaction of aldedyes or ketones with chloromethyl phenyl sulfone to give epoxysulfones with ee up to 83% (Eq. 14.60) [113].

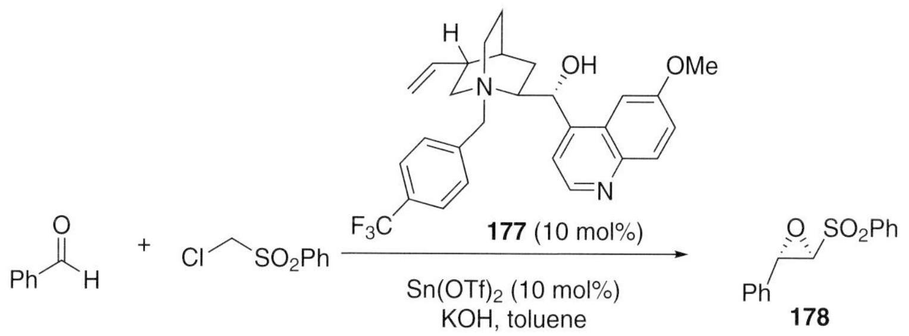

$$\text{Ph-CHO} + \text{Cl-CH}_2\text{SO}_2\text{Ph} \xrightarrow[\text{KOH, toluene}]{\textbf{177} (10 \text{ mol\%}) \atop \text{Sn(OTf)}_2 (10 \text{ mol\%})} \underset{\substack{\textbf{178}\\66\%, 83\% \text{ ee}}}{\text{Ph}\overset{\text{O}}{\diagdown}\text{SO}_2\text{Ph}}$$

(14.60)

## 14.4 Aziridinations

Aziridines are the nitrogen analogues of epoxides and exhibit similar reaction patterns as electrophilic reagents. They undergo highly regio- and stereo-selective transformations and, therefore, are useful building blocks in organic synthesis. In addition, aziridines may exhibit antitumor or antibiotic activity or other biological properties, which makes them attractive synthetic targets in their own right. The aziridination reactions can be divided into two categories: aziridination of olefins via nitrene and aziridination of imines with carbon-transfer reagents (Scheme 14.4).

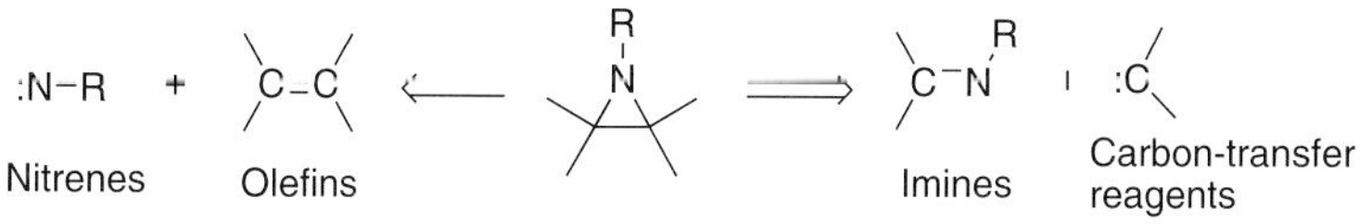

**Scheme 14.4** Synthesis of aziridines from olefins and imines.

### 14.4.1 Aziridination of Alkenes

#### 14.4.1.1 Metal-Catalyzed/Mediated Aziridination

In 1983, Groves reported the stiochiometric aziridation of alkenes with a manganese-imido complex generated *in situ* from an isolable (prophyrin)manganese-nitrido intermediate [114]. Later, nitrodo complexes of manganese salen [115], ruthenium prophyrin [116], and ruthenium salen [117] were used as the reagents for aziridination of alkenes. Mansuy *et al.* in 1984 documented that azirdination of olefins can be achieved with (N-(p-toluenesulfonyl)imino))phenyliodinane (PhI=NTs) using Fe(III)- or Mn(III)-porphyrins as catalysts [118]. The methodology was further extended by Evans' group to the reaction of both electron-rich and electron-deficient olefins with copper(I) or copper (II) catalysts [119]. Recently, Halfen et al. described that the non-heme iron(II) complexes are efficient catalysts for the aziridination of olefins by PhI=NTs (Eq. 14.61) [120]. These systems provide aziridines in moderate to good yields and only a small excess of olefin is required for optimum reactivity.

$$Ph\diagup\diagdown + PhINTs \xrightarrow[\substack{CH_2Cl_2,\ 25\ °C}]{\substack{(Me_5dien)Fe(CF_3SO_3)_2 \\ \text{or } (^iPr_3TACN)Fe(CF_3SO_3)_2 \\ \text{catalyst/PhINTs/olefin} \\ (1:20:100-500)}} Ph\diagup\underset{N}{\triangle}^{Ts} + PhI$$

**179** → **180** 62–95%

Me$_5$dien = 1,1,4,7,7-pentamethyldiethylenetriamine
$^i$Pr$_3$TACN = 1,4,7-triisopropyl-1,4,7-triazacyclononane

(14.61)

A facile, high yielding, and stereospecific method for olefin aziridination using trichloroethylsulfamate, $H_2NSO_3CH_2CCl_3$ **182**, as a nitrogen source was reported by Bios et al. (Eq. 14.62) [121]. In the presence of 1–2 mol% $Rh_2(CF_3CONH_2)_4$, PhI (OAc)$_2$ was utilized as the oxidant to promote N-atom transfer reactions. A wide range of alkenes are suitable substrates.

$$\underset{R^2}{\overset{R^1}{\diagdown}}\diagup{R^3} + H_2NSO_3CH_2CCl_3 \xrightarrow[\substack{PhI(OAc)_2,\ MgO \\ C_6H_6,\ 0\ °C}]{\substack{Rh_2(CF_3CONH_2)_4 \\ (1-2\ \text{mol\%})}} \underset{R^2}{\overset{R^1}{\diagdown}}\underset{R^3}{\overset{NSO_3CH_2CCl_3}{\diagup}}$$

**181** + **182** (1.1 equiv) → **183** 65–95%

(14.62)

Doyle et al. reported a dirhodium (II) carprolacatamate-catalyzed aziridination of olefins. They found that the use of p-toluenesulfonamide, N-bromosuccinimide, and potassium carbonate readily furnishes aziridines in up to 95% yields under mild conditions. The catalyst loading could be reduced to 0.01 mol% (Eq. 14.63) [122].

**184** = Rh$_2$(cap)$_4$

dihydronaphthalene $\xrightarrow[\substack{TsNH_2,\ NBS,\ K_2CO_3}]{Rh_2(cap)_4}$ **185** (NTs-aziridine)

(14.63)

1.0 mol% Rh$_2$(cap)$_4$, 95% yield
0.1 mol% Rh$_2$(cap)$_4$, 84% yield

Very recently, Kim et al. developed an efficient protocol for copper-catalyzed olefin aziridination using 5-methyl-2-pyridinesulfonamide or 2-pyridinesulfonyl azide as the nitrenoid source. The 2-pyridyl group significantly facilitates aziridination, suggesting that the reaction is driven by the favorable formation of a pyridyl-coordinated nitrenoid intermediate [123].

Lebel *et al.* reported that *N*-tosyloxycarbamates could be used as a source of metal nitrenes in the rhodium-catalyzed aziridination reactions (Eq. 14.64) [124]. The reaction proceeded at room temperature using a rhodium catalyst and an excess of potassium carbonate. They also developed the intermolecular aziridination of styrenes with trichloroethyl *N*-tosyloxycarbamates using pyridine copper complexes as catalysts (Eq. 14.65) [125].

$$186 \xrightarrow[\text{acetone, 25 °C}]{\text{K}_2\text{CO}_3 \text{ (7 equiv)} \atop \text{Rh}_2(\text{OAc})_4 \text{ (5 mol\%)}} 187 \quad 62\text{-}79\% \tag{14.64}$$

$$188 + 189 \xrightarrow[\text{benzene, 25 °C}]{\text{K}_2\text{CO}_3 \text{ (10 equiv)} \atop \text{Cu(pyridine)}_4(\text{BF}_4)_2 \text{ (10 mol\%)}} 190 \quad 51\text{-}68\% \tag{14.65}$$

Chloramine-T [126], bromamine-T [127], *N,N*-dichloro-*p*-toluenesulfonamide [128] and azides [129] have also been used as the source of nitrenes for metal-catalyzed aziridination reactions.

The enantioselective aziridination of alkenes via nitrene transfer was achieved with copper diimine [130] or copper bis(oxazoline) catalysts [131]. Recently, a one-pot procedure for copper-catalyzed PhI(OAc)$_2$-mediated asymmetric alkene aziridianation has been reported. With 5 mol% of copper/bis(oxazoline) catalyst, the reaction of alkene with 4-nitrobenzenesulfonamide afforded the desired product in 94% yield with 75% ee (Eq. 14.66) [132]. Xu *et al.* documented a highly enantioselective aziridination of chalcones catalyzed by CuOTf/1,8-bisoxazolidinylanthrace (AnBox) with up to 99% ee (Eq. 14.67) [133].

$$191 \xrightarrow[\text{PhI(OAc)}_2, \text{benzene}]{21 \text{ (6 mol\%)} \atop [\text{Cu(CH}_3\text{CN)}_4]\text{ClO}_4 \text{ (5 mol\%)}} 192 \tag{14.66}$$

43-95%, 32-72% ee

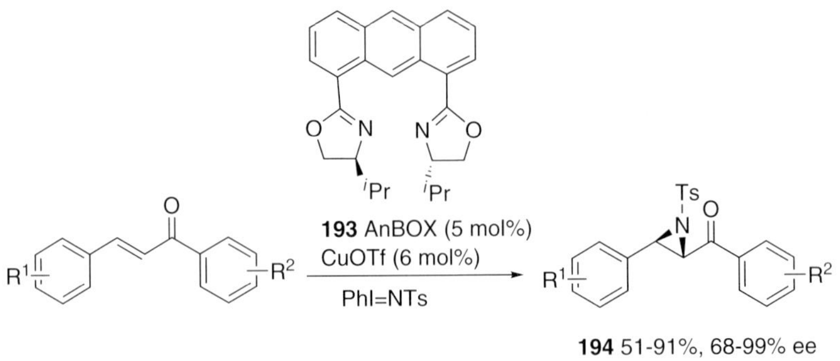

**716** | *14 Cyclopropanation, Epoxidation and Aziridination Reactions*

$$\text{193 AnBOX (5 mol\%)}$$

R¹-ArCH=CH-C(O)-Ar-R² + PhI=NTs → **194** (Ts-aziridine) **51-91%, 68-99% ee**

with **193 AnBOX (5 mol%)**, CuOTf (6 mol%)

(14.67)

Ding *et al.* reported that a diimine derivative, derived from D-manitol, is an effective chiral ligand for the copper-catalyzed asymmetric aziridination of olefin derivatives with PhI=NTs as the nitrene source, affording the corresponding N-sulfonyl azirindine derivatives in good to excellent yields with up to 99% ee (Eq. 14.68) [134].

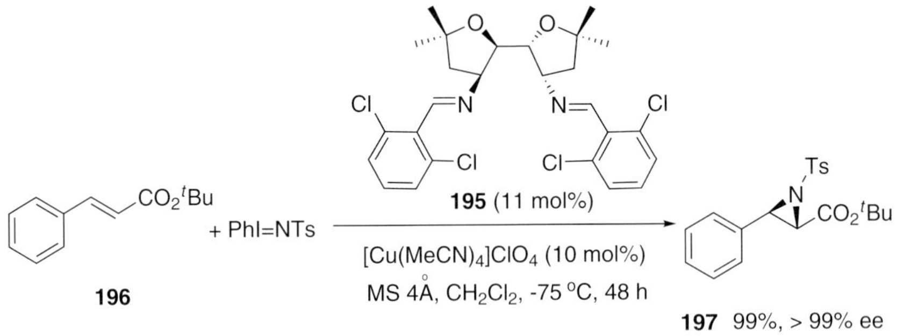

**196** PhCH=CH-CO$_2^t$Bu + PhI=NTs → **197** 99%, > 99% ee

**195** (11 mol%), [Cu(MeCN)$_4$]ClO$_4$ (10 mol%), MS 4Å, CH$_2$Cl$_2$, -75 °C, 48 h

(14.68)

### 14.4.1.2 Metal-Free Aziridination

Che *et al.* reported an aryl iodide mediated aziridination of alkenes with N-aminophthalimide under metal-free conditions (Eq. 14.69) [135]. This method probably involves *in situ* generation of an aryl-$\lambda^3$-iodane species from oxidation of the aryl iodide with *m*-CPBA. Reaction of this intermediate with N-aminophthalimide produces N-benzoyloxyaminophthalimide as the active specie that undergoes C=C bond addition and cyclization to give aziridines. Recently, Minakata found that *tert*-butyl hypoiodite is a powerful reagent for synthesis of aziridines from olefins and sulfonamides (Eq. 14.70) [136].

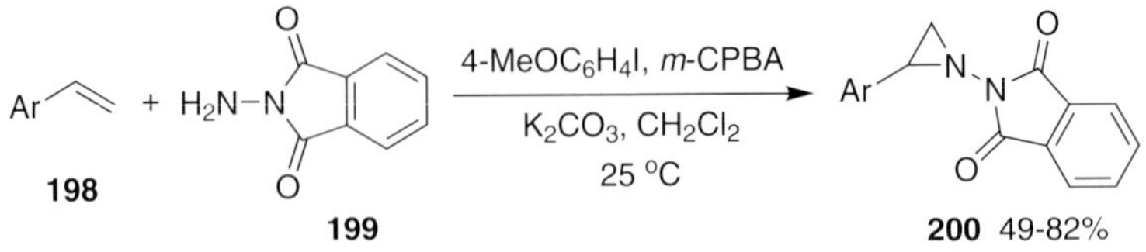

Ar-CH=CH$_2$ + H$_2$N-N(phthalimide) → Ar-aziridine-N-phthalimide

**198** + **199** → **200** 49-82%

4-MeOC$_6$H$_4$I, *m*-CPBA, K$_2$CO$_3$, CH$_2$Cl$_2$, 25 °C

(14.69)

$$R-\underset{\underset{NH_2}{\overset{\overset{O}{\parallel}}{S}}}{\overset{O}{\parallel}} + \underset{R^2}{\overset{R^1}{\diagdown}}=\underset{R^3}{\diagup} \xrightarrow[\text{MeCN, RT}]{{}^tBuOCl, NaI} \underset{R^2 \quad R^3}{\overset{\overset{SO_2R}{|}}{\underset{}{\triangle}}}^{R^1} \quad (14.70)$$

**201**    **202**            **203** 58-82%

Another useful approach to aziridines is the reaction of the N-N-ylide (aminimine) with Michael acceptors. In 1980, Ikeda et al. reported that aminimide could directly iminate chalcones to form azirdines [137]. Shi et al. found that the aminimide, generated in situ from tertiary amine and O-mesitylenesulfonylhydroxylamine (MSH) in the presence of a base, was an effective NH-transfer reagent for the aziridination of chalcones. The amine promoter in this aziridination process could be used in a catalytic amount and a reasonable level of enantioselectivity can be obtained when chiral tertiary amine is used (Eq. 14.71) [138]. The reaction of O-(diphenylphosphinyl)hydroxylamine with tertiary amine was used by Armstrong for the in situ generation of aminimine for the aziridination of enones. When quinine was employed as the chiral promoter, moderate enantioselectivities were achieved [139].

**204** (+)-Troger's Base

$$Ar\underset{205}{\diagup}\overset{O}{\diagdown}Ar' \xrightarrow[\text{CH}_3\text{CN/CH}_2\text{Cl}_2 \text{ (2:1)}]{\text{MSH (2.0 equiv)} \atop \text{CsOH·H}_2\text{O (3.0 equiv)}} Ar\underset{206}{\diagup}\overset{\overset{H}{\underset{N}{\triangle}}\;\overset{O}{\parallel}}{\diagdown}Ar' \quad (14.71)$$

50-93%, 55-67% ee

Organocatalytic reactions that involve a combination of catalytic iminium and enamine intermediates have been developed rapidly recently. Córdova et al. found that chiral amine **207** could catalyze the aziridination of α, β-unsaturated aldehydes with acetyl hydroxyamines to give the corresponding 2-formylaziridines in good yields with good diastereoselectivities and up to 99% ee (Eq. 14.72) [140].

$$R\underset{208}{\diagup}\overset{O}{\diagdown}H + \underset{209}{\overset{R^1}{\underset{H}{N}}\diagdown OAc} \xrightarrow[\text{CHCl}_3]{207 \text{ (20 mol\%)}} R\underset{210}{\overset{\overset{R^1}{\underset{N}{\triangle}}}{\diagup}}\overset{O}{\diagdown}H \quad (14.72)$$

**210** 54-78%
4:1 - 10:1 dr, 84-99% ee

## 14.4.2
### Aziridination of Imines

The transformation of imines into aziridines remains less developed than the analogous transformation of aldehydes into epoxides. The major two means of

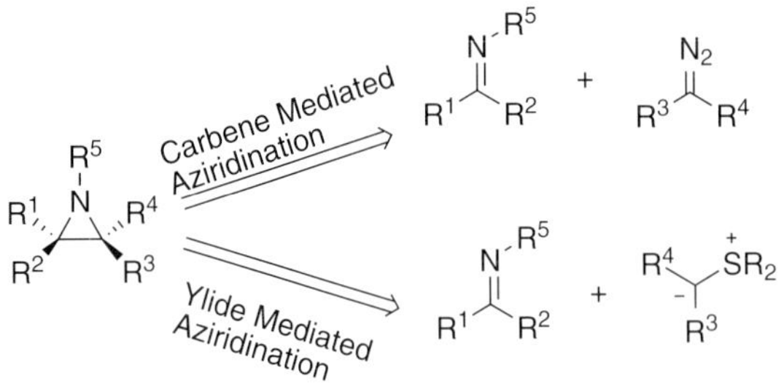

# 14 Cyclopropanation, Epoxidation and Aziridination Reactions

**Scheme 14.5** Carbene and ylide-mediated aziridination.

access to aziridines from imines are the reactions mediated by carbenes (carbenoids) or ylides (Scheme 14.5).

### 14.4.2.1 With Carbenes

The reaction of carbenes with Schiff bases to afford aziridines is well known. A recent example reported by Aggarwal was the addition of trimethylsilyldiazomethane to N-sulfonylimines to afford aziridines in good yields with high *cis* stereoselectivities (Eq. 14.73) [141].

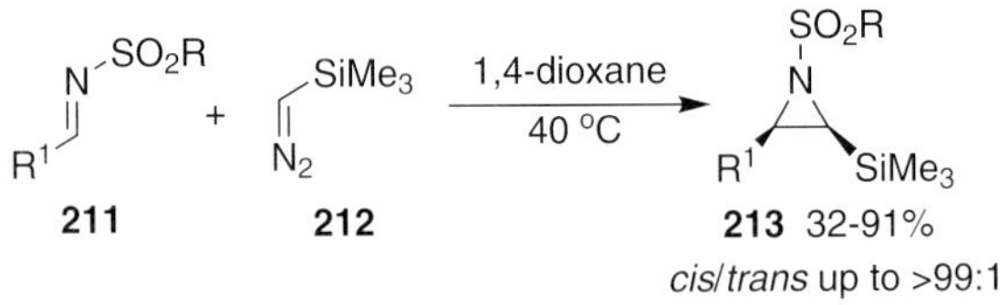

$$\text{211} + \text{212} \xrightarrow{\text{1,4-dioxane}, 40\,°C} \text{213} \quad 32-91\% \quad cis/trans \text{ up to } >99:1 \tag{14.73}$$

The formation of aziridines upon transition metal-catalyzed decomposition of diazo compounds in the presence of C=N double bonds is also well established [142]. The first asymmetric aziridination of imines was reported by Jacobsen (Eq. 14.74) [143], in which ethyl diazoacetates reacted with N-arylaldimines in the presence of Ph-BOX/copper(I) hexafluorophosphate to give N-arylaziridine carboxylates. Though the diastereoselectivities of the reaction were acceptable, the enantioselectivities were low.

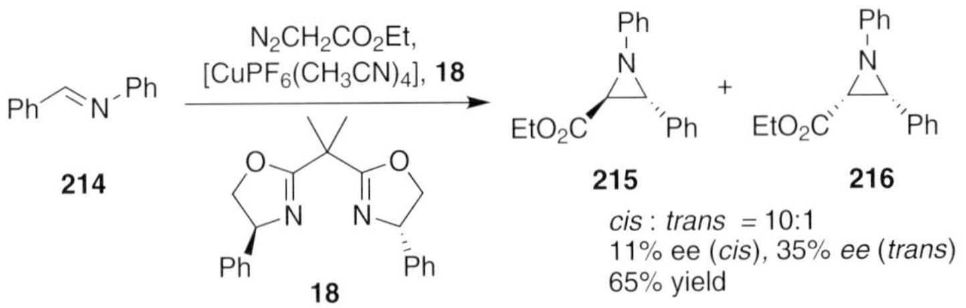

215 cis : trans = 10:1
11% ee (*cis*), 35% ee (*trans*)
65% yield

(14.74)

Many Lewis acids were found to be efficient catalysts for aziridination of imines [144]. Wulff *et al.* reported that the axially chiral boron Lewis acids VAPOL could promote the aziridination of N-benzhydrylimines with ethyl diazoacetate (Eq. 14.75).

This reaction provided aziridines with very high cis diastereoselectivities and up to 99% ee [145].

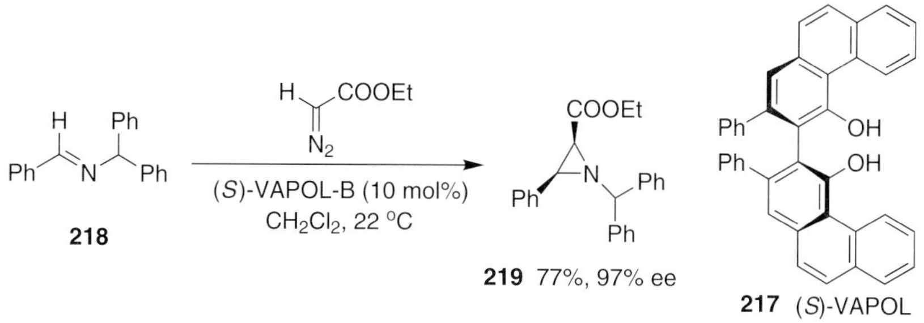

$$\text{218} \xrightarrow[\text{CH}_2\text{Cl}_2,\ 22\ ^\circ\text{C}]{\text{(S)-VAPOL-B (10 mol\%)}} \text{219}\ 77\%,\ 97\%\ ee \qquad \text{217 (S)-VAPOL} \tag{14.75}$$

Brønsted acid-catalyzed aziridination of imines with diazo compound was also reported by Johnston recently (Eq. 14.76) [146].

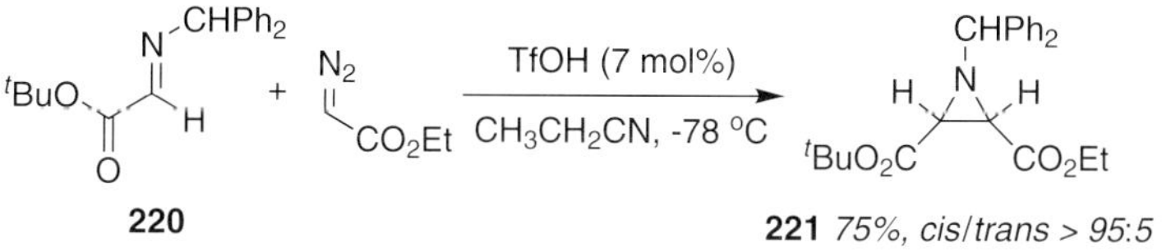

$$\text{220} + \text{N}_2\text{CHCO}_2\text{Et} \xrightarrow[\text{CH}_3\text{CH}_2\text{CN},\ -78\ ^\circ\text{C}]{\text{TfOH (7 mol\%)}} \text{221}\ 75\%,\ cis/trans > 95:5 \tag{14.76}$$

### 14.4.2.2 With Ylides

The azirdination of alkenes with ylides has been summarized systematically in 1997 and 2007 [147]. Dai *et al.* reported the sulfide-catalyzed aziridintaions in 1996. Based on the reaction of a sulfide with cinnamyl bromide, followed by deprotonation of the resulting sulfonium salt to form an ylide, the reaction of the ylide with an imine yielded an aziridine (Eq. 14.77) [148]. Asymmetric aziridination with moderate to high ee was successfully realized using the chiral sulfonium ylide derived from D-camphor (Eq. 14.78). Saito *et al.* described the synthesis of aziridines via a chiral sulfide **228** with moderate diastereoselectivities and good enantioselectivities (Eq. 14.79) [149].

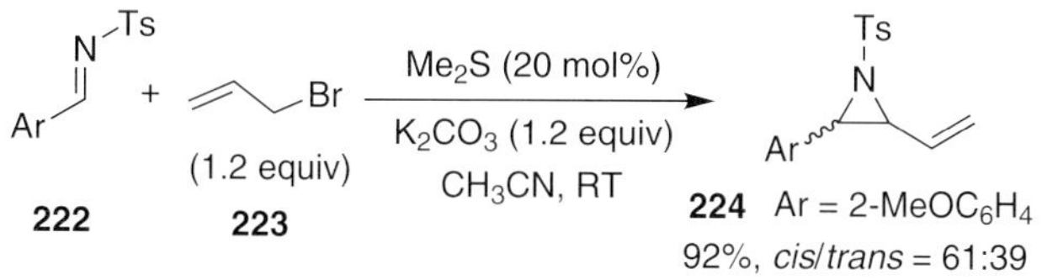

$$\text{222} + \text{223 (1.2 equiv)} \xrightarrow[\text{CH}_3\text{CN, RT}]{\text{Me}_2\text{S (20 mol\%)} \atop \text{K}_2\text{CO}_3\ (1.2\ \text{equiv})} \text{224}\ \text{Ar} = 2\text{-MeOC}_6\text{H}_4 \tag{14.77}$$

92%, cis/trans = 61:39

$$\text{225} + \text{226} \xrightarrow[\text{CH}_2\text{Cl}_2,\ \text{RT}]{\text{Cs}_2\text{CO}_3} \text{227} \tag{14.78}$$

85%, 77.5% ee

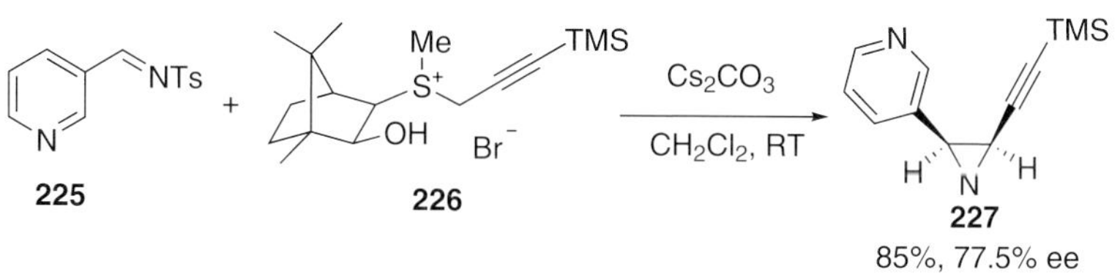

## 14 Cyclopropanation, Epoxidation and Aziridination Reactions

$$(14.79)$$

**229** + **230** (3.0 equiv) → **231** 99%
trans/cis = 75:25, 92%ee (trans)

Conditions: **228** (1.0 equiv), K$_2$CO$_3$ (3 equiv), CH$_3$CN

Vinylaziridnes are important subunits in a number of biologically active compounds and versatile intermediates for the synthesis of nitrogen-containing compounds. Tang et al. reported that optically active cis-2-substituted vinylaziridines could be synthesized by the reaction of N-tert-butylsulfinylimines with telluronium ylides with excellent diastereoselectivities in good to excellent yields (Eq. 14.80) [150].

$$(14.80)$$

**233** → **234** R = alkyl, aryl; R$^1$ = H, Ph, TMS
56-98%, up to 98% de, 30:1 (cis/trans)

Conditions: 1) LiHMDS, 2) $^t$Bu-S(O)-N=CHR **232**

Aggarwal et al. have developed a catalytic asymmetric aziridination of imines via sulfur ylides generated by the reaction of a metallocarbene with a sulfide. The carbenoid was generated by the decomposition of phenyl diazo compounds in the presence of a suitable transition metal salt. Phenyl N-tosylhydrazone salts were also effective as a carbene source. For example, a range of imines were aziridinated in good yields with moderate diastereoselectivities and high enantioselectivities by using N-tosylhydrazone salt as the carbene source in the presence of a catalytic amount of chiral sulfide (Eq. 14.81) [151].

$$(14.81)$$

**236** + **60** → **237** 50-72%
dr = 2:1 - 8:1
73-98% ee (trans)

Conditions: **235** (20 mol%), Rh$_2$(OAc)$_4$ (1 mol%), BnEt$_3$N$^+$Cl$^-$ (10 mol%), 1,4-dioxane, 40 °C

## 14.5
## Experimental: Selected Procedures

### 14.5.1
### Synthesis of Compound 14 with Catalyst 12 [19]

To a flame-dried and Ar-filled 5 mL-vial (Vial **A**) was added a solution of **12** (112.4 mg, 0.30 mmol, 0.25 equiv.) in freshly distilled $CH_2Cl_2$ (1.0 mL), followed by the addition of neat $ZnEt_2$ (37.0 mg, 31 μL, 0.30 mmol, 0.25 equiv). The solution was stirred at room temperature for 1 h. Upon cooling to 0 °C under Ar, $CH_2I_2$ (80.4 mg, 24 μL, 0.30 mmol, 0.25 equiv) was added dropwise, and then the solution was stirred at 0 °C for 0.5 h.

Simultaneously, to another flame-dried and Ar-filled 5 mL-vial (Vial **B**) was added freshly distilled $CH_2Cl_2$ (1.0 mL), followed by neat $ZnEt_2$ (148.2 mg, 123 μL, 1.2 mmol, 1.0 equiv). Upon cooling to −78 °C under Ar, $CH_2I_2$ (642.8 mg, 193 μL, 2.4 mmol, 2.0 equiv) was added dropwise. After the resulting mixture was stirred at this temperature for 1 h (a white precipitate formed), ethyl methoxyacetate (EMA) (141.8 mg, 141 μL, 1.2 mmol, 1.0 equiv) was added. The reaction mixture was warmed to −50 °C and a homogeneous solution was formed, the reaction mixture was then cooled to −78 °C.

To the third flame-dried and Ar-filled 5 ml-vial (Vial **C**) was added $I_2$ (152.4 mg, 0.6 mmol, 0.5 equiv). Vacuum was applied for 10 s and then $CH_2Cl_2$ (0.5 mL) was added. After the mixture was cooled to 0 °C, neat $ZnEt_2$ (37.0 mg, 31 μL, 0.30 mmol, 0.25 equiv) was added. After stirring at 0 °C for 1 h, the reaction mixture was cooled to −78 °C under Ar. The contents of Vial **A** and Vial **B** were then transferred to vial **C** via a cannula, followed by the addition of **13** ($R^1$ = Ph, $R^2$ = H, $R^3$ = Me) (247.5 mg, 1.2 mmol, 1.0 equiv). After warming to −40 °C and stirring at this temperature for 48 h, the reaction mixture was poured into hexanes (100 mL) and filtered. The filtrate was concentrated and purified by chromatography (silica gel, hexanes : ether = 6 : 1, v/v) to afford **14** as colorless liquid (261.0 mg, 99% yield, 93% ee).

### 14.5.2
### Synthesis of Compound 27 with Catalyst 24 [27]

A mixture of $CuPF_6(CH_3CN)_4$ (7.7 mg, 0.021 mmol), trisoxazoline **24** (9.0 mg, 0.023 mmol), **25** ($R^1$ = Ph, $R^2$ = H) (0.2 mL, 1.9 mmol, 5 equiv) in ethyl acetate (1 mL) was stirred at room temperature for 2 h. The resulting mixture was heated to 40 °C and then 3 Å MS (200 mg) was added. Into this solution was injected **26** (78 mg, 0.42 mmol) in 2 mL of ethyl acetate via a syringe pump within 6 h. After the reaction was complete (monitored by TLC), the mixture was filtered rapidly through a glass funnel with a thin layer of silica gel and eluted with dichloromethane. The filtrate was concentrated under reduced pressure and the residue was purified by flash chromatography (silica gel, petroleum ether/ethyl acetate) to afford the desired product **27** (Yield: 112 mg (92%), 92% ee).

### 14.5.3
**Synthesis of Compound 56 with Sulfonium salt 54 [37]**

To a stirred suspension of sulfonium salt **54** (R = TMS)(236 mg, 0.6 mmol) and α,β-unsaturated compound ($R^1$ = Ph, $R^2$ = COOMe) (0.5 mmol) in 2 mL of THF was added $^t$BuOK (202 mg, 1.8 mmol) in one portion at −78 °C. After stirring for 4 h at −78 °C, the reaction mixture was passed through a short silica gel column, which was eluted with ethyl acetate. After concentration of the elution, the residue was purified by flash column chromatography to afford **56** ($R^1$ = Ph, $R^2$ = COOMe, R = TMS) (Yield: 85%. 97% ee).

### 14.5.4
**Synthesis of Compound 79 with Catalyst 76 [45]**

The catalyst **76** (0.2 equiv) followed by **77** ($R^1$ = O$t$-Bu) (1.0 equiv) and **78** ($R^2$ = Ph, $R^3$ = H) (1.2 equiv) was added to a stirred solution of $Cs_2CO_3$ (1.2 equiv) in MeCN (0.25 M) and stirred at 80 °C for 24 h. The reaction was quenched with 1 M aqueous HCl solution and extracted three times with $Et_2O$ or EtOAc. The combined organic phases were washed with a saturated aqueous solution of $NaHCO_3$, dried ($MgSO_4$) and concentrated under reduced pressure. The residue was purified by flash column chromatography to afford **79** ($R^1$ = $^t$BuO, $R^2$ = Ph) (Yield: 96%. 86% ee).

### 14.5.5
**Synthesis of Compound 90 with Catalyst 88 [50]**

The enyne **91** (0.10–0.50 mmol) in $CH_2Cl_2$ (1 mL) was added to a mixture of **90** (2 mol%) in $CH_2Cl_2$ (2 mL) and the mixture was stirred for 5 min at 23 °C. The resulting mixture was filtered through $SiO_2$ and the solvent was evaporated to give **92**, which was purified by column chromatography (EtOAc/hexane mixtures).

### 14.5.6
**Synthesis of Compound 107 with (+)-104 [57]**

A 500-mL, 1-neck round-bottom flask equipped with a Teflon-coated magnetic stir bar was oven dried and then fitted with a serum cap and flushed with $N_2$. The flask was charged with 200 mL of dry $CH_2Cl_2$ and cooled in a −23 °C bath (dry ice/$CCl_4$). Then the following liquids were added sequentially via syringe while stirring in the cooling bath: 5.94 mL (5.68 g, 20 mmol) of titanium tetraisopropoxide; 3.43 mL (4.12 g, 20 mmol) of L(+)-diethyl tartrate, stirred 5 min before the next addition; 3.47 mL (3.08 g, 20 mmol) of geraniol; and, finally, about 11 mL of a dichloromethane solution (3.67 M in TBHP) containing about 40 mmol (2 equiv) of anhydrous *tert*-butyl hydroperoxide (TBHP).

The resulting homogeneous solution was then stored overnight in the freezer at about −20 °C in the sealed reaction vessel. Then the flask was placed in a −23 °C bath (dry ice/$CCl_4$) and 50 mL of 10% aqueous tartaric acid solution was added while stirring; the aqueous layer solidified. After 30 min, the cooling bath was removed and stirring was continued at room temperature for 1 h or until the aqueous layer became clear. After separation of the aqueous layer, the organic layer was washed once with water, dried ($Na_2SO_4$), and concentrated to afford a colorless oil with an odor revealing contamination by TBHP.

This oil was diluted with 150 mL of ether, and the resulting solution was cooled in an ice bath, and then 60 mL of 1 N NaOH solution was added. This produced a two-phase mixture which was stirred at 0 °C for 0.5 h. The ether phase was washed with brine, dried ($Na_2SO_4$), and concentrated to give 4.24 g of clear oil. Chromatography on silica gel afforded 2.6 g (77%) **107**. Analysis of this material as the MTPA ester gave an enantiomeric excess (ee) of >95% whereas analysis of the derived epoxy acetate by using Eu(hfbc)$_3$ chiral shift reagent gave 94% ee.

### 14.5.7
**Synthesis of Compound 120 with catalyst 118 [64a]**

A 50-mL round-bottomed flask equipped with a magnetic stir bar and a rubber septum was charged with 1, 2-dihydronaphthalene (1.36 g, 10 mol), 4-PPNO (170 mg, 10 mol%), **119a** (X = OMe) (72 mg, 1 mol%), and chlorobenzene (10 mL). The mixture was purged of air, cooled in an ice bath, and stirred under positive $N_2$ pressure. Precooled, purged, buffered NaOCl solution (30 mL, 1.8 equiv) was added via syringe, and the mixture was stirred for 2 h. At this point, **119b** (X = t-Bu) (37 mg, 0.5 mol%) in chlorobenzene (1 mL) was added, and the mixture was stirred for an additional 2 h. Heptane (10 mL) and Celite (0.5 g) were added with stirring, and the mixture was filtered through Celite. The filtrate was washed with water (2 × 30 mL) and then saturated aqueous $NaHCO_3$ (30 mL). The organic layer was dried ($Na_2SO_4$), filtered, and concentrated. The crude product was purified by flash chromatography on silica gel (EtOAc/hexanes 1 : 7). Epoxide **120** was isolated as a crystalline solid, mp 45–46 °C (0.784 g, 52% yield, 98% ee).

### 14.5.8
**Synthesis of Compound 141 with Catalyst 139 (General Procedure) [78]**

Aqueous $Na_2$-(EDTA) ($1 \times 10^{-4}$ M, 10 mL) and a catalytic amount of tetrabutylammonium hydrogen sulfate were added to a solution of *trans*-stilbene (0.18 g, 1 mmol) in acetonitrile (15 mL) with vigorous stirring at 0 °C. A mixture of oxone (3.07 g, 5 mmol) and sodium bicarbonate (1.3 g, 15.5 mmol) was pulverized, and a small portion of this mixture was added to the reaction mixture to bring the pH to >7. After 5 min, ketone **1** (0.77 g, 3 mmol) was added portionwise over a period of 1 h. Simultaneously, the rest of the oxone and sodium bicarbonate was added portionwise over 50 min. After completion of the addition of ketone **1**, the reaction mixture was stirred for another 1 h at 0 °C, diluted with water (30 mL), and extracted with hexanes (4 × 40 mL). The combined extracts were washed with brine, dried ($Na_2SO_4$), filtered, concentrated, and purified by flash chromatography [the silica gel was buffered with 1% triethylamine solution in hexane; hexane-ether (1 : 0 to

50 : 1 v/v) was used as the eluent] to afford *trans*-stilbene oxide as white crystals (0.149 g, 73% yield, 95.2% ee).

### 14.5.9
### Synthesis of Compound 157 with Catalyst 155 [89]

The catalyst **155** (0.05 mmol, 10 mol%) was added at room temperature to a solution of the aldehyde **156** (0.5 mmol) in $CH_2Cl_2$ (1.0 mL) followed by the addition of 35% $H_2O_2$ (aq.) (63 mg, 0.65 mmol, 1.3 equiv). After 4 h, the crude reaction mixture was passed through a silica gel FC column (ether/pentane) to give oxirane-2-carbaldehyde **157**.

### 14.5.10
### Synthesis of Compound 162 with Catalyst 161 [96]

To a mixture of enone **153** (0.10 mmol) and chiral quaternary ammonium salt **161** (6 mg, 0.003 mmol, 3 mol%) in toluene (0.3 mL) was added 13% aqueous sodium hypochlorite (NaOCl, 0.15 mL) and the mixture was stirred for several hours at 0 °C under Ar atmosphere. The resulting mixture was diluted with water and the organic phase was separated. The aqueous phase was then extracted with $CH_2Cl_2$ (3 times). The combined organic extracts were dried over $Na_2SO_4$. The solvents were evaporated and the residual oil was purified by flash column chromatography on silica gel (ethyl acetate/hexane as eluant) to afford the corresponding epoxy ketone **162**.

### 14.5.11
### Synthesis of Compound 174 with Sulfide 172 [107]

To a 5-mL round-bottomed flask fitted with a $N_2$ balloon and containing a magnetic stirrer bar were added sequentially sulfide **172** (12.5 mg, 0.05 mmol), anhydrous $CH_3CN$ (0.8 mL), rhodium(II) acetate dimer (4.5 mg, 0.01 mmol), benzyltriethylammonium chloride (11.4 mg, 0.05 mmol), aldehyde (1.0 mmol) and tosylhydrazone sodium salt **173** (148 mg, 1.0 mmol). The reaction mixture was stirred at 40 °C for 24 h, then cooled to RT and tosylhydrazone sodium salt **173** (148 mg, 1.0 mmol) was added. The mixture was stirred at 40 °C for a further 24 h. Work-up consisted of sequential addition to the reaction mixture of water (1.0 mL) and ethyl acetate (1.0 mL). The separated aqueous layer was washed with ethyl acetate (2 × 2.0 mL) and the combined organic phases dried over $MgSO_4$, filtered and concentrated *in vacuo*. The crude product was analyzed by $^1H$ NMR to determine the diastereomeric ratio and then purified by flash chromatography to afford the corresponding epoxide **174**.

### 14.5.12
### Synthesis of Compound 190 with Trichloroethyl N-tosyloxycarbamates 188 [125]

$Cu(pyridine)_4(BF_4)_2$ (28 mg, 0.050 mmol) was suspended in benzene (5 mL) in a 20-mL scintillation vial under air. $K_2CO_3$ (692 mg, 5.00 mmol) and the olefin **189**

(2.50 mmol) were successively added. The heterogeneous solution was stirred for 15 min at 25 °C and trichloroethyl N-tosyloxycarbamates **188** (182 mg, 0.50 mmol) was added in one portion. The solution was stirred at 25 °C overnight. Ether (15 mL) was added and the resulting mixture was filtered. The solid was washed 4 times with 10 mL of ether. The combined filtrate was concentrated under reduced pressure. The residue was purified by flash chromatography on silica gel using ether/hexanes as eluent to afford **190** (51–68% yield). The silica gel was pre-treated with 1% $Et_3N$/hexanes.

## 14.5.13
### Synthesis of Compound 210 with Catalyst 207 [140]

To a stirred solution of catalyst **207** (20 mol%) and **209** ($R^1$ = Cbz) (1.2 equiv, 0.30 mmol) in chloroform (1.0 mL) at 40 °C was added α,β-unsaturated aldehyde **208** (R = nPr) (1.0 equiv, 0.25 mmol). The reaction was vigorously stirred for 0.5 h. Next, the reaction was directly loaded upon a silica-gel column and immediate chromatography (pentane/EtOAc-mixtures or toluene/EtOAc-mixtures) furnished **203** ($R^1$ Cbz, R = n-Pr) (78% yield, 96% ee). The enantiomeric excess was determined by HPLC on Daicel Chiralpak OJ with iso-hexane/i-PrOH (90/10) as the eluent.

## 14.5.14
### Synthesis of Compound 219 with (S)-VAPOL Catalyst 217 [145]

The (S)-VAPOL-B catalyst from 54 mg (0.10 mmol) of (S)-VAPOL **217** was dissolved in 1 mL of toluene and transferred via syringe to a 10-mL flame-dried flask with stir bar at room temperature. To this flask was first added 115 µL (1.1 mmol) of ethyl diazoacetate by syringe. After 5 min stirring 271 mg (1.00 mmol) of imine **218** (R = Ph) in 1 mL of toluene was added via syringe pump addition over 3 h. After additional stirring for 2 h the reaction contents were diluted with 10 mL of diethyl ether and washed twice with 20 mL portions of brine. The organic layer was dried over $MgSO_4$, gravity filtered, and concentrated by rotary evaporation to give the crude aziridine as an off-white solid. The solid crude reaction mixture was purified by column chromatography to give aziridine **219** (R = Ph) as a white solid (275 mg) in 77% isolated yield. An optical purity of 97% ee was determined by HPLC analysis using a chiralcel OD column with 9/1 hexanes/2-propanol as the eluent, flow rate = 1.0 mL min$^{-1}$.

## Abbreviations

| | |
|---|---|
| Ac | acetyl |
| Bn | benzyl |
| Boc | tert-butyloxycarbonyl |
| Bu | butyl |
| DCM | dichloromethane |

| | |
|---|---|
| DIC | N, N′-diisopropylcarbodiimide |
| DMAP | 4-dimethylaminopryidine |
| dr | diastereomeric ratio |
| ee | enantiomeric excess |
| LDA | lithium diisopropylamide |
| LiHMDS | lithium hexamethyldisilazide |
| LiTMP | lithium 2,2,6,6-tetramethylpiperidin-1-ide |
| Mes | mesityl |
| MS | molecular sieve |
| TBHP | *tert*-butyl hydroperoxide |
| THF | tetrahydrofuran |
| TMS | trimethylsilyl |

**Acknowledgments**

We are grateful for financial support from the Chinese Academy of Sciences and the Natural Sciences Foundation of China for our research in this area.

**References**

1 (a) Donaldson, W.A. (2001) *Tetrahedron*, **57**, 8589–8267; (b) Maas, G. (2004) *Chem. Soc. Rev.*, **33**, 183–190; (c) de Meijere, A. (2003) *Chem. Rev.*, **103**, 931–932; (d) Lebel, H., Marcoux, J.F., Molinaro, C. and Charette, A.B. (2003) *Chem. Rev.*, **103**, 977–1055.

2 (a) Shi, Y. (2004) *Acc. Chem. Res.*, **37**, 488–496; (b) Yang, D. (2004) *Acc. Chem. Res.*, **37**, 497–505; (c) Aggarwal, V.K. and Winn, C.L. (2004) *Acc. Chem. Res.*, **37**, 611–620; (d) Adolfsson, H. and Balan, D. (2006) *Aziridines and Epoxides in Asymmetric Synthesis* (ed. A.K. Yudin), Wiley-VCH, Weinheim, Chapter 6.

3 (a) Muller, P. and Fruit, C. (2003) *Chem. Rev.*, **103**, 2905–2919; (b) Sweeney, J.B. (2006) *Aziridines and Epoxides in Asymmetric Synthesis* (ed. A.K. Yudin), Wiley-VCH, Weinheim, Chapter 4.

4 Simmons, H.E. and Smith, R.D. (1958) *J. Am. Chem. Soc.*, **80**, 5323–5324.

5 (a) Wittig, G. and Wingler, F. (1964) *Chem. Ber.*, **97**, 2146–2164; (b) Furukawa, J., Kawabata, N. and Nishimura, J. (1966) *Tetrahedron Lett.*, **7**, 3353–3354; (c) Denmark, S.E. and Edwards, J.P. (1991) *J. Org. Chem.*, **56**, 6974–6981; (d) Piers, E. and Coish, O.D. (1995) *Synthesis*, 47–55.

6 Molander, G.A. and Etter, J.B. (1987) *J. Org. Chem.*, **52**, 3942–3944.

7 Maruoka, K., Fukutani, Y. and Yamamoto, H. (1985) *J. Org. Chem.*, **50**, 4412–4414.

8 Closs, G.L. and Moss, R.A. (1964) *J. Am. Chem. Soc.*, **86**, 4042–4053.

9 Takai, K., Toshikawa, S., Inoue, A. and Kokumai, R. (2003) *J. Am. Chem. Soc.*, **125**, 12990–12991.

10 Tsai, C.C., Hsieh, I.L., Cheng, T.T., Tsai, P.K., Lin, K.W. and Yan, T.H. (2006) *Org. Lett.*, **8**, 2261–2263.

11 (a) Charette, A.B. and Juteau, H. (1994) *J. Am. Chem. Soc.*, **116**, 2651–2652; (b) Charette, A.B., Juteau, H., Lebel, H. and Molinaro, C. (1998) *J. Am. Chem. Soc.*, **120**, 11943–11953.

12 Charette, A.B., Molinaro, C. and Brochu, C. (2001) *J. Am. Chem. Soc.*, **123**, 12168–12175.

13 Imai, N., Sakamoto, K., Takahashi, H. and Kobayashi, S. (1994) *Tetrahedron Lett.*, **35**, 7045–7048.

14 (a) Denmark, S.E., Christenson, B.L., Coe, D.M. and O'Connor, S.P. (1995) *Tetrahedron Lett.*, **36**, 2215–2218; (b) Denmark, S.E. and O'Connor, S.P. (1997) *J. Org. Chem.*, **62**, 3390–3401.

15 Aggarwal, V.K., Fang, G.Y. and Meek, G. (2003) *Org. Lett.*, **5**, 4417–4420.

16 Concellón, J.M., Rodríguez-Solla, H. and Simal, C. (2007) *Org. Lett.*, **9**, 2685–2688.

17 (a) Concellón, J.M., Rodríguez-Solla, H. and Gómez, C. (2002) *Angew. Chem. Int. Ed.*, **41**, 1917–1919; (b) Concellón, J.M., Rodríguez-Solla, H., Méjica, C. and Blanco, E.G. (2007) *Org. Lett.*, **9**, 2981–2984.

18 Song, Z., Lu, T., Hsung, R.P., Al-Rashid, Z.F., Ko, C. and Tang, Y. (2007) *Angew. Chem. Int. Ed.*, **46**, 4069–4072.

19 Du, H., Long, J. and Shi, Y. (2006) *Org. Lett.*, **8**, 2827–2829.

20 (a) Long, J., Yuan, Y. and Shi, Y. (2003) *J. Am. Chem. Soc.*, **125**, 13632–13633; (b) Long, J., Du, H., Li, K. and Shi, Y. (2005) *Tetrahedron Lett.*, **46**, 2737–2740.

21 Charette, A.B., Janes, M.K. and Lebel, H. (2003) *Tetrahedron: Asym.*, **14**, 867–872.

22 Aggarwal, V.K., de Vicente, J. and Bonnert, R.V. (2001) *Org. Lett.*, **3**, 2785–2788.

23 Adams, L.A., Aggarwal, V.K., Bonnert, R.V., Bressel, B., Cox, R.J., Shepherd, J., de Vicente, J., Walter, M., Whittingham, W.G. and Winn, C.L. (2003) *J. Org. Chem.*, **68**, 9433–9440.

24 Ye, T. (1998) *Modern Catalytic Methods for Organic Synthesis with Diazo Compounds: From Cyclopropanes to Ylides*, Wiley & Sons, New York.

25 Nozaki, H., Moriuti, S., Takaya, H. and Noyori, R. (1966) *Tetrahedron Lett.*, **7**, 5239–5244.

26 Evans, D.A., Woerpel, K.A., Hinman, M.M. and Faul, M.M. (1991) *J. Am. Chem. Soc*, **113**, 726–728.

27 Xu, Z.H., Zhu, S.N., Sun, X.L., Tang, Y. and Dai, L.X. (2007) *Chem. Commun.*, 1960–1963.

28 Nishiyama, H., Itoh, Y., Matsumoto, H., Park, S.B. and Itoh, K. (1994) *J. Am. Chem. Soc.*, **116**, 2223–2224.

29 Lo, W.-C., Chen, C.-M., Cheng, K.-F. and Mak, T.C.W. (1997) *Chem. Commun.*, 12051206.

30 Munslow, I.J., Gillespie, K.M., Deeth, R.J. and Scott, P. (2001) *Chem. Commun.*, 1638–1639.

31 Miller, J.A., Jin, W. and Nguyen, S.T. (2002) *Angew. Chem. Int. Ed.*, **41**, 2953–2956.

32 Chen, Y. and Zhang, X.P. (2007) *J. Am. Chem. Soc.*, **129**, 12074–12705.

33 Niino, T., Toganoh, M., Andrioletti, B. and Furuta, H. (2006) *Chem. Commun.*, 4335–4337.

34 Hu, W., Timmons, D.J. and Doyle, M.P. (2002) *Org. Lett.*, **4**, 901–904.

35 Kanchiku, S., Suematsu, H., Matsumoto, K., Uchida, T. and Katsuki, T. (2007) *Angew. Chem. Int. Ed.*, **46**, 3889–3891.

36 Thompson, J.L. and Davies, H.M.L. (2007) *J. Am. Chem. Soc.*, **129**, 6090–6091.

37 (a) Ye, S., Huang, Z.-Z., Xia, C.-A., Tang, Y. and Dai, L.-X. (2002) *J. Am. Chem. Soc.*, **124**, 2432–2433; (b) Deng, X.M., Cai, P., Ye, S., Sun, X.L., Liao, W.W., Li, K., Tang, Y., Wu, Y.D. and Dai, L.X. (2006) *J. Am. Chem. Soc.*, **128**, 9730–9740.

38 (a) Aggarwal, V.K., Alonso, E., Fang, G., Ferrara, M., Hynd, G. and Porcelloni, M. (2001) *Angew. Chem. Int. Ed.*, **40**, 1433–1436; (b) Aggarwal, V.K. and Grange, E. (2006) *Chem. Eur. J.*, **12**, 568–575.

39 Mamai, A. and Madalengoitia, J.S. (2000) *Tetrahedron Lett.*, **41**, 9009–9014.

40 Kakei, H., Sone, T., Sohtome, Y., Matsunaga, S. and Shibasaki, M. (2007) *J. Am. Chem. Soc.*, **129**, 13410–13411.

41 Kunz, R.K. and MacMillan, D.W.C. (2005) *J. Am. Chem. Soc.*, **127**, 3240–3241.

42 (a) Hartikka, A. and Arvidsson, P.I. (2007) *J. Org. Chem.*, **72**, 5874–5877; (b) Hartikka, A., Slosarczyk, A.T. and Arvidsson, P.I. (2007) *Tetrahedron: Asym.*, **18**, 1403–1409.

43 Tang, Y., Ye, S. and Sun, X.L., (2005) *Synlett.*, 2720–2730.

44 Liao, W.W., Li, K. and Tang, Y. (2003) *J. Am. Chem. Soc.*, **125**, 13030–13031.

45 (a) Papageorgiou, C.D., de Dios, M.A.C., Ley, S.V. and Gaunt, M.J. (2004) *Angew. Chem. Int. Ed.*, **43**, 4641–4644;
(b) Johansson, C.C.C., Bremeyer, N., Ley, S.V., Owen, D.R., Smith, S.C. and Gaunt, M.J. (2006) *Angew. Chem. Int. Ed.*, **45**, 6024–6028.

46 Fishcer, E.O. and Dötz, K.H. (1970) *Chem. Ber.*, **103**, 1273–1278.

47 Harvey, D.F. and Sigano, D.M. (1996) *Chem. Rev.*, **96**, 271–288.

48 (a) Barluenga, J., Lopez, S., Trabanco, A.A., Fernandez-Acebes, A. and Florez, J. (2000) *J. Am. Chem. Soc.*, **122**, 8145–8154;
(b) Barluenga, J., Suarez-Sobrino, A.L., Tomas, M., Garcia-Granda, S. and Santiago-Garcia, R. (2001) *J. Am. Chem. Soc.*, **123**, 10494–10501; (c) Barluenga, J., Aznar, F., Gutierrez, I. and Martin, J.A. (2002) *Org. Lett.*, **4**, 2719–2722;
(d) Barluenga, J., Dieguez, A., Rodriguez, H. and Fananas, F.J. (2005) *Angew. Chem. Int. Ed.*, **44**, 126–128; (e) Barluenga, J., Fernandez-Rodriguez, M.A., Garcia-Garcia, P., Aguilar, E. and Merino, I. (2005) *Chem. Eur. J.*, **12**, 303–313;
(f) Barluenga, J., Andina, F., Aznar, F. and Valdes, C. (2007) *Org. Lett.*, **9**, 4143–4146.

49 Barluenga, J., Dieguez, A., Rodriguez, H. and Fananas, F.J. (2005) *Angew. Chem. Int. Ed.*, **44**, 126–128.

50 (a) Nieto-Oberhuber, C., Lopez, S., Munoz, M.P., Jimenez-Nunez, E., Bunuel, E., Cardenas, D.J. and Echavarren, A.M. (2006) *Chem. Eur. J.*, **12**, 1694–1702; (b) Lopez, S., Herrero-Gomez, E., Perez-Galan, P., Nieto-Oberhuber, C. and Echavarren, A.M. (2006) *Angew. Chem. Int. Ed.*, **45**, 6029–6032.

51 Nishino, F., Miki, K., Kato, Y., Ohe, K. and Uemura, S. (2003) *Org. Lett.*, **5**, 2615–2617.

52 Rautenstrauch, V. (1984) *J. Org. Chem.*, **49**, 950–952.

53 Marco-Contelles, J. and Soriano, E. (2007) *Chem. Eur. J.*, **13**, 1350–1357 and references cited therein.

54 Miki, K., Ohe, K. and Uemura, S. (2003) *J. Org. Chem.*, **68**, 8505–8513.

55 Tenaglia, A. and Marc, S. (2006) *J. Org. Chem.*, **71**, 3569–3575.

56 Johansson, M.J., Gorin, D.J., Staben, S.T. and Toste, F.D. (2005) *J. Am. Chem. Soc.*, **127**, 18002–18003.

57 (a) Katsuki, T. and Sharpless, K.B. (1980) *J. Am. Chem. Soc.*, **102**, 5974–5976;
(b) Katsuki, T. (1999) *Comprehensive Asymmetric Catalysis*, vol. 2 (eds E.N. Jacobsen, A. Pfaltz and H. Yamamoto), Springer, Berlin, Chapter 18. 1.

58 (a) Matsumoto, K., Sawada, Y., Saito, B., Sakai, K. and Katsuki, T. (2005) *Angew. Chem. Int. Ed.*, **44**, 4935–4939; (b) Sawada, Y., Matsumoto, K., Kondo, S., Watanabe, H., Ozawa, T., Suzuki, K., Saito, B. and Katsuki, T. (2006) *Angew. Chem. Int. Ed.*, **45**, 3478–3480; (c) Sawada, Y., Matsumoto, K. and Katsuki, T. (2007) *Angew. Chem. Int. Ed.*, **46**, 4559–4561;
(d) Arends, I.W.C.E. (2006) *Angew. Chem. Int. Ed.*, **45**, 6250–6252.

59 (a) Sheng, M.N. and Zajacek, J.G. (1970) *J. Org. Chem.*, **35**, 1839–1843; (b) Sharpless, K.B. and Michaelson, R.C. (1973) *J. Am. Chem Soc.*, **95**, 6136–6137.

60 Michaelson, R.C., Palermo, R.E. and Sharpless, K.B. (1977) *J. Am. Chem. Soc.*, **99**, 1992–1993.

61 Zhang, W., Basak, A., Kosugi, Y., Hoshino, Y. and Yamamoto, H. (2005) *Angew. Chem. Int. Ed.*, **44**, 4389–4391.

62 (a) Zhang, W. and Yamamoto, H. (2007) *J. Am. Chem. Soc.*, **129**, 286–287; (b) Makita, N., Hoshino, Y. and Yamamoto, H. (2003) *Angew. Chem. Int. Ed.*, **42**, 941–943.

63 Srinivasan, K., Michaud, P. and Kochi, J.K. (1986) *J. Am. Chem. Soc*, **108**, 2309–2320.

64 (a) Zhang, W., Loebach, J.L., Wilson, S.R. and Jacobsen, E.N. (1990) *J. Am Chem. Soc.*, **112**, 2801–2803; (b) Irie, R., Noda, K., Ito, Y., Matsumoto, N. and Katsuki, T. (1990) *Tetrahedron Lett.*, **31**, 7345.

65 (a) Chen, K. and Que, L. Jr, (1999) *Chem. Commun.*, 1375–1376. (b) Bukowski, M.R., Comba, P., Lienke, A., Limberg, C., de Laorden, C.L., Mas-Ballesté, R., Merz, M. and Que, L. Jr, (2006) *Angew. Chem. Int. Ed.*, **45**, 3446–3449.

66 White, M.C., Doyle, A.G. and Jacobsen, E.N. (2001) *J. Am. Chem. Soc.*, **123**, 7194–7195.

67 Gelalcha, F.G., Bitterlich, B., Anilkumar, G., Tse, M.k. and Beller, M. (2007) *Angew. Chem. Int. Ed.*, **46**, 7293–7296.

68 (a) Groves, J.T., Bonchio, M., Carofiglio, T. and Shalyaev, K. (1996) *J. Am. Chem. Soc.*, **118**, 8961–8962; (b) Higuchi, T., Othake, H. and Hirobe, M. (1989) *Tetrahedron Lett.*, **30**, 6545–6548; (c) Ohtake, H., Higuchi, T. and Hirobe, M. (1992) *Tetrahedron Lett.*, **33**, 2521–2524.

69 Tse, M.K., Klawonn, M., Bhor, S., Döbler, C., Anilkumar, G., Hugl, H., Mägerlein, W. and Beller, M. (2005) *Org. Lett.*, **7**, 987–990.

70 (a) Bhor, S., Anilkumar, G., Tse, M.K., Klawonn, M., Döbler, C., Bitterlich, B., Grotevendt, A. and Beller, M. (2005) *Org. Lett.*, **7**, 3393–3396; (b) Berkessel, A., Kaiser, P. and Lex, J. (2003) *Chem. Eur. J.*, **9**, 4746–4756; (c) Nishiyama, H., Shimida, T., Itoh, H., Sugiyama, H. and Motoyama, Y. (1997) *Chem. Commun.*, 1863–1864; (d) Tse, M.K., Döbler, C., Bhor, S., Klawonn, M., Mägerlein, W., Hugl, H. and Beller, M. (2004) *Angew. Chem. Int. Ed.*, **43**, 5255–5260.

71 Colladon, M., Scarso, A., Sgarbossa, P., Michelin, R. and Strukul, G. (2006) *J. Am. Chem. Soc.*, **128**, 14006–14007.

72 (a) Bougauchi, M., Watanabe, S., Arai, T., Sasai, H. and Shibaski, M. (1997) *J. Am. Chem. Soc.*, **119**, 2329–2330; (b) Matsunaga, S., Kinoshita, T., Okada, S., Harada, S. and Shibasaki, M. (2004) *J. Am. Chem. Soc.*, **126**, 7559–7570.

73 Kakei, H., Tsuji, R., Ohshima, T. and Shibasaki, M. (2005) *J. Am Chem. Soc.*, **127**, 8962–8963.

74 Curci, R., Fiorentino, M. and Serio, M.R. (1984) *J. Chem Soc, Chem. Commun.*, 155–156.

75 Yang, D., Yip, Y.-C., Tang, M.-W., Wang, M.K., Zheng, J.-H. and Cheung, K.-K. (1996) *J. Am. Chem. Soc.*, **118**, 491–492.

76 (a) Yang, D., Wang, X.C., Wong, M.K., Yip, Y.C. and Tang, M.W. (1996) *J. Am. Chem. Soc.*, **118**, 11311–13312; (b) Yang, D., Wong, M.K., Yip, Y.C., Wang, X.C., Tang, M.W., Zheng, J.H. and Cheung, K.K. (1998) *J. Am. Chem. Soc.*, **120**, 5943–5952; (c) Yang, D. (2004) *Acc. Chem. Res.*, **37**, 497–505.

77 (a) Shi, Y. (2004) *Acc. Chem. Res.*, **37**, 488–496; (b) Wang, Z.-X., Tu, Y., Frohn, M., Zhang, J.-R. and Shi, Y. (1997) *J. Am. Chem. Soc.*, **119**, 11224–11235.

78 Wu, X.Y., She, X. and Shi, Y. (2002) *J. Am. Chem. Soc.*, **124**, 8792–8793.

79 Tian, H., She, X., Shu, L., Yu, H. and Shi, Y. (2000) *J. Am. Chem. Soc.*, **122**, 11551–11552.

80 (a) Tian, H., She, X., Xu, J. and Shi, Y. (2001) *Org. Lett.*, **3**, 1929–1931; (b) Tian, H., She, X., Yu, H., Shu, L. and Shi, Y. (2002) *J. Org. Chem.*, **67**, 2435–2446; (c) Goeddel, D., Shu, L., Yuan, Y., Wong, O.A., Wang, B. and Shi, Y. (2006) *J. Org. Chem.*, **71**, 1715–1717; (d) Wong, O.A. and Shi, Y. (2006) *J. Org. Chem.*, **71**, 3973–3976.

81 Burke, C.P. and Shi, Y. (2006) *Angew. Chem. Int. Ed.*, **45**, 4475–4478.

82 (a) Hanquet, G., Lusinchi, X. and Millet, P. (1987) *Tetrahedron Lett.*, **28**, 6061–6064; (b) Biscoe, M.R. and Breslow, R. (2005) *J. Am. Chem. Soc.*, **127**, 10812–10813.

83 Aggarwal, V.K. and Wang, M.F. (1996) *Chem. Commun.*, 191–192.

84 Page, P.C.B., Farah, M.M., Buckley, B.R. and Blacker, A.J. (2007) *J. Org. Chem.*, **72**, 4424–4430.

85 Page, P.C.B., Buckley, B.R., Heaney, H. and Blacker, A.J. (2005) *Org. Lett.*, **7**, 375–377.

86 Page, P.C.B., Buckley, B.R. and Blacker, A.J. (2004) *Org. Lett.*, **6**, 1543–1546.

87 Adamo, M.F.A., Aggarwal, V.K. and Sage, M.A. (2000) *J. Am. Chem. Soc.*, **122**, 8317–8318; Aggarwal, V.K., Lopin, C. and

Sandrinelli, F. (2003) *J. Am. Chem. Soc.*, **125**, 7596–7601.
88 (a) Lattanzi, A. (2005) *Org. Lett.*, **7**, 2579–2582; (b) Li, Y., Liu, X., Yang, Y. and Zhao, G. (2007) *J. Org. Chem.*, **72**, 288–291.
89 Marigo, M., Franzen, J., Poulsen, T.B., Zhuang, W. and Jorgensen, K.A. (2005) *J. Am. Chem. Soc.*, **127**, 6964–6965.
90 Erkkilä, A., Pihko, P.M. and Clarke, M.-R. (2007) *Adv. Synth. Catal.*, **349**, 802–806.
91 Wang, X. and List, B. (2008) *Angew. Chem. Int. Ed*, **47**, 1119–1122.
92 Hashimoto, T. and Maruoka, K. (2007) *Chem. Rev.*, **107**, 5656–5682.
93 Helder, R., Hummelen, J.C., Laaane, R.W.P.M., Wiering, J.S. and Wynberg, H. (1976) *Tetrahedron Lett.*, **17**, 1831–1834.
94 (a) Lygo, B. and Wainwright, P.G. (1998) *Tetrahedron Lett.*, **39**, 1599–1602; (b) Corey, E.J. and Zhang, E.-Y. (1999) *Org. Lett.*, **1**, 1287–1290.
95 Jew, S., Lee, J.-H., Jeong, B.-S., Yoo, M.-S., Kim, M.-J., Lee, Y.-J., Lee, J., Choi, S., Lee, K., Lah, M.S. and Park, H. (2005) *Angew. Chem. Int. Ed.*, **44**, 1383–1385.
96 Ooi, T., Ohara, D., Tamura, M. and Maruoka, K. (2004) *J. Am. Chem. Soc.*, **126**, 6844–6845.
97 Peris, G., Jakobsche, C.E. and Miller, S.J. (2007) *J. Am. Chem. Soc.*, **129**, 8710–8711.
98 (a) McGarrigle, E.M., Myers, E.L., Illa, O., Shaw, M.A., Riches, S.L. and Aggarwal, V.K., (2007) *Chem. Rev.*, **107**, 5841–5883; (b) Aggarwal, V.K. and Winn, C.L. (2004) *Acc. Chem. Res.*, **37**, 611–620.
99 Johnson, A.W. and LaCount, R.B. (1958) *Chem. Ind. (London)*, 1440–1441.
100 Furukawa, N., Okano, K. and Fujihara, H. (1987) *Nippon Kagaku Kaishi*, 1353–1358.
101 Jurukawa, N., Sugihara, Y. and Fujihara, H. (1989) *J. Org. Chem.*, **54**, 4222–4224.
102 Zhou, Y.G., Li, A.-H., Hou, X.-L. and Dai, L.-X. (1996) *Chem. Commun.*, 1353–1354.
103 Li, A.-H., Dai, L.-X., Hou, X.-L., Huang, Y.-Z. and Li, F.-W. (1996) *J. Org. Chem.*, **61**, 489–493.
104 Davoust, M., Brière, J.-F., Jaffrès, P.-A. and Metzner, P. (2005) *J. Org. Chem.*, **70**, 4166–4169.
105 Aggarwal, V.K., Rahman, H.A., Jones, R.V.H., Lee, H.Y. and Reid, B.D. (1994) *J. Am. Chem. Soc.*, **116**, 5973–5974.
106 Fulton, J.R., Aggarwal, V.K. and Vicente, J. (2005) *Eur. J. Org. Chem.*, 1479–1492.
107 Aggarwal, V.K., Alonso, E., Hynd, G., Lydon, K.M., Palmer, M.J., Porcelloni, M. and Studley, J.R. (2001) *Angew. Chem. Int. Ed.*, **40**, 1430–1433.
108 Zhou, Z.L., Shi, L.-L. and Huang, Y.-Z. (1990) *Tetrahedron Lett.*, **31**, 7657–7660.
109 (a) Brière, J.F., Takada, H. and Metzner, P. (2005) *Phosphorus, Sulfur Silicon Relat. Elem.*, **180**, 965–968; (b) Ou, W.-H. and Huang, Z.-Z. (2005) *Synthesis*, 2857–2860.
110 (a) Kiyooka, S.I. and Shahid, K.A. (2000) *Tetrahedron: Asym.*, **11**, 1537–1542; (b) Achard, T.J.R., Belokon, Y.N., Ilyin, M., Moskalenko, M., North, M. and Pizzato, F. (2007) *Tetrahedron Lett.*, **48**, 2965–2969.
111 Corey, E.J. and Choi, S. (1991) *Tetrahedron Lett.*, **32**, 2857–2860.
112 (a) Humelen, J.C. and Wynberg, H. (1978) *Tetrahedron Lett.*, **19**, 1089–1092; (b) Aggarwal, V.K., Hynd, G., Picoul, W. and Vasse, J.-L. (2002) *J. Am. Chem. Soc.*, **124**, 9964–9965; (c) Arai, S., Tokumaru, K. and Aoyama, T. (2004) *Tetrahedron Lett.*, **45**, 1845–1848 and references cited therein.
113 Arai, S. and Shioiri, T. (2002) *Tetrahedron*, **58**, 1407–1413 and references cited therein.
114 Groves, J.T. and Takahashi, T. (1983) *J. Am. Chem. Soc.*, **105**, 2073–2074.
115 Du Bois, J., Tomooka, C.S., Hong, J. and Carreira, E.M. (1997) *Acc. Chem. Res.*, **30**, 364–372.
116 Leung, S.K.Y., Huang, J.S., Liang, J.L., Che, C.M. and Zhou, Z.Y. (2003) *Angew. Chem. Int. Ed*, **42**, 340–343.
117 Man, W.L., Lam, W.W.Y., Yiu, S.M., Lau, T.C. and Peng, S.M. (2004) *J. Am. Chem. Soc.*, **126**, 15336–15337.
118 Mansuy, D., Mahy, J.-P., Dureault, A., Bedi, G. and Battioni, P. (1984) *J. Chem. Soc, Chem. Commun.*, 1161–1163.

119 Evans, D.A., Faul, M.M. and Bilodeau, M.T. (1991) *J. Org. Chem.*, **56**, 6744–6746.

120 Klotz, K.L., Slominski, L.M., Hull, A.V., Gottsacker, V.M., Mas-Balleste, R., Que, L. and Halfen, J.A. (2007) *Chem. Comm.*, 2063–2065.

121 Guthikonda, K. and Du Bois, J. (2002) *J. Am. Chem. Soc.*, **124**, 13672–13673.

122 Catino, A.J., Nichols, J.M., Forslund, R.E. and Doyle, M.P. (2005) *Org. Lett.*, **7**, 2787–2790.

123 Han, H., Par, S.B., Kim, S.K. and Chang, S. (2008) *J. Org. Chem.*, **73**, 2862–2870.

124 Lebel, H., Huard, K. and Lectard, S. (2005) *J. Am. Chem. Soc.*, **127**, 14198–14199.

125 Lebel, H., Lectard, S. and Parmentier, M. (2007) *Org. Lett.*, **9**, 4797–4800.

126 (a) Albone, D.P., Aujla, P.S., Taylor, P.C., Challener, S. and Derrck, A.M. (1998) *J. Org. Chem.*, **63**, 9569–9571; (b) Simkhovich, L. and Grross, Z. (2001) *Tetrahedron Lett.*, **42**, 8089–8092.

127 (a) chanda, B.M., Vyas, R. and Bedekar, A.V. (2001) *J. Org. Chem.*, **66**, 30–34; (b) Gao, G.Y., Harden, J.D. and Zhang, X.P. (2005) *Org. Lett.*, **7**, 3191–3193; (c) Antunes, A.M.M., Bonifacio, V.D.B., Nascimento, S.C.C., Lobo, A.M., Branco, P.S. and Prabhakar, S. (2007) *Tetrahedron.*, **63**, 7009–7017.

128 Han, J.L., Li, Y.F., Zhi, S.J., Pan, Y., Timmons, C. and Li, G.G. (2006) *Tetrahedron Lett.*, **47**, 7225–7228.

129 (a) Li, Z., Quan, R.W. and Jacobsen, E.N. (1995) *J. Am. Chem. Soc.*, **117**, 5889–5890; (b) Omura, K., Uchida, T., Irie, R. and Katsuki, T. (2004) *Chem. Commun.*, 2060–2061; (c) Gao, G.Y., Jones, J.E., Vyas, R., Harden, J.D. and Zhang, X.P. (2006) *J. Org. Chem.*, **71**, 6655–6658.

130 Li, Z., Conser, K.R. and Jacobsen, E.N. (1993) *J. Am. Chem. Soc.*, **115**, 5326–5327.

131 (a) Evans, D.A., Faul, M.M., Bilodeau, M.T., Anderson, B.A. and Barnes, D.M. (1993) *J. Am. Chem. Soc.*, **115**, 5328–5329; (b) Evans, D.A., Faul, M.M. and Bilodeau, M.T. (1991) *J. Org. Chem*, **56**, 6744–6746.

132 Kwong, H.-L., Liu, D., Chan, K.-Y., Lee, C.-S., Huang, K.-H. and Che, C.-M. (2004) *Tetrahedron Lett.*, **45**, 3965–3968.

133 Xu, J., Ma, L. and Jiao, P. (2004) *Chem. Commun.*, 1616–1617.

134 Wang, X.S. and Ding, K.L. (2006) *Chem. Eur. J.*, **12**, 4568–475.

135 Li, J., Chan, P.W.H. and Che, C.M. (2005) *Org. Lett.*, **7**, 5801–5804.

136 Minakata, S., Morino, Y., Oderaotoshi, Y. and Komatsu, M. (2006) *Chem. Commun.*, 3337–3339.

137 Ikeda, I., Machii, Y. and Okahara, M. (1980) *Synthesis*, 650–651.

138 Shen, Y.M., Zhao, M.X., Xu, J.X. and Shi, Y. (2006) *Angew. Chem. Int. Ed.*, **45**, 8005–8008.

139 Armstrong, A., Baxter, C.A., Lamont, S.G., Pape, A.R. and Wincewicz, R. (2007) *Org. Lett.*, **9**, 351–353.

140 Vesely, J., Ibrahem, I., Zhao, G.L., Rios, R. and Córdova, A. (2007) *Angew. Chem. Int. Ed.*, **46**, 778.

141 Aggarwal, V.K. and Ferrara, M. (2000) *Org. Lett.*, **2**, 4107–4110.

142 Doyle, M.P., Hu, W. and Timmons, D.J. (2001) *Org. Lett.*, **3**, 933–935.

143 Hansen, K.B., Finney, N.S. and Jacobsen, E.N. (1995) *Angew. Chem., Int. Ed.*, **34**, 676–678.

144 (a) Casarrubios, L., Pérez, J.A., Brookhart, M. and Tempeton, J.L. (1996) *J. Org. Chem.*, **61**, 8358–8359; (b) Zhu, Z. and Espenson, J.H. (1996) *J. Am. Chem. Soc.*, **118**, 9901–9907; (c) Sengupta, S. and Mondal, S. (2000) *Tetrahedron Lett.*, **41**, 6245–6248.

145 (a) Wulff, W.D. and Antilla, J.C. (1999) *J. Am. Chem. Soc.*, **121**, 5099–5100; (b) Su, Y., Rabalakos, C., Mitchell, W.D. and Wulff, W.D. (2005) *Org. Lett.*, **7**, 367–369.

146 Williams, A.L. and Johnston, J.N. (2004) *J. Am. Chem. Soc.*, **126**, 1612–1613.

147 (a) Li, A.H., Dai, L.X. and Aggarwal, V.K. (1997) *Chem. Rev.*, **97**, 2341–2372; (b) McGarrigle, E.M., Myers, E.L., Illa, O., Shaw, M.A., Riches, S.L. and Aggarwal, V.K. (2007) *Chem. Rev.*, **107**, 5841–5883.

148 Li, A.-H., Dai, L.-X. and Hou, X.-L. (1996) *J. Chem. Soc., Perkin Trans. 1*, 867–868.

**149** Saito, T., Sakairi, M. and Akiba, D. (2001) *Tetrahedron Lett.*, **42**, 5451–5454.

**150** Zheng, J.C., Liao, W.W., Sun, X.X., Sun, X.L., Tang, Y., Dai, L.X. and Deng, J.G. (2005) *Org. Lett.*, **7**, 5789–5792.

**151** Aggarwal, V.K., Alonso, E., Fang, G., Ferrara, M., Hynd, G. and Porcelloni, M. (2001) *Angew. Chem. Int. Ed.*, **40**, 1433–1436.

# 15
# Cyclization of Cyclopropane- or Cyclopropene-Containing Compounds

*Junliang Zhang and Yuanjing Xiao*

## 15.1
## Introduction

Three-membered carbocycles, important versatile building blocks for organic chemistry, have unique highly strained structural and electronic properties which give rise to unusually high reactivity and readily undergo many interesting and unique transformations. Many methods have been developed for the synthesis of various kinds of three-membered carbocyles [1]. This chapter concerns recent advances in the cyclization reaction of cyclopropane and cyclopropene-containing compounds, including cyclopropanes, vinylcyclopropanes, methylenecyclopropanes, vinylidenecyclopropanes and cyclopropenes. Typical examples using these reactions in synthetic applications will also be discussed. Some overlap with existing reviews covering related topics is unavoidable, these reviews will be mentioned and listed in the corresponding sections.

## 15.2
## Cycloisomerization Reactions

### 15.2.1
### Cycloisomerization of Vinylcyclopropanes and Their Hetero-Analogues

#### 15.2.1.1 Cycloisomerization of Vinylcyclopropanes to Cyclopentenes
The vinylcyclopropane–cyclopentene rearrangement represents one of the most typical transformations of vinylcyclopropanes [3]. The proposed mechanism for transition metal-mediated vinylcyclopropane–cyclopentene is depicted in Scheme 15.1. Oxidative addition of the cyclopropyl ring forms a vinyl metallacyclobutane which then undergoes rearrangement to metallacyclohexene followed by reductive elimination to give the cyclopentene product. The reaction pathway via vinylcyclopropane M-$\eta^2$ **5** to M-$\eta^3$ **6** and then cyclopentene M-$\eta^2$ **7** cannot be ruled out.

**Scheme 15.1** Plausible reaction pathway for vinylcyclopropane–cyclopentene rearrangement.

This rearrangement can occur under pyrolytic conditions [3] and in the presence of transition metals and Lewis acid [4]. Metal-free rearrangement of simple vinylcyclopropanes (i.e., without strong polar substituents) generally requires high temperatures and tends to produce mixtures of products. For example, Hudlicky demonstrated that bicyclic vinylcyclopropanes **9** undergo vinylcyclopropane–cyclopentene rearrangement under pyrolytic conditions or in the presence of a rhodium complex as catalyst but give the products **10** and **11** with different diasteroselectivity. Later, he successfully applied this Rh(I)-catalyzed transformation for the total synthesis of hirsutene **14** (Scheme 15.2) [5].

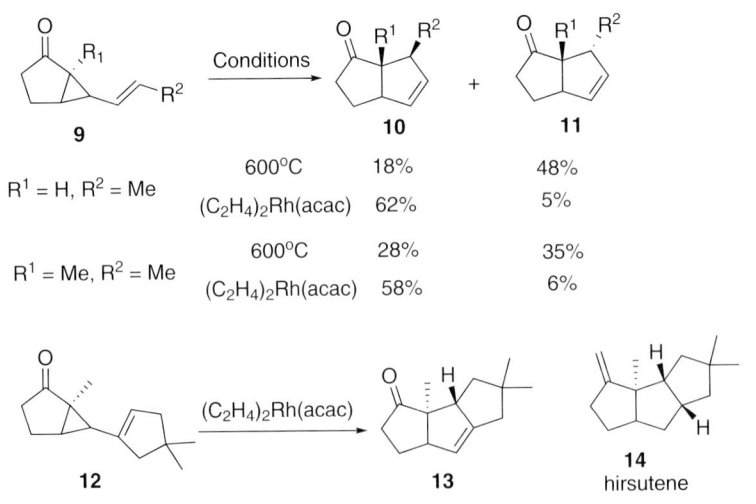

**Scheme 15.2** Vinylcyclopropane–cyclopentene rearrangement under pyrolytic conditions or under the catalysis of a rhodium complex.

Recently, Ni(0)-catalyzed transformation of vinylcyclopropanes without an activating group has been developed. Suginome and Ito observed that 1-(1-cyclopropylvinyl)benzene can rearrange smoothly into 1-cyclopentenylbenzene in toluene at 90 °C in the presence of Ni(acac)$_2$/tricyclohexylphosphine with a catalytic amount of

**Scheme 15.3** Nickel/PCy$_3$-catalyzed simple vinylcyclopropane rearrangement.

DIBAL-H(Scheme 15.3) [6]. Recently, a better catalyst system Ni(COD)$_2$/IPr was explored by Louie and Murakami [7], which makes the reaction occur at lower temperature and gives better yields of the products with complete retention of stereochemistry (Scheme 15.4).

**Scheme 15.4** [Ni(COD)$_2$]/IPr-catalyzed simple vinylcyclopropane rearrangement.

Hiroi demonstrated that activated optically active VCPs **17** and **19** rearrange into nonracemic cyclopentene derivatives under the catalysis of transition metal complexes such as Pd, Pt, or Ni catalysts. The stereoselectivity of this rearrangement depends on the catalyst used, providing the best results with Pd catalysts (Scheme 15.5) [8].

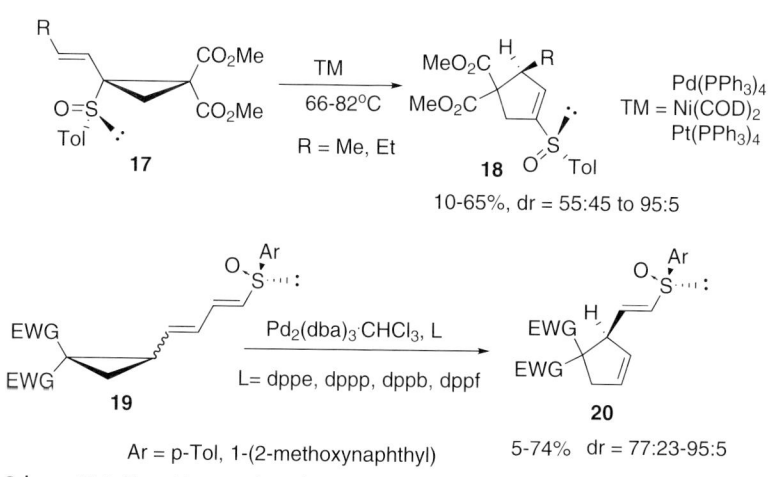

**Scheme 15.5** Transition-metal-catalyzed rearrangement of activated optically active VCPs.

This kind of vinylcyclopropane–cyclopentene rearrangement could also be catalyzed by a Lewis acid. Corey disclosed that the cycloisomerization reaction of substrate **21** to product **22** proceeded smoothly in the presence of $Et_2AlCl$ (Scheme 15.6) [9].

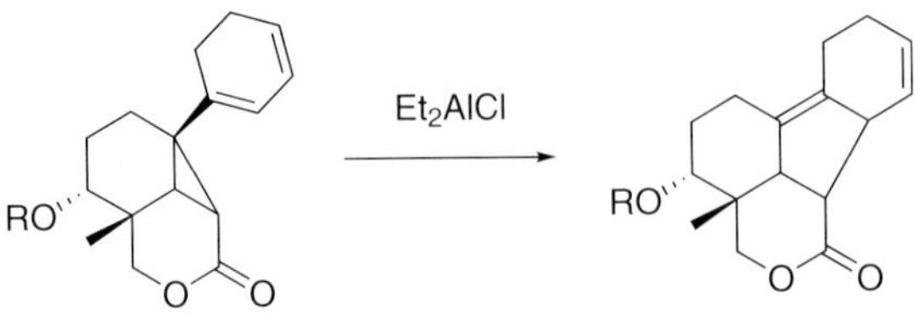

**Scheme 15.6** $Et_2AlCl$-catalyzed vinylcyclopropane–cyclopentene rearrangement.

The donor–acceptor-substituted vinylcyclopropanes **23** and **25** revealed a remarkably high stereoselectivity of the rearrangement, leading to cyclopentene derivatives **24** and **26** in excellent yields (Scheme 15.7) [10].

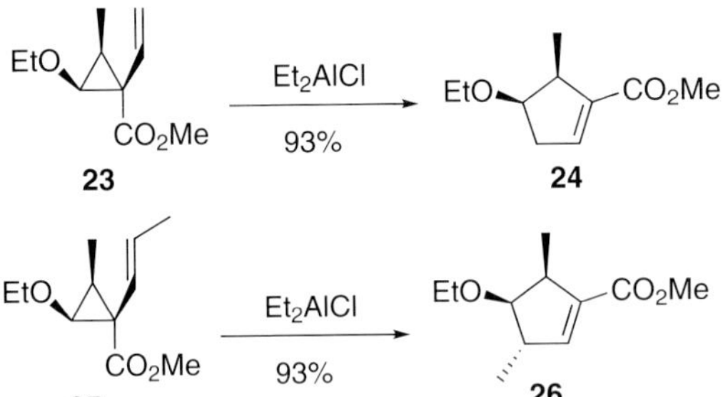

**Scheme 15.7** $Et_2AlCl$-catalyzed rearrangemt of D–A VCPs.

When these diethylaluminum chloride-assisted ring enlargements were performed with enantioenriched vinylcyclopropanes **27** and **29**, the former precursor **27** almost completely lost its chiral information (only 9% ee left) during the transformation into **28**, whereas **29** was converted into cyclopentene derivatives **30** without decrease in the enantiopurity, indicating that the additional cyclic ring is essential to keep the chiral information during the rearrangement reaction (Scheme 15.8) [11].

It was found that the Lewis acid $SnCl_4$ can be employed as catalyst in the ring enlargement of **31** to the tricyclic product **32** (Scheme 15.9) [12].

### 15.2.1.2 Cycloisomerization of Hetero-Analogues of VCPs to Five-Membered Heterocycles

Johnson reported a mild Ni(0)-catalyzed rearrangements of 1-acyl-2-vinylcyclopropanes **33** to afford dihydrofurans **34** in good to excellent yields (Scheme 15.10) [13]. Oxidative cleavage of the cyclopropane ring results in a (π-allyl)metal species **35** with a

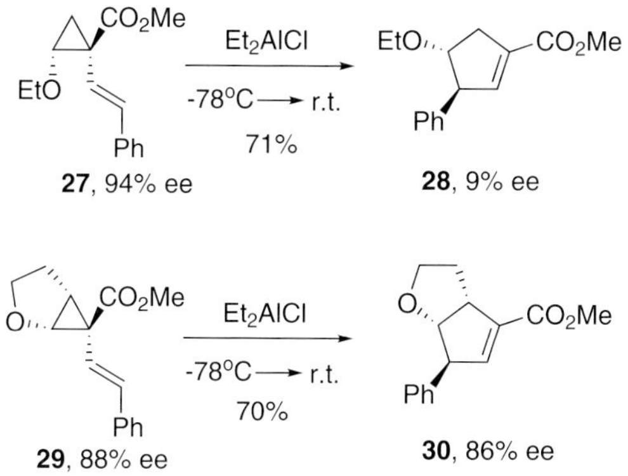

**Scheme 15.8** Et$_2$AlCl-catalyzed rearrangemt of optically pure D–A VCPs.

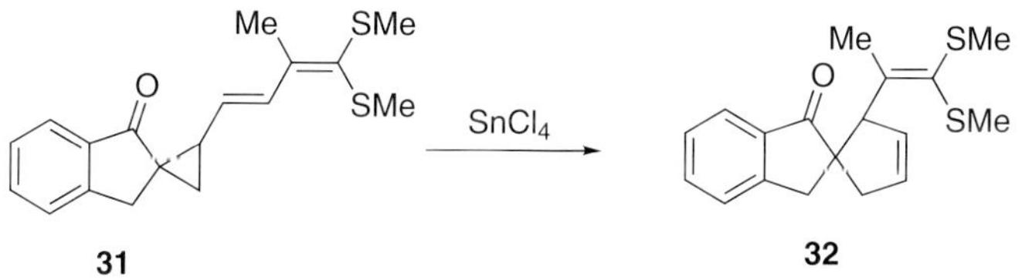

**Scheme 15.9** SnCl$_4$-catalyzed vinylcyclopropane–cyclopentene rearrangement.

pendant enolate which undergoes reductive elimination to afford the final dihydrofuran products. The optically pure substrate **36** (88% ee), treated with a catalytic amount of Ni(COD)$_2$/Ph$_3$P, furnishes the corresponding optically pure dihydrofuran product *trans*-**37** (88%ee) in 94% yield as a single diastereomer.

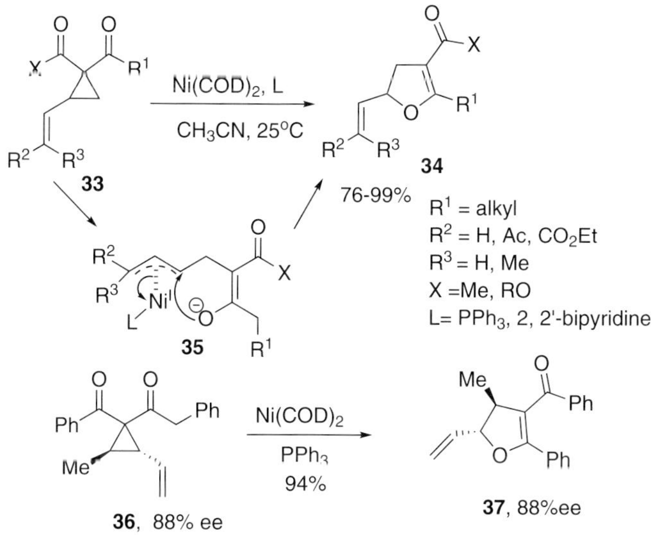

R$^1$ = alkyl
R$^2$ = H, Ac, CO$_2$Et
R$^3$ = H, Me
X = Me, RO
L = PPh$_3$, 2, 2'-bipyridine

**Scheme 15.10** Ni(0)-catalyzed rearrangements of 1-acyl-2-vinylcyclopropanes.

Cylopropyl silyl ketones **38** can undergo similar rearrangement to 5-silyl-2,3-dihydrofuran derivatives in up to 99% yields in the presence of trimethylsilyl trifluoromethanesulfonate, regardless of the substituents on the cyclopropane ring or silicon atom (Scheme 15.11) [14].

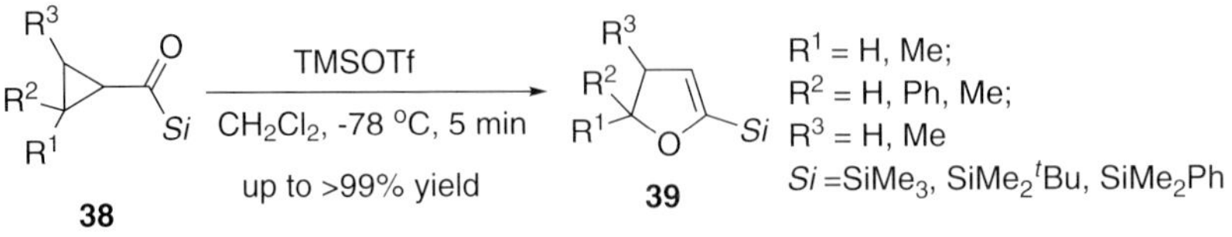

**Scheme 15.11** TMSOTf-catalyzed rearrangements of cylopropyl silyl ketones.

Chang demonstrated that cyclopropyl thioimidate **42**, generated from the reaction of substituted cyclopropyl thioamides **40** with methyl iodide, could undergo rearrangement to give pyrrolothiomethylimidates **41**. The iodide, generated in the reaction pathway, mediated the cleavage of the cyclopropyl ring of **42** (Scheme 15.12) [15].

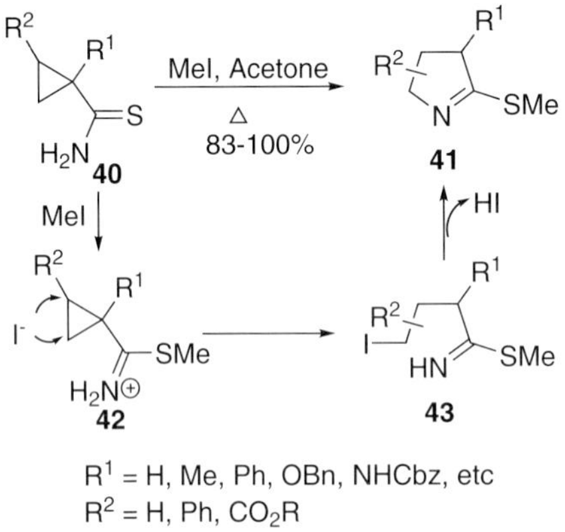

**Scheme 15.12** Synthesis of pyrrolothiomethylimidates via rearrangement.

## 15.2.2
### Cycloisomerization of Vinyl-, Allenyl-, and Alkynylcyclopropanes

Trost recently demonstrated an efficient asymmetric cycloisomerization of vinylcyclopropanols **44** into vinylcyclobutanones **45** in the presence of catalytic amounts of Pd$_2$(dba)$_3$ and chiral bidentate phosphine ligand (R,R)-**46** (Scheme 15.13) [16]. This transformation takes place via the extremely facile Wagner–Meerwein-type ring expansion of a cyclopropylmethyl cation, which is stabilized in the form of a π-allylpalladium intermediate **47**.

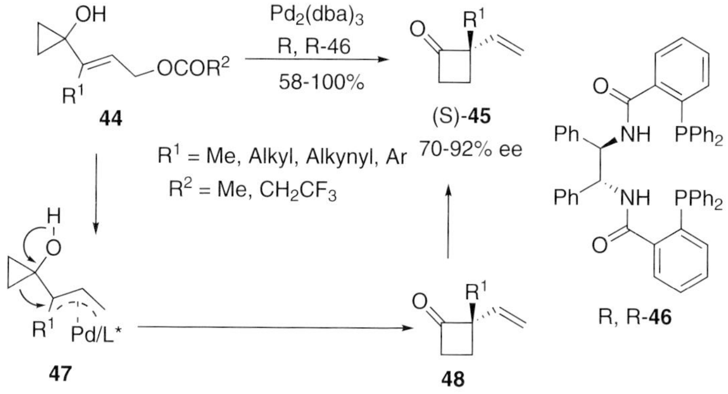

## 15.2 Cycloisomerization Reactions

**Scheme 15.13** Asymmetric cycloisomerization of vinylcyclopropanols under chiral palladium complex catalysis.

Allenylcyclopropanes **49** undergo the enlargement rearrangement reaction to 3-alkylidenecyclopent-1-enes **50/51** catalyzed by Rh-complexes, the regioselectivity could be controlled by the ligand of the catalyst (Scheme 15.14) [17].

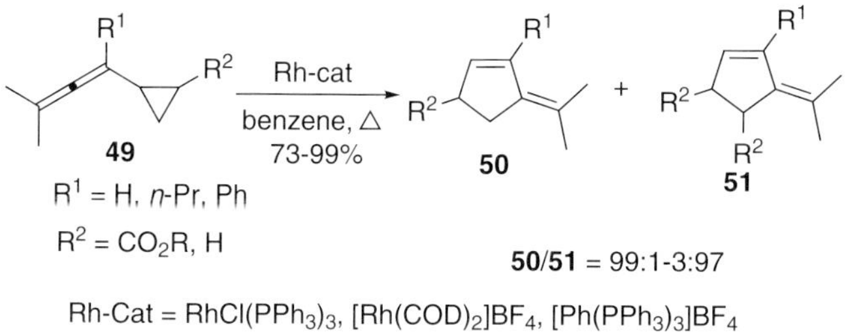

**Scheme 15.14** Rh(I)-catalyzed rearrangement of allenylcyclopropanes.

Iwasawa reported a $Co_2(CO)_8$-mediated rearrangement reaction of 1-(alk-1-ynyl)cyclopropanols **52** into (E)-2-alkylidenecyclobutanones **54** in moderate to excellent yields. In contrast, The TBS-protected 1-(alk-1-ynyl)cyclopropanols ether **55** underwent a different rearrangement to give conjugated cyclopentenones **56** or **57** under the same reaction conditions. The stereochemistry of the substrate affects the reaction, the cis isomer gives a better yield and high diastereoselectivity (Scheme 15.15) [18].

Recently, Toste developed an Au(I)-catalyzed version of this 1-(alk-1-ynyl)cyclopropanols to 2-alkylidenecyclobutanones enlargement cycloisomerization reaction (Scheme 15.16) [19]. Initially, the cationic gold complex binds the triple bond of **58** to form the intermediate **59**. The final product was afforded via the successive Wagner–Meerwein-type ring expansion followed by hydrolysis.

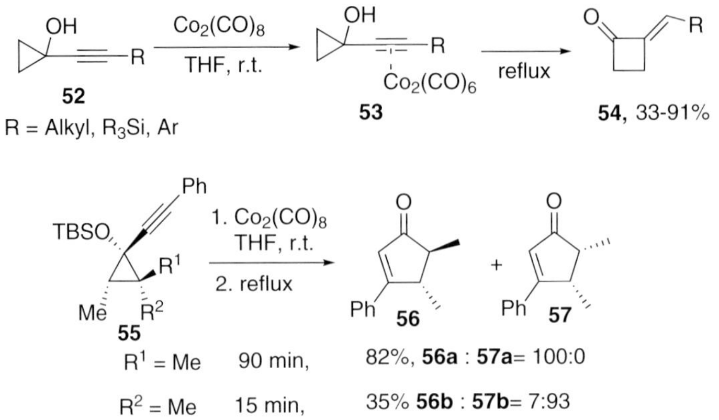

**Scheme 15.15** Co$_2$(CO)$_8$-mediated rearrangement reaction of 1-(alk-1-ynyl)cyclopropanols.

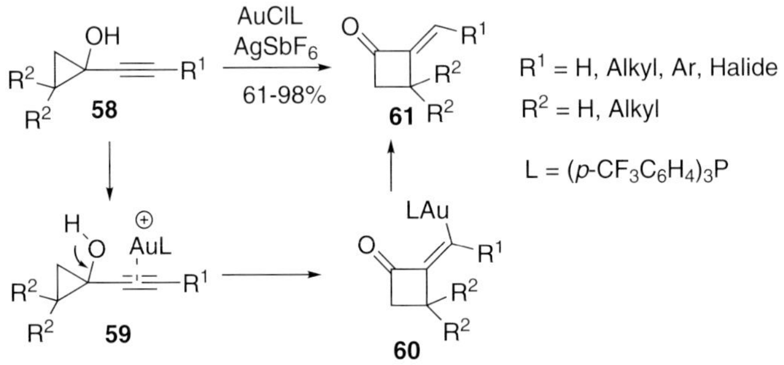

**Scheme 15.16** Au(I)-catalyzed ring enlargement of 1-(alk-1-ynyl) cyclopropanols to 2-alkylidenecyclobutanones.

## 15.2.3
### Cycloisomerization of Other Cyclopropanes

3,3-Sigmatropic rearrangements of divinylcyclopropanes to cycloheptadiene derivatives are among the most important reactions for construction of seven-membered carbocycles [20]. For example, Nakamura developed an efficient route for the construction of cycloheptadiene **64** from the divinylcyclopropane **63**, which was easily prepared from the cyclopropene **62** by the addition of the copper reagent and subsequent cross-coupling with alkenyl iodide under the catalysis of a palladium complex (Scheme 15.17) [21].

One of the vinyl groups of divinylcyclopropanes can be replaced by a hetero double bond to afford the corresponding heterocyclic compounds. For example, the 2-vinylcyclopropyl methanol **65** was oxidized by Dess–Martin periodinane to give vinylcyclopropyl aldehyde, which undergoes 3,3-sigmatropic rearrangement to 2,5-dihydrooxepine product **66**, which may be easily converted into the corresponding cyclopentenes **67** via [1,3] ring contraction in the presence of EtAlCl$_2$ (Scheme 15.18) [22].

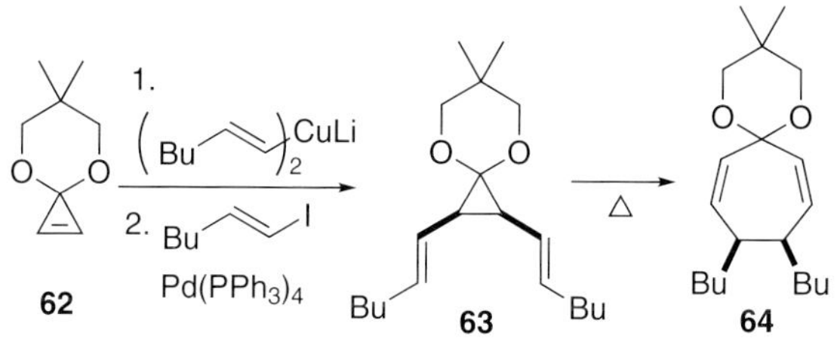

## 15.2 Cycloisomerization Reactions

**Scheme 15.17** Synthesis of cycloheptadiene from cyclopropene.

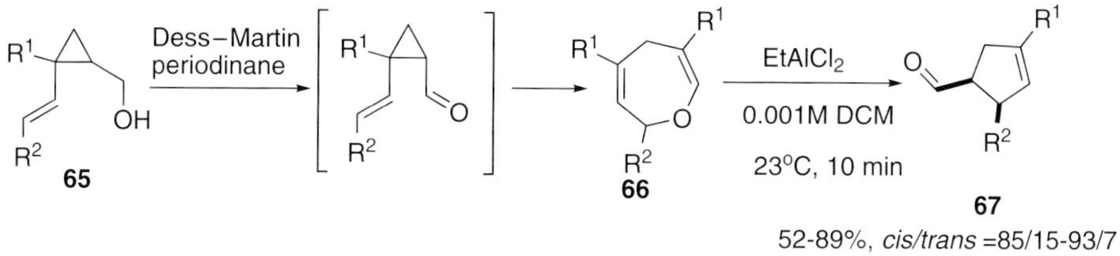

**Scheme 15.18** Synthesis of cyclopentenes from 2-vinylcyclopropyl methanol.

Analogously, vinylcyclopropyl carboxylic esters **65** were hydrolyzed by KOSiMe$_3$ and then reacted with (PhO)$_2$P(O)N$_3$ to generate the carboxylic azide which readily undergoes rearrangement to afford vinylcyclopropanyl isocyanates and 3, 3-sigmatropic rearrangement to yield 1H-azepin-2(3H)-ones **69** (Scheme 15.19) [23].

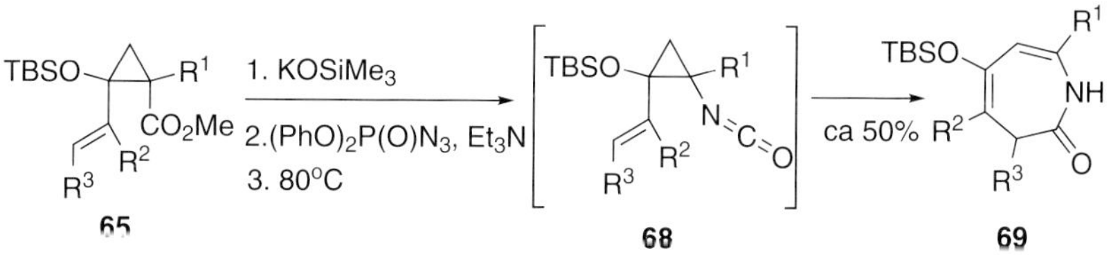

**Scheme 15.19** Synthesis of 1H-azepin-2(3H)-ones from vinylcyclopropyl carboxylic esters.

Widenhoefer demonstrated that the cycloisomerization of alkene-VCPs **70** initialized by hydrosilylation under catalysis of a palladium (II) complex produced alkenylcyclopropanes **71** in 54–83% yields with high stereoselectivity (Scheme 15.20) [24]. The catalytic cycle starts from the silapalladition of the terminal olefin moiety of **70**, providing intermediate **72**, which undergoes alkene insertion via 5-exo-trig cyclization and subsequent β-carbon elimination to produce the homoallyl palladium species **75**. The catalytic cycle would go into the next run after hyride transfer from silanes to the product.

A rhodium-catalyzed cycloisomerization of N-allylated bicyclo[1.1.0]butylalkylamines **76** was recently reported by Wipf [25]. These reactions provide pyrrolidines **77** and azepines **78** with high levels of stereo- and regiocontrol, which depend on the nature of the Rh(I) ligands (Scheme 15.21). Under catalysis of the

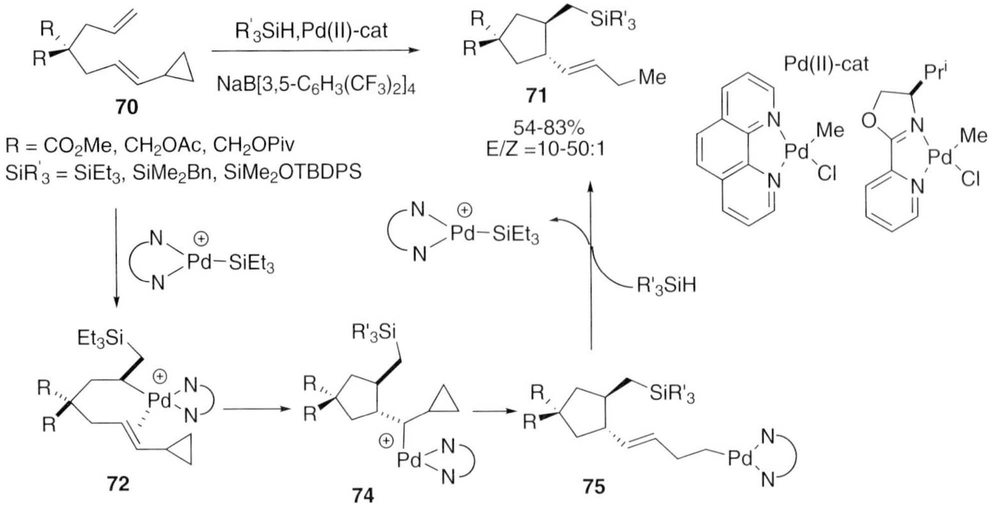

**Scheme 15.20** Palladium-catalyzed hydrosilylation of alkene-VCPs.

[Rh(CH$_2$=CH$_2$)$_2$Cl]$_2$/Ph$_3$P system, the reaction gives pyrrolidines **77** as the major products in 55–75% yields, whereas azepines **78** were produced in the presence of [Rh(CO)$_2$Cl]$_2$. The controllable regioselectivity comes from the regioselective formation of the homoallyl Rh-carbenes **80** and **81**, which were generated from the same intermediate **79** through oxidative addition of Rh(I) across the central σ-bond in bicyclo[1.1.0]butane.

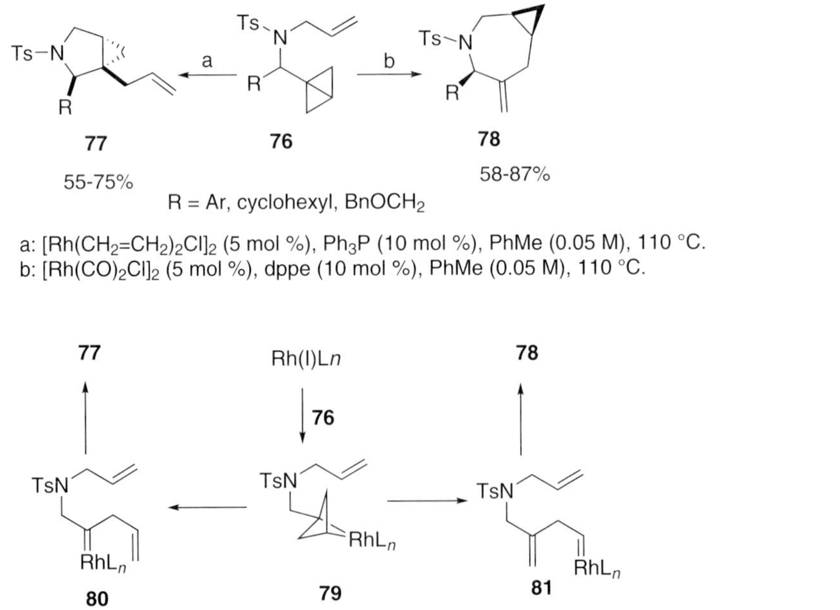

a: [Rh(CH$_2$=CH$_2$)$_2$Cl]$_2$ (5 mol %), Ph$_3$P (10 mol %), PhMe (0.05 M), 110 °C.
b: [Rh(CO)$_2$Cl]$_2$ (5 mol %), dppe (10 mol %), PhMe (0.05 M), 110 °C.

**Scheme 15.21** Rhodium-catalyzed cycloisomerization of N-allylated bicyclo[1.1.0]butylalkylamines.

## 15.2.4
### Cycloisomerization of Methylene- and Alkylidenecyclopropanes

Very recently, Fürstner [26] and Shi [27] independently observed an efficient synthesis of cyclobutenes **83** via the Pt(II) or Pd(II)-catalyzed enlargement reaction of alkylidenecyclopropanes **82**. In both cases, a clean 1, 2-deuterium shift with >97% D incorporation into the cyclobutene product was observed. In the $PtCl_2$/CO system, the reaction mechanism was suggested as cycle A in Scheme 15.22. The electrophilic addition of the platinum catalyst to the double bond of **82** generates the cyclopropylmethyl cation **84**, which undergoes a Demjanov rearrangement to afford the zwitterionic cyclobutyl species **85** or the carbenoid platinum complex **86**. The subsequent 1, 2-hydride shift provides the cyclobutyl cation **87**, which regenerates the catalyst and affords the final product via elimination of the metal. An alternative mechanism proposed by Shi is shown in Cycle B of Scheme 15.22. Regioselective halopalladation of MCP **82** with $PdBr_2$, generated *in situ* from the reaction between $Pd(OAc)_2$ and $CuBr_2$, gives the cyclopropylpalladium complex **88**. After β-hydride elimination and reversed-regioselective insertion, the cyclopropylmethylpalladium species **90** was formed. Subsequent α-bromide elimination gave the carbenoid species **91**, which undergoes ring enlargement to afford the final cyclobutene product.

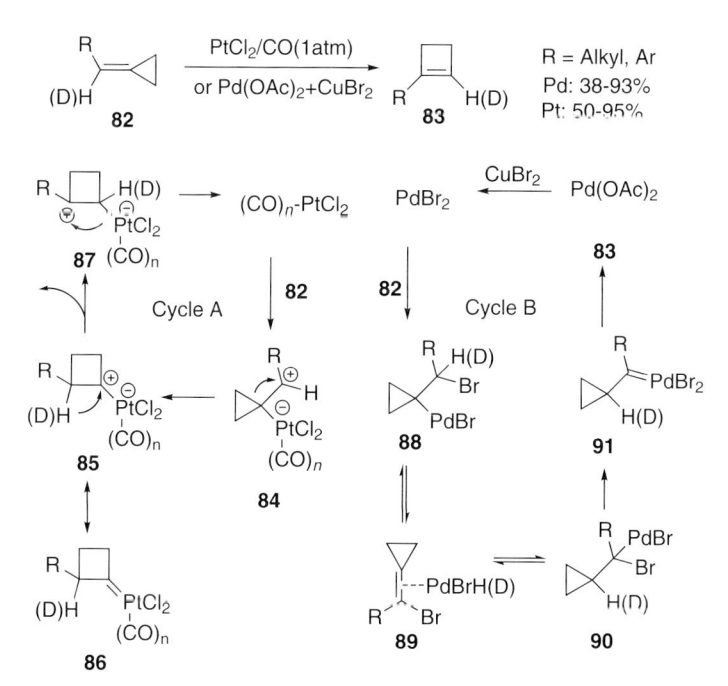

**Scheme 15.22** Transition-metal-catalyzed cycloisomerization of alkylidenecyclopropanes.

### 15.2.4.1 Cycloisomerization of Acyl Methylene- and Alkylidenecyclopropanes

An efficient synthesis of various heterocycles from the same starting material acetyl methylene- and alkylidenecyclopropanes under various controlled reaction conditions was disclosed by Ma (Scheme 15.23) [28]. Cycloisomerization of **92** under a catalytic amount of NaI afforded 3-alkylidene-2, 3-dihydrofurans **93**, which can be converted into 2, 3, 4-trisubstituted furans **94** via aromatization under acidic conditions. 2, 3, 4-Trisubstituted furans **94** could also be prepared from **92** directly but under the co-catalyst $PdCl_2(CH_3CN)_2$/NaI. 3-Methylene-2, 3-dihydrofurans **96** was given as the major product under the catalysis of the Pd(0) species, similarly, 3-methylene-2, 3-dihydrofurans **96** isomerize to the corresponding furans **97** upon treatment with 3 M HCl in THF. In contrast to the Pd(0)-catalyzed cycloisomerization of **92**, the reaction of acyl-MCPs under the catalysis of Pd(II) complex gave 4$H$-pyrans **95**. ($Z$)-1-(2-(chloromethylene)-1-(phenylsulfonyl)cyclopropyl)ethanone **98** isomerizes to 5-chloro-2-methyl-3-(phenylsulfonyl)-4$H$-pyran **99** in 76% yield in the presence of Pd(II) cpmplexes, in which a 1,2-chloride shift was observed in this transformation.

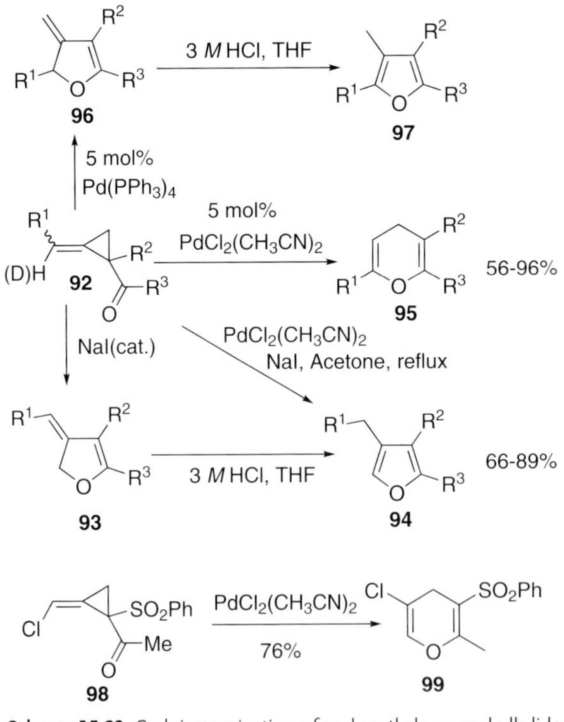

**Scheme 15.23** Cycloisomerization of acyl methylene- and alkylidenecyclopropanes.

A mechanistic rationale for the Pd(0)-catalyzed cycloisomerization of acyl-MCPs into furans is proposed in Scheme 15.24. The direct oxidative addition of Pd(0) species into the distal C—C bond of **92** results in the formation of palladacyclobutane **100**, which undergoes rearrangement to palladium enolate **101**. The resulting

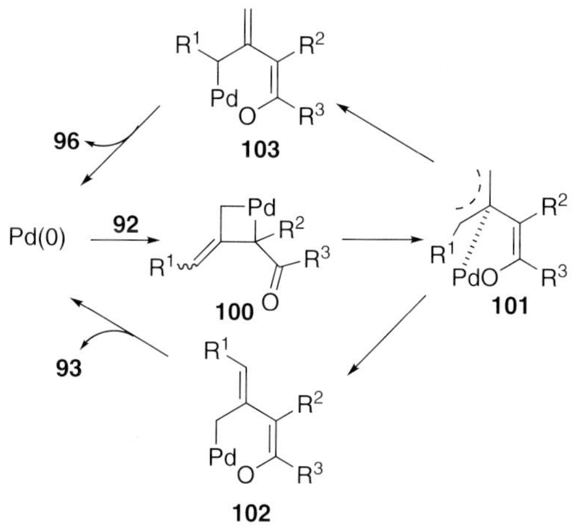

**Scheme 15.24** Plausible reaction pathway for Pd(0)-catalyzed cycloisomerization of acyl alkylidenecyclopropanes.

intermediate **101** generates palladaheterocyles **102** and **103** via η3–η1 rearrangement, followed by reductive elimination to give the final furan product and regenerate the Pd(0) species.

For the Pd(II)-catalyzed cycloisomerization of **92** to pyrans, there are two possible reaction pathways a and b. Path a begins with regioselective halopalladation of the exo double bond of **92** forming the cyclopropylmethylpalladium species **104**, which gives palladium enolate **105** via regioselective β-carbon elimination and enolization. Subsequent regioselective insertion of the double bond of **105** into the O–Pd bond affords the key intermediate – the dihydropyranyl palladium species **106**, from which different reaction pathways are followed according to the substitutents. When $R^4$ = H, tandem β-H elimination and hydropalladation with reversed regioselectivity occurs to give the dihydropyranyl palladium species **108**, followed by β-halide elimination to give the 4H-pyran **95**. The 2H-pyran **114** was formed when $R^4$ = alkyl, experiencing a tandem β-$H^a$ elimination, hydropalladation with a reversed regioselectivity, and β-halide elimination. When $R^4$ = chloride, direct β-halide elimination was observed to give the 1,2-chloride shift product 4H-pyran **99**. The key intermediate **106** can be formed via an alternative reaction pathway b: The coordination of Pd(II) species to the exo double bond of **92**, making it electrophilic enough to trigger an intramolecular cyclization to afford the cyclopropyl oxocarbenium species **110**. Subsequent ring expansion produces cation **111** which, in turn, gives the key intermediate **106**.

Ma also disclosed that intermolecular cyclization of acyl alkylidenecyclopropanes with amines provides an efficient route to the synthesis of 2, 3, 4-trisubstituted pyrroles. Two possible reaction pathways are depicted in Scheme 15.26 [29]. One pathway begins with the initial formation of the alkylidenecyclopropyl imine **119** by condensation of the corresponding ketones with amines which, in turn, undergoes Cloke rearrangement to give 3-alkylidene-2, 3-dihydro-1H-pyrrole **120**. Subsequent aromatization of **120** gives the final pyrrole product **118**. The other pathway starts from the nucleophilic attack of the amine at the less sterically hindered carbon center of the three-membered ring and thereby under cleavage of the distal bond to afford intermediate **115** (path b). The intramolecular condensation between amine and ketone and subsequent aromatization produces pyrrole **118**.

### 15.2.4.2 Cycloisomerization via a Cascade Process

A rhodium-catalyzed tandem C–H activation/cycloisomerization of MCP **121**, containing a vinylpyridine, to produce the cycloheptene **125** was disclosed by Fürstner in 2007 (Scheme 15.27) [30]. The formation of an *E*-configured exocyclic C=C bond is consistent with a mechanistic scenario comprising a C–H activation (forming **122**)/hydrometallation (forming **123**)/β-carbon elimination (forming **124**)/reductive elimination sequence directed by the basic nitrogen atom of the pyridine ring. This interpretation is also strongly supported by the deuterium-labeling experiment.

Fürstner also found that **126** with an aldehyde functionality could undergo a similar tandem transformation leading to cyclohept-4-en-1-ones **127** (Scheme 15.28) [30]. Syn-addition of the resulting Rh–H species, generated by the insertion of the catalyst into the C–H bond of the aldehyde group of **126**, to the adjacent double bond provides metallacycle **128** as the primary product. After the necessary rotation to eclipse the C–Rh bond with the C–C bond of the cyclopropane, the intermediate **129**

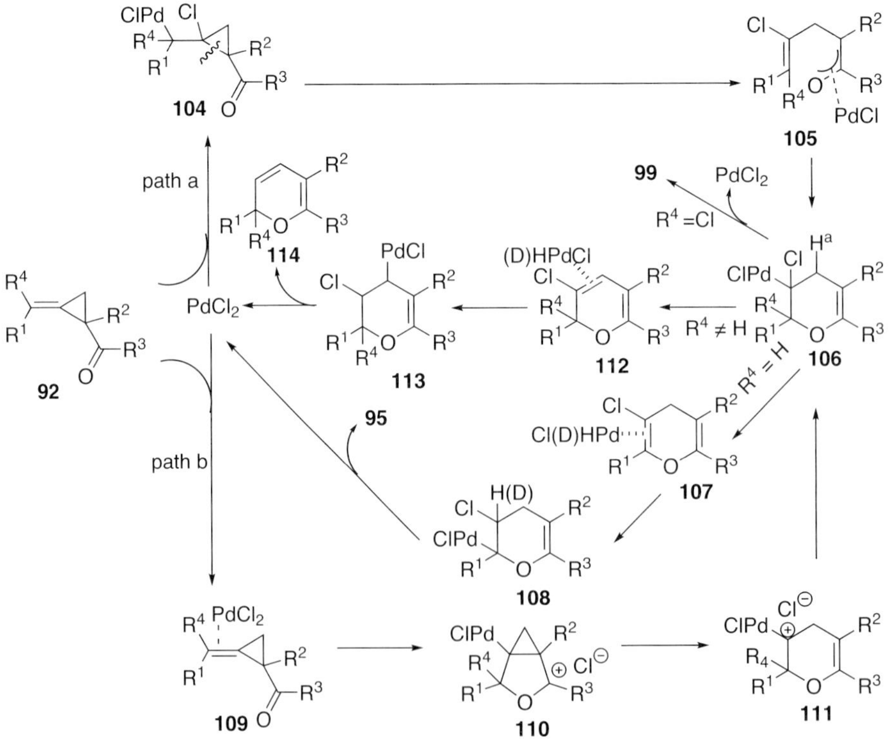

**Scheme 15.25** Plausible reaction pathway for Pd(II)-catalyzed cycloisomerization of acyl alkylidenecyclopropanes.

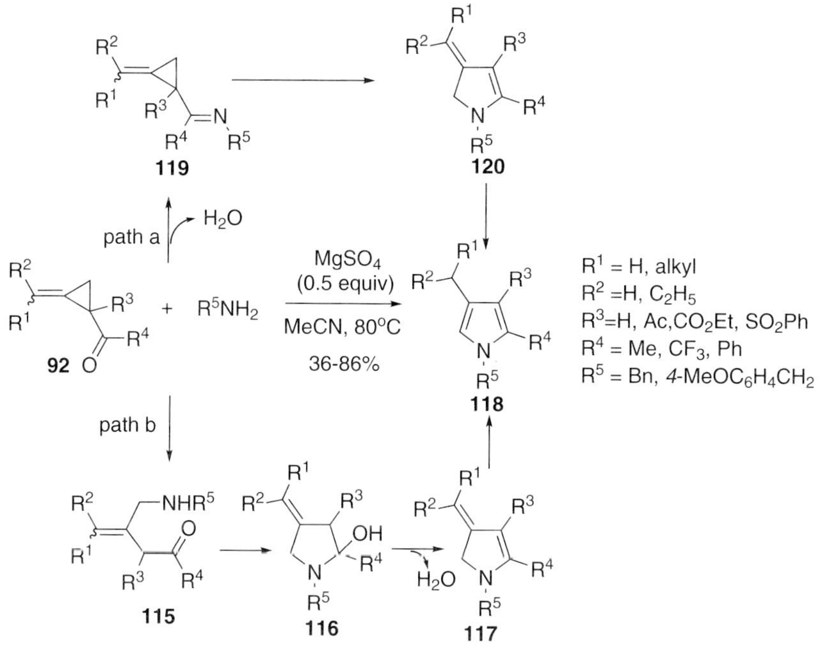

**Scheme 15.26** Intermolecular cyclization of acyl alkylidenecyclopropanes with amines.

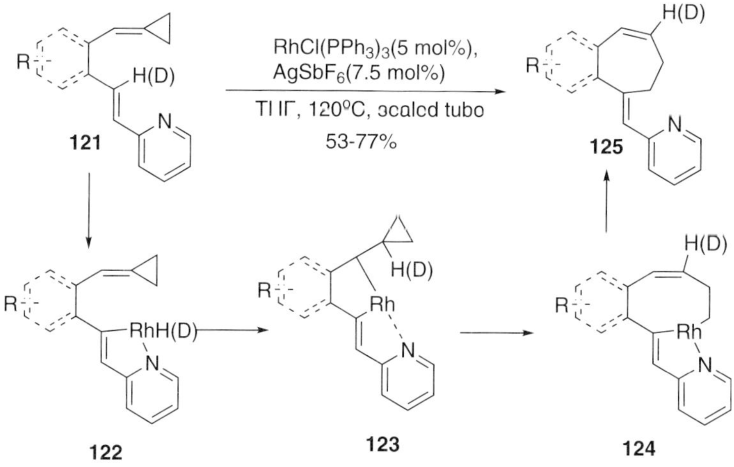

**Scheme 15.27** Rhodium-catalyzed tandem C–H activation/cycloisomerization of MCP with vinylpyridine.

undergoes a ring expansion to furnish rhodacyclooctenone **130**. Subsequent reductive elimination leads to the final product **127**.

An analogous rhodium-catalyzed hydroacylation of VCP **131** into cyclooctenones **132** as the major product along with a trace amount of cyclopropylcyclopentanones

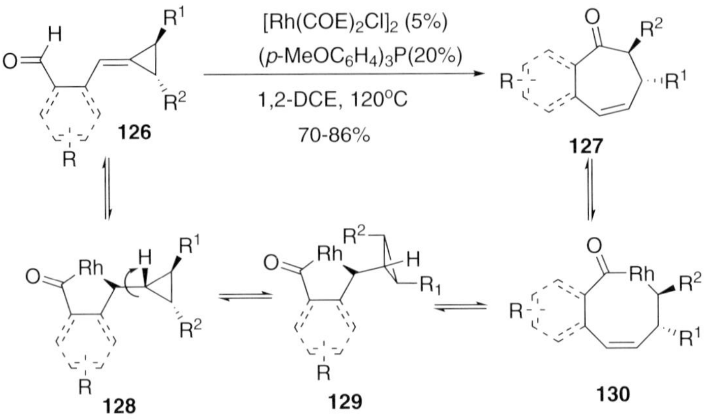

Scheme 15.28 Rhodium-catalyzed hydroacylation of MCP with aldehyde.

**133** was reported by Shair in 2000 (Scheme 15.29) [31]. The mechanism of this transformation starts from the oxidative addition of Rh(I) species to the C—H bond of aldehyde **131**, which undergoes 6-endo-trig hydrorhodation to produce the rhodacyclohexanone **135**. Subsequent β-carbon elimination produces the rhodacyclononenone **136**, followed by reductive elimination to give the cyclooctenone **132** as the major product. The trace of the minor product **133** was given by the direct reductive elimination of **135**.

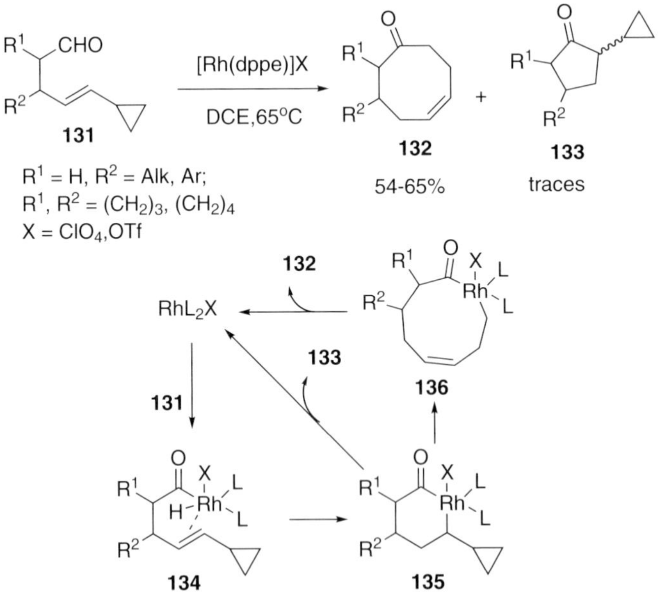

Scheme 15.29 Rhodium-catalyzed hydroacylation of VCP with aldehyde.

## 15.2.5
### Cycloisomerization of Vinylidenecyclopropanes

Shi [32] reported an efficient Lewis acid catalyzed cycloisomerization of arylvinylidene-cyclopropanes **137**, with three substituents at the 1, 2-positions of the cyclopropane, to provide easy access to 6a*H*-benzo[c]fluorine derivatives via a double intramolecular Friedel–Crafts reaction under mild reaction conditions in good to excellent yields (Scheme 15.30). The mechanism of this transformation is rationalized as the initial formation of the zwitterionic intermediate **139** through Lewis acid coordination of the double bond of the allene moiety resulting in the formation of the π-allyl cation **140a/b**. Intramolecular Friedel–Crafts reaction with either the $R^1$ or $R^2$ group gives the cyclic zwitterionic intermediate **141** followed by allylic rearrangement to produce intermediate **142**. A second stereoselective intramolecular Friedel–Crafts reaction with the $R^3$ group occurs to produce **143** with syn-configuration. Subsequently, deprotonation and [1,3]-H shift give the intermediate **146** followed by release of the Lewis acid to afford the final product.

## 15.2.6
### Cycloisomerization of Cyclopropenes

#### 15.2.6.1 Cycloisomerization of 3-alkoxylcarbonylcyclopropenes to Furans
Nefedov disclosed a first example of Cu(I)-catalyzed cycloisomerization of 3-alkoxylcarbonylcyclopropenes to furans via a vinylcarbenoid species **148**, generated by the ring opening of cyclopropenes (Scheme 15.31) [33]. The generated carbenoid **148** was trapped by active olefin such as norbornadiene (nbd) to produce the vinylcyclopropane adduct **149** in good yield. Treatment with with other alkenes failed to afford the cyclopropanation product, but led instead to the formation of 2-alkoxyfurans **152** via a cascade process involving isomerization/cyclization/reductive elimination.

Later, a few other groups observed a similar rearrangement of cyclopropenes into furans as a side reaction in the Rh(II) complexes-catalyzed cyclopropenation reaction. The formation of the carbenoid intermediate during the cycloisomerization was proved by the rearrangement of cyclopropene **153** under the catalysis of $Rh_2pfb_4$. The cyclopentylideneacetate **155** was produced as the major product along with 7% yield of furan **157**. It was proposed that the major product **155** was generated from the rhodium vinylcarbene species *E*-**154** via intramolecular C–H insertion at the tethered methoxyalkyl group. In contrast, the Z-isomer of the rhodium vinylcarbene species *E*-**154**, with the ester group oriented close to the rhodium center is able to cycloisomerize to the furan **157** via rapid electrocyclic cyclization into rhodapyran **156** instead of the intramolecular C–H insertion, followed by reductive elimination (Scheme 15.32) [34].

#### 15.2.6.2 Cycloisomerization of 3-acylcyclopropenes to Furans
Similar to the 3-alkoxylcarbonylcyclopropenes, the corresponding 3-acylcyclopropenes could undergo the same transformation to furans via a vinylcarbenoid species, generated by the ring opening of cyclopropenes in the presence of a transition-metal

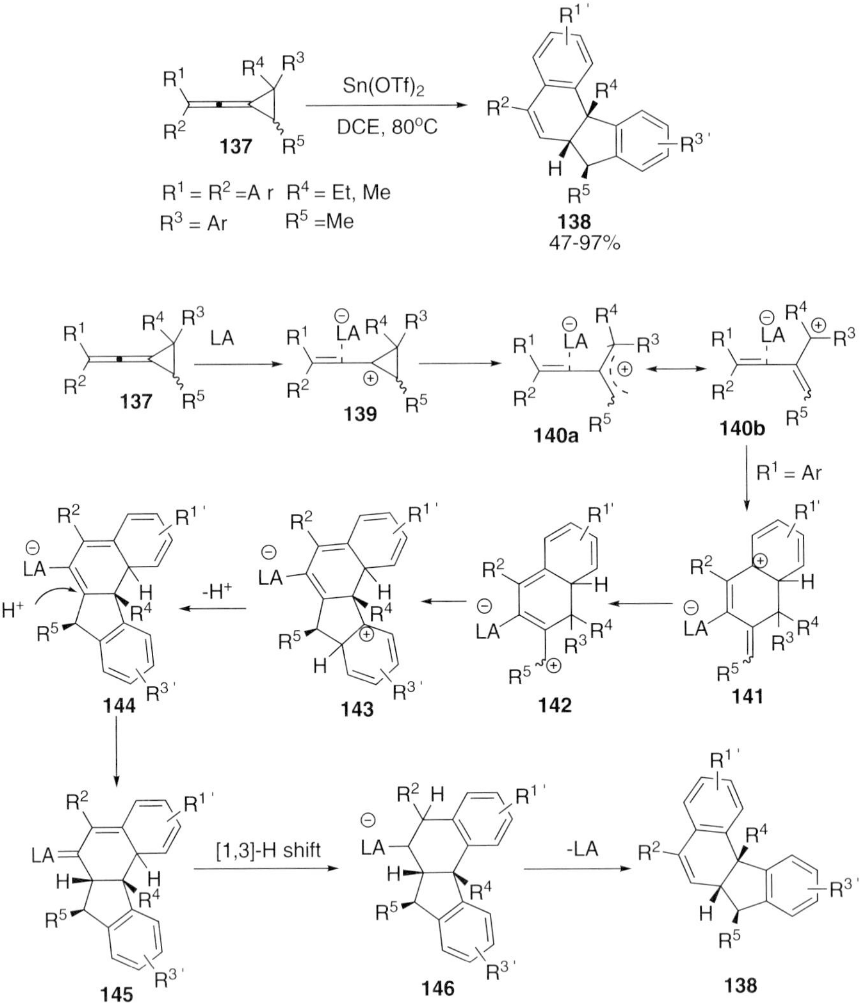

**Scheme 15.30** Sn(OTf)$_2$-catalyzed cycloisomerization of VCPs.

catalyst. Padwa reported that furan **160**, generated from cyclopropene **158** catalyzed by Rh(II) acetate via the vinylcarbenoid species **159**, rapidly underwent a second intermolecular cyclopropanation with **159** to give the reactive oxabicyclohexene **161**, which was followed by facile Cope rearrangement to afford the oxabicyclic compound **162** as the major product (Scheme 15.33) [35].

Padwa [35] also reported that Rh(I) complexes can catalyze the cycloisomerization of 3-acylcyclopropenes to 2,5-disubsituted furans instead of 2,4-disubstituted products. The proposed mechanistic rationale is depicted in Scheme 15.34. The direct regioselective oxidative addition of rhodium to the C1−C3 bond of cyclopropene **163** produces the metallacyclobutene species **164**. Subsequent enolization and reductive elimination give the 2, 5-disubstituted furan **166**. Alternatively, **164** can be generated

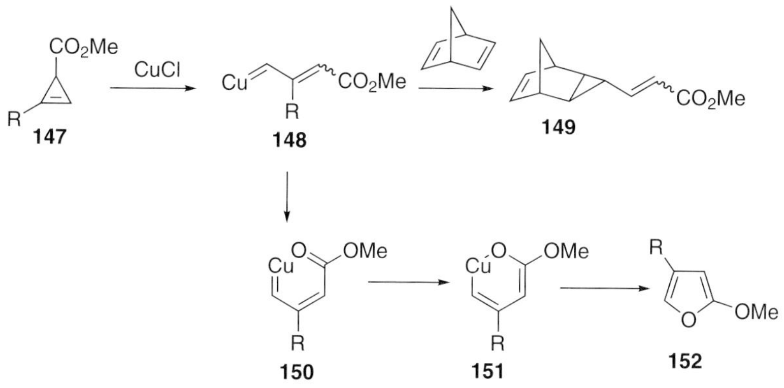

**Scheme 15.31** Cu(I)-catalyzed cycloisomerization of 3-alkoxycarbonylcyclopropenes.

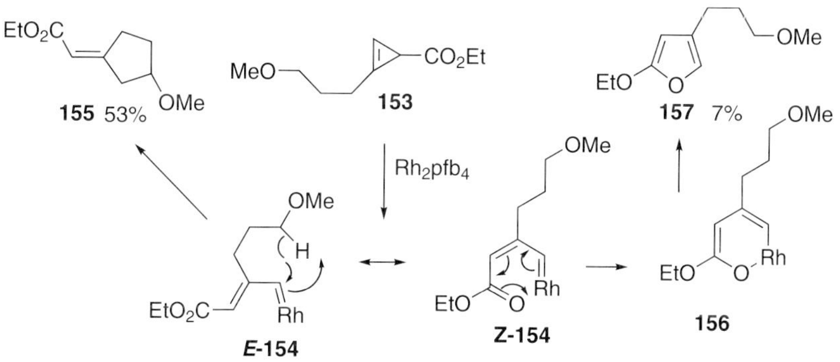

**Scheme 15.32** Rh(II)-catalyzed cycloisomerization of 3-alkoxylcarbonylcyclopropenes.

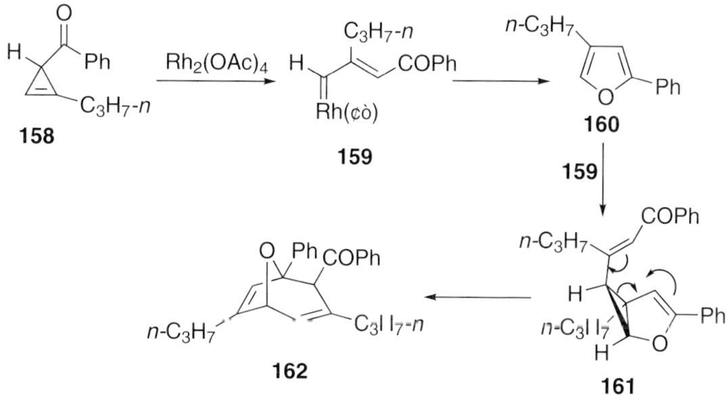

**Scheme 15.33** Rh(II)-catalyzed cycloisomerization of 3-acylcyclopropenes.

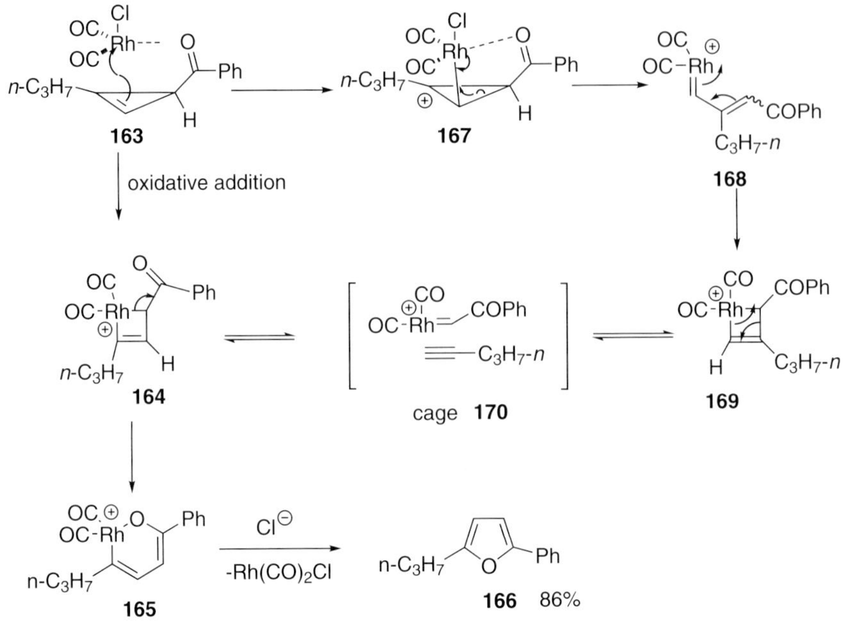

**Scheme 15.34** Rh(I)-catalyzed regioselective cycloisomerization of 3-acylcyclopropenes.

from the [2 + 2] cycloaddition of the two components of cage **170**, arising from the reverse [2 + 2] cycloaddition of the metallacyclobutene species **169**, which is generated from the vinylcarbenoid species **168**.

Ma [36] demonstrated an alternative way to control the regioselectivity of this cycloisomerization by employing different transition metal catalysts instead of the different oxidation states of the Rh-complexes developed by Padwa. The reaction of cyclopropene **171** produced 2, 3, 5-trisubstituted furan **179** in the presence of $PdCl_2$, while 2, 3, 5-trisubstituted furan **175** was produced under the catalysis of CuI. The opposite regioselectivity resulted from the different initial halometallation in the two different catalytic cycles (Scheme 15.35). The Pd(II)-catalyzed process begins with the regioselective chloropalladation of **171** to generate the cyclopropylpalladium species **176**, in which the Pd atom is at the less substituted carbon atom. Subsequent β-carbon elimination and enolization produces the palladium enolate **177**. The endo-oxapalladation of **177** and subsequent β-halide elimination of the resulting dihydrofuryl species **178** affords the final furan product **179**. In contrast, the copper-catalyzed reaction begins with iodocupration with the opposite regiochemistry, producing the cyclopropyl cuprate **172**, in which the metal resides at the more substituted carbon atom. As in the Pd(II)-catalyzed case, the following steps involve formation of the copper enolate **173** via β-carbon elimination and enolization, followed by endo-oxacupration and β-halide elimination to produce the 2, 3, 4-trisubstituted furan **175** with opposite regiochemistry. The reason for the opposite regiochemistry in the halometallation is not quite clear, but the more electrophilic $PdCl_2$ coordinates the double bond of cyclopropene to facilate the formation of

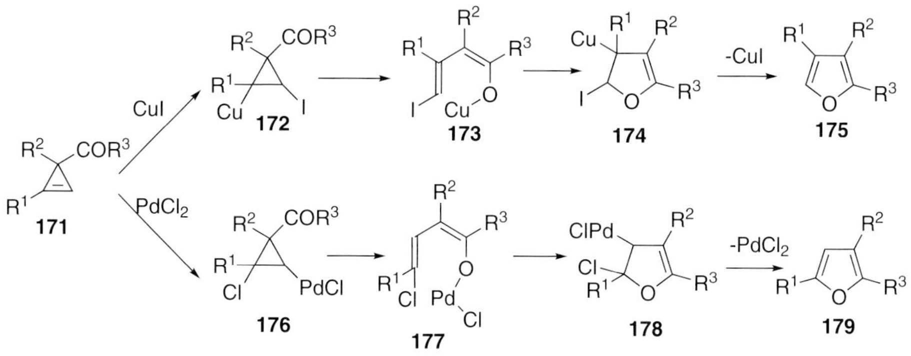

**Scheme 15.35** Regioselectivity control in transition-metal-catalyzed cycloisomerization of 3-acylcyclopropenes.

intermediate **176** and the good nucleophilic iodide of CuI may help the opposite regioselective addition.

### 15.2.6.3 Cycloisomerization of Cyclopropenyl Imines to Pyrroloheterocycles

It is not surprising that cyclopropenyl imines, aza-analogues of acylcyclopropenes, can undergo cycloisomerization to pyrroles in the presence of transition metal catalysts. Gevorgyan demonstrated a transition-metal-catalyzed cycloisomerization of **180** to pyrroloheterocycles, in which the regioselectivity of cycloisomerization can be controlled by employment of different transition metal complexes (Scheme 15.36) [37]. Under the catalysis of the Rh(I)-complex, the reaction of **180** produces 1, 3, 5-trisubstituted indolizines **182** via the rhodium vinylcarbenoid species **181**, which is consistent with the case in Scheme 15.33. In contrast, complementary ring cleavage of the cyclopropene leading to **183** was observed when the Cu(I) complex was employed as the catalyst, which is in turn consistent with the result demonstated by Ma

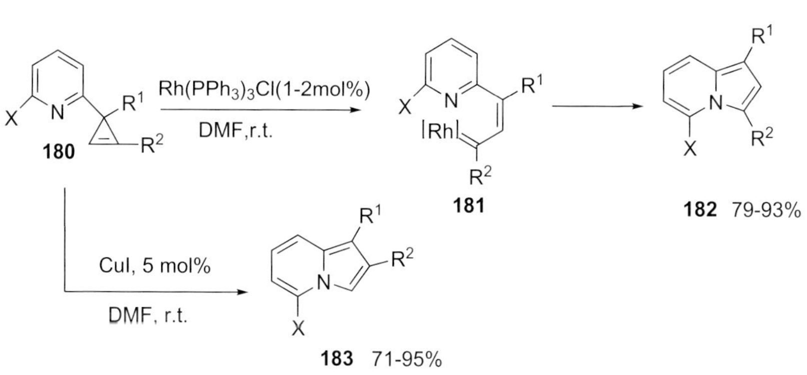

**Scheme 15.36** Regioselectivity control in transition metal-catalyzed cycloisomerization of 3-(pyridin-2-yl)cyclopropenes.

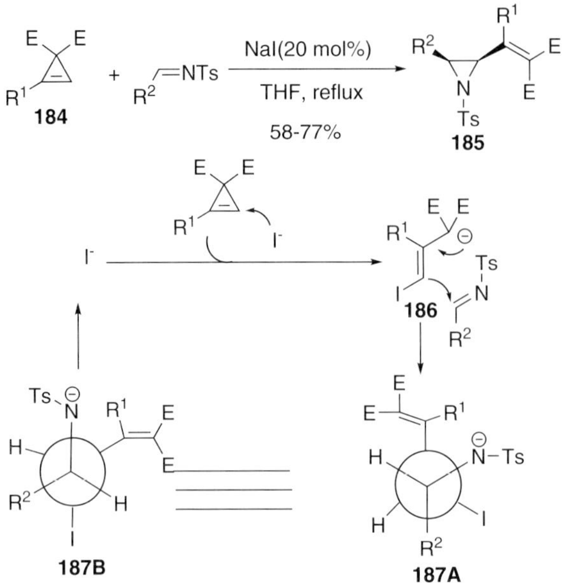

**Scheme 15.37** NaI-catalyzed [1 + 2] aziridination reaction of cyclopropenes with activated imines.

## 15.3
## Cycloaddition Reactions

### 15.3.1
### [1 + 2] Cycloaddition Reaction

Nucleophilic attack of iodide on the less substituted carbon of activated cyclopropene results in the formation of the allylic anion, via ring opening, which can be trapped by electrophiles [38, 39]. Ma reported a novel NaI-catalyzed [1 + 2] aziridination reaction of activated cyclopropenes **184** with activated imines affording polyfunctionalized *cis*-vinyl aziridines **185**, in which cyclopropenes act as the 1-carbon component [39]. The mechanistic rationale was proposed as Scheme 15.37. Iodide regioselectively attacks the less substituted carbon of the activated cyclopropane to form a stereodefined allylic carbanion intermediate **186**, which reacts with the activated imine to produce intermediate **187A**. Subsequent cyclization affords the cycloadduct and regenerates iodide.

### 15.3.2
### [2 + m + n] Cycloaddition

Cyclopropenes, MCPs and VCPs have a double bond inside, at the same time, the high strain energy of the cyclopropane (in MCP and cyclopropenes) make the double bond quite active, thus, these compounds can serve as the two-carbon component in the [2 + m + n] cycloaddition in the presence of transition metal complexes. Such advances have been nicely summarized in a recent review [40].

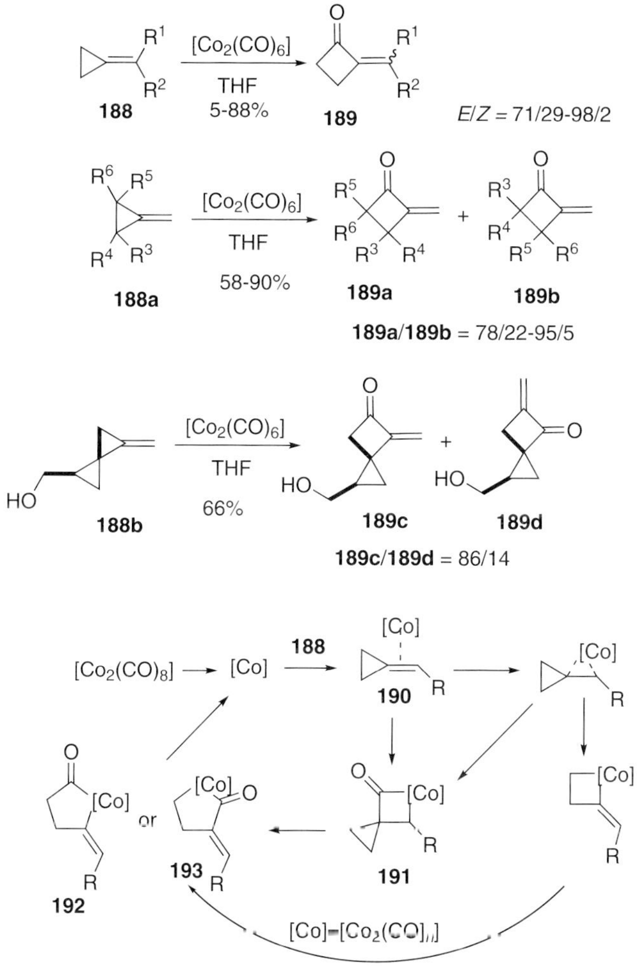

**Scheme 15.38** Cobalt-mediated or catalyzed [3 + 1] carbonylation of MCPs.

## 15.3.3
### [3 + 1] cycloaddition of MCPs

Meijere developed a new cobalt-mediated or catalyzed [3 + 1] carbonylative cocyclization of methylenecyclopropanes **188** to give 2-alkylidenecyclobutanones **189** under mild conditions (Scheme 15.38) [41]. It is worth pointing out that only the cyclopropane ring with methylene can undergo the carbonylative cyclization, that is, the 4-(hydroxymethyl)methylenespiropentane **189b** reacts with [Co$_2$(CO)$_8$] to produce the regioisomeric methylenespiro[2.3]hexanones **189c**/**189d** in a ratio of 86 : 14 in 66% yields, in which the cyclopropane without methylene was kept inert. The author proposed that the exchange of one or two CO ligands of the [Co$_2$(CO)$_8$] complex with a methylenecyclopropane forms the alkene cobalt complex **190**, which can undergo two possible reaction pathways. One is oxidative addition of the C=C bond and CO

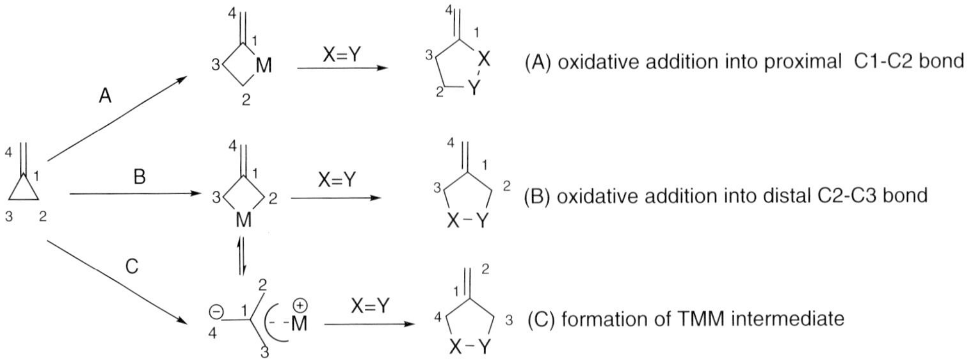

**Scheme 15.39** Three different reaction pathways for transition metal catalyzed [3 + 2] cycloadditions of MCPS.

insertion to give the cobaltaspiro[2.3]hexanone **191**, which was followed by (cyclopropylmethyl)metal-to-homoallylmetal rearrangement. Subsequent reductive elimination of the resulting intermediate **192/193** yields the cycloadduct. The other is a sequentially oxidative addition of the proximal bond of MCP and CO insertion of the resulting cobaltacyclobutane to give the final product.

### 15.3.4
### [3 + 2] Cycloaddition Reaction

#### 15.3.4.1 [3 + 2] Cycloaddition Reaction of Methylene- and Alkylidene-Cyclopropanes

**Transition-Metal as the Catalyst**  There are three different cyclopropyl ring cleavage modes in transition-metal-catalyzed [3 + 2] cycloadditions involving an MCP unit as a three-carbon component, that is, oxidative addition of the transition metal to the proximal (A) or distal (B) C−C bond or formation of a TMM intermediate (C) (Scheme 15.39) [42]. Transition-metal-catalyzed intermolecular and intramolecular [3 + 2] cycloadditions of MCPs with various unsaturated compounds, including alkenes [43], alkynes [44], carbon dioxide [45], and ketenimines [46], have been developed, pioneered by Noyori [47], Binger [48], and Trost [49]. Intramolecular versions of this kind of reaction have also been developed by several groups [50, 53].

Recently, Yamamoto demonstrated that palladium-catalyzed [3 + 2] cycloadditions of MCPs with aldehydes and imines afford 3-methylene-tetrahydrofurans and pyrrolidines **194**, respectively (Scheme 15.40) [51]. Almost no diastereoselectivity was observed in these reactions. It is proposed that oxidative addition of palladium complexes to the distal bond of MCP produces the 3-methylene-palladacyclobutane **195**, which is followed by formal [3 + 2] cycloaddition to yield the final product.

It is not surprising to find that [3 + 2] cycloaddition of MCPs with 1, 2-diazines **198** produces the corresponding cycloadducts **199** in the presence of Pd complexes, in which 1, 2-diazines **198** act as two-atom components as imines. Subsequent dehydrogenation of **199** followed by aromatization affords pyrrolopyridazines **200** in moderate to good yields (Scheme 15.41) [52].

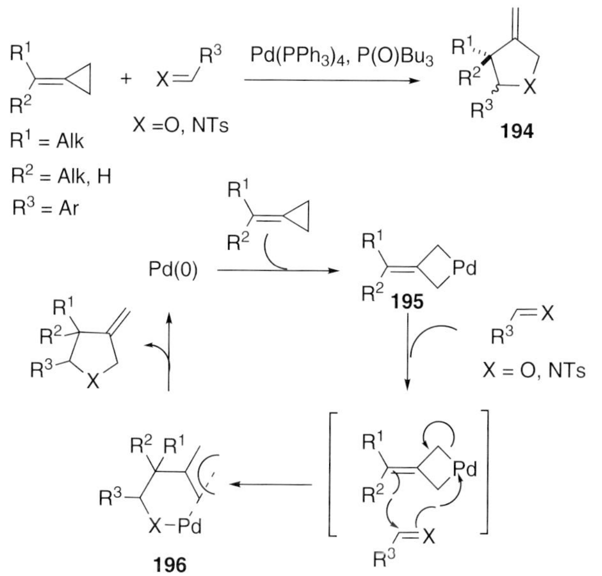

**Scheme 15.40** Palladium-catalyzed [3 + 2] cycloadditions of MCPs with aldehydes and imines.

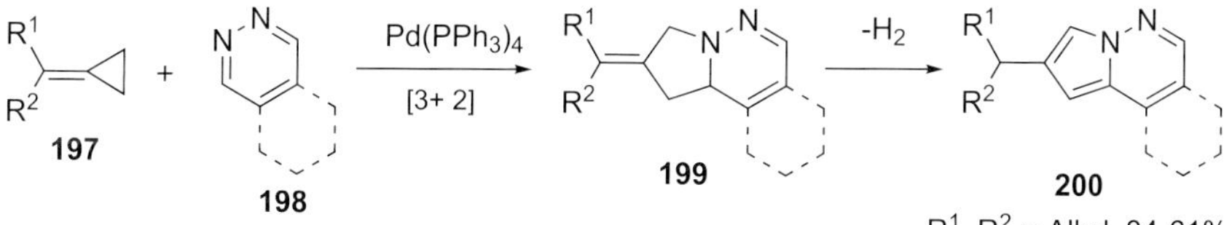

**Scheme 15.41** Palladium-catalyzed [3 + 2] cycloadditions of MCPs with 1,2-diazines.

Intramolecular [3 + 2] cycloadditions of MCPs to alkynes via the TMM intermediate **203** could happen in the presence of Ni- or Pd-complexes (Scheme 15.42) [53]. It is proposed that under transition-metal catalysis, alkyne-tethered MCPs **201** with an exo-methylene double bond, and the isomeric precursors **202** with an internal olefin moiety give the same TMM intermediate **203**, which in turn undergoes intramolecular [3 + 2] cycloaddition with the intramolecular C—C triple bond to give the corresponding bicyclic products **204** in moderate to excellent yields.

The first-generation Grubbs catalyst also showed high efficiency in catalyzing this type of intramolecular [3 + 2] cycloaddition of **202b** [54], but the mechanism of this transformation is believed to be different from the Ni- and Pd-catalyzed case. A reasonable mechanism is that the reaction proceeds via cyclometallation and subsequent β carbon elimination and reductive elimination (Scheme 15.43).

Recently, Mascareñas reported a first example of Pd(0)-catalyzed intramolecular [3 + 2] cycloaddition of alkene-MCPs **207**, providing a rapid and practical way to bicyclo[3.3.0]octane systems **208** containing up to three stereocenters

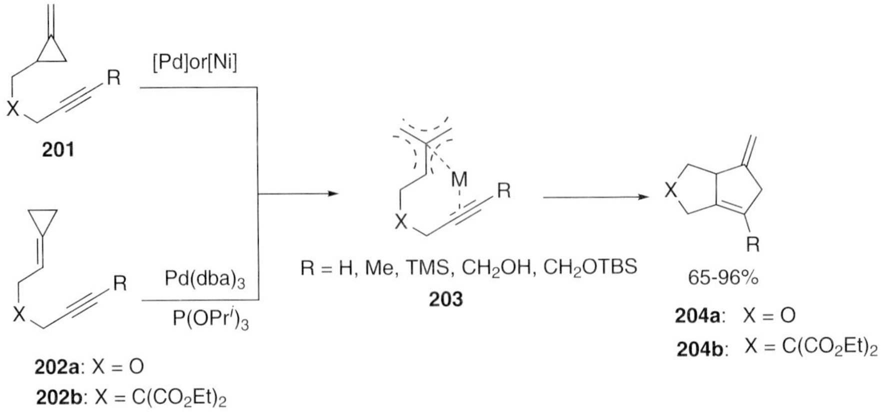

**Scheme 15.42** Pd(0) or Ni(0)-catalyzed intramolecular [3 + 2] cycloaddition of MCPs to alkynes.

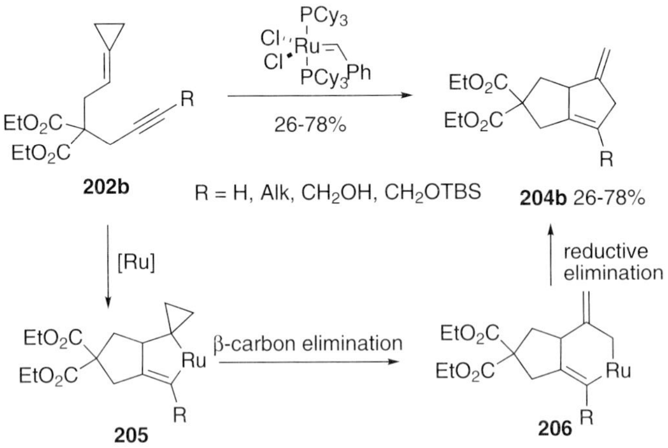

**Scheme 15.43** Grubbs catalyst-catalyzed intramolecular [3 + 2] cycloaddition of MCPs to alkynes.

(Scheme 15.44) [55]. The same author also demonstrated that the allene-MCP systems **209** can undergo similar transformation to give the corresponding bicyclo[3.3.0]octanes **210a** and **210b** with high diastereoselectivity [56].

**Lewis acid as the Catalyst** In 2002, Lautens reported a novel approach to five-membered heterocyclic compounds via a formal [3 + 2] cycloaddition of mono-activated methylenecyclopropanes (MCPs) with active aldimines in the presence of $MgI_2$ or MAD. The reaction of MCP **211** with active aldimines produced methylenepyrrolidine **212** in good yields as a mixture of diastereomers (diastereoselctivity up to >20: 1) mediated by $MgI_2$. Later, the same author reported an asymmetric catalyzed

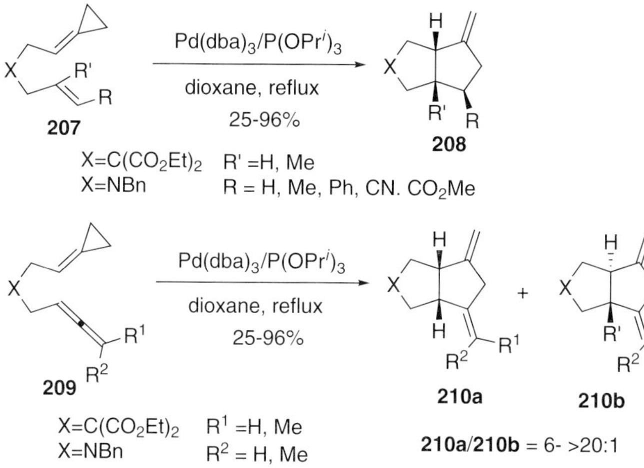

**Scheme 15.44** Pd(0)-catalyzed intramolecular [3 + 2] cycloaddition of alkene- and allene-MCPs.

version of this transformation and the ee values were up to 86%. Complementary regiochemistry was realized when MAD was used as catalyst to give pyrrolidines **213** in 35–92% yields with up to >20 : 1 diastereoselectivity (Scheme 15.45) [57]. When using MCP **211b** as the substrate to run the reaction under the same reaction conditions, it is interesting to find that six-membered hetereocycles **214** were produced in good yields. The nucleophilic attack of I$^-$ at the monoactivated MCP **211** generates the enolate **215**, which undergoes 1,2-addition with active aldimines to give the intermediate **216**, followed by nucleophilic substitution to give the 5 membered heterocycles. The 6-membered heterocycles **214** were produced via γ-addition followed by cyclization.

Shi [58] demonstrated that a Lewis-acid catalyzed formal [3 + 2] cycloaddition of MCP **197** to aldehydes or imines gave 2-substituted 3-methylene-tetrahydrofuran or 3-methylene-pyrrolide **218** with a regiochemistry complementary to that of the Pd-catalyzed version (Scheme 15.40) in the presence of BF$_3$·OEt$_2$ at −25 °C (Scheme 15.46). It is interesting to find that if the reaction was run at room temperature, the same Lewis acid catalyzed the reaction of MCP **197a** with at least one benzene ring, with aldehyde or imines to give the indenes **219** and/or the 6-membered heterocycles **220**. The addition of MCP **197** to the Lewis acid-activated aldehyde gives the cyclopropylcarbinyl cation **221** and its resonance-stabilized zwitterionic intermediate **222** and **223**, respectively (Scheme 15.47) [58]. The homoallylic rearrangement of cations **221** and **223** generates **227** and **224** which can further afford the interconvertible intermediates **226** and **225**, respectively, in the presence of Lewis acid. The tandem intramolecular Friedel–Crafts reaction from **226** produces the indene **219** (for arylaldehydes). The intramolecular cyclization of cation **223** gives cation **228**, which can be viewed as a nonclassical carbenium cation, followed by isomerization to give the 6-membered heterocycles.

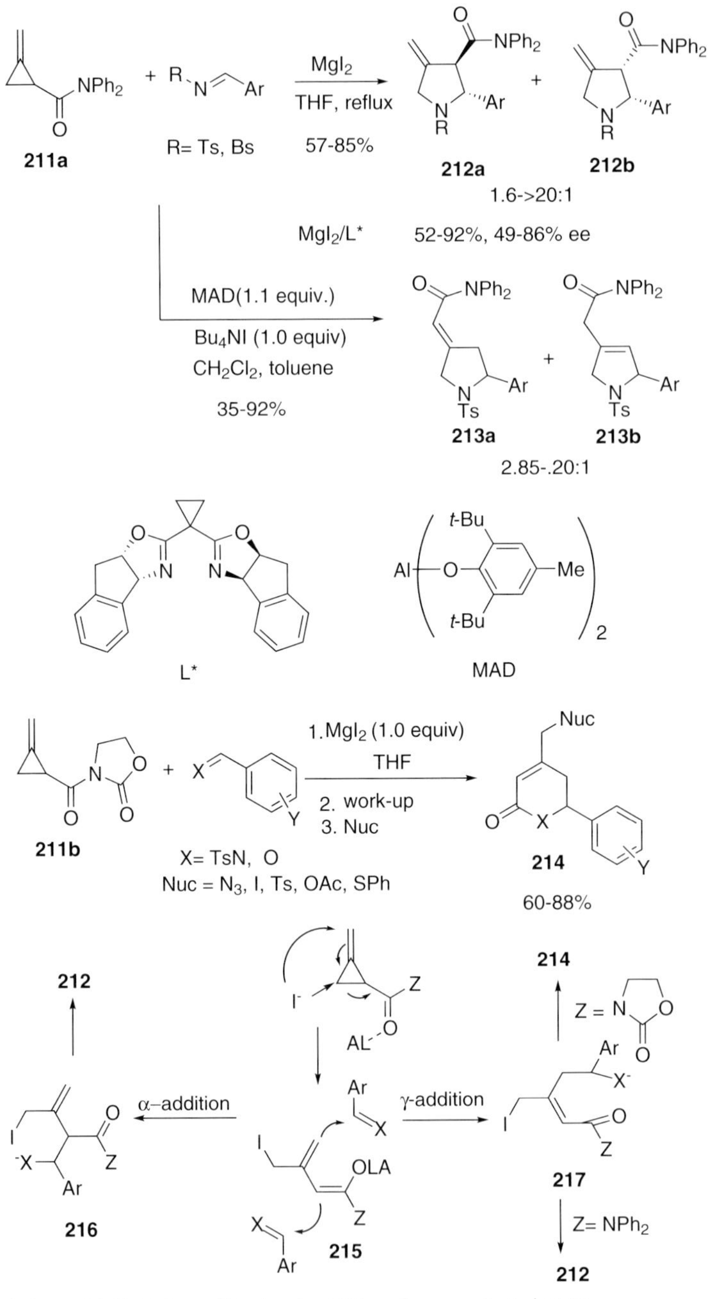

**Scheme 15.45** MgI$_2$-mediated cycloaddition of monoactivated MCPs.

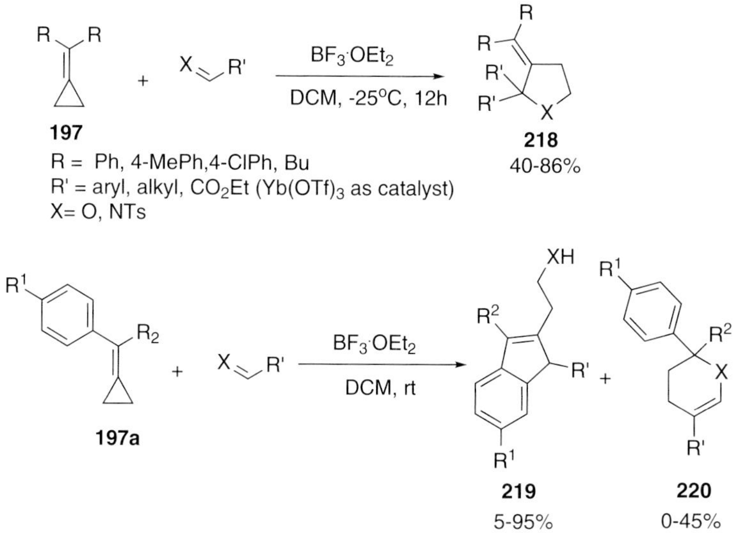

**Scheme 15.46** BF$_3$·OEt$_2$-catalyzed cyclization of MCPs with aldehydes or imines.

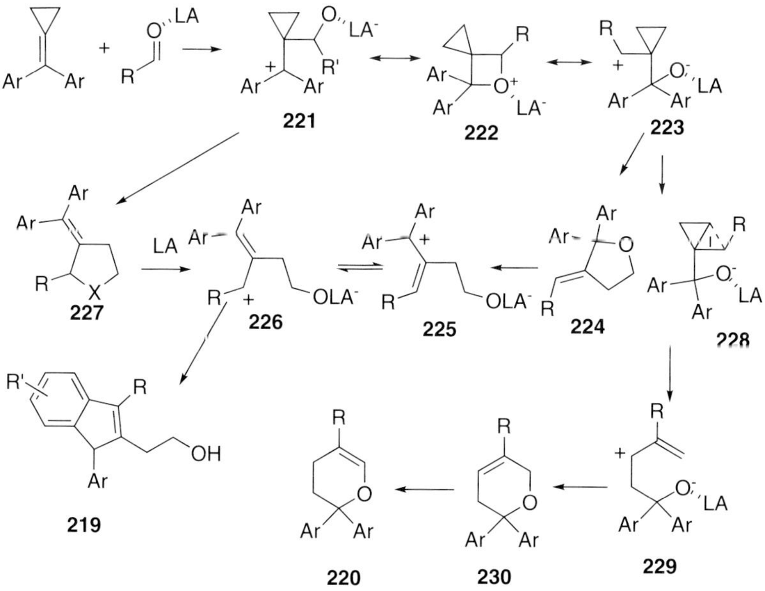

**Scheme 15.47** Plausible reaction pathway for BF$_3$·OEt$_2$-catalyzed cyclization of MCPs with aldehydes.

### 15.3.4.2 [3 + 2] Cycloaddition Reaction of Vinylidenecyclopropanes

Vinylidenecyclopropanes **231** can react with activated imines under the catalysis of a Lewis acid, as reported by Shi recently (Scheme 15.48) [59]. The reaction selectivity depends on the electronic nature of both the activated imines and the Ar$^1$ or Ar$^2$

of **231**. When the $R^7$ group is an electron-donating group, the reaction undergoes formal [4 + 2] cycloaddition to give 1, 2, 3, 4-tetrahydroquinoline **233**, in which vinylidenecyclopropanes act as the 2-C component. Whereas pyrrolidines **232** were produced via the [3 + 2] cycloaddition, in which VCPs act as the 3-C component. The mechanistic rationale is proposed by the author: the imines were activated by $BF_3\cdot OEt_2$ which is attacked by the central carbon of the allene moiety to produce allylic cations **234** and **235**. Intermediate **236** undergoes a ring-opening reaction to give a new allylic cation **236** followed by cyclization to afford the [3 + 2] cycloadducts when $R^7$ is an electron-withdrawing group. Intermediate **237** undergoes Friedel–Crafts reaction and subsequent protonation to give the [4 + 2] cycloadducts **233** when $R^7$ is an electron-donating group.

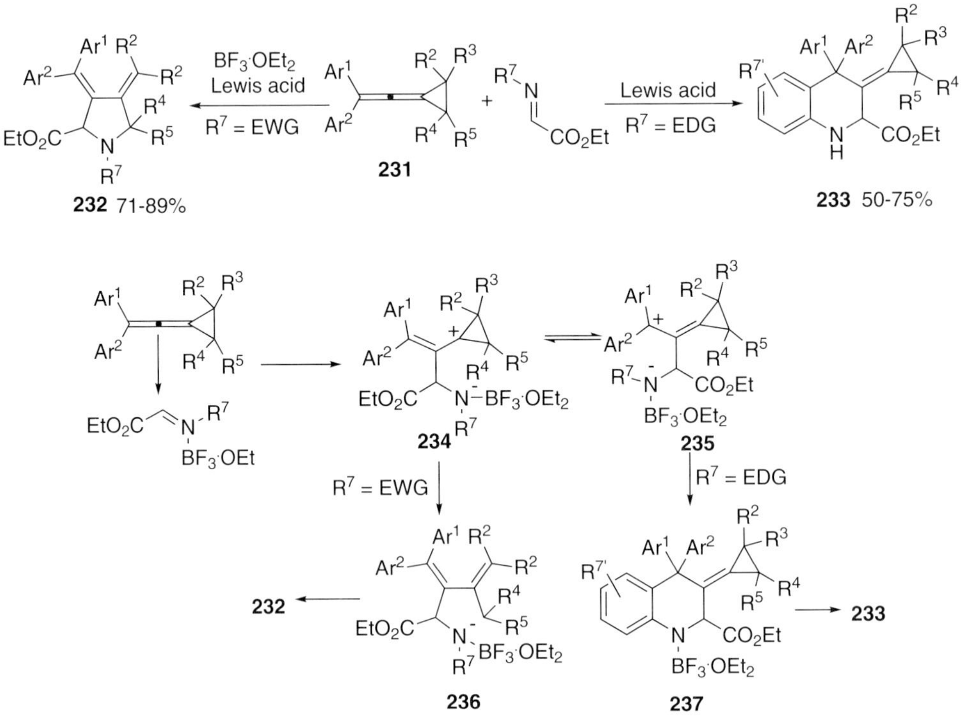

**Scheme 15.48** $BF_3\cdot OEt_2$-catalyzed cyclization of arylvinylidenecyclopropanes with imines.

Shi further demonstrated that vinylidenecyclopropanes **238** and ethyl glyoxylate **239a** can undergo analogous [3 + 2] cycloaddition to produce the tetrahydrofuran derivatives **240** in 77–99% yields in the presence of $BF_3\cdot OEt_2$, whereas if diethyl ketomalonate **239b** was employed instead of ethyl glyoxylate **239a**, the reaction gave 3, 6-dihydropyrans **241** instead of the anticipated tetrahydrofurans via [4 + 2] cycloaddition in 30–66% yields under the catalysis of $Nd(OTf)_3$. The mechanism for forming tetrahydrofuran is analogous to that of activated imines. The reaction mechanism of [4 + 2] cycloaddition of **238** with diethyl ketomalonate **239b** is believed to proceed via a cascade ring-opening rearrangement to triene **245**, followed by ene reaction forming **246** and then cyclization (Scheme 15.49) [60].

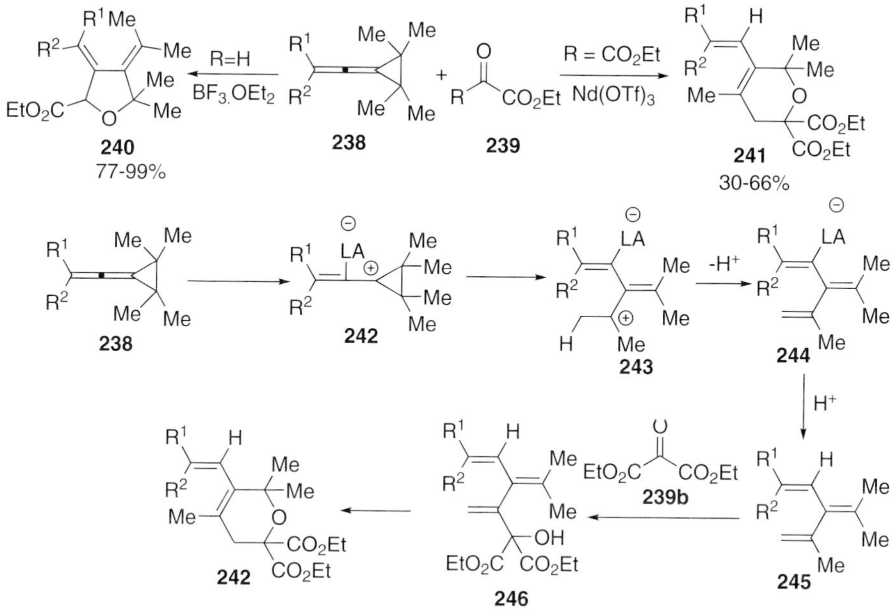

**Scheme 15.49** Lewis acid-catalyzed cyclization of vinylidenecyclopropanes with glyoxylate or diethyl ketomalonate.

A Brønsted acid TfOH-mediated [3 + 2] cycloaddition reaction of diarylvinylidenecyclopropanes with nitriles was also disclosed by Shi recently (Scheme 15.50) [61]. The reactions normally give 3,4-dihydro-2H-pyrrole derivatives **250** in 58–94% yields leaving the allene moiety untouched, except in the case of **247**, which has a strong electron-donating methoxy group at the para position of the benzene ring, giving a different type of 3,4-dihydro-2H-pyrrole derivative **248** via cleavage of the distal bond of the cyclopropane ring (Scheme 15.50).

### 15.3.4.3 [3 + 2] Cycloaddition Reaction of Activated Cyclopropanes

In 2005, Johnson [62] demonstrated that $Sn(OTf)_2$-catalyzed formal [3 + 2] cycloaddition of aldehydes with donor–acceptor (D–A) cyclopropanes **256** bearing a malonyl diester acceptor group and a carbon-based resonance donor substituent, leading to 2,5-disubstituted tetrahydrofurans **258** with high diastereoselectivity (Scheme 15.51). In the same year, he developed an asymmetric version of this reaction. The reaction of the deuterium labeled substrate **D-256a** with benzaldehyde gives a mixture of diastereomer **D-258a** and **D-258b** with a ratio of 96 : 4, from which he proposed an unusual $S_N2$ process, where the aldehyde acts as a nucleophile inverting the stereochemistry at the activated C-2 carbon of the cyclopropane (forming intermediate **259**).

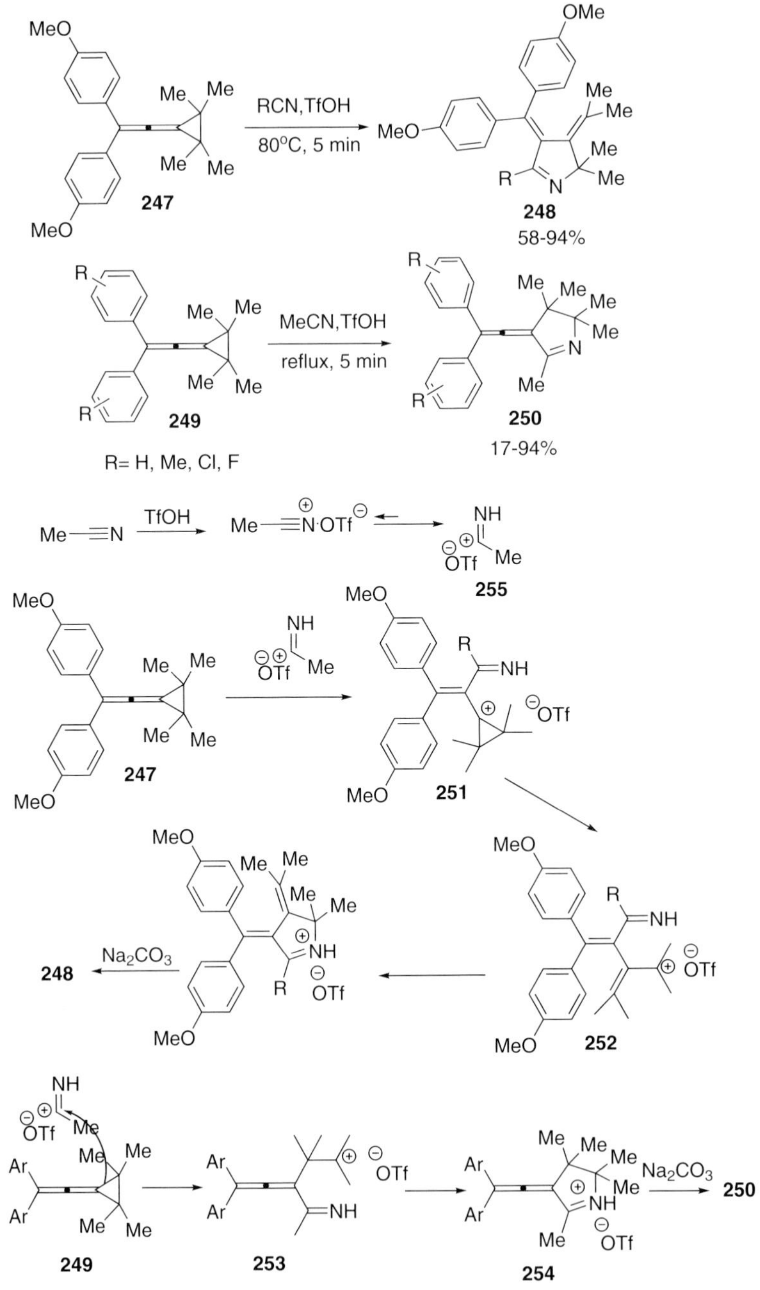

**Scheme 15.50** TfOH-mediated [3 + 2] cycloaddition reaction of diarylvinylidenecyclopropanes with nitriles.

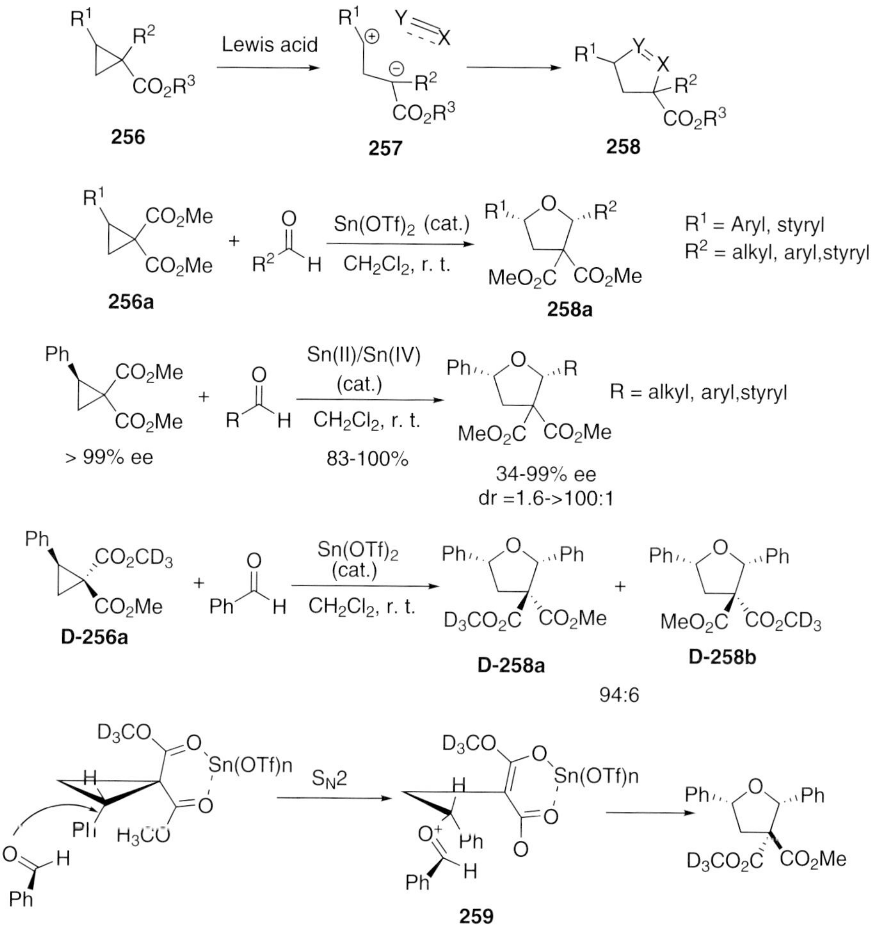

**Scheme 15.51** Sn(OTf)$_2$-catalyzed formal [3 + 2] cycloaddition of aldehydes with D-A cyclopropanes.

Imines can also react as the 2-atom component with activated D–A cyclopropanes **256a** under Lewis acid catalysis. In 2006, Tang developed a Sc(OTf)$_3$-catalyzed formal [3 + 2] cycloaddition of imines with 2-substituted cyclopropane-1,1-dicarboxylates to produce highly substituted pyrrolidines **260** in up to 94% yield and with up to 30/1 diastereoselctivity (Scheme 15.52) [63].

In 1999, Carreira [64] developed a novel approach to the construction of the spiroring system **262** in good to excellent yields via a MgI$_2$-catalyzed formal [3 + 2] cycloaddition of imines to activated cyclopropane **261**, in which the ratio of the two diastereomers is up to 98 : 2. A potential mechanism for this reaction is that the iodide effects ring opening of the cyclopropane via nucleophilic attack furnishing enolate **263**, which would react with the starting imine to furnish intermediate **264**, followed by alkylative cyclization to give the final cycloadduct (Scheme 15.53).

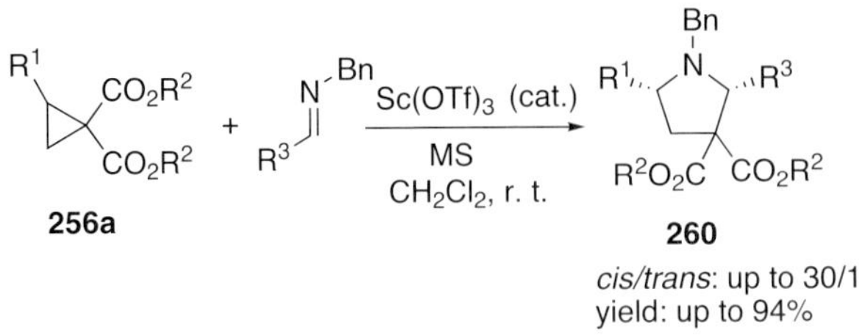

**Scheme 15.52** Sc(OTf)$_3$-catalyzed formal [3 + 2] cycloaddition of imines with 2-substituted cyclopropane-1,1-dicarboxylates.

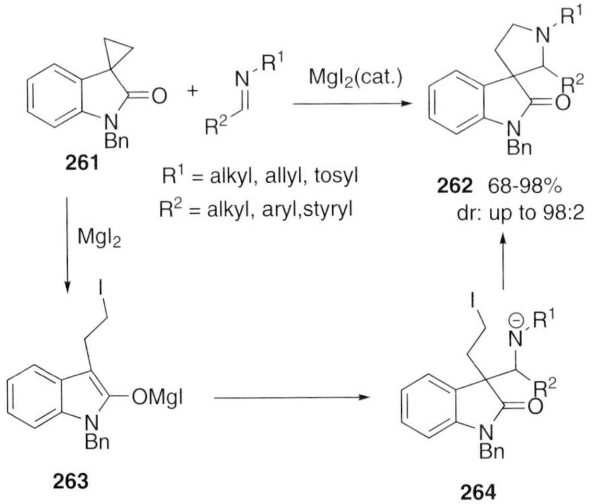

**Scheme 15.53** MgI$_2$-catalyzed [3 + 2] cycloaddition of imines with spirocyclopropane.

This MgI$_2$-mediated formal [3 + 2] cycloaddition of **265** amd **266** has been successfully employed by Carreira to assemble compound **267**, the core of the alkolid spirotryprostatin B in 2005 (Scheme 15.54) [65]. He also demonstrated the synthetic application of this MgI$_2$-mediated formal [3 + 2] cycloaddition to the total synthesis of (±)-strychnofoline, in which the two key important starting materials were activated cyclopropane **268** and cyclic imine **269** (Scheme 15.54) [66].

The intramolecular [3 + 2] cycloaddition reaction of oxime ether tethered cyclopropane dicarboxylate leading to substituted pyrrolo-isoxazolidines with high diastereoselectivity (up to >99 : 1) in good to excellent yields in the presence of a catalytic amount of Yb(OTf)$_3$ was disclosed by Kerr (Scheme 15.55) [67]. It is interesting to find that the Z-oxime ether of Z-**271a** was treated with Yb(OTf)$_3$ to afford the diastereomer *cis*-**272a** exclusively, while the E isomer of E-**271a** cyclized to the trans product *trans*-**272a**.

Furthermore, the author showed that a simple reversal of the addition order of catalyst and substrate results in the formation of two discrete diastereomers in a highly selective and predictable manner (Scheme 15.56) [67]. A reasonable mechanistic rationale for this dramatic effect of the order of the addition of substrate and

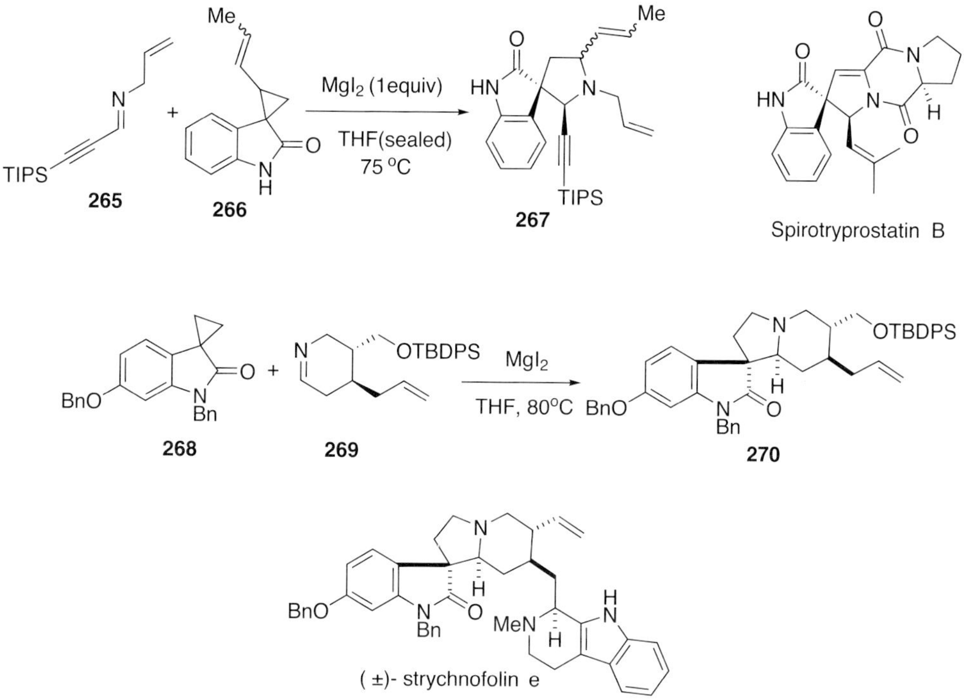

**Scheme 15.54** Synthetic application of MgI$_2$-mediated formal [3 + 2] cycloaddition in the total synthesis of (±)-strychnofoline.

catalyst was proposed. Treatment of the cyclopropyl-alkoxylamine **273** with catalytic Yb(OTf)$_3$ affords isoxazolidine **274** with complete conversion. Subsequently, **274** reacts with the subsequently added aldehyde to generate the intermediate Z-oxyiminium species **275**, which undergoes Manich reaction to give the final *cis* adduct. The *trans*-adducts are formed through the Manich reaction of the intermediate E-oxyiminium species **276**, which was generated by intramolecular attack of the N-atom of the oximes at the alkyl substituted position of the cyclopropanes.

De Meijere reported GaCl$_3$-catalyzed formal [3 + 2] cycloadditions of 2-arylcyclopropane-1,1-dicarboxylates **256** to diazenes **277**. The reaction proceeds with complete regioselectivity to furnish pyrazolidine-1, 2, 3, 3-tetracarboxylates **278** exclusively; however, the *cis*-configured azo compound N-phenyltriazolinedione gave the two possible regioisomeric pyrazolidines in ratios varying from 1:1.5 to 1:3 (Scheme 15.57) [68].

Silyl enol ethers can also be the 2-carbon component of [3 + 2] cycloadditions. In 2006, Takashu and Iharaa showed that silyl enol ethers **279** and 2-(*p*-methoxyphenyl) cyclopropyl phenyl ketone **280** undergo [3 + 2] cycloaddition in the presence of a catalytic amount of triflic imide (Tf$_2$NH) to give functionalized cyclopentanes **281** in 62–94% yields (Scheme 15.58) [69]. Moreover, cascade multicomponent [4 + 2]/[3 + 2] cycloadditions of α, β-unsaturated carbonyl compounds, 2-siloxydienes and

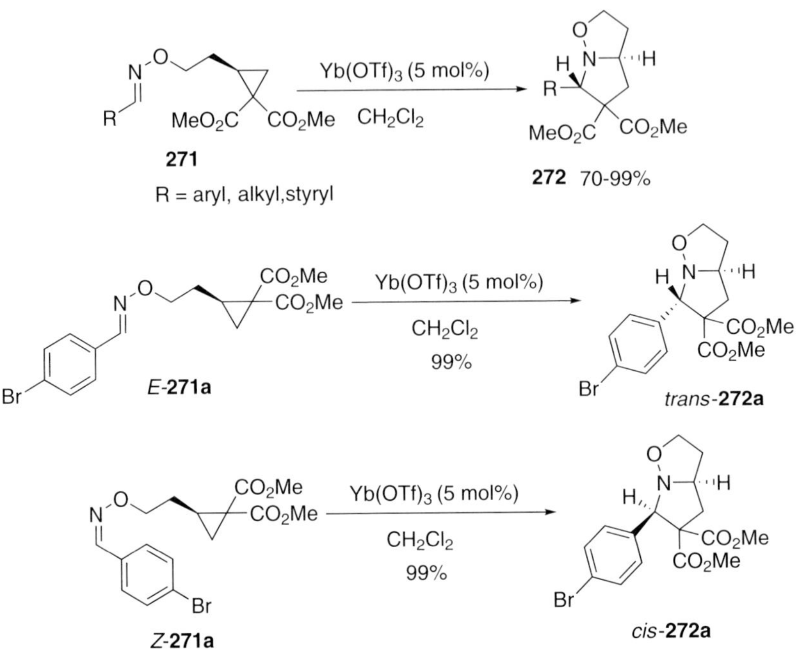

**Scheme 15.55** Yb(OTf)$_3$ –catalyzed intramolecular [3 + 2] cycloaddition reaction of oxime ether tethered cyclopropane diesters.

D–A cyclopropanes **280** to form highly-functionalized bicyclo[4.3.0]nonanes **282** have been disclosed.

In 2007, Pagenkopf reported a novel TMSOTf catalyzed [3 + 2] annulation reaction between various indoles and the 2-alkoxycyclopropanoate esters **283**. Both high efficiency and complete stereochemical control were observed in some cases with this annulation process (Scheme 15.59) [70]. Attack on the oxo-carbenium ion intermediate **285** by indole occurred from the indole Re or Si face and anti to the ester substituent. Two intermediates **286a** and **296b** undergo cyclization to give the corresponding products **284**. In this transformation, one of the non-enolizable stereocenters in the original cyclopropane (the R$^1$-connected carbon center) could induce the relative chemistry of up to four new stereocenters.

A formal [3 + 2] cycloaddition reaction between *tert*-butyldiphenylsilylmethyl-substituted cyclopropyl phenyl/alkyl ketones **287** and aryl acetylenes in the presence of TiCl$_4$ leading to cyclopentene derivatives **288** with high regio- and stereoselectivity was explored by Yadav (Scheme 15.60) [71]. Aryl acetylenes with electron-donating substituents reacted better than those substituted with electron-withdrawing groups, indicating the reaction might be a stepwise process, that is, the initial attack of a terminal acetylenic carbon center at the positive end of the cyclopropane dipole **289**, which was generated in the presence of TiCl$_4$. Subsequent interception of the resulting intermediate **290** would produce the final product.

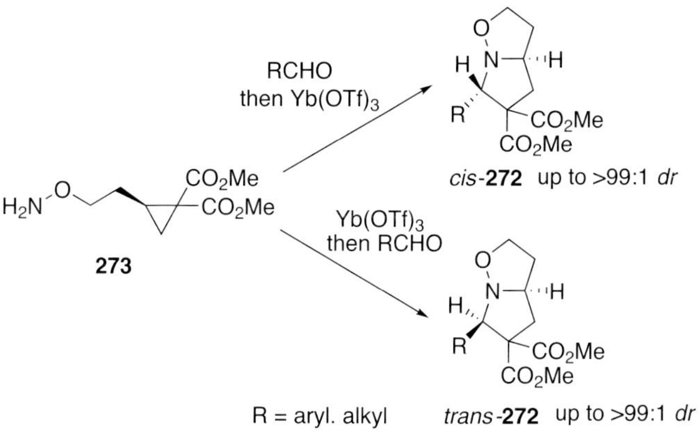

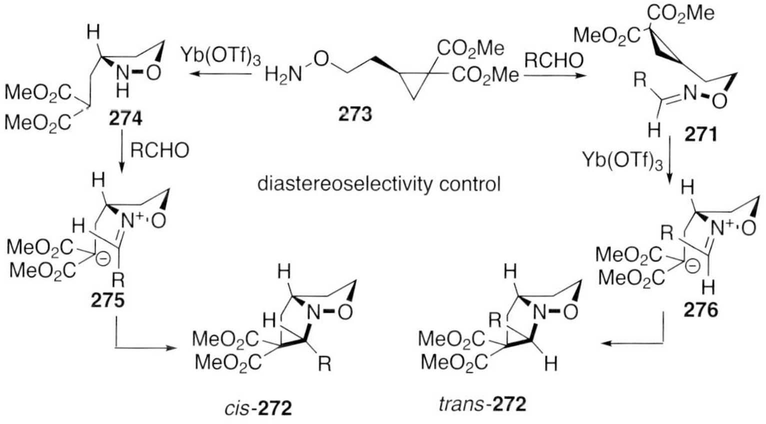

**Scheme 15.56** Stereoselectivity control in the Yb(OTf)$_3$-catalyzed [3 + 2] cycloaddition reaction.

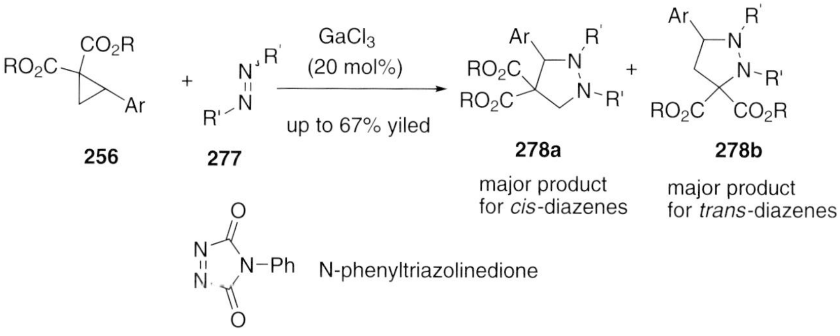

**Scheme 15.57** GaCl$_3$-catalyzed formal [3 + 2] cycloadditions of 2-arylcyclopropane-1,1-dicarboxylates with diazenes.

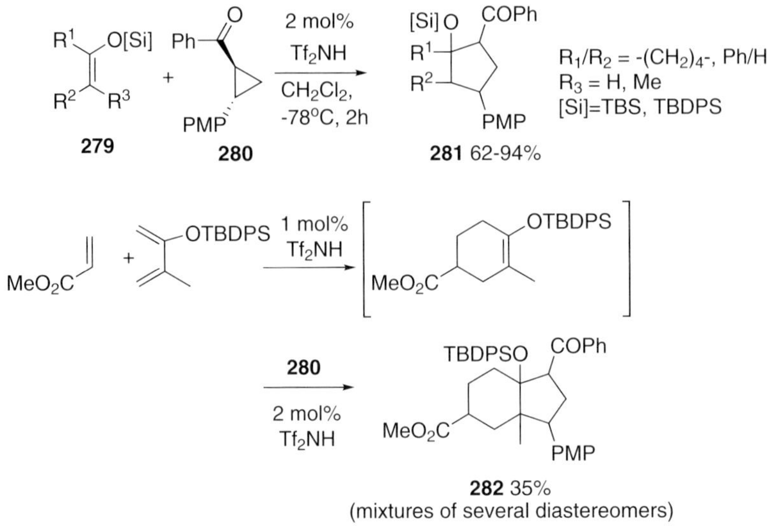

**Scheme 15.58** Tf$_2$NH-catalyzed [3 + 2] cycloaddition of acyl cyclopropanes with silyl enol ethers.

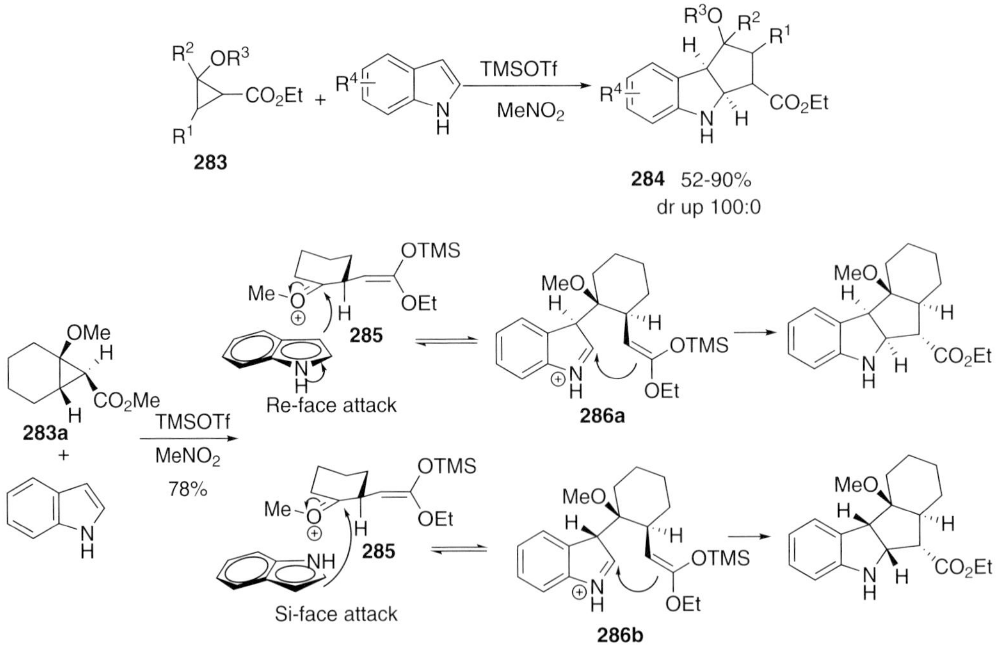

**Scheme 15.59** TMSOTf-mediated [3 + 2] annulation reaction of 2-alkoxycyclopropane carboxylates with indoles.

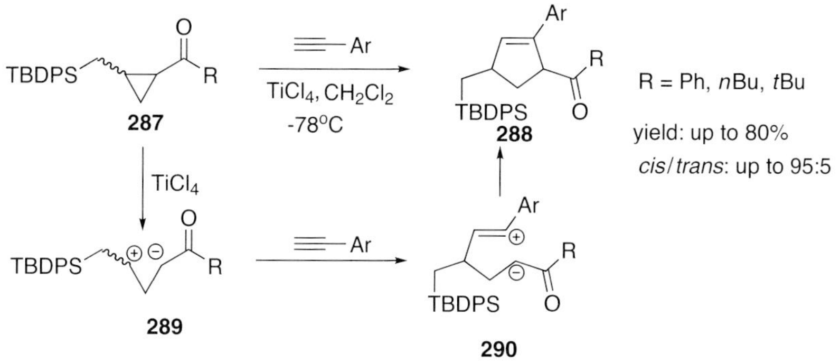

## 15.3 Cycloaddition Reactions

**Scheme 15.60** TiCl$_4$-catalyzed [3 + 2] cycloaddition reaction between *tert*-butyldiphenylsilylmethyl-substituted cyclopropyl phenyl/alkyl ketones.

In the presence of Lewis acid, D–A cyclopropanes can react with nitriles to give five-membered heterocycles. For example, in 2003, Pagenkopf reported a TMSOTf-mediated formal [3 + 2] cycloaddition reaction between donor–acceptor cyclopropane **291** and nitriles to afford 3, 4-dihydro-2H-pyrrole cycloaddition products **292** in good to excellent yields (Scheme 15.61) [72]. TMSOTf also effectively mediates the dehydration of the 3, 4-dihydro-2H-pyrrole cycloaddition products **294** and tautomerization to afford pyrroles **295** in moderate to excellent overall yield. Therefore, highly substituted pyrroles could be furnished by this cost-effective and regiospecific cascade reaction from the corresponding active donor–acceptor cyclopropanes **293** with nitriles.

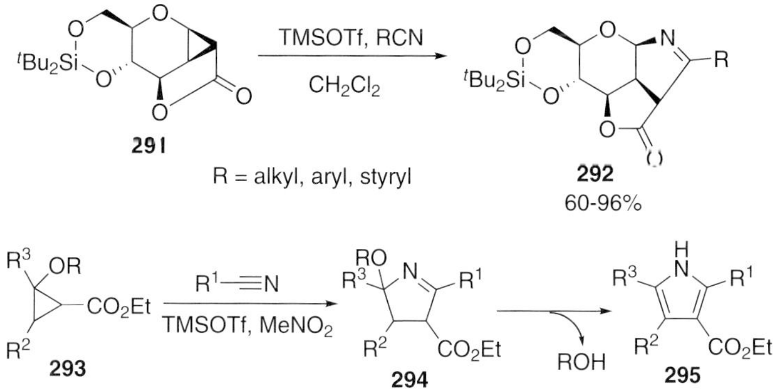

**Scheme 15.61** TMSOTf-mediated [3 + 2] cycloaddition of D–A cyclopropanes with nitriles.

2-(Iodomethyl)cyclopropane-1,1-dicarboxylates **296** can react as a 3-carbon component with olefins via a radical process. Treatment of 1-diethylphosphonyl- or 1-phenylsulfonyl-2-(iodomethyl)cyclopropane-1-carboxylate **296** with Et$_3$B produces an unsymmetrical homoallyl radical species **298** that gives functionalized cyclopentane derivatives **297** with high stereoselectivity through iodine atom transfer [3 + 2] cycloaddition reaction with alkenes (Scheme 15.62) [73].

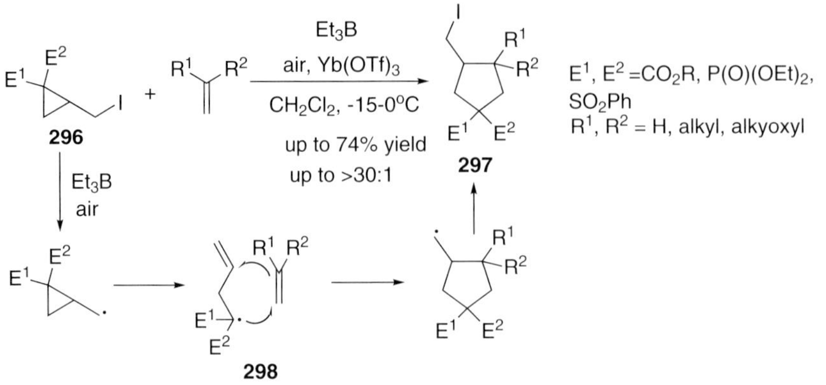

**Scheme 15.62** Iodine atom transfer [3 + 2] cycloaddition reaction of 2-(Iodomethyl)cyclopropane-1-carboxylate with alkenes.

### 15.3.4.4 [3 + 2] Cycloaddition of VCPs

VCPs **299** undergo a formal [3 + 2] cycloaddition with electron-deficient olefins in the presence of a palladium catalyst, yielding the cyclopentanes **300** in 23–89% yields (Scheme 15.63) [74]. The proposed mechanistic rationale involves initial oxidative addition of VCPs to form a zwitterionic π-allyl palladium species **301**. After undergoing Michael addition to electron-deficient olefins, a new zwitterionic π-allyl palladium species **302** was formed which, upon subsequent intramolecular nucleophilic substitution, produces the final product.

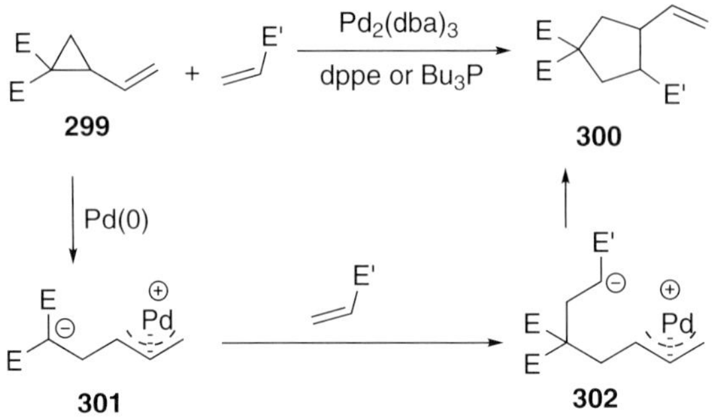

**Scheme 15.63** Pd(0)-catalyzed [3 + 2] cycloaddition of activated VCPs with electron-deficient olefins.

Vinyl-cyclopropanes normally act as a 5-carbon component in [5 + 2] or [5 + 2 + 1] cycloaddition reactions (see Sections 15.3.10 and 15.4.1), but in some special cases they can also act as a 3-carbon component in the [3 + 2] cycloaddition reaction. For example, Yu [75] recently reported a Rh(I)-catalyzed intramolecular [3 + 2] cycloaddition of trans-VCP-enes **303** yielding 5,5-fused bicycles **304** with a yield of 49–83% (Scheme 15.64). A chirality transfer phenomenon was found in this transformation, that is, optically pure 5, 5-fused bicycle (+)-**304a** could be prepared from optically active (+)-**303a** without loss of the enantiopurity. It is very interesting to find that the cycloaddition of the cis isomer of VCP-ene *cis*-**303b** gives the 5,7-fused bicycle **305** instead of the 5,5-fused bicycle via [5 + 2] cycloaddition. A reasonable mechanistic rationale for this transformation is that the oxidative cleavage of the cyclopropyl ring of *trans*-**303** produces a π-allyl Rh(III) species **306a**, in which the alkene chain stays at the opposite side of the π-allyl moiety. The alkene insertion affords the intermediate **307a**. Subsequent stereochemistry-favored reductive elimination of the two proximal carbons C1′ and C3 of **307a** gives the [3 + 2] cycloadduct. The cis isomer forms intermediate **306b** followed by alkene insertion to produce intermediate **307b**. The C1′ and C5 are close in proximity which means that subsequent reductive elimination between C1′ and C5 is favored to yield the [5 + 2] cycloadduct.

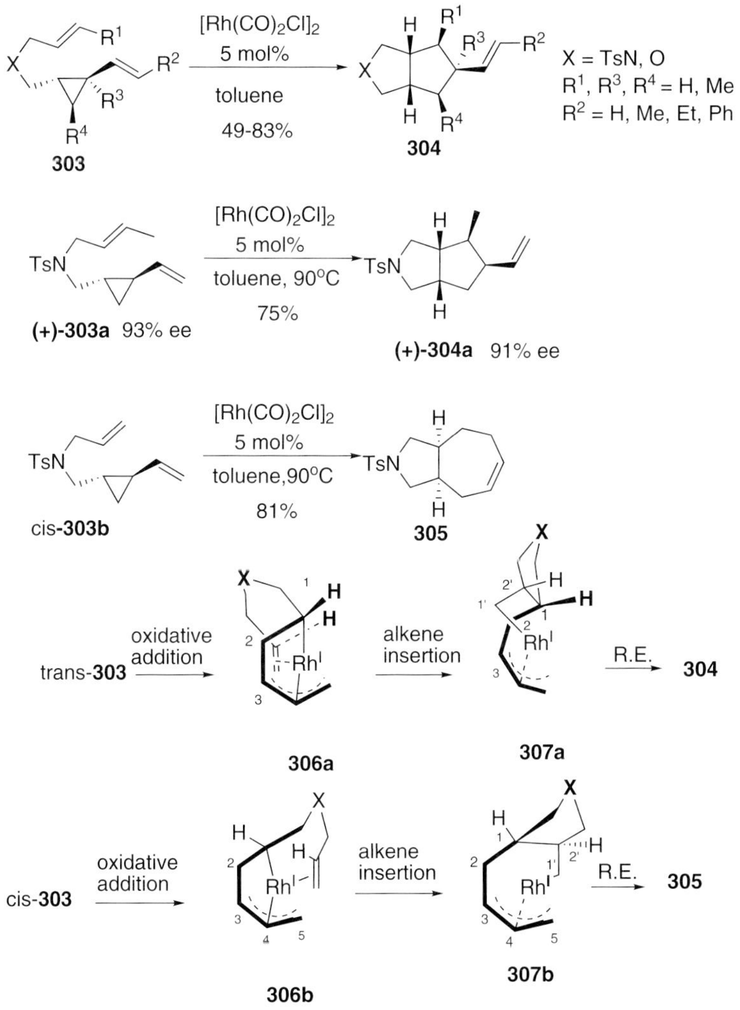

**Scheme 15.64** Rh(I)-catalyzed intramolecular cycloaddition of VCP-enes.

## 15.3.5
## [3 + 3] Cycloaddition of Activated Cyclopropanes

Kerr discovered a formal [3 + 3] cycloaddition of nitrones [76] with cyclopropane-1,1-dicarboxylates **256a** under the catalysis of Lewis acids such as Yb(OTf)$_3$ and MgI$_2$ to give the uncommon tetrahydro-1,2-oxazines **308** (Scheme 15.65) [77]. When using Yb (OTf)$_3$ as catalyst, this [3 + 3] cycloaddition reaction is highly regio- and stereo-selective yielding the cis adducts exclusively [77a], while the reaction with MgI$_2$ gave the product with moderate to high diasteroselectivity [77b].

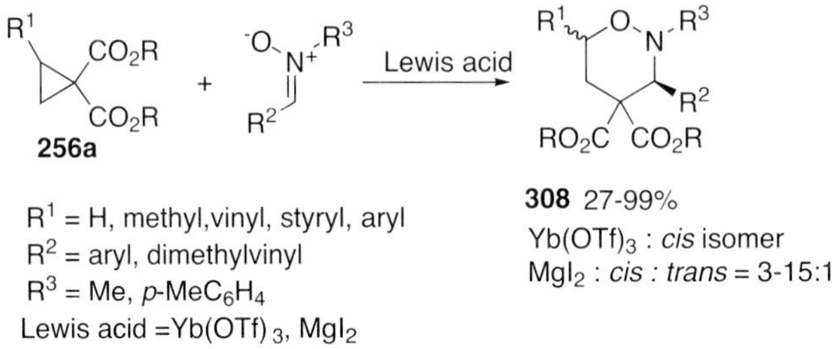

$R^1$ = H, methyl, vinyl, styryl, aryl
$R^2$ = aryl, dimethylvinyl
$R^3$ = Me, p-MeC$_6$H$_4$
Lewis acid = Yb(OTf)$_3$, MgI$_2$

**308** 27–99%
Yb(OTf)$_3$ : cis isomer
MgI$_2$ : cis : trans = 3–15:1

**Scheme 15.65** Lewis acid-catalyzed [3 + 3] cycloaddition of cyclopropane-1,1-dicarboxylates with nitrones.

Kerr also developed a three-component version of this transformation, in which the nitrones were formed *in situ* by reaction of the corresponding hydroxylamines with aldehydes. This strategy has been applied as the key step in the total synthesis of the natural products (+)-phyllantidine [78] and nakadomarin A [79] (Scheme 15.66). The three-component cycloaddition of N-hydroxy(4-methoxyphenyl)methanamine, aldehyde **309**, 2-vinyl cyclopropane diesters **310** under the catalysis of Yb(OTf)$_3$ produces **311**, which is then converted to the final natural product (+)-phyllantidine. The approach to Nakadomarin A starts from cycloaddition of 2-phenyl 1,1-cyclopropane dicarboxylates **256b**, aldehyde **312** and N-hydroxybenzyl amine in toluene in the presence of Yb(OTf)$_3$ to give **313**, which is converted to the corresponding core of nakadomarin A **314**.

Kerr also studied the cycloaddition of nitrones **315** with optically pure 2-substituted cyclopropane-1, 1-dicarboxylates **256a** and found that the ee of the adducts is temperature and $R^3$ dependent. Optically active 2-substituted cyclopropane-1, 1-dicarboxylates **256a** are shown to racemize under typical reaction conditions (i.e., 110 °C) (Scheme 15.67) [80].

Sibi reported the first enantioselective version of this reaction [81]. In the presence of Ni(ClO$_4$)$_2$ and chiral bisoxazoline ligand **320**, enantiomerically pure tetrahydrooxazines **319** could be prepared from racemic cyclopropyldicarboxylates **318** and nitrones in excellent yields (up to 99%) and high enantioselectivity (71–99% ee). In contrast to Yb(OTf)$_3$, this chiral catalyst gives very poor cis/trans diastereoselectivity, but affords excellent enantioselectivity for both diastereomers (Scheme 15.68).

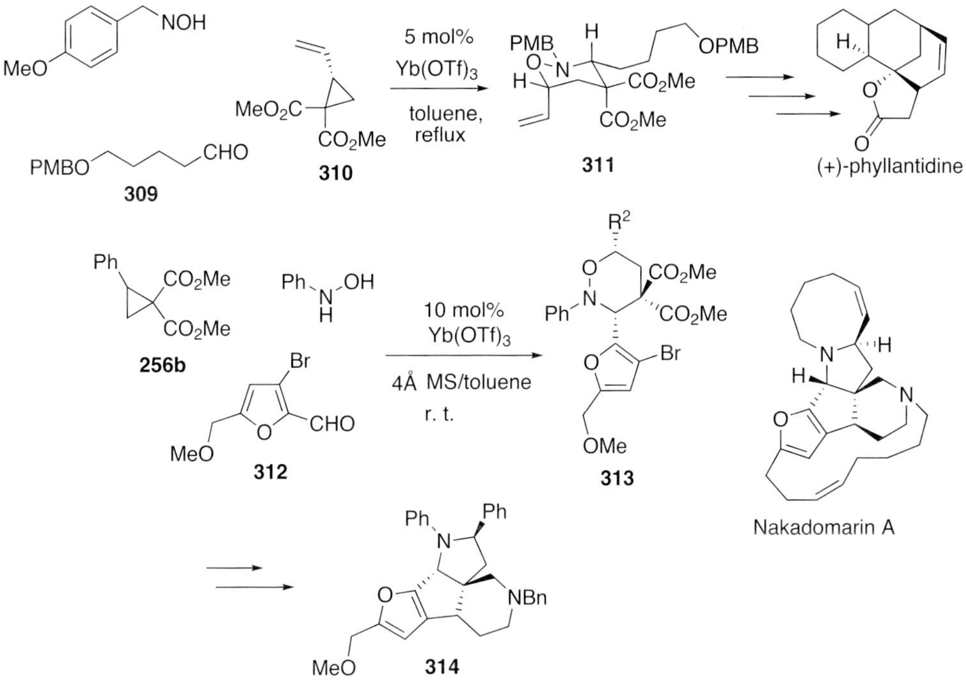

**Scheme 15.66** Synthetic application of three-component [3 + 2] cycloaddition reaction in total synthesis.

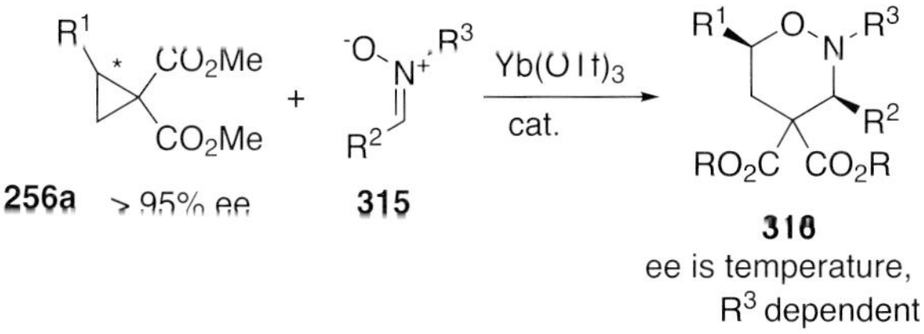

**Scheme 15.67** [3 + 3] Cycloaddition of optically active 2-substituted cyclopropane-1, 1-dicarboxylates.

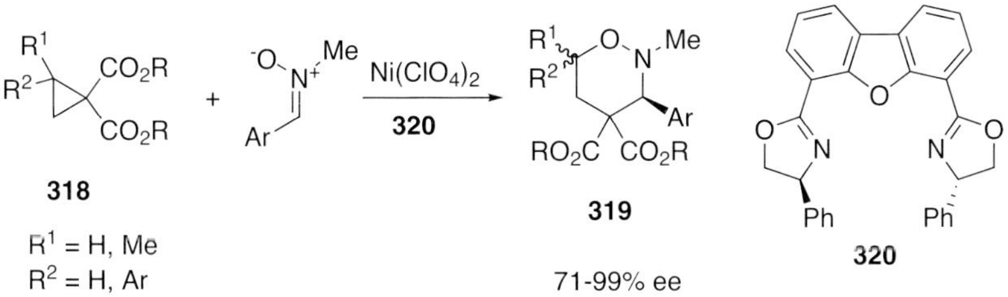

**Scheme 15.68** Ni(ClO$_4$)$_2$/$C_2$-symmetric bisoxazoline **320** catalyzed enantioselective [3 + 3] cycloaddition.

Recently, Tang [82] has applied pseudo-$C_3$-symmetric trisoxazoline **321** as the chiral ligand for this Ni(II)-catalyzed enantioselective [3 + 3] cycloaddition reaction. In contrast to bisoxazoline **320**, this chiral catalyst gave high diastereoselecivty (cis/trans up to 13/1) and high enantioselectivity (80–97% ee). It is noteworthy that this reaction can be employed for the kinetic resolution of 2-substituted cyclopropane-dicarboxylates to give optically active cyclopropanes with excellent ee values (Scheme 15.69). It is quite interesting to note that the optically active cyclopropanes prepared by this kinetic resolution could react with nitrones to afford the corresponding optically active tetrahydrooxazines in the presence of a catalytic amount of Ni(ClO$_4$)$_2$ without chiral ligand. Both enantiomers could be prepared from the racemic 2-substituted cyclopropane 1, 1-dicarboxylate by direct enantioselective cycloaddition in the presence of a catalytic amount of Ni(ClO$_4$)$_2$/**321** or by kinetic resolution followed by cycloaddition with nitrones.

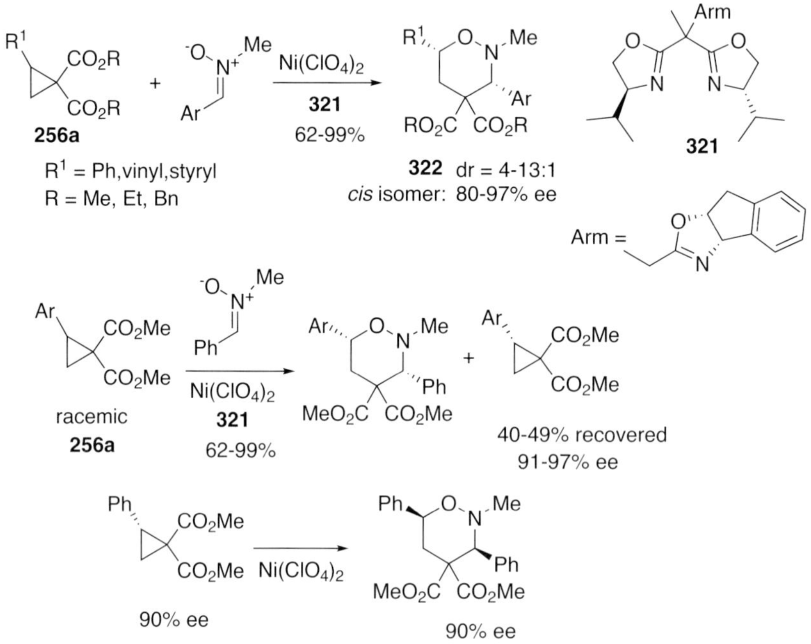

**Scheme 15.69** Ni(ClO$_4$)$_2$/pseudo-$C_3$-symmetric trisoxazoline **321** catalyzed enantioselective [3 + 3] cycloaddition.

A Ni(ClO$_4$)$_2$-catalyzed [3 + 3] cycloaddition of aromatic azomethine imines **323** with cyclopropane-1,1-dicarboxylates **256a** was achieved by Charette (Scheme 15.70) [83]. This methodology provides access to the unique tricyclic dihydroquinoline derivatives **324** with dr up to 6.6 : 1. Complete retention of stereogenic information at the carbon 2 of the cyclopropane was observed, thus a stepwise mechanism consisting of nucleophilic opening of the cyclopropane followed by diastereoselective ring closure reaction was proposed.

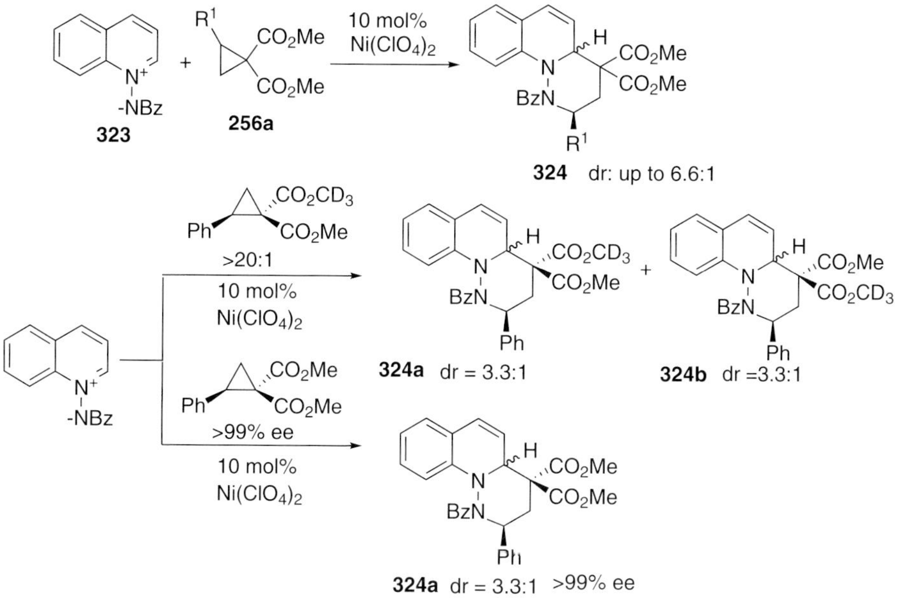

**Scheme 15.70** Ni(ClO$_4$)$_2$-catalyzed [3 + 3] cycloaddition of 1,1-cyclopropane diesters with aromatic azomethine imines.

## 15.3.6
## [3 + 4] Cycloaddition

Cyclopropanes can act as a 3-carbon component of [3 + 4] cycloaddition under transition metal or Lewis acid catalysis. Recently, Mascareñas reported a Pd(0)-catalyzed intramolecular [4 + 3] cycloaddition of readily accessible diene-MCPs **325** to furnish 5,7-fused bicyclic systems **326** and 5,5-fused bicycles **327** with high selectivity, in which the nature of the ligands affects the ratio of the 5,7-fused and 5,5-fused bicycles(Scheme 15.71) [84]. Moderate enantioselectivity (64% ee) was achieved when using chiral L1 as the chiral ligand. The mechanism for this transformation is depicted in Scheme 15.71. The formation of the TMM intermediate and subsequent insertion of one of the double bonds of the dienes affords a palladacyclohexane intermediate **328**, which undergoes two possible reaction pathways: direct reductive elimination affording the 5,5-fused bicycles or π-allylic rearrangement yielding the palladacyclooctane **329**, followed by reductive elimination to give the 5,7-fused bicycles.

Ivanova [85] reported a Lewis acid catalyzed formal [4 + 3] cycloaddition between 2-aryl cyclopropane-1,1-dicarboxylates **330** and 1,3-diphenylisobenzofuran **331** under mild reaction conditions to produce two diastereoisomeric cycloadducts **332** in a combined yield of 84–92% and with up to 86/14 daistereoselectivity (Scheme 15.72). The predominant formation of the less stable exo isomer suggests a concerted mechanism with orbital control of the stereochemical course of the reaction.

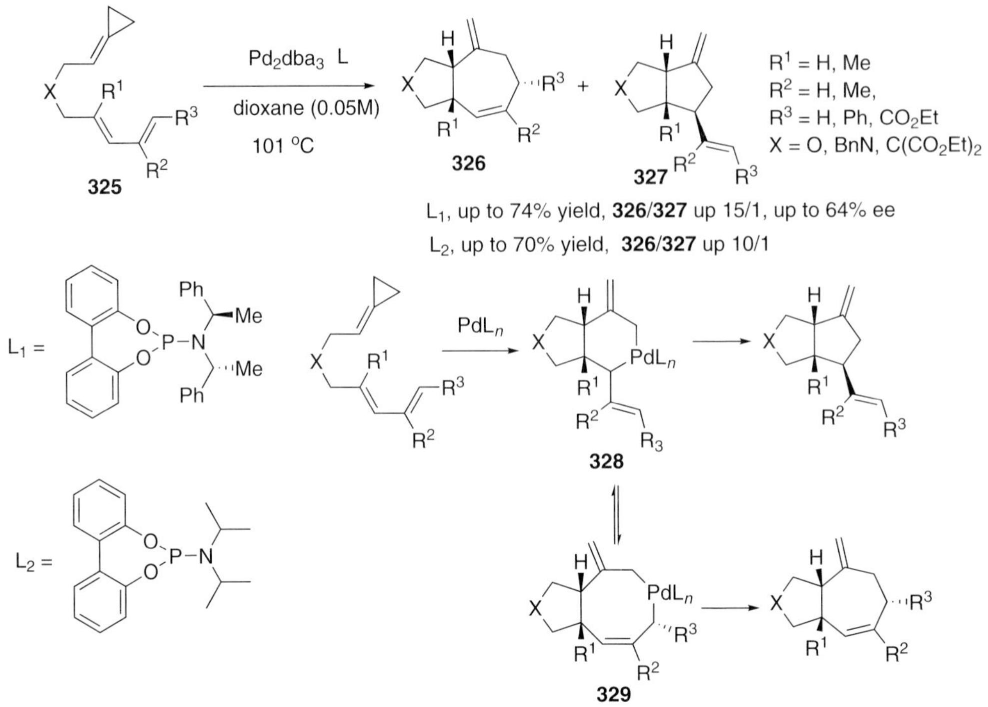

**Scheme 15.71** Pd(0)-catalyzed intramolecular [4 + 3] cycloaddition of diene-MCPs.

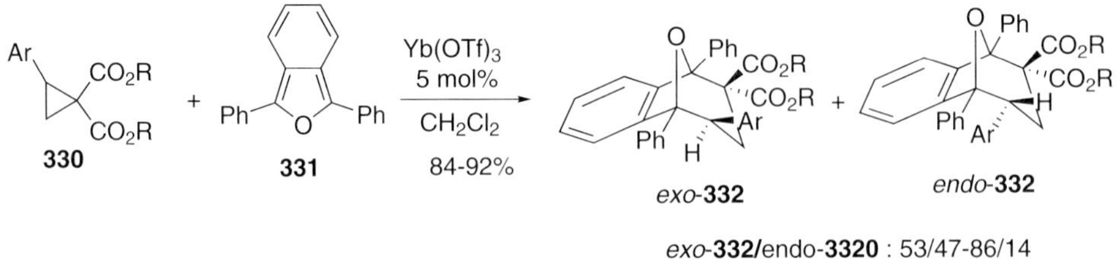

**Scheme 15.72** Lewis acid catalyzed formal [4 + 3] cycloaddition between cyclopropane diesters and 1,3-diphenylisobenzofuran.

### 15.3.7
### [4 + 1] Cycloaddition

De Meijere developed a formal [4 + 1] cycloaddition of MCPs **333** with Fisher carbenes **334**, in which the MCPS act as the 4-carbon component, yielding cyclopentenones **335** in 28–58% yields with up to 93:7 diastereoselectivity (Scheme 15.73) [86]. The mechanism of this reaction was rationalized through the initial formation of chromacyclobutane **336** via [2 + 2] cycloaddition, followed by β-carbon elimination to produce cyclopentane **337**, which after migratory insertion of CO and subsequent reductive elimination affords cyclopentenone **338**. The latter then isomerizes into the final more stable cyclopentenones **335**.

**Scheme 15.73** [4 + 1] Cycloaddition of MCPs with Fisher carbenes **334**.

A Rh(I)-catalyzed formal [4 + 1] cycloaddition of 1-(1-alkynyl)cyclopropyl ketones **339** with CO was developed by Zhang (Scheme 15.74) [87]. The reaction proceeded in DCE at 70 °C, providing 1, 3, 5-trisubstituted 5, 6-dihydrocyclopenta[c]furan-4-ones **340** in good to excellent yields with traces of trisubstituted furans **341** as by-product. This reaction is believed to proceed via initial regioselective oxidative addition of 1-(1-alkynyl)cyclopropyl ketones **339** to produce rhodacyclobutane **342**, which upon cyclization gives rhodabicycles **343**. Migratory insertion of CO into **343** forms rhodium species **344** followed by subsequent reductive elimination to give the 1, 3, 5-trisubstituted 5, 6-dihydrocyclopenta[c]furan-4-ones **340**. Trace amount of by-product furan **341** is generated via β-H elimination of species **343** before CO migratory insertion and subsequent reductive elimination.

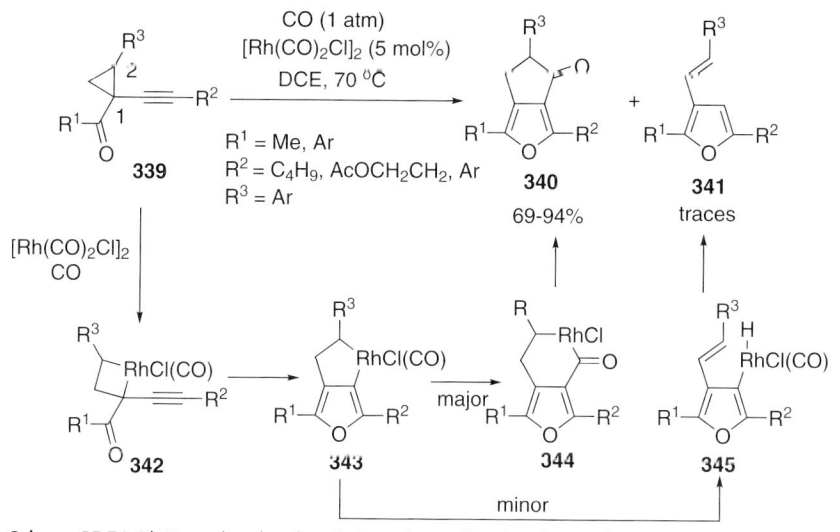

**Scheme 15.74** Rh(I)-catalyzed carbonylation of 1-(1-alkynyl)cyclopropyl ketones.

## 15.3.8
### [4 + 2] Cycloaddition of 1-(1-Alkynyl)cyclopropyl Ketones

Zhang [88] reported a Au(I)-catalyzed [4 + 2] cycloaddition of 1-(1-alkynyl)cyclopropyl ketones **346** with various modified indoles, aldehydes, imines and silyl enol ether **347** to produce furan fused 6-membered carbo- and heterocycles **348** in good to excellent yields. When $R^3$ is not H, the diastereoselectivity of this reaction is from 1:1 to 4:1. The reaction is believed to proceed via the oxocarbenium **349**, generated from the **346** mediated by the Au(I) complex which, upon ring expansion, gives the furanyl gold species **350**. The furanyl gold species **350** have separated charge and can serve as 1,4-dipoles to react with dipolarophiles, providing the [4 + 2] cycloadducts (Scheme 15.75).

## 15.3.9
### [5 + 1] Cycloaddition

It is reported by de Meijere [89] that VCPs can undergo formal [5 + 1] cycloaddition with CO in the presence of transition metal complexes, the VCPs acting as the 4-carbon component, producing cyclohexenones **352** in 16–95% yields. The transition metal complex can be stoichiometric octacarbonyldicobalt or a catalytic amount of Co- and Rh-complexes (Scheme 15.76). It is noteworthy that the reaction of aryl- and vinyl-substituted VCPs proceeds smoothly only in the presence of stoichiometric octacarbonyldicabalt, in contrast, the simple vinylcyclopropane without other substituents failed to give the product. This transformation starts from the formation of the cobaltacyclopropane species **354**, followed by ring expansion to produce cobaltacyclohexene **355**. The latter undergoes CO insertion and reductive elimination to give the cyclohexenones **352**.

Taber reported a regioselective [5 + 1] cycloaddition of 2-substituted VCPs **356** with the substituent on the cyclopropane ring by employment of the iron pentacarbonyl complex $Fe(CO)_5$ [90], providing highly substituted cyclohexenones **357** and the minor regioisomer **358** in moderate to good yields with diastereoselectivity up to 16.6:1 (Scheme 15.77) [91]. The reaction may proceed via the initial formation of metallacyclopropane **359**, which can undergo ring expansion through cleavage of bond a or b to give the ferracyclohexene species **360** and **363**. Subsequent CO insertion of **360** and reductive elimination gives the nonconjugated cyclohexenones **362** which readily isomerize to the more stable conjugated cyclohexenones **357** in the presence of DBU. The minor regioisomer **358** is produced from the ferracyclohexene species **363** by a similar process. This reaction has been successfully employed by Taber as the key step in the total synthesis of (+)-delobanone from the VCP **364** [92].

During the study of the cycloisomerization reaction of cyclopropenyl ketones and esters in the presence of $[Rh(CO)_2Cl]$, Liebeskind [93] found that the reaction of cyclopropenyl esters **365** favored the pyrone products **369/370** via formal [5 + 1] cycloaddition, where the regioselectivity depends on the size of the $R^2$ and $R^3$ groups. In contrast, the reaction of cyclopropenyl ketones giving pyrones needs higher CO

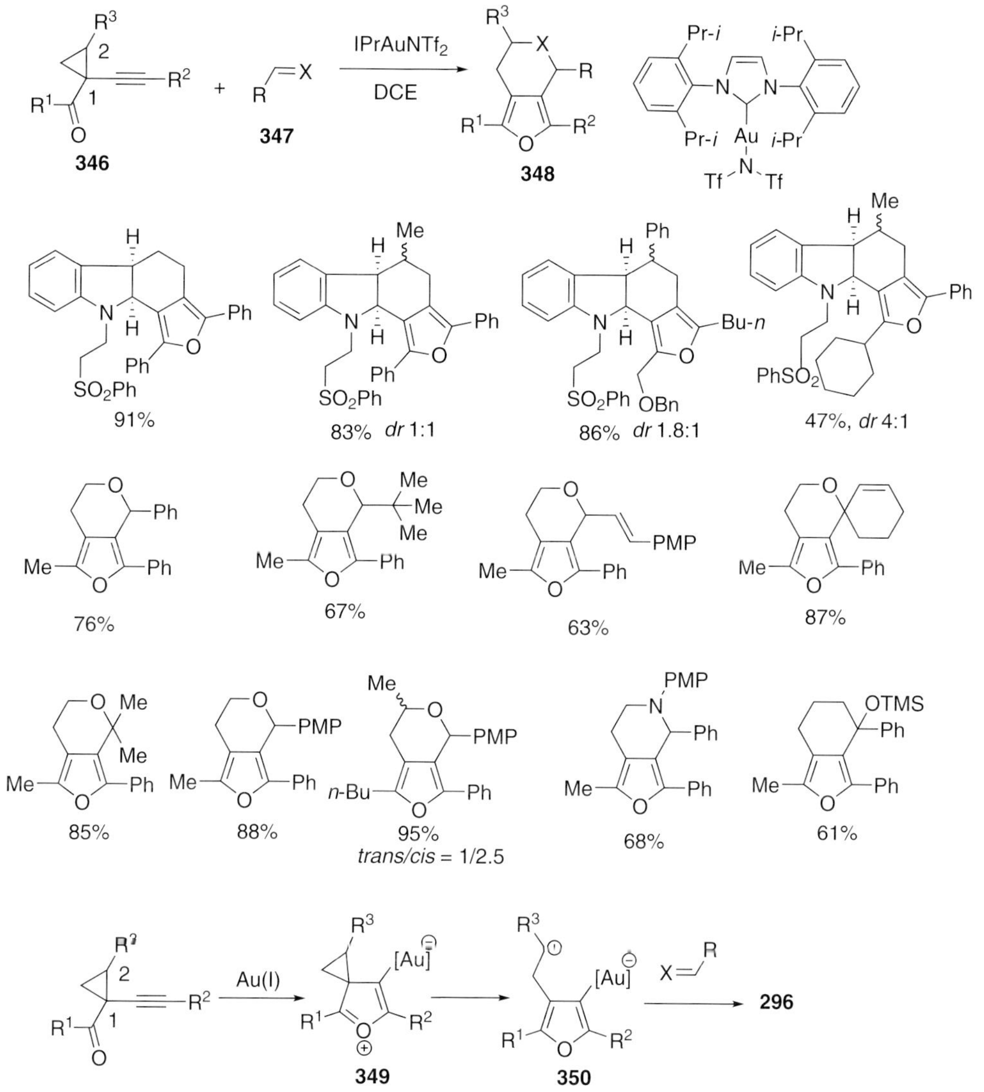

**Scheme 15.75** Au(I)-catalyzed [4 + 2] cycloaddition of 1-(1-alkynyl)cyclopropyl ketones.

pressure to suppress the side reaction to form furans. The reaction is believed to proceed via the initial formation of the tertiary cyclopropyl cation **366** followed by ring expansion to produce rhodacyclobutene **367**, which equilibrates with its regioisomer **368** via a reverse [2 + 2] cycloaddition and anti-regioselective [2 + 2] cycloaddition pathway. CO insertion of **367** and **368**, and subsequent enolization and reductive elimination gives the pyrones **369** and **370**, respectively. Pyrone **369** will be the major regioisomer when $R^3 = H$ because the 1, 3-disubstituted rhodacyclobutene **368** is more stable than its 1,2-disubstituted analogue **367** (Scheme 15.78).

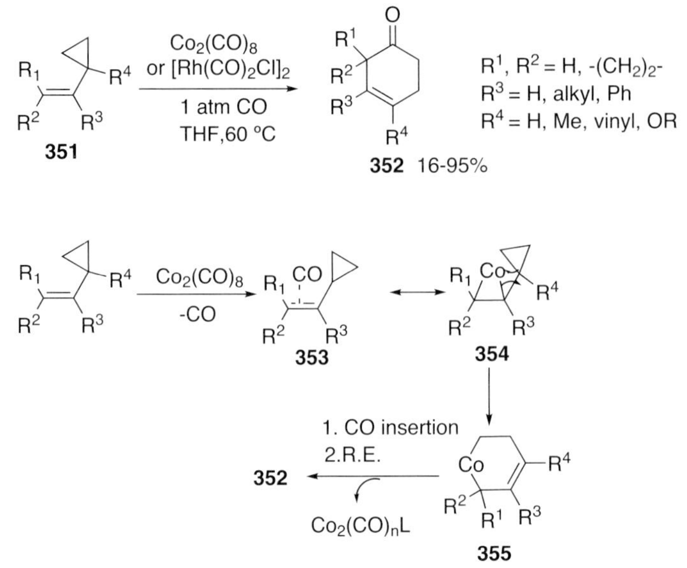

**Scheme 15.76** Co$_2$(CO)$_8$-mediated or Rh(I)-catalyzed formal [5 + 1] cycloaddition of VCPs with CO.

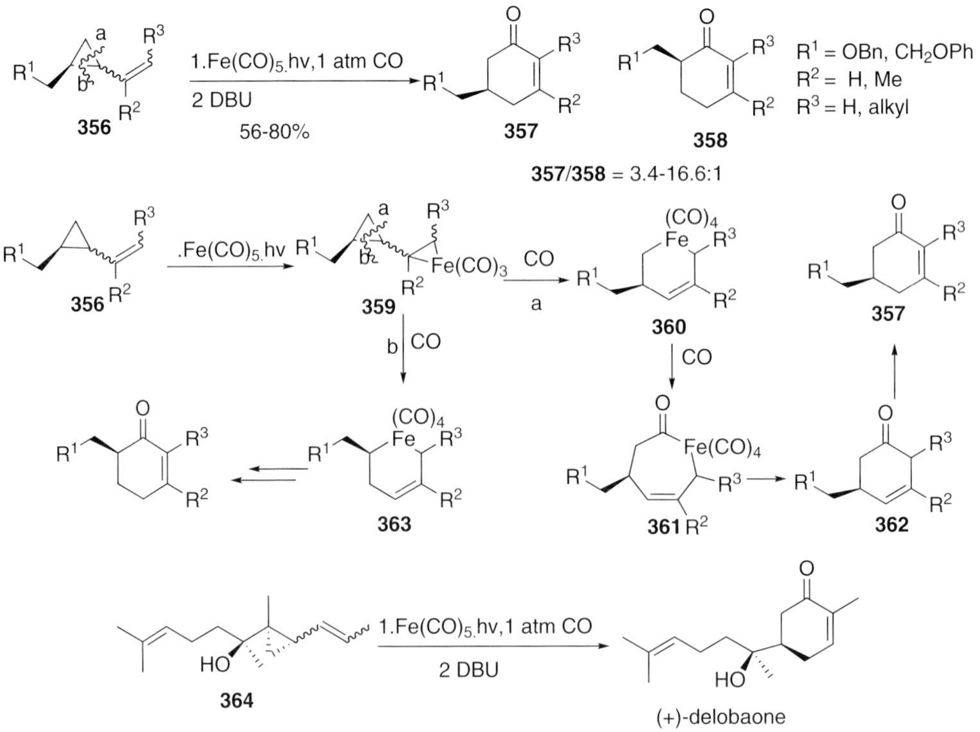

**Scheme 15.77** Fe(CO)$_5$-catalyzed regioselective [5 + 1] cycloaddition of 2-substituted VCPs.

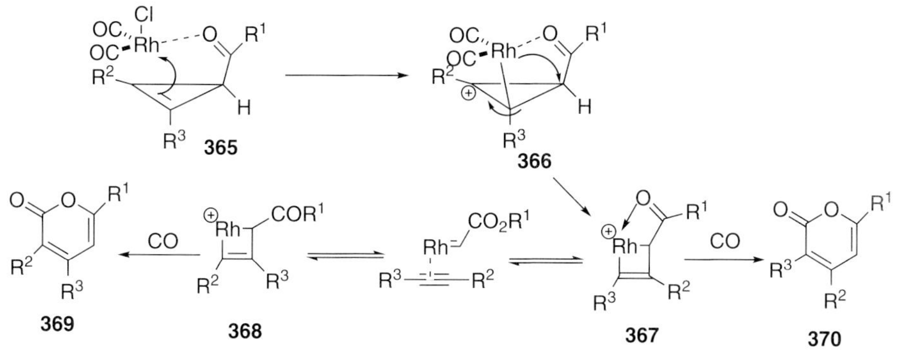

**Scheme 15.78** Rh(I)-catalyzed regioselective [5 + 1] cycloaddition of cyclopropenyl ketones and esters.

It was demonstrated that arylcyclopropenes **371** can undergo carbonylative cycloisomerization (formal [5 + 1] cycloaddition) to form the naphthols **372** in 35–76% combined yields with up to 100:0 diastereroselectivities in the presence of substoichiometric and even catalytic amounts of Cr and Mo carbonyl complexes (Scheme 15.79) [94]. The regioselectivity comes from the regioselective oxidative addition of the bond of the cyclopropenes forming the metallacyclobutenes **373a** and **373b**, which undergo subsequent CO insertion and cyclization.

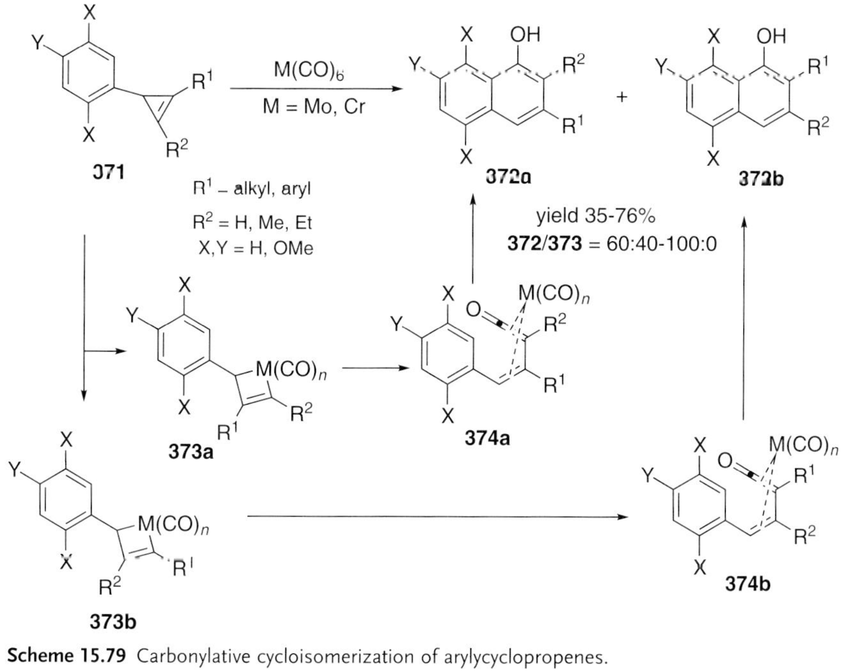

**Scheme 15.79** Carbonylative cycloisomerization of arylycyclopropenes.

## 15.3.10
## [5 + 2] Cycloaddition of VCPs

### 15.3.10.1 Intramolecular [5 + 2] Cycloaddition of VCPs

Wender [95] in 1995 developed the first example of Rh(I)-catalyzed intramolecular [5 + 2] cycloaddition of vinylcyclopropanes(VCPs) **375** with alkynes, leading to 5,7-fused bicycles **376** in moderate to excellent yields (Scheme 15.80). This transformation was designed only as a homologue of the Diels–Alder, but more important than this, it truly opens a field in which VCPs acts as the 5-carbon component in a series of [5 + m] or [5 + m + n] cycloadditions. After developing the first generation catalyst by combination of Wilkinson's catalyst Rh(PPh$_3$)$_3$Cl with AgOTf [95], Wender found that [Rh(CO)$_2$Cl] [96] and the rhodium arene complex **379** [97] are better catalysts, especially the latter, which can efficiently catalyze intramolecular [5 + 2] cycloaddition of both alkyne-and alkene-tethered VCPs **377** under mild conditions and suppresses the undesired olefin isomerization observed in the former catalyzed case. Other cationic rhodium complexes such as Rh(DIPHOS)(CH$_2$Cl$_2$)$_2$SbF$_6$ and [Rh(dppb)Cl]$_2$/AgSbF$_6$ were later developed by Gilbertson [98] and Zhang [97], respectively.

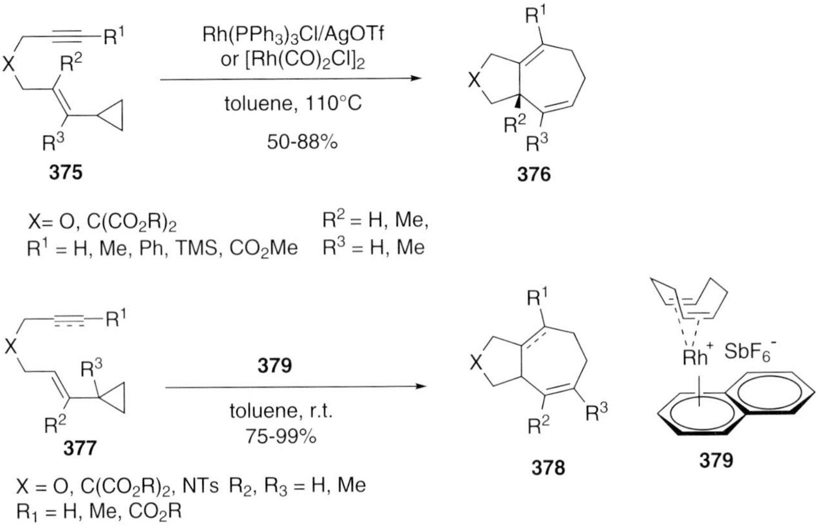

**Scheme 15.80** Rh(I)-catalyzed intramolecular [5 + 2] cycloaddition of vinylcyclopropanes (VCPs) to alkynes.

Instead of the expensive Rh complex, CpRu(NCCH$_3$)$_3$]PF$_6$ was explored by Trost as an efficient catalyst capable of enabling [5 + 2] cycloaddition of alkyne-VCPs **380** into 5,7-fused bicycles **381** in yields of 73–95% under mild conditions (Scheme 15.81) [100].

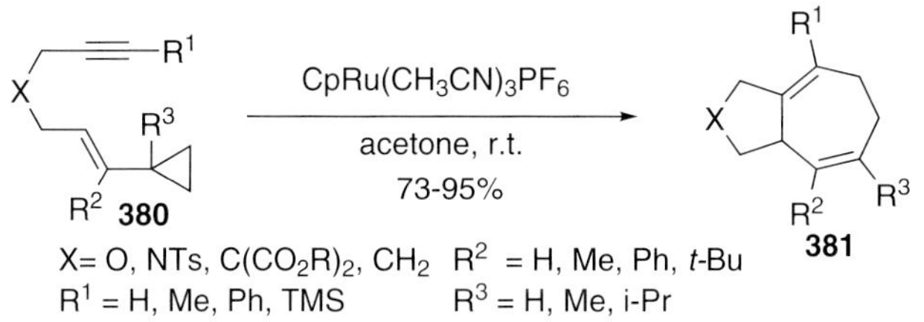

**Scheme 15.81** CpRu(NCCH$_3$)$_3$PF$_6$-catalyzed intramolecular [5 + 2] cycloaddition of vinylcyclopropanes (VCPs) to alkynes.

Trost [101] further showed that CpRu(NCCH$_3$)$_3$PF$_6$ catalyzed the intramolecular cycloaddition of alkyne-VCPs **382** with one hydroxy or TMS ether group forming 5,7-fused bicycles **383**, in which the stereochemistry of the newly generated bridged carbon stereocenter is cis to the adjacent hydroxy group, as a result of the inside alkoxy effect (the hydroxy group may coordinate with Ru, see intermediate **384**) (Scheme 15.82).

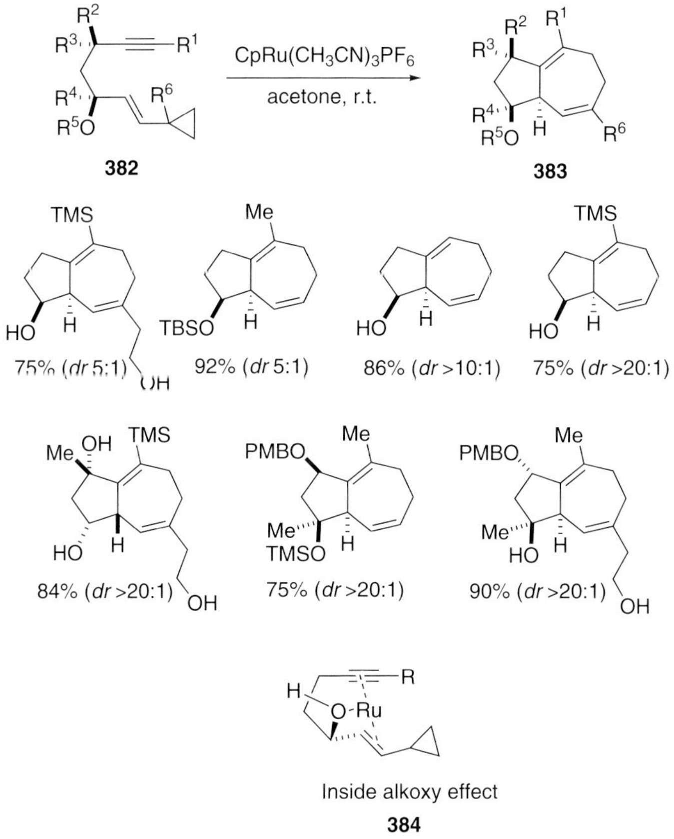

**Scheme 15.82** Inside alkoxy effect controlling the stereochemistry.

This Ru-catalyzed [5 + 2] cycloaddition can be extended to bicyclic substrates **385**, providing an efficient route to the complex tricyclic skeleton **386** in high yields and with good diastereoselectivities (Scheme 15.83) [100b,102].

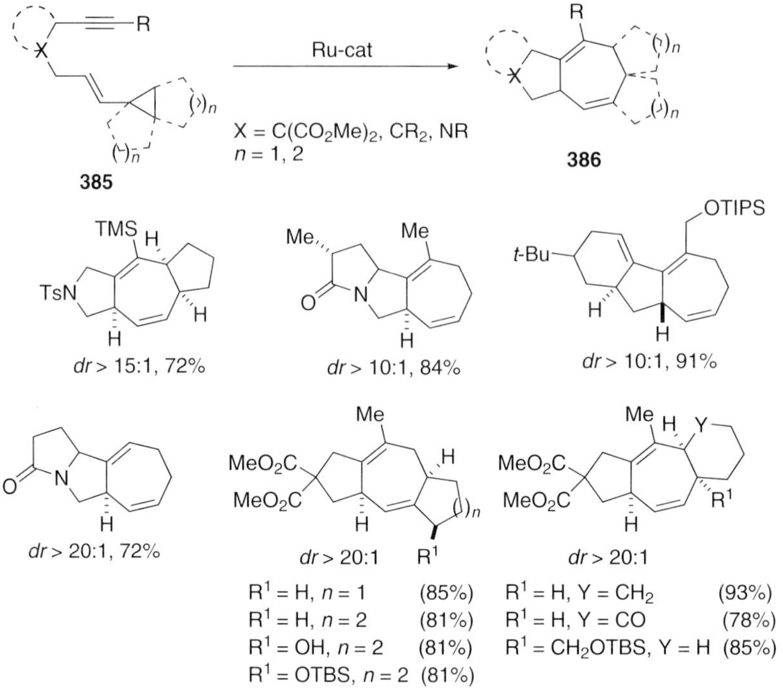
R
Ru-cat
X = C(CO2Me)2, CR2, NR
n = 1, 2
385
386
TMS
TsN
dr > 15:1, 72%
Me
dr > 10:1, 84%
OTIPS
t-Bu
dr > 10:1, 91%
dr > 20:1, 72%
MeO2C
MeO2C
dr > 20:1
R1 = H, n = 1 (85%)
R1 = H, n = 2 (81%)
R1 = OH, n = 2 (81%)
R1 = OTBS, n = 2 (81%)
dr > 20:1
R1 = H, Y = CH2 (93%)
R1 = H, Y = CO (78%)
R1 = CH2OTBS, Y = H (85%)

**Scheme 15.83** Ru-catalyzed [5 + 2] cycloaddition of bicyclic substartes **385**.

Similarly to alkyne-VCPs, it was found by Wender that alkene-VCPs **387** can also undergo intramolecular [5 + 2] cycyloaddition to give *cis*-5, 7- and *trans*-6,7-fused bicycles **388** in yields of 70–94%. It is noteworthy that the combined catalyst system ($Rh(PPh_3)_3Cl$/AgOTf) gives a better result than the usually more efficient catalyst $[Rh(CO)_2Cl]_2$ which did not catalyze this reaction (Scheme 15.84) [103]. In contrast, employment of the Ru complex $CpRu(NCCH_3)_3]PF_6$ for cycloaddition of alkene-VCP **387a** failed to provide the desired 5, 7-fused bicycle, instead the Ru complex **389** was isolated in good yields when the catalyst loading was 0.5 equiv to the substrate [104].

In order to reduce organic solvent waste and make the work-up simpler, [5 + 2] cycloaddition performed in aqueous media has been successfully developed by employment of a Rh catalyst modified by introduction of a water-soluble sulfonic acid group to the phenyl ring of the phosphine ligand (Scheme 15.85) [105].

An enantioselective access to the optically active 5, 7-fused bicycles **393a–e** from alkene-VCPs **392a–e** was achieved by employment of $[Rh((R)\text{-BINAP})]SbF_6$ as the chiral Rh complex (Scheme 15.86) [106]. t is noteworthy that the enantioselectivity of this transformation depends strongly on the substitution pattern of both VCP and the

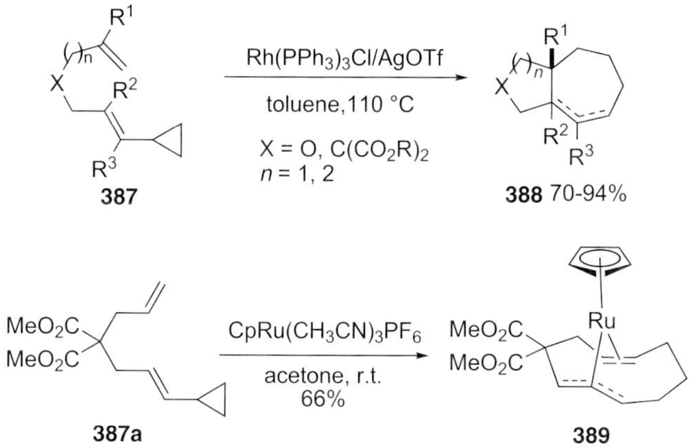

**Scheme 15.84** Cyclization of alkene-VCPs.

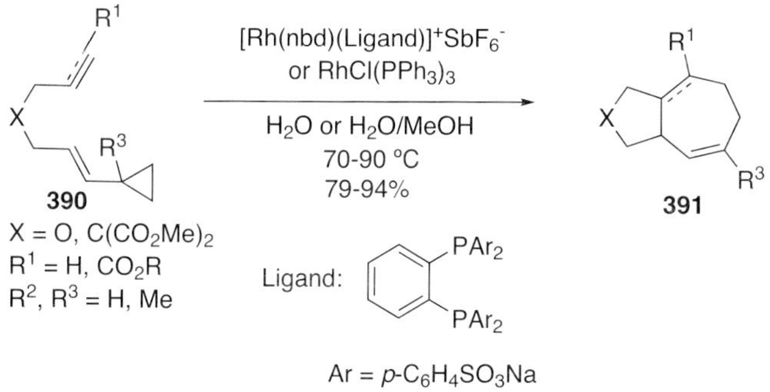

**Scheme 15.85** Modified Rh complex for the [5 + 2] cycloaddition in aqueous media.

tethered functional group. Either introduction of at least one substitutent (R1 or R2) to the substrate or changing the tethered functional group from malonate to TsN will dramatically improved the enantioselectivity, whereas asymmetric [5 + 2] cycloaddition of alkyne-VCPs gives the the corresponding cycloadducts with only moderate ees.

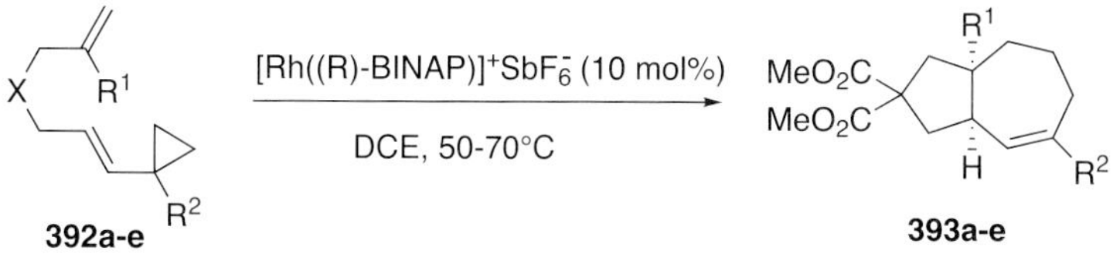

**393a**: X= C(CO$_2$Me)$_2$, R$^1$ = R$^2$ = H        73% (ee 52%)
**393b**: X= C(CO$_2$Me)$_2$, R$^1$ = Me, R$^2$ = H    72% (ee 95%)
**393c**: X= C(CO$_2$Me)$_2$, R$^1$ = CH$_2$OBn, R$^2$ = H   80% (ee 99%)
**393d**: X= C(CO$_2$Me)$_2$, R$^1$ = H, R$^2$ = Me    92% (ee 95%)
**393e**: X= TsN, R$^1$ = R$^2$ = H                    90% (ee 96%)

**Scheme 15.86** Enantioselective [5 + 2] cycloaddition of alkene-VCPs.

Furthermore, similar to alkene-VCPs, allene-VCPs **394** have also been successfully employed as the substrate in the Rh complex catalyzed [5 + 2] cycloaddition, providing an easy approach (70–93% yields) to 5,7-fused bicycles **395** with an exocyclic double bond in the seven-membered ring. A complete chirality transfer phenomenon was observed in the cycloaddition of optically active allene-VCPs (Scheme 15.87) [107].

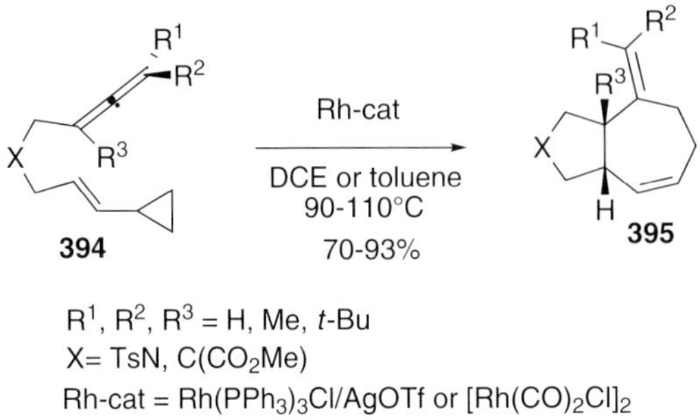

R1
R2
R3
X
394
Rh-cat
DCE or toluene
90-110°C
70-93%
R1
R2
R3
X
H
395
R1, R2, R3 = H, Me, t-Bu
X= TsN, C(CO2Me)
Rh-cat = Rh(PPh3)3Cl/AgOTf or [Rh(CO)2Cl]2


**Scheme 15.87** Rh(I)-catalyzed [5 + 2] cycloaddition of allene-VCPs.

A general mechanism for [5 + 2] cycloaddition of alkyne-, alkene- and allene-VCPs is proposed in Scheme 15.88. It is believed that there are two possible reaction pathways forming the intermediate 5,8-fused rhodabicycle **398** which undergoes reductive elimination to produce the cycloadduct, that is, either from the metallacyclohexene **396** by double or triple carbon–carbon bond insertion or from the metallacyclopentene **397** via cyclopropyl methyl metal to home allyl metal rearrangement. Metallacyclohexene **396** was generated by the oxidative cleavage of the cyclopropane ring. Wender and Houk preferred the former pathway for the $[Rh(CO)_2Cl]_2$-catalyzed reaction according to their computational studies [108]. In contrast, it is believed that the Ru-catalyzed case proceeded preferentially via the latter reaction pathway [100]. Recent studies by Yu, Wender and Houk showed the reductive elimination step of intermediate **364** is the determining step of the whole reaction and the reactivity of the one generated from the alkene-VCP is the lowest among three alkene-, alkyne- and allene-VCPs [109].

Regiochemistry in this transformation often depends on the catalyst used and the stereochemistry of the VCPs (Scheme 15.89) [110]. For example, [5 + 2] cycloaddition of *trans*-alkyne-VCPs **399** under the catalysis of $Rh(PPh_3)_3OTf$ gives 5,7-fused bicycles **400** via mode a cleavage, while $[Rh(CO)_2Cl]_2$ catalyst affords **401** with a complementary regiochemistry via mode b cleavage. For the *cis*-alkyne-VCPs **402**, the regioisomers **403** and **404** with complementary regiochemistry are produced, in which **403** was always produced as the major or only product. Different from the Rh-catalyzed case, the $CpRu(NCCH_3)_3]PF_6$ catalyzed cycloaddition of *trans*-cyclopropanes occurs with low regioselectivity.

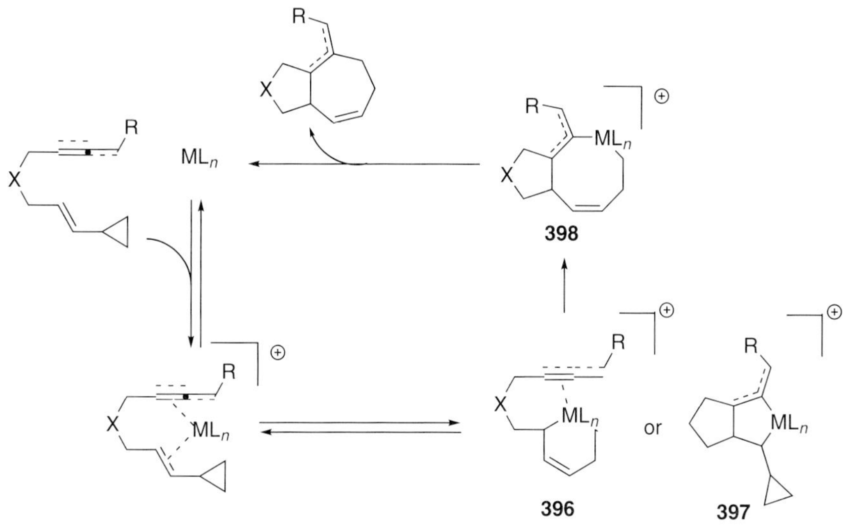

**Scheme 15.88** Plausible general mechanism of [5 + 2] cycloaddition of alkyne-, alkene- and allene-VCPs.

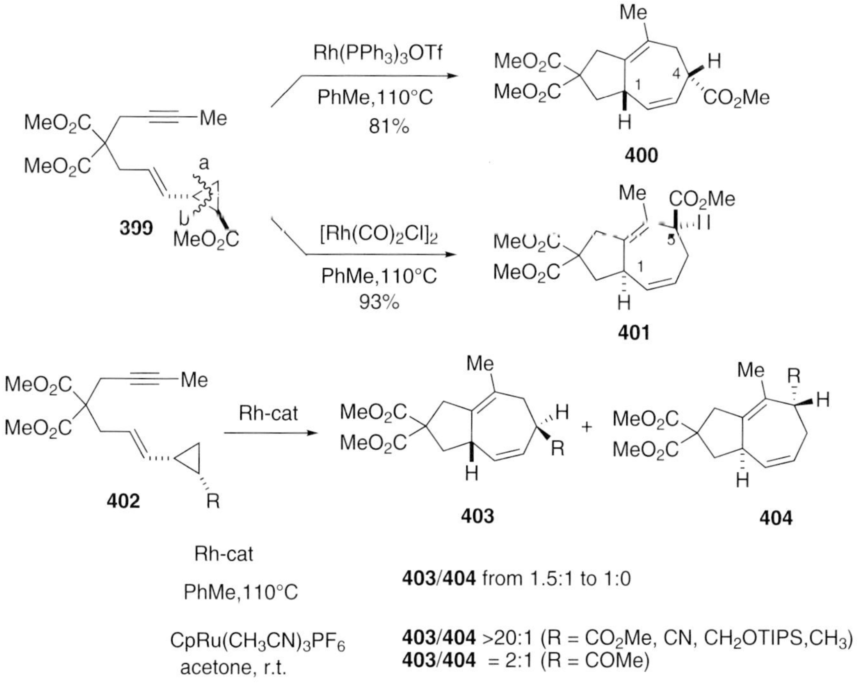

**Scheme 15.89** Regiochemistry and stereochemistry in Rh(I)-catalyzed alkyne-VCPs.

403/404 from 1.5:1 to 1:0

403/404 >20:1 (R = $CO_2Me$, CN, $CH_2OTIPS$, $CH_3$)
403/404 = 2:1 (R = COMe)

Synthetic application of the [5 + 2] cycloaddition of alkyne and allene tethered VCPs to the asymmetric total synthesis of several natural products has been demonstrated by several groups (Scheme 15.90). For example, a Rh-catalyzed [5 + 2] cycloaddition reaction of **405** was employed by Wender for the highly selective assembly of the key bicyclic skeleton of (+)-dictamnol **406** [111]. (+)-Aphanamol I [112] is another target molecule studied by Wender, of which the key transformation is Rh(I)-catalyzed [5 + 2] cycloaddition of the tetrasubstituted allene-VCPs **407** leading to bicyclic **408**. The Rh(I)-catalyzed cycloaddition of **409** to **410** has been employed by Martin in the enantioselective total synthesis of Tremulenediol A and Tremulenolide A [113]. Very recently, Trost realized the total synthesis of (−)-pseudolaric acid B [114] by employment of Ru complexes-catalyzed cycloaddition of optically active alkyne-VCP **411** as the key transformation (Scheme 15.91).

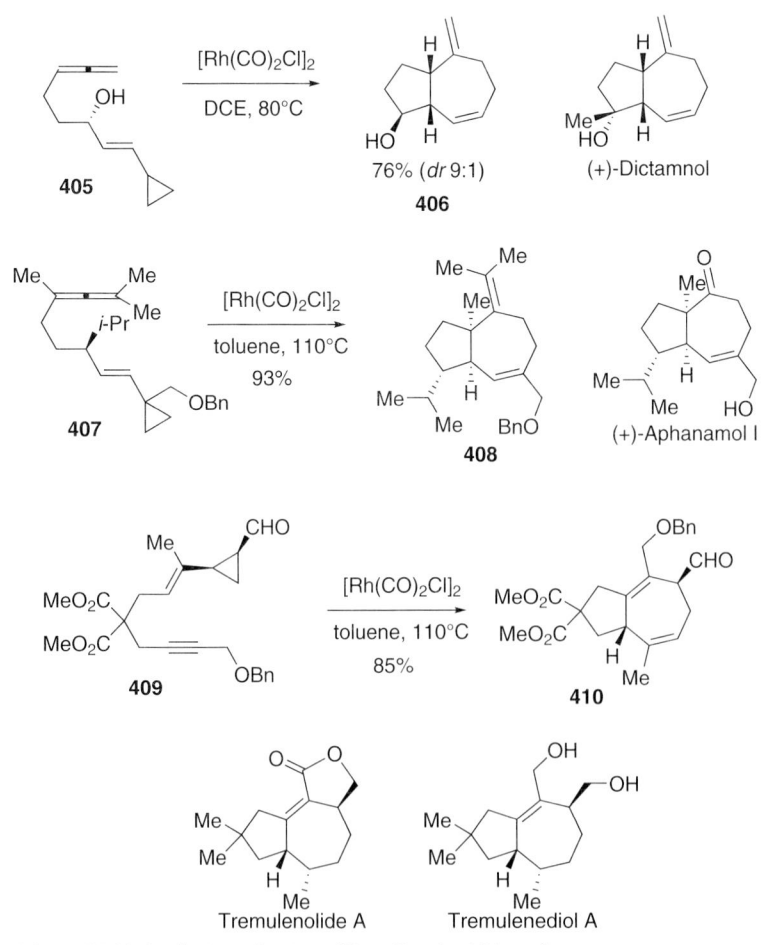

**Scheme 15.90** Synthetic application of [5 + 2] cycloaddition of alkyne- or allene-VCPs to asymmetric total synthesis of natural products.

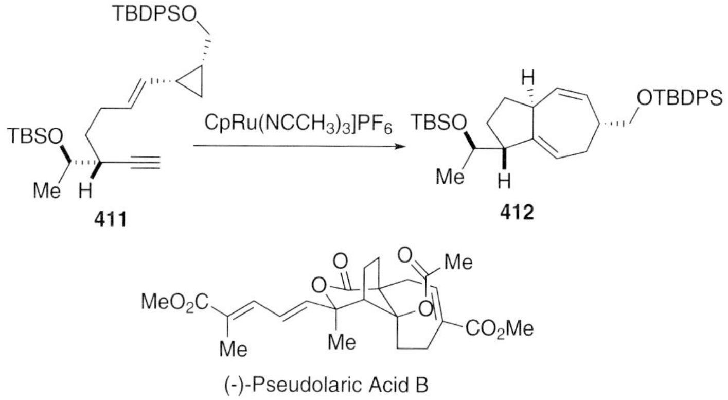

(-)-Pseudolaric Acid B

**Scheme 15.91** Synthetic application in total synthesis of (−)-Pseudolaric Acid B.

### 15.3.10.2 Intermolecular [5 + 2] Cycloaddition of VCPs with Alkynes and Allenes

Wender [115] demonstrated the first intermolecular version of [5 + 2] cycloaddition by employment of siloxy or alkoxy vinylcyclopropanes **413** as the three-carbon component and alkynes as the corresponding two-carbon component. Various functional groups on the alkyne are tolerated and the cycloadducts **414** were given in good to excellent yields (Scheme 15.92).

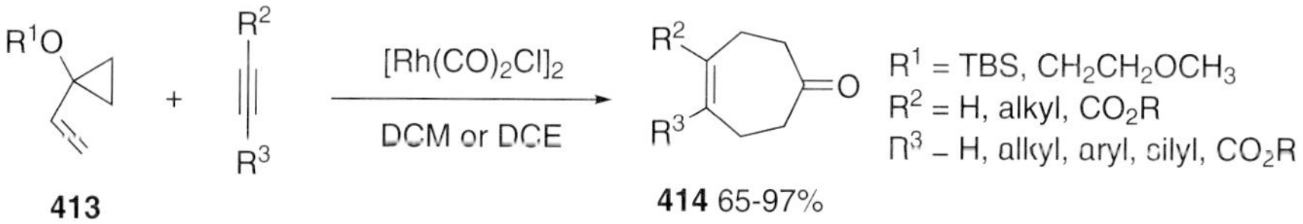

**Scheme 15.92** Rh(I)-catalyzed intermolecular [5 + 2] cycloaddition of siloxy- or alkoxy-vinylcyclopropanes with alkynes.

The proposed mechanistic rationale is shown in Scheme 15.93. Similar to the intramolecular case, there are two possible reaction pathways. The first is oxidative addition of the vinylcyclopropane ring producing the interconvertible rhodacyclohexene **415** and π-allyl rhodium complex **416**, followed by alkyne insertion. The reductive elimination of the resulting rhodacyclooctene **417** gives the final product and regenerates the Rh species. The other is initiated with cyclometallation to form rhodacyclopentene **418**, followed by β-carbon elimination to give the same key intermediate **417**.

Activated allenes **419** with a conjugated alkynyl, alkenyl or cyano group can be employed as the two-carbon component instead of alkynes and, reacting with vinylcyclopropanes, provide alkylidenecycloheptanones **420** as mixtures of two E/Z isomers. It is noteworthy that nonactivated allenes do not undergo this kind of intermolecular [5 + 2] cycloaddition (Scheme 15.94) [116].

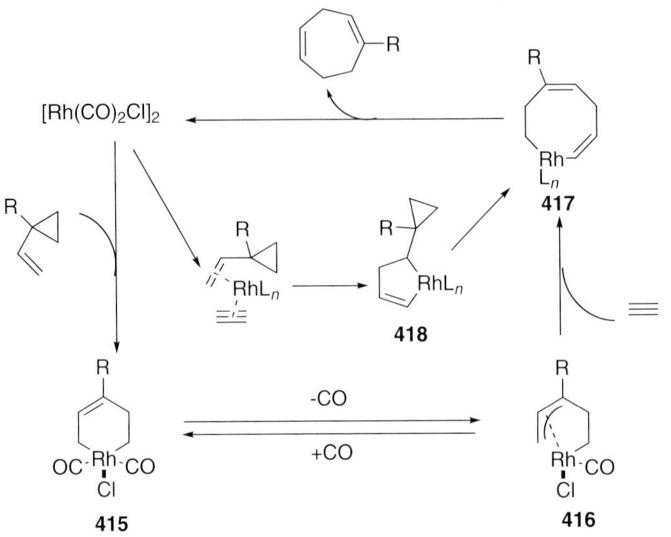

**Scheme 15.93** Plausible reaction pathway of Rh(I)-catalyzed intermolecular [5 + 2] cycloaddition.

**Scheme 15.94** Rh(I)-catalyzed intermolecular [5 + 2] cycloaddition of alkoxy-vinylcyclopropanes with activated allenes.

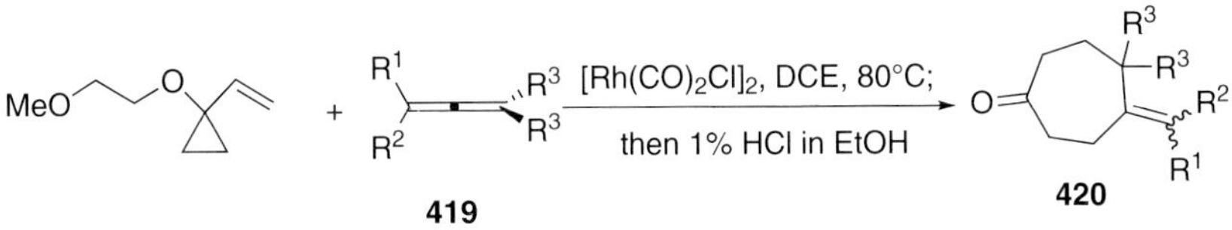

$R^1$ = ≡—R, CN
$R^2$ = H, n-Bu, $CH_2CO_2Et$
$R^3$ = Me, -$(CH_2)_5$-

22-99% (dr from 2.5:1 to 1:2.5)

The efficiency of cycloaddition reactions between simple alkyl-substituted VCPs **421** and terminal alkynes is strongly dependent on the $R^1$ group at C-1. A large and bulky substituent $R^1$ at C-1 and incorporation of $R^2$ and $R^3$ at the vinyl part make the cycloaddition reaction proceed smoothly. A dramatic solvent effect was observed in this cycloaddition [117]. Higher yields and a faster reaction rate were realized by employment of 2, 2, 2-trifluoroethanol (TFE) as co-solvent, which was rationalized through stabilization of the reaction intermediates described above (Scheme 15.95)[117].

A cascade [5 + 2]/[4 + 2] cycloaddition of vinylcyclopropanes, enynes and activated dienophiles was recently disclosed by Wender [118], providing fused bicyclic or tricyclic rings **423** in good to excellent yields. It was found that the addition of all three components at once could prevent decomposition of the resulting dienes from cycloaddition of VCPs and enynes and thus improve the yields (Scheme 15.96).

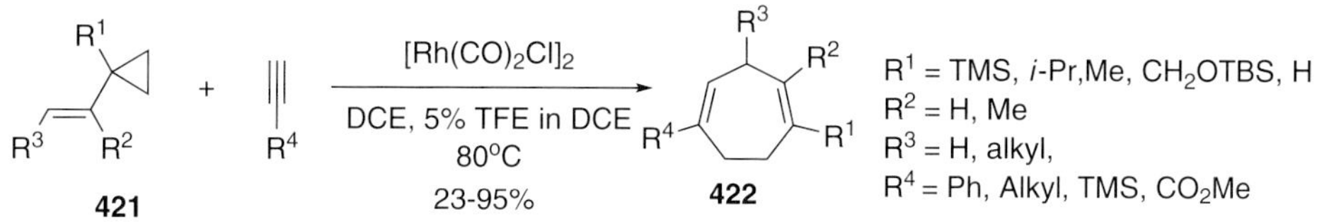

**Scheme 15.95** Rh(I)-catalyzed [5 + 2] cycloaddition of simple alkyl-substituted VCPs with terminal alkynes.

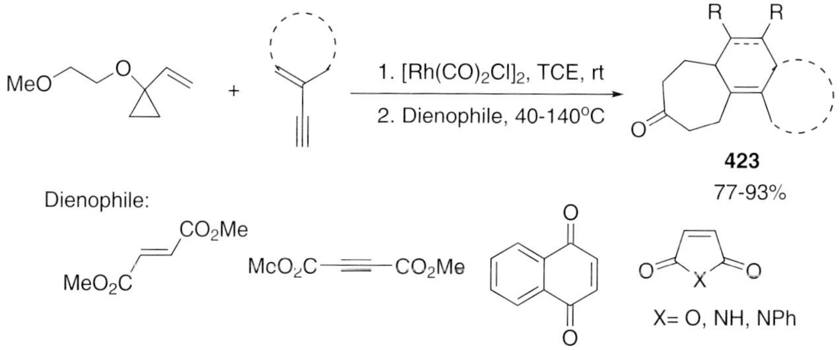

**Scheme 15.96** Cascade [5 + 2]/[4 + 2] cycloaddition of vinylcyclopropanes, enynes and activated dienophiles.

Wender further demonstrated that a hetero-[5 + 2] cycloaddition of VCP analogue cyclopropyl imines **424** with dimethyl acetylenedicarboxylate affords dihydroazepines **425** in 68–95% yields in the presence of a Rh complex. Since the imines could be formed *in situ* from the corresponding aldehydes **426** and amines, an efficient one-pot imination/aza-[5 + 2] cycloaddition strategy was then developed (Scheme 15.97) [119].

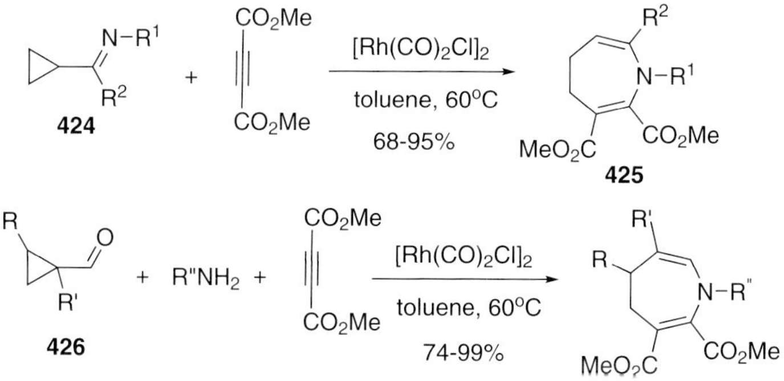

**Scheme 15.97** Rh(I)-catalyzed hetero-[5 + 2] cycloaddition of VCP analogue cyclopropyl imines with dimethyl acetylenedicarboxylate.

## 15.4
**Multicomponent Cyclization Reactions**

### 15.4.1
**[5 + 2 + 1] Cycloaddition of Alkene-VCPs**

Wender [120] developed the first example of transition-metal-catalyzed three-component [5 + 2 + 1] cycloaddition of alkoxyvinylcyclopropane, activated alkyne **427**, and CO. The reaction first gives the cyclooctadienone **431** which rapidly undergoes transannular condensation and hydrolysis to afford the bicyclo[3.3.0]octenone **428** in good to excellent yields and with excellent regioselectivity. The CO insertion of the resulting metallacyclooctadiene **429** from intermolecular [5 + 2] cycloaddition of VCP with alkyne, followed by CO insertion (forming intermediate **430**) and subsequent reductive elimination are believed to be the possible reaction pathway (Scheme 15.98).

Similarly to the case of [5 + 2] cycloaddition, activated alkynyl-allene **432** can be employed as substrate in this [5 + 2 + 1] cycloaddition reaction of alkoxyvinylcyclopropane and CO, providing a mixture of cyclooctadiones **433** and its transannular aldol product hydroxybicyclo[3.3.0]octanone **434** (ratio 1:2.2) in 88%. A lower pressure of CO (1 atm) could improve the selectivity (1: 2.2) but decrease the yield (62%) (Scheme 15.99) [116].

A Rh(I)-catalyzed [(5 + 2) + 1] cycloaddition of alkene-VCPs **435** with CO was recently developed by Yu [121]. Various substituted 5/8 or 6/8 fused bicyclic cyclooctenones **436** were produced in 29–92% yields with high diastereoselectivity. The E/Z geometry of the C=C bonds of the VCPs affects the cis/trans stereochemistry of the bicyclic products, that is, the *E*-isomer ene-VCP gives the trans isomer of the cycloadduct (Scheme 15.100).

Yu [122] has also developed a tandem [(5 + 2) + 1] cycloaddition/transannular aldol condensation reaction of alkene-VCPs **438** with CO to construct the fused tricyclic skeletons and their heteroatom-imbedded analogues **439**. Meanwhile, he has also successfully applied this method to the total synthesis of natural products. Both (±)-hirsutene and (±)-1-desoxhypnophilin have been synthesized from the same key synthon **441**, which is easily produced by this tandem reaction of alkene-VCP **440** and CO (Scheme 15.101).

### 15.4.2
**[5 + 1 + 2 + 1] Cycloaddition of Alkene-VCPs**

Wender [123] developed a Rh(I)-catalyzed [5 + 1 + 2 + 1] cycloaddition of alkoxy-VCPs, CO and terminal alkynes producing cyclononadienone **443** which, upon enolization and 6-π electron cyclization and elimination, affords the final product hydroxyindanones **444** (Scheme 15.102).

### 15.4.3
**[3 + 2 + 2] Cycloaddition of MCPs**

A Ni-catalyzed intermolecular [3 + 2 + 2] cyclization of bulky terminal alkynes with methylene-cyclopropanes with electron-withdrawing groups at the C=C double bond

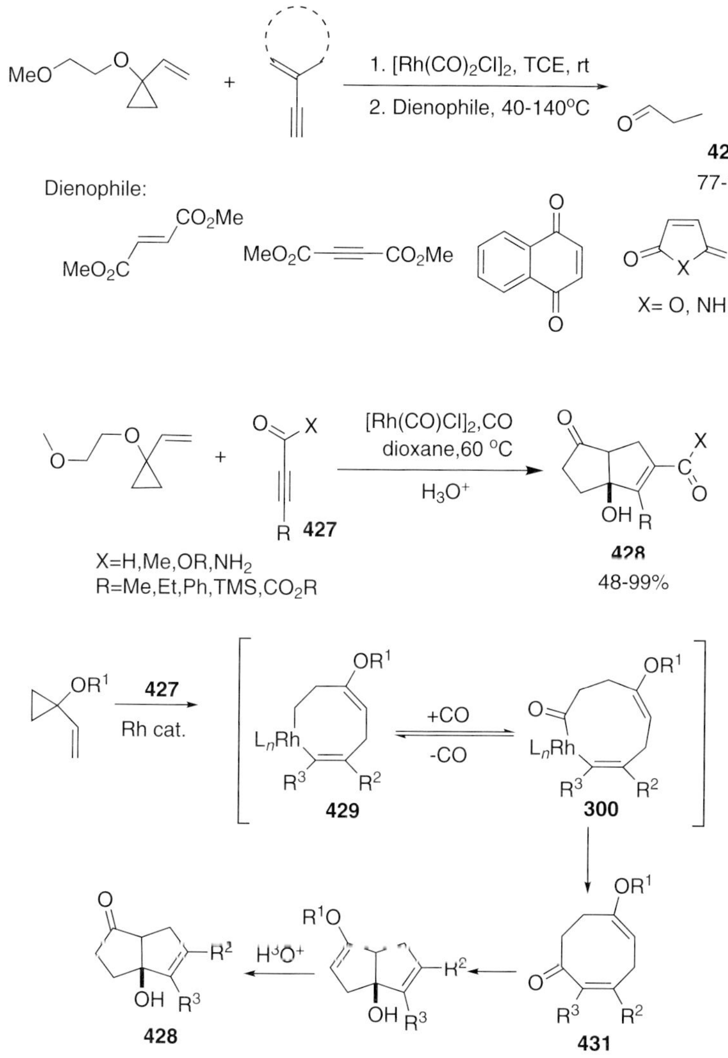

**Scheme 15.98** Cascade [5 + 2 + 1] cycloaddition/transannular condensation of alkoxyvinylcyclopropane, activated alkyne and CO.

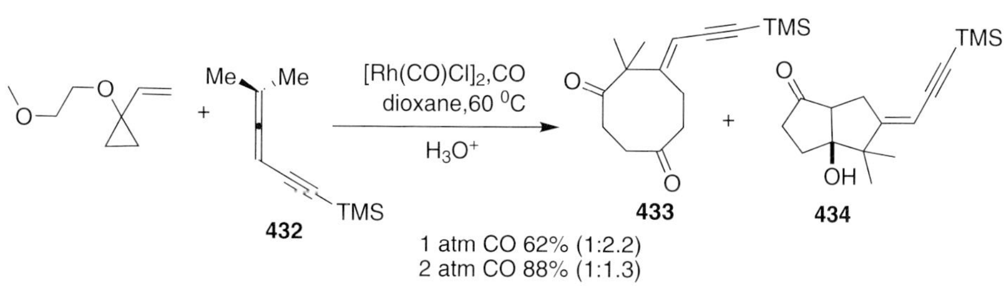

**Scheme 15.99** Cascade [5 + 2 + 1] cycloaddition/transannular condensation of alkoxyvinylcyclopropane, activated allenes and CO.

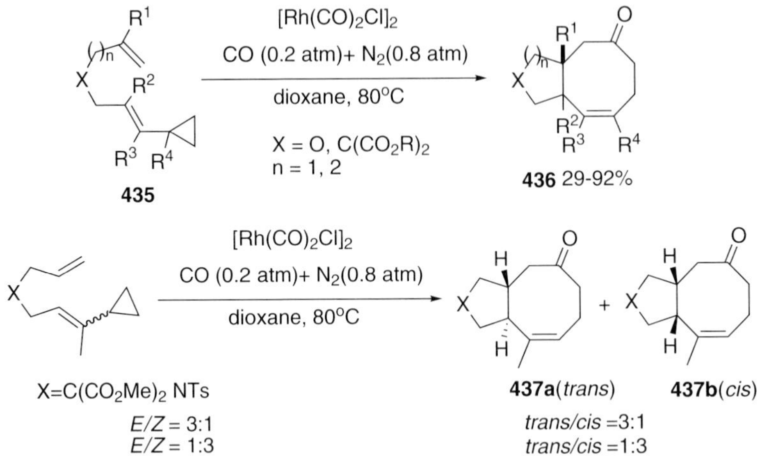

**Scheme 15.100** Rh(I)-catalyzed [(5 + 2) + 1] cycloaddition of alkene-VCPs with CO.

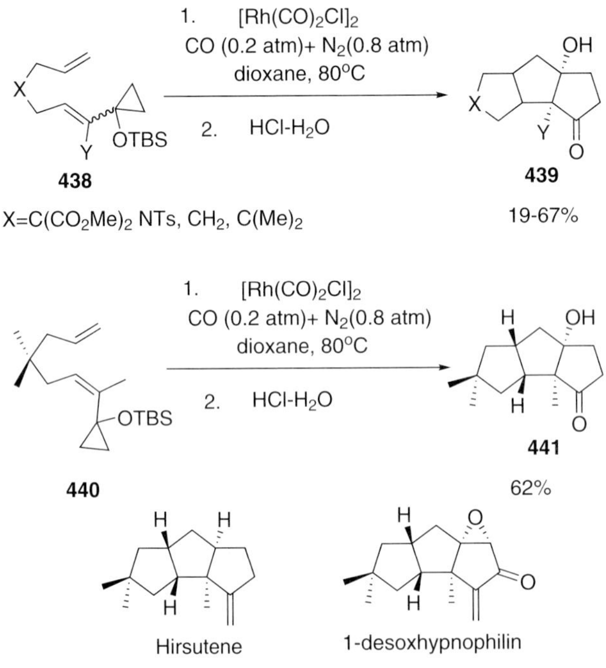

**Scheme 15.101** Tandem [(5 + 2) + 1] cycloaddition/transannular aldol condensation and synthetic application.

producing cycloheptadienes **445** was explored by Saito (Scheme 15.103) [124]. It is noteworthy that the presence of an electron-deficient substituent at the MCP was crucial for the success of this cyclization. The reaction may be initiated by formation of the nickelacyclopentadiene species **446** from two molecules of alkynes, which

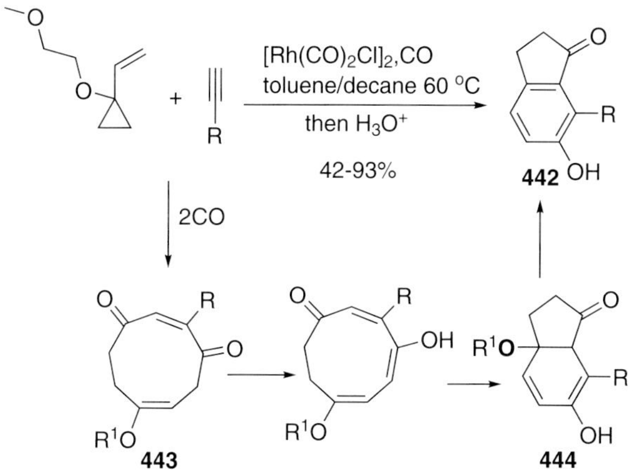

**Scheme 15.102** Rh(I)-catalyzed [5 + 1 + 2 + 1] cycloaddition of alkoxy-VCPs, CO and terminal alkynes.

could then undergo two possible pathways. One is alkene insertion of MCP into **446** and subsequent β-carbon elimination to give nickelamethylenecyclooctadiene **447**. The other way to form intermediate **447** is by direct insertion of the proximal bond of MCPs. The reductive elimination of intermediate **447** gives the final product.

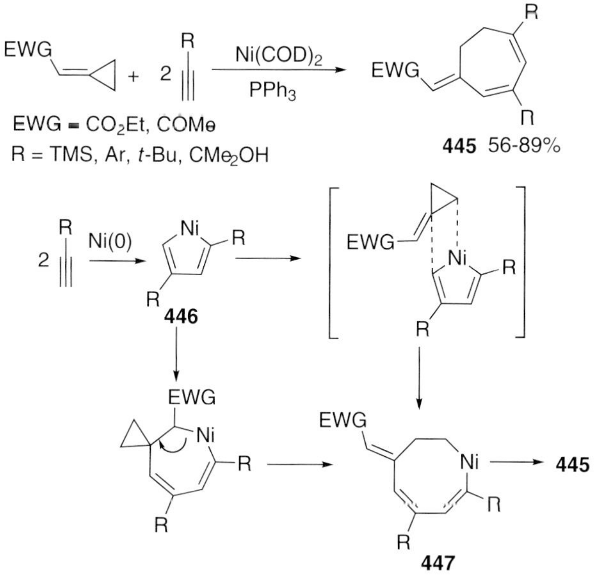

**Scheme 15.103** Ni-catalyzed intermolecular [3 + 2 + 2] cyclization.

## 15.4.4
## [(3 + 3) + 1] Cycloaddition

Chung [125] reported a Rh(I)-catalyzed carbonylative [3 + 3 + 1] cycloaddition of vinyl-biscyclopropanes **448** and **450** to give the corresponding seven-membered-ring products **449** and **451** in reasonable to high yields. The latter also gives a by-product **452** in 11–34% yields via rearrangement (Scheme 15.104). Oxidative addition of a rhodium complex into the C–C bond of the fused cyclopropane and subsequent ring expansion of the other cyclopropane produces intermediate **454** or **456** which, upon CO insertion and reductive elimination, gives the cycloadduct **449** or **451**, respectively. The by-product **452** was generated via β-H elimination from the intermediate **456** and reductive elimination of the resulting intermediate.

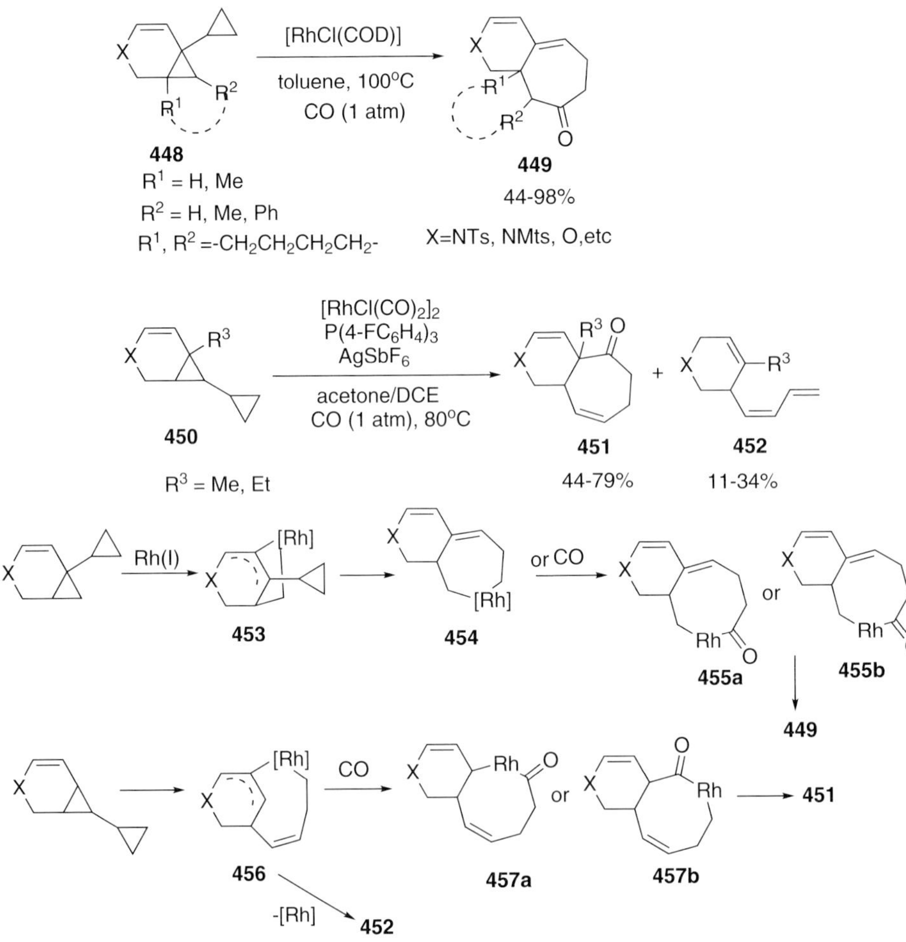

**Scheme 15.104** Rh(I)-catalyzed carbonylative [3 + 3 + 1] cycloaddition of vinyl biscyclopropanes **448** or **450**.

## 15.5
## Cyclization Involving Miscellaneous and Cascade Reactions

### 15.5.1
### Miscellaneous Reactions Involving [4 + 2] Cycloaddition

De Meijere reported a cascade Heck-type arylation/[4 + 2] cycloaddition of allenylcyclopropanes **458** with dienophiles under the catalysis of a Pd complex, providing vinylcyclohexenes **459a** and their diastereomer **459b** in a ratio ranging from 1.4 : 1 to 5 : 1 (Scheme 15.105) [126]. The mechanism of this reaction is initiated by the arylpalladation of the allenyl moiety in **458** to form a $\eta^1$-allyl palladium species **460** and/or a $\eta^1$-allyl palladium intermediate **461**. The latter undergoes β-carbon elimination to produce the intermediate **462** which, upon subsequent β-H elimination, gives the unstable trienes **463**. Diels–Alder cycloaddition of the trienes **463** with activated dienophiles produces a mixture of two diastereomers **459a** and **459b**.

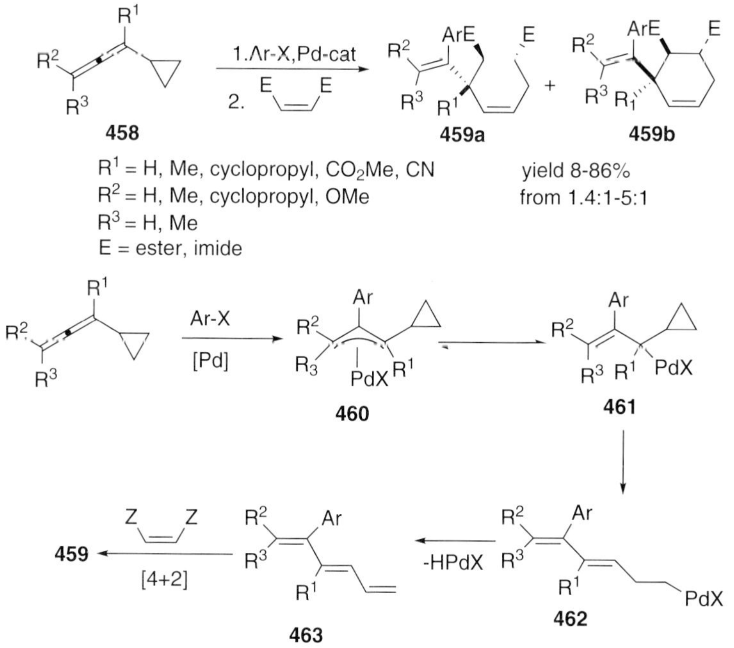

**Scheme 15.105** Cascade Heck-type arylation/[4 + 2] cycloaddition of allenylcyclopropanes **458** with dienophiles.

De Meijere [127] found that Heck-type reaction of cyclopropylidenecyclopropane **464** with iodobenzene could produce an unstable diene **465** under catalysis of a palladium complex, which readily undergoes [4 + 2] cycloaddition with α,β-unsaturated enoates to afford **466**. If vinyl iodide was used instead of iodobenzene, the unstable triene **467** would be afforded, which could be trapped by dimethyl maleate at room temperature to give the diene **468**. Compound **469** could be afforded in moderate yield via a cascade Heck-type reaction and double [4 + 2] cycloaddition (Scheme 15.106).

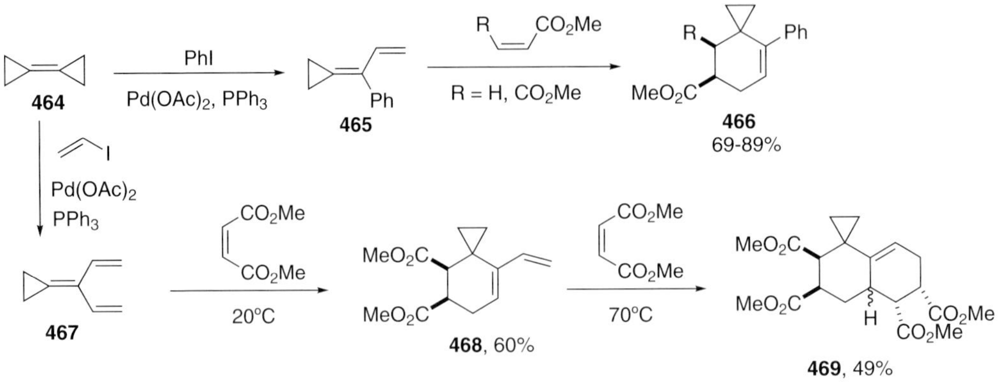

**Scheme 15.106** Cascade Heck-type arylation/[4 + 2] cycloaddition of bicyclic MCPs with dienophiles.

During the study of the [4 + 2] cycloaddition of 1-(1-alkynyl)cyclopropyl ketones **470**, Zhang found a different reaction outcome with ethyl enol ether, providing substituted bicyclo[3. 2. 0]heptane derivatives **471** in 48–98% yields, instead of the anticipated furans after treatment of the reaction mixture by TsOH (20 mol%) in acetone/$H_2O$ (7/1, 2 mM) (Scheme 15.107) [128]. The mechanism of this transformation is believed to proceed via initial formation of oxocarbenium intermediate **472** through 5-endo-dig cyclization of the carbonyl group onto the Au-activated C−C triple bond. Subsequent [4 + 2] cycloaddition of the oxocarbenium intermediate **472** with enol ether would form the bridged bicycle **473**. Ring rearrangement leads to formation of **475**, followed by elimination of Au catalyst to produce the bicyclic enone **476**. Treatment of **476** with Brønsted acid gives the cation **478**, which could be trapped by $H_2O$ to give **479**. Subsquent EtOH elimination affords the bicyclo[3. 2. 0]heptane derivatives **471**.

## 15.5.2
### Cyclization Involving a Cascade Reaction

Schmalz developed a gold-catalyzed cascade reaction of 1-(1-alkynyl)cyclopropyl ketones **480** with nucleophiles which provides efficient access to highly substituted furans **481** under mild conditions in up to 96% yield (Scheme 15.108) [129]. The reaction starts from the formation of a primary gold-chelated intermediate **482**, which could be attacked by the nucleophile in a regioselective homo-Michael-type addition to generate a gold-enolate intermediate **483**. Cycloisomerization of this Au-enolate affords the furanyl gold complex **484** which undergoes protonation to give the final product.

Kilburna [130] developed a $SmI_2$-mediated 6-exo cyclization of methylenecyclopropyl ketone **485** leading to the cycloheptane derivative **486** in excellent yield; the stereochemistry of the starting material favors suitable conformations for the cyclization (Scheme 15.109). Such cyclizations can be incorporated into cascade reaction sequences, providing routes to polycyclic products **488** with natural product-like structures.

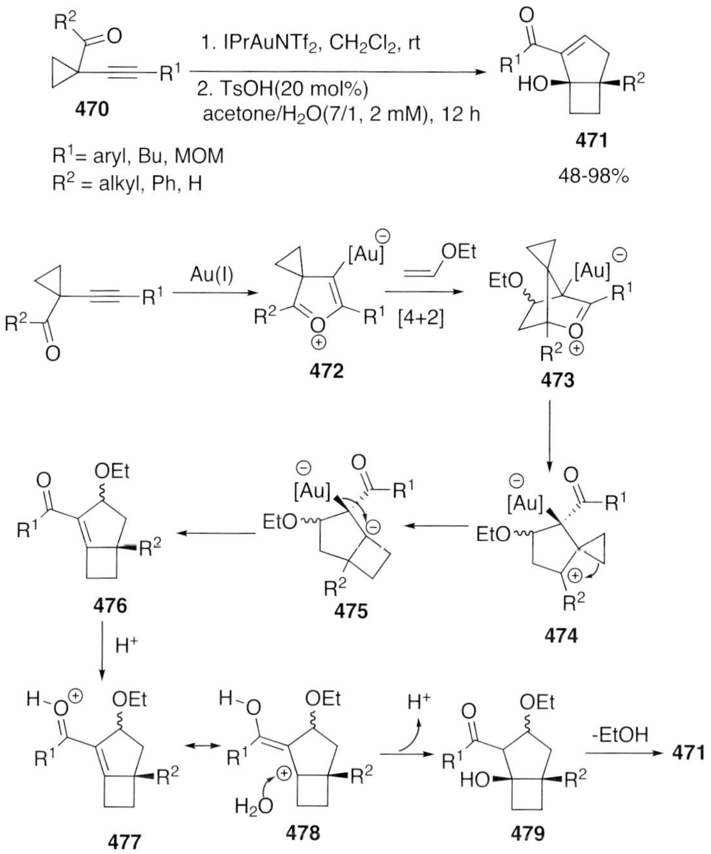

**Scheme 15.107** Au(I)-catalyzed cyclization of 1-(1-alkynyl) cyclopropyl ketones with ethyl vinyl ether.

An $SnCl_4$-mediated tandem ring-opening/double cyclization process of the doubly activated cyclopropanes **489** leading to the furoquinoline derivatives **490** was recently explored by Zhang and Liu [131]. This new strategy is very powerful due to its high chemo- and regioselectivity, high efficiency, operational simplicity, and the ready availability of a wide range of substrates from cheap starting materials. The reaction is believed to proceed via initial formation of the Sn-enolate **491** through nucleophilic attack-triggered ring opening of cyclopropane which, upon cyclization, produces 2, 3-dihydronfuran **492**. Subsequent intramolecular Friedel–Crafts reaction results in formation of the intermediate **493** followed by protonation and dehydration (Scheme 15.110).

Murakami [132] developed a rhodium catalyzed carbonylation reaction of spiropentanes **495** to cyclopentenones **496** in 56–84% yields, which involves two different types of carbon–carbon bond cleavage processes. Initial oxidative addition of the rhodium complex to the bond between C1 and C2 forms the spirocyclic rhodacy-

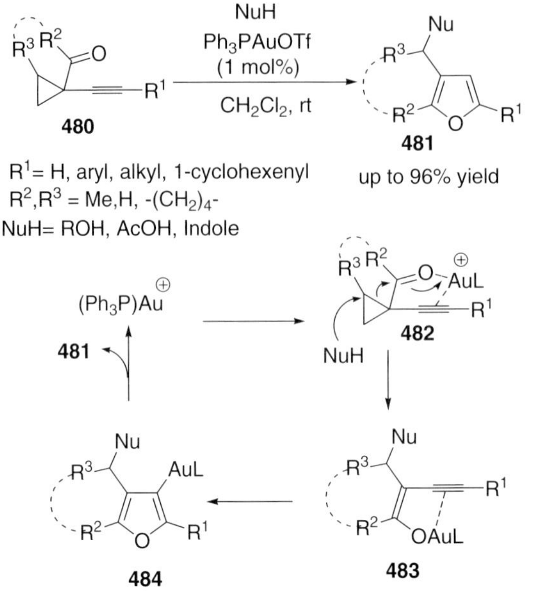

**Scheme 15.108** Au(I)-catalyzed cyclization 1-(1-alkynyl)cyclopropyl ketones with nucleophiles.

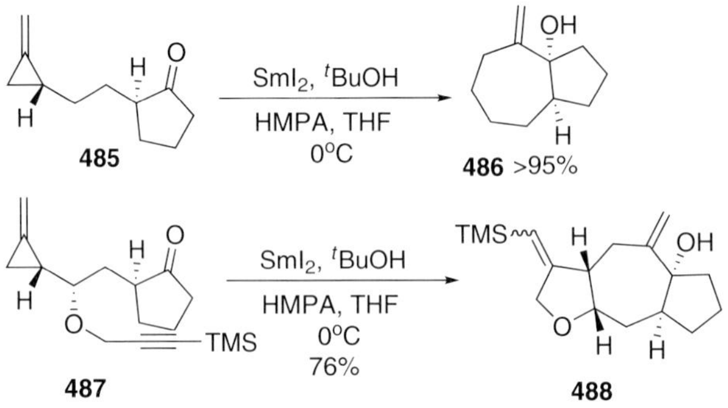

**Scheme 15.109** SmI$_2$-mediated 6-exo cyclization of methylenecyclopropyl ketone.

clobutane **497** which, upon CO insertion and β-carbon elimination, produces rhodacyclohexanone **499**. Subsequent reductive elimination furnishes 3-methylene-cyclopentanone **500**, which finally isomerizes into cyclopentenones **496** under the reaction conditions (Scheme 15.111).

A concise synthesis of multifunctionalized seven-membered carbocycles **503** has been achieved by application of the sequential Kulinkovich cyclopropanation of acetal-containing esters **501** and the Lewis acid mediated a cascade ring opening/cyclization of the resulting cyclopropyl silyl ethers **502** (Scheme 15.112) [133].

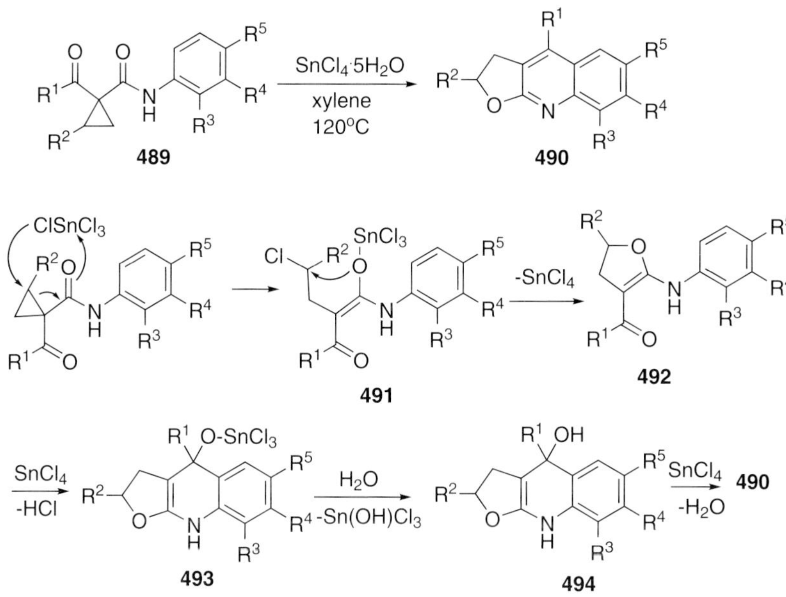

**Scheme 15.110** SnCl$_4$-mediated tandem ring-opening/double cyclization process of the doubly activated cyclopropanes.

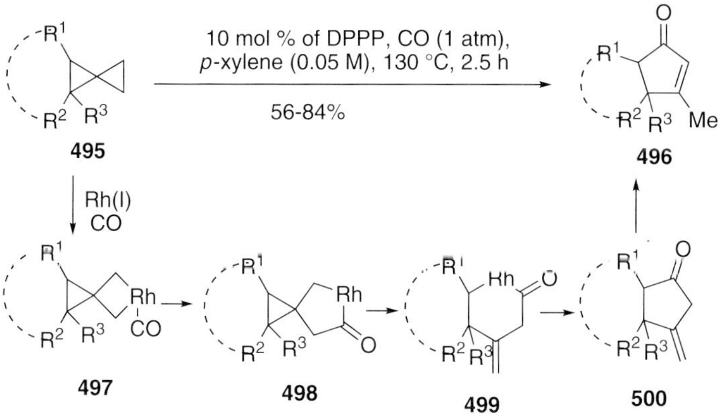

**Scheme 15.111** Rhodium-catalyzed carbonylation reaction of spiropentanes.

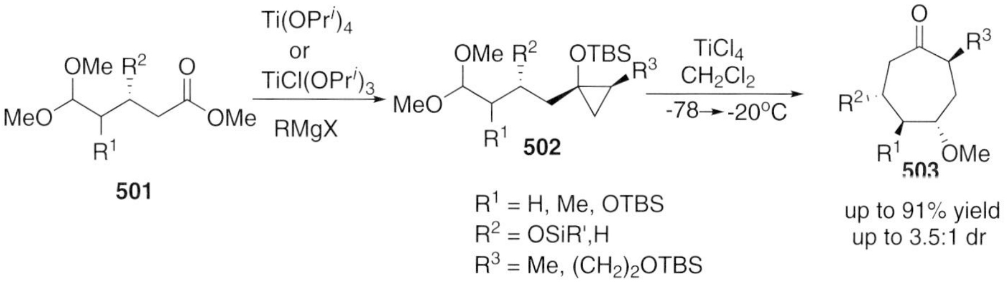

**Scheme 15.112** Concise synthesis of multifunctionalized seven-membered carbocycles.

## 15.6
## Experimental: Selected Procedures

### 15.6.1
### Synthesis of Compound 16 via Vinylcyclopropane–Cyclopentene Rearrangement (Scheme 15.4) [7a]

An oven-dried two-neck round-bottomed flask equipped with a magnetic stirring bar and gas-line adapter was evacuated and filled with $N_2$. A (dry and degassed) pentane solution of diyne was added via syringe. To the stirring solution, a solution of Ni(COD)$_2$ and IPr (in dry and degassed pentane) was added. The dark greenish-black reaction mixture was stirred for 10 h until complete consumption of starting material was observed by GC. The contents of the reaction vessel were then concentrated and purified by silica gel column chromatography.

### 15.6.2
### Regioselective Cycloisomerization of Cyclopropenyl Ketone 171 (Scheme 15.35) [37]

#### 15.6.2.1 Isomerization under Catalysis of CuI
To a solution of 1-acetyl-2-butylcycloprop-2-enecarboxylic acid ethyl ester **171a** (105 mg, 0.5 mmol) (105 mg, 0.5 mmol) in 2 mL of $CH_3CN$ was added CuI (4.7 mg, 5 mol%) and the reaction mixture was then stirred at 80 °C till complete consumption of **171a** (9 h). After evaporation, the residue was purified by column chromatography on silica gel (petroleum ether/diethyl ether 10 : 1) to afford 2-methyl-4-butylfuran-3-carboxylic acid ethyl ester **175a** (96 mg, 91%) as a liquid.

#### 15.6.2.2 Isomerization under Catalysis of PdCl$_2$(CH$_3$CN)$_2$
To a solution of 1-acetyl-2-butylcycloprop-2-enecarboxylic acid ethyl ester **171a** (105 mg, 0.5 mmol) in 2 mL of $CHCl_3$ was added PdCl$_2$(CH$_3$CN)$_2$ (6.5 mg, 5 mol %) and the reaction mixture was then stirred at reflux till complete consumption of **171a** (18 h). After evaporation, the residue was purified by column chromatography on silica gel (petroleum ether/diethyl ether 10 : 1) to afford 2-methyl-5-butylfuran-3-carboxylic acid ethyl ester **179a** (96 mg, 91%) as a liquid.

### 15.6.3
### Synthesis of Compound 194 via [3 + 2] Cycloaddition Reaction of Alkylidenecycloproapnes with Aldehyde (Scheme 15.40) [[51]a]

Aldehyde (1.5 mmol) and alkylidenecycloproapne (0.5 mmol) were added to a mixture of [Pd(PPh$_3$)$_4$] (11.5 mg, 0.01 mmol) and tributylphosphane oxide (4.4 mg, 0.02 mmol) under argon in a pressure vial. The reaction mixture was heated at 120 °C for 5 h, and then filtered through a silica-gel column using ethyl acetate as an eluent. After concentration, the residue was purified by column chromatography (silica gel) to afford compound **194**.

### 15.6.4
**Stereoselective Synthesis of 2, 5-*cis* or 2, 5-*trans*-pyrrolo-isoxazolidines 272 (Scheme 15.56) [67]**

#### 15.6.4.1 Synthesis of 2, 5-*cis*-pyrrolo-isoxazolidines *cis*-272
To a solution of alkoxylamine-tethered cyclopropane **273** (1.0 equiv) in $CH_2Cl_2$ (0.2 M) was added $Yb(OTf)_3$ (5 mol%). After stirring for 30 min at room temperature, the aldehyde (1.1 equiv.) was then added and the reaction mixture stirred until complete consumption of **273**. After concentration, the residue was purified by column chromatography (silica gel) to afford compound *cis*-**272**.

#### 15.6.4.2 Synthesis of 2, 5-*trans*-pyrrolo-isoxazolidines *trans*-272
To a solution of oxime ether tethered cyclopropane dicarboxylate *E*-**271** (1.0 equiv.) in dry $CH_2Cl_2$ (0.1 M) was added $Yb(OTf)_3 \cdot xH_2O$ (5 mol%). The mixture was stirred overnight at room temperature and monitored by TLC for completion. The reaction mixture was diluted with $CH_2Cl_2$, washed successively with $H_2O(1\times)$, brine ($1\times$), dried over $MgSO_4$ and concentrated *in vacuo*. The crude product was pre-absorbed onto silica and purified by flash column chromatography (elution EtOAc/hexanes).

### 15.6.5
**Synthesis of Optically Active Tetrahydrooxazines 322 (Scheme 15.69) [82]**

A mixture of $Ni(ClO_4)_2 \cdot 6H_2O$ (0.040 mmol) and the trisoxazoline **321** (0.044 mmol) in dimethoxyethane (1 mL) was stirred at 50 °C for 2 h under nitrogen. The mixture was then cooled to room temperature and added to the cyclopropane diester **256** (0.44 mmol) with a syringe. Activated 4 Å molecular sieves (100 mg) were added to the resulting solution. The mixture was stirred at $-30$ °C for 30 min, and then the nitrone (0.20 mmol) was added. After the reaction was complete (monitored by TLC), the mixture was passed rapidly through a glass funnel with a thin layer (20 mm) of silica gel (300–400 mesh), which was then washed with $CH_2Cl_2$ (50 mL). The filtrate was concentrated under reduced pressure, and the residue was purified by flash chromatography to afford the product.

### 15.6.6
**Synthesis of 5, 7-fused Bicycles 326 via Intramolecular [3 + 4] Cycloaddition (Scheme 15.71) [84]**

Triene **325** (91 mg, 0.25 mmol) was added to a suspension of $Pd_2(dba)_3$ (13.7 mg, 6 mol%) and phosphoramidite $L_2$ (18.9 mg, 24 mol%) in dry, freshly distilled and carefully deoxygenated dioxane (4 mL, 50 mM). The resulting mixture was introduced into a preheated bath and refluxed for 3 h. After cooling to room temperature, the mixture was diluted with hexanes (8 mL) and filtered through a short pad of silica gel eluting with 10% EtOAc/hexanes. The filtrate was concentrated and purified by flash chromatography (3% $Et_2O$/hexanes) to give compound **326**.

### 15.6.7
### [4 + 2] Cycloaddition of 1-(1-Alkynyl)cyclopropyl Ketones 346 with Aldehyde (Scheme 15.75) [88]

To a solution of a cyclopropyl ketone (0.1 mmol, 1 equiv) and a ketone or an aldehyde (0.2 mmol, 2 equiv) in anhydrous $ClCH_2CH_2Cl$ (2 mL) was added a solution of $IPrAuNTf_2$ in $ClCH_2CH_2Cl$ (0.05 M, 0.1 mL, 5 mol%). The reaction mixture was stirred at 50 °C until the cyclopropyl ketone was completely consumed (monitored by TLC). $Et_3N$ (0.1 mL) was added to quench the reaction, and the mixture was then concentrated. The resulting residue was purified through silica gel flash column chromatography (eluents: EtOAc and hexanes) to yield the desired dihydropyran-fused furan.

### 15.6.8
### Asymmetric Rh(I)-Catalyzed Intramolecular [5 + 2] Cycloaddition of Alkene-VCPs 392 (Scheme 15.86) [106]

Alkene-vinylcyclopropane **392b** (17.6 mg, 66.1 μmol), $[((R)\text{-BINAP})Rh(MeOH)_2]^+$ $SbF_6^-$ (6.6 μmol, 6.8 mg), and 1,2-dichloroethane (1.32 mL to make [VCP] = 0.05 M) were added to an oven-dried 10-mL vial. The reaction mixture was heated at 70 °C for 25 h. The crude product mixture was concentrated by rotary evaporation, and the crude product was purified by flash column chromatography eluting with 10% $Et_2O$/pentane to yield 12.7 mg (72%, >95% ee) of product as a colorless oil.

### 15.6.9
### [(5 + 2) + 1] Cycloaddition of Alkene-VCPs 435 (Scheme 15.100) [121]

$[Rh(CO)_2Cl]_2$ (5 mol% to the substrate) was charged in a base-washed oven-dried Schlenk flask under an atmosphere of nitrogen, and then a solution of the ene-VCP substrate in degassed dioxane (0.05 M) was added. The solution was bubbled with the mixed CO (0.2 atm CO + 0.8 atm $N_2$) for 5 min. The reaction mixture was stirred at 80 °C under the mixed CO atmosphere until TLC indicated completion of the reaction. After being cooled to room temperature, the mixture was concentrated and the residue was purified by flash column chromatography with silica gel to afford the cycloaddition products.

### Abbreviations

| | |
|---|---|
| Ac | acetyl |
| Bn | benzyl |
| Boc | *tert*-butyloxycarbonyl |
| Cy | cyclohexyl |
| DBU | 1, 5-diazabicyclo[5.4.0]undec-5-ene |
| DCE | dichloroethane |

| | |
|---|---|
| DCM | dichloromethane |
| EWG | electron-withdrawing group |
| MCP | methylenecyclopropane |
| MCR | multicomponent reaction |
| Mes | mesityl |
| MOM | methoxymethyl |
| Piv | pivaloyl |
| PMB | *p*-methoxybenzyl |
| RCM | ring-closing metathesis |
| rt | room temperature |
| TBAF | tetrabutylammonium fluoride |
| TBDPS | *tert*-butyldiphenylsilyl |
| TBS | *tert*-butyldimethylsilyl |
| TES | triethylsilyl |
| Tf | triflyl, trifluoromethylsulfonyl |
| TFE | 2, 2, 2-trifluoroethanol |
| THF | tetrahydrofuran |
| TMEDA | $N,N, N',N'$-tetramethylethylenediamine |
| TMS | trimethylsilyl |
| VCP | vinylcyclopropane |

## Acknowledgments

Financial support from the National Natural Science Foundation of China (20 702 015) and Shanghai Municipal Committee of Sciences and Technology (07pj14039) is greatly appreciated. This work was also sponsored by the Shanghai Shuguang Program (07SG27) and the Shanghai Leading Academic Discipline Project (B409). We also want to thank Mr. Hongyin Gao, Lu Liu, Feng liu and Wanxiang zhao for their structure drawing.

## References

1. (a) Walsh, R. (2005) *Chem. Soc. Rev.*, **34**, 714–732; (b) Yu, M. and Pagenkopf, B.L. (2005) *Tetrahedron*, **61**, 321–329; (c) Baird, M.S. (2003) *Chem. Rev.*, **103**, 1271–1275; (d) Lebel, H., Marcoux, J.-F., Molinaro, C. and Charette, A.B. (2003) *Chem. Rev.*, **103**, 977–1050; (e) Sekiguchi, A. and Lee, V.Y. (2003) *Chem. Rev.*, **103**, 1429–1447.
2. (a) Brandi, A. and Goti, A. (1998) *Chem. Rev.*, **98**, 589–635; (b) Ohkita, M., Nishida, S. and Tsuji, T. (1995) *Chemistry of the Cyclopropyl Group*, (Ed. Rapport, Z.) **2**, 261–318; (c) Chen, K.-C. and Lee, G.-A. (2006) *Huaxue*, **64**, 73; (d) Rubin, M., Rubina, M. and Gevorgyan, V. (2006) *Synthesis*, 1221–1226.
3. (a) Wang, S.C. and Tantillo, D.J. (2006) *J. Organomet. Chem.*, **691**, 4386–4392; (b) Hudlicky, T., Kutchan, T.M. and Naqvi, S.M. (1985) in *Organic Reactions*, vol. 33 (ed. A.S. Kende), (series book), 247–335.
4. Baldwin, J.E. (2003) *Chem. Rev.*, **103**, 1197–1212.

5 (a) Hudlicky, T., Koszyk, F.J., Kutchan, T.M. and Sheth, J.P. (1980) *J. Org.Chem.*, **45**, 5020–5027; (b) Hudlicky, T., Kutchan, T.M., Wilson, S.R. and Mao, D.T. (1980) *J. Am.Chem. Soc.*, **102**, 6351–6353.

6 Suginome, M., Matsuda, T., Yoshimoto, T. and Ito, Y. (2002) *Organometallics*, **21**, 1537–1539.

7 (a) Zuo, G. and Louie, J. (2004) *Angew. Chem., Int. Ed.*, **43**, 2277–2279; (b) Miura, T., Sasaki, T., Harumashi, T. and Murakami, M. (2006) *J. Am. Chem. Soc.*, **128**, 2516–2517.

8 (a) Hiroi, K. and Arinaga, Y. (1994) *Tetrahedron Lett.*, **35**, 153–156; (b) Hiroi, K., Yoshida, Y. and Kaneko, Y. (1999) *Tetrahedron Lett.*, **40**, 3431–3434; (c) Hiroi, K., Suzuki, Y., Kaneko, Y., Kato, F., Abe, I. and Kawagishi, R. (2000) *Polyhedron*, **19**, 525–528.

9 (a) Corey, E.J. and Myers, A.G. (1984) *Tetrahedron Lett.*, **25**, 3559–3562; (b) Corey, E.J. and Kigoshi, H. (1991) *Tetrahedron Lett.*, **32**, 5025–5028.

10 Davies, H.M.L. and Hu, B. (1992) *J. Org. Chem.*, **57**, 3186–3190.

11 Davies, H.M.L., Ahmed, G., Calvo, R.L., Churchill, M.R. and Churchill, D.G. (1998) *J. Org. Chem.*, **63**, 2641–2645.

12 Satyanarayana, J., Rao, M.V.B., Ila, H. and Junjappa, H. (1996) *Tetrahedron Lett.*, **37**, 3565–3568.

13 Bowman, R.K. and Johnson, J.S. (2006) *Org. Lett.*, **8**, 573–576.

14 Honda, M., Naitou, T., Hoshino, H., Takagi, S., Segi, M. and Nakajima, T. (2005) *Tetrahedron Lett.*, **46**, 7345–7348.

15 Chang, R.K., DiPardo, R.M. and Kuduk, S.D. (2005) *Tetrahedron Lett.*, **46**, 8513–8516.

16 Trost, B.M. and Yasukata, T. (2001) *J. Am. Chem. Soc.*, **123**, 7162–7163.

17 Hayashi, M., Ohmatsu, T., Meng, Y.-P. and Saigo, K. (1998) *Angew. Chem., Int. Ed.*, **37**, 837–839.

18 Iwasawa, N., Matsuo, T., Iwamoto, M. and Ikeno, T. (1998) *J. Am. Chem. Soc.*, **120**, 3903–3914.

19 Markham, J.P., Staben, S.T. and Toste, F.D. (2005) *J. Am. Chem. Soc.*, **127**, 9708–9709.

20 (a) Piers, E. (1991) *Comprehensive Organic Synthesis*, vol. 5 (eds B.M. Trost and I. Fleming), Pergamon Press, Oxford, p. 971, Chapter 8.2; (b) Hudlicky, T., Fan, R., Reed, J.W. and Gadamasetti, K.G. (1992) *Org. React.*, **41**, 1–133.

21 Kubota, K., Isaka, M. and Nakamura, E. (1996) *Heterocycles*, **42**, 565–575.

22 Nasveschuk, C.G. and Rovis, T. (2005) *Angew. Chem. Int. Ed.*, **44**, 3264–3267.

23 Böttcher, G. and Reissig, H.-U. (2000) *Synlett*, 725–727.

24 Wang, X., Stankovich, S.Z. and Widenhoefer, R. (2002) *Organometallics*, **21**, 901–905.

25 Walczak, M.A.A. and Wipf, P. (2008) *J. Am. Chem. Soc.*, **130**, 6924–6925.

26 Fürstner, A. and Aissa, C. (2006) *J. Am. Chem. Soc.*, **128**, 6306–6307.

27 Shi, M., Liu, L.-P. and Tang, J. (2006) *J. Am. Chem. Soc.*, **128**, 7430–7431.

28 (a) Ma, S. and Zhang, J. (2003) *Angew. Chem., Int. Ed.*, **42**, 183–187; (b) Ma, S., Lu, L. and Zhang, J. (2004) *J. Am. Chem. Soc.*, **126**, 9645–9660.

29 Lu, L., Chen, G. and Ma, S. (2006) *Org. Lett.*, **8**, 835–838.

30 Aïssa, C. and Fvrstner, A. (2007) *J. Am. Chem. Soc.*, **129**, 14836–14837.

31 Aloise, A.D., Layton, M.E. and Shair, M.D. (2000) *J. Am. Chem. Soc.*, **122**, 12610–12611.

32 Xu, G.-C., Liu, L.-P., Lu, J.-M. and Shi, M. (2005) *J. Am. Chem. Soc.*, **127**, 14552–14553.

33 Tomilov, Y.V., Shapiro, E.A., Protopopova, M.N., Ioffe, A.I., Dolgii, I.E. and Nefedov, O.M. (1985) *Izv. Akad. Nauk SSSR Ser. Khim.*, 631.

34 (a) Müller, P., Pautex, N., Doyle, M.P. and Bagheri, V. (1990) *Helv. Chim. Acta*, **73**, 1233–1241; (b) Müller, C. and Gränicher, C., (1993) *Helv. Chim. Acta*, **76**, 521–534; (c) Müller, P. and Gränicher, C., (1995) *Helv. Chim. Acta*, **78**, 129–144.

35 Padwa, A., Kassir, J.M. and Xu, S.L. (1991) *J. Org. Chem.*, **56**, 6971–6972.
36 Ma, S. and Zhang, J. (2003) *J. Am. Chem. Soc.*, **125**, 12386–12387.
37 Chuprakov, S. and Gevorgyan, V. (2007) *Org. Lett.*, **9**, 4463–4466.
38 Ma, S., Zhang, J., Cai, Y. and Lu, L. (2003) *J. Am. Chem. Soc.*, **125**, 13954–13955.
39 Ma, S., Zhang, J., Lu, L., Jin, X., Cai, Y. and Hou, H. (2005) *Chem. Commun.*, 909–911.
40 For a recent review of [2 + 2 + 1] cycloaddition, see: Rubin, M., Rubina, M. and Gevorgyan, V. (2007) *Chem. Rev.*, **107**, 3117–3179.
41 Kurahashi, T. and de Meijere, A. (2005) *Angew. Chem., Int. Ed.*, **44**, 7881–7884.
42 Nakamura, I. and Yamamoto, Y. (2002) *Adv. Synth. Catal.*, **344**, 111–129.
43 (a) de Meijere, A., Nüske, H., El-Sayed, M., Labahn, T., Schroen, M. and Bräse, S. (1999) *Angew. Chem., Int. Ed.*, **38**, 3669–3672; (b) de Meijere, A. and Kozhushkov, S.I. (2000) *Eur. J. Org. Chem.*, 3809–3822; (c) Binger, P., Wedemann, P., Kozhushkov, S.I. and de Meijere, A. (1998) *Eur. J. Org. Chem.*, 113–119; (d) Kawasaki, T., Saito, S. and Yamamoto, Y. (2002) *J. Org. Chem.*, **67**, 4911–4915.
44 Yamago, S. and Nakamura, E. (1988) *J. Chem. Soc., Chem. Commun.*, 1112–1113.
45 (a) Inoue, Y., Hibi, T., Satake, M. and Hashimoto, H. (1979) *J. Chem. Soc., Chem. Commun.*, 982; (b) Binger, P. and Weintz, H.-J. (1984) *Chem. Ber.*, **117**, 654–665.
46 Weintz, H.-J. and Binger, P. (1985) *Tetrahedron Lett.*, **26**, 4075–4078.
47 Noyori, R., Odagi, T. and Takaya, H. (1970) *J. Am. Chem. Soc.*, **92**, 5780–5781.
48 (a) Binger, P., Doyle, M. and Benn, R. (1983) *Chem. Ber.*, **116**, 1–10; (b) Binger, P. and Schuchardt, U. (1981) *Chem. Ber.*, **114**, 3313–3324.
49 Trost, B.M. (1986) *Angew. Chem., Int. Ed.*, **25**, 1–20.
50 Corlay, H., Lewis, R.T., Motherwell, W.B. and Shipman, M. (1995) *Tetrahedron.*, **51**, 3303–3318.
51 (a) Nakamura, I., Oh, B.H., Saito, S. and Yamamoto, Y. (2001) *Angew. Chem., Int. Ed.*, **40**, 1298–1300; (b) Oh, B.-H., Nakamura, I., Saito, S. and Yamamoto, Y. (2001) *Tetrahedron Lett.*, **42**, 6203–6205.
52 Siriwardana, A.I., Nakamura, I. and Yamamoto, Y. (2004) *J. Org. Chem.*, **69**, 3202–3204.
53 (a) Lautens, M., Ren, Y. and Delanghe, P.H.M. (1994) *J. Am. Chem. Soc.*, **116**, 8821–8822; (b) Lautens, M. and Ren, Y. (1996) *J. Am. Chem. Soc.*, **118**, 9597–9605; (c) Delgado, A., Rodrigues, J.R., Castedo, L. and Mascareñas, J.L. (2003) *J. Am. Chem. Soc.*, **125**, 9282–9283.
54 López, F., Delgado, A., Rodrígues, J.R., Castedo, L. and Mascareñas, J.L. (2004) *J. Am. Chem. Soc.*, **126**, 10262–10263.
55 Gulías, M., García, R., Delgado, A., Castedo, L. and Mascareñas, J.L. (2006) *J. Am. Chem. Soc.*, **128**, 384–385.
56 Trillo, B., Gulías, M., López, F., Castedo, L. and Mascareñas, J.L. (2006) *Adv. Synth. Catal.*, **348**, 2381–2384.
57 (a) Lautens, M. and Han, W. (2002) *J. Am. Chem. Soc.*, **124**, 6312–6316; (b) Taillier, C., Bethuel, Y. and Lautens, M. (2007) *Tetrahedron*, **63**, 8469–8477; (c) Taillier, C. and Lautens, M. (2007) *Org. Lett.*, **9**, 591–593.
58 (a) Shi, M., Xu, B. and Huang, J.-W. (2004) *Org. Lett.*, **6**, 1175–1178; (b) Shi, M. and Xu, B. (2003) *Tetrahedron Lett.*, **44**, 3839–3842; (c) Liu, L.-P. and Shi, M. (2007) *Tetrahedron*, **63**, 4535–4542.
59 Lu, J.-M. and Shi, M. (2007) *Org. Lett.*, **9**, 91805–91808.
60 Lu, J.-M. and Shi, M. (2008) *J. Org. Chem.*, **73**, 2206–2210.
61 Li, W. and Shi, M. (2008) *J. Org. Chem.*, **73**, 4151–4154.
62 (a) Pohlhaus, P.D. and Johnson, J.S. (2005) *J. Org. Chem.*, **70**, 1057–1059; (b) Pohlhaus, P.D. and Johnson, J.S. (2005) *J. Am. Chem. Soc.*, **127**, 16014–16015.
63 Kang, Y.-B., Tang, Y. and Sun, X.-L. (2006) *Org. Biomol. Chem.*, **4**, 299–301.

64 Alper, P.B., Meyers, C., Lerchner, A., Siegel, D.R. and Carreira, E.M. (1999) *Angew. Chem., Int. Ed.*, **38**, 3186–3189.
65 Marti, C. and Carreira, E.M. (2005) *J. Am. Chem. Soc.*, **127**, 11505–11515.
66 Lerchner, A. and Carreira, E.M. (2006) *Chem. Eur. J.*, **12**, 8208–8219.
67 Jackson, S.K., Karadeolian, A., Driega, A.B. and Kerr, M.A. (2008) *J. Am. Chem. Soc.*, **130**, 4196–4201.
68 Korotkov, V.S., Larionov, O.V., Hofmeister, A., Magull, J. and de Meijer, A. (2007) *J. Org. Chem.*, **72**, 7504–7510.
69 Takasu, K., Nagao, S. and Iharaa, M. (2006) *Adv. Synth. Catal.*, **348**, 2376–2380.
70 Bajtos, B., Yu, M., Zhao, H. and Pagenkopf, B.L. (2007) *J. Am. Chem. Soc.*, **129**, 9631–9634.
71 Yadav, V.K. and Sriramurthy, V. (2004) *Angew. Chem., Int. Ed.*, **43**, 2669–2671.
72 Ming, Yu and Pagenkopf*, Brian L. (2003) *J. Am. Chem. Soc.*, **125**, 8122–8123.
73 (a) Kitagawa, O., Fujiwara, H. and Taguchi, T. (2001) *Tetrahedron Lett.*, **42**, 2165–2167; (b) Kitagawa, O., Miyaji, S., Sakuma, C. and Taguchi, T. (2004) *J. Org. Chem.*, **69**, 2607–2610.
74 Shimizu, I., Ohashi, Y. and Tsuji, J. (1985) *Tetrahedron Lett.*, **26**, 3825–3828.
75 Jiao, L., Ye, S. and Yu, Z.-X. (2008) *J. Am. Chem. Soc.*, **130**, 7178–7179.
76 Cardona, F. and Goti, A. (2005) *Angew. Chem., Int. Ed.*, **44**, 7832–7835.
77 (a) Young, I.S. and Kerr, M.A. (2003) *Angew. Chem., Int. Ed.*, **42**, 3023–3026; (b) Ganton, M.D. and Kerr, M.A. (2004) *J. Org. Chem.*, **69**, 8554–8557.
78 Carson, C.A. and Kerr, M.A. (2006) *Angew. Chem., Int. Ed.*, **45**, 6560–6563.
79 (a) Young, I.S., Williams, J.L. and Kerr, M.A. (2005) *Org. Lett.*, **7**, 953–955; (b) Young, I.S. and Kerr, M.A. (2007) *J. Am. Chem. Soc.*, **129**, 1465–1469.
80 Sapeta, K. and Kerr, M.A. (2007) *J. Org. Chem.*, **72**, 8597–8599.
81 Sibi, M.P., Ma, Z. and Jasperse, C.P. (2005) *J. Am. Chem. Soc.*, **127**, 5764–5765.
82 Kang, Y.-B., Sun, X.-L. and Tang, Y. (2007) *Angew. Chem., Int. Ed.*, **46**, 3918–3921.
83 Perreault, C., Goudreau, S.R., Zimmer, L.E. and Charette, A.B. (2008) *Org. Lett.*, **10**, 689–692.
84 Gulías, M., Durán, J., López, F., Castedo, L. and Mascareñas, J.L. (2007) *J. Am. Chem. Soc.*, **129**, 11026–11027.
85 Ivanova, O.A., Budynina, E.M., Grishin, Y.K., Trushkov, I.V. and Verteletskii, P.V. (2008) *Angew. Chem., Int. Ed.*, **47**, 1107–1110.
86 Kurahashi, T., Wu, Y.-T., Meindl, K., Rühl, S. and de Meijere, A. (2005) *Synlett*, 805–808.
87 Zhang, Y., Chen, Z., Xiao, Y. and Zhang, J. (2008), *Chem. Eur. J.*, **15**, 5208–5211.
88 Zhang, G., Huang, X., Li, G. and Zhang, L. (2008) *J. Am. Chem. Soc.*, **130**, 1814–1815.
89 de Meijere, A. and Kurahashi, T. (2005) *Synlett*, 2619–2622.
90 (a) Khusnutdinov, R.I. and Dzhemilev, U.M. (1994) *J. Organomet. Chem.*, **471**, 1–18; (b) Sarel, S. (1978) *Acc. Chem. Res.*, **11**, 204–211; (c) Schultze, M.M. and Gockel, U. (1996) *Tetrahedron Lett.*, **37**, 357–358; (d) Schultze, M.M. and Gockel, U. (1996) *J. Organomet. Chem.*, **525**, 155–158.
91 (a) Taber, D.F., Kanai, K., Jiang, Q. and Bui, G. (2000) *J. Am. Chem. Soc.*, **122**, 6807–6808; (b) Taber, D.F., Joshi, P.V. and Kanai, K. (2004) *J. Org. Chem.*, **69**, 2268–2271.
92 Taber, D.F., Bui, G. and Chen, B. (2001) *J. Org. Chem.*, **66**, 3423–3426.
93 Cho, S.H. and Liebeskind, L.S. (1987) *J. Org. Chem.*, **52**, 2631–2634.
94 (a) Semmelhack, M.F., Ho, S., Steigerwald, M. and Lee, M.C. (1987) *J. Am.Chem. Soc.*, **109**, 4397–4399; (b) Semmelhack, M.F., Ho, S., Cohen, D., Steigerwald, M., Lee, M.C., Lee, G., Gilbert, A.M., Wulff, W.D. and Ball, R.G. (1994) *J. Am. Chem. Soc.*, **116**, 7108–7122.
95 Wender, P.A., Takahashi, H. and Witulski, B. (1995) *J. Am. Chem. Soc.*, **117**, 4720–4721.
96 Wender, P.A. and Sperandio, D. (1998) *J. Org. Chem.*, **63**, 4164–4165.

97 Wender, P.A. and Williams, T.J. (2002) *Angew. Chem., Int. Ed.*, **41**, 4550–4553.
98 Gilbertson, S.R. and Hoge, G.S. (1998) *Tetrahedron Lett.*, **39**, 2075–2078.
99 Wang, B., Cao, P. and Zhang, X. (2000) *Tetrahedron Lett.*, **41**, 8041–8044.
100 (a) Trost, B.M., Toste, F.D. and Shen, H. (2000) *J. Am. Chem. Soc.*, **122**, 2379–2380; (b) Trost, B.M., Shen, H.C., Horne, D.B., Toste, F.D., Steinmetz, B.G. and Koradin, C. (2005) *Chem. Eur. J.*, **11**, 2577–2590.
101 Trost, B.M., Shen, H.C., Schulz, T., Koradin, C. and Schirok, H. (2003) *Org. Lett.*, **5**, 4149–4151.
102 Trost, B.M. and Shen, H.C. (2001) *Angew. Chem., Int. Ed.*, **40**, 2313–2316.
103 (a) Wender, P.A., Husfeld, C.O., Langkopf, E., Love, J.A. and Pleuss, N. (1998) *Tetrahedron*, **54**, 7203–7220; (b) Wender, P.A., Husfeld, C.O., Langkopf, E. and Love, J.A. (1998) *J. Am. Chem. Soc.*, **120**, 1940–1941.
104 Trost, B.M. and Toste, F.D. (2001) *Angew. Chem., Int. Ed.*, **40**, 1114–1116.
105 Wender, P.A., Love, J.A. and Williams, T.J. (2003) *Synlett*, 1295–1298.
106 Wender, P.A., Haustedt, L.O., Lim, J., Love, J.A., Williams, T.J. and Yoon, J.-Y. (2006) *J. Am. Chem. Soc.*, **128**, 6302–6303.
107 Wender, P.A., Glorius, F., Husfeld, C.O., Langkopf, E. and Love, J.A. (1999) *J. Am. Chem. Soc.*, **121**, 5348–5349.
108 Yu, Z.-X., Wender, P.A. and Houk, K.N. (2004) *J. Am. Chem. Soc.*, **126**, 9154–9155.
109 Yu, Z.-X., Cheong, P.H.-Y., Liu, P., Legault, C.Y., Wender, P.A. and Houk, K.N. (2008) *J. Am. Chem. Soc.*, **130**, 2378–2379.
110 (a) Wender, P.A. and Dyckman, A.J. (1999) *Org. Lett.*, **1**, 2089–2092; (b) Trost, B.M. and Shen, H.C. (2000) *Org. Lett.*, **2**, 2523–2525.
111 Wender, P.A., Fuji, M., Husfeld, C.O. and Love, J.A. (1999) *Org. Lett.*, **1**, 137–139.
112 Wender, P.A. and Zhang, L. (2000) *Org. Lett.*, **2**, 2323–2326.
113 Ashfeld, B.L. and Martin, S.F. (2005) *Org. Lett.*, **7**, 4535–4537.
114 Trost, B.M., Waser, J. and Meyer, A. (2007) *J. Am. Chem. Soc.*, **129**, 14556–14557.
115 (a) Wender, P.A., Rieck, H. and Fuji, M. (1998) *J. Am. Chem. Soc.*, **120**, 10976–10977; (b) Wender, P.A., Dyckman, A.J., Husfeld, C.O. and Scanio, M.J.C. (2000) *Org. Lett.*, **2**, 1609–1611.
116 Wegner, H.A., de Meijere, A. and Wender, P.A. (2005) *J. Am. Chem. Soc.*, **127**, 6530–6531.
117 Wender, P.A., Barzilay, C.M. and Dyckman, A.J. (2001) *J. Am. Chem. Soc.*, **123**, 179–180.
118 Wender, P.A., Gamber, G.G. and Scanio, M.J.C. (2001) *Angew. Chem., Int. Ed.*, **40**, 3895–3897.
119 Wender, P.A., Pedersen, T.M. and Scanio, M.J.C. (2002) *J. Am. Chem. Soc.*, **124**, 15154–15155.
120 Wender, P.A., Gamber, G.G., Hubbard, R.D. and Zhang, L. (2002) *J. Am.Chem. Soc.*, **124**, 2876–2877.
121 Wang, Y., Wang, J., Su, J., Huang, F., Jiao, L., Liang, Y., Yang, D., Zhang, S., Wender, P.A. and Yu, Z.-X. (2007) *J. Am. Chem. Soc.*, **129**, 10060–10061.
122 Jiao, L., Yuan, C. and Yu, Z.-X. (2008) *J. Am. Chem. Soc.*, **130**, 4421–4430.
123 Wender, P.A., Gamber, G.G., Hubbard, R.D., Pham, S.M. and Zhang, L. (2005) *J. Am. Chem. Soc.*, **127**, 2836–2837.
124 Saito, S., Masuda, M. and Komagawa, S. (2004) *J. Am. Chem. Soc.*, **126**, 10540–10541.
125 Kim, S.Y., Lee, S.I., Choi, S.Y. and Chung, Y.K. (2008) *Angew. Chem. Int. Ed.*, **47**, 4914–4917.
126 (a) de Meijere, A., Schelper, M., Knoke, M., Yucel, B., Sünneman, H.W., Scheurich, R.P. and Arve, L. (2003) *J. Organomet. Chem.*, **687**, 249–255; (b) Knoke, M. and de Meijere, A. (2003) *Synlett*, 195–198; (c) Knoke, M. and de Meijere, A. (2005) *Eur. J. Org. Chem.*, 2259–2268.
127 (a) Bräse, S. and de Meijere, A. (1995) *Angew. Chem., Int. Ed.*, **34**, 2545–2547; (b) Nüske, H., Bräse, S., Kozhushkov, S.I., Noltemeyer, M., ElSayed, M. and de

Meijere, A. (2002) *Chem. Eur. J.*, **8**, 2350–2369.

**128** Li, G., Huang, X. and Zhang, L. (2008) *J. Am. Chem. Soc.*, **130**, 6944–6945.

**129** Zhang, J. and Schmalz, H.-G. (2006) *Angew. Chem., Int. Ed.*, **45**, 6704–6707.

**130** Underwood, J.J., Hollingworth, G.J., Horton, P.N., Hursthousea, M.B. and Kilburna, J.D. (2004) *Tetrahedron Lett.*, **45**, 2223–2225.

**131** Zhang, Z., Zhang, Q., Sun, S., Xiong, T. and Liu, Q. (2007) *Angew. Chem., Int. Ed.*, **46**, 1726–1729.

**132** Matsuda, T., Tsuboi, T. and Murakami, M. (2007) *J. Am. Chem. Soc.*, **129**, 12596–12597.

# 16
# Transition Metal-Catalyzed Ring Expansion Cyclization Reactions

*Masahiro Yoshida and Yoshimitsu Nagao*

## 16.1
### Introduction

The ring expansion reaction is a valuable methodology for the synthesis of a variety of substituted ring compounds. Pinacol or semipinacol rearrangement is one of the most common and the historical process in which cyclic vicinal diols or 2-hetero-substituted cyclic alcohols are converted to the ring expanded products by acid-catalyzed 1,2-rearrangement (Scheme 16.1) [1]. As another methodology, ring expansion using a transition metal complex has been explored. In this reaction, a cyclic alcohol having a carbon–carbon multiple bond is activated by coordination to the transition metal, which causes a 1,2-rearrangement of the ring to give the ring expanded product. In the last 10 years, various kinds of reactions and their application have been intensively reported. This chapter focuses on a recent overview of the 1,2-migration reactions of cyclic compounds using transition metals, mainly palladium, as catalyst.

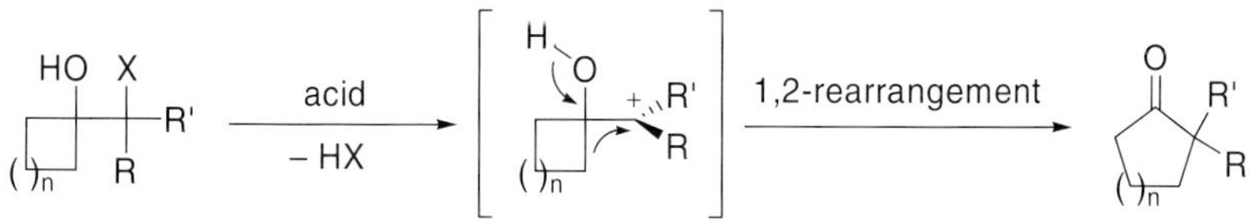

**Scheme 16.1** Acid-catalyzed ring expansion of 2-hetero atom-substituted cyclic alcohols.

## 16.2
### Palladium-Catalyzed Ring Expansions

#### 16.2.1
##### Reactivities of Alkenyl Substrates

The ring expansion reaction of vinylcyclobutanol derivatives **1** by palladium, which was developed by Clark in 1985, is a useful reaction for the construction of substituted

*Handbook of Cyclization Reactions. Volume 2.*
Edited by Shengming Ma
Copyright © 2010 WILEY-VCH Verlag GmbH & Co. KGaA, Weinheim
ISBN: 978-3-527-32088-2

five-membered ring systems [2]. The reaction is triggered by a release in the strain of the four-membered ring systems to afford the 2-methylidenecyclopentanones **4** via the intermediates **2** and **3** (Scheme 16.2) [3].

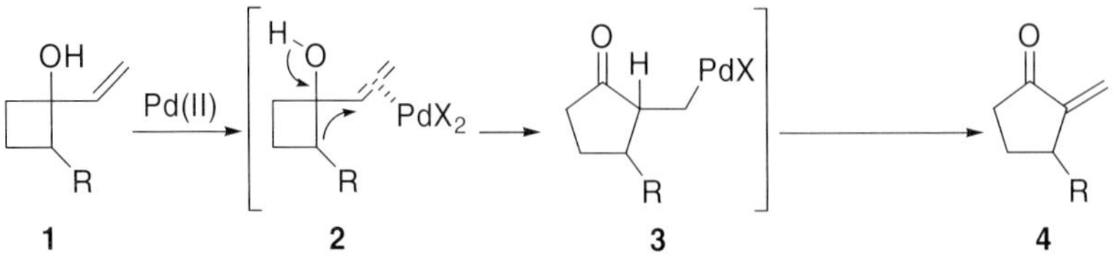

**Scheme 16.2** Palladium-promoted ring expansion of vinylcyclobutanols.

Application to the total synthesis of trichothecane-type sesquiterpenoids was reported (Scheme 16.3). Vinylcyclobutanol **6**, which was prepared from the cyclobutanone **5**, reacted with PdCl$_2$(MeCN)$_2$ to afford the ring expanded compounds **7**. Total synthesis of scirpene was achieved after several sequences of modifications from **7** [4].

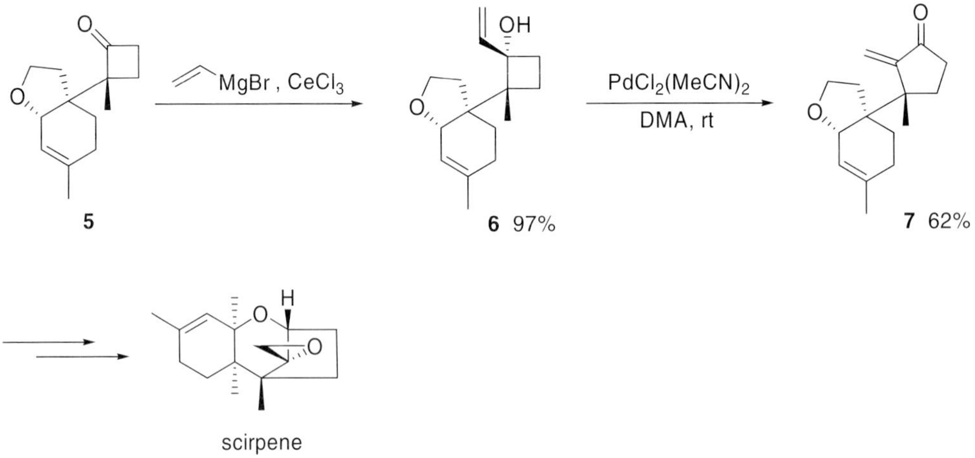

**Scheme 16.3** Application to the total synthesis of scirpene.

Similarly, the synthesis of 4-deoxyverrucarol has been accomplished by utilizing the ring expansion reaction from **9** to **10** as the key step (Scheme 16.4) [5]. In the reaction, the desired product **10** was obtained in 90% yield when Pd(OAc)$_2$ was used.

A sequential ring expansion–insertion process was accomplished by the reaction of isopropenyl-substituted cyclobutanol **11** (Scheme 16.5) [6, 7]. In this case, coordination of the isopropenyl group to a palladium(II) complex followed by the ring expansion of the cyclobutanol ring generated the intermediate **13**, which undergoes cyclization to construct the steroidal tetracyclic compound in a single step. The obtained products were *cis*- and *trans*-naphthohydrindans **14** and **16**, in

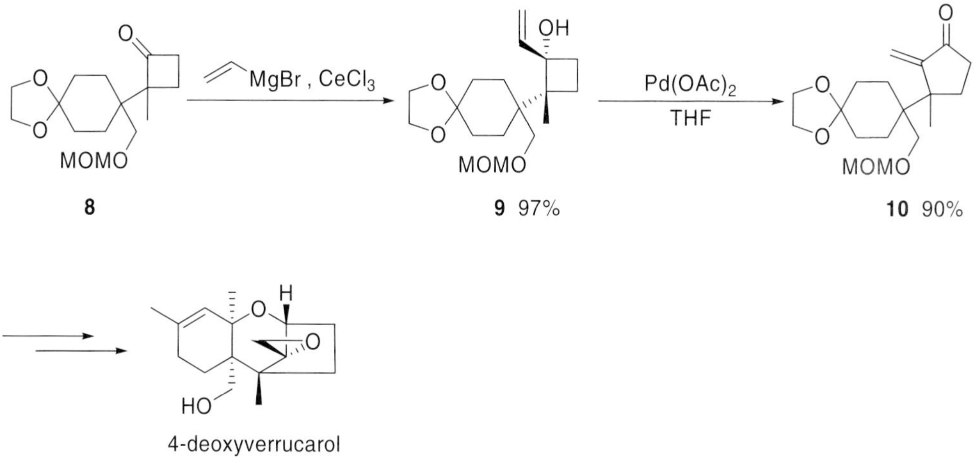

**Scheme 16.4** Application to the synthesis of 4-deoxyverrucarol.

accordance with their olefin-isomerized products **15** and **17**, and it was clear that the diastereoselectivity is controlled by the proper choice of solvent. The cis-fused products **14** and **15** were predominantly produced in dichloroethane, the trans-fused products **16** and **17** were yielded selectively when a mixture of THF and HMPA was used as the solvent. It is presumed that the stereoselectivity of the products depends on the conformation of the isopropenyl group during the 1,2-migration step. In a non-polar solvent such as dichloroethane, the ring expansion reaction would proceed via the intermediate **18** leading to the cis products, in which palladium would be associated with both the olefin and the hydroxy group. In contrast, in the case of a polar solvent such as HMPA, the reaction would take place via **19** to afford the trans products, in which palladium would be associated with only the olefin because the solvent coordinates with palladium as a ligand.

The total synthesis of (+)-equilenin utilizing this ring expansion reaction has been accomplished (Scheme 16.6) [7]. In the synthesis, the optically active cyclobutanone **22** was produced by the cascade asymmetric epoxidation–ring expansion reaction of the cyclopropylidene derivative **20** using a chiral (salen)Mn(III) complex **21** [8]. After introduction of the isopropenyl group, the resulting substrate **23** was predominantly converted to *trans*-naphthohydrindan **24** by the palladium-promoted reaction in HMPA-THF. The product **24** was further transformed to (+)-equilenin in an additional few steps.

This palladium-promoted ring expansion methodology was also applied to other strained ring systems. A strained bicyclic 1-alkenylcyclopentanol **26** underwent 1,2-rearrangement to fenchone by treatment with a catalytic amount of $PdCl_2(PPh_3)_2$ (Scheme 16.7) [9]. In the reaction, coordination of the alkene moiety in **26** to the palladium(II) complex triggers the 1,2 rearrangement to form the intermediate **28**, which affords the product by reaction with the resulting HCl.

The reaction of propenoylcyclopentanol **29** in the presence of $Pd(PPh)_4$ with P(o-tolyl)$_3$ produced the ring expanded hydroxycyclohexanone **32** (Scheme 16.8) [10]. It

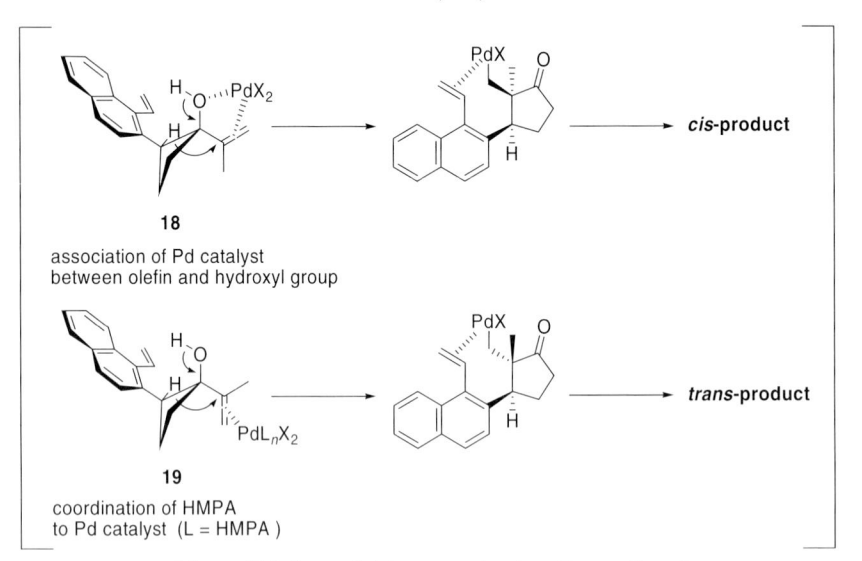

**Scheme 16.5** Sequential ring expansion–insertion reaction of isopropenyl-substituted cyclobutanol.

was found that the combination of the palladium complex with Lewis basic phosphine enhanced the reactivity. As a speculative mechanism it was proposed that the oxidative addition of the hydroxy group of **29** to palladium followed by the conjugate addition of the phosphine to the propenyl moiety generates the complex **31**, which causes a 1,2-migration thus affording the ring expanded product **32**. The reaction also proceeded in the absence of phosphine or palladium catalyst although the reactivity was decreased. In these reactions, it is expected that the corresponding activated intermediate **33** or **34** was formed in the migration step.

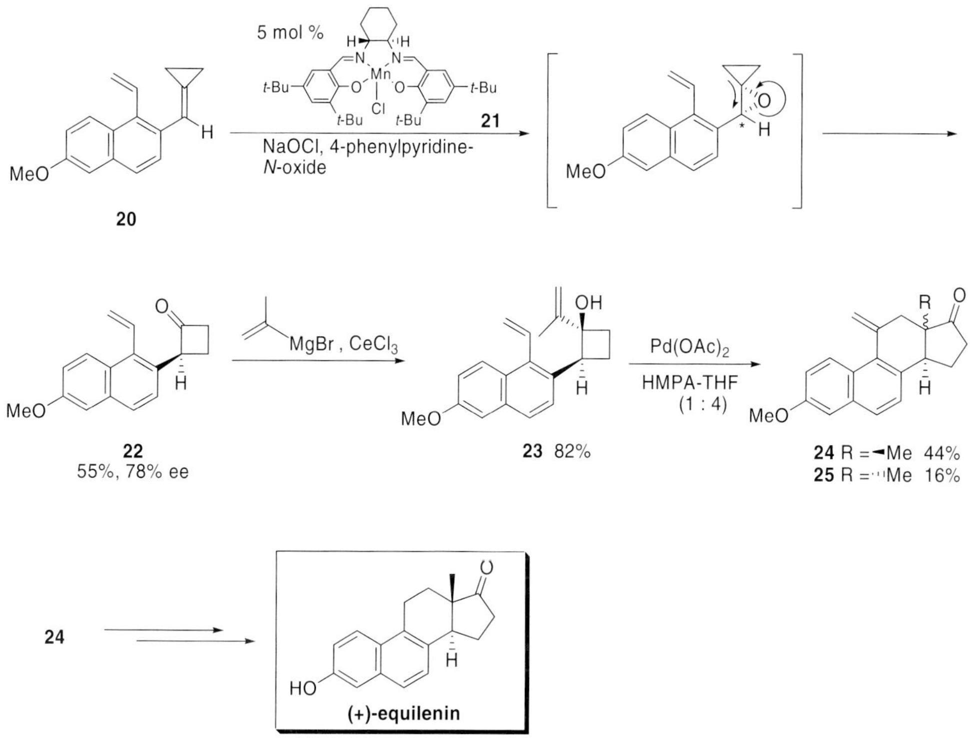

**Scheme 16.6** Application to the total synthesis of (+)-equilenin.

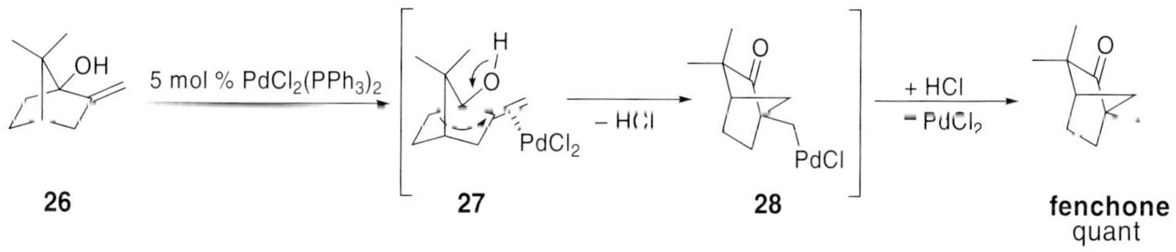

**Scheme 16.7** Ring expansion of bicyclic 1-alkenylcyclopentanol.

As an application of this rearrangement, steroidal ring transformation has been achieved (Scheme 16.9) [10]. Thus, the propenyl group was initially introduced into estrone **35** by reaction with the allene **36**, and the resulting propenylcyclopentanol **37** was converted to the ring expanded product **38** in a stereospecific manner by the palladium-catalyzed reaction. The observed stereospecificity would indicate that the rearrangement proceeds via the intermediate **39**. Although rearrangements of the steroidal D-ring are typically performed by employing bases and Lewis acids [11], such a neutral palladium-catalyzed reaction has never been reported.

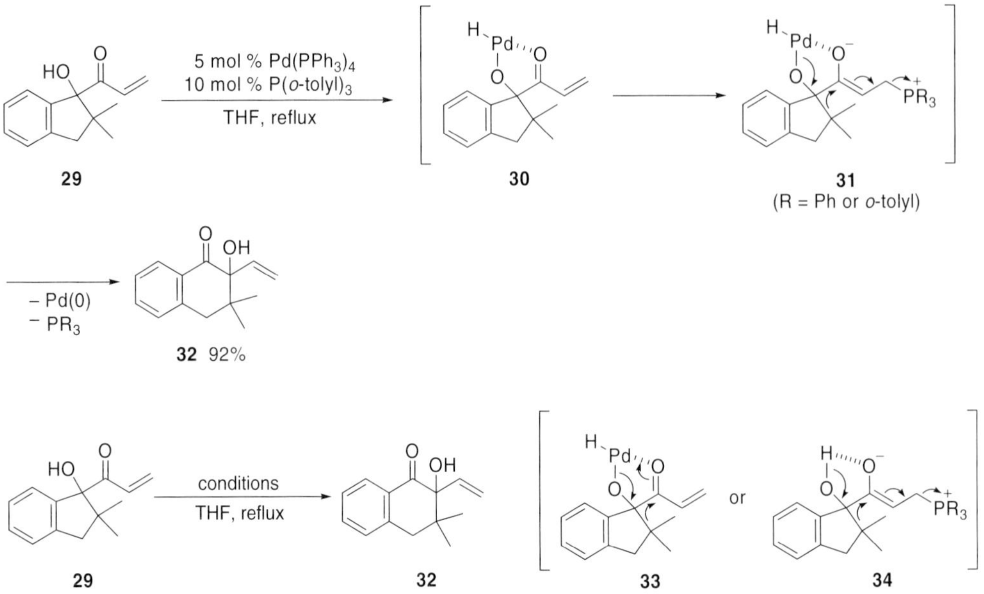

**Scheme 16.8** Palladium- and phosphine-catalyzed ring expansion of propenoylcyclopentanol.

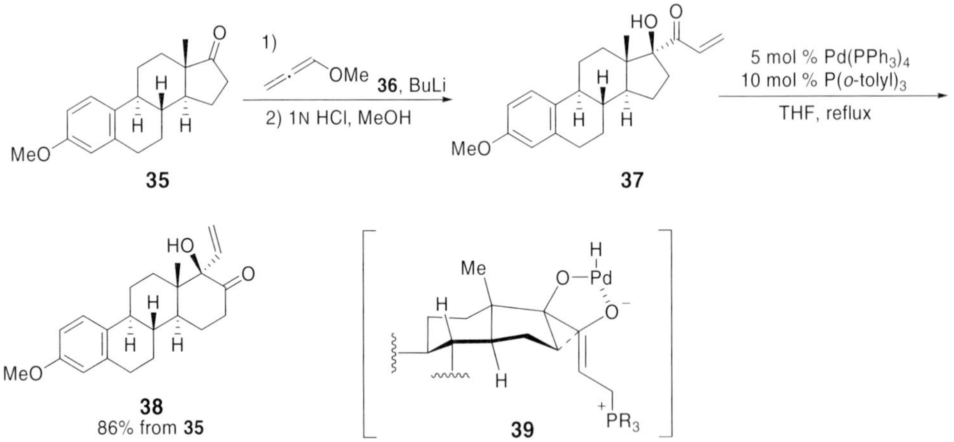

| conditions | yield |
|---|---|
| 5 mol % PdCl$_2$(MeCN)$_2$ | 26% |
| 10 mol % P(o-tolyl)$_3$ | 43% |

**Scheme 16.9** Steroidal ring transformation of an estrone derivative.

An enantioselective ring expansion process has been performed by employing a chiral palladium catalyst (Scheme 16.10) [12]. When the three-membered ring substrate **40**, having an allylic carbonate moiety, was subjected to reaction with Pd$_2$(dba)$_3$·CHCl$_3$ and chiral phosphine ligand **41** [13] the optically active vinylcyclobutanone **44** was obtained in high enantiomeric purity. It is presumed that both the π-allylpalladium species **42** and **43** are formed as the reaction intermediates via π–σ–π interconversion, in which the rearrangement of complex **43** leads to the observed **44**. Similarly, the corresponding vinylcyclopentanone **46** was successively prepared in good enantioselectivity from the reaction of the substrate **45** which has a cyclobutane ring. The reaction provides a useful methodology for the construction of a quaternary carbon stereocenter in an enantioselective manner.

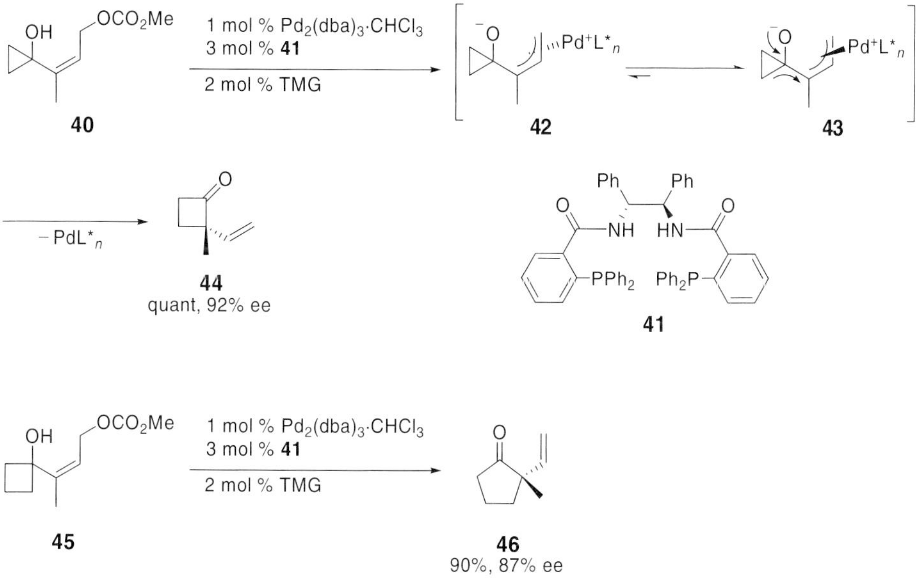

**Scheme 16.10** Enantioselective ring expansion process using a chiral palladium catalyst.

In contrast to the predominant migration of a secondary or tertiary carbon atom in most ring expansion reactions, a primary carbon was selectively rearranged in the oxidative reaction of vinylcyclobutanol **47** (Scheme 16.11) [14]. When the bicyclic cyclobutanol **47** was subjected to reaction with 10 mol% Pd(OAc)$_2$ under atmospheric pressure of oxygen, cyclopentanone **51** was produced as the sole product. Palladium alcoholate **48** is initially formed in the reaction, and then β-carbon elimination generates the intermediate **49**. Intramolecular carbopalladation of **49** followed by β-hydrogen elimination of the resulting **50** provides the ring expanded product **51**. Tricyclic ketone **56** was obtained in 92% yield in the case of the phenyl-substituted substrate **52**. In the reaction, an intermediate **53**, produced by β-carbon elimination, undergoes intramolecular cyclization with the phenyl ring to form the intermediate **55**. Elimination of β-hydrogen in the molecule **55** produces the product **56**.

As another type of palladium promoted ring expansion, conversion of aryl-substituted methylenecyclopropanes to cyclobutenes was reported (Scheme 16.12) [15]. When the substrate **57** was treated with a catalytic amount of Pd(OAc)$_2$ and CuBr$_2$, cyclobutene **62** was obtained in 93% yield. The reactive palladium species is proposed

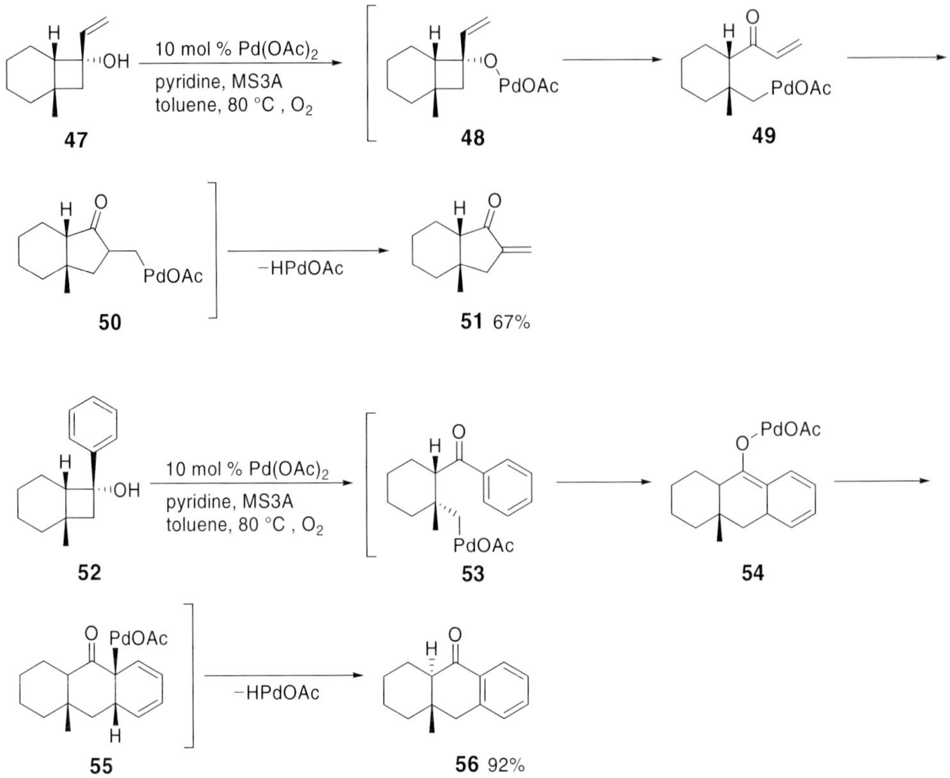

**Scheme 16.11** Migration of a primary carbon in the reaction of vinyl- and phenyl-substituted cyclobutanols.

to be PdBr$_2$, which causes a bromopalladation with methylenecyclopropane to afford the intermediate **58**. After transformation to the palladium carbenoid **61** via β-hydrogen elimination followed by hydropalladation with reversed regioselectivity (from **58** to **60**), 1,2-rearrangement of **61** generates the product **62**. A deuterium labeling experiment using **63** provided the corresponding cyclobutene **64** with the deuterium at the 2-position. The result supports that the β-hydrogen elimination–hydropalladation process from **58** to **60** surely occurs.

## 16.2.2
### Reactivities of Dienyl Substrates

It is known that several types of palladium-promoted reactions of allenes and 1.3-dienes produce the π-allylpalladium intermediates, which offer useful methodology for making carbon–carbon and carbon–heteroatom bonds [16]. Utilization of allenes and 1.3-dienes in the palladium-catalyzed ring expansion reaction has also been examined extensively.

The first report on the ring expansion reaction of allenyl-subsituted substrates appeared in 1997. Allenyl-substituted cyclobutanol **65** reacted with iodotoluene in the

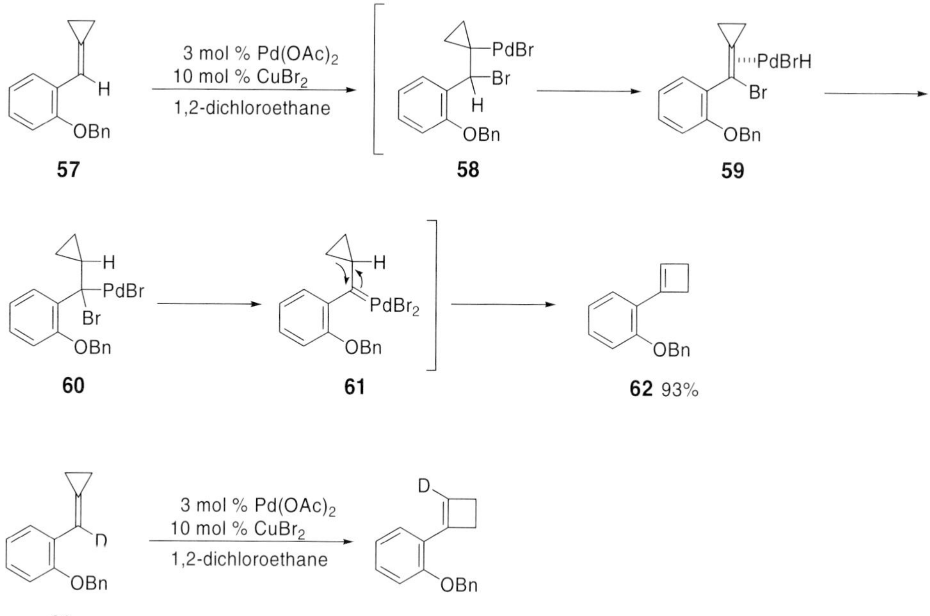

**Scheme 16.12** Ring expansion of methylenecyclopropanes.

presence of 10 mol% Pd(PPh$_3$)$_4$ with K$_2$CO$_3$ to afford the tolyl-substituted cyclopentanone **69** (Scheme 16.13) [17]. In the reaction, regioselective carbopalladation of allene with the resulting arylpalladium species takes place to give the π-allylpalladium intermediate **67**, which causes a 1,2-migration of the cyclobutane ring yielding the ring expanded compound **68**. The compound is further isomerized to the thermodynamically stable product **69**.

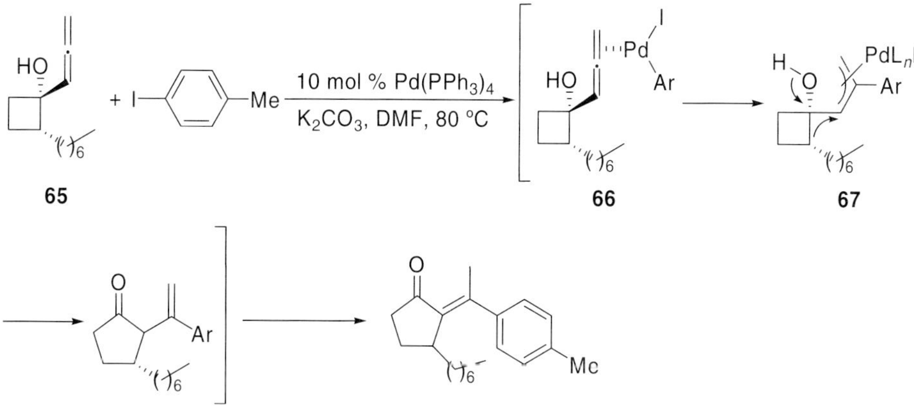

**Scheme 16.13** Ring expansion of an allenylcyclobutanol derivative in the presence of iodotoluene.

The ring expansion of allenylcyclobutanols was successfully applied to the synthesis of bicyclic medium-size ring compounds by the intramolecular cyclization–ring expansion reactions (Scheme 16.14) [17]. When the substrate **70** which has a Z-alkenyl iodide group was subjected to reaction with 10 mol% Pd(PPh$_3$)$_4$ and Ag$_2$CO$_3$, the bicyclic product **71** containing a seven-membered ring was obtained in 63% yield. Similarly, eight-membered ring products **73–75** were produced as a mixture from the reaction of **72**.

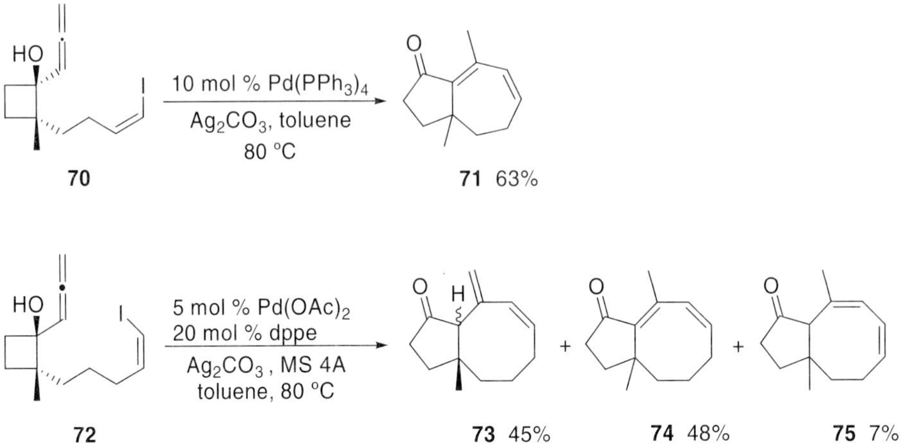

**Scheme 16.14** Intramolecular cyclization–ring expansion reaction of allenylcyclobutanols.

The reaction of allenylcyclobutanol **76** with a methyl group at the 1-position of the allenyl moiety, gave the cyclopentanone **77** and the diastereomer **78** (Scheme 16.15) [18]. The ratio of the products was influenced by the reaction temperature, the diastereoselectivity for **77** being increased when the reaction was carried out at high temperature. In comparison with the selective production of the *trans*-cyclopentanone **80** as a sole product by the reaction of the substrate **79**, the reaction of the diastereomer **81** exclusively provides the cis-product **82**. It is clear from these results that the ring expansion reaction proceeds in a stereospecific manner. The plausible mechanism for the observed stereospecificity is as follows (Scheme 16.16) [18]. The stereochemistry of the product would be reflected by two transition states **84** and **87**, which are each formed by the insertion of two conformers **83** or **86** into the aryl palladium species, followed by formation of the π-allylpalladium complex. Equilibrium between **84** and **87** would cause the stereospecific production of **85** by shifting to the more stable **84**. Improvement in diastereoselectivities at the higher reaction temperature supports the equilibrium-controlled mechanism.

A heterocyclic ring expansion reaction was reported by using hydroxy allenylisoindolinone **89** as the substrate (Scheme 16.17) [19]. When **89** was subjected to reaction with *p*-iodotoluene in the presence of 5 mol% Pd(PPh$_3$)$_4$ with K$_2$CO$_3$, the ring expanded product **91** was obtained in 90% yield via the formation of the π-allylpalladium intermediate **90**. Furthermore, synthesis of the tetracyclic compound **93** was achieved by an intramolecular carbopalladation–ring expansion reaction using the *o*-iodophenyl-introduced substrate **92**.

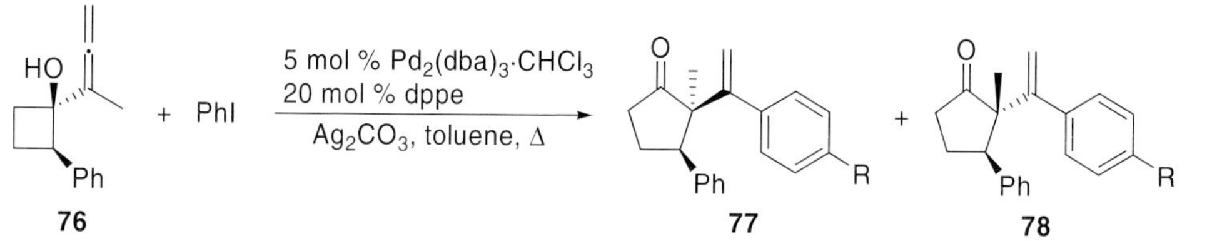

## 16.2 Palladium-Catalyzed Ring Expansions

| temp | 77 : 78 | total yields |
|---|---|---|
| 60 °C | 60 : 40 | 80% |
| 80 °C | 88 : 12 | 70% |
| reflux | 97 : 3 | 57% |

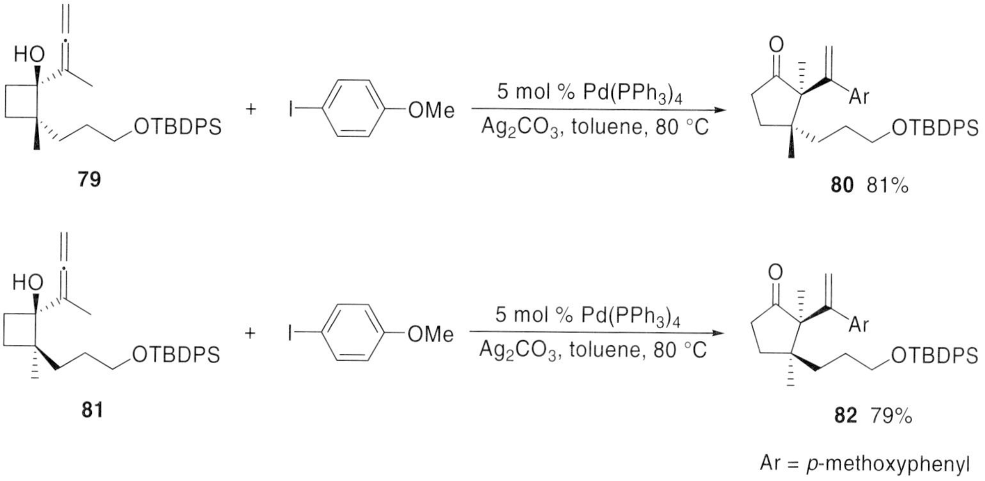

Ar = p-methoxyphenyl

**Scheme 16.15** Stereospecific ring expansion of allenylcyclobutanols.

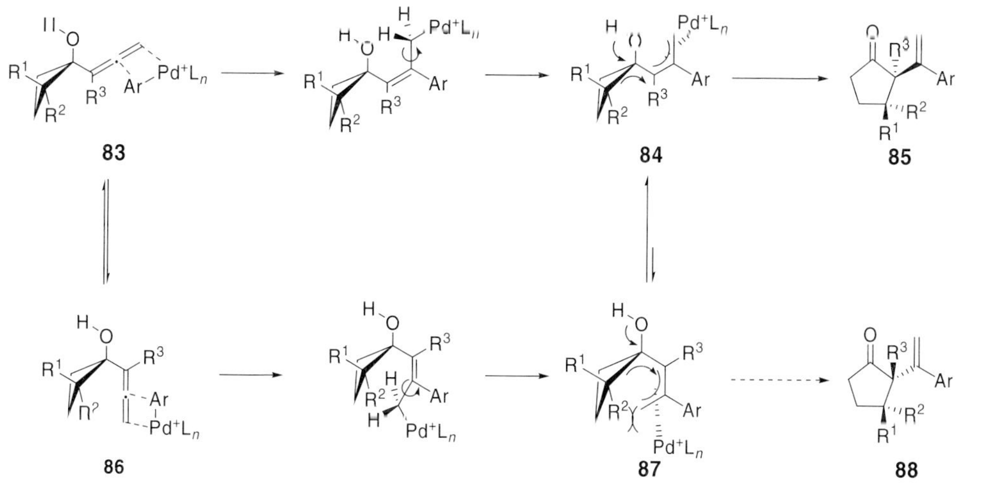

**Scheme 16.16** A plausible mechanism for the stereospecificity.

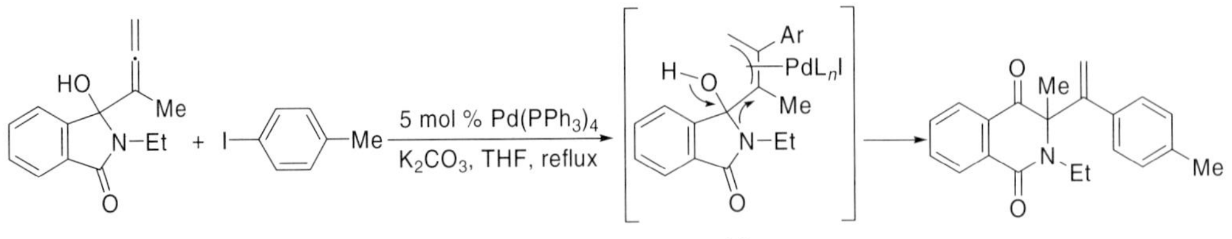


89 90 91 90%

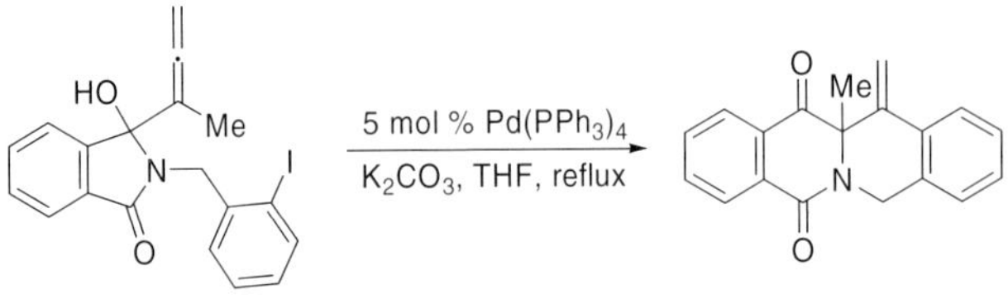


92 93 72%

**Scheme 16.17** Heterocyclic ring expansion using hydroxy allenylisoindolinones.

Allenyl alcohol **94**, which contains a methoxy group at the allenyl moiety, reacted with iodobenzene to give the corresponding isoquinolone **95** in 61% yield (Scheme 16.18) [20]. The reaction of **94** with vinyl bromide also proceeded uneventfully, producing the vinyl-substituted product **96** in 67% yield. The intramolecular reaction using **97** produced the methoxy-substituted tetracyclic compound **98** although the yield was low.

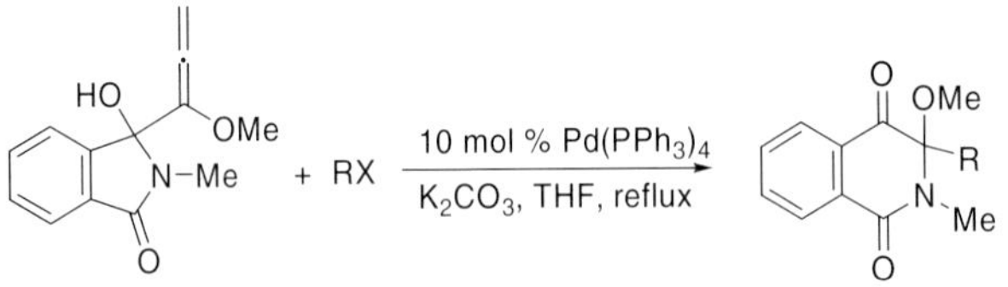


94

| RX | product | yield |
|---|---|---|
| iodobenzene | **95** R = Ph | 61% |
| vinyl bromide | **96** R = vinyl | 67% |

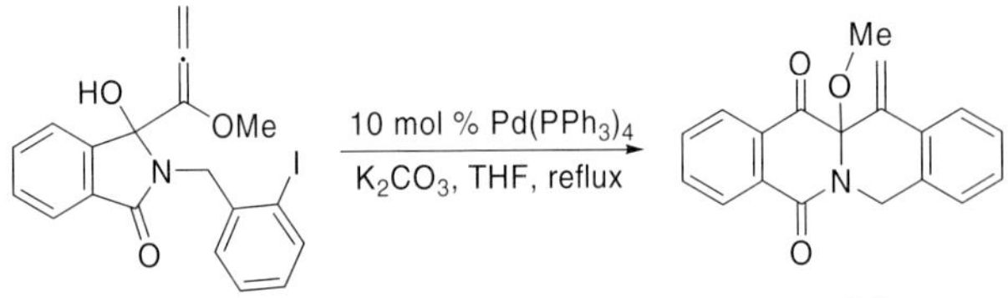


97 98 37%

**Scheme 16.18** Ring expansion of methoxy-substituted allenylisoindolinones.

Similarly, the substituted isochromanone derivative **100** was prepared in high yield by the reaction of hydroxyallenylphthalan **99** with iodotoluene in the presence of 5 mol% $Pd(PPh_3)_4$ with $K_2CO_3$ (Scheme 16.19) [21].

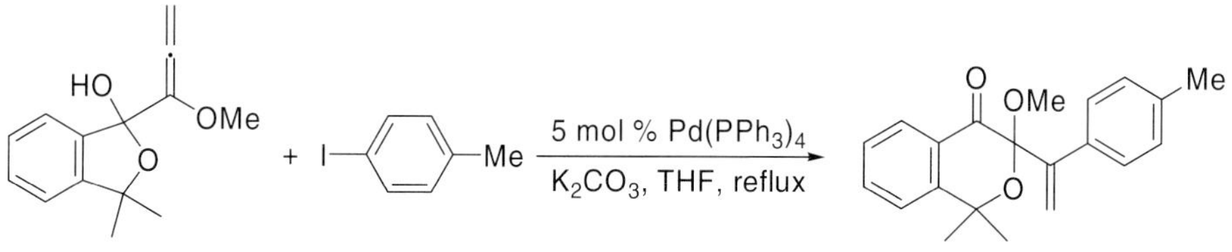

**Scheme 16.19** Ring expansion of hydroxyallenylphthalan.

While ring expansion reactions using allenic compounds, as described above, are initiated by a carbopalladation, a hydropalladation–ring expansion process occurred in the reaction of hydroxyallenylisoindolinone **94** (Scheme 16.20) [22]. Thus, when the substrate **94** was treated with 5 mol% Pd(PPh$_3$)$_4$ and 10 mol% P(o-tolyl)$_3$, the vinyl-substituted isoquinolone **103** was obtained in 93% yield. The proposed mechanism is that oxidative addition of the hydroxy group in the substrate to palladium forms the complex **101**, which causes intramolecular regioselective hydropalladation to generate the π-allylpalladium intermediate **102**. Successive 1,2-rearrangement from **102** provides the ring expanded product **103**. The reaction was applied to the synthesis of naphthoquinone **105** and isochromanone **106** by the reaction of allenylindanone **104** and allenylphthalan **99**.

Enantioselective hydropalladation–ring expansion reaction of allenylcyclobutanols has been achieved by employing a chiral palladium catalyst (Scheme 16.21) [23]. Optically active vinylcyclopentanone **110** was produced in high enantiomeric excess when the substrate **107** was subjected to reaction with a catalytic amount of Pd$_2$(dba)$_3$·CHCl$_3$ and the chiral ligand **41**. With benzoic acid and triethylamine as the additives, a good yield and enantioselectivity were observed. As the reaction mechanism, it is expected that the hydropalladation of the allene in **108** generated *in situ* affords the π-allylpalladium **109**. The enantioface-selective 1,2-rearrangement from **109** gives the optically active product **110**. As an application of this methodology, the chiral hetero-spiro compound **113** was synthesized via the enantioselective ring expansion of **111** followed by the ring-closing metathesis of the resulting product **112**.

Compared to the diversity of allenic compounds in the transition-metal-catalyzed ring expansion reaction, utilization of 1,3-dienes has not been extensively examined. Only one example has been reported concerning the palladium-catalyzed ring expansion reaction of 1,3-dienylcyclobutanols with aryl iodides (Scheme 16.22) [24]. When the substrate **114** was treated with iodobenzene in the presence of 10 mol% Pd(PPh$_3$)$_4$ and Ag$_2$CO$_3$, phenyl-substituted alkenylcyclopentanone **118** having a (Z)-olefinic geometry was selectively furnished in 83% yield. In the reaction, regioselective insertion of the 1,3-dienyl moiety in **114** into the arylpalladium complex affords the allylpalladium species **116**. After the complex is transformed to the π-allylpalladium intermediate **117**, a 1,2-rearrangement occurs to produce the ring expanded product **118**. The trans-product **120** was obtained diastereoselectively when trans-substituted cyclobutanol **119** was

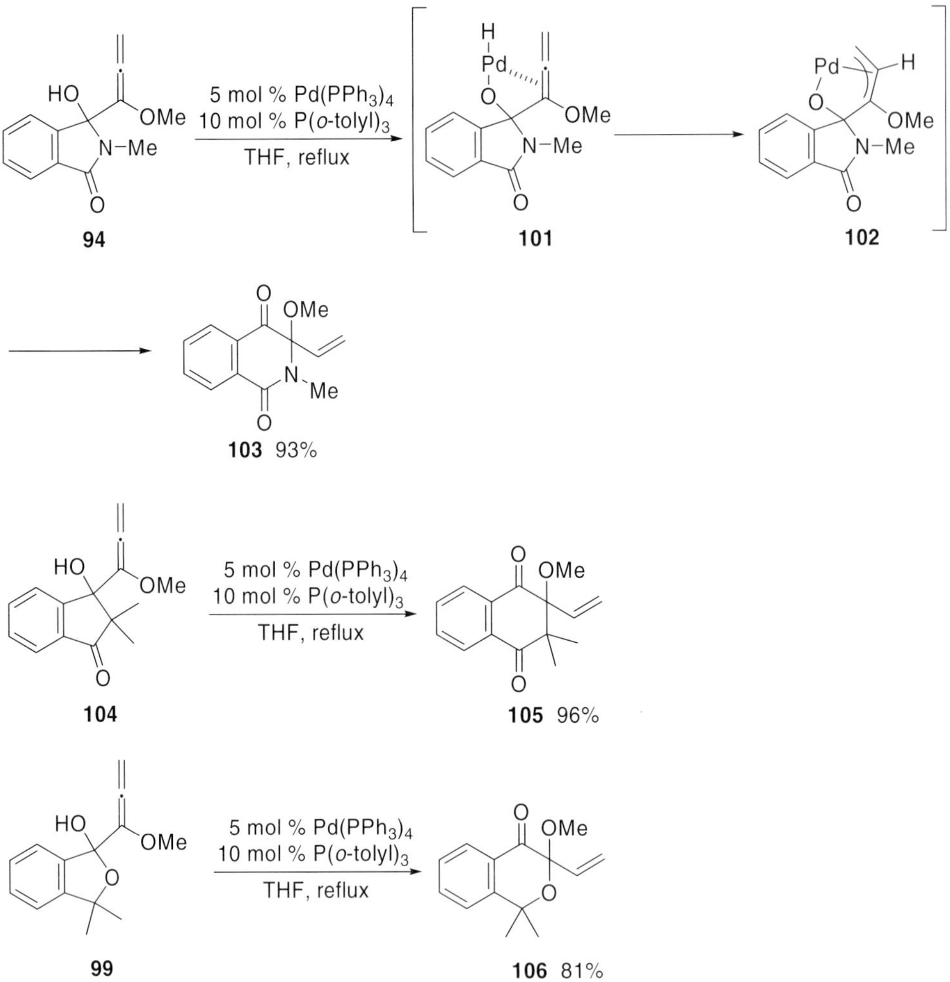


**94** **101** **102**

**103** 93%

**104** **105** 96%

**99** **106** 81%

**Scheme 16.20** Hydropalladation–ring expansion reaction of a hydroxyallenylisoindolinone and its related compounds.

subjected to the same reaction. The substrate **121** having a cis stereochemistry was converted to the corresponding diastereomeric cis-product **122** exclusively. It is ascertained from these results that the ring expansion reaction proceeded in a stereospecific manner. No reaction occurred when the substrate **123** having a (*E*)-1,3-butadienyl group was used.

Mechanistic rationalizations for the observed (Z)-specificity and diastereospecificity are described as follows (Scheme 16.23) [24]. Coordination of both the olefin and the hydroxy group within the substrate to arylpalladium initially forms the intermediate **124**, in which the dienyl group is located at the less hindered site to avoid a steric repulsion with the substituents $R^1$ and $R^2$. Regio- and diastereoselective

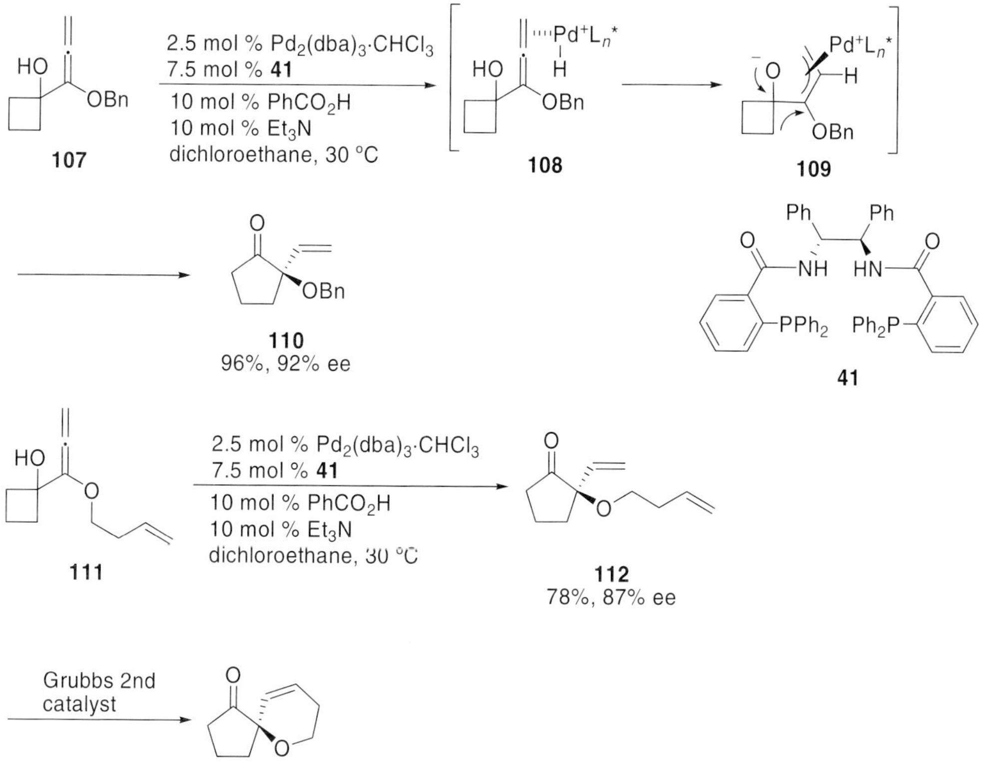

**Scheme 16.21** Enantioselective hydropalladation–ring expansion reaction of allenylcyclobutanols.

insertion into palladium gives allylpalladium **125**, which is successively transformed to π-allylpalladium **126**. The complex **126** interconverts by π–σ–π isomerization to **127**, which further causes the concerted 1,2-rearrangement of the cyclobutane ring on the opposite face of the palladium to produce the product **128** in a stereospecific manner. A reason for the low reactivity of the (E)-substrate **123** could be that the chelation-controlled insertion of the intermediate **129** is difficult because the distance between the hydroxy group and the olefin becomes longer than that of the (Z)-intermediate **124**.

### 16.2.3
### Reactivities of Alkynyl Substrates

Since the first report on the palladium-catalyzed ring expansion reaction of alkynylcyclobutanols in 1987 [25], several examples of the use of the alkynyl-substituted compounds have been reported, especially in the last 10 years.

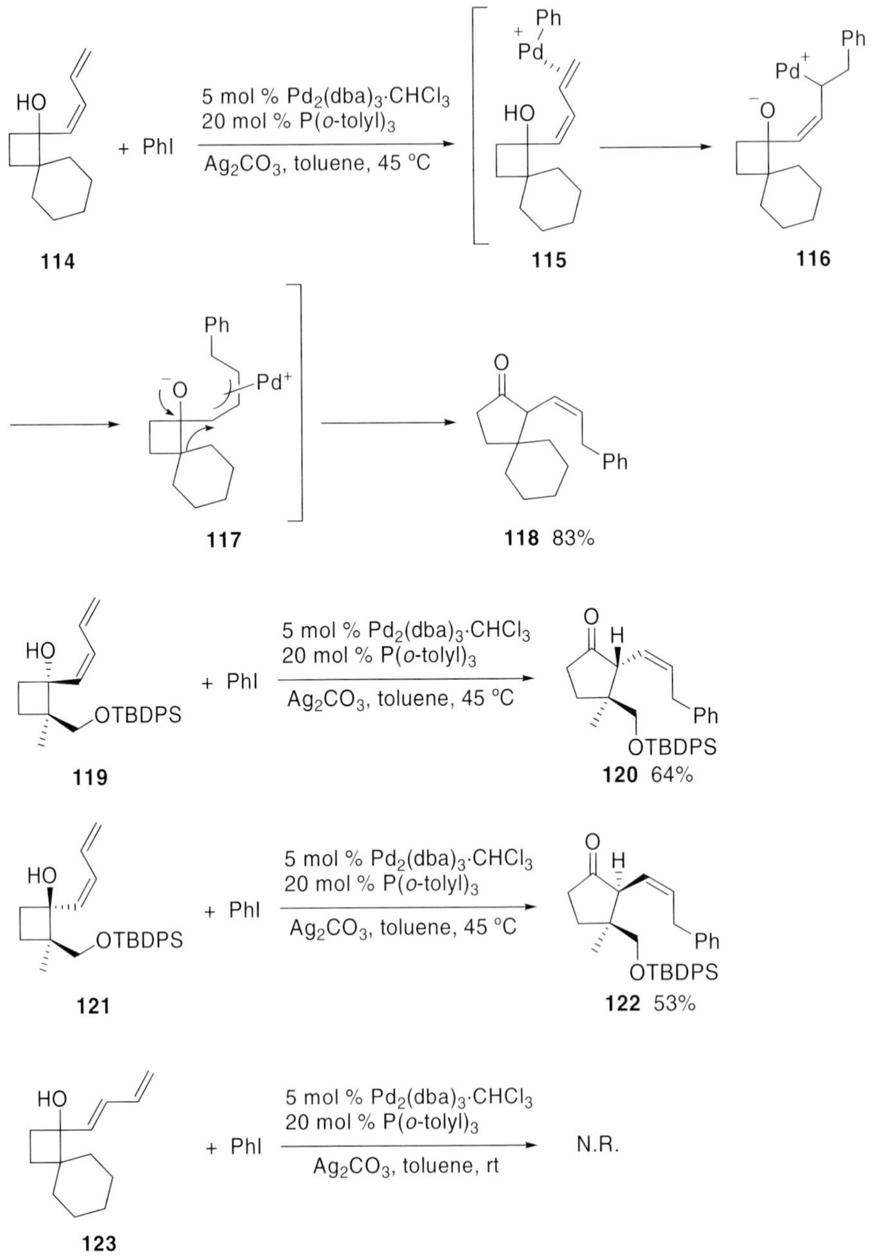

Scheme 16.22 Ring expansion of 1,3-dienylcyclobutanols.

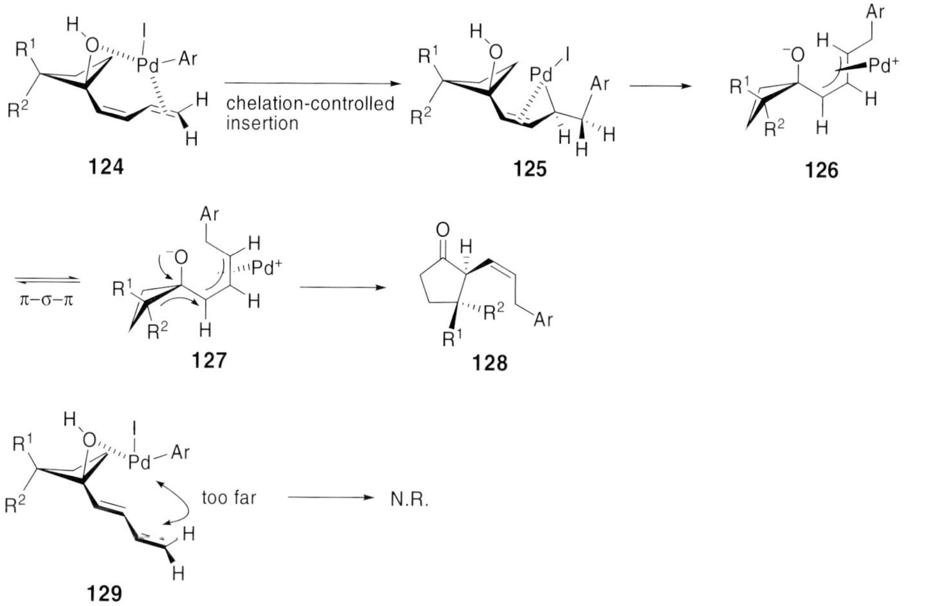

**Scheme 16.23** Mechanistic rationalization for the stereospecificity.

Synthesis of 2-alkylidenecyclopentanone **133** has been accomplished by a palladium-catalyzed reaction of alkynylcyclobutanol **130** with iodobenzene (Scheme 16.24) [26, 27]. In the reaction, the regioselective carbopalladation of the alkyne to the arylpalladium affords the alkenylpalladium intermediate **131**. The 1,2-migration of **131** by release of the ring strain forms the palladacyclohexanone **132**, which causes subsequent reductive elimination to give the ring expanded product **133**. It is noteworthy that the process appears to involve a novel ring expansion process via the palladacycle intermediate.

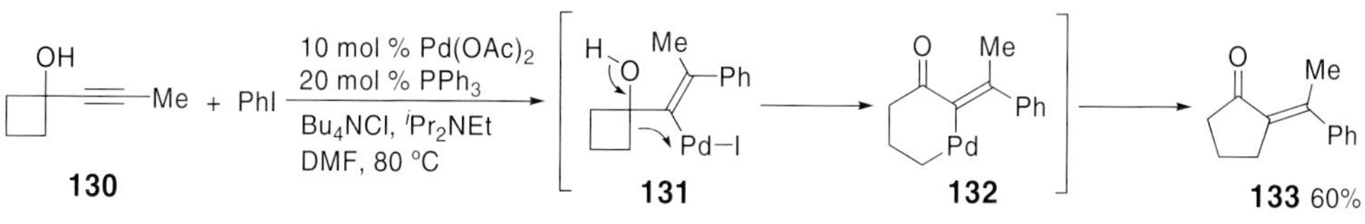

**Scheme 16.24** Ring expansion of an alkynylcyclobutanol in the presence of iodobenzene.

It is well known that propargylic compounds exhibit versatile reactivity in the presence of a palladium complex, and the reactions make up an important class of palladium-catalyzed reactions [28]. The key step in the reactions is the formation of the π-propargyl/allenylpalladium complex by facile elimination of the leaving group, which reacts further with nucleophilic reactants to afford a variety of substituted products. Application of the propargylic reactivity to the ring expansion reaction has also been reported. When treatment of cyclobutanol **134** containing a propargyl carbonate moiety with *p*-cresol in the presence of 5 mol% Pd$_2$(dba)$_3$·CHCl$_3$ and dppe was carried out, the phenoxyalkenyl-substituted cyclopentanone **138** was obtained in 99% yield (Scheme 16.25) [29]. As a plausible mechanism, the palladium catalyst initially promotes decarboxylation of **134** to yield the allenylpalladium species **135**. The complex is regarded as a π-propargylpalladium intermediate **136**, which undergoes nucleophilic attack by *p*-cresol to form the π-allylpalladium intermediate **137**. Finally, ring expansion reaction of **137** gives the cyclopentanone **138**.

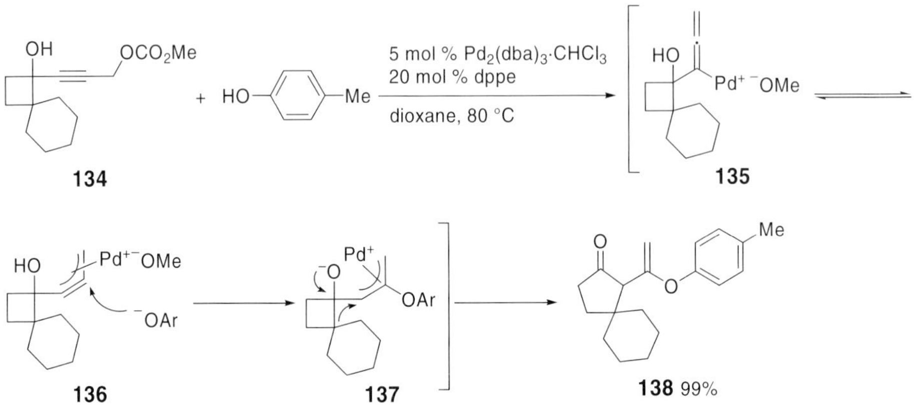

**Scheme 16.25** Ring expansion of a cyclobutanol containing a propargyl carbonate moiety in the presence of *p*-cresol.

When the trans-substituted substrate **139** was subjected to the reaction, the *trans*-cyclopentanone **141** was produced as a single diastereomer (Scheme 16.26) [29]. On the other hand, α,β-unsaturated cyclopentanone **145** was selectively produced by the reaction of the diastereomeric cis-substrate **142**. The stereochemical outcome in the reaction can be explained in terms of the different conformation of the π-allylpalladium complex during the ring expansion step. In the case of the trans-substrate **139**, the ring expansion process would proceed via **140**, a more stable conformer, to give **141**. Similarly, when the cis-substrate **142** is employed, the corresponding product **144** is produced via the intermediate **143**. The olefinic moiety of the product is readily isomerized to the thermodynamically stable **145** due to the steric repulsion in **144**.

Naphthalimide **147** also reacted as a nucleophile with the substrate **146** to afford the corresponding substituted cyclopentanone **148** in 58% yield (Scheme 16.27) [[29] b]. When sodium methoxide was used as a nucleophile in the reaction of **146**, only solvolysis of the carbonate moiety occurred. However, the desired reaction proceeded by carrying out the reaction using propargyl bromide **149** to produce the methoxy-introduced product **150**. In these ring expansion reactions using propargylic compounds, various substituted cyclopentanones can be synthesized along with the formation of a carbon–oxygen bond or a carbon–nitrogen bond.

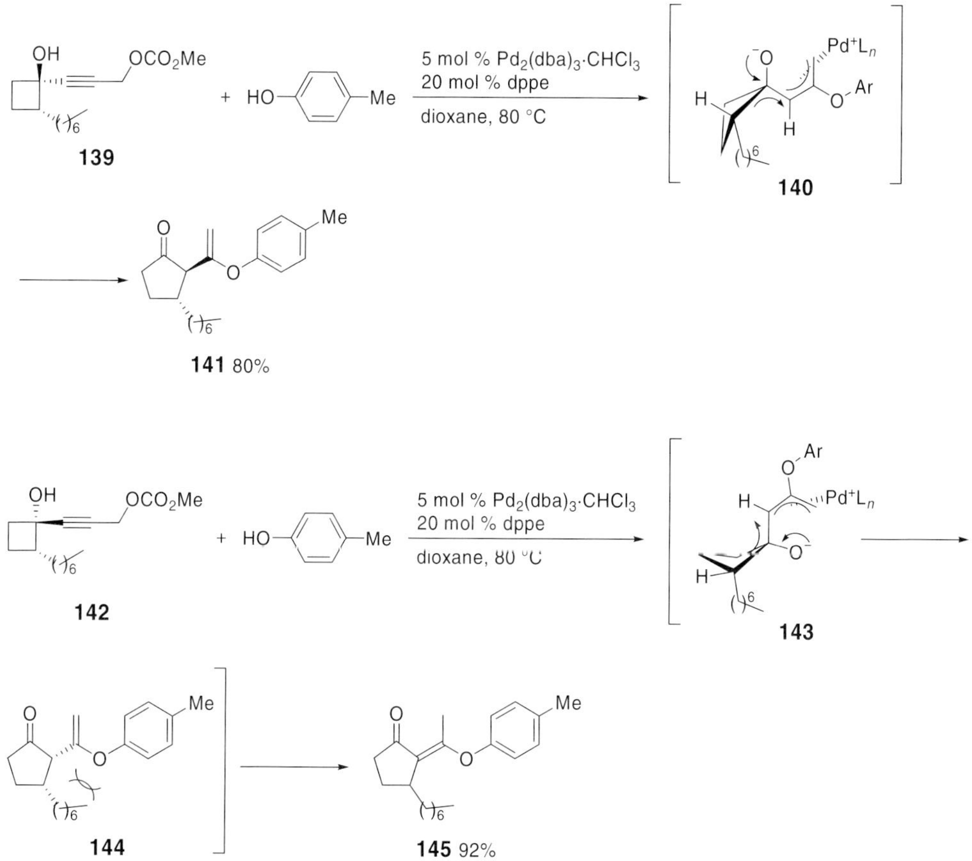

**Scheme 16.26** Stereospecific ring expansion of propargylic cyclobutanols.

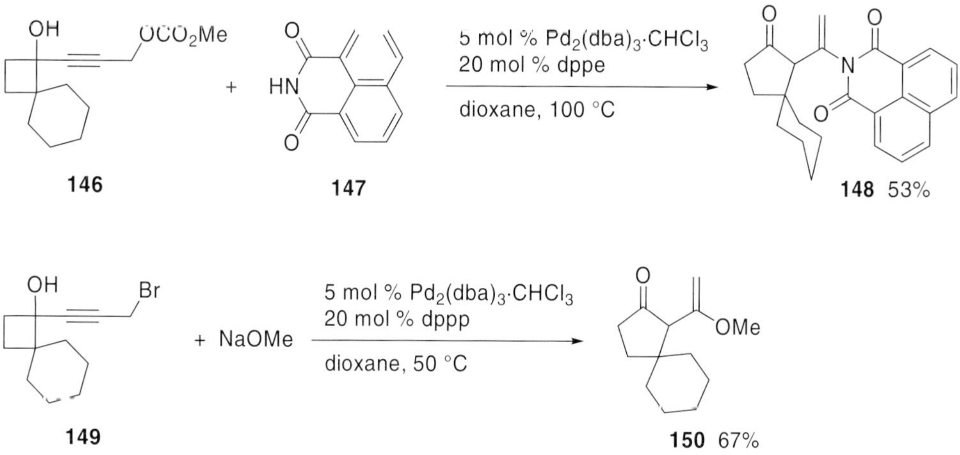

**Scheme 16.27** Ring expansion of propargylic cyclobutanols in the presence of an imide or an alkoxide.

## 16.3
## Ring Expansions with Other Transition Metals

### 16.3.1
### Ruthenium-Catalyzed Reactions

Utilization of other transition metals in ring expansion reactions had not been extensively examined in contrast with the multiplicity of palladium-promoted reactions. In 2001, it was reported that a ruthenium complex catalyzes the ring expansion reaction of allenylcyclobutanols with α,β-unsaturated carbonyl compounds (Scheme 16.28) [30, 31]. Thus, the substituted cyclopentanone **156** was obtained in 84% yield by reaction of the substrate **151** and methyl vinyl ketone with 10 mol% CpRu(MeCN)$_3$PF$_6$. In the reaction, the ruthenacycles **152** or **153** are initially formed, which induce the 1,2-migration of the four-membered ring after transformation to the π-allylruthenium intermediate **154**. Finally, protonation of the resulting ruthenium enolate **155** affords the product. The corresponding cyclopentanone **157** was produced in 90% yield when acrolein was used as an α,β-unsaturated carbonyl compound.

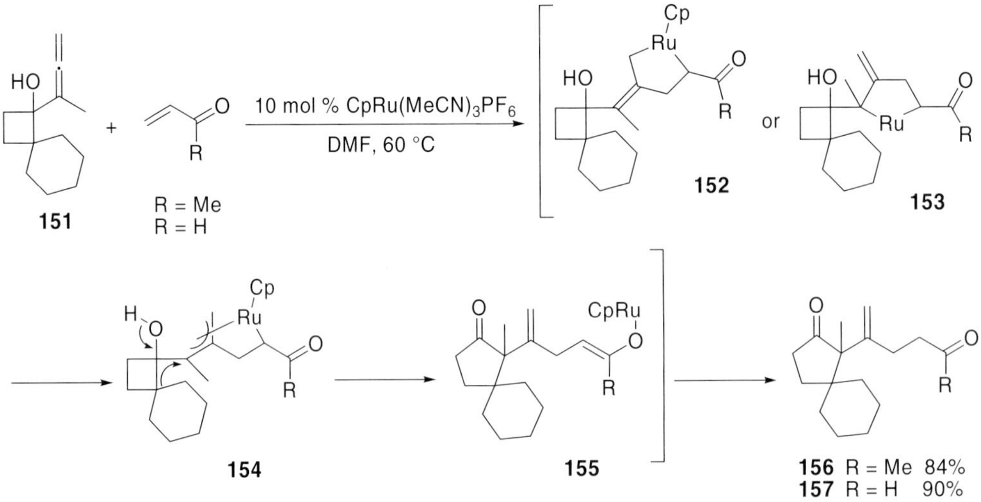

**Scheme 16.28** Ruthenium-catalyzed ring expansion of allenylcyclobutanols in the presence of α,β-unsaturated carbonyl compounds.

The ruthenium-catalyzed ring expansion of alkynylcyclobutanols was also reported (Scheme 16.29) [32]. The reaction of the substrate **158** with methyl vinyl ketone in the presence of 10 mol% CpRu(PPh$_3$)$_2$Cl with 15 mol% CeCl$_3$ provided the cyclopentanone **162** in 68% yield. As the mechanistic rationalization, the ruthenium catalyst initially coordinates to the alkyne moiety to afford the complex **159**, which successively induces a 1,2-rearrangement of the cyclobutane ring to give the vinylruthenium complex **160**. The insertion of methyl vinyl ketone into **160**, followed by protonolysis of the resulting ruthenium enolate **161**, produces the ring expanded product **162**.

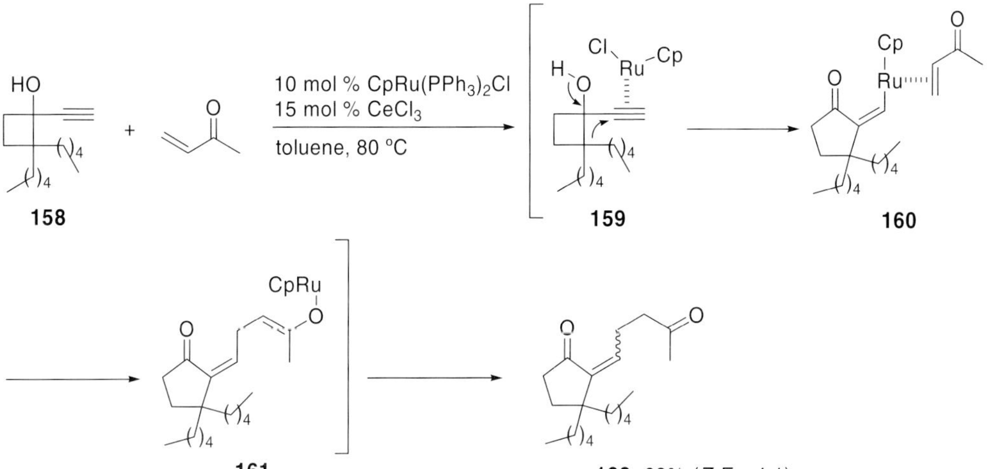

**Scheme 16.29** Ruthenium-catalyzed ring expansion of alkynylcyclobutanols.

It was found that the olefinic stereochemistries of the products are dependent on the ruthenium catalyst used in the reaction (Scheme 16.30). In contrast to the predominant formation of the (Z)-product **164** in the reaction of **163** with CpRu(PPh$_3$)$_2$Cl, the (E)-product **165** was selectively obtained when CpRu(MeCN)$_3$PF$_6$ was used. It is predicted that acid-catalyzed isomerization of the initially formed **164** proceeds in the case of the phosphine-free CpRu(MeCN)$_3$PF$_6$ condition, leading to the thermodynamically stable **165**.

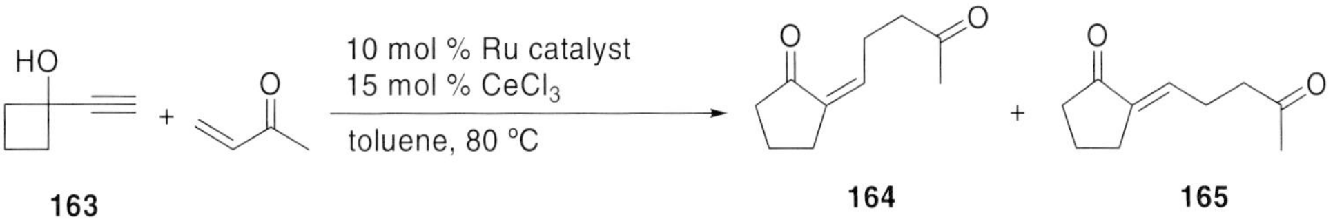

| Ru catalyst | 164 : 165 | total yields |
|---|---|---|
| CpRu(PPh$_3$)$_2$Cl | 2.9 : 1 | 58% |
| CpRu(MeCN)$_3$PF$_6$ | 1 : 16 | 52% |

**Scheme 16.30** Control of the olefinic stereochemistry in ruthenium-catalyzed ring expansion.

An unexpected ruthenium-catalyzed dimerization–ring expansion reaction occurred to give the product **172** when alkynylcyclobutanol **166** was only treated with 10 mol% CpRu(MeCN)$_3$PF$_6$ (Scheme 16.31) [33]. It is expected that the key intermediate is a ruthenacycle **168**, which is formed by coordination of the ruthenium catalyst with two molecules of acetylenylcyclobutanol. There is an equilibrium between the complex **168** and a zwitterionic intermediate **169**, and ring rearrangement of **169** followed by ring opening of the resulting ruthenacycle **170** to form an alkenyl ruthenium hydride **171**. Finally, reductive elimination of ruthenium produces the ring expanded dimer **172**.

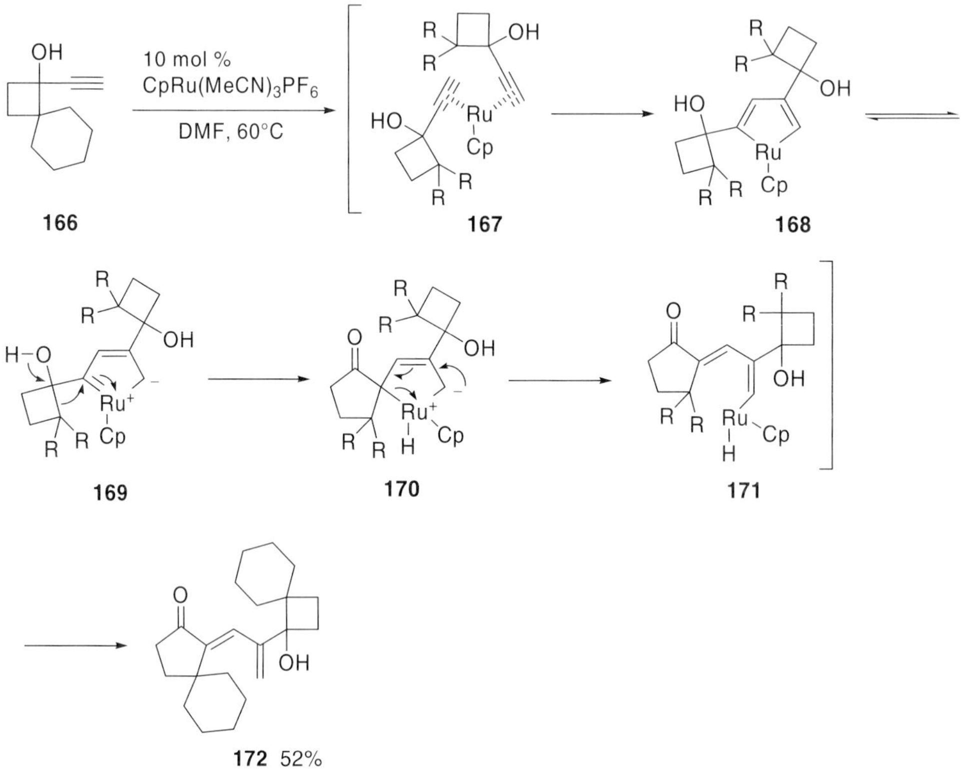

**Scheme 16.31** Ruthenium-catalyzed dimerization–ring expansion reaction.

## 16.3.2
### Gold-Catalyzed Reactions

Since gold complexes turned out to be excellent catalysts for the electrophilic activation of alkynes, extensive study on the gold-catalyzed reaction has been performed [34]. As an example of the reactions, gold-catalyzed ring expansion of alkynylcycloalkanols was reported (Scheme 16.32) [35]. Treatment of the alkynylcyclopropanol **173** with 1 mol% Au[P(*p*-CF$_3$C$_6$H$_4$)$_3$]Cl and AgSbF$_6$ selectively produced the (*E*)-alkylidenecyclobutanone **176** in 99% yield. In the mechanism, coordination of the alkyne moiety to the cationic gold species to give **174** induces the 1,2-rearrangement of the four-membered ring giving the intermediate **175**, which was further subjected to protonolysis with retention of the olefinic geometry to afford the product **176**. The corresponding cyclopentanone **178** was obtained by the same reaction of the alkynylcyclobutanol **177**.

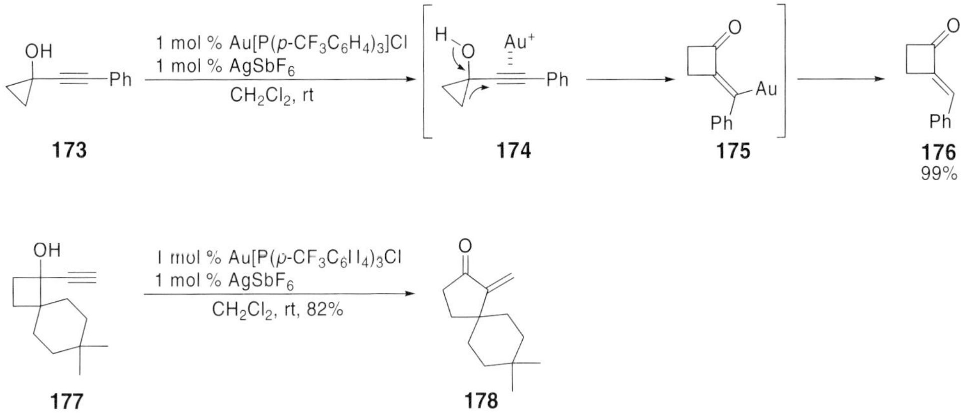

**Scheme 16.32** Gold-catalyzed ring expansion of an alkynylcycloalkanol.

Gold-catalyzed transformation of cyclobutanols having an allenyl or a propargyl ester moiety was also examined (Scheme 16.33) [36]. In the presence of 5 mol% Au(PPh$_3$)OTf, allenylcyclobutanol **107** was converted to the vinyl-substituted cyclopentanone **110** in 85% yield. The reactive oxonium intermediate **179** is formed by the reaction of the allenyl moiety with gold, and ring rearrangement successively occurs to give **110** via the intermediate **180**. When the reaction of the substrate **181**, containing a propargyl acetate, with the gold catalyst was carried out, the ring expanded product **184** was obtained in 96% yield. In the reaction, coordination of the alkyne to gold causes [3.3]-sigmatropic rearrangement to yield *O*-acetyl allene **182**. Further transformation of **182** with gold yields the oxonium intermediate **183**, which triggers the ring expansion to give **184**.

## 16.3.3
### Reactions with Fischer-Carbene Complexes

A variety of transformations have been developed for the reaction of transition metal-carbene complexes [37]. Among them, a ring expansion reaction of alkynylcyclobutenols with Fischer carbene complexes has been reported (Scheme 16.34) [38]. Treatment of alkynylcyclobutenol **185** with the chromium-carbene complex **186** led

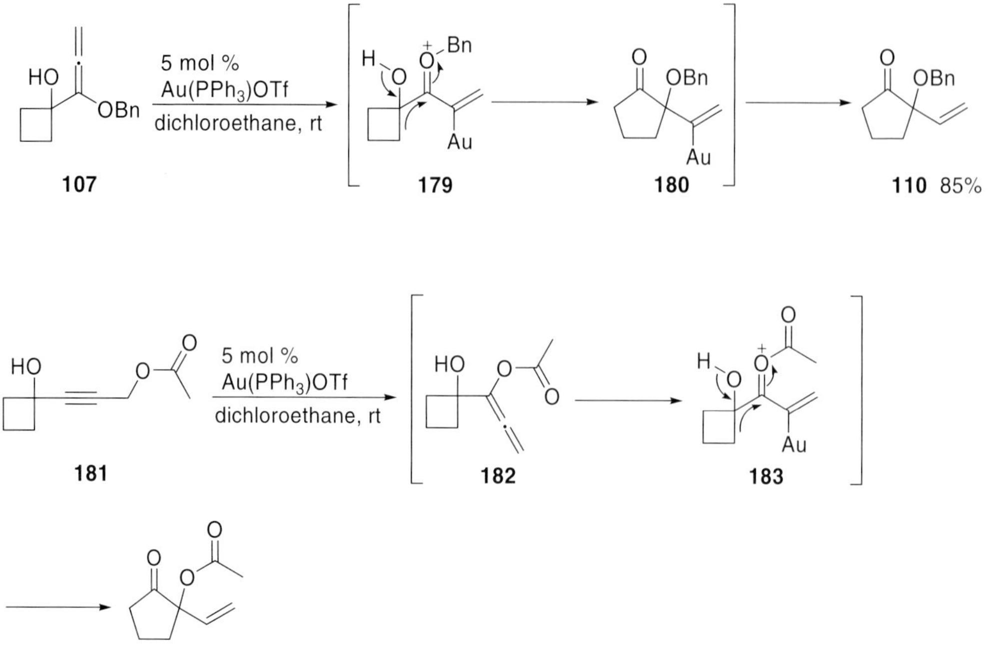


107 179 180 110 85%

181 182 183

184 96%

**Scheme 16.33** Gold-catalyzed ring expansion of the cyclobutanols having an allenyl or a propargylic ester moiety.

to the ring expanded product **190** in 72% yield. In the reaction, the vinylcarbene intermediate **187**, which is generated from **185** and **186**, induces the 1,2-migration of the four-membered ring to yield the intermediate **188**. Reductive elimination of **188**, followed by hydrolysis of the resulting enol ether **189**, affords the product **190**.

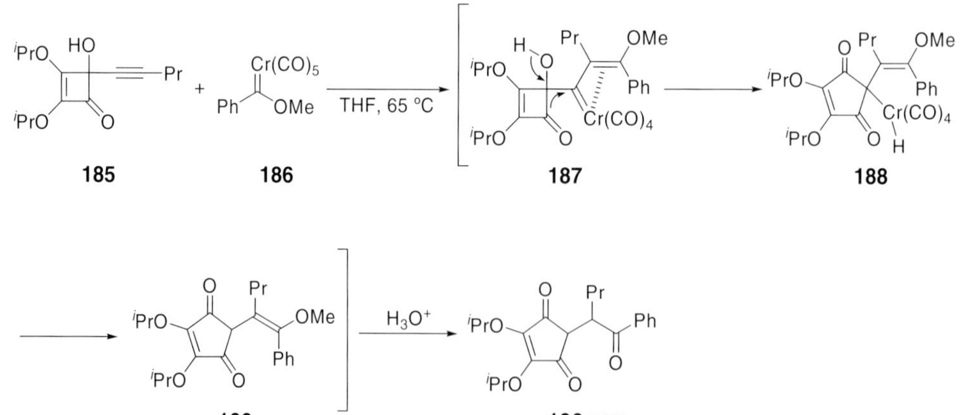


185 186 187 188

189 190 72%

**Scheme 16.34** Ring expansion reaction of an alkynylcyclobutenol with chromium-carbene complex.

## 16.4
## Conclusion

During the last 10 years, transition-metal-catalyzed ring expansion reactions have been realized to be one of the useful methodologies for the construction of substituted four- to six-membered ring compounds. Wide scope for synthetic applications including stereocontrolled or cascade reactions is possible by the specific design of functionalized substrates and the choice of catalyst. It can be predicted that broader applications will be developed in the future.

## 16.5
## Experimental: Selected Procedures

### 16.5.1
### General

Materials were obtained from commercial suppliers and used without further purification except when otherwise noted. Solvents were dried and distilled according to standard protocols. The phrase "residue upon work-up" refers to the residue obtained when the organic layer was separated and dried over anhydrous $MgSO_4$ and the solvent was evaporated under reduced pressure.

### 16.5.2
### Synthesis of Compound 10 (Scheme 16.4) [5]

To a stirred solution of **9** (1.28 g, 3.92 mmol) in THF (150 mL) under argon was added $Pd(OAc)_2$ (1.3 g, 5.8 mmol) at room temperature, and stirring was continued for 8 h at the same temperature. The reaction mixture was passed through a short pad of silica gel with $Et_2O$ as eluent. The filtrate was concentrated under reduced pressure, and the residue was chromatographed on silica gel with hexane–AcOEt (90 : 10 v/v) to give **10** (1.15 g, 90%) as colorless prisms.

### 16.5.3
### Synthesis of Compounds 24 and 25 (Scheme 16.6) [7]

To a stirred solution of the isopropenylcyclobutanol **23** (80.2 mg, 0.273 mmol) in HMPA-THF (1 : 4) (5 mL) was added $Pd(OAc)_2$ (73.6 mg, 0.328 mmol) at room temperature and the stirring was continued for 5 h at the same temperature. The resulting solution was diluted with water and extracted with AcOEt. The combined extracts were washed with saturated aqueous NaCl. The residue upon work-up was chromatographed on silica gel with hexane–AcOEt (98 : 2 v/v) as eluent to give the naphthohydrindans **24** and **25** (**24** : **25** = 73 : 27, 47.6 mg, 60%) as a colorless oil.

### 16.5.4
**Synthesis of Compound 32 (Scheme 16.8) [10]**

A mixture of hydroxypropenoylindan **29** (31.8 mg, 0.147 mmol), Pd(PPh$_3$)$_4$ (8.5 mg, 0.007 mmol), and P(o-tolyl)$_3$ (3.9 mg, 0.015 mmol) in THF (1.5 mL) was refluxed under argon atmosphere for 7 h. The reaction mixture was evaporated *in vacuo* to afford a crude product, which was purified by column chromatography on silica gel with *n*-hexane–AcOEt (10:1) to give vinyltetralone **32** (29.3 mg, 92%) as white crystals.

### 16.5.5
**Synthesis of Compound 71 (Scheme 16.14) [17]**

A slurry of the allenylcyclobutanol **70** (27.8 mg, 0.094 mmol), Ag$_2$CO$_3$ (50.0 mg, 0.181 mmol), and Pd(PPh$_3$)$_4$ (10.9 mg, 9.4 μmol) in toluene (50 mL) was stirred for 16 h at 80 °C. The reaction mixture was diluted with water and extracted with Et$_2$O. The combined extracts were washed with saturated aqueous NaCl solution. The residue upon work-up was chromatographed on silica gel with hexane–AcOEt (99:1 v/v) as eluent to give the bicyclic compound **71** (10.1 mg, 63%) as a colorless oil.

### 16.5.6
**Synthesis of Compound 91 (Scheme 16.17) [19]**

A mixture of allenylisoindolinone **89** (150 mg, 0.66 mmol), Pd(PPh$_3$)$_4$ (40 mg, 0.033 mmol), 4-iodotoluene (0.43 g, 1.98 mmol), and K$_2$CO$_3$ (0.11 g, 0.66 mmol) in anhydrous THF (5 mL) was refluxed under a nitrogen atmosphere for 26 h. The reaction mixture was quenched with saturated NH$_4$Cl and then extracted with ether. The organic layer was washed with brine, dried over MgSO$_4$, and filtered. The filtrate was evaporated *in vacuo* to afford the crude product, which was purified by flash column chromatography on silica gel with ether–hexane (1:3) to give 4-oxo-1-isoquinolone **91** (190 mg, 90% yield) as a pale yellow oil.

### 16.5.7
**Synthesis of Compound 105 (Scheme 16.20) [22]**

A mixture of allenylisoindanone **104** (122 mg, 0.50 mmol), Pd(PPh$_3$)$_4$ (29.0 mg, 0.025 mmol), and P(o-tolyl)$_3$ (15.0 mg, 0.050 mmol) in anhydrous THF (5 mL) was refluxed under a nitrogen atmosphere for 9 h. The reaction mixture was evaporated *in vacuo* to afford a crude product, which was purified by flash column chromatography on silica gel with hexane–AcOEt (50:1) to give naphthoquinone **105** (117 mg, 96% yield) as colorless prisms.

### 16.5.8
**Synthesis of Compound 118 (Scheme 16.22) [24]**

To a stirred solution of the (Z)-1,3-dienylcyclobutanol **114** (35.7 mg, 186 μmol) in toluene (0.74 mL) were added iodobenzene (56.8 mg, 278 μmol), Pd$_2$(dba)$_3$·CHCl$_3$

(9.6 mg, 9.3 μmol), P(o-tolyl)$_3$ (11.3 mg, 37.2 μmol) and Ag$_2$CO$_3$ (103 mg, 372 μmol) at room temperature, and stirring was continued for 4 h at 45 °C. After filtration of the reaction mixture using AcOEt with a small amount of the mixture of silica gel and Celite, followed by evaporation of the eluate, the residue was chromatographed on silica gel with hexane–AcOEt (98 : 2 v/v) as eluent to give the cyclopentanone **118** (41.3 mg, 83%) as a colorless oil.

### 16.5.9
**Synthesis of Compound 141 (Scheme 16.26) [29]**

A slurry of the cyclobutanol **139** (35.8 mg, 0.127 mmol), p-cresol (16.4 mg, 0.152 mmol), Pd$_2$(dba)$_3$·CHCl$_3$ (6.6 mg, 6.4 μmol), and dppe (10.1 mg, 25.4 μmol) in dioxane (3 mL) was stirred for 1 h at 80 °C. After evaporation of the solvent, the residue was chromatographed on silica gel with hexane–AcOEt (98 : 2 v/v) as eluent to afford the cyclopentanone **141** (32.1 mg, 80%) as a colorless oil.

### 16.5.10
**Synthesis of Compound 156 (Scheme 16.28) [30]**

To a stirred solution of the allenylcyclobutanol **151** (61.8 mg, 0.321 mmol) and methyl vinyl ketone (33.8 mg, 0.482 mmol) was added [CpRu(MeCN)$_3$]PF$_6$ (14.0 mg, 0.0321 mmol) at room temperature in a sealed tube. After being stirred for 0.5 h at 60 °C, the reaction mixture was filtered through Celite, then washed with water and extracted with Et$_2$O. The combined organic layer was washed with saturated aqueous NaCl, dried over anhydrous MgSO$_4$ and concentrated under reduced pressure. The residue was purified by chromatography on silica gel with hexane–AcOEt (85 : 15 v/v) as eluent to give the cyclopentanone **156** (70.4 mg, 84%) as a colorless oil.

### 16.5.11
**Synthesis of Compound 162 (Scheme 16.29) [32]**

To a stirred solution of ethynylcyclobutanol **158** (47.7 mg, 202 μmol) in toluene (1.0 mL) were added methyl vinyl ketone (70.0 mg, 1010 μmol), anhydrous CeCl$_3$ (10.0 mg, 40.4 μmol) and CpRu(PPh$_3$)$_2$Cl (14.7 mg, 20.2 μmol) at room temperature, and stirring was continued for 2 h at 80 °C. After filtration of the reaction mixture using AcOEt with a small amount of the mixture of silica gel and a Celite pad, followed by evaporation of the elute, the residue was chromatographed on silica gel with hexane–AcOEt (92: 8 v/v) as eluent to give the cyclopentanones **(Z)-162** (33.4 mg, 54%) and **(E)-162** (8.8 mg, 14%) as a colorless oil.

### Abbreviations

| | |
|---|---|
| Ac | acetyl |
| Ar | aryl |
| Bn | benzyl |

| | |
|---|---|
| Cp | cyclopentadienyl |
| dba | dibenzylideneacetone |
| DMA | $N,N$-dimethylacetamide |
| dppe | 1,2-bis(diphenylphosphino)ethane |
| dppp | 1,3- bis(diphenylphosphino)propane |
| HMPA | hexamethylphosphoramide |
| L | ligand |
| MOM | methoxymethyl |
| TBDPS | *tert*-butyldiphenylsilyl |
| Tf | trifluoromethanesulfonyl |
| TMG | trimethylene glycol |

## Acknowledgments

A part of these studies was supported by a Grant-in-Aid for Scientific Research on Priority Areas (A) (2) (No. 13 029 085) from the Ministry of Education, Culture, Sports, Science, and Technology, Japan and for Scientific Research (B) (2) (No. 16 390 008) from the Japan Society for the Promotion of Science, and by a Pilot Research Grant (2004–2006) from The University of Tokushima, Japan.

## References

1 For an overview of Pinacol-type rearrangements including related reactions, see: (a) Hanson, J.R. (1991) *Comprehensive Organic Synthesis*, vol. 3 (eds B.M. Trost, I. Fleming, and G. Pattenden), Pergamon, Oxford, pp. 705–719; (b) Ricknorn, B. *ibid*, pp. 721–732 and pp. 733–775;(c) Covency, D. *ibid*, pp. 777–801.

2 Clark, G.R. and Thiensathit, S. (1985) *Tetrahedron Lett.*, **26**, 2503–2506.

3 For other examples of the palladium-catalyzed ring expansion reactions of alkenylcyclobutanols, see: (a) Demuth, M., Pandey, B., Wietfeld, B., Said, H. and Viader, J. (1988) *Helv. Chim. Acta*, **71**, 1392–1398; (b) de Almeida Barbosa, L.-C. and Mann, J. (1990) *J. Chem. Soc., Perkin Trans. 1*, 177–178; (c) Kim, S., Uh, K.H., Lee, S. and Park, J.H. (1991) *Tetrahedron Lett.*, **32**, 3395–3396; (d) Nemoto, H., Nagamochi, M. and Fukumoto, K. (1993) *J. Chem. Soc., Perkin Trans. 1*, 2329–2332; (e) Nemoto, H., Nagamochi, M., Ishibashi, H. and Fukumoto, K. (1994) *J. Org. Chem.*, **59**, 74–79; (f) Nemoto, H., Shiraki, M. and Fukumoto, K. (1994) *Synlett*, 599–600; (g) Nemoto, H., Miyata, J. and Fukumoto, K. (1996) *Tetrahedron*, **52**, 10363–10374; (h) Kocovský, P., Dunn, V., Gogoll, A. and Langer, V. (1999) *J. Org. Chem.*, **64**, 101–119; (i) Hegedus, L.S. and Ranslow, P.B. (2000) *Synthesis*, 953–958.

4 Nemoto, H., Takahashi, E. and Ihara, M. (1999) *Org. Lett.*, **1**, 517–520.

5 (a) Nemoto, H., Miyata, J. and Ihara, M. (1999) *Tetrahedron Lett.*, **40**, 1933–1936; (b) Miyata, J., Nemoto, H. and Ihara, M. (2000) *J. Org. Chem.*, **65**, 504–512.

6 Nemoto, H., Miyata, J., Yoshida, M., Raku, N. and Fukumoto, K. (1997) *J. Org. Chem.*, **62**, 7850–7851.

7 (a) Nemoto, H., Yoshida, M., Fukumoto, K. and Ihara, M. (1999) *Tetrahedron Lett.*, **40**,

907–910; (b) Yoshida, M., Ismail, M.A.-H., Nemoto, H. and Ihara, M. (2000) *J. Chem. Soc., Perkin Trans. 1*, 2629–2635.

8 Jacobsen, E.N., Zhang, W., Muci, A.R., Ecker, J.R. and Deng, L. (1991) *J. Am. Chem. Soc.*, **113**, 7063–7064.

9 García Martínez, A., Teso Vilar, E., García Fraile, A., de la Moya Cerero, S. and Lora Maroto, B. (2005) *Tetrahedron Lett.*, **46**, 3509–3511.

10 Nagao, Y., Tanaka, S., Ueki, A., Kumazawa, M., Goto, S., Ooi, T., Sano, S. and Shiro, M. (2004) *Org. Lett.*, **6**, 2133–2136.

11 (a) Creary, X., Inocencio, P.A., Underiner, T.L. and Kostromin, R. (1985) *J. Org. Chem.*, **50**, 1932–1938; (b) Bischofberger, N. and Walker, K.A.M. (1985) *J. Org. Chem.*, **50**, 3604–3609; (c) Schor, L., Seldes, A.M.S. and Gros, E.G. (1990) *J. Chem. Soc., Perkin Trans. 1*, 163–166.

12 Trost, B.M. and Yasukata, T. (2001) *J. Am. Chem. Soc.*, **123**, 7162–7163.

13 Trost, B.M., Vranken, D.L. and Bingel, C. (1992) *J. Am. Chem. Soc.*, **114**, 9327–9343.

14 Nishimura, T., Ohe, K. and Uemura, S. (2001) *J. Org. Chem.*, **66**, 1455–1465.

15 Shi, M., Liu, L.-P. and Tang, J. (2006) *J. Am. Chem. Soc.*, **128**, 7430–7431.

16 Tsuji, J. (2004) *Palladium Reagents and Catalysts: New Perspectives for the 21st Century*, John Wiley& Sons, Chichester, pp. 519–542.

17 Nemoto, H., Yoshida, M. and Fukumoto, K. (1997) *J. Org. Chem.*, **62**, 6450–6451.

18 (a) Yoshida, M., Sugimoto, K. and Ihara, M. (2000) *Tetrahedron Lett.*, **41**, 5089–5092; (b) Yoshida, M., Sugimoto, K. and Ihara, M. (2002) *Tetrahedron*, **58**, 7839–7846.

19 (a) Jeong, I.-Y. and Nagao, Y. (1998) *Tetrahedron Lett.*, **39**, 8677–8680; (b) Nagao, Y. (2002) *Yakugaku Zasshi*, **122**, 1–27; (c) Nagao, Y. and Sano, S. (2003) *J. Synth. Org. Chem. Jpn.*, **61**, 1088–1098.

20 Nagao, Y., Tanaka, S., Ueki, A., Jeong, I.-Y., Sano, S. and Shiro, M. (2002) *Synlett*, 480–482.

21 Jeong, I.-Y., Shiro, M. and Nagao, Y. (2000) *Heterocycles*, **52**, 85–89.

22 Nagao, Y., Ueki, A., Asano, K., Tanaka, S., Sano, S. and Shiro, M. (2002) *Org. Lett.*, **4**, 455–457.

23 Trost, B.M. and Xie, J. (2006) *J. Am. Chem. Soc.*, **128**, 6044–6045.

24 Yoshida, M., Sugimoto, K. and Ihara, M. (2004) *Org. Lett.*, **6**, 1979–1982.

25 (a) Liebeskind, L.S., Mitchell, D. and Foster, B.S. (1987) *J. Am. Chem. Soc.*, **109**, 7908–7910; (b) Mitchell, D. and Liebeskind, L.S. (1990) *J. Am. Chem. Soc.*, **112**, 291–296; (c) Liebeskind, L.S. and Bombrun, A. (1994) *J. Org. Chem.*, **59**, 1149–1159.

26 (a) Larock, R.C. and Reddy, C.K. (2000) *Org. Lett.*, **2**, 3325–3327; (b) Larock, R.C. and Reddy, C.K. (2002) *J. Org. Chem.*, **67**, 2027–2033.

27 The same ring expansion reaction was reported in 2003: Wei, L.-M., Wei, L.-L., Pan, W.-B. and Wu, M.-J. (2003) *Tetrahedron Lett.*, **44**, 595–597.

28 (a) Tsuji, J. (2004) *Palladium Reagents and Catalysts: New Perspectives for the 21st Century*, John Wiley & Sons, Chichester, pp. 543–563; (b) Tsuji, J. and Mandai, T. (1995) *Angew. Chem.*, **107**, 2830–2854; (1995) *Angew. Chem. Int. Ed.*, **34**, 2589–2612.

29 (a) Yoshida, M., Nemoto, H. and Ihara, M. (1999) *Tetrahedron Lett.*, **40**, 8583–8586; (b) Yoshida, M., Komatsuzaki, Y., Nemoto, H. and Ihara, M. (2004) *Org. Biomol. Chem.*, **2**, 3099–3107.

30 Yoshida, M., Sugimoto, K. and Ihara, M. (2001) *Tetrahedron Lett.*, **42**, 3877–3880.

31 For an overview of the ruthenium-catalyzed reactions, see: Trost, B.M., Frederiksen, M.U. and Rudd, M.T., (2005) *Angew. Chem.*, **117**, 6788–6825; (2005) *Angew. Chem. Int. Ed.*, **44**, 6630–6666.

32 Sugimoto, K., Yoshida, M. and Ihara, M. (2006) *Synlett*, 1923–1927.

33 Yoshida, M., Sugimoto, K. and Ihara, M. (2003) *ARKIVOC*, **xiii**, 35–44.

34 For reviews on gold-catalyzed reactions, see: (a) Krause, N. and Hoffmann-Röder, A. (2005) *Org. Biomol. Chem.*, **3**, 387–391; (b) Hashmi, A.S.K. and Hutchings, G.

(2006) *Angew. Chem.*, **118**, 8064–8105; (2006) *Angew. Chem. Int. Ed.*, **45**, 7896–7936.

35 Markham, J.P., Staben, S.T. and Toste, F.D. (2005) *J. Am. Chem. Soc.*, **127**, 9708–9709.

36 Yeom, H.-S., Yoon, S.-J. and Shin, S. (2007) *Tetrahedron Lett.*, **48**, 4817–4820.

37 For an overview on the reactivity of metal-carbenes: Dorwald, F.Z. (1999) *Metal Carbenes in Organic Synthesis*, Wiley–VCH, Weinheim.

38 Zora, M., Herndon, J.W., Li, Y. and Rossi, J. (2001) *Tetrahedron*, **57**, 5097–5107.

# 17
# Hydrometallation-Initiated Cyclization Reactions
*Isamu Matsuda*

## 17.1
## Introduction

Transition metals have demonstrated their importance in organic synthesis during recent decades [1]. Their ability to act as catalysts is also important from the viewpoint of sustainable chemistry. Hydrogen molecule and other hydride species are used as conventional reagents for reducing a carbon–carbon or carbon–heteroatom bond with the aid of a transition metal catalyst. Hydro-metallation is regarded as a fundamental step in this reaction. The carbon–metal species, regarded as a key intermediate in these reactions, would be expected to show high reactivity toward another component such as the species derived from the oxidative addition of aryl halide to a low valent metal species [2] or transmetallation with another carbanion source.

On the other hand, substituted carbocycles represent one of the most prevalent structural features of naturally occurring and biologically active compounds. For this reason, considerable effort has been directed toward the development of new methods for the synthesis of functionalized carbocycles. In this chapter, we focus on cyclization initiated by hydrometallation to a carbon–carbon unsaturated bond among many aspects on transition-metal catalyzed approaches.

## 17.2
## Self-Feeding Cyclization

### 17.2.1
### Intramolecular Hydrosilylation to a Double Bond

Many transition-metal catalysts are applicable to both intramolecular and intermolecular hydrosilylation. Exploration of the oxidation protocol of a carbon–silicon bond disclosed a novel route for transforming a carbon–carbon multiple bond to an alkanol which proceeds by a hydrosilylation protocol [3]. Thus, a new synthetic route

*Handbook of Cyclization Reactions. Volume 2.*
Edited by Shengming Ma
Copyright © 2010 WILEY-VCH Verlag GmbH & Co. KGaA, Weinheim
ISBN: 978-3-527-32088-2

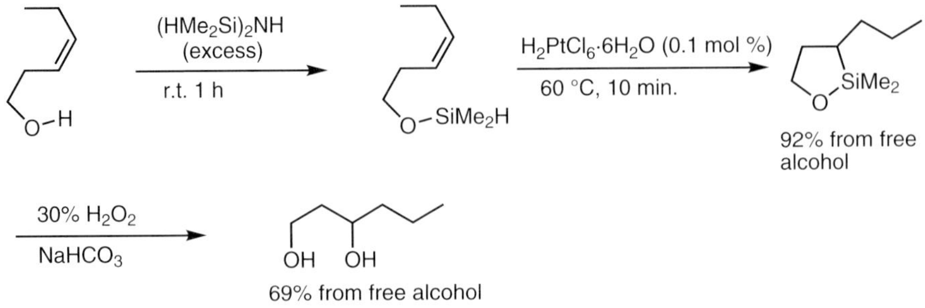

**Scheme 17.1** Regio-controlled route to hexane-1,3-diol.

to 1,3-diols and 2-aminoalcohols was explored by Tamao [4–6]. It is completed through a couple of sequential steps, starting from allylic or homoallylic alcohols and allylic amines, in which intramolecular hydrosilylation plays an important role. For example, (Z)-3-hexen-1-ol is transformed to hexane-1,3-diol in 69% overall yield, as shown in Scheme 17.1 [4]. The most important feature of this protocol is that oxidation of a Si–C moiety proceeds with the retention of the configuration at the cleaved carbon atom. An endo-trig type of ring-closure is observed exclusively in the reaction of 2-methyl-1-phenylpropen-1-ol to form the cyclic compound **1** with syn-selectivity (syn : anti = 6.7 : 1) (Eq. 17.1) [5a].

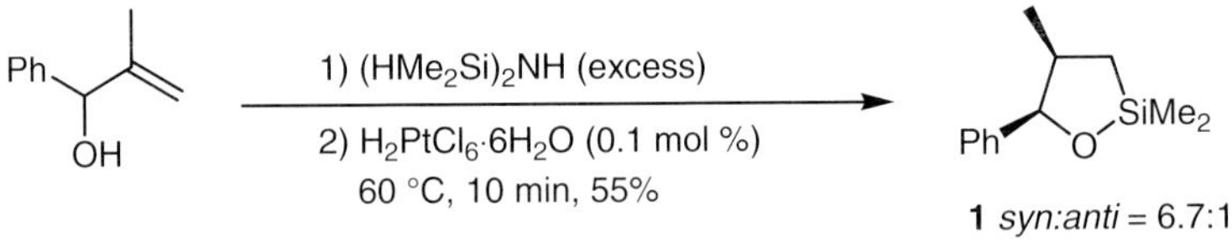

$$(17.1)$$

General trends for cyclization are summarized as follows: (i) An exo-trig mode is exclusive or predominant in the reaction of the internal type olefin moieties included in allylic or homoallylic alcohol derivatives (allyl/exo and homoallyl/exo). (ii) With the terminal olefins, an endo process occurs highly selectively in allylic cases (allyl/endo), while an exo mode is observed predominantly in homoallylic alcohols (homoallyl/exo). (iii) An endo mode occurs exclusively in reactions of vinylidene type allylic and homoallylic alcohols (allyl/endo and homoallyl/endo) [4]. The resultant cyclic silylether is converted to the corresponding diol regiospecifically [5].

In contrast to the ring-closure of allylic alcohols, which depends on the structure of the respective substrate, exo-trig closure occurs uniformly in the hydrosilylation of aminosilanes derived from allylic amines. It is also necessary to use Pt[{(CH$_2$=CH)Me$_2$Si}$_2$O]$_2$**2** instead of H$_2$PtCl$_6$·6H$_2$O as a catalyst (Eq. 17.2) [6]. The same catalyst may be applied to the synthesis of 1-aza-2,4-disilacyclobutane (Eq. 17.3) [7].

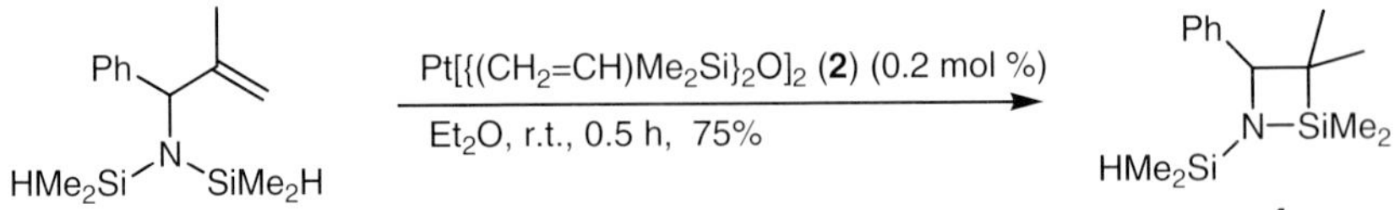

$$(17.2)$$

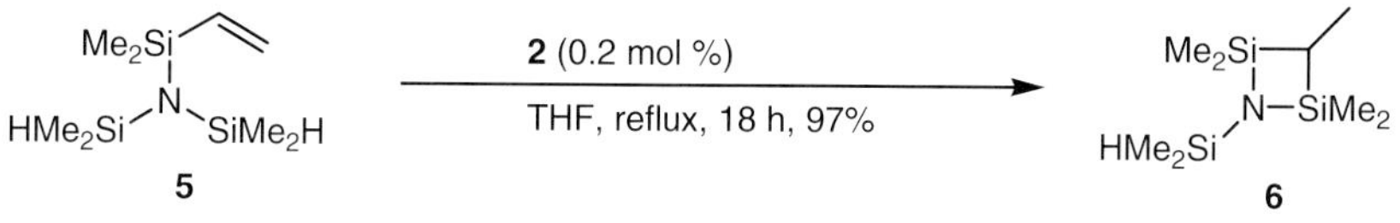

## 17.2 Self-Feeding Cyclization

(17.3)

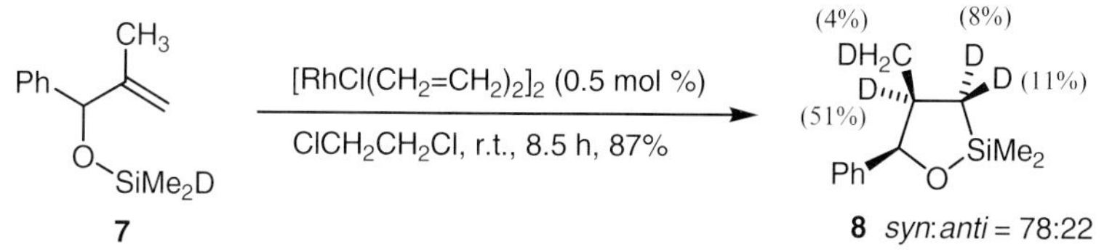

(17.4)

[RhCl(CH$_2$=CH$_2$)$_2$]$_2$ is also active as a catalyst precursor for intramolecular hydrosilylation. When Rh-catalyst is used, endo-trig type cyclization occurs exclusively in the reactions of **7** and **9** (Eq. 17.4 and Scheme 17.2). The result of **9** contrasts sharply with the exclusive exo-trig mode in the catalysis of **2**. It is concluded, from the deuterium labeled experiments, that the silylmetallation mechanism contributes in the Pt case and the hydrometallation mechanism in the Rh case [8].

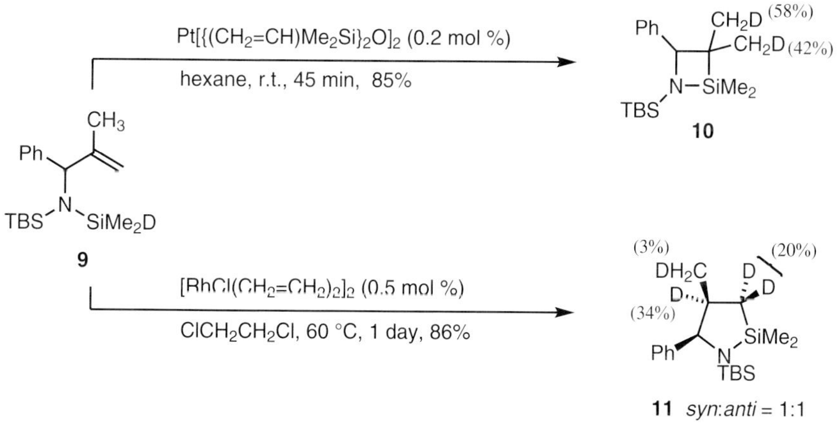

**Scheme 17.2** Clear metal-dependency in cyclization modes of **9**.

Intramolecular hydrosilylation occupies its own position as a reliable tool for the synthesis of obtusenyne [9], jatrophatrione, citlalitrione [10], and leucascandrolide A [11], as shown in Schemes 17.3–17.5.

In addition to hydroxy and amino groups, a carboxylic acid group can become a directing group to introduce a hydroxy group to a remote position through an intramolecular hydrosilylation and oxidation protocol. Optically active 2-cyclopentene carboxylic acid and N-BOC-3,4-dehydroproline are successfully transformed to the corresponding 2-hydroxycyclopentane carboxylic acid, and **20**, respectively (Scheme 17.6) [12].

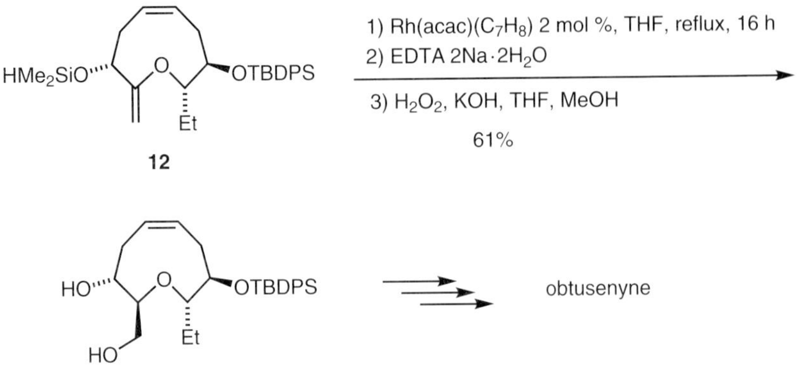
HMe2SiO
OTBDPS
Et
1) Rh(acac)(C7H8) 2 mol %, THF, reflux, 16 h
2) EDTA 2Na·2H2O
3) H2O2, KOH, THF, MeOH
61%
12
HO
OTBDPS
HO
Et
obtusenyne

**13** (*anti*:*syn* = >95:<5)

**Scheme 17.3** Application of intramolecular hydrosilylation for the synthesis of obtusenyne.

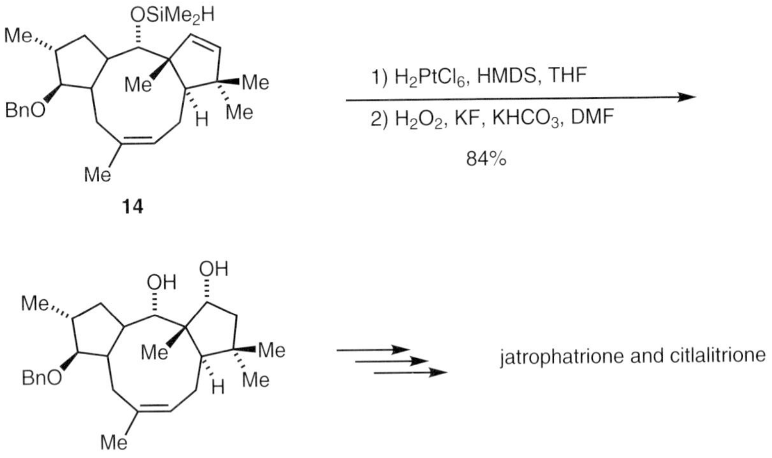
OSiMe2H
Me
BnO
Me
Me
H
Me
Me
1) H2PtCl6, HMDS, THF
2) H2O2, KF, KHCO3, DMF
84%
14
OH
OH
Me
BnO
Me
Me
H
Me
Me
jatrophatrione and citlalitrione

**15**

**Scheme 17.4** Application of intramolecular hydrosilylation for the synthesis of jatrophatrione and citlalitrione.

Although the reaction rate is quite slow, syn-selective asymmetric intramolecular hydrosilylation of **21** is attained to give **22** with 87% ee by pretreating $[RhCl(CH_2{=}CH_2)_2]_2$ with (*R,R*)-DIOP or (*R*)-BINAP (Eq. 17.5) [13].

When isolated $[Rh(S,S)\text{-chiraphos}]_2(ClO_4)_2$ is used as a catalyst, the reaction proceeds in a wide range of substrates at a far higher rate than that using the similar species formed *in situ* [14]. Olefin insertion into a Si–Rh bond is proposed for intramolecular hydrosilylation on the basis of these results and deuterium labeling experiments, however, it is quite difficult to explain consistently all the data at this stage without isolation of any intermediate. Enantioselectivity and deuterium distribution in the products are sharply affected, depending on the substrates, ligands, and solvents [15]. A carbonyl group, as well as carbon–carbon multiple bonds, is also

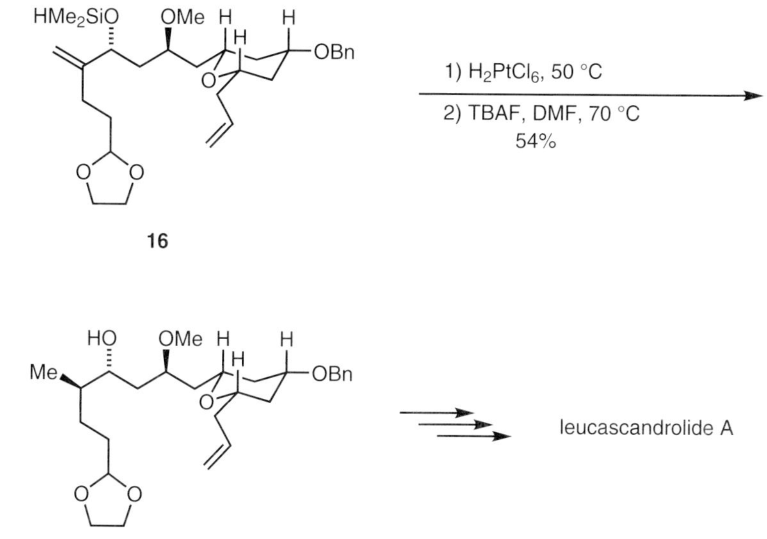

**Scheme 17.5** Application of intramolecular hydrosilylation for the synthesis of leucascandrolide A.

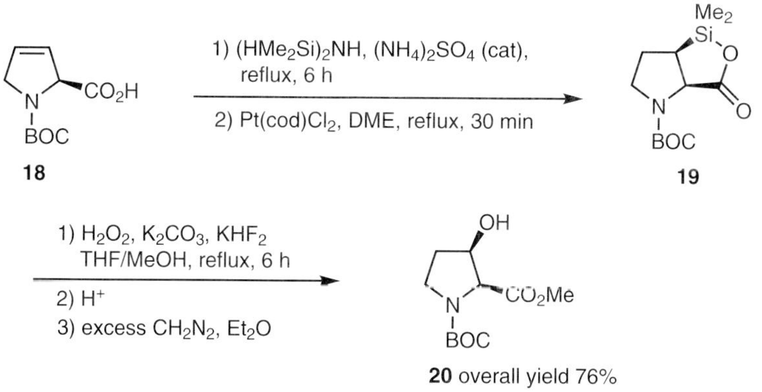

**Scheme 17.6** Regio-defined hydroxylation of olefin unit of **18**.

susceptible to intramolecular hydrosilylation, in which asymmetric induction is attained in good to excellent enantiomeric excess yield (Eq. 17.6) [16].

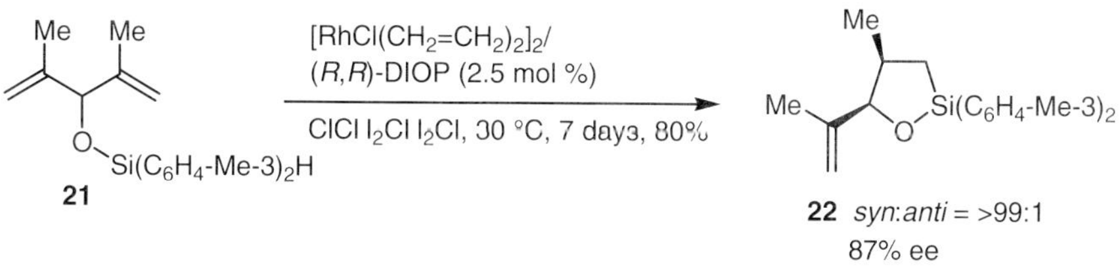

(17.5)

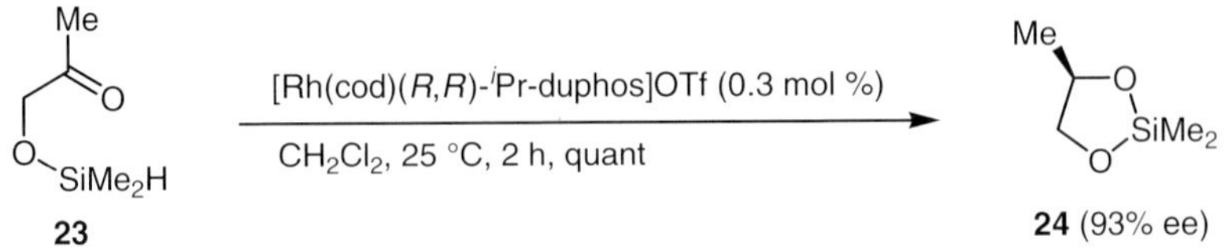

## 17 Hydrometallation-Initiated Cyclization Reactions

$$\text{23} \xrightarrow[\text{CH}_2\text{Cl}_2,\ 25\ °\text{C},\ 2\ \text{h, quant}]{[\text{Rh(cod)}(R,R)\text{-}^i\text{Pr-duphos}]\text{OTf (0.3 mol \%)}} \text{24 (93\% ee)} \quad (17.6)$$

Formation of products similar to **1** and **24** can be carried out in a one-pot operation using a dihydrosilane with any of an allylic alcohol, α,β-unsaturated aldehyde, or α-ketoaldehyde [17]. Consecutive intramolecular hydrosilylation using dihydrosilane was also demonstrated in the synthesis of a chiral spirosilane with high diastereo- and enantioselectivity (Eq. 17.7) [18].

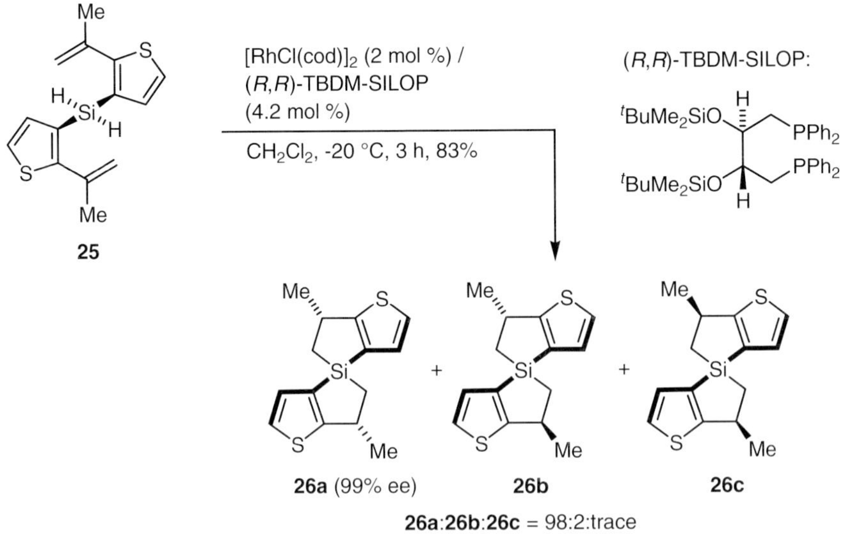

$$\text{25} \xrightarrow[\text{CH}_2\text{Cl}_2,\ -20\ °\text{C},\ 3\ \text{h, 83\%}]{\substack{[\text{RhCl(cod)}]_2\ (2\ \text{mol \%}) / \\ (R,R)\text{-TBDM-SILOP} \\ (4.2\ \text{mol \%})}} \text{26a} + \text{26b} + \text{26c}$$

**26a** (99% ee)  **26b**  **26c**

**26a:26b:26c** = 98:2:trace

(17.7)

Palladium complex **27**, which was designed for olefin polymerization [19], also works as a catalyst precursor for intramolecular hydrosilylation to form silacyclohexane rings selectively, in good yields, from **28** or **29** (Scheme 17.7) [20a]. It is known that an almost 1 : 1 mixture of six and seven-membered silacycles is given in the reaction of **29** with the assistance of $H_2PtCl_6$ [20b]. The catalytic behavior is explained by a similar mechanism to that proposed in the intermolecular version [21].

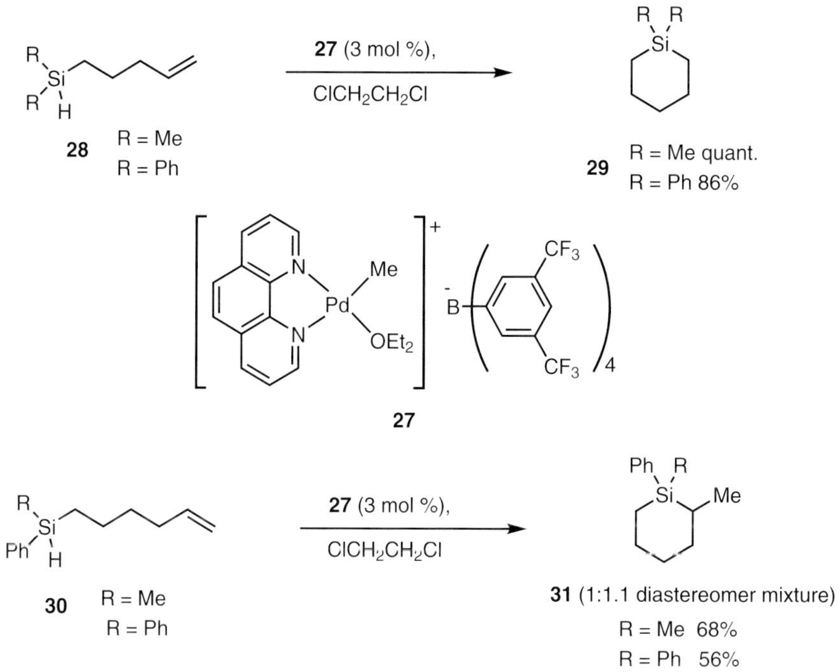

**Scheme 17.7** Pd-Catalyzed intramolecular hydrosilylation of **28** and **30**.

## 17.2.2
## Intramolecular Hydrosilylation to a C−C Triple Bond

### 17.2.2.1 syn-Addition

A similar protocol has been successfully extended to 3-decyn-1-ol in which intramolecular hydrosilylation occurs exclusively in an exo-dig mode with syn-addition to the C−C triple bond to give **33** (Eq. 17.8) [22].

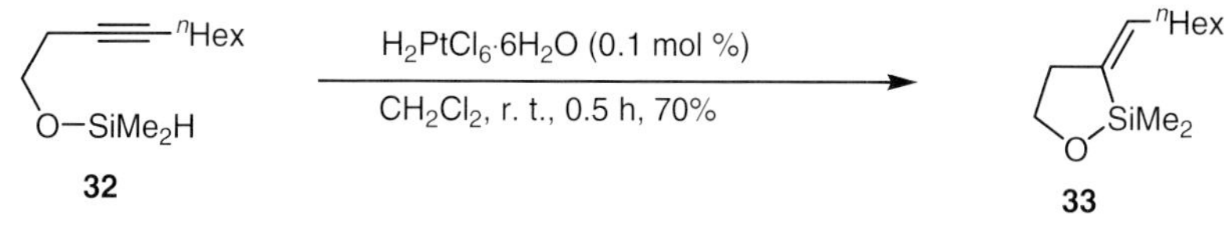

(17.8)

Detailed study of this cyclization and the subsequent oxidation demonstrates sufficient reliability for regiochemistry and stereochemistry in more complex models. A sequential transformation from a homopropargylic alcohol was utilized as a key transformation in the synthesis of (−)- and (+)-membrenone C **36** (Scheme 17.8) [23].

**Scheme 17.8** Application of intramolecular hydrosilylation for the synthesis of membrenone-C.

Since alkenylsilanes can be used as a partner in a coupling reaction with an aryl halide [24], the products formed from intramolecular hydrosilylation of alkynylsilanes can be used to construct designed trisubstituted alkenes [25].

### 17.2.2.2 Anti-Addition

(Z)-1-(o-Dimethylsilylphenyl)but-1-en-3-ynes **38**, which are readily derived from (Z)-1-(o-bromophenyl)but-1-en-3-ynes **37**, react to give 1,1-dimethyl-2-methylidenesilachromens **39**, with the aid of a catalytic amount of $H_2PtCl_6 \cdot 6H_2O$. The reaction proceeds in a regioselective (6-exo-dig) and stereoselective (E-selective) manner (Scheme 17.9) [26].

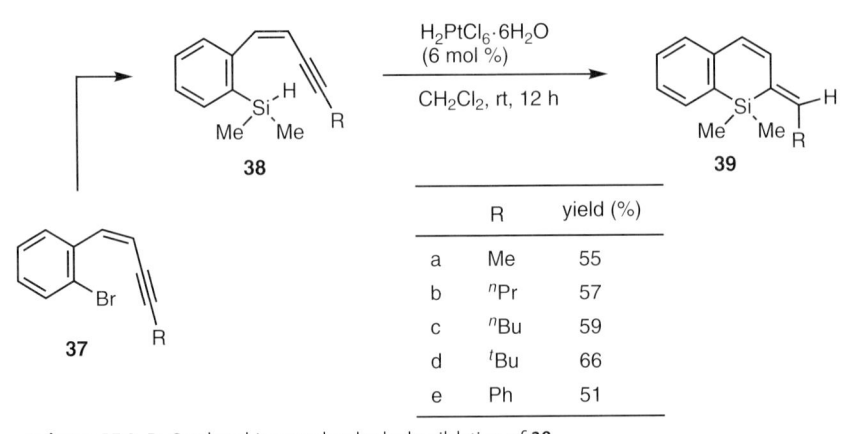

|   | R | yield (%) |
|---|---|---|
| a | Me | 55 |
| b | $^nPr$ | 57 |
| c | $^nBu$ | 59 |
| d | $^tBu$ | 66 |
| e | Ph | 51 |

**Scheme 17.9** Pt-Catalyzed intramolecular hydrosilylation of **38**.

Lewis acids such as $AlCl_3$ can also catalyze intramolecular hydrosilylation of alkynylsilanes to form silacycles. Anti-addition to a C–C triple bond is observed uniformly in all substrates, while 6-endo, 6-exo, and 7-exo cyclization occur specifically, corresponding to the structure of the starting substrates [27].

## 17.2 Self-Feeding Cyclization

When 3-pentynyloxydiisopropylsilane **40** was treated with a catalytic amount of RuHCl(CO)(P$^i$Pr$_3$)$_2$, **41** (5-exo-syn cyclization) was formed exclusively, whereas **42** was formed as the result of 5-exo-anti cyclization with [RuCl$_2$(C$_6$H$_5$)]$_2$ as the catalyst (Scheme 17.10).

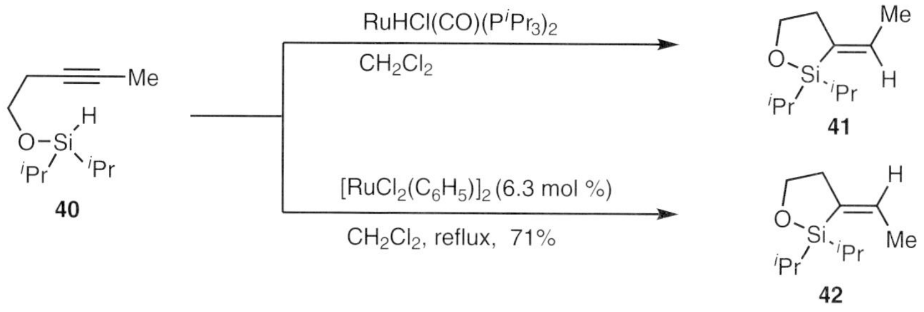

**Scheme 17.10** Syn- and anti-addition in Ru-catalyzed intramolecular hydrosilylation.

Less hindered 3-octynyloxydimethylsilane reacts to give oligomers **44** under similar conditions (Eq. 17.9). This product can be subjected to cross-coupling with alkenyl halides under the catalysis of a Pd complex [28]. The same catalyst provides **46** through an intramolecular hydrosilylation of **45** (Eq. 17.10). The product was obtained as oligomers which show sufficient reactivity for cross-coupling with iodoarenes [29].

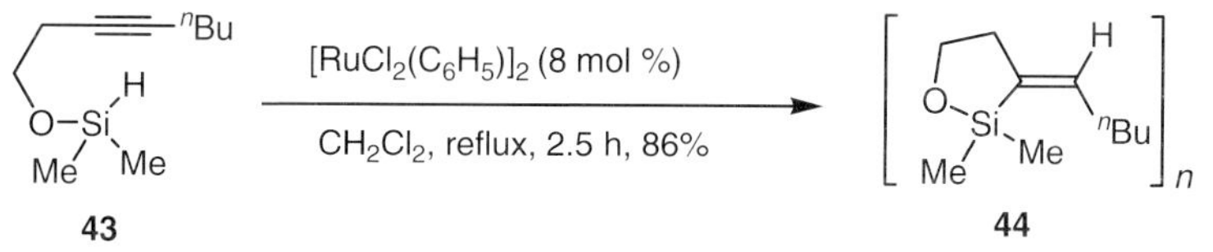

(17.9)

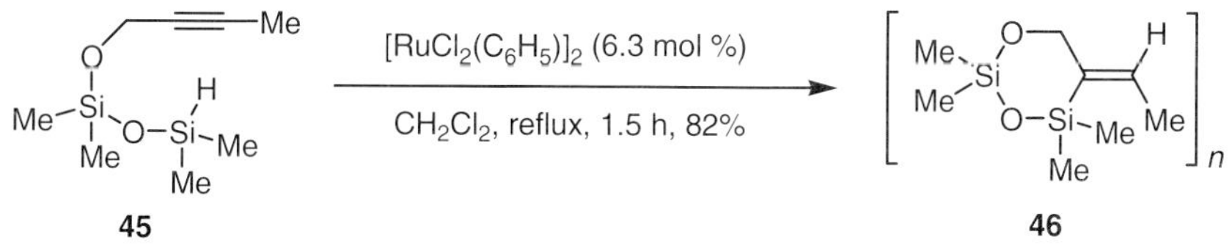

(17.10)

A complementary regio-chemistry for anti-addition to a C-C triple bond is realized by catalysis with [Cp*Ru(CH$_3$CN)$_3$]PF$_6$ for intermolecular hydrosilylation of 1-phenyl-2-butyn-1-ol **47** (Eq. 17.11) [30].

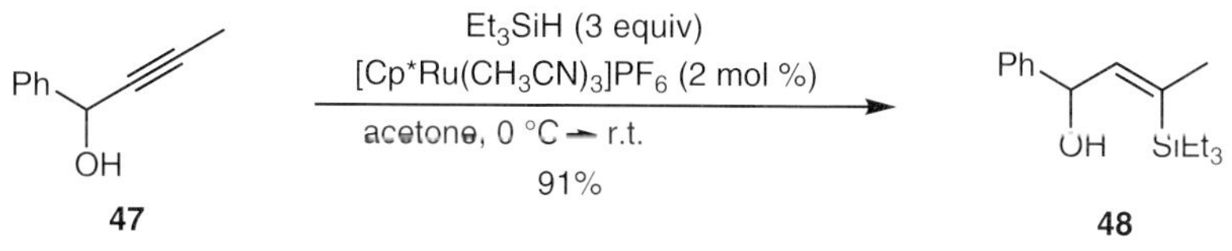

(17.11)

When this catalyst is applied to the intramolecular hydrosilylation of homopropargylic and bishomopropargylic alcohol derivatives **49**, cyclization proceeds selectively in the endo-dig mode. It is notable that seven-membered cyclic products **51** ($n = 2$) are formed selectively from bishomopropargylic alcohols **49** ($n = 2$), even though the exo-dig process would form the more favorable six-membered ring (Scheme 17.11) [31].

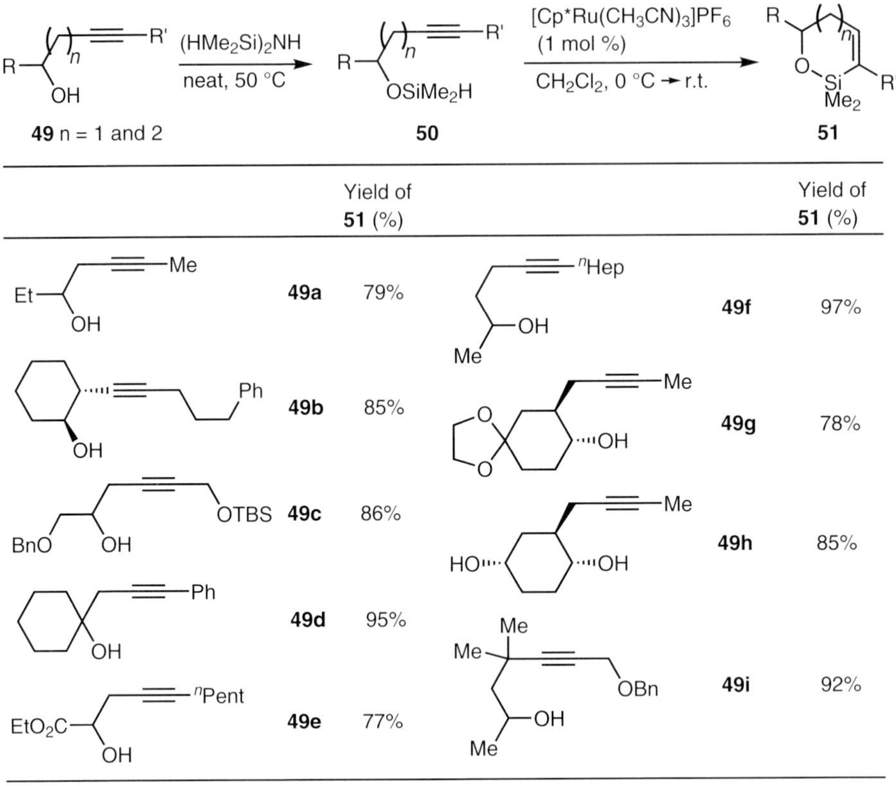

Scheme 17.11 Anti-endo-dig type intramolecular hydrosilylation.

Trost proposed a pathway to afford the anti-endo-dig product on the basis of a theoretical study: (i) insertion of a C—C triple bond into Ru—H to give a $\eta^2$-vinylruthenium intermediate is concerted with oxidative addition of the Si—H to the Ru-complex, (ii) a counterclockwise rotation around the resultant C—C double bond of the intermediate occurs during silylmigration from the Ru to the $sp^2$-carbon, and (iii) the anti-endo-dig product is formed by reductive elimination [32]. Modification of the oxidation conditions toward the vinylsilyl moiety makes possible an alkyne hydrosilylation-oxidation strategy to form the carbonyl functionality. Application of this method to **52**, which is readily constructed from 1-tetradecyne and 1-buten-3-ol, realizes a short synthesis of the piperidine alkaloid, (+)-spectaline **55** (Scheme 17.12) [33].

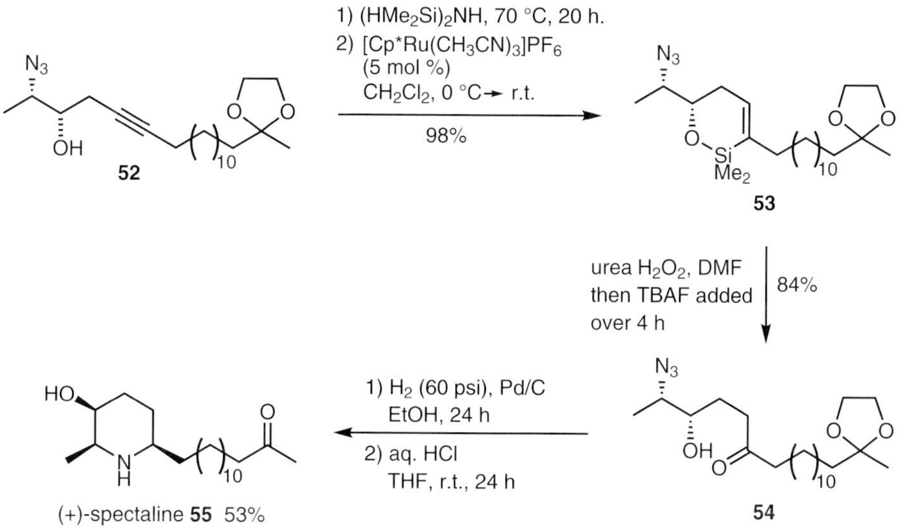

**Scheme 17.12** Ru-Catalyzed intramolecular hydrosilylation approach for the synthesis of (+)-spectaline **55**.

## 17.2.3
### Intramolecular Version of Silylformylation of Alkynes and Alkenes

In silylformylation of 1-alkynes, the internal sp carbon was formylated exclusively, while silylformylation of internal alkynes gives rise to lower regioselectivity. An sp-carbon bearing a bulkier substituent seems to be preferred to that with a less bulky substituent at the formylation site. For example, 2-hexyne and 1-phenylpropyne gave the corresponding products with a mixture of the regioisomers, respectively [34]. On the basis of a general nature in the inter-molecular silylformylation, intramolecular hydrosilylation, and cyanosilylation [35], it seems reasonable to assume that alkynyldiorganylsilanes could react with CO in an intramolecular fashion. Thus, regio-defined formylation of internal acetylenic bonds is achieved by means of intramolecular silylformylation. Reaction of 4-heptynyldimethylsilane **56** is shown in Scheme 17.13 as a typical example. **56** readily incorporates CO to give aldehyde (Z)-**57** in 83% yield with the aid of $Rh_4(CO)_{12}$. According to Baldwin's rule, this cyclization pattern may be categorized as the 5-exo-dig mode if the rule is applicable to a reaction involving organometallic species. This simple operation does not give any endo-dig mode cyclization product **58** or intermolecular silylformylation of **56**. However, alkynylsilane **56** behaves as a hydrosilane in the intermolecular syn-silylformylation with phenylacetylene to give **59** in 86% yield (Scheme 17.13) [36].

Other examples of syn-exo-dig cyclic silylformylation are summarized in Table 17.1. Alkynes **60**, **62**, and **64** afforded **61**, **63**, and **65**, respectively, as the sole product of CO incorporation. Regardless of substituents, both alkyl and aryl acetylenes undergo cyclization, clearly discriminating two sp-carbons. Intramolecular cyclization is particularly effective with terminal alkynes **66** [36], **68** [36], and **70** [37] containing three or four methylene units between the acetylenic and silyl moieties.

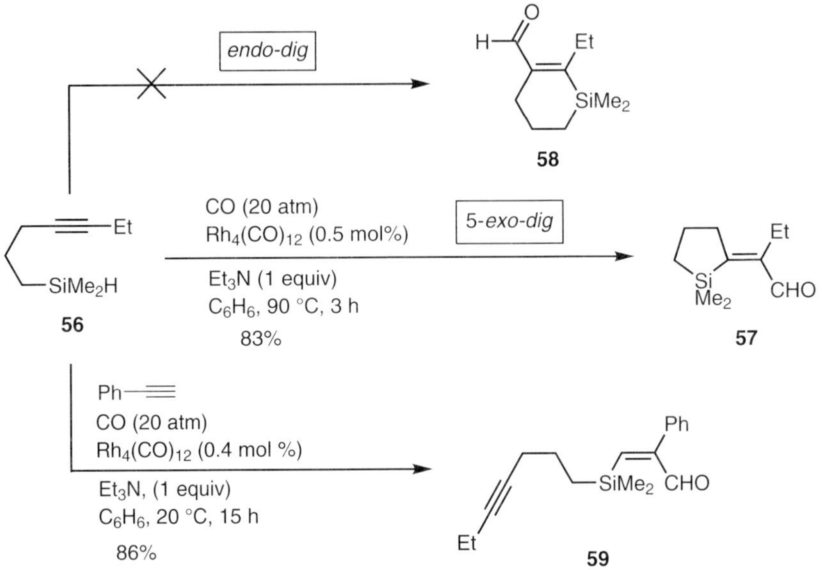

**Scheme 17.13** Intra- and inter-molecular silylformylation of **56**.

But-3-ynyl(methyl)phenylsilane does not give any positive result for CO incorporation. The regioselectivity of the silylformylation is completely reversed from that in the standard silylformylation. A bulky *tert*-butyl group in **70**($R^1 = {}^tBu$) plays an important role for diastereoselective formation of **71** with trans-geometry between phenyl(Si) and the *tert*-butyl group, whereas the methyl group in **70** ($R^1 = Me$) is a spectator of the diastereomeric discrimination [37].

A hydrosilyloxy moiety also works as a directing group for regio-control in the intramolecular silylformylation of both 1-alkynes and internal alkynes (Table 17.2) [38].

This type of intramolecular silylformylation is applicable to 4-silyloxy-1,6-diynes **84** which give **85** in good yields. The remaining alkynyl moiety in **85** can further undergo intermolecular silylformylation (Eq. 17.12) [39].

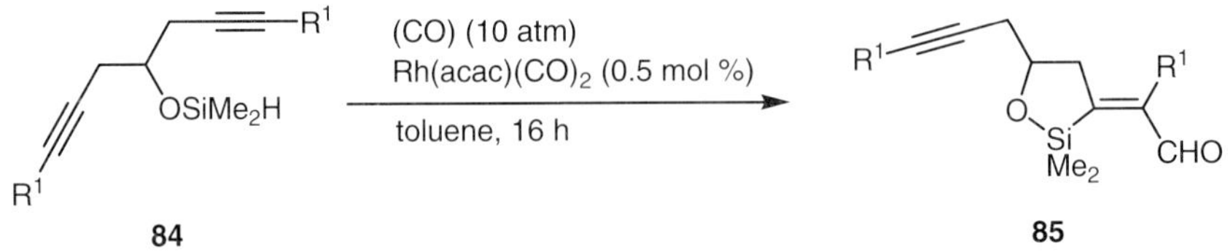

|   | $R^1$ | temperature (°C) | Yield (%) |
|---|-------|------------------|-----------|
| a | H     | 25               | 98        |
| b | Me    | 60               | 82        |

(17.12)

**Table 17.1** [Rh]-catalyzed intramolecular silylformylation of alkynylsilanes[a].

| No. | Alkynylsilanes | Catalyst[b] | Products | Yield (%) | Ref. |
|---|---|---|---|---|---|
| 1 | 60 | A | 61 | 79 | [36] |
| 2 | 62 | A | 63 | 71 | [36] |
| 3 | 64 | A | 65 | 69 | [36] |
| 4 | 66, $R^1$ = Me | A | 67 | 56 | [36] |
| 5 | 66, $R^1$ = Ph | B | 67 | 37[c] | [36] |
| 6 | 68, $R^1$ = Me | B | 69 | 56[c] | [36] |
| 7 | 68, $R^1$ = Ph | B | 69 | 49[c] | [36] |
| 8 | 70, $R^1$ = Me | B | 71 | 73[c] | [37] |
| 9 | 70, $R^1$ = $^t$Bu | B | 71 | 70[c] | [37] |

[a]Reactions were conducted on a 1 mmol scale of an alkynylsilane in $C_6H_6$ solution with 0.5 mol% of catalyst and an equimolar amount of $Et_3N$ under CO pressure (20 atm) for 3 h at 90 °C unless otherwise noted.
[b]A; $Rh_4(CO)_{12}$, B;.

[c]conditions: catalyst (1 mol%), CO (20 atm), $CH_2Cl_2$, 40 °C, 24 h.

**Table 17.2** [Rh]-catalyzed intramolecular silylformylation of ω-(hydrodimethylsilyloxy)alkynes [38].

| No. | Alkynes | Catalyst[a] | Products | Yield (%)[c] |
|---|---|---|---|---|
| 1 | **72** | ($^t$BuNC)$_4$RhCo(CO)$_4$ | **73** | 42 |
| 2 | | Rh$_2$Co$_2$(CO)$_{12}$ | | (46) |
| 3 | | Rh(acac)(CO)$_2$ | | (62) |
| 4 | **74** | ($^t$BuNC)$_4$RhCo(CO)$_4$ | **75** | (81) |
| 5 | | Rh(acac)(CO)$_2$ | | (73) |
| 6 | **76** | Rh$_4$(CO)$_{12}$, Et$_3$N[b] | **77** | 77 |
| 7 | **78** | ($^t$BuNC)$_4$RhCo(CO)$_4$ | **79** | 66 |
| 8 | | Rh$_2$Co$_2$(CO)$_{12}$ | | (84) |
| 9 | | Rh(acac)(CO)$_2$ | | (93) |
| 10 | | Rh$_4$(CO)$_{12}$, Et$_3$N[b] | | 69 |
| 11 | **80** | ($^t$BuNC)$_4$RhCo(CO)$_4$ | **81** | 69 |
| 12 | | Rh$_2$Co$_2$(CO)$_{12}$ | | (92) |
| 13 | | Rh(acac)(CO)$_2$ | | (94) |
| 14 | **82** | ($^t$BuNC)$_4$RhCo(CO)$_4$ | **83** | 73 |
| 15 | | Rh$_2$Co$_2$(CO)$_{12}$ | | (80) |
| 16 | | Rh(acac)(CO)$_2$ | | (95) |

[a] Reactions were conducted on a 0.5 mmol scale of an alkyne in toluene solution with 0.5 mol% of catalyst under CO pressure (10 atm) at 65–70 °C for 3–23 h unless otherwise noted.
[b] Reactions were conducted on a 2.8 mmol scale of an alkyne in C$_6$H$_6$ solution with 0.25 mol% of catalyst and 2.9 mmol of Et$_3$N under CO pressure (20 atm) at 85 °C for 3 h.
[c] Isolated yield. The value in parentheses is GC yield.

A similar approach using an aminosilyl directing group is possible. In fact, alkynylamine **86** reacts with CO to give **87** under the catalysis of Rh(acac)(CO)$_2$ (Eq. 17.13). Since 3-aminosilylalkenals derived from other alkynylamines are too unstable to be isolated, the corresponding reaction mixture was converted to the

## 17.2 Self-Feeding Cyclization

**Table 17.3** [Rh]-catalyzed intramolecular silylformylation of bis(silyl)aminoalkynes [40].

| No. | Substrates | Catalyst | Products[a] | Yield (%) |
|---|---|---|---|---|
| 1 | **88** | Rh(acac)(CO)$_2$ | **89** | 52 |
| 2 |  | ($^t$BuNC)$_4$RhCo(CO)$_4$ |  | 80 |
| 3 |  | Rh$_2$Co$_2$(CO)$_{12}$ |  | 75 |
| 4 | **90** | Rh(acac)(CO)$_2$ | **91** | 71 |
| 5 |  | ($^t$BuNC)$_4$RhCo(CO)$_4$ |  | 87 |
| 6 | **92** | Rh(acac)(CO)$_2$ | **93** | 90 |
| 7 | **94** | Rh(acac)(CO)$_2$ | **95** | 84 |
| 8 | **96** | Rh(acac)(CO)$_2$ | **97** | 91 |
| 9 |  | ($^t$BuNC)$_4$RhCo(CO)$_4$ |  | 78 |

[a]Reactions were conducted on a 0.5 mmol scale of an alkyne in toluene solution with 0.5 mol% of catalyst under CO pressure (10 atm) at 60 °C for 14 h. The resulting mixture was poured into a solution of NaBH$_4$ (10 equiv) in MeOH at 0 °C to form the final products.

stable alcohols by reduction with NaBH$_4$ in MeOH without isolation of the silylformylated product (Table 17.3) [40].

(17.13)

Although silylformylation of simple alkenes does not proceed under the standard conditions for alkynes, the first example of the silylformylation for a C—C double bond was successfully attained in the reaction of **98** under CO pressure [41]. Since the most serious competitive reaction is intramolecular hydrosilylation with the 5-exo-trig mode, a key point for selective silylformylation is an appropriate choice of the combination of substituents $R^1$ and R in **98**. The presence of $R^1$ (except H) is required for advantageous selectivity leading to silylformylation. Because of the instability of **99** the structure of **99** is identified as **100** which is obtained by reduction of the formyl group in **99** followed by acetylation. An additional feature of this silylformylation is cis-selectivity with respect to the oxasilacyclopentane ring in **99** (Scheme 17.14) [41].

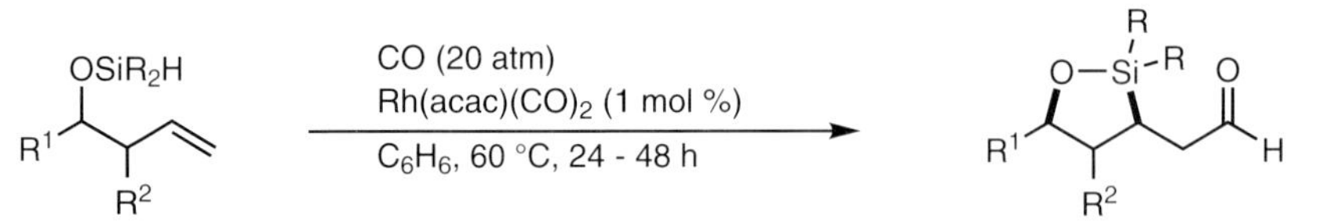

|   | $R^1$ | $R^2$ | R | Yield (%) of **100** | dr |
|---|---|---|---|---|---|
| a | Me | H | Ph | 67 | 4.5:1 |
| b | allyl | H | Ph | 64 | 4:1 |
| c | $^i$Pr | H | Ph | 79 | 6:1 |
| d | $^t$BuMe$_2$SiO(CH$_2$)$_2$ | H | Ph | 60 | 4:1 |
| e | Ph | H | $^i$Pr | 54 | 7:1 |
| f | $^i$Pr | Me | Ph | 71 | 10:1 |

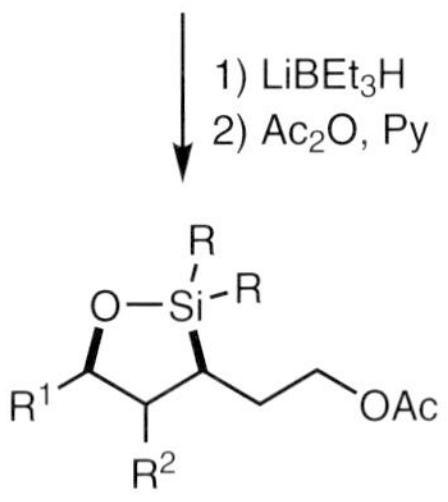

**Scheme 17.14** Rh-catalyzed intramolecular silylformylation of **98**.

## 17.3
## Cyclization Tied by Outsourcing Reagents

### 17.3.1
### Hydrosilane

#### 17.3.1.1 Diynes
Since it was revealed that a Pd(0) complex generates a Pd—H species in acetic acid, cycloisomerization of linear diynes, enynes, and dienes on the basis of this knowledge has become a powerful tool for the construction of complex frameworks in synthetic organic chemistry [42]. In the reaction of a hydrosilane with 1,6-diynes a hydrosilane behaves as a hydrogen-atom donor and the silyl moiety is converted to a silyl acetate, as shown in Eq. 17.14 [43]. Later the system was modified by replacing acetic acid with formic acid to enhance the activity for the synthesis of more complex

molecules with P(2-furyl)$_3$ as the ligand (Eq. 17.15) [44].

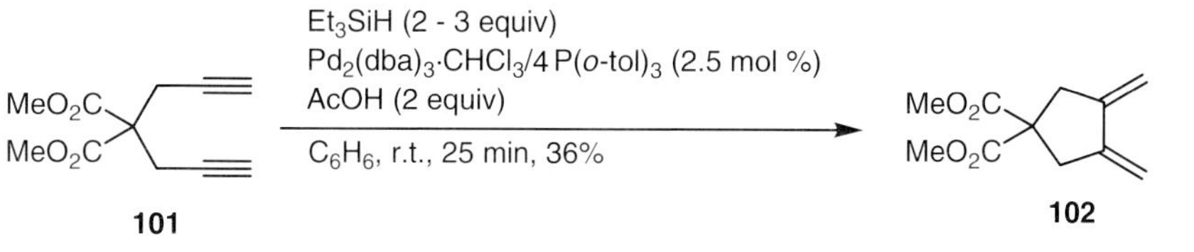

$$\text{101} \xrightarrow[\text{C}_6\text{H}_6,\ \text{r.t.,}\ 25\ \text{min,}\ 36\%]{\substack{\text{Et}_3\text{SiH (2 - 3 equiv)} \\ \text{Pd}_2(\text{dba})_3 \cdot \text{CHCl}_3/4\ \text{P}(o\text{-tol})_3\ (2.5\ \text{mol \%}) \\ \text{AcOH (2 equiv)}}} \text{102}$$

(17.14)

$$\text{103} \xrightarrow[\text{toluene, 80 °C, 20 min, 87\%}]{\substack{\text{Et}_3\text{SiH (2 equiv)} \\ \text{Pd}_2(\text{dba})_3 \cdot \text{CHCl}_3/4\ \text{P}(2\text{-furyl})_3\ (2.5\ \text{mol \%}) \\ \text{HCOOH (2 equiv)}}} \text{104}$$

Ar = 2,6-diMeO-4-MeC$_6$H$_2$

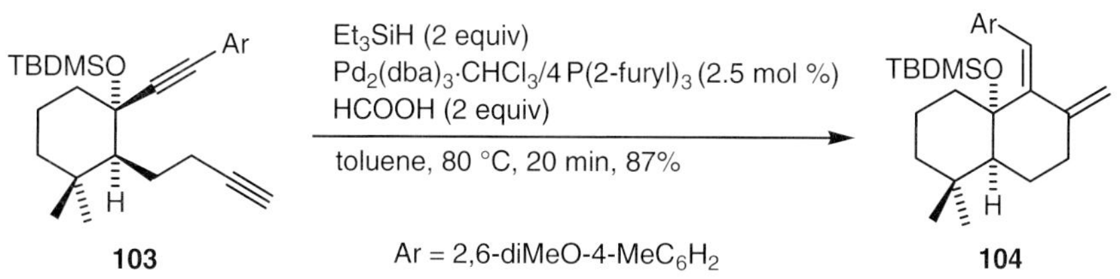

(17.15)

A hydrosilane was successfully incorporated into 1,7-diynes to form six-membered cyclic compounds in the presence of Ni(0) catalyst (Scheme 17.15) [45]. The silyl group in the products may be converted to a hydroxy group or an aryl group.

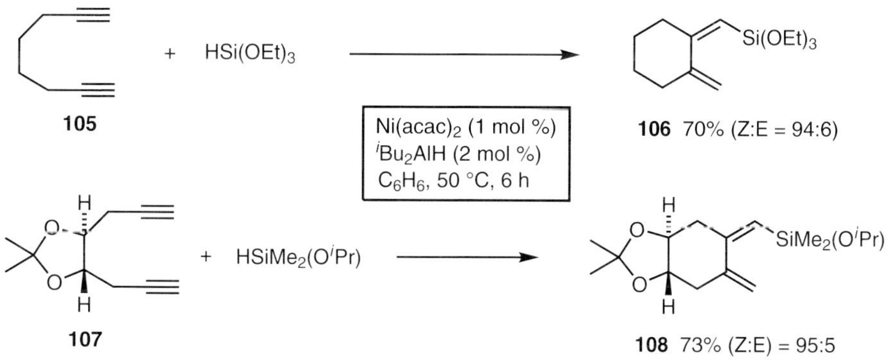

105 + HSi(OEt)$_3$ → 106 70% (Z:E = 94:6)

Ni(acac)$_2$ (1 mol %)
$^i$Bu$_2$AlH (2 mol %)
C$_6$H$_6$, 50 °C, 6 h

107 + HSiMe$_2$(O$^i$Pr) → 108 73% (Z:E) = 95:5

**Scheme 17.15** Ni-catalyzed silylative annulation of 1,7-diynes.

The inefficiency of the Ni(0) catalyst for 1,6-diynes is overcome by using RhCl(PPh$_3$)$_3$ or RhH(SiR$_3$)(Cl)(PPh$_3$)$_3$ as catalyst in place of Ni(0). Since acetylenic triple bonds are relatively susceptible to a transition-metal complex, it is imperative in the reaction of 1,6-heptadiyne to suppress the side reaction in which two acetylenic bonds react independently with hydrosilanes and cyclocodimerization occurs between two starting substrates. These issues may be addressed by adopting the slow addition of a solution containing the starting substrate and the hydrosilane to a solution of the catalyst. Thus, reproducible results are obtained in the reactions using RhH(SiR$_3$)(Cl)

(PPh$_3$)$_2$ as a catalyst. E-selectivity was observed with a moderate yield of the desired product (Eq. 17.16) [46].

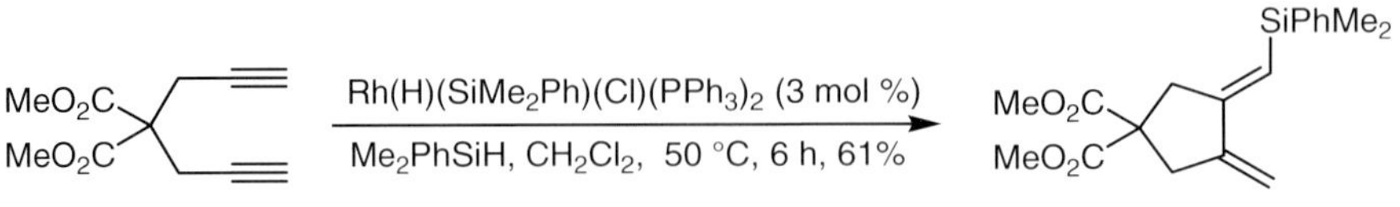

$$\text{101} \xrightarrow[\text{Me}_2\text{PhSiH, CH}_2\text{Cl}_2,\ 50\ °\text{C, 6 h, 61\%}]{\text{Rh(H)(SiMe}_2\text{Ph)(Cl)(PPh}_3)_2\ (3\ \text{mol \%})} \text{109}$$

(17.16)

Use of the Pt(II) complex **110** together with B(C$_6$F$_5$)$_3$ as the catalyst system dramatically improves product selectivity, reaction rate, and Z-selectivity in a wide range of substrates, as shown in Scheme 17.16. The catalyst is effective in the reaction

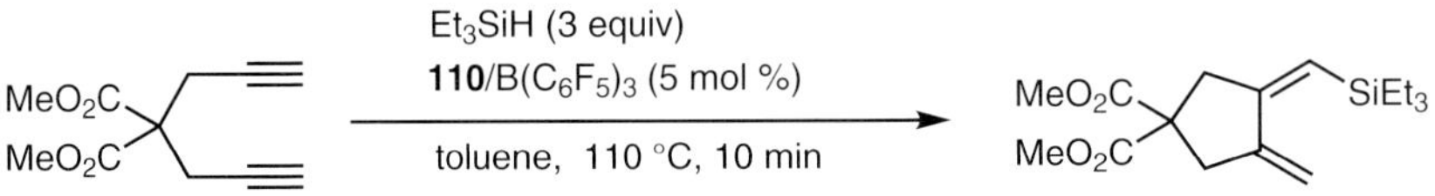

$$\text{101} \xrightarrow[\text{toluene, 110 °C, 10 min}]{\text{Et}_3\text{SiH (3 equiv)}\quad \text{110/B(C}_6\text{F}_5)_3\ (5\ \text{mol \%})} \text{111}\ 82\%\ (>30:1)$$

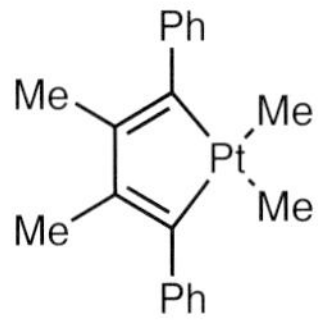

**110**

**112**
R = Piv 76% (29:1)
R = Bn 86% (>30:1)
R = TBDMS 80% 23:1

**113** 77% (26:1)

**114**
R = Ph 84% (25:1)
R = SO$_2$Me 74% (29:1)
R = CONMe$_2$ 80% (>30:1)
R = COMe 65% (29:1)

**115** 72% (30:1)

**116** 78% (12:1)

**117** 36% (21:1)

**118** 43% (>30:1)

**119** 56% (9:1)

**120** 77% (28:1)

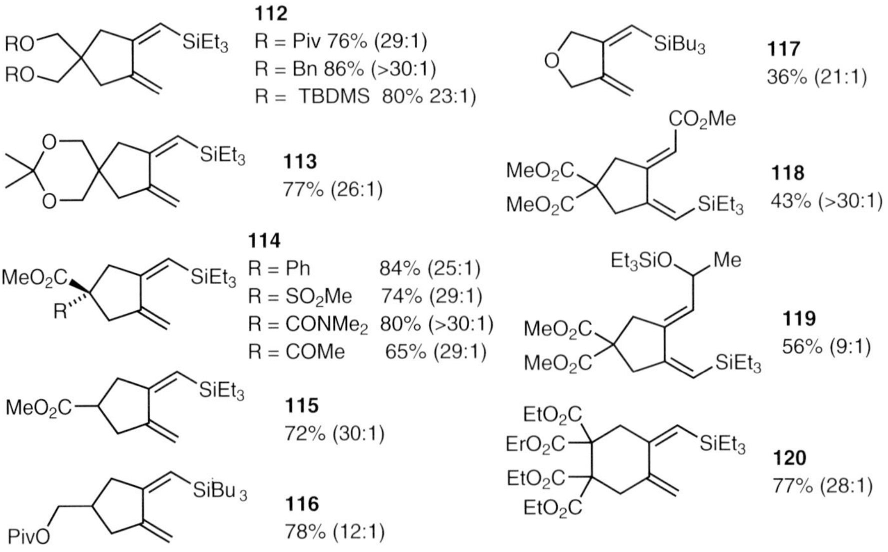

**Scheme 17.16** Pt(II)-catalyzed silylative annulation of 1,6- and 1,7-diynes.

of 1,6-diynes with a methoxycarbonyl group or acetyl group on the terminal sp-carbon. Its effectiveness has also been demonstrated for the reaction of 4,4,5,5-tetraethoxycarbonyl-1,7-octadiyne with $Et_3SiH$ to give **120** [47].

In contrast to these examples, this Pt(II) catalyst system is not applicable to diynes having two internal acetylenic moieties since separate hydrosilylation of each C—C triple bond is performed in preference to cyclization. The formation of disilylated compound is efficiently suppressed to give dialkylidenecyclopentane derivatives with sufficient Z-selectivity by catalysis with $[Rh(binap)(cod)]BF_4$ combined with the slow addition of a hydrosilane into a solution of the diyne and catalyst. Similar products listed below are uniformly obtained from the corresponding diyne and a hydrosilane in moderate yield and with good diastereoselectivity, as shown in Scheme 17.17 [48].

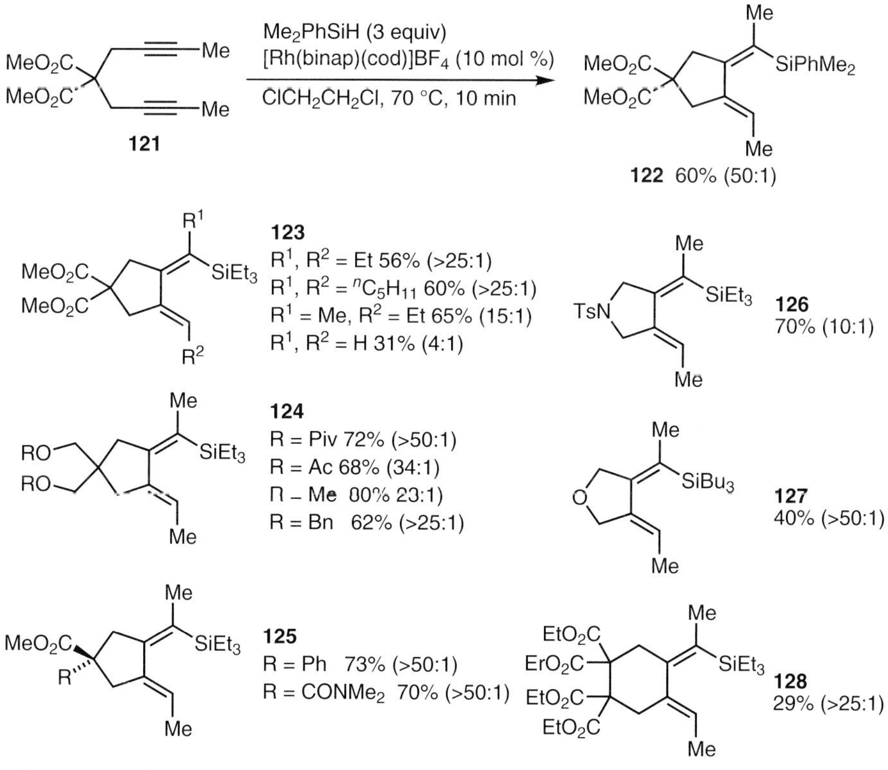

**Scheme 17.17** Rh-catalyzed silylative annulation of 1,6- and 1,7-diynes.

Although no intermediate has been isolated in these reactions, a rationale for the formation of products is shown in Scheme 17.18. The metallated alkylidene ring **132** would be formed by insertion of a C—C triple bond of **130** into the M—Si bond of **129**.

Oxidative addition of another hydrosilane with M and reductive elimination gives **134** and regenerates the M−Si species (Scheme 17.18).

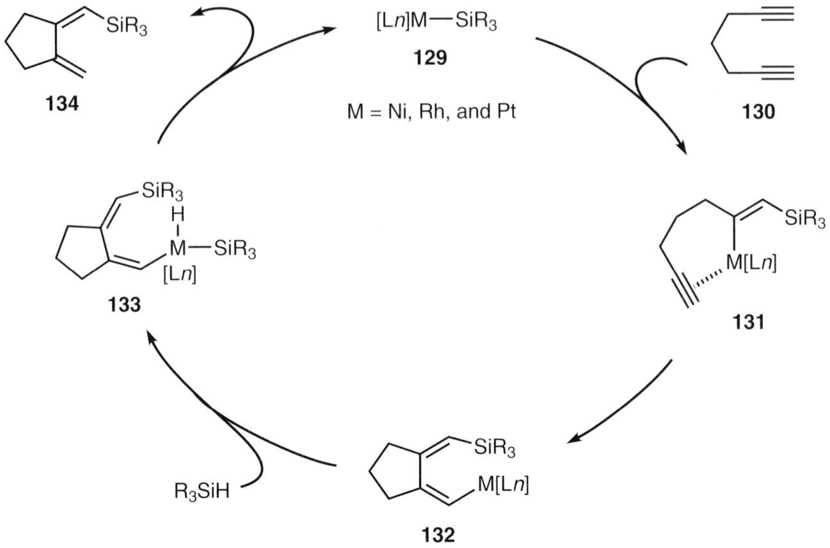

**Scheme 17.18** Rationale for the silylative annulation of 1,6-heptadiyne.

[Pd(η$^3$-C$_3$H$_5$)(cod)]PF$_6$ also catalyzes an analogous cyclization, however, the rate is quite low and the regiochemistry in the reaction of **135** (Eq. 17.17) and 4-oxa-1,6-octadiyne is the reverse of that observed in the reactions catalyzed by other metal species. These contrasting results in regiochemistry can be explained by supposing the steps to be different from Scheme 17.18: (i) a less hindered C−C triple bond inserts into the Pd−H species and (ii) subsequent annulation forms the Pd–vinyl species which interacts with another molecule of a hydrosilane to give the product [49].

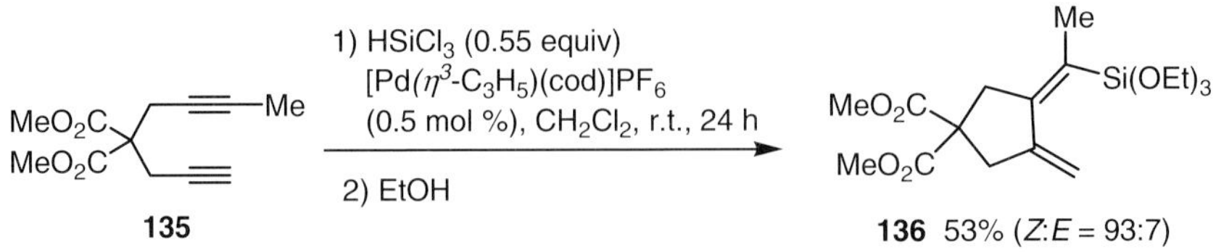

$$(17.17)$$

$^n$Bu$_3$SnH [50], Me$_3$SiSnMe$_3$ [51], and 1,3-dimethyl-2-dimethylphenylsilyl-2-bora-1,3-diazacyclopentane [52] is applicable to the cyclization of 1,6-diynes with the aid of a Pd catalyst. These starting substrates play the role of strings like R$_3$SiH in the diyne cyclization.

**Table 17.4** Rh-catalyzed hydrogenative annulation of 1,6-diynes.

|   | X | R¹ | R² | Ligand | Yield (%) |
|---|---|----|----|--------|-----------|
| a | H$_2$C | Ph | Ph | BINAP | 89 |
| b | (MeO$_2$C)$_2$C | Ph | Ph | BINAP | 85 |
| c | (MeO$_2$C)$_2$C | Ph | Me | BIPHEP | 68 |
| d | (MeO$_2$C)$_2$C | Ph | Me$_3$Si | BIPHEP | 51 |
| e | (MeO$_2$C)$_2$C | Me | Me | BINAP | 73 |
| f | cyclohexane-1,3-dione-2,2-diyl | Ph | Ph | BIPHEP | 90 |
| g | 2,2-dimethyl-1,3-dioxane-4,6-dione-5,5-diyl | Ph | Ph | BIPHEP | 79 |
| h | O | Ph | Ph | BINAP | 78 |
| i | TsN | Me | Me | BINAP | 62 |

Recently, it has been demonstrated that molecular hydrogen participates in the reductive cyclization of 1,6-diynes **137** under the catalysis of Rh(cod)$_2$OTf/*rac*-BINAP or BIPHEP (Table 17.4). The formation of **138** is elucidated by either one of two catalytic cycles, the hydrometallation mechanism (cycle B) and the oxidative coupling mechanism (cycle A), as shown in Scheme 17.19 [53]. It is quite difficult to discriminate unambiguously between the two cycles.

### 17.3.1.2 Enynes

Enynes also readily cyclize when a hydrosilane is added in the presence of a Rh complex. In fact *N*-benzyl-*N*-propargylallylamine reacts with Me$_2$PhSiH to give the expected product even under CO pressure (20 atm) in the presence of Rh$_4$(CO)$_{12}$ or Rh(CO)$_2$(acac) as a catalyst (conditions a in Scheme 17.20) [54]. The same transformation was also achieved under atmospheric pressure of CO in hexane at room temperature (conditions b in Scheme 17.20), when identical substrates are

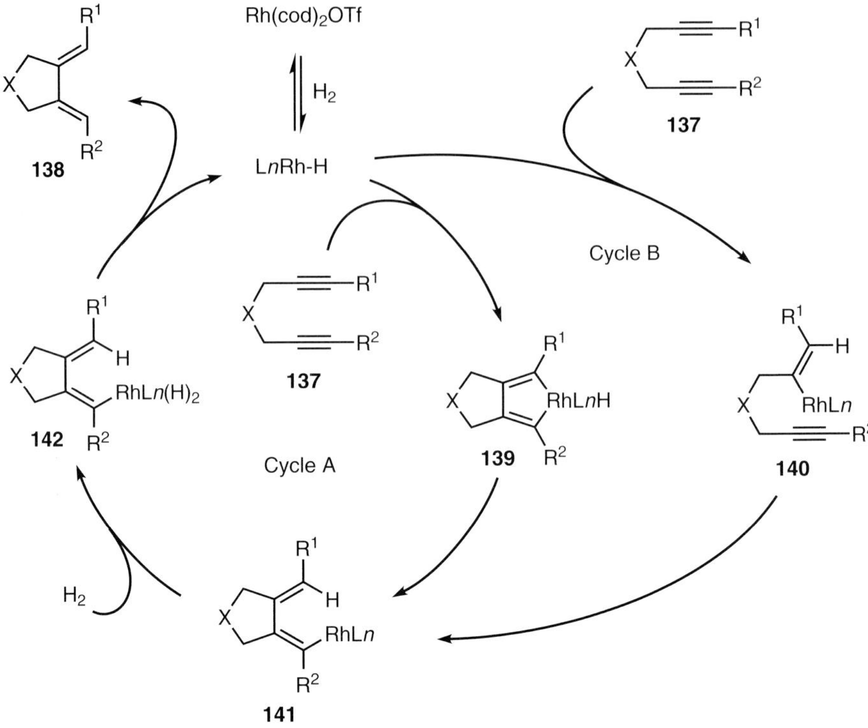

Scheme 17.19 Proposed mechanism for the hydrogenative annulation of 1,6-diynes.

reacted with catalysis by $Rh_4(CO)_{12}$. Extension of this type of cyclization to a wide range of enynes and hydrosilanes under the optimized conditions in Scheme 17.20 has been reported [55]. The benzyldimethylsilyl group in the products can be successfully utilized for further cross coupling [56]. Rhodium complexes modified by a carbene ligand **156** may also be used as catalyst for a similar cyclization [57]. The catalytic cycle here also seems to proceed similarly to Scheme 17.18: the formation of an M−SiR$_3$ species by the interaction of rhodium catalyst with a hydrosilane, insertion of a C−C triple bond into a Rh−Si bond to form the carbocycle with a metallated alkyl group instead of the metallated vinylidene group of **132** in Scheme 17.18, subsequent oxidative addition of another hydrosilane to the rhodium metal, and reductive elimination of a product to regenerate the Rh−Si species (Scheme 17.20).

Asymmetric induction for this protocol has been realized in a good enantiomeric excess with a moderate chemical yield by using a mixture of [Rh(cod)$_2$]SbF$_6$/(R)-BIPHEMP as a catalyst precursor (Eq. 17.18) [58]. Higher enantioselectivity is attained by replacement of [Rh(cod)$_2$]SbF$_6$/(R)-BIPHEMP with [Rh(cod)$_2$]BF$_4$/(R)-SDP [59]. These results are summarized in Table 17.5.

a) CO (20 atm)
Rh$_4$(CO)$_{12}$ (0.5 mol %)
C$_6$H$_6$, 70 °C, 14 h, 88%

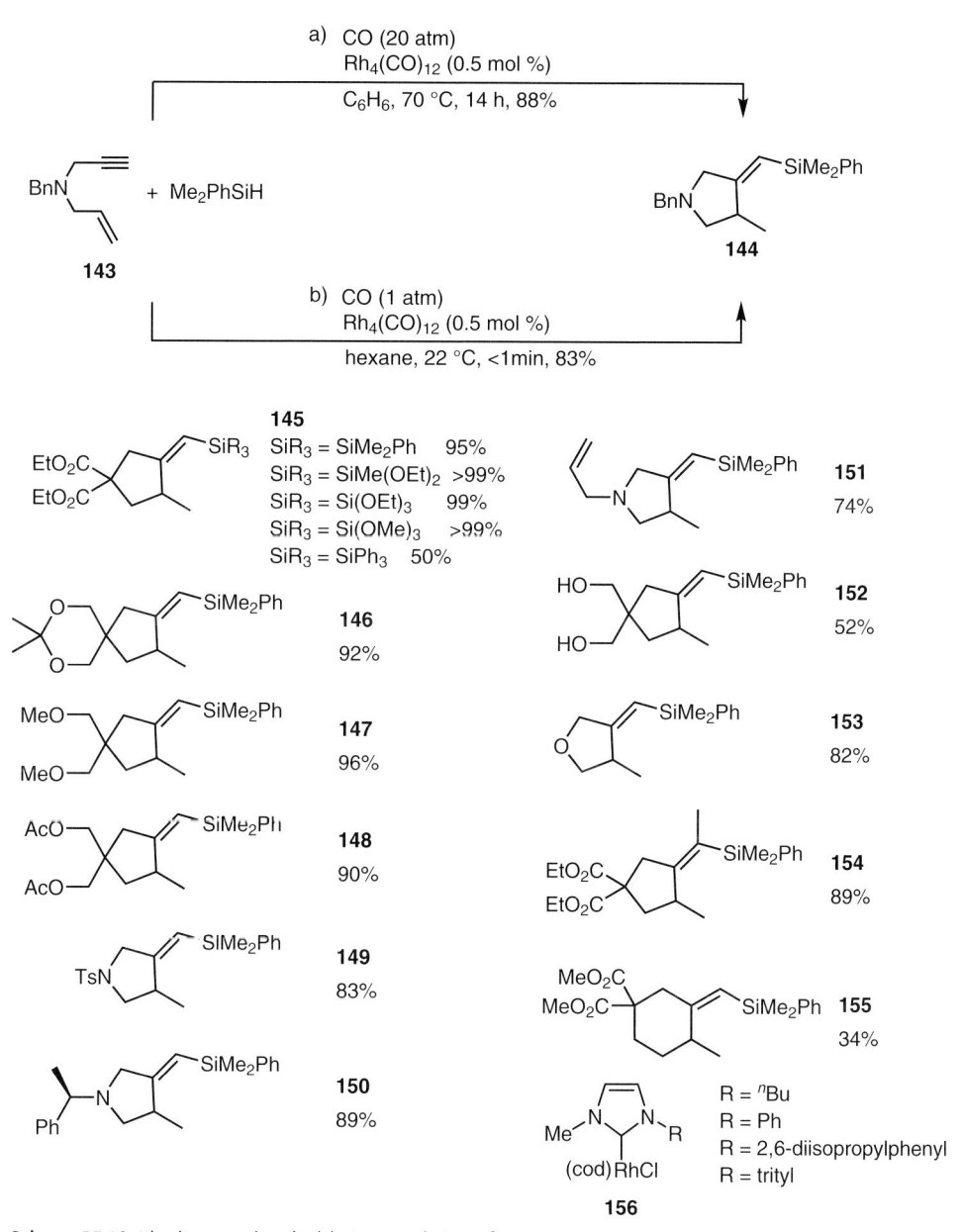

b) CO (1 atm)
Rh$_4$(CO)$_{12}$ (0.5 mol %)
hexane, 22 °C, <1min, 83%

**143** BnN-enyne + Me$_2$PhSiH → **144**

**145** SiR$_3$ = SiMe$_2$Ph 95%
SiR$_3$ = SiMe(OEt)$_2$ >99%
SiR$_3$ = Si(OEt)$_3$ 99%
SiR$_3$ = Si(OMe)$_3$ >99%
SiR$_3$ = SiPh$_3$ 50%

**146** 92%

**147** 96%

**148** 90%

**149** 83%

**150** 89%

**151** 74%

**152** 52%

**153** 82%

**154** 89%

**155** 34%

**156**
R = $^n$Bu
R = Ph
R = 2,6-diisopropylphenyl
R = trityl

**Scheme 17.20** Rhodium-catalyzed silylative annulation of enynes.

**Table 17.5** Rhodium-catalyzed asymmetric induction in silylated annulation of 1,6-enynes in the presence of $R_3SiH$.

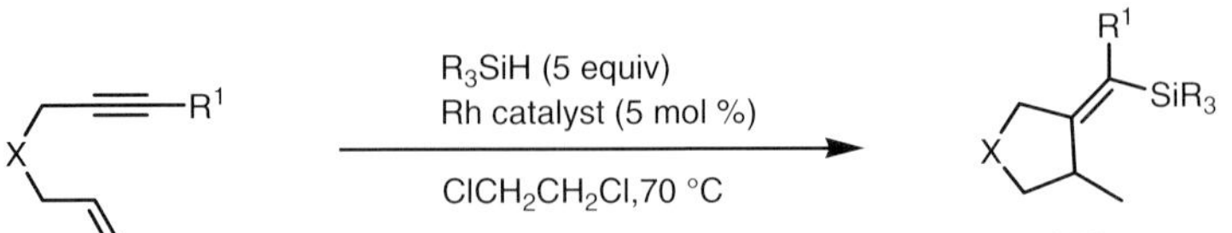


157 → 158

| No. | X | R¹ | $R_3SiH$ | Catalyst[a] | Yield (%) | ee (%) | Ref. |
|---|---|---|---|---|---|---|---|
| 1 | $(MeO_2C)_2C$ | H | $Et_3SiH$ | A | 72 | 94 | [59] |
| 2 | $(MeO_2C)_2C$ | Me | $Et_3SiH$ | A | 50 | 95 | [59] |
| 3 | $(EtO_2C)_2C$ | H | $(EtO)_3SiH$ | A | 93 | 92 | [59] |
| 4 | $(EtO_2C)_2C$ | Me | $(EtO)_3SiH$ | A | 67 | 96 | [59] |
| 5 | $(MeOCH_2)_2C$ | Me | $Et_3SiH$ | B | 65 | 80 | [58] |
| 6 | $(AcOCH_2)_2C$ | Me | $Et_3SiH$ | B | 58 | 83 | [58] |
| 7 | $(EtC(=O))_2C$ | Me | $Et_3SiH$ | B | 48 | 87 | [58] |
| 8 | $(NC)_2C$ | H | $Et_3SiH$ | A | 50 | 99.5 | [59] |
| 9 | TsN | H | $Et_3SiH$ | A | 60 | 98 | [59] |
| 10 | TsN | H | $Et_2MeSiH$ | A | 75 | 98 | [59] |
| 11 | TsN | H | $(EtO)_3SiH$ | A | 75 | 98 | [59] |
| 12 | TsN | Me | $MePh_2SiH$ | B | 73 | 80 | [58] |
| 13 | TsN | Me | $(EtO)_3SiH$ | A | 93 | 92 | [59] |

[a] A; $[Rh(cod)_2BF_4/(R)\text{-SDP}$, B; $[Rh(cod)_2]SbF_6/(R)\text{-BIPHEMP}$.

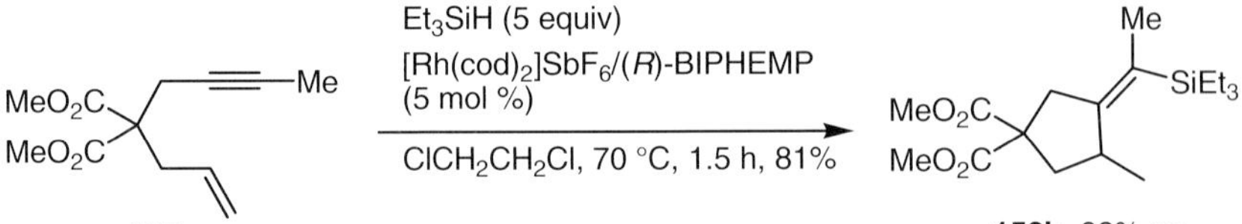


157b

158b 92% ee

(17.18)

In the case of a Pd-catalyst, the regiochemistry is different from the result obtained in the reactions with Rh-catalyst (Eq. 17.19) [60].

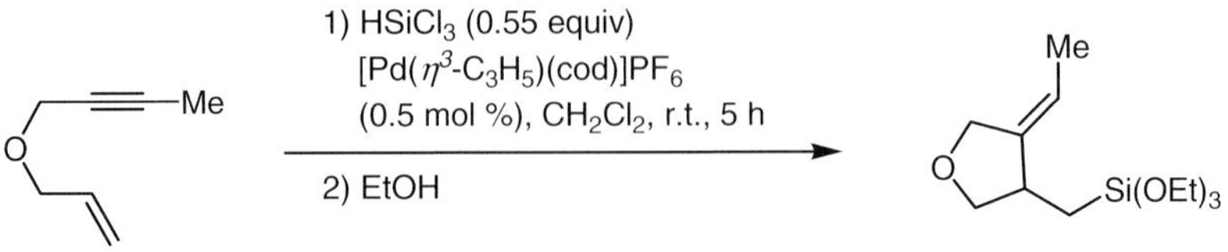


**159** 74% (dr = 4:1)

**157** X = O, R¹ = Me

(17.19)

Though Group 3 metals possess high Lewis acidity and have extreme instability in air, $Cp^*_2YCH_3 \cdot THF$ works as a catalyst for the cyclization of enynes whose structure requires relatively bulky substituents on the sp-carbon and the allylic carbon to realize high conversion and high diastereoselectivity (Eqs. 17.20 and 17.21). The results are explained by the sequential process shown in Scheme 17.21: formation of the Y—H species, selective syn-insertion of an alkyne moiety into the Y—H bond to form **165**, subsequent cyclic carbometallation to form **167** via **166**, and σ-bond metathesis of **167** with $PhSiH_3$ affords the final product **168** [61].

$^nBu_3SnH$ [62], $Me_3SiSn^nBu_3$ [63], and catecholborane [64] may also be used as a component for the cyclization of enynes.

When a combination of $Rh(cod)_2OTf$ with rac-BINAP or BIPHEP is adopted as a catalyst, molecular hydrogen can be incorporated directly in the cyclization of 1,6-enynes to give alkylidene cyclopentane rings, **169** (Table 17.6) [53]. With a chiral ligand, such as (R)-BINAP or (R)-Cl,MeO-BIPHEP, enantiomeric face selection is attained to give **170** with excellent enantiomer excess [65].

Allenyne **171** is also applicable for the formation of the 2-silylvinylalkylidenecyclopentane framework **172** in which combination of $Rh(acac)(CO)_2$ and a trialkoxysilane is crucial for acceptable yield and selectivity (Table 17.7) [66].

### 17.3.1.3 Dienes

Following Marks' work [67] on the intramolecular cyclization of 1,5-hexadienes with a hydrosilane, it has been demonstrated that some complexes of Group 3 metals possess high catalytic ability for the hydrosilylative cyclization of dienes. For example, 1,6-heptadiene reacts with phenylsilane to give 1-silylmethyl-2-methylcyclopentane as a major product (54% yield, trans : cis = 85 : 15) in the presence of 1 mol% of $Cp^*_2NdCH(SiMe)_2$ at room temperature [68]. 1,5-Hexadiene also cyclizes to give a mixture of (silylmethyl)cyclopentane and silacycloheptane (4 : 1) in 95% yield with the aid of 3 mol% of $La[N(SiMe_3)_2]_3$ [69]. The high performance of the yttrium complex, $Cp^*_2YCH_3 \cdot THF$, has been demonstrated in silylative cyclization for a wide range of

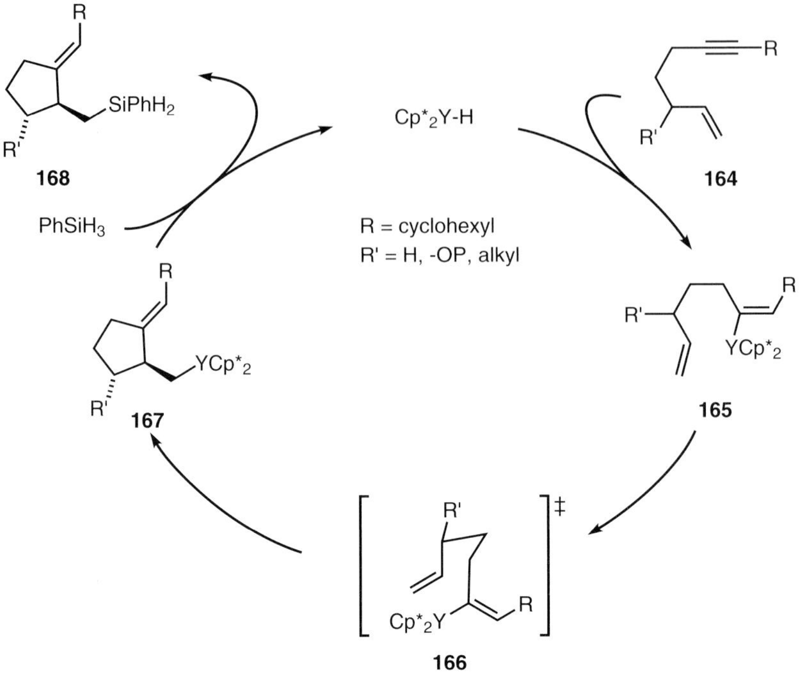

**Scheme 17.21** Plausible catalytic cycle for yttrium-catalyzed reaction of a hydrosilane with a 1,6-enynes.

**Table 17.6** Rhodium-catalyzed hydrogenative annulation of **169**.

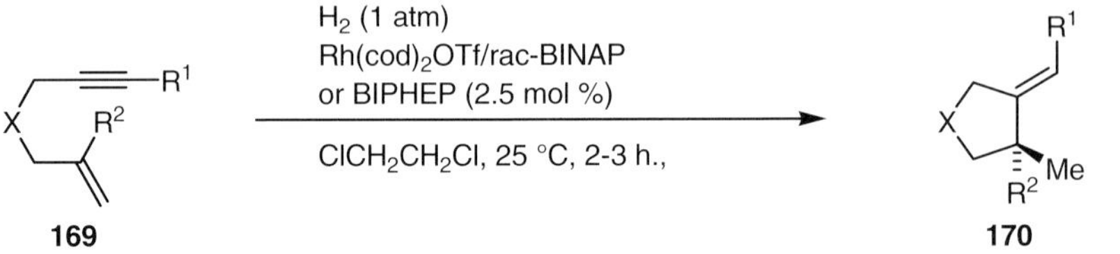

$H_2$ (1 atm), Rh(cod)$_2$OTf/rac-BINAP or BIPHEP (2.5 mol %), ClCH$_2$CH$_2$Cl, 25 °C, 2–3 h.

| | X | R$^1$ | R$^2$ | Ligand | Yield of 170 (%) |
|---|---|---|---|---|---|
| a | H$_2$C | Ph | H | BIPHEP | 65 |
| b | (MeO$_2$C)$_2$C | H | H | BIPHEP | 49 |
| c | (MeO$_2$C)$_2$C | Me | H | BINAP | 75 |
| d | (MeO$_2$C)$_2$C | Ph | H | BIPHEP | 89 |
| e | (MeO$_2$C)$_2$C | Ph | Me | BINAP | 80 |
| f | O | Ph | H | BIPHEP | 82 |
| g | TsN | Me | H | BINAP | 79 |
| h | TsN | Ph | H | BIPHEP | 91 |

**Table 17.7** Rhodium-catalyzed silylative annulation of **171**.

| | Z | R | Yield of 172 (%) |
|---|---|---|---|
| a | TsN | Me | 80 |
| b | TsN | $^n$Bu | 79 |
| c | TsN | Ph | 50 |
| d | O | Me | 63 |
| e | O | Ph | 54 |
| f | $(EtO_2C)_2C$ | Ph | 49 |

1,5-dienes **173** and 1,6-dienes **175** to give the 1,2-disubstituted cyclopentane **174** (Scheme 17.22) and 1,3-disubstituted cyclohexanes **176** (Eq. 17.22), respectively, with the trans-isomer as the major product. In these reactions, the formation of Cp*$_2$YH·THF through σ-bond metathesis with a hydrosilane is regarded as the trigger of the catalytic cycle. The resultant Y—H species adds to the less hindered double bond in anti-Markovnikov manner. However, two organosilane products are isolated in 99% crude yield as a 7:1 mixture in the reaction of o-allylstyrene. The minor product is the cyclic organosilane resulting from initial yttrium hydride addition at the more hindered vinyl group followed by cyclization and silane trapping [70].

| | R | yield (%) | diastereomer ratio |
|---|---|---|---|
| a | H | 100 | - |
| b | OTBDMS | 75 | 4:1 |
| c | Ph | 74 | 9:1 |
| d | OCPh$_3$ | 71 | 24:1 |
| e | OBn | 76 | 3.5:1 |

**Scheme 17.22** Yttrium-catalyzed silylative cyclization of 1,6-dienes.

The results for 1,6-dienes **177**, **179**, and **181** with defined relative stereochemistry clearly show that annulation most likely proceeds exclusively through a chair-like transition structure, avoiding unfavorable interaction between the catalyst and the pre-existing ring. The diastereoselectivity of the annulations was determined after oxidation of **178**, **180**, and **182** which includes diastereoisomers arising from the silicon stereocenter. Such remarkable diastereoselectivity is not observed in the case of cyclopentane derivatives (Eqs. 17.23, 17.24, and 17.25) [71].

Since the catalytic potential of Cp*$_2$YCH$_3$·THF is insufficient for silylative cyclization of 1,1-disubstituted olefins, the more active **183** was introduced. It works well

## 17.3 Cyclization Tied by Outsourcing Reagents

as a catalyst for cyclization of more bulky olefins, as shown in Table 17.5. When **183** is applied as a catalyst to the reactions to construct fused ring systems, exclusive cis-ring fusion is observed in the relative stereochemistry between two carbon skeletons [72]. This point is quite different to the result with Cp*$_2$YCH$_3$·THF that gives trans-fused bicyclic-systems (*vide infra*) selectively.

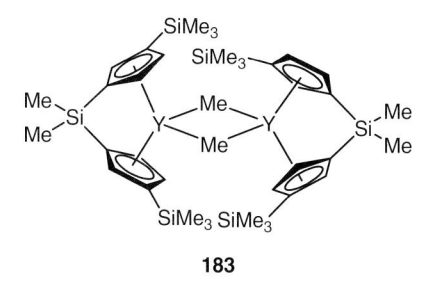

**183**

**187**

The usefulness of this protocol is demonstrated by the synthesis of the precursor **180** of (±)-Epilupinine (Scheme 17.23) [73]. Despite this successful demonstration, much improvement in catalytic efficiency [72a, 74], and the possibility for modification of the complex with the chiral ligand **187** [75], the majority of organic chemists do not use this protocol because of the strong Lewis acidity and extreme instability of catalyst. Molecular hydrogen is used as a reducing agent for reductive cyclization of a limited type of 1,5-dienes and 1,6-dienes catalyzed by a Ln complex [76].

**Scheme 17.23** Yttrium-catalyzed synthesis of epilupinine **186**.

Cationic Pd(II) complexes carrying a weakly coordinating counterion, [(phen)PdMe(Cl)]BAr$_4$ (Ar = 3,5-bis(trfluoromethyl)phenyl), are becoming promising candidates as a catalyst for the above protocol. These complexes display high activity with respect to olefin insertion and catalyze the polymerization of α-olefins [77]. Complex **27** catalyzes olefin hydrosilylation through cleavage of Si—H bonds via a low-energy σ-bond metathesis pathway [21]. In fact, complex **27** catalyzes the reaction of **188** with an R$_3$SiH to give the cyclized product in good yield with high trans-selectivity. Since isolated **27** decomposes easily in solution at room temperature, the corresponding catalytically active species can be generated *in situ* from a 1 : 1 mixture of (phen)PdMe$_2$ and HBAr$_4$ (Eq. 17.26) [78c] or (phen)PdMeCl **190** and NaBAr$_4$ in ClCH$_2$CH$_2$Cl at 0 °C. Typical precursors are illustrated below [78].

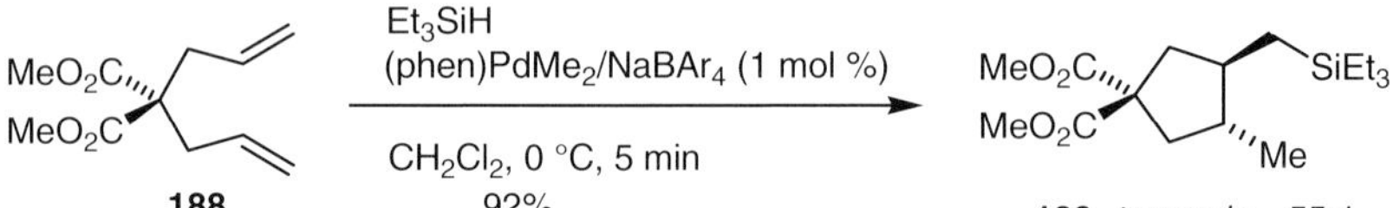

$$\text{MeO}_2\text{C} \diagup \diagdown \xrightarrow[\text{CH}_2\text{Cl}_2,\ 0\ °\text{C},\ 5\ \text{min}]{\substack{\text{Et}_3\text{SiH} \\ (\text{phen})\text{PdMe}_2/\text{NaBAr}_4\ (1\ \text{mol}\ \%)}} \text{MeO}_2\text{C} \diagup \text{SiEt}_3$$

**188**  92%  **189** trans:cis = 55:1

(17.26)

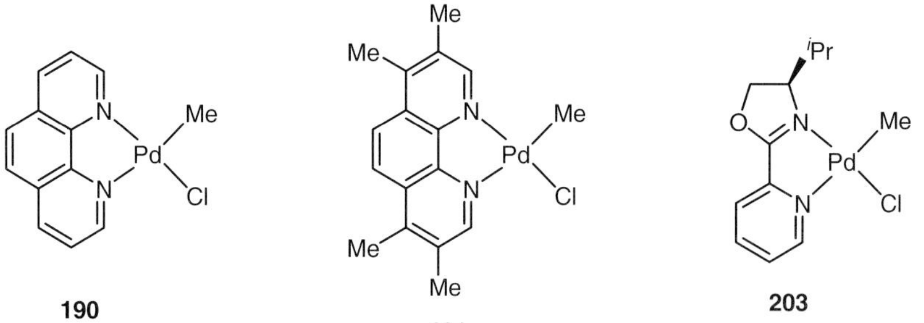

**190**   **191**   **203**

Using any of these catalyst systems the protocol is applicable to the combination of 1,6-dienes or 1,7-dienes with Me$_3$SiH, Et$_3$SiH, Me$_2$PhSiH, $^t$BuMe$_2$SiH, Ph$_3$SiH, or BnMe$_2$SiH. Some of these results are summarized in Table 17.8. It is noteworthy that more than one electron-withdrawing group should be involved in the starting dienes for smooth completion of reaction, except for a few examples. A remarkable acceleration of rate and increase in yield for this cyclization was observed when a mixture of (Me$_4$-phen)Pd(Me)Cl **191**/NaBAr$_4$ in ClCH$_2$CH$_2$Cl was used as catalyst [79].

According to $^1$H NMR spectroscopic data at low temperature, a Pd—Si species **193** and two cationic species **196** and **198** regarded as the intermediates shown in Scheme 17.24 were characterized by tracing a stoichiometric mixture of [(phen)PdMe(NCAr)]BAr$_4$ (**192**), Et$_3$SiH, and **188** in CD$_2$Cl$_2$. On the basis of this fact, Widenhoefer proposed the catalytic cycle of this transformation as depicted in Scheme 17.24. Silylation of **192** with Et$_3$SiH forms the silyl palladium complex **193**. Displacement of the nitrile ligand in **193** with one of the double bonds of **188** would form the silyl palladium olefin species **194** of which coordinated double bond inserts into the Pd—Si bond to form the coordinatively unsaturated palladium alkyl

**Table 17.8** [(Phen)PdMe]X-catalyzed silylative cyclization of 1,6- and 1,7-dienes [78c].

| Diene | Product | Yield (%) dr | Diene | Product | Yield (%) dr |
|---|---|---|---|---|---|
| (diene, E=CO₂Me) | (cyclopentane with SiMe₃, Me; E=CO₂Me) | 80<br>54:1 | (diene, RO-, R=COMe) | (cyclopentane with SiEt₃, Me, RO-; R=COMe) | 99<br>30:1 |
| (diene, E=CO₂Me) | (cyclopentane with SiEt₃, Et; E=CO₂Me) | 88<br>10:1 | (diene, RO-, R=Me) | (cyclopentane with SiEt₃, Me, RO-; R=Me) | 92<br>>50:1 |
| (diene, E=CO₂Me) | (cyclopentane with SiEt₃, E; E=CO₂Me) | 51<br>>50:1 | (diene, E=CO₂Me) | (cyclopentane with SiEt₃, Me, E; E=CO₂Me) | 86<br>16:5:4:1 |

(Continued)

**Table 17.8** (Continued)

| Diene | Product | Yield (%) dr | Diene | Product | Yield (%) dr |
|---|---|---|---|---|---|
| (diene with E, E, Me, Me substituents) E = CO₂Et | (cyclopentane with SiEt₃, Me, E, E, Me, Me) E = CO₂Et | 69 44:1 | (N-allyl diene) R = COCF₃ | (pyrrolidine with SiEt₃, Me) R = COCF₃ | 87 33:1 |
| (diene with E, E, Me) E = CO₂Et | (cyclopentane with SiEt₃, Me, E, E, Me) E = CO₂Et | 81 1:1 | (diene with E, E) E = CO₂Et | (cyclohexane with SiEt₃, Me, E, E) E = CO₂Et | 93 31:1 |

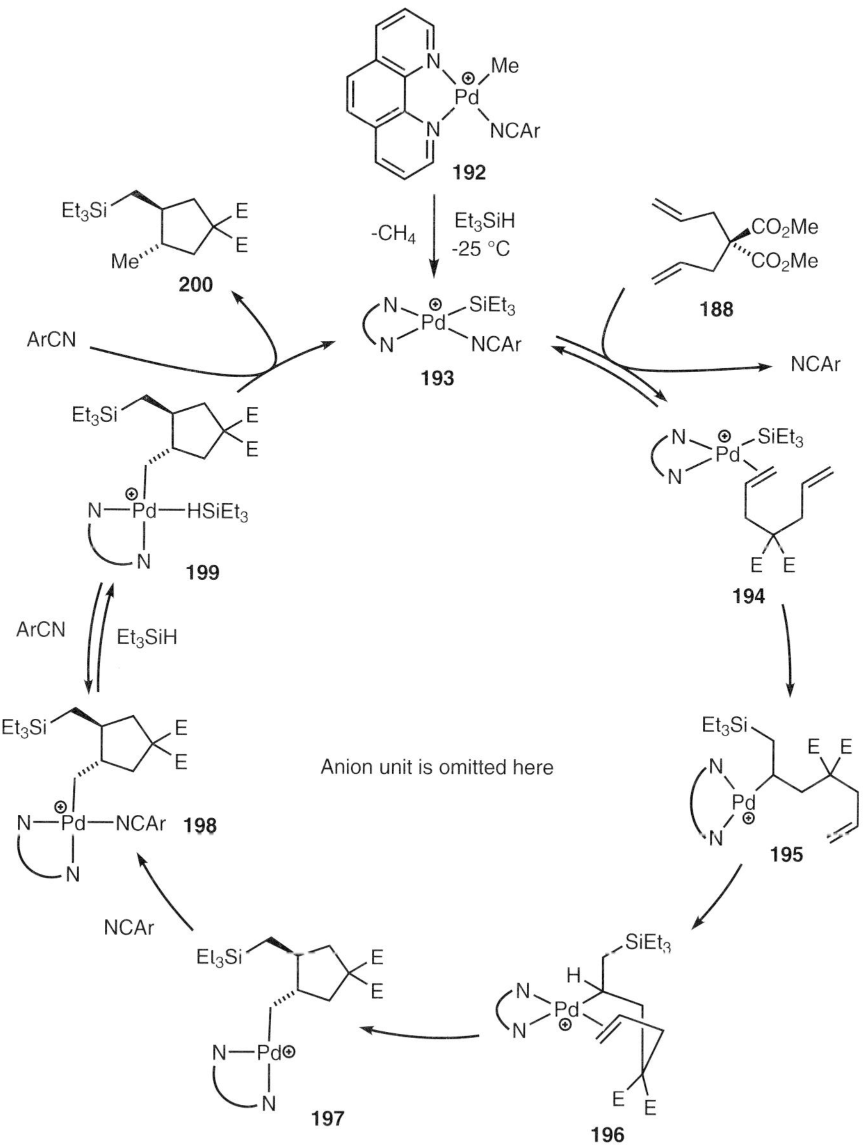

**Scheme 17.24** Catalytic pathway for Pd-catalyzed reaction of Et₃SiH with a 1,6-diene.

intermediate **195**. Immediate coordination of the pendant olefin of **195** would form the observed alkyl olefin chelate complex **196**. β-Migratory insertion of coordinated olefin into the Pd–C bond of **196** to generate the coordinatively unsaturated palladium cyclopentylmethyl complex **197**, followed by ligand capture with ArCN, would form the observed **198**. Associative silylation of **198** via the palladium–silane intermediate **199** would release the carbocycle **200** and regenerate the palladium silyl complex **193**[80]. In addition, the trans-isomer is formed almost exclusively (>97%) in the silylative cyclization of 1,6-diene **201** including the cycloalkenyl moiety as a diene component (Eq. 17.27), while a 1:1 mixture of diastereomers is obtained in the reaction of a less hindered 1,6-diene (see Table 17.8). These contrasting results suggest that the process of silylpalladation is reversible under these conditions [81].

$$\text{201} \xrightarrow[\text{ClCH}_2\text{CH}_2\text{Cl, r.t., 20 min.}]{\text{Et}_3\text{SiH, 190/NaBAr}_4 \text{ (5 mol \%)}} \text{202 (dr = 43:1)}$$

(17.27)

Asymmetric induction in the transformation of **188** to **189** using **203** as a catalyst precursor has been realized in excellent chemical yield with a moderate enantiomeric excess under conditions similar to those in Eq. 17.28 [82]. While the presence of about 20 equiv of $H_2O$ in a similar reaction changed the major product to **204**. When **188** and $Et_3SiH$ are subjected to the silylative cyclization conditions with a 1:1 mixture of **203** and $NaBAr_4$ in the presence of about 20 equiv of water, **204** was obtained in good yield and diastereoselectivity with low enantiomeric excess, despite the concomitant formation of **205** (Eq. 17.28) [83].

$$\text{188} \xrightarrow[\text{ClCH}_2\text{CH}_2\text{Cl, 50 °C, 1 h, 88\%}]{\text{Et}_3\text{SiH (2 eq), H}_2\text{O (20 eq), 203/NaBAr}_4 \text{ (5 mol \%)}} \text{204} + \text{205}$$

E = CO₂Me, **204** (81:19) **205**, de >95%, ee 38%

(17.28)

Although $Et_3SiH$ gives an excellent result for the silylative cyclization of 1,6-dienes, the $Et_3Si$ group is not suitable for conventional oxidative transformation to the hydroxy group. Thus, benzhydryldimethylsilane is chosen as the supplier of a silyl group for the corresponding two-step transformation of 1,6-dienes. As a result, 93% enantiomeric excess was observed at best in the reaction of **188** to form the congener of **189** and the resultant silyl group in the product was readily transformed to the hydroxy group [84].

When the $[PdCl(\eta^3\text{-}C_3H_5)]_2/P(o\text{-Tol})_3$ system was used as a catalyst, the reaction of o-allylstyrene with $Cl_3SiH$ afforded 1-methyl-2-(trichlorosilylmethyl)indan as a major product (87% selectivity) [85]. Its formation is explained by hydropalladation with Markovnikov type selection to the double bond of the styrene moiety rather than a silylpalladation mechanism. This is quite in contrast to the result with the $Cp^*_2YCH_3 \cdot THF$ catalyst where *anti*-Markovnikov type addition to the allylic double bond precedes addition to the styrenic one (Scheme 17.25) [71].

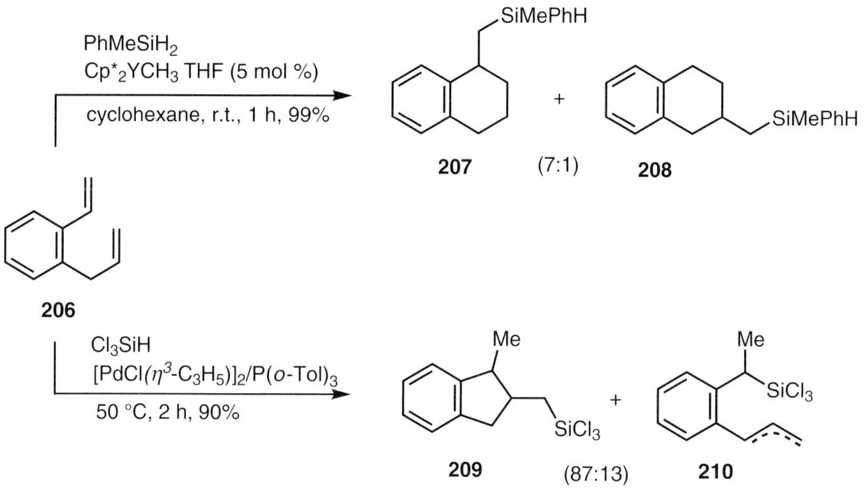

**Scheme 17.25** Product distribution depending on the metal-catalyst in the silylative cyclization of **206**.

### 17.3.1.4 Compounds Involving Three C—C Unsaturated Bonds

Cp*$_2$YCH$_3$·THF is also an effective catalyst for the reaction of 4-vinyl-1,6-dienes with a hydrosilane to form bicyclo[3.3.0]octane frameworks (Scheme 17.26), which suggests 5-exo-trig cyclization is favored over the 6-exo-trig cyclization. The most notable feature of this catalyst is that two cyclopentane rings are fused in a trans-manner (*vide supra*), which is thermodynamically less favored. The generation of a single trans-fused diastereomer **212** is explained by supposing that the pseudo axial orientation **215** prefers the trimethylsilyloxy substituent, as shown in Scheme 17.27. A trans-decalin framework is constructed by a similar procedure with triene **229**. The trans-juncture between two 6-membered rings of **218** is proved as the corresponding hydroxy form derived from **218** via an oxidation process (Scheme 17.28) [71].

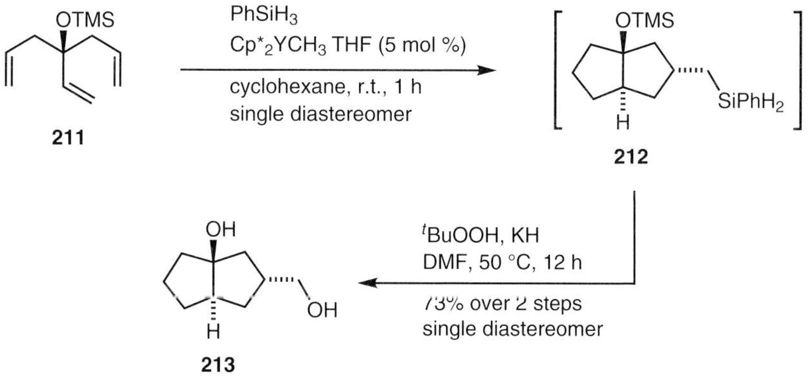

**Scheme 17.26** Cp*$_2$YCH$_3$·THF-catalyzed silylative cyclization of a triene **211**.

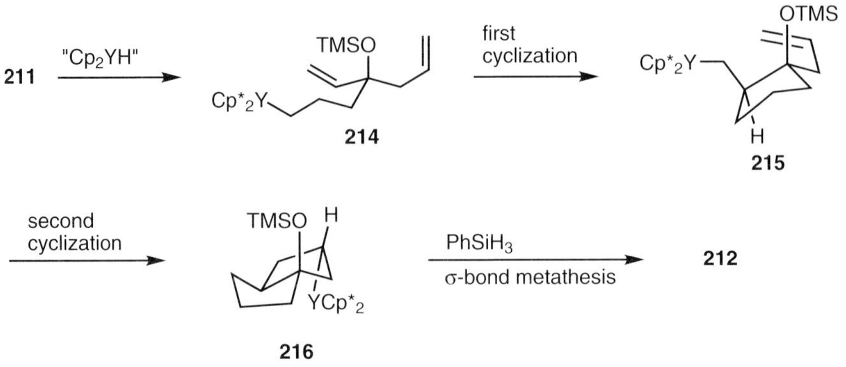

**Scheme 17.27** Explanation for stereochemical course for yttrium-catalyzed cascade cyclization.

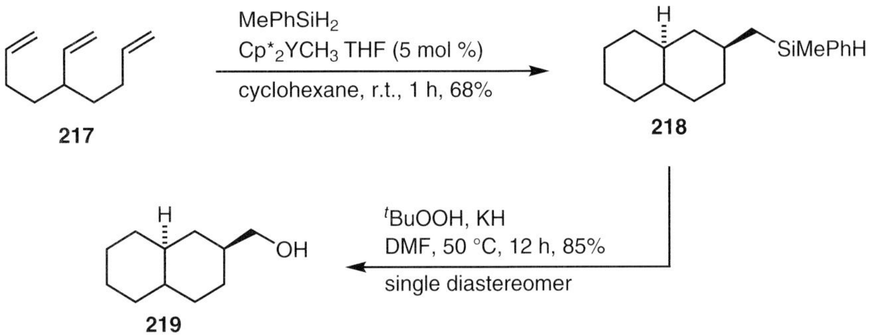

**Scheme 17.28** Cp*$_2$YCH$_3$·THF-catalyzed silylative cyclization of **217**.

Although a stoichiometric quantity is required to obtain a reasonable yield, Cp*$_2$YCH$_3$·THF works well in the consecutive cyclization of a dienyne unit with PhSiH$_3$, in which ring fusion is also trans and the diastereochemistry of the product reflects that of the starting dienyne (Eqs. 17.29 and 17.30). Additionally, the larger the bulkiness of the substituent on the allylic carbon, the higher the diastereoselectivity [86].

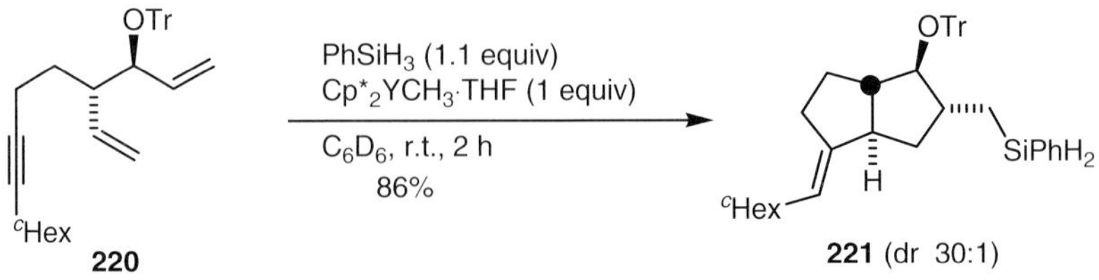

(17.29)

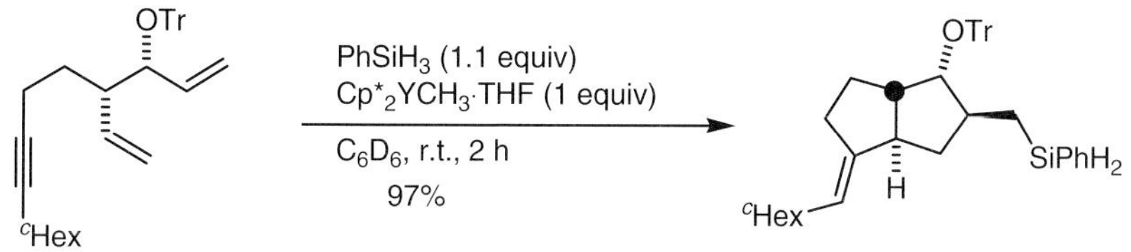


**222** **223** (dr = >60:1)

(17.30)

Another type of cascade cyclization of 1,6,11-undecatrienes with $Et_3SiH$ is attained by the catalysis of a 1 : 1 mixture of (phen)Pd(Me)Cl and $NaBAr_4$, as shown in Eqs. 17.31 and 17.32. In these reactions, selective formation of cascade type products **225** and **227** indicates that insertion of the pendant olefin into the Pd–C bond in an intermediate like **228** is significantly faster than the reaction of **228** with free silane [81].

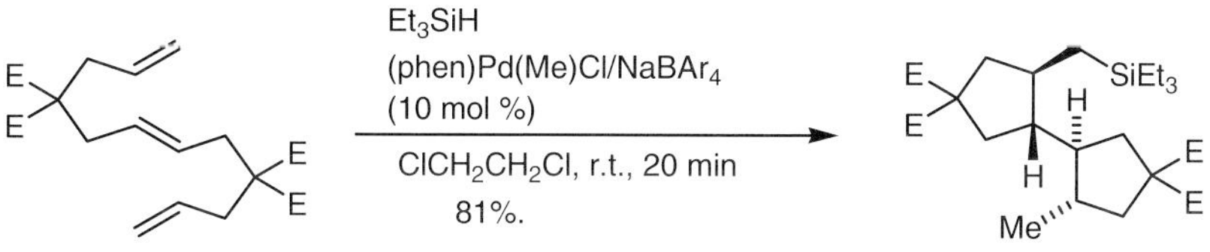


**224** E = $CO_2Me$

**225** (dr = >25:1)

(17.31)

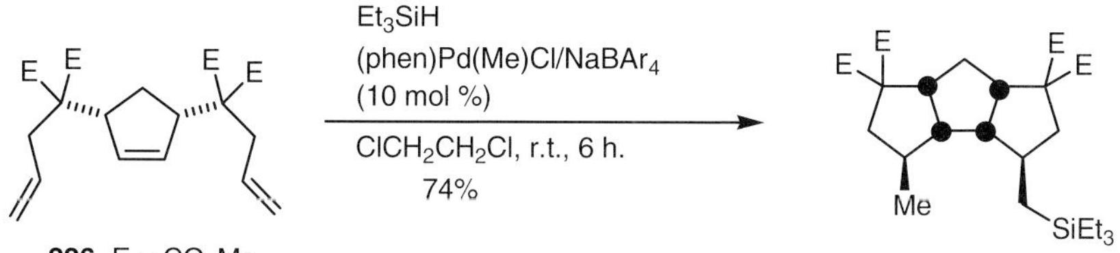


**226** E = $CO_2Me$

**227** (dr = 20:1)

(17.32)

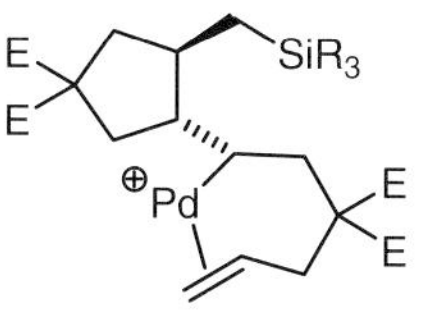


**228** E =$CO_2Me$

6(*E*)-Dodec-6-ene-1,11-diyne **229** reacts with 1 equiv of $Me_2PhSiH$ in the presence of 1 mol% of $Rh(acac)(CO)_2$ in toluene to give a dumbbell type compound **230** as a result of consecutive cyclization triggered by silylrhodation, while **231** is formed selectively from the (*Z*)-isomer (Eqs. 17.33 and 17.34). The exclusive formation of two diastereoisomers can be rationalized by supposing the stereospecific generation of the alkyl–[Rh](H) species followed by the stereospecific insertion of the pendant acetylene moiety into the alkyl–[Rh] bond [87].

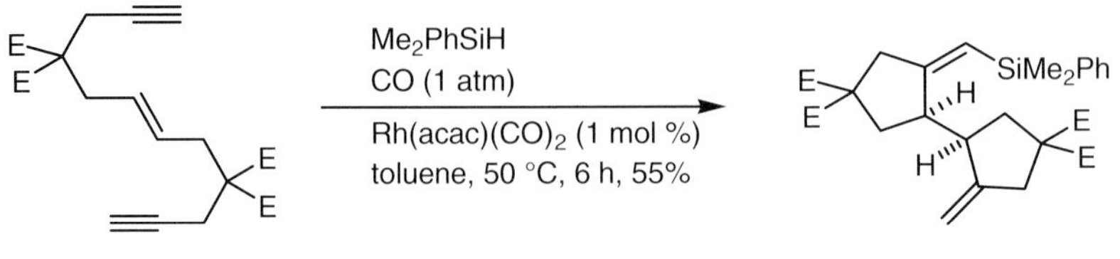

**229-(E)**  E = CO$_2$Et

Me$_2$PhSiH
CO (1 atm)
⟶
Rh(acac)(CO)$_2$ (1 mol %)
toluene, 50 °C, 6 h, 55%

**230**

(17.33)

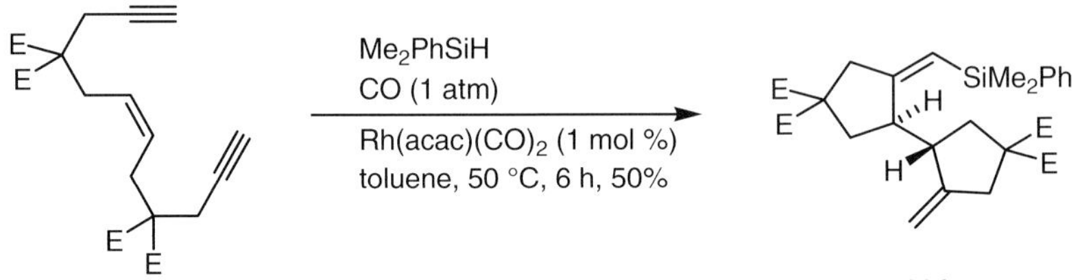

**229-(Z)**  E = CO$_2$Et

Me$_2$PhSiH
CO (1 atm)
⟶
Rh(acac)(CO)$_2$ (1 mol %)
toluene, 50 °C, 6 h, 50%

**231**

(17.34)

Some limited 1,6,11-triynes form a tricycle framework such as **233** and **234** in rhodium-catalyzed reactions with a hydrosilane, which is constructed by iterative cyclization triggered by insertion of a terminal C−C triple bond into the Rh−Si bond and β-elimination of the Rh−H species (**233**) or Rh−Si species (**234**) from the common intermediate as the final step, although the formation of **234** is also explained by intramolecular cyclotrimerization of three acetylenic bonds in **232** (Eq. 17.35) [88].

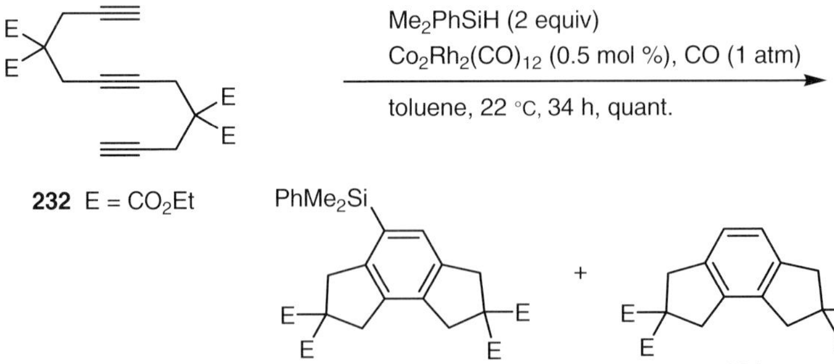

**232** E = CO$_2$Et

Me$_2$PhSiH (2 equiv)
Co$_2$Rh$_2$(CO)$_{12}$ (0.5 mol %), CO (1 atm)
⟶
toluene, 22 °C, 34 h, quant.

**233**  +  **234**

(17.35)

A Pd−H species derived from [Pd(η$^3$-C$_3$H$_5$)Cl]$_2$ and formic acid works well as a catalyst to construct tricyclic frameworks from enediynes (Eq. 17.36), in which two hydrogen atoms originate from formic acid [89].

$$\text{235} \xrightarrow[\text{DMF, 80 °C, 1 h}]{\substack{[Pd(\eta^3\text{-}C_3H_5)Cl]_2/4\ PPh_3 \\ (5\ mol\ \%) \\ HCOOH\ (2\text{-}3\ equiv)}} \text{236}\quad 72\% \quad (17.36)$$

### 17.3.1.5 Carbonyl Compounds Tethered by a C–C Unsaturated Bond

Oxidative coupling to form metallacyclopentanes in a stoichiometric reaction has been known as a short-cut tool achieving intramolecular carbocyclization of linear molecules including two unsaturated bonds or γ,δ-unsaturated carbonyl compounds. In due course, it was revealed that a hydrosilane works well as a reducing agent via a σ-bond metathesis with a metallacyclopentane intermediate. A report on the use of $Cp_2Ti(PMe_3)_2$ to form oxatitanabicyclooctanes [90] prompted many researchers to apply this knowledge to the design of a catalytic cycle. For example, reductive cyclization of γ,δ-unsaturated carbonyl compounds has been realized, as shown in Eqs. 17.37 and 17.38 [91].

$$\text{237} \xrightarrow[\text{2) HCl/acetone, 3 h}]{\substack{1)\ Ph_2SiH_2 \\ Cp_2Ti(PMe_3)_2\ (10\ mol\ \%) \\ toluene, -20\ °C, 2\ h}} \text{238} \quad (17.37)$$

$$\text{239} \xrightarrow[\text{pentane, r.t., 0.5 h, 73\%}]{\substack{(EtO)_3SiH \\ Cp_2Ti(PMe_3)_2\ (20\ mol\ \%)}} \text{240} \quad (17.38)$$

A similar transformation is also realized by a nickel catalyst, prepared from a 1:2 mixture of $Ni(cod)_2$ and $PBu_3$, for the synthesis of some heterocycles including azacycles which may be used immediately as an intermediate for the synthesis of alkaloids (Eqs. 17.39 and 17.40). In fact, this reaction was incorporated as the key step in a total synthesis of allopumiliotoxin 267 A (**247**) (Scheme 17.29) [92].

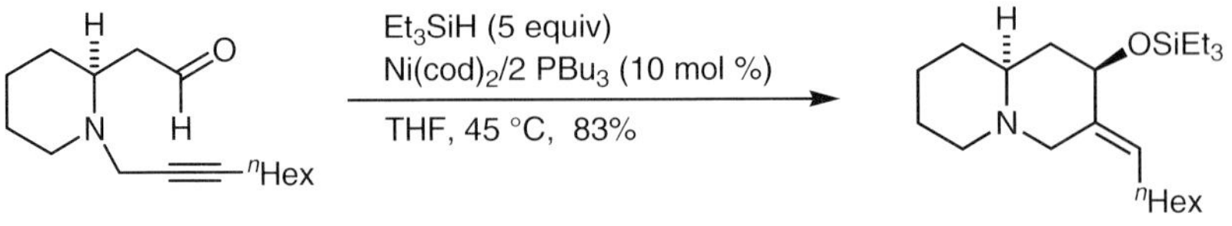


**241** **242** (dr = >97:<3)

(17.39)

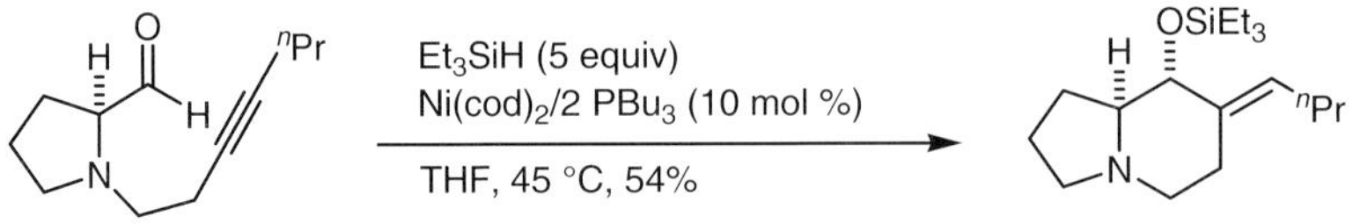


**243** **244** (dr = <3:>97)

(17.40)

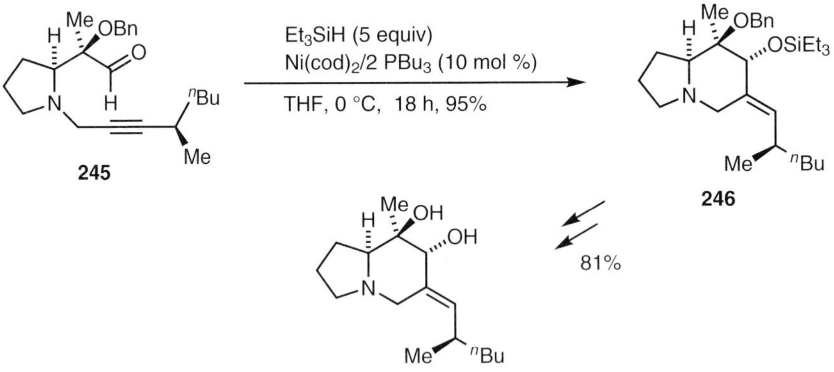


**245** **246** **247**

allopumiliotoxin 267A

**Scheme 17.29** Application of Ni-catalyzed silylative cyclization for the synthesis of allopumiliotoxin-267 A.

When **248** was used as the ligand instead of $PBu_3$, the reactions seemed to proceed via a catalytic cycle including insertion of a C–C triple bond into the Ni–H bond [93].

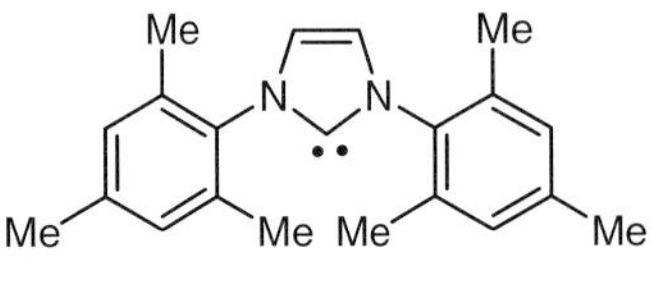


**248**

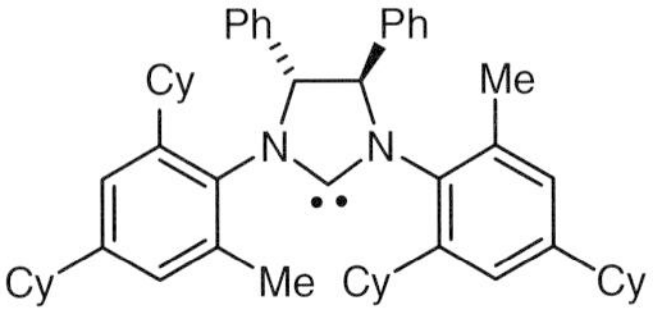


**249**

Mechanism aside, a catalyst system modified by **248** or **249** is effective for silylative endo cyclization of ynals with terminal or internal alkynes to give a macrocycle. Asymmetric induction accompanied by the macrocyclization was realized by modification of the catalyst with **249**, despite moderate enantiomer excess [94].

On the other hand, cyclization of compounds including a 1,3-diene and a carbonyl moiety in remote positions has been developed. The knowledge that the Ni–C species possesses sufficient nucleophilicity toward a carbonyl carbon is skillfully combined with the use of a hydrosilane as a hydrogen supplying agent in the cyclization of ynals. Aldehydes including the 1,3-diene moiety react with triethylsilane in the presence of an Ni(0) species derived from Ni(cod)$_2$ and 2 equiv of PPh$_3$ under mild conditions to give substituted cycloalkanes (Scheme 17.30). The formation of these products is rationalized according to Scheme 17.31 [95]. First, the Si–Ni–H species is formed by oxidative addition of Et$_3$SiH to the Ni(0) species derived from Ni(cod)$_2$ and PPh$_3$. The subsequent insertion of **250** (R = H) into the Ni–H bond produces **252**, which is immediately converted to the π-allylnickel complex **252**. The π-allylnickel moiety in **253** reacts with the tethered carbonyl group to give **254**, from which product **251** is formed by reductive elimination accompanied by the regeneration of the Ni(0) species. Although the relative stereochemistry is unambiguously determined, it is not clear why the relative configuration between the connecting carbons fluctuates, depending on the formed ring size. Presumably the less hindered side would be chosen in the transition state for cyclization [96]. Formal total synthesis of (−)-elaeokanine C has been successfully demonstrated despite insufficient stereocontrol in the key silylative cyclization [96]. Some trials for asymmetric induction give a clue for much improvement [97].

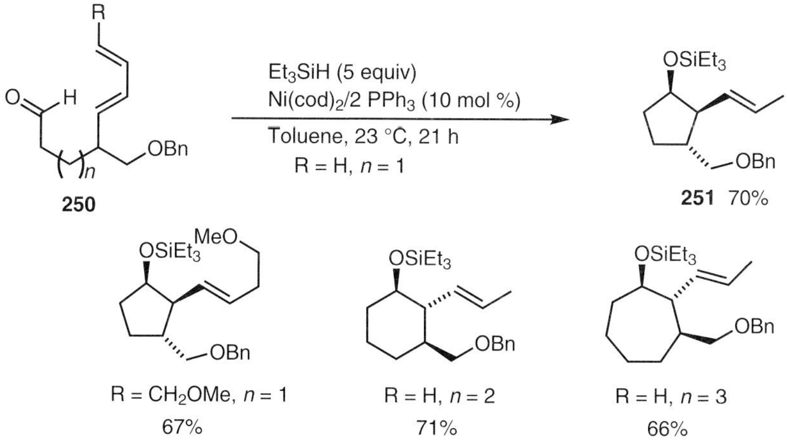

**Scheme 17.30** Ni-catalyzed silylative cyclization of dienals.

In relation to these protocols, reductive cyclization of alkynals [98], alkynones [98b], aldehydes tethered by a conjugated diene [99], and epoxides involving an alkyne moiety [100] have been successfully performed to give the corresponding products

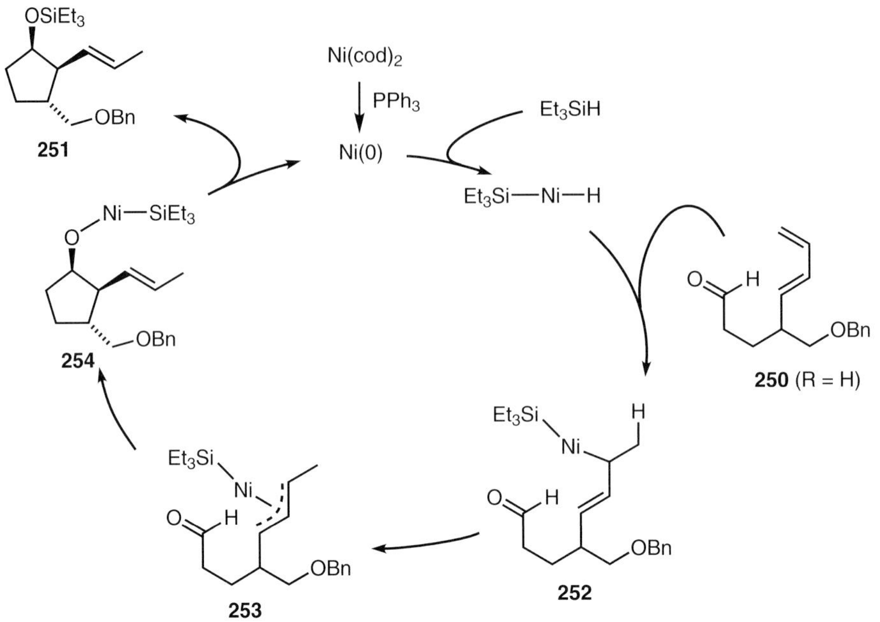

**Scheme 17.31** Proposed catalytic cycle for Ni-catalyzed silylative cyclization of **250**.

with assistance of Ni(cod)$_2$/PPh$_3$/Et$_2$Zn, Ni(acac)$_2$/BEt$_3$, or Ni(cod)$_2$/PMe$_2$Ph/BEt$_3$ catalyst systems.

Rhodium-catalyzed silylative cyclization of alkynals is difficult to achieve under CO pressure (1–20 atm), since the silylformylation of the alkyne moiety proceeds selectively [34]. In contrast to the results with alkynals, silylative cyclization of allenyl-carbonyls proceeds selectively to give cycloalkanols in the reactions with Et$_3$SiH under CO pressure (20 atm) with the aid of Co$_2$Rh$_2$(CO)$_{12}$ (Table 17.9) [101]. The identical transformation is carried out by using Me$_3$SiSn$^n$Bu$_3$ instead of Et$_3$SiH under the catalysis of [Pd($\eta^3$-C$_3$H$_5$)Cl]$_2$ [102].

In contrast to the reaction with a hydrosilane under CO gas, alkynals **257** react smoothly with hydrogen molecule to give cyclopentane rings **258** with high enantiomer excess under the catalysis of Rh(cod)$_2$OTf with (R)-Cl,MeO-BIPHEP under mild conditions (Table 17.10). The presence of 2-naphthoic acid (5 mol%) is crucial to attain high conversion of alkynals reproducibly [103].

### 17.3.1.6 Aldehydes Tethered by an α,β-Unsaturated Carbonyl Group

Methyl 5-(E)-oxo-2-hexenoate **259** captures an equimolar amount of Et$_3$SiH with catalysis by 1 mol% of RhCl(PPh$_3$)$_3$ or RhH(PPh$_3$)$_4$ to give the cyclic compound **260** in good yield. The stereochemical outcome of the reaction is not altered when the Z-isomer of **259** is employed as a substrate. It is observed that the catalyst precursor used significantly influences the diastereoselectivity of this cyclization: RhCl(PPh$_3$)$_3$

## 17.3 Cyclization Tied by Outsourcing Reagents

**Table 17.9** $Co_2Rh_2(CO)_{12}$-catalyzed silylative cyclization of allenyl-carbonyls **255**.

Conditions: $Et_3SiH$ (2 equiv), $Co_2Rh_2(CO)_{12}$ (1 mol %), CO (20 atm), $Et_2O$, 70 °C; **255** → **256**

| | Z | R | n | Yield of 256 (%) |
|---|---|---|---|---|
| a | TsN | H | 1 | 74 |
| b | TsN | H | 2 | 61 |
| c | TsN | Me | 1 | 54 |
| d | TsN | Et | 1 | 56 |
| e | TsN | Me | 2 | 54 |
| f | $(MeO_2C)_2C$ | H | 1 | 68 |
| g | $(MeO_2C)_2C$ | Me | 1 | 56 |
| h | O | H | 2 | 68 |

**Table 17.10** Rh-catalyzed asymmetric hydrogenative cyclization of alkynal **257**.

Conditions: $H_2$ (1 atm), $Rh(cod)_2OTf/(R)$-Cl,MeO-BIPHEP (5 mol %), 2-naphthoic acid (5 mol), $ClCH_2CH_2Cl$, 45 °C; **257** → **258**

| | X | Y | $R^1$ | $R^2$ | Yield of 258 (%) | ee (%) |
|---|---|---|---|---|---|---|
| a | $p$-BrBnN | O=C | Ph | H | 85 | 97 |
| b | BnN | O=C | Me | H | 83 | 99 |
| c | BnN | $H_2C$ | $^cPr$ | H | 67 | 95 |
| d | TsN | $H_2C$ | Ph | $-(CH_2)_5-$ | 99 | 91 |
| e | TsN | $H_2C$ | Me | $-(CH_2)_5-$ | 83 | 90 |
| f | TsN | $H_2C$ | $^cPr$ | Me | 77 | 91 |
| g | TsN | $H_2C$ | Me | Me | 85 | 96 |
| h | TsN | $H_2C$ | H | Me | 96 | 96 |
| I | TsN | $Me_2C$ | Me | H | 63 | 98 |
| j | O | $H_2C$ | Ph | Me | 71 | 93 |
| k | O | $H_2C$ | $^n$Hex | Me | 86 | 99 |
| l | O | $H_2C$ | Me | $-(CH_2)_4-$ | 73 | 99 |

shows a moderate syn-selectivity while RhH(PPh$_3$)$_4$ shows predominant formation of the anti-form (Eq. 17.41) [104a,b].

$$\text{259} \xrightarrow[\text{toluene, 50 °C, 16 h}]{\substack{\text{Et}_3\text{SiH (2.1 equiv)} \\ \text{[Rh] (1 mol \%)}}} \text{260}$$

RhCl(PPh$_3$)$_3$    81% (*syn:anti* = 3:1)
PhH(PPh$_3$)$_4$    81% (*syn:anti* = 1:11)

(17.41)

Some other results are summarized in Table 17.11. In this type of transformation the olefin moiety of **259** inserts into the Rh—H bond of the Si—Rh—H species formed by oxidative addition of Et$_3$SiH to a Rh(I) complex. The resultant rhodium enolate attacks the pendant formyl group and subsequent reductive elimination of Rh(I) species releases the cyclic product [104]. This scenario is imaged from that proposed in intermolecular aldol couplings [105]. If the process of aldol cyclization is controlled according to a 6-membered transition state, stereochemical outcome in the products reflects the ratio of the geometrical isomer in the enolate intermediate.

Later, this protocol was extended to another type of model **261** by using a 1:1 mixture of Cu(OAc)$_2$·H$_2$O and DPPF as a catalyst system in THF, in which high syn-diastereoselectivity (>95%) is realized, regardless of the substrate (Table 17.12). In addition, asymmetric cyclization is successfully attained (62–83% ee) by replacing DPPF with a chiral ligand, such as (S)-BINAP, (S)-MeO-BIPHEP, (R)-3,5-xyl-MeO-BIPHEP, and (S)-SEGPHOS [106].

When this catalyst system is applied to compounds including an amide moiety **261**, the process suffers from low yield because of attenuated electrophilicity of the α,β-unsaturated amide. This issue is surmounted by replacement of Cu(OAc)$_2$·H$_2$O with either Co(acac)$_2$ or CoCl$_2$/PCy$_2$Ph with the concomitant use of 2 equiv Et$_2$Zn as a reducing agent (Scheme 17.32) [107].

Similar intramolecular aldol coupling is performed by Co(dpm)$_2$ catalyst, in which the relative stereochemistry of the product is selectively syn (Scheme 17.33 Path **A**) [108]. A reaction of 6-oxo-1-phenylhept-2(*E*)-en-1-one **266b** with PhSiH$_3$ is also catalyzed by In(OAc)$_3$ (10 mol%) under reflux conditions in EtOH to give **266b** in 77% yield [109]. The same product was isolated in the reaction of **265b** with a stoichiometric amount of $^n$Bu$_2$ISnH after the subsequent hydration under acidic conditions [110]. Reductive aldol cyclization in the presence of In(OAc)$_3$ or $^n$Bu$_2$ISnH is applicable to ketones tethered by an enone moiety.

Hydrogen molecule is also oxidatively added to a low-valent metal to give a M—H species which makes it possible to form a metal enolate supposed in the transformations shown in Eq. 17.41 and Path **A** in Scheme 17.33. In fact, the identical transformation is attained by atmospheric pressure of H$_2$ molecule instead of PhSiH$_3$.

17.3 Cyclization Tied by Outsourcing Reagents | 887

Table 17.11 RhH(PPh$_3$)$_4$-catalyzed cyclization of 6-oxo-2-hexenonate derivatives [104b].

| No. | Substrate | Product | Yield (%) syn : anti | No | Substrate | Product | Yield (%) syn : anti |
|---|---|---|---|---|---|---|---|
| 1 | | | 62<br>6.4 : 1 | 5 | | | 81 complex mixture |
| 2 | | | 59<br>1 : 2 | 6 | | | 69<br>1 : 20 |
| 3 | | | 68<br>1 : 2.5 | 7 | | | 65<br>1 : 3 |
| 4 | | | 61<br>1 : 11 | 8 | | | 68<br>2.5 : 1 |

**Table 17.12** Cu-Catalyzed intramolecular reductive aldol type cyclization.

**261** → **262**

1) HMe$_2$SiOSiMe$_2$H (1 equiv)
Cu(OAc)$_2$·H$_2$O/DPPF (5 mol %)
THF, r.t., 13 - 29 h
2) MeOH/H$^+$

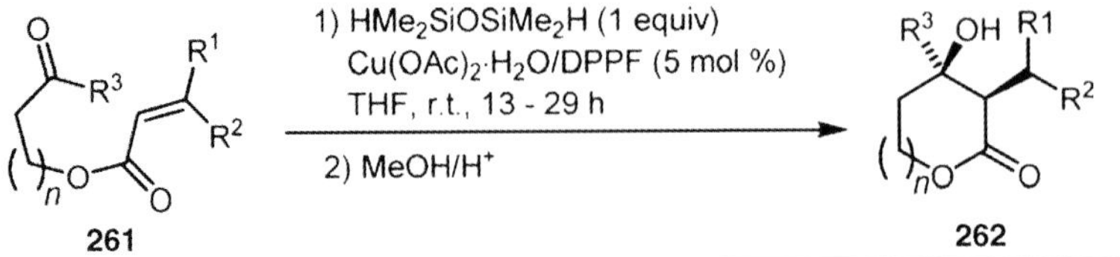

|   | R$^1$ | R$^2$ | R$^3$ | n | Yield of 262 (%) |
|---|---|---|---|---|---|
| a | H | H | Ph | 0 | 69 |
| b | H | Me | Ph | 0 | 72 |
| c | H | $^i$Bu | Ph | 0 | 65 |
| d | H | PhCH$_2$CH$_2$ | Ph | 0 | 60 |
| e | H | H | Ph | 1 | 71 |
| f | H | Me | Ph | 1 | 50 |
| g | H | Ph | Ph | 1 | 47 |
| h | H | 4-MeO–C$_6$H$_4$ | Me | 1 | 61 |
| I | H | 4-Cl–C$_6$H$_4$ | Me | 1 | 65 |
| j | H | 2-furyl | Me | 1 | 63 |
| k | H | $^i$Bu | Me | 1 | 73 |
| l | H | PhCH$_2$CH$_2$ | Me | 1 | 61 |
| m | Me | Me | Me | 1 | 60 |

The addition of 30 mol% of KOAc to the catalyst system composed of Rh(cod)$_2$OTf and P(p-CF$_3$C$_6$H$_4$)$_3$ is crucial to make **266** for direct selective aldol type cyclization with sufficient conversion (Scheme 17.33 Path **B**) [111]. This protocol is applicable to some keto-enones by a slight adjustment in the catalyst system, although it is necessary to suppress the formation of the 1,4-hydrogenated product inevitably

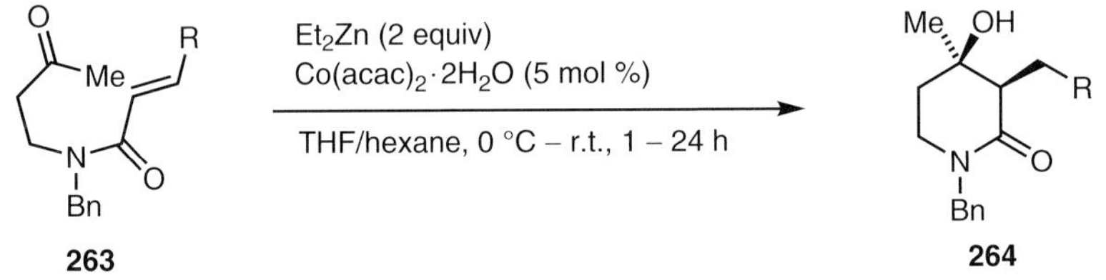

**263** → **264**

Et$_2$Zn (2 equiv)
Co(acac)$_2$·2H$_2$O (5 mol %)
THF/hexane, 0 °C – r.t., 1 – 24 h

|   | R | yield of 264 (%) | syn:anti |
|---|---|---|---|
| a | H | 88 | 9:1 |
| b | $^i$Pr | >99 | >19:1 |
| c | Ph | 97 | >19:1 |
| d | 2-furyl | >99 | >19:1 |

**Scheme 17.32** Cu-catalyzed intramolecular reductive aldol cyclization.

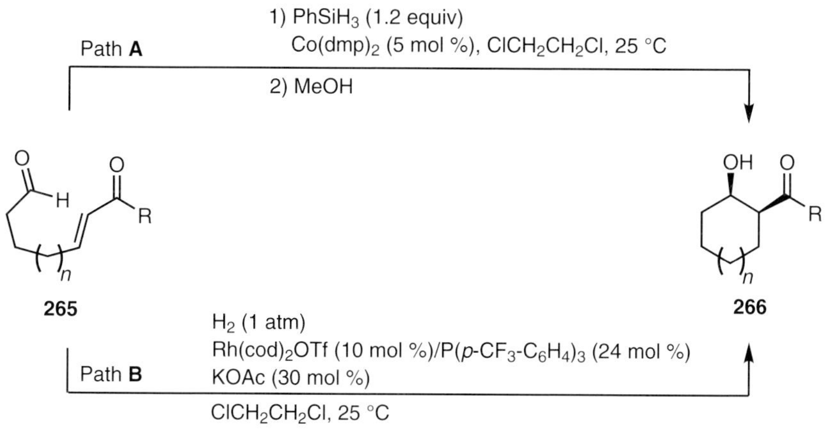

|   | R | n | Product 266 Path A yield (%) | Path B yield (%) (syn:anti) |
|---|---|---|---|---|
| a | Ph | 0 | 70 | 71 (24:1) |
| b | Ph | 1 | 87 | 89 (10:1) |
| c | Me | 1 | 38 | 65 (1:5) |
| d | 2-naphthyl | 1 | 68 | 90 (10:1) |
| e | 2-furyl | 1 | 75 | 70 (6:1) |
| f | 2-thiophenyl | 1 | 73 | 76 (19:1) |
| g | Ph | 2 | 35 | - - |

**Scheme 17.33** Co- and Rh-catalyzed reductive intramolecular aldol cyclization.

accompanied with 8 to 20% yield in this reaction (Eq. 17.42) [112].

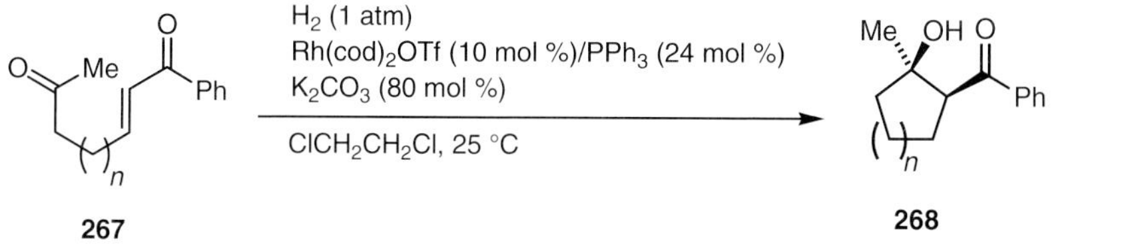

**a** n = 1  75% (syn;anti = >95:5)
**b** n = 2  72% (syn;anti = >95:5)

(17.42)

Stryker's reagent, [(Ph$_3$P)CuH]$_6$ [113] is also reliable for this type of transformation with prior formation of the *syn*-form when it is used as a reagent in stoichiometric quantity [114]. The amount of [(Ph$_3$P)CuH]$_6$ is reduced to 10 mol% by using polymethylhydrosiloxane as a reductant. However, the concomitant formation of the over-reduction or dehydration product is a serious issue in this type of model **269** (Eq. 17.43) [115].

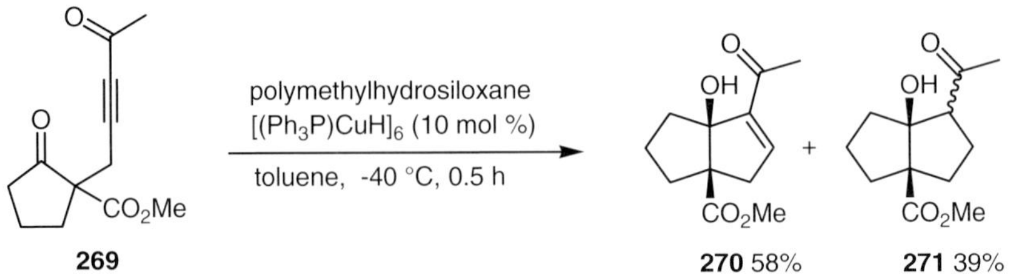

(17.43)

**269** → **270** 58% + **271** 39%

Conditions: polymethylhydrosiloxane, [(Ph$_3$P)CuH]$_6$ (10 mol %), toluene, −40 °C, 0.5 h

A combination of Co(dmp)$_2$ and PhSiH$_3$ is effective for the cyclization via 1,4-addition of PhSiH$_3$ and subsequent intramolecular Michael-type coupling. In this case, the relative stereochemistry is regulated to the anti-form (Scheme 17.34) [108]. Yields of cyclized products are slightly improved, when In(OAc)$_3$ is used in place of Co(dmp)$_2$ as catalyst [109]. The cyclization along with a similar concept is demonstrated by the reactions with a stoichiometric amount of Li($^s$Bu)$_3$BH [116], [(Ph$_3$P)CuH]$_6$ [117], or $^n$Bu$_2$ISnH [110].

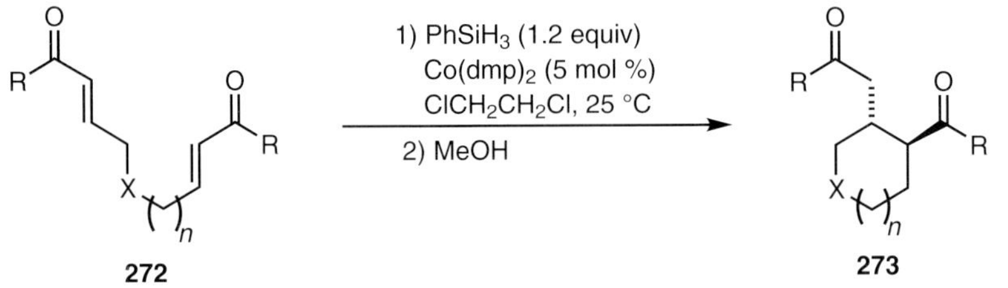

**272** → **273**

Conditions: 1) PhSiH$_3$ (1.2 equiv), Co(dmp)$_2$ (5 mol %), ClCH$_2$CH$_2$Cl, 25 °C; 2) MeOH

|   | X | R | n | yield of **273** (%) |
|---|---|---|---|---|
| a | CH$_2$ | Ph | 0 | 62 |
| b | CH$_2$ | Ph | 1 | 73 |
| c | O | Ph | 1 | 63 |
| d | CH$_2$ | 2-furyl | 1 | 52 |

**Scheme 17.34** Co-catalyzed reductive intramolecular Michael type cyclization.

## 17.4
## Cascade Reactions Incorporating CO

### 17.4.1
### Annulative Silylcarbonylation

#### 17.4.1.1 1,6-Diynes

Silylformylation of 1-alkynes is a quite reliable tool to give 2-substituted 3-silylpropenals, unless the molar ratio of 1-alkyne to R$_3$SiH deviates from 1:1. When less reactive $^{t}BuMe_2SiH$ is used as a starting substrate in the silylformylation of phenylacetylene, the reaction must be carried out at 25 °C to obtain **274** in acceptable yields. Operation under forcing conditions (100 °C, 16 h) does not enhance the formation of **274**, but rather results in a moderate yield (49%) of **274** with concomitant formation of **275** (23%) and **276** (3%). The yields of cyclopentenones, **275** and **276**, increase with a slight decrease in **274**, by using 2 equiv phenylacetylene under CO pressure (Eq. 17.44) [34b].

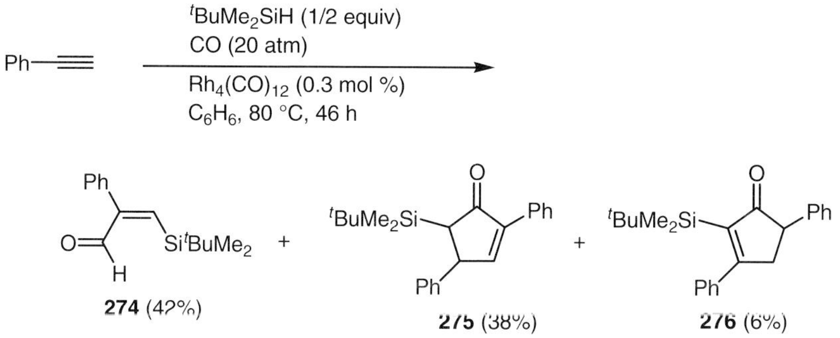

$$Ph-\!\!\equiv\;\xrightarrow[\substack{Rh_4(CO)_{12}\ (0.3\ mol\ \%)\\ C_6H_6,\ 80\ °C,\ 46\ h}]{\substack{^{t}BuMe_2SiH\ (1/2\ equiv)\\ CO\ (20\ atm)}}$$

**274** (42%) + **275** (38%) + **276** (6%)

(17.44)

Although congeners of **275** and **276** are formed in the reaction of phenylacetylene with either $Me_2PhSiH$ or $Et_2MeSiH$ (phenylacetylene : silane ≈ >2 : 1) under CO pressure (20 atm), the congener of **275** is readily protodesilylated during purification to give 2,4-diphenylcyclopent-2-enone, which is identical to the product obtained by the protodesilylation of **275**. When phenylacetylene is replaced by 1-hexyne in the silylformylation using less than 0.5 equiv silanes, homologues of **276** become the major cyclocarbonylation products. In particular, various types of carbonylation products are isolated in the reaction of 1-hexyne with 0.5 equiv $^{t}BuMe_2SiH$ under more forcing conditions than that in the silylformylation (Eq. 17.45) [34b].

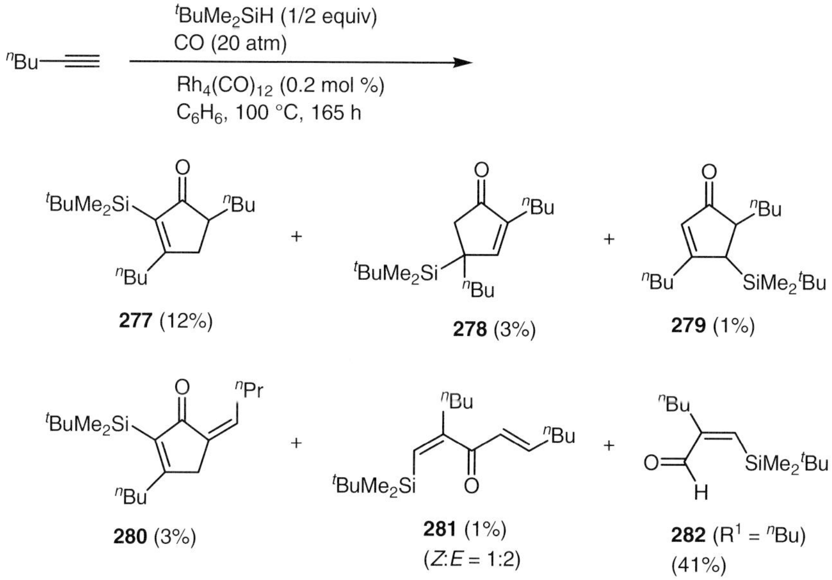

$$^{n}Bu-\!\!\equiv\;\xrightarrow[\substack{Rh_4(CO)_{12}\ (0.2\ mol\ \%)\\ C_6H_6,\ 100\ °C,\ 165\ h}]{\substack{^{t}BuMe_2SiH\ (1/2\ equiv)\\ CO\ (20\ atm)}}$$

**277** (12%) + **278** (3%) + **279** (1%)

**280** (3%) + **281** (1%) (Z:E = 1:2) + **282** ($R^1 = {}^{n}Bu$) (41%)

(17.45)

The construction of these cyclopentenone derivatives may be consistently explained by the continuous insertion of two alkynes into the Rh–Si bond to form **284** via **283**. The subsequent insertion of CO into the Rh–C bond of **284** gives **285**, which cyclizes to form **286**. The final products **275**, **276**, **277**, **278**, **279**, or **280** are derived from **287**, which represents some isomers generated by prototropy or silatropy from **286** and the subsequent oxidative addition of $R_3SiH$ to Rh metal (Scheme 17.35) [34b]. The formation of divinyl ketone **281** is described by a reverse order of insertion into the Rh–C bond of **283**. Insertion of CO into the Rh–C bond of **283** forms **288** which is the key intermediate in the silylformylation. Subsequent insertion of 1-alkyne into the Rh–C(acyl) bond of **288** provides **289**, which is linked to **281** via the interaction with another molecule of $R_3SiH$ (Scheme 17.35). The rate of this path is far slower than that of the normal silylformylation (Eq. 17.45) [34b].

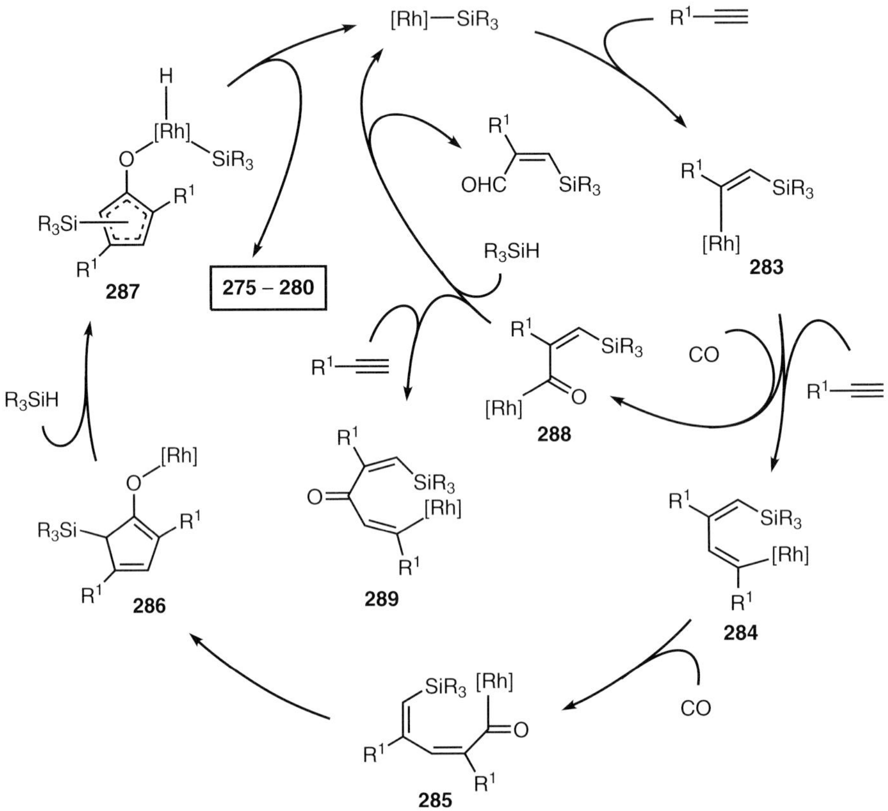

**Scheme 17.35** Proposed catalytic cycle for Rh-catalyzed annulative silylcarbonylation.

The selectivity for the formation of cyclopentenones is not improved in the intermolecular reactions. 1,6-Heptadiyne, **130**, however, reacts in benzene, with 2 equiv of $^tBuMe_2SiH$ to give two bicyclo[3.3.0]octenones **290** and **291** in 14% and 63% yield, respectively at 95 °C under CO (20 atm) pressure [118]. When benzene was replaced with acetonitrile, **291** was obtained as the sole product in 60% yield under similar conditions. On the other hand, the reaction in acetonitrile at 25 °C gives the normal silylformylation product **292** as a major component (59%) with the concomitant formation of **291** (20%) (Scheme 17.36) [118].

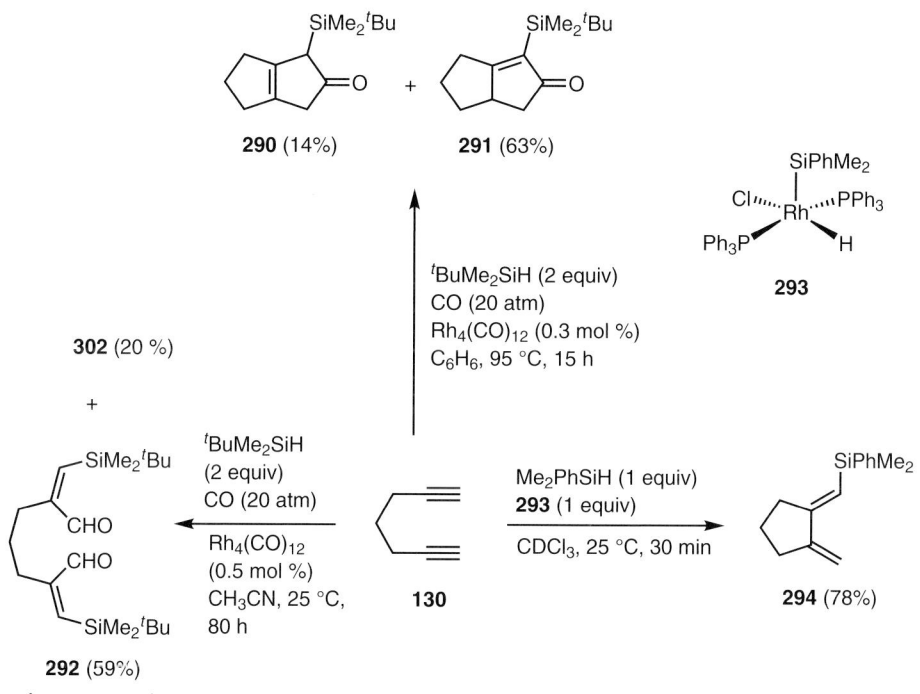

**Scheme 17.36** Rh-catalyzed reactions of **130** with a hydrosilane.

The fact that the normal silylformylation product **292** becomes a major product in the reaction at 25 °C suggests that the activation energy to form a bicyclooctenone framework is higher than that to form **292**. The construction of bicyclo[3.3.0] octenone **290** and **291** seems to be performed by the intermediacy of **131** and **132** shown in Scheme 17.18 by analogy with the catalytic pathway for the intermolecular formation of cyclopentenones (Scheme 17.35). The idea of this type of ring-closing triggered by silylrhodation of an alkyne moiety is supported by the fact that **130** reacts with Me$_2$PhSiH to give **294** under a nitrogen atmosphere in the presence of complex, **293** (1 equiv) which is prepared *in situ* by the interaction of RhCl(PPh$_3$)$_3$ with 1 equiv of Me$_2$PhSiH [46].

This unique [2 + 2 + 1] cyclocoupling may be extended to a variety of 1,6-diyne units (Table 17.13). All types of 1,6-diynes **137** (R$^1$ = R$^2$ = H) listed here react uniformly to give cyclocarbonylation products in moderate to good yields, although the product distribution depends on the substrate used. The most obvious point is that the thermodynamically less stable **295** is formed as a major product in the reaction of **101** and its ethoxy congener under conditions similar to the reaction of

**Table 17.13** Rh-catalyzed silylative cyclocarbonylation of 1,6-diynes with $^t$BuMe$_2$SiH.

**137** (R$^1$ = R$^2$ = H) → **295** + **296**

Conditions: $^t$BuMe$_2$SiH (2 eq), CO pressure, [Rh]

| No. | X in 1,6-diyne | Conditions[a] | Solvent | Yield (%) of products 295 | Yield (%) of products 296 | Ref. |
|---|---|---|---|---|---|---|
| 1 | (MeO$_2$C)$_2$C | A | C$_6$H$_6$ | 58 | 6 | [118] |
| 2 | (MeO$_2$C)$_2$C | A | CH$_3$CN | 70 | 14 | [118] |
| 3 | (EtO$_2$C)$_2$C | B | toluene | 82 | 0 | [120b] |
| 4 | (EtO$_2$C)(Me)C | B | toluene | 70 | 0 | [120b] |
| 5 | (EtO$_2$C)(H)C | B | toluene | 48 | 15 | [120b] |
| 6 | (HOCH$_2$)$_2$C | C | toluene | 0 | 71[b] | [120a] |
| 7 | ($^t$BuMe$_2$SiOCH$_2$)$_2$C | C | toluene | 0 | 95[b] | [120a] |
| 8 | (AcOCH$_2$)$_2$C | C | toluene | 0 | 72[b] | [120a] |
| 9 | (AcOCH$_2$)(H)C | B | toluene | 73 | 0 | [120b] |
| 10 | p-TsN | A | C$_6$H$_6$ | 18[c] | 51 | [118] |
| 11 | PhCH$_2$N | A | C$_6$H$_6$ | 31[d] | 9 | [118] |
| 12 | PhCH$_2$N | A | CH$_3$CN | 72[d] | 8 | [118] |
| 13 | PhCH$_2$N | D | toluene | 70[d] | 18 | [120b] |
| 14 | CH$_2$=CHCH$_2$N | D | toluene | 56[d] | 22 | [120b] |
| 15 | O | A | C$_6$H$_6$ | 0[e] | 3 | [118] |
| 16 | O | A | CH$_3$CN | 2[f] | 35 | [118] |
| 17 | O | B | toluene | 27 | 22 | [120b] |

[a] A: CO (20 atm), Rh$_4$(CO)$_{12}$ (0.5 mol%), 95 °C, 14 h. B: CO (50 atm), Rh(acac)(CO)$_2$ (2 mol%), 120 °C, 5–14 h. C: CO (50 atm), ($^t$BuNC)$_4$RhCo(CO)$_4$, 65 °C, 10 h. D: CO (15 atm), Rh(acac)(CO)$_2$ (2 mol%), 50 °C, 12 h.
[b] This value means the yield of **297**.
[c] A bicyclooctenone, **298**, was isolated as an additional product in 5% yield.
[d] The yield of **298**.
[e] A bicyclooctenone, **299** was isolated as an additional product in 6% yield.
[f] A bicyclooctenone, **299** was isolated as an additional product in 2% yield.

**297**, **298**, **299**

1,6-heptadiyne **130**. The isolated **295** {X = (MeO$_2$C)$_2$C or (EtO$_2$C)$_2$C} is readily converted to **296** under the catalysis of RhCl$_3$·3H$_2$O in ethanol [118, 119]. It is quite difficult to isolate **295** as a single species, since this type of isomerization proceeds partially during purification through a pad of silica gel. Bicyclooctadienones **297** are isolated as the sole product in the reaction of **137** (R$^1$ = R$^2$ = H) under forcing conditions (entries 6, 7, and 8 in Table 17.13). The *p*-tosyl derivative of dipropargylamine preferentially forms **296** as a major product (entry 10 in Table 17.13) [118], while benzyl and allyl groups preferentially form **298** (entries 11–14 in Table 17.13) [118, 120]. The products corresponding to **296** (X=CH$_2$=CHCH$_2$N) and **298** (X=CH$_2$=CHCH$_2$N) were ill-defined by speculation [[55]a]. Bicyclic product **299** (X=O) is also isolated along with **295** (X=O) and **296** (X=O) in the reaction of dipropargyl ether (entries 15 and 16 in Table 17.13) [118]. Formation of **299** (X=O) suggests participation of a process involving silatropy in the catalytic cycle. 2,8-Decadiyne is also carbonylated to give the analogue of **295** in 43% yield under relatively low CO pressure (5 atm) [118]. It is difficult at present to specify the factor determining the product, since prototropy, silatropy, or dehydrogenation from the intermediate may vary the product distribution, depending on the structure of the starting diyne **137** in addition to the thermodynamic instability of the primary product.

4,4-Dimethyl-1,6-heptadiyn-3-ol **300** reacts with $^t$BuMe$_2$SiH to give a mixture of **301** and **302** (Eq. 17.46) [120]. A careful analysis of the reaction mixture, however, shows the product is composed of a complex mixture of regioisomers and diastereomers, including **301** and **302** [121]. This observation suggests that it is difficult to differentiate between the two acetylenic moieties in **300** at the silylrhodation step.

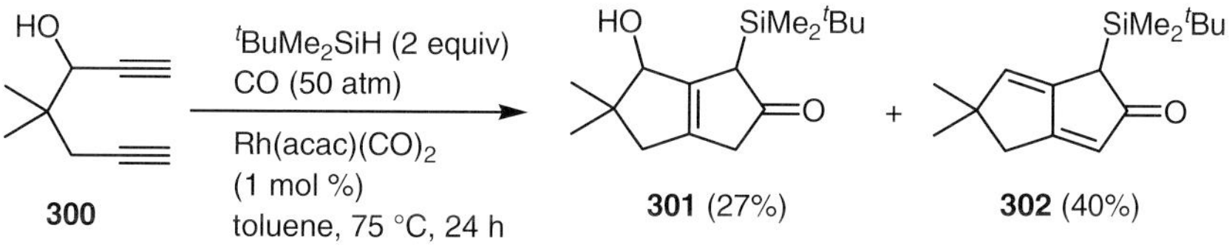

(17.46)

In contrast to rhodium catalyst which gives bicyclo[3.3.0]octenone frameworks from **137** (R$^1$ = R$^2$ = H), Ru$_3$(CO)$_{12}$ gives catechol derivatives **303** from the identical 1,6-diyne units under severe conditions (Scheme 17.37) [122].

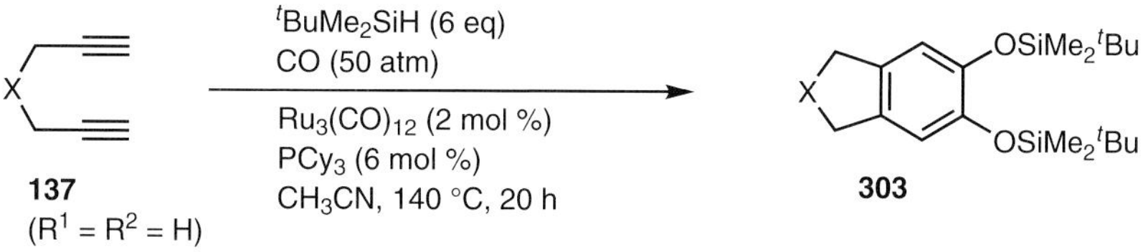

|   | X | Yield (%) |
|---|---|---|
| a | Ac$_2$C | 71 |
| b | O | 40 |
| c | *p*-TsN | 45 |

**Scheme 17.37** Ru-catalyzed CO incorporation into 1,6-diynes.

### 17.4.1.2 1,6-Enynes

Rhodasilylation and rhodavinylation of two acetylenic bonds involved in the same molecule at an early stage of cyclization is proposed to explain the formation of bicyclo [3.3.0]octenone **295** from 1,6-diyne **137** in a one-pot operation. If this similar cyclization pattern would proceed in the reaction of 1,6-enyne, this simple operation could provide a new tool for the construction of bicyclo[3.3.0] frameworks. When **157** {X = (MeO$_2$C)$_2$C, R$^1$ = H} was subjected to the reaction with $^t$BuMe$_2$SiH under conditions similar to the case of 1,6-diyne, aldehyde **304** was unexpectedly obtained in 34% yield as the sole product (Eq. 17.47) [54].

$$\begin{array}{c} \text{MeO}_2\text{C} \\ \text{MeO}_2\text{C} \end{array} \xrightarrow[\substack{\text{Rh}_4(\text{CO})_{12}\ (0.5\ \text{mol}\ \%) \\ \text{C}_6\text{H}_6,\ 90\ °\text{C},\ 14\ \text{h} \\ 34\%}]{\substack{^t\text{BuMe}_2\text{SiH}\ (1.2\ \text{equiv}) \\ \text{CO}\ (20\ \text{atm})}} \begin{array}{c} \text{MeO}_2\text{C} \\ \text{MeO}_2\text{C} \end{array}\!\!\!\!\!\!\!\!\!\!\!\!\!\!\!\!\!\!\!\!\!\!\!\!\!\!\!\!\!\!\!\!\!\!\!\!\!\!\!\!\!\!\!\!\!\!\!\!\!\!\!\!\!\!\!\!\!\!\!\!\!\!\!\!\!\!\!\text{SiMe}_2{}^t\text{Bu} / \text{CHO}$$

**157** X = (MeO$_2$C)$_2$C
R$^1$ = H

**304**

(17.47)

This result shows clearly a possibility of continuous insertion of alkyne and alkene into Rh—Si and Rh—C= bonds. When more reactive hydrosilane is used to enhance the conversion of **157** {X = (MeO$_2$C)$_2$C, R$^1$ = H}, the most serious side reaction is annulative hydrosilylation to give **306** in this type of transformation. This new protocol for the formylation of an olefinic bond is optimized in a normal concentration of substrates (0.5–0.1 mol l$^{-1}$) by adopting the following conditions: (i) use of Me$_2$PhSiH as a hydrosilylation agent, (ii) mixing the starting substrates in a solvent saturated by CO, and (iii) high CO pressure (>20 atm) [54]. Additional modification of the reaction conditions also provides higher yields and higher selectivity of **305**: (i) operation with highly dilute reactants (0.02 mol l$^{-1}$) and (ii) addition of 10 mol% of P(OEt)$_3$ to the starting mixture [55]. Thus, **305**, which was formed by annulative silylformylation was isolated as the sole product in the reaction of 1,6-enyne, **157** (R$^1$ = H), with a molar equivalent of Me$_2$PhSiH under CO pressure (mode 1 in Scheme 17.38) (Table 17.14) [54, 55]. N-Benzyl-N-propargylallylamine, **143** is an exceptional example in which CO is not incorporated under the same reaction conditions. Pyrrolidine derivative **144** was the sole product isolated [54].

When this reaction was carried out under nitrogen atmosphere or CO atmosphere, a cyclopentane **306** was formed selectively within a few minutes at 25 °C (mode 2 in Scheme 17.38) [54, 55]. 1,6-Enyne **157** (R$^1$ = H) was used as a typical model for the Pauson–Khand reaction (mode 3 in Scheme 17.38) [123], whereas formation of **308** was completely suppressed under the conditions of mode 1 in Scheme 17.38. It is noteworthy that simple alkyne silylformylation product **307** is not detected. This is quite different from the silylformylation of 1-penten-4-yne under similar conditions. These results can be explained by a pathway similar to the reaction of 1,6-diynes: (i) sequential insertion of the acetylenic and olefinic moieties into the Rh—Si bond, and (ii) subsequent interaction of CO and Me$_2$PhSiH with the resultant intermediate to give **305**. The fact that **306** is formed selectively under CO atmosphere supports elucidation of the stepwise insertion of the acetylenic and olefinic moieties involved in the same molecule into the Rh—Si bond.

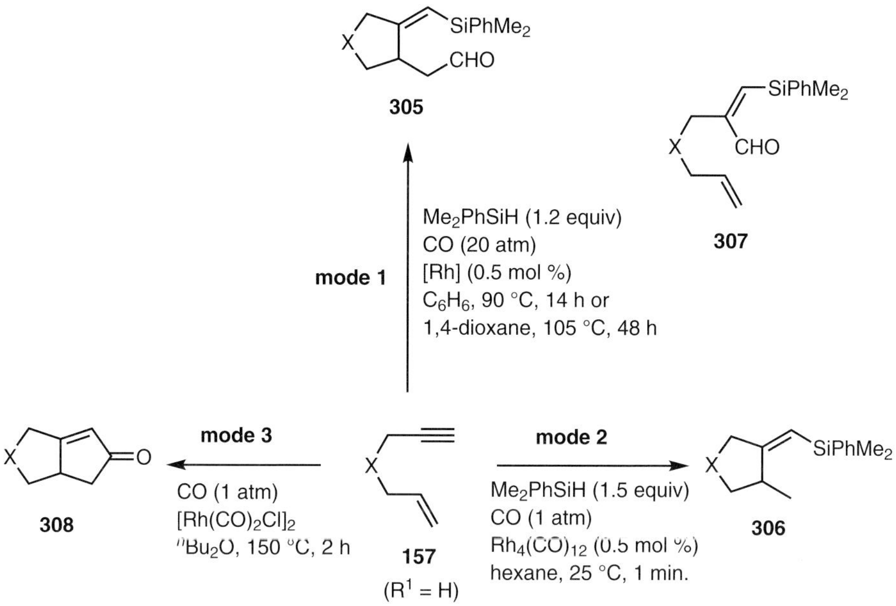

**Scheme 17.38** Reaction modes catalyzed by a Rh-complex.

**Table 17.14** Annulative silylformylation of **157** ($R^1$ = H) with $Me_2PhSiH$.

| No. | X in 1,6-enyne | Conditions[a] | Solvent | Yield (%) of 305 | Ref. |
|---|---|---|---|---|---|
| 1 | $(MeO_2C)_2C$ | A | $C_6H_6$ | 85[b] | [54] |
| 2 | $(MeO_2C)_2C$ | B | $C_6H_6$ | 89 | [54] |
| 3 | $(MeO_2C)_2C$ | C | 1,4-dioxane | 98 | [55b] |
| 4 | $(EtO_2C)_2C$ | C | 1,4-dioxane | 91 | [55b] |
| 5 | 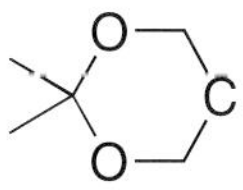 (dioxane-C) | A | $C_6H_6$ | 76 | [54] |
| 6 | (dioxane-C) | B | $C_6H_6$ | 89 | [54] |
| 7 | (dioxane-C) | C | 1,4-dioxane | 89 | [55b] |
| 8 | $(MeOCH_2)_2C$ | C | 1,4-dioxane | 83 | [55b] |
| 9 | $(AcOCH_2)_2C$ | C | 1,4-dioxane | 91 | [55b] |
| 10 | TsN | A | $C_6H_6$ | 59 | [54] |
| 11 | TsN | B | $C_6H_6$ | 74 | [54] |
| 12 | TsN | C | 1,4-dioxane | 85 | [55b] |
| 13 | MsN | C | 1,4-dioxane | 56 | [55b] |
| 14 | BnN | A | $C_6H_6$ | 88[c] | [54] |
| 15 | BnN | B | $C_6H_6$ | 78[c] | [54] |
| 16 | O | A | $C_6H_6$ | 37 | [54] |
| 17 | O | B | $C_6H_6$ | 37 | [54] |

[a]A: $Rh_4(CO)_{12}$ 0.5 mol%, 90 °C, 14 h. B: $Rh(acac)(CO)_2$ 0.5 mol%, 90 °C, 14 h. C: $Rh_4(CO)_{12}$ 0.5 mol %, $P(OEt)_3$ 10 mol%, 105 °C, 45 h.
[b]CO pressure was 36 atm. When the reaction was carried out under 20 atm of CO pressure, 74% of **305** was isolated accompanied with 11% of **306**.
[c]No **305** (X = BnN) was obtained. The yield of **144**.

4,4-Dimethyl-5-trimethylsilyloxyhept-1-en-6-yne **307** and its positional isomer 4,4-dimethyl-3-trimethylsilyloxyhept-1-en-6-yne **309** react with $Me_2PhSiH$ to give the corresponding aldehydes, **308** and **310**, respectively, as a mixture of two diastereomers under conditions similar to mode 1 in Scheme 17.38 (Eqs. 17.48 and 17.49) [54]. In the reaction of **309**, an appreciable amount of **311** is formed concomitantly. This fact implies that the trimethylsilyloxy group on the allylic carbon seems to retard the cyclopentane annulation step.

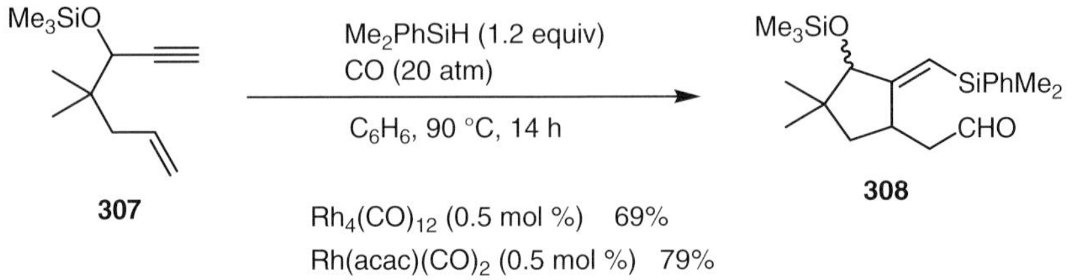

(17.48)

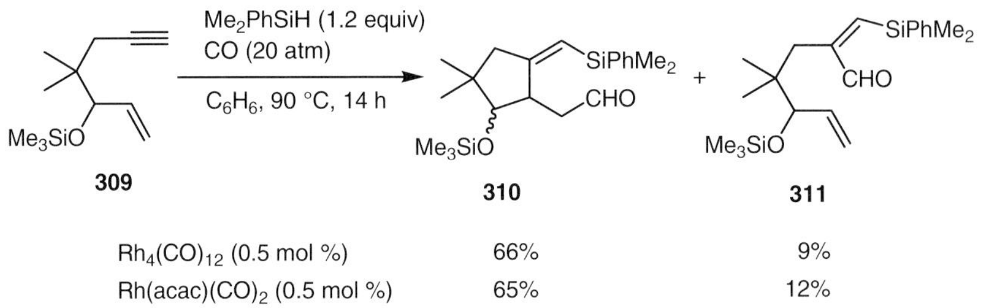

(17.49)

It is notable that the reaction of **157b** with $Me_2PhSiH$ gives **312** as the sole cyclic compound; less than 3% of the standard silylformylation product is formed (Eq. 17.50) [54]. The outcome is rationalized by intervention of **313** in which intramolecular coordination of an olefinic moiety dominates the orientation of silylrhodation toward the alkyne moiety in **157b** ($E = CO_2Me$).

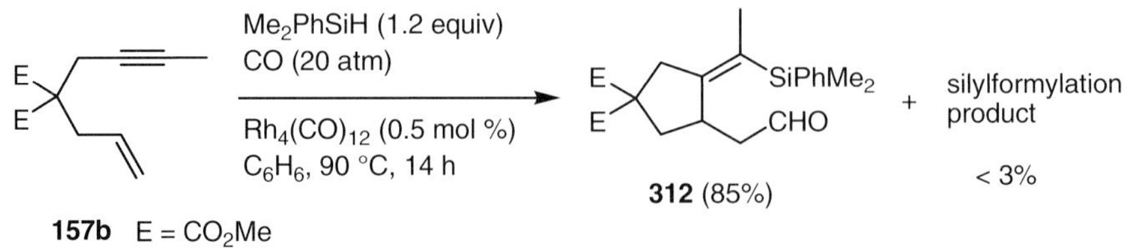

(17.50)

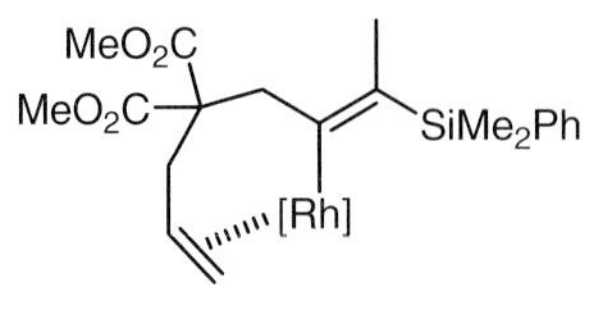

## 17.4.2
### Silylative Lactonization of Alkynols

When a 1 : 1 mixture of 1 equiv each of propargylic alcohol, **314**, Me$_2$PhSiH, and Et$_3$N is heated under CO pressure (20 atm) in the presence of Rh$_4$(CO)$_{12}$ (0.5 mol%), silylformylation proceeds to give **315** in 52% yield accompanied by 43% yield of β-lactone **316** (Scheme 17.39) [124].

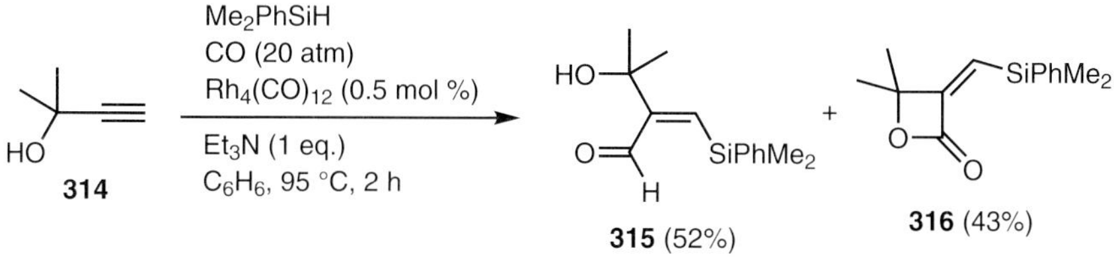

| | hydrosilane | base | yield (%) | |
|---|---|---|---|---|
| | | | 315 congener | 316 congener |
| a | Me$_2$PhSiH | Pyridine | 67 | 0 |
| b | Me$_2$PhSiH | TMEDA | 32 | 59 |
| c | Me$_2$PhSiH | DBU (0.1 equiv) | 0 | 81 |
| d | Et$_3$SiH | Et$_3$N | 6 | 64 |
| e | $^i$Pr$_3$SiH | Et$_3$N | 0 | 33 |
| f | $^t$BuMe$_2$SiH | Et$_3$N | 0 | 96 |

**Scheme 17.39** Rh-catalyzed silylative β-lactone formation from **314**.

Selectivity for the formation of **316** is dramatically improved by use of either a stronger base, such as 1,8-diazabicyclo[5.4.0]undec-7-ene (DBU), (0.1 equiv of DBU is sufficient for selective formation of **316**) instead of Et$_3$N or a bulkier silane such as $^t$BuMe$_2$SiH. On the basis of this preliminary information, this protocol was extended to some types of acetylenic alcohols, as shown in Table 17.15 [124]. Spiro type β-lactone **320**, **322**, and **324** are also obtained in good yields by this improved method. In the reaction of less substituted 3-butyn-2-ol **317**, the combined use of $^t$BuMe$_2$SiH and 0.1 equiv of DBU is required for selective formation of β-lactone **318** [124a]. Homopropargyl-type alcohols **325**, **327**, **329**, and **331** are readily converted to lactone rings **326**, **328**, **330**, and **332**, respectively, in which the lactone ring formation is much easier than in the case of propargylic alcohols. δ-Lactones **334** and **336** are also selectively constructed from **333** and **335**, respectively, by applying a similar modification [124b].

The presence of 2–3 equiv of H$_2$O or an alcohol does not affect the product pattern in the standard silylformylation toward a C—C triple bond. While **326** is derived

**Table 17.15** Silylative lactonization of alkynylalcohols [124].

| No. | Alkynyl alcohol | Hydrosilane | Base | Product[a] | Yield (%) |
|---|---|---|---|---|---|
| 1 | **317** | Me$_2$PhSiH | Et$_3$N | **318** | 15 |
| 2 |  | Me$_2$PhSiH | DBU |  | 54 |
| 3 |  | $^t$BuMe$_2$SiH | DBU |  | 79 |
| 4 | **319** | $^t$BuMe$_2$SiH | DBU | **320** | 68 |
| 5 | **321** | Me$_2$PhSiH | DBU | **322** | 85 |
| 6 | **323** | Me$_2$PhSiH | DBU | **324** | 86 |
| 7 | **325** | Me$_2$PhSiH | Et$_3$N | **326** | 86 |
| 8 | **327** | $^t$BuMe$_2$SiH | DBU | **328** | 84 |
| 9 | **329** | Me$_2$PhSiH | Et$_3$N | **330** | 87 |
| 10 |  | $^t$BuMe$_2$SiH | DBU |  | 83 |

(*Continued*)

## Table 17.15 (Continued)

| No. | Alkynyl alcohol | Hydrosilane | Base | Product[a] | Yield (%) |
|-----|-----------------|-------------|------|------------|-----------|
| 11  | 331             | Me$_2$PhSiH | Et$_3$N | 332     | 87        |
| 12  | 331             | $^t$BuMe$_2$SiH | DBU | 332    | 82        |
| 13  | 333             | $^t$BuMe$_2$SiH | Et$_3$N | 334  | 84        |
| 14  | 335             | $^t$BuMe$_2$SiH | DBU | 336    | 73        |

[a] Reactions were carried out on a 2 mmol scale in C$_6$H$_6$ solution with 0.25 mol% of Rh$_4$(CO)$_{12}$ and equimolar amounts of an alkynyl alcohol, a hydrosilane, and a base (except DBU of which content was 10 mol% for the completion of reactions) under CO pressure (20 atm) at 100 °C for 2 h.

from **325** without a base under conditions similar to silylformylation. This fact implies strongly the existence of intramolecular coordination of a hydroxy group located in an appropriate position to the rhodium center and an additional role of base in the product determination step. This point is quite different from the preceding butenolide formation that proceeds through incorporation of 2 equiv CO into an acetylenic triple bond in the presence of Rh$_4$(CO)$_{12}$ [125]. On the other hand, propargylic alcohols **337** lead almost selectively lead to **338** in the presence of Rh$_4$(CO)$_{12}$ (0.5 mol%) under synthetic gas pressure (CO/H$_2$ = 1/1, 40 atm) (Scheme 17.40) [126].

Hydrorhodation to a C—C triple bond seems to be the first step in these reactions. The transformation of **325** to **326** (entry 7 in Table 17.15) resembles the palladium-catalyzed carbonylation of **325** to form α-methylene-γ-butyrolactone, in which an alkoxycarbonylpalladium complex plays an important role for cyclization of homopropargyl alcohols [127].

From this evidence, selective formation of lactones in Table 17.15 can be explained by supposing the presence of **340** ($n=1$, 2, or 3), which plays a pivotal role in the differentiation between intramolecular nucleophilic attack of the hydroxy group and reductive elimination of silylformylation products like **315**. Thus, addition of a base is believed to accelerate the conversion of **340** ($n=1$, 2, or 3) to rhodate anion **341** ($n=1$, 2, or 3). The formation of β-lactone and δ-lactone frameworks can be explained by identical logic. This is supported by the fact that the introduction of a strong base such as DBU is advantageous for the selective formation of a lactone framework. The

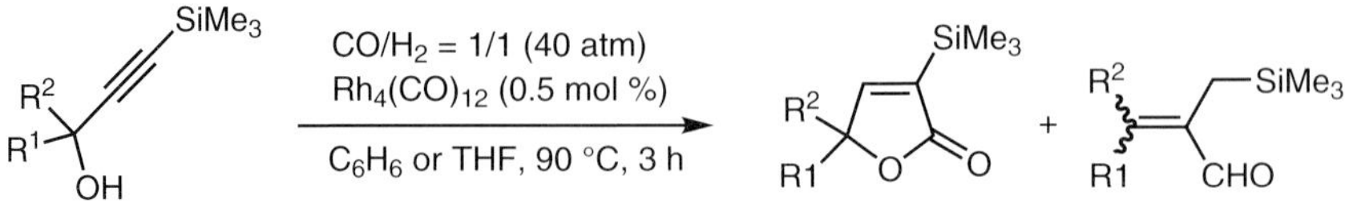

## Scheme 17.40 Rh-catalyzed dihydrofuranone formation from 337.

Reagents: CO/H$_2$ = 1/1 (40 atm), Rh$_4$(CO)$_{12}$ (0.5 mol %), C$_6$H$_6$ or THF, 90 °C, 3 h. Substrate **337** → **338** + **339**.

| | R$^1$ | R$^2$ | Yield (%) of 338 | 338:339 |
|---|---|---|---|---|
| a | −(CH$_2$)$_4$− | | 82 | 95:5 |
| b | −(CH$_2$)$_5$− | | 95 | 99:1 |
| c | Et | Et | 83 | 98:2 |
| d | Me | Ph | 88 | 97:3 |
| e | Ph | Ph | 85 | 94:6 |
| f | H | Oct | 72 | 96:4 |
| g | H | $^i$Pr | 74 | 93:7 |
| h | H | Ph | 95 | 96:4 |

reactivity of $^t$BuMe$_2$SiH and of Me$_2$PhSiH is almost comparable in this lactone formation. Formation of esters is not observed in the intermolecular silylformylation of a mixture containing 1-alkyne and an alcohol [124a]. The facts also support that an appreciable energy gain is expected by the intramolecular coordination shown in **340** and **341** in the formation of lactone rings.

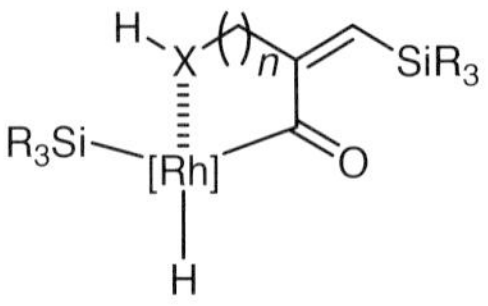

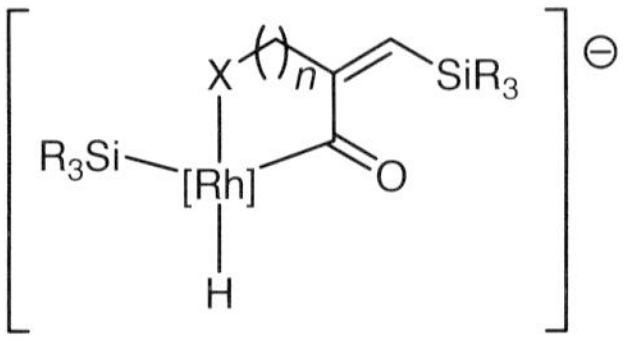

**340** X = O (n = 1, 2, and 3)
**342** X = NR$^1$ (n = 1, 2, 3, and 4)

**341** X = O (n = 1, 2, and 3)
**343** X = NR$^1$ (n = 1, 2, 3, and 4)

### 17.4.3
### Silylative Carbamoylation of Alkynes

A protocol similar to the previous section is applicable to the selective formation of lactams from alkynylamines, since an amino group involved in alkynylamines seems to be suitable for intramolecular coordination, as shown in **342** and **343**. Propargylic amines **344** with two substituents on the propargyl carbon react readily with an equimolar amount of Me$_2$PhSiH to give β-lactam **345** as the sole product in the presence of Rh$_4$(CO)$_{12}$ (0.3 mol%) and DBU (0.1 equiv) (Scheme 17.41) [128a]. Protection of the amino group as *p*-tosylamide is required for selective formation of the β-lactam ring. When $^t$BuMe$_2$SiH is used together with DBU, protection of the amino group is not necessary for the selective formation of **347** (Scheme 17.42) [128a].

|   | R$^1$ | Yield (%) |
|---|---|---|
| a | Me | 80 |
| b | Me$_2$CHCH$_2$– | 65 |

**Scheme 17.41** Rh-catalyzed β-lactam formation.

|   | Z | Yield (%) |
|---|---|---|
| a | *p*-Ts | 81 |
| b | Bn | 98 |
| c | H | 64 |

**Scheme 17.42** Rh-catalyzed spiro-type β-lactam formation.

In contrast to the results of **344** and **346**, the expected β-lactam is obtained as a minor product in the reaction of sterically less constrained propargylic amine under similar conditions [128a]. In the presence of DBU the amino group locating in the 'just fitting' position participates immediately for the formation of energetically more stable lactam rings under silylformylation conditions (Table 17.16) [128b]. While, in the absence of DBU, the lactam formation suffers from lack of reproducibility in yields and loss of Z-selectivity in the silylmethylene moiety [129]. The formation of lactam rings is elucidated by the intermediacy of **342** and **343**.

**Table 17.16** Rh$_4$(CO)$_{12}$-catalyzed construction of lactam frameworks [128b].

| No. | Alkynylamine | | Product[a] | Yield (%) |
|---|---|---|---|---|
| 1 | 348 | R$^1$ = H | 349 | 98 |
| 2 | | R$^1$ = Ph | | 84 |
| 3 | 350 | | 351 | 75 |
| 4 | 352 | | 353 | 68 |
| 5 | 354 | | 355 | 97 |
| 6 | 356 | n = 1 | 357 | 66 |
| 7 | | n = 2 | | 55 |

[a] Reactions were carried out on a 1.6 mmol scale of an alkynylamine and $^t$BuMe$_2$SiH in C$_6$H$_6$ solution with 0.25 mol% of Rh$_4$(CO)$_{12}$ and 10 mol% of DBU under CO pressure (20 atm) at 100 °C for 2 h.

## 17.5
## Conclusion

The hydrometallation protocol has been applied to the construction of diverse types of cyclic compounds. Among many candidates as a precursor for the formation of the M−H species a hydrosilane has occupied the central position in this protocol, since it has been demonstrated that the silyl group involved in Si−C(sp$^3$) and Si−C(sp$^2$)

moieties can be replaced by a hydroxy group or an additional carbon unit. Intramolecular hydrosilylation has been accepted as a powerful tool for constructing stereo-defined 1,3-alkanediols or trisubstituted alkenes. Hydrosilanes have also played an important role as a trigger and string for the annulation in the intermolecular reactions of 1,6- and 1,7-dienes, enynes, and diynes with the catalysis of a transition-metal complex. Moreover, the combined use of a hydrosilane and CO has disclosed a new short-cut for the construction of functionalized cyclic compounds. However, the following work is strongly required in the future: (i) exploration of more active and specific catalysts and (ii) exploration of a new concept for the formation of M−H species.

## 17.6
## Experimental: Selected Procedures

### 17.6.1
### Intramolecular Hydrosilylation of 9 Catalyzed by 2 [8]

To a mixture of silylamine **9** (311 mg, 0.97 mmol) and hexane (1.9 mL) was added **2** (0.25 M xylene solution, 19 μL, $4.9 \times 10^{-3}$ mmol, 0.5 mol%) at room temperature under argon. After 45 min, disappearance of the olefin was confirmed by $^1$H NMR spectroscopy. Bulb-to-bulb distillation (bath temperature 95–115 °C/0.3 mmHg) of the reaction mixture gave the four-membered-ring product **10** (264 mg, 85% yield).

### 17.6.2
### Intramolecular Hydrosilylation of 9 Catalyzed by [RhCl(CH$_2$=CH$_2$)$_2$]$_2$ [8]

A mixture of **9** (156 mg, 0.50 mmol), dichloroethane (1.0 mL), and [RhCl(CH$_2$=CH$_2$)$_2$]$_2$ (3.9 mg, $9.9 \times 10^{-3}$ mmol) was stirred at 70 °C under nitrogen for 1 day. Disappearance of olefin was confirmed by $^1$H NMR spectroscopy. Bulb-to-bulb distillation (bath temperature 180 –195 °C/0.9 mmHg) of the reaction mixture gave the four-membered-ring product **11** (137 mg, 86% yield).

### 17.6.3
### Intramolecular Hydrosilylation of 34 to Form 35 [23b]

Crude **34**, derived from the corresponding dihydroxy form, (0.219 g, 0.99 mmol), and tetramethyldisilazane (3.5 mL, 19.7 mmol) was dissolved in THF (4 mL), and H$_2$PtCl$_6$ (0.05 M in THF, 0.1 mL, $4.95 \times 10^{-3}$ mmol) was added. The yellow solution was heated to 75 °C. After 5 h the hydrosilylation was judged complete by IR analysis (Si H, 2121 cm$^{-1}$). The mixture was diluted with ether and filtered through Celite 521. The mother liquor was concentrated under reduced pressure to afford 0.323 g (97%) of **35** as a yellow oil that was used in the oxidation without further purification.

## 17.6.4
### Synthesis of Spectaline 55 from 52 [33]

A mixture of alkyne **52** (50 mg, 0.127 mmol) and 1,1,3,3-tetramethyldisilazane (85 mg, 0.635 mmol) was heated to 70 °C for 20 h. The neat mixture was cooled to room temperature and concentrated *in vacuo* (ca. 1 Torr) to remove volatile organic species. After 30 min under vacuum, the residue was taken up in $CH_2Cl_2$ (0.3 mL) and treated with solid $[Cp^*Ru(CH_3CN)_3]PF_6$ (3.2 mg, 0.0064 mmol) at 0 °C. The solution was allowed to warm to room temperature over 2 h and was then filtered through florisil (ca. 3 cm), eluting with ether (5 mL), to afford 56 mg (98%) of spectroscopically homogeneous cyclic vinylsilane **53** without additional purification.

The azido-siloxane **53** (7.1 mg, 0.016 mmol) and solid urea-hydrogen peroxide adduct (6 mg, 0.064 mmol) were taken up in DMF (0.16 mL) and treated at 0 °C with TBAF (63 µL, 0.063 mmol, 1.0 M in THF) dropwise via a syringe pump over 4 h. The mixture was then allowed to warm to room temperature and was stirred for 10 h. At this time, TLC analysis indicated incomplete consumption of the starting vinylsilane, and so additional urea-hydrogen peroxide adduct (6 mg, 0.064 mmol) was added, followed by additional TBAF (63 µL, 0.063 mmol, 1.0 M in THF) via syringe over 4 h as before. The mixture was then allowed to warm to room temperature and was stirred for 20 h. The mixture was diluted with water (5 mL) and extracted with AcOEt (3 × 4 mL). The organic layers were dried over $Na_2SO_4$ and concentrated *in vacuo*. The residue was purified on a silica gel column (eluent: pet. ether/AcOEt: 7/1 − 4/1) to afford 5.2 mg (84%) of the desired ketone **54**.

The keto-azide (2.4 mg, 0.0058 mmol) was taken up in EtOH (0.5 mL) and treated with 10% Pd/C (1 mg). The mixture was placed under 60 psi of $H_2$ and stirred for 24 h. The mixture was filtered through Celite, washing with additional EtOH (2 mL). The filtrate was concentrated *in vacuo* and taken up in THF (0.5 mL). The solution was treated with aqueous HCl (0.05 mL, 2.0 M solution) and stirred for 24 h at room temperature. The mixture was then quenched by addition of aqueous NaOH (0.5 mL, 1.0 M solution). The mixture was extracted with AcOEt (2 × 3 mL), dried over $Na_2SO_4$, and concentrated *in vacuo*. The residue was purified on a column of basic alumina (eluent 100 : 1 $CHCl_3$ : EtOH), affording 1.0 mg (53%) of (+)-spectaline **55**.

## 17.6.5
### $Rh_4(CO)_{12}$-Catalyzed Intramolecular Silylformylation of 56 [36]

A glass tube fitted with a stirring bar was charged with $Rh_4(CO)_{12}$ (0.0042 g, 0.0056 mmol) and benzene (5 mL) saturated with CO. The tube was put in a 100 mL stainless steel autoclave. The reactor was pressurized by CO to 20 atm. The contents were stirred for 5 min at ambient temperature and the pressurized CO was purged in a hood. To this tube were added subsequently **56** (0.168 g, 1.09 mmol) in $C_6H_6$ (1 mL) and $Et_3N$ (0.119 g, 1.18 mmol) through a syringe needle at the same temperature. The reactor was pressurized again by CO to 20 atm. The content was stirred for 3 h at 90 °C and cooled to ambient temperature. After excess of CO was purged in a hood, the reaction mixture was concentrated under reduced pressure. The residual oily liquid was purified by column chromatography on silica gel using hexane/AcOEt (97/3) as an eluent to give an oily liquid. The subsequent bulb-to-bulb distillation of this oil gave pure **57** (0.164 g, 83%) as a colorless liquid: bp 110 °C/10 Torr.

## 17.6.6
### Reaction of 101 with Et₃SiH to form (Z)-111 [47b]

Toluene (20 mL) and Et$_3$SiH (150 μL, 1.0 mmol) were added sequentially to a mixture of **110** (12 mg, 0.025 mmol), B(C$_6$F$_5$)$_3$ (12 mg, 0.023 mmol), and **101** (105 mg, 0.50 mmol) at 0 °C. The resulting orange solution was heated at 110 °C for 10 min, cooled to room temperature, and concentrated under vacuum. Chromatography of the residue {Al$_2$O$_3$; hexane/AcOEt (24/1)} gave (Z)-**111** (128 mg, 82%) as a faintly pink oil.

## 17.6.7
### Rh₄(CO)₁₂-Catalyzed Silylative Annulation of 143 with Me₂PhSiH [55b]

A reaction vessel equipped with a stirring bar and a CO inlet, was charged with Rh$_4$(CO)$_{12}$ (3.8 mg, 0.005 mol, 0.5 mol%). After purging the vessel with CO, hexane (1.0 mL) was added to dissolve the catalyst. Me$_2$PhSiH (68 mg, 0.5 mmol) was added via a syringe. After stirring for 5 min at ambient temperature, the reaction mixture was then cannulated into a 5-mL round-bottomed flask containing a solution of **143** (238 mg, 1 mmol), Me$_2$PhSiH (138 mg, 1 mmol) in hexane (1.5 mL) via CO pressure without stirring. The resulting mixture was stirred for less than 1 min, and the reaction mixture was submitted to GC analysis. After the reaction was complete, all volatiles were removed under reduced pressure, and the crude product was purified by column chromatography on silica gel to give **144** (83% yield) as a viscous colorless liquid.

## 17.6.8
### [Pd]-Catalyzed Silylative Annulation of 188 to 189 [78c]

Dichloromethane (10 mL) and triethylsilane (230 mg, 2.0 mmol) were added sequentially via syringe to a mixture of **188** (100 mg, 0.47 mmol), (phen)PdMe$_2$ (7 mg, 0.022 mmol), and NaBAr$_4$ (24 mg, 0.024 mmol) at 0 °C and stirred for 5 min to form a dark brown solution. Evaporation of solvent and chromatography (hexane/AcOEt: 12/1) gave **189** (42 mg, 92%) as a colorless oil.

## 17.6.9
### Silylative Cascade Cyclization of 217 with the Catalysis of Cp*₂YCH₃·THF [71]

In a nitrogen atmosphere glove box Cp$^*$$_2$YCH$_3$·THF (5 mol%) was dissolved in cyclohexane (0.5 M). Then methylphenylsilane (31 mg, 0.25 mmol) and **217** (35 mg,

0.23 mmol) were added sequentially in the same flask. The reaction mixture was initially bright yellow and faded to light yellow over 1 h after which no **217** was observed by gas chromatography. The reaction mixture was removed from the glove box and filtered through a small plug of Florisil to remove the catalyst prior to concentration by rotary evaporation. The crude product was purified by flash chromatography. Kugelrohr distillation provided **218** (43 mg, 0.16 mmol, 68% yield, dr = 12.8 : 1).

### 17.6.10
**Silylative Cyclization of 245 to Form 246 [92b]**

PBu$_3$ (63 mL, 0.25 mmol) was added dropwise by syringe to a solution of Ni(cod)$_2$ (34 mg 0.12 mmol) in THF (20 mL) at rt under an atmosphere of argon. After 5 min, Et$_3$SiH (500 μL, 3.13 mmol) was added dropwise. The mixture was cooled to 0 °C with an ice bath and solution of **245** (222 mg, 0.63 mmol) in THF (5 mL) was added dropwise. The reaction mixture was then stirred at 0 °C for 20 h. Upon the completion of the reaction by TLC analysis, the mixture was diluted with Et$_2$O, washed with saturated NaHCO$_3$, brine, dried over Na$_2$SO$_4$, and filtered. After concentration under reduced pressure, the residue was purified by column chromatography on silica gel (CH$_2$Cl$_2$/MeOH/NH$_3$: 92/8/0.1) to give **246** (280 mg, 95%) as a colorless oil.

### 17.6.11
**Rh-Catalyzed Aldol Type Cyclization of 259 [104b]**

Triethylsilane (0.86 g, 7.4 mmol, 2.1 equiv) was added slowly to a stirred solution of methyl (*E*)-6-oxo-hexenoate (*E*)-**259** and RhCl(PPh$_3$)$_3$ (48 mg, 0.05 mmol, 1 mol%) in anhydrous, degassed toluene (20 mL) at ambient temperature. The resulting solution was heated for 16 h at 50 °C and then cooled to room temperature. The reaction mixture was diluted with 2 M aqueous sodium hydroxide (10 mL) and extracted with dichloromethane (60 mL). The combined organic extracts were dried over MgSO$_4$, filtered and the filtrate concentrated under reduced pressure. The resulting crude oil was purified by flash column chromatography eluting with P.E. 30–40 °C/AcOEt (90 : 10) to afford the product **260** (0.73 g, 81%) as a 3 : 1 mixture of cis- and trans-diastereomers as a colorless oil. An identical procedure was followed when using RhH(PPh$_3$)$_4$ (I).

### 17.6.12
**Rh$_4$(CO)$_{12}$-Catalyzed Bicyclocarbonylation of 130 [118]**

A glass tube containing a stirring bar, Rh$_4$(CO)$_{12}$ (0.011 g, 0.014 mmol) and benzene (8 mL) was put in a 100 mL stainless steel autoclave. The reactor was pressurized by CO to 20 atm and the content was stirred for 15 min at ambient temperature to dissolve Rh$_4$(CO)$_{12}$. CO was then purged in a hood and into this tube were added subsequently $^t$BuMe$_2$SiH (0.641 g, 5.52 mmol) in C$_6$H$_6$ (2 mL) and **130** (0.281 g, 3.05 mmol) in C$_6$H$_6$ (2 mL) through a syringe needle at ambient temperature. The reactor was pressurized by CO to 20 atm again. The contents were stirred for 15 h at 95 °C and cooled to ambient temperature. After excess of CO was purged in a hood, the reaction mixture was concentrated under reduced pressure. The residual oily liquid was chromatographed on silica gel using hexane/AcOEt (49/1–48/2) as an eluent to give **290** (0.105 g, 14%) and **291** (0.454 g, 63%, bp 100 °C/0.4 mmHg) as an oily liquid, respectively.

### 17.6.13
### $Rh_4(CO)_{12}$-Catalyzed Silylcarbonylation of 157 {X = $(MeO_2C)_2C$, $R^1$ = H} (Scheme 17.38 Mode 1) [54]

A glass tube containing a stirring bar, $Rh_4(CO)_{12}$ (0.004 g, 0.005 mmol) and benzene (8 mL) was put in a 100 mL stainless steel autoclave. The reactor was pressurized by CO to 20 atm and the contents were stirred for 15 min at ambient temperature to dissolve $Rh_4(CO)_{12}$. CO was then purged in a hood and to this tube were added subsequently $Me_2PhSiH$ (0.152 g, 1.12 mmol) in $C_6H_6$ (1 mL) and **157** (0.212 g, 1.01 mmol) in $C_6H_6$ (1 mL) through a syringe needle at ambient temperature. The reactor was pressurized by CO to 36 atm again. The contents were stirred for 14 h at 90 °C and then cooled to ambient temperature. After excess of CO was purged in a hood, the reaction mixture was concentrated under reduced pressure. The residual oily liquid was chromatographed on silica gel using hexane/AcOEt (49/1–48/2) as an eluent to give **305** {X = $(MeO_2)_2C$} (0.378 g, 85%) as an oily liquid.

### 17.6.14
### $Rh_4(CO)_{12}$-Catalyzed Silyl Carbonylation of 314 to Form 316 [124a]

A glass tube fitted with a stirring bar was charged with $Rh_4(CO)_{12}$ (0.004 g, 0.005 mmol) and $C_6H_6$ (3 mL) saturated with CO. The tube was put in a 100 mL stainless steel autoclave. The reactor was cooled to an external temperature of −78 °C with an acetone–dry ice bath under an atmosphere of CO. To this tube were added subsequently $^tBuMe_2SiH$ (0.124 g, 1.06 mmol) in $C_6H_6$ (1 mL), **314** (0.097 g, 1.15 mmol) in $C_6H_6$ (1 mL), and $Et_3N$ (0.105 g, 1.04 mmol) through a syringe needle at the same temperature. The reactor was pressurized by CO to 20 atm. The contents were stirred for 2 h at 100 °C and cooled to ambient temperature. After excess CO was purged in a hood, the reaction mixture was concentrated under reduced pressure. The residue was chromatographed on silica gel using hexane/AcOEt (48/2) as an eluent to give an oily liquid. The subsequent bulb-to-bulb distillation of this oil gave pure **316** (0.211 g, 96%) as a colorless oil which solidified during storage: bp 81 °C/0.11 mmHg, mp 36.8–38.5 °C.

### 17.6.15
### Rh-Catalyzed Silylcarbamoylation of 346 to Form β-Lactam 347 [128b]

A glass tube fitted with a stirring bar was charged with $Rh_4(CO)_{12}$ (0.003 g, 0.004 mmol) and $C_6H_6$ (5 mL) saturated with CO. The tube was put in a 100 mL stainless steel autoclave. The reactor was cooled to an external temperature of $-78\,^\circ$C with an acetone–dry ice bath under an atmosphere of CO. To this tube were added subsequently $^t$BuMe$_2$SiH (0.183 g, 1.57 mmol) in C$_6$H$_6$ (1 mL), *N*-benzyl-*N*-1-ethynylcyclohexylamine **346b** (0.335 g, 1.57 mmol) in C$_6$H$_6$ (1 mL), and DBU (0.013 g, 0.009 mmol) in C$_6$H$_6$ (1 mL) through a syringe needle at the same temperature. The reactor was pressurized with CO to 20 atm. The contents were stirred for 4 h at 100 $^\circ$C and cooled to ambient temperature. After excess CO was purged in a hood, the reaction mixture was concentrated under reduced pressure. The residue was chromatographed on silica gel using hexane/AcOEt (20/1) as an eluent to give **347b** (0.550 g, 98%) as a pale yellow oily liquid.

## Abbreviations

| | |
|---|---|
| Ac | acetyl |
| acac | acetylacetone |
| Ar | aryl |
| BINAP | 2,2′-bis(diphenylphosphino)-1,1′-binaphthyl |
| BIPHEMP | 6,6′-bis-(diphenylphosphino)-2,2′-dimethylbiphenyl |
| BIPHEP | 2,2′-bis(diphenylphosphino)-1,1′-biphenyl |
| BOC | *tert*-butoxycarbonyl |
| Bn | benzyl |
| chiraphos | 2,3-bis(diphenylphosphino)butane |
| cod | 1,5-cyclooctadiene |
| Cp | cyclopentadienyl |
| Cp* | 1,2,3,4,5-pentamethylcyclopentadienyl |
| Cy | cyclohexyl |
| DBU | 1,8-diazabicyclo[5.4.0]undec-7-ene |
| DIOP | 2,3-*O*-isopropylidene-2,3-dihydroxy-1,4-bis(diphenylphosphino)butane |
| DPPF | 1,1′-bis(diphenylphosphino)ferrocene |
| dba | dibenzylideneacetone |
| dpm | dipivaloylmethane or 2,2,6,6-teramethylpentan-3,5-dione |
| phen | 1,10-phenanthroline |
| Piv | pivaloyl |
| SDP | 7,7′-bis(diphenylphosphino)-1,1′-spirobiindane |
| SEGPHOS | 5,5′-bis(diphenylphosphino)-4,4′-bi-1,3-benzodioxole |
| TBAF | tetrabutylammonium fluoride |
| TBDMS | *tert*-butyldimethylsilyl |
| TMEDA | *N*,*N*,*N*′,*N*′-tetramethylethylenediamine |
| TMS | trimethylsilyl |
| Tr | triphenylmethyl |
| Ts | *p*-toluenesulfonyl |

# References

1. Hegedus, L.S. (1994) *Transition Metals in the Synthesis of Complex Organic Materials*, University Science Books, USA.
2. (a) Heck, R.F. (1985) *Palladium Reagents in Organic Synthesis*, Academic Press, USA; (b) Beletskaya, I.P. and Cheprakov, A.V. (2000) *Chem. Rev.*, **100**, 3009–3066.
3. (a) Tamao, K., Ishida, N. and Kumada, M. (1983) *J. Org. Chem.*, **48**, 2120–2122; (b) Tamao, K., Ishida, N., Tanaka, T. and Kumada, M. (1983) *Organometallics*, **2**, 1694–1696; (c) Tamao, K. (1988) *J. Synth. Org. Chem. Jpn.*, **46**, 861–878; (d) Fleming, I., Henning, R. and Plaut, H. (1984) *J. Chem. Soc. Chem. Commun.*, 29–31; (e) Fleming, I. and Sanderson, P.E.J. (1987) *Tetrahedron Lett.*, **28**, 4229–4232; (f) Fleming, I., Barbero, A. and Walter, D. (1997) *Chem. Rev*, **97**, 2063–2192; (g) Jones, G.R. and Landais, Y. (1996) *Tetrahedron*, **52**, 7599–7662.
4. Tamao, K., Nakajima, T., Sumiya, R., Arai, H. and Ito, Y. (1986) *Tetrahedron Lett.*, **27**, 3377–3380.
5. (a) Tamao, K., Nakajima, T., Sumiya, R., Arai, H., Higuchi, N. and Ito, Y. (1986) *J. Am. Chem. Soc.*, **108**, 6090–6093; (b) Tamao, K., Yamauchi, T. and Ito, Y. (1987) *Tetrahedron Lett.*, **28**, 171–174; (c) Tamao, K., Nakagawa, Y., Arai, H., Higuchi, N. and Ito, Y. (1988) *J. Am. Chem. Soc.*, **110**, 3712–3714.
6. Tamao, K., Nakagawa, Y. and Ito, Y. (1990) *J. Org. Chem.*, **55**, 3438–3439.
7. Yoon, K. and Son, D.Y. (1999) *Org. Lett.*, **3**, 423–425.
8. Tamao, K., Nakagawa, Y. and Ito, Y. (1993) *Organometallics*, **12**, 2297–2308.
9. Curtis, N.R. and Holmes, A.B. (1992) *Tetrahedron Lett.*, **33**, 675–678.
10. Jang, J., Long, Y.O. and Paquette, L.A. (2003) *J. Am. Chem. Soc.*, **125**, 1567–1574.
11. (a) Kozmin, S.A. (2001) *Org. Lett.*, **3**, 755–758; (b) Wang, Y., Janjic, J. and Kozmin, S.A. (2005) *Pure Appl. Chem.*, **77**, 1161–1169.
12. Sibi, M.P. and Christensen, J.W. (1995) *Tetrahedron Lett.*, **36**, 6213–6216.
13. Tamao, K., Tohma, T., Inui, N., Nakayama, O. and Ito, Y. (1990) *Tetrahedron Lett.*, **31**, 7333–7336.
14. Bergens, S.H., Noheda, P., Whelan, J. and Bosnich, B. (1992) *J. Am. Chem. Soc.*, **114**, 2121–2128.
15. (a) Bergens, S.H., Noheda, P., Whelan, J.W. and Bosnich, B. (1992) *J. Am. Chem. Soc.*, **114**, 2128–2135; (b) Wang, X. and Bosnich, B. (1994) *Organometallics*, **13**, 4131–4133.
16. Burk, M.J. and Feaster, J.E. (1992) *Tetrahedron Lett.*, **33**, 2099–2102.
17. Wang, X., Ellis, W.W. and Bosnich, B. (1996) *Chem. Commun.*, 2561–2563.
18. Tamao, K., Nakamura, K., Ishii, H., Yamaguchi, S. and Shiro, M. (1996) *J. Am. Chem. Soc.*, **118**, 12469–12470.
19. Rix, F.C., Brookhart, M. and White, P.S. (1996) *J. Am. Chem. Soc.*, **118**, 4746–4764.
20. (a) Widenhoefer, R.A., Krzyzanowska, B. and Webb-Wood, G. (1998) *Organometallics*, **17**, 5124–5127; (b) Swisher, J.V. and Chon, H. (1974) *J. Organomet. Chem.*, **69**, 83–91.
21. LaPointe, A.M. and Brookhart, M. (1997) *J. Am. Chem. Soc.*, **119**, 906–917.
22. Tamao, K., Maeda, K., Tanaka, T. and Ito, Y. (1988) *Tetrahedron Lett.*, **29**, 6955–6956.
23. (a) Marshall, J.A. and Yanik, M.M. (2000) *Org. Lett.*, **2**, 2173–2175; (b) Marshall, J.A. and Ellis, K.C. (2003) *Org. Lett.*, **5**, 1729–1732.
24. (a) Tamao, K., Kobayashi, K. and Ito, Y. (1989) *Tetrahedron Lett.*, **30**, 6051–6054; (b) Hiyama, T. (1998) in *Metal-Catalyzed Cross-Coupling Reactions* (eds F. Diedrich and P.J. Stang), Wiley-VCH, Germany, Chapter 10; (c) Hiyama, T. and Shirakawa, E. (2002) *Top. Curr. Chem.*, **219**, 61–85.
25. (a) Denmark, S.E. and Pan, W. (2001) *Org. Lett.*, **3**, 61–64; (b) Denmark, S. and Sweis, R. (2002) *Acc. Chem. Res.*, **35**, 835–846;

(c) Denmark, S.E. and Sweis, R.F. (2002) *Chem. Pharm. Bull.*, **50**, 1531–1541.
26 Sashida, H. and Kuroda, A. (1999) *Synthesis*, 921–923.
27 Sudo, T., Asao, N. and Yamamoto, Y. (2000) *J. Org. Chem.*, **65**, 8919–8923.
28 Denmark, S.E. and Pan, W. (2002) *Org. Lett.*, **4**, 4163–4166.
29 Denmark, S.E. and Pan, W. (2003) *Org. Lett.*, **5**, 1119–1122.
30 Trost, B.M., Ball, Z.T. and Jöge, T. (2003) *Angew. Chem. Int. Ed.*, **42**, 3415–3418.
31 Trost, B.M. and Ball, Z.T. (2003) *J. Am. Chem. Soc.*, **125**, 30–31.
32 Chung, L., Wu, Y., Trost, B.M. and Ball, Z.T. (2003) *J. Am. Chem. Soc.*, **125**, 11578–11582.
33 Trost, B.M., Ball, Z.T. and Laemmerhold, K.M. (2005) *J. Am. Chem. Soc.*, **127**, 10028–10038.
34 (a) Matsuda, I., Ogiso, A., Sato, S. and Izumi, Y. (1989) *J. Am. Chem. Soc.*, **111**, 2332–2333; (b) Matsuda, I., Fukuta, Y., Tsuchihashi, T., Nagashima, H. and Itoh, K. (1997) *Organometallics*, **16**, 4327–4345; (c) Ojima, I., Ingallina, P., Donovan, R.J. and Clos, N. (1991) *Organometallics*, **10**, 38–41; (d) Ojima, I., Tzamarioudaki, M.T. and Tsai, C. (1994) *J. Am. Chem. Soc.*, **116**, 3643–3644.
35 Suginome, M., Kinugasa, H. and Ito, Y. (1994) *Tetrahedron Lett.*, **35**, 8635–8638.
36 Monteil, F., Matsuda, I. and Alper, H. (1995) *J. Am. Chem. Soc.*, **117**, 4419–4420.
37 Aronica, L.A., Caporusso, A.M., Salvadori, P. and Alper, H. (1999) *J. Org. Chem.*, **64**, 9711–9714.
38 Ojima, I., Vidal, E., Tzamarioudaki, M. and Matsuda, I. (1995) *J. Am. Chem. Soc.*, **117**, 6797–6798.
39 Bonafoux, D. and Ojima, I. (2001) *Org. Lett.*, **3**, 1303–1305.
40 Ojima, I. and Vidal, E.S. (1999) *Organometallics*, **18**, 5103–5107.
41 Leighton, J.L. and Chapman, E. (1997) *J. Am. Chem. Soc.*, **119**, 12416–12417.
42 (a) Trost, B.M. (1998) *Chem. Eur. J.*, **4**, 2405–2412; (b) Trost, B.M. and Kirsche, M.J. *Synlett*, 1–16.
43 Trost, B.M. and Lee, D.C. (1988) *J. Am. Chem. Soc.*, **110**, 7255–7258.
44 Trost, B.M., Freitz, F.J. and Watkins, W.J. (1996) *J. Am. Chem. Soc.*, **118**, 5146–5147.
45 (a) Tamao, K., Kobayashi, K. and Ito, Y. (1989) *J. Am. Chem. Soc.*, **111**, 6478–6480; (b) Tamao, K., Kobayashi, K. and Ito, Y. (1992) *Synlett*, 539–546.
46 (a) Muraoka, T., Matsuda, I. and Itoh, K. (1998) *Tetrahedron Lett.*, **39**, 7325–7328; (b) Muraoka, T., Matsuda, I. and Itoh, K. (2002) *Organometallics*, **21**, 3650–3660.
47 (a) Madine, J.W., Wang, X. and Widenhoefer, R.A. (2001) *Org. Lett.*, **3**, 385–388; (b) Wang, X., Chakrapani, H., Madine, J.W., Keyerleber, M.A. and Widenhoefer, R.A. (2002) *J. Org. Chem.*, **67**, 2778–2788.
48 Liu, C. and Widenhoefer, W.A. (2002) *Organometallics*, **21**, 5666–5673.
49 Uno, T., Wakayanagi, S., Sonoda, Y. and Yamamoto, K. (2003) *Synlett*, 1997–2000.
50 Lautens, M., Smith, N.D. and Ostrovsky, D. (1997) *J. Org. Chem.*, **62**, 8970–8971.
51 Gréau, S., Radetich, B. and RajanBabu, T.V. (2000) *J. Am. Chem. Soc.*, **122**, 8579–8580.
52 (a) Onozawa, S., Hatanaka, Y. and Tanaka, M. (1997) *Chem. Commun.*, 1229–1230; (b) Onozawa, S., Hatanaka, Y., Choi, N. and Tanaka, M. (1997) *Organometallics*, **16**, 5389–5391.
53 Yang, H. and Krische, M.J. (2004) *J. Am. Chem. Soc.*, **126**, 7875–7880.
54 Fukuta, Y., Matsuda, I. and Itoh, K. (1999) *Tetrahedron Lett.*, **49**, 4703–4706.
55 (a) Ojima, I., Donovan, R.J. and Shay, W.R. (1992) *J. Am. Chem. Soc.*, **114**, 6580–6582; (b) Ojima, I., Vu, A.T., Lee, S., McCullagh, J.V., Moralee, A.C., Fujiwara, M. and Hoang, T.H. (2002) *J. Am. Chem. Soc.*, **124**, 9164–9174; (c) Park, K., Jung, I., Kim, S. and Chung, Y. (2003) *Org. Lett.*, **5**, 4967–4970.
56 Denmark, S.E. and Liu, J.H. (2007) *J. Am. Chem. Soc.*, **129**, 3737–3744.
57 Park, K., Kim, S., Son, S. and Chung, Y. (2003) *Eur. J. Org. Chem.*, 4341–4345.

58 Chakrapani, H., Liu, C. and Widenhoefer, R.A. (2003) *Org. Lett.*, **5**, 157–159.
59 Fan, B., Xie, J., Li, S., Wang, L. and Zhou, Q. (2007) *Angew. Chem. Iint. Ed.*, **46**, 1275–1277.
60 Wakayanagi, S., Shimamoto, T., Chimori, M. and Yamamoto, K. (2005) *Chem. Lett.*, **34**, 160–162.
61 (a) Molander, G.A. and Retsch, W.H. (1997) *J. Am. Chem. Soc.*, **119**, 8817–8825; (b) Molander, G.A. and Corrette, C.P. (1999) *J. Org. Chem.*, **64**, 9697–9703.
62 Lautens, M. and Mancuso, J. (2000) *Org. Lett.*, **2**, 671–673.
63 (a) Mori, M., Hirose, T., Wakamatsu, H., Imakuni, N. and Sato, Y. (2001) *Organometallics*, **20**, 1907–1909; (b) Sato, Y., makuni, N.I., Hirose, T., Wakamatsu, H. and Mori, M. (2003) *J. Organomet. Chem.*, **687**, 392–402; (c) Lautens, M. and Mancuso, J. (2002) *Synlett*, 394–398.
64 Kinder, R.E. and Widenhoefer, R.A. (2006) *Org. Lett.*, **8**, 1967–1967.
65 Jang, H., Hughes, F.W., Gong, H., Zhang, J., Brodbelt, J.S. and Kirsche, M.J. (2005) *J. Am. Chem. Soc.*, **127**, 6174–6175.
66 Shibata, T., Kadowaki, S. and Takagi, K. (2004) *Organometallics*, **23**, 4116–4120.
67 Jeseke, G., Lauke, H., Mauermann, H., Swepston, P.N., Schumann, H. and Marks, T.J. (1985) *J. Am. Chem. Soc.*, **107**, 8091–8103.
68 Onozawa, S., Sakakura, T. and Tanaka, M. (1994) *Tetraheron Lett.*, **35**, 8177–8180.
69 Horino, Y. and Livinghouse, T. (2004) *Organometallics*, **23**, 12–14.
70 Molander, G.A. and Nichols, P.J. (1995) *J. Am. Chem. Soc.*, **117**, 4415–4416.
71 Molander, G.A., Nichols, P.J. and Noll, B.C. (1998) *J. Org. Chem.*, **63**, 2292–2306.
72 Molander, G.A. and Dowdy, E.C. (1998) *J. Org. Chem.*, **63**, 3386–3396.
73 Molander, G.A. and Nichols, P.J. (1996) *J. Org. Chem.*, **61**, 6040–6043.
74 (a) Molander, G.A. and Schmitt, M.H. (2000) *J. Org. Chem.*, **65**, 3767–3770; (b) Molander, G.A. and Romero, J.A.C. (2005) *Tetrahedron*, **61**, 2631–2643.
75 Muci, A.R. and Bercaw, J.E. (2000) *Tetrahedron Lett.*, **41**, 7609–7612.
76 (a) Piers, W.E. and Bercaw, J.E. (1990) *J. Am. Chem. Soc.*, **112**, 9406–9407; (b) Molander, G.A. and Hoberg, J.O. (1992) *J. Am. Chem. Soc.*, **114**, 3123–3125.
77 (a) Rix, F.C. and Brookhart, M. (1995) *J. Am. Chem. Soc.*, **117**, 1137–1138; (b) Johnson, L.K., Killian, C.M. and Brookhart, M. (1995) *J. Am. Chem. Soc.*, **117**, 6414–6415; (c) Johnson, L.K., Mecking, S. and Brookhart, M. (1996) *J. Am. Chem. Soc.*, **118**, 267–268; (d) Mecking, S., Johnson, L.K., Wang, L. and Brookhart, M. (1998) *J. Am. Chem. Soc.*, **120**, 888–899.
78 (a) Widenhoefer, R.A. and DeCarli, M.A. (1998) *J. Am. Chem. Soc.*, **120**, 3805–3806; (b) Stengone, C.N. and Widenhoefer, R.A. (1999) *Tetrahedron Lett.*, **40**, 1451–1454; (c) Widenhoefer, R.A. and Stengone, C.N. (1999) *J. Org. Chem.*, **64**, 8681–8692.
79 Widenhoefer, R.A. and Vadehra, A. (1999) *Tetrahedron Lett.*, **40**, 8499–8502.
80 (a) Perch, N.S. and Widenhoefer, R.A. (2001) *Organometallics*, **20**, 5251–5253; (b) Perch, N.S. and Widenhoefer, R.A. (2004) *J. Am. Chem. Soc.*, **126**, 6332–6346.
81 Wang, X., Chakrapani, H., Stengone, C.N. and Widenhoefer, R.A. (2001) *J. Org. Chem.*, **66**, 1755–1760.
82 (a) Perch, N.S. and Widenhoefer, R.A. (1999) *J. Am. Chem. Soc.*, **121**, 6960–6961; (b) Perch, N.S., Pei, T. and Widenhoefer, R.A. (2000) *J. Org. Chem.*, **65**, 3836–3845.
83 Perch, N.S., Kisanga, P. and Widenhoefer, R.A. (2000) *Organometallics*, **19**, 2541–2548.
84 (a) Pei, T. and Widenhoefer, R.A. (2000) *Org. Lett.*, **2**, 1469–1471; (b) Pei, T. and Widenhoefer, R.A. (2000) *Tetrahedron. Lett.*, **41**, 7597–7600; (c) Pei, T. and Widenhoefer, R.A. (2001) *J. Org. Chem.*, **66**, 7639–7645.
85 Uozumi, Y., Tsuji, H. and Hayashi, T. (1998) *J. Org. Chem.*, **63**, 6137–6140.
86 Molander, G.A. and Retsch, W.H. (1998) *J. Org. Chem.*, **63**, 5507–5518.

87 Ojima, I., McCullagh, J.V. and Shay, W.R. (1996) *J. Organomet. Chem.*, **521**, 421–423.
88 Ojima, I., Vu, A.T., McCullagh, J.V. and Kinoshita, A. (1999) *J. Am. Chem. Soc.*, **121**, 3230–3231.
89 Oh, C., Rhim, C., Kang, J., Kim, A., Park, B. and Seo, Y. (1996) *Tetrahedron Lett.*, **37**, 8875–8878.
90 Hewlett, D.F. and Whitby, R. (1990) *J. Chem. Soc. Chem. Commun.*, 1684–1686.
91 (a) Kablaoui, N.M. and Buchwald, S.L. (1996) *J. Am. Chem. Soc.*, **118**, 3182–3191; (b) Crowe, W.E. and Rachita, M.J. (1995) *J. Am. Chem. Soc*, **117**, 6787–6788.
92 (a) Tang, X. and Montgomery, J. (1999) *J. Am. Chem. Soc.*, **121**, 6098–6099. (b) Tang, X. and Montgomery, J. (2000) *J. Am. Chem. Soc.*, **122**, 6950–6954.
93 Mahandru, G.M., Liu, G. and Montgomery, J. (2004) *J. Am. Chem. Soc.*, **126**, 3698–3699.
94 (a) Knapp-Reed, B., Mahandru, G.M. and Montgomery, J. (2005) *J. Am. Chem. Soc.*, **127**, 13156–13157; (b) Chaulagain, M.R., Sormunen, G.J. and Montgomery, J. (2007) *J. Am. Chem. Soc.*, **129**, 95686–9569.
95 (a) Sato, Y., Takimoto, M., Hayashi, K., Katsuhara, T., Takagi, K. and Mori, M. (1994) *J. Am. Chem. Soc.*, **116**, 9771–9772; (b) Sato, Y., Takimoto, M., Hayashi, K., Katsuhara, T., Takagi, K. and Mori, M. (2000) *J. Am. Chem. Soc.*, **122**, 1624–1634.
96 (a) Sato, Y. and Mori, M. (1997) *Tetrahedron Lett.*, **38**, 3931–3934; (b) Sato, Y. and Mori, M. (1998) *Tetrahedron*, **54**, 1153–1168.
97 (a) Sato, Y., Saito, N. and Mori, M. (2000) *J. Am. Chem. Soc.*, **122**, 2371–2372; (b) Sato, Y., Saito, N. and Mori, M. (2002) *J. Org. Chem.*, **67**, 9310–9317.
98 (a) Montgomery, J. (2004) *Angew. Chem. Int. Ed.*, **43**, 3890–3908; (b) Montgomery, J., Oblinger, E. and Savchenki, A.V. (1997) *J. Am. Chem. Soc.*, **119**, 4911–4920; (c) Montgomery, J., Chevliakov, M.V. and Brielmann, H.L. (1997) *Tetrahedron*, **83**, 16449–16462.
99 Shibata, K., Kimura, M., Shimada, M. and Tamaru, Y. (2001) *Org. Lett.*, **3**, 2181–2183.
100 (a) Morinaro, C. and Jamison, T.F. (2003) *J. Am. Chem. Soc.*, **125**, 8076–8077; (b) Woodin, K.S. and Jamison, T.F. (2007) *J. Org. Chem.*, **72**, 7451–7454.
101 Kang, S., Hong, Y., Lee, J., Kim, W., Lee, I. and Yu, C. (2003) *Org. Lett.*, **5**, 2813–2816.
102 Kang, S., Ko, B., Lim, Y. and Jung, J. (2002) *Angew. Chem. Int. Ed.*, **41**, 343–345.
103 Rhee, J. and Krische, M.J. (2006) *J. Am. Chem. Soc.*, **128**, 10674–10675.
104 (a) Emiabata-Smith, D., McKillop, A., Mills, C., Motherwell, W.B. and Whitehead, A.J. (2001) *Synlett*, 1302–1304; (b) Freiría, M., Whitehead, A.J., Tocher, D.A. and Motherwell, W.B. (2004) *Tetrahedron*, **60**, 2673–2692; (c) Freiría, M., Whitehead, A.J. and Motherwell, W.B. (2005) *Synthesis*, 3079–3084.
105 Matsuda, I., Takahashi, K. and Sato, S. (1990) *Tetrahedron Lett.*, **31**, 5331–5334.
106 Lam, H. and Joensuu, P.M. (2005) *Org. Lett.*, **7**, 4225–4228.
107 (a) Lam, H., Murray, G.J. and Firth, J.D. (2005) *Org. Lett.*, **7**, 5743–5746; (b) Lam, H., Joensuu, P.M., Murray, G.J., Fordyce, E.A.F., Prieto, O. and Luebers, T. (2006) *Org. Lett.*, **8**, 3729–3732.
108 (a) Baik, T., Liuis, A.N., Wang, L. and Krische, M.J. (2001) *J. Am. Chem. Soc.*, **123**, 5112–5113; (b) Wang, L., Jang, H., Roh, Y., Lynch, V., Schultz, A.J., Wang, X. and Krische, M.J. (2002) *J. Am. Chem. Soc.*, **124**, 9448–9453; (c) Hudleston, R.R. and Krische, M.J. (2003) *Synlett*, 12–21.
109 Miura, K., Yamada, Y., Tomita, M. and Hosomi, A. (2004) *Synlett*, 1985–1989.
110 Suwa, T., Nishino, K., Miyatake, M., Shibata, I. and Baba, A. (2000) *Tetrahedron Lett.*, **41**, 3403–3406.
111 (a) Jang, H., Huddleston, R.R. and Krische, M.J. (2002) *J. Am. Chem. Soc.*, **124**, 15156–15157; (b) Jang, H. and Krische, M.J. (2004) *Eur. J. Org. Chem.*, 3953–3958.
112 Huddleston, R.R. and Krische, M.J. (2003) *Org. Lett.*, **5**, 1143–1146; (b) Koech, P.K. and Krische, M.J. (2004) *Org. Lett.*, **6**, 691–694.

113 (a) Mahoney, W.S., Brestensky, D.M. and Stryker, J.M. (1988) *J. Am. Chem. Soc.*, **110**, 291–293; (b) Brestensky, D.M. and Stryker, J.M. (1989) *Tetrahedron Lett.*, **30**, 5677–5680.

114 Chiu, P., Szeto, C., eng, Z. and Cheng, K. (2001) *Org. Lett.*, **3**, 1901–1903.

115 Chiu, P. and Leung, S. (2004) *Chem. Commun.*, 2308–2309.

116 Yoshii, E., Hori, K., Nomura, K. and Yamaguchi, K. (1995) *Synlett*, 568–570.

117 Kamenecka, T.M., Overman, L.E. and Ly Sakata, S.K. (2002) *Org. Lett.*, **4**, 79–82.

118 Matsuda, I., Ishibashi, H. and Ii, N. (1995) *Tetrahedron Lett.*, **36**, 241–244.

119 Ojima, I., Fracchiolla, D.A., Donovan, R.J. and Banerji, P. (1994) *J. Org. Chem.*, **59**, 7594–7595.

120 (a) Ojima, I., Kass, D.F. and Zhu, J. (1996) *Organometallics*, **15**, 5191–5195; (b) Ojima, I., Zhu, J., Vidal, S.E. and Kass, D.F. (1998) *J. Am. Chem. Soc.*, **120**, 6690–6697.

121 Fukuta, Y. (2001) Dissertation in Nagoya University, *Study on Silylative Cyclocarbonylation of 1,6-Diynes and Enynes*, 17.

122 Chatani, N., Fukumoto, Y., Ida, T. and Murai, S. (1993) *J. Am. Chem. Soc.*, **115**, 11614–11615.

123 Kobayashi, T., Koga, Y. and Narasaka, K. (2001) *J. Organomet. Chem.*, **624**, 73–87.

124 (a) Matsuda, I., Ogiso, A. and Sato, S. (1990) *J. Am. Chem. Soc.*, **112**, 6120–6121; (b) Matsuda, I. and Niikawa, N. Unpublished results.

125 (a) Mise, T., Hong, P. and Yamazaki, H. (1983) *J. Org. Chem.*, **48**, 238–242; (b) Doyama, K., Joh, T., Onitsuka, K., Shiohara, T. and Takahashi, S. (1987) *J. Chem. Soc. Chem. Commun.*, 649.

126 Fukuta, Y., Matsuda, I. and Itoh, K. (2001) *Tetrahedron Lett.*, **42**, 1301–1304.

127 (a) Murray, T.F. and Norton, J.R. (1979) *J. Am. Chem. Soc.*, **101**, 4107–4119; (b) Murray, T.F., Samsel, E.G., Varma, V. and Norton, J.R. (1981) *J. Am. Chem. Soc.*, **103**, 7520–7528; (c) Tsuji, Y., Kondo, T. and Watanabe, Y. (1987) *J. Mol. Cat.*, **40**, 295–304.

128 (a) Matsuda, I., Sakakibara, J. and Nagashima, H. (1991) *Tetrahedron Lett.*, **32**, 7431–7434; (b) Matsuda, I., Tsuchihashi, T. and Takeuchi, K. Unpublished results.

129 Ojima, I., Machnik, D., Donovan, R.J. and Mneimne, O. (1996) *Inorg. Chim. Acta*, **251**, 299–307.

# 18
# Catalytic Dipolar Cycloadditions of Alkynes with Azides and Nitrile Oxides
*Valery V. Fokin*

## 18.1
## Introduction

1,3-Dipolar cycloaddition reactions provide direct access to a wide range of useful heterocyclic systems. Many of the resulting heterocycles are found in natural products, man-made drugs, agrochemicals, and materials. As "fusion" processes, dipolar cycloadditions are atom economical, and the variety of readily accessible dipoles and dipolarophiles make these transformations particularly suitable for the synthesis of structurally and functionally diverse collections of compounds. This large family of reactions has been a subject of intensive research, most notably by Rolf Huisgen and coworkers, whose work led to the formulation of the general concept of 1,3-dipolar cycloadditions in 1958 [1]. Since then, dipolar cycloaddition chemistry has found extensive applications in organic synthesis and has been a subject of several reviews [2, 3].

Although cycloaddition reactions of azides and olefins have been widely utilized in organic synthesis, the similar azide–alkyne cycloaddition has received far less attention. This may be due to the lack of reactivity of organic azides with the carbon–carbon triple bond as well as a certain degree of azidophobia among synthetic organic chemists. Indeed, even though organic azides have been used in synthesis for over 100 years, their utility has been often limited to the facile introduction of the amino group into organic molecules [4]. Other facets of their unique reactivity remained largely unexplored until relatively recently. The surge of interest in the applications of organic azides in the studies of biological systems may be attributed to the pioneering work of Bertozzi and colleagues on the development of small-molecule chemical reporters. Recognizing the nearly bio-orthogonal properties of azides and the ease of their introduction into the biological molecules, Bertozzi exploited their reaction with phosphines (the modified Staudinger ligation) in metabolic oligosaccharide engineering studies [5, 6]. Around the same time, Sharpless and colleagues introduced click chemistry as a set of "near-perfect" bond-forming reactions useful for rapid assembly of molecules with desired function [7]. Click transformations are easy to perform, give rise to their intended products in very high yields with little or no byproducts, work well under many conditions (usually especially well in water), and are unaffected by the nature of the groups being connected to each other. The "click" name is meant to signify that with the use of these methods, joining molecular fragments is as easy as "clicking" together the two pieces of a buckle. The buckle works no matter what is attached to it, as long as its two pieces can reach each other, and the components of the buckle can make a connection only with each other. The potential of organic azides as highly energetic, yet very selective functional groups in organic synthesis was highlighted, and their dipolar cycloadditions with olefins and alkynes were placed among the reactions fulfilling the click criteria. However, the inherently low reaction rates of the azide–alkyne cycloaddition did not make it very useful in the click context, and its potential was not revealed until the discovery of the copper catalysis in 2002. The copper-catalyzed reaction was discovered simultaneously and independently by the groups of Meldal in Denmark [8] and Fokin and Sharpless in the United States [9]. Meldal used the tranformation for the synthesis of peptidotriazoles in organic solvents starting from alkynylated amino acids attached to solid supports, whereas the Scripps groups immediately turned to aqueous systems and devised a straightforward and practical procedure for the covalent "stitching" of virtually any fragments containing an azide and an alkyne functionality, noting the broad utility and versatility of the novel process "for those organic synthesis endeavors which depend on the creation of covalent links between diverse building blocks" [9]. Within months, the applications of this reaction began to appear across the chemical disciplines, making it the most known and, arguably the most useful, click reaction discovered to date. Indeed, since its discovery, the copper-catalyzed azide–alkyne cycloaddition, CuAAC, has firmly established its leading position among the most reliable means for the covalent assembly of complex molecules. It has enabled a number of applications in synthesis, medicinal chemistry, molecular biology, and materials science. Unsurprisingly, it is often equated with click chemistry, whereas it is in fact only a representative of the click family of reactions.

Numerous applications of the CuAAC reaction reported during the last several years have been regularly reviewed [10–15], and this chapter is not intended to cover them comprehensively [16]. Instead, it focuses on the fundamental aspects of the CuAAC and related processes and provides practitioners with general guidelines for choosing the appropriate reaction conditions, catalysts, and ligands suitable for their specific applications. The applications discussed below have been chosen for illustrative purposes, and many related examples can be found in the current literature.

In 2005, another efficient catalytic azide–alkyne cycloaddition, catalyzed by ruthenium complexes, was reported [17]. This sister reaction, termed RuAAC (ruthenium-catalyzed azide–alkyne cycloaddition), provides ready access to the complementary 1,5-disubstituted triazole isomers, as well as to fully substituted 1,2,3-triazoles, since it also engages internal alkynes in the cycloaddition. The second part of the chapter is devoted to the ruthenium-catalyzed cycloadditions.

## 18.2
### Azide–Alkyne Cycloaddition: Basics

The fundamental thermal reaction, involving terminal or internal alkynes (Scheme 18.1A, 1,3-dipolar azide–alkyne cycloaddition, AAC), has been known for over a century (the first 1,2,3-triazole was synthesized by Michael from phenyl azide and diethyl acetylenedicarboxylate in 1893) [18], and was most thoroughly investigated by Huisgen and coworkers in the 1950s–70s in the course of their studies of the family of 1,3-dipolar cycloaddition reactions [19–25]. Although the reaction is highly exothermic (about 50–65 kcal mol$^{-1}$), its high activation barrier (25–26 kcal mol$^{-1}$ for methyl azide and propyne [26]) results in exceedingly low reaction rates for unactivated reactants, even at elevated temperature. Furthermore, since the difference in the HOMO–LUMO energy levels for both azides and alkynes are of similar magnitude, both dipole–HOMO- and dipole–LUMO-controlled pathways operate in these cycloadditions. As a result, a mixture of regioisomeric 1,2,3-triazole products is usually formed when an alkyne is unsymmetrically substituted. Copper catalysts

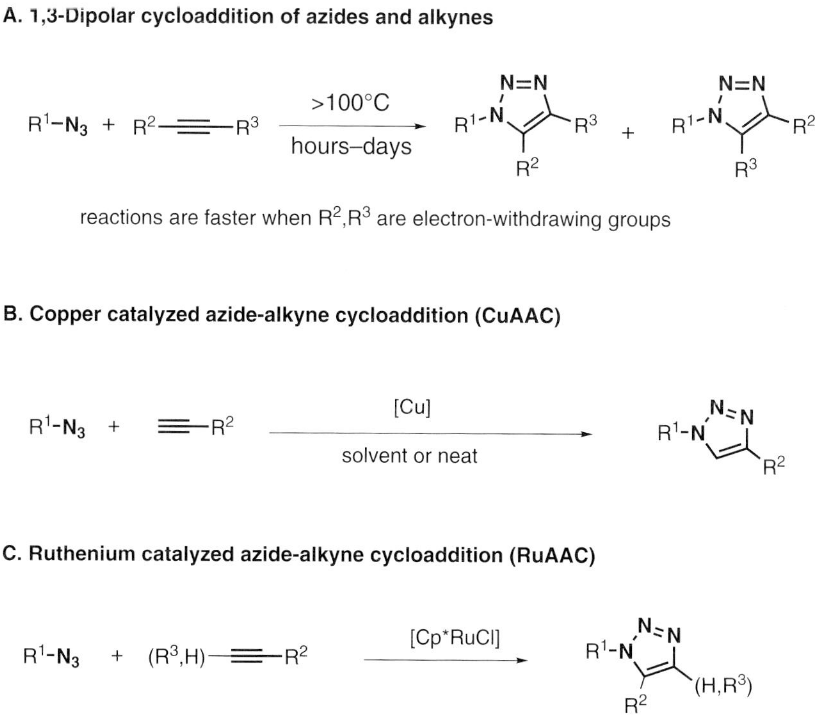

**A. 1,3-Dipolar cycloaddition of azides and alkynes**

reactions are faster when R$^2$,R$^3$ are electron-withdrawing groups

**B. Copper catalyzed azide-alkyne cycloaddition (CuAAC)**

**C. Ruthenium catalyzed azide-alkyne cycloaddition (RuAAC)**

**Scheme 18.1** Thermal cycloaddition of azides and alkynes usually requires prolonged heating and results in mixtures of both 1,4- and 1,5-regioisomers (A), whereas CuAAC produces only 1,4-disubstituted-1,2,3-triazoles at room temperature in excellent yields (B). The RuAAC reaction proceeds with both terminal and internal alkynes and gives 1,5-disubstituted and fully, 1,4,5-trisubstituted-1,2,3-triazoles.

(Scheme 18.1B) accelerate the reaction of azides with terminal alkynes [8, 9] by approximately $10^7$ relative to the uncatalyzed version [26], making it conveniently fast at and even below room temperature. The reaction is not significantly affected by the steric and electronic properties of the groups attached to the azide and alkyne reactive centers. For example, primary, secondary, and even tertiary, electron-deficient and electron-rich, aliphatic, aromatic, and heteroaromatic azides usually react well with variously substituted terminal alkynes. The reaction proceeds well in most protic and aprotic solvents, including water, and is unaffected by most organic and inorganic functional groups, therefore all but eliminating the need for protecting group chemistry [27]. The 1,2,3-triazole heterocycle has the advantageous properties of high chemical stability (in general, being inert to severe hydrolytic, oxidizing, and reducing conditions, even at high temperature), strong dipole moment (4.8–5.6 D), aromatic character, and hydrogen-bond-accepting ability [28, 29]. Thus it can interact productively in several ways with biological molecules and can serve as a hydrolytically-stable replacement for the amide bond [30–32]. Compatibility of the CuAAC reaction with a broad range of functional groups and reaction conditions [33–36] has made it broadly useful across the chemical disciplines, and its applications include the synthesis of biologically active compounds, the preparation of conjugates to proteins and polynucleotides, the synthesis of dyes, the elaboration of known polymers and the synthesis of new ones, the creation of responsive materials, and the covalent attachment of desired structures to surfaces [10, 37–40].

The ruthenium-catalyzed version of the azide–alkyne cycloaddition (RuAAC, Scheme 18.1C) was an important addition to the family of catalytic dipolar cycloaddition reactions. Similarly to the CuAAC, the ruthenium-catalyzed process is regioselective, but when terminal alkynes are used the opposite regioselectivity is observed and 1,5-disubstituted triazoles are obtained. In sharp contrast to the CuAAC process, both terminal and internal alkynes participate in the ruthenium-catalyzed reaction, so the latter allows preparation of the fully substituted triazoles. The RuAAC reaction proceeds best in organic solvents (1,2-dichloroethane, toluene, dioxane, tetrahydrofuran, and benzene have been used successfully)[17, 41–43] at temperatures ranging from ambient to 140 °C (under microwave irradiation) [44]. The commonly used catalysts are readily available and stable $Cp^*RuCl(PPh_3)_2$ and $Cp^*RuCl(COD)$ complexes; $[Cp^*RuCl]_4$ catalyst is more active but less stable, and was found to be particularly effective for the cycloadditions of aryl azides at elevated temperature under microwave irradiation [44].

## 18.3
## Copper-Catalyzed Cycloadditions

### 18.3.1
### Catalysts and Ligands

A wide range of experimental conditions for the CuAAC have been employed since its discovery, underscoring the robustness of the process and its compatibility with most functional groups, solvents, and additives, regardless of the source of the catalyst. The most commonly used protocols and their advantages and limitations are discussed below, and the representative experimental procedures are described at the end of this chapter.

Different copper(I) sources can be utilized in the reaction. Copper(I) salts (iodide, bromide, chloride, acetate) and coordination complexes (such as $[Cu(CH_3CN)_4]PF_6$ [9], $[Cu(POMe_3)_3]Br$ [45], $[Cu(PPh_3)_3]Br$ [46, 47]) can be used directly. The latter were favored for the reactions in organic solvents, in which cuprous salts have limited solubility. (Although it should be noted that solubility of the reactants and catalysts is not strictly required. As evidenced by a number of examples, triazole products are often obtained in excellent yield and purity from reagents whose solubility is very limited in the reaction medium).

Copper(II) salts and coordination complexes are not competent catalysts, and reports describing Cu(II)-catalyzed cycloadditions [48, 49] are not accurate. Copper (II) is, of course, a well-known oxidizing agent for organic compounds [50]. Alcohols, amines, aldehydes, thiols, phenols, and carboxylic acids may be oxidized by the cupric ion, reducing it to the catalytically active copper(I) species in the process. Especially relevant is the family of oxidative acetylenic couplings catalyzed by the cupric species [51], with the venerable Glaser coupling [52, 53] being the most studied example. Since terminal acetylenes are necessarily present in the CuAAC reaction, their oxidation is an inevitable side reaction which would, in turn, produce the needed catalytically active copper(I) species.

Depending on the ligand environment, Cu(I) can be readily oxidized to the catalytically inactive Cu(II) species. The standard potential of the $Cu^{2+}/Cu^+$ couple is 159 mV, but can vary widely, depending on the solvent and the ligands coordinated to the metal. As an oxidant itself, Cu(II) can mediate oxidative alkyne coupling reactions, thereby impairing the efficiency of the cycloaddition and resulting in the formation of undesired byproducts. In fact, when the cycloaddition is performed in organic solvents using copper(I) halides as catalysts, the conditions originally reported by the group of Meldal [8], the reaction is plagued by the formation of oxidative coupling byproducts **4a–d** unless the alkyne is bound to a solid support (**5**, Scheme 18.2A and B). If the alkyne is present in solution and the azide is immobilized on the resin, only traces of the desired triazole product are formed. The above side reactions are a consequence of the thermodynamic instability of the Cu(I) oxidation state. Therefore, when copper(I) catalyst is used directly, whether by itself or in conjunction with amine ligands, exclusion of oxygen may be helpful to prevent these side reactions.

The use of a reducing agent, most commonly sodium ascorbate, introduced by Fokin and coworkers [9], is a convenient and practical alternative to oxygen-free conditions. Its combination with a copper(II) salt, such as the readily available and stable copper(II) sulfate pentahydrate or copper(II) acetate, has become the method of choice for preparative synthesis of 1,2,3-triazoles. Water appears to be an ideal solvent capable of supporting copper(I) acetylides in their reactive state, especially when they are formed *in situ*. "The aqueous ascorbate" procedure often furnishes triazole products in nearly quantitative yield and over 90% purity, does not require

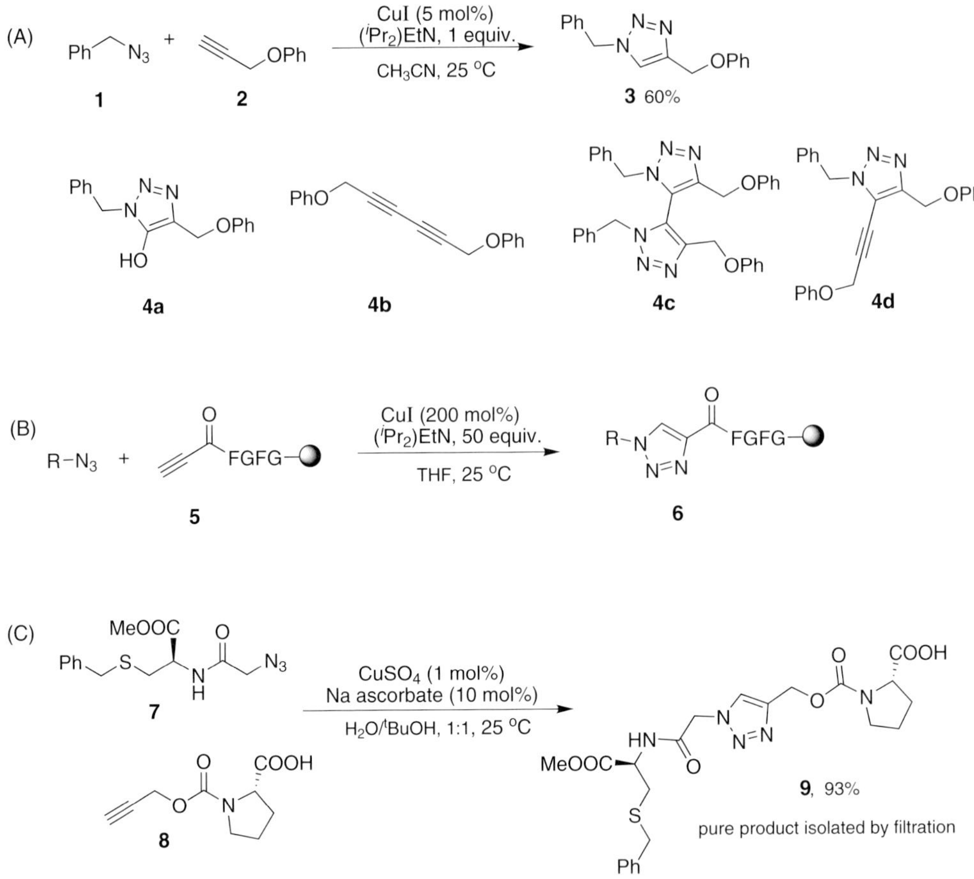

**Scheme 18.2** **A:** Oxidative coupling byproducts in the CuAAC reactions catalyzed by copper(I) salts; **B:** CuAAC with immobilized alkyne avoids the formation of the oxidative byproducts but requires large excess of the catalyst; reactions with immobilized azide fail; **C:** solution-phase CuAAC in the presence of sodium ascorbate.

any ligands, nor does the reaction mixture need to be protected from oxygen (Scheme 18.2C). Of course, copper(I) salts can also be used in combination with ascorbate, wherein it converts any oxidized copper(II) species back to the catalytically active +1 oxidation state.

The reaction can also be catalyzed by elemental copper, thus further simplifying the experimental procedure – a small piece of copper metal (wire or turning) is all that is added to the reaction mixture, followed by shaking or stirring for 12–48 h [9, 26, 54]. Aqueous alcohols (methanol, ethanol, *tert*-butanol), tetrahydrofuran, and dimethylsulfoxide can be used as solvents. Cu(II) sulfate may be added to accelerate the reaction; however, this is not necessary in most cases, as copper oxides and carbonates (the patina), which are normally present on the surface of the copper metal, are sufficient to initiate the catalytic cycle.

Although the procedure based on the copper metal requires longer reaction times when performed at ambient temperature, it usually provides access to very pure triazole products with low levels of copper contamination. Alternatively, the reaction can be performed under microwave irradiation at elevated temperature (120–160 °C), reducing the reaction time to 10–30 min [54–60].

The copper metal procedure is also very simple experimentally and is particularly convenient for high-throughput synthesis of compound libraries for biological screening. The reaction is very selective, and triazole products are generally isolated in >85–90% yields, and can often be submitted for screening directly. When required, trace quantities of copper remaining in the reaction mixture can be removed with an ion-exchange resin or by using solid-phase extraction techniques. Other heterogeneous copper(0) and copper(I) catalysts, such as copper nanoclusters, which are easily obtained and are stable [61], and copper/cuprous oxide nanoparticles [62], have also shown excellent catalytic activity.

Chemical transformations used in the synthesis of bioconjugates impose additional demands on the efficiency and selectivity. They must be exquisitely chemoselective, biocompatible, and fast. Despite the experimental simplicity and efficiency of the ascorbate procedure, the reaction is often simply not fast enough when the concentrations of the reactants are low. Tris(benzyltriazolyl)methyl amine ligand **10** (TBTA, Figure 18.1), introduced soon after the CuAAC discovery [63, 64], has been shown to significantly accelerate the CuAAC reaction and stabilize its +1 oxidation state. In addition, it appears to sequester copper ions, thereby preventing damage to the biological molecules. After its utility was demonstrated by efficient attachment of 60 alkyne-containing fluorescent dye molecules to the azide-labeled cowpea mosaic virus, [35, 65] it has been used numerous times in bioconjugation studies with nucleic acids [66, 67], proteins [34, 68], *E. Coli* bacteria [33, 36, 69], and mammalian cells [70, 71]. It preserved the integrity of biological targets and was essential for successful labeling of whole cells, proteins from tissue lysates, viruses, and nucleic

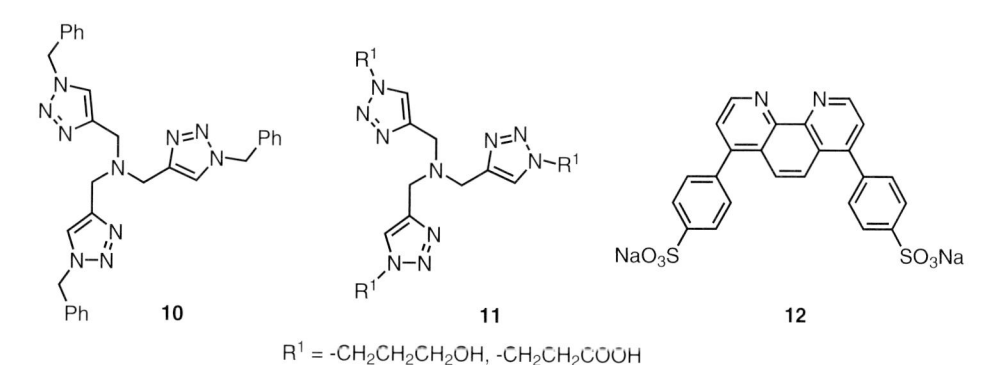

$R^1$ = -CH$_2$CH$_2$CH$_2$OH, -CH$_2$CH$_2$COOH

**Figure 18.1** (1,2,3-Triazolyl)methyl amine (TBTA), its water-soluble analogs, and sulfonated bathophenanthroline ligands for the CuAAC.

acids. It should be noted that TBTA has low solubility in water and, although complete solubility does not appear to be necessary for successful conjugation, hydroxylated TBTA analogs such as **11** can be used when higher ligand solubility is desired [72]. Other polydentate trimethylamine derivatives containing benzimidazole and benzothiazole "side-arm" substituents have also been recently reported, although they have not yet been used as widely as TBTA [73, 74]. As an alternative to tetradentate azole-containing ligands for CuAAC, the sulfonated bathophenanthroline **12** has been identified as another water-soluble commercially-available ligand [75]. Although it is kinetically competent, it adjusts the redox potential of the copper(I)/(II) couple, making the catalyst more susceptible to oxidation, thus requiring rigorous exclusion of oxygen from the reaction mixture.

### 18.3.2
### CuAAC with *In Situ* Generated Azides

Although organic azides are generally stable and safe compounds, those of low molecular weight can spontaneously decompose and, therefore, could be difficult to handle. This is especially true for small molecules with several azide functionalities that would be of much interest for the generation of polyfunctionalized structures. In these cases, CuAAC can be performed in a one-pot two-step sequence, whereby an *in situ* generated organic azide is immediately consumed in a reaction with a copper acetylide (Schemes 18.3 and 18.4). Alkyl azides can be readily obtained from the corresponding halides or arylsulfonates by reaction with sodium azide (Scheme 18.5). The latter does not interfere with the cycloaddition reaction even if it is used in excess.

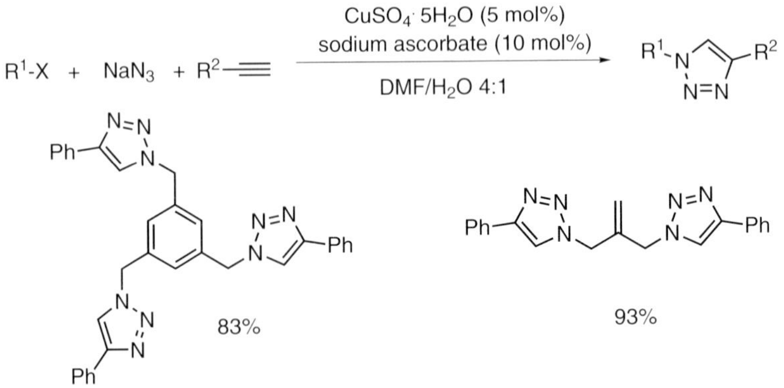

**Scheme 18.3** One-pot synthesis of triazoles from alkyl halides.

Similarly, aryl and vinyl azides can be accessed in one step from aryl and vinyl halides via a copper-catalyzed reaction with sodium azide in the presence of a catalytic amount of L-proline (Scheme 18.4) [76]. In this fashion, a range of 1,4-disubstituted 1,2,3-triazoles can be prepared in excellent yields [77–79]. This reaction sequence can be performed at elevated temperature under microwave irradiation, reducing the reaction time to 10–30 min [54].

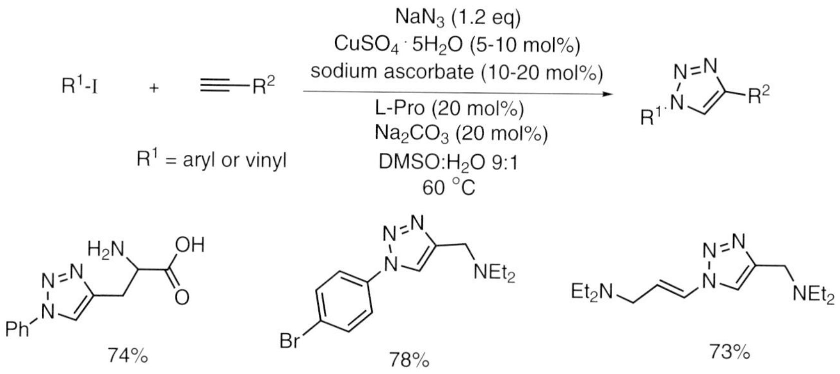

**Scheme 18.4** One-pot synthesis of triazoles from aryl halides.

Anilines can also be converted to aryl azides by the reaction with *tert*-butyl nitrite and azidotrimethylsilane [80]. The resulting azides can be submitted to the CuAAC conditions without isolation, furnishing triazole products in excellent yields. Microwave heating further improves the reaction, significantly reducing the reaction time [60].

In a recent example, Pfizer chemists developed a continuous flow process wherein a library of 1,4-disubstituted 1,2,3-triazoles was synthesized in a copper flow reactor from alkyl halides, sodium azide, and terminal acetylenes (Scheme 18.5) [81]. Organic azides were generated *in situ* and were immediately captured with acetylenes under the copper catalysis conditions simply by passing the reaction mixture through a heated copper tubing reactor (0.75 mm id). This process eliminates both the handling of organic azides and the need for additional copper catalyst and permits the facile preparation of numerous triazoles in a continuous flow process.

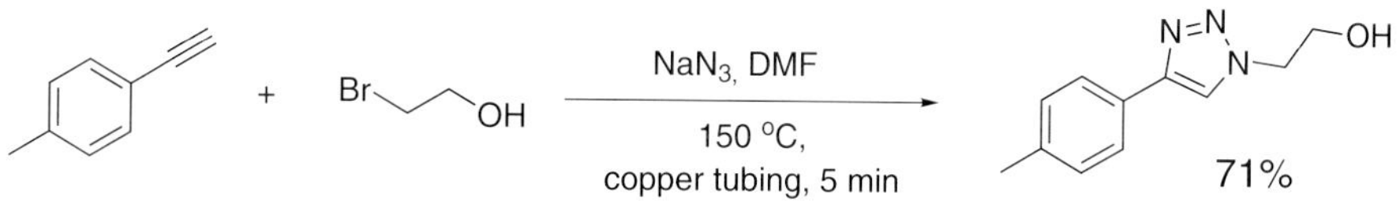

**Scheme 18.5** CuAAC in a copper flow reactor.

## 18.3.3
### Mechanistic Aspects of the CuAAC

Intermediacy of copper(I) acetylides in CuAAC was postulated early on based on the lack of reactivity of internal alkynes under copper catalysis conditions. While the history of copper(I) acetylides dates back as far as Glaser's discovery of the oxidative dimerization of phenylacetylide of copper in 1869, the precise nature of the reactive alkynyl copper species is not well understood. The chief complications are the tendency of copper species to form polynuclear clusters and the facility of the ligand exchange at the copper center, hence the involvement of multiple equilibria in the reactions involving organocopper intermediates.

The initial computational treatment of the CuAAC focused on the possible reaction pathways between the isolated, mononuclear copper(I) acetylides and organic azides (propyne and methyl azide were chosen for simplicity) [26]. The key bond-making steps are shown in Scheme 18.6. Formation of Cu-acetylide **14** (step A) was calculated to be exothermic by 11.7 kcal mol$^{-1}$, consistent with the well-known facility of this step which probably occurs through a π-alkyne copper complex intermediate. π-Coordination of alkyne to copper is calculated to bring the p$K_a$ of the terminal proton down by about 10 units, bringing it into the proper range to be deprotonated in an aqueous medium. A concerted 1,3-dipolar cycloaddition of azide to the Cu-acetylide has a high calculated barrier (23.7 kcal mol$^{-1}$), thus the metal must play an additional role. In the proposed sequence, the azide is activated by coordination to copper (step B), forming intermediate **15**. This ligand exchange step is nearly thermoneutral computationally (2.0 kcal mol$^{-1}$ uphill when L is water). The key bond-forming event takes place in the next step, when **15** is converted to the unusual 6-member copper metallacycle **16**. This step is endothermic by 12.6 kcal mol$^{-1}$ with a calculated barrier of 18.7 kcal mol$^{-1}$, which corresponds roughly to the observed rate

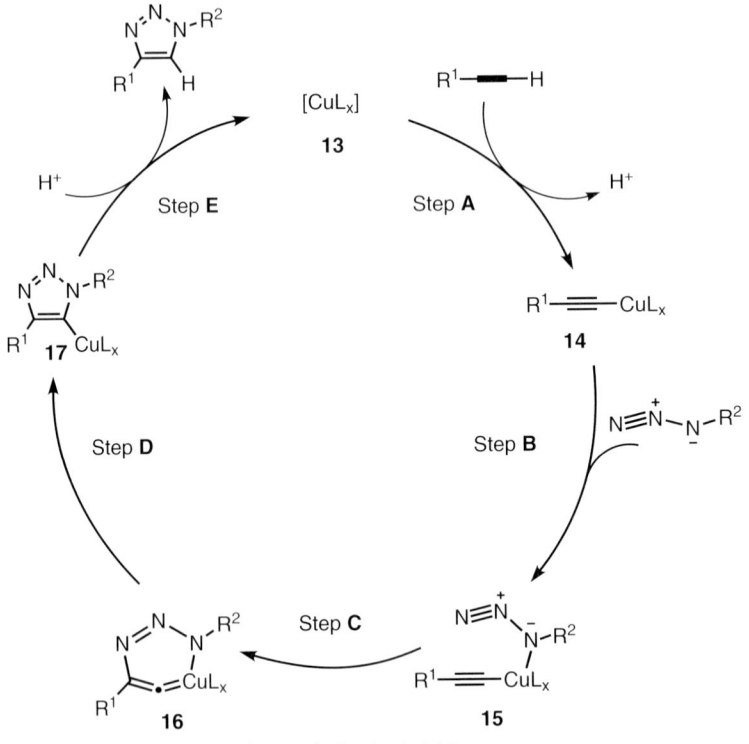

**Scheme 18.6** Proposed catalytic cycle for the CuAAC.

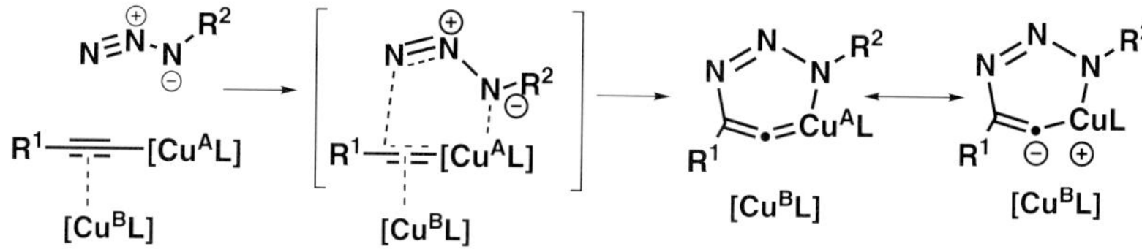

**Scheme 18.7** Introduction of a second copper(I) atom may favorably influence the energetic profile of the reaction (bottom). L = H$_2$O.

increase and is considerably lower than the barrier for the uncatalyzed reaction (approximately 26.0 kcal mol$^{-1}$), thus accounting for the significant rate acceleration accomplished by Cu(I). The CuAAC reaction is therefore not a true concerted cycloaddition, and its regiospecificity is explained by binding of both azide and alkyne to copper prior to C–N bond formation. From **16**, the barrier for ring contraction, which forms the triazolyl-copper derivative **7**, is quite low, 3.2 kcal mol$^{-1}$. Proteolysis of **17** releases the triazole product, thereby completing the catalytic cycle.

The density functional theory (DFT) investigation described above was soon followed by a study of the kinetics of the copper-mediated reaction between benzyl azide and phenylacetylene, revealing that with catalytic Cu(I) concentrations under saturating conditions (rate independent of [alkyne]), the reaction was found to be *second order* in metal [82].

$$\text{rate} = k[\text{alkyne}]^0[\text{azide}]^{0.2\pm0.1}[\text{Cu}]^{2.0\pm0.1}$$

The second order in the catalyst was also observed in the CuAAC reactions in the presence of tris(benzimidazolyl)methyl amine ligands [73]. This dependence on Cu(I) is not unreasonable since most copper(I) acetylides are highly aggregated species, engaging in various σ- and π-interactions [83–85]. The extent of polynucleation has a significant effect on the stability, properties, and catalytic activity of copper (I) complexes. More recent DFT (B3LYP) studies of the reaction between dinuclear copper(I) acetylide complexes and an organic azide revealed a substantial further drop in the activation barrier already reported for the mono-copper ensemble [86, 87].

Two transition states containing a second copper(I) center were located (Scheme 18.7). The two transition states differed only in the spectator ligand on the second copper atom (Figure 18.2). In the first, it was an acetylide (**TS-di$^{\text{alkyne}}$**), and in the second the spectator ligand was a chloride (**TS-di$^{\text{chloride}}$**). In both **TS-di$^{\text{alkyne}}$** and **TS-di$^{\text{chloride}}$** the second copper atom, Cu$^B$, forms a bond with C$^1$. The distances are calculated at 1.93 Å and 1.90 Å for **TS-di$^{\text{alkyne}}$** and **TS-di$^{\text{chloride}}$**, respectively. This short distance indicates a strong interaction between the second metal atom and the reacting copper acetylide [88].

The interaction of the Cu$^B$ with the proposed intermediates in the triazole-forming sequence and with the transition state indicates that a second copper atom is involved in the key steps of the CuAAC process. To compare the reactivity of the dinuclear copper complexes to their mono-copper analog, the transition state energy was

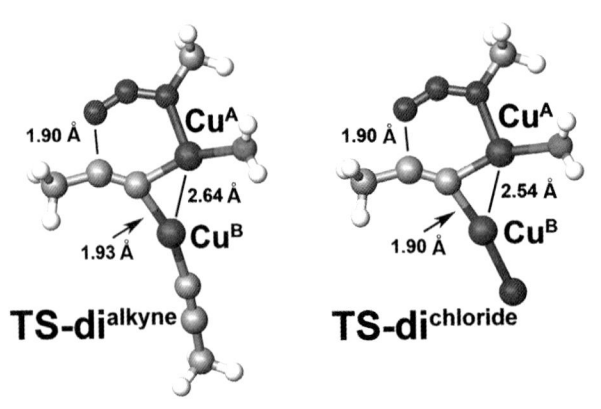

**Figure 18.2** Interaction of the second copper atom with the transition state of the metallacycle **16** formation step.

compared relative to the azide and dicopper species in isolation. The geometries of dinuclear copper acetylide species **18**$^{alkyne}$ and **18**$^{chloride}$ (Figure 18.3) were found to be similar. The reacting acetylide is σ-coordinated to Cu$^A$, while the second metal center, Cu$^B$X (where X is acetylide or chloride), exhibits a stronger interaction with C$^1$, with possible interactions between Cu$^B$ and C$^2$ and Cu$^A$. Interestingly, the μ$^2$ mode of coordination of the alkyne was not observed unless two identical spectator ligands were present on both copper atoms (e.g., [MeCC(CuCl)$_2$]$^-$). The barrier for the addition of methyl azide to the dinuclear copper acetylide complex **18**$^{alkyne}$ was calculated at 12.9 kcal mol$^{-1}$. The corresponding barrier for the chloride analog **18**$^{chloride}$ was found to be even lower, 10.5 kcal mol$^{-1}$. Compared to the mononuclear copper acetylides, for which an overall barrier of 17 kcal mol$^{-1}$ was calculated, the reactivity of the dinuclear complexes is expected to be several orders of magnitude higher.

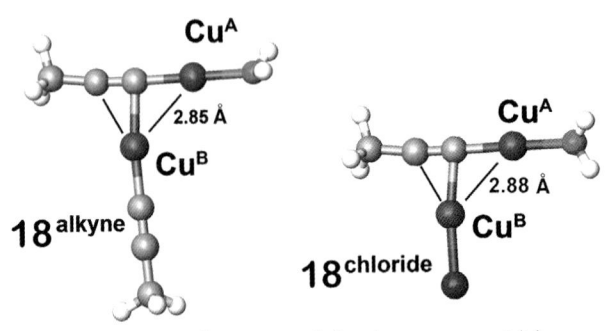

**Figure 18.3** Optimized structures of dinuclear copper acetylides.

Furthermore, a similar effect of the second copper center was observed in the six-membered intermediates **19**$^{alkyne}$ and **19**$^{chloride}$ (Figure 18.4). Both transformations of **18** to **19** were found to be endothermic, by 6 and 3.6 kcal mol$^{-1}$ for **19**$^{alkyne}$ and **19**$^{chloride}$, respectively – substantially lower endothermicity than for the corresponding mono-copper transformation (11.2 kcal mol$^{-1}$).

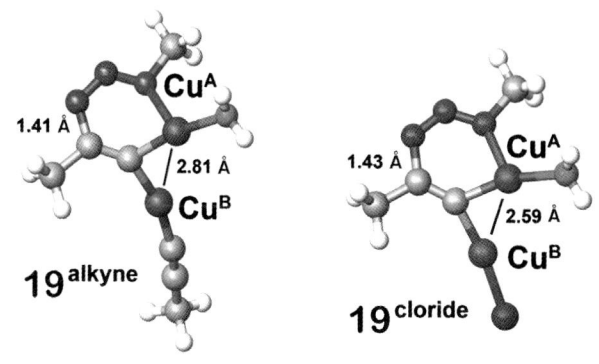

**Figure 18.4** Optimized structures of the dinuclear analogs of the metallacycle **19** (a dinuclear analog of the metallacycle **16** from Scheme 18.6.

### 18.3.4
### Reactions of Sulfonyl Azides

Reaction of sulfonyl azides with terminal alkynes under the copper catalysis conditions is an interesting exception in the CuAAC family. Depending on the conditions and reagents, it can result in the formation of different products (Scheme 18.8).

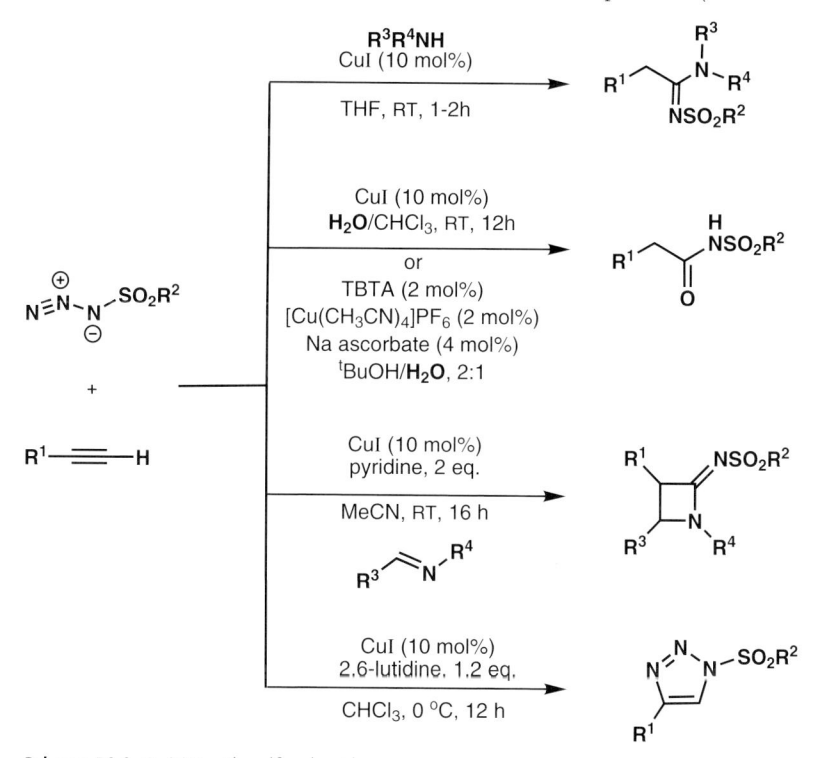

**Scheme 18.8** CuAAC with sulfonyl azides.

Chang reported that N-sulfonyl azides are converted to N-sulfonyl amidines when the reaction is conducted in the presence of amines [89]. In aqueous conditions, N-acyl sulfonamides are the major products [90, 91].

Although the catalysis initially appears to follow the same pathway as in reactions with other azides, the cuprated triazole intermediate **20** (Scheme 18.9, the N-sulfonylated analog of **17** in Scheme 18.6), can undergo a ring-chain isomerization to form the cuprated diazoimine **21** which, upon the loss of a molecule of dinitrogen, furnishes the N-sulfonyl ketenimine **22** [92]. Alternatively, copper(I) alkynamide **23** can be generated with a concomitant elimination of dinitrogen and, after protonation, would again generate a reactive N-sulfonyl ketonimine species **12**. In addition to amines and water, the latter can be trapped with imines, furnishing N-sulfonyl azetidinimines (Scheme 18.8) [92]. However, when the reaction is performed in chloroform in the presence of 2,6-lutidine, N-sulfonyl triazoles are obtained in good yields [93].

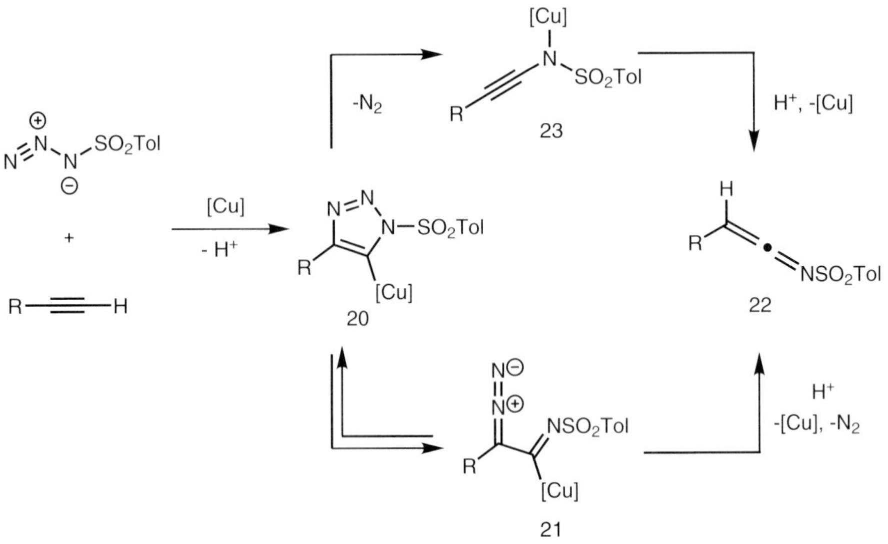

**Scheme 18.9** Possible pathways leading to the ketenimine in the CuAAC with sulfonyl azides.

## 18.3.5
### Examples of Application of the CuAAC Reaction

#### 18.3.5.1 Synthesis of Compound Libraries for Biological Screening

The CuAAC process performs well in most common laboratory solvents and usually does not require protection from oxygen and water (in fact, aqueous solvents are commonly used and, in many cases, result in cleaner isolated products), making it an ideal tool for the synthesis of libraries for initial screening as well as focused sets of compounds for structure–activity profiling. The lack of byproducts and high conversions often allow screening of the reaction mixtures without further purification.

When necessary, traces of copper can be removed by solid phase extraction utilizing a metal-scavenging resin or simple filtration through a plug of silica gel.

In a study aimed at the discovery of inhibitors of human fucosyltransferase, CuAAC was used to link 85 azides to the GDP-derived alkyne **24** with excellent yields (Scheme 18.10) [94]. The library was screened directly, and a nanomolar inhibitor **25** was identified. Testing the purified hit compound **25** against several glycosyltransferases and nucleotide binding enzymes revealed that it was the most potent and selective inhibitor of human α-1,3-fucosyltransferase VI at that time.

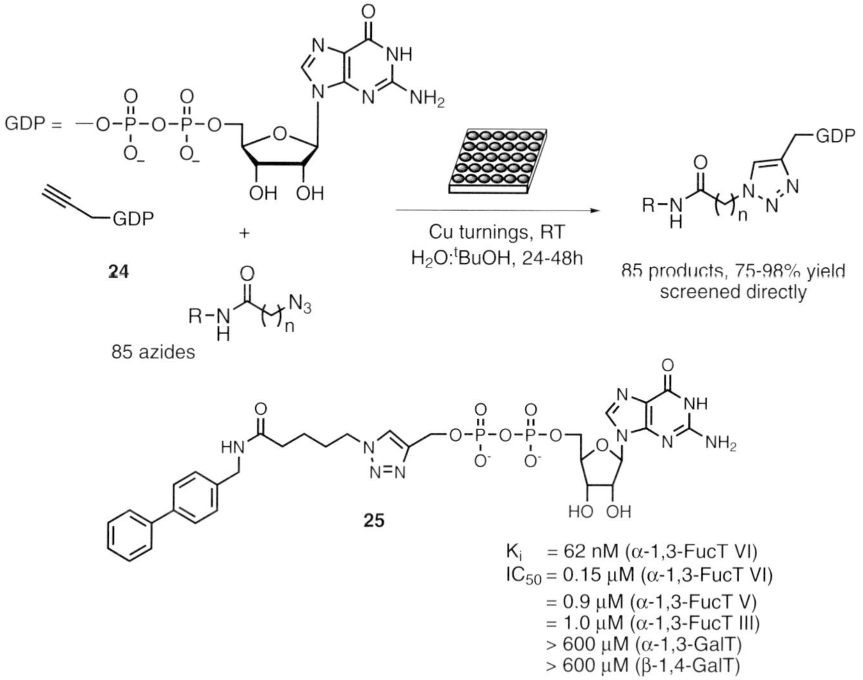

**Scheme 18.10** Synthesis and direct screening of fucosyl transferase inhibitors.

In another example, a novel family of potent HIV-1 protease inhibitors was discovered using the CuAAC reaction [95]. A focused library of azide-containing fragments was united with a diverse array of functionalized alkyne-containing building blocks (Scheme 18.11). After the direct screening of the crude reaction products, a lead structure **26** with $K_i = 98$ nM was identified. Optimization of both azide and alkyne fragments was equally facile (compound **27**, $K_i = 23$ nM). Further functionalization of the triazole at C-5 gave a series of compounds with increased activity, exhibiting $K_i$ values as low as 8 nM (compound **28**).

### 18.3.5.2 Synthesis of Glycoconjugates

Many cell-surface recognition events, which result in cellular adhesion, inflammation and immune surveillance, are mediated by complex carbohydrate interactions. It is therefore not surprising that development of synthetic architectures that bear

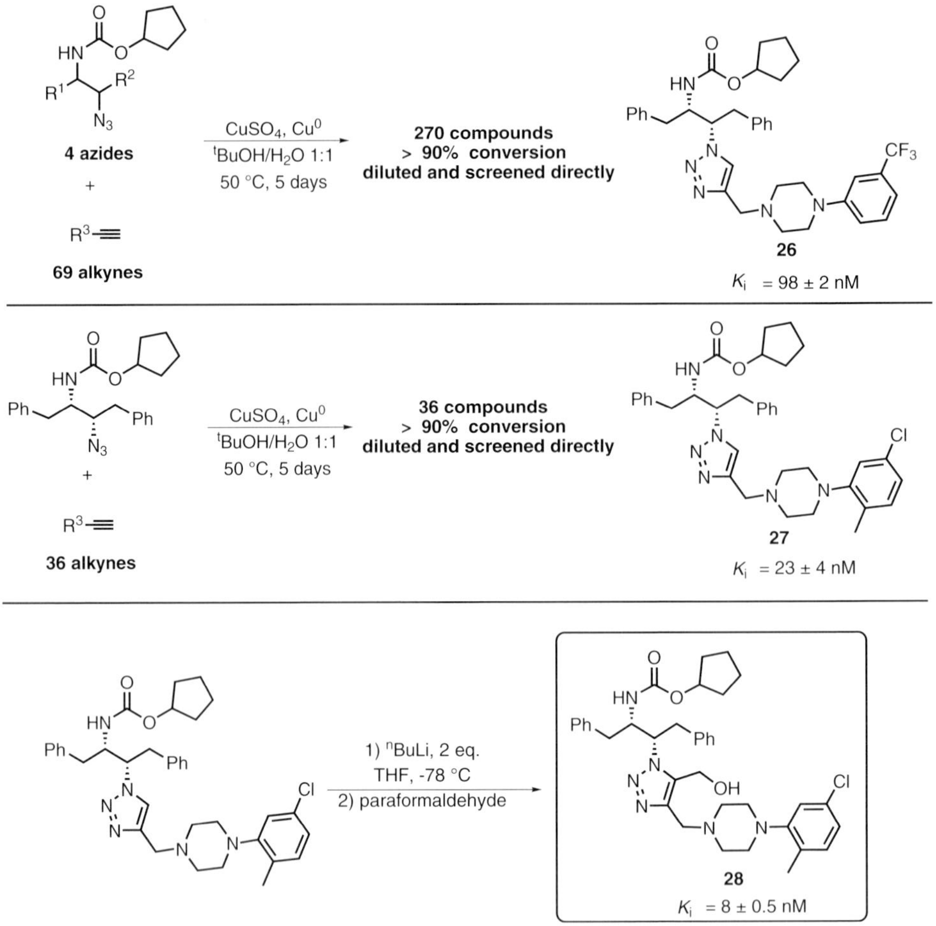

**Scheme 18.11** Novel series of inhibitors of HIV-1 protease via CuAAC.

multiple carbohydrate units has received much attention. CuAAC with organic soluble catalysts $(Ph_3P)_3CuBr$ and $(EtO)_3PCuI$ in combination with diisopropyl ethyl amine were used by Santoyo-Gonzalez and coworkers to synthesize a number of polyvalent (di- to hepta-) mannosylated ligands (Figure 18.5) [45, 96]. Under microwave irradiation, the reactions were generally complete in less than 30 min, and polyvalent products were isolated in high yield. C-glycoside clusters [97], neoglycopolymers [98, 99], and virus-glycopolymer conjugates [65] have been reported utilizing CuAAC alone or in combination with atom-transfer polymerization techniques.

### 18.3.5.3 Modification and Biological Profiling of Natural Products

Many bioactive natural products have limited therapeutic potential because of the narrow activity–toxicity window (the minimum therapeutically useful concentration and the concentration at which side effects make treatment impractical). Therefore, modification of potentially useful bioactive natural products with the goal of

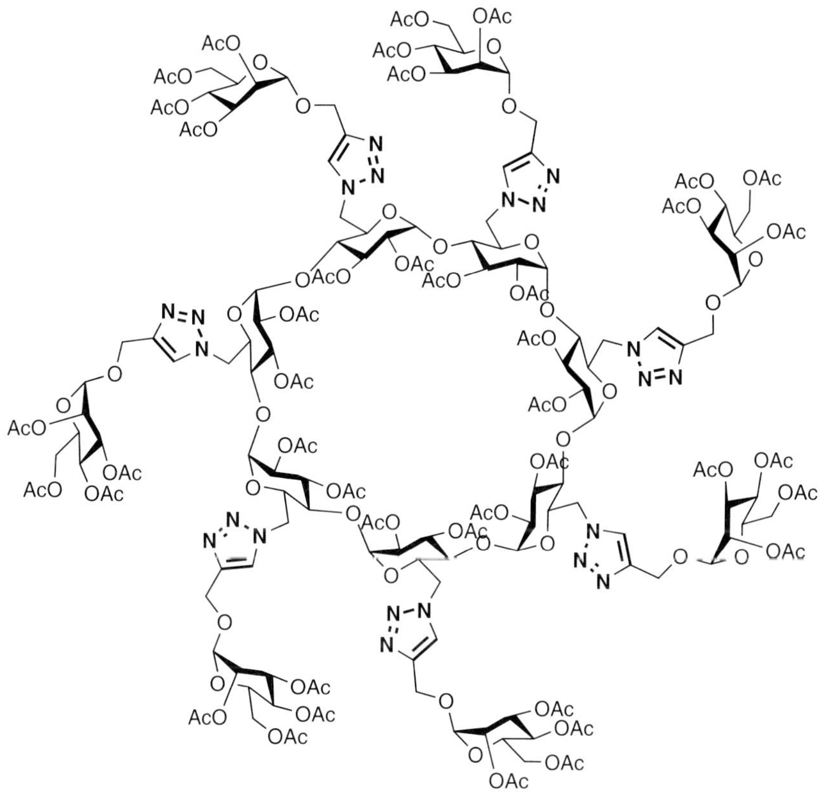

**Figure 18.5** Heptamannosylated cyclodextrin derivative synthesized by CuAAC in one step from the corresponding heptaazide.

improving their therapeutic index is a viable approach. Ideally, such modifications should be accomplished on the molecule itself to avoid protection/deprotection steps. In reality, most chemical transformations are not compatible with a range of functional groups present in the parent compound. The high selectivity and fidelity of the CuAAC make it a good candidate for the last-step derivatization of complex bioactive molecules. This approach was demonstrated by Walsh and Lin, who utilized CuAAC for the preparation of an analog library of glycopeptide antibiotic tyrocidine [100], a nonribosomal peptide which is believed to be targeting the bacterial lipid bilayer. Although a potent antibiotic, it can cause hemolysis at concentrations close to MIC. First, 13 head-to-tail cyclic tyrocidine derivatives containing 1 to 3 propargylglycines were generated from linear peptide N-acetyl cysteamine peptides by action of a thioesterase domain excised from tyrocidine synthetase (Scheme 18.12). The resulting cyclic peptides were then conjugated to 21 azido sugars at less than 1 mM concentration, resulting in a library of 247 isolated compounds. Two most active derivatives, **29** and **30**, identified by antibacterial and hemolysis assays were purified and further profiled, revealing a sixfold improvement in the therapeutic index

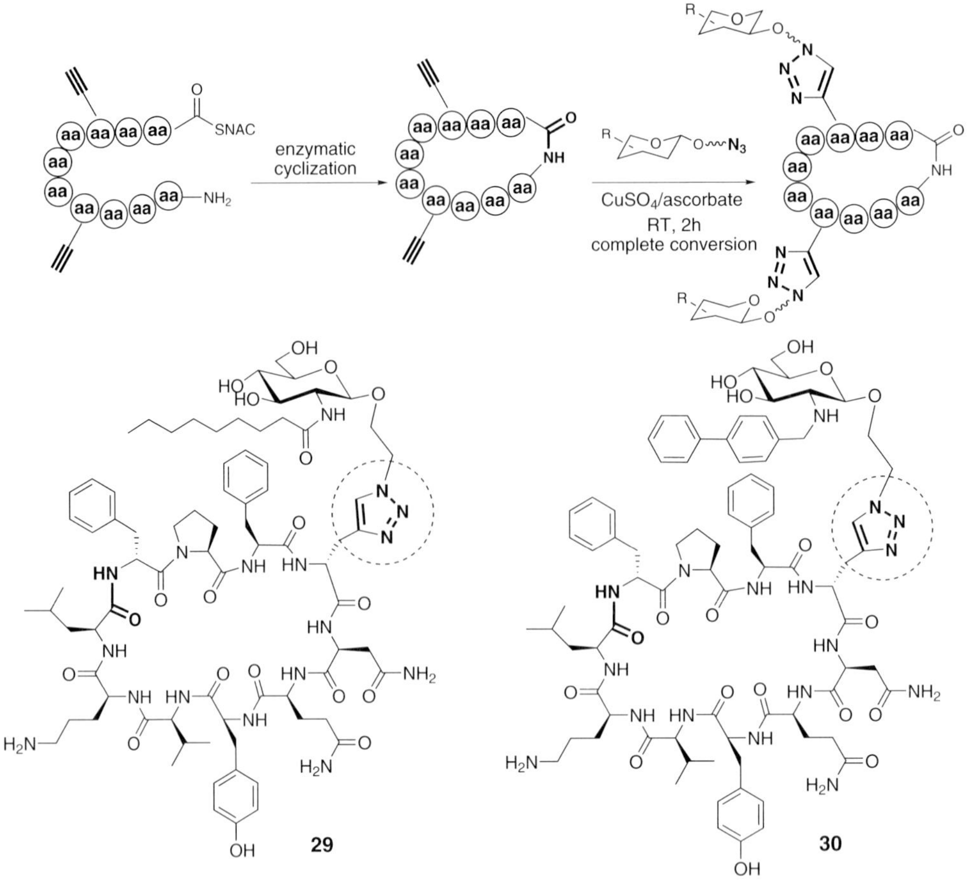

**Scheme 18.12** Chemoenzymatic functionalization of antibiotic tyrocidine.

compared to the wild-type tyrocidine. The toxicity of different reagents that can be used for the CuAAC was also examined, and neither ascorbate, Cu(II), or TBTA caused a hemolytic reaction, thus allowing direct assaying of the analog library.

## 18.3.6
### Copper-Catalyzed Synthesis of Isoxazoles

Although the uncatalyzed 1,3-dipolar cycloaddition of nitrile oxides to acetylenes has been known for a long time, its applications to the synthesis of isoxazoles are scarce. In the reported procedures the yields of isoxazole products are often quite low, side reactions result in impurities, and both regioisomers are often obtained [101]. Furthermore, nitrile oxides are not very stable themselves and dimerize readily, especially when electron-rich substituents are present. In the presence of a copper(I) catalyst, 3,5-disubstituted isoxazoles can be readily synthesized from aldehydes and alkynes in a one-pot three-step process, whereupon nitrile oxide intermediates are generated *in situ* and further react with copper(I) acetylides without isolation [102]. In this procedure, an aldehyde is first converted to the corresponding aldoxime via the reaction with hydroxylamine. Without isolation, the aldoxime is transformed to the corresponding nitrile oxide using 1.05 equiv of chloramine-T trihydrate, which acts both as a halogenating agent and as a base [103]. In the presence of a catalytic amount of copper(I), obtained from Cu metal and copper(II) sulfate, the *in situ* generated nitrile oxide undergoes a stepwise addition to a copper(I) acetylide at ambient temperature, furnishing the 3,5-disubstituted isoxazole (Scheme 18.13).

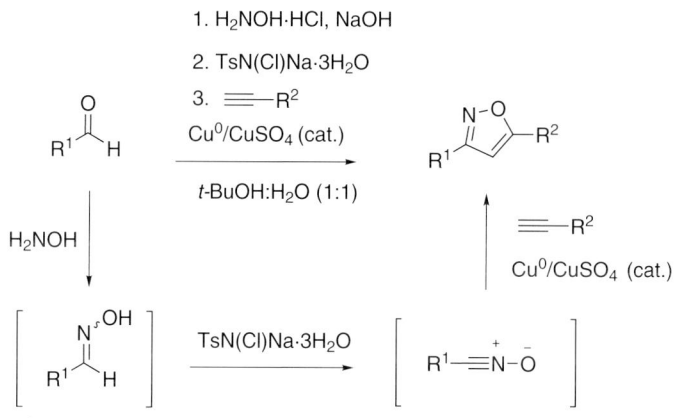

**Scheme 18.13** One-pot synthesis of 3,5-disubstituted isoxazoles from aldehydes, hydroxylamine, and terminal alkynes.

Nitrile oxides generated from a range of aliphatic aldehydes readily participate in this transformation. The 3,5-disubstituted isoxazole products are obtained in moderate to good yields after a simple filtration or aqueous work up. Trace amounts of toluenesulfonamide and unreacted acetylene are easily removed by recrystallization or by passing the product through a short plug of silica gel.

The same sequence can be applied to aromatic aldehydes, affording the products in 60–76% yields. The yields obtained in this one-pot procedure compare favorably to other reported syntheses of isoxazoles, especially in cases with alkenyl, activated aryl or heteroaryl groups, such as, for example, α-furyl. Table 18.1 lists several representative examples.

## 18.4
## Ruthenium-Catalyzed Cycloadditions

### 18.4.1
### Ruthenium-Catalyzed Azide-Alkyne Cycloaddition

The broadness, reliability, and selectivity of the CuAAC process notwithstanding, the reaction is not capable of producing 1,5-disubstitued 1,2,3-triazoles and of engaging

Table 18.1 Isoxazoles prepared from aldehydes, hydroxyl amine, and terminal alkynes.

| Entry | Aldehyde | Acetylene | Isoxazole | Yield (%) |
|---|---|---|---|---|
| 1 | cinnamaldehyde | phenylacetylene | 3-styryl-5-phenyl-isoxazole | 72 |
| 2 | propanal | phenylacetylene | 3-ethyl-5-phenyl-isoxazole | 73 |
| 3 | pentafluorobenzaldehyde | phenylacetylene | 3-(pentafluorophenyl)-5-phenyl-isoxazole | 68 |
| 4 | pentafluorobenzaldehyde | propargyl alcohol | 3-(pentafluorophenyl)-5-(hydroxymethyl)-isoxazole | 64 |
| 5 | 2-furaldehyde | 4-(propargyloxy)acetophenone | 3-(2-furyl)-5-[(4-acetylphenoxy)methyl]-isoxazole | 68 |
| 6 | pivaldehyde | 4-methoxyphenyl propargyl ether | 3-tert-butyl-5-[(4-methoxyphenoxy)methyl]-isoxazole | 57 |
| 7 | diphenylacetaldehyde | 4-(propargyloxy)acetophenone | 3-(diphenylmethyl)-5-[(4-acetylphenoxy)methyl]-isoxazole | 61 |

internal alkynes in catalysis. This agrees well with theoretical studies, which predict that the formation of 1,5-regioisomers is disfavored by over 15 kcal mol$^{-1}$ and that copper(I) acetylides are required intermediates [26]. Among the available methods are the reactions of stabilized phosphonium ylides [104–109] or enamines [28, 110] with aryl azides and the addition of magnesium acetylides to azides [111]. However, these methods have considerable limitations.

In a process complementary to the CuAAC, pentamethylcyclopentadienyl ruthenium chloride [Cp*RuCl] complexes catalyze the cycloaddition of azides with terminal alkynes, regioselectively leading to 1,5-disubstituted 1,2,3-triazoles [17]. Furthermore, and in stark contrast to the CuAAC, its sister ruthenium-catalyzed process, RuAAC, readily engages internal alkynes in catalysis, providing access to fully substituted 1,2,3-triazoles [43].

[Cp*RuCl] complexes, for example, [Cp*RuCl]$_4$, Cp*RuCl(PPh$_3$)$_2$, and Cp*RuCl (COD), are especially active and selective catalysts, producing the triazole products in excellent yields. Both bis(triphenylphosphine) **31** and cyclooctadiene **32** catalysts are readily available and stable. Both complexes can be conveniently obtained by treatment of [Cp*RuCl]$_4$ precursor with excess of PPh$_3$ or cyclooctadiene, respectively.

The [Cp*RuCl]-catalyzed cycloadditions proceed well in a variety of aprotic organic solvents including tetrahydrofuran, dioxane, toluene, benzene, dimethylformamide, and 1,2-dichloroethane. Performing the reaction in protic solvents (MeOH, i-PrOH) gives reduced yields and causes formation of byproducts. Nevertheless, the presence of adventitious water or protic functional groups in the reactants usually does not affect the performance of the catalysts; in other words, the solvents employed need not be scrupulously dried. The cycloadditions with Cp*RuCl(PPh$_3$)$_2$ can be carried out at temperatures ranging from ambient to 110 °C. The reaction does not appear to be very sensitive to atmospheric oxygen. Indeed, the cycloaddition of benzylazide and Ph$_2$C(OH)C≡CH in the presence of 1 mol% of Cp*RuCl(PPh$_3$)$_2$ in benzene also proceeded smoothly, even when the reaction was performed without exclusion of oxygen (4 h under reflux, >99% conversion).

The cyclooctadiene ligand in Cp*RuCl(COD) catalyst is more labile than phosphines and is displaced more readily than phosphine ligands in the bis(triphenylphosphine) complex. This is evidenced by the activity of the former catalyst, even at room temperature. This feature of the Cp*RuCl(COD) catalyst is particularly advantageous for the reactions involving internal alkynes or aryl azides. The deactivation of the catalyst via the formation of the tetraazadiene complex [43] is common for aryl azides, but it is minimized at room temperature, and the triazoles are obtained in good to excellent yields and with excellent regioselectivity. Examples of the triazoles prepared by the RuAAC process are shown in Table 18.2.

## 18.4.2
### Ruthenium-Catalyzed Synthesis of Isoxazoles

As described above, copper(I) acetylides react regioselectively with nitrile oxides generating 3,5-disubstiuted isoxazoles [26]. However, there are no reported methods for generating the regiocomplementary 3,4-disubstiuted isomers. In fact, even when thermal cycloadditions of nitrile oxides with alkynes are successful, they favor the formation of the 3,5-disubstituted isomer (Scheme 18.1). Furthermore, examples of reactions of nitrile oxides with internal alkynes are limited to a handful of highly activated alkynes (e.g., acetylene dicarboxylate and related electron-deficient acetylenes). Unactivated, electron-rich, or sterically hindered acetylenes usually fail to react altogether.

**Table 18.2** [Cp*RuCl]-catalyzed cycloadditions of azides and internal alkynes.

| Entry | Product | Yield (%) | Entry | Product | Yield (%) |
|---|---|---|---|---|---|
| 1 | | 80 | 6 | | 69 |
| 2 | | 81 | 7 | | 93 |
| 3 | | 72 | 8 | | 71 |
| 4 | | 76 | 9 | | 97 |
| 5 | | 83 | 10 | | 79 |

[Cp*RuCl] catalysts fill in this gap: both 3,4-di- and 3,4,5-tri-substituted isoxazoles can be obtained with excellent regioselectivity at room temperature by a ruthenium-catalyzed cycloaddition of nitrile oxides (generated *in situ* from hydroximoyl chlorides by treatment with Et$_3$N) and terminal (Table 18.3) or internal (Table 18.4) alkynes, respectively [112]. For example, phenylacetylene **34** and 4-chloro-*N*-hydroxybenzimidoyl chloride **33** give only the 3,5-disubstituted regioisomer in the absence of a Ru-catalyst. In contrast, in the presence of 5 mol% of Cp*RuCl(COD), the 3,4-isomer **35** was formed almost exclusively (Scheme 18.14).

## 18.4.3
### Mechanistic Studies of the Ruthenium-Catalyzed Cycloadditions

Participation of both terminal and internal alkynes in ruthenium catalysis suggests that ruthenium acetylides are not involved in the catalytic cycle. This assumption is supported experimentally. Since different Ru(II) complexes, such as Cp*RuCl

**Table 18.3** Ru-catalyzed reactions of nitrile oxides with terminal alkynes.

| Entry | Product | Yield (%) |
|---|---|---|
| 1 | isoxazole with 4-Cl-C6H4 at C3, Ph at C4 | 86 |
| 2 | isoxazole with 4-OMe-C6H4 at C3, 4-Me-C6H4 at C4 | 87 |
| 3 | isoxazole with 4-OMe-C6H4 at C3, 1-hydroxycyclohexyl at C4 | 67 |
| 4 | isoxazole with 4-OMe-C6H4 at C3, CH2-(thiomorpholine-1,1-dioxide) at C4 | 77 |
| 5 | isoxazole with Ph at C3, styryl at C4 | 87 |
| 6 | isoxazole with 2,2-dimethyl-chromene (OAc) substituent, styryl at C4 | 93 |
| 7 | isoxazole with CH2CH2CN at C3, CH2Ph at C4 | 72 |
| 8 | isoxazole with Et at C3, CH2NHTs at C4 | 85 |
| 9 | isoxazole with styryl at C3, CH2NHTs at C4 | 93 |

(PPh$_3$)$_2$, Cp*RuCl(COD), Cp*RuCl(NBD) and [Cp*RuCl]$_4$ are competent catalysts, the neutral [Cp*RuCl] is the proposed catalytically active species. This hypothesis is supported by the observations that (i) [Cp*RuBr] and [Cp*RuI] complexes are significantly less active catalysts, (ii) [Cp*Ru]$^+$ cationic complexes obtained by the removal of the chloride with Ag$^+$ are devoid of catalytic activity altogether, and (iii) chelating diphosphines, such as bis(diphenylphosphino)ethane, completely deactivate the catalyst.

Catalytic cyclotrimerization of alkynes catalyzed by [Cp*RuCl] and similar complexes is, of course, well known [113–118], and the current mechanistic proposal is

**Table 18.4** Isoxazoles produced from internal alkynes (**a** and **b** are regioisomers of the product isoxazole).

| Entry | Products | Yield (%) | |
|---|---|---|---|
| | | a | b |
| 1 | 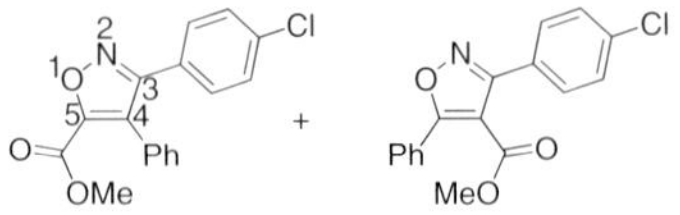 (isoxazole with Cl-C6H4, Ph, CO2Me) + regioisomer | 71 | 5 |
| 4 | 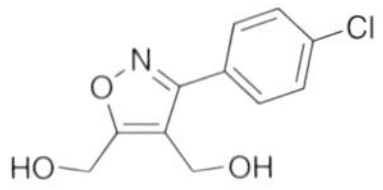 (isoxazole with Cl-C6H4, CH2OH, CH2OH) | 78 | — |
| 5 | 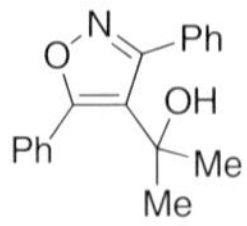 (isoxazole with Ph, Ph, C(Me)2OH) | 83 | — |
| 6 | 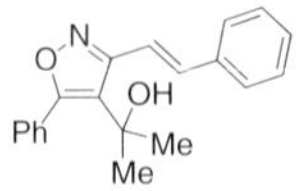 (isoxazole with Ph, styryl, C(Me)2OH) | 99 | — |
| 7 | 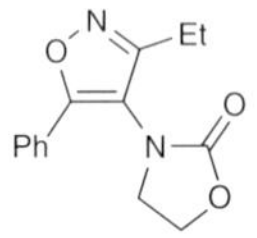 (isoxazole with Ph, Et, oxazolidinone) | 39 | — |

that RuAAC represents a simple case that shunts off the usual alkyne oligomerization sequence, as schematically outlined in Scheme 18.15 [43]. The displacement of the spectator ligands (step A) produces the activated complex **36**, which is converted, via the oxidative coupling of an alkyne and an azide (step B) to the ruthenacycle **37**. This step controls the regioselectivity of the overall process. The new C−N bond is formed between the more nucleophilic and less sterically demanding carbon and the electrophilic terminal nitrogen of the azide. The metallacycle intermediate then undergoes reductive elimination (step C) releasing the aromatic triazole product and regenerating the catalyst (step D).

[CpRuCl] complexes, such as CpRuCl(PPh$_3$)$_2$ also catalyze the reaction, albeit much less efficiently and not regioselectively. The higher catalytic activity of

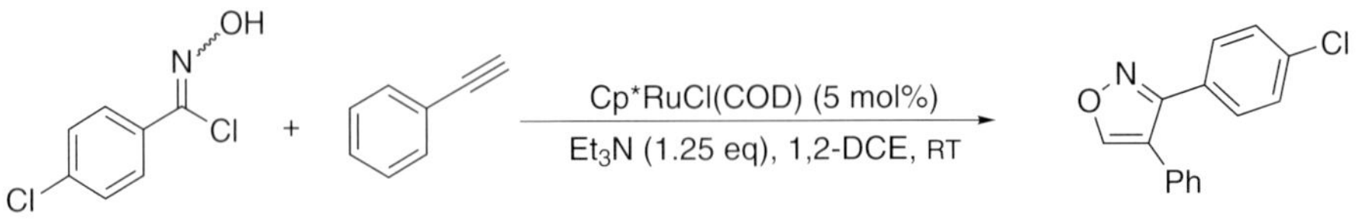

**Scheme 18.14** Ruthenium-catalyzed synthesis of isoxazoles.

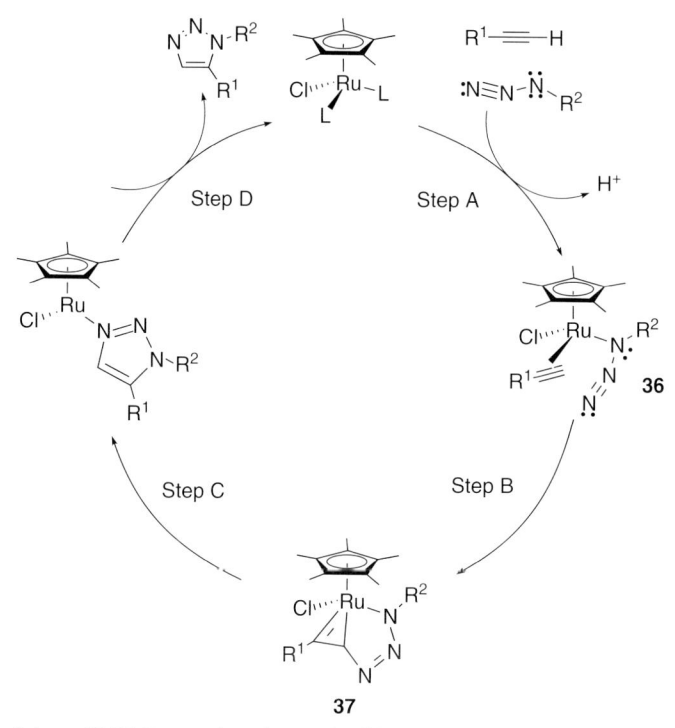

**Scheme 18.15** Proposed catalytic cycle of the RuAAC reaction.

[Cp*RuCl] catalysts can be attributed to the lability of the bystander ligands in such systems (thus facilitating the formation of the activated complex **36**) and the more sterically demanding nature of the Cp* ligand (which, in turn, facilitates the reductive elimination, step C, in the catalytic cycle).

The proposed reaction mechanism was examined computationally employing DFT calculations using the B3LYP hybrid functional. In this computational study, methyl azide and propyne were used as the model reactants with [CpRuCl] as catalyst. Methyl azide can, in principle, coordinate to the metal center via either N-1 or N-3. Both modes of coordination are known, although the former is observed far more commonly [119]. The alkyne can also coordinate to the metal center in π-fashion in two distinct orientations. Therefore, four activated azide:[Ru]:alkyne complexes, **PCA**, **PCB**, **PCC** and **PCD**, are possible, as shown in Figure 18.6. **PCA** and **PCC**

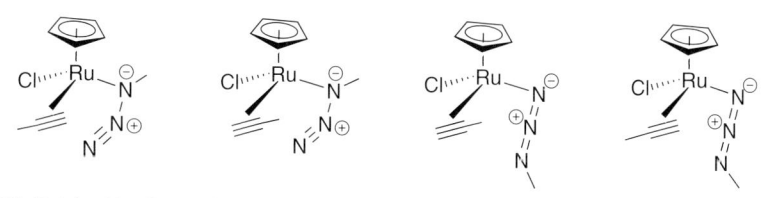

**PCA** (0.0 kcal/mol)   **PCB** (0.8 kcal/mol)   **PCC** (1.0 kcal/mol)   **PCD** (0.7 kcal/mol)

**Figure 18.6** Structures and computed energies of the activated complexes.

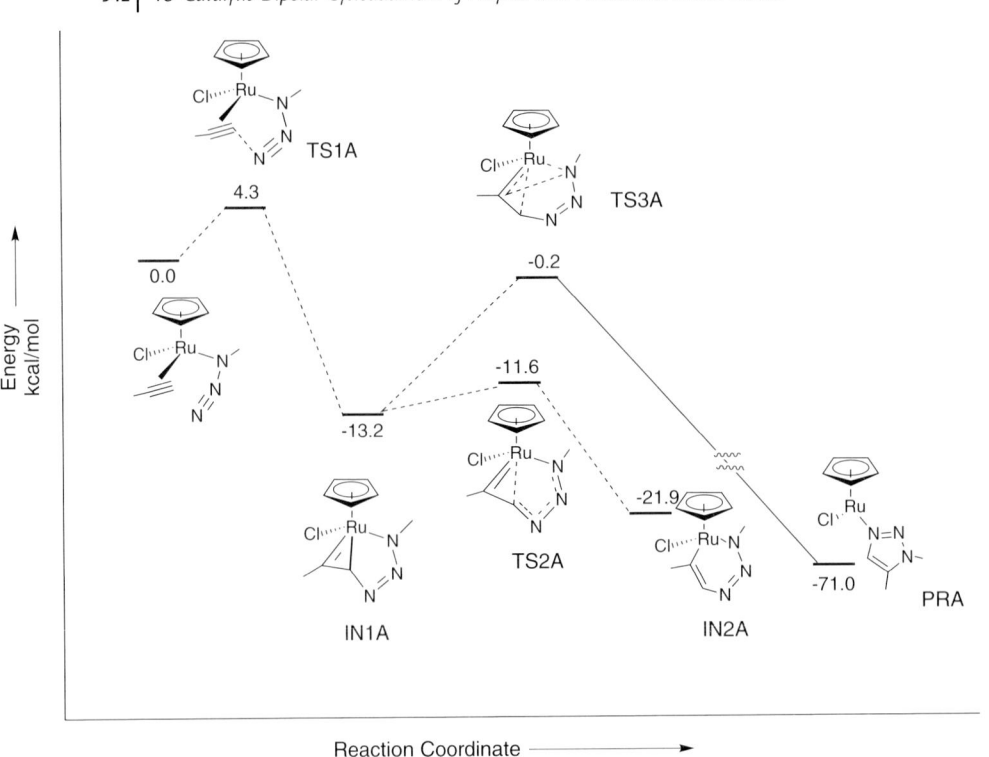

**Figure 18.7** Schematic representation (energy vs. reaction coordinate) of the reaction of organic azides and alkynes catalyzed by [CpRuCl].

complexes lead to the 1,5-disubstituted triazole product, whereas **PCB** and **PCD** result in the formation of the 1,4-regioisomer. The computed energies of the complexes are within 1 kcal mol$^{-1}$.

The energy profile for the reaction of **PCA** is illustrated in Figure 18.7. The first step, oxidative coupling of methyl azide and propyne, results in the formation of the six-membered ruthenacycle **IN1A**. This step is exothermic by 13.2 kcal mol$^{-1}$, and the calculated barrier is only 4.3 kcal mol$^{-1}$. **IN1A** can undergo reductive elimination via transition state **TS3A** (the rate-determining step, 13 kcal mol$^{-1}$ activation energy) forming the metal-triazole complex **PRA**, or can isomerize to a more stable ruthenacycle **IN2A** with a very low barrier of 1.6 kcal mol$^{-1}$ (**TS2A**). In other words, the facility and the outcome of the overall process is influenced by the relative energies and ease of interconversion of intermediates **IN1A** and **IN2A** but, in any case, the oxidative addition step is irreversible under the reaction conditions, and the activation energy of the rate-determining step is reduced from 24 kcal mol$^{-1}$ for the uncatalyzed process [26] to 13 kcal mol$^{-1}$, accounting for the observed rate acceleration. Pathway B, leading to the 1,4-regioisomer (see Supporting Information for complete details) is disfavored by about 3 kcal mol$^{-1}$. Further examination of the structural features of the species involved in the oxidative coupling steps (Figure 18.8) reveals the steric

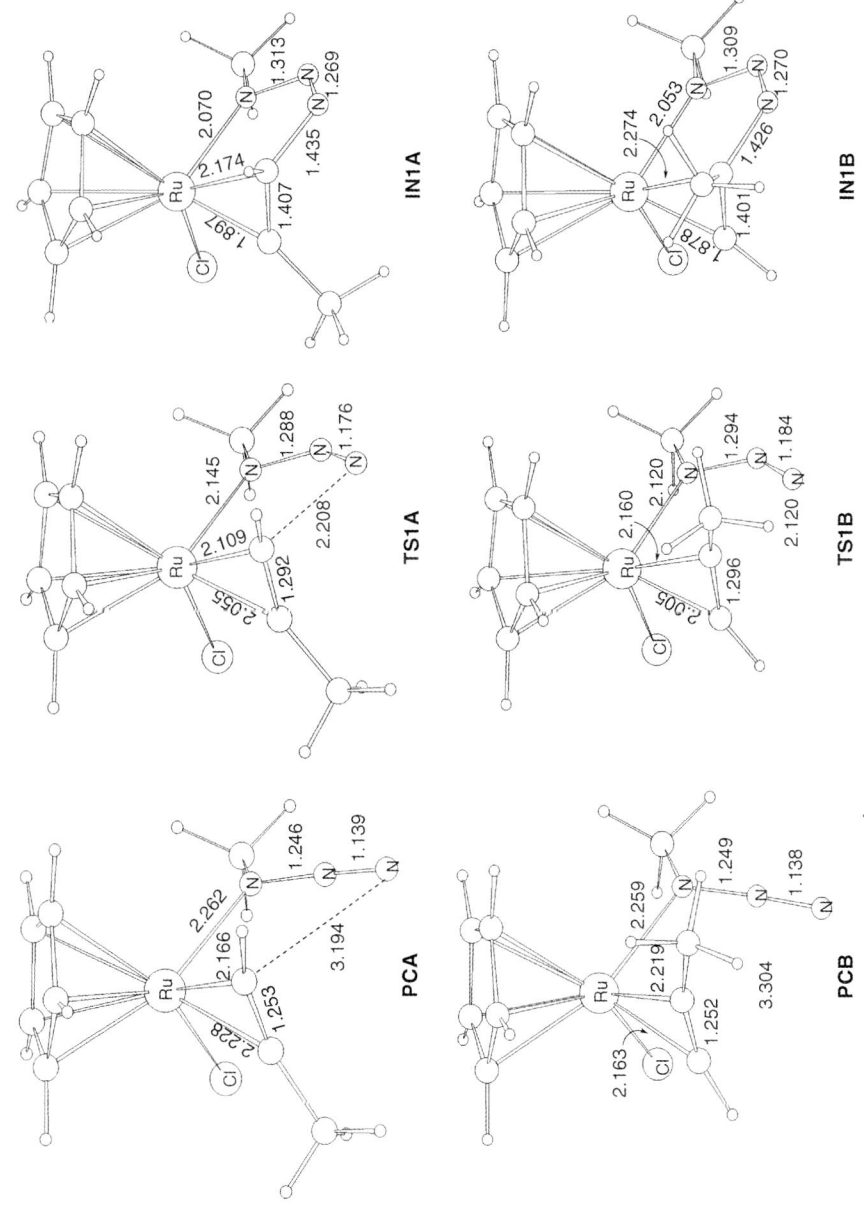

**Figure 18.8** Selected structural parameters (Å) calculated for species involved in the oxidative coupling steps of pathways A and B (from **PCA** and **PCB**, respectively).

repulsion between the methyl substituent of the propyne and the hydrogens on the Cp ring in the metallacycle intermediate **IN1B**, whereas it is avoided in the intermediate **IN1A**. The Ru–C$_\beta$ bond distance in intermediate **IN1B** is 2.274 Å, longer than the Ru–C$_\beta$ distance in **IN1A**, supporting the steric repulsion argument. Clearly, these steric interactions are amplified in the [Cp*RuCl] complexes, explaining the high regioselectivity of those catalytic systems. Pathways beginning from the azide coordinated to ruthenium via the terminal nitrogen (starting from **PCC** and **PCD**) were also evaluated, revealing much higher activation barriers for the formation of even the first intermediate, 20.9 and kcal mol$^{-1}$, respectively.

The computational results are in general agreement with the observed performance of the CpRuCl(PPh$_3$)$_2$ catalyst: the reactions are much slower than with the Cp* analog, the regioselectivity does not exceed about 85:15 (in favor of the 1,5-regioisomer), and the yields are modest at best. Higher sensitivity of the catalysis to the steric demands of the azide substituent are also explained by this model: σ-coordination of the azide to the ruthenium atom via its N-1 nitrogen is a required event for the subsequent oxidative addition step, the stability of the complex **PCA** (and the strength of this interaction) is directly affected by the size of the azide substituent.

It appears that the oxidative coupling step is, in essence, a nucleophilic attack of the coordinated alkyne on the terminal nitrogen of the coordinated azide, whose electrophilicity is further increased by coordination to the ruthenium center. Both steric and electronic factors favor the pathway starting from the **PCA** activated complex species which leads to the 1,5-regioselectivity observed experimentally.

## 18.5
### Experimental: Selected Procedures

### 18.5.1
#### Copper-Catalyzed Cycloadditions

*General Procedure 1, "aqueous ascorbate" method for the synthesis of 1,2,3-triazoles* [9, 26]. 17-Ethynyl estradiol (888 mg, 3 mmol) and (S)-3-azidopropane-1,2-diol (352 mg, 3 mmol) were suspended in 12 mL of 1:1 water/*tert*-butanol mixture. Sodium ascorbate (0.3 mmol, 300 μL of freshly prepared 1 M solution in water) was added, followed by copper(II) sulfate pentahydrate (7.5 mg, 0.03 mmol, in 100 μL of water). The heterogeneous mixture was stirred vigorously overnight, at which point it cleared and TLC analysis indicated complete consumption of the reactants. The reaction mixture was diluted with 50 mL of water, cooled in ice, and the white precipitate was collected by filtration. After washing with cold water (2 × 25 mL), the precipitate was dried under vacuum to afford 1.17 g (94%) of pure product as an off-white powder.

*General Procedure 2, copper metal-catalyzed synthesis of 1,2,3-triazoles* [9, 26]. Phenylacetylene (2.04 g, 20 mmol) and 2,2-bis(azidomethyl)propane-1,3-diol (1.86 g, 10 mmol) were dissolved in 1:2 *tert*-butanol/water mixture (50 mL). About 1 g of copper metal turnings were added, and the reaction mixture was stirred for 24 h, after which time TLC analysis indicated complete consumption of the starting materials.

Copper was removed, and the white product was filtered off, washed with water, and dried to yield 3.85 g (98%) of pure bis-triazole product.

*General Procedure 3, copper-catalyzed synthesis of isoxazoles* [102]. *trans*-Cinnamaldehyde (2.6 g, 20 mmol) was added to the solution of hydroxylamine hydrochloride (1.46 g, 21 mmol) in 80 mL of 1 : 1 *t*-BuOH : $H_2O$. To this was added NaOH (0.84 g, 21 mmol), and after stirring for 30 min at ambient temperature, TLC analysis indicated that oxime formation was complete. Chloramine-T trihydrate (5.9 g, 21 mmol) was added in small portions over 5 min, followed by $CuSO_4 \cdot 5H_2O$ (0.15 g, 0.6 mmol) and copper turnings (ca. 50 mg). 1-Ethynylcyclohexene (2.23 g, 21 mmol) was added, the pH was adjusted to about 6 by addition of a few drops of 1 M NaOH, and stirring was continued for another 6 h. The reaction mixture was poured into ice/water (150 mL), and 10 mL of dilute $NH_4OH$ was added to remove all copper salts. The product was collected by filtration, redissolved, and passed through a short plug of silica gel (ethyl acetate : hexanes 1 : 6, $R_F = 0.6$) affording 3.6 g (72%) of 5-cyclohex-1-enyl-3-styrylisoxazole as an off-white solid, m.p. 138–139 °C.

## 18.5.2
### Ruthenium-Catalyzed Cycloadditions

*General Procedure 4, Cp\*RuCl(PPh$_3$)$_2$ catalyzed reaction* [43]. A solution of 1-ethynylcyclohexanol (51 μL, 0.40 mmol) and ethyl 2-(2-azidoacetamido)-3-hydroxypropanoate (86 mg, 0.40 mmol) in 0.5 mL of dioxane was added to Cp*RuCl(PPh$_3$)$_2$ (6.4 mg, 0.008 mmol) dissolved in 2.5 mL of dioxane. The vial was purged with nitrogen, sealed and heated in an oil bath at 60 °C for 12 h, at which point TLC and LC-MS analyses indicated complete consumption of the alkyne and the azide starting materials. The mixture was absorbed onto silica gel and chromatographed with hexanes : ethyl acetate (1 : 1) to remove non-polar impurities, followed by ethyl acetate to elute the product, which was isolated as a pale yellow oil. (Note: in all reactions the alkyne was added first, followed by the addition of the azide, or azide and alkyne were dissolved in the reaction solvent and added to the solution of the catalyst).

*General Procedure 5, Cp\*RuCl(COD) catalyzed reactions* [43]. Cp*RuCl(COD) (4.0 mg, 0.010 mmol) was added to a tube with septa cap. The tube was sealed then evacuated and filled with nitrogen three times. 5 mL of toluene (degassed for 1 h with nitrogen purge) was added, followed by 2-methyl-4-phenylbut-3-yn-2-ol (80 mg, 0.50 mmol) and 1-azido-4-methylbenzene (67 mg, 0.50 mmol). The reaction was stirred at room temperature for 30 min, at which time TLC analysis indicated complete consumption of the starting materials. The mixture was absorbed onto silica gel and chromatographed with 4 : 1 hexanes : ethyl acetate, then 2 : 1 hexanes : ethyl acetate to afford the pure product as a white solid. (Note: in all reactions the alkyne was added first, followed by the addition of the azide, or azide and alkyne were dissolved in the reaction solvent and added to the solution of the catalyst; azide should not be added first).

*General Procedure 6, Ruthenium-catalyzed synthesis of isoxazoles* [112]. 4-Chloro-*N*-hydroxybenzimidoyl chloride (208 mg, 1.1 mmol) and phenylacetylene (110 μL, 1.0 mmol) were combined in a 20 mL screw top vial purged with dry nitrogen.

Degassed 1,2-dichloroethane (10 mL) was added followed by Cp*RuCl(COD) (19 mg, 0.05 mmol) and triethylamine (176 µL, 1.25 mmol) at room temperature and the vial was capped. After 10 h, the reaction mixture was passed through a silica gel plug eluting with $CH_2Cl_2$. The resulting solution was concentrated and flashed over silica gel (hexanes to 10 : 1 hexanes : ethyl acetate) to provide the indicated compound as a yellow oil (219 mg, 86%).

## References

1 Huisgen, R. (1963) *Angew. Chem. Int. Ed.*, **2**, 565–598.
2 Padwa, A. (ed). (1984) *1,3-Dipolar Cycloaddition Chemistry*, John Wiley & Sons, New York.
3 Padwa, A. and Pearson, W.H. (eds) (2002) *Synthetic Applications of 1,3 Dipolar Cycloaddition Chemistry Toward Heterocycles and Natural Products*, John Wiley & Sons, New York.
4 *Chemistry of the Azido Group*. Patai S., Ed. New York: John Wiley & Sons, 1971.
5 Saxon, E. and Bertozzi, C.R. (2000) *Science*, **287**, 2007–2010.
6 Hang, H.C., Yu, C., Kato, D.L., and Bertozzi, C.R. (2003) *Proc. Natl. Acad. Sci. USA*, **100**, 14846–14851.
7 Kolb, H.C., Finn, M.G., and Sharpless, K.B. (2001) *Angew. Chem. Int. Ed.*, **40**, 2004–2021.
8 Tornøe, C.W., Christensen, C., and Meldal, M. (2002) *J. Org. Chem.*, **67**, 3057–3062.
9 Rostovtsev, V.V., Green, L.G., Fokin, V.V., and Sharpless, K.B. (2002) *Angew. Chem. Int. Ed.*, **41**, 2596–2599.
10 Bock, V.D., Hiemstra, H., and van Maarseveen, J.H. (2006) *Eur. J. Org. Chem.*, 51–68.
11 Fokin, V.V. (2007) *ACS Chem. Biol.*, **2**, 775–778.
12 Moses, J.E. and Moorhouse, A.D. (2007) *Chem. Soc. Rev.*, **36**, 1249–1262.
13 Johnson, J.A., Koberstein, J.T., Finn, M.G., and Turro, N.J. (2008) *Macromol. Rapid Commun.*, **29**, 1052–1072.
14 Lutz, J.-F. and Zarafshani, Z. (2008) *Adv. Drug Delivery Rev.*, **60**, 958–970.
15 Tron, G.C., Pirali, T., Billington, R.A., Canonico, P.L., Sorba, G., and Genazzani, A.A. (2008) *Med. Res. Rev.*, **28**, 278–308.
16 For a continuously updated list of applications of CuAAC and a compehensive list of reaction conditions, the reader is referred to http://www.scripps.edu/chem/fokin/cuaac.
17 Zhang, L., Chen, X., Xue, P., Sun, H.H.Y., Williams, I.D., Sharpless, K.B., Fokin, V.V., and Jia, G. (2005) *J. Am. Chem. Soc.*, **127**, 15998–15999.
18 Michael, A. (1893) *J. Prakt. Chem.*, **48**, 94.
19 Huisgen, R. and Blaschke, H. (1965) *Chem. Ber.*, **98**, 2985–2997.
20 Huisgen, R., Knorr, R., Moebius, L., and Szeimies, G. (1965) *Chem. Ber.*, **98**, 4014–4021.
21 Huisgen, R., Moebius, L., and Szeimies, G. (1965) *Chem. Ber.*, **98**, 1138–1152.
22 Huisgen, R., Szeimies, G., and Moebius, L. (1966) *Chem. Ber.*, **99**, 475–490.
23 Huisgen, R., Szeimies, G., and Moebius, L. (1967) *Chem. Ber.*, **100**, 2494–2507.
24 Huisgen, R. (1984) in *1,3-Dipolar Cycloaddition Chemistry*, vol. 1 (ed. A. Padwa), John Wiley & Sons, New York, pp. 1–176.
25 Huisgen, R. (1989) *Pure Appl. Chem.*, **61**, 613–628.
26 Himo, F., Lovell, T., Hilgraf, R., Rostovtsev, V.V., Noodleman, L., Sharpless, K.B., and Fokin, V.V. (2005) *J. Am. Chem. Soc.*, **127**, 210–216.
27 Wu, P. and Fokin, V.V. (2007) *Aldrichim. Acta*, **40**, 7–17.
28 Tomé, A.C. (2004) in *Science of Synthesis: Houben-Weyl Methods of Molecular*

*Transformations* (eds R.C. Storr and T.L. Gilchrist), vol. 13: Five-Membered Hetarenes with Three or More Heteroatoms, Thieme, Stuttgart, pp. 415–601.

29 Krivopalov, V.P. and Shkurko, O.P. (2005) *Russ. Chem. Rev.*, **74**, 339–379.

30 Brik, A., Muldoon, J., Lin, Y.-c., Elder, J.H., Goodsell, D.S., Olson, A.J., Fokin, V.V., Sharpless, K.B., and Wong, C.-h. (2003) *ChemBioChem*, **4**, 1246–1248.

31 Brik, A., Alexandratos, J., Lin, Y.-C., Elder, J.H., Olson, A.J., Wlodawer, A., Goodsell, D.S., and Wong, C.-H. (2005) *ChemBioChem*, **6**, 1167–1169.

32 Tam, A., Arnold, U., Soellner, M.B., and Raines, R.T. (2007) *J. Am. Chem. Soc.*, **129**, 12670–12671.

33 Link, A.J. and Tirrell, D.A. (2003) *J. Am. Chem. Soc.*, **125**, 11164–11165.

34 Speers, A.E., Adam, G.C., and Cravatt, B.F. (2003) *J. Am. Chem. Soc.*, **125**, 4686–4687.

35 Wang, Q., Chan, T.R., Hilgraf, R., Fokin, V.V., Sharpless, K.B., and Finn, M.G. (2003) *J. Am. Chem. Soc.*, **125**, 3192–3193.

36 Link, A.J., Vink, M.K.S., and Tirrell, D.A. (2004) *J. Am. Chem. Soc.*, **126**, 10598–10602.

37 Kolb, H.C. and Sharpless, K.B. (2003) *Drug Discov. Today*, **8**, 1128–1137.

38 Wang, Q., Chittaboina, S., and Barnhill, H.N. (2005) *Lett. Org. Chem.*, **2**, 293–301.

39 Goodall, G.W. and Hayes, W. (2006) *Chem. Soc. Rev.*, **35**, 280–312.

40 Graham, D. and Enright, A. (2006) *Curr. Org. Syn.*, **3**, 9–17.

41 Majireck, M.M. and Weinreb, S.M. (2006) *J. Org. Chem.*, **71**, 8680–8683.

42 Oppilliart, S., Mousseau, G., Zhang, L., Jia, G., Thuery, P., Rousseau, B., and Cintrat, J.-C. (2007) *Tetrahedron*, **63**, 8094–8098.

43 Boren, B.C., Narayan, S., Rasmussen, L.K., Zhang, L., Zhao, H., Lin, Z., Jia, G., and Fokin, V.V. (2008) *J. Am. Chem. Soc.*, **130**, 8923–8930.

44 Rasmussen, L.K., Boren, B.C., and Fokin, V.V. (2007) *Org. Lett.*, **9**, 5337–5339.

45 Perez-Balderas, F., Ortega-Munoz, M., Morales-Sanfrutos, J., Hernandez-Mateo, F., Calvo-Flores, F.G., Calvo-Asin, J.A., Isac-Garcia, J., and Santoyo-Gonzalez, F. (2003) *Org. Lett.*, **5**, 1951–1954.

46 Wu, P., Feldman, A.K., Nugent, A.K., Hawker, C.J., Scheel, A., Voit, B., Pyun, J., Frechet, J.M.J., Sharpless, K.B., and Fokin, V.V. (2004) *Angew. Chem. Int. Ed.*, **43**, 3928–3932.

47 Malkoch, M., Schleicher, K., Drockenmuller, E., Hawker, C.J., Russell, T.P., Wu, P., and Fokin, V.V. (2005) *Macromolecules*, **38**, 3663–3678.

48 Reddy, K.R., Rajgopal, K., and Kantam, M.L. (2006) *Synlett*, 957–959.

49 Fukuzawa, S., Shimizu, E., and Kikuchi, S. (2007) *Synlett*, 2436–2438.

50 Nigh, W.G. (1973) in *Oxidation in Organic Chemistry*, vol. **Part B** (ed. W.S. Trahanovsky), Academic Press, NewYork, pp. 1–95.

51 Siemsen, P., Livingston, R.C., and Diederich, F. (2000) *Angew. Chem. Int. Ed.*, **39**, 2632–2657.

52 Glaser, C. (1869) *Chem. Ber.*, **2**, 422.

53 Glaser, C. (1870) *Ann. Chem. Pharm.*, **154**, 159.

54 Appukkuttan, P., Dehaen, W., Fokin, V.V., and Van der Eycken, E. (2004) *Org. Lett.*, **6**, 4223–4225.

55 Ermolat'ev, D., Dehaen, W., and Van der Eycken, E. (2004) *QSAR Comb. Sci.*, **23**, 915–918.

56 Khanetskyy, B., Dallinger, D., and Kappe, C.O. (2004) *J. Comb. Chem.*, **6**, 884–892.

57 Bouillon, C., Meyer, A., Vidal, S., Jochum, A., Chevolot, Y., Cloarec, J.-P., Praly, J.-P., Vasseur, J.-J., and Morvan, F. (2006) *J. Org. Chem.*, **71**, 4700–4702.

58 Appukkuttan, P. and Van der Eyeken, E. (2008) *Eur. J. Org. Chem.*, 1133–1155.

59 Lucas, R., Neto, V., Bouazza, A.H., Zerrouki, R., Granet, R., Krausz, P., and Champavier, Y. (2008) *Tetrahedron Lett.*, **49**, 1004–1007.

60 Moorhouse, A.D. and Moses, J.E. (2008) *Synlett*, 2089–2092.

61 Pachon, L.D., van Maarseveen, J.H., and Rothenberg, G. (2005) *Adv. Synth. Cat.*, **347**, 811–815.

62 Molteni, G., Bianchi, C.L., Marinoni, G., Santo, N., and Ponti, A. (2006) *New J. Chem.*, **30**, 1137–1139.

63 Chan, T.R., Hilgraf, R., Sharpless, K.B., and Fokin, V.V. (2004) *Org. Lett.*, **6**, 2853–2855.

64 Chan, T.R. (2005) Ph.D. Dissertation, The Scripps Research Institute, *Diss. Abstr. Int.*, B 2006, **66** (8), 4224.

65 Gupta, S.S., Raja, K.S., Kaltgrad, E., Strable, E., and Finn, M.G. (2005) *Chem. Comm.*, 4315–4317.

66 Weller, R.L. and Rajski, S.R. (2005) *Org. Lett.*, **7**, 2141–2144.

67 Burley, G.A., Gierlich, J., Mofid, M.R., Nir, H., Tal, S., Eichen, Y., and Carell, T. (2006) *J. Am. Chem. Soc.*, **128**, 1398–1399.

68 Speers, A.E. and Cravatt, B.F. (2004) *Chem. Biol.*, **11**, 535–546.

69 Beatty, K.E., Xie, F., Wang, Q., and Tirrell, D.A. (2005) *J. Am. Chem. Soc.*, **127**, 14150–14151.

70 Dieterich, D.C., Link, A.J., Graumann, J., Tirrell, D.A., and Schuman, E.M. (2006) *Proc. Natl. Acad. Sci. USA*, **103**, 9482–9487.

71 Sawa, M., Hsu, T.-L., Itoh, T., Sugiyama, M., Hanson Sarah, R., Vogt Peter, K., and Wong, C.-H. (2006) *Proc. Natl. Acad. Sci. USA*, **103**, 12371–12376.

72 El-Sagheer, A.H. and Brown, T. (2008) *Current Protocols in Nucleic Acid Chemistry* 4.33.1–4.33.21.

73 Rodionov, V.O., Presolski, S.I., Diaz, D.D., Fokin, V.V., and Finn, M.G. (2007) *J. Am. Chem. Soc.*, **129**, 12705–12712.

74 Rodionov, V.O., Presolski, S.I., Gardinier, S., Lim, Y.-H., and Finn, M.G. (2007) *J. Am. Chem. Soc.*, **129**, 12696–12704.

75 Lewis, W.G., Magallon, F.G., Fokin, V.V., and Finn, M.G. (2004) *J. Am. Chem. Soc.*, **126**, 9152–9153.

76 Zhu, W. and Ma, D. (2004) *Chem. Comm.*, 888–889.

77 Feldman, A.K., Colasson, B., and Fokin, V.V. (2004) *Org. Lett.*, **6**, 3897–3899.

78 Chittaboina, S., Xie, F., and Wang, Q. (2005) *Tetrahedron Lett.*, **46**, 2331–2336.

79 Kacprzak, K. (2005) *Synlett*, 943–946.

80 Barral, K., Moorhouse, A.D., and Moses, J.E. (2007) *Org. Lett.*, **9**, 1809–1811.

81 Bogdan, A.R. and Sach, N.W. (2009) *Adv. Synth. Cat.*, **351**, 849–854.

82 Rodionov, V.O., Fokin, V.V., and Finn, M.G. (2005) *Angew. Chem. Int. Ed.*, **44**, 2210–2215.

83 Vrieze, K. and van Koten, G. (1987) in *Comprehensive Coordination Chemistry*, vol. **2**, Pergamon, Oxford, pp. 189–245.

84 Mykhalichko, B.M., Temkin, O.N., and Mys'kiv, M.G. (2001) *Russ. Chem. Rev.*, **69**, 957–984.

85 Xie, X., Auel, C., Henze, W., and Gschwind, R.M. (2003) *J. Am. Chem. Soc.*, **125**, 1595–1601.

86 Ahlquist, M. and Fokin, V.V. (2007) *Organometallics*, **26**, 4389–4391.

87 Straub, B.F. (2007) *Chem. Comm.*, 3868–3870.

88 Chui, S.S.Y., Ng, M.F.Y., and Che, C.-M. (2005) *Chem. Eur. J.*, **11**, 1739–1749.

89 Bae, I., Han, H., and Chang, S. (2005) *J. Am. Chem. Soc.*, **127**, 2038–2039.

90 Cho, S.H., Yoo, E.J., Bae, I., and Chang, S. (2005) *J. Am. Chem. Soc.*, **127**, 16046–16047.

91 Cassidy, M.P., Raushel, J., and Fokin, V.V. (2006) *Angew. Chem. Int. Ed.*, **45**, 3154–3157.

92 Whiting, M. and Fokin, V.V. (2006) *Angew. Chem. Int. Ed.*, **45**, 3157–3161.

93 Yoo, E.J., Ahlquist, M., Kim, S.H., Bae, I., Fokin, V.V., Sharpless, K.B., and Chang, S. (2007) *Angew. Chem. Int. Ed.*, **46**, 1730–1733.

94 Lee, L.V., Mitchell, M.L., Huang, S.-J., Fokin, V.V., Sharpless, K.B., and Wong, C.-H. (2003) *J. Am. Chem. Soc.*, **125**, 9588–9589.

95 Whiting, M., Tripp, J.C., Lin, Y.C., Lindstrom, W., Olson, A.J., Elder, J.H., Sharpless, K.B., and Fokin, V.V. (2006) *J. Med. Chem.*, **49**, 7697–7710.

96 Perez-Balderas, F., Morales-Sanfrutos, J., Hernandez-Mateo, F., Isac-Garcia, J., and

Santoyo-Gonzalez, F. (2009) *Eur. J. Org. Chem.*, 2441–2453.

97 Dondoni, A. and Marra, A. (2006) *J. Org. Chem.*, **71**, 7546–7557.

98 Ladmiral, V., Mantovani, G., Clarkson, G.J., Cauet, S., Irwin, J.L., and Haddleton, D.M. (2006) *J. Am. Chem. Soc.*, **128**, 4823–4830.

99 Geng, J., Mantovani, G., Tao, L., Nicolas, J., Chen, G., Wallis, R., Mitchell, D.A., Johnson, B.R.G., Evans, S.D., and Haddleton, D.M. (2007) *J. Am. Chem. Soc.*, **129**, 15156–15163.

100 Lin, H. and Walsh, C.T. (2004) *J. Am. Chem. Soc.*, **126**, 13998–14003.

101 Lang, S.A. and Lin, Y.I. (1984) in *Comprehensive Heterocyclic Chemistry*, vol. VI/4B (eds A.R. Katritzky and C.W. Rees), Pergamon, Oxford, pp. 1–130.

102 Hansen, T.V., Wu, P., and Fokin, V.V. (2005) *J. Org. Chem.*, **70**, 7761–7764.

103 Hassner, A. and Rai, K.M. (1989) *Synthesis*, **1**, 57–59.

104 Labbe, G., Ykman, P., and Smets, G. (1969) *Tetrahedron*, **25**, 5421.

105 Ykman, P., Labbe, G., and Smets, G. (1970) *Tetrahedron Lett.*, 5225.

106 Ykman, P., Labbe, G., and Smets, G. (1971) *Tetrahedron*, **27**, 5623.

107 Ykman, P., Labbe, G., and Smets, G. (1971) *Tetrahedron*, **27**, 845.

108 Ykman, P., Smets, G., Mathys, G., and Labbe, G. (1972) *J. Org. Chem.*, **37**, 3213.

109 Ykman, P., Labbe, G., and Smets, G. (1973) *Tetrahedron*, **29**, 195–198.

110 Fioravanti, S., Pellacani, L., Ricci, D., and Tardella, P.A. (1997) *Tetrahedron-Asym.*, **8**, 2261–2266.

111 Krasinski, A., Fokin, V.V., and Sharpless, K.B. (2004) *Org. Lett.*, **6**, 1237–1240.

112 Grecian, S. and Fokin, V.V. (2008) *Angew. Chem. Int. Ed.*, **47**, 8285–8287.

113 Yamamoto, Y., Kitahara, H., Ogawa, R., Kawaguchi, H., Tatsumi, K., and Itoh, K. (2000) *J. Am. Chem. Soc.*, **122**, 4310–4319.

114 Yamamoto, Y., Takagishi, H., and Itoh, K. (2002) *J. Am. Chem. Soc.*, **124**, 28–29.

115 Kirchner, K., Calhorda, M.J., Schmid, R., and Veiros, L.F. (2003) *J. Am. Chem. Soc.*, **125**, 11721–11729.

116 Yamamoto, Y., Arakawa, T., Ogawa, R., and Itoh, K. (2003) *J. Am. Chem. Soc.*, **125**, 12143–12160.

117 Ura, Y., Sato, Y., Shiotsuki, M., Kondo, T., and Mitsudo, T. (2004) *J. Mol. Catal. A-Chem.*, **209**, 35–39.

118 Yamamoto, Y., Kinpara, K., Saigoku, T., Takagishi, H., Okuda, S., Nishiyama, H., and Itoh, K. (2005) *J. Am. Chem. Soc.*, **127**, 605–613.

119 Cenini, S., Gallo, E., Caselli, A., Ragaini, F., Fantauzzi, S., and Piangiolino, C. (2006) *Coord. Chem. Rev.*, **250**, 1234–1253.

# 19
# Electrophilic Cyclizations

*Félix Rodríguez and Francisco J. Fañanás*

## 19.1
## Introduction

Electrophilic cyclizations are those reactions in which an external electrophilic reagent activates the double (or triple) bond of an alkene (or alkyne) favoring a subsequent intramolecular addition of an internal nucleophile (see Scheme 19.1). These processes result in the formation of cyclic products (carbocycles from carbon-centered nucleophiles and heterocycles from heteroatom-centered nucleophiles). The electrophilic cyclization reactions initially provide compounds with a functionality derived from the original electrophilic reagent which may be further transformed.

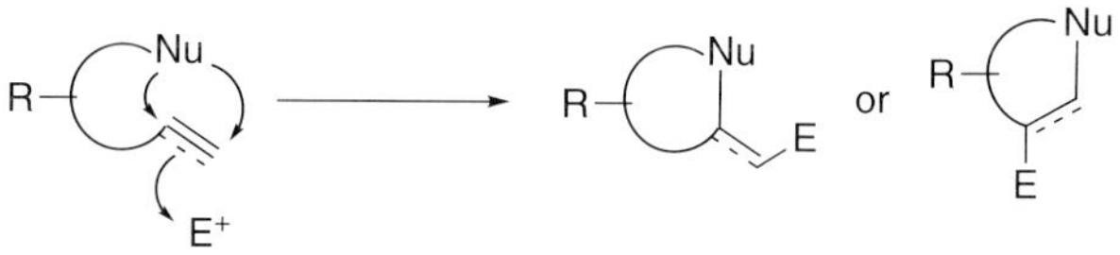

**Scheme 19.1** Concept of the electrophilic cyclization.

The work by Bougault at the turn of the 20th century on the iodocyclization of unsaturated acids may be considered the birth of electrophilic cyclization reactions [1]. Since then, the interest in this type of reaction has undergone incredible growth, especially over the last decades. This growth is exemplified by the number of reviews on this subject that have appeared [2]. In the light of this substantial coverage, it is not the goal of this chapter to present a complete or comprehensive review of electrophilic cyclizations, but rather a selective and illustrative one. The major focus will be given to those more recent applications (mainly from the year 2000 to date) with real potential for organic synthesis. The reader is referred to the earlier reviews for full mechanistic information and details of historic interest.

*Handbook of Cyclization Reactions. Volume 2.*
Edited by Shengming Ma
Copyright © 2010 WILEY-VCH Verlag GmbH & Co. KGaA, Weinheim
ISBN: 978-3-527-32088-2

Numerous electrophiles have been used to accomplish electrophilic cyclizations. Probably, the most common are the halonium ions ($I^+$, $Br^+$, and $Cl^+$) generated from the molecular form or from more complex reagents. Other electrophiles frequently used are selenium and several metallic salts. In this context, the use of metallic complexes as catalysts in electrophilic cyclizations is nowadays one of the most important research fields in organic chemistry (see Chapter 10 in this book). Although other electrophiles such as sulphur or a simple proton can be used to activate the double or triple bond, in this review we will focus on halocyclizations and selenocyclizations.

## 19.2
## Mechanism, Regio- and Stereoselectivity

The general mechanism for the electrophile-initiated cyclofunctionalization is shown in Scheme 19.2. The process is initiated by reaction of the π-system with the electrophile to generate an activated intermediate. Although several structures may be proposed for this intermediate, most cyclization reactions are best rationalized by considering the cyclic "onium" ion **A** or the π-complex **B**. Both intermediates **A** and **B** present two reactive centers for the subsequent intramolecular nucleophilic attack leading to *exo*- or *endo*-products.

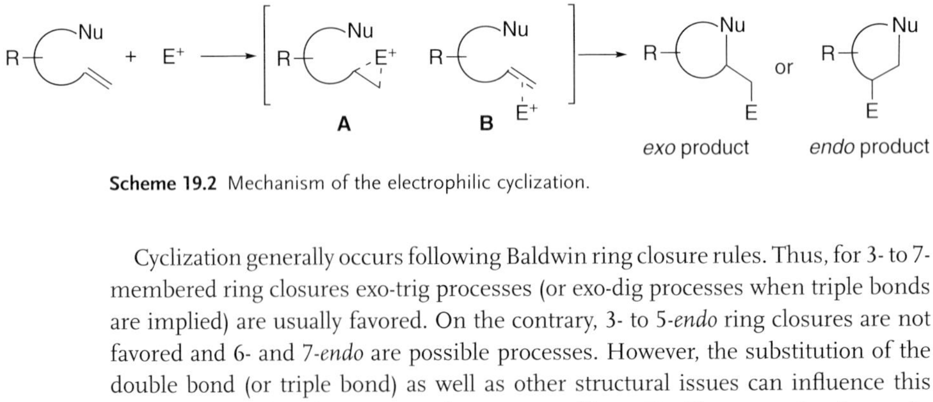

**Scheme 19.2** Mechanism of the electrophilic cyclization.

Cyclization generally occurs following Baldwin ring closure rules. Thus, for 3- to 7-membered ring closures exo-trig processes (or exo-dig processes when triple bonds are implied) are usually favored. On the contrary, 3- to 5-*endo* ring closures are not favored and 6- and 7-*endo* are possible processes. However, the substitution of the double bond (or triple bond) as well as other structural issues can influence this regiochemistry by electronic and/or steric effects [3]. The example shown in Scheme 19.3 illustrates the difficulties in obtaining a single regioisomer with some substrates [4].

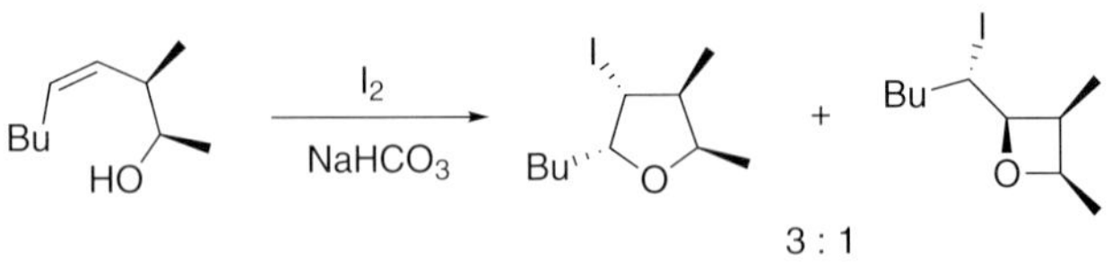

**Scheme 19.3** Mixture of regioisomers obtained in an electrophilic cyclization.

Another interesting feature of the electrophilic cyclization reactions, regarding the steresoselectivity of the processes, is that most of them proceed through a stereospecific anti addition across the π-bond. So, if a stereospecific anti addition is assumed, when the starting material contains stereogenic centers and the cyclization process generates a new stereogenic center, only two diastereoisomers of the final product can be formed. However, depending on the structure of the substrate and/or the reaction conditions, a single diastereoisomer of the final cyclic product may be obtained [5]. Illustrative examples are shown in Scheme 19.4 [6].

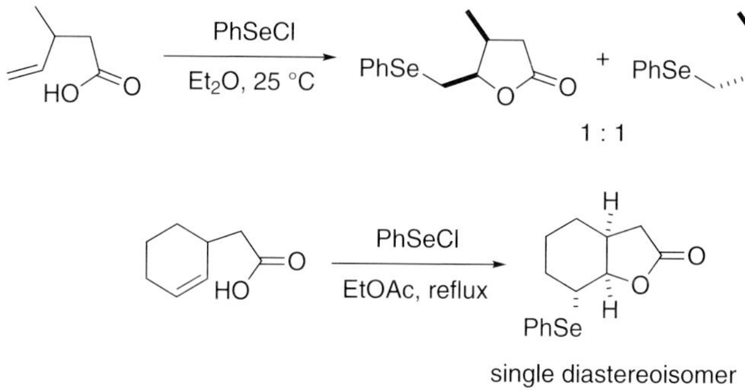


**Scheme 19.4** Stereoselectivity in electrophilic cyclization reactions.

## 19.3
## Halocyclizations

Due to the huge number of works reported in the literature dealing with halocyclization reactions, it is difficult to cover all of them and the reader is referred to early reviews for comprehensive reports [2]. In our opinion, nowadays the most important research lines in this field are the development of highly stereoselective processes and their application in the synthesis of relevant or natural products. Hence, in this chapter we will focus on these issues.

As shown in Scheme 19.5, halocyclization reactions may be performed with iodonium, bromonium, and chloronium ions [7]. Not only iodine but also electrophilic iodine reagents such as ICl, *N*-iodosuccinimide, bis(collidine)iodine(I) perchlorate and Barluenga's reagent $IPy_2BF_4$ [8] have been successfully used in iodocyclization reactions. Bromocyclizations have been mainly carried out by using

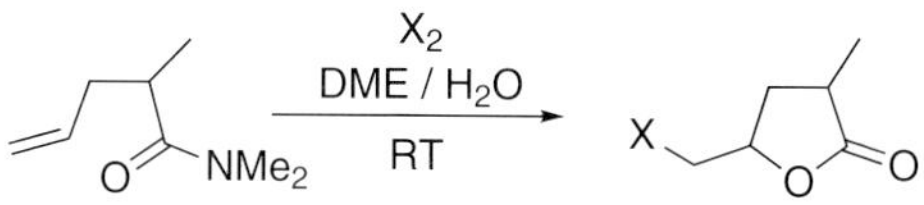


X = I, 85% (*cis* / *trans*: 90 / 10)
X = Br, 75% (*cis* / *trans*: 99 / 1)
X = Cl, 60% (*cis* / *trans*: 60 / 40)

**Scheme 19.5** Halocyclization reactions with iodine, bromine or chlorine.

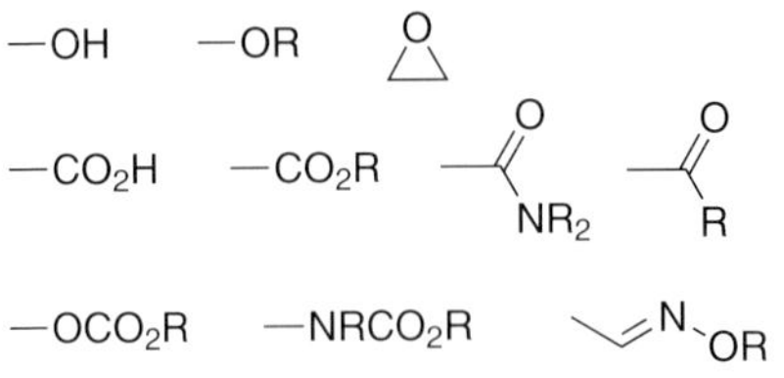

**Figure 19.1** Functional groups used as oxygen-centered nucleophiles.

bromine and N-bromosuccinimide as electrophiles and chlorine is the most useful reagent for chlorocyclization reactions.

### 19.3.1
### Oxygen-Centered Nucleophiles

Halocyclization reactions involving oxygen nucleophiles have been studied more extensively than reactions involving any other nucleophiles. Some examples of functional groups used as oxygen nucleophiles that will be discussed in the next sections are shown in Figure 19.1.

#### 19.3.1.1 Cyclization of Alcohols and Ethers

**Intramolecular Addition to Alkenes** The synthesis of cyclic ethers by the use of the hydroxy functional group as nucleophile is, probably, the most simple and widely reported halocyclization reaction.

Recent efforts in this area are focused on the development of stereoselective processes. The stereoselectivity in these cyclization reactions is dictated by the facial differentiation of the olefinic double bond. The stereoselective halocyclizations of hydroxy-alkenes have been mostly limited to substrate-controlled reactions where the pre-existing stereogenic centers control the chirality of the newly formed stereocenters. Thus, for example, fluorine has been shown to be a highly efficient *syn*-stereodirecting group in iodocyclization processes when alcohols containing a fluorine atom at the allylic position are used as starting materials [9] (Scheme 19.6).

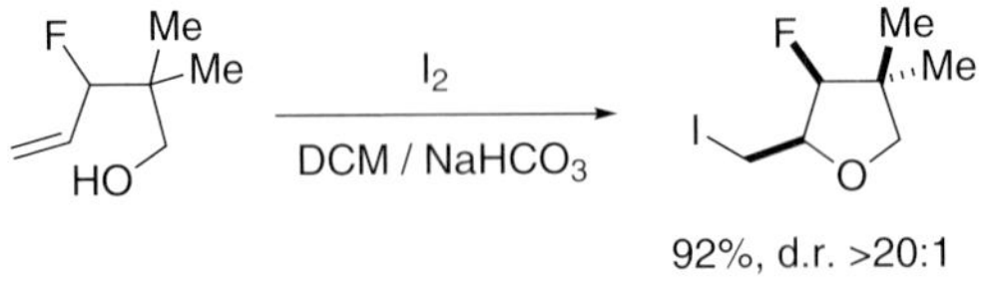

**Scheme 19.6** Fluorine-directed diastereoselective iodocyclization.

Another illustrative example of a substrate-controlled halocyclization reaction is shown in Scheme 19.7. In this reaction, a complex cyclic terpene-derived compound is obtained as a single diastereoisomer (single enantiomer) by using the reagent $IPy_2BF_4$ as the iodonium source [10].

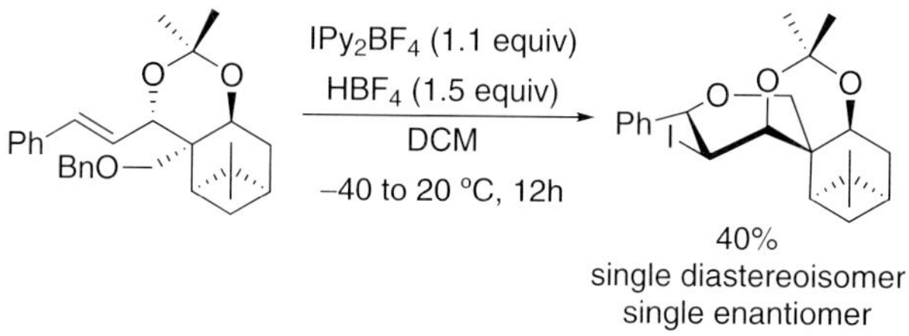

**Scheme 19.7** Diastereoselective iodocyclization promoted by IPy$_2$BF$_4$.

A substrate-controlled iodocyclization reaction has been applied as the key step in the total synthesis of (+)-lasonolide A [11]. The synthesis of the required cis-2,6-disubstituted tetrahydropyran was achieved by treatment of the alkenol precursor (91% ee) with I$_2$ in the presence of K$_2$CO$_3$. Under these conditions a 27 : 1 mixture of cis:trans isomers was obtained (Scheme 19.8).

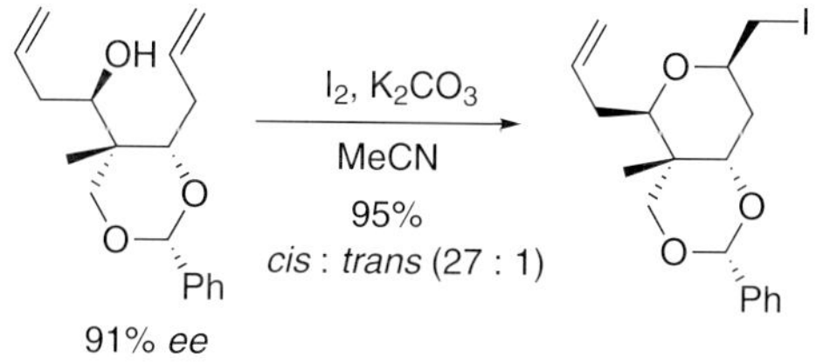

**Scheme 19.8** Diastereoselective synthesis of a *cis*-2,6-disubstituted tetrahydropyran required in the total synthesis of (+)-lasonolide A.

Considering the high levels of diastereocontrol in the halocyclization reactions when starting from compounds containing chiral centers, an interesting application of these processes is the desymmetrization of cyclohexa-1,4-dienes [12]. An example is shown in Scheme 19.9.

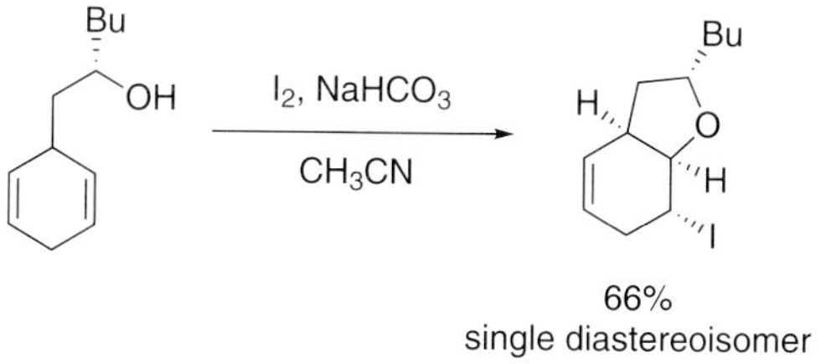

**Scheme 19.9** Desymmetrization of a cyclohexa-1,4-diene derivative.

As we have seen in previous examples, the substrate-controlled halocyclization reaction allows the synthesis of diastereoisomerically pure cyclic ethers. Obviously, following this strategy, enantiomerically pure compounds may be available from enantiomerically pure starting materials. Much more appealing is the enantioselective halocyclization reaction of achiral substrates. To achieve this end, reagent-controlled versions of the halocyclization reaction are required. In this context, some salen-metal

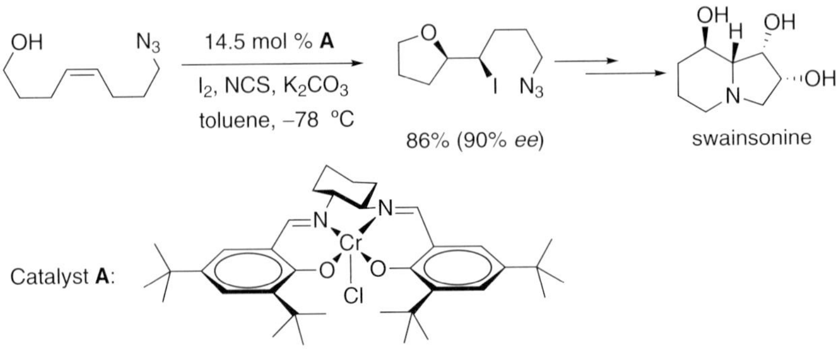

**Scheme 19.10** Asymmetric iodocyclization, synthetic application to swainsonine.

complexes have been reported as efficient catalysts for the iodocyclization of γ-hydroxy-*cis*-alkenes [13] (Scheme 19.10). The synthetic value of this reaction has been demonstrated in the synthesis of the indolizidine alkaloid swainsonine.

The halocyclization reaction of alkenol derivatives has been used extensively in synthesis and many biologically important compounds have been made by applying this reaction as the key step. The mild reaction conditions allow the use of this reaction on structurally complex substrates. For example, as shown in Scheme 19.11, a NIS-mediated iodoetherification reaction has been used in the assemblage of the G-ring of azaspiracid-1 [14].

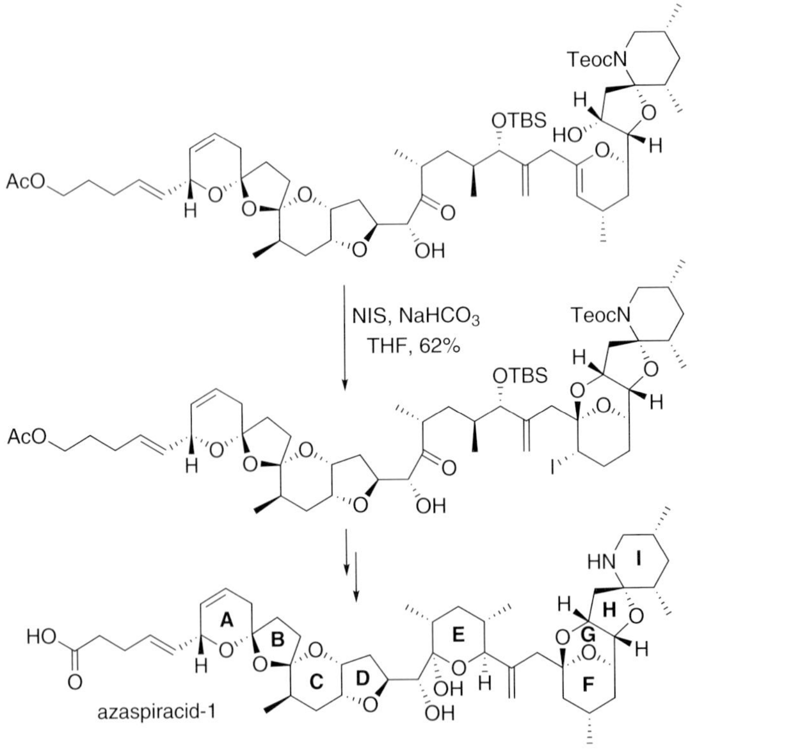

**Scheme 19.11** Application of the iodocyclization reaction in the total synthesis of azaspiracid-1.

**Intramolecular Addition to Alkynes and Allenes** A variety of oxygen-containing heterocycles have been obtained by the halocyclization of acetylenic alcohols. The reaction of 3-alkyne-1,2-diols with $I_2$ in the presence of $NaHCO_3$ provides a very convenient synthesis of 3-iodofurans [15] (Scheme 19.12). The reaction proceeds through an initial electrophilic cyclization followed by a dehydration reaction. Other important structures such as benzo[b]furans [16] and chromenes [17], that are prevalent in a wide variety of biologically active natural and unnatural compounds, have been synthesized from methoxyether derivatives through iodocyclization reactions (Scheme 19.12). In these cases the oxygen of the methoxy group acts as the nucleophile. After nucleophilic attack on the activated iodonium intermediate, a facile removal of the methyl group via $S_N2$ displacement by the chloride (or iodide) anion present in the reaction medium leads to the final products. Interestingly, all these iodine-containing products can be further elaborated to a wide range of compounds using subsequent palladium-catalyzed processes.

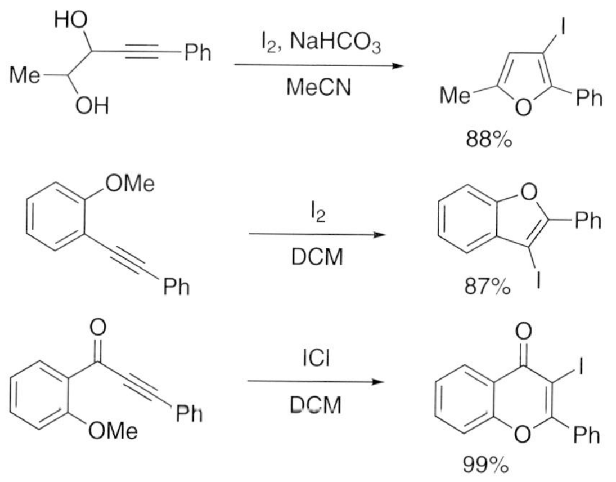

**Scheme 19.12** Synthesis of furans, benzo[b]furans and chromenes by iodocyclization from alkynol derivatives.

Chiral γ-substituted allenamides have been shown to undergo efficient N-iodosuccinimide-mediated cyclizations to give highly functionalized dihydrofurans [18]. These reactions proceed rapidly and without loss of stereochemistry (Scheme 19.13). The vinyl iodide functionality present in the final dihydrofuran derivatives allows a range of functionalization by metal-catalyzed coupling chemistry. In the example in Scheme 19.13, a typical Suzuki coupling is shown.

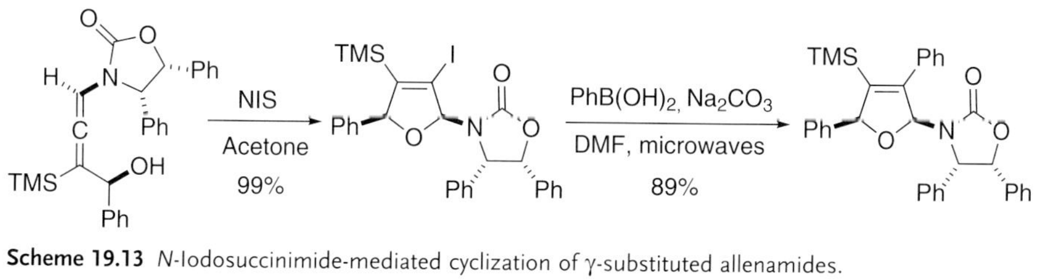

**Scheme 19.13** N-Iodosuccinimide-mediated cyclization of γ-substituted allenamides.

**Intramolecular Addition to Enyne Derivatives** The bromoallene motif occurs in a variety of important natural products. An interesting strategy to access this structure is the biomimetic-type bromoetherification reaction of an enyne derivative. This reaction has recently been applied to the synthesis of (−)-kumausallene [19] and (+)-laurallene [20]. As shown in Scheme 19.14, 2,4,4,6-tetrabromocyclohexanedione (TBCD) is used as the bromonium source. The main drawback of these processes is that mixtures of two isomeric bromoallenes are obtained and a final separation step is required.

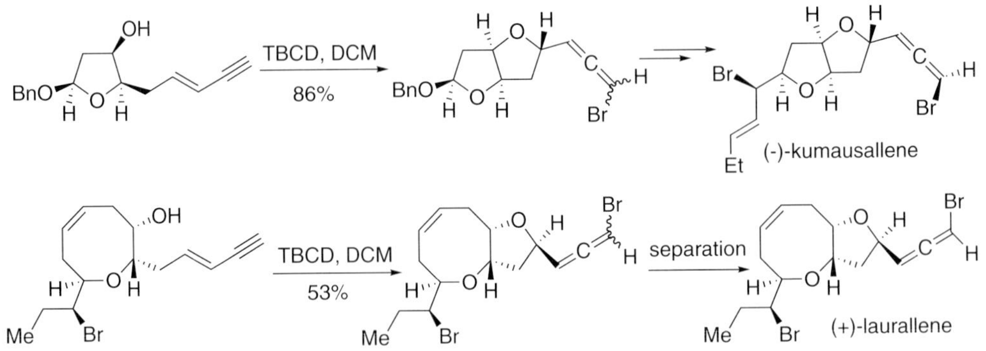

**Scheme 19.14** Bromoallenes by means of bromoetherification reactions of enynes.

### 19.3.1.2 Cyclization of Carboxylic Acids and Carboxylic Acid Derivatives

In this section we will focus on halocyclization reactions where the nucleophile is a carboxylic acid or a carboxylic acid derivative (esters or amides). The final products of these processes are lactones, in all cases (Scheme 19.15). As shown, these reactions proceed through coordination of the electrophile to the double bond to form the expected cyclic "onium" intermediate, followed by an intramolecular nucleophilic attack of the carbonylic oxygen atom leading to a new "onium" intermediate that further evolves, in general through a hydrolysis step, to give the final lactone product. When amides are used as starting materials, a competitive nucleophilic attack of the nitrogen atom, to give lactam derivatives, is sometimes observed.

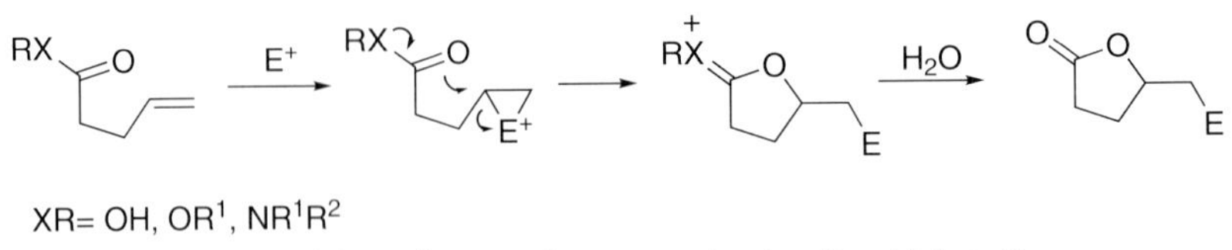

XR= OH, OR$^1$, NR$^1$R$^2$

**Scheme 19.15** Electrophilic cyclization of unsaturated carboxylic acid derivatives.

**Intramolecular Addition to Alkenes** Halolactonization reactions of unsaturated carboxylic acid derivatives have received sustained attention and have become a very reliable strategy in synthetic design. As in the case of the haloetherification reactions discussed previously, recent advances in halolactonization processes are related to

## 19.3 Halocyclizations

the development of stereoselective reactions and their application in the synthesis of relevant products. In this section we will focus on these issues.

Most work on stereocontrolled halolactonization reactions has been in the field of substrate-controlled processes where an existing chiral moiety in the substrate controls the outcome of the reaction. This type of stereochemical control has been systematically studied and several general trends may be inferred [5]. For example, when the starting unsaturated acid derivative is substituted at the allylic position (1,2-asymmetric induction), 1,2-*trans*-configured products are usually obtained. Thus, 3-methyl-4-pentenoic acid reacts with iodine to give the corresponding *trans*-iodomethyllactone derivatives as the major products [5a] (Scheme 19.16). However, the presence of hydroxy (or alkoxy) substituents [21] or fluorine atoms [9] at the allylic position reverses the configuration to give 1,2-*cis*-products (Scheme 19.16).

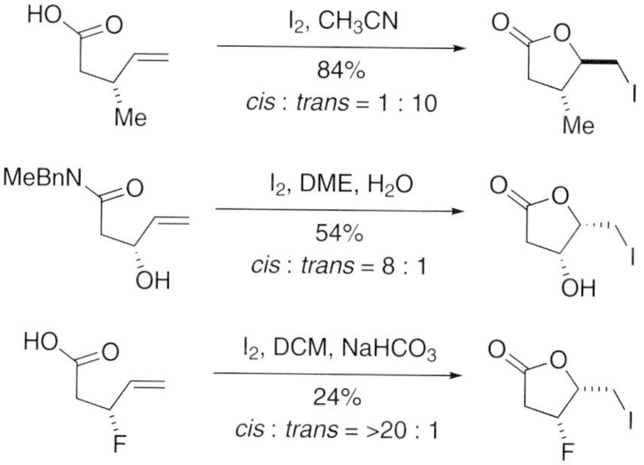

**Scheme 19.16** Diastereoselective iodolactonization: 1,2-asymmetric induction.

While 1,2-asymmetric induction is the most profound in halolactonization reactions, the presence of substituents at other locations different from the allylic position within the unsaturated acid derivative may also induce asymmetry, namely 1,3- and 1,4-asymmetric induction. The pioneering work of Bartlett in this field shows that, in general, 1,3-asymmetric induction produces the 1,3-disubstituted cis-isomer as the major product. However, almost no control is observed by substituents beyond the allylic and homoallylic positions (1,4-asymmetric induction and further) [5f] (Scheme 19.17).

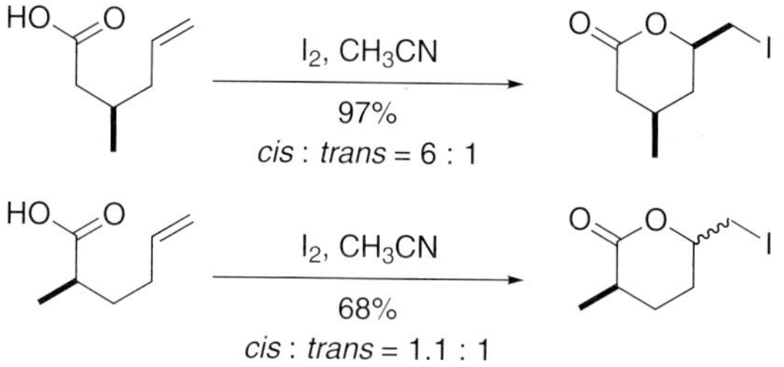

**Scheme 19.17** Diastereoselective iodolactonization; 1,3-, and 1,4-asymmetric induction.

Optically active lactone derivatives are available by using chiral auxiliaries. In this context, stereoselective halolactonization reactions with unsaturated amides derived from chiral amines have largely been reported [22]. However, in general, modest enantiomeric excesses have been reached. A different strategy to achieve enantioselective halolactonization reactions involves the use of chiral halo-reagents (reagent-controlled halolactonization). In order to get a chiral environment around the halonium ion, its coordination to a chiral amine is the strategy most widely applied [23]. However, the enantiomeric excesses reported till now are not high enough for practical asymmetric synthesis. However, further optimized versions of these interesting reagent controlled reactions are expected. Scheme 19.18 shows a comparative study of the effectiveness of a chiral amine used as auxiliary (substrate control) or as a ligand (reagent control). The induction is smaller in the substrate-controlled experiment and, interestingly, the major enantiomer is the opposite of that obtained in the reagent-controlled experiment.

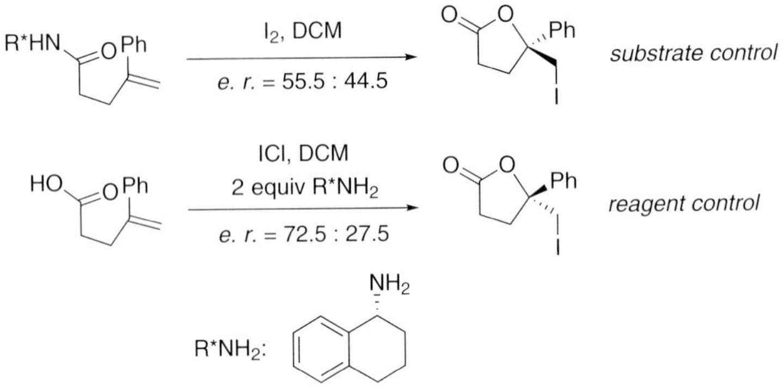

**Scheme 19.18** Enantioselective iodolactonization, substrate and reagent control.

As demonstrated in this section, the halocyclization reaction is a powerful tool in organic synthesis for the construction of lactones and so several natural products have been synthesized using this reaction. Illustrative examples are the synthesis of fumonisin $B_2$ [24] and (−)-cinatrin B [25] shown in Scheme 19.19. In both cases a highly diastereoselective iodolactonization reaction is used as the key step.

**Intramolecular Addition to Alkynes and Allenes** Halolactonization of alkynoic acid derivatives has been much less investigated and employed in synthetic organic chemistry than the halolactonization reaction of alkenoic acids summarized in the previous section [26]. Nevertheless, it has found utility in the preparation of interesting compounds such as α-pyrone and isocoumarin derivatives [27]. Illustrative examples of halolactonization reactions of alkynoic acid derivatives are shown in Scheme 19.20. Thus, the reaction of o-(1-alkynynyl)benzoate or (Z)-2-alken-4-ynoate derivatives with iodine in dichloromethane leads to the expected iodine-containing isocoumarin and α-pyrone derivatives. The vinyl iodide functionality present in the final products allows a range of functionalization by metal-catalyzed coupling chemistry. Typical Sonagashira and Suzuki reactions are shown in Scheme 19.20.

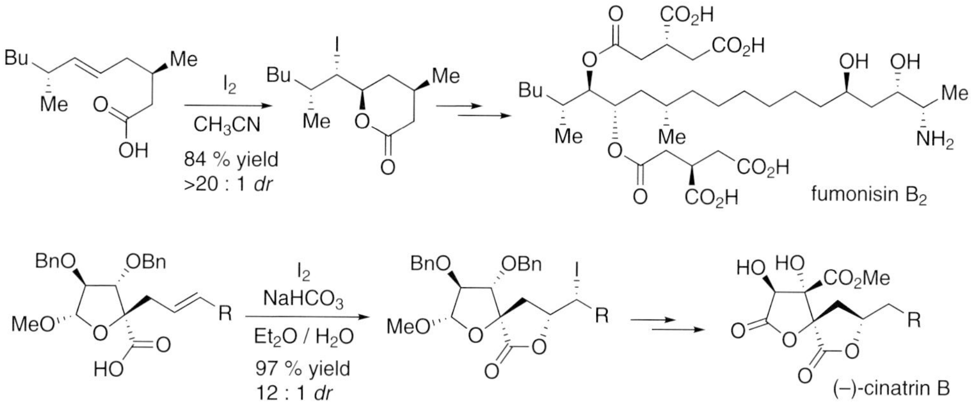

**Scheme 19.19** Application of the iodolactonization reaction in the total synthesis of fumonisin B$_2$ and (−)-cinatrin B.

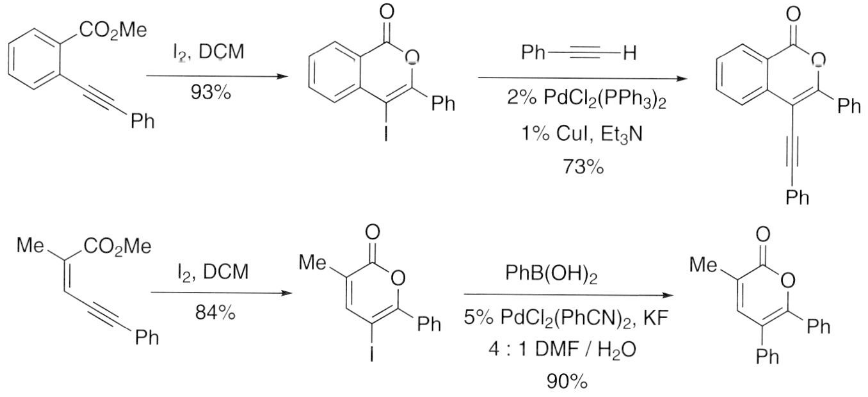

**Scheme 19.20** Synthesis of isocoumarins and α-pyrones through iodolactonization reactions and further functionalization by coupling reactions.

In a related process, phosphaisocoumarins have been prepared under mild conditions by the reaction of *o*-(1-alkynyl)phenylphosphonates with I$_2$ or ICl [28]. The resulting iodides can be further elaborated using palladium-catalyzed coupling reactions.

The regioselective halolactonization reaction of 2,3-allenoic acids or their ester derivatives provides a simple and convenient strategy for the synthesis of interesting 4-halobutenolides [29]. These reactions can be efficiently performed by treating the 2,3-allenoate derivative with iodine in aqueous acetonitrile (Scheme 19.21).

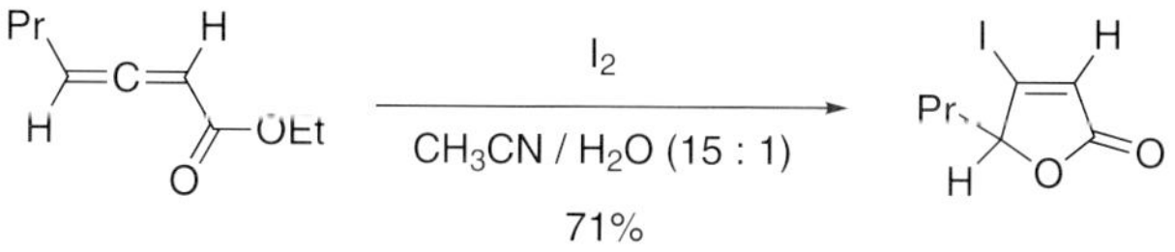

**Scheme 19.21** Iodolactonization reaction of a 2,3-allenoate derivative to give a 4-iodobutenolide.

**Other Cyclizations From Amides** In the previous sections, some examples of the cyclization of amides to give lactone derivatives have been shown. In all these cases, after the cyclization reaction the amine moiety is lost during the work-up of the reaction. However, in those amides where the nitrogen atom is part of the tether linking the carbonyl group and the double bond, the halocyclization reaction proceeds by nucleophilic attack of the carbonylic oxygen to give nitrogen-containing heterocycles. For example, oxazoline derivatives have been obtained by applying this approach [30] (an example is shown in Scheme 19.22). A similar strategy has been used for the synthesis of pyrano[2,3-b]quinolines derivatives [31] (an example is shown in Scheme 19.22). In this case, the nitrogen atom of the initial amide is not part of the tether but a very favorable final aromatization reaction, to form the quinoline ring, keeps the nitrogen in the final product.

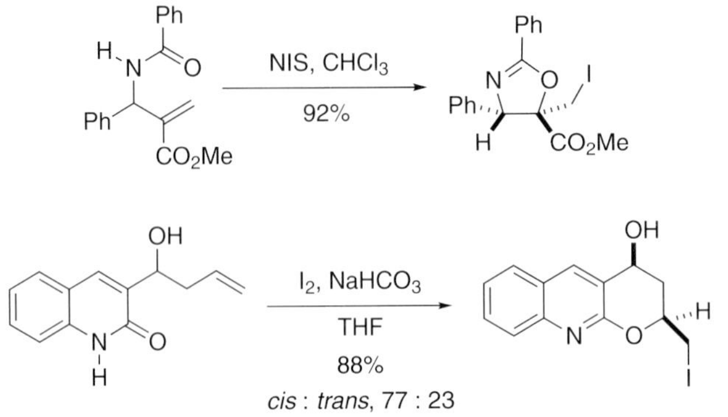

**Scheme 19.22** Oxazoles and pyrano[2,3-b]quinolines from amides by iodocyclization reactions.

### 19.3.1.3 Cyclization of Carbonates and Carbamates

A useful strategy to construct the *syn*-1,3-diol moiety is the iodine-induced carbonate cyclization methodology originally published by Bartlett *et al.* [5e] and later improved by Smith and Duan [32]. A recent example where a homoallylic *tert*-butyl carbonate is reacted with *N*-iodosuccinimide in acetonitrile to give the corresponding iodocarbonate derivative in high yield as a single diastereoisomer is shown in Scheme 19.23 [33].

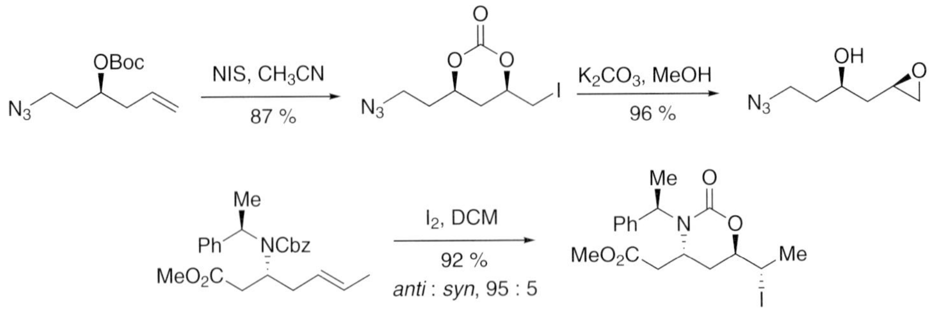

**Scheme 19.23** 1,3-diol and 1,3-aminoalcohol derivatives by iodocyclization reactions.

Further deprotection of the new carbonate moiety leads to a *syn*-epoxy alcohol used in the total synthesis of the atorvastatin side-chain. The iodocyclization reaction proceeds through nucleophilic attack of the carbonate carbonyl group on the cyclic iodonium intermediate and subsequent *tert*-butyl fragmentation, likely by releasing isobutene and a proton.

Similarly, the 1,3-aminoalcohol moiety is available from homoallylic *tert*-butyl carbamates or from the corresponding *N*-Cbz substrates. For example, this strategy was used in the total synthesis of sperabillins B and D [34]. As shown in Scheme 19.23, in this case the iodocyclization reaction proceeds in high yield to give the *anti*-1,3-aminoalcohol derivative as the major isomer.

### 19.3.1.4 Cyclization of Aldehydes and Ketones

The reaction of *o*-alkynylbenzaldehydes or ketones with different alcohols or carbon-based nucleophiles in the presence of the reagent $IPy_2BF_4$ gives functionalized 4-iodo-1*H*-isochromenes in a regioselective manner. When alkynes and alkenes are used as nucleophiles, regioselective benzannulation reactions take place to form 1-iodonaphthalenes and 1-naphthyl ketones respectively [35]. Representative examples of these processes are shown in Scheme 19.24.

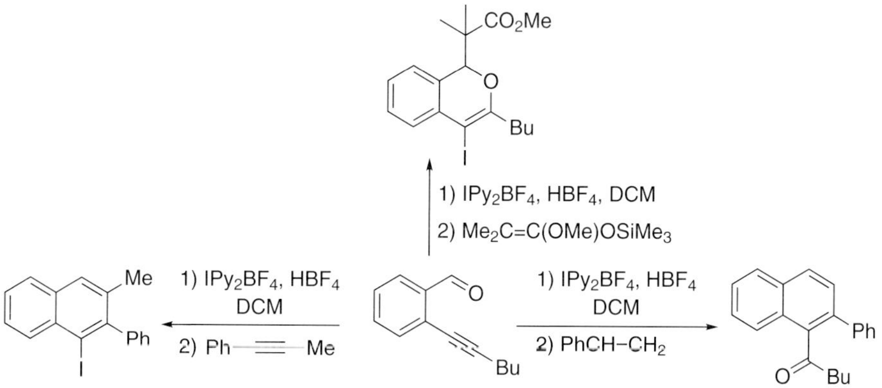

**Scheme 19.24** Synthesis of 4-iodo-1*H*-isochromenes, 1-iodonaphtalenes and 1-naphthyl ketones from *o*-alkynylbenzaldehydes by iodocyclization reactions promoted by $IPy_2BF_4$.

These reactions involve an initial interaction of the iodonium ion with the alkyne, assisted by the neighboring carbonyl group that affords the key benzo[c]pyrilium cation. Nucleophilic attack on this reactive species gives the resulting 1*H*-isochromene derivatives. More complex mechanisms, involving the participation of polycyclic intermediates, are proposed for the reaction with alkynes and alkenes as nucleophiles (Scheme 19.25).

Furthermore, this chemistry was implemented for 3-alkynylpyrrole-2-carboxaldehydes and related heterocyclic systems that eventually lead to substituted indole, benzofuran and benzothiophene skeletons and to other heterocyclic derivatives [36].

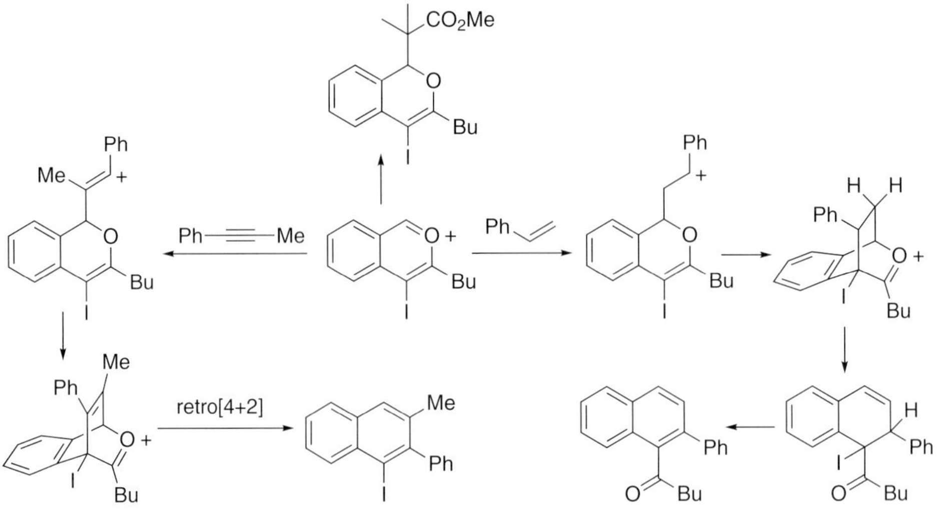

**Scheme 19.25** Proposed reaction pathways accounting for the formation of 4-iodo-1H-isochromenes, 1-iodonaphtalenes and 1-naphthyl ketones from a common isobenzopyrilium cation.

An interesting synthesis of 4-iodo-3-furanones from 2-alkynyl-2-siloxy carbonyl compounds promoted by iodonium ions has been developed recently [37]. The process involves a tandem sequence initiated by interaction of the iodonium ion with the alkyne, assisted by the neighboring carbonyl group, to give an oxonium ion intermediate. Subsequent 1,2-shift gives the final product through a formal α-ketol rearrangement (Scheme 19.26).

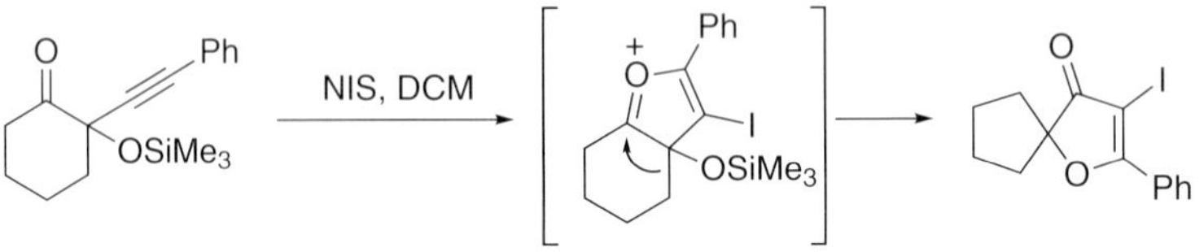

**Scheme 19.26** Synthesis of 4-iodo-3-furanones from 2-alkynyl-2-siloxy carbonyl compounds promoted by iodonium ions.

### 19.3.1.5 Cyclizations with Other Oxygen-Centered Nucleophiles

The electrophilic cyclization of propargylic oxirane compounds with iodine offers an efficient and straightforward route to substituted iodofurans [38]. The presence of the iodide functional group on the furan ring provides an opportunity for further functionalization by conventional palladium-catalyzed cross-coupling reactions. The iodocyclization reaction proceeds through attack of the oxirane oxygen on the iodonium intermediate followed by elimination of a proton (Scheme 19.27).

4-Iodoisoxazoles are easily available by reaction of Z-O-methyl oximes of 2-alkyn-1-one derivatives in the presence of ICl [39]. Further functionalization of the

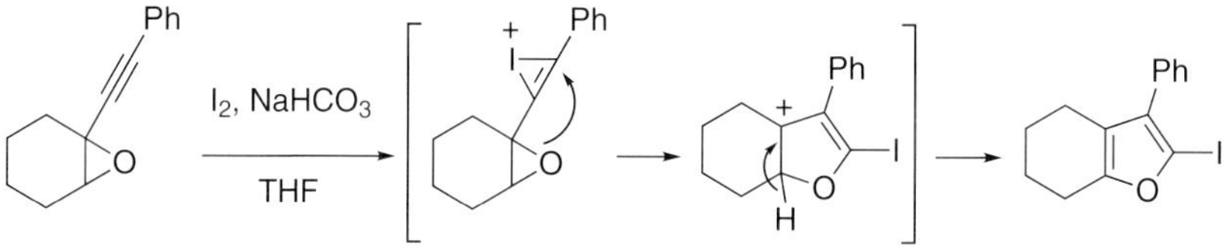

**Scheme 19.27** Iodonium-induced cyclization of propargylic oxiranes.

resulting iodine-containing products by palladium-catalyzed reactions yields 3,4,5-trisubstituted isoxazoles, including valdecoxib, a highly potent COX-2 inhibitor (Scheme 19.28).

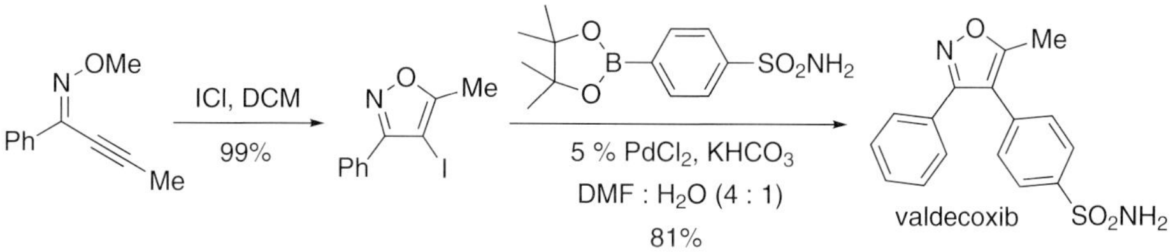

**Scheme 19.28** Synthesis of 4-iodoisoxazoles by iodocyclization reactions, synthesis of valdecoxib.

### 19.3.2
### Nitrogen-Centered Nucleophiles

The halocyclization reaction of unsaturated compounds involving nitrogen nucleophiles is an important strategy to access nitrogen heterocycles. The most useful functional groups used as nitrogen nucleophiles are shown in Figure 19.2.

#### 19.3.2.1 Cyclization of Amines

**Intramolecular Addition to Alkenes** The intramolecular haloamination of double bonds has been widely used to access many interesting nitrogen-containing heterocycles [40]. As in the haloetherification reaction, the most important issue in haloamination reactions is the development of regio- and stereo-selective processes. The substrate-controlled stereoselective reactions represent by far the major route used till now to introduce new stereogenic centers. Thus, many examples of haloamination reactions of chiral substrates or substrates attached to a chiral auxiliary have

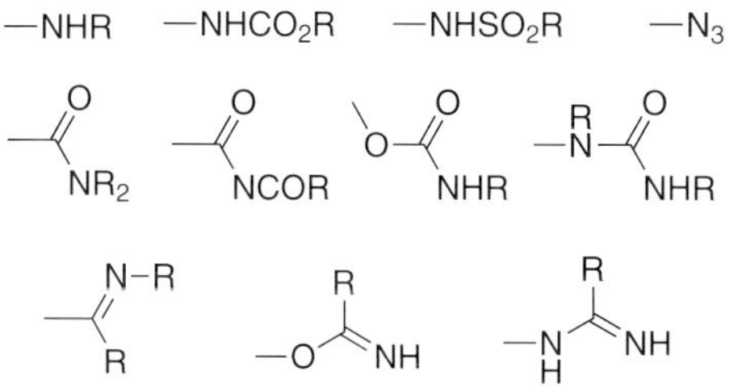

**Figure 19.2** Functional groups used as nitrogen-centered nucleophiles.

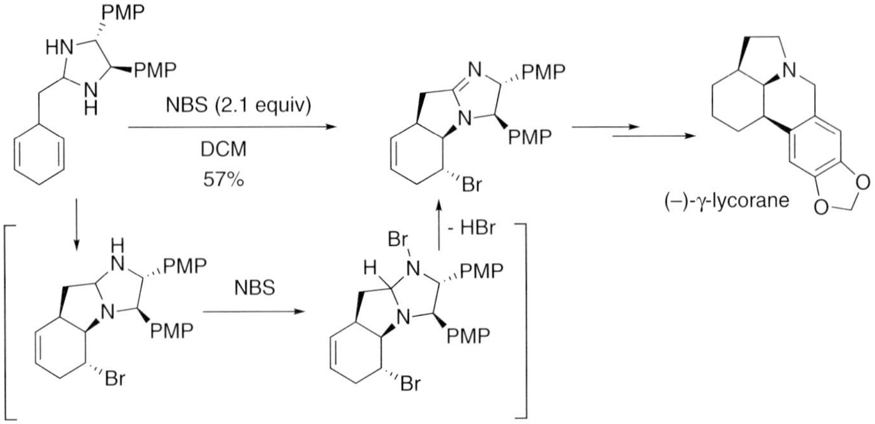

**Scheme 19.29** Desymmetrization of two olefins in 1,4-cyclohexadiene by asymmetric bromoamination.

been published. A representative example is shown in Scheme 19.29. In this process an asymmetric desymmetrization of two olefins in 1,4-cyclohexadiene using chiral 1,2-diamines is achieved [41]. Interestingly, this reaction proceeds through a conventional bromoamination reaction followed by *in situ* oxidation of the aminal unit. The method has been applied to a concise synthesis of (−)-γ-lycorane.

The intramolecular haloamination of double bonds has also been applied as a key step in the total synthesis of the alkaloid (+)-lyconadin A [42]. As shown in Scheme 19.30, the iodoamination reaction provides the tetracyclic skeleton of the natural product in very high yield and as a single isomer.

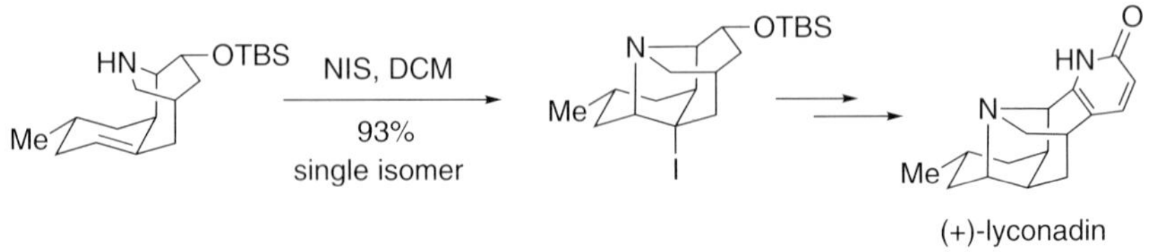

**Scheme 19.30** Application of the iodoamination reaction in the total synthesis of (+)-lyconadin A.

**Intramolecular Addition to Alkynes** The iodocyclization of o-alkynylanilines or some of their *N*-protected forms with the reagent IPy$_2$BF$_4$ affords 3-iodoindole derivatives [43] (Scheme 19.31). The versatile iodine functionality can be transformed using organometallic chemistry to synthesize more complex molecules. Moreover, this reaction can also be applied to the solid-phase synthesis of indoles.

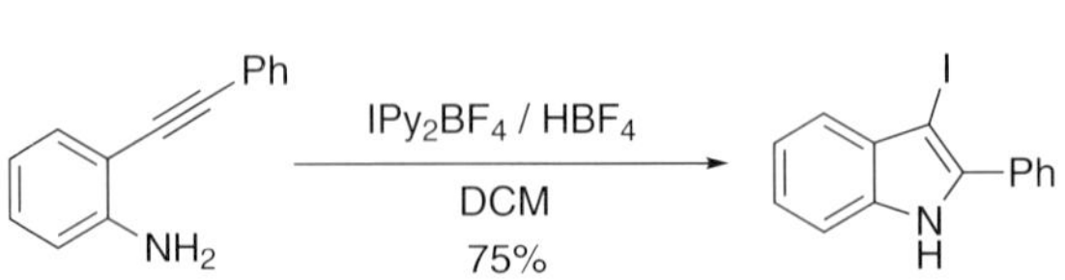

**Scheme 19.31** Iodocyclization of anilines to give 3-iodoindoles by reaction with IPy$_2$BF$_4$.

### 19.3.2.2 Cyclization of Amides

In general, the halocyclization of unsaturated amides leads to the corresponding lactone derivatives by nucleophilic attack of the carbonylic oxygen on the halonium intermediate, followed by a hydrolysis step (see Section 19.3.1.2). However, in some cases, nucleophilic attack of the amide nitrogen is observed and so lactams are obtained [44]. In order to get lactams instead of lactones it is necessary to increase the nucleophilicity of the nitrogen atom. Two different ways have been used to achieve this. The first involves the introduction of an electron-withdrawing group on the nitrogen atom that enhances the amide proton acidity. In this context, carbamates, N-acyl amides, sulfonamides, and so on, in the presence of a base, have been used [45]. The second is the modification of the nitrogen hybridization by conversion of the amide function into an imidate, since the nonbonding nitrogen electrons become more nucleophilic [46]. Thioimidates have also been used as amide equivalents [47]. Illustrative examples of these two strategies are shown in Scheme 19.32. Thus, an asymmetric iodolactamization induced by a chiral oxazolidine auxiliary is presented [48]. The other example shows the synthesis of a chiral γ-lactam from 3(S)-hydroxy-4-pentenamide which served as the intermediate in the synthesis of (−)-slaframine [49].

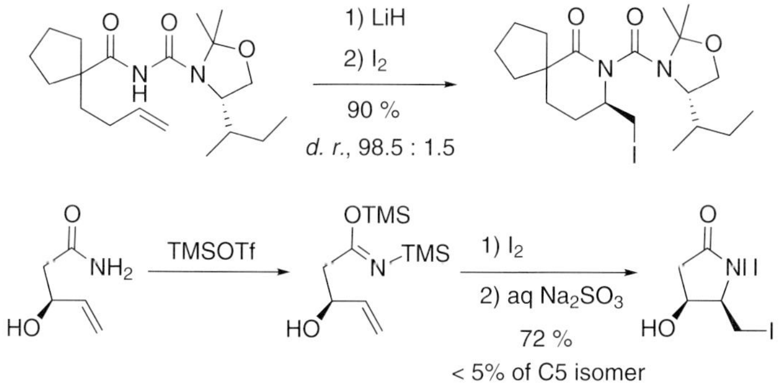

**Scheme 19.32** Two different strategies to access lactams from amides through halolactamization reactions.

### 19.3.2.3 Cyclization of Carbamates, Amidines, Ureas and Isoureas

The halocyclization of carbamate derivatives usually proceeds through nucleophilic attack of the carbonyl oxygen on the activated halonium intermediate to give finally the corresponding diol derivatives (see Section 19.3.1.3). However, nitrogen attack becomes predominant or exclusive by introduction of an electron-withdrawing group (for example a tosyl group) on the nitrogen atom [50] (Scheme 19.33). The regiochemistry of the reaction can also be controlled by the appropriate choice of basic metallic reagent [45d].

The halocyclization of urea derivatives leads to the formation of the corresponding diamine derivative, as shown in the example of Scheme 19.33 [51]. As before, the base used in this process may direct the reaction to O-cyclized or N-cyclized products [45d].

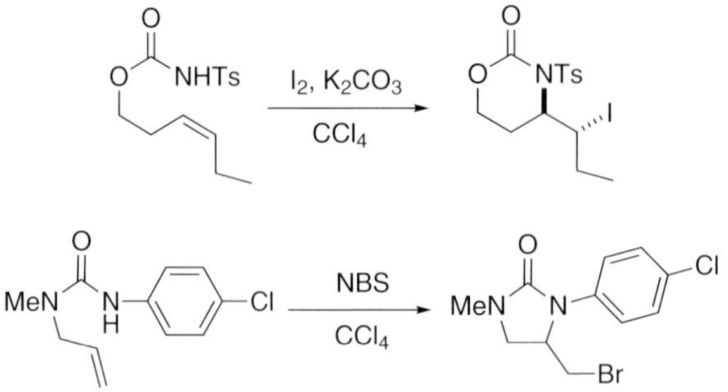

**Scheme 19.33** Halocyclization of carbamates and ureas.

Other nitrogen-containing functional groups that provide a nitrogen nucleophile in halocyclization reactions are acetimidates [52], amidines [53], isoureas [54] and isothioureas [55]. Representative examples are shown in Scheme 19.34.

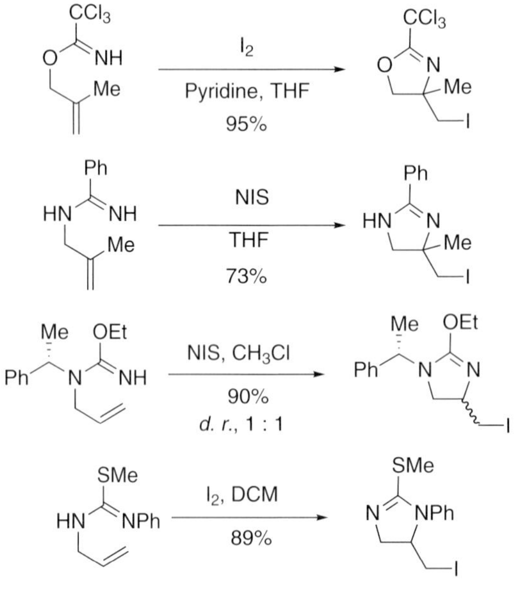

**Scheme 19.34** Halocyclization of acetimidates, amidines, isoureas and isothioureas.

### 19.3.2.4 Cyclization of Imines

*tert*-Butylimines of *o*-(1-alkynyl)benzaldehydes react under mild conditions in the presence of $I_2$ or ICl to give isoquinoline derivatives [56]. The reaction proceeds through nucleophilic attack of the imine nitrogen on the iodonium intermediate to give an isoquinolinium salt that further evolves to produce the final isoquinoline derivatives and a *tert*-butyl cation, which generates isobutylene (Scheme 19.35).

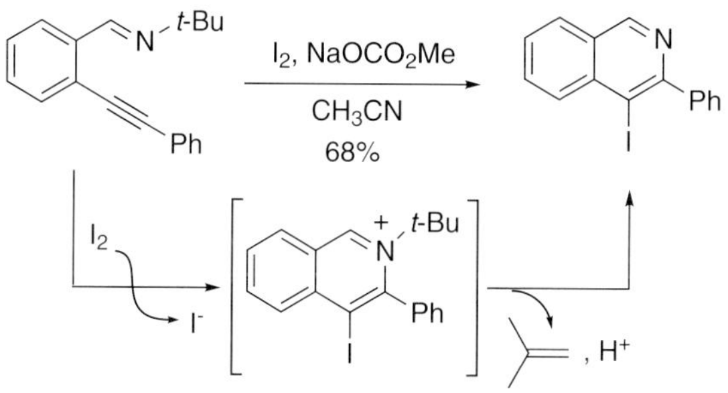

Scheme 19.35 Iodocyclization of iminoalkynes to give isoquinolines.

### 19.3.2.5 Cyclization of Azides

Isoquinoline derivatives are also available from 2-alkynyl benzyl azides by reaction with iodine [57]. The reaction proceeds by initial coordination of the iodonium ion to the triple bond followed by nucleophilic attack of the azide and subsequent elimination of nitrogen and a proton (Scheme 19.36).

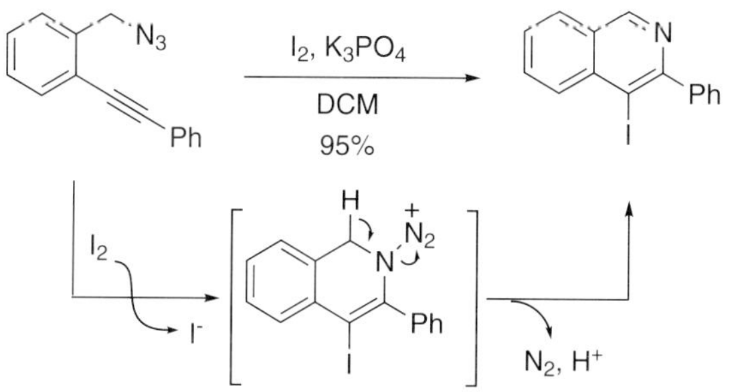

Scheme 19.36 Iodocyclization of 2-alkynyl benzyl azides to give isoquinolines.

### 19.3.3
### Sulfur- and Selenium-Centered Nucleophiles

3-Halobenzo[b]thiophenes have been synthesized from o-ethynylphenyl sulfides by halocyclization reactions using iodine, bromine or N-bromosuccinimide as electrophiles [58]. The resulting halothiophenes can be further elaborated using palladium-catalyzed coupling and/or metallation techniques. Interestingly, this strategy has been used for the synthesis of tubulin binding agents. The analogous benzo[b]selenophenes have been prepared from the corresponding o-ethynylphenyl selenides by applying the same approach [59]. Representative examples of both processes are shown in Scheme 19.37. These reactions proceed, as expected, by initial coordination of the iodonium or bromonium ion to the triple bond followed by nucleophilic attack of sulfur or selenium to give a cationic intermediate that undergoes facile removal of the alkyl group (benzyl or methyl groups in Scheme 19.37) via an $S_N2$ displacement by the counterion iodide or bromide generated *in situ* during the cyclization.

Iodine also induces the cyclization of sulfur-containing acetylenic alcohols to either ketone- or vinylic iodide-containing benzothiophenes [60].

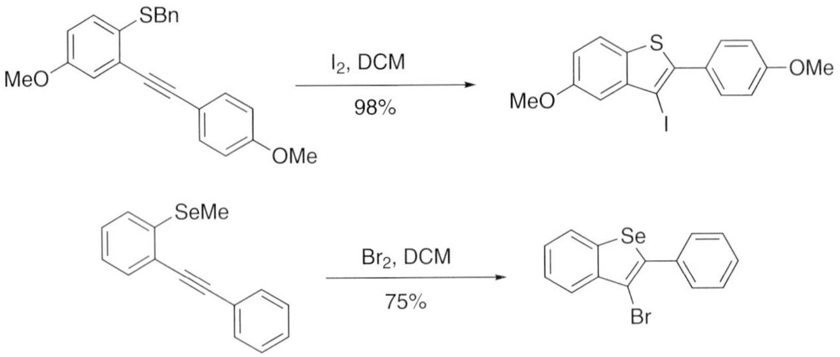

**Scheme 19.37** Halocyclization of *o*-ethynylphenyl sulfides or selenides to give benzo[*b*]thiophenes or benzo[*b*]selenophenes.

## 19.3.4
## Carbon-Centered Nucleophiles

The development of new strategies to make carbon–carbon bonds is without any doubt one of the main challenges in organic synthesis. In this context the halocarbocyclization of alkenyl or alkynyl groups is an attractive option. In contrast to the well-known halocyclization of oxygen- and nitrogen-centered nucleophiles, summarized in previous sections, the reaction of substrates with a carbon atom as the intramolecular nucleophile is less common. The most useful functional groups acting as carbon nucleophiles are active methine groups, aryl groups and alkynes. Hence, in the following sections we will focus on the reactivity of compounds containing these carbon-centered nucleophiles.

### 19.3.4.1 Cyclization of Compounds Containing Active Methine Groups

The iodocarbocyclization reaction of malonate derivatives having unactivated alkenyl groups proceeds in a highly stereospecific manner through the *trans*-addition of iodine and the malonate enolate to the carbon–carbon double bond in the presence of titanium alkoxides [61]. In this reaction, iodine is an appropriate electrophile to activate the alkene, while the titanium complex acts as a basic reagent to enhance the nucleophilicity of the malonate moiety through the formation of a titanium enolate. The asymmetric version of this reaction has been successfully developed by using the chiral titanium alkoxide Ti(TADDOLate)$_2$ [62] (Scheme 19.38). Furthermore, this iodocarbocyclization reaction is not limited to malonate derivatives and other active methine compounds have provided excellent results when performing the process in the presence of titanium tetrachloride, triethylamine and iodine [63].

The above commented iodocarbocyclization reaction can also be applied to malonate derivatives having a carbon–carbon triple bond. This strategy is of practical interest for assembly of the (*E*)-2-iodomethylenecyclopentane scaffold [64]. The scope of this cyclization was expanded showing that an alternative carbotitanation of the alkyne followed by iodolysis of the resulting carbon–titanium bond renders the (Z)-isomers [65]. In any case, these reactions proceed through a 5-*exo-dig* cyclization mode.

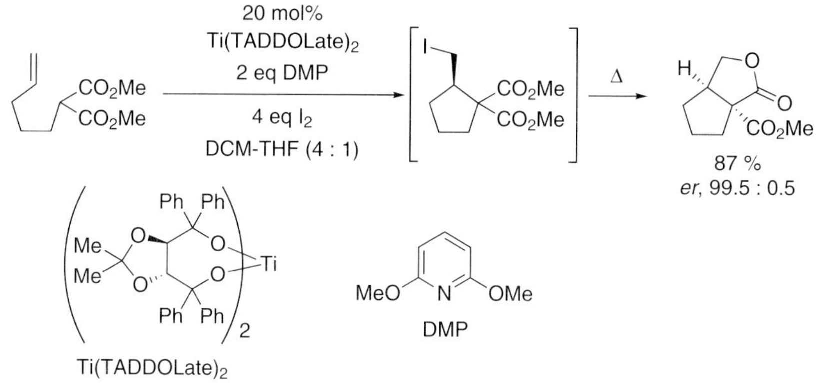

**Scheme 19.38** Catalytic asymmetric iodocarbocyclization by means of titanium enolates.

Iodocarbocyclization reactions through 6-*endo-dig* processes have been observed in the reaction of 2-(2-ethynyl)benzylmalonate derivatives [66]. The more elusive 5-*endo-dig* mode has been recently accomplished [67] (Scheme 19.39). In this work, cyclizations involving terminal and substituted (alkyl, aryl, Br, I) alkynes were accessed.

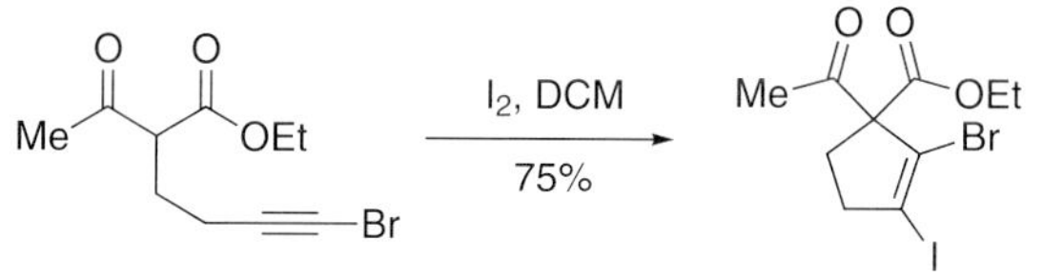

**Scheme 19.39** Iodocarbocyclization reaction of an alkyne derivative through a 5-endo-dig process.

### 19.3.4.2 Alkynes as Nucleophiles

A single example of an interesting cyclization of an alkyne onto another alkyne mediated by IPy$_2$BF$_4$ has been published [68] (Scheme 19.40). The overall process comprises two different carbon–carbon bond-forming reactions, namely, an alkyne–alkyne coupling and a Friedel–Crafts-like ring closure.

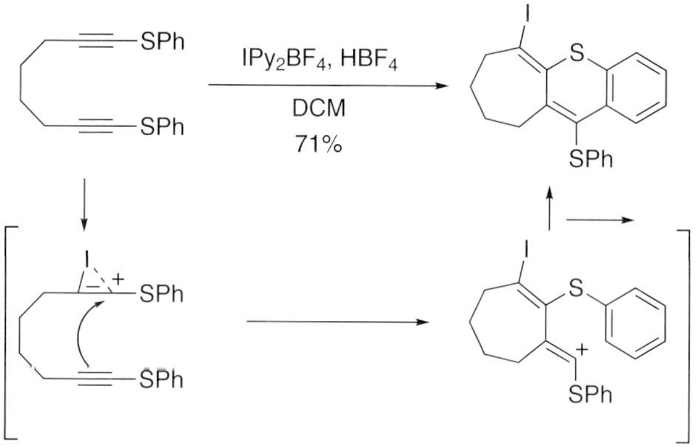

**Scheme 19.40** Intramolecular cyclization of diynyl sulfide derivatives upon reaction with IPy$_2$BF$_4$.

### 19.3.4.3 Aryl Groups as Nucleophiles

The intramolecular iodoarylation reaction of alkenes has been implemented in a fast and flexible access to tetrahydronaphthalene derivatives [69] and the heterocyclic core of both chromanes and tetrahydroquinoline derivatives [70]. Substituted allyl aryl ethers and N-allyl-N-sulfonyl anilines undergo a diastereoselective ring-closing process in the presence of the iodinating reagent $IPy_2BF_4$, giving rise to the corresponding heterocyclic compound (Scheme 19.41).

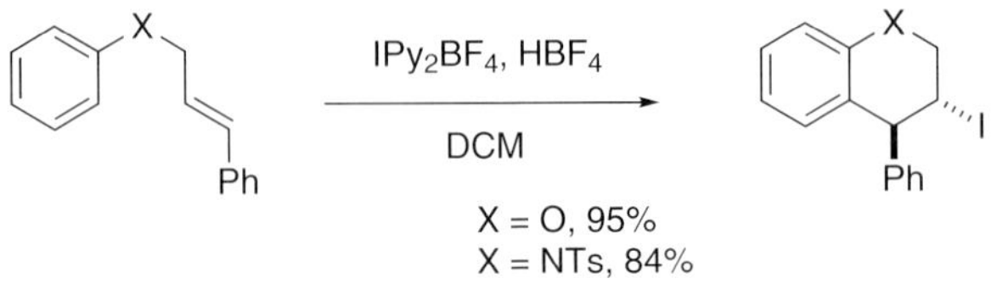

X = O, 95%
X = NTs, 84%

**Scheme 19.41** Iodonium-promoted synthesis of chromanes and tetrahydroquinoline derivatives.

The reagent $IPy_2BF_4$ has also been used successfully in the cyclization of arylalkynes [69]. The versatility of the method in the preparation of interesting benzofused heterocycles [71] and complex fused polycyclic aromatics [72] has been established (Scheme 19.42).

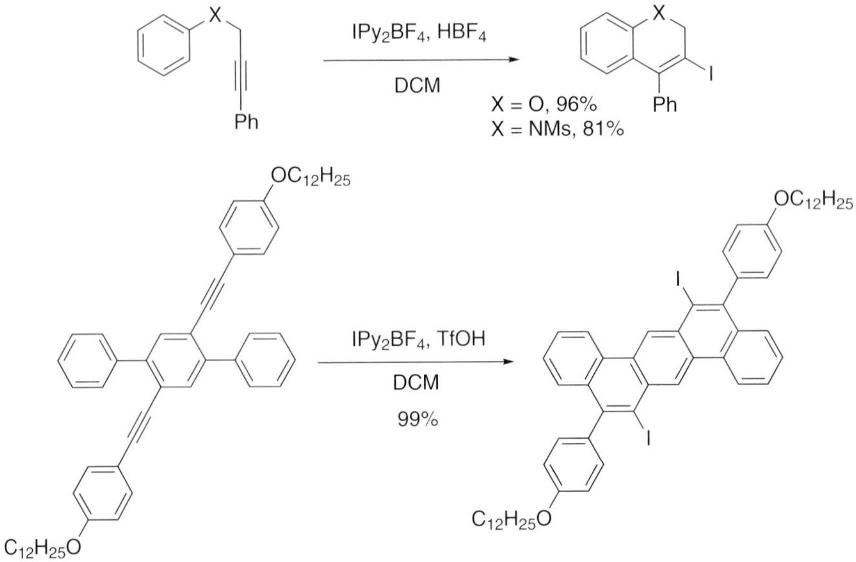

X = O, 96%
X = NMs, 81%

**Scheme 19.42** $IPy_2BF_4$-promoted synthesis of benzofused heterocycles and polycyclic aromatics.

## 19.4 Selenocyclizations

Since the first example of selenolactonization described by Campos and Petragnani in 1960 [73], organoselenium-induced cyclization reactions have been widely expanded in organic synthesis, mainly due to the extraordinary versatility of the carbon–selenium bond for further elaboration. As will be described in the following sections, depending on the internal nucleophile used, many carbo- and heterocyclic ring systems have been prepared by this strategy.

The most common reagents used in selenocyclization reactions are the commercially available phenylselenyl chloride and bromide. In some cases, the presence of the halide counterion has been shown to be a problem as competitive nucleophilic ring opening of the selenonium intermediate between the internal nucleophile and the halide have been observed. Hence, several selenium-reagents free of the nucleophilic counterion, such as N-phenylselenophthalimide (NPSP), N-phenylselenosuccinimide (NPSS), phenylselenyl hexafluoroantimoniate (PhSeSbF$_6$), phenylselenyl hexafluorophosphate (PhSePF$_6$), phenylselenyl p-toluensulfonate (PhSeOTs), and phenylselenyl triflate (PhSeOTf) have been used. Several methods involve the oxidative cleavage of diphenyldiselenide with different oxidation reagents (m-nitrobenzene sulfonyl peroxide, ammonium nitrate, ammonium cerium (IV) nitrate and iodosobenzene diacetate) to form in situ phenylselenyl m-nitrobenzene sulfonate, nitrate, and acetate respectively. (2,4,6-Triisopropylphenyl)selenyl bromide (TIPPSeBr) and a highly electrophilic selenium-copper complex, PhSeCN·Cu(OTf)$_2$ have also been used.

Cyclization of unsaturated compounds mediated by electrophilic selenium-compounds is very well documented in the literature and the growing interest in selenocyclofunctionalization is demonstrated in some excellent reviews [74]. As previously emphasized, it is not the goal of this chapter to present a complete or comprehensive review of the issue, but rather a selective and illustrative one. The major focus will be on recent progress in the area. The reader is referred to the earlier reviews for full details on selenocyclofunctionalization reactions.

## 19.4.1
### Oxygen-Centered Nucleophiles

Selenocyclization reactions involving oxygen nucleophiles have been widely studied and most of the functional groups shown in Figure 19.1 have been used in this type of reaction. Some of the most representative will be discussed in the following sections.

#### 19.4.1.1 Cyclization of Alcohols

**Intramolecular Addition to Alkenes** Selenium-mediated cyclizations of alkenol derivatives leading to cyclic ethers are well established reactions as convenient pathways in the synthesis of natural products and related compounds. In particular, this reaction has been mainly applied to the synthesis of furan and pyran derivatives through 5-endo, 5-exo, 6-endo, and 6-exo cyclizations [75]. Less common are 4-exo cyclizations to give oxetane derivatives [76], and 7-exo and 7-endo cyclizations to get oxepane derivatives [77] (Scheme 19.43).

Substituents on the unsaturated alcohols have considerable influence on the stereoselectivity of the seleno-etherification reactions. Systematic studies on this issue have been performed [78]. Also, it has been shown recently that the cyclization

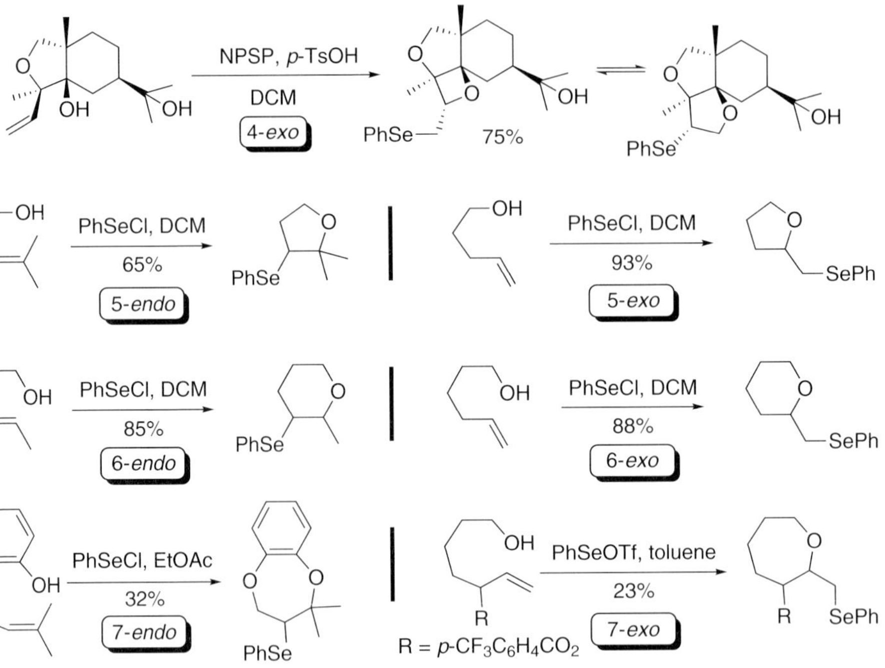

**Scheme 19.43** Typical selenocyclizations of unsaturated alcohols.

can be dramatically influenced and directed by counterion and additives to the phenylselenyl electrophile, or by internal coordination with hydroxy moieties [79].

In recent years, the asymmetric syntheses of cyclic ethers by selenocyclization reactions have become of special interest [80]. Illustrative examples of substrate-controlled reactions, where the pre-existing stereogenic centers control the chirality of the newly formed stereocenters, are shown in Scheme 19.44. Thus, for example, the

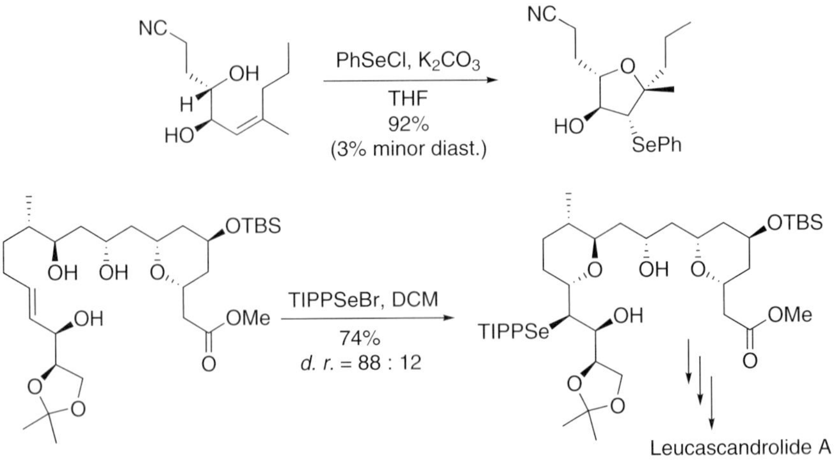

**Scheme 19.44** Substrate-controlled selenocyclization processes in the stereoselective synthesis of furan and pyran derivatives.

tetrahydrofuran moiety of amphidinolides X and Y has been synthesized recently through a highly selective selenocyclization reaction [81]. Another interesting example of a substrate-controlled selenocyclization process is that described in the total synthesis of leucascandrolide A [82]. This reaction is the first example of a stereoselective synthesis of a 2,6-trans-disubstituted tetrahydropyran mediated by a bulky selenium electrophile.

Some chiral electrophilic selenium reagents have been prepared recently and their application in reagent-controlled asymmetric synthesis has attracted the attention of several groups. These reagents include chiral aryl and terpene-derived selenides [83]. Two representative examples of stereoselective selenoetherification reactions are shown in Scheme 19.45. In the first reaction, a chiral ferrocenyl selenide is used to get a furan derivative in good yield and diastereoselectivity [84]. In a similar way, chiral sulfur-containing selenides have been used successfully in the asymmetric synthesis of several cyclic ethers, including furan and pyran derivatives [85] (see the second reaction in Scheme 19.45).

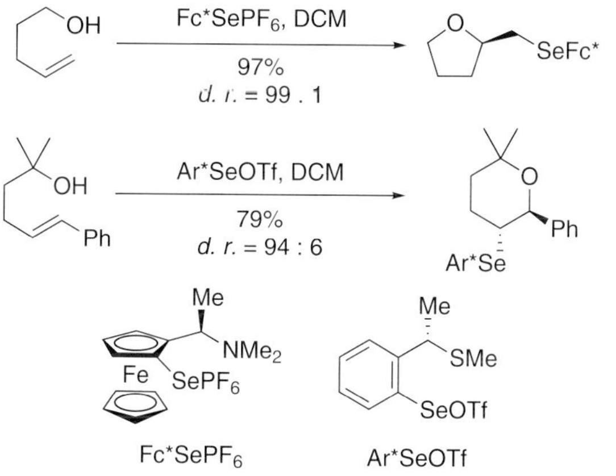

**Scheme 19.45** Reagent-controlled selenocyclization processes in the stereoselective synthesis of furan and pyran derivatives (the absolute configuration in the final products has not been determined).

Selenoetherification reactions have been wisely applied to the solid-phase synthesis of libraries of benzopyran derivatives, a structural motif found in numerous natural products [86]. The strategy is based on the use of a polystyrene-based selenyl bromide which is capable of loading substrates through electrophilic cyclization reactions [87]. The loading of o-prenylated phenol derivatives through a 6-endo-trig cyclization gives the corresponding resin-bound benzopyrans linked to to the resin through a selenoether. Such resin can be then elaborated to a variety of natural and designed structures using the functionalities of the aromatic ring. Finally, the desired benzopyran derivative can be released from the solid support by oxidation of the selenide to the corresponding selenoxide which undergoes spontaneous *syn* elimination (Scheme 19.46).

**Intramolecular Addition to Alkynes and Allenes**  Selenoetherification reactions are not limited to compounds containing a carbon–carbon double bond and so alkyne and allene derivatives have also been used in this type of cyclizations. Representative

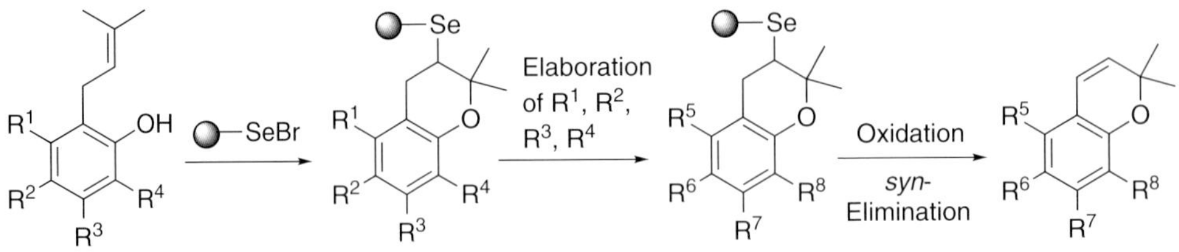

**Scheme 19.46** Solid-phase loading, elaboration and cleavage of benzopyran derivatives.

examples are shown in Scheme 19.47. Selenium-containing benzo[b]furans are easily obtained from 2-(alkynyl)anisole derivatives in a reaction similar to that performed with iodonium-reagents [16a]. Allenyl carbinols also react in a 5-*endo-trig* selenoetherification process to give dihydrofuran derivatives [88].

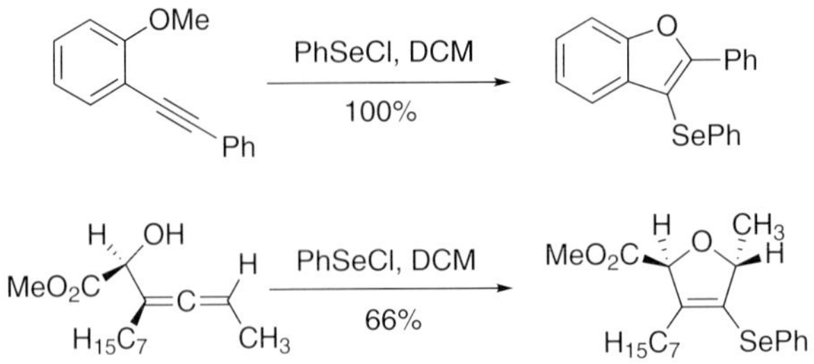

**Scheme 19.47** Selenoetherification reactions of alkyne and allene derivatives.

### 19.4.1.2 Cyclization of Aldehydes and Ketones

Selenoetherification reactions can be performed not only with alkenol derivatives as starting materials but also with 1,3-dicarbonyl derivatives [89]. These compounds react through their enol form in the presence of the electrophilic selenium-reagent *N*-phenylselenophthalimide (NPSP) and a catalytic amount of iodine (or alternatively, trace amounts of tin tetrachloride) to give the corresponding cyclic ether (Scheme 19.48). Interestingly, when the reaction is carried out in the presence of zinc diiodide or stoichiometric amounts of tin tetrachloride the carbocyclic product resulting from the nucleophilic attack of the active methylene group is formed.

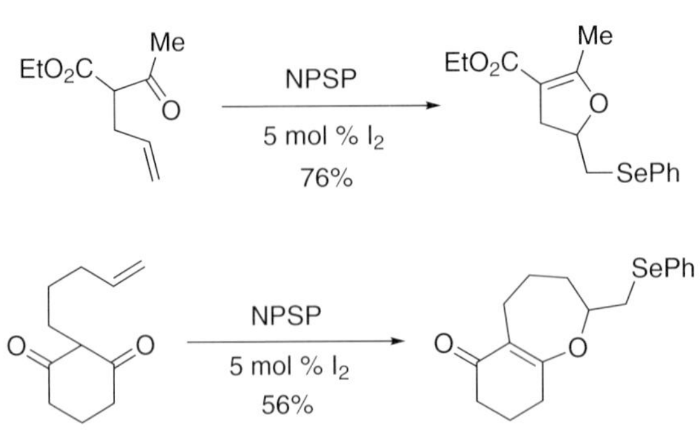

**Scheme 19.48** Selenoetherification reactions from 1,3-dicarbonylic compounds.

A conceptually interesting and synthetically useful reaction is the organoselenium-mediated trapping of a hemiacetal by a suitably disposed carbon–carbon double bond [90]. This reaction has been applied in natural product synthesis [91] (Scheme 19.49). The process has also been wisely applied to the synthesis of spiroacetal derivatives by using alkenyl hydroxyketones as starting materials [92] (Scheme 19.49).

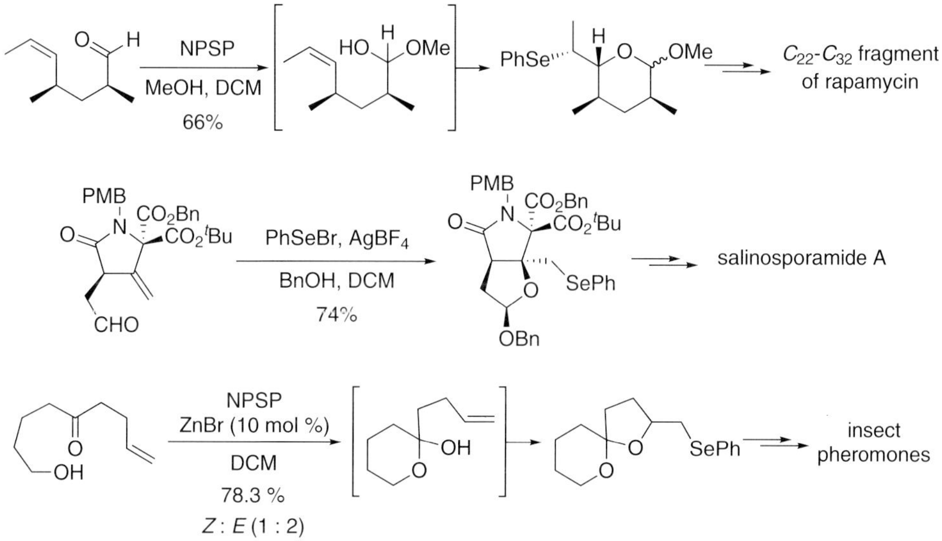

**Scheme 19.49** Selenocyclization reactions by hemiacetal trapping.

### 19.4.1.3 Cyclization of Carboxylic Acids and Carboxylic Acid Derivatives

**Intramolecular Addition to Alkenes** The selenocyclization reaction of unsaturated carboxylic acids or carboxylic acid derivatives (esters or amides) is a powerful reaction in organic chemistry for the synthesis of lactone derivatives [2d, e, 74d]. In particular, this reaction has been applied to the synthesis of γ- and δ-lactones (Scheme 19.50). The syntheses of several natural products and analogues include a selenolactonization step [93]. The selenolactonization reactions follow a mechanism analogous to that described in Section 19.3.1.2 for the halolactonization reactions (see also Scheme 19.15) [94]. When amides are used as starting materials, a competitive nucleophilic attack of the nitrogen atom, to give lactam derivatives, is sometimes observed.

As shown in the last two examples of Scheme 19.50, the diastereoselective synthesis of lactone derivatives can be achieved by both substrate- and reagent-controlled strategies. In the substrate-controlled reactions the pre-existing stereogenic centers control the chirality of the newly formed stereocenters. On the other hand, the reagent-controlled reactions are performed with chiral electrophilic selenium reagents [83–85].

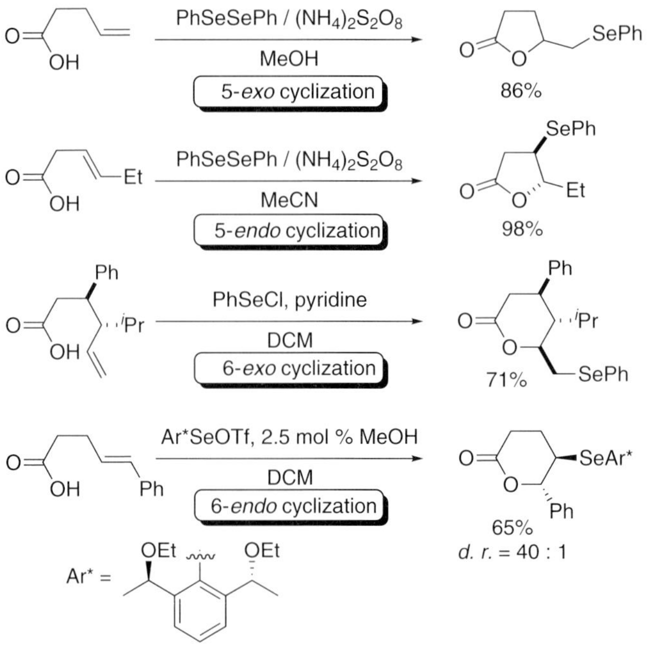

**Scheme 19.50** Selenolactonization reactions.

Recent advances in the field of selenolactonization reactions are related to the use of catalytic selenium electrophiles [95] and the development of Lewis base-catalyzed selenolactonization processes [96] (Scheme 19.51). The first method supposes the reaction of butenoic acids in the presence of 5 mol % of diphenyl diselenide and 1.05 equiv of [bis(trifluoroacetoxy)iodo]benzene in acetonitrile to afford the corresponding butenolide derivatives. This reaction proceeds through a tandem selenocyclization–elimination sequence. In the second process, significant rate enhancements are observed in a typical selenolactonization reaction when a catalytic amount

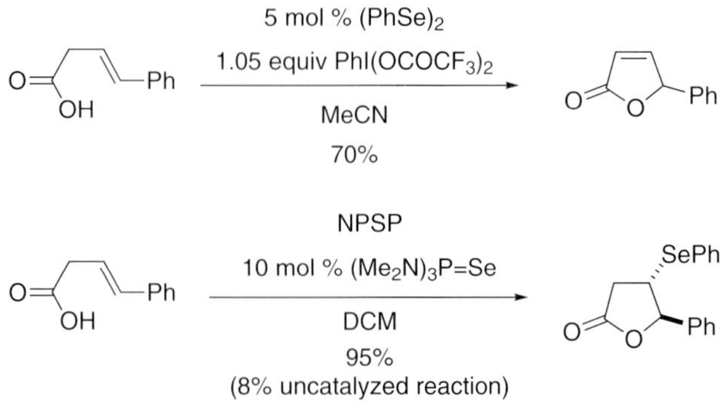

**Scheme 19.51** Catalytic selenolactonization reactions. Use of catalytic selenium electrophiles and Lewis base catalyzed selenolactonization processes.

of a Lewis base (hexamethylphosphoramide or the corresponding thio- or selenophosphoramide analogues) is added.

**Intramolecular Addition to Alkynes** Selenolactonization reactions can also be carried out with compounds containing a carbon–carbon triple bond instead of a double bond. Thus, this reaction has been recently applied in the synthesis of isocoumarins and α-pyrones from *o*-(1-alkynyl)benzoates and (*Z*)-2-alken-4-ynoates, respectively [27] (Scheme 19.52).

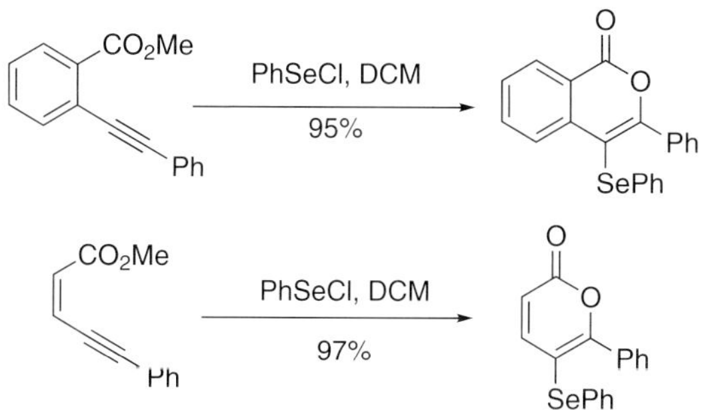


**Scheme 19.52** Synthesis of isocoumarins and α-pyrones through selenolactonization reactions.

**Other Cyclizations From Amides** 2-Oxazoline and 1,3-oxazine derivatives can also be obtained from amides [97] (Scheme 19.53). In these cases, the nitrogen atom of the starting amide is part of the tether linking the carbonyl group and the double bond. The selenocyclization reaction proceeds by nucleophilic attack of the carbonylic oxygen on the activated double bond.

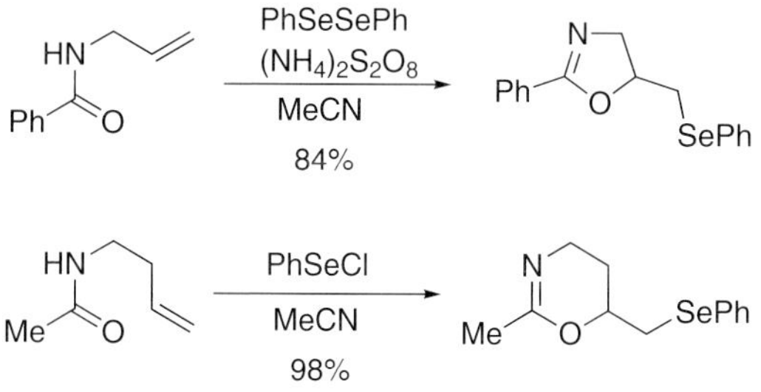


**Scheme 19.53** Synthesis of 2-oxazoline and 1,3-oxazine derivatives through selenolactonization reactions of amides.

## 19.4.2
## Nitrogen-Centered Nucleophiles

The selenocyclization reaction of unsaturated compounds involving nitrogen nucleophiles is an important strategy to access nitrogen heterocycles. Functional groups that can be used as nitrogen nucleophiles are shown in Figure 19.2. However, in the next sections we will refer only to the most useful ones.

### 19.4.2.1 Cyclization of Amines

The cyclization of alkenylamines has been mainly performed with amines bearing an electron-withdrawing group at the nitrogen atom due to the undesired side reactions that, in general, occur when the reaction is carried out with primary alkenylamines [2d].

The selenocyclization reaction of alkenylamines is an interesting strategy to access pyrrolidine or piperidine derivatives [98] (Scheme 19.54). As shown in Scheme 19.54, high levels of stereoselectivity are achieved when an hydroxy substituent is placed at the allylic position, allowing the synthesis of biologically relevant products such as *trans*-3-hydroxy-2-(hydroxymethyl)piperidine obtained by simple oxidation of the initially formed selenide.

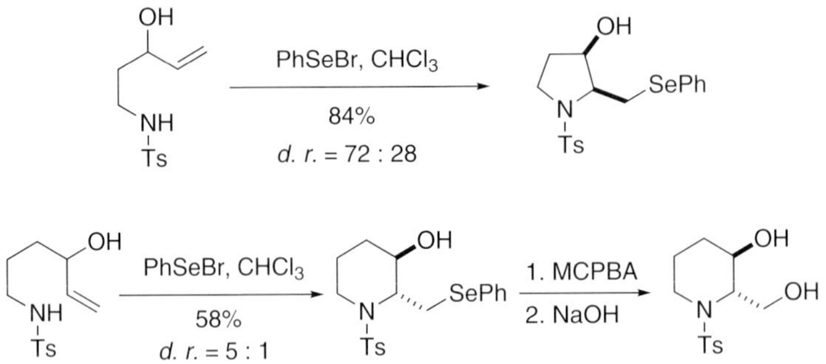

**Scheme 19.54** Synthesis of pyrrolidine or piperidine derivatives by selenium induced cyclization of alkenylamines.

A very interesting synthetic application of the selenocyclization of amines has been recently developed by Ley and coworkers to establish the important 3a-hydroxypyrrolo[2,3-b]indole core [99]. This method is based on a two-step selenocyclization–oxidative deselenation sequence of tryptophan derivatives (Scheme 19.55). This reaction has been used in the total synthesis of okaramine C [100] and the central amino acid component of chloptosin [101].

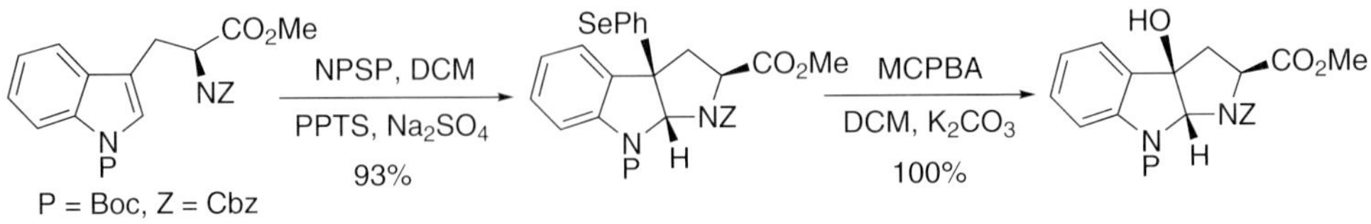

**Scheme 19.55** Selenocyclization-oxidative deselenation sequence used to elaborate the 3a-hydroxypyrrolo[2,3-b]indole core.

### 19.4.2.2 Cyclization of Amides and Imidates

The selenocyclization of alkenyl amides gives rise to either iminolactones or lactams, depending on the structure of the alkenyl amide [102]. Formation of iminolactones or lactams results, respectively, from the oxygen or nitrogen attack on the transient episelenonium ion. Representative examples of both possibilities are shown in Scheme 19.56.

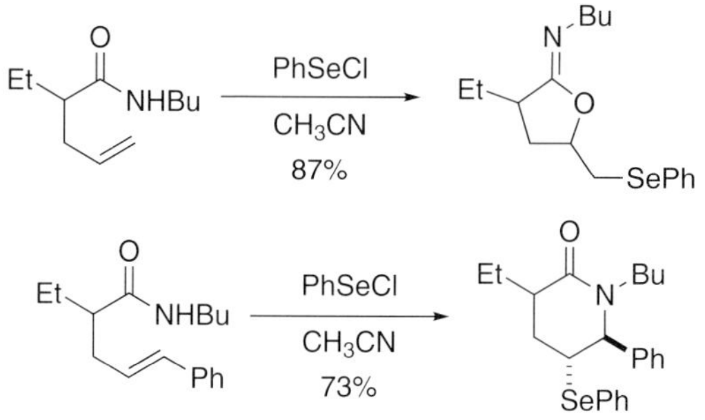

**Scheme 19.56** Selenocyclization of amides. Formation of iminolactones or lactams.

In order to increase the nucleophilicity of the nitrogen atom and so get lactams instead of iminolactones, imidates have been used instead of amides [102a]. Thus, in general, the reaction of imidates leads to the exclusive formation of lactams by the attack of the nitrogen atom on the transient episelenonium ion. A representative example is shown in Scheme 19.57.

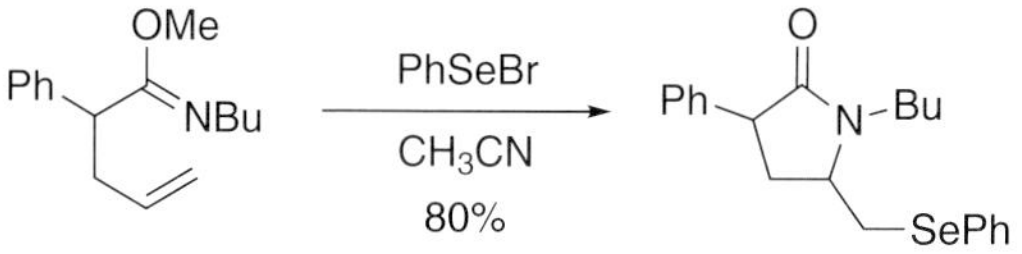

**Scheme 19.57** Selenocyclization of imidates. Formation of lactams.

### 19.4.2.3 Cyclization of Imines and Oximes

Alkenimines react with PhSeBr in dichloromethane to produce cyclic pyrrolinium or piperidinium salts. Further reduction of the crude iminium salts with NaBH$_4$ affords the corresponding cyclic pyrrolidines and piperidines [103] (Scheme 19.58).

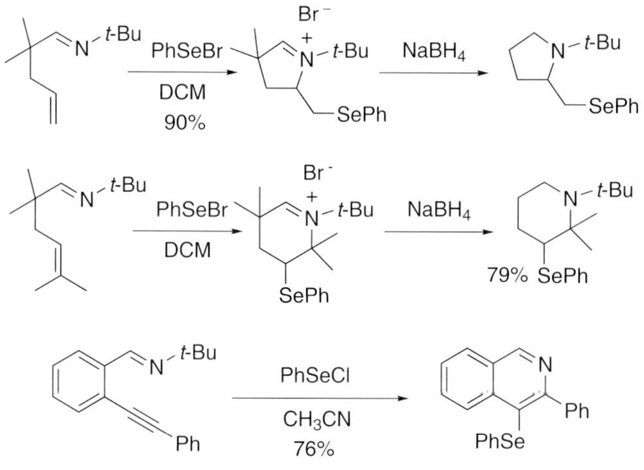

**Scheme 19.58** Selenocyclization of imines.

The *tert*-butylimines of o-(1-alkynyl)benzaldehydes and analogous pyridinecarbaldehydes have been cyclized under mild conditions in the presence of PhSeCl to give the corresponding selenium-containing isoquinoline and naphthyridine derivatives [56] (Scheme 19.58). This cyclization follows a mechanism analogous to that of the halocyclization reaction (see Section 19.3.2.4).

Alkenyl oximes react in the presence of selenium reagents to give the corresponding nitrone derivatives [104] (Scheme 19.59). In some cases either the oxygen or the nitrogen atom can act as the nucleophile to afford 1,2-oxazine or cyclic nitrone derivatives, respectively [105].

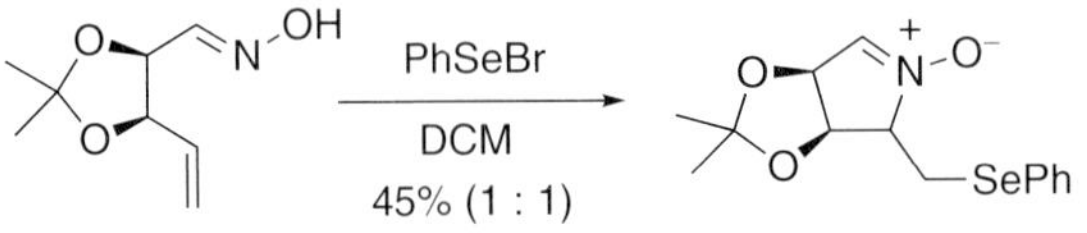

**Scheme 19.59** Selenocyclization of oximes.

## 19.4.3
### Sulfur- and Selenium-Centered Nucleophiles

3-Selenobenzo[*b*]thiophenes and 3-selenobenzo[*b*]selenophenes have been synthesized from o-ethynylphenyl sulfides and selenides, respectively [58b, 59]. These processes are similar to the halocyclization reactions described in Section 19.3.3 (see also Scheme 19.37).

## 19.4.4
### Carbon-Centered Nucleophiles

Selenocyclization reactions with carbon-centered nucleophiles are less explored processes than the corresponding halocyclization reactions described previously (see Section 19.3.4).

Interesting examples of this kind of cyclization are shown in Scheme 19.60. Thus, propargylic anilines are treated with PhSeBr in acetonitrile in the presence of

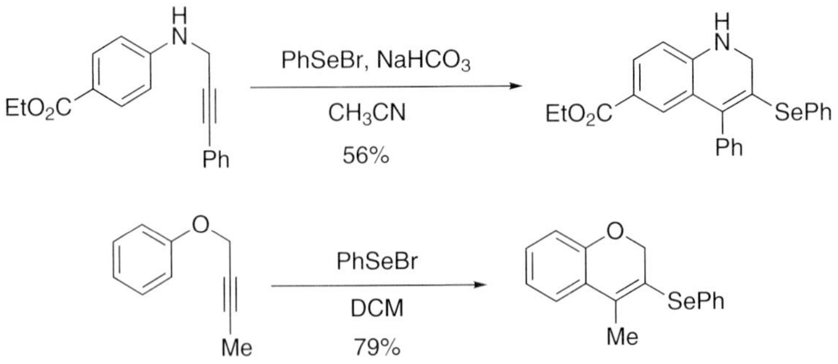

**Scheme 19.60** Selenocyclization of propargylic anilines and propargylic aryl ethers to give quinolines and 2*H*-benzopyrans respectively.

NaHCO$_3$ to give the corresponding 3-substituted quinoline derivatives [71b]. In a related process, substituted 2H-benzopyrans are easily obtained from the corresponding propargylic aryl ethers [71c].

## 19.5
## Conclusion and Perspectives

The electrophilic cyclization reactions have become valuable tools in organic synthesis for the construction of carbo- and heterocycles from easily available starting materials containing carbon–carbon double or triple bonds. An interesting feature of halocyclization and selenocyclization reactions is that a halogen or selenium is introduced in the final product, allowing further functionalization. This includes the generation of new carbon–carbon and carbon–heteroatom bonds from carbon–halogen or carbon–selenium bonds. As these reactions occur, in general, with high stereochemical control they have been widely used in the synthesis of various natural products. An aspect that requires further investigation is the development of new asymmetric halo- and seleno-cyclization reactions by using chiral reagents. As shown in this chapter, some progress in this field has been made by the discovery of new chiral selenium compounds. However, the asymmetric reagent-controlled halocyclization reaction is still in its infancy and the development of catalytic versions of this reaction is a real challenge. In this context, a recent work from Ishihara and coworkers should be remarked [106]. Key to the reaction is a chiral phosphoramidite reagent that activates an iodine atom from N-iodosuccinimide, transfers it enantioselectivity to a linear substrate and induces cyclization (Scheme 19.61).

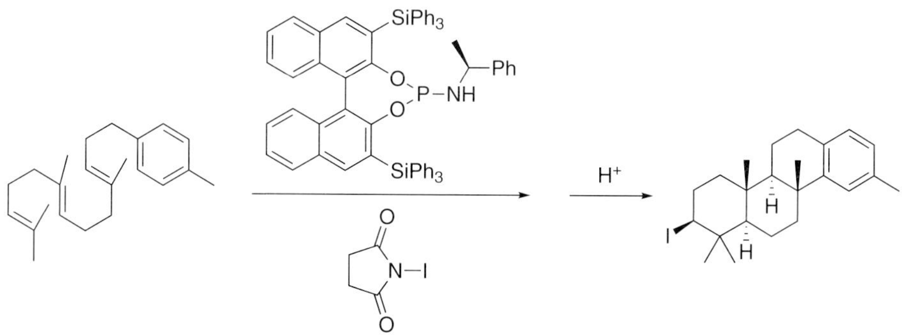

**Scheme 19.61** Biomimetic chiral halocyclization.

In conclusion, the electrophilic cyclization reactions discovered a century ago are not only important tools in current organic synthesis but also future exciting discoveries are awaited.

## 19.6
## Experimental: Selected Procedures

### 19.6.1
### Halocyclization Reactions

#### 19.6.1.1 IPy$_2$BF$_4$-Promoted Intramolecular Addition of Anilines to Alkynes (Scheme 19.31)

IPy$_2$BF$_4$ (0.41 g, 1.1 mmol, 1.1 equiv) was dissolved in dry CH$_2$Cl$_2$ (10 mL) and stirred for 5 min at room temperature. The solution was cooled in a cryocool apparatus, then tetrafluoroboric acid (137 µL, 54% solution in diethyl ether, 1.0 mmol, 1.0 equiv) was added. After 10 min the corresponding 2-ethynylaniline was added, and the solution was stirred until disappearance of the starting aniline. The reaction mixture was poured into 100 g of crushed ice and vigorously stirred, allowing the temperature to rise to room temperature. The organic layer was washed with a 5% aqueous solution of Na$_2$S$_2$O$_3$ (50 mL), dried over sodium sulfate, and concentrated. The product was purified by filtration through a short column of deactivated silica gel with hexane/ethyl acetate as eluent.

### 19.6.2
### Selenocyclization Reactions

#### 19.6.2.1 PhSeCl-Promoted Intramolecular Addition of Ethers to Alkynes (Scheme 19.47)

To a solution of 0.25 mmol of the alkyne and CH$_2$Cl$_2$ (5 mL) was added 0.375 mmol of PhSeCl. The mixture was flushed with argon and allowed to stir at 25 °C for 6 h. The reaction mixture was washed with 20 mL of water and extracted with diethyl ether. The combined ether layers were dried over anhydrous Na$_2$SO$_4$ and concentrated under vacuum to yield the crude product, which was further purified by flash chromatography on silica gel with ethyl acetate/hexanes as the eluent.

### Acknowledgments

Financial support was provided by the MEC. We are also indebted to all our collaborators and in particular to Professor José Barluenga for his invaluable assistance.

### References

1 Bougault, M.J. (1904) *C. R. Acad. Sci.*, **139**, 864–867.
2 (a) Dowle, M.D. and Davies, D.I. (1979) *Chem. Soc. Rev.*, **8**, 171–197; (b) Bartlett, P.A. (1984) *Asymmetric Synthesis*, vol. 3 (ed. J.D. Morrison), Academic Press, San Diego, chapter 6; (c) Cardillo, G. and Orena, M. (1990) *Tetrahedron*, **46**, 3321–3408; (d) Harding, K.E. and Tiner, T.H. (1991) *Comprehensive Organic*

Synthesis, vol. 4 (ed. B.M. Trost), Pergamon Press, New York, pp. 363–421; (e) Robin, S. and Rousseau, G. (1998) *Tetrahedron*, **54**, 13681–13736; (f) Ranganathan, S., Muraleedharan, K.M., Vaish, N.K. and Jayaraman, N. (2004) *Tetrahedron*, **60**, 5273–5308; (g) Larock, R.C. (2005) *Acetylene Chemistry: Chemistry, Biology, and Material Science* (eds F. Diederich, P.J. Stang and R.R. Tykwinski), Wiley-VCH, New York, pp. 51–99; (h) French, A.N., Bissmire, S. and Wirth, T. (2004) *Chem. Soc. Rev.*, **33**, 354–362.

**3** (a) Snider, B.B. and Johnston, M.I. (1985) *Tetrahedron Lett.*, **26**, 5497–5500; (b) Bravo, F. and Castillón, S. (2001) *Eur. J. Org. Chem.*, 507–516; (c) Robin, S. and Rousseau, G. (2002) *Eur. J. Org. Chem.*, 3099–3114; (d) Van de Weghe, P., Bourg, S. and Eustache, J. (2003) *Tetrahedron*, **59**, 7365–7336.

**4** Barks, J.M., Knight, D.W., Seaman, C.J. and Weingarten, G.G. (1994) *Tetrahedron Lett.*, **35**, 7259–7262.

**5** For stereochemical studies, see: (a) Bartlett, P.A. and Myerson, J. (1978) *J. Am. Chem. Soc.*, **100**, 3950–3952; (b) Collum, D.B., McDonald, J.H. III, and Still, W.C. (1980) *J. Am. Chem. Soc.*, **102**, 2118–2120; (c) Bartlett, P.A. and Jernstedt, K.K. (1980) *Tetrahedron Lett.*, **21**, 1607–1610; (d) Rollinson, S.W., Amos, R.A. and Katzenellenbogen, J.A. (1981) *J. Am. Chem. Soc.*, **103**, 4114–4125; (e) Bartlett, P.A., Meadows, J.D., Brown, E.G., Morimoto, A. and Jernstedt, K.K. (1982) *J. Org. Chem.*, **47**, 4013–4018; (f) Bartlett, P.A., Richardson, D.P. and Myerson, J. (1984) *Tetrahedron*, **40**, 2317–2327. (g) Toru, T., Fujita, S. and Maekawa, E. (1985) *J. Chem. Soc., Chem. Commun.*, 1082–1083. (h) Tomoskozi, I., Gruber, L. and Gulacsi, E. (1985) *Tetrahedron Lett.*, **26**, 3141–3144; (i) Hoffmann, R.W., Stürmer, R. and Harms, K. (1994) *Tetrahedron Lett.*, **35**, 6263–6266; (j) Macritchie, J.A., Peakman, T.M., Silcock, A. and Willis, C.L. (1998) *Tetrahedron Lett.*, **39**, 7415–7418. See also reference [2h].

**6** Clive, D.L.J., Russell, C.G., Chittattu, G. and Singh, A. (1980) *Tetrahedron*, **36**, 1399–1408.

**7** Tamaru, Y., Mizutani, M., Furukawa, Y., Kawamura, S.I., Yoshida, Z., Yanagi, K. and Minobe, M. (1984) *J. Am. Chem. Soc.*, **106**, 1079–1085.

**8** Barluenga, J., González, J.M., Campos, P.J. and Asensio, G. (1985) *Angew. Chem. Int. Ed. Engl.*, **24**, 319–320.

**9** Tredwell, M., Luft, J.A.R., Schuler, M., Tenza, K., Houk, K.N. and Gouverneur, V. (2008) *Angew. Chem. Int. Ed.*, **47**, 357–360.

**10** Barluenga, J., Alvarez-Pérez, M., Rodríguez, F., Fañanás, F.J., Cuesta, J.A. and García-Granda, S. (2003) *J. Org. Chem.*, **68**, 6583–6586.

**11** Kang, S.H., Kang, S.Y., Kim, C.M., Choi, H., Jun, H., Lee, B.M., Park, C.M. and Jeong, J.W. (2003) *Angew. Chem. Int. Ed.*, **42**, 4779–4782.

**12** Butters, M., Elliot, M.C., Hill-Cousins, J., Paine, J.S. and Walker, J.K.E. (2007) *Org. Lett.*, **9**, 3635–3638.

**13** (a) Kang, S.H., Lee, S.B. and Park, C.M. (2003) *J. Am. Chem. Soc.*, **125**, 15748–15749; (b) Kwon, H.Y., Park, C.M., Lee, S.B., Youn, J. and Kang, S.H. (2008) *Chem. Eur. J.*, **14**, 1023–1028.

**14** (a) Nicolaou, K.C., Koftis, T.V., Vyskocil, S., Petrovic, G., Tang, W., Frederick, M.O., Chen, D.Y., Li, Y., Ling, T. and Yamada, Y.M.A. (2006) *J. Am. Chem. Soc.*, **128**, 2859–2872; (b) Nicolaou, K.C., Frederick, M.O., Petrovic, G., Cole, K.P. and Loizidou, E.Z. (2006) *Angew. Chem. Int. Ed.*, **45**, 2609–2615.

**15** Bew, S.P. and Knight, D.W. (1996) *Chem. Commun.*, 1007–1008.

**16** (a) Yue, D., Yao, T. and Larock, R.C. (2005) *J. Org. Chem.*, **70**, 10292–10296; (b) Yao, T., Yue, D. and Larock, R.C. (2005) *J. Org. Chem.*, **70**, 9985–9989; (c) Arcadi, A., Cacchi, S., di Giuseppe, S., Fabrizi, G. and Marinelli, F. (2002) *Org. Lett.*, **4**, 2409–2412.

17 Zhou, C., Dubrovsky, A.V. and Larock, R.C. (2006) *J. Org. Chem.*, **71**, 1626–1632.

18 Hyland, C.J.T. and Hegedus, L.S. (2006) *J. Org. Chem.*, **71**, 8658–8660.

19 Evans, P.A., Murthy, V.S., Roseman, J.D. and Rheingold, A.L. (1999) *Angew. Chem. Int. Ed.*, **38**, 3175–3177.

20 Crimmins, M.T. and Tabet, E.A. (2000) *J. Am. Chem. Soc.*, **122**, 5473–5476.

21 Takahata, H., Uchida, Y. and Momose, T. (1994) *Tetrahedron Lett.*, **35**, 4123–4124.

22 Representative examples: (a) Bradbury, R.H., Major, J.S., Oldham, A.A., Rivett, J.E., Roberts, D.A., Slater, A.M., Timms, D. and Waterson, D. (1990) *J. Med. Chem.*, **33**, 2335–2342; (b) Najdi, S., Reichlin, D. and Kurth, M.J. (1990) *J. Org. Chem.*, **55**, 6241–6244; (c) Price, M.D., Kurth, M.J. and Schore, N.E. (2002) *J. Org. Chem.*, **67**, 7769–7773.

23 (a) Grossman, R.B. and Trupp, R.J. (1998) *Can. J. Chem.*, **76**, 1233–1237; (b) Cui, X.-L. and Brown, R.S. (2000) *J. Org. Chem.*, **65**, 5653–5658; (c) Haas, J., Piguel, S. and Wirth, T. (2002) *Org. Lett.*, **4**, 297–300; (d) Wang, M., Gao, L.X., Mai, W.P., Xia, A.X., Wang, F. and Zhang, S.B. (2004) *J. Org. Chem.*, **69**, 2874–2876; (e) Haas, J., Bissmire, S. and Wirth, T. (2005) *Chem. Eur. J.*, **11**, 5777–5785; (f) Garnier, J.M., Robin, S. and Rousseau, G. (2007) *Eur. J. Org. Chem.*, 3281–3291. For an example where the acid forms a complex with a chiral titanium reagent, see: (g) Kitagawa, O., Hanano, T., Tanabe, K., Shiro, M. and Taguchi, T. (1992) *J. Chem. Soc., Chem. Commun.*, 1005–1007.

24 Shi, Y., Peng, L.F. and Kishi, Y. (1997) *J. Org. Chem.*, **62**, 5666–5667.

25 Cuzzupe, A.N., Di Florio, R. and Rizzacasa, M.A. (2002) *J. Org. Chem.*, **67**, 4392–4398.

26 Larock, R.C. (1999) *Comprehensive Organic Transformations*, Wiley-VCH, New York, p. 1896.

27 (a) Bellina, F., Biagetti, M., Carpita, A. and Rossi, R. (2001) *Tetrahedron*, **57**, 2857–2863; (b) Biagetti, M., Bellina, F., Carpita, A., Stabile, P. and Rossi, R. (2002) *Tetrahedron*, **58**, 5023–5038; (c) Rossi, R., Carpita, A., Bellina, F., Stabile, P. and Mannina, L. (2003) *Tetrahedron*, **59**, 2067–2081; (d) Yao, T. and Larock, R.C. (2003) *J. Org. Chem.*, **68**, 5936–5942.

28 Peng, A. and Ding, Y. (2004) *Org. Lett.*, **6**, 1119–1121.

29 (a) Marshall, A., Wolf, M.A. and Wallace, E.M. (1997) *J. Org. Chem.*, **62**, 367–371; (b) Ma, S., Shi, Z. and Yu, Z. (1999) *Tetrahedron Lett.*, **40**, 2393–2396; (c) Ma, S., Shi, Z. and Yu, Z. (1999) *Tetrahedron*, **55**, 12137–12148. (d) Fu, C. and Ma, S. (2005) *Eur. J. Org. Chem.*, 3942–3945.

30 Galeazzi, R., Martelli, G., Mobbili, G., Orena, M. and Rinaldi, S. (2004) *Org. Lett.*, **6**, 2571–2574.

31 Singh, M.K., Chandra, A., Singh, B. and Singh, R.M. (2007) *Tetrahedron Lett.*, **48**, 5987–5990.

32 Duan, J.J. and Smith, A.B. III, (1993) *J. Org. Chem.*, **58**, 3703–3711.

33 George, S. and Sudalai, A. (2007) *Tetrahedron Lett.*, **48**, 8544–8546.

34 Davies, S.G., Haggitt, J.R., Ichihara, O., Kelly, R.J., Leech, M.A., Mortimer, A.J.P., Roberts, P.M. and Smith, A.D. (2004) *Org. Biomol. Chem.*, **2**, 2630–2649.

35 (a) Barluenga, J., Vázquez-Villa, H., Ballesteros, A. and González, J.M. (2003) *J. Am. Chem. Soc.*, **125**, 9028–9029; (b) Barluenga, J., Vázquez-Villa, H., Ballesteros, A. and González, J.M. (2003) *Org. Lett.*, **5**, 4121–4123; (c) Yue, D., Della Cá, N. and Larock, R.C. (2006) *J. Org. Chem.*, **71**, 3381–3388.

36 (a) Barluenga, J., Vázquez-Villa, H., Ballesteros, A. and González, J.M. (2005) *Adv. Synth. Catal.*, **347**, 526–530; (b) Barluenga, J., Vázquez-Villa, H., Merino, I., Ballesteros, A. and González, J.M. (2006) *Chem. Eur. J.*, **12**, 5790–5905.

37 Crone, B. and Kirsch, S.F. (2007) *J. Org. Chem.*, **72**, 5435–5438.

38 Xie, Y., Liu, X., Wu, L., Han, Y., Zhao, L., Fan, M. and Liang, Y. (2008) *Eur. J. Org. Chem.*, 1013–1018.

39 Waldo, J.P. and Larock, R.C. (2007) *J. Org. Chem.*, **72**, 9643–9647.

40 For some recent illustrative examples, see: (a) Kitagawa, O., Suzuki, T. and Taguchi, T. (1998) *J. Org. Chem.*, **63**, 4842–4845; (b) Davies, S.G., Nicholson, R.L., Price, P.D., Roberts, P.M. and Smith, A.D. (2004) *Synlett*, 901–903; (c) Kim, J.H., Curtis-Long, M.J., Seo, W.D., Ryu, Y.B., Yang, M.S. and Park, K.H. (2005) *J. Org. Chem.*, **70**, 4082–4087; (d) Fiorelli, C., Marchioro, C., Martelli, G., Monari, M. and Savoia, D. (2005) *Eur. J. Org. Chem.*, 3987–3993; (e) Faltz, H., Bender, C. and Liebscher, J. (2006) *Synthesis*, 2907–2922; (f) Morino, Y., Hidaka, I., Oderaotoshi, Y., Komatsu, M. and Minakata, S. (2006) *Tetrahedron*, **62**, 12247–12251; (g) Minakata, S., Morino, Y., Oderaotoshi, Y. and Komatsu, M. (2006) *Org. Lett.*, **8**, 3335–3337; (h) Diaba, F., Ricou, E. and Bonjoch, J. (2007) *Org. Lett.*, **9**, 2633–2636.

41 Fujioka, H., Murai, K., Ohba, Y., Hirose, H. and Kita, Y. (2006) *Chem. Commun.*, 832–834.

42 Beshore, D.C. and Smith, A.B. III, (2007) *J. Am. Chem. Soc.*, **129**, 4148–4149.

43 (a) Barluenga, J., Trincado, M., Rubio, E. and González, J.M. (2003) *Angew. Chem. Int. Ed.*, **42**, 2406–2409. For subsequent related approaches to indoles, see: (b) Amjad, M. and Knight, D.W. (2004) *Tetrahedron Lett.*, **45**, 539–541; (c) Yue, D. and Larock, R.C. (2004) *Org. Lett.*, **6**, 1037–1040; (d) Yue, D., Yao, T. and Larock, R.C. (2006) *J. Org. Chem.*, **71**, 62–69.

44 Hu, T., Liu, K., Shen, M., Yuan, X., Tang, Y. and Li, C. (2007) *J. Org. Chem.*, **72**, 8555–8558.

45 For selected examples, see: (a) Biloski, A.J., Wood, R.D. and Ganem, B. (1982) *J. Am. Chem. Soc.*, **104**, 3233–3235; (b) Rajendra, G. and Miller, M.J. (1987) *J. Org. Chem.*, **52**, 4471–4477; (c) Boeckman, R.K. Jr and Cornnell, B.T. (1995) *J. Am. Chem. Soc.*, **117**, 12368–12369; (d) Fujita, M., Kitagawa, O., Suzuki, T. and Taguchi, T. (1997) *J. Org. Chem.*, **62**, 7330–7335.

46 Knapp, S. (1996) *Advances in Heterocyclic Natural Product Synthesis*, vol 3, JAI Press, Greenwich, pp. 57–98.

47 (a) Kano, S., Yokomatsu, T., Iwasawa, H. and Shibuya, S. (1987) *Heterocycles*, **26**, 359–362; (b) Takahata, H., Takamatsu, T. and Yamazaki, T. (1989) *J. Org. Chem.*, **54**, 4812–4822; (c) Takahata, H., Yamazaki, K., Takamatsu, T., Yamazaki, T. and Momose, T. (1990) *J. Org. Chem.*, **55**, 3947–3950.

48 Shen, M. and Li, C. (2004) *J. Org. Chem.*, **69**, 7906–7909.

49 Knapp, S. and Gibson, F.S. (1992) *J. Org. Chem.*, **57**, 4802–4809.

50 Hirama, M., Iwashita, M., Yamazaki, Y. and Ito, S. (1984) *Tetrahedron Lett.*, **25**, 4963–4964.

51 Balko, T.W., Brinkmeyer, R.S. and Terando, N.H. (1989) *Tetrahedron Lett.*, **30**, 2045–2048.

52 Cardillo, G., Orena, M., Porzi, G. and Sandri, S. (1982) *J. Chem. Soc., Chem. Commun.*, 1309–1311.

53 Hunt, P.A., May, C. and Moody, C.J. (1988) *Tetrahedron Lett.*, **29**, 3001–3002.

54 Bruni, E., Cardillo, C., Orena, M., Sandri, S. and Tomasini, C. (1989) *Tetrahedron Lett.*, **30**, 1679–1682.

55 Creeke, P.I. and Mellor, J.M. (1989) *Tetrahedron Lett.*, **30**, 4435–4438.

56 (a) Huang, Q., Hunter, J.A. and Larock, R.C. (2001) *Org. Lett.*, **3**, 2973–2976; (b) Huang, Q., Hunter, J.A. and Larock, R.C. (2002) *J. Org. Chem.*, **67**, 3437–3444.

57 Fischer, D., Tomeba, H., Pahadi, N.K., Patil, N.T. and Yamamoto, Y. (2007) *Angew. Chem. Int. Ed.*, **46**, 4764–4766.

58 (a) Flynn, B.L., Verdier-Pinard, P. and Hamel, E. (2001) *Org. Lett.*, **3**, 651–654; (b) Yue, D. and Larock, R.C. (2002) *J. Org. Chem.*, **67**, 1905–1909.

59 Kesharwani, T., Worlikar, S.A. and Larock, R.C. (2006) *J. Org. Chem.*, **71**, 2307–2312.

60 Hessian, K.O. and Flynn, B.L. (2003) *Org. Lett.*, **5**, 4377–4380.

61 (a) Inoue, T., Kitagawa, O., Oda, Y. and Taguchi, T. (1996) *J. Org. Chem.*, **61**, 8256–8263; (b) Kitagawa, O., Inoue, T. and Taguchi, T. (1996) *Rev. Heteroat. Chem.*, **15**, 243–262. For a related synthetic application of this reaction, see: (c) Siegel, D.R. and Danishefsky, S.J. (2006) *J. Am. Chem. Soc.*, **128**, 1048–1049.

62 Inoue, T., Kitagawa, O., Saito, A. and Taguchi, T. (1997) *J. Org. Chem.*, **62**, 7384–7389.

63 Kitagawa, O., Suzuki, T., Inoue, T., Watanabe, Y. and Taguchi, T. (1998) *J. Org. Chem.*, **63**, 9470–9475.

64 Kitagawa, O., Inoue, T., Hirano, K. and Taguchi, T. (1993) *J. Org. Chem.*, **58**, 3106–3112.

65 Kitagawa, O., Suzuki, T., Inoue, T. and Taguchi, T. (1998) *Tetrahedron Lett.*, **39**, 7357–7360.

66 Bi, H.-P., Guo, L.-N., Duan, X.-H., Gou, F.-R., Huang, S.-H., Liu, X.-Y. and Liang, Y.-M. (2007) *Org. Lett.*, **9**, 397–400.

67 Barluenga, J., Palomas, D., Rubio, E. and González, J.M. (2007) *Org. Lett.*, **9**, 2823–2826.

68 Barluenga, J., Romanelli, G.P., Alvarez-García, L.J., Llorente, I., González, J.M., García-Rodríguez, E. and García-Granda, S. (1998) *Angew. Chem. Int. Ed.*, **37**, 3136–3139.

69 Barluenga, J., González, J.M., Campos, P.J. and Asensio, G. (1988) *Angew. Chem. Int. Ed. Engl.*, **27**, 1546–1547.

70 Barluenga, J., Trincado, M., Rubio, E. and González, J.M. (2004) *J. Am. Chem. Soc.*, **126**, 3416–3417.

71 (a) Barluenga, J., Trincado, M., Marco-Arias, M., Ballesteros, A., Rubio, E. and González, J.M. (2005) *Chem. Commun.*, 2008–2010. For related cyclizations using other sources of iodonium ions, see: (b) Zhang, X., Campo, M.A., Yao, T. and Larock, R.C. (2005) *Org. Lett.*, **7**, 763–766; (c) Worlikar, S.A., Kesharwani, T., Yao, T. and Larock, R.C. (2007) *J. Org. Chem.*, **72**, 1347–1353.

72 Goldfinger, M.B., Crawford, K.B. and Swager, T.M. (1997) *J. Am. Chem. Soc.*, **119**, 4578–4593. For related cyclizations using other sources of iodonium ions, see: Yao, T., Campo, M.A. and Larock, R.C. (2005) *J. Org. Chem.*, **70**, 3511–3517.

73 de Moura Campos, M. and Petragnani, N. (1960) *Chem. Ber.*, **93**, 317–320.

74 (a) Paulmier, C. (1986) *Selenium Reagents and Intermediates in Organic Synthesis*, Pergamon Press, Oxford, pp. 229–255; (b) Liotta, D. (ed.) (1987) *Organoselenium Chemistry*, Wiley, New York; (c) Tiecco, M. (2000) *Topics in Current Chemistry: Organoselenium Chemistry* (ed. T. Wirth), Springer, Heidelberg, pp. 7–54; (d) Petragnani, N., Stefani, H.A. and Vaduga, C.J. (2001) *Tetrahedron*, **57**, 1411–1448.

75 (a) Vukićević, R., Konstantinović, S. and MIhailović, M.L. (1991) *Tetrahedron*, **47**, 859–865; (b) Konstantinović, S., Bugarčić, Z., Milosavljević, S., Schroth, G. and MIhailović, M.L. (1992) *Liebigs Ann. Chem.*, 261–268.

76 (a) Prangé, T., Rodríguez, M.S. and Suárez, E. (2003) *J. Org. Chem.*, **68**, 4422–4431; (b) Van de Weghe, P., Bourg, S. and Eustache, J. (2003) *Tetrahedron*, **59**, 7365–7376.

77 (a) Xiang, L. and Kozikowski, A.P. (1990) *Synlett*, 279–281; (b) Williams, R.M. and Cushing, T.D. (1990) *Tetrahedron Lett.*, **44**, 6325–6328; (c) Inoue, H., Murata, S. and Suzuki, T. (1994) *Liebigs Ann. Chem.*, 901–909.

78 See for example: (a) Reitz, A.B., Nortey, S.O., Maryanoff, B.E., Litta, D. and Monahan, R. (1987) *J. Org. Chem.*, **52**, 4191–4202; (b) Lipshutz, B.H. and Barton, J.H. (1992) *J. Am. Chem. Soc.*, **114**, 1084–1086; (c) Mihelich, E.D. and Hite, G.A. (1992) *J. Am. Chem. Soc.*, **114**, 7318–7319; (d) Lipshutz, B.H. and Gross, T. (1995) *J. Org. Chem.*, **60**, 3572–3573; (e) Landais, Y., Planchenault, D. and Weber, V. (1995) *Tetrahedron Lett.*, **36**, 2987–2990; (f) Landais, Y. and Planchenault, D. (1995) *Synlett*, 1191–1193; (g) Andrey, O., Glanzmann, C., Landais, Y. and Parra-Rapado, L. (1997)

*Tetrahedron*, **53**, 2835–2854; (h) Andrey, O., Ducry, L., Landais, Y., Planchenault, D. and Weber, V. (1997) *Tetrahedron*, **53**, 4339–4352; (i) Tiecco, M., Testaferri, L. and Santi, C. (1999) *Eur. J. Org. Chem.*, 797–803.

79. Khokhar, S.S. and Wirth, T. (2004) *Angew. Chem. Int. Ed.*, **43**, 631–633.

80. (a) Browne, D.M. and Wirth, T. (2006) *Curr. Org. Chem.*, **10**, 1893–1903; (b) Zhu, C. and Huang, Y. (2006) *Curr. Org. Chem.*, **10**, 1905–1920; (c) Braga, A.L., Lüdtke, D.S. and Vargas, F. (2006) *Curr. Org. Chem.*, **10**, 1921–1938.

81. Rodríguez-Escrich, C., Olivella, A., Urpí, F. and Vilarrasa, J. (2007) *Org. Lett.*, **9**, 989–992.

82. Fettes, A. and Carreira, E.M. (2003) *J. Org. Chem.*, **68**, 9274–9283.

83. (a) Tomoda, S. and Iwaoka, M. (1988) *Chem. Lett.*, 1895–1898. (b) Deziel, R., Goulet, S., Grenier, L., Bordeleau, J. and Bernier, J. (1993) *J. Org. Chem.*, **58**, 3619–3621; (c) Fujita, K., Iwaoka, M. and Tomoda, S. (1994) *Chem. Lett.*, 923–926; (d) Nishibayashi, Y., Singh, J.D., Uemura, S. and Fukuzawa, S. (1994) *Tetrahedron Lett.*, **35**, 3115–3118; (e) Wirth, T. (1995) *Angew. Chem. Int. Ed. Engl.*, **34**, 1726–1728; (f) Back, T.G. and Dyck, B.P. (1994) *Chem. Commun.*, 515–516. (g) Déziel, R., Malenfant, E., Thibault, C., Fréchette, S. and Gravel, M. (1997) *Tetrahedron Lett.*, **38**, 4753–4756; (h) Fragale, G., Neuburger, M. and Wirth, T. (1998) *Chem. Commun.*, 1867–1868; (i) Tiecco, M., Testaferri, L., Bagnoli, L., Marini, F., Temperini, A., Tomassini, C. and Santi, C. (2000) *Tetrahedron Lett.*, **41**, 3241–3245; (j) Ścianowski, J., Rafiński, Z. and Wojtczak, A. (2006) *Eur. J. Org. Chem.*, 3216–3225.

84. Takada, H., Nishibayashi, Y. and Uemura, S. (1999) *J. Chem. Soc., Perkin Trans. 1*, 1511–1516.

85. Tiecco, M., Testaferri, L., Marini, F., Sternativo, S., Bagnoli, L., Santi, C. and Temperini, A. (2001) *Tetrahedron: Asym.*, **12**, 1493–1502.

86. (a) Nicolaou, K.C., Pfefferkorn, J.A. and Cao, G.-Q. (2000) *Angew. Chem. Int. Ed.*, **39**, 734–739; (b) Nicolaou, K.C., Cao, G.-Q. and Pfefferkorn, J.A. (2000) *Angew. Chem. Int. Ed.*, **39**, 739–743; (c) Nicolaou, K.C., Pfefferkorn, J.A., Roecker, A.J., Cao, G.-Q., Barluenga, S. and Mitchell, H.J. (2000) *J. Am. Chem. Soc.*, **122**, 9939–9953; (d) Nicolaou, K.C., Pfefferkorn, J.A., Mitchell, H.J., Roecker, A.J., Barluenga, S., Cao, G.-Q., Affleck, R.L. and Lilling, J.E. (2000) *J. Am. Chem. Soc.*, **122**, 9954–9967; (e) Nicolaou, K.C., Pfefferkorn, J.A., Barluenga, S., Mitchell, H.J., Roecker, A.J. and Cao, G.-Q. (2000) *J. Am. Chem. Soc.*, **122**, 9968–9976.

87. Nicolaou, K.C., Pastor, J., Barluenga, S. and Winssinger, N. (1998) *Chem. Commun.*, 1947–1948.

88. (a) Marshall, J.A. and Wang, X. (1990) *J. Org. Chem.*, **55**, 2995–2996; (b) Ma, S., Pan, F., Hao, X. and Huang, X. (2004) *Synlett*, 85–88; (c) Chen, G., Fu, C. and Ma, S. (2006) *Tetrahedron*, **62**, 4444–4452.

89. (a) Jackson, W.P., Ley, S.V. and Morton, J.A. (1980) *J. Chem. Soc., Chem. Commun.*, 1028–1029. (b) Jackson, W.P., Ley, S.V. and Whittle, A.J. (1980) *J. Chem. Soc., Chem. Commun.*, 1173–1174; (c) Ley, S.V., Lygo, B., Molines, H. and Morton, J.A. (1982) *J. Chem. Soc., Chem. Commun.*, 1251–1252; (d) Ley, S.V., Lygo, B. and Molines, H. (1984) *J. Chem. Soc., Perkin Trans. I*, 2403–2405.

90. Current, S. and Sharpless, K.B. (1978) *Tetrahedron Lett.*, **19**, 5075–5078.

91. See for example: (a) Anderson, J.C., Ley, S.V. and Marsden, S.P. (1994) *Tetrahedron Lett.*, **35**, 2087–2090; (b) Endo, A. and Danishefsky, S.J. (2005) *J. Am. Chem. Soc.*, **127**, 8298–8299.

92. (a) Ley, S.V. and Lygo, B. (1982) *Tetrahedron Lett.*, **23**, 4625–4628; (b) Doherty, A.M., Ley, S.V., Lygo, B. and Williams, D.J. (1984) *J. Chem. Soc., Perkin Trans. I*, 1371–1377.

93. See for example: (a) Kozikowski, A.P. and Lee, J. (1990) *J. Org. Chem.*, **55**, 863–870; (b) Bennett, F., Knight, D.W. and Fenton,

G. (1991) *J. Chem. Soc., Perkin Trans. 1*, 133–140.

94 For a recent study on the mechanism of this reaction see: Denmark, S.E. and Edwards, M.G. (2006) *J. Org. Chem.*, **71**, 7293–7306.

95 Browne, D.M., Niyomura, O. and Wirth, T. (2007) *Org. Lett.*, **9**, 3169–3171.

96 Denmark, S.E. and Collins, W.R. (2007) *Org. Lett.*, **9**, 3801–3804.

97 See for example: (a) Chretien, F. and Chapleur, Y. (1988) *J. Org. Chem.*, **53**, 3615–3617; (b) Engman, L. (1991) *J. Org. Chem.*, **56**, 3425–3430; (c) Tiecco, M., Testaferri, L., Tingoli, M. and Marini, F. (1993) *J. Org. Chem.*, **58**, 1349–1354; (d) Tingoli, M., Tiecco, M., Testaferri, L. and Temperini, A. (1998) *Synth. Commun.*, **28**, 1769–1778.

98 (a) Toshimitsu, A., Terao, K. and Uemura, S. (1986) *J. Org. Chem.*, **51**, 1724–1729; (b) Tiecco, M., Testaferri, L., Tingoli, M., Bartoli, D. and Balducci, R. (1990) *J. Org. Chem.*, **55**, 429–434; (c) Cooper, M.A. and Ward, A.D. (1992) *Tetrahedron Lett.*, **33**, 5999–6002; (d) Cooper, M.A. and Ward, A.D. (1994) *Tetrahedron Lett.*, **35**, 5065–5068; (e) Berthe, B., Outurquin, F. and Paulmier, C. (1997) *Tetrahedron Lett.*, **38**, 1393–1396; (f) Jones, A.D., Knight, D.W., Redfern, A.L. and Gilmore, J. (1999) *Tetrahedron Lett.*, **40**, 3267–3270.

99 Ley, S.V., Cleator, E. and Hewitt, P. (2003) *Org. Biomol. Chem.*, **1**, 3492–3494.

100 Hewitt, P., Cleator, E. and Ley, S.V. (2004) *Org. Biomol. Chem.*, **2**, 2415–2417.

101 Hong, W.-X., Chen, L.-J., Zhong, C.-L. and Yao, Z.-J. (2006) *Org. Lett.*, **8**, 4919–4922.

102 (a) Toshimitsu, A., Terao, K. and Uemura, S. (1987) *J. Org. Chem.*, **52**, 2018–2026; (b) Izumi, T. and Morishita, N. (1994) *J. Heterocycl. Chem.*, **31**, 145–152; (c) Diederich, M. and Nubbemeyer, U. (1996) *Chem. Eur. J.*, **2**, 894–900; (d) Sudau, A. and Nubbemeyer, U. (1998) *Angew. Chem. Int. Ed. Engl.*, **37**, 1140–1143; (e) Sudau, A., Münch, W., Nubbemeyer, U. and Bats, J.W. (2000) *J. Org. Chem.*, **65**, 1710–1720.

103 (a) De Kimpe, N. and Boelens, M. (1993) *J. Chem. Soc., Chem Commun.*, 916–918. (b) De Smaele, D. and De Kimpe, N. (1995) *J. Chem. Soc., Chem Commun.*, 2029–2030.

104 (a) Hall, A., Meldrum, K.P., Therond, P.R. and Wightman, R.H. (1997) *Synlett*, 123–125. See also: (b) Grigg, R., Hadjisoteriou, M., Kennewell, P. and Markandu, J. (1992) *J. Chem. Soc., Chem Commun.*, 1537–1538.

105 (a) Tiecco, M., Testaferri, L., Tingoli, M., Bagnoli, L. and Marini, F. (1993) *J. Chem. Soc., Perkin Trans. 1*, 1989–1993. (b) Tiecco, M., Testaferri, L., Bagnoli, L., Purgatorio, V., Temperini, A., Marini, F. and Santi, C. (2001) *Tetrahedron: Asym.*, **12**, 3297–3304.

106 Sakakura, A., Ukai, A. and Ishihara, K. (2007) *Nature*, **445**, 900–903. See also references [13] and [23].

# 20
# Cyclizations Based on C—H Activation
*David A. Capretto, Zigang Li, and Chuan He*

## 20.1
## Introduction

Past studies on C—H activation chemistry have focused on mechanistic understanding, and are consequently well explored [1]. Since these studies in the middle and late twentieth century, the field of C—H activation chemistry has re-emerged with a new focus on developing catalytic processes to facilitate the synthesis of complex molecules [2, 3]. As the transformations have increased in difficulty and complexity, the range of catalytic systems has become increasingly broad, with the use of different metals, ligands, supporting salts, and other additives continuing to push the chemistry in new directions. If the current rate of success is any indication of the transformations yet to be discovered, C—H activation chemistry will remain at the forefront for many years to come.

Although some isolated examples of stoichiometric C—H activation were seen earlier, the field started to expand at a more rapid pace in the 1960s. The advances that started this expansion of chemistry include, but are not limited to, the aryl activation of azobenzene by cobalt and nickel, Chatt and Davidson's oxidative addition of an aryl C—H bond of naphthalene across a ruthenium center, and platinum reactivity reported by Shaw, Garnett, Hodges, and Shilov [4]. These works and other early examples have been discussed in great detail with interesting historical perspectives in previous reviews [5].

Murai's report of the ruthenium-catalyzed ortho-alkylation of aromatic ketones with olefins in 1993 is considered a major advance in the field of C—H activation due to the high efficiency and selectivity of the reaction [2]. This work was influenced by two earlier examples that use similar concepts. The first, published in the mid-1980s by Lewis, achieved the ethylation of a phenol through an ortho-metallation pathway utilizing a ruthenium phosphine complex [6]. The second, published three years later by Jordan, coupled propene with alpha-picoline utilizing a zirconium complex [7]. Both of these examples made use of metal coordination to a heteroatom: in the ruthenium case the phenolic oxygen, and in the zirconium case the picolinic

*Handbook of Cyclization Reactions. Volume 2.*
Edited by Shengming Ma
Copyright © 2010 WILEY-VCH Verlag GmbH & Co. KGaA, Weinheim
ISBN: 978-3-527-32088-2

# 20 Cyclizations Based on C–H Activation

Table 20.1 A variety of substrates are amenable to Murai's C–C bond formation strategy.

| Substrate 1 | Substrate 2 | Catalyst | Product | Yield (%) |
|---|---|---|---|---|
| α-tetralone | =Si(OEt)₃ | RuH₂(CO)(PPh₃)₃ | α-tetralone with CH₂CH₂Si(OEt)₃ ortho substituent | 100 |
| cyclohex-1-enyl tBu ketone | =Si(OEt)₃ | RuH₂(CO)(PPh₃)₃ | cyclohex-1-enyl tBu ketone with Si(OEt)₃ substituent | 96 |
| N-tBu aryl imine with CF₃ | =Si(OEt)₃ | Ru₃(CO)₁₂ | ortho-alkylated N-tBu aryl imine with CF₃ and CH₂CH₂Si(OEt)₃ | 75 |
| 2-(pent-4-enyl)pyridine | | RhCl(PPh₃)₃ | 2-(cyclopentylidenemethyl)pyridine | 93 |

Conditions for each reaction can be found in Ref. [5b].

nitrogen. While lacking in generality and efficiency, these two systems demonstrated the concept of chelation-assistance that was crucial to the success of Murai's system.

This initial success led to study of other reactions by Murai. As shown in Table 20.1, ortho-alkylation was amenable to different substrates including non-aromatic olefins and imines. By switching to a rhodium catalyst, the intramolecular version of this reaction proceeded with good yield. Later studies involving deuterium exchanges and $^{13}$C kinetic isotope effects supported the mechanism shown in Scheme 20.1a. This chemistry was also shown to be amenable to building polymers, as shown in Gupta and Weber's use of ruthenium to build polymeric structures in Scheme 20.1b [8].

When considering the cyclization of substrates via C–H activation, there are two main routes that achieve the transformation, as shown in Scheme 20.2. In the first case, a metal inserts into the C–H bond, forming an activated C–M species. This active C–M species is then able to attack a nearby bond (usually tethered to the molecule), making a new bond and, hence, a cyclic structure. In the second scenario, the metal interacts with a functional group to form a reactive species, which then goes on to attack a C–H bond, forming the cyclic structure. While both cases lead to the same result, the C–H activation occurs first only in the first case. In this first case, the moiety denoted "X" can be a functional group with good leaving abilities, or a species that can be reduced without being a leaving group (e.g., an olefin). In the second case, the "Y" moiety can be anything that is efficient at interacting with the metal of choice.

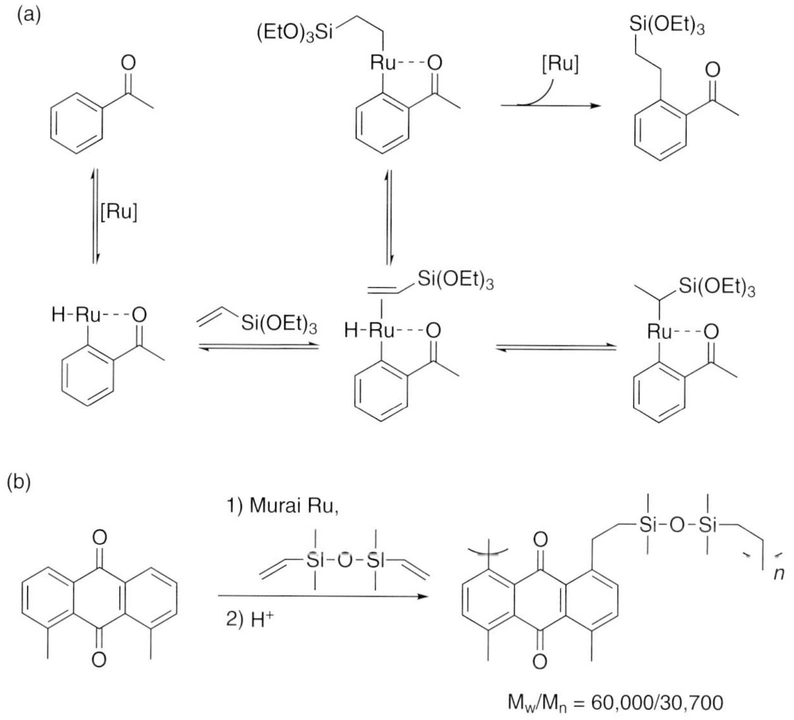

**Scheme 20.1** (a) Proposed mechanism for aryl–olefin coupling reaction. (b) Using Murai chemistry to build polymeric structures containing silicon.

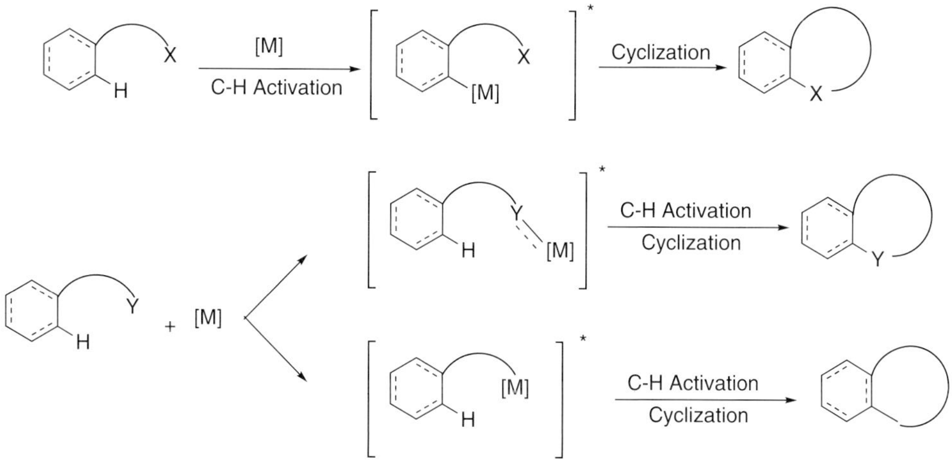

**Scheme 20.2** The concept of cyclization by C—H activation.

This can range from another proton in the case of some palladium cyclizations all the way to carbene and nitrene sources, which interact with rhodium, ruthenium, and silver.

In this chapter, the aim is to present and discuss the advances in the cyclization of substrates by C—H activation from 2000 to 2008. The chapter is divided based on the type of bond activated, then further divided by metal. Due to the popularity of C—H activation, numerous books and review articles on assorted aspects of this chemistry have been published with great frequency. Pertinent review articles for each type of transformation are included in their respective sections. We do not intend to cover all works in this review, and consequently some studies may not be presented.

## 20.2
## Aryl C—H Bond Activation

### 20.2.1
### Rhodium

Rhodium systems exhibited true promise in aryl C—H activations in Chung's report of a rhodium(I) system that formed a C—C bond through the intermolecular ortho-alkylation of aromatic imines [9]. This was followed by a report from Ellman and Bergman in 2001 on the annulation of alkenyl-substituted heterocycles shown in Scheme 20.3 [10]. In this reaction, the catalyst is a modification of Wilkinson's rhodium(I) catalyst. In later studies, they were able to both isolate the catalyst-bound intermediate and gain an in-depth understanding of the mechanism of the reaction, which is shown in Scheme 20.4 [11].

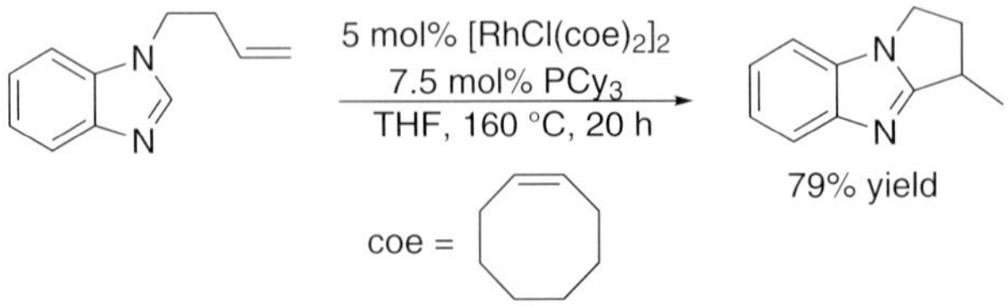

**Scheme 20.3** Aryl C—H bond activation by a rhodium-based catalyst.

Since then, Ellman and Bergman have broadened the scope of this cyclization [12]. In 2004, they made the aromatic imine cyclization enantioselective, the first transformation of its kind [13]. One set of optimized conditions (Scheme 20.5) led to quantitative yields of product in less than 2 h with ee values above 85%. Ligand studies revealed that the backbone of the phosphine was responsible for chiral induction; reversing the stereochemistry of the P—N bond changed neither the ee nor the configuration of the resulting product. Further substrate studies showed that a slight decrease in yield was complemented with improved ee if the reaction temperature was lowered to 50 °C (and even room temperature in one case).

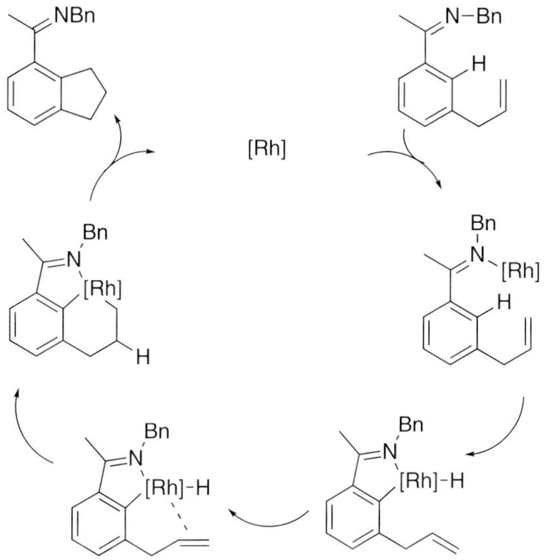

**Scheme 20.4** Mechanism for rhodium-based C—H activation.

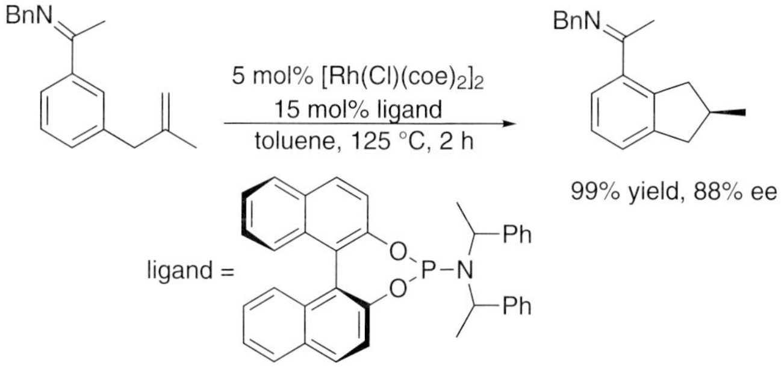

**Scheme 20.5** Highly enantioselective aryl C—H bond activation by rhodium phosphine complexes.

Realizing the strength of this transformation, they moved to applying it in natural product synthesis. They were able, in 2005, to synthesize a fragment of the natural product (+)-lithospermic acid, which has potent anti-HIV activity [14]. Shown in Scheme 20.6, a chiral imine was first attached to the substrate in order to guide the olefin insertion in a diastereoselective manner. This step was required for the reaction to proceed stereoselectively. The amine was added (which forms an imine), the substrate cyclized, and the imine hydrolyzed to give back the aldehyde functionality. The rhodium-supporting ligand used in this case was slightly different

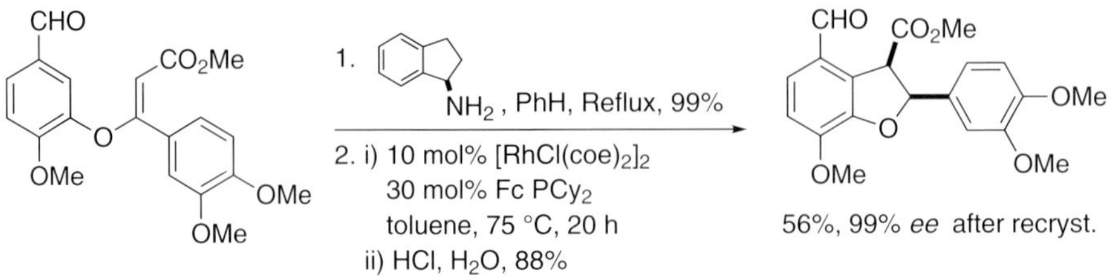

**Scheme 20.6** Enantioselective synthesis of (+)-lithospermic acid fragment by rhodium cyclization.

from that used in their original report (ferrocene-based phosphine). Overall, (+)-lithospermic acid was synthesized in 10 steps with a 5.9% overall yield. This is the first reported synthesis of this natural product. This strategy should be amenable to building other testable analogues due to its simple chemistry.

The group further reported, in two papers, the synthesis of tricyclic mescaline analogues using catalytic rhodium [15]. In 2006, the synthesis of bicyclic indole structures that are biologically active protein kinase C inhibitors was reported [16]. The protein kinase C transformation is easily performed achirally by the use of Wilkinson's catalyst in the same strategy used to make lithospermic acid (through formation of imine, cyclization, and hydrolysis, see Scheme 20.6). Hoping to make the reaction stereospecific, they tailored both the rhodium-supporting ligand and the imine directing group. The best case yielded product in 61% yield with 90% ee. Active c-Jun N-terminal kinase inhibitors were also synthesized using the same type of reaction in 2007 [17].

In 2007, Aissa and Furstner reported a rhodium(I)-catalyzed tandem reaction involving a C−H activation followed by cycloisomerization [18]. They were successful with two different types of transformations, as shown in Scheme 20.7. In both cases, phosphine-supported rhodium complexes gave the best yields. In the pyridine case a silver hexafluoroantiminate additive was required for clean conversion. Deuterium labeling studies support a mechanism in which rhodium activates the vinylic C−H bond, forming a metallacycle that is stabilized by interaction with the pyridine nitrogen.

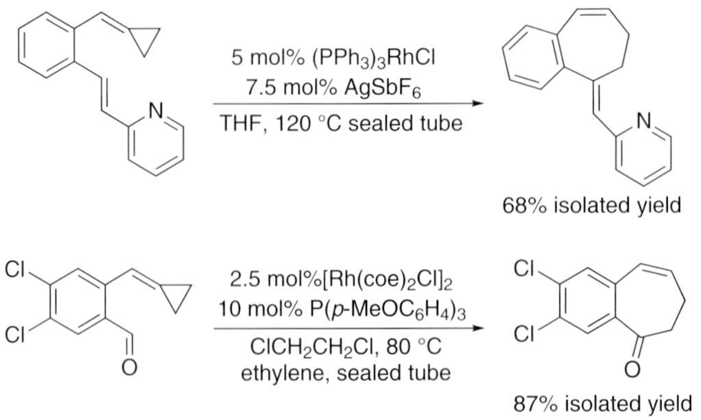

**Scheme 20.7** Rhodium-catalyzed tandem C−H activation/cycloisomerization reaction.

## 20.2.2
## Palladium

Palladium is currently the most reliable metal for aryl C—H bond activation thanks to its versatility and the ability to accomplish a broad range of transformations. There are a number of recent reviews on palladium-catalyzed chemistry. These reviews have covered several specific aspects, including dioxygen-based oxidation of organic substrates, synthesis of heterocycles through π-interactions, and heterocycle synthesis through oxidative additions [19]. More general reviews have focused on similar topics, with equally thorough treatment of palladium [20]. There are a number of different book volumes that describe both the long history and recent aspects of palladium chemistry [21]. In addition, the Sanford, Yu, Shi, and Daugalis groups have made many recent contributions towards expanding the substrate scope in aryl activations [22].

One general transformation that palladium is widely known for is the Heck reaction. The classic intramolecular Heck reaction uses a Pd(0) source which inserts into an aryl–halogen bond in order to add a pendant olefin. The oxidative Heck or Fujiwara–Moritani reaction is an improvement on the normal Heck reaction, which utilizes a Pd(II) source that inserts into an aryl–hydrogen bond, eliminating the need for a halogenated aryl ring (Scheme 20.8). Both have been used extensively in the past to build cyclic systems, and the limits of the oxidative Heck reaction continue to be explored.

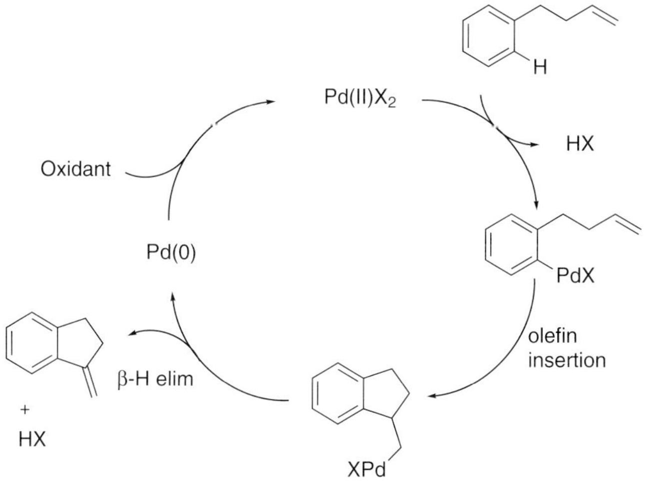

**Scheme 20.8** The Fujiwara–Moritani (oxidative Heck) transformation.

In 2004, Stoltz reported use of the oxidative Heck reaction to form functionalized benzofuran derivatives [23]. The aryl rings must be very electron-rich (at least two methoxy substitutions), but both benzofuran and dihydrobenzofuran rings with a quaternary carbon center could be synthesized in good yields. Cumulenes have also recently been cyclized using a similar strategy [24]. Stoltz and coworkers also reported the palladium-catalyzed annulations of indoles in 2003 [25]. Shown in Scheme 20.9, this reaction uses molecular oxygen as the oxidant and progresses with good yields. In the same year, Broggini and coworkers published their work on similar chemistry with the formation of β-carbolinone structures [26].

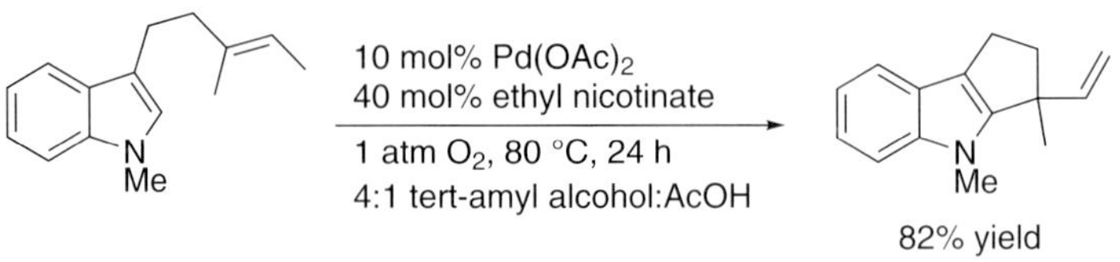

**Scheme 20.9** The annulation of indoles utilizing palladium.

Through the exploration of palladium migrations, Larock has made many contributions to cyclization by aryl C–H bond activations, which have been recently reviewed [27]. In some cases, oxidative Heck chemistry was used to achieve the transformations. The first report of this work came in 2003, describing the use of a 1,4-palladium migration to form a variety of polycyclic structures [28a]. An aryl-halide species was employed to generate the original aryl-palladium species. Aryl C–H bond activation causes migration of the palladium-halogen species, which then activates another aryl C–H bond to form a new C–C bond, leading to a fused cycle. Soon after this report, alkyl to aryl migration was discovered, broadening the scope of polycyclic structures formed [28b].

In 2006, Larock and coworkers reported on the expansion in substrate type and mechanistic studies of an aryl coupling reaction different from those just discussed [29]. The original report, published in 2005, described the formation of carbazoles through a nitrogen-directed vinylic to aryl migration shown in Scheme 20.10 [30]. Based on that reaction, they hypothesized that they should be able to synthesize substituted indoles in the same way. A vinyl to aryl migration in an N-allyl aniline derivative should put the palladium in place to perform an intermolecular Heck reaction. Although the substrate scope is limited, the reaction shown in Scheme 20.11 was successful under the original conditions. They also synthesized dibenzofurans with a substrate scope similar to the carbazoles, even with Pd–O coordination known to be weaker than Pd–N coordination in this type of catalyst system. It should be noted that with unsymmetrical alkynes, isomeric ratios varied

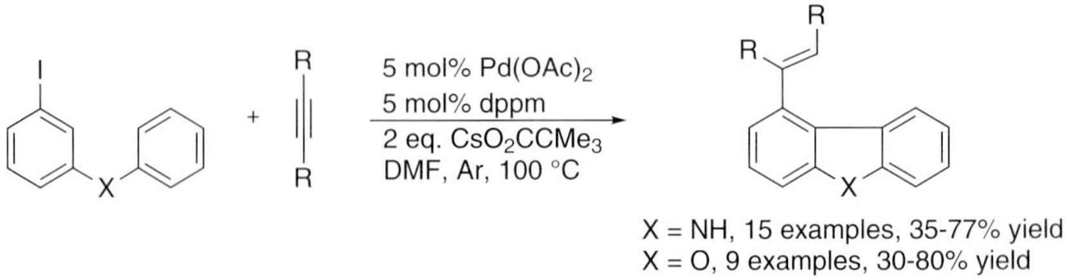

**Scheme 20.10** A vinylic to aryl palladium migration to form carbazoles and dibenzofurans.

from 7 : 1 to 15 : 1 in favor of the product leading to less steric hindrance around the aryl ring. Mechanistically, deuterium labeling studies strongly support the existence of the hypothesized palladium migration.

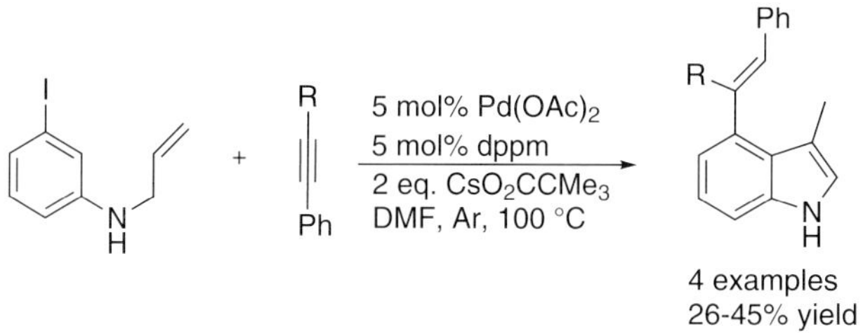

**Scheme 20.11** Substituted indoles through palladium migration.

In terms of general C–C bond formation through aryl C–H activation, a number of complex structures have been designed. A number of groups have built functionalized carbazole structures, in one case using Pd(OAc)$_2$ to couple aryl rings towards the synthesis of the carbazole alkaloid carbazomadurin A [31]. The combination of this chemistry with Buchwald–Hartwig coupling allowed the generation of carbazoles from separate aryl rings in one pot [32]. In 2005, Lautens and coworkers synthesized complex heterocycles through the formation of both an alkyl–aryl and aryl–heteroaryl bond in one pot, shown in Scheme 20.12 [33]. A variety of heterocyclic systems were formed through this reaction type, which coupled indoles, azaindoles, pyrroles and pyrazoles with aryl iodides. By varying the nitrogen substitution on the nitrogen-based rings, tetracyclic ring systems that contained six, seven, and even eight-membered rings were formed.

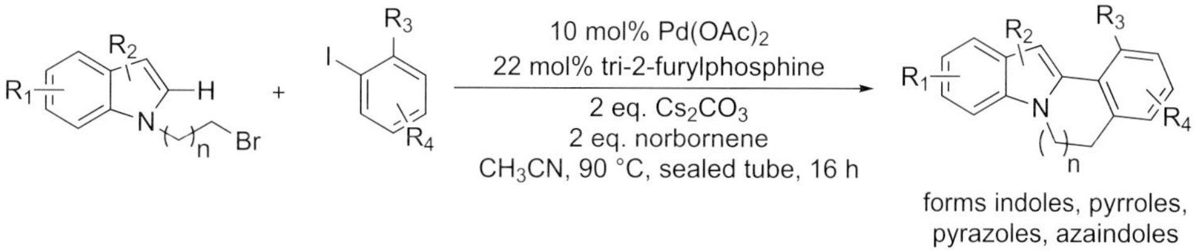

**Scheme 20.12** The synthesis of complex heterocycles through palladium-catalyzed reactivity.

Stoltz and coworkers have used palladium to accomplish natural product syntheses. They reported the total synthesis of (+)-Dragmacidin F utilizing a palladium catalyst in 2004, shown in Scheme 20.13 [34]. The reaction has 74% yield and the overall synthesis takes 19 steps from their starting point.

Another popular reaction type seen in palladium cyclization chemistry is the tandem reaction, in which multiple different bonds are formed in one step. This chemistry has been used to cyclize substrates such as bromoenynes and diaryl anilines [35]. Intermolecular reactions can also lead to cyclized products, which Wolfe and Bertrand demonstrated in the coupling of aryl bromides with cyclopentene

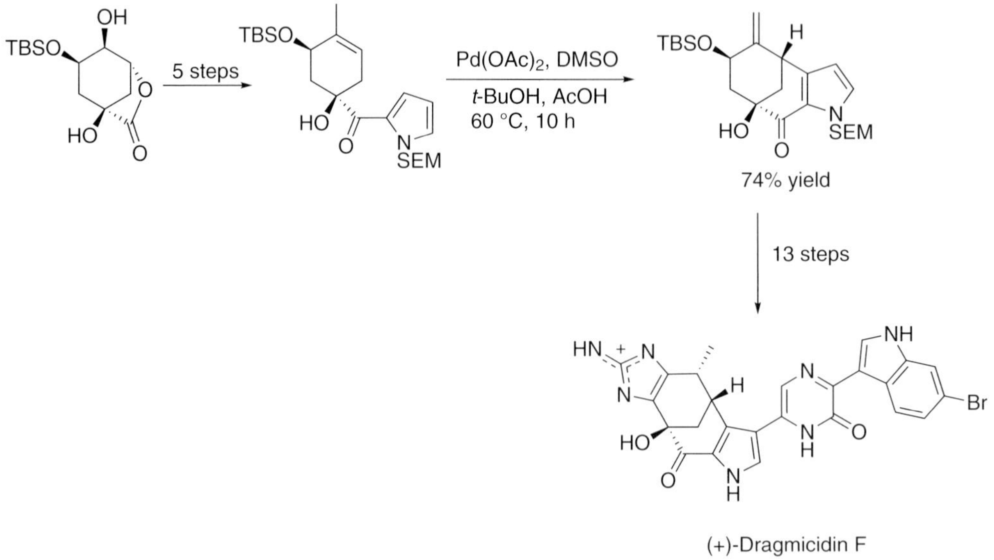

**Scheme 20.13** Palladium-assisted synthesis of Dragmicidin F.

derivatives to form tricyclic systems that contained a cyclobutane ring fused to both an aryl ring and a cyclopentane ring, shown in Scheme 20.14 [36]. Another case is shown in Scheme 20.15; this is an interesting reaction type due to the multifused ring systems that can be formed from relatively simple substrates.

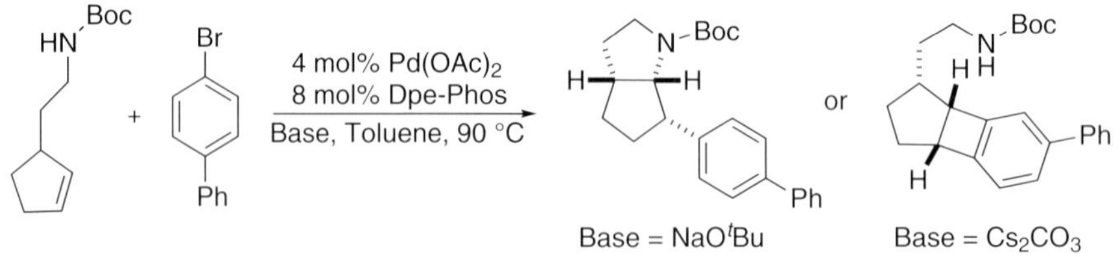

**Scheme 20.14** Palladium-catalyzed tandem reactivity.

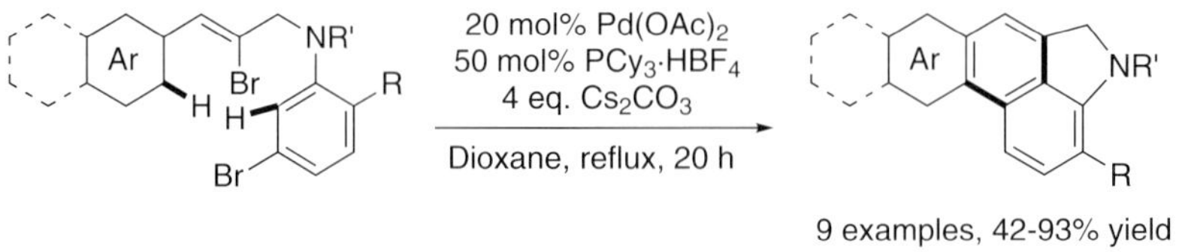

**Scheme 20.15** Formation of fused ring systems utilizing palladium.

Alkynes are popular in palladium-catalyzed cyclizations for their ability to oxidatively add across a C—H bond to form a vinylic palladium species. Groups have recently capitalized on this chemistry to form polycyclic structures, with a popular product being the oxindole species [37]. Zhu and coworkers reported a notable use of this chemistry in 2007 [38]. By generating the active phenyl alkyne species needed for cyclization, they were able to combine Sonogashira coupling with oxidative cyclization reaction in one pot. Shown in Scheme 20.16, as long as the two aryl species are added sequentially, the addition can be controlled to form the highly functionalized oxindole products.

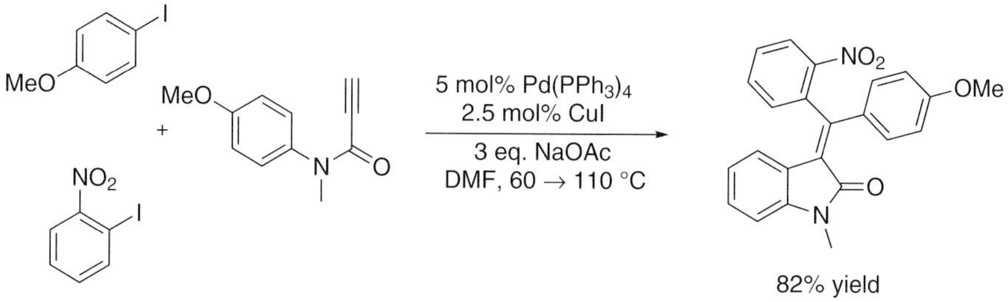

**Scheme 20.16** One-pot Sonagashira coupling and palladium cyclization to form oxindoles.

There are a few examples of C−N bond formation through aryl activation. Larock has contributed to C−N bond formation using his migration chemistry [39]. In 2007, he reported an aryl to imidoyl migration that yielded fluoren-9-one and xanthen-9-one structures. The reaction conditions were the same as in earlier cases (see Scheme 20.9), and mechanistic studies supported a mechanism similar to the other migrations. It should be noted that the mechanisms supported both in this case and in the carbazole formations (see above) include the formation of an unprecedented palladium(IV) hydride species.

In 2005, Buchwald reported the palladium-catalyzed production of carbazoles through an intramolecular C−N bond formation in 2-phenylacetanilides [40]. The catalyst is the readily available Pd(OAc)$_2$ with Cu(OAc)$_2$ as the catalytic oxidant. The reaction will take place in air, although pure O$_2$ leads to slightly higher yields. They postulate that the reaction proceeds via ortho-metallation, a pathway that has precedent and would seem to fit for this type of reaction. One notable result of the substrate scope shows that when the 2-phenyl moiety is substituted regioselectivity is excellent ($\geq 16:1$).

Shi and coworkers used a different strategy to produce carbazoles in early 2008 [41]. In the first step of the reaction, they performed a biaryl coupling of N-acetyltetrahydroquinoline with simple arenes to produce biaryl species. Exposure of the biaryl to Buchwald coupling conditions led to intramolecular C−N bond formation, yielding carbazoles. Even better, they could do this in a single pot; the second transformation proceeded by simply adding additional palladium catalyst. Besides the obvious usefulness of this being a one-pot reaction, the arene couplings proceed without need for halogenation. The catalyst was Pd(OAc)$_2$ with copper(II) triflate as the oxidant, and the reaction proceeds under 1 atm of dioxygen.

In mid-2007, Hiroya and coworkers reported the formation of indazoles from hydrazones shown in Scheme 20.17 [42]. The optimized reaction conditions involved Pd(OAc)$_2$ as the catalytic source, DMSO as a solvent and a temperature of 50 °C.

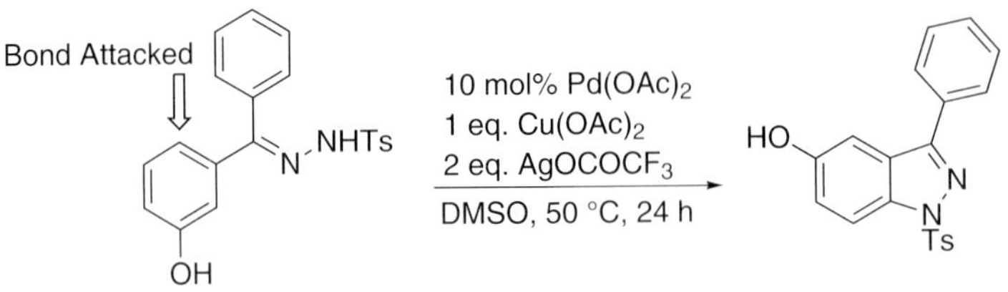

**Scheme 20.17** The formation of indazoles from hydrazones with palladium.

Stoichiometric amounts of both Cu(OAc)$_2$ and silver trifluoroacetate were also needed. The reaction could proceed by eliminating one of the copper or silver species; however catalyst loadings of palladium needed to be increased to see comparable yields.

Observing the reaction scheme, one may note that if the two aryl rings are substituted differently, two different products can be formed, depending on which ring is attacked. Further substrate analysis revealed that in the case of meta-substitution on both aryl rings, an electron-donating substituent led to high selectivity for C—N bond formation on the more electron-rich ring. For substrates with only one aryl ring carrying a functional group, the selection of which aryl ring was attacked depended on the position of that group. In the case of meta-substitution, the more electron-withdrawing the substituent, the more likely it was that the unsubstituted ring was used to make the C—N bond. In the case of para-substitution, the unsubstituted ring was always favored to form the C—N bond. In some cases (OH, NH$_2$, and NO$_2$ substitutions), complete regioselectivity was observed. The trends seen in the substrate analysis would support an aryl ring palladation, which the authors postulate to be the case. A more electron-rich aryl ring would be more likely to be attacked by Pd(OAc)$_2$. It would be interesting to see exactly what role both copper and silver play, and whether or not the roles are the same, that is, re-oxidizing the Pd(0) back to Pd(II).

## 20.2.3
## Gold

In recent years, gold has emerged as an interesting metal for catalysis, and consequently has enjoyed a "renaissance." Due to this, reviews have covered a broad range of gold chemistry; however, only a few selected examples will be covered in this chapter [43].

Our group has reported on some aryl C—H bond activations utilizing gold in recent years. In 2003, we reported the formation of 3-chromanols from (phenoxymethyl)oxiranes using simple gold(III) chloride through a cyclic alkylation pathway [44]. This and a similar reaction reported by us are shown in Scheme 20.18 [45]. The intramolecular formation of an aryl–alkene bond, shown in Scheme 20.19, was realized in 2004 [46]. This reaction can be performed both at room temperature and in the absence of solvent as long as the substrates are liquids. Then, in 2006, we reported the formation of dihydrobenzofurans from aryl allyl ethers catalyzed by simple gold-based phosphine complexes [47].

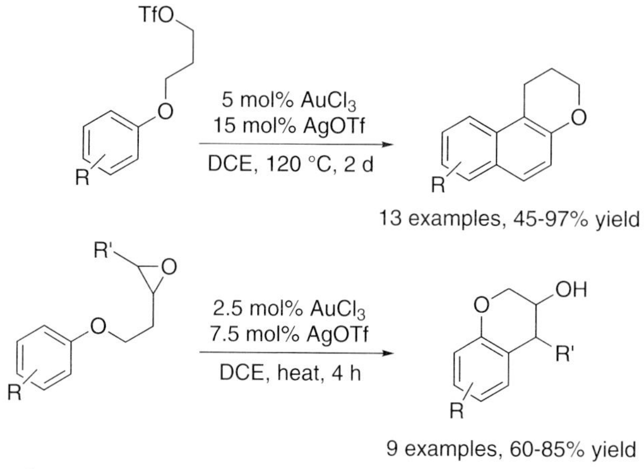

**Scheme 20.18** Gold-based aryl C—H activation.

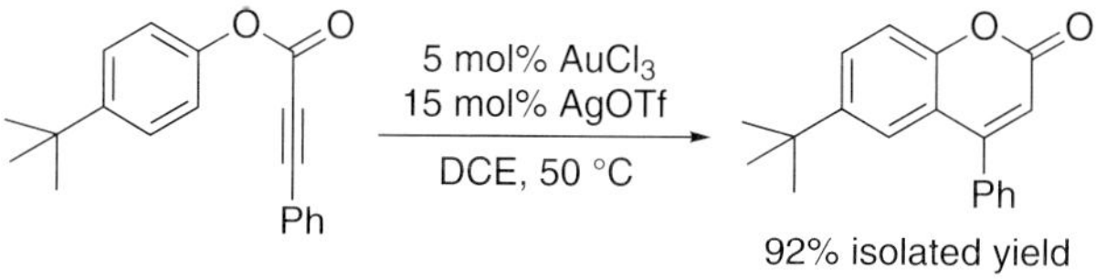

**Scheme 20.19** Intramolecular aryl-alkene bond formation utilizing gold.

Other groups have also reported on some aryl C—H activation reactions utilizing different gold salts. In 2006, Nolan and coworkers reported an intramolecular hydroarylation reaction utilizing a gold(I)-N-heterocyclic carbene catalyst shown in Scheme 20.20, with an expansion of scope and mechanistic study a year later [48]. Other types of transformations achieved by gold include the formation of quinolines, annulation of phenols with dienes, [4 + 3] and [3 + 2] cyclizations, and the hydroarylation of allenes [49].

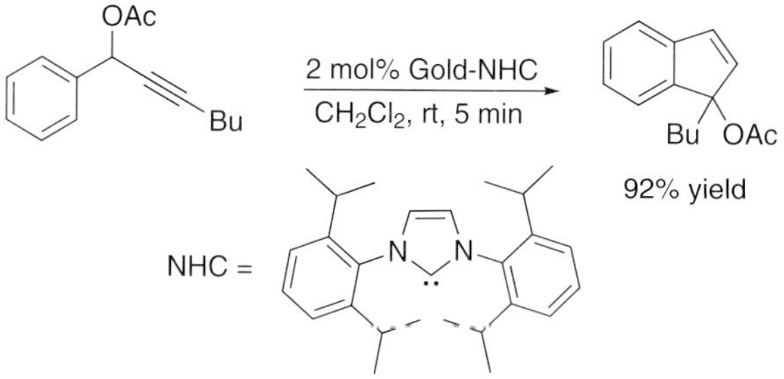

**Scheme 20.20** The cyclization of propargylic substrates utilizing a gold N-heterocyclic carbene system.

## 20.2.4
### Other Metals

Other assorted metals have also contributed to aryl C–H activation chemistry, albeit in lesser amounts. The cyclization of propargylic aryl ethers to form chromenes, dihydroquinolines, and coumarins with $PtCl_4$ was reported by Sames and coworkers [50]. The reaction is sterospecific, and mechanistic studies favor the activation of the alkyne by platinum, followed by attack from the benzene ring to yield the cyclic structure. All products are in the 6-endo form, which does not support a Heck-type mechanism. Sames also reported the hydroarylation of a variety of arene–alkene substrates using $RuCl_3$ with silver trifluoromethylsulfonate as an additive [51]. The substrate scope is broad, with tricyclic systems being formed either with fused aryl ring systems or cyclic alkenes. In addition, chiral ethers were cyclized without racemization.

Yi reported the synthesis of tricyclic nitrogen-based ring systems using a ruthenium carbonyl catalyst [52]. Mechanistic studies support the formation of a cationic ruthenium acetylide species, which adds to the nitrogen by hydroamination, generating a vinyl ruthenium. Subsequent ortho-C–H bond activation followed by another alkyne insertion leads to product formation. In a later work, they expanded the substrate scope to include coupling of arylamines and further supported their proposed mechanism with more detailed studies [53].

Rhenium does not have much catalytic chemistry associated with it, though one notable example is an $sp^3$-carbon activation reported by Hartwig [54]. However, Takai has recently reported some rhenium-based aryl C–H bond activations in the formation of indenes and phthalimindines [55]. Both cases utilize the dirhenium system $[ReBr(CO)_3(thf)]_2$ in catalytic quantities. Exposing a mixture of an aryl aldimine with an alkyne in refluxing toluene leads to the formation of indenes with an amino substitution. Para-substitution on the aldimine was tolerated as long as it was electron donating and ortho-substitution did not give high yields no matter what the functional group. The use of unsymmetric alkynes led to the formation of a single isomer. If an isocyanate instead of an alkyne was reacted with the aldimine, phthalimide derivatives were formed in good yields, regardless of substitution location or electronic nature. Competition studies of alkyne versus isocyanide addition showed the formation of the phthalimide was faster, which shows preference for a polar molecule. This behavior is actually the reverse of that seen with ruthenium and rhodium catalysts of similar reactivity. A commentary on this work and some other rhenium chemistry was published by Horino in 2007, and the reader is encouraged to reference this for further information [56].

Takai later reported a one-pot procedure utilizing this chemistry [57]. In one pot, the aldimine was first formed by the ruthenium-catalyzed reaction of an acetylene with aniline. After complete formation of the aldimine species, the rhenium catalyst was added along with ethyl acrylate to generate the indene derivative, with the aniline being regenerated as a by-product of the reaction.

## 20.3
## Alkyl C—H Bond Activation

### 20.3.1
### C—C Bond Formation

#### 20.3.1.1 Rhodium

Rhodium is a versatile metal for achieving alkyl C—H activation, whether it is forming a C—C or C—N bond. Dirhodium systems are extremely successful in sp$^3$ C—H activation, especially in carbene and nitrene insertion chemistry, which has undergone rapid growth in the last five years. In all these cases, the dirhodium structure has been shown to be crucial in the reactivity of these complexes.

A good introduction to the commonality of dirhodium carbene chemistry is the work of Jung and coworkers [58]. In 2001, they reported the cyclization shown in Scheme 20.21. The catalyst was simple dirhodium tetraacetate, which has been the most robust and reliable of the dirhodium compounds. The product formed is a five-membered ring, which is the preferred ring size for this transformation. The reaction, although simple, shows the power of this catalyst system, with a γ-lactam being formed. In all but a few cases yields are greater than 90%, and no additional products are added to the reaction besides the substrate and catalyst.

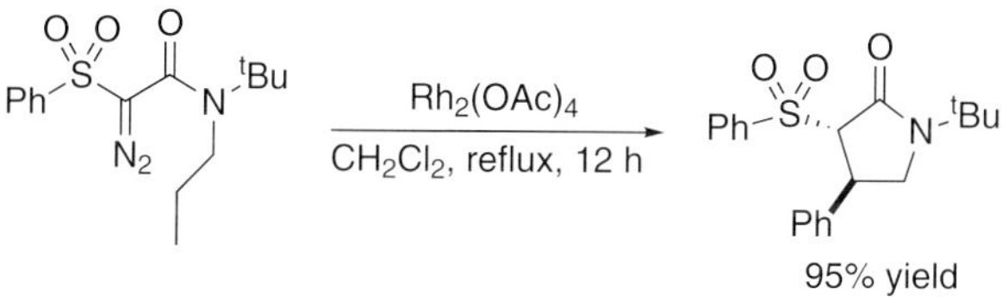

**Scheme 20.21** Basic dirhodium carbene chemistry to create cyclic species.

An interesting use of dirhodium carbene cyclization chemistry was reported by Hashimoto and coworkers in 2001 [59]. With the use of a substrate with two diazo linkages, a double C—H insertion occurred and yielded a spirocyclic compound. The product was formed in 78% yield with 80% ee, with recrystallization giving optically pure spirocycle. The other enantiomer could also be produced by changing the stereochemistry of the rhodium catalyst.

In 2002, Brown and coworkers reported the synthesis of some bicyclic structures utilizing Rh$_2$(OAc)$_4$ [60]. While not the overall aim of the report, the substrates cyclized in high yields to give furofuranone products. A similar report catalyzed the formation of cyclopentane derivatives for studies towards the synthesis of triquinane skeletons [61].

In 2007, Taber and Tian reported the synthesis of α-aryl cyclopentanones using a rhodium catalyst [62]. For this reaction, the order of substrate addition was essential; the diazo species was added to the rhodium catalyst mixture. Reversal of addition led to increased dimerization of the diazo species. Yields were generally good, with the lowest yielding substrate containing a para-methyl substitution. Only a slightly electron-donating moiety seems to have a deleterious effect.

The Afonso group has published some interesting contributions to the rhodium-catalyzed reactions. Their first report, in 2003, examined the cyclization of α-dizao-acetates and α-diazo-acetimides with phosphoryl functionalities to form lactams and lactones [63]. In the case of lactams, varying the substituents on the nitrogen substrate greatly affected both the distribution of products (β-lactams versus γ-lactams) and the conformations of those products. Substrates with a dibenzyl substitution on the nitrogen yielded β-lactam in 81% yield overall, with 95% of the trans-isomer. Switching one benzyl group to a *tert*-butyl group led to a complete reversal in the configuration of the double bond, with 89% overall yield and 88% cis-isomer. They attribute this change to the rigidity of the *tert*-butyl group forcing the benzyl moiety in the six-membered transition state into an axial position, as shown in Scheme 20.22. Electronic effects were shown to play a role in both the size of the lactam formed as well as the distribution of isomers. Using chiral rhodium catalysts, modest ees were later achieved [64].

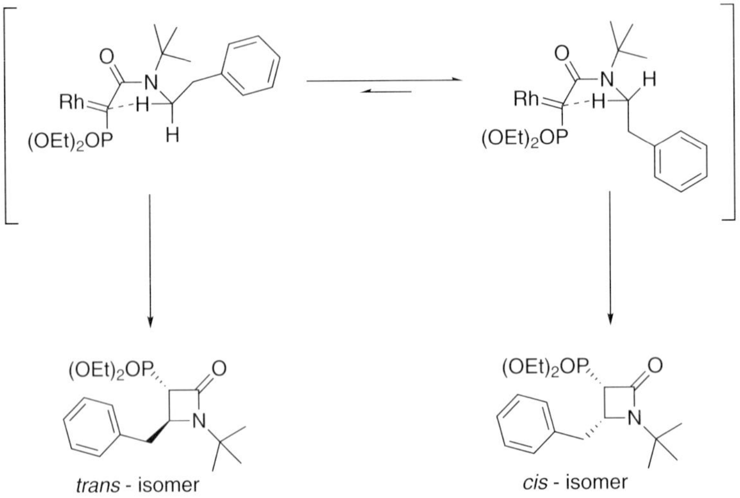

**Scheme 20.22** Functional group control of isomeric products in dirhodium carbene chemistry.

Almost simultaneously, the same cyclization chemistry was reported in a wet ionic liquid, with yields either comparable to or better than the original report [65]. Since the catalyst is completely soluble in the ionic liquid, they could extract out the product, add more substrate, and reuse the catalyst up to six times with no decrease in yield along the way. The reaction was then further extended to take place in water [66]. Just like in the case of ionic liquids, the catalyst was reused up to eleven times before the conversion began to suffer from lower yields. In any sort of wet solvent, the diazo compound should be expected to add water to the carbene center, resulting in simple alcohol formation. They hypothesize that the extremely hydrophobic substrate interacts with the rhodium to create an equally hydrophobic intermediate. With the intermediate being so hydrophobic, the water is not able to access the rhodium-carbene species. They went on to support this hypothesis by showing that substrates with more hydrophilic moieties (n-propoxycarbonyl, for example) gave exclusive alcohol formation. They also showed that by using a more hydrophobic rhodium source, they could further increase the yield and selectivity of cyclized product versus alcohol, with one case yielding 97% cyclized product.

In 2004, Tabber and Joshi published an interesting mechanistic study of intramolecular C—H insertion [67]. By using a thoughtfully designed substrate, they were able to see how catalyst composition affected product formation. Seen in Scheme 20.23, there are a variety of possible products that can be formed, whose distribution can tell about the rate of rhodium carbene formation, ratio of C—H insertion to β-hydride elimination, chemoselectivity, and diastereoselectivity. With regard to the ratio of elimination versus insertion, both electronic and steric effects of the catalyst were observed. In an electronic sense, a more positively charged rhodium center leads to an early transition state and hence more elimination product, which is observed. The steric effect dictates that if the rhodium species is too large, the system arranges itself to proceed by β-hydride elimination. The observed chemoselectivities are attributed to the catalyst polarizability; a methine carbon (C—H) will be more electron rich, and a more polarizable catalyst will interact more favorably with that carbon versus a methylene carbon, which is more electron poor. The diastereoselectivity, which is a result of insertion into a $CH_2$ bond, is believed to be governed by what they call the "tightness" of the transition state, which they defined as "the C—C distance at the point of commitment."

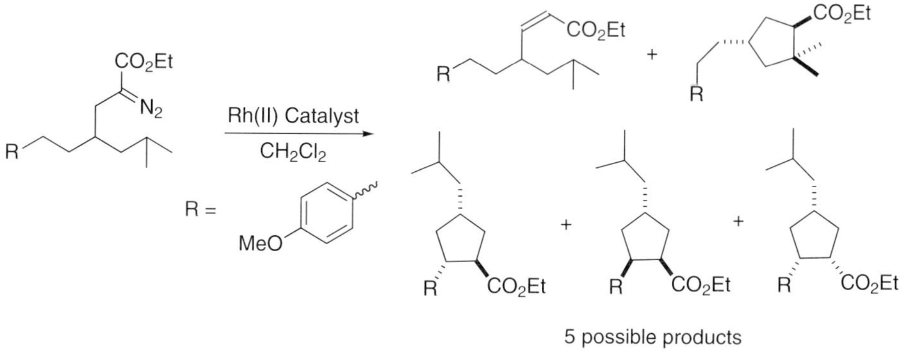

**Scheme 20.23** Understanding mechanism by looking at different products for rhodium cyclization.

In another mechanistic study, Clark and coworkers looked at the products formed with some diazo ketone species [68]. With the initial substrate they used, they were able to isolate a cyclic acetal species, which formed as a result of an abnormal C—H insertion reaction. They further explored this anomalous species, showing that rhodium catalysts with strongly electron-withdrawing ligands increased the relative amount of the acetal. They also revealed that if a substituent is placed on the tether connecting the ether moiety with the diazo functionality, an increase in the amount of C—H insertion product versus cyclopropanation is observed.

What is interesting about this study is that the results they observe suggest a mechanism very different from that commonly accepted for C−H insertion. The formation of the acetal would proceed through a hydride transfer to form an oxenium ion, which could form either the acetal or the normal C−H insertion product. Could the commonly accepted mechanism for this transformation be the wrong one? Using deuterium-labeled substrates, they concluded that the formation of the normal C−H insertion product and the acetal proceeded via different mechanisms; with the commonly accepted mechanism occurring for the normal insertion and with a hydride transfer mechanism occurring for the acetal formation.

It was stated before, and can be seen experimentally, that examination of the rhodium-carbene cyclization products shows an overwhelming favorability for the formation of five-membered ring products. There have been two reported cases in which six-membered products are formed. The first report of this transformation came in 2007 by Novikov and John, who formed six-membered rings using carbethoxy diazosulfones and sulfonates using $Rh_2(OAc)_4$ at room temperature [69]. After synthesizing the substrates in two to three steps they were subjected to the conditions for cyclization. The diazo substrate had to be added to the catalyst mixture over 5 h, and yields were generally low. Later in the same year, Du Bois and coworkers reported an improvement in the same chemistry [70]. Besides expanding the substrate scope, they showed that *in situ* generated iodonium ylides would also cyclize into six-membered products (Scheme 20.24). Yields were better in most cases compared to Novikov, and this method eliminates the formation of unstable diazo compounds. Base and molecular sieves were essential to the success of the reaction. The authors also note that the sparing solubility of the catalyst and iodine source actually facilitates the reaction; if the reaction is tested in a different solvent in which these materials are soluble, the yields decrease significantly.

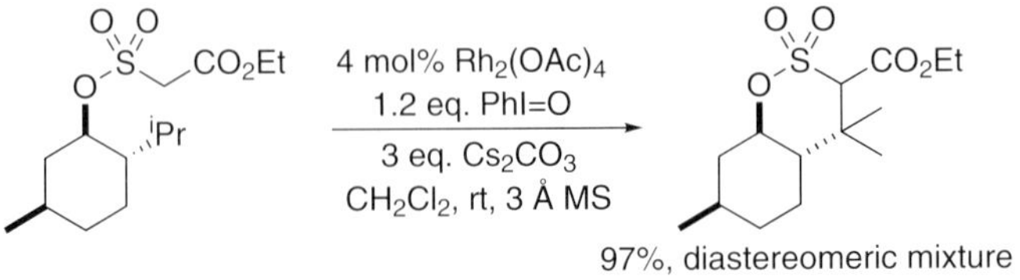

**Scheme 20.24** Formation of six-membered rings by a dirhodium catalyst.

### 20.3.1.2 Palladium

The activation of alkyl C−H bonds by palladium is not as ubiquitous as aryl C−H bond activation, but significant advances have been made, especially in the last five years [22t–v,71].

A very unique palladium-catalyzed C−C bond formation that is able to activate an $sp^3$-hybridized C−H bond to cyclize and form heterocycles was reported by Knöchel [72]. This chemistry has some precedent in cyclization, but it is extremely limited in scope due to the difficulty in activating a strong C−H bond. The reaction reported by Knöchel utilizes a pyrrole moiety to generate the reactivity shown in Scheme 20.25. If the pyrrole ring is unsymmetrically substituted with a phenyl group, a tetracyclic system is formed.

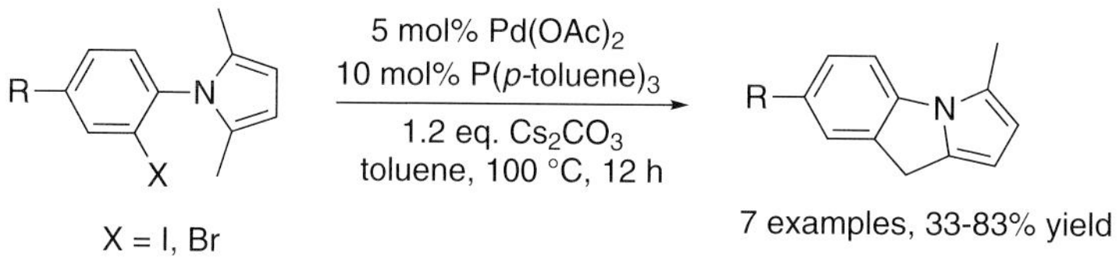

$$R\text{—Ar(X)—N(pyrrole)} \xrightarrow[\substack{1.2 \text{ eq. } Cs_2CO_3 \\ \text{toluene, } 100\,^\circ C, 12\,h}]{\substack{5\text{ mol\% } Pd(OAc)_2 \\ 10\text{ mol\% } P(p\text{-toluene})_3}} R\text{—fused pyrrole product}$$

X = I, Br    7 examples, 33–83% yield

**Scheme 20.25** sp³ C—H bond activation to form heterocycles.

Heterocycles generated by the cleavage of an sp³ C—H bond were also reported by Fagnou [73]. In his case the starting materials are aryl alkyl ethers with a bromo-substitution on the 2-position of the aryl ring. Exposure of this system to a palladium catalyst generates the benzofuran derivatives in high yields. They believe the reaction goes by the formation of a palladium aryl bond with displacement of the bromide, followed by activation of the C—H bond. Computational studies strongly support this mechanism as opposed to palladium activation of the C—H bond first; the Pd(IV)-hydride species gave a $\Delta G$ value of almost 50 kcal mol$^{-1}$. The formation of ω-unsaturated-α-cyano-ketones through a palladium acetate-mediated cyclization was reported in 2003 by Liu and coworkers [74]. The formation of a fused methylcyclopentane takes place under mild conditions and only forms a single stereoisomer in each case.

Palladium, like rhodium, has also aided in the synthesis of complex natural products. In 2002, Sames and coworkers reported the synthesis of the core of teleocidin B4, a complex natural product [75]. Two of the steps in the formation of the core utilized palladium and are shown in Scheme 20.26.

### 20.3.1.3 Other Metals

In 2004, Sames reported the first cross-coupling of an sp³ C—H bond with an alkene that has an amide functionality using an iridium catalyst [76]. This reaction shows good functional group tolerance, with the coupling proceeding smoothly with a silyl ether substitution. In addition, the reaction is stereospecific.

The stereochemical aspects of intramolecular C—H insertion reactions of iron carbenes was studied by Helquist [77]. Iron carbenes are well-known species, and have previously been used in the synthesis of natural products. Their studies, which included an isotope effect measurement, strongly support a concerted insertion into a C—H bond. Most of the racemic substrates were cyclized with >90% diastereos-electivity as long as neighboring functionalities were bulky enough. The same group later used this information for the total synthesis of sterpurene.

An isolated example of a gold(I)-catalyzed hyroacylation reaction was reported by Skouta and Li in 2007 [78]. Using a gold(I)-phosphine complex, the annulatation of a variety of salicylaldehydes with different aryl acetylenes occurred, yielding isoflavanone

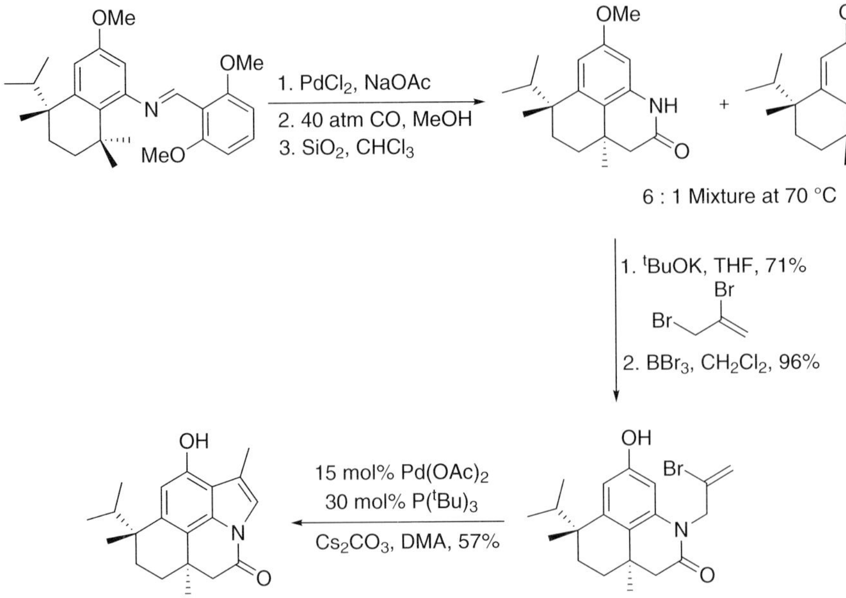

**Scheme 20.26** Synthesis of the teleocidin B4 core.

core structures. While the proposed mechanism involves the activation of the aldehydic C−H bond, the authors do note that a gold carbine-based mechanism may be more operative, especially considering the literature precedent.

### 20.3.2
### C−N Bond Formation

C−N bond formations continue to be both highly valued and highly studied in C−H activation chemistry thanks to the abundance of natural products containing nitrogen functionalities and heterocycles. The great potential in C−N bond formation chemistry was realized with a report from Breslow, who, in the early 1980s demonstrated both inter- and intramolecular nitrene transfer using both dirhodium tetraacetate and an iron porphyrin catalyst, which is shown in Scheme 20.27 [79]. However, ways to generate C−N bonds in a facile, enantioselective manner are still few and far between. In the last five years, great strides have been made in this area, and further efforts will undoubtedly lead to new results.

Because of its impact and importance, many recent and thorough reviews have been published on catalytic C−N bond formations. In 2003, Muller and Fruit published a general review on nitrene transfers into C−H bonds, which covers much of the relevant chemistry through 2002 [3]. This review is a nice introduction to nitrene transfers in general, complete with historical background. A more specific mini-review on intramolecular aminations was published in 2005 by Davies and Long, and covers a good amount of cyclization chemistry from roughly 2002 to mid-2005 with a focus on rhodium systems [80]. Davies also published a minireview in

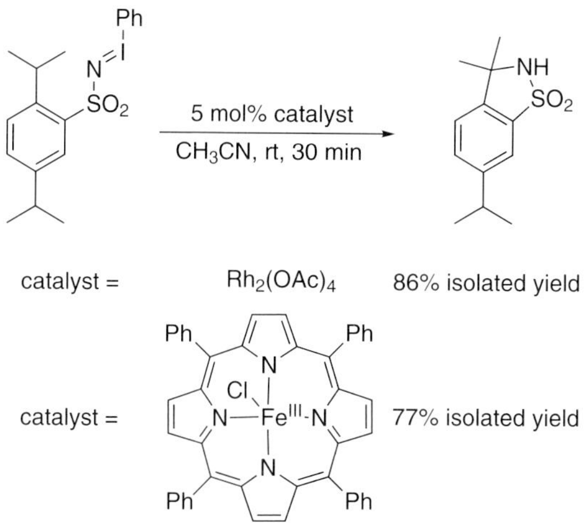

**Scheme 20.27** Breslow's initial nitrene transfer reactivity.

2008 which covers nitrene transfer chemistry in some detail, including applications toward the synthesis of natural products [81].

### 20.3.2.1 Rhodium

The Du Bois group at Stanford University has made many recent advances in rhodium-mediated C—N bond formations. In 2001, they reported the cyclization of carbamates to form oxazolidinones, shown in Scheme 20.28 [82]. This initial work led to many subsequent publications involving substrate expansions, catalyst optimizations, and natural product strategies. Later reports achieved the cyclization of different sulfamate esters to form 1,3-difunctionalized amines, propargylic amines, and N,O acetals [83]. In addition, they designed a new catalyst system $Rh_2(esp)_2$, shown in Figure 20.1, which has increased reactivity [84].

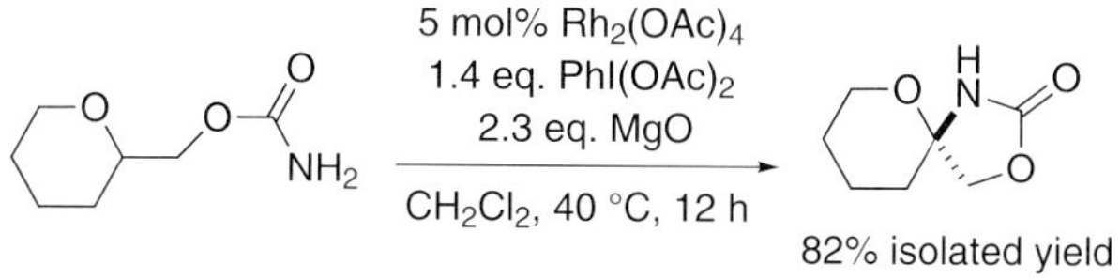

**Scheme 20.28** Carbamate cyclization through nitrene transfer by a dirhodium catalyst.

In 2004, Fruit and Muller reported the generation of cylic sulfamidates from aromatic sulfamate esters utilizing a rhodium catalyst [85]. By using the supporting ligand show in Figure 20.2, modest enantioselectivity was garnered with catalyst loadings as low as 2 mol%. If cyclization of the sulfamate with an allylic functionality was attempted, only intramolecular aziridination took place, yielding a seven-membered ring structure with lower ee values.

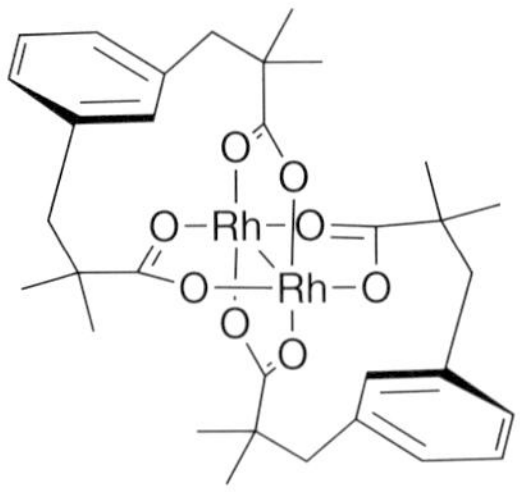

**Figure 20.1** Rh$_2$(α,α,α′,α′-tetramethyl-1,3-benzenedipropionate)$_2$ (Rh$_2$(esp)$_2$).

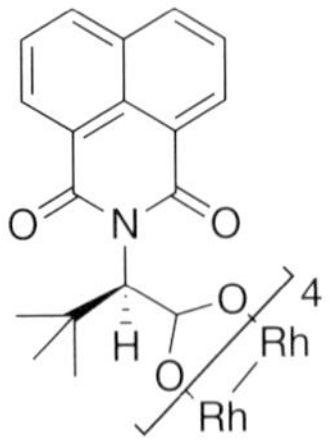

[Rh$_2${(S)-nttl}$_4$]

**Figure 20.2** N-1,8-naphthoyl-t-leucine supported di-rhodium catalyst.

In 2005, Du Bois and Wehn reported an extension of the same C—H amination strategy [86]. They used simple Rh$_2$(OAc)$_4$ to generate cyclic sulfamidates with functionalities different from those previously reported. They also generated the seven-membered aziridine structures, and extended this chemistry by performing a nucleophilic ring-opening addition of alcohols to form eight-membered rings. No ee was observed, but in their case it was not needed, as the cyclic sulfamidates were used to generate more C—C and C—N bonds utilizing a nickel cross-coupling strategy. In 2006, Du Bois reported some success in the cyclization of urea and guanidine derivatives [87]. They were able to cyclize a variety of carbamate and guanidine starting materials with good to excellent yields and retention of stereochemistry. In all cases, the trichloroethoxysulfonyl (Tces) moiety was required; the free amine substrates did not yield any recognizable products.

Lebel and coworkers in 2005 reported an extremely useful synthesis of oxazolidinones from N-tosyloxycarbamates [88]. In most reported aziridinations and C—H bond insertions, hypervalent iodine species are used as the nitrogen sources, with a by-product being stoichiometric amounts of iodobenzene. Lebel first tried to avoid this by the use of an azidoformate, which would leave only gaseous nitrogen as the by-product. When this failed, different alkoxy substituents were attempted, with the tosyl group giving the highest conversion. Yields were good for a variety of substrates, with activation of both benzylic and aliphatic C—H bonds, including 64% yield for the cyclization of a secondary aliphatic C—H bond. Allylic N-tosyloxycarbamates were converted to aziridines as well. Both reaction types occurred with complete retention of stereochemistry, further proving the usefulness of this method.

In 2006, Davies and Reddy reported a new rhodium system capable of performing enantioselective intramolecular amidations [89]. The rhodium catalyst, shown in Scheme 20.29, is based on an adamantylglycine framework. The intramolecular amination substrates were the same ones used originally by Lebel in 2005, with good yields and ee values.

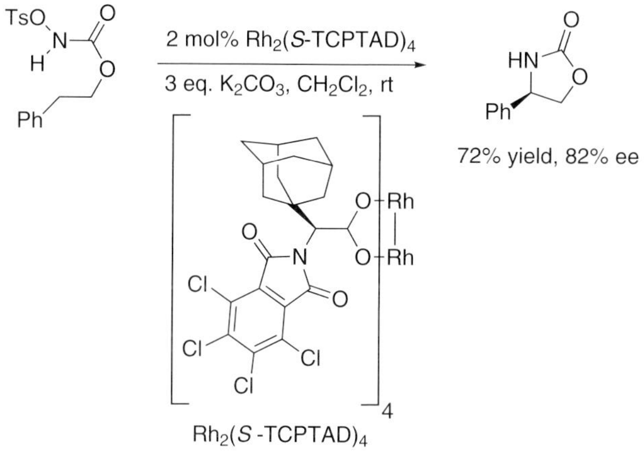

**Scheme 20.29** Enantioselective intramolecular C—H amination by a dirhodium system.

Rhodium-based alkyl activation has been used towards the synthesis of complex natural product scaffolds. In 2003, Du Bois reported a stereoselective synthesis of (−)-tetrodoxin, an interesting guanidium poison [90]. Two stereoselective steps in the natural product synthesis utilized dirhodium and involved both carbene and nitrene insertions, shown in Scheme 20.30. In 2007, Conrad and Du Bois reported some attempts to use C—H amination strategies to make the natural product aconitine [91]. This is a complex molecule, with a distinctive diterpenoid frame containing 19 carbons. Using $Rh_2(OAc)_4$, tsulfamate esters cyclized to form N,O-acetal species. By switching to the $Rh_2(esp)_2$ catalyst developed in their laboratory and shown in Figure 20.1, they were able to regioselectively form a C—N bond in a complex precursor for the aconitine structure, shown in Scheme 20.31. Subsequent synthetic steps on this molecule towards the natural product unfortunately led to messy product mixtures, and they determined that the strategy put forth was not amenable to constructing the ring system. Even with this drawback, these two approaches show the synthetic power of a well-established catalytic system.

### 20.3.2.2 Palladium

Shi and coworkers reported an interesting case of palladium-catalyzed chemistry in 2006 [92]. Using only phenylethylamines, they synthesized trisubstituted pyrrole

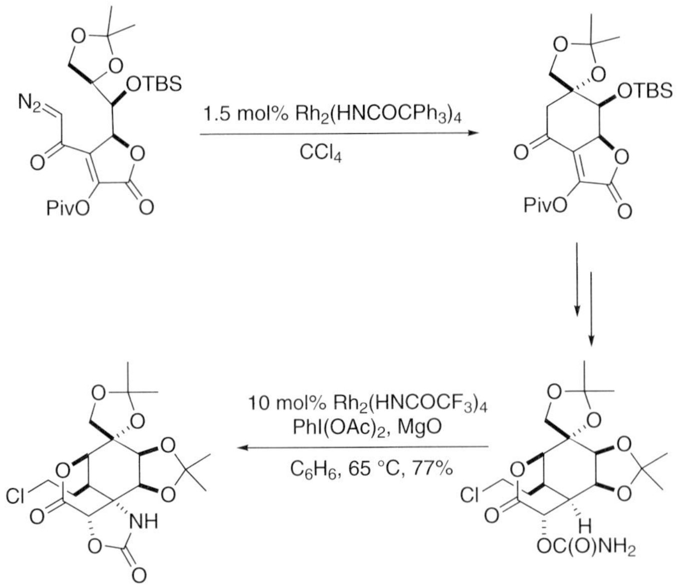

**Scheme 20.30** Rhodium-assisted total synthesis of (−)-tetrodoxin.

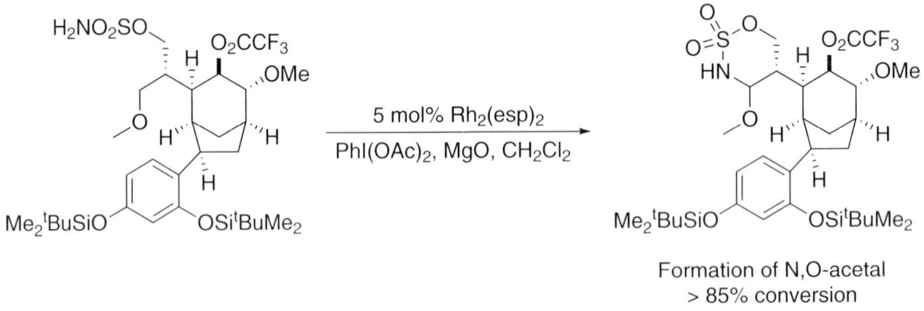

**Scheme 20.31** Use of rhodium towards the synthesis of complex *N,O*-acetal structures.

derivates using a palladium catalyst with 1 equiv of Cu(OAc)$_2$. This process, shown in Scheme 20.32, involves the breaking of a remarkable twelve different bonds (6 N−H, 2 C−N, 4 C−H) and the formation of five new bonds (2 C−C, 3 C−N).

White and Fraunhoffer reported the first example of a catalytic allylic C−H amination reaction shown in Scheme 20.33 [93]. The catalyst is an interesting palladium system with a sulfoxide supporting ligand. Yields are good for all substrates, and if a chiral starting material is used, diastereoselectivities of 6 : 1 or better in most cases are observed. They demonstrated the utility of this reaction by converting the products to *syn*-1,2-amino alcohols, which are widely seen in natural products. This method provides a more facile way to generate these functionalities, which should be useful for future syntheses.

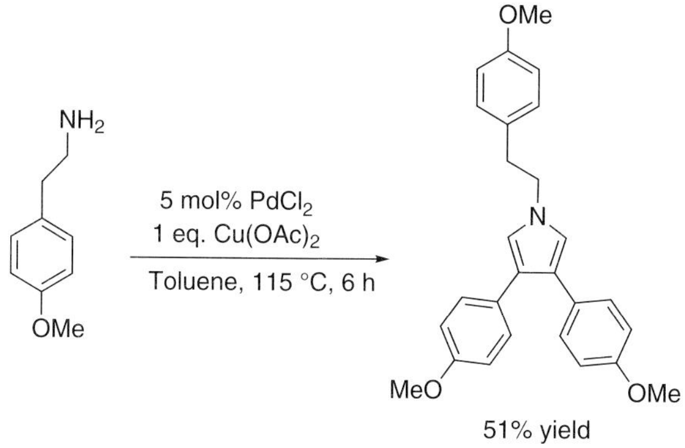

**Scheme 20.32** The formation of five new bonds in one pot with a palladium catalyst.

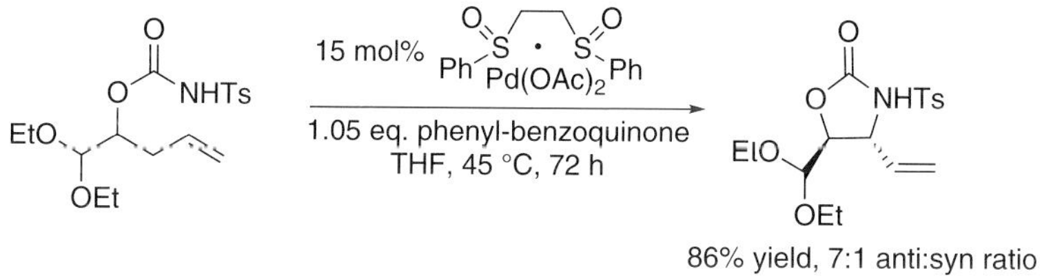

**Scheme 20.33** Allylic intramolecular C—H amination with a palladium catalyst.

### 20.3.2.3 Other Metals

The Che group at the University of Hong Kong has made important contributions to metal-catalyzed nitrene transfer chemistry. Extensive studies of ruthenium porphyrin systems in the 1990s led to isolation of a variety of nitrene-bound ruthenium systems. These systems showed activity for nitrene transfer, both in the aziridination of double bonds as well as in the C—H activation of simple alkanes [94]. These works laid the foundation for many reports that followed.

In 2002, his group reported the successful intramolecular cyclization of sulfamate esters with two ruthenium porphyrin systems shown in Figure 20.3 [95]. Their typical procedure involved 2 equiv of PhI(OAc)$_2$ as oxidant, with Al$_2$O$_3$ as a basic additive. Reactions were performed at 40 °C and in dichloromethane. Catalyst loadings were as low as 0.1 mol%, giving the highest turnover number (300) seen for this type of

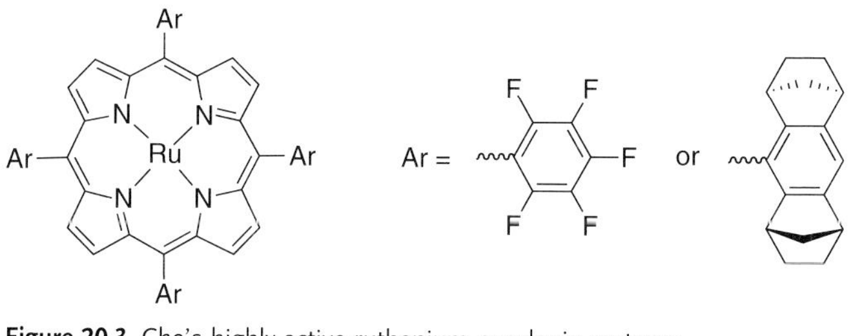

**Figure 20.3** Che's highly active ruthenium porphyrin systems.

reaction. By making the ruthenium porphyrin chiral, enantioselectivity was achieved. In this case, the equivalents of oxidant and base were slightly different, and solvent was switched to benzene. Lower reaction temperature increased the ee while it decreased the overall yield of product. This chemistry is very impressive, with all optimized ee values greater than 80%. Later, they reported the use of a manganese (III) Schiff-base complex that was successful in the cyclization of sulfamate esters [96]. The best improvement of this system over Che's previously reported systems is the use of a cheaper metal and a ligand system that is more easily synthesized and modified when compared to porphyrin frameworks.

Silver has shown some promise in both intra- and intermolecular C—N bond formation reactions. In 2004, we reported on the intramolecular amidation of saturated C—H bonds with a silver-terpyridine catalyst that was previously shown to be active in aziridination [97]. The catalyst and general reaction are shown in Scheme 20.34. In all cases, a stoichiometric amount of PhI(OAc)$_2$ was required to generate the active nitrene species. Both carbamate and sulfamate species cyclized. Sulfamate esters formed six-membered ring products, something rhodium has some difficulty in achieving. The substrate scope is fairly large, and includes activation of both benzylic and aliphatic C—H bonds with good yields. In addition, the reaction was stereoselective. Later, it was discovered that a silver-bathophenanthroline catalyst was able to perform the same reactions at 50 °C versus 80 °C for the terpyridine catalyst with higher yields of cyclized product [98]. This new system also mediates intermolecular C—H aminations.

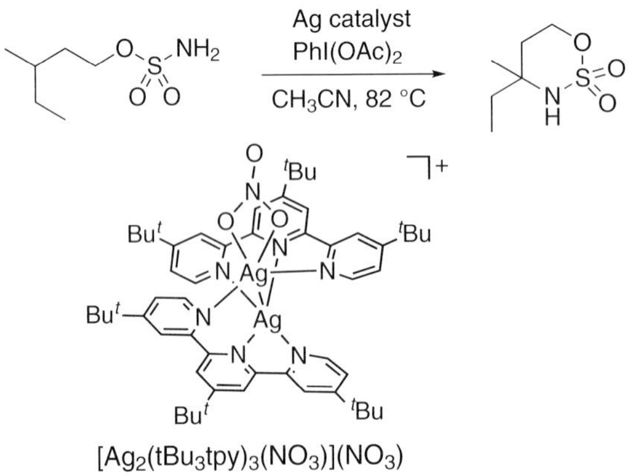

**Scheme 20.34** Intramolecular amidation by C—H activation with a silver catalyst.

## 20.3.3
### C—O Bond Formation

One recently reported reaction by Hashmi and coworkers utilizes gold to form isochromenes at room temperature [99]. This reaction is not easily controlled, with one by-product showing the formation of eight new bonds based on a dimerization of the substrates.

The activation of an sp$^3$ C—H group to form a C—O bond in the hydroxylation of L-valine was reported in 2001 by Sames [100]. They were able to achieve this process with catalytic amounts of $K_2PtCl_4$ along with stoichiometric amounts of $CuCl_2$ as an oxidant. They showed that this reaction was in fact not a radical process, and a variety of amino acids were intramolecularly hydroxylated to form cyclic species. The selectivity for ring formation was not good; nonetheless, this transformation is rare and hence valuable.

As shown in this chapter, the development of facile C—H activation strategies has allowed the laboratory synthesis of complex structures. The demand for more efficient and economic syntheses of both simple and complex molecules will undoubtedly lead to new methodologies in the future.

## 20.4
## Experimental: Selected Procedures

### 20.4.1
### Experimental Protocol for Reaction Shown in Scheme 20.4

In a water-free environment (dry box), a solution of [RhCl(coe)$_2$]$_2$ (5 mol%), chiral phosphine ligand (15 mol%), and imine were mixed in toluene (solution 0.1 M in imine). The mixture was stirred in a 125 °C oil bath for 2 h. The reaction was cooled, concentrated, and the product hydrolyzed by addition of aqueous 1 N HCl with stirring for 3 h. The reaction mixture was extracted three times with ethyl acetate, dried, filtered, and concentrated. The mixture was purified by dry loading onto a silica gel column followed by elution to yield pure product [12].

### 20.4.2
### Experimental Procedure for Reaction Shown in Scheme 20.13

A flame-dried Schlenk tube with a stir bar was cooled under a nitrogen atmosphere. Added to the tube were aryl bromide (1.2 equiv), Pd(OAc)$_2$ (4 mol%), dpe-phos (8 mol%) and Cs$_2$CO$_3$ (2.3 equiv). To this mixture was added the N-protected amine (1.0 equiv) in dioxane (4 mL of solvent for every 1 mmol amine) via a syringe. This mixture was heated to 100 °C with stirring. The reaction was monitored by GC, and when complete was cooled to room temperature. Saturated aqueous ammonium chloride (1 mL) and ethyl acetate (1 mL) were added to the mixture. The layers were separated, and the water layer was further washed with additional ethyl acetate. The combined organic layers were dried with Na$_2$SO$_4$, filtered, and concentrated. The crude product was purified on silica gel by flash chromatography [36].

### 20.4.3
### Experimental Protocol for Reaction Shown in Scheme 20.19

In a round-bottom flask with a septum top and in the absence of light, 0.02 mmol AgBF$_4$ was added to a solution of [(IPr)AuCl] (0.02 mmol) in anhydrous dichloromethane (35 mL). The solution became cloudy, and a solution of the propargylic acetate (1 mmol) in anhydrous dichloromethane (5 mL) was added through the septum. Reaction completion was monitored by TLC, and upon completion, the solvent was removed. The crude reaction mixture was dissolved in pentane, filtered through a Celite pad, and evaporated. Next, the mixture was purified on silica gel by flash chromatography [48a].

### 20.4.4
**Experimental Protocol for Reaction Shown in Scheme 20.23**

Over a period of roughly 1 min, a solution of diazosulfonate (1.4 mmol) in 10 mL dichloromethane was added to a refluxing solution of $Rh_2(O_2CR)_4$ (0.02 equiv) in 15 mL dichloromethane. An additional 3 mL dichloromethane ensured quantitative transfer of the diazosulfonate. The resulting green mixture was refluxed for another 60 min and then cooled to room temperature. The mixture was filtered through a Celite pad, using extra dichloromethane to wash both the reaction flask and Celite pad. The filtrate was concentrated and the mixture was purified by silica gel flash chromatography [70].

### 20.4.5
**Experimental Protocol for Reaction Shown in Scheme 20.24**

A mixture of the starting aryl bromide (1 mmol), $Pd(OAc)_2$ (5 mol%), tri(p-tolyl)phosphine (10 mol%), and $Cs_2CO_3$ (1.2 mmol) was heated at 110 °C in toluene (5 mL) in a sealed tube under a nitrogen atmosphere for 12 h. The reaction was cooled to room temperature, 10 mL of water was added, and the reaction was extracted three times with diethyl ether. The combined organic fractions were washed with a brine solution, dried over sodium sulfate, and concentrated. The crude mixture was purified by flash chromatography (silica gel, 10 : 1 hexane/diethyl ether) to yield the pure product as white solid in 83% yield [72].

### 20.4.6
**Experimental Protocol for Reaction Shown in Scheme 20.28**

To a solution of the substrate shown in Scheme 20.28 (0.5 mmol) in dichloromethane (10 mL) was added 3 equiv of $K_2CO_3$ and $Rh_2(S\text{-TCPTAD})_4$ (2 mol%). The mixture was stirred at room temperature for 4 h. The mixture was filtered, concentrated *in vacuo*, and purified by flash chromatography on silica gel [89].

### 20.4.7
**Experimental Protocol for Reaction Shown in Scheme 20.33**

To prepare the catalyst, silver nitrate (0.01 mmol) and *tert*-$Bu_3$tpy (0.01 mmol) were mixed for 10 min in $CH_3CN$. This mixture was then added to a mixture of the sulfamate ester (0.2 mmol) and $PhI(OAc)_2$ (0.28 mmol) in $CH_3CN$. The reaction mixture was stirred at 82 °C in a sealed tube for 6–12 h. During this time, the reaction mixture changed from a pale yellow to dark brown. The mixture was cooled, concentrated *in vacuo*, and purified by flash chromatography on silica gel (3 : 1 hexanes/ethyl acetate) [97].

## Abbreviations

| | |
|---|---|
| $^t$Bu | *tert*-butyl |
| DMSO | dimethylsulfoxide |
| TBS | tri-butyl silyl ether protecting group |
| OAc | acetate |
| SEM | 2-(trimethylsilyl)ethoxymethyl protecting group |
| Dpe-phos | bis(2-diphenylphosphinophenyl)ether |
| Cy | cyclohexyl |
| Dppm | bis[diphenylphosphino]methane |
| Ph | phenyl |
| Coe | cyclooctene |
| Fc | ferrocene |
| TFA | trifluoroacetate |
| DIPEA | $N,N'$-diisopropylethylamine |
| Tpy | terpyridine |
| Bn | benzyl |
| Ts | tosyl |
| Boc | *tert*-butyloxycarbonyl |
| $^i$Pr | isopropyl |

## References

1 Shilov, A.E. and Shul'pin, G.B. (1997) *Chem. Rev.*, **97**, 2879–2932.

2 Murai, S., Kakiuchi, F., Sekine, S., Tanaka, Y., Kamatani, A., Sonoda, M. and Chatani, N. (1993) *Nature*, **366**, 529–531.

3 (a) Dick, A.R. and Sanford, M.S. (2006) *Tetrahedron*, **62**, 2439–2463; (b) Davies, H.M.L. and Beckwith, R.E.J. (2003) *Chem. Rev.*, **103**, 2861–2904; (c) Muller, P. and Fruit, C. (2003) *Chem. Rev.*, **103**, 2905–2919.

4 (a) Klinman, J.P. and Dubeck, M. (1963) *J. Am. Chem. Soc.*, **85**, 1544–1545; (b) Horiie, S. and Muruhashi, S. (1960) *Bull. Chem. Soc. Jpn.*, **23**, 247–249; (c) Chatt, J. and Davidson, J.M. (1965) *J. Chem. Soc.*, 843–855; (d) Chency, A.J. and Shaw, B.L. (1972) *J. Chem. Soc., Dalton Trans.*, **754**, 860–865; (e) Garnett, J.L. and Hodges, R.J. (1967) *J. Am. Chem. Soc.*, **89**, 4546–4547; (f) Tyabin, M.B., Shilov, A.E. and Shteinman, A.A. (1971) *Dokl. Akad. Nauk. SSSR*, **198**, 380–383.

5 (a) Crabtree, R.H. (1985) *Chem. Rev.*, **85**, 245–269; (b) Kakiuchi, F. and Murai, S. (2002) *Acc. Chem. Res.*, **35**, 826–834.

6 Lewis, L.N. and Smith, J.F. (1986) *J. Am. Chem. Soc.*, **108**, 2728–2735.

7 Jordan, R.F. and Taylor, D.F. (1989) *J. Am. Chem. Soc.*, **111**, 778–779.
8 Gupta, S.K. and Weber, W.P. (2000) *Macromolecules*, **33**, 108–114.
9 Jun, C.-H., Hong, J.-B., Kim, Y.-H. and Chung, Y.-K. (2000) *Angew. Chem. Int. Ed.*, **39**, 3440–3442.
10 Tan, K.L., Bergman, R.G. and Ellman, J.A. (2001) *J. Am. Chem. Soc.*, **123**, 2685–2686.
11 (a) Tan, K.L., Bergman, R.G. and Ellman, J.A. (2002) *J. Am. Chem. Soc.*, **124**, 3202–3203; (b) Wiedemann, S.H., Lewis, J.C., Ellman, J.A. and Bergman, R.G. (2006) *J. Am. Chem. Soc.*, **128**, 2452–2462.
12 Thalji, R.K., Ahrendt, K.A., Bergman, R.G. and Ellman, J.A. (2001) *J. Am. Chem. Soc.*, **123**, 9692–9693.
13 Thalji, R.K., Ellman, J.A. and Bergman, R.G. (2004) *J. Am. Chem. Soc.*, **126**, 7192–7193.
14 O'Malley, S.J., Tan, K.L., Watzke, A., Bergman, R.G. and Ellman, J.A. (2005) *J. Am. Chem. Soc.*, **127**, 13496–13497.
15 (a) Ahrendt, K.A., Bergman, R.G. and Ellman, J.A. (2003) *Org. Lett.*, **5**, 1301–1303; (b) Thalji, R.K., Ahrendt, K.A., Bergman, R.G. and Ellman, J.A. (2005) *J. Org. Chem.*, **70**, 6775–6781.
16 Wilson, R.M., Thalji, R.K., Bergman, R.G. and Ellman, J.A. (2006) *Org. Lett.*, **8**, 1745–1747.
17 Rech, J.C., Yato, M., Duckett, D., Ember, B., LoGrasso, P.V., Bergman, R.G. and Ellman, J.A. (2007) *J. Am. Chem. Soc.*, **129**, 490–491.
18 Aissa, C. and Furstner, A. (2007) *J. Am. Chem. Soc.*, **129**, 14836–14837.
19 (a) Stahl, S.S. (2004) *Angew. Chem. Int. Ed.*, **43**, 3400–3419; (b) Zeni, G. and Larock, R.C. (2004) *Chem. Rev.*, **104**, 2285–2310; (c) Zeni, G. and Larock, R.C. (2006) *Chem. Rev.*, **106**, 4644–4680.
20 (a) Wolfe, J.P. and Thomas, J.S. (2005) *Curr. Org. Chem.*, **9**, 625–655; (b) Alberico, D., Scott, M.E. and Lautens, M. (2007) *Chem. Rev.*, **107**, 174–238; (c) Fairlamb, I.J.S. (2007) *Annu. Rep. Sect. B (Org. Chem.)*, **103**, 68–89; (d) Nakamura, I. and Yamamoto, Y. (2004) *Chem. Rev.*, **104**, 2127–2198.
21 (a) Li, J.J. and Gribble, G.W. (2000) *Palladium in Heterocyclic Chemistry. A Guide for the Synthetic Chemist*, vol. 20, Pergamon (An Imprint of Elsevier), Amsterdam; (b) Negishi, E. and de Meijere, A. (2002) *Handbook of Organopalladium Chemistry for Organic Synthesis*, John Wiley and Sons; (c) Dyker, G. (2005) *Handbook of C–H Transformations: Applications in Organic Synthesis*, Wiley; (d) Tsuji, J. (1997) *Palladium Reagents and Catalysts: Innovations in Organic Synthesis*, John Wiley and Sons.
22 (a) Hull, K.L., Lanni, E.L. and Sanford, M.S. (2006) *J. Am. Chem. Soc.*, **128**, 14047–14049; (b) Kalyani, D., Dick, A.R., Anani, W.Q. and Sanford, M.S. (2006) *Org. Lett.*, **8**, 2523–2526; (c) Deprez, N.R., Kalyani, D., Krause, A. and Sanford, M.S. (2006) *J. Am. Chem. Soc.*, **128**, 4972–4973; (d) Desai, L.V., Malik, H.A. and Sanford, M.S. (2006) *Org. Lett.*, **8**, 1141–1144; (e) Kalyani, D. and Sanford, M.S. (2005) *Org. Lett.*, **7**, 4149–4152; (f) Dick, A.R., Kampf, J.W. and Sanford, M.S. (2005) *J. Am. Chem. Soc.*, **127**, 12790–12791; (g) Kalyani, D., Deprez, N.R., Desai, L.V. and Sanford, M.S. (2005) *J. Am. Chem. Soc.*, **127**, 7330–7331; (h) Dick, A.R., Hull, K.L. and Sanford, M.S. (2004) *J. Am. Chem. Soc.*, **126**, 2300–2301; (i) Chen, X., Li, J.-J., Hao, X.-S., Goodhue, C.E. and Yu, J.-Q. (2006) *J. Am. Chem. Soc.*, **128**, 78–79; (j) Olafs Daugulis, V.G.Z. (2005) *Angew. Chem. Int. Ed.*, **44**, 4046–4048; (k) Shabashov, D. and Daugulis, O. (2007) *J. Org. Chem.*, **72**, 7720–7725; (l) Chiong, H.A. and Daugulis, O. (2007) *Org. Lett.*, **9**, 1449–1451; (m) Lazareva, A. and Daugulis, O. (2006) *Org. Lett.*, **8**, 5211–5213; (n) Shabashov, D. and Daugulis, O. (2006) *Org. Lett.*, **8**, 4947–4949; (o) Chiong, H.A., Pham, Q.-N. and Daugulis, O. (2007) *J. Am. Chem. Soc.*, **129**, 9879–9884; (p) Zaitsev, V.G. and Daugulis, O. (2005) *J. Am. Chem.*

Soc., **127**, 4156–4157; (q) Wan, X., Ma, Z., Li, B., Zhang, K., Cao, S., Zhang, S. and Shi, Z. (2006) *J. Am. Chem. Soc.*, **128**, 7416–7417; (r) Cai, G., Fu, Y., Li, Y., Wan, X. and Shi, Z. (2007) *J. Am. Chem. Soc.*, **129**, 7666–7673; (s) Yang, S., Li, B., Wan, X. and Shi, Z. (2007) *J. Am. Chem. Soc.*, **129**, 6066–6067; (t) Hull, K.L., Anani, W.Q. and Sanford, M.S. (2006) *J. Am. Chem. Soc.*, **128**, 7134–7135; (u) Giri, R., Maugel, N., Li, J.-J., Wang, D.-H., Breazzano, S.P., Saunders, L.B. and Yu, J.-Q. (2007) *J. Am. Chem. Soc.*, **129**, 3510–3511; (v) Chen, X., Goodhue, C.E. and Yu, J.-Q. (2006) *J. Am. Chem. Soc.*, **128**, 12634–12635.

23 Zhang, H., Ferreira, E.M. and Stoltz, B.M. (2004) *Angew. Chem. Int. Ed.*, **43**, 6144–6148.

24 Furuta, T., Asakawa, T., Iinuma, M., Fujii, S., Tanaka, K. and Kan, T. (2006) *Chem. Comm.*, 3648–3650.

25 Ferreira, E.M. and Stoltz, B.M. (2003) *J. Am. Chem. Soc.*, **125**, 9578–9579.

26 Abbiati, G., Beccalli, E.M., Broggini, G. and Zoni, C. (2003) *J. Org. Chem.*, **68**, 7625–7628.

27 Ma, S. and Gu, Z. (2005) *Angew. Chem. Int. Ed.*, **44**, 7512–7517.

28 (a) Campo, M.A., Huang, Q., Yao, T., Tian, Q. and Larock, R.C. (2003) *J. Am. Chem. Soc.*, **125**, 11506–11507; (b) Huang, Q., Fazio, A., Dai, G., Campo, M.A. and Larock, R.C. (2004) *J. Am. Chem. Soc.*, **126**, 7460–7461.

29 Zhao, J. and Larock, R.C. (2006) *J. Org. Chem.*, **71**, 5340–5348.

30 Zhao, J. and Larock, R.C. (2005) *Org. Lett.*, **7**, 701–704.

31 (a) Sridharan, V., Martín, M.A. and Menéndez, J.C. (2006) *Synlett*, 2375–2378; (b) Knolker, H.-J. and Knoll, J. (2003) *Chem. Comm.*, 1170–1171.

32 Bedford, R.B. and Betham, M. (2006) *J. Org. Chem.*, **71**, 9403–9410.

33 (a) Bressy, C., Alberico, D. and Lautens, M. (2005) *J. Am. Chem. Soc.*, **127**, 13148–13149; (b) Blaszykowski, C., Aktoudianakis, E., Alberico, D., Bressy, C., Hulcoop, D.G., Jafarpour, F., Joushaghani, A., Laleu, B. and Lautens, M. (2008) *J. Org. Chem.*, **73**, 1888–1897.

34 Garg, N.K., Caspi, D.D. and Stoltz, B.M. (2004) *J. Am. Chem. Soc.*, **126**, 9552–9553.

35 (a) Ohno, H., Yamamoto, M., Iuchi, M. and Tanaka, T. (2005) *Angew. Chem. Int. Ed.*, **44**, 5103–5106; (b) Ohno, H., Iuchi, M., Fujii, N. and Tanaka, T. (2007) *Org. Lett.*, **9**, 4813–4815.

36 Bertrand, M.B. and Wolfe, J.P. (2007) *Org. Lett.*, **9**, 3073–3075.

37 (a) Tang, S., Peng, P., Pi, S.-F., Liang, Y., Wang, N.-X. and Li, J.-H. (2008) *Org. Lett.*, **10**, 1179–1182; (b) Pinto, A., Neuville, L., Retailleau, P. and Zhu, J. (2006) *Org. Lett.*, **8**, 4927–4930; (c) Guo, L.-N., Duan, X.-H., Liu, X.-Y., Hu, J., Bi, H.-P. and Liang, Y.-M. (2007) *Org. Lett.*, **9**, 5425–5428.

38 Pinto, A., Neuville, L. and Zhu, J. (2007) *Angew. Chem. Int. Ed.*, **46**, 3291–3295.

39 Zhao, J., Yue, D., Campo, M.A. and Larock, R.C. (2007) *J. Am. Chem. Soc.*, **129**, 5288–5295.

40 Tsang, W.C.P., Zheng, N. and Buchwald, S.L. (2005) *J. Am. Chem. Soc.*, **127**, 14560–14561.

41 Li, B.-J., Tian, S.-L., Fang, Z. and Shi, Z.-J. (2008) *Angew. Chem. Int. Ed.*, **47**, 1115–1118.

42 Inamoto, K., Saito, T., Katsuno, M., Sakamoto, T. and Hiroya, K. (2007) *Org. Lett.*, **9**, 2931–2934.

43 Hashmi, A.S.K. (2007) *Chem. Rev.*, **107**, 3180–3211.

44 Shi, Z. and He, C. (2004) *J. Am. Chem. Soc.*, **126**, 5964–5965.

45 Shi, Z. and He, C. (2004) *J. Am. Chem. Soc.*, **126**, 13596–13597.

46 Shi, Z. and He, C. (2004) *J. Org. Chem.*, **69**, 3669–3671.

47 Reich, N., Yang, C.-G., Shi, Z. and He, C. (2006) *Synlett*, 1278–1280.

48 (a) Marion, N., Díez-González, S., de Frémont, P., Noble, A.R. and Nolan, S.P. (2006) *Angew. Chem. Int. Ed.*, **45**, 3647–3650; (b) Marion, N., Carlqvist, P., Gealageas, R., de Frémont, P., Maseras, F.

and Nolan, S.P. (2007) *Chem. Eur. J.*, **13**, 6437–6451.

49 (a) Liu, X.-Y., Ding, P., Huang, J.-S. and Che, C.-M. (2007) *Org. Lett.*, **9**, 2645–2648; (b) Nguyen, R.-V., Yao, X. and Li, C.-J. (2006) *Org. Lett.*, **8**, 2397–2399; (c) Gorin, D.J., Dube, P. and Toste, F.D. (2006) *J. Am. Chem. Soc.*, **128**, 14480–14481; (d) Lian, J.-J., Chen, P.-C., Lin, Y.-P., Ting, H.-C. and Liu, R.-S. (2006) *J. Am. Chem. Soc.*, **128**, 11372–11373; (e) Peng, L., Zhang, X., Zhang, S. and Wang, J. (2007) *J. Org. Chem.*, **72**, 1192–1197; (f) Watanabe, T., Oishi, S., Fujii, N. and Ohno, H. (2007) *Org. Lett.*, **9**, 4821–4824.

50 (a) Pastine, S.J., Youn, S.W. and Sames, D. (2003) *Org. Lett.*, **5**, 1055–1058; (b) Pastine, S.J., Youn, S.W. and Sames, D. (2003) *Tetrahedron*, **59**, 8859–8868.

51 Youn, S.W., Pastine, S.J. and Sames, D. (2004) *Org. Lett.*, **6**, 581–584.

52 Yi, C.S., Yun, S.Y. and Guzei, I.A. (2005) *J. Am. Chem. Soc.*, **127**, 5782–5783.

53 Yi, C.S. and Yun, S.Y. (2005) *J. Am. Chem. Soc.*, **127**, 17000–17006.

54 Chen, H. and Hartwig, J.F. (1999) *Angew. Chem. Int. Ed.*, **38**, 3391–3393.

55 (a) Kuninobu, Y., Kawata, A. and Takai, K. (2005) *J. Am. Chem. Soc.*, **127**, 13498–13499; (b) Kuninobu, Y., Tokunaga, Y., Kawata, A. and Takai, K. (2006) *J. Am. Chem. Soc.*, **128**, 202–209.

56 Horino, Y. (2007) *Angew. Chem. Int. Ed.*, **46**, 2144–2146.

57 Kuninobu, Y., Nishina, Y. and Takai, K. (2006) *Org. Lett.*, **8**, 2891–2893.

58 Yoon, C.H., Zaworotko, M.J., Moulton, B. and Jung, K.W. (2001) *Org. Lett.*, **3**, 3539–3542.

59 Takahashi, T., Tsutsui, H., Tamura, M., Kitagaki, S., Nakajima, M. and Hashimoto, S. (2001) *Chem. Comm.*, 1604–1605.

60 Swain, N.A., Brown, R.C.D. and Bruton, G. (2002) *Chem. Comm.*, 2042–2043.

61 Sengupta, S. and Mondal, S. (2003) *Tetrahedron Lett.*, **44**, 6405–6408.

62 Taber, D.F. and Tian, W. (2007) *J. Org. Chem.*, **72**, 3207–3210.

63 Gois, P.M.P. and Afonso, C.A.M. (2003) *Eur. J. Org. Chem.*, **2003**, 3798–3810.

64 Gois, P.M.P., Candeias, N.R. and Afonso, C.A.M. (2005) *J. Mol. Catal. A: Chem.*, **227**, 17–24.

65 Gois, P.M.P. and Afonso, C.A.M. (2003) *Tetrahedron Lett.*, **44**, 6571–6573.

66 (a) Candeias, N.R., Gois, P.M.P. and Afonso, C.A.M. (2005) *Chem. Comm.*, 391–393; (b) Candeias, N.R., Gois, P.M.P. and Afonso, C.A.M. (2006) *J. Org. Chem.*, **71**, 5489–5497.

67 Taber, D.F. and Joshi, P.V. (2004) *J. Org. Chem.*, **69**, 4276–4278.

68 Clark, J.S., Dossetter, A.G., Wong, Y.-S., Townsend, R.J., Whittingham, W.G. and Russell, C.A. (2004) *J. Org. Chem.*, **69**, 3886–3898.

69 John, J.P. and Novikov, A.V. (2007) *Org. Lett.*, **9**, 61–63.

70 Wolckenhauer, S.A., Devlin, A.S. and Du Bois, J. (2007) *Org. Lett.*, **9**, 4363–4366.

71 (a) Desai, L.V., Hull, K.L. and Sanford, M.S. (2004) *J. Am. Chem. Soc.*, **126**, 9542–9543; (b) Giri, R., Chen, X. and Yu, J.-Q. (2005) *Angew. Chem. Int. Ed.*, **44**, 2112–2115; (c) Giri, J.L.R., Lei, J.-G., Li, J.-J., Wang, D.-H., Chen, X., Naggar, I.C., Guo, C., Foxman, B.M. and Yu, J.-Q. (2005) *Angew. Chem. Int. Ed.*, **44**, 7420–7424; (d) Wang, D.-H., Hao, X.-S., Wu, D.-F. and Yu, J.-Q. (2006) *Org. Lett.*, **8**, 3387–3390; (e) Giri, R., Wasa, M., Breazzano, S.P. and Yu, J.-Q. (2006) *Org. Lett.*, **8**, 5685–5688; (f) Shabashov, D. and Daugulis, O. (2005) *Org. Lett.*, **7**, 3657–3659; (g) Zaitsev, V.G., Shabashov, D. and Daugulis, O. (2005) *J. Am. Chem. Soc.*, **127**, 13154–13155.

72 (a) Hongjun Ren, P.K. (2006) *Angew. Chem. Int. Ed.*, **45**, 3462–3465; (b) Ren, H., Li, Z. and Knochel, P. (2007) *Chem. Asian J.*, **2**, 416–433.

73 Lafrance, M., Gorelsky, S.I. and Fagnou, K. (2007) *J. Am. Chem. Soc.*, **129**, 14570–14571.

74 Kung, L.-R., Tu, C.-H., Shia, K.-S. and Liu, H.-J. (2003) *Chem. Comm.*, 2490–2491.

75 Dangel, B.D., Godula, K., Youn, S.W., Sezen, B. and Sames, D. (2002) *J. Am. Chem. Soc.*, **124**, 11856–11857.

76 DeBoef, B., Pastine, S.J. and Sames, D. (2004) *J. Am. Chem. Soc.*, **126**, 6556–6557.

77 (a) Ishii, S., Zhao, S. and Helquist, P. (2000) *J. Am. Chem. Soc.*, **122**, 5897–5898; (b) Ishii, S., Zhao, S., Mehta, G., Knors, C.J. and Helquist, P. (2001) *J. Org. Chem.*, **66**, 3449–3458.

78 Skouta, R. and Li, C.-J. (2007) *Angew. Chem. Int. Ed.*, **46**, 1117–1119.

79 Breslow, R. and Gellman, S.H. (1983) *J. Am. Chem. Soc.*, **105**, 6728–6729.

80 Davies, H.M.L. and Long, M.S. (2005) *Angew. Chem. Int. Ed.*, **44**, 3518–3519.

81 Davies, H.M.L. and Manning, J.R. (2008) *Nature*, **451**, 417–424.

82 Espino, C.G. and Du Bois, J. (2001) *Angew. Chem. Int. Ed.*, **40**, 598–600.

83 (a) Espino, C.G., Wehn, P.M., Chow, J. and Du Bois, J. (2001) *J. Am. Chem. Soc.*, **123**, 6935–6936; (b) Fleming, J.J., Fiori, K.W. and Du Bois, J. (2003) *J. Am. Chem. Soc.*, **125**, 2028–2029; (c) Fiori, K.W., Fleming, J.J. and Du Bois, J. (2004) *Angew. Chem. Int. Ed.*, **43**, 4349–4352; (d) Wehn, P.M., Lee, J. and Du Bois, J. (2003) *Org. Lett.*, **5**, 4823–4826.

84 Espino, C.G., Fiori, K.W., Kim, M. and Du Bois, J. (2004) *J. Am. Chem. Soc.*, **126**, 15378–15379.

85 (a) Fruit, C. and Müller, P. (2004) *Helv. Chim. Acta*, **87**, 1607–1615; (b) Fruit, C. and Muller, P. (2004) *Tetrahedron: Asym.*, **15**, 1019–1026.

86 Wehn, P.M. and Du Bois, J. (2005) *Org. Lett.*, **7**, 4685–4688.

87 Kim, M., Mulcahy, J.V., Espino, C.G. and Du Bois, J. (2006) *Org. Lett.*, **8**, 1073–1076.

88 Lebel, H., Huard, K. and Lectard, S. (2005) *J. Am. Chem. Soc.*, **127**, 14198–14199.

89 Reddy, R.P. and Davies, H.M.L. (2006) *Org. Lett.*, **8**, 5013–5016.

90 Hinman, A. and Du Bois, J. (2003) *J. Am. Chem. Soc.*, **125**, 11510–11511.

91 Conrad, R.M. and Du Bois, J. (2007) *Org. Lett.*, **9**, 5465–5468.

92 Wan, X., Xing, D., Fang, Z., Li, B., Zhao, F., Zhang, K., Yang, L. and Shi, Z. (2006) *J. Am. Chem. Soc.*, **128**, 12046–12047.

93 Fraunhoffer, K.J. and White, M.C. (2007) *J. Am. Chem. Soc.*, **129**, 7274–7276.

94 (a) Au, S.-M., Fung, W.-H., Cheng, M.-C., Che, C.-M. and Peng, S.-M. (1997) *Chem. Comm.*, 1655–1656; (b) Zhou, X.-G., Yu, X.-Q., Huang, J.-S. and Che, C.-M. (1999) *Chem. Comm.*, 2377–2378; (c) Au, S.-M., Zhang, S.-B., Fu, W.-H., Yu, W.-Y., Che, C.-M. and Cheung, K.-K. (1998) *Chem. Comm.*, 2677–2678; (d) Au, S.-M., Huang, J.-S., Yu, W.-Y., Fung, W.-H. and Che, C.-M. (1999) *J. Am. Chem. Soc.*, **121**, 9120–9132; (e) Yu, X.-Q., Huang, J.-S., Zhou, X.-G. and Che, C.-M. (2000) *Org. Lett.*, **2**, 2233–2236; (f) Huang, J.-S., Sun, X.-R., Leung, S.K.-Y., Cheung, K.-K. and Che, C.-M. (2000) *Chem. Eur. J.*, **6**, 334–344.

95 (a) Liang, J.-L., Yuan, S.-X., Huang, J.-S., Yu, W.-Y. and Che, C.-M. (2002) *Angew. Chem. Int. Ed.*, **41**, 3465–3468; (b) Liang, J.-L., Yuan, S.-X., Huang, J.-S. and Che, C.-M. (2004) *J. Org. Chem.*, **69**, 3610–3619.

96 Zhang, J., Hong Chan, P.W. and Che, C.-M. (2005) *Tetrahedron Lett.*, **46**, 5403–5408.

97 Cui, Y. and He, C. (2004) *Angew. Chem. Int. Ed.*, **43**, 4210–4212.

98 Li, Z., Capretto, D.A., Rahaman, R. and He, C. (2007) *Angew. Chem. Int. Ed.*, **46**, 5184–5186.

99 Hashmi, A.S.K., Schäfer, S., Wölfle, M., Diez Gil, C., Fischer, P., Laguna, A., Blanco, M.C. and Gimeno, M.C. (2007) *Angew. Chem. Int. Ed.*, **46**, 6184–6187.

100 Dangel, B.D., Johnson, J.A. and Sames, D. (2001) *J. Am. Chem. Soc.*, **123**, 8149–8150.

# 21
# Friedel–Crafts Type Cyclizations
*Ryuji Hayashi and Gregory R. Cook*

## 21.1
## Introduction

More than 130 years after its inception, the Friedel–Crafts (FC) reaction (electrophilic aromatic substitution) still remains one of the most important methods for C–C bond formation and is the subject of continued investigation today (Scheme 21.1) [1]. In general, because nucleophilicity of aromatic compounds is relatively low compared to other typical nucleophiles, FC reactions traditionally require severe conditions such as the use of stoichiometric amounts of strong Lewis acids, excess amounts of arenes, and high temperatures [2]. The original conditions have been replaced by milder and more environmentally friendly procedures. Although great efforts have been made in catalytic FC reactions, there remain several challenges. In particular, FC reactions with electron-deficient arenes are a continuing problem with few solutions. Therefore, new and -efficient methods to activate electrophiles for FC reactions under catalytic and mild conditions are highly significant.

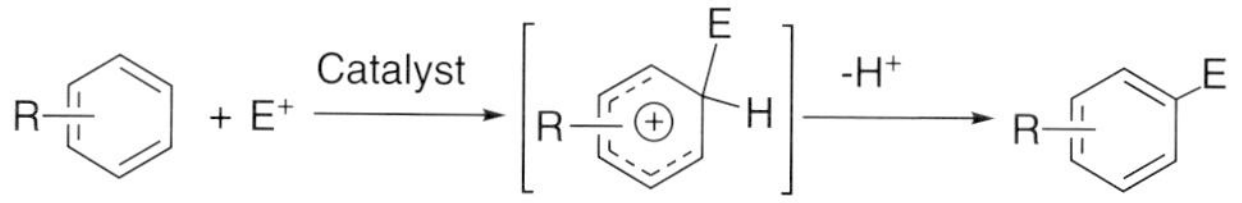

**Scheme 21.1** Friedel–Crafts reaction.

The synthesis of aromatic polyfunctionalized compounds, including benzocycles, has been carried out by a number of research groups due to their wide scope of applicability in the agrochemical as well as pharmaceutical fields [3]. The ring-closing Friedel–Crafts (RCFC) cyclization reaction is one of the most widely utilized methods in order to synthesize benzocycles (Scheme 21.2). This chapter deals with the recent development (in the last 10 years) of intramolecular FC cyclization reactions to construct a variety of cyclic molecules.

*Handbook of Cyclization Reactions. Volume 2.*
Edited by Shengming Ma
Copyright © 2010 WILEY-VCH Verlag GmbH & Co. KGaA, Weinheim
ISBN: 978-3-527-32088-2

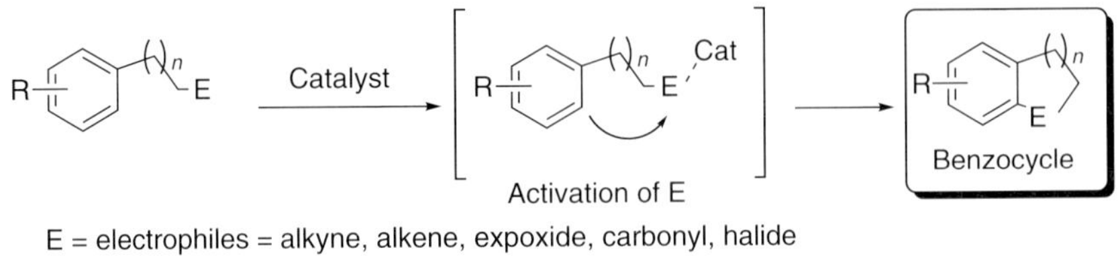


**Scheme 21.2** Synthesis of benzocycles via Friedel–Crafts reactions.

## 21.2
## Classic Intramolecular Friedel–Crafts Cyclizations

### 21.2.1
### RCFC Reactions via Activations of Alkyl Halides

The original FC reaction dealt with the substitution of aromatic compounds with an alkyl halide by using Lewis acids, intramolecular FC reactions via activation of an alkyl halide [4] by Lewis acids are still one of the major methods for ring constructions. For example, Heidelbaugh and coworkers demonstrated a convenient synthesis of 7-halo-1-indanones and 8-halo-1-tetralones (Scheme 21.3) [5]. Homologation of the 2-halobenzoic acids **1** was followed by cyclization to the indanone products **3**. Classic conditions, where stoichiometric amounts of $AlCl_3$ and high temperature were utilized, resulted in the efficient RCFC for substrate **2** to afford **3**. According to this protocol, gram quantities of the product **3** were prepared.

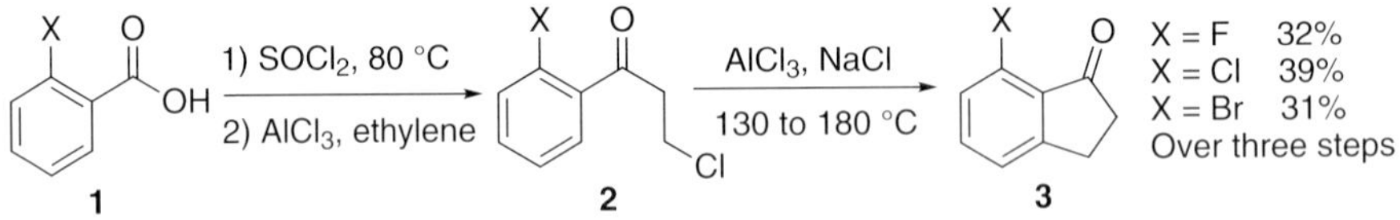


1 2 3

**Scheme 21.3** A convenient synthesis of halogen-substituted benzocycles.

In contrast to classic conditions for the FC reaction of an alkyl halide: use of high temperature and a stoichiometric amount of Lewis acid, our group has demonstrated remarkably mild and efficient RCFC under catalytic conditions [6]. We discovered that In(III) salts were highly effective catalysts for the RCFC of simple allylic halides and arenes **4**. In(III) salts appeared to be the most general and possess unique halophilic properties as a Lewis acid for this reaction. Even for deactivated arenes possessing chloride and fluoride substituents, reaction proceeded smoothly when activated by $InCl_3$ (Scheme 21.4).

### 21.2.2
### RCFC Reactions of Carboxylic Acid Derivatives

Acid chlorides are some of the most common types of electrophiles utilized for the RCFC reactions. The acid chloride **7** can be easily prepared from the carboxylic acid **6**,

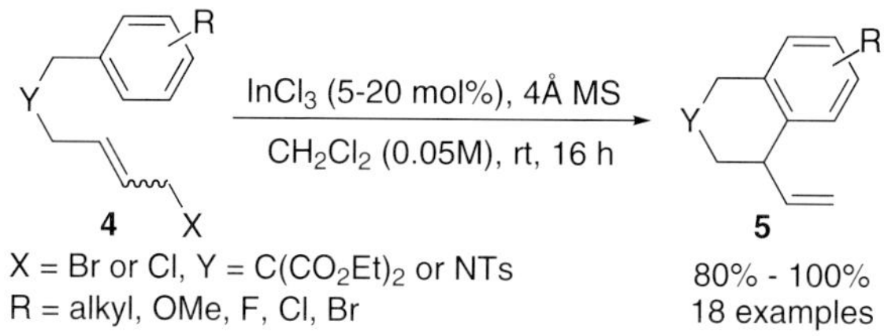

## 21.2 Classic Intramolecular Friedel–Crafts Cyclizations

**Scheme 21.4** Catalytic synthesis of benzocycles.

4: X = Br or Cl, Y = C(CO$_2$Et)$_2$ or NTs
R = alkyl, OMe, F, Cl, Br

InCl$_3$ (5-20 mol%), 4Å MS
CH$_2$Cl$_2$ (0.05M), rt, 16 h

5: 80% - 100%, 18 examples

and **7** can undergo RCFC reaction to afford the cyclized product **8** (Scheme 21.5). This two-step process has been used by many research groups and a variety of cyclic compounds have been synthesized including 5-carbocycles [7], 6-carbocycles [8] medium-sized rings [9], and heterocycles [10].

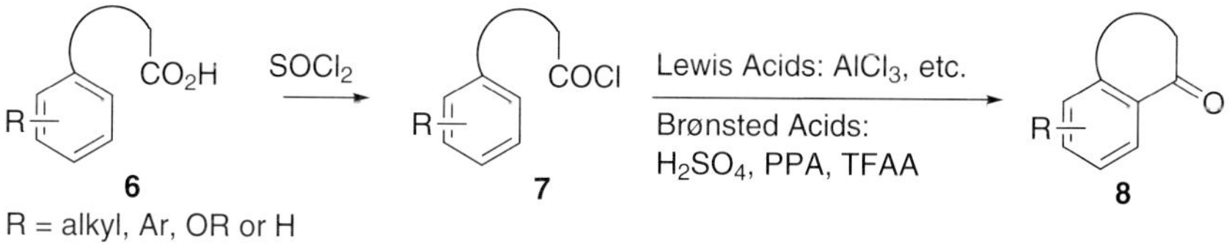

**Scheme 21.5** Two-step synthesis of benzocycles from carboxylic acids.

6 (CO$_2$H) → SOCl$_2$ → 7 (COCl) → Lewis Acids: AlCl$_3$, etc. / Brønsted Acids: H$_2$SO$_4$, PPA, TFAA → 8
R = alkyl, Ar, OR or H

There are many examples of direct RCFC reactions of carboxylic acids **6** to synthesize the benzocycles **8** (Scheme 21.6). Reagents including PPA [11], TFAA [12], and H$_2$SO$_4$ [13] are known to be very effective for this transformation, and other [14] examples can be found in the literature. Even though very high temperatures are required, substoichiometric amounts of Lewis acids are known to catalyze the RCFC reaction of carboxylic acids [15]. Similarly, direct RCFC of ester substrates has also been reported [16].

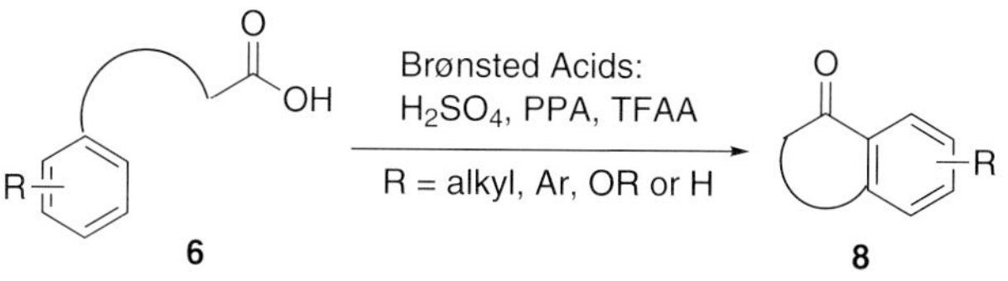

**Scheme 21.6** Direct RCFC reaction of carboxylic acids.

6 → Brønsted Acids: H$_2$SO$_4$, PPA, TFAA; R = alkyl, Ar, OR or H → 8

In some cases, typical conditions for the RCFC reaction of carboxylic acids or acid chlorides are not effective. The Yoshikawa group showed an elegant solution for such a situation (Scheme 21.7). For example, the direct RCFC reaction of the acid **9** failed in the presence of H$_3$PO$_4$ or trifluoroacetic anhydride. An attempt to synthesize the

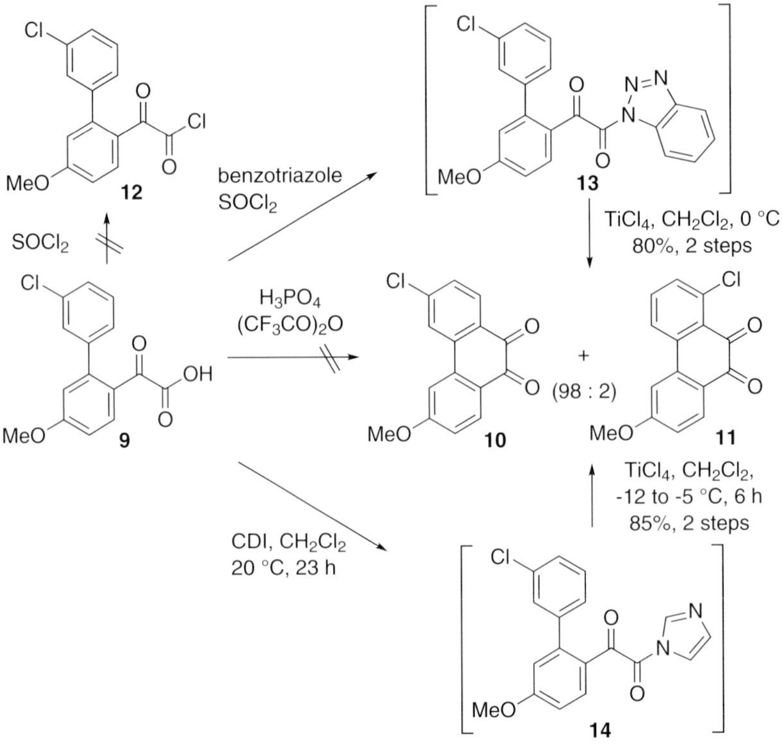

**Scheme 21.7** Heterocycles as a leaving group for RCFC reactions.

corresponding acid chloride **12** from the carboxylic acid **9** resulted in the formation of an unidentified by-product. In contrast, the treatment of the same acid **9** with thionyl chloride in the presence of benzotriazole produced smoothly the corresponding benzotriazolide **13**. The resulting benzotriazolide, stable only up to about 20 °C, rapidly underwent a RCFC reaction upon activation of the carbonyl group by TiCl$_4$ Lewis acid to produce the desired product **10** in 80% yield (2 steps from **9**). Three percent of the regioisomer **11** was formed. Due to the safety concern about the use of benzotriazole on a large scale, they also examined the use of 1,1′-carbonyldiimidazole (DCI) as an alternative reagent. Similarly, the phenanthrenequinone **10** was obtained in an improved yield (85%) with a slightly better selectivity. They applied this method to a 20 kg scale and 77% yield was obtained [17].

The combination of chlorotrimethylsilane and indium(III) chloride as a special catalyst has been utilized in RCFC reactions. For these reactions, it was pointed out that neither chlorotrimethylsilane nor indium(III) chloride was effective alone and the reactions proceeded only when both were combined. This notable activation of Lewis acidity of chlorosilane by indium(III) chloride was later emphasized in various reactions [18]. The Baba group applied this activation of Lewis acids to intramolecular FC reactions between arene and carboxylic acids (Scheme 21.8) [19].

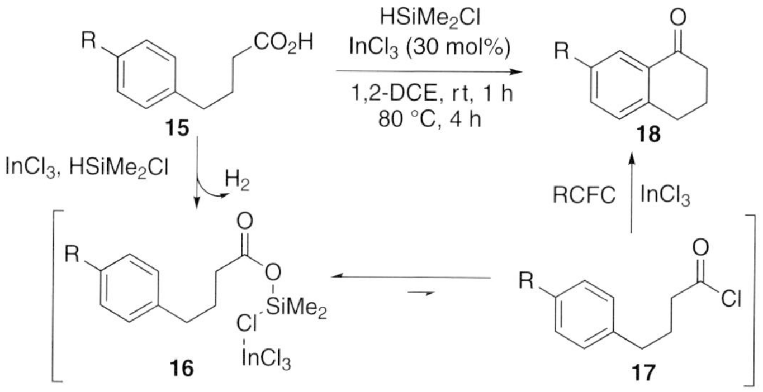

**Scheme 21.8** Reactions of acids with special Lewis acids.

## 21.3
## RCFC Reactions via Activation of Carbonyl Derivatives and Alcohols

### 21.3.1
### RCFC Reactions of Aldehydes and Ketones (1,2-Addition)

There are many literature reports dealing with the RCFC reaction via 1,2-addition to aldehydes [20] and ketones [21]. For example, the Wipf group utilized RCFC in the enantioselective synthesis of the AB-ring system of tetrazomine (Scheme 21.9) [22]. Swern oxidation of the substrate **19** to the sensitive α-formyl amide **20** followed by treatment with *p*-TsOH in dioxane provided the desired isoquinoline **21** in 75% overall yield as a single diastereomer.

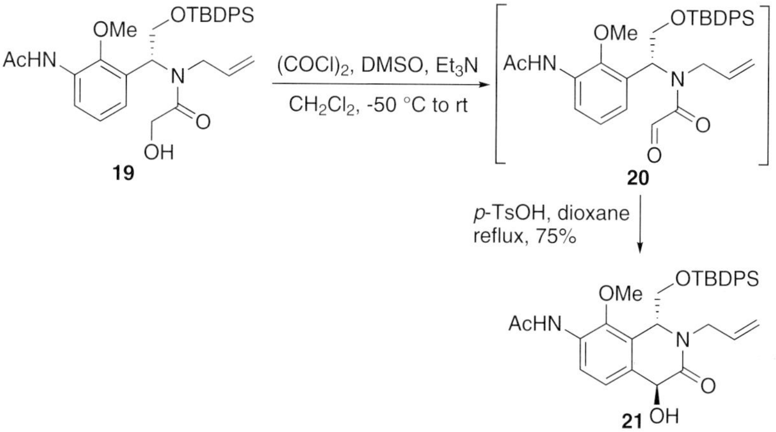

**Scheme 21.9** RCFC reactions as a key step in the total synthesis of tetrazomine.

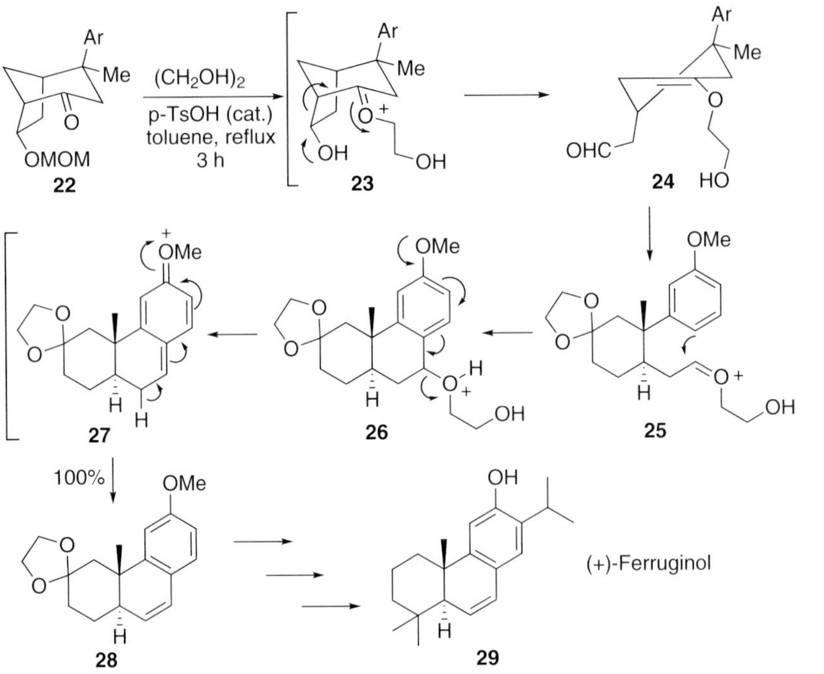

**Scheme 21.10** RCFC reaction as a key step in the total synthesis of (−)-ferruginol.

The Ogasawara group demonstrated a concise route to (+)-ferruginol via a tandem retro-aldol, intramolecular FC reaction (Scheme 21.10) [23]. Ethylene glycol was essential to attain a cyclization of **22** to **28**. In the absence of ethylene glycol, the reaction proceeded very slowly and was accompanied by a mixture of by-products. Accordingly, as exemplified by **22**, the reaction commenced with concomitant ethylene glycol-mediated cleavage of the MOM ether and ketalization of the ketone functionality, leading to the activated intermediate **23** which subsequently collapsed to the aldehyde **24** by a retro-aldol reaction. In the presence of ethylene glycol, the reaction proceeded further to generate the activated oxonium intermediate **25** which underwent an RCFC reaction to afford the product **28** via intermediates **26** and **27**. From **28**, the (+)-ferruginol **29** was synthesized in several steps.

## 21.3.2
### RCFC Reactions of Imines (1,2-Addition)

In order to synthesize nitrogen-containing benzocycles, RCFC reactions of imines can provide such an important class of molecules [24]. For example, Mizukami reported the synthesis of medium-sized cyclic amines via an RCFC reaction of **30** bearing an acetylene dicobalt moiety (Scheme 21.11) [25].

The Hsung group showed a practical approach to the Pictet–Spengler cyclization of arene-ynamides **32** and they synthesized a variety of polycyclic compounds utilizing

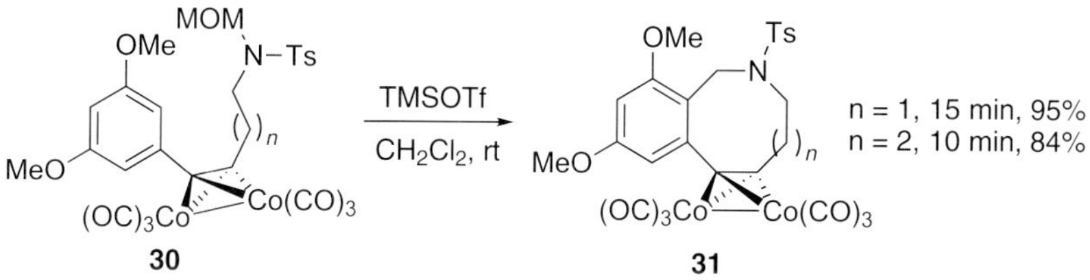

**Scheme 21.11** Synthesis of 8- and 9-membered cyclic amines via RCFC reaction.

this methodology [26]. For the arene-ynamide **32**, the application of 1 mol% HNTf$_2$ at room temperature for only a few minutes provided the cyclized product **35** in excellent yield. They hypothesized that the Pictet–Spengler cyclization took place via the highly reactive keteniminum ion **34** which underwent RCFC reaction (Scheme 21.12).

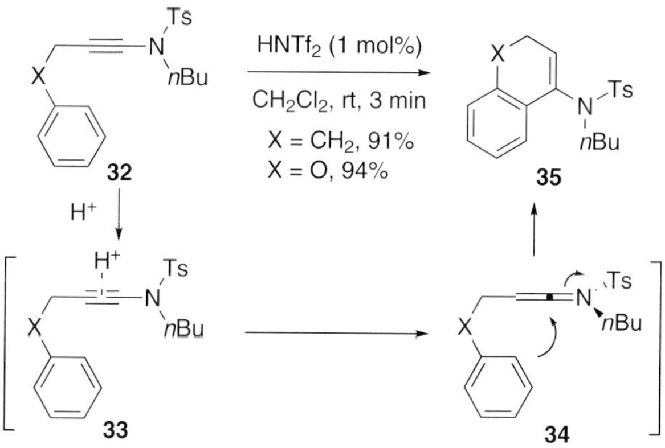

**Scheme 21.12** Acid-catalyzed RCFC Reaction of arene-ynamides.

In addition to the RCFC reaction of arene ynamides, the cyclization of indole-ynamides was also demonstrated. The indole-ynamide **35** was treated with 15 mol% *para*-nitrobenzenesulfonic acid (PNBSA) and the product **36** was obtained in 67% yield. From **36**, 11-desbromoarborescidine C **37** was synthesized in several steps (Scheme 21.13).

### 21.3.3
### RCFC Reactions of Acetals

RCFC reaction of acetal derivatives [27] has been utilized in the synthesis of a variety of natural and unnatural products. The Lee group reported macro-C-glycosidation in the total synthesis of kendomycin [28]. They initially attempted to cyclize **38** under a

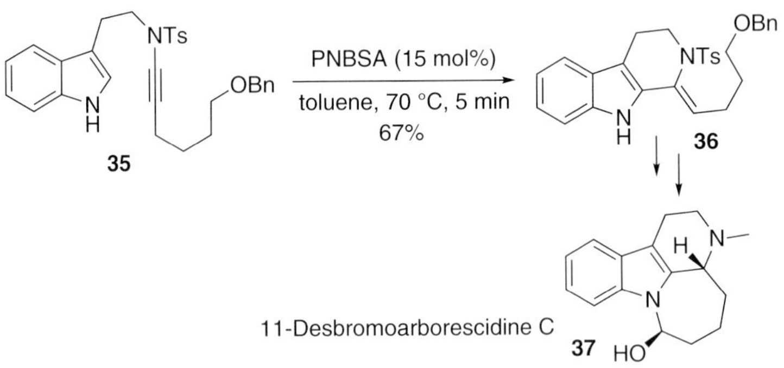

**Scheme 21.13** Total synthesis of 11-desbromoarborescidine C.

variety of FC conditions but were unsuccessful. Only hydrolysis of the anomeric acetate was observed. The reaction with the phenol compound **39**, in contrast, smoothly underwent RCFC reaction via the hemiacetal intermediate **40** to obtain the desired macrocycle **41** as a single isomer in 40–70% yields (Scheme 21.14). From **41**, kendomycin **42** was synthesized in a few steps.

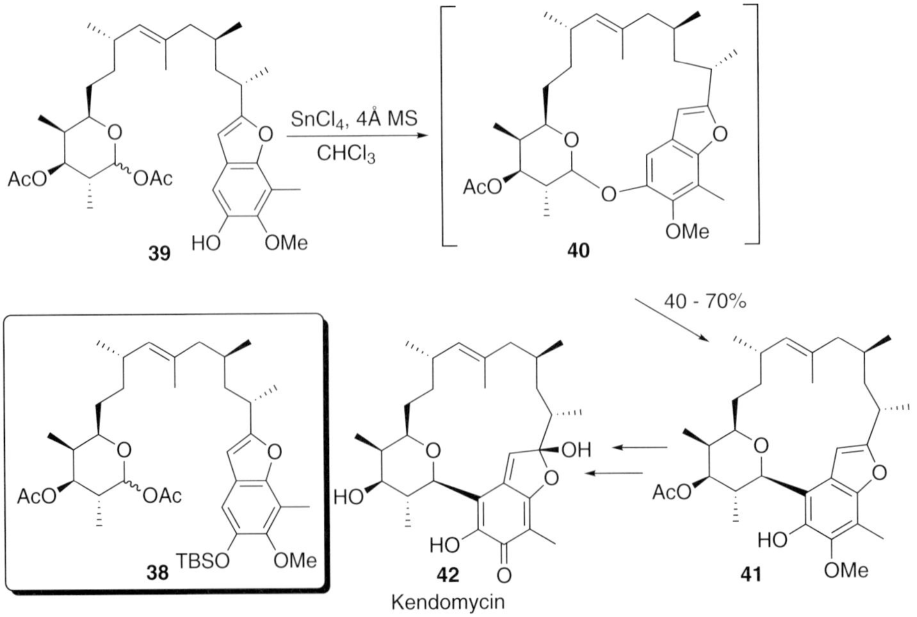

**Scheme 21.14** Total synthesis of kendomycin.

## 21.3.4
### RCFC Reactions via Conjugate Additions

RCFC reactions via conjugate additions have been utilized to construct a variety of cyclic compounds[29]. The Trauner group showed a practical and atom-economical RCFC reaction activated by triflation of α,β-unsaturated ketones [30]. For the α,β-unsaturated cyclic ketone substrate **42**, the RCFC triflation reaction proceeded well with highly nucleophilic arenes to afford the *cis*-fused product **43** (Scheme 21.15). Interestingly, for substrate **44** under the same conditions, the 1,6-addition product **45** was produced in 77% yield. The reaction was also amenable to the formation of indole products.

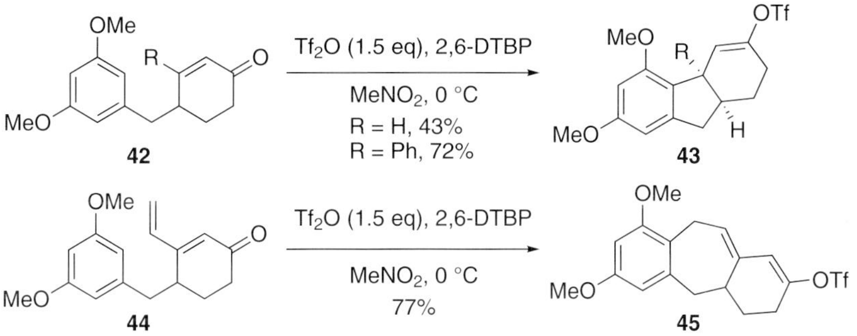

**Scheme 21.15** RCFC triflations.

A unique RCFC 1,6-conjugate addition was used in the total synthesis of (−)-salviasperanol (Scheme 21.16) [31]. The substrate **46** was treated with excess BF$_3$-etherate and a stoichiometric quantity of ethanethiol in CH$_2$Cl$_2$ at room temperature for 12 h to provide the dienone product **48** in 94% yield via the intermediate **47**. Formation of the vinyl sulfide intermediate **47** took less than 1 h and the key intermediate **47** was isolated and characterized.

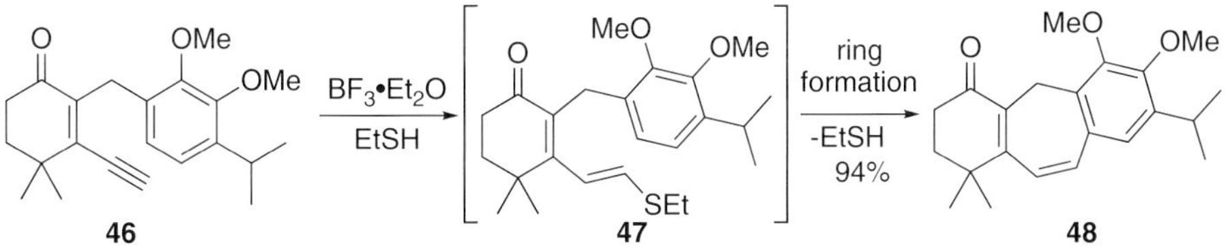

**Scheme 21.16** RCFC reaction as a key step in the total synthesis of (−)-salviasperanol.

## 21.3.5
### S$_N$1 Type Reactions and Direct Substitution of Alcohols

This section concerns the electrophilic substitution via carbocations (S$_N$1 type reactions) and direct electrophilic substitution via activation of alcohols [32]. The

Nishii and Tanabe groups reported an interesting chirality exchange reaction involving a single step chirality transfer from sp$^3$ central chirality to axial chirality. Benzannulation of the optically active compound **49** afforded the chiral biaryl compound **52** with an excellent level of stereoinduction [33]. Their proposed mechanism suggested that TiCl$_4$ first chelated with the oxygen and chlorine of the substrate **49** to generate a rigid intermediate **50**. The chlorine on the ortho-position oriented itself at the backside of the chelation face due to steric repulsion. Elimination of the hydroxy group by TiCl$_4$ gave the cationic intermediate **51** while sufficiently maintaining the conformation by freezing the free rotation of bonds *a* and *b*. The conjugation between the cyclopropylmethyl cation and the benzene ring system could contribute to prevention of the free rotation of bond *b* bearing an accessory chlorine-substituted phenyl group in **51**. Finally, ring-opening of cyclopropane and RCFC reaction took place sequentially to give product **52** in excellent yield and with the retention of optical activity (Scheme 21.17).

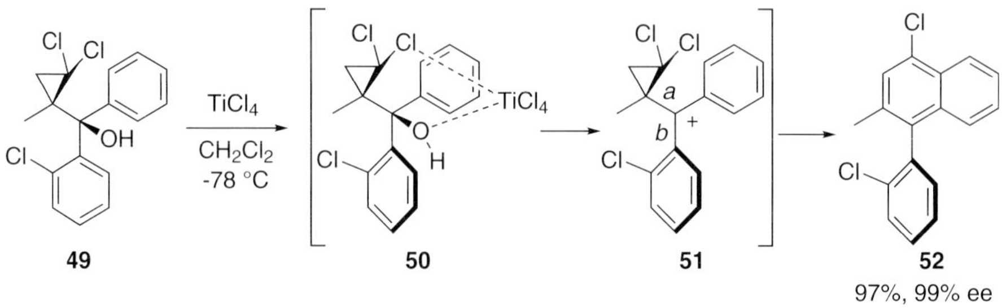

**Scheme 21.17** Chirality exchange via benzannulations.

The Rawal group used an acid-catalyzed RCFC reaction of alcohols in the facile synthesis of the indeno-tetrahydropyridine core of Haouamine A [34]. The alcohol **53** was treated with triflic acid and cyclization proceeded smoothly to afford the indenotetrahydropyridine **54** in 71% yield (Scheme 21.18).

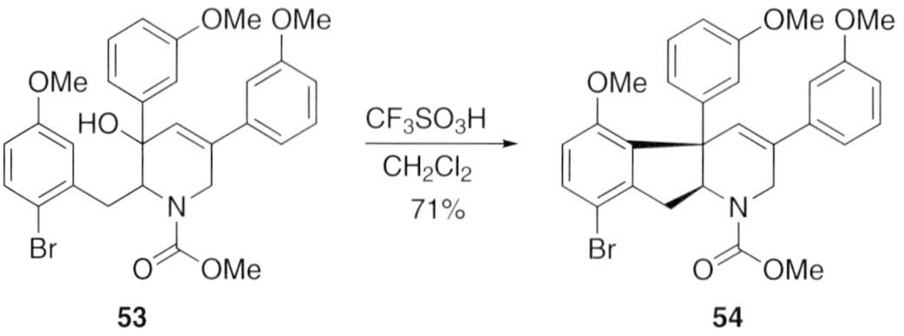

**Scheme 21.18** RCFC reaction in the synthesis of haouamine A.

The Suginome group has demonstrated the stereoselective synthesis of benzooxadecaline frameworks via three-component cascade reactions [35]. Reaction of substrate **55** in the presence of TiCl$_4$ via sequential incorporation of two different aldehydes provided the interemediate **56**. Cationic cascade cyclization of the intermediate **56** afforded the cationic intermediate **57**, and RCFC reaction yielded the tricyclic *trans*-1,2-benzooxadecaline **58** framework stereoselectively with high yield (Scheme 21.19).

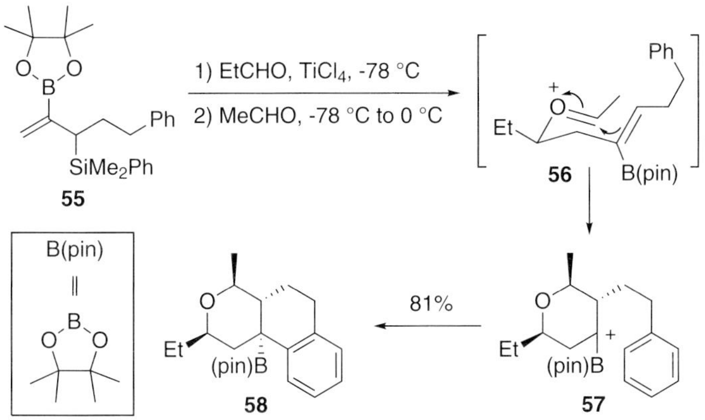

**Scheme 21.19** RCFC in three-component reactions.

## 21.3.6
### RCFC Reactions of Epoxide Derivatives

RCFC reactions of epoxides [36] can produce cyclic chiral alcohols. For example, Pericàs showed an efficient synthesis of chiral tetrahydrobenzoxepinol via epoxide ring-opening reactions (Scheme 21.20) [37]. When the epoxy ether **59** was treated with $BF_3$-$OEt_2$, a fast reaction took place, and the enantiomerically pure medium-size ring product **60** was isolated in 80% yield.

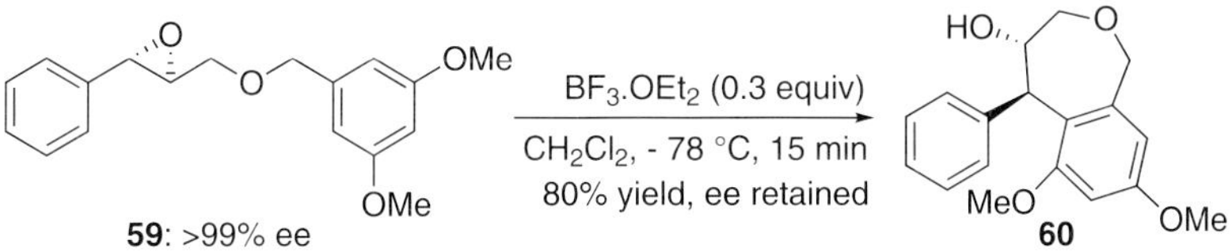

**Scheme 21.20** Ring-opening of epoxides.

In addition, the He group showed the effectiveness of Au(III) salts in promoting the ring opening of oxiranes by tethered electron-rich arenes [38]. From their control experiments performed in the absence of catalyst and in the presence of a variety of Lewis acids, both in catalytic and in stoichiometric amounts, the critical role played by the Au(III) catalyst in the mechanistic cycle was clearly verified. As shown in Scheme 21.21, highly stereospecific ring opening of the epoxide **61** was performed to afford the corresponding chromanol **62** in 82% yield as a single enantiomer, suggesting that this reaction must be proceeding via an $S_N2$-type mechanism.

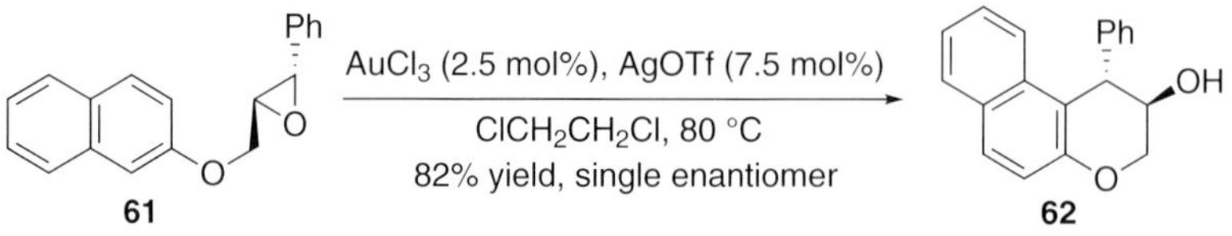

**Scheme 21.21** Au(III)-catalyzed epoxide ring-openings.

Similarly, the Branchaud group reported the RCFC reaction of episulfonium ion intermediates [39]. Treatment of the enantiomerically enriched hydroxysulfide **63** or **64** with BF$_3$·Et$_2$O led to the enantiomerically enriched product **66** in 35% yield, or **67** in 76% yield with retention of the stereochemistry via the episulfonium ion intermediate **65** (Scheme 21.22).

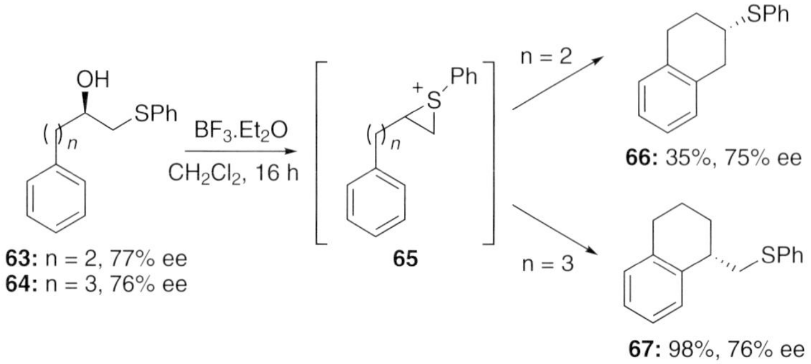

**Scheme 21.22** RCFC reactions of episulfonium ions.

## 21.4
## RCFC Reactions of C—C π Bonds

### 21.4.1
### RCFC Reactions of Alkenes

There have been numerous reports regarding RCFC reactions of alkenes in the presence of stoichiometric amounts of Brønsted acids including H$_2$SO$_4$ [40], TFA [41], PPA [42], and others [43]. The activation of double bonds by transition metals [44] has also become a popular protocol for initiating FC reactions.

2,2-Difluorovinyl ketone **68** has been uniquely utilized in RCFC reactions [45]. Despite fluorine's electron-withdrawing inductive effect, through donation of its unshared electron pair to the vacant p-orbital of the α-carbon, fluorine has a stabilizing effect on α-carbocations [46]. In the presence of TMSOTf, the 2,2-difluorovinyl ketone bearing an aryl group **68** underwent RCFC reaction via the α-fluorine stabilized carbocation intermediate **69** to afford the cyclic product **70** in high yield (Scheme 21.23).

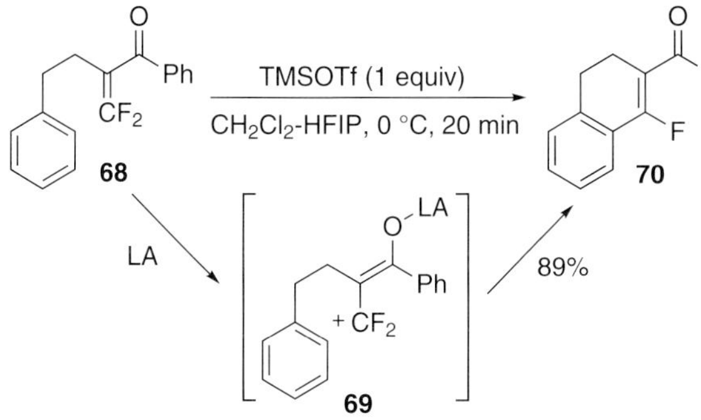

## 21.4 RCFC Reactions of C—C π Bonds | 1037

**Scheme 21.23** RCFC reactions via fluorocarbocations.

Additionally, RCFC reaction via activation of the 1,1-difluoro-1-alkene by a Pd(II) complex has been reported [47]. Under Pd(II) catalysis, in the absence of $BF_3\text{-}OEt_2$, reaction of substrate **71** produced only a trace amount of the cyclic product **75**. However, treatment of the substrate **71** with both cationic Pd(II) catalyst and $BF_3\text{-}OEt_2$ resulted in the formation of cyclic ketone **75** (89%) and fluoroarene **77**. The fluoroarene **77** could be generated via β-fluorine and β-hydrogen elimination from the cyclic intermediate **73**. Because of the high affinity of boron for fluorine it was suggested that $BF_3\text{-}OEt_2$ worked in concert with the palladium complex to accelerate the β-fluorine elimination from the intermediate **73** and regenerate the active cationic catalyst, $PdL_4(BF_4)_2$ (Scheme 21.24).

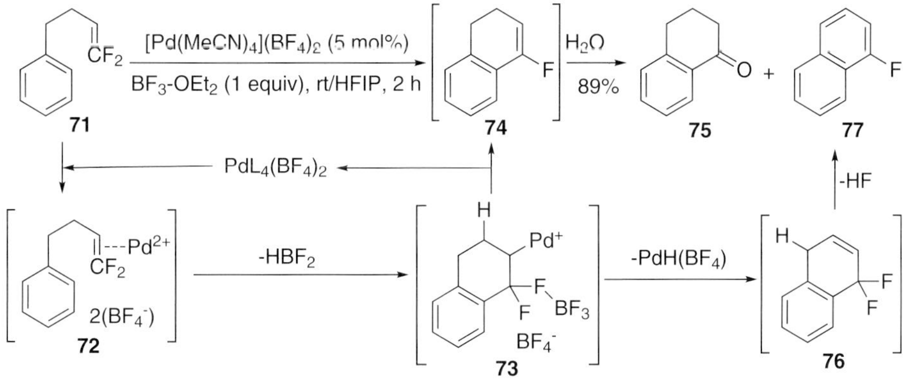

**Scheme 21.24** RCFC reactions via activation of 1,1-difluoro-1-alkene.

Balme and coworkers reported the RCFC reaction of aryl dienes catalyzed by Sc(OTf)$_3$ [48]. In the presence of 20 mol% Sc(OTf)$_3$, the 1,3-bis-exo cyclic diene **78** was smoothly cyclized to afford the tricyclic product **82** in 96% yield. Possible mechanisms could be the electrophilic activation of the ethylene intermediate **79** or the carbocationic intermediate **80** induced by the metal triflate (Scheme 21.25).

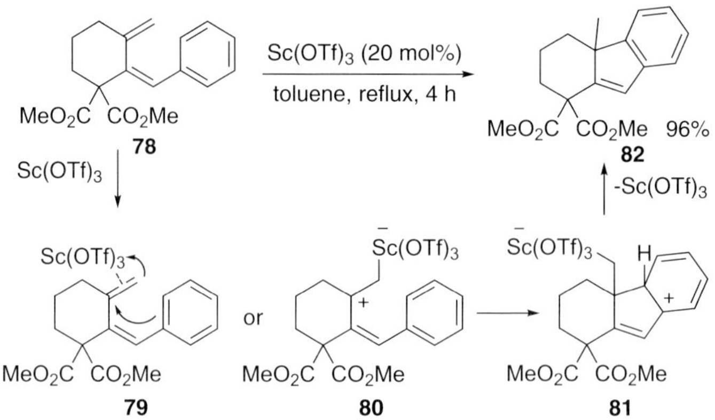

**Scheme 21.25** Sc(OTf)$_3$ catalyzed RCFC of 1,3-bisdienes.

Sames and coworkers demonstrated an efficient and practical synthesis of a variety of cyclic molecules including hydronaphthalenes, hydrobenzofurans, hydrobenzopyrans, hydroquinolines, and hydrocarbazoles via RCFC reactions [49]. Under RuCl$_3$/AgOTf catalysis, **83** smoothly underwent tandem cyclization via hydroarylation to provide the tricyclic product **84** in excellent yield (Scheme 21.26).

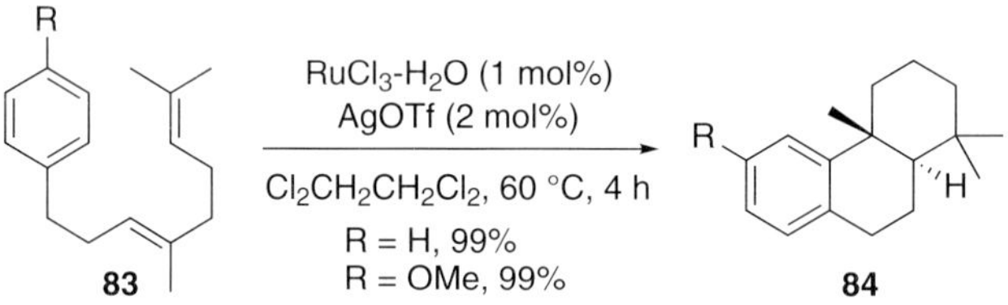

**Scheme 21.26** Hydroarylations catalyzed by Ru(III).

The Widenhoefer group addressed an efficient synthesis of tetrahydrocarbazoles and tetrahydro-β-carbolinones [50]. Cyclization of the indole-alkene substrate **85** under Pd catalysis yielded the cyclization/carboalkoxylation product **86** in good yield. This reaction proceeded by a chemoselective 6-exo-trig cyclization in the presence of stoichiometric amounts of Cu(II) salts as an oxidant for the final Pd(0) species. Under Pt(II) catalysis, 6-exo-trig cyclization of an unactivated double bond took place to afford the product **87** (Scheme 21.27). Mechanistically D-labeling experiments suggested that Pt(II) activates only a C—C double bond, ruling out C—H activation of the indole as a possibility.

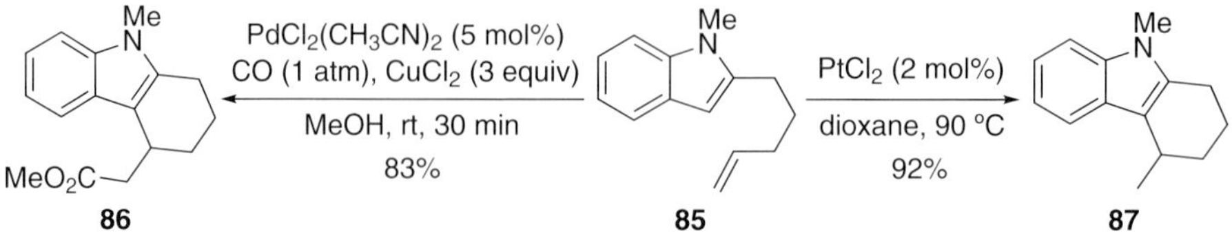

**Scheme 21.27** Metal-catalyzed RCFC reaction of unactivated C—C double bonds.

## 21.4.2
### RCFC Reactions of Alkynes

The use of alkynes as electrophiles in RCFC reactions has garnered a great deal of interest in the last few years since this can be an atom-economical replacement for the intramolecular Mizoroki–Heck coupling reaction (no halogen atom needed). A variety of Brønsted acids have been utilized for RCFC reactions of arenes with alkynes [51]. The Kozmin group developed catalytic acid-promoted carbocyclizations of siloxyalkynes to yield a range of substituted tetralones and cyclohexenones [52]. Treatment of the siloxyalkyne **88** with 10 mol% $Tf_2NH$ generated the highly reactive ketenium ion intermediate **90** that was readily trapped by nucleophilic arenes to afford the product **89** in 86% yield (Scheme 21.28).

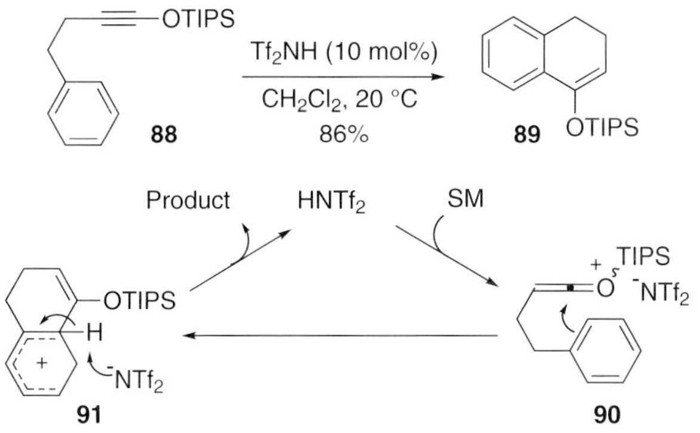

**Scheme 21.28** Acid-catalyzed RCFC reactions via ketenium ions.

Recently, many late transition metals have been utilized to activate C–C triple bonds. The high electron affinity of electrophilic late transition metals for π-bonds and their excellent tolerance toward many hard functional groups suggested transition metal salts to be potentially effective catalysts for the activation of alkynes. Especially, cationic Pt(II) and Ru(II) salts were found to be active in promoting reactions via activation of alkynes. For example, in 1996, Ru-vinylidine complexes were utilized for RCFC reactions of alkyne with arene [53]. Another example from the Murai group showed that a catalytic amount of $GaCl_3$ was a very effective catalyst for the RCFC reaction of alkyne **92** (Scheme 21.29) [54].

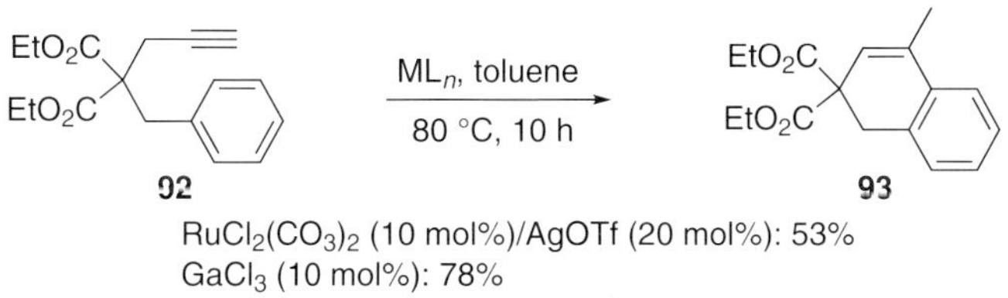

$RuCl_2(CO_3)_2$ (10 mol%)/AgOTf (20 mol%): 53%
$GaCl_3$ (10 mol%): 78%

**Scheme 21.29** Late transition metal-catalyzed RCFC reactions of alkyne.

The Nishizawa group accomplished RCFC reactions of alkyne with arene under remarkably mild conditions [55]. They utilized only 0.1–0.2 mol% Hg(OTf)$_2$-(TMU)$_3$ catalyst and a variety of both carbo- and heterocyclic compounds were synthesized at ambient temperature in acetonitrile. The catalytic activity was high, reaching up to 1000 turnovers (Scheme 21.30).

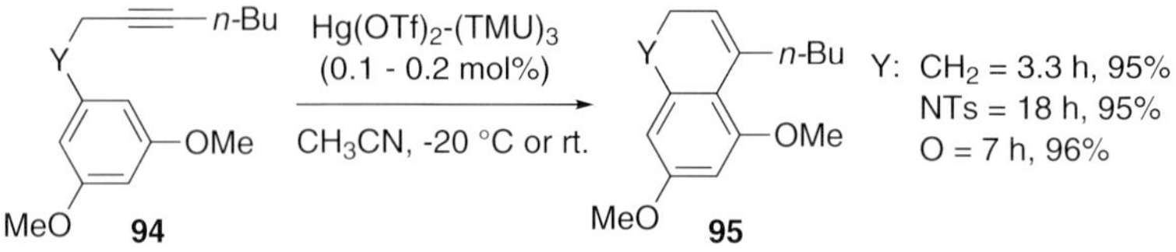

**Scheme 21.30** Hg(OTf)$_2$-(TMU)$_3$ catalyzed RCFC reaction under mild conditions.

The Fürstner group extended this protocol to the synthesis of 10-halophenanthrenes and the total synthesis of the aporphine alkaloid [56]. Interestingly, treatment of the halogen-substituted alkyne-arene substrate **96** with a catalytic amount of AuCl gave the product **98** via 1,2-migration of the Au-vinylidine intermediate **97**, whereas a catalytic amount of InCl$_3$ yielded the 10-halophenanthrene product **99** (Scheme 21.31). The Soriano group reported detailed computational studies for these transformations [57]. Moreover, under Au catalysis, similar reactions can be accomplished under milder conditions [58].

Additionally, a combination of ionic liquid and metal triflate dramatically enhanced the catalytic activities in the intramolecular FC alkynylation reactions (Scheme 21.32) [59].

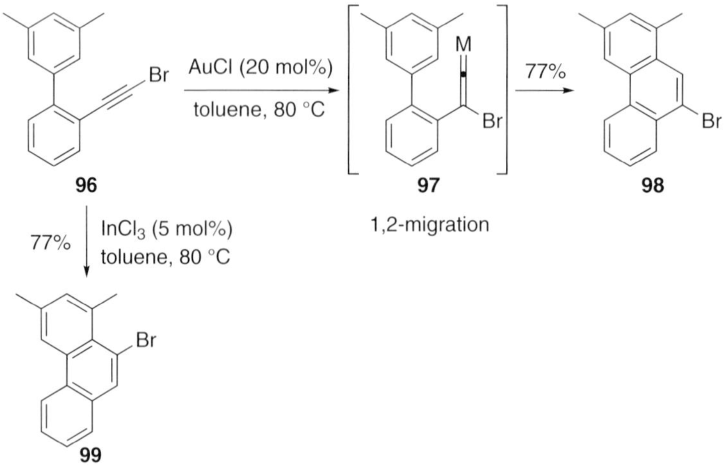

**Scheme 21.31** Synthesis of heteroarenes by Au(I) and In(III) Catalysis.

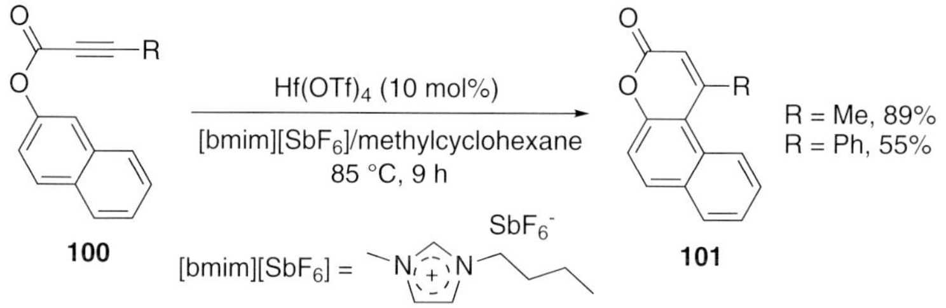

## 21.4 RCFC Reactions of C—C π Bonds

**Scheme 21.32** RCFC reactions in the presence of ionic liquid.

A variety of quinolines were synthesized by $BF_3$-$OEt_2$ catalyzed RCFC reaction of allenyl cations [60]. Treatment of the propargyl silyl ether **102** with 20 mol% $BF_3$-$OEt_2$ generated the cationic intermediates **103** and **104**. Aromatic substitution of allenyl cations proceeded to produce the cyclized product **105** as the major regioisomer. Similarly, this protocol was applied to the synthesis of medium-sized rings **107** and **108** (Scheme 21.33). Additionally, direct intramolecular FC cyclization of arenes with propargylic alcohols was reported very recently [61].

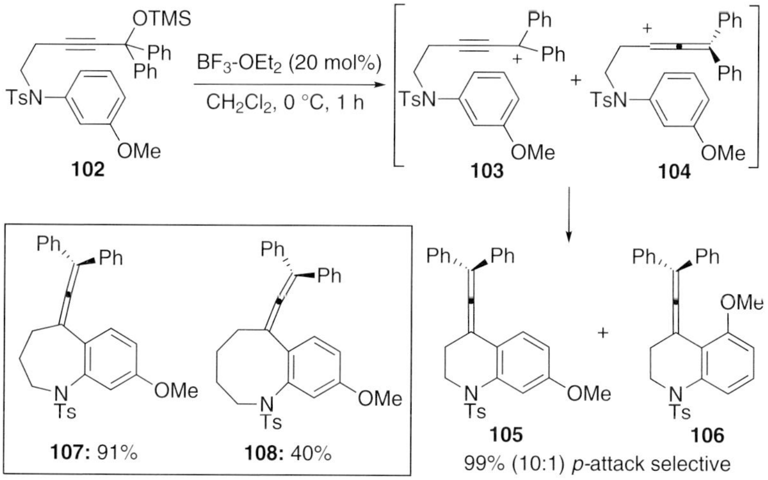

**Scheme 21.33** B(III)-catalyzed RCFC reaction of allenyl cations.

The Fürstner group has also demonstrated the intramolecular aromatic substitution of both carbo- and heteroaromatic compounds with alkynes under $PtCl_2$ catalysis [62]. Reactions proceeded efficiently to produce polyaromatic compounds in good yield, as exemplified in Scheme 21.34.

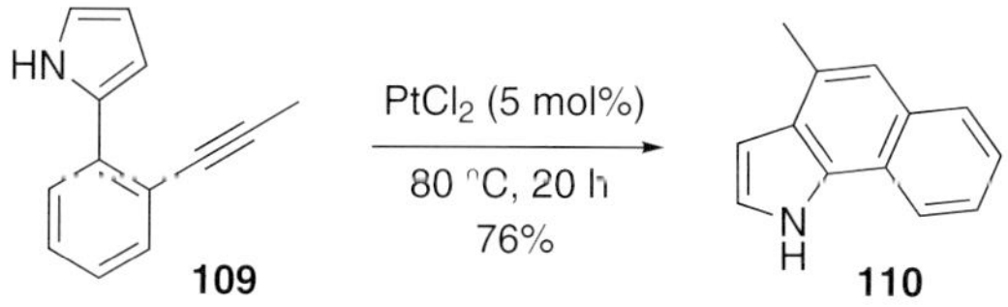

**Scheme 21.34** Pt(II)-catalyzed RCFC reaction of alkyne with pyrrole.

## 21.5
## Miscellaneous Reactions

The Hegedus group reported intramolecular FC acylations of chromium-carbene-complex-derived ketenes [63]. Photolysis of the Fischer-carbene complex **111** generated the intermediate **112** which has ketene-like reactivity. Under Lewis acid catalysis, intramolecular FC acylation of the ketene intermediate **112** proceeded to afford the cyclic product **113** (Scheme 21.35).

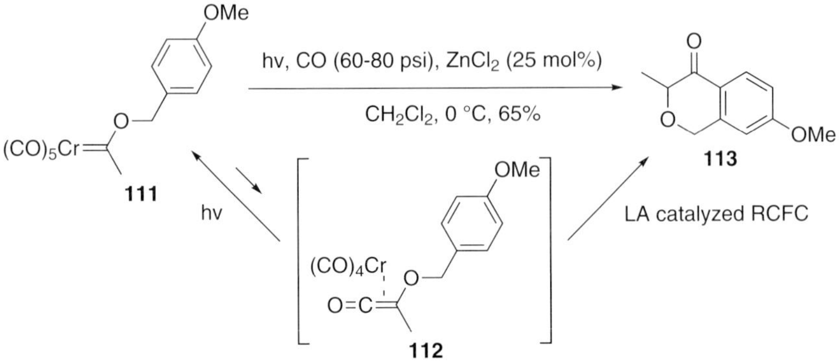

**Scheme 21.35** RCFC reaction of Fischer-carbene.

The Fillion group demonstrated a one-pot synthesis of tetrahydrofluorenones via thermal Diels–Alder/Lewis acid-catalyzed FC cyclizations [64]. The Diels–Alder reaction of the Meldrum's acid **114** and the butadiene generated *in situ* from butadiene sulfone **115** took place to give the intermediate **116**. Subsequently $BF_3$-$OEt_2$ catalyzed the FC cyclization to yield tetrahydrofluorenone **117** in good yield (Scheme 21.36). Similarly, when using substituted dienes such as isoprene, 74% yield of products **118** and **119** was obtained with high regioselectivity. The Fillion group also showed a unique approach for the synthesis of bicyclo[6.1.0]nonane-9-carboxylic acids via RCFC reactions [65].

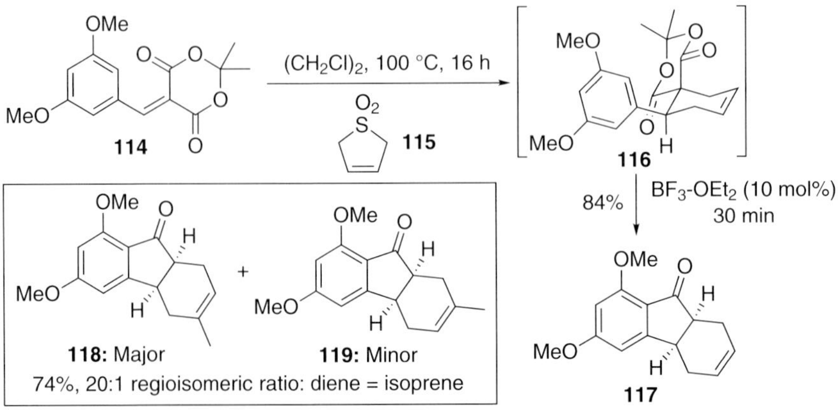

**Scheme 21.36** One-pot synthesis of tetrahydrofluorenones.

The Shi group reported Lewis acid-catalyzed rearrangement of multi-substituted arylvinylidenecyclopropanes [66, 67]. In the presence of 10 mol% Sn(OTf)$_2$, the arylvinylidenecyclopropane **120** transformed into the tetracyclic product **121**. The possible mechanism of the rearrangement is shown in Scheme 21.37. First, initial coordination of the substrate **122** to the Lewis acid gave zwitterionic intermediate **123**, a vinyl group stabilized the cyclopropyl cationic intermediate, which resulted in the formation of the cyclopropane ring-opened intermediate **124**, or the resonance stabilized intermediate **125** by the aromatic group. Then, RCFC reaction with either the aromatic R$^1$ or R$^2$ group produced the cyclized intermediate **126**. This afforded the intermediate **127** via allylic rearrangement, which was followed by another sterically demanding RCFC reaction with the aromatic group to produce the cyclized intermediate **128** with syn-configuration. Deprotonation of **128** afforded the corresponding intermediate **129** and the addition of the released proton produced the intermediate **130**. The 1,3-proton shift along with the release of the Lewis acid afforded the corresponding product **132**.

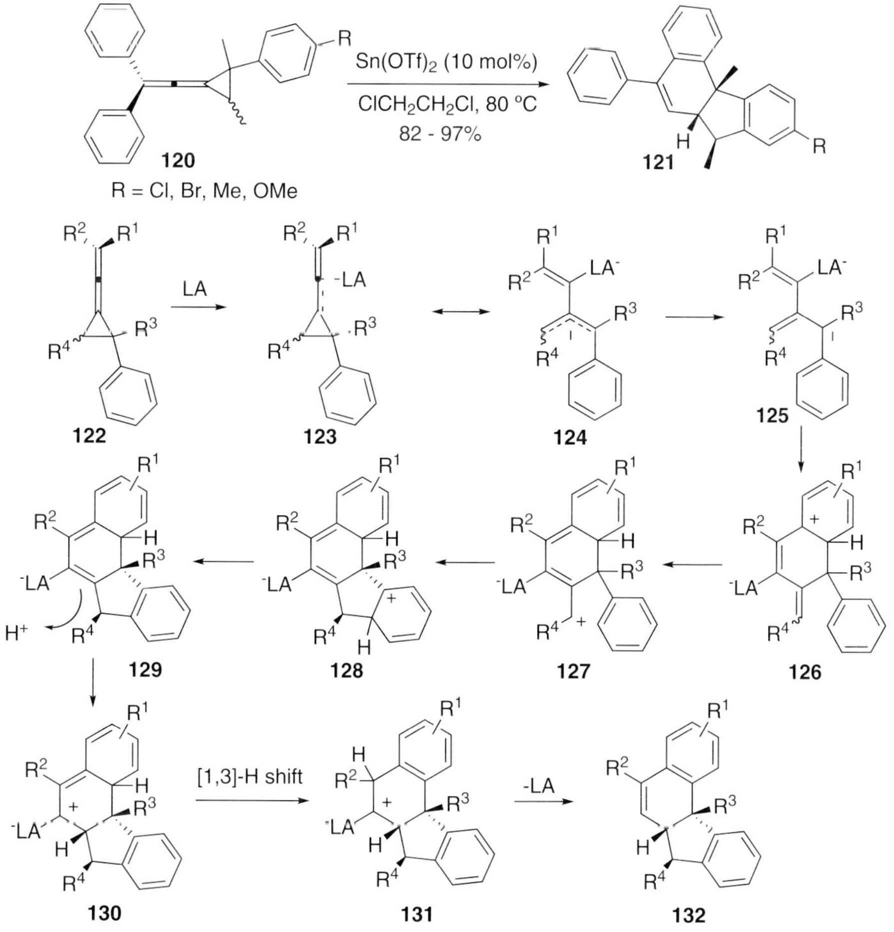

**Scheme 21.37** Acid-catalyzed rearrangement of arylvinylidenecyclopropanes.

## 21.6
## Catalytic Enantioselective RCFC Reactions

The Evans group showed enantioselective intramolecular FC cyclizations of α,β-unsaturated 2-acyl imidazoles catalyzed by chiral Lewis acid complexes [68]. For the 2-acyl imidazole tethered with indole **133**, the chiral bis(oxazolinyl)pyridine-scandium(III) triflate Lewis acid complex **134** catalyzed the efficient RCFC reaction to afford the cyclic heterocycle **135** in excellent yield as well as enantiomeric excess (Scheme 21.38).

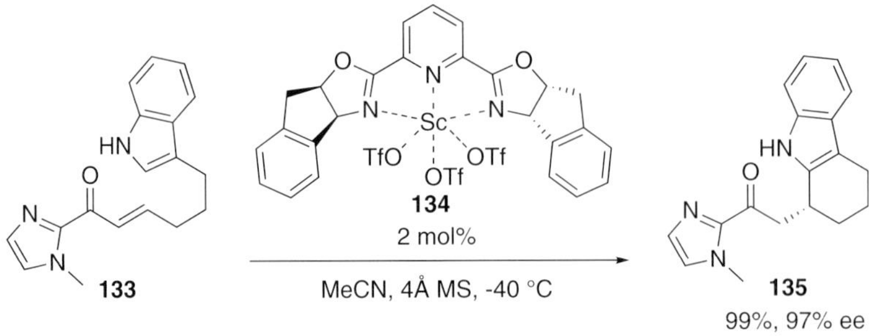

**Scheme 21.38** Chiral Lewis acid-catalyzed RCFC reaction via Michael addition.

Xiao reported an enantioselective organocatalytic RCFC alkylation of indoles via Michael addition [69]. In the presence of the heterocyclic catalyst [70] **137**, FC cyclization of the α,β-unsaturated aldehyde tethered with a variety of indoles yielded the tricyclic product **138** in moderate to excellent yields and high ee was observed (Scheme 21.39).

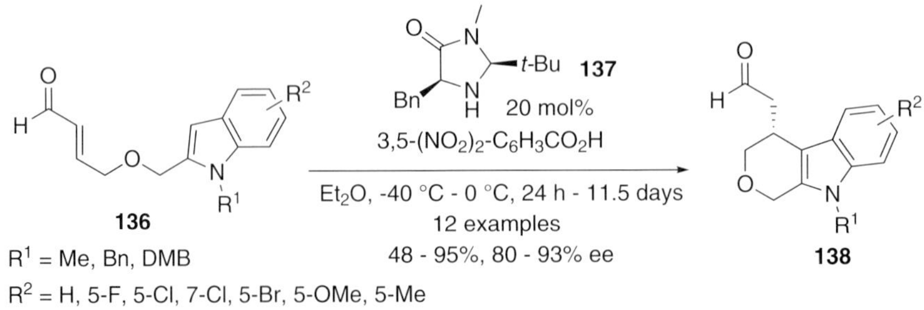

$R^1$ = Me, Bn, DMB
$R^2$ = H, 5-F, 5-Cl, 7-Cl, 5-Br, 5-OMe, 5-Me

**Scheme 21.39** Enantioselective organocatalytic RCFC reaction of indole.

As discussed previously, in order to synthesize tetrahydroisoquinoline derivatives, the Pictet–Spengler reaction is one of the most useful protocols. There have been some reports regarding diastereoselective ring-closing Pictet–Spengler cyclizations, however, stoichiometric amounts of acid catalysts were required for these protocols [71]. In 2004, the Jacobsen group reported catalytic enantioselective ring-closing Pictet–Spengler reactions of arenes with indoles via a hydrogen-bonding chiral thiourea catalyst [72]. For example, the imine **139** tethered with indole smoothly underwent intramolecular FC cyclization to produce the tetrahydroisoquinoline product **141** in 70% yield and 93% ee in the presence of 10 mol% thiourea catalyst **140** (Scheme 21.40).

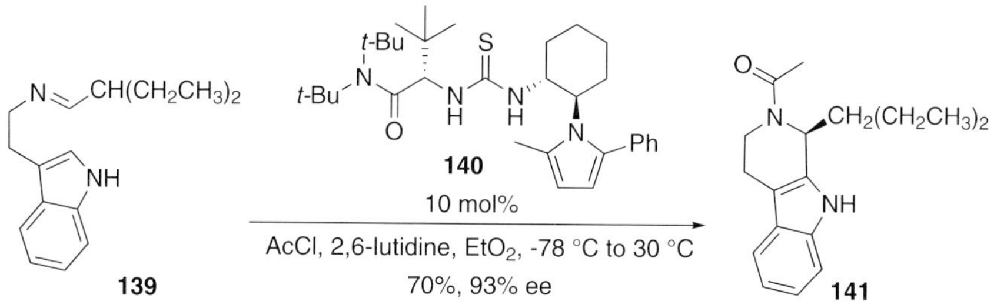

**Scheme 21.40** Catalytic enantioselective ring-closing Pictet–Spengler reaction.

Additionally, Bandini, Umani-Ronchi and coworkers demonstrated a metal-catalyzed RCFC reaction to synthesize both chiral tetrahydro-β-carboline and tetrahydroisoquinoline derivatives [73]. In the presence of a catalytic amount of Trost ligand **143** and Pd$^0$, the substrate **142** underwent RCFC reaction to yield the product **144a** in good yield with high ee. Additionally, when R$^1$ was substituted, they succeeded in synthesizing the enantiomerically pure cyclic product with a tert-chiral center **144b** in high ee. When R$^2$ was substituted, enantio- and diastereo-selective RCFC proceeded to afford the desired product **144c** in moderate yield with high dr and ee (Scheme 21.41).

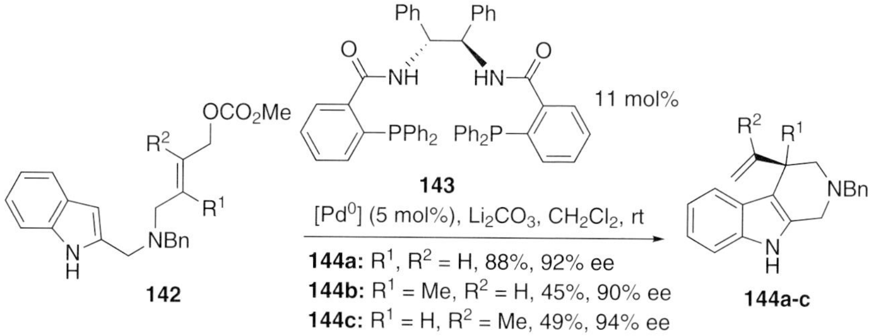

**Scheme 21.41** Pd-catalyzed Pictet-Spengler RCFC reactions of indoles.

In 2004, the Widenhoefer group showed one example of an enantioselective RCFC reaction of indole with unactivated olefin [50a]. They utilized a catalytic amount of the chiral Pt-BIPHEP complex **146** for the reaction of **145** and FC cyclization took place to afford the cyclic product **147** in 85% yield with 69% ee (Scheme 21.42).

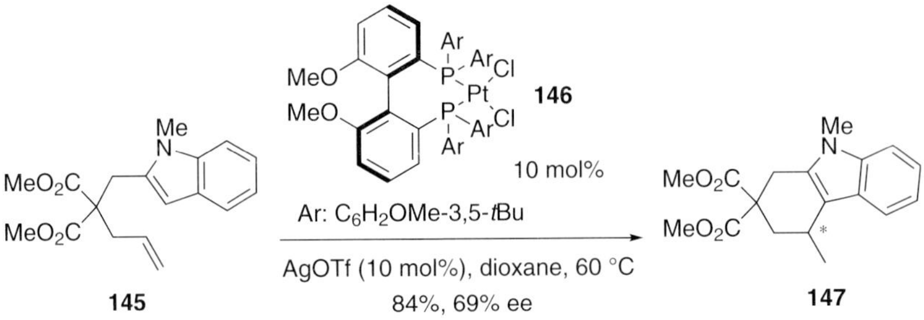

**Scheme 21.42** Pt-catalyzed ring-closing alkylation of indoles with olefins.

Three years later, Widenhoefer and Liu reported catalytic enantioselective RCFC reactions of indoles with allenes [74] and they obtained higher ee than in the previous result (Scheme 21.43). In the presence of catalytic amounts of chiral Au-BIPHEP complex **149**, the allene substrate **148** underwent intramolrcular FC alkylation to produce the cyclic products **150** in high yield and up to 92% ee (Scheme 21.43). They also succeeded in synthesizing the enantiomerically pure 7-membered ring.

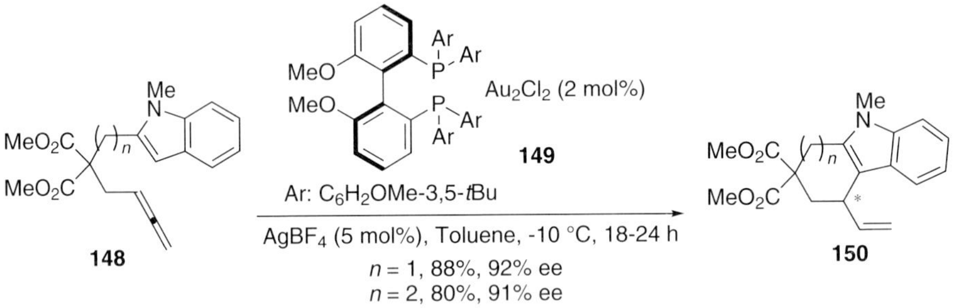

**Scheme 21.43** Au-catalyzed enantioselective RCFC reactions of indoles with allenes.

## 21.7
## Conclusion

This chapter has dealt with ring constructions via intramolecular FC cyclization reactions categorized by a variety of electrophiles reported over the last decade. Products from FC cyclization reactions can provide a variety of aromatic polyfunctionalized compounds. Importantly, many of these compounds have been applied in the field of pharmaceutical as well as agricultural studies. Even though the FC reaction has been studied for more than 130 years, there are still several limitations, particularly in its severe reaction conditions. In particular, FC reactions with an electron-deficient arene has been a very tough problem with few solutions. Therefore, more practical protocols are still highly desired, and as a consequence, further improvements of FC reactions can open the door to better synthesis in the future.

## 21.8
### Experimental: Selected Procedures

### 21.8.1
#### In(III)-Catalyzed RCFC Reaction for Synthesis of 5

Substrate **4** (0.1 mmol) and $CH_2Cl_2$ (2 mL, 0.05 M) was placed in a small screw-cap scintillation vial equipped with a magnetic stir bar. Powdered 4 Å molecular sieves (50 mg) and Lewis acid catalyst (0.01 mmol) were added. The reaction was allowed to stir at room temperature. Upon completion of the reaction (usually 16 h), the mixture was loaded onto a silica gel column directly and eluted using 10:1 hexane–ethyl acetate solution to afford the cyclic product **5**.

### 21.8.2
#### RCFC Reaction of Arene-Ynamides for Synthesis of 35

To a 0.05 M solution of an appropriate ynamide in $CH_2Cl_2$ was added $HNTf_2$ (0.01–0.05 equiv) at room temperature. The resulting solution was stirred at room temperature and the reaction progress was monitored using TLC analysis. When the starting material was completely consumed, the reaction mixture was quenched with $Et_3N$. Removal of the solvent *in vacuo* led to a crude material that was purified using silica gel flash column chromatography (gradient eluent: EtOAc in hexane).

### 21.8.3
#### Ru(III)-Catalyzed Hydroarylation

A mixture of $RuCl_3 \cdot H_2O$ (1 mol%) and AgOTf (2 mol%) in DCE was stirred vigorously for 1 h. To the resulting solution was added the substrate **83** in DCE (0.2 M). After full conversion, the solvent was evaporated and the residue was purified by column chromatography on silica gel to give the corresponding product.

### 21.8.4
#### Acid-Catalyzed RCFC Reaction via Ketenium Ions

Under $N_2$ atmosphere, a flame-dried air-free flask was charged with $CH_2Cl_2$ (25 mL) and siloxy alkyne **88** (0.31 mmol). The solution was treated with $HNTf_2$ in $CH_2Cl_2$ (0.153 M, 0.2 mL, 0.1 equiv) dropwise as an orange color developed. After stirring at room temperature for 1 h, the reaction mixture was treated with diisopropylethylamine (0.1 mL) and stirred for a further 0.5 h. The resulting solution was then diluted with hexanes (20 mL), washed with HCl (1 M, 10 mL), saturated aqueous $NaHCO_3$ (10 mL), $H_2O$ (10 mL) and brine (10 mL), dried ($MgSO_4$), filtered and concentrated. The residue was purified by silica gel flash chromatography (hexanes:ethyl acetate = 10 : 1) to afford the silyl enol ether **89**.

### 21.8.5
### Enantioselective Organocatalytic RCFC of Indole

The indolyl α,β-unsaturated aldehyde **136** was dissolved in ether (0.1 M) and then cooled to constant temperature (−40 °C). After 20 min, 20 mol% (2S, 5S)-5-benzyl-2-*tert*-butyl-3-methyl-imidazolidin-4-one **137** and 20 mol% 3,5-dinitrobenzoic acid were added in one portion. The resulting mixture was stirred at −40 °C until complete consumption of the indolyl α, β-unsaturated aldehyde as determined by TLC. The reaction mixture was then passed through a cold silica gel plug with $Et_2O$ and then concentrated. The resulting residue was purified by silica gel chromatography to afford the title compounds.

### 21.8.6
### Thiourea-Catalyzed Pictet–Spengler Reaction

Molecular sieves (250 mg, 3 Å spherical) were flame-dried *in vacuo* and cooled to 23 °C under nitrogen. Dichloromethane (1.25 mL), then tryptamine (40 mg, 0.25 mmol, 1.0 equiv.) were added. 2-Ethylbutanal (32 mL, 0.275 mmol, 1.0 equiv) was added dropwise by syringe to the suspension, which was allowed to stand at 23 °C for 7 h and swirled occasionally to ensure mixing of the contents. The resulting solution was filtered by cannula transfer to a flame-dried round-bottomed flask. The desiccant was rinsed twice with dichloromethane (5 mL) and the rinses were combined with the filtrate by cannula transfer. The solution was concentrated *in vacuo*, yielding the imine as a pale brown oil, which was dissolved in diethyl ether (5.0 mL). Catalyst **140** (6.7 mg, 0.013 mmol, 5 mol%) was added, and the solution cooled to −78 °C in a dry ice/acetone bath. 2,6-Lutidine (29 mL, 0.25 mmol, 1.0 equiv), then acetyl chloride (18 mL, 0.25 mmol, 1.0 equiv), were added dropwise by syringe. The mixture was stirred at −78 °C for 5 min, then warmed to −30 °C and stirred for 22 h. The resulting heterogeneous mixture was allowed to warm to 23 °C and concentrated *in vacuo*. The residue was purified by column chromatography on silica gel, eluting with 20% ethyl acetate in dichloromethane, yielding the product **141**.

### 21.8.7
### Asymmetric Pd-Catalyzed Pictet–Spengler RCFC

The reaction was carried out at room temperature under nitrogen atmosphere. A solution of $[Pd_2dba_3]\oplus CHCl_3$ (3.6 mmol, 5 mol%) and the chiral ligand **143** (11 mol%) in anhydrous $CH_2Cl_2$ (1.0 mL) was stirred until the solution color turned from deep red to orange (about 30 min). Then, the carbonate **142** (0.07 mmol, 1 equiv) dissolved in 0.5 mL of $CH_2Cl_2$ and $Li_2CO_3$ (2 equiv) were added. The resulting reaction mixture slowly turned yellow. The reaction was stirred overnight (complete consumption of **142** as judged by TLC) and then quenched with water (4 mL) and extracted with EtOAc. The combined organics were dried over $Na_2SO_4$ and concentrated under reduced pressure. The crude was purified by passage through a pad of silica gel and the enantiomeric excess determined through HPLC analysis with chiral column. The C/N regiochemistry of the cyclization was determined on the reaction crude by HPLC and confirmed after flash chromatography.

### 21.8.8
### Au-Catalyzed RCFC of Indole with Allenes

A mixture of the catalyst **149** (2 mol%) and $AgBF_4$ (5 mol%) in toluene (0.2 mL) was stirred at $-20\,°C$ for 10 min. To this a solution of **148** (41 mg, 0.13 mmol) in toluene (0.3 mL) was added via syringe. The resulting mixture was maintained at $-10\,°C$ for 17 h. Column chromatography of the reaction mixture (hexanes:EtOAc = 10:1 → 5:1) gave **150** (71 mg, 87%) as a pale yellow oil.

## Abbreviations

| | |
|---|---|
| Ac | acetyl |
| Ar | arene |
| Bn | benzyl |
| DCE | 1,2-dichloroethane |
| DCI | 1,1′-carbonyldiimidazole |
| E | electrophile |
| LA | Lewis acid |
| MOM | methoxymethyl |
| PNBSA | *p*-nitrobenzenesulfonic acid |
| PPA | poly(phosphoric acid) |
| RCFC | ring-closing Friedel–Crafts |
| TBDPS | *tert*-butyldiphenylsilyl |
| TBS | *tert*-butyldimethylsilyl |
| TFAA | trifluoroacetic anhydride |
| TMS | trimethylsilyl |
| Ts | *p*-toluene sulfonyl |
| TsOH | *p*-toluenesulfonic acid |

## Acknowledgment

We are grateful to the ND EPSCoR, North Dakota State University, the NDSU Center for Protease Research, the National Science Foundation and the National Institutes of Health for support of our research programs. RH acknowledges North Dakota State University and ND EPSCoR for graduate fellowship support.

## References

1. (a) Friedel, C. and Crafts, J.M.C.R. (1877) *Hebd. Seances Acad. Sci.*, **84**, 1392–1393.
2. (a) Olah, G.A. (1963) *Friedel-Crafts and Related Reactions*, Wiley, New York; (b) Olah, G.A. (1973) *Friedel-Crafts Chemistry*, Wiley, New York; (c) Roberts, R.M. and Khalaf, A.A. (1984) *Friedel-Crafts Alkylation Chemistry. A Century of Discovery*, Marcel Dekker, New York; (d) Olah, G.A., Krishnamurti, R. and Prakash, G.K.S. (1991) *Comprehensive Organic Synthesis*, 1st edn, vol. 3, Pergamon, New York, pp. 293–339; (e) Bandini, M., Melloni, A. and Umani-Ronchi, A. (2004) *Angew. Chem. Int. Ed.*, **43**, 550–556.
3. Bandini, M., Emer, E., Tommasi, S. and Umani-Ronchi, A. (2006) *Eur. J. Org. Chem.*, 3527–3544.
4. (a) Song, Y.S., Kong, S.D., Khan, S.A., Yoo, B.R. and Jung, I.N. (2001) *Organometallics*, **20**, 5586–5590; (b) González-Gmez, J.C., Santana, L. and Uriarte, E. (2002) *Synthesis*, 43–46; (c) Palma, A., Agredo, J.S., Carrillo, C., Kouznetsov, V., Stashenko, E., Bahsas, A. and Amaro-Luis, J. (2002) *Tetrahedron*, **58**, 8719–8727; (d) Kurteva, V.B., Santos, A.G. and Afonso, C.A.M. (2004) *Org. Biomol. Chem.*, **2**, 514–523; (e) Dupont, R. and Cotelle, P. (1998) *Tetrahedron Lett.*, **39**, 8457–8460.
5. Nguyen, P., Coupuz, E., Heidelbaugh, T.M., Chow, K. and Garst, M.E. (2003) *J. Org. Chem.*, **68**, 10195–10198.
6. Hayashi, R. and Cook, G.R. (2007) *Org. Lett.*, **9**, 1311–1314. For a similar cyclization of 6-acetoxy-4-alkenyl arenes, see: Ma, S. and Zhang, J., (2002) *Tetrahedron Lett.*, **43**, 3435–3438;Ma, S. and Zhang, J. (2003) *Tetrahedron*, **59**, 6273–6283.
7. (a) Sengupta, S. and Mondal, S. (1999) *Tetrahedron Lett.*, **40**, 3469–3470; (b) Uchikawa, O., Fukatsu, K., Tokunoh, R., Kawada, M., Matsumoto, K., Imai, Y., Hinuma, S., Kato, K., Nishikawa, H., Hirai, K., Miyamoto, M. and Ohkawa, S. (2002) *J. Med. Chem.*, **45**, 4222–4239; (c) terWiel, M.K.J., van Delden, R.A., Meetsma, A. and Feringa, B.L. (2003) *J. Am. Chem. Soc.*, **125**, 15076–15086; (d) Xiao, X. and Cushman, M. (2005) *J. Org. Chem.*, **70**, 6496–6498; (e) Sharma, A.K., Subramani, A.V. and Gorman, C.B. (2007) *Tetrahedron*, **63**, 389–395; (f) Zhao, C., Zhang, Y. and Ng, M. (2007) *J. Org. Chem.*, **72**, 6364–6371; (g) Reim, S., Lau, M. and Langer, P. (2006) *Tetrahedron Lett.*, **47**, 6903–6905.
8. (a) Su, J. and Wulff, W.D. (1998) *J. Org. Chem.*, **63**, 8440–8447; (b) Vicario, J.L., Badía, D., Domínguez, E. and Carrillo, L. (2000) *Tetrahedron; Asym.*, **11**, 1227–1237; (c) Ragot, J.P., Prime, M.E., Archibald, S.J. and Taylor, R.J.K. (2000) *Org. Lett.*, **2**, 1613–1616; (d) Tagat, J.R., McCombie, S.W., Nazareno, D.V., Boyle, C.D., Kozlowski, J.A., Chackalamannil, S., Josien, H., Wang, Y. and Zhou, G. (2002) *J. Org. Chem.*, **67**, 1171–1177; (e) Mulrooney, C.A., Li, X., DiVirgilio, E.S. and Kozlowski, M.C. (2003) *J. Am. Chem. Soc.*, **125**, 6856–6857; (f) Liu, F., Zha, H. and Yao, Z. (2003) *J. Org. Chem.*, **68**, 6679–6684; (g) Miller, S.A. and Bercaw, J.E. (2004) *Organometallics*, **23**, 1777–1789; Quesada, E., Stockley, M. and Taylor, R.J.K. (2004) *Tetrahedron Lett.*, **45**, 4877–4881; (h) Liu, F., Hu, T. and Yao, Z. (2005) *Tetrahedron*, **61**, 4971–4981; (i) Aprahamian, I., Preda, D.V., Bancu, M., Belanger, A.P., Sheradsky, T., Scott, L.T. and Rabinovitz, M. (2006) *J. Org. Chem.*, **71**, 290–298; (j) Wehlan, H., Jezek, E., Lebrasseur, N., Pavé, G., Roulland, E., White, A.J.P., Burrows, J.N. and Barrett, A.G.M. (2006) *J. Org. Chem.*, **71**, 8151–8158; (k) Dahl, B.J. and Branchaud, B.P. (2006) *Org. Lett.*, **8**, 5841–5844; (l) Xiao, X., Morrell, A., Fanwick, P.E. and Cushman, M. (2006) *Tetrahedron*, **62**, 9705–9712.
9. (a) Lee, T.B.K., Tebben, A.J., Weiberth, F.J., Wong, G.S.K. and George, S.K. (1998) *Syn. Commun.*, **28**, 747–751; (b) Othman, M., Pigeon, P. and Decroix, B. (1998)

J. Heterocyclic Chem., **35**, 1429–1433; (c) Chavan, S.P., Thakkar, M. and Kalkote, U.R. (2007) *Tetrahedron Lett.*, **48**, 535–537; (d) Ślusarczyk, M., De Borggraeve, W.M., Toppet, S. and Hoornaert, G.J. (2007) *Eur. J. Org. Chem.*, 2987–2994.

10 (a) Othman, M., Pigeon, P. and Decroix, B. (1998) *J. Heterocyclic Chem.*, **35**, 1371–1375; (b) Kajiro, H., Mitamura, S., Mori, A. and Hiyama, T. (1998) *Synlett*, 51–52; (c) Locke, A.J. and Richards, C.J. (1999) *Organometallics*, **18**, 3750–3759; (d) Legrand, A., Rigo, B., Gautret, P., Hénichart, J.-P. and Couturier, D. (1999) *J. Heterocyclic. Chem.*, **36**, 1263–1270; (e) Hénichart, J.-P. and Couturier, D. (2000) *J. Heterocyclic. Chem.*, **37**, 215–227; (f) Akuédu, R., Ebrik, S.A.A., Witczak-Legrand, A., Fasseur, D., Ghammarti, S.E., Couturier, D., Decroix, B., Othman, M., Debacker, M. and Rigo, B. (2002) *Tetrahedron*, **58**, 9239–9247; (g) Bourry, A., Akué-Gédu, R., Hénichart, J.-P., Sanz, G. and Rigo, B. (2004) *Tetrahedron Lett.*, **45**, 2097–2101; (h) Andreu, I., Cabedo, N., Fabis, F., Cortes, D. and Rault, S. (2005) *Tetrahedron*, **61**, 8282–8287; Ebrik, S.A.A., Legrand, A., Rigo, B. and Couturier, D. (1999) *J. Heterocyclic Chem.*, **36**, 997–1000; (i) Eggers, K., Fyles, T.M. and Montoya-Pelaez, P.J. (2001) *J. Org. Chem.*, **66**, 2966–2977; (j) Field, J.E., Hill, T.J. and Vankataraman, D. (2003) *J. Org. Chem.*, **68**, 6071–6078; (k) Hanessian, S., Papeo, G., Angiolini, M., Fettis, K., Beretta, M. and Munro, A. (2003) *J. Org. Chem.*, **68**, 7204–7218; (l) Vicario, J.L., Badía, D. and Carrillo, L. (2003) *Tetrahedron; Asym.*, **14**, 489–495.

11 (a) Loh, T.-P. and Hu, Q.-Y. (2001) *Org. Lett.*, **3**, 279–281; (b) Banerjee, A.K., Vera, W. and Laya, M.S. (2004) *Syn. Commun.*, **34**, 2301–2308; (c) Walters, R.S., Ho, D.M. and Pascal, R.A. (2005) *Tetrahedron Lett.*, **46**, 6487–6489; (d) Maertens, F., Toppet, S., Hoornaert, G.J. and Compernolle, F. (2005) *Tetrahedron*, **61**, 1715–1722; (d) Maertens, F., Bogaert, A.V., Compernolle, F. and Hoornaert, G.J. (2004) *Eur. J. Org. Chem.*, 4648–4656.

12 (a) Boger, D.L., Hong, J., Hikota, M. and Ishida, M. (1999) *J. Am. Chem. Soc.*, **121**, 2471–2477; (b) Piggott, M.J. and Wege, D. (2006) *Tetrahedron*, **62**, 3550–3556; (c) Lei, X. and Porco, J.A. (2006) *J. Am. Chem. Soc.*, **128**, 14790–14791.

13 (a) Luxenburger, A. (2003) *Tetrahedron*, **59**, 6045–6049; (b) Basavaiah, D. and Reddy, R.M. (2001) *Tetrahedron Lett.*, **42**, 3025–3027; (c) Lee, J. and Hong, J. (2004) *J. Org. Chem.*, **69**, 6433–6440.

14 (a) Esteban, G., López-Sánchez, M.A., Martinez, M.E. and Plumet, J. (1998) *Tetrahedron*, **54**, 197–212; (b) Lan, K. and Shan, Z.-X. (2007) *Syn. Commun.*, **37**, 2171–2177; (c) Fei, Z. and McDonald, F.E. (2007) *Org. Lett.*, **9**, 3547–3550.

15 (a) Cui, D.-M., Kawamura, M., Shimada, S., hayashi, T. and Tanaka, M. (2003) *Tetrahedron Lett.*, **44**, 4007–4010; (b) Cui, D.-M., Zhang, C., Kawamura, M. and Shimada, S. (2004) *Tetrahedron Lett.*, **45**, 1741–1745.

16 (a) Patil, M.L., Borate, H.B., Ponde, D.E. and Deshpande, V.H. (2002) *Tetrahedron*, **58**, 6615–6620; (b) Kalinin, A.V., Reed, M.A., Norman, B.H. and Snieckus, V. (2003) *J. Org. Chem.*, **68**, 5992–5999; (c) Fillion, E. and Fishlock, D. (2003) *Org. Lett.*, **5**, 4653–4656; (d) Fillion, E., Fishlock, D., Wilsily, A. and Goll, J.M. (2005) *J. Org. Chem.*, **70**, 1316–1327; (e) Tran, Y.S. and Kwon, O. (2005) *Org. Lett.*, **7**, 4289–4291; (f) Fillion, E. and Fishlock, D. (2005) *J. Am. Chem. Soc.*, **127**, 13144–13145.

17 (a) Yoshikawa, N., Doyle, A., Tan, L., Murry, J.A., Akao, A., Kawasaki, M. and Sato, K. (2007) *Org. Lett.*, **9**, 4103–4106; (b) Katritzky, A.R., Jiang, R. and Suzuki, K. (2005) *J. Org. Chem.*, **70**, 4993–5000.

18 (a) Yang, L., Xu, L.-W. and Xia, C.-G. (2007) *Tetrahedron Lett.*, **48**, 1599–1603; (b) Onishi, Y., Ito, T., Yasuda, M. and Baba, A. (2002) *Tetrahedron*, **58**, 8227–8235; (c) Onishi, Y., Ogawa, D., Yasuda, M. and Baba, A. (2002) *J. Am. Chem. Soc.*, **124**, 13690–13691; (d) Yasuda, M., Yamasaki, S., Onishi, Y. and Baba, A. (2004) *J. Am. Chem. Soc.*, **126**, 7186–7187; (e) Yasuda, M., Onishi, Y., Ueba, M., Miyai, T. and Baba, A. (2001)

*J. Org. Chem.*, **66**, 7741–7744; (f) Yasuda, M., Saito, T., Ueba, M. and Baba, A. (2004) *Angew. Chem. Int. Ed.*, **43**, 1414–1416; (g) Saito, T., Yasuda, M. and Baba, A. (2005) *Synlett*, 1737–1739; (h) Saito, T., Nishimoto, Y., Yasuda, M. and Baba, A. (2006) *J. Org. Chem.*, **71**, 8516–8522.

19 Baru, S.A., Yasuda, M. and Baba, A. (2007) *Org. Lett.*, **9**, 405–408.

20 (a) Hanessian, S., Mauduit, M., Demont, E. and Talbot, C. (2002) *Tetrahedron*, **58**, 1485–1490; (b) Nakahara, S. and Kubo, A. (2003) *Heterocycles*, **12**, 2717–2725; (c) Yadav, J.S., Reddy, B.V.S. and Padmavani, B. (2004) *Synthesis*, 405–408; (d) Yadav, J.S., Reddy, B.V.S., Parimala, G. and Krishnam Raju, A. (2004) *Tetrahedron Lett.*, **45**, 1543–1546; (e) Maurin, C., Bailly, F. and Cotelle, P. (2005) *Tetrahedron*, **61**, 7054–7058; (f) Klumpp, D.A., Kindelin, P.J. and Li, A. (2005) *Tetrahedron Lett.*, **46**, 2931–2935; (g) Hanessian, S., Talbot, C., Mauduit, M., Saravanan, P. and Gone, J.R. (2006) *Heterocycles*, **67**, 205–206; (h) Womark, G.B., Angeles, J.G., Fanelli, V.E. and Heyer, C.A. (2007) *J. Org. Chem.*, **72**, 7046–7049.

21 (a) Hanna, I., Michaut, V. and Ricard, L. (2001) *Tetrahedron Lett.*, **42**, 231–234; (b) Peng, X., She, X., Su, Y., Wu, T. and Pan, X. (2004) *Tetrahedron Lett.*, **45**, 3283–3285; (c) Abid, M., Spaeth, A. and Török, B. (2006) *Adv. Synth. Catal.*, **348**, 2191–2198.

22 Wipf, P. and Hopkins, C.R. (2001) *J. Org. Chem.*, **66**, 3133–3139.

23 Nagata, H., Miyazawa, N. and Ogasawara, K. (2001) *Org. Lett.*, **3**, 1737–1740.

24 (a) Seong, M.R., Song, H.N. and Kim, J.N. (1998) *Tetrahedron Lett.*, **39**, 7101–7104; (b) Katritzky, A.R., Cobo-Domingo, J., Yang, B. and Steel, P.J. (1999) *Tetrahedron; Asym.*, **10**, 255–263; (c) Casimir, J.R., Tourwé, D., Iterbeke, K., Guichard, G. and Briand, J.-P. (2000) *J. Org. Chem.*, **65**, 6487–6492; (d) Sánchez, J.D., Ramos, M.T. and Avendaño, C. (2001) *J. Org. Chem.*, **66**, 5731–5735; (e) Auricchio, S., Magnani, C. and Truscello, A.M. (2002) *Eur. J. Org. Chem.*, 2411–2416; (f) Yang, J., Che, X., Dang, Q., Wei, Z., Gao, S. and Bai, X. (2005) *Org. Lett.*, **7**, 1541–1543; (g) Zhuang, W., Haell, R.G. and Jørgensen, K.A. (2005) *Org. Biomol. Chem.*, **3**, 2566–2571; (h) Hilt, G., Galbiati, F. and Harms, K. (2006) *Synthesis*, 3575–3584.

25 Mizukami, M., Saito, H., Higuchi, T., Imai, M., Bando, H., Kawahara, N. and Nagumo, S. (2007) *Tetrahedron Lett.*, **48**, 7228–7231.

26 Zhang, Y., Hsung, R.P., Zhang, X., Huang, J., Slafer, B.W. and Davis, A. (2005) *Org. Lett.*, **7**, 1047–1050.

27 (a) Kao, C.-L., Yen, S.Y. and Chern, J.-W. (2000) *Tetrahedron Lett.*, **41**, 2207–2210; (b) Fearnley, S.P. and Tidwell, M.W. (2002) *Org. Lett.*, **4**, 3797–3798; (c) López, F., Castedo, L. and Mascareñas, J.L. (2005) *Org. Lett.*, **7**, 287–290; (d) Solorio, D.M. and Jennings, M.P. (2007) *J. Org. Chem.*, **72**, 6621–6623; (e) Ma, S. and Zhang, J. (2002) *Tetrahedron Lett.*, **43**, 3435–3438; (f) Ma, S. and Zhang, J. (2003) *Tetrahedron*, **59**, 6273.

28 Yuan, Y., Men, H. and Lee, C. (2004) *J. Am. Chem. Soc.*, **126**, 14720–14721.

29 (a) Liu, H.-J. and Tran, D.D.-P. (1999) *Tetrahedron Lett.*, **40**, 3827–3830; (b) Basavaiah, D., Bakthadoss, M. and Reddy, G.J. (2001) *Synthesis*, 919–927; (c) Yim, H.-K., Liao, Y. and Wong, H.N.C. (2003) *Tetrahedron*, **59**, 1877–1884; (d) Yamazaki, S., Morikawa, S., Iwata, Y., Yamamoto, M. and Kuramoto, K. (2004) *Org. Biomol. Chem.*, **2**, 3134–3138; (e) Chin, C.-L., Tran, D.D.-P., Shia, K.-S. and Liu, H.-J. (2005) *Synlett*, 417–420; (f) Angeli, M., Bandini, M., Garelli, A., Piccinelli, F., Tommasi, S. and Umani-Ronchi, A. (2006) *Org. Biomol. Chem.*, **4**, 3291–3296; (g) Cruz, M.C., Jiménez, F., Delgado, F. and Tamariz, J. (2006) *Synlett*, 749–755; (h) Yin, W., Ma, Y., Xu, J. and Zhao, Y. (2006) *J. Org. Chem.*, **71**, 4312–4315; (i) Li, W.-D.Z. and Wang, X.-W. (2007) *Org. Lett.*, **9**, 1211–1214; (j) Behenna, D.C., Stockdill, J.L. and Stoltz, B.M. (2007) *Angew. Chem. Int. Ed.*, **46**, 4077–4080; (k) Bandini, M., Melloni, A., Tommasi, S. and Umani-Ronchi, A. (2005)

*Synlett*, 1199–1222; (l) Agnusdei, M., Bandini, M., Melloni, A. and Umani-Ronchi, A. (2003) *J. Org. Chem.*, **68**, 7126–7129.
30 Grundl, M.A., Kaster, A., Beaulieu, E.D. and Trauner, D. (2006) *Org. Lett.*, **8**, 5429–5432.
31 Majetich, G., Zou, G. and Grove, J. (2008) *Org. Lett.*, **10**, 85–87.
32 (a) Kokubo, K., Koizumi, T., Yamaguchi, H. and Oshima, T. (2001) *Tetrahedron Lett.*, **42**, 5025–5028; (b) Galland, J.-C., Dias, S., Savignac, M. and Genêt, J.-P. (2001) *Tetrahedron*, **57**, 5137–5148; (c) Kouznetsov, V., Palma, A., Rozo, W., Stashenko, E., Bahsas, A. and Amaro-Luis, J. (2002) *Syn. Commun.*, **32**, 2965–2971; (d) Mandal, P.K., Cabell, L.A. and MacMurray, J.S. (2005) *Tetrahedron Lett.*, **46**, 3715–3718; (e) Williams, C.M., Mander, L.N., Bernhardt, P.V. and Willis, V.C. (2005) *Tetrahedron*, **61**, 3759–3769; (f) Oyanagi, T., Sakurai, Y. and Yamamura, K. (2005) *Heterocycles*, **65**, 2791–2798; (g) Basavaiah, D. and Reddy, K.R. (2007) *Org. Lett.*, **9**, 57–60; (h) Lim, H.N., Ji, S.-H. and Lee, K.-J. (2007) *Synthesis*, 2454–2460; (i) Langer, P. (1999) *Chem. Commun.*, 1217–1218; (j) Mühlthau, F., Schuster, O. and Bach, T. (2005) *J. Am. Chem. Soc.*, **127**, 9348–9349; (k) Hauser, F.M. and Ganguly, D. (2000) *J. Org. Chem.*, **65**, 1842–1849; (l) Chandrasekhar, S. and Reddy, M.V. (2000) *Tetrahedron*, **56**, 1111–1114; (m) McMills, M.C., Wright, D.L. and Weekly, R.M. (2002) *Syn. Commun.*, **32**, 2417–2425; (n) Chandrasekhar, S., Reddy, N.R., Reddy, M.V., Jagannadh, B., Nagaraju, A., Sankara, A.R. and Kunwar, A.C. (2002) *Tetrahedron Lett.*, **43**, 1885–1888; (o) Chandrasekhar, S., Mohanty, P.K., Harikishan, K. and Sasmal, P.K. (1999) *Org. Lett.*, **1**, 877–878; (p) Zhai, H., Luo, S., Ye, C. and Ma, Y. (2003) *J. Org. Chem.*, **68**, 8268–8271; (q) Hu, H. and Zhai, H. (2003) *Synlett*, 2129–2130; (r) Ma, Z. and Zhai, H. (2007) *Synlett*, 161–163; Ma, Z., Hu, H., Xiong, W. and Zhai, H. (2007) *Tetrahedron*, **63**, 7523–7531.
33 Nishii, Y., Wakasugi, K., Koga, K. and Tanabe, Y. (2004) *J. Am. Chem. Soc.*, **126**, 5358–5359.
34 Smith, N.D., Hayashida, J. and Rawal, V.H. (2005) *Org. Lett.*, **7**, 4309–4312.
35 Suginome, M., Ohmori, Y. and Ito, Y. (2001) *Chem. Commun.*, 1090–1091.
36 (a) Ho, W.-B. and Broka, C. (2000) *J. Org. Chem.*, **65**, 6743–6748; (b) Nagumo, S., Miyoshi, I., Akita, H. and Kawahara, N. (2002) *Tetrahedron Lett.*, **43**, 2223–2226; (c) Mühlthau, F. and Bach, T. (2005) *Synthesis*, 3428–3436.
37 Islas-González, G., Benet-Buchholz, J., Maestro, M.A., Riera, A. and Pericàs, M.A. (2006) *J. Org. Chem.*, **71**, 1537–1544.
38 (a) Shi, Z. and He, C. (2004) *J. Am. Chem. Soc.*, **126**, 5964–5965; (b) Shi, Z. and He, C. (2004) *J. Org. Chem.*, **69**, 3669–3671.
39 (a) Branchaud, B.P. and Blanchette, H.S. (2002) *Tetrahedron Lett.*, **43**, 351–353; (b) Philippe, N., Denivet, F., Vasse, J.-L., Santos, J.S.O., Levacher, V. and Dupas, G. (2003) *Tetrahedron*, **59**, 8049–8056.
40 (a) Palma, A., Barajas, J.J., Kouznetsov, V.V., Stashenko, E., Bahsas, A. and Amaro-Luis, J. (2004) *Synlett*, 2721–2724; (b) Zubkov, F.I., Nikitina, E.V., Kouznetsov, V.V. and Duarte, L.D.A. (2004) *Eur. J. Org. Chem.*, 5064–5074; (c) Kouznetsov, V.V., Zubkov, F.I., Kikitina, E.V. and Duarte, L.D.A. (2004) *Tetrahedron Lett.*, **45**, 1981–1984; (d) Méndez, L.Y.V. and Kouznetsov, V.V. (2007) *Tetrahedron Lett.*, **48**, 2509–2512.
41 (a) Ma, S. and Zhang, J. (2002) *Tetrahedron Lett.*, **43**, 3435–3438; (b) Ma, S. and Zhang, J. (2003) *Tetrahedron*, **59**, 6273–6283.
42 (a) Vicario, J.L., Badía, D., Domínguez, E., Crespo, A. and Carrillo, L. (1999) *Tetrahedron; Asym.*, **10**, 1947–1959; (b) Yépez, A.F., Palma, A., Stashenko, E., Bahsas, A. and Amaro-Luis, J.M. (2006) *Tetrahedron Lett.*, **47**, 5825–5828.
43 (a) Harmata, M., Hong, X. and Barnes, C.L. (2004) *Org. Lett.*, **6**, 2201–2203; (b) Wang, Y., Wu, J. and Xia, P. (2006) *Syn. Commun.*, **36**, 2685–2698; (c) Yang, S.-M. and Fang, J.-M. (2007) *Tetrahedron*, **63**, 1421–1428.

44 (a) Lomberget, T., Bouyssi, D. and Balme, G. (2005) *Synthesis*, 311–319; (b) Doherty, S., Knight, J.G., Smyth, C.H., Harrington, R.W. and Clegg, W. (2007) *Organometallics*, **26**, 5961–5966.

45 (a) Ichikawa, J., Kaneko, M., Yokota, M., Itonaga, M. and Yokoyama, T. (2006) *Org. Lett.*, **8**, 3167–3170; (b) Ichikawa, J., Jyono, H., Kudo, T., Fujiwara, M. and Yokota, M. (2005) *Synthesis*, 39–46.

46 Smart, B.E. (1994) *Organofluorine Chemistry, Principles and Commercial Applications* (eds E. Banks, B.E. Smart and J.C. Tatlow), Plenum Press, New York.

47 Yokota, M., Fujita, D. and Ichikawa, J. (2007) *Org. Lett.*, **9**, 4639–4642.

48 Lomberget, T., Bentz, E., Bouyssi, D. and Balme, G. (2003) *Org. Lett.*, **5**, 2055–2057.

49 Youn, S.W., Pastine, S.J. and Sames, D. (2004) *Org. Lett.*, **6**, 581–584.

50 (a) Liu, C., Han, X., Wang, X. and Widenhoefer, R.A. (2004) *J. Am. Chem. Soc.*, **126**, 3700–3731; (b) Liu, C. and Widenhoefer, R.A. (2004) *J. Am. Chem. Soc.*, **126**, 10250–10251.

51 GowriSankar, S., Lee, k.Y., Lee, C.G., Kim, J.N. (2004) *Tetrahedron Lett.*, **45**, 6141–6146.

52 Zhang, L. and Kozmin, S.A. (2004) *J. Am. Chem. Soc.*, **126**, 10204–10205.

53 Merlic, C.A. and Pauly, M.E. (1996) *J. Am. Chem. Soc.*, **118**, 11319–11320.

54 (a) Chatani, N., Inoue, H., Ikeda, T. and Murai, S. (2000) *J. Org. Chem.*, **65**, 4913–4918; (b) Inoue, H., Chatani, N. and Murai, S. (2002) *J. Org. Chem.*, **67**, 1414–1417.

55 Nishizawa, M., Takao, H., Yadav, V.K., Imagawa, H. and Sugihara, T. (2003) *Org. Lett.*, **5**, 4563–4565.

56 (a) Fürstner, A. and Mamane, V. (2003) *Chem. Commun.*, 2112–2113; (b) Mamane, V., Hannen, P. and Fürstner, A. (2004) *Chem. Eur. J.*, **10**, 4556–4575.

57 Soriano, E. and Marco-Contelles, J. (2006) *Organometallics*, **25**, 4542–4553.

58 Nevado, C. and Echavarren, A.A. (2005) *Chem. Eur. J.*, **11**, 3155–3164.

59 Song, C.E., Jung, D., Choung, S.Y., Roh, E.J. and Lee, S. (2004) *Angew. Chem. Int. Ed.*, **43**, 6183–6185.

60 Ishikawa, T., Manabe, S., Aikawa, T., Kudo, T. and Saito, S. (2004) *Org. Lett.*, **6**, 2361–2364.

61 Huang, W., Shen, Q., Wang, J. and Zhou, X. (2008) *J. Org. Chem.*, **73**, 1586–1589.

62 Fürstner, A. and Mamane, V. (2002) *J. Org. Chem.*, **67**, 6264–6267.

63 Bueno, A.B., Moser, W.H. and Hegedus, L.S. (1998) *J. Org. Chem.*, **63**, 1462–1466.

64 Fillion, E., Dumas, A.M. and Hogg, S.A. (2006) *J. Org. Chem.*, **71**, 9899–9902.

65 Fillion, E. and Beingessner, R.L. (2003) *J. Org. Chem.*, **68**, 9485–9488.

66 Xu, G.-C., Liu, L.-P., Lu, J.-M. and Shi, M. (2005) *J. Am. Chem. Soc.*, **127**, 14552–14553.

67 (a) Wang, B.-Y., Jiang, R.-S., Li, J. and Shi, M. (2005) *Eur. J. Org. Chem.*, 4002–4008; (b) Shi, M. and Lu, J.-M. (2006) *J. Org. Chem.*, **71**, 1920–1923; (c) Lu, J.-M. and Shi, M. (2006) *Org. Lett.*, **8**, 5317–5320; (d) Zhang, Y.-P., Lu, J.-M., Xu, G.-C. and Shi, M. (2007) *J. Org. Chem.*, **72**, 509–516.

68 (a) Evans, D.A., Fandrick, K.R. and Song, H.-J. (2005) *J. Am. Chem. Soc.*, **127**, 8942–8943; (b) Evans, D.A., Fandrick, K.R., Song, H.-J., Scheidt, K.A. and Xu, R. (2007) *J. Am. Chem. Soc.*, **129**, 10029–10030.

69 Li, C.-F., Liu, H., Liao, J., Cao, Y.-J., Liu, X.-P. and Xiao, W.-J. (2007) *Org. Lett.*, **9**, 1847–1850.

70 Paras, N.A. and MacMillan, D.W.C. (2001) *J. Am. Chem. Soc.*, **123**, 4370–4371.

71 (a) Gremmen, C., Willemse, B., Wanner, M.J. and Koomen, G.-J. (2000) *Org. Lett.*, **2**, 1955–1958; (b) Silveira, C.C., Felix, L.A., Braga, A.L. and Kaufman, T.S. (2005) *Org. Lett.*, **7**, 3701–3705.

72 Taylor, M.S. and Jacobsen, E.N. (2004) *J. Am. Chem. Soc.*, **126**, 10558–10559.

73 Bandini, M., Melloni, A., Piccinelli, F., Sinisi, R., Tommasi, S. and Umani-Ronchi, A. (2006) *J. Am. Chem. Soc.*, **128**, 1424–1425.

74 Liu, C. and Widenhoefer, R.A. (2007) *Org. Lett.*, **9**, 1935–1938.

# 22
# Macrolactones and Macrolactams
*Chang-Liang Sun, Bi-Jie Li, and Zhang-Jie Shi*

In the universal natural and synthetic world, macrocycles are unique and important structural scaffolds. Generally, cyclic structural units containing $n$ atoms ($n > 12$) are considered as macrocycles. Among such macrocyclic structures, macrolactone and macrolactam exist widely, exhibiting broad biological activities and playing important roles in synthetic and natural drug families. In this chapter, we describe general and efficient methods to construct macrolactones and macrolactams and their applications in organic syntheses. Since the formation of macrolactones and macrolactams through C–C formation, such as the RCM reaction, and coupling reactions, will be discussed in other chapters, this chapter focuses only on the formation of macrolactones and macrolactams through C–X (X=O or N) construction.

## 22.1
## Macrolactones

### 22.1.1
### Introduction

Macrolactone, which contains a ring with twelve or more members and one or more ester linkages, is one of the most important frameworks in natural products and synthetically useful compounds in organic chemistry [1]. Three main factors make macrolactonization rather complicated: (i) various functional groups on the precursor must be protected properly; (ii) the self-curl of the precursor is an entropy-decreasing process, which is disfavored thermodynamically; (iii) possible intermolecular reaction may compete with the desired macrolactonization. Thus, the challenge is to develop reagents to activate either or both of the two ends of the precursor to approach macrolactonization. Certainly, the conditions should be mild enough to ensure a good yield and no influence on the other functional groups and chiral centers.

Great progress has been made in the chemical synthesis of macrolactones by the development of various efficient methods for ring closure of ω-hydroxy acids or their activated derivatives [1, 2]. Recently, several effective C–C bond-forming transformations, such as transition metal-promoted coupling and olefin metathesis, have been widely applied to the synthesis of macrocyclic compounds. However, macrolactonization is still the most popular and direct method for producing macrolactone frameworks by developing effective and practical methods to construct the ester linkage. This chapter mainly introduces the methods of macrolactonization from ω-hydroxy carboxylic acids and their applications in the synthesis of natural products.

Numerous methods have been reported for syntheses of macrolactones; however, there are only a few reactions that are widely used in total syntheses of natural products, such as the Corey–Nicolaou lactonization via the S-pyridyl ester, Masamune lactonization via thiol ester activation, Mukaiyama lactonization via the onium salt, Yamaguchi lactonization via a mixed-anhydride, Keck–Steglich lactonization through DCC/DMAP·HCl activation, Mitsunobu substitution by alcohol activation, and Shiina lactonization via benzoic anhydride. In this chapter, we introduce these methods according to their mechanisms and applications in total syntheses of natural products.

## 22.1.2
### Macrolactonization via Formation of an Acyl–Oxygen Bond

Direct formation of an acyl–oxygen bond toward a macrolactone is a straightforward method to build up the desired scaffolds. Mechanistically, the key step in these transformations is the formation of the acyl–oxygen bond, generally by activating the carboxyl group or the hydroxy group or both. These reactions were developed very early but are still broadly applied for macrolactonization.

#### 22.1.2.1 Corey–Nicolaou Macrolactonization

**General** Corey–Nicolaou macrolactonization, first developed in 1974 [3], proved to be one of the most efficient and widely used macrolactonization methods. In their reports, a series of ω-hydroxy acids were efficiently cyclized by first converting them to the corresponding 2-pyridinethiol esters, which were further added slowly to refluxing benzene. Slow addition is required to maintain a highly dilute condition, keeping the undesired intermolecular esterification product to a minimum.

The use of the 2-pyridinethiol ester method for the lactonization of molecules bearing multiple functionality was first illustrated by the synthesis of (±)-zaralenone (Scheme 22.1) [4]. The 2-pyridinethiol ester of the protected (±)-hydroxy acid derivative **1** was prepared in benzene at 25 °C and subjected to cyclization in benzene under reflux to afford (±)-zaralenone **2** in 75% isolated yield after removal of the protecting groups. Over the past three decades, several modifications of this method have been developed including: (i) the use of silver perchlorate or silver tetrafluoroborate to activate the 2-pyridinethiol ester by complexation [5a] and (ii) the development and use of other bis-heterocyclic disulfide reagents [5b].

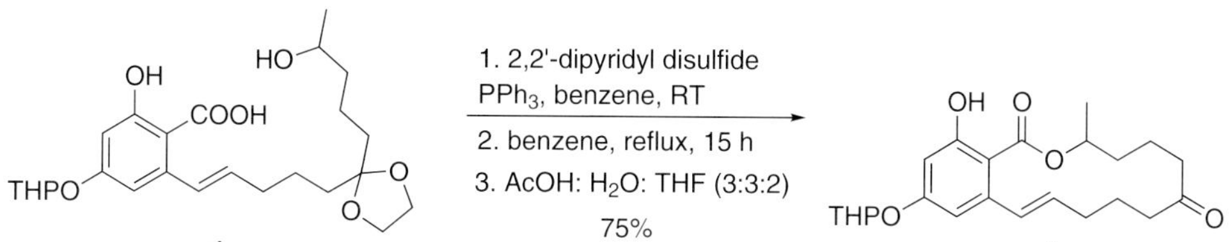

Scheme 22.1 Corey–Nicolaou macrolactonization step in the synthesis of (±)-zaralenone.

**Mechanism** This macrolactonization is considered to proceed via a stepwise process (Scheme 22.2). 2-Pyridinethiol ester undergoes an intramolecular proton transfer to afford a dipolar intermediate, in which the carbonyl group is held by hydrogen bonding. The intramolecular attack of the alkoxide anion on the activated carbonyl group followed by collapse of the tetrahedral intermediate yields the lactone along with 2-pyridinethione [6].

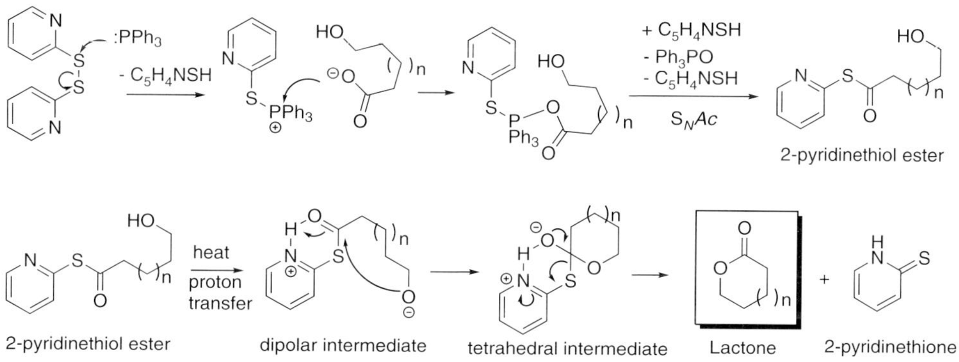

Scheme 22.2 Mechanism of Corey–Nicolaou macrolactonization.

**Applications** Due to its generality with regard to the ring size as well as its outstanding functional group tolerance, Corey–Nicolaou macrolactonization was frequently applied to total synthesis [7, 8]. Nicolaou and coworkers applied the modified Corey–Nicolaou macrolactonization to the construction of the medium cyclic lactone ring in brevetoxin A [7]. The dihydroxy dicarboxylic acid substrate **3** was first converted to bis-2-pyridinethiol ester which gave the bis-lactonization product **4** upon heating in toluene at low substrate concentration (Scheme 22.3).

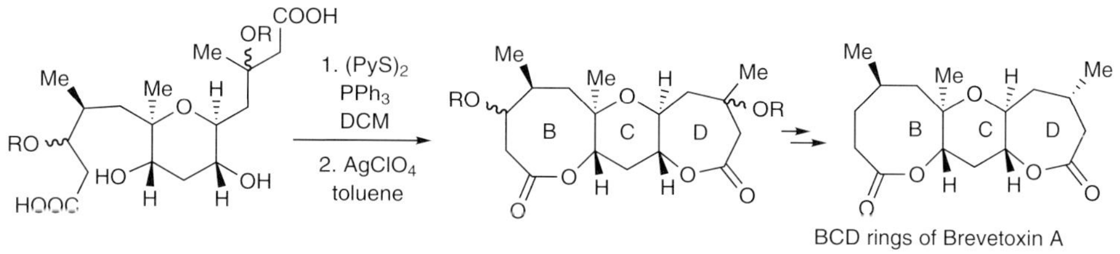

Scheme 22.3 Corey–Nicolaou macrolactonization applied in the construction of BCD ring system of brevetoxin A.

## 22 Macrolactones and Macrolactams

The first total synthesis of the ichthyotoxic marine natural product (−)-aplyolide A **7** was accomplished by Stenstrom and coworkers in 2001 [8]. This natural product contains a 16-membered lactone ring, four (Z)-double bonds and a stereogenic center. Among the numerous macrolactonization methods tested, starting from the precursor **6**, only Corey–Nicolaou macrolactonization gave the desired product while others gave mainly diolide (Scheme 22.4).

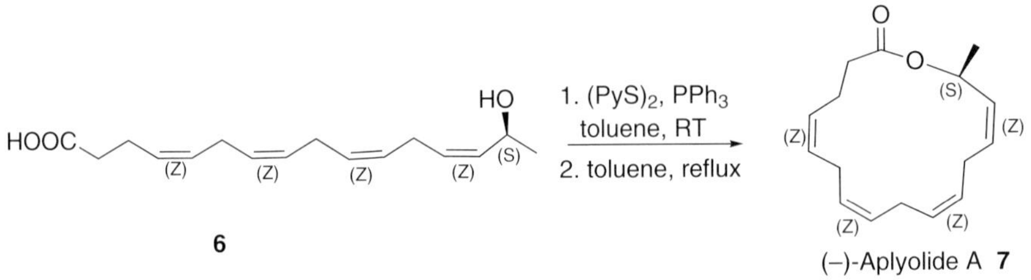

**Scheme 22.4** Corey–Nicolaou macrolactonization applied in the synthesis of (−)-aplyolide A.

### 22.1.2.2 Masamune Macrolactonization

**General** In 1981, Masamune employed the *S-tert*-butyl ester as an intermediate in the total synthesis of 6-deoxyerythronolide B **10**. During this synthesis, the thiol ester was activated and transformed to the desired polyoxygenated macrolactone **9** by a metal-promoted cyclization (Scheme 22.5) [9]. This transformation to produce macrolactone from the linear precursor was called Masamune lactonization.

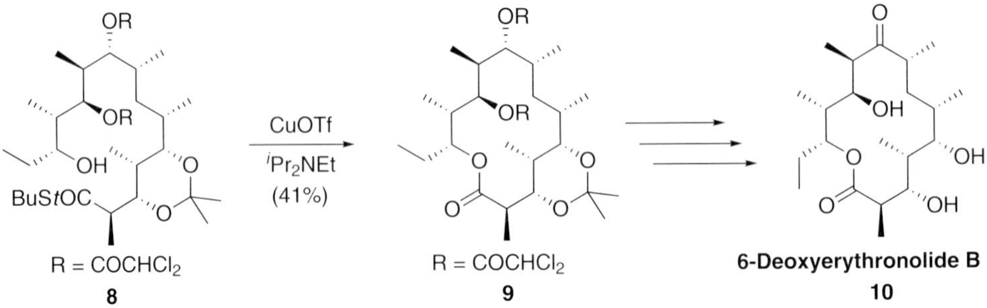

**Scheme 22.5** Masamune lactonization applied in the synthesis of 6-deoxyerythronolide B.

**Mechanism** The *tert*-butyl thiol ester is relatively stable and can survive under either mild acidic or alkaline conditions. The heavy metals used as thiophilic cations, including Hg(II), Ag(I), Cu(I) and Cu(II), are generally used to activate the thiol esters [10]. Different thioates can be transformed into lactones promoted by a properly selected thiophilic metal. Generally, $^i$Pr$_2$NEt serves as the base to quench the proton from the hydroxy group. The mechanism for this macrolactonization was hypothesized as shown in Scheme 22.6.

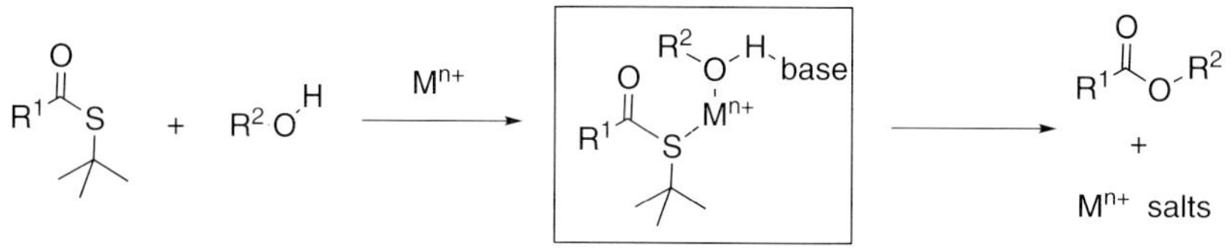

**Scheme 22.6** Mechanism of Masamune lactonization.

**Applications** There are only a few examples of total synthesis with the application of this method. Besides 6-deoxyerythronolide B [9], Masamune also successfully introduced his method of macrolactonization into the synthesis of macrolide antibiotics I methymycin **13** (Scheme 22.7) [11].

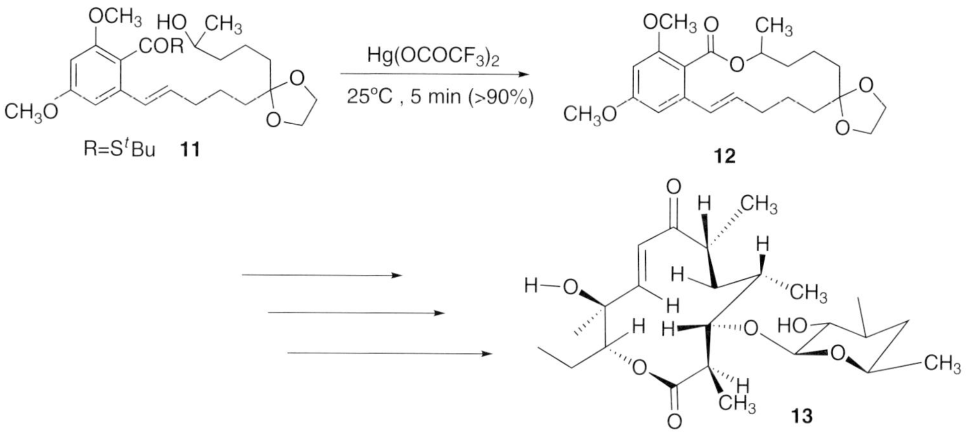

**Scheme 22.7** Masamune lactonization applied in the synthesis of macrolide antibiotics I methymycin.

### 22.1.2.3 Mukaiyama Macrolactonization

**General** Since 1976, Mukaiyama has developed a new method for the preparation of carboxylic esters or lactones in the presence of tertiary amines by using onium salts such as 2-halopyridinium, benzoxazolium, benzothiazolium, and pyridinium salts, which functioned as activating reagents for acids [12]. After the generation of the activated onium salts from ω-hydroxy acids, spontaneous lactonization proceeded smoothly to produce the corresponding lactones. Bartlett and coworkers first applied this method to the preparation of brefeldin A **16** (Scheme 22.8) [13a]. Other groups also succeeded in the total synthesis of complex molecules employing this useful reagent [13b, 14, 15].

**Mechanism** Mukaiyama macrolactonization is initiated by nucleophilic substitution by acid in 2-halopyridinium with the assistance of tertiary amines. After the formation of ester, the acyl C—O is activated and can accept intramolecular attack of

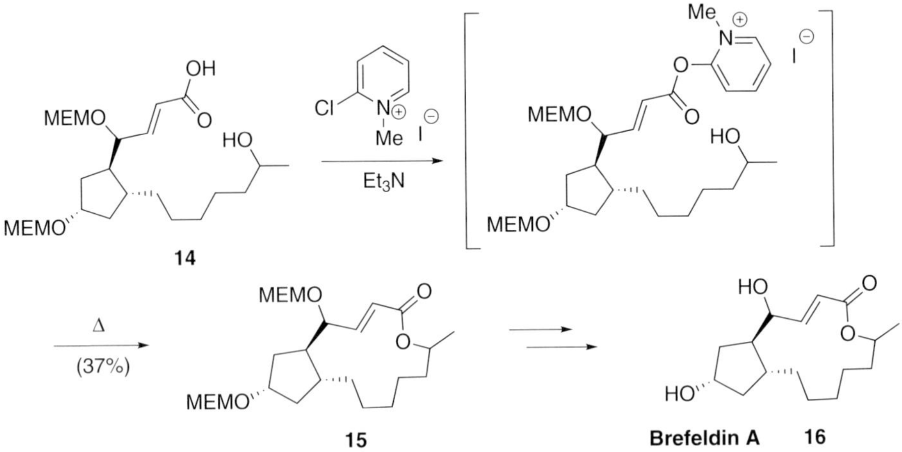

**Scheme 22.8** Mukaiyama macrolactonization applied in the preparation of brefeldin A.

the alcohol to form the desired lactones (Scheme 22.9). This method was also applied to lactamization when using amines as the nucleophiles.

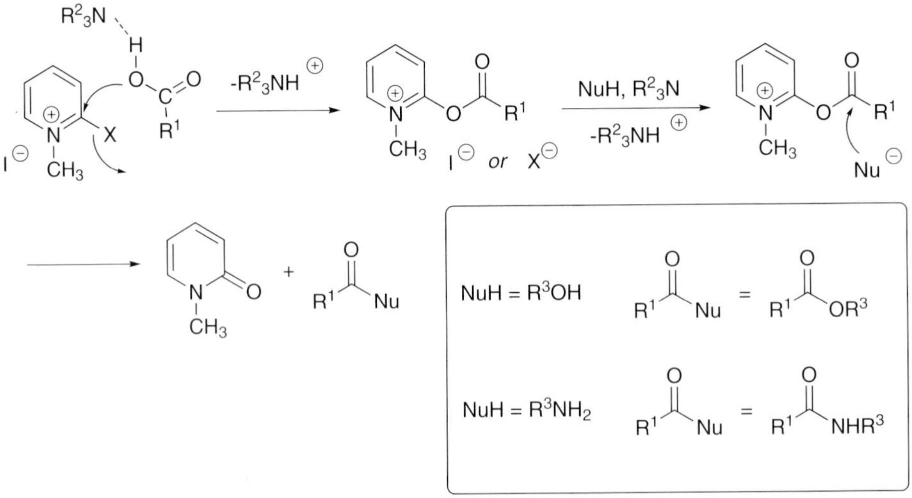

**Scheme 22.9** Mechanism of Mukaiyama macrolactonization.

**Applications** Since many functional groups survive well under this condition, many research groups successfully introduced this method into their total syntheses of complex molecules. For example, avermectin $A_{1a}$ **19** was synthesized by Danishefsky in 1988 (Scheme 22.10) [14] and the total synthesis of antiparasitic agent Avermectin $B_{1a}$ was accomplished by White in 1995 with the use of this method [15].

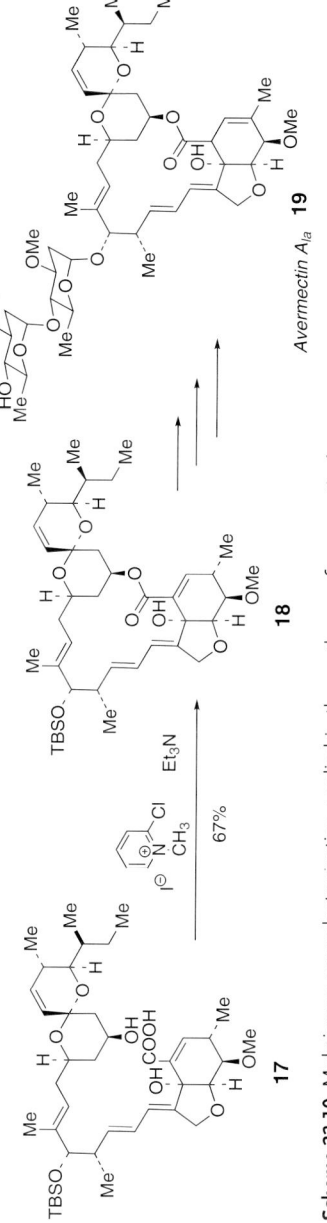

**Scheme 22.10** Mukaiyama macrolactonization applied in the syntheses of avermectin A$_{1a}$.

## 22.1.2.4 Yamaguchi Macrolactonization

**General** In 1979, a new and efficient method for the preparation of esters and lactones was developed by Yamaguchi and coworkers. They found that 2,4,6-trichlorobenzoyl chloride was an efficient reagent to produce the mixed anhydride with ω-hydroxy acid in the presence of triethylamine [16]. After removal of the triethylamine hydrochloride, the resulting mixed anhydride was transformed into the desired lactones by slow addition to a highly diluted solution of DMAP in toluene under mild conditions. The procedure is quite simple and easy to handle. It is important to note that Yamaguchi macrolactonization tolerates various different functional groups, and even a very acid-sensitive substrate, which can be rapidly decomposed by trace amounts of HCl, can be efficiently cyclized without observation of any decomposed product (Scheme 22.11). During this transformation, a highly dilute condition is required to avoid intermolecular coupling. Generally, benzene and toluene are used as efficient solvents and usually several equivalents of DMAP are used as the additive.

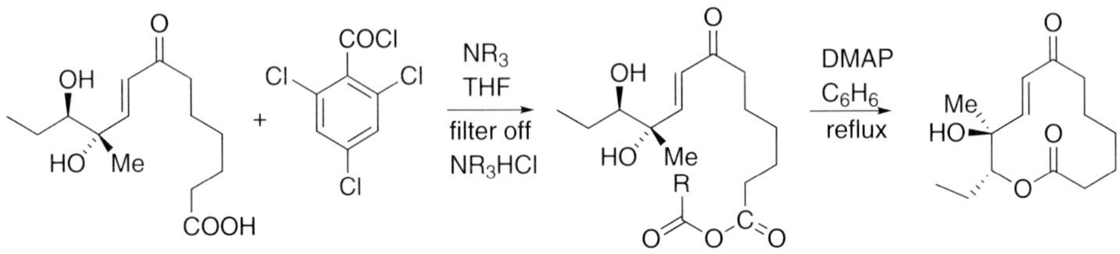

**Scheme 22.11** First discovery of Yamaguchi macrolactonization.

**Mechanism** During this transformation, a common acylation of the carboxylic acid would give the mixed anhydride as a key intermediate. DMAP, a known catalyst for acyl transfer reactions, catalyzed the transfer acylation of alcohol moiety with the mixed anhydride (Scheme 22.12).

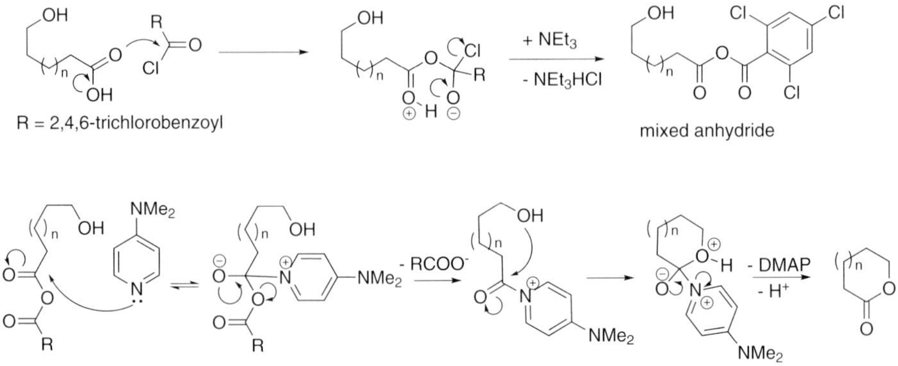

**Scheme 22.12** Mechanism of Yamaguchi macrolactonization.

**Applications** Yamaguchi macrolactonization is one of the most popular acyl C–O formation methods in the total syntheses of many natural macrolactones. For example, the stereocontrolled total synthesis of (−)-Macrolactin A **25** was accomplished by Marino and coworkers featuring a key step of the Yonemitsu modification of the Yamaguchi macrolactonization [17, 18]. The preformed mixed anhydride was added in one portion to a highly dilute solution of DMAP in toluene. Then removal of the protecting groups under acidic conditions furnished the 24-membered macrolide (Scheme 22.13).

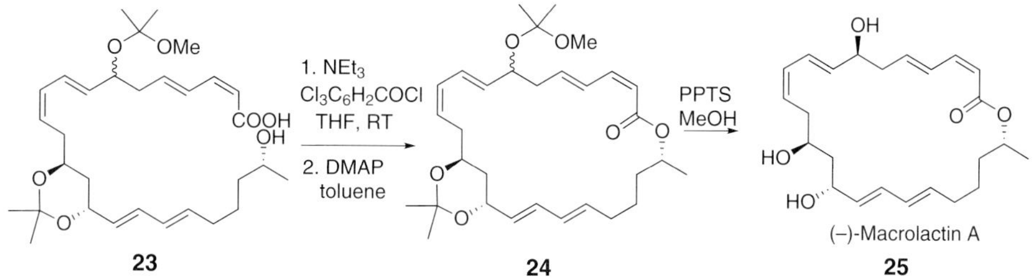

**Scheme 22.13** Yamaguchi macrolactonization applied in the synthesis of (−)-macrolactin A.

The enantioselective total synthesis of oleandolide, the aglycon of the macrolide antibiotic Oleandomycin **28**, was carried out by Panek's group [19]. The key step involved a modified Yamaguchi macrolatonization method. The desired 14-membered lactone was obtained in 94% isolated yield without observation of any undesired 12-membered lactone (Scheme 22.14). The origin of the excellent selectivity was ascribed to the strong conformational preference induced by the large substituent at C9.

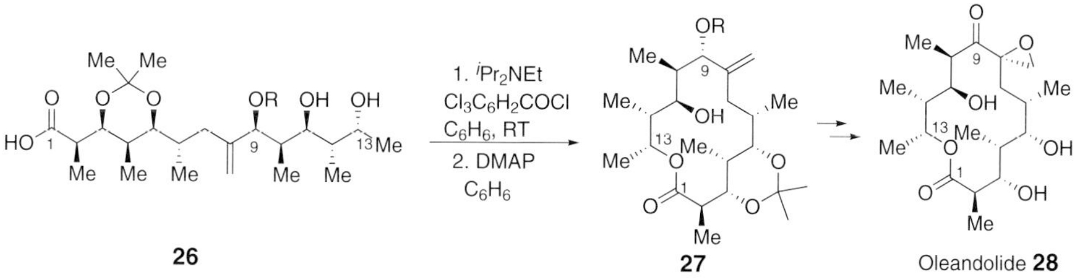

**Scheme 22.14** Yamaguchi macrolactonization applied in the synthesis of macrolide antibiotic oleandomycin.

### 22.1.2.5 Keck Macrolactonization

**General** In the early 1980s, total syntheses of several very complex macrolide antibiotics were accomplished. In 1985, Keck and Boden developed a novel method to facilitate direct macrolactionizaiton via the activated esters derived from the ω-hydroxy acid substrates, which were generated *in situ* without the need for isolation [20]. They initially used the conditions of the Steglich esterification

(DMAP/DCC) [21] for the formation of macrolactones. Unfortunately, the macrolatonization did not proceed even in the presence of excess reagents. With much effort, they found that a proton source such as the hydrochloride salt of dimethylamino pyridine (DMAP·HCl) was required to facilitate the crucial proton-transfer step. In the presence of this proton source, the macrocyclizations occurred in good to excellent yields. This transformation was named the Keck macrolactonization (Scheme 22.15).

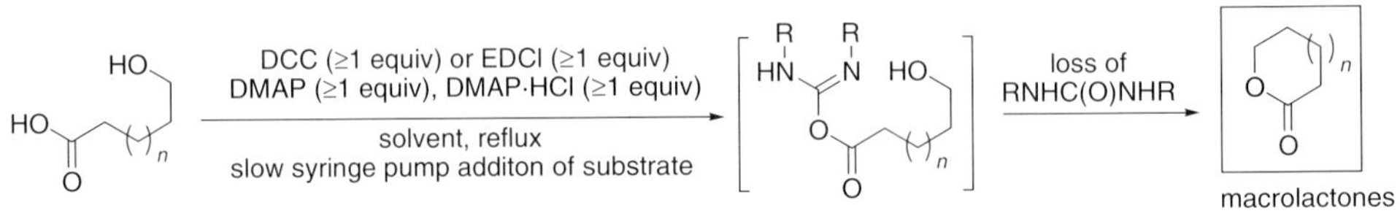

**Scheme 22.15** Keck macrolactonization.

This transformation was proposed to proceed through the DCC-activated ester, which underwent further nucleophilic substitution, promoted by DMAP·HCl. The general features of this transformation are: (i) similar to other macrolactonization procedures the reaction requires high-dilution conditions ($\leq 0.03$ M); (ii) the substrate is usually dissolved in an aprotic solvent and added to refluxing solution of the reagents via a syringe pump over several hours; (iii) the activating agent is DCC or EDCl that prevents the trace amount of water from destroying the activated acyl derivative; (iv) the carbodiimide reagent is typically used in several-fold excess to ensure high conversion of the starting material; (v) the use of DMAP·HCl prolongs the lifetime of the activated intermediate and suppresses the formation of the undesired by-product. The most important modification of the Keck macrolactonization utilizes polymer-supported DCC to simplify the work-up [22].

**Mechanism** The general mechanism is described in Scheme 22.16.

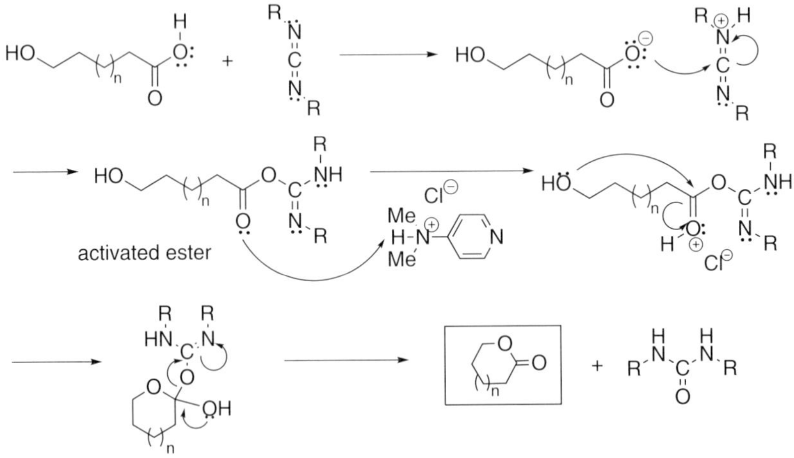

**Scheme 22.16** General mechanism of Keck macrolactonization.

**Applications** Keck and Boden have successfully applied this macrolactionization protocol in their synthesis of colletodiol **31** in high yield (Scheme 22.17) [23]. This transformation showed high efficiency and excellent selectivity.

**Scheme 22.17** Keck macrolactonization applied in the synthesis of colletodiol.

### 22.1.2.6 Shiina Macrolactonization

**General** Recently, Shiina and coworkers reported a novel mixed-anhydride method for the preparation of macrolactones as well as medium-sized ring compounds. They further applied this new methodology in the preparation of several macrocyclic molecules. This reaction could be promoted by an acidic or basic activator, such as Lewis acids, DMAP, or DMAPO. Ricinoleic acid lactone **33** was first synthesized by the combination of 4-trifluoromethylbenzoic anhydride (TFBA) with Lewis acid catalysts [24], while the aleuritic acid lactone **35** was alternatively synthesized using 2-methyl-6-nitrobenzoic anhydride (MNBA) with basic additives (Scheme 22.18) [25].

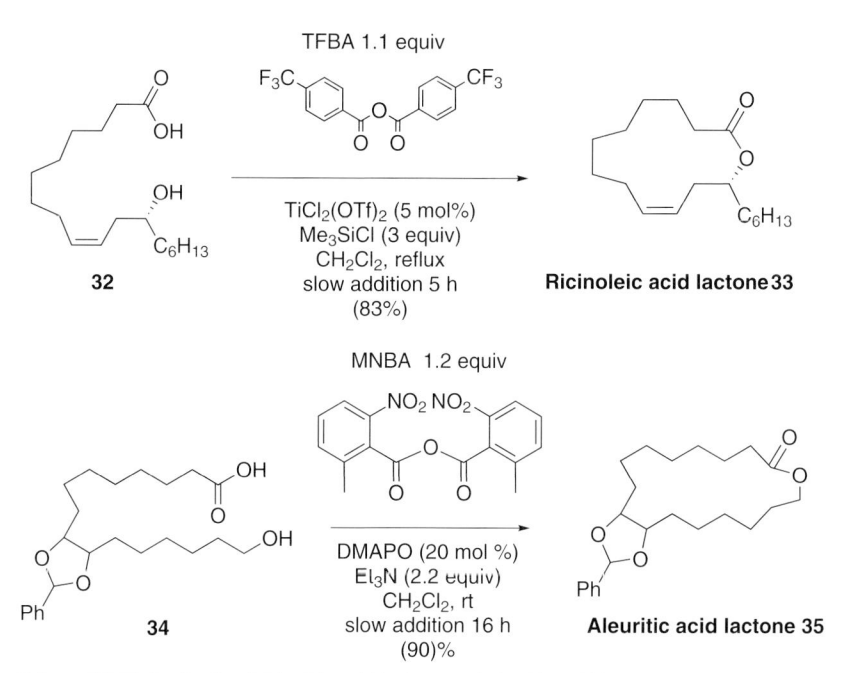

**Scheme 22.18** Synthesis of ricinoleic acid lactone and aleuritic acid lactone.

**Mechanism** An effective method for syntheses of carboxylic esters from nearly equimolar amounts of silyl carboxylates and alkyl silyl ethers via mixed-anhydrides was developed by employing a substituted benzoic anhydride and a Lewis acid catalyst [24a]. During this transformation, it was indicated that the following successive reactions would lead to the formation of carboxylic esters: (i) the initial formation of the mixed-anhydride from benzoic anhydride and silyl carboxylates with the promotion of Lewis acid, and (ii) the alcoholysis of the mixed anhydrides by alkyl silyl ethers with the assistance of a Lewis acid (Scheme 22.19).

**Applications** A variety of molecules [26] were successfully synthesized according to the substituted benzoic anhydride method under acidic or basic conditions, including the medium-sized lactones, especially 8 or 9-membered rings, such as cephalosporolide D [27], antimycin $A_{3b}$ [28]. (Scheme 22.20)

### 22.1.3
### Macrolactonization Via Formation of an Alkyl C—O Bond

Mechanistically, the key step of this type of macrolactonization is the formation of an alkyl C—O bond. Therefore, most activators react on the hydroxy groups rather than carboxyl groups. Thus, both oxygen atoms in the ester linkage come from the carboxyl group. This type of macrolactionization mainly refers to the Mitsunobu reaction, and includes some other methods, which have been used infrequently.

#### 22.1.3.1 Mitsunobu Reaction

**General** In 1967, Mitsunobu first reported the efficient esterification of secondary alcohols with carboxylic acids in the presence of diethyl azodicarboxylate (DEAD) and $Ph_3P$ [29]. Later, the procedure was applied to synthesize optically active amines, azides, ethers, thioethers, and even alkanes from the alcohols. DEAD and DIAD were most frequently used as the azodicarboxylate reagents. This method can also be used to construct macrolactam. Furthermore, Mukaiyama developed an important variant of the Mitsunobu reaction by the preparation of inverted *tert*-alkyl carboxylates from chiral tertiary alcohols via alkoxydiphenylphosphines, which were formed *in situ* by using 2,6-dimethyl-1,4- benzoquinone [30].

**Mechanism** In this protocol, the carboxylate anion reacts as a nucleophile to attack the alcohol derivatives, which is regarded as an $S_N2$ reaction. Thus, the product undergoes complete inversion of configuration. Many other nucleophiles also undergo this reaction and the corresponding products can be produced efficiently (Scheme 22.21) [31].

**Applications** The Mitsunobu reaction has been widely applied in the syntheses of natural products. In 1984, Mitsunobu demonstrated the potential of the lactonization in the alternative total synthesis of colletodiol **38** (Scheme 22.22) [32]. Meyers recently

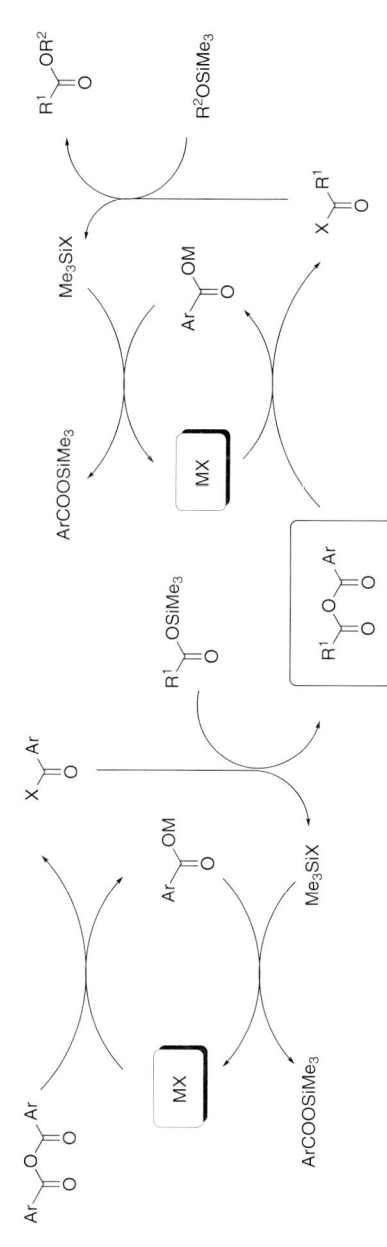

**Scheme 22.19** Mechanism of Shiina macrolactonization.

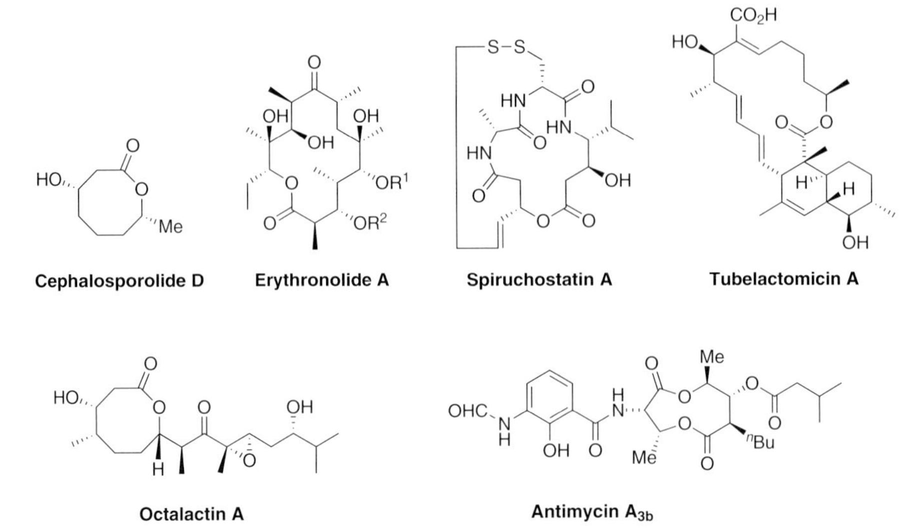
Cephalosporolide D
Erythronolide A
Spiruchostatin A
Tubelactomicin A
Octalactin A
Antimycin A3b

**Scheme 22.20** Molecules successfully synthesized using Shiina macrolactonization.

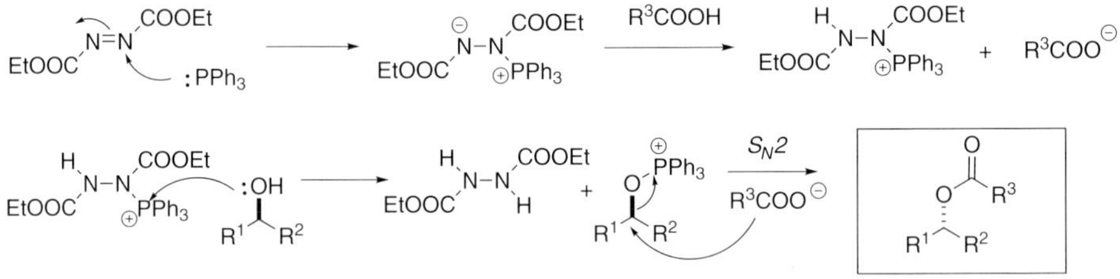
R3COOH
SN2

**Scheme 22.21** Mechanism of Mitsunobu reaction.

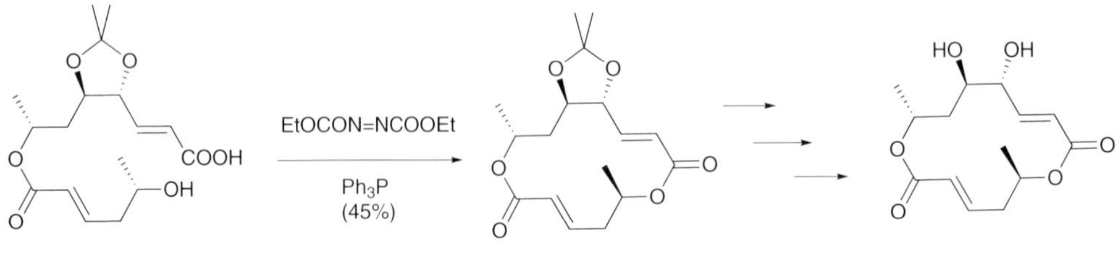
EtOCON=NCOOEt
Ph3P
(45%)

36 37 colletodiol **38**

**Scheme 22.22** Mitsunobu reaction applied in the synthesis of colletodiol.

accomplished the total synthesis of Griseoviridin **41** utilizing this lactone formation reaction (Scheme 22.23) [33].

#### 22.1.3.2 Other Methods

In 1989, Matsuyama and coworkers reported a useful method for the synthesis of macrolactones using (ω-carboxyalkyl)sulfonium salts. Base-catalyzed intramolecular

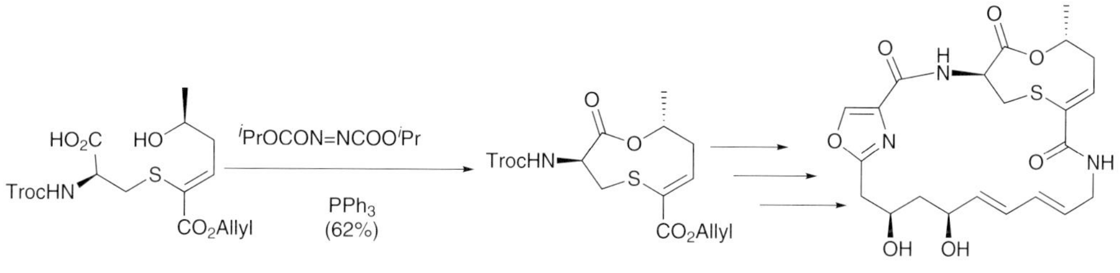

**Scheme 22.23** Mitsunobu reaction applied in the synthesis of griseoviridin.

cyclization of (ω-carboxyalkyl)sulfonium salts gave simple macrolactones in high yields under highly dilute conditions (Scheme 22.24) [34].

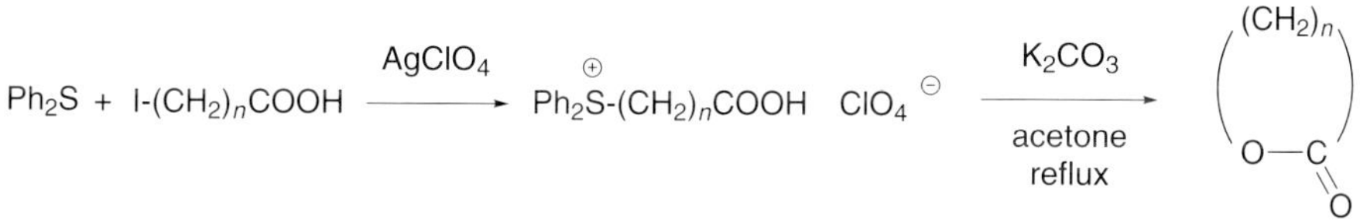

**Scheme 22.24** Synthesis of macrolactones using (ω-carboxyalkyl)sulfonium salts.

In 1990, Karim and coworkers developed an efficient method for macrolactonization. The intramolecular nucleophilic displacement of chloride from the highly electrophilic chloro ketone moiety by a remote carboxylate nucleophile resulted in the clean formation of the 11-membered keto lactone (Scheme 22.25). Relatively high substrate concentrations (up to 18 mM) could be employed without formation of dimeric or oligomeric by-products. The slow mixation of substrate and base was not required [35a].

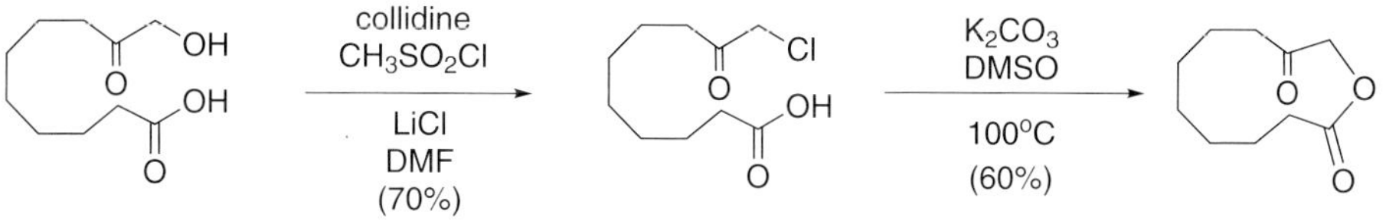

**Scheme 22.25** An efficient method for macrolactonization developed by Karim and coworkers.

Very recently, White and coworkers developed a new macrolactonization method via hydrocarbon oxidation. Bis-sulfoxide/BQ were used as serial ligands to promote efficient Pd-catalyzed allylic C−H activation. Linear ω-alkenoic acids react to furnish 14- to 19-membered alkyl and aryl macrolides with a range of diverse functionalities. Using the allylic C−H oxidative macrolactonization, a novel 17-membered bis (indolyl)maleimide macrolide **43** was readily formed in 58% isolated yield (Scheme 22.26) [35b].

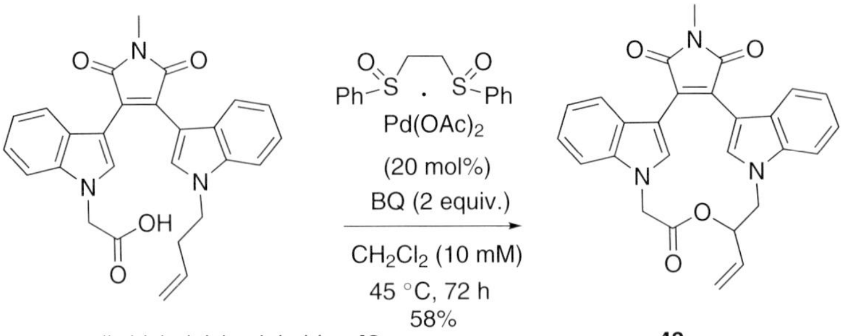

**Scheme 22.26** The synthesis of bis(indolyl)maleimide macrolide using the allylic C—H oxidative macrolactonization.

Using Rh catalyzed C—H bond functionalization strategy involving an unprecedented Rh-catalyzed hydroacylation of ketones, Dong and coworkers were able to realize an enantioselective approach to lactones [35c]. Keto-aldehyde was converted to the corresponding seven-membered lactone ring in a high yield and with high enantioselectivity (Scheme 22.27). Although those new developments showed a good potential to construct macrolactones, only limited examples have been demonstrated in natural product syntheses.

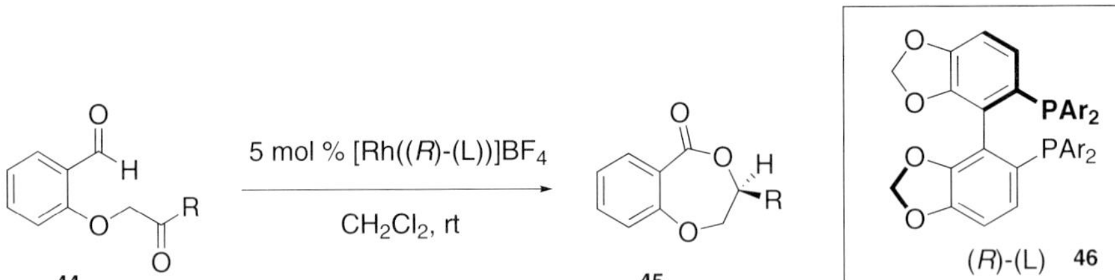

**Scheme 22.27** Macrolactonization using Rh catalyzed C—H bond functionalization strategy.

## 22.1.4
### Macrolactonization in Biosystems

Recently, biological processes have been broadly developed and used in organic syntheses. The application of biocatalysts to organic synthesis is a well-defined field now, several reviews [36] and books [37] have already been published. The synthetic application of biocatalysts to the preparation of chiral synthons or special molecular structures can be carried out in whole cells of animals [38], plants [39], or microorganisms [40]. In these cases, the complicated mechanism of the enzymes present in the cells can be utilized without disrupting the integrity of the membrane or as a crude homogenate [41]. However, most biocatalysts proceed in a water medium. Besides the advantages of biocatalysts in a water medium, the newly-developed methods of enzyme catalysis in organic media have many unique properties, such as the high solubility of nonpolar substrates, an equilibrium favoring the product side, depressing by-products caused by water and giving an easy isolation and purification of products. Generally, the available enzymes are traditionally divided into six classes [42] (oxidoreductases, transferases, hydrolases, lyases, isomerases, and lipases) according to the specific type of reaction. Although there is no specific type of enzyme for lactonization or esterification, hydrolytic enzymes such as esterases, lipases, and proteases may, generally, catalyze the reverse reaction, such as esterification and amidation, with a low concentration condition. In order to carry out the esterification process of preparative significance smoothly, the hydrolysis and other by-products have to be suppressed by optimizing the reaction conditions. Based on the facts above and references reported, macrolactonization can be achieved via biocatalysts and many examples have been reported.

In 1993, Mori reported his syntheses of several macrolides of insect pheromones [43]. Compared with chemical methods, the enzyme-catalyzed macrolactonization favored the formation of twelve-membered rings and beyond (Scheme 22.28). When dealing with eleven-membered rings, the only products are dimers and trimers.

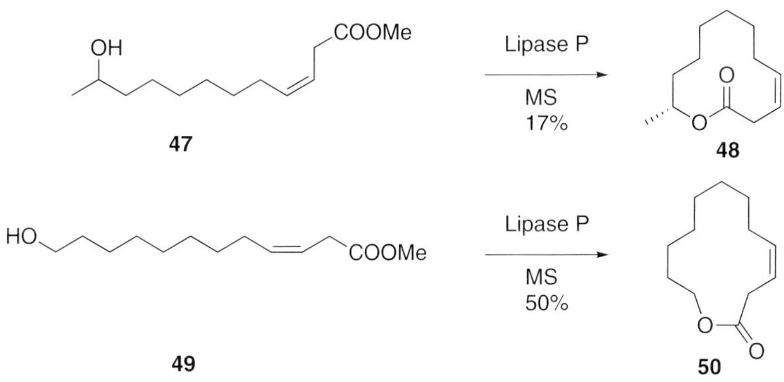

**Scheme 22.28** Enzyme-catalyzed macrolactonization.

In 1993, Mamdapur used porcine pancreatic lipase (PPL) as the biocatalyst in the lactonization in his synthesis of a 1,4-dienic macrolide pheromone of cucujid grain beetles [44]. The reaction could be carried out in benzene to provide the target compound (78%) along with the unconverted starting material (Scheme 22.29).

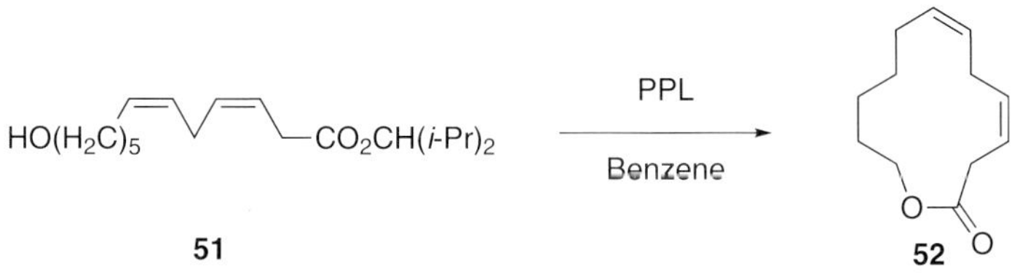

**Scheme 22.29** PPL as the biocatalyst in the lactonization in the synthesis of 1,4-dienic macrolide.

## 22.2
## Macrolactams

### 22.2.1
### Introduction

The term macrolactam refers to those compounds having a twelve-membered or larger ring system, which is linked by one or more amide bond. Like macrolactone, macrolactam is another of the most important scaffolds in natural products and synthetic compounds. Therefore, the development of efficient methods of macrolactam synthesis continues to be an active research area in organic chemistry. Generally, the methods of macrolactam synthesis fall into two categories, depending on how the macro-ring is formed. One is direct amidation to form the macro-ring, and the other is ring closure of an amide bond-containing molecule to form the macrolactam. Both of these strategies require the synthesis of the amide bond. Therefore, efficient methods of amide bond formation are crucial for the success of macrolactam synthesis.

The amide bond-forming methods have been largely stimulated by peptide chemistry since the amide bond is the key functionality in peptides and proteins. It is also an important component of many natural products and drugs [45]. The development in peptide chemistry has also resulted in great progress in the amidation methods.

Direct condensation of carboxylic acid and amine to form amide requires very high temperature (160–180 °C), which is obviously impractical in total synthesis [46]. Thus, preactivation of the carboxylic acid moiety to an active acylation reagent is necessary before reaction with the amine. In general, the active acylation reagents include acyl halides, acyl azides, acylimidazoles, anhydrides, esters, and so on (Scheme 22.30). Excellent reviews and monographs on this topic have already appeared [47].

$$\text{RCOOH} \xrightarrow[\text{activation step}]{\text{activating reagent}} \text{RCOX} \xrightarrow[\text{aminolysis step}]{\text{R'NH}_2} \boxed{\text{RCONHR'}}$$

**Scheme 22.30** Preactivation of the carboxylic acid moiety into an active acylation reagent.

Below we discuss various methods of macrolatam synthesis in detail. There are two subsections. Detailed amidation methods are introduced in the first while the second deals with different ring closing methods of the amide bond-containing molecule. While a comprehensive compilation of the literature on amidation methods and their application in macrolactam synthesis is not easy to achieve, we mainly introduce those methods that are frequently used and, if appropriate, several examples in total synthesis are given.

## 22.2.2
### Formation of Macrolactams Via Direct Amidation

This section discusses several mild amidation methods and, if possible, their application in macrolactam synthesis. The amidation methods are classified according to the activated forms of carboxylic acids such as acyl halides, acylimidazoles, anhydrides, esters, and so on.

#### 22.2.2.1 Acyl Halides

Acyl halides, such as acyl chloride, acyl bromide and acyl fluoride, can be easily prepared from carboxylic acid. Among the acyl halides, acyl chloride is the most frequently used while the use of acyl bromide and fluoride is rare.

The corresponding acyl chlorides can be obtained by simply treating carboxylic acids with thionyl chloride ($SOCl_2$), phosphorus trichloride ($PCl_3$), phosphorus oxychloride ($POCl_3$), phosphorus pentachloride ($PCl_5$) or oxalyl chloride (($COCl)_2$) (Scheme 22.31) [48a–c]. In some cases, a combination of $PPh_3$ and $CCl_4$ was also applicable [48d]. In addition, many acyl chlorides are commercially available. Acyl fluoride could be prepared by the reaction of carboxylic acid and cyanuric fluoride (Scheme 22.32) [49].

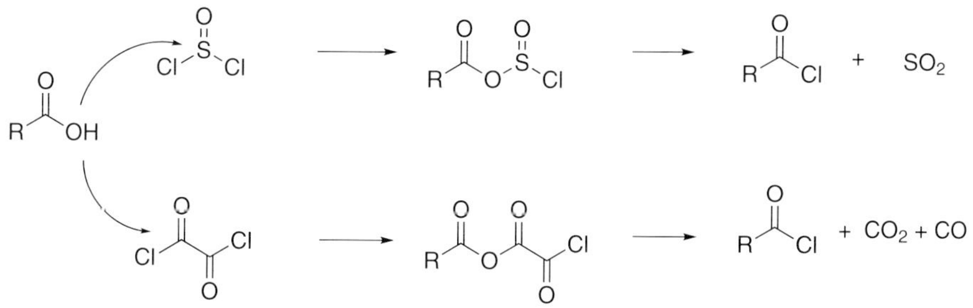

**Scheme 22.31** Carboxylic acids reacting with chlorides.

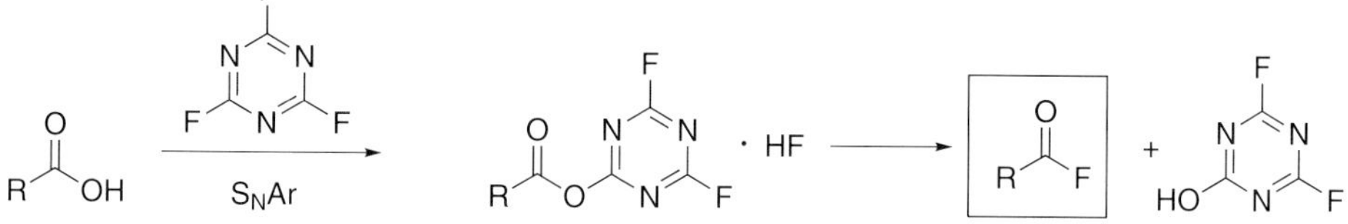

**Scheme 22.32** Acyl fluorides prepared by the reaction of carboxylic acids and cyanuric fluoride.

The amide bond was formed by the reaction of the acyl halide with amine. Typically, the addition of base was helpful to accelerate the reaction. Pyridine or $N,N$-dimethylaminopyridine (DMAP) is commonly used. The acyl-transfer intermediate was responsible for the rate acceleration (Scheme 22.33) [50]. Both the acyl chloride and acyl fluoride reacted in the same way.

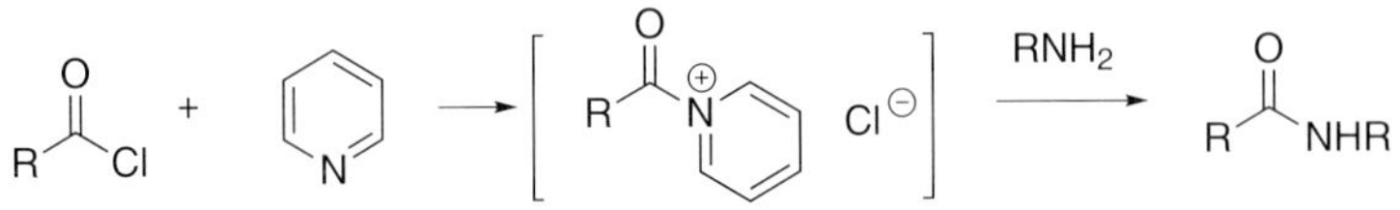

# 22 Macrolactones and Macrolactams

**Scheme 22.33** The amide bond formed by the reaction of the acyl halide with amine.

However, acyl halides have a limited application in macrolactam synthesis, due to the inherent nature of the reaction. Racemization, cleavage of protecting groups, and other side reactions often accompanied the amidation process. These drawbacks are not compatible with natural product synthesis.

### 22.2.2.2 Acylimidazoles

Carbonyl diimidazole (CDI) can reacted with carboxylic acid to form a reactive acylation intermediate. This method allows one-pot amide bond formation. With the help of the basic imidazole generated *in situ*, the acylimidazole reacts readily with amine to form the desired amide (Scheme 22.34) [51].

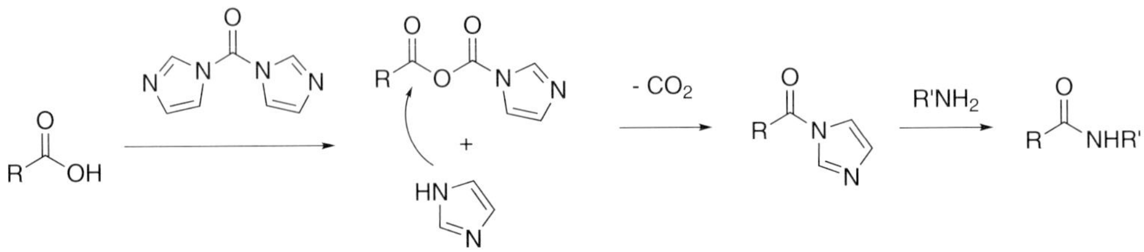

**Scheme 22.34** CDI reacted with carboxylic acid to form reactive acylation intermediate.

### 22.2.2.3 Anhydrides

Anhydrides react smoothly with amine to give amide. Generally, anhydrides can be divided into two kinds: symmetric anhydrides and mixed anhydrides.

**Symmetric Anhydrides** Symmetric anhydride can be prepared by simply heating the carboxylic acid at elevated temperature. Addition of dehydrating reagents such as dicyclohexyl carbodiimide (DCC) allows the preparation to be carried out under much milder conditions (Scheme 22.35) [52].

The preformed anhydride reacts readily with amine to form the amide. The major drawback of this method is that half of the acid is wasted during this transformation (Scheme 22.36). In macrolactam synthesis, this is obviously impractical.

**Mixed Anhydrides** Methods using mixed anhydrides have been developed in order to overcome the uneconomical problem of symmetric anhydrides. Mixed anhydrides could be obtained by incorporating a second cheap and readily available acid. However, another problem of regioselectivity control emerged during the aminolysis. This problem was solved by choosing an appropriate second carboxylic acid moiety. For example, mixed pivalic anhydrides react with amine to form only one amide

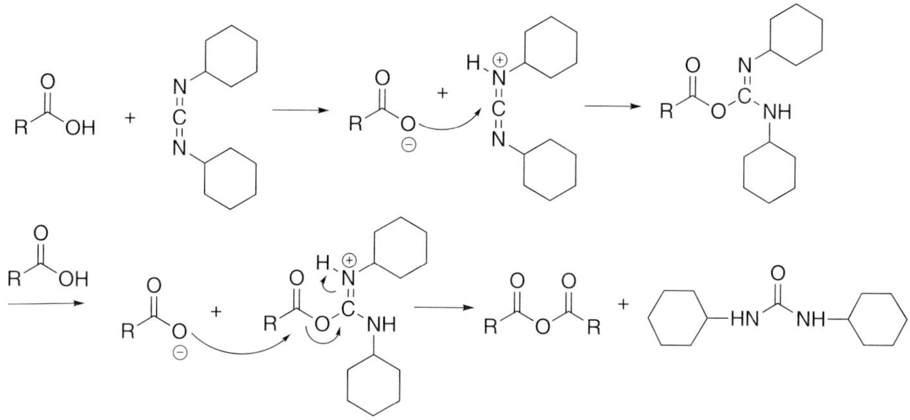

**Scheme 22.35** Preparation process of symmetric anhydrides.

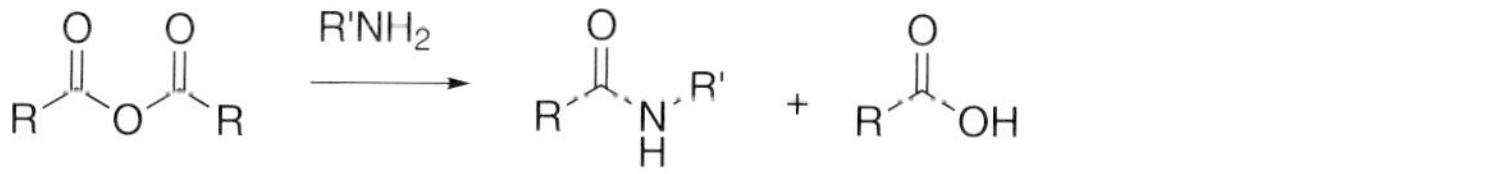

**Scheme 22.36** The anhydride preformed reacts readily with amine to form the amide.

selectively. The high level of regioselectivity is due to the steric hindrance. Carbonic anhydrides also give high regioselectivity, which arises from the different reactivity between the two acyl C—O bonds (Scheme 22.37) [53].

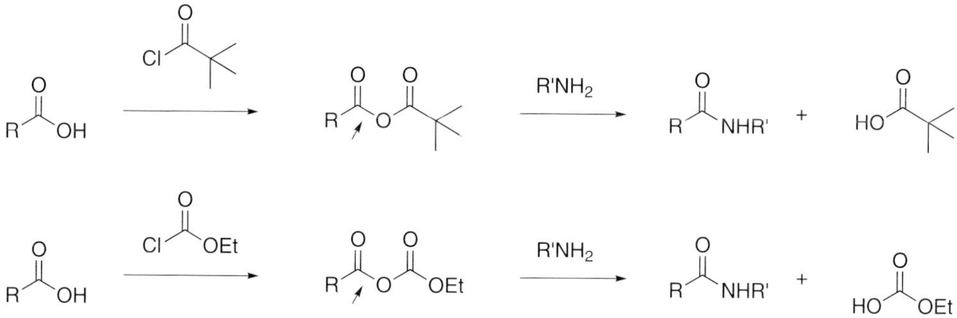

**Scheme 22.37** Mixed anhydride methods.

Conceptually, the reactive intermediates resulting from the condensation of carboxylic acids with other acids or their derivatives can also be regarded as mixed anhydrides. These acids and their derivatives include ethoxyacetylene, boronic acid derivatives, carbodiimides, phosphoric acid derivatives, and so on.

Ethoxyacetylene reacts with carboxylic acids, converting them to masked anhydrides, which readily undergo aminolysis (Scheme 22.38) [54].

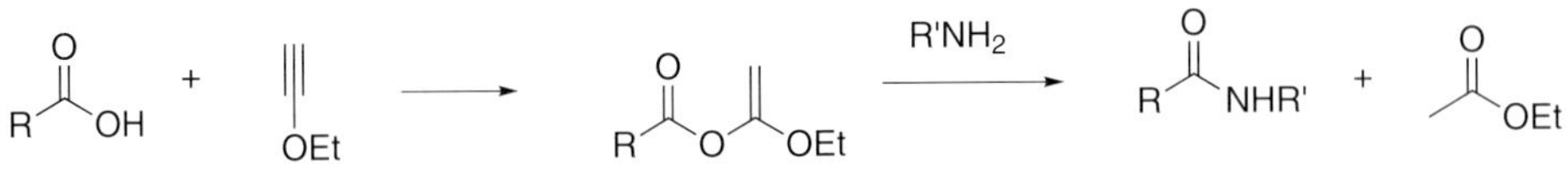

**Scheme 22.38** Ethoxyacetylene used for masked anhydrides.

Acyloxyboron species, generated from carboxylic acids and boron reagents, also react smoothly with amines to give the amides [55a,b]. A catalytic reaction was also developed. For example, arylboronic acids with electron-withdrawing groups such as 3,5-bis-(trifluoromethyl) phenylboronic acid and 3,4,5-trifluorophenylboronic acid catalyze the amidation in refluxing toluene while the use of 2-iodophenylboronic acid allows amidation to proceed at room temperature [55c,d]. In the proposed mechanism, an acyloxyboron species is considered as the key intermediate (Scheme 22.39).

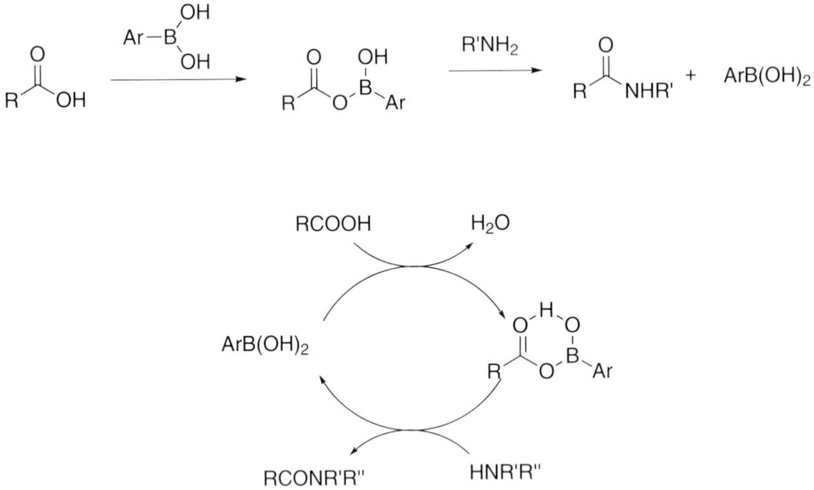

**Scheme 22.39** Proposed mechanism using acyloxyboron species.

Carbodiimides, such as dicyclohexyl carbodiimide (DCC), diisopropyl carbodiimide (DIC) and 1-ethyl-3-(3-dimethyl amino)carbodiimide HCl salt (EDC or WSC-HCl) are frequently used for the activation of carboxylic acids [56]. The carbodiimide first reacts with carboxylic acid to afford the O-acylisourea mixed anhydride, which subsequently undergoes aminolysis to give the amidation product. The method can be performed in one pot (Scheme 22.40).

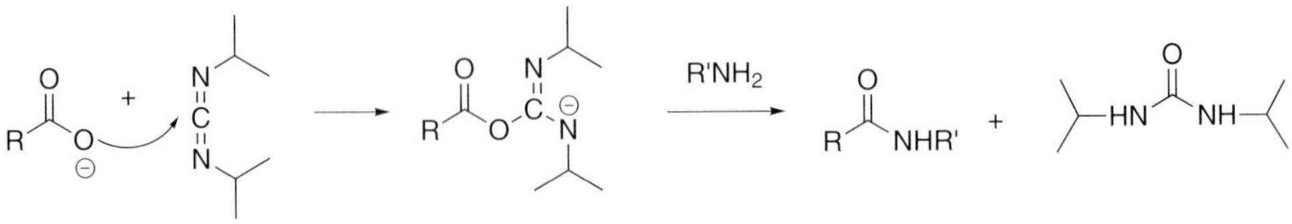

**Scheme 22.40** DCC, DIC, EDC and WSC-HCl used for the activation of carboxylic acids.

In some cases, acyl transfer to form the unproductive N-acylurea was observed. This could be suppressed by the addition of DMAP and HOBt. These nucleophiles reacted with O-acylisourea faster than the intramolecular acyl transfer. The formed intermediate was still reactive enough towards amine to yield the amide (Scheme 22.41) [57].

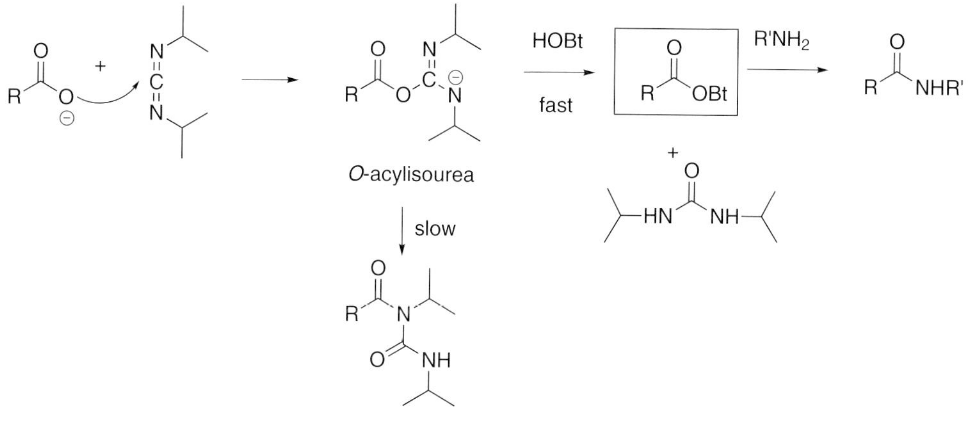

**Scheme 22.41** The addition of DMAP and HOBt suppressing unproductive N-acylurea.

In synthetic applications, symmetric anhydrides are seldom employed in total synthesis due to the difficulty in their preparation and the waste of half of the acid. Mixed anhydrides and their extended analogues are used more frequently.

Luzopeptins have attracted much interest because of their potency as inhibitors of HIV replication *in vitro* at non-cytotoxic doses. In Ciufolini's approach, by simple exposure of the starting material to EDCI/HOAt at 0 °C (EDCI = 1-ethyl-3-[3-(dimethylanimo)propyl]-carbodiimide), dimerization to form macrolactam **54** was achieved, albeit in a moderate yield, together with an undesired monomer **55**. After subsequent transformation of **56**, the desired macrolactam **57** was obtained (Scheme 22.42) [58].

### 22.2.2.4 Esters

By choosing different alcohols, the esters formed can be very reactive due to enhanced electrophilicity of the carbonyl center. Their higher reactivity allows the reaction with amine to proceed under mild conditions (Scheme 22.43) [59]. The activated ester can be used either *in situ* or after isolation. Accordingly, there are two general strategies: a multi-step strategy and a one-pot strategy.

**Multi-Step Strategy** For this strategy, the active esters are prepared in advance and isolated before use and their further reaction with amine gives the desired amides.

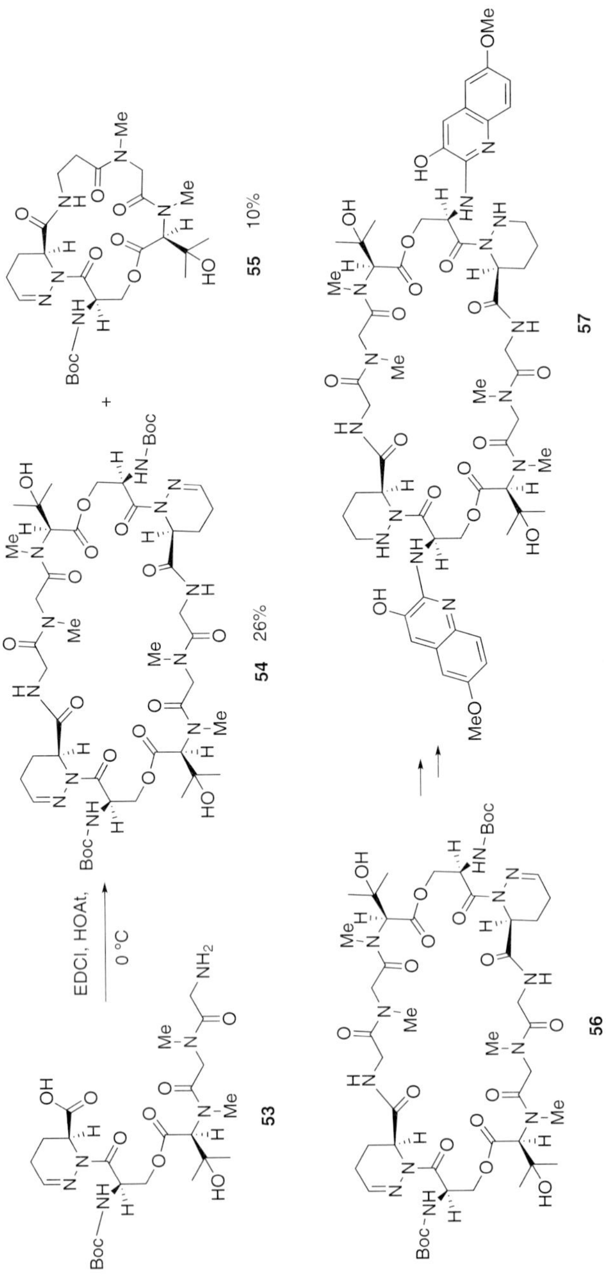

**Scheme 22.42** Macrolactamization used in the synthesis of Luzopeptins.

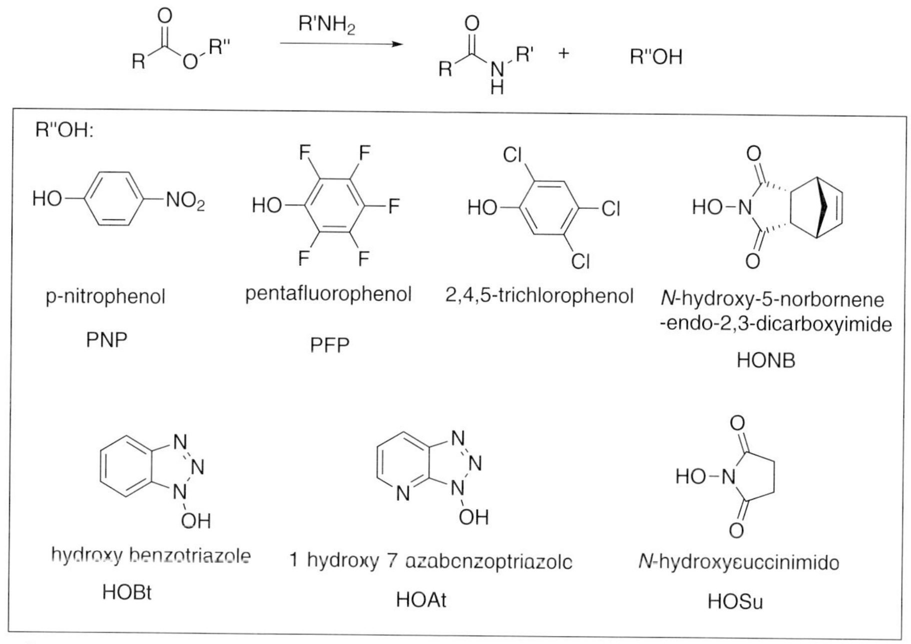

## 22.2 Macrolactams

$$R-C(=O)-O-R'' \xrightarrow{R'NH_2} R-C(=O)-NH-R' + R''OH$$

R''OH:
- p-nitrophenol (PNP)
- pentafluorophenol (PFP)
- 2,4,5-trichlorophenol
- N-hydroxy-5-norbornene-endo-2,3-dicarboxyimide (HONB)
- hydroxy benzotriazole (HOBt)
- 1-hydroxy-7-azabenzotriazole (HOAt)
- N-hydroxysuccinimide (HOSu)

**Scheme 22.43** Different alcohols used to form the ester intermediates.

Hirama reported the convergent synthesis of maduropeptin chromophore aglycon **60**. The 15-membered macrolactam is formed by slow addition of the isolated azido-PFP ester pentafluorophenyl (PFP) ester **59** to excess triphenylphosphine in THF–H$_2$O (30 : 1) at 45 °C through the intermediacy of the corresponding primary amine (66% yield) (Scheme 22.44) [60].

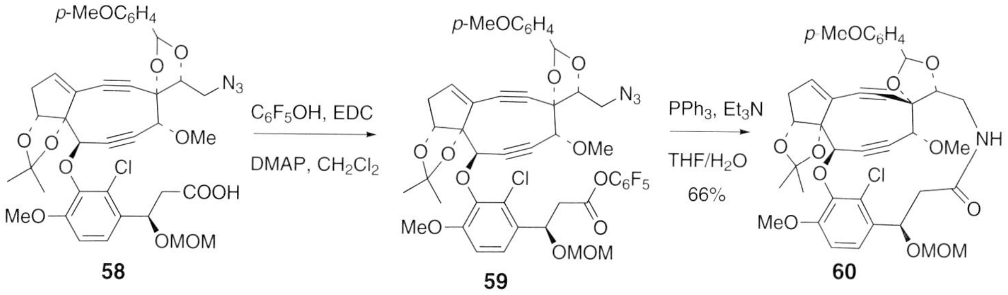

**Scheme 22.44** Multiple steps strategy used in the synthesis of maduropeptin chromophore aglycon.

**One-Pot Strategy** For the one-pot strategy, the active ester is prepared *in situ* and reacts directly with the amine without isolation. In the method to activate carboxylic acid with DCC and HOBt (Scheme 22.52), the reactive intermediate is, indeed, an activated ester. Well-defined catalysts that possess the phenol moiety have been developed and proved particularly effective for amide couplings. According to their nature, they could be classified as phosphonium, uronium, and immonium salts.

Due to its simplicity in handling, mild reaction conditions, and high efficiency in amidation, this strategy has been widely applied in total synthesis. Benzotriazol-1-yl-oxytris-(dimethylamino)-phosphonium hexafluorophosphate (BOP), also called Castro's reagent, was among the earliest examples of these HOBt-based phosphonium salt reagents. The one-pot coupling procedure was quite simple. By mixing the carboxylic acid and amine with BOP and a base, the desired amide was formed, usually in high yield. Mechanistically, the carboxylic acid first reacted with BOP to generate an activated acylphosphonium species and $BtO^-$. Then $BtO^-$ reacted with the activated acylphosphonium species to produce a reactive Bt ester, which finally underwent aminolysis (Scheme 22.45) [61].

In the enantioselective total synthesis of Ustiloxin D **65**, Wandless applied two steps of amide coupling promoted by phosphonium salts to couple an amino acid first and then close up the macrolactam ring (Scheme 22.46) [62].

Tamandarins A and B are a class of marine natural cyclodepsipeptides with structures and biological activities closely related to those of the didemnins. In Joullie's approach, the final steps featured the use of a different amide coupling method to construct this cyclopeptide (Scheme 22.47) [63].

During the total synthesis of (−)-Stevastelin Chida and coworkers used diethyl phosphorocyanidate (DEPC) as a key reaction reagent to accomplish the synthesis of macrolactam **72** (Scheme 22.48) [64].

Another class of catalyst is uronium-based. This kind of catalyst, such as O-(1H benzotriazol-1-yl)-N,N,N,N-tetramethyluronium hexafluorophosphate (HBTU) and its tetrafluoroborate analogue TBTU, is very efficient in amide-forming reactions. The counterion has no influence on the reactivity. Both the mechanism and the coupling procedure are similar to those of phosphonium species (Scheme 22.49) [65].

In Nicolaou's total synthesis of Diazonamide A, attempts to accomplish macrolactam formation from **74** encountered high difficulty. Among various methods tested, HATU-2,4,6-collidine was the only reagent combination, which was able to form compound **75** in a moderate yield (Scheme 22.50) [66].

The synthesis of two epothilone analogues, 15(S)-aza-12,13-desoxyepothilone B and the epimeric 15(R)-aza-12,13-desoxyepothilone B, was achieved in Danishefsky's laboratory. The macrolactam **77** was synthesized by using HATU. Subsequent transformation easily gave the desired targets (Scheme 22.51) [67].

In Nicolaou's total synthesis of Halipeptins A and D, two amide bond-forming reactions were used. First, PyAOP was used to synthesize the amide **81** through intermolecular amidation. Subsequently, the macrolactam **83** was obtained by using HATU as the amidation reagent (Scheme 22.52) [68].

In the total synthesis of the potent sistone deacetylase inhibitor FR235222, uronium salts were also used in the crucial macrocyclization step. The unprotected linear tetrapeptide was cyclized by exposure to HATU (2 equiv) and DIEA (2 equiv) in a very dilute solution. The cyclized product **85** was obtained in 68% yield after purification by HPLC (Scheme 22.53) [69].

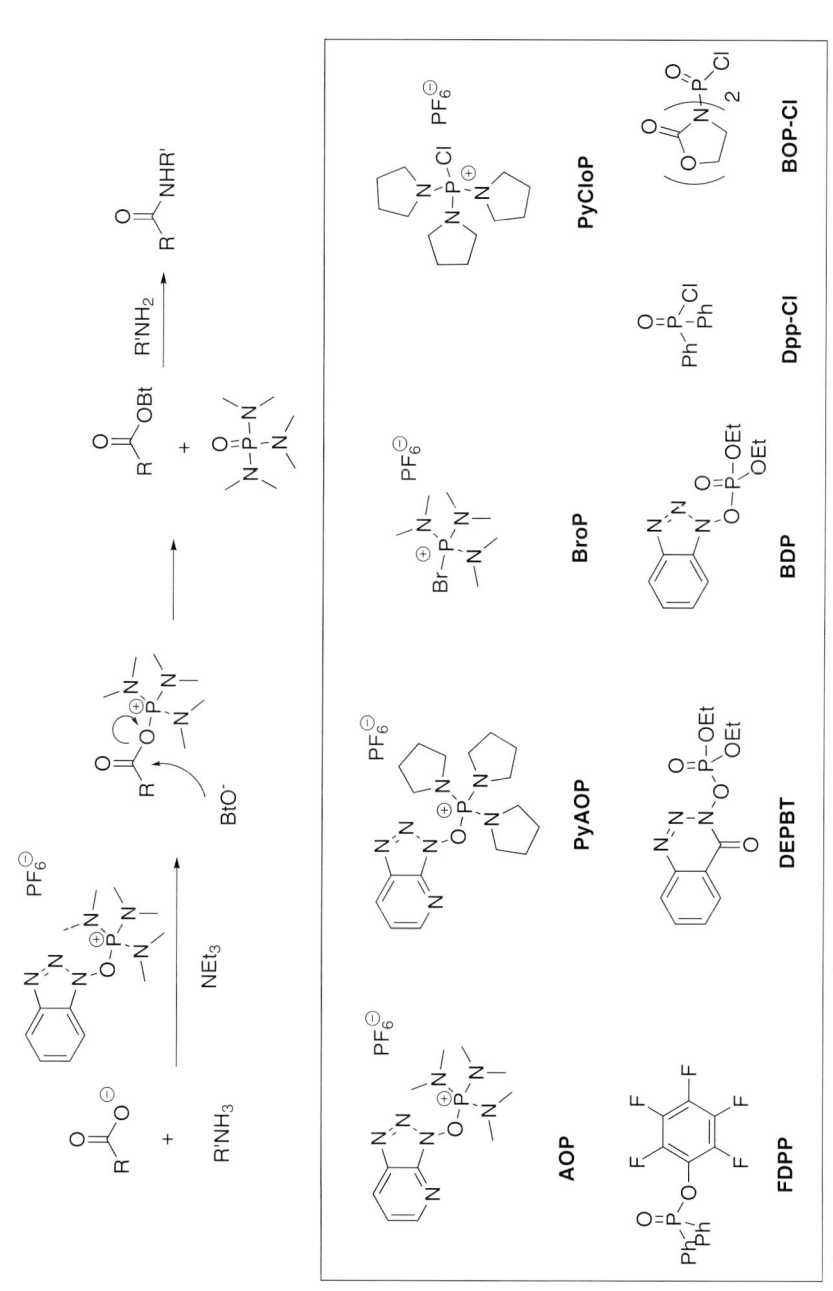

Scheme 22.45 Castro's reagent and its analogues used in amidation.

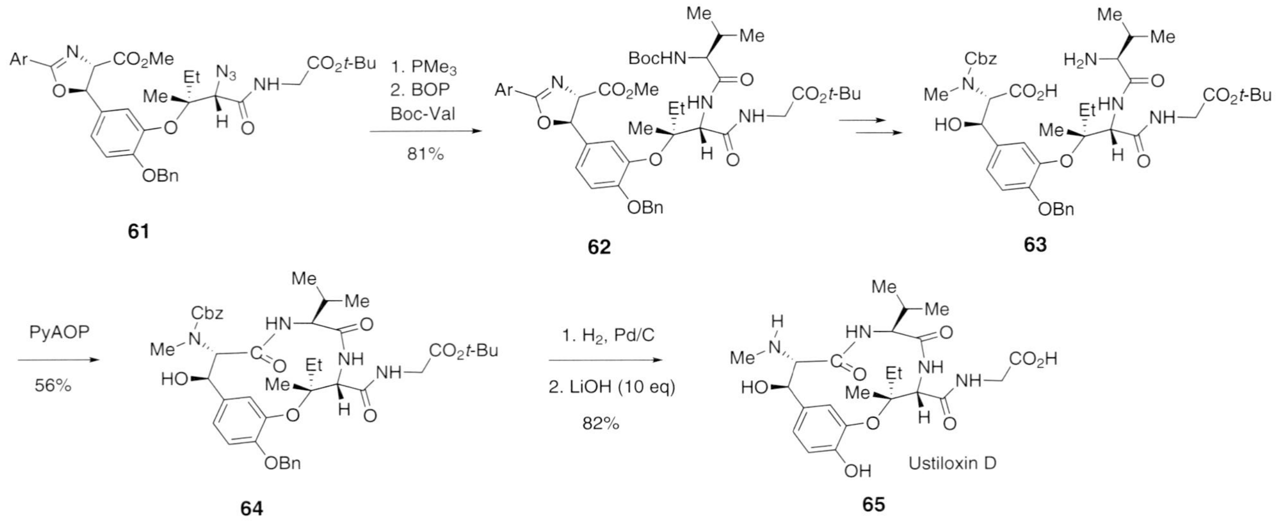

**Scheme 22.46** Two steps strategy used in the synthesis of Ustiloxin D.

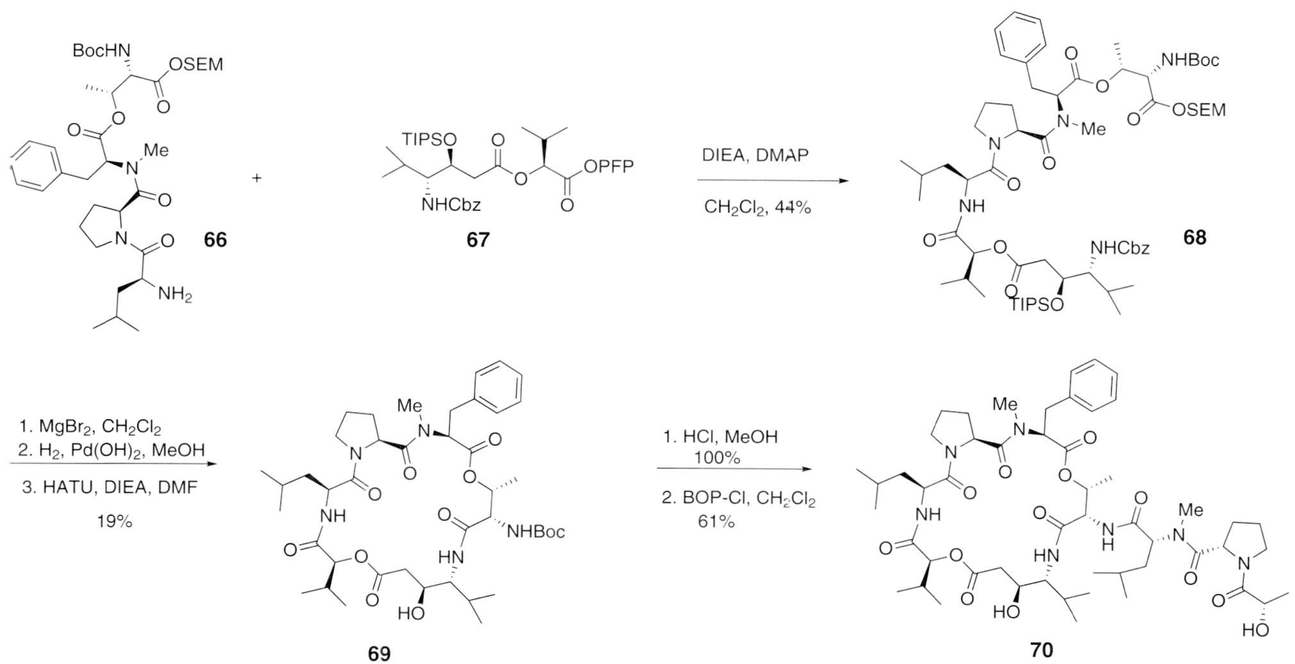

**Scheme 22.47** Different amide coupling methods to construct Tamandarins A and B.

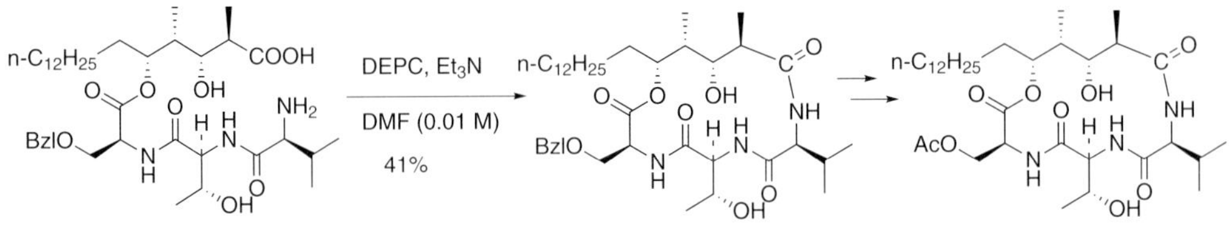


**71** **72** stevastelin B1 **73**

**Scheme 22.48** DEPC as a key reaction reagent used in the total synthesis of (−)-stevastelin B.

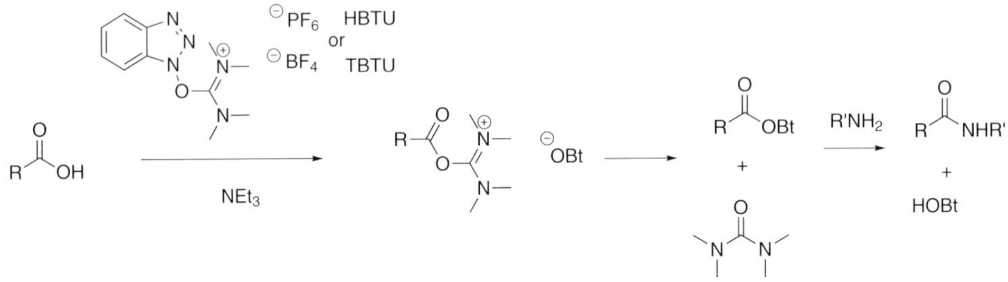


**Scheme 22.49** Mechanism and the coupling procedure of uronium-based catalysts.

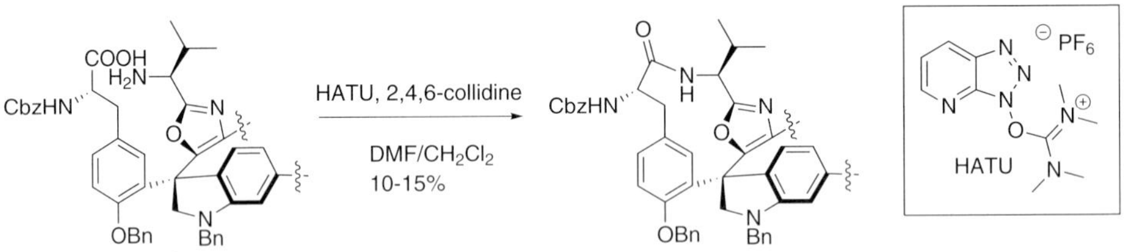


**74** **75**

**Scheme 22.50** HATU and 2,4,6-collidine reagents used in the total synthesis of diazonamide A.

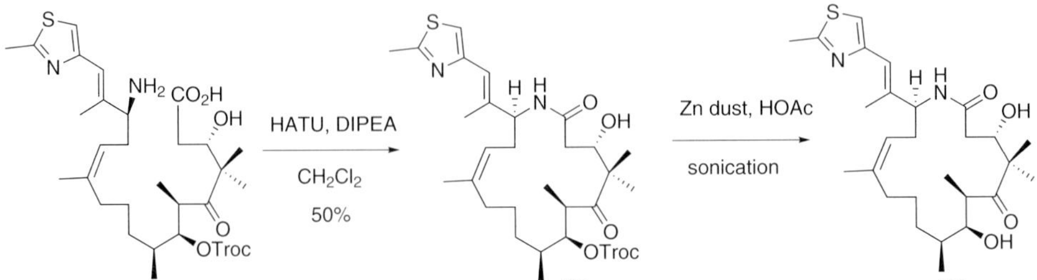


**76** **77** **78**

**Scheme 22.51** The synthesis of 15(*S* and *R*)-aza-12,13-desoxyepothilone B by using HATU.

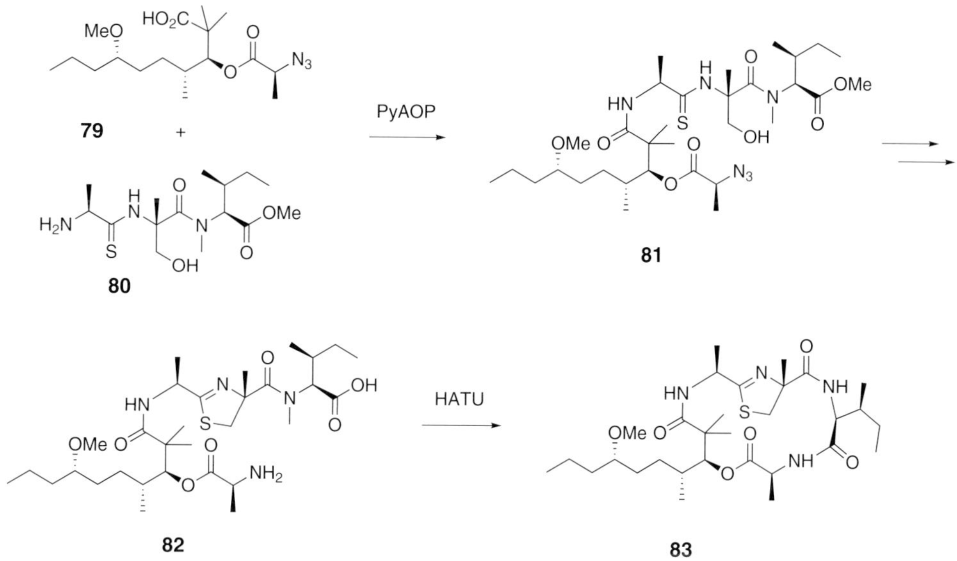

**Scheme 22.52** Two amide bond-forming reactions used in the total synthesis of Halipeptins A and D.

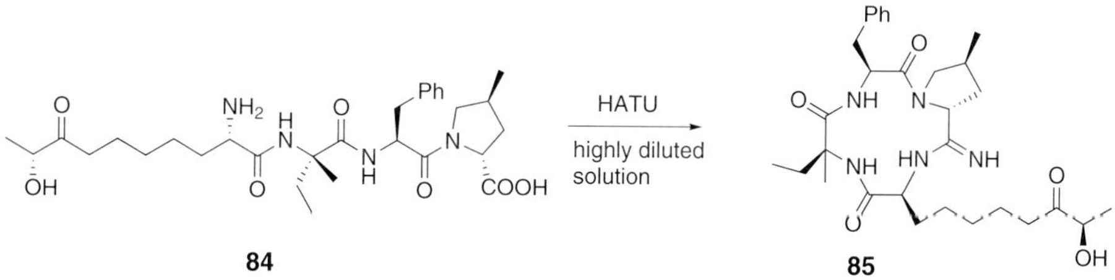

**Scheme 22.53** Uronium salts used in the synthesis of potent sistone deacetylase inhibitor FR235222.

The third class of amidation catalyst is ammonium salt. An excellent example is Mukaiyama's reagent (2-chloro-1-methylpyridinium iodide). In the presence of a tertiary amine, Mukaiyama's reagent first reacts with carboxylic acid to give an activated pyridinium ester, which further undergoes aminolysis. Usually, pyridinium compounds with tetrafluoroborate and hexachloroantimonate counterions have better solubility (Scheme 22.54) [70].

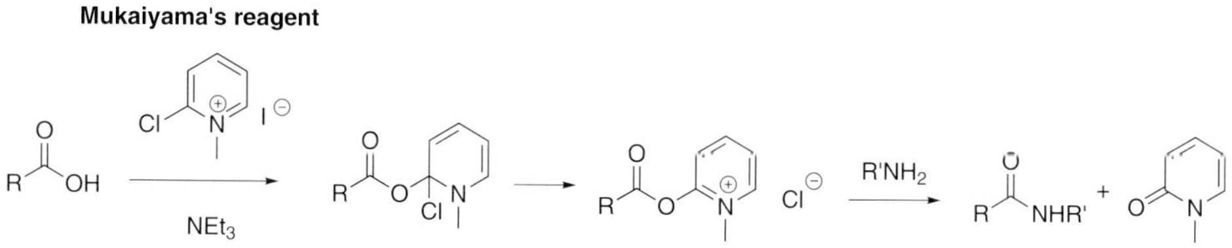

**Scheme 22.54** Mukaiyama's reagent used to give an activated pyridinium ester intermediate.

Smith III reported the first total synthesis of (+)-Thiazinotrienomycin E, a member of a novel class of cytotoxic ansamycin antibiotics. The unstable amino acid, formed after the removal of the alloc group and without purification, was slowly added to a mixture of 2-chloro-1-methylpyridinium iodide (Mukaiyama salt) and TEA in toluene. The macrocyclic lactam **87** was obtained in 61% yield for the two steps (Scheme 22.55) [71].

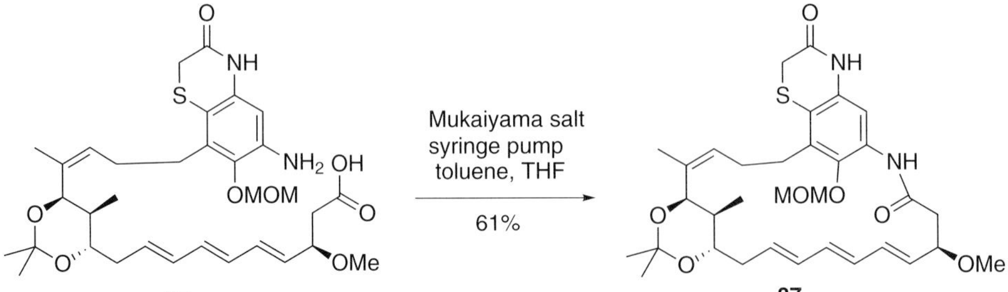

**Scheme 22.55** Mukaiyama salts used in the total synthesis of (+)-thiazinotrienomycin E.

## 22.2.3
### Formation of Macrolactam via Aryl C–N Bond Formation

Very recently, Buchwald–Hartwig amidation has also been applied to construct macrolactam. Reblastatin **90**, isolated in 2000 from a benzenoid ansamycin-like cell cycle inhibitor by Takatsu and coworkers, was synthesized by Panek. Subjection of the acyclic skeleton to Buchwald's amidation conditions smoothly provided the desired macrolactam **89** in 83% yield (Scheme 22.56) [72].

## 22.3
### Conclusion and Perspectives

Macrolactones or macrolides, which are different from medium or small lactones in many aspects, are among the most important structures in natural products and bioactive compounds. Many methods of macrolactonizations by the formation of ester linkage have been developed in the past several decades. Direct lactonization, an intramolecular version of esterification, has been relatively well studied and introduced as a powerful tool and widely applied in organic synthesis. However, drawbacks of the developed methods exist, which limit their applications. Undoubtedly, the development of highly efficient and relatively universal methods for macrolactonization is of great significance, not only in fundamental organic chemistry, but also in the medical and pharmaceutical industries.

Since the first specific method of macrolactonization, Corey–Nicolaou macrolactonization, more and more methods of macrolactonization have been developed for different synthetic environments, different precursors, and different targets. Several of these methods have become very popular with organic chemists for the

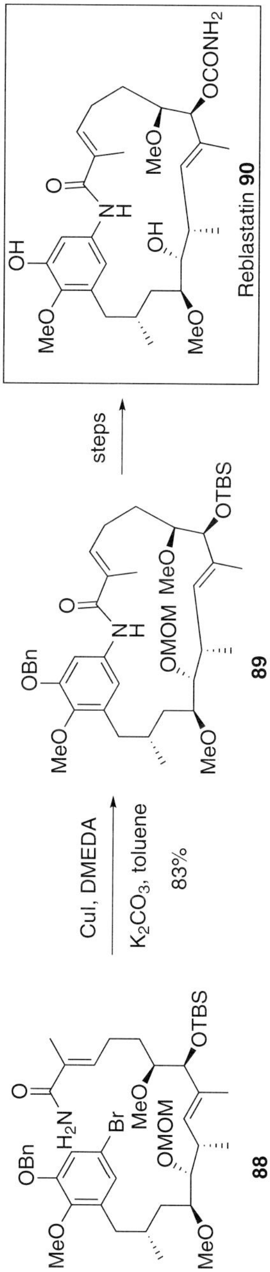

**Scheme 22.56** Buchwald–Hartwig amidation applied in the synthesis of Reblastatin.

total synthesis of numerous natural product molecules. However, almost all the methods focused on how to activate the two ends of the acid precursors and most of the macrolactonization methods were used as the ring-closing steps. Actually, the macrolactonization steps were often less efficient in many synthetic cases. More and more chemists designed new strategies for macrolactones using other more efficient and general ring-closing methods with pre-existing ester linkages. These newly-developed ring-closing methods enriched the arsenal of organic chemists and the strategies in organic synthesis. Furthermore, the biocatalysts – enzymes – have also been useful and powerful tools for macrolactonization accompanying the fast development of biochemistry.

Macrolactam features the amide functionality. Therefore, the synthesis of macrolactam relies heavily on amide bond-forming methods. The amidation reaction goes back to the beginning of organic chemistry. However, methodologies to form an amide bond in macrolactam have been speeded up only in the last two decades. More general amide coupling methodologies that allow efficient access to complex macrolactam synthesis are still highly desired.

After their discovery, the so-called "onium salts" have gradually taken the place of traditional carbodiimide and active ester-based methods. HOBt and HOAt-based uronium, phosphonium and immonium salts stand out as very efficient. Other reagents, however, are seldom used except for very special cases. Depending on the specific circumstances of natural product synthesis, the chemists will have many choices from different methodologies and strategies. In most cases, chemists have to screen these methods in order to find the most suitable one.

The long-standing interest in macrolactam synthesis calls for the development of more and more efficient amide-coupling methods. The development of amidation methods, in its turn, can shorten the steps of macrolactam synthesis. The development of efficient amidation methods as well as macrolactam synthesis will continue to be a research area of active interest.

## 22.4
### Experimental: Selected Procedures

### 22.4.1
### Synthesis of Compound 7 through Corey-Nicolaou Macrolactonization [8]

The ω-hydroxy acid **6** (0.5 mmol), 2,2′-dipyridyl disulfide (165 mg, 0.75 mmol) and triphenylphosphine (197 mg, 0.75 mmol) were dissolved in dry and oxygen-free xylene under argon and the resulting mixture was stirred at 25 °C for 5 h. The reaction mixture containing the 2-pyridinethiol ester was diluted with 10 mL of dry and oxygen-free xylene and the resulting solution was added slowly from a mechanically driven syringe over 15 h to 100 mL of dry xylene under reflux and argon conditions. Refluxing was continued for an additional 10 h. The solvent was removed under reduced pressure and the ether-soluble part of the residue was subjected to preparative layer chromatography on silica gel (10 % ether in pentane for development) to afford pure lactones, which were purified further by recrystallization from hexane.

## 22.4.2
**Synthesis of Brefeldin A 16 [13a]**

A solution of 173 mg (0.51 mmol) of hydroxy acid **14** and 0.56 mL (4.1 mmol) of triethylamine in 40 mL of methylene chloride was added via syringe pump to a refluxing solution of 508 mg (2.36 mmol) of 2-chloro-l-methylpyridinium tetrafluoroborate in 37 mL of $CH_2Cl_2$ and 12 mL of acetonitrile over an 8 h. The solvent was removed under reduced pressure and the residue was dissolved in ethyl acetate, washed with water, dilute $H_3PO_4$, saturated $NaHCO_3$ and brine, dried, filtered and concentrated to give 170 mg of a dark oil. This material was chromatographed (ether/pentane, 1:1) to give 60 mg (37%) of a mixture of C-15 epimers macrolactones **15**.

## 22.4.3
**Synthesis of Avermectin $A_1$ 19 [14]**

A solution of 33 mg (0.045 mmol) of the ω-hydroxy acid **17** in 10 mL of $CH_3CN$ was added very slowly over 2 h to a refluxing solution of TEA (50 mL, 0.358 mmol) and 43 mg (0.16 mmol) of 2-chloro-*N*-methylpyridinium iodide. After the addition was complete, the reaction was cooled to room temperature and 2 mL of $CH_2Cl_2$ was added, and the organic phase was washed once with 10 mL of saturated bicarbonate solution, dried over anhydrous $MgSO_4$, filtered, and concentrated at reduced pressure. Chromatography gave 19.8 mg (62%) of conjugated $A_1$, aglycon $C_{13}$ TBS in a crystalline form, which was identical in all respects with an authentic sample prepared by degradation of Avermectin $A_1$.

## 22.4.4
**Synthesis of (−)-Macrolactin A 25 [17, 18]**

A solution of ester **23** (294 mg, 539 μmol) 2:1 methanol/1 M KOH (12 mL) was heated to 40 °C for 3 h. The methanol was evaporated under reduced pressure and the resulting residue was poured into diethyl ether (50 mL). A saturated aqueous solution of ammonium chloride (30 mL) was added and the organic layer was isolated. The aqueous layer was extracted with diethyl ether (3 × 15 mL). The combined organic layers were dried over anhydrous sodium sulfate and filtered through Celite. Triethylamine (1 mL) was added to the filtrate before removing the solvent under reduced pressure. To a solution of the triethylamine salt in THF (5.4 mL) was added 2,4,6-trichlorobenzoyl chloride (126 μL, 810 μmol) at room temperature. The reaction mixture was allowed to stir at room temperature for 12 h before removing the THF under reduced pressure. The residue was re-dissolved in toluene (54 mL) and added quickly to a rapidly stirred solution of DMAP (395 mg, 3.24 mmol) in toluene (162 mL). After stirring at room temperature for 1.5 h, the solvent was removed under reduced pressure and the residue was chromatographed on silica gel (gradient elution, 9:1 to 85:15 hexane/ethyl acetate). The crude protected lactone was dissolved in wet methanol and treated with PPTS (cat. amount). After stirring at room temperature for 1 h the solvent was removed under reduced pressure and the resulting oil was purified by HPLC (C18, 7:3 methanol/water) to give 57 mg (26%) of

Macrolactin A and 22 mg (10%) of 7-epimacrolactin A for a combined chemical yield of 36% over 4 steps.

### 22.4.5
### Synthesis of Oleandolide 28 [19]

To a solution of azeotropically dried hydroxy acid (5.0 mg, 7.87 μmol) in benzene (0.5 mL) at room temperature was added $^i$Pr$_2$NEt (41 μL, 0.236 mmol) and 2,4,6-trichlorobenzoyl chloride (26 μL, 0.157 mmol). The reaction solution was stirred at room temperature for 13 h before it was diluted by the addition of 8 mL of benzene and treated with DMAP (38 mg, 0.315 μmol). The resulting white reaction mixture was allowed to stir vigorously at room temperature for 4 h before being quenched with saturated aqueous NH$_4$Cl solution (20 mL) and then extracted with EtOAc (2 × 15 mL). The organic layers were combined and dried over MgSO$_4$, filtered, and concentrated *in vacuo*. The crude product was flash chromatographed on silica gel (10% EtOAc in hexanes as eluent) to afford the desired macrocycle (4.7 mg, 97% yield) as white crystalline residue.

### 22.4.6
### Synthesis of Colletodiol 31 [23]

To a solution of dicyclohexylcarbodiimide (288.9 mg, 1.4 mmol), 4-(dimethylamino)pyridine (171 mg, 1.4 mmol), and 4-(dimethyamino)pyridine hydrochloride (222 mg, 1.4 mmol) in ethanol-free chloroform (20 mL) under reflux was added, via syringe pump, a solution of ω-hydroxy acid (99 mg, 0.29 mmol) in ethanol-free chloroform (1 mL) over a period of 16 h with stirring. The residual contents of the syringe and needle were rinsed into a tared flask and concentrated to give 16 mg of recovered starting material. The solution was cooled to room temperature and quenched by the addition of methanol (3 mL) and 10 drops of acetic acid. The solution was transferred to a 100-mL round-bottom flask and concentrated to a volume of 5 mL, diluted with ether, and filtered through a pad of Celite. The solvent was removed under reduced pressure, and the product was isolated by HPLC, collecting 8 mL fractions and eluting with hexanes (80 mL), 5% EtOAc/hexanes (80 mL), 10% EtOAc/hexanes (120 mL), and 20% EtOAc/hexanes (200 mL). The product was found in fractions 40–49 and concentrated to give 65 mg (82%) of a colorless oil.

### 22.4.7
### Synthesis of Compound Aleuritic Acid Lactone 35 [25b]

#### 22.4.7.1 Method A
To a solution of 2-methyl-6-nitrobenzoic anhydride (MNBA) (157.8 mg, 0.456 mmol) and DMAP (111.4 mg, 0.912 mmol) in dichloromethane (135.8 mL) at room temperature was slowly added a solution of ω-hydroxy acid **34** (149.2 mg, 0.380 mmol) in dichloromethane (84.6 mL) with a mechanically driven syringe

over a 16.5 h period. After addition of the solution, the reaction mixture was additionally stirred for 1 h at room temperature. The reaction mixture was concentrated to about 20 mL by evaporation of the solvent under reduced pressure and then saturated aqueous NaHCO$_3$ was added at 0 °C. The usual work-up and purification of the mixture by TLC on silica gel afforded 128.2 mg (90%) of protected *erythro*-aleuritic macrolactone, 4.4 mg (1.5%) of diolide and 1.8 mg (0.4%) of triolide.

#### 22.4.7.2  Method B

To a solution of MNBA (148.7 mg, 0.432 mmol), triethylamine (80.1 mg, 0.792 mmol) and DMAPO (9.9 mg, 0.072 mmol) in dichloromethane (151.0 mL) at room temperature was slowly added a solution of ω-hydroxy acid **34** (141.3 mg, 0.360 mmol) in dichloromethane (108.0 mL) with a mechanically driven syringe over a 16 h period. After addition of the solution, the reaction mixture was additionally stirred for 1 h at room temperature. The same treatment of the reaction mixture as above afforded 120.8 mg (90%) of the macrolactone, 4.9 mg (1.8%) of diolide and 0.7 mg (0.2%) of triolide.

### 22.4.8
### Synthesis of Maduropeptin Chromophore Aglycon 60 [60]

To a solution of the substrate **58** (409 μmol), DMAP (500 mg, 4.09 mmol) and EDC·HCl (391 mg, 2.05 mmol) in CH$_2$Cl$_2$ (20 mL) was added a separate solution of pentafluorophenol (377 mg, 2.05 mmol) in CH$_2$Cl$_2$ (4 mL) at 0 °C. After being stirred for 2.5 h at room temperature, the reaction mixture was poured into saturated aqueous NH$_4$Cl and extracted with hexane. The organic layer was washed with brine, dried over Na$_2$SO$_4$, and concentrated. The residue was purified by flash column chromatography to obtain the activated ester.

To a solution of PPh$_3$ (4.89 g, 18.6 mmol) and Et$_3$N (52 μL, 373 μmol) in THF (60 mL) and water (2.8 mL) was added a solution of the activated ester (347.7 mg, 373 μmol) in THF (25 mL). The reaction mixture was stirred for 3 h at 45 °C. After cooling, the reaction mixture was concentrated. The residue was purified by flash column chromatography.

### 22.4.9
### Preparation of Macrocyclic Lactam 70 [63]

To a mixture of the macrocycle amine salt **69** (9 mg, 0.01 mmol) and side chain (6.1 mg, 0.015 mmol) in CH$_2$Cl$_2$ (0.50 mL) at 0 °C was added BOP (8.4 mg, 0.015 mmol) and NMM (6 μL, 0.05 mmol). After 30 min at 0 °C, the reaction was stirred at room temperature overnight. The reaction mixture was treated with NaCl solution (2 mL, saturated) and extracted with EtOAc (2 × 10 mL). The organic layers were washed with 10% HCl (5 mL), 5% NaHCO$_3$ (5 mL), and NaCl (5 mL, saturated), dried (Na$_2$SO$_4$), filtered, and concentrated. The crude oil (8 mg, 61%) was purified by HPLC.

## 22.4.10
### Preparation of Macrocyclic Lactam 85 [69]

The cyclization was performed in solution at a concentration of $7.7 \times 10^{-5}$ M with HATU (18.8 mg, 0.05 mmol) and DIEA (11 µL, 0.062 mmol) in DCM. The solution was stirred at 4 °C for 1 h and then allowed to warm to room temperature for 1 h. The solvent was removed under reduced pressure. The crude cyclopeptide was purified by HPLC.

## 22.4.11
### Preparation of Macrocyclic Lactam 87 [71]

A solution of the amino acid **86** in THF-toluene (2 : 3, 5 mL) was treated with TEA (0.02 mL, 0.143 mmol). The resultant yellow solution was then added over a 2 h period via a syringe pump to a suspension of 2-chloro-1-methylpyridinium iodide (30 mg, 0.117 mmol) and TEA (0.03 mL, 0.215 mmol) in toluene (5 mL). After addition, the mixture was stirred for an additional 1 h. The resultant solid was then removed by filtration and the yellow filtrate was concentrated *in vacuo*. Flash chromatography (ethyl acetate/hexanes, 3 : 1) provided the corresponding macrolactam.

**Abbreviations**

| | |
|---|---|
| Ac | acetyl |
| BOP | (1*H*-benzo[d][1,2,3]triazol-1-yloxy)tris(dimethylamino)phosphonium hexafluorophosphate |
| BQ | *p*-benzoquinone |
| CDI | carbonyl diimidazole |
| DCC | dicyclohexylcarbodiimide |
| DCM | dichloromethane |
| DEAD | diethyl azodicarboxylate |
| DEPC | diethyl phosphorocyanidate |
| DIAD | diisopropyl azodicarboxylate |
| DIEPA | *N,N*-diisopropylethylamine |
| DMAP | *N,N*-4-dimethylaminopyridine |
| DMAPO | *N,N*-4-dimethylaminopyridine *N*-oxide |
| EDC | 1-ethyl-3-(3-dimethylaminopropyl)carbodiimide |
| EDCl | 1-ethyl-3-(3-dimethylaminopropyl)carbodiimide hydrochloride |
| HATU | *O*-(7-azabenzotriazol-1-yl)-*N,N,N',N'*-tetramethyluronium hexafluorophosphate |
| HBTU | *O*-(1H benzotriazol-1-yl)-*N,N,N',N'*-tetramethyluronium hexafluorophosphate |
| HOAt | 1-hydroxy-7-azabenzotriazole |
| HOBt | hydroxy benzotriazole |
| HWE | Horner–Wadsworth–Emmons reaction (olefination) |

| | |
|---|---|
| MNBA | 2-methyl- 6-nitrobenzoic anhydride |
| NMM | *N*-methylmorpholine |
| NMP | *N*-methyl-2-pyrrolidinone |
| PFP | pentafluorophenyl |
| PPL | porcine pancreatic lipase |
| PPTS | pyridinium *p*-toluenesulfonate |
| RCM | ring-closing metathesis |
| TBTU | *O*-(1H benzotriazol-1-yl)-*N,N,N′,N′*-tetramethyluronium tetrafluoroborate |
| TEA | triethylamine |
| TFBA | 4-trifluoromethylbenzoic anhydride |
| THF | tetrahydrofuran |

## Acknowledgments

Support by a starter grant from Peking University and grants from the National Sciences Foundation of China (No. 20542001, 20521202, 20672006) and the National Basic Research Program of China (973 Program: 2009CB825300) are gratefully acknowledged.

## References

1 (a) Lukacs, G. and Ohno, M. (1990) *Recent Progress in the Chemical Synthesis of Antibiotics*, Springer-Verlag, Berlin; (b) Omura, S. (2002) *Macrolide Antibiotics*, 2nd edn, Academic Press, San Diego, CA.

2 (a) Nicolaou, K.C. (1977) *Tetrahedron*, **33**, 683–710; (b) Masamune, S., Bates, G.S. and Corcoran, J.W. (1977) *Angew. Chem.*, **89**, 602–624; (1977) *Angew. Chem. Int. Ed. Engl.*, **16**, 585–607; (c) Paterson, I. and Mansuri, M.M. (1985) *Tetrahedron*, **41**, 3569–3624; (d) Mulzer, J. (1991) *Comprehensive Organic Synthesis*, vol. 6 (eds B.M. Trost and I. Fleming), Pergamon Press, Oxford, pp. 323–380; (e) Rousseau, G. (1995) *Tetrahedron*, **51**, 2777–2849; (f) Norcross, R.D. and Paterson, I. (1995) *Chem. Rev.*, **95**, 2041–2114; (g) Nakata, T., (2002) *Macrolide Antibiotics*, 2nd edn (ed. S. Omura), Academic Press, San Diego, CA, pp. 181–284; (h) Parenty, A., Moreau, X. and Campagne, J.-M. (2006) *Chem. Rev.*, **106**, 911–939.

3 Corey, E.J. and Nicolaou, K.C. (1974) *J. Am. Chem. Soc.*, **96**, 5614–5616.

4 Corey, E.J., Nicolaou, K.C. and Melvin, L.S. Jr. (1975) *J. Am. Chem. Soc.*, **97**, 653–654.

5 (a) Gerlach, H. and Thalmann, A. (1974) *Helv. Chim. Acta*, **57**, 2661–2663; (b) Corey, E.J. and Brunelle, D.J. (1976) *Tetrahedron Lett.*, **17**, 3409–3412.

6 Behinpour, K., Hopkins, A. and Williams, A. (1981) *Tetrahedron Lett.*, **22**, 275–278.

7 (a) Nicolaou, K.C., Bunnage, M.E., McGarry, D.G., Shi, S., Somers, P.K., Wallace, P.A., Chu, X.-J., Agrios, K.A., Gunzner, G.L. and Yang, Z. (1999) *Chem. Eur. J.*, **5**, 599–617; (b) ISasaki, T., Inoue, M. and Hirama, M. (2001) *Tetrahedron Lett.*, **42**, 5299–5303; (c) Andrus, M.B. and Shih, T.-L. (1996) *J. Org. Chem.*, **61**, 8780–8785.

8 Hansen, T.V. and Stenstrom, Y. (2001) *Tetrahedron: Asym.*, **12**, 1407–1409.

9 Masamune, S., Hirama, M., Mori, S., Ali, S.A. and Garvey, D.S. (1981) *J. Am. Chem. Soc.*, **103**, 1568–1571.

10 (a) Masamune, S., Hayase, Y., Schilling, W., Chan, W.K. and Bates, G.S. (1977) *J. Am. Chem. Soc.*, **99**, 6756–6758; (b) Masamune, S., Kamata, S. and Schilling, W. (1975) *J. Am. Chem. Soc.*, **97**, 3515–3516.

11 (a) Masamune, S., Kim, C.U., Wilson, K.E., Spessard, G.O., Georghiou, P.E. and Bates, G.S. (1975) *J. Am. Chem. Soc.*, **97**, 3512–3513; (b) Masamune, S., Yamamoto, H., Kamata, S. and Fukuzawa, A. (1975) *J. Am. Chem. Soc.*, **97**, 3513–3514.

12 (a) Mukaiyama, T. Usui, M. Shimada, E. and Saigo, K. (1975) *Chem. Lett.*, 1045–1048; (b) Mukaiyama, T., Usui, M. and Saigo, K. (1976) *Chem. Lett.*, 49–50; (c) Saigo, K., Usui, M., Kikuchi, K., Shimada, E. and Mukaiyama, T. (1977) *Bull. Chem. Soc. Jpn.*, **50**, 1863–1866; (d) Mukaiyama, T. (1979) *Angew. Chem.*, **91**, 798–812; (1979) *Angew. Chem. Int. Ed. Engl.*, **18**, 707–721.

13 (a) Bartlett, P.A. and Green, F.R. III (1978) *J. Am. Chem. Soc.*, **100**, 4858–4865; (b) Ley, S.V., Armstrong, A., Diéz-Martén, D. Ford, M.J., Grice, P., Knight, J.G., Kolb, H.C., Madin, A., Marby, C.A., Mukherjee, S., Shaw, A.N., Slawin, A.M.Z., Vile, S., White, A.D., Williams, D.J. and Woods, M. (1991) *J. Chem. Soc., Perkin Trans. 1*, 667–692.

14 Danishefsky, S.J., Armistead, D.M., Wincott, F.E., Selnick, H.G. and Hungate, R. (1989) *J. Am. Chem. Soc.*, **111**, 2967–2980.

15 White, J.D., Bolton, G.L., Dantanarayana, A.P., Fox, C.M.J., Hiner, R.N., Jackson, R.W., Sakuma, K. and Warrier, U.S. (1995) *J. Am. Chem. Soc.*, **117**, 1908–1939.

16 Inaga, J., Hirata, K., Saeki, H., Katsuki, T. and Yamaguchi, M. (1979) *Bull. Chem. Soc. Jpn.*, **52**, 1989–1993.

17 Marino, J.P., McClure, M.S., Holub, D.P., Comasseto, J.V. and Tucci, F.C. (2002) *J. Am. Chem. Soc.*, **124**, 1664–1668.

18 Hikota, M., Tone, H., Hirota, K. and Yonemitsu, O. (1990) *Tetrahedron*, **46**, 4613–4328.

19 Hu, T., Takenaka, N. and Panek, J.S. (2002) *J. Am. Chem. Soc.*, **124**, 12806–12815.

20 Boden, E.P. and Keck, G.E. (1985) *J. Org. Chem.*, **50**, 2394–2395.

21 Neises, B. and Steglich, W. (1978) *Angew. Chem*, **90**, 556–558; (1978) *Angew. Chem. Int. Ed. Engl*, **17**, 522–524.

22 Keck, G.E., Sanchez, C. and Wager, C.A. (2000) *Tetrahedron Lett.*, **41**, 8673–8676.

23 Keck, G.E., Boden, E.P. and Wiley, M.R. (1989) *J. Org. Chem.*, **54**, 896–906.

24 (a) Shiina, I. (2004) *Tetrahedron*, **60**, 1587–1599; (b) Shiina, I. and Mukaiyama, T. (1994) *Chem. Lett.*, 677–680; (c) Shiina, I., Miyoshi, S., Miyashita, M. and Mukaiyama, T. (1994) *Chem. Lett.*, 515–518. See also: (d) Mukaiyama, T., Shiina, I. and Miyashita, M. (1992) *Chem. Lett.*, 625–628; (e) Mukaiyama, T., Miyashita, M. and Shiina, I. (1992) *Chem. Lett.*, 1747–1750; (f) Mukaiyama, T., Izumi, J., Miyashita, M. and Shiina, I. (1993) *Chem. Lett.*, 907–910; (g) Miyashita, M., Shiina, I. and Mukaiyama, T. (1993) *Chem. Lett.*, 1053–1054; (h) Miyashita, M., Shiina, I., Miyoshi, S. and Mukaiyama, T. (1993) *Bull. Chem. Soc. Jpn.*, **66**, 1516–1527; (i) Miyashita, M., Shiina, I. and Mukaiyama, T. (1994) *Bull. Chem. Soc. Jpn.*, **67**, 210–215; (j) Shiina, I., Miyashita, M., Nagai, M. and Mukaiyama, T. (1995) *Heterocycles*, **40**, 141–148.

25 (a) Shiina, I., Kubota, M., Oshiumi, H. and Hashizume, M. (2004) *J. Org. Chem.*, **69**, 1822–1830; (b) Shiina, I., Kubota, M. and Ibuka, R. (2002) *Tetrahedron Lett.*, **43**, 7535–7539; (c) Shiina, I., Ibuka, R. and Kubota, M. (2002) *Chem. Lett.*, 286–287; (d) Shiina, I. (2005) *J. Synth. Org. Chem. Jpn.*, **63**, 2–17. See also: (e) Shiina, I. and Kawakita, Y. (2004) *Tetrahedron*, **60**, 4729–4933.

26 (a) Shiina, I., Hashizume, M., Yamai, Y., Oshiumi, H., Shimazaki, T., Takasuna, Y. and Ibuka, R. (2005) *Chem. Eur. J.*, **11**, 6601–6608; (b) Shiina, I., Oshiumi, H.,

Hashizume, M., Yamai, Y. and Ibuka, R. (2004) *Tetrahedron Lett.*, **45**, 543–547; (c) Shiina, I., Katoh, T. and Hashizume, M. (2006) Abstracts of Papers, 86th National Meeting of the Chemical Society of Japan, Chiba, vol. 2, 4L428; (d) Shiina, I., Katoh, T., Komiyama, Y., Sasaki, A., Suzuki, R., Hitomi, S. and Fukui, H. (2006) Abstracts of Papers, 48th Symposium on the Chemistry of Natural Products, Sendai, p. 34; (e) Shiina, I., Oshiumi, H., Hashizume, M., Yamai, Y. and Ibuka, R. (2004) *Tetrahedron Lett.*, **45**, 543–547; (f) Shiina, I., Katoh, T. and Hashizume, M. (2006) Abstracts of Papers, 86th National Meeting of the Chemical Society of Japan, Chiba, vol. 2 4L428; (g) Shiina, I., Katoh, T., Komiyama, Y., Sasaki, A., Suzuki, R., Hitomi, S. and Fukui, H. (2006) Abstracts of Papers, 48th Symposium on the Chemistry of Natural Products, Sendai, p. 34.

27 (a) Shiina, I., Fukuda, Y., Ishii, T., Fujisawa, H. and Mukaiyama, T. (1998) *Chem. Lett.*, 831–832; (b) Shiina, I., Fujisawa, H., Ishii, T. and Fukuda, Y. (2000) *Heterocycles*, **52**, 1105–1123.

28 (a) Wu, Y. and Yang, Y.-Q. (2006) *J. Org. Chem.*, **71**, 4296–4301; (b) Nishii, T., Suzuki, S., Yoshida, K., Arakaki, K. and Tsunoda, T. (2003) *Tetrahedron Lett.*, **44**, 7829–7832, and references cited therein.

29 (a) Mitsunobu, O. and Yamada, M. (1967) *Bull. Chem. Soc. Jpn.*, **40**, 2380–2382; (b) Mitsunobu, O., Yamada, M. and Mukaiyama, T. (1967) *Bull. Chem. Soc. Jpn.*, **40**, 935–939.

30 (a) Mukaiyama, T., Shintou, T. and Fukumoto, K. (2003) *J. Am. Chem. Soc.*, **125**, 10538–10539; (b) Shintou, T., Fukumoto, K. and Mukaiyama, T. (2004) *Bull. Chem. Soc. Jpn.*, **77**, 1569–1579.

31 (a) Grochowski, E., Hilton, B.D. and Michejda, C.J. (1982) *J. Am. Chem. Soc.*, **104**, 6876–6877; (b) Ahn, C., Correia, R. and DeShong, P. (2002) *J. Org. Chem.*, **67**, 1751–1753; (c) Adam, W., Narita, N. and Nishizawa, Y. (1984) *J. Am. Chem. Soc.*, **106**, 1843–1845; (d) Varasi, M., Walker, K.A.M. and Maddox, M.L. (1987) *J. Org. Chem.*, **52**, 4235–4238; (e) Hughes, D.L., Reamer, R.A., Bergan, J.J. and Grabowski, E.J. (1988) *J. Am. Chem. Soc.*, **110**, 6487–6491.

32 Tsutsui, H. and Mitsunobu, O. (1984) *Tetrahedron Lett.*, **25**, 2163–2166.

33 (a) Dvorak, C.A., Schmitz, W.D., Poon, D.J., Pryde, D.C., Lawson, J.P., Amos, R.A. and Meyers, A.I. (2000) *Angew. Chem.*, **112**, 1730–1732; (2000) *Angew. Chem. Int. Ed.*, **39**, 1664–1666; (b) Meyers, A.I., Lawson, J., Amos, R.A., Walker, D.G. and Spohn, R.F. (1982) *Pure Appl. Chem.*, **54**, 2537–2544; (c) Moreau, X. and Campagne, J.-M. (2003) *J. Org. Chem.*, **68**, 5346–5350, and references cited therein.

34 Matsuyama, H., Nakamura, T. and Kamigata, N. (1989) *J. Org. Chem.*, **54**, 5218–5223.

35 (a) Karim, M.R. and Sampson, P. (1990) *J. Org. Chem.*, **55**, 598–605; (b) Fraunhoffer, K.J., Prabagaran, N., Sirois, L.E. and White, M.C. (2006) *J. Am. Chem. Soc.*, **128**, 9032–9033; (c) Shen, Z.M., SKhan, Z.H.A. and Dong, V.M. (2008) *J. Am. Chem. Soc.*, **130**, 2916–2917.

36 (a) Sariaslani, F.S. and Rosazza, J.P.N. (1984) *Enzyme Microb. Technol.*, **6**, 241–252; (b) Yamada, H. and Shimizu, S. (1988) *Angew. Chem.*, **100**, 640–662; (1988) *Angew. Chem. Int. Ed. Engl.*, **27**, 622–642; (c) Gramatica, P. (1988) *Chim. Oggi*, **6**, 17; (1989), **7**, 43; (1989), **11**, 9; (d) Crout, D.H. and Christen, M. (1989) *Modern Synthetic Methods*, vol. 5 (ed. R. Scheffold), Springer, Berlin, pp. 1–114.

37 (a) Jones, J.B., Sih, C.J. and Perlman, D. (eds) (1976) *Applications of Biochemical Systems in Organic Chemistry*, Wiley, New York; (b) Tramper, J., van der Plas, H.C. and Linko, P. (eds) (1985) Biocatalysts in Organic Syntheses, *Studies in Organic Chemistry*, Elsevier, Amsterdam, p. 22; (c) Davies, H.G., Green, R.H., Kelly, D.R. and Roberts, S.M. (eds) (1989) *Biotransformations in Preparative Organic Chemistry*, Academic Press, London; (d) Abramowicz, D.A. (ed.) (1990)

*Biocatalysis*, Van Nostrand Reinhold, New York.
38 Mizrahi, A. (1986) *Process Biochem.*, 108–112; (b) Arathoon, W.R. and Birch, J.R. (1986) *Science*, **232**, 1390–1395.
39 (a) Kutney, J.P. (1990) *Nat. Prod. Rep.*, **7**, 85–103; (b) Kutney, J.P. (1991) *Synlett*, 11–19; (c) Suga, T. and Hirata, T. (1990) *Phytochemistry*, **29**, 2393–2406; (d) CibaFoundations Symposium (1988) *Applications of Plant Cells and Tissue Cultures*, vol. 137, John Wiley & Sons, Chichester, pp. 228–238.
40 (a) Tamm, C. (1962) *Angew. Chem.*, **74**, 225–242; (1962) *Angew. Chem. Int. Ed. Engl.*, **1**, 178–195; (b) Kieslich, K. (1976) *Microbial Transformations of Non Steroid Cyclic Compounds*, Thieme, Stuttgart; (c) Demain, A.L. (1981) *Science*, **214**, 987–995; (d) Rosazza, J.P.N. (ed.) (1982) *Microbial Transformations of Bioactive Compounds*, vols. 1 and 2, CRC Press, Boca Raton, FL; (e) Kieslich, K. (1986) *Arzneim.-Forsch.*, **36**, 774, 888, 1006.
41 Yamada, H. and Shimizu, S. (1988) *Angew. Chem.*, **100**, 640–660; (1988) *Angew. Chem. Int. Ed. Engl.*, **27**, 622–642.
42 (1973) *Enzyme Nomenclature*, Elsevier, Amsterdam.
43 Mori, K. and Tomioka, H. (1992) *Liebigs Ann. Chem.*, 1011–1017.
44 Pawar, A.S., Chattopadhyay, S. and Chattopadhyay, A. (1993) *J. Org. Chem.*, **58**, 7535–7536.
45 Ghose, A.K., Viswanadhan, V.N. and Wendoloski, J.J. (1999) *J. Comb. Chem.*, **1**, 55–68.
46 Jursic, S. and Zdravkovski, Z. (1993) *Synth. Commun.*, **23**, 2761–2770.
47 (a) Montalbetti, A.G.N. and Falque, V. (2005) *Tetrahedron*, **61**, 10827–10852; (b) Humphrey, J.M. and Chamberlin, A.R. (1997) *Chem. Rev.*, **97**, 2243–2266.
48 (a) Chu, W., Tu, Z., McElveen, E., Xu, J., Taylor, M., Luedtke, R.R. and Mach, R.H. (2005) *Bioorg. Med. Chem.*, **13**, 77–87; (b) Adams, R. and Ulrich, L.H. (1920) *J. Am. Chem. Soc.*, **42**, 599–611; (c) Pearson, A.J. and Roush, W.R. (eds) (1999) *Handbook of Reagents for Organic Synthesis: Activating Agents and Protecting Groups*, Wiley, New York, p. 333; (d) Lee, J.B. (1966) *J. Am. Chem. Soc.*, **88**, 3440–3441.
49 (a) Carpino, L.A., Beyermann, M., Wenschuh, H. and Bienert, M. (1996) *Acc. Chem. Res.*, **29**, 268–274; (b) Pryor, K.E., Shipps, G.W. Jr., Skyler, D.A. and Rebek, J. Jr. (1998) *Tetrahedron*, **54**, 4107–4124.
50 Ragnarsson, U. and Grehn, L. (1998) *Acc. Chem. Res.*, **31**, 494–501.
51 Klausner, Y. and Bodansky, M. (1974) *Synthesis*, 549–559.
52 Mikolajczyk, M. and Kielbasinski, P. (1981) *Tetrahedron*, **37**, 233–284.
53 (a) Wittenberger, S.J. and McLaughlin, M.A. (1999) *Tetrahedron Lett.*, **40**, 7175–7178; (b) Chu, W., Tu, Z., McElveen, E., Xu, J., Taylor, M., Luedtke, R.R. and Mach, R.H. (2005) *Bioorg. Med. Chem.*, **13**, 77–87.
54 (a) Broekema, R., Van Der Werf, S. and Arens, J.F. (1958) *Recl. Trav. Chim.*, **77**, 258–259; (b) Sheehan, J.C. and Hlavka, J.J. (1958) *J. Org. Chem.*, **23**, 635–636.
55 (a) Pelter, A., Levitt, T.E. and Nelsoni, P. (1970) *Tetrahedron*, **26**, 1539–1544; (b) Ishihara, K., Ohara, S. and Yamamoto, H. (2000) *Macromolecules*, **33**, 3511–3513; (c) Ishihara, K., Ohara, S. and Yamamoto, H. (1996) *J. Org. Chem.*, **61**, 4196–4197; (d) Al-Zoubi, R.M. Marion, O. and Hall, D.G. (2008) *Angew. Chem.*, **120**, 2918–2921; (2008) *Angew. Chem. Int. Ed.*, **47**, 2876–2879.
56 (a) Sheehan, J.C. and Hess, G.P. (1955) *J. Am. Chem. Soc.*, **77**, 1067–1068; (b) Sheehan, J., Cruickshank, P. and Boshart, G. (1961) *J. Org. Chem.*, **26**, 2525–2528.
57 Windridge, G.C. and Jorgensen, E.C. (1971) *J. Am. Chem. Soc.*, **93**, 6318–6319.
58 Ciufolini, M.A., Valognes, D. and Xi, N. (2000) *Angew. Chem.*, **112**, 2612–2614; (2000) *Angew. Chem. Int. Ed.*, **39**, 2493–2495.
59 (a) Kisfaludy, L., Schon, I., Szirtes, T., Nyeki, O. and Low, M. (1974) *Tetrahedron*

*Lett.*, **15**, 1785–1786; (b) Fujino, M., Kobayashi, S., Obayashi, M., Fukuda, T., Shinagawa, S. and Nishimura, O. (1974) *Chem. Pharm. Bull.*, **22**, 1857–1863; (c) Carpino, L.A. (1993) *J. Am. Chem. Soc.*, **115**, 4397–4398.

60 Komano, K., Shimamura, S., Inoue, M. and Hirama, M. (2007) *J. Am. Chem. Soc.*, **129**, 14184–14186.

61 (a) Castro, B., Dormoy, J.R., Evin, G., and Selve, C. (1975) *Tetrahedron Lett.*, **16**, 1219–1220; (b) Castro, B., Dormoy, J.-R., Dourtoglou, B., Evin, G., Selve, C. and Ziegler, J.-C. (1976) *Synthesis*, 751–752.

62 Tanaka, H., Sawayama, A.M. and Wandless, T.J. (2003) *J. Am. Chem. Soc.*, **125**, 6864–6865.

63 Adrio, J., Cuevas, C., Manzanares, I. and Joullie, M.M. (2007) *J. Org. Chem.*, **72**, 5129–5138.

64 Kurosawa, K., Nagase, T. and Chida, N. (2002) *Chem. Comm.*, 1280–1281.

65 (a) Dourtoglou, V., Ziegler, J.-C. and Gross, B. (1978) *Tetrahedron Lett.*, **19**, 1269–1272; (b) Knorr, R., Trzeciak, A., Bannwarth, W. and Gillessen, D. (1989) *Tetrahedron Lett.*, **30**, 1927–1930; (c) Pettit, G.R. and Taylor, S.R. (1996) *J. Org. Chem.*, **61**, 2322–2327.

66 Nicolaou, K.C., Lizos, D.E., Kim, D.W., Schlawe, D., de Noronha, R.G. Longbottom, D.A., Rodriquez, M., Bucci, M. and Cirino, G. (2006) *J. Am. Chem. Soc.*, **128**, 4460–4470.

67 Stachel, S.J., Chappell, M.D., Lee, C.B., Danishefsky, S.J., Chou, T.-C., He, L. and Horwitz, S.B. (2000) *Org. Lett.*, **2**, 1637–1639.

68 Nicolaou, K.C., Hao, J., Reddy, M.V., Rao, P.B., Rassias, G., Snyder, S.A., Huang, X., Chen, D.Y.-K., Brenzovich, W.E., Giuseppone, N., Giannakakou, P. and Brate, A. (2004) *J. Am. Chem. Soc.*, **126**, 12897–12906.

69 Knorr, R., Trzeciak, A., Bannwarth, W. and Gillessen, D. (1989) *Tetrahedron Lett.*, **30**, 1927–1930.

70 (a) Bald, E., Saigo, K. and Mukaiyama, T. (1975) *Chem. Lett.*, 1163–1166; (b) Huang, H., Iwasawa, N. and Mukaiyama, T. (1984) *Chem. Lett.*, 1465–1466; (c) Xu, J.C. and Li, P. (1999) *Tetrahedron Lett.*, **40**, 8301–8304.

71 Smith, A.B. III and Wan, Z. (1999) *Org. Lett.*, **1**, 1491–1494.

72 Wrona, I.E., Gabarda, A.E., Evano, G. and Panek, J.S. (2005) *J. Am. Chem. Soc.*, **127**, 15026–15027.

# 23
# Free Radical Cyclization Reactions
*Jake Zimmerman, Amanda Halloway, and Mukund P. Sibi*

## 23.1
### Radical Cyclizations: General Definition

Free radical cyclizations are an invaluable tool for the construction of various complex natural and nonnatural target molecules [1]. There are several advantages offered by these types of bond-forming reactions including, but not limited to, mild conditions, functional group tolerability, the possibility of tandem (or cascade) reactions, and control of regio- and stereochemistry in the products. Radical cyclization reactions have been a technology that synthetic chemists have used widely over the past forty years and several reviews have been written on this topic [2]. Therefore, it is the goal of this chapter to concentrate on the past ten years and to include topics such as asymmetric radical cyclizations, which have emerged only in the past decade.

A simple definition of a radical cyclization reaction is the formation of a cyclic structure via the addition of a radical (unpaired electron) to an acceptor. There are three main steps in this process: (i) generation of the initial radical, (ii) cyclization onto the acceptor, and (iii) termination of the product radical. Scheme 23.1 gives a graphic representation for the different pathways of a free radical cyclization. Once the radical is formed an exo or endo cyclization is possible (**2** to **3** or **4**). The ring size in the product, substituents on the radical precursor, and hybridization of the atom on the acceptor, will dictate whether the exo or endo process is favored (path a or b). For instance, in the classical 5-hexenyl radical, the 5-exo trig will be favored over the 6-endo trig [1a]. The preferences for exo or endo processes will be discussed in more detail throughout this review as this can be dependent on several factors. Once the cyclization has taken place, there are different termination pathways for the resulting radical (**3** or **4**). Scheme 23.1 lists four separate possibilities: (i) reduction, (ii) atom or group transfer, (iii) oxidation, or (iv) further additions (i.e., a tandem or cascade reaction). Each of these methods is discussed in some detail.

*Handbook of Cyclization Reactions. Volume 2.*
Edited by Shengming Ma
Copyright © 2010 WILEY-VCH Verlag GmbH & Co. KGaA, Weinheim
ISBN: 978-3-527-32088-2

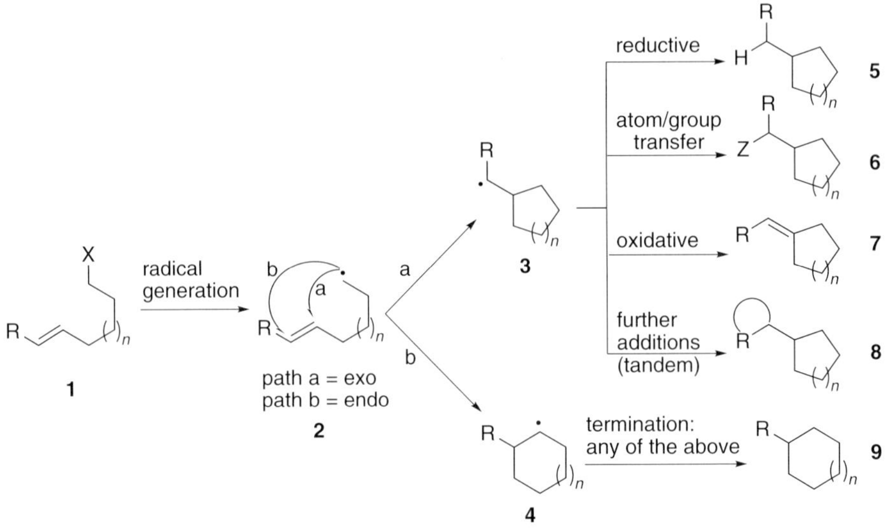

**Scheme 23.1** Radical cyclization pathways.

## 23.1.1
### Formation of a Single Ring

One major advantage of a free radical method is that both carbocyclic and heterocyclic structures are easily accessible using this technology. Scheme 23.2 shows four different monocyclic radical cyclization pathways. Equation (1) displays a prototypical method for accessing a carbocycle via a nucleophilic radical conjugate addition onto an α,β-unsaturated ester [3]. There are many other reports of reactions such as these in the literature [4]. A similar strategy can be used to make lactones. In this case the reactivity of the radical and acceptor are reversed as the radical is the more electrophilic partner while the alkene is the nucleophilic species. Using an acyclic iodo-ester **12**, a 5-exo cyclization gives the δ-lactone **13** in good yields. These reactions are not as simple as they appear due to the fact that the ester geometry favors a trans-orientation and, therefore, if these are performed under reductive conditions, the simple, acyclic, reduced compound is the major product. Oshima and coworkers showed that, indeed, these reactions are synthetically feasible using water as the solvent under atom-transfer conditions (Eq. (2)) [5]. Another more common radical strategy to form lactones is the Ueno–Stork reaction [2j]. This method typically involves a halo acetal such as **14** (Eq. (3)) as the radical precursor [6]. The geometry for radical cyclization is more suitable for facile reaction rates. A simple oxidation of cyclic acetal **15** using Jones' reagent affords the lactone. Nitrogen-containing heterocycles such as amides and pyrrolines can also be accessed through radical cyclization processes. Li and coworkers devised a novel route to access electrophilic nitrogen-centered radicals that will cyclize onto simple alkenes yielding a variety of N-phenyl lactams (Eq. (4)) [7].

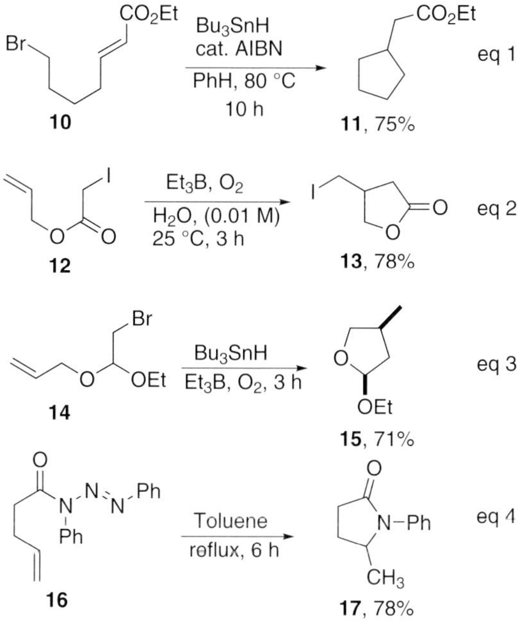

**Scheme 23.2** Single-ring radical cyclizations.

## 23.1.2
## Tandem Cyclizations

The construction of polycyclic molecules is a popular area of research in synthetic organic chemistry. The ability to carry out multiple C—C bond-forming reactions in a single transformation to form a multiple ring system is highly desirable and these processes have been called tandem, domino, or cascade reactions. Free radical cascade reactions are among one of the best ways to achieve such transformations [8]. Using these cascade reactions, the number of separate organic manipulations of a potential target molecule will be reduced and, therefore, less time, solvent, and, ultimately, cost are needed for the entire process. To achieve such sophisticated transformations, the starting material must be assembled properly in order to allow a reasonable reaction coordinate to be followed. Based on recent reviews in the area of this topic [8], this section will focus on literature reports from 2001 to 2007.

Murphy and Patro used a tandem radical cyclization strategy to synthesize an advanced intermediate towards the total synthesis of aspidospermidine (Scheme 23.3) [9]. The cascade process begins with a 5-exo cyclization of the aryl radical **19** onto the cyclohexene to provide radical intermediate **20**. This is then trapped by the azide to generate **21** followed by loss of nitrogen to give the desired tetracyclic product **23** in 40% yield. The reaction was carried out under reductive conditions using AIBN as the radical initiator and tristrimethylsilylsilane (TTMSS) as the hydrogen atom source in refluxing benzene. One of the key features of this tandem process is the selectivity of the silyl radical for iodide atom abstraction in the

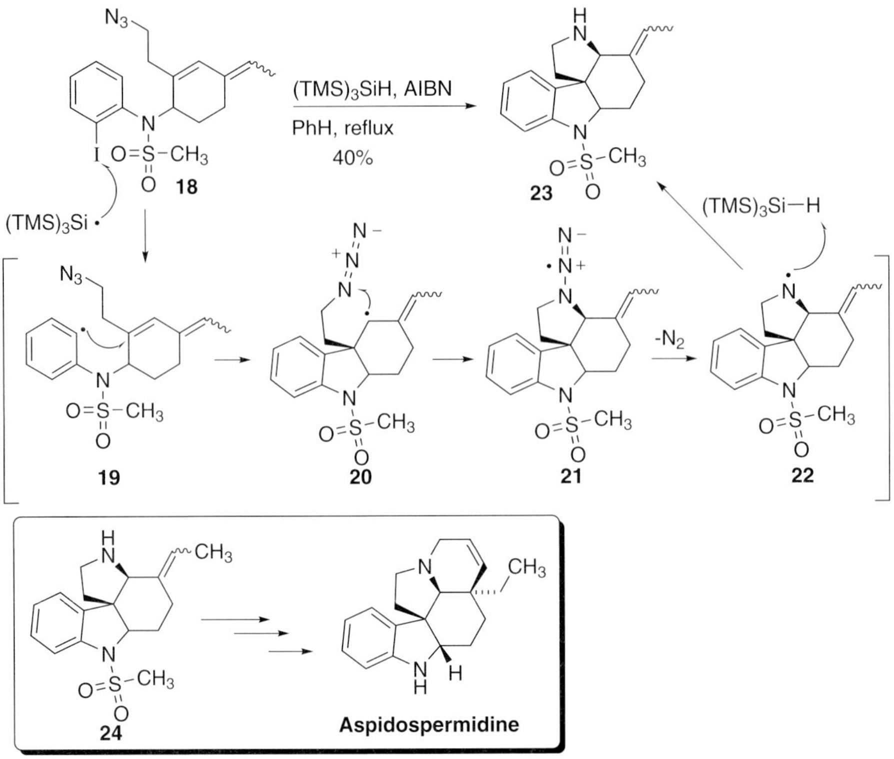

**Scheme 23.3** Tandem cyclization in the synthesis of aspidospermidine.

presence of the alkyl azide. The formal total synthesis was then completed from compound **23** in four steps to give aspidospermidine.

Ihara and coworkers reported a more non-conventional cascade radical cyclization strategy. They utilized an electrochemical radical process to synthesize a variety of bi- and tricyclic scaffolds (Scheme 23.4) [10]. The mechanism for this process begins with cathodic reduction of the vinyl bromide **27**. The corresponding radical intermediate **28** then cyclizes onto the olefin of the allylic side chain followed by a second ring closure via an addition of the primary radical **29** to the α,β-unsaturated enone **30**. They did a thorough study to find an optimal catalyst for the tandem radical process; [Ni(teta)](ClO$_4$)$_2$ (teta = 5,5,7,12,12,14-hexamethyl-1,4,8,11-tetraazacyclotetradecane) was chosen as the catalyst for this reaction [11]. It was also observed that reaction times were decreased at elevated temperatures as the conversion of **27** to **30** was completed in half the time when the reaction was performed at 60 °C (20 h) versus room temperature (48 h). A significant by-product in these reactions was the monocyclized product (around 35% in some cases). This is the only product observed when there is a substituent on the β-carbon of the enone olefin.

Vinyl radical cyclizations have also been reported in the synthesis of propellanes [12]. This tandem reaction is quite unique due to the exclusive formation of

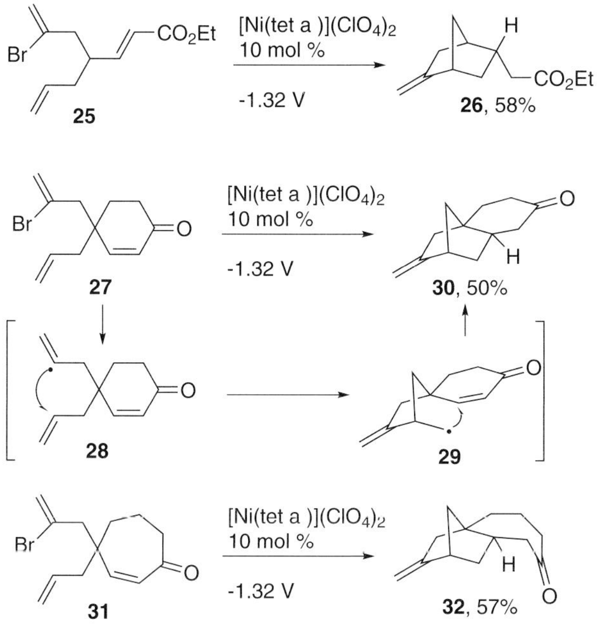

**Scheme 23.4** Tandem electroreductive radical cyclization.

6-endo products. This cascade process is carried out under classical tributyltin hydride/AIBN conditions (Scheme 23.5). Initially, the tin radical adds to the terminal alkyne yielding a vinyl radical **35**, which undergoes cyclization onto the alkene (**35** to **36**). Overall, this process is a 6-endo cyclization, however, it is believed to go through a cyclopropyl methyl intermediate (via a 5-exo cyclization) that rearranges into the tertiary radical intermediate **36** [13]. Next, a 5-exo-dig process yields **37** followed by hydrogen atom abstraction from tributyltin hydride to give the tricyclic structure **38**. Finally, treatment with $SiO_2$ and aqueous HCl removes the silicon and tin moieties. This methodology was utilized as a crucial step in the synthesis of modhephene, one of only few known natural products with the propellane skeleton.

Another common way to execute a cascade radical cyclization process is to devise an acyclic polyene structure such that, upon radical generation, multiple radical additions take place yielding bi-, tri-, and even tetracyclic molecules. These polyene cyclization reactions often begin with homolysis of traditional alkyl halides or acyl selenides. More recently it has been reported that homolytic bond cleavage of epoxides using titanocene chloride can result in tandem radical cyclizations of these polyene substrates [14]. Fernandez-Mateos and coworkers reported a very recent example of this technology [15]. Their strategy was unique in the sense that the terminal cyclization was performed on a nitrile as opposed to the more traditional alkene or alkyne (Scheme 23.6). The mechanism for the formation of the cyclic product is proposed to start with a 6-endo-trig cyclization of tertiary radical **45** followed by two more 6-endo processes. Very interestingly, a 4-exo cyclization onto the nitrile terminates the cascade process. A final treatment with aqueous acid

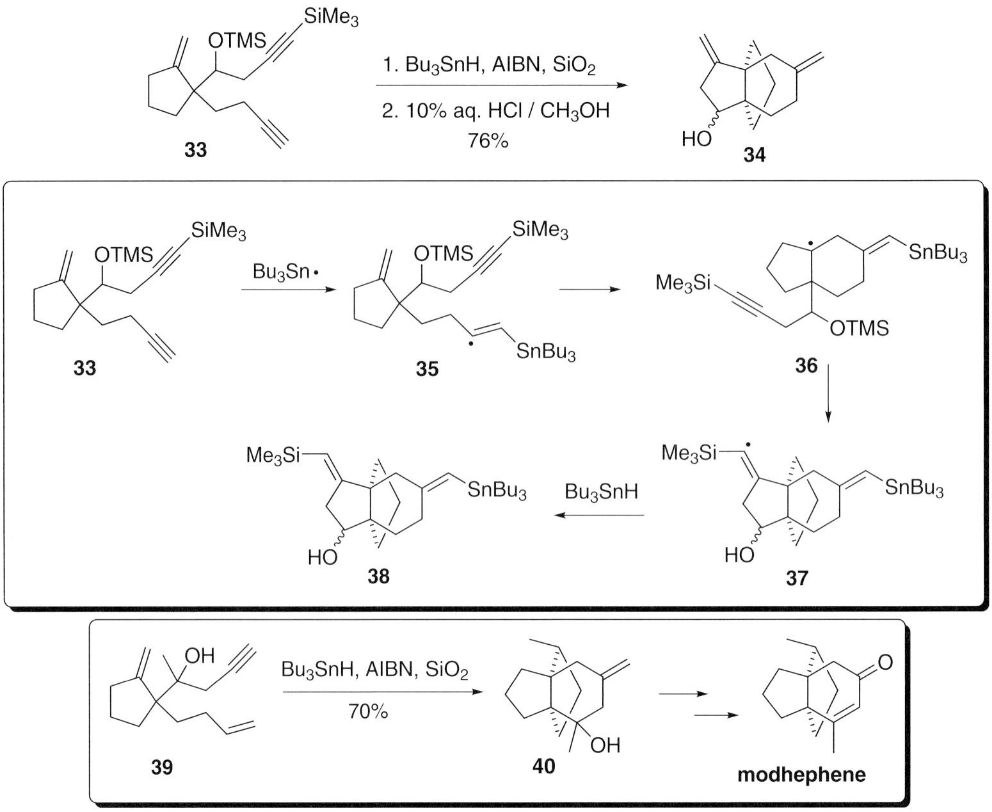

**Scheme 23.5** Tandem cyclizations in the synthesis of propellanes.

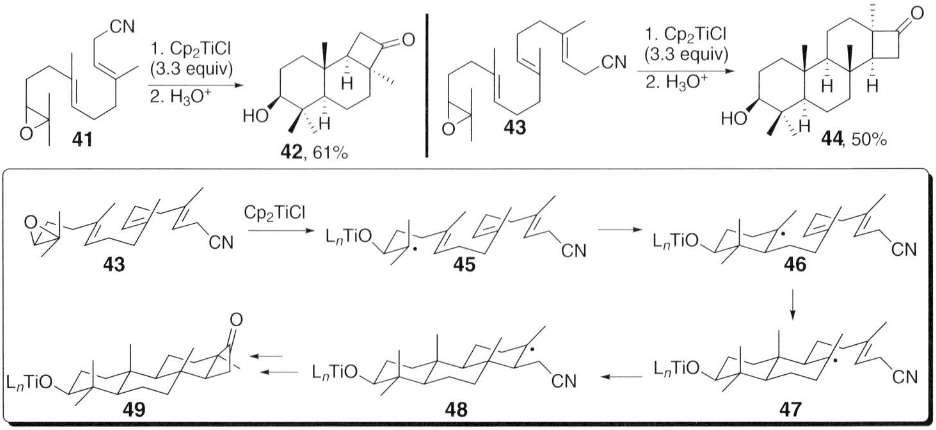

**Scheme 23.6** Titanocene chloride-mediated radical cascade.

hydrolyzed the imine and yields the tetracyclic ketone **44**. It should be noted that the cyclization was observed to go to completion or not at all. The stereochemical outcome of the reaction was established by X-ray crystallography and is explained by the chair-like transition states shown below.

A novel domino radical cyclization was reported for the synthesis of spirocyclic indole structures. This method uses copper(I) chloride as the catalyst for this atom-transfer radical cascade transformation (Scheme 23.7) [16]. Initially, a chlorine atom abstraction from the trichloroacetyl group takes place followed by a 5-exo cyclization of the electrophilic radical onto the indole ring, forming the spirocyclic radical intermediate **52**. A 6-endo-trig ring closure gives **53**, which, after chlorine atom transfer, yields the product **54** and regenerates the copper catalyst. The reactions were conducted using 0.4 equiv of CuCl and 0.8 equiv of TMEDA (tetramethyl ethylene diamine) in methylene chloride. It was observed that long reaction times were needed when the reaction was carried out at room temperature and, therefore, reflux temperatures were used. The efficiency for this procedure was only moderate, 20–39% isolated yields were reported. The researchers noted that a common by-product was the intermediate spirocyclic compound where only the initial 5-exo cyclization took place.

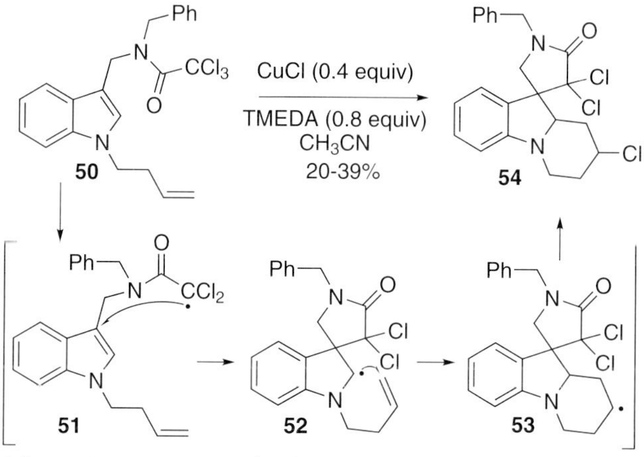

**Scheme 23.7** Copper(I)-catalyzed atom-transfer tandem cyclization.

## 23.2
## Different Methods for Radical Cyclizations

### 23.2.1
### Reductive Cyclizations

Arguably one of the most common strategies used in radical cyclization processes is the use of a reducing agent, which usually plays a significant role as a chain carrying species. One of the major issues using reductive conditions is the possibility for reduction of the radical prior to cyclization. This can be overcome by adjusting the concentrations of the hydrogen atom source so that cyclization will be favorable and the chain process will remain efficient. Tin hydrides have been a key player in these reductive radical cyclizations for nearly half a century now [17]. There are numerous reports of these processes [1, 2] and it is the purpose of this section to highlight specific examples from the recent literature.

### 23.2.1.1 Tin Hydride

The classical Ueno–Stork reaction is an efficient method for the preparation of cyclic acetals and, upon oxidation, the corresponding lactones. Recently, Friestad and coworkers reported that haloacetals of α- and β-hydroxyhydrazones undergo 5- and 6-exo cyclizations to give aminosugar derivatives (Scheme 23.8) [18]. The hydrazone substrates were prepared from the condensation of the corresponding hydroxyaldehydes with hydrazine. Treatment of the hydroxyhydrazones with N-iodosuccinimide and ethyl vinyl ether yielded the cyclization precursors (**55** and **57**). The cyclization reactions were conducted under standard reductive conditions using tributyltin hydride and AIBN in benzene at reflux. Both iodo and bromo haloacetals were examined with the iodo derivatives giving slightly better yields. The effects on the stereoselectivity for the cyclization reaction using chiral substrates derived from enantiomerically pure hydroxyaldehydes was also studied.

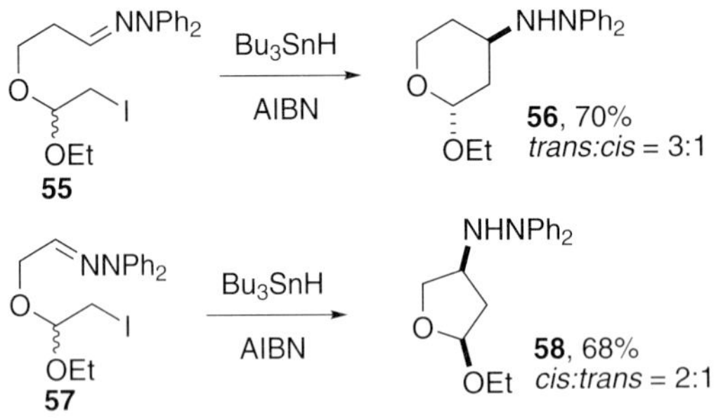

**Scheme 23.8** Cyclizations onto hydrazone acceptors.

Other radical precursors, such as nitro groups, can be used in radical cyclizations under classical tin hydride conditions. Jahn and Rudakov showed that treatment of nitro ethers, such as **59**, with tributyltin hydride and AIBN at reflux provided efficient access to a class of tetrahydrofuran lignans via a 5-exo cyclization pathway (Scheme 23.9) [19].

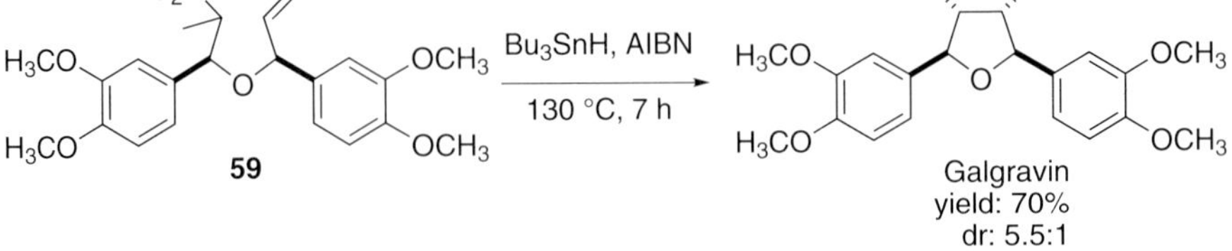

**Scheme 23.9** Synthesis of galgravin.

A recent report by Clive and coworkers describes the radical carbocyclizations of O-trityl oximes [20]. The products for these reactions are oximes of cyclic ketones and, therefore, are synthons for the corresponding amines and ketones. A key advantage in this technology is the fact that the acceptor π-bond is retained in the product. The same research group had previously shown a similar cyclization process for the preparation of lactones [21]. Extension of the methodology to carbocyclization (i.e., no heteroatom in the cyclization fragment) was unsuccessful. A recent modification, outlined in Scheme 23.10, however, shows that indeed these cyclizations are general for the all-carbon substrates. The mechanism for this process involves a 5- or 6-exo ring closure onto the oxime carbon followed by fragmentation and loss of a trityl radical. Subsequent tautomerization yields the cyclic oximes. A major discovery in the development of the new reaction conditions was the observation that the build-up of the stable trityl radical was leading to chain collapse and, therefore, a reagent was needed that would remedy this problem. Diphenyl diselenide, in the presence of tributyltin hydride, will form phenyl selenol (PhSeH). Literature reports show that hydrogen atom abstraction for a variety of carbon-centered radicals from PhSeH is several orders of magnitude faster than with $Bu_3SnH$ [22]. Benzene selenol, then, should lead to efficient reduction of the trityl radical and overall a more efficient chain process. Addition of 0.2 equiv of PhSeSePh was found to be optimal. It was also found that a nitrogen base ($i$-$Pr_2NEt$, Hunig's base) was needed to facilitate the final tautomerization. Overall, this process is very general as a variety of substrates give moderate to excellent yields.

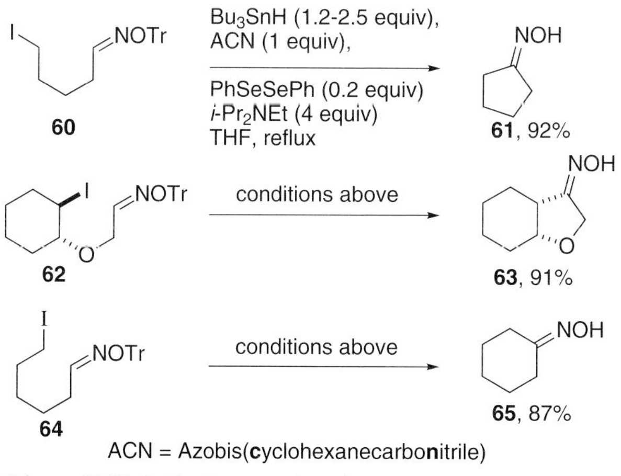

**Scheme 23.10** Cyclization onto O-trityl oximes.

The preparation of nitrogen-containing heterocycles is an important area of research as these motifs are commonly found in biologically important molecules. Ishibashi *et al.* surveyed radical cyclizations onto enamides that produced 5–8 membered ring heterocycles [23]. Two representative examples are shown in Scheme 23.11. The position of the acyl side chain had a dramatic effect on the outcome of the reaction. For instance, a 6-endo cyclization was favored over a 5-exo pathway when the radical precursor (halogen) is not part of the enamide acyl group (**68** to **69**). This is a very unusual observation, as it is well known that 5-exo processes are generally preferred over 6-endo ring closures. This endo-selective methodology was further applied to a tandem radical reaction giving an efficient synthetic route to the cylindricine class of natural products.

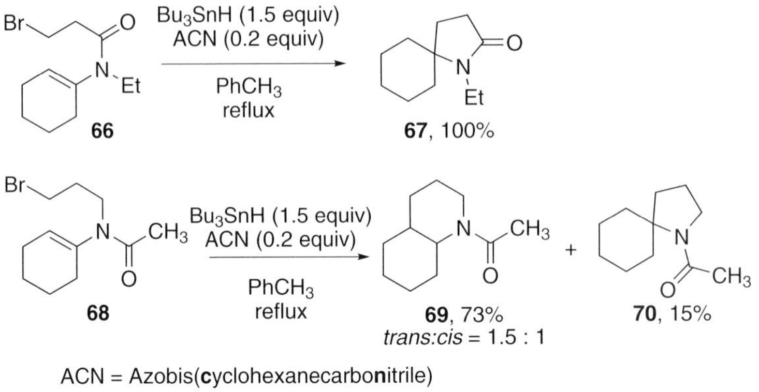

ACN = Azobis(**c**yclohexanecarbo**n**itrile)
**Scheme 23.11** Cyclization of enamides.

Although they are very efficient hydrogen atom sources, tin hydrides suffer from toxicity and difficulty in post-reaction purification. Because of these problems researchers have explored several alternatives to tin and some specific examples will be discussed below.

### 23.2.1.2 Silicon

Organosilanes emerged as one of the best alternatives for alkyltin hydrides although other organometallic hydrides such as organogermanium [24] and organomercury hydrides [25] have been studied. Arguably, the most routinely used organosilicon hydride in reductive radical cyclizations is tris(trimethylsilyl)silane (TTMSS) which was originally reported by Gilman and coworkers in 1965 [26] but its potential as a hydrogen atom source and chain carrier in radical reactions lay dormant for more than two decades [27]. In general, TTMSS is a less reactive hydrogen atom donor towards carbon-centered radicals. This is most likely due to the slightly stronger Si—H (79 kcal mol$^{-1}$ in TTMSS) versus the Sn—H bond (74 kcal mol$^{-1}$ in Bu$_3$SnH) [28]. The major advantage of using silanes in place of tin is the lower toxicity of the silyl by-products [29] and the simplification of reaction purification. It should be noted that multiple reviews that cover silanes in radical reactions have appeared in the literature during the last decade [30].

In a recent report, researchers were examining an intramolecular conjugate addition radical cyclization process for the preparation of 2,4-disubstituted piperidines [31]. A dramatic increase in diastereoselectivity was observed when Bu$_3$SnH was replaced with TTMSS (Scheme 23.12). This increase in stereoselectivity was not general for all cyclization precursors, but when the 2-substituent was secondary or benzylic, better drs were observed. The proposed explanation was based on the slower hydrogen atom transfer from TTMSS to the resulting secondary radical after cyclization. Due to the longer lifetime of this radical intermediate, a 1,5-hydrogen atom transfer can take place in the cis-isomer. This leads to a Smiles-type rearrangement through a radical addition to the aromatic ring of the tosylate followed by sulfur dioxide extrusion. Therefore, the minor diastereomer is ultimately converted into a different product. Although the selectivity improves under these conditions, the yield suffers due to the alternate reaction pathway for the cis-compound.

**Scheme 23.12** TTMSS as a hydrogen atom source.

### 23.2.1.3 Germanium

Alkylgermanium hydrides represent another popular alternative for tin hydrides. Germanium is a Group 14 element, like tin and silicon, which makes it a good candidate as an efficient hydrogen atom source. Organogermanium hydrides have been known for over 20 years [32] but only in the last 10 to 12 years have they been used in synthetic radical processes [33]. Reports of low toxicity for organogermanium compounds represent another major benefit for using these reagents in radical processes [34]. Several germanium hydride derivatives have been prepared and studied, including tris(trimethylsilyl)germanium hydride, a derivative of the silicon counterpart (TTMSS). The rate of hydrogen atom abstraction by primary radicals from $(TMS)_3GeH$ is nearly identical to $Bu_3SnH$ ($3.1 \times 10^6 \, M^{-1} \, s^{-1}$ at 25 °C and $2.4 \times 10^6 \, M^{-1} \, s^{-1}$ at 25 °C, respectively). Due to its high cost, however, $(TMS)_3GeH$, is not a reasonable replacement for tributyltin hydride. A more affordable organogermanium hydride is $Bu_3GeH$, but the rates of H-atom abstraction are significantly less than that for $Bu_3SnH$ (primary radical abstraction is approximately 24 times slower from $Bu_3GeH$ than from $Bu_3SnH$). The slower rate for reduction of primary radicals by $Bu_3GeH$ can be beneficial for cyclization reactions in that the radical precursor will not be prematurely reduced prior to cyclization. An example of an organogermanium hydride-mediated cyclization for the preparation of lactams was reported by Bowman and coworkers (Scheme 23.13) [35]. The cyclization of **75** to **76** using $Bu_3GeH$ was much more efficient than the corresponding $Bu_3SnH$ conditions (67% and 29% yields, respectively). Another marked benefit noted in this report was the superior stability of $Bu_3GeH$ over $Bu_3SnH$ as the germanium hydride reagent was stable for several weeks

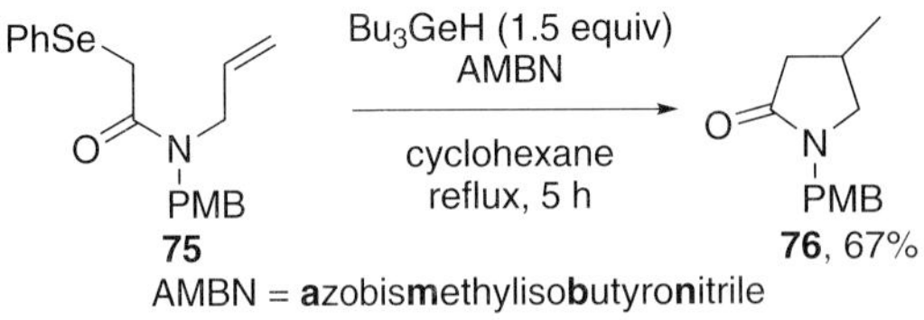

**Scheme 23.13** Organogermanium hydrides as a hydrogen atom source.

in CDCl$_3$ while the tributyltin hydride showed significant decomposition within one day.

#### 23.2.1.4 Indium

Hydrides of indium(III) salts have received significant attention as both radical initiators and hydrogen atom sources in a variety of free radical processes over the past decade. Indium hydrides are easily prepared via the transmetallation of indium (III) trihalides with metal hydrides or silanes. Initially, hydride sources such as Bu$_3$SnH [36], NaBH$_4$ [37], and DIBAL-H [38] were used in the preparation of the indium hydride reagent, but due to the strong hydridic nature of these reagents, problems with substrate compatibility were an issue.

In 2004, Baba and coworkers reported that triethylsilane was suitable for preparing HInCl$_2$ from InCl$_3$ [39]. This was a major advantage since Et$_3$SiH is a very mild hydride source and no unwanted side reactions should take place. The mechanism for the indium hydride-mediated radical cyclizations of enyne **77** is illustrated in Scheme 23.14. A mixture of InCl$_3$ and triethylsilane in acetonitrile at 0 °C generates an equilibrium mixture of HInCl$_2$. The initiator used for this process is triethylborane,

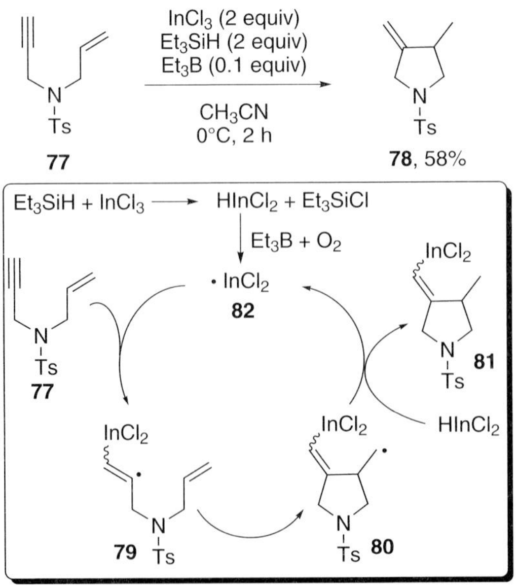

**Scheme 23.14** Indium hydride as a H-atom source.

which is known to generate ethyl radicals upon exposure to oxygen. An ethyl radical abstracts a hydrogen atom from indium hydride forming an indium radical that selectively adds to the terminal end to the alkyne, generating the radical intermediate **79**. A 5-exo cyclization yields **80**, followed by a hydrogen atom transfer from $HInCl_2$ to yield **81** and the chain-carrying indium radical (**82**). An acidic work-up gives the pyrrolidine product **78**. There have been several other reports using indium species such as $In(OAc)_3$ [40], In(I) [41], and In(0) [42] in reductive radical cyclizations.

### 23.2.1.5 Phosphorous Acid

Phosphorus reagents have been receiving significant attention as plausible replacements for tributyltin hydride in reductive radical reactions. In particular, hypophosphorous acid and its closely related derivatives are of particular interest for many research groups [43]. Some of the major benefits that this class of reagents offer are, water solubility and stability and ease of separation from organic products after reaction. Due to the nature of these compounds researchers have been able to show that these radical cyclizations can be carried out in water [44] or more traditional organic solvents.

Renaud and coworkers reported the use of dimethyl phosphite in a radical cyclization process. This method is based upon Curran's alkenyl radical translocation/cyclization process that was developed using traditional $Bu_3SnH$ conditions [45]. The proposed reaction mechanism is illustrated in Scheme 23.15 [46]. First, a phosphorus radical adds to the terminal alkyne, generating the vinyl radical intermediate **84**. Next, a 1,5-hydrogen atom transfer followed by a 5-exo cyclization takes place to generate the cis-bicyclic product **86**. In this report, several five-membered ring formations proceeded with good to excellent yields. Spirocyclic products were also prepared in good efficiency.

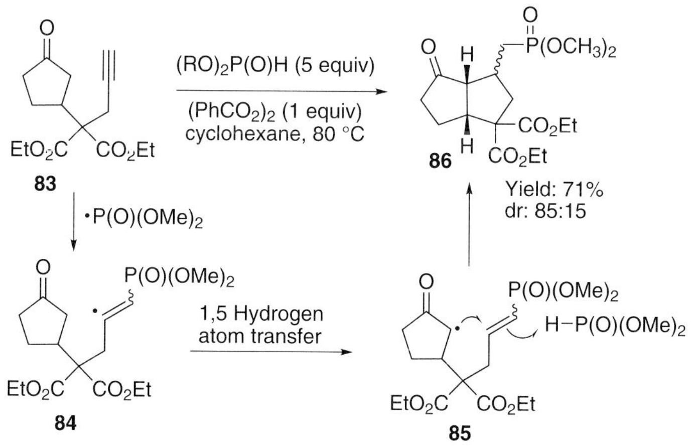

**Scheme 23.15** Dimethyl phosphite in radical cyclizations.

Diethyl phosphine oxide (DEPO) is another phosphorus-based alternative for alkyltin hydrides. DEPO has been used in radical cyclizations that yield spirocyclic indole products (Scheme 23.16) [47]. A major advantage for this methodology is the fact that the reactions are carried out in water. Because of the aqueous conditions, a water-soluble initiator was needed (V-501). An extension of this method by the same research group was used as a key step in the synthesis of horsfiline, a natural product with potential anti-cancer properties.

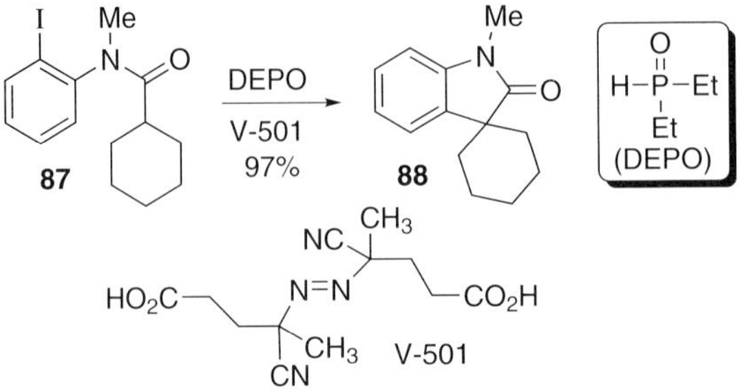

Scheme 23.16 Diethyl phosphine oxide-mediated cyclizations.

### 23.2.1.6 Thiols

Thiyl radicals are easily prepared from the corresponding thiol (or disulfide), and, therefore, thiols have been used in reductive radical cyclizations for many years [48]. It is well known that thiyl radicals will add reversibly to alkenes and alkynes, generating carbon-centered radicals that can undergo further reactions such as cyclization. Examples of these cyclization reactions are numerous and include examples of subsequent ring closures onto acceptors such as alkenes and hydrazones [49].

As described in the previous section, a phosphorus radical addition/translocation/cyclization method was developed for the synthesis of 5-membered rings. A similar strategy was developed using thiols rather than phosphites (Scheme 23.17) [50]. The mechanism for the thiol-mediated reaction is nearly identical to the phosphorus-mediated reaction. A thiyl radical adds to the terminal alkyne followed by a 1,5-hydrogen atom transfer and then subsequent cyclization onto the resulting alkene to give a variety of cyclic and bicyclic structures **90**. It was observed that two equivalents of initiator (AIBN) and thiophenol (under syringe pump addition) in refluxing *tert*-butanol were optimal conditions for this method. The efficiency of this reaction is also quite interesting as thiophenol is a powerful reducing agent and only trace amounts of the uncyclized product was observed. One drawback for the reaction is that when hydrogen atom transfer is slow the resulting vinyl radical will add to the phenyl ring of the thiophenol moiety, resulting in benzothiophene by-products. This result is consistent with previous reports using tin hydride methodology [51].

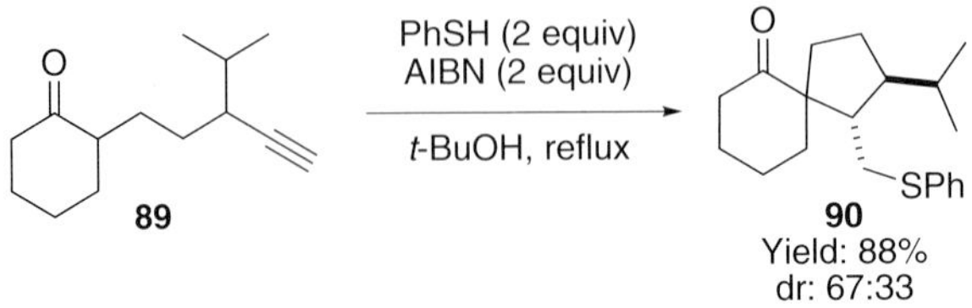

Scheme 23.17 Thiol-mediated cyclization.

Thiols can also be used in the direct formation of acyl radicals starting from simple aldehydes. Researchers used this strategy for preparing 2-substituted cyclopent- and cyclohexanones using an odorless thiol, *tert*-butyldodecanethiol (Scheme 23.18) [52]. Only catalytic amounts of the thiol and initiator (AIBN or V-40) were needed to obtain good to excellent yields of the cyclic adducts. The proposed mechanism for the reaction is believed to start with thiyl radical formation followed by a hydrogen atom abstraction of the aldehyde giving the intermediate acyl radical **92**. Next, an intramolecular radical cyclization (5- or 6-exo) followed by a reduction of the resulting radical yields the target product **93** and regenerates the thiyl radical. The method was quite general as both electron deficient and unactivated alkenes worked as acceptors although it was noted that α,β-unsaturated olefins gave slightly higher yields.

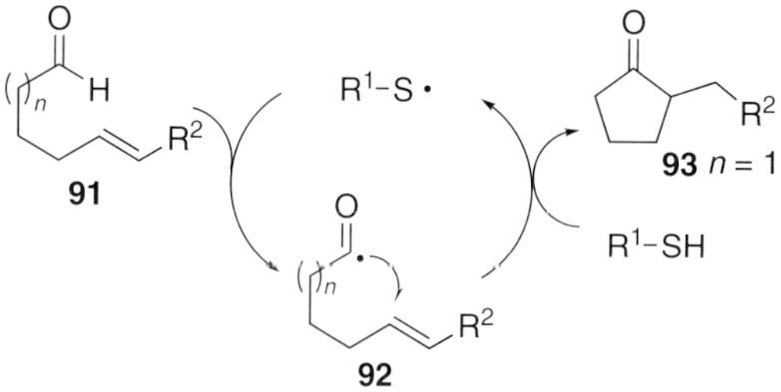

**Scheme 23.18** Thiol-catalyzed acyl radical cyclization.

## 23.2.2
## Atom-Transfer Cyclizations

One major disadvantage for reductive radical cyclizations is that one ultimately loses functionality in the end product. For instance, in Scheme 23.1 the cyclization precursor contains an alkyl halide and an alkene as the acceptor while the product contains neither functional group. Atom or group transfer methods, however, allow for the retention of the radical precursor atom (or group) in the product, which ultimately leaves a handle for further synthetic modification of the product. The intramolecular version of this method, atom-transfer radical cyclization (ATRC), has been studied for many years and provides access to various cyclic structures [53].

### 23.2.2.1 Copper
One of the most successful ATRC methods is the copper(I)-catalyzed reaction of 2,2,2-trichloro carbonyl compounds. These are oxidative atom-transfer reactions that involve a redox process between copper(I) and copper (II) intermediates. Two recent reviews on this topic have been published and, therefore, a discussion on more recent reports will follow [54].

Yang and coworkers reported the chlorine atom-transfer cyclization reactions of α-chloro β-keto esters [55]. They were able to show that both mono and tandem radical cyclization reactions proceeded efficiently using a Cu(I)-bipyridine (or bisoxazoline) complex. Scheme 23.19 outlines the proposed mechanism for the transformation.

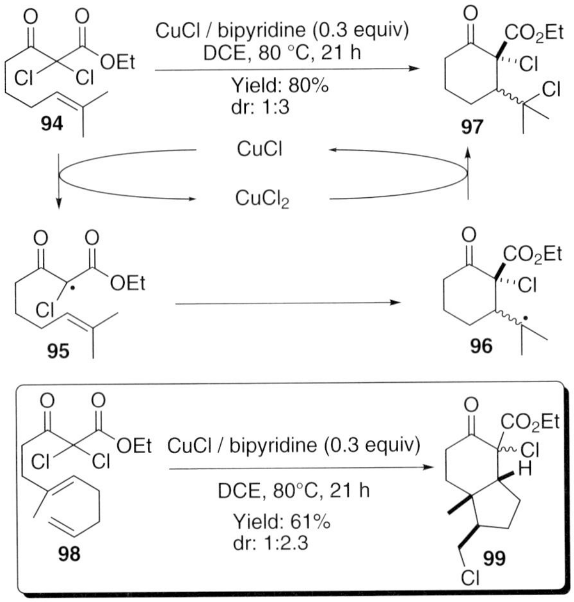

**Scheme 23.19** Copper(I)-catalyzed ATRC reaction.

Initially CuCl abstracts a chlorine atom from **94** to generate intermediate **95** and $CuCl_2$. A 6-exo cyclization yields **96** which abstracts a chlorine atom from $CuCl_2$ giving the desired product **97** and regenerating the catalyst (CuCl). The product contained a mixture of diastereomers (about 1 : 3 in the example shown) which was explained by the fact that the carbonyls could be weakly chelated by the $CuCl_2$ Lewis acid or the carbonyls could be orientated in an anti-fashion so as to minimize both dipole and steric repulsions. The reaction can be carried out at both room temperature or in refluxing dichloroethane, depending on what catalyst system is used. Copper(I)-bisoxazoline complexes were found to be efficient at room temperature while the bipyridine complex was best at reflux. A tandem process using the same catalyst system was also reported as **99** was formed in 61% yield with a diastereomeric ratio of 1 to 2.3.

The use of trichloroacetates in Cu(I)-catalyzed ATRC reactions to afford lactones is well known, however, the application towards target synthesis is not as well documented. Quayle and collaborators used this technology as a strategy to approach the synthesis of steganacin derivatives [56]. The reaction proceeds with good diastereoselectivity, yielding the threo-isomer **101** in a 19 : 1 (threo : erythro) ratio (Scheme 23.20). The researchers discovered, however, that upon extended thermolysis conditions, the product would undergo equilibration to afford a 3 : 2 (threo : erythro) ratio. Although the equilibration was a concern, when product **101** was subjected to solvolysis conditions in benzyl alcohol, a single threo-isomer of **102** was isolated in 85% yield. This intermediate **102** was used as a key building block in the synthesis of steganone derivatives.

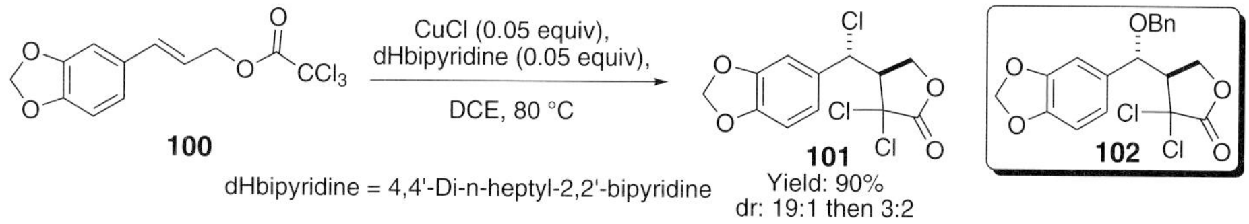

## 23.2 Different Methods for Radical Cyclizations

**Scheme 23.20** Copper(I)Cl-mediated atom-transfer cyclization.

Reagents for 100 → 101: CuCl (0.05 equiv), dHbipyridine (0.05 equiv), DCE, 80 °C. Yield: 90%, dr: 19:1 then 3:2. dHbipyridine = 4,4'-Di-n-heptyl-2,2'-bipyridine. Compound 102 shown with OBn group.

### 23.2.2.2 Ruthenium

Ruthenium complexes represent another class of transition metal catalysts for mediating ATRC reactions [57]. Arguably one of the most popular ruthenium catalysts is Grubb's catalyst **104**, however, it is most notably recognized as an olefin metathesis catalyst and not as an atom-transfer catalyst. A report by Snapper and coworkers showed that catalyst **104** successfully promoted Kharasch reactions [58]. Grubbs also reported that **104** catalyzed atom-transfer radical polymerization reactions [59]. An ATRC reaction using Grubbs' catalyst **104** was later reported by Quayle et al. (Scheme 23.21) [60]. The cinnamyl ester **103** was treated with a 5 mol% quantity of the catalyst and refluxed for 3 h to yield an isomerically pure γ-lactone product **105** in 54% yield. Thus demonstrating that the Grubbs catalyst can be used to initiate ATRC reactions and that only low catalyst loading is necessary to produce effective chemical conversions.

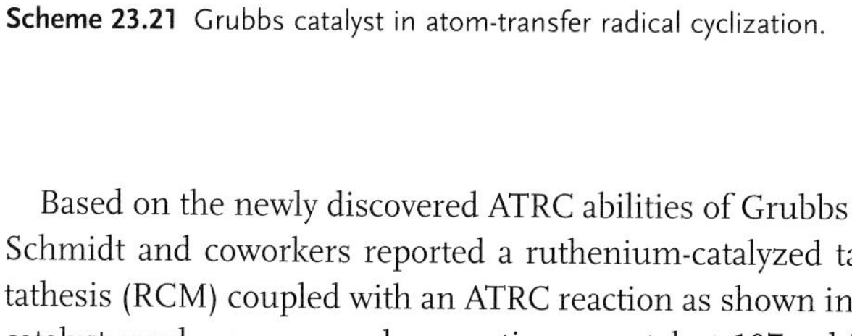

**Scheme 23.21** Grubbs catalyst in atom-transfer radical cyclization.

103 + catalyst 104 (5 mol%), toluene, reflux → 105, 54%.

Based on the newly discovered ATRC abilities of Grubbs catalyst by Quayle et al., Schmidt and coworkers reported a ruthenium-catalyzed tandem ring-closing metathesis (RCM) coupled with an ATRC reaction as shown in Scheme 23.22 [61]. The catalyst used was a second generation precatalyst **107**, which had been previously noted for its accelerated activity in metathesis reactions [62]. Interestingly enough, this catalyst was able to mediate both steps in this tandem sequence when it was previously believed only a year earlier that these two pathways were competing [63]. The result was a 61% yield of the product **108** with excellent diastereoselectivity in the ATRC step. This good stereocontrol has been previously demonstrated with copper-catalyzed ATRC reactions [64].

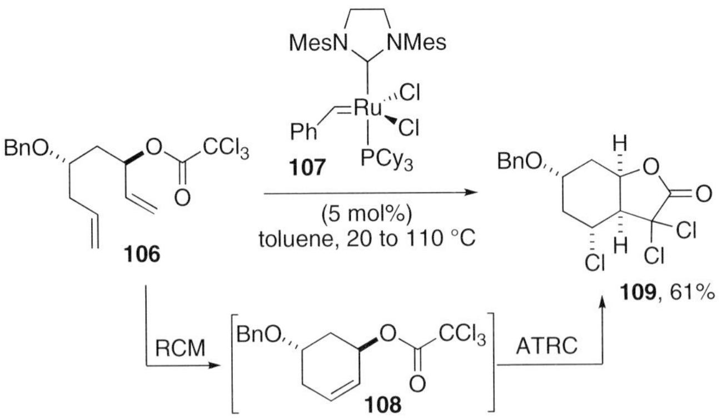

**Scheme 23.22** Tandem RCM and ATRC reaction using Grubbs catalyst.

### 23.2.2.3 Organotin

Until recently, halogen atom-transfer cyclizations had not been reported in 8-endo radical cyclizations of α-carbonyl radicals. A study by Wang and colleagues, however, recently reported that such cyclizations were successfully completed using bis(tributyltin) as the radical initiator [65]. As shown below in Scheme 23.23, bis(tributyltin) is capable of initiating radical cyclization reactions, both in the presence and absence of $BF_3-Et_2O$. Reaction with **110** in Scheme 23.23(A) was conducted at room temperature in order to prevent rearrangement products and required the use of a Lewis acid to enhance the radical reactivity and promote product selectivity, as had been shown in previous studies [66]. The reaction yielded the product **111** in pure form in 80% yield. Reaction in Scheme 23.23(B) was carried out under higher temperatures and without the Lewis acid additive. These conditions favored the formation of the rearrangement product **112** (80% yield). The mechanism for the rearranged lactone product is thought to go through a bicyclic intermediate **113** where an iodide anion ring opens the cyclic intermediate to yield the γ-lactone product **112**.

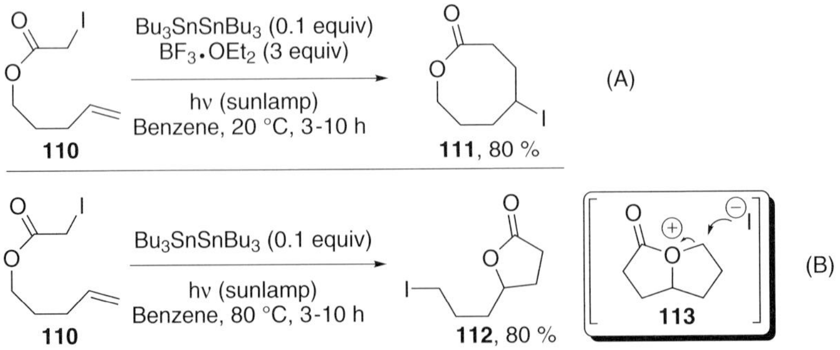

**Scheme 23.23** Hexabutylditin-mediated ATRC reactions.

### 23.2.2.4 Peroxides

Peroxides represent another avenue for atom-transfer radical cyclizations. Traditionally atom-transfer reactions have been carried out using sunlamp irradiation in the presence of hexaalkyldistannanes, as shown by Curran *et al.* [67]. Due to the toxicity of tin by-products and the inability to remove such by-products, there has been a need for a new method of synthesis [68]. Peroxides such as dilauroyl peroxide (DLP) function as radical initiators and generate alkyl radicals from the corresponding alkyl iodides or bromides [69]. Renaud and coworkers have successfully shown that DLP-mediated iodine atom-transfer reactions can be carried out very efficiently, as shown in Scheme 23.24 [70]. The α-iodoacetates **114** as well as other substrates were shown to undergo radical cyclization reactions using DLP as the radical initiator with similar yields to Curran's classical organotin method [68d].

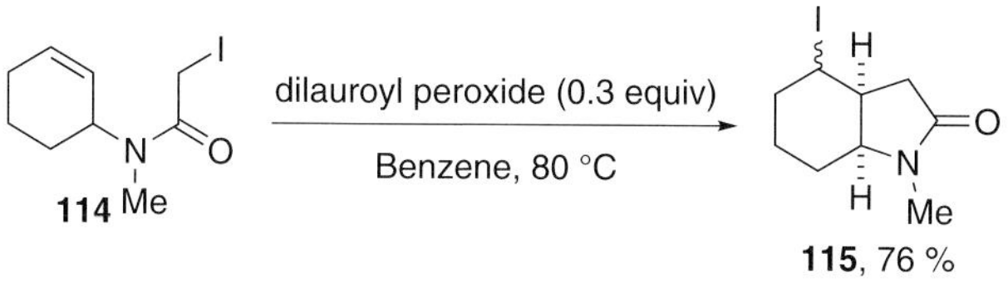

**Scheme 23.24** Dilauroyl peroxide-mediated ATRC reaction.

## 23.2.3
### Manganese Triacetate-Mediated Cyclization

Manganese triacetate is an extraordinary reagent for the selective formation of highly electrophilic radicals, which can then be employed in a variety of transformations, especially cyclizations. There are several comprehensive reviews on the use of manganese triacetate in radical chemistry and thus only a few informative and recent applications are presented here [71].

The most common use of manganese triacetate is radical cyclization under oxidative conditions. The intermediate electrophilic radical generated using manganese triacetate adds readily to alkenes and the resultant product radical after cyclization can be effectively trapped to introduce additional functional groups. Snider and coworkers have reported that radicals can be trapped using azides (Scheme 23.25) [72]. Treatment of **116** with manganese triacetate in methanol in the presence of sodium azide furnishes the cyclized product **117** in high yield. Similarly, compound **118** undergoes cyclization and trapping to give a mixture of isomers **119** and **120** in good yield and modest diastereoselectivity.

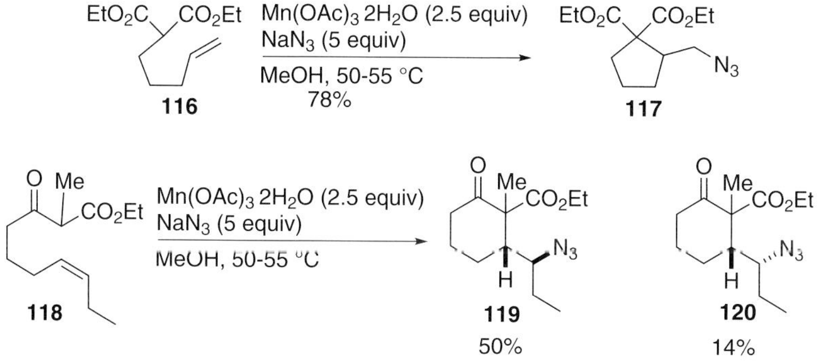

**Scheme 23.25** Mn(OAc)$_3$-mediated radical cyclization followed by azide trapping.

Ishibashi and coworkers [73] have developed a novel manganese triacetate-mediated tandem cyclization as the key step in the total synthesis of 3-demethoxyerythratidinone (Scheme 23.26). Treatment of the thiomethylacetamide **121** with a mixture of manganese triacetate and copper triflate generates the α-carbonyl radical which cyclizes (5-endo trig) on the alkene to form a tricyclic radical intermediate. This radical is further oxidized to a cation, which is then trapped by the electron-rich aromatic ring (Friedel–Crafts reaction). This compound **122** was converted to the target using standard transformations. It is important to note that the 5-endo trig cyclization does not proceed in the absence of a thiomethyl substituent and copper triflate is necessary for the reaction course shown (copper acetate gives alternate products).

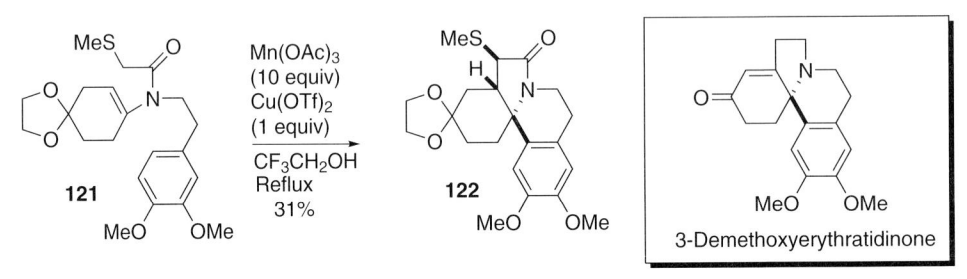

**Scheme 23.26** Effect of co-catalyst on Mn(OAc)$_3$-mediated radical cyclization.

As part of a program on devising efficient approaches to the total synthesis of oroidin alkaloid natural products, Chen and coworkers have evaluated manganese triacetate-mediated radical cyclizations [74]. A few model cyclizations are illustrated in (Scheme 23.27). Cyclization of **123** using manganese triacetate/acetic acid at 60 °C

**Scheme 23.27** Mn(OAc)$_3$-mediated cyclization – oroidin alkaloids.

gave a mixture of diastereomeric products in modest yield and selectivity. In this example a 5-exo cyclization is followed by a 6-endo cyclization. In the second example (**126** to **127** and **128**) where a chlorine atom blocks the second 6-endo pathway, a spirocyclic compound **127** and a monocyclic compound **128** are formed in high yield in nearly equal amounts. The authors present cyclizations using a variety of substrates in this work.

### 23.2.4
### Samarium Diiodide-Mediated Cyclization

Samarium iodide is a versatile reagent that is used extensively in radical chemistry. Samarium iodide or samarium metal, are excellent single electron transfer reagents. Additionally, one can use excess samarium reagent to carry out a second electron transfer to convert the product radical into an anion and thereby extend the utility of the process. There are multiple reviews on the use of samarium iodide in radical reactions [75, 76]. This section will highlight reactions from the recent literature.

Riessig and coworkers [77] have carried out an interesting radical cyclization using samarium iodide (Scheme 23.28). The reaction involves the treatment of ketone **129** with excess samarium iodide. This results in the initial formation of a ketyl radical which cyclizes on the aromatic ring. The resulting radical is further reduced to an anion, which is protonated to yield the product **130** in low yield. It is noteworthy that four contiguous stereocenters are established in a single operation. Furthermore, the product has structural features of a steroid.

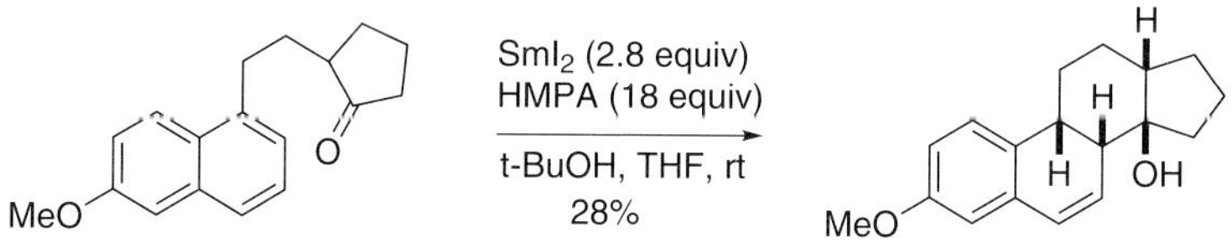

**Scheme 23.28** Samarium diiodide-mediated radical cyclization.

Samarium iodide can also transfer a single electron to a carbon–halogen bond generating a radical. St. Jean and coworkers have reported an interesting reaction in which a diiodide is used as a starting material (Scheme 23.29) [78]. Treatment of the diiodide **131** with a mixture of excess $SmI_2$ and a catalytic amount of $NiI_2$ in the

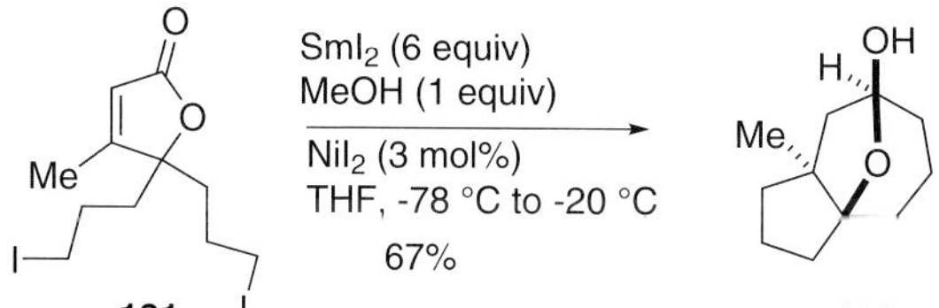

**Scheme 23.29** Tandem radical reactions using $SmI_2$.

presence of 1 equiv of methanol furnished the tricyclic alcohol **132** in good yield. The reaction begins with the formation of a C-centered radical, which adds conjugatively to the butenolide. The resultant radical intermediate is reduced to an enolate by the excess reagent, which is protonated by methanol to form the saturated lactone. The remaining C—I bond is then converted to a carbanion, which then adds on the lactone carbonyl to furnish the product. The impact of the nature of the protic additive and its stoichiometry on reaction outcome has been determined.

Curran and coworkers have compared the efficiency of samarium-mediated radical cyclizations with Heck cyclizations in model studies towards the synthesis of penitrem, a molecule with potent neurotoxic activity [79]. Treatment of iodoarene **133** with $SmI_2$/HMPA and in the presence of acetone furnishes **134** in modest yield (Scheme 23.30). This is an example of a radical/polar crossover reaction. The aryl radical formed from reaction between the C—I bond and $SmI_2$ cyclizes onto the cyclobutene ring and subsequently the resultant radical is reduced to the anion. This anionic intermediate is trapped by acetone to furnish **134** wherein two new chiral centers are established during the cyclization/trapping process. Product **135** is formed due to inefficient trapping by acetone. The authors have investigated the cyclization reaction using tinhydride and under Heck conditions.

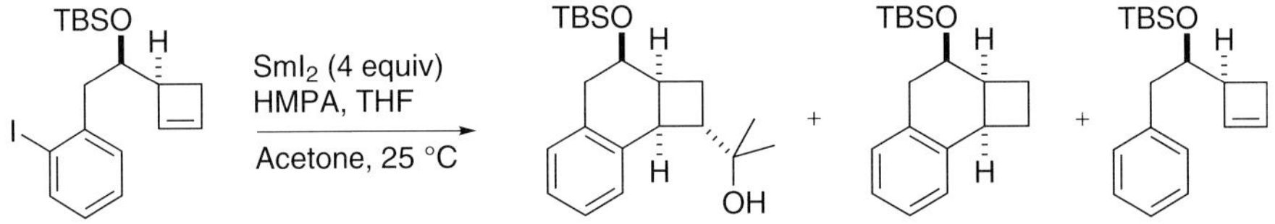

**Scheme 23.30** Radical-polar crossover reaction using $SmI_2$.

## 23.2.5
### Titanium (III)-Mediated Cyclization

Metal-mediated radical chemistry has had and continues to enjoy a lot of popularity. In this context, the application of titanocene(III) chloride in radical chemistry has received much attention. Since the original report of Nugent and Rajanbabu on the generation of radical from epoxides using Ti(III) reagents, this field has seen much activity and there are many reviews [80]. This section illustrates recent examples of cyclization reactions using Ti(III) reagents.

Epoxides undergo ring opening in the presence of Ti(III) reagents to form the more substituted titanoyloxy radical, which undergoes further reactions. Roy and coworkers have employed this methodology extensively for the synthesis of tetrahydrofuran natural products (Scheme 23.31) [81]. Reaction of alkynyl epoxide **137** with $Cp_2TiCl$ forms the radical intermediate **138**, which undergoes 6-exo-dig cyclization to furnish **141**. Alternatively, a 5-exo-trig cyclization can be carried out under different reaction conditions leading to a different class of natural products. Reaction of **143** with Cp$_2$TiCl gave furano lignan natural products **144** as the major isomer. A variety of starting materials with different Ar$^1$ and Ar$^2$ groups have been investigated for accessing different natural products. Alternatively, reaction of **143** with Cp$_2$TiCl and I$_2$ at 60 °C gave access to furofurano lignan natural products **142**.

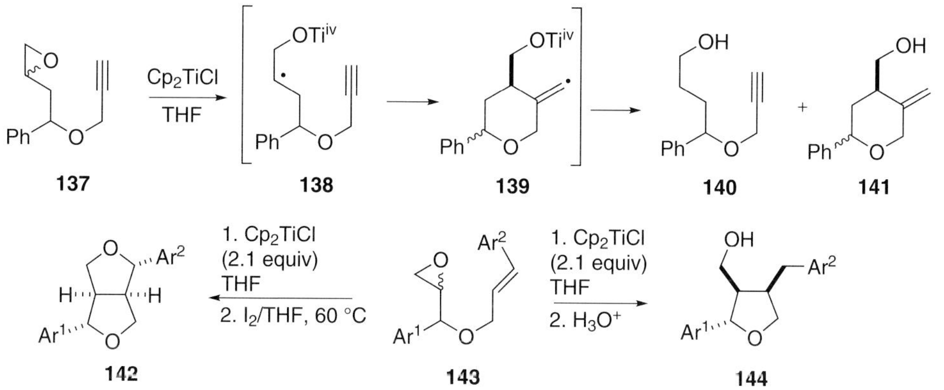

**Scheme 23.31** Ti(III)-mediated radical cyclizations.

Justicia and co-workers have developed an efficient synthesis of α-ambrinol using Ti(III)-mediated radical cyclization (Scheme 23.32) [82]. Treatment of **145**, which is readily prepared from geranylacetone, with Cp$_2$TiCl gave the cyclohexane **146** in good yield as the sole product. The process involves a highly selective 6-endo-trig cyclization and since the reaction is carried out under anhydrous conditions an alkene is formed as the product. The conversion of **146** to α-ambrinol requires three steps and proceeds in high overall yield.

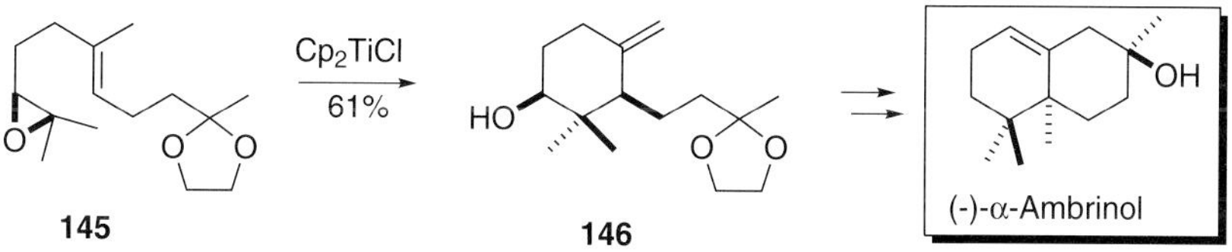

**Scheme 23.32** Synthesis of ambrinol using Ti(III)-mediated radical cyclization.

Cyclizations leading to four-membered rings (4-exo pathway) are slow and thus not suitable for efficient radical chain propagation. Gansauer and coworkers have offered a nice solution to this problem using titanocene reagents (Scheme 23.33) [83]. Treatment of the epoxide **147** with different titanium catalysts (**149**–**151**) in the presence of manganese and collidine hydrochloride gave the 4-exo cyclization product **148** as a mixture of diastereomers. The catalysts shown vary in terms of steric bulk. Catalysts **149** and **150** containing ligands that are less bulky, provide the diastereomeric products in excellent yield and modest selectivity. In contrast, **151** with a large *tert*-butyl group gave a low yield of the products with a slightly improved selectivity. The authors have carried out cyclizations using chiral catalysts.

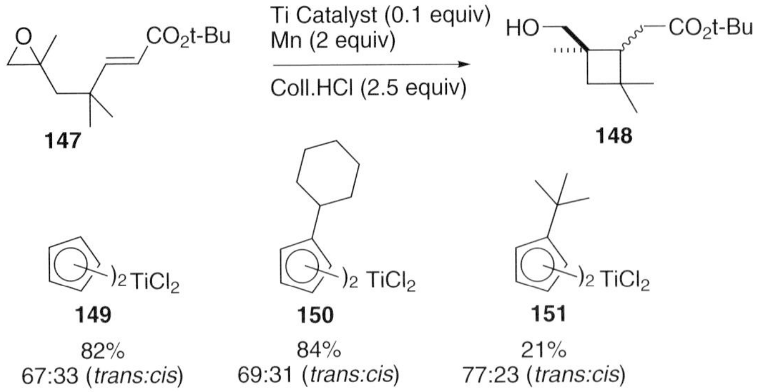

Scheme 23.33 Impact of Ti(III) catalysts on radical cyclization.

## 23.2.6
### Photochemical Cyclizations

Radicals can be generated under photolysis conditions. There are a large number of examples that utilize photolysis conditions to carry out radical cyclizations [84]. A few recent and selected examples of radical cyclizations using photolysis conditions are detailed below.

Narasaka and coworkers have made seminal contributions to the use of light energy for radical reactions and an example is shown in Scheme 23.34 [85]. The authors have investigated cyclizations of alkenyl acylphosphonates under photochemical and thermolysis conditions. Treatment of **152** with 300 nm light in benzene provides the cyclic product **156** in 93% yield. The product can be formed via two different pathways. Carbon–phosphorus bond homolysis from the photoactivated intermediate produces the radical pair intermediate **153**. Acyl radical cyclization followed by radical trapping provides **156** (path a). Alternatively, addition of the photochemically generated diphenyl phosphinoyl radical to **152** provides the intermediate **154** (path b). Cyclization of the alkyl radical onto the acyl phosphonate furnishes **155** which collapses to the product. Crossover experiment has established that the reaction proceeds via path b. The authors have compared the photochemical reactions with thermolytic ones.

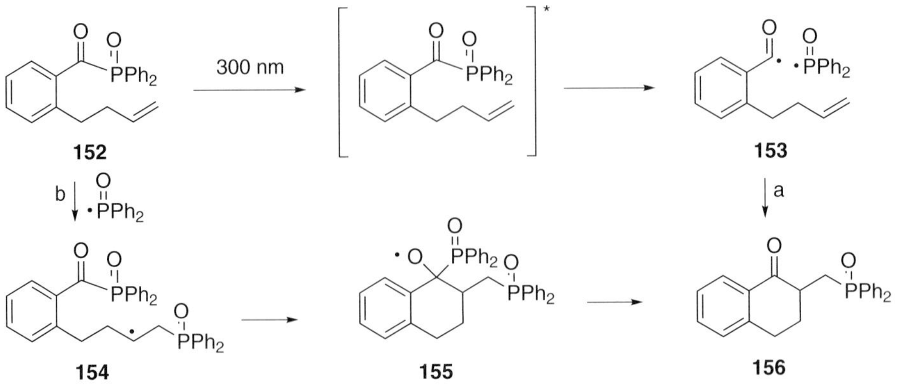

Scheme 23.34 Photochemical cyclizations of alkenyl acylphosphonates.

## 23.2 Different Methods for Radical Cyclizations

As part of a study on the synthesis of angularly fused cyclopentanoids, Mattay and coworkers have reported an intramolecular tandem fragmentation-radical anion cyclization by photochemical electron transfer of tricyclic α-cyclopropyl ketones (Scheme 23.35) [86]. Photolysis (254 nm) of a dry acetonitrile solution of ketone **157** with lithium perchlorate as an additive furnished the 6-endo-dig product **159** and the non-cyclized product **160**. The reaction did not provide any 5-exo-dig cyclization product. The reaction proceeds by ring opening of the cyclopropyl ring to provide the distonic-γ-keto radical anion **158**, which undergoes radical cyclization after protonation.

**Scheme 23.35** Photochemical electron transfer in radical cyclization.

Perfluoroalkyl iodides undergo cleavage of the C—I bond on photolysis with near-UV light. Ogawa and coworkers have evaluated atom-transfer cyclizations of dienes, diynes, and enynes using perfluoroalkyl iodides under photolytic conditions (Scheme 23.36) [87]. Reaction of diallyl ether **160** with $C_{10}F_{21}I$ under photolysis gave the cyclized product **163** as a mixture of cis/trans-isomers in modest yield. A similar sequence with 1,6-heptadiene **162** also furnished the corresponding cyclized product **164** in low yield. In both cases the non-cyclized product was also obtained.

161 = O
162 = CH$_2$

163 = O 58%; *cis*:*trans* = 75:25    165 9%
164 = CH$_2$ 36%; *cis*:*trans* = 85:15    166 40%

**Scheme 23.36** Photochemical tandem radical cyclizations using perfluoroalkyl iodides.

Diphenyl diselenide on photolysis with 300 nm light undergoes homolysis to produce phenylselenyl radical with high efficiency. Ogawa and coworkers have developed a novel four-component coupling reaction to form cyclopentanes (Scheme 23.37) [88]. Addition of the phenylselenyl radical formed by photolysis to **167** at the less substituted carbon produces a vinyl radical which then adds selectively to **168** producing a nucleophilic tertiary α-alkoxy radical. This radical adds to the electrophilic terminus of *tert*-butylacrylate **169** which then undergoes a 5-exo-trig cyclization to produce an α-selenyl radical which completes the chain by abstracting a phenylselenyl group from diphenyl diselenide to furnish **171** and **172** in 47% yield (91 : 9). Several by-products were also formed in minor amounts. The reactivity of each component is carefully chosen such that the reaction proceeds efficiently and attests to the type of complex reactions that can be carried out using radical intermediates.

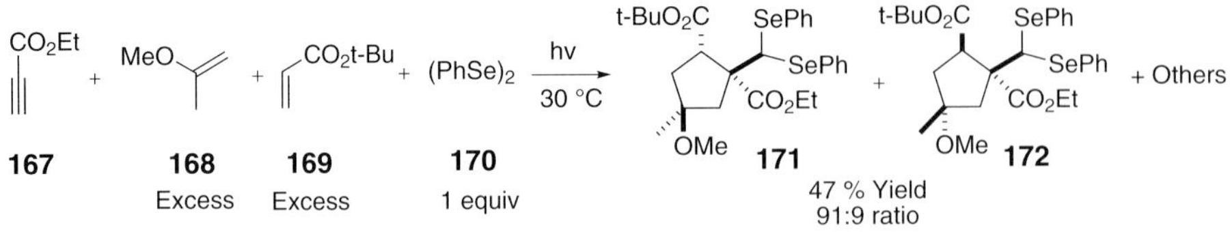

**Scheme 23.37** Multicomponent radical reactions.

## 23.3
## Annulation Reactions

Free radical chemistry is well suited to form multiple carbon–carbon bonds in a single operation. One such process is the annulation reaction. The annulation reaction in its simplistic form is defined as a ring-forming process in which two molecular fragments are united with the formation of two new bonds (Scheme 23.38). The reaction entails the addition of a radical to a neutral, leading to a product radical (**173** to **174**), which then cyclizes to form a ring (**174** to **175**). This section focuses on recent reports and readers should consult a review for additional information.

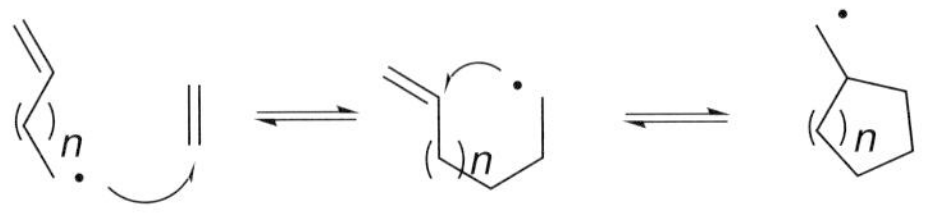

**Scheme 23.38** Basic outline for an annulation reaction.

Curran and coworkers have made seminal contributions to the development of radical annulation reactions. One area where annulation reactions have made significant impact is in the development of efficient methods for the synthesis of very promising cancer therapeutics, camptothecin and analogs [89]. The formation of a tetracyclic camptothecin analog via an annulation reaction is illustrated in Scheme 23.39. Treatment of **177** using hexabutylditin as a radical initiator as well as a chain carrier under photolytic conditions starts the annulation reaction with the partner isocyanide **176**. The reaction provides two isomeric products depending on the substitution pattern of the annulation partner.

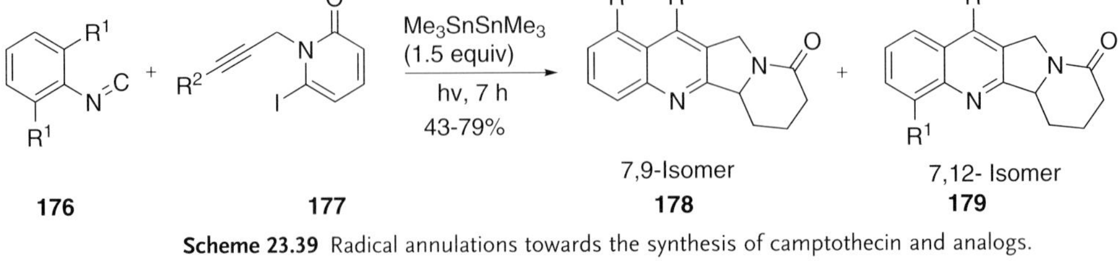

**Scheme 23.39** Radical annulations towards the synthesis of camptothecin and analogs.

Taguchi and coworkers [90] have investigated an atom-transfer [3 + 2] annulation sequence, which involves the ring opening of a cyclopropylmethyl radical (Scheme 23.40). Treatment of the disubstituted cyclopropane **180** with triethylborane initiates the reaction by forming the cyclopropylmethyl radical which undergoes rapid ring opening to form a stable homoallyl radical. The resultant tertiary electrophilic radical adds to **181** on the less substituted alkene carbon resulting in a product radical which then cyclizes to form the five-membered ring. The reaction cycle is completed by iodine atom transfer from the starting material to form the product in modest yield. The reaction is facilitated by the addition of a lanthanide Lewis acid. The product is formed with very high selectivity.

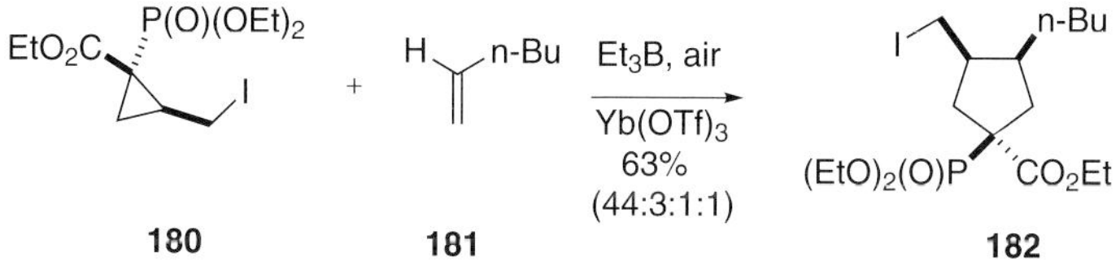

**Scheme 23.40** [3 + 2] Annulations.

Renaud and coworkers have developed a [3 + 2] annulation process leading to cyclopentanes in which radical termination is carried out using an azide (Scheme 23.41) [91]. The process involves treatment of **183** with triethyl borane which forms an electrophilic homoallyl tertiary radical. This radical then adds to 1-octene at the less substituted end and the resulting intermediate radical cyclizes to form another tertiary radical. This intermediate then reacts with phenylsulfonyl azide which then transfers an azido group to terminate the reaction and furnish the product **185**.

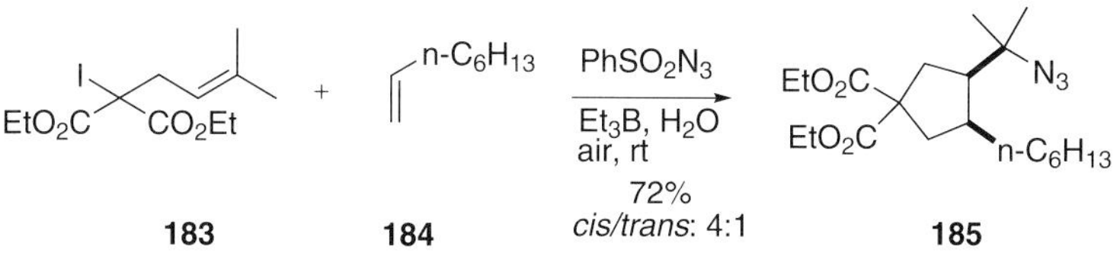

**Scheme 23.41** [3 + 2] Annulations with azide trapping.

An interesting [5 + 1] annulation process to form cyclic sulfones has been reported by Brasalau and coworkers (Scheme 23.42) [92]. The process involves the generation of a sulfur radical (thiophenyl) which then initiates the reaction by addition to the less substituted alkene of **186** yielding intermediate **188**. The intermediate primary radical reacts with sulfur dioxide to form **189**, which then cyclizes (6-endo trig) to yield the product sulfone **187** in high yield and regenerate the

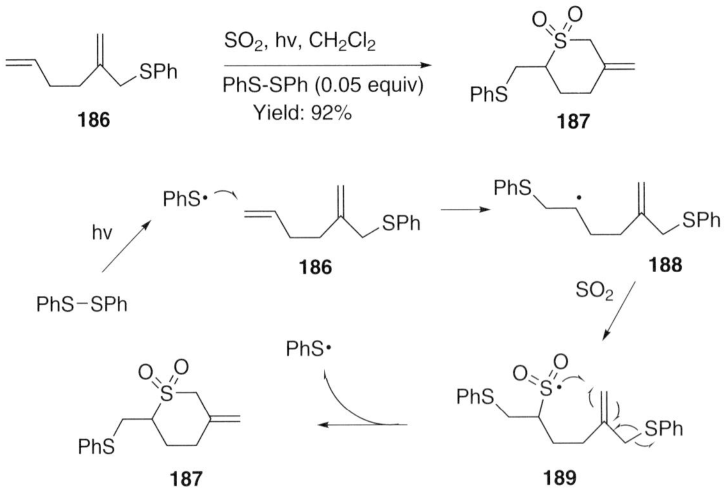

**Scheme 23.42** [5 + 1] Radical annulations using sulfur dioxide.

thiophenyl radical to continue the chain. The authors report formation of sulfones of different ring sizes ($n + 1$ annulations).

Oshima and coworkers have developed a novel [3 + 2] annulation reaction leading to the formation of substituted pyrrolidines (Scheme 23.43) [93]. The annulating reagent is a protected N-chloroamine **190** (Bs = benzene sulfonyl). The N-centered radical is formed by the initiator triethyl borane on reaction with **190**. The radical then adds to the styrene with a high degree of regiocontrol to form the intermediate **193**. Cyclization followed by chlorine atom transfer from the starting material provides the product **192**. It is important to note that four contiguous chiral centers are established in this annulation reaction. The annulation has broad substrate (different alkenes) and reagent scope (different 3 atom components, amines).

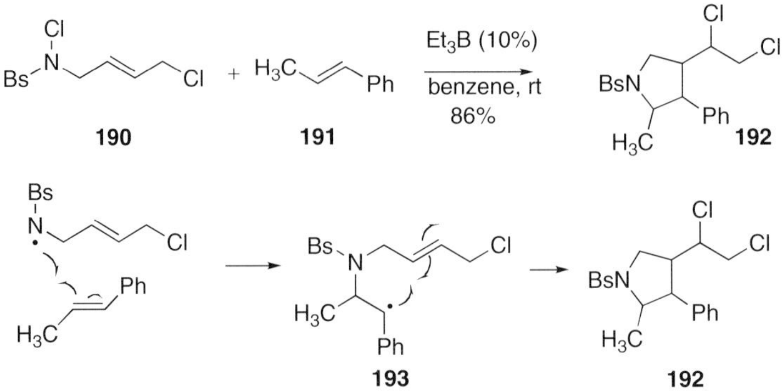

**Scheme 23.43** [3 + 2] Annulations leading to pyrrolidines.

## 23.4
### Diastereoselective Radical Cyclizations

The control of absolute stereochemistry in radical reactions has been extensively investigated using chiral auxiliaries [94]. A variety of both inter- and intramolecular reactions have been developed. The diastereoselectivity in the cyclization reactions reported in the past five years has been controlled by the native facial bias of the blocking group but at times also with the assistance of Lewis acids.

Pedrosa and coworkers have investigated the use of perhydro-1,3-benzoxazines derived from 8-aminomenthol as chiral auxiliaries in radical cyclizations (Scheme 23.44) [95]. Treatment of the amide **194** under tin hydride conditions generates a mixture of cyclic lactams in good yield and modest diastereoselectivity. The cyclization proceeds exclusively in the 5-exo mode. The major isomer **195** can be readily converted to enantioenriched 3-methylpyrrolidine **197**. The authors have demonstrated that the methodology can be readily extended to the synthesis of 3,4-disubstituted pyrrolidines.

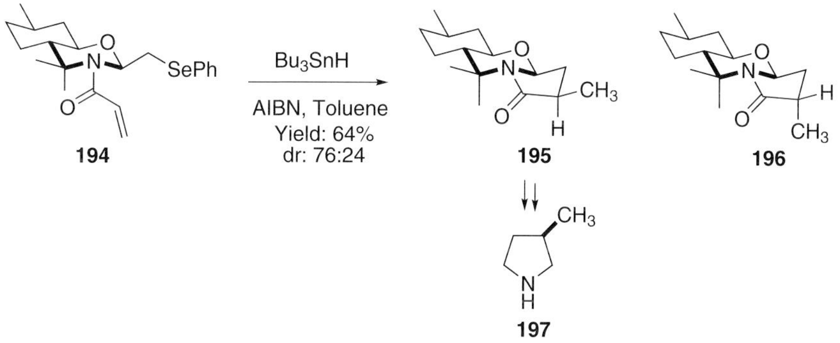

**Scheme 23.44** Perhydro-1,3-benzoxazine auxiliary in radical cyclization.

Sulfoxides are excellent chiral auxiliaries in radical cyclizations. Malacria and coworkers have investigated vinyl radical addition to alkylidine bissulfoxides (Scheme 23.45) [96]. Cyclization of **198** under tin hydride conditions gave the cyclized product **199** in good yield and diastereoselectivity. The impact of Lewis acid additives on diastereoselectivity has also been investigated.

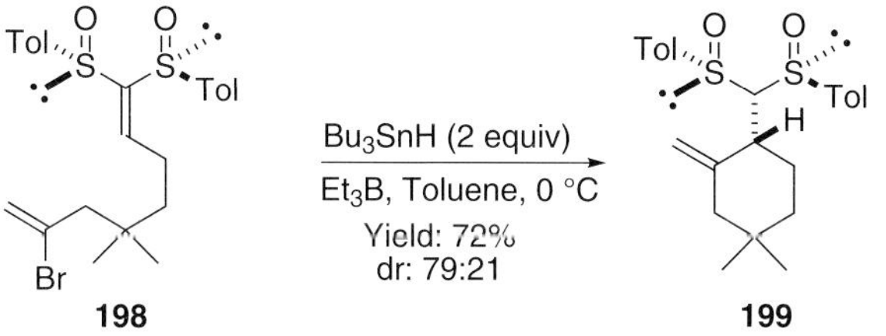

**Scheme 23.45** Radical cyclizations using chiral sulfoxides.

As noted above, chiral vinyl sulfoxides can control stereochemistry during radical cyclizations. Lee and coworkers have examined the impact of sulfur chirality and alkene geometry on diastereoselective radical cyclizations leading to substituted tetrahydrofurans (Scheme 23.46) [97]. Treatment of vinyl sulfoxide **200** with tin hydride gave the tetrahydrofuran **201** in high yield and with near perfect stereocontrol. Cyclization of the Z-isomer of **200** also gave the product **201** in high yield and selectivity. The absolute stereochemistry of the sulfoxides had a small impact with the matched case providing near perfect stereocontrol. These results show the preference for the formation of the cis-2,5-disubstituted product.

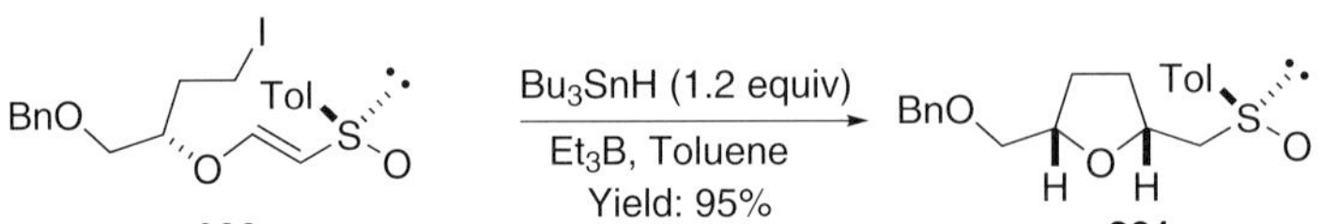

**Scheme 23.46** Use of chiral sulfoxides in the synthesis of tetrahydrofurans.

A highly diastereoselective atom-transfer radical cyclization using chiral oxazolidinone auxiliary has been reported by Yang and coworkers (Scheme 23.47) [98]. The process involves reaction of **202** with triethylborane in the presence of ytterbium triflate as a Lewis acid. The cyclized product **203** was obtained in high yield as a single isomer. Catalytic amounts of the Lewis acid additive were also effective in the cyclization with minimal impact on selectivity or efficiency. The Lewis acid additive chelates the carbonyls such that the benzyl group can provide optimal face shielding (**204**) and also enhances the electrophilicity of the α-carbonyl radical.

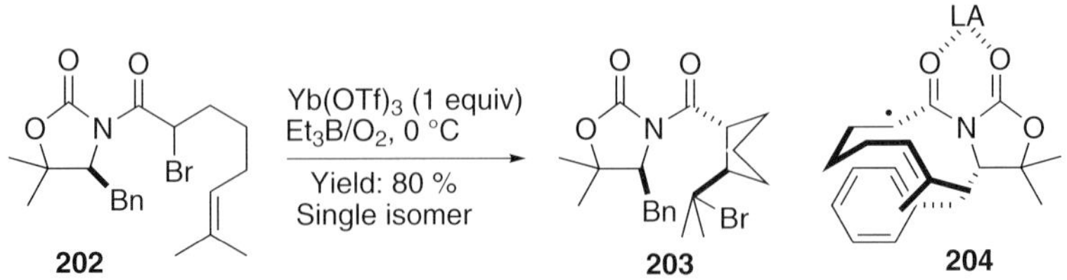

**Scheme 23.47** Chiral oxazolidinones in ATRC reactions.

Enholm and coworkers have reported a 5-hexenyl radical cyclization using readily available carbohydrate-derived chiral auxiliaries (Scheme 23.48) [99]. 5-exo cyclization of the isosorbide hexose-derived chiral ester **205** under reductive conditions and using magnesium bromide as a Lewis acid gave the cyclized product **206** in high yield and with near perfect diastereoselectivity. The use of an alternate carbohydrate-derived chiral auxiliary for the cyclization reaction as well as screening of various Lewis acids and reaction conditions have been carried out.

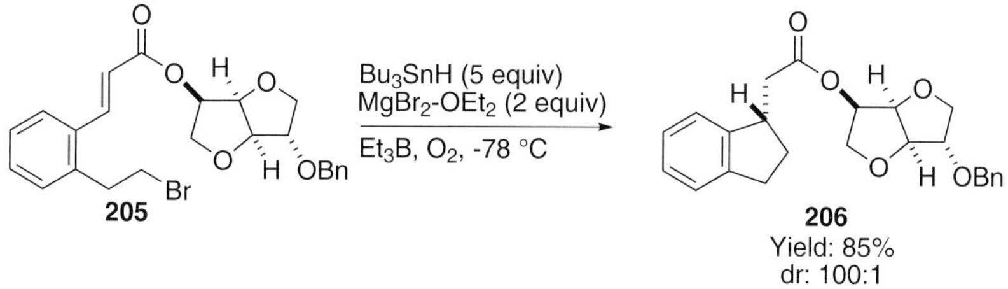

**Scheme 23.48** Carbohydrate-based auxiliaries in radical cyclization.

Diastereoselective tandem cyclization mediated by manganese triacetate under oxidative conditions has been investigated by Valdivia and coworkers (Scheme 23.49) [100]. The reaction provides access to the bicyclic skeleton of podolactones. Treatment of the chiral β-ketoester **207** with a mixture of manganese triacetate and copper acetate gave a mixture of bicyclic compounds **208** and **209** in modest yield and diastereoselectivity. The diastereoselectivity could be improved (6.5 : 1) by carrying out the reaction in methanol at 0 °C.

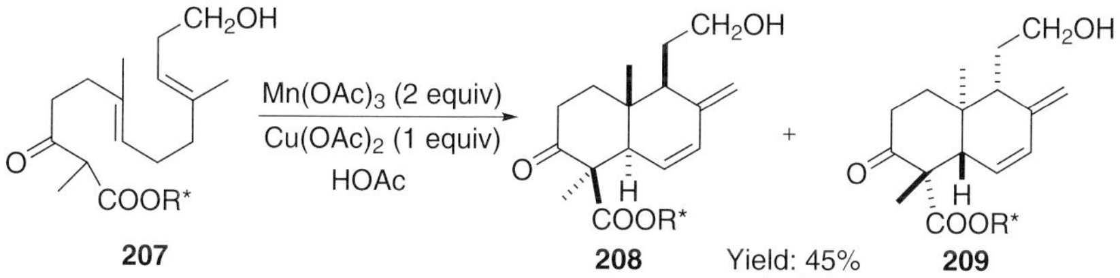

**Scheme 23.49** Mn(OAc)$_3$-mediated diastereoselective radical cyclizations.

## 23.5
## Enantioselective Radical Cyclizations

There are several reports on enantioselective radical cyclization reactions using chiral Lewis acids as well as organocatalysts. Different types of radical cyclizations have been investigated. Yang and coworkers have reported a highly enantioselective atom-transfer radical cyclization reaction mediated by chiral Lewis acids [101]. Advantages of these enantioselective ATRC include installing multiple chiral centers and retaining a halogen atom in the product, which allows for further functionalization. The atom-transfer radical cyclization of unsaturated β-keto ester **210** using Mg(ClO$_4$)$_2$ and chiral ligand **211** is shown in Scheme 23.50. Toluene as a solvent generally gave higher enantioselectivities than CH$_2$Cl$_2$ (see entries 1 and 2). Both 5-exo and 6-exo (not shown) cyclizations proceeded uneventfully. The addition of activated 4 Å molecular sieves led to enhancement in ees and allowed the use of substoichiometric amounts of the chiral Lewis acid (entry 3). The molecular sieves are thought to act as a drying agent: the addition of 1.0 equiv of water drastically reduces the selectivity and cyclization rate of **210** (see entries 2 and 4). Catalytic loading of the chiral Lewis acid showed nearly identical efficiency to the use of stoichiometric amounts of chiral Lewis acid with respect to both chemical yields and enantioselectivities (compare entries 2 and 3).

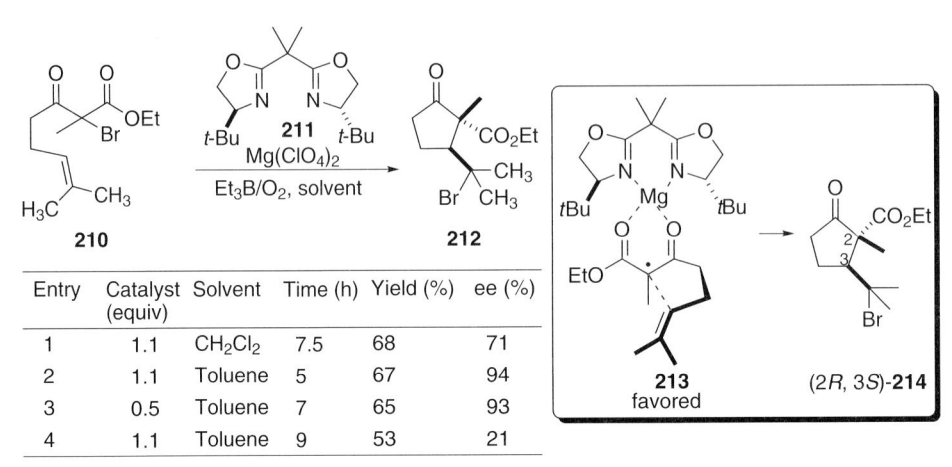

| Entry | Catalyst (equiv) | Solvent | Time (h) | Yield (%) | ee (%) |
|---|---|---|---|---|---|
| 1 | 1.1 | CH$_2$Cl$_2$ | 7.5 | 68 | 71 |
| 2 | 1.1 | Toluene | 5 | 67 | 94 |
| 3 | 0.5 | Toluene | 7 | 65 | 93 |
| 4 | 1.1 | Toluene | 9 | 53 | 21 |

**Scheme 23.50** Enantioselective atom-transfer cyclization.

The model shown in Scheme 23.50 has been used to explain the high selectivity in the cyclization. Due to the steric bulk of the *tert*-butyl groups of bisoxazoline ligand **211**, *re*-face cyclization is favored over *si*-face cyclization. Transition state **213** results in the lowest overall steric interaction and leads to product **214** with (2R, 3S) configuration where the ester group on C2 and the alkyl group on C3 are trans to one another.

Yang and coworkers have reported the first Lewis acid-catalyzed enantioselective atom-transfer tandem cyclization reactions (Scheme 23.51). Tandem radical

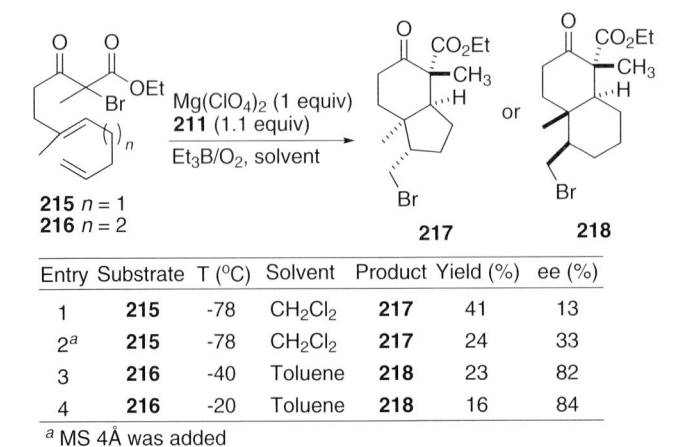

| Entry | Substrate | T (°C) | Solvent | Product | Yield (%) | ee (%) |
|---|---|---|---|---|---|---|
| 1 | 215 | -78 | CH$_2$Cl$_2$ | 217 | 41 | 13 |
| 2[a] | 215 | -78 | CH$_2$Cl$_2$ | 217 | 24 | 33 |
| 3 | 216 | -40 | Toluene | 218 | 23 | 82 |
| 4 | 216 | -20 | Toluene | 218 | 16 | 84 |

[a] MS 4Å was added

**Scheme 23.51** Tandem enantioselective radical cyclizations using atom-transfer additions.

cyclization reactions are noteworthy since they provide access to highly functionalized polycyclic compounds with multiple stereocenters. Enantioselective tandem cyclization of **215** using Mg(ClO$_4$)$_2$ and chiral ligand **211** in CH$_2$Cl$_2$ gave poor ees (entry 1). Molecular sieves slightly increased the ee but reduced the chemical yield (entry 2). Substrate **216** in toluene at higher temperatures gave good enantioselectivities but poor yields were still obtained (entries 3 and 4).

Chiral Lewis acids prepared from Yb(OTf)$_3$ and chiral ligands were investigated for the cyclization of substrate **215** (Scheme 23.52) [102]. A chiral Lewis acid complex, **220**/Yb in CH$_2$Cl$_2$, gave 60% yield of **217** with 66% ee (entry 1). Addition of 4 Å molecular sieves gave nearly complete reversal of enantiofacial selectivity in the tandem radical cyclization along with increase in the reduced product (compare entries 1 and 2). Toluene as solvent gave low yields of **217** and ees compared to those with methylene chloride (entry 3).

| Entry | Ligand | Solvent | t (h) | Yield (%) [**219**] | Yield (%) [**217**] | ee (%) |
|---|---|---|---|---|---|---|
| 1 | 220 | CH$_2$Cl$_2$ | 15 | 23 | 60 | 66 |
| 2[a] | 220 | CH$_2$Cl$_2$ | 13 | 68 | 11 | -56 |
| 3 | 221 | Toluene | 10 | 64 | 17 | -43 |

[a] MS 4Å was used

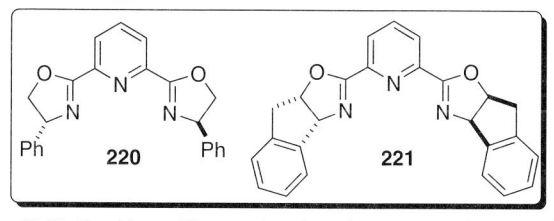

**Scheme 23.52** Ytterbium triflate-mediated tandem enantioselective cyclization.

A model for the stereochemical outcome for the tandem radical cyclization in the presence of the [Yb(Ph-pybox)(OTf)$_3$] (pybox = 2,6-bis(2-oxazolin-2-yl)pyridine) is shown in Scheme 23.53. The reactive complex **222** has an octahedral geometry (with one triflate still bound to the metal) where *re*-face cyclization is favored due to the steric interactions of the substrate and the ligand's phenyl groups. The 6-endo cyclization takes place via a chair-like transition state to yield a tertiary radical **223** followed by a ring flip and a 5-exo cyclization (**224** to **225** to **217**). The primary radical in **225** then abstracts a bromine atom from **215** to yield (2*R*, 3*S*, 4*S*, 5*S*)-**217**.

**Scheme 23.53** Stereochemical model for Yb(OTf)$_3$-mediated asymmetric radical cyclization.

Nishida and coworkers have carried out cyclizations using chiral aluminum Lewis acids derived from BINOLs and Me$_3$Al (Scheme 23.54) [103]. Initial formation of a vinyl radical is followed by a 5-exo or 6-exo (for $n = 1$ or 2) cyclization where the face selection is controlled by the chiral Lewis acid. Cyclization of ester **226** at $-78\,^\circ$C gave the cyclopentane in good yield and modest selectivity. Similarly, reaction with ester **226** (n=2) at 0 $^\circ$C led to cyclohexane formation. Lower yields of the six-membered ring products are due to difficulty in 6-exo cyclizations. A stoichiometric amount of **227** provided cyclized products in low ees. When 4 equiv of **227** were used, however, cyclized products (R)-**228** and (R)-**229** were obtained in 36% and 48% ee respectively. Replacement of the ester with the Weinreb amide led to smooth cyclization to furnish (S)-**230** in 26% ee. The low ee in this experiment is due to background reaction. The nature of the carboxylic substituent impacts selectivity in the cyclization. The esters upon complexation are oriented in an s-trans-fashion whereas the Weinreb amides adopt an s-cis-conformation.

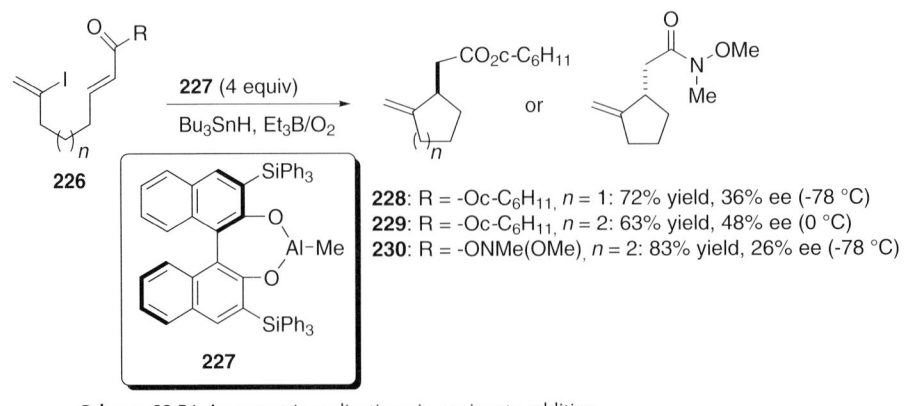

**228**: R = -Oc-C$_6$H$_{11}$, n = 1: 72% yield, 36% ee (-78 °C)
**229**: R = -Oc-C$_6$H$_{11}$, n = 2: 63% yield, 48% ee (0 °C)
**230**: R = -ONMe(OMe), n = 2: 83% yield, 26% ee (-78 °C)

**Scheme 23.54** Asymmetric cyclization via conjugate addition.

## 23.5 Enantioselective Radical Cyclizations

Cyclizations of α-bromo-N-allyl amides and sulfonamides with triethylborane as a radical initiator and further reduction of the cyclized radical with tin hydride has been reported by Hiroi and coworkers (Scheme 23.55) [104]. Various Lewis acids were explored and titanium tetraisopropoxide emerged to be superior to either triethylaluminum or magnesium triflate. Among the substrates, less bulky substituents on nitrogen resulted in better reaction efficiency, with larger substituents like 2,4,6-triisopropylphenyl- and 1-naphthalenesulfonyl leading to reduced products along with recovered starting materials. It was observed that substrate **231** (with *p*-tolyl sulfonyl substituent) was ideal in terms of both reaction efficiency and enantioselectivity. The products obtained possessed trans-geometry at the newly formed C−C bond. The highest selectivity was obtained with ligand **234** for the trans-product **232**.

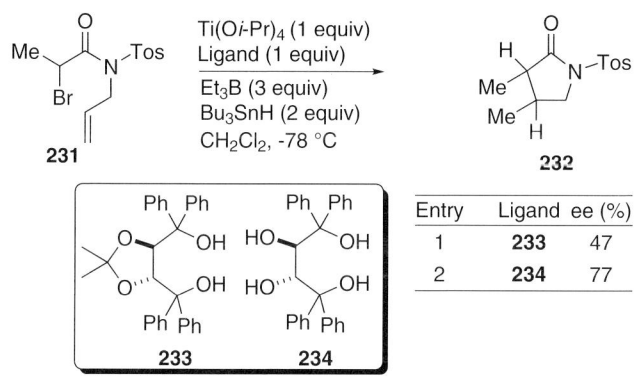

| Entry | Ligand | ee (%) |
|---|---|---|
| 1 | 233 | 47 |
| 2 | 234 | 77 |

**Scheme 23.55** Cyclization of α-bromo-N-allylamides and sulfonamides.

Curran and coworkers have discovered a unique cyclization procedure in which they showed that axial chirality could be transferred to a new stereocenter with retention of chirality (Scheme 23.56) [105]. The required substrates *M* or *P*-**235** were either prepared from the chiral pool or by racemic synthesis followed by preparative chiral HPLC separation. Treatment of *M* or *P*-**235** to conditions shown in Scheme 23.56 gave the products (*R*) or (*S*)-**236** in good yields and high ees. The high ees are due to the almost complete absence of racemization of radical intermediates **237** or **238**. This is in turn related to the efficiency of the aryl radical addition to the olefin. The intermediate **237** obtained from *M*-**235a** has to rotate around the aryl-nitrogen bond in order for the proper overlap required for cyclization to occur. If the cyclization is not efficient, there is a possibility of the bond rotation going further, leading to **238** and hence to racemization. These factors are borne out in the examples presented. It was observed that higher ees are obtained when $R_E$ is phenyl: the delocalization (and hence stabilization) provided by the carbonyl group becomes less important due to the delocalization provided by conjugation with the phenyl ring. This allows *M*-**235b** to react faster furnishing higher ees. Curran and coworkers have another report on asymmetric cyclization of transient atropisomers [106].

An interesting intramolecular radical cyclization followed by enantioselective hydrogen atom transfer has recently been reported (Scheme 23.57) [107]. This reaction

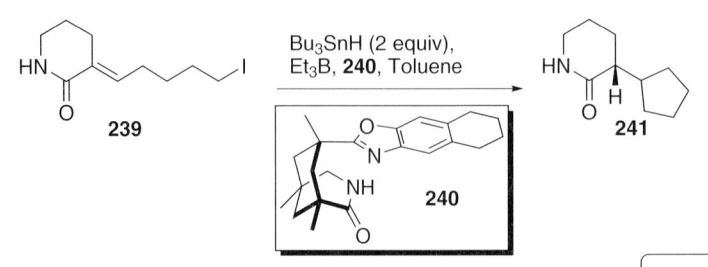

**Scheme 23.56** Memory of chirality in cyclization.

| Entry | Precursor | $R_1$ | $R_E$ | $R_Z$ | X | Yield (%) | ee (%) |
|---|---|---|---|---|---|---|---|
| 1 | M-235a | Me | Me | H | I | 73 | 89 (R) |
| 2 | P-235a | Me | Me | H | I | 70 | 85 (S) |
| 3 | M-235b | Me | Ph | H | I | 75 | 94 (R) |
| 4 | P-235b | Me | Ph | H | I | 73 | 92 (S) |

is carried out in the presence of a chiral complexing agent **240**, which can hydrogen bond to an appropriate acceptor. The authors studied the enantioselectivity of the reductive cyclization of 3-(ω-iodoalkylidene)piperidin-2-ones **239**. The observed selectivity was determined to be dependent on three factors. First, the temperature

| Entry | T (°C) | Equiv (**240**) | Et$_3$B (mol%) | Yield (%) | ee (%) |
|---|---|---|---|---|---|
| 1 | 25 | - | 50 | 83 | - |
| 2 | -10 | 2.5 | 10 | 79 | 55 |
| 3 | -78 | 1.0 | 20 | 91 | 40 |
| 4 | -78 | 2.5 | 50 | 84 | 41 |
| 5 | -78 | 2.5 | 20 | 81 | 84 |

**Scheme 23.57** Hydrogen bonding as a chiral controller in enantioselective H-atom transfer.

needed to be very low ($-78\,^\circ$C). Secondly, lower amounts of radical initiator (Et$_3$B) gave improved enantioselectivities (entries 4 and 5). Another interesting reaction parameter was the need for a large excess of the chiral source **240** (entries 3 and 5). A proposed model for the observed stereochemistry is shown in Scheme 23.57. The *re*-face is shielded by the tetrahydronaphthalene moiety and the enantioselective hydrogen atom transfer proceeds from the more accessible *si*-face of the prochiral radical **242**.

## 23.6
## Applications of Radical Cyclizations in Natural Product Synthesis

Due to their limited availability from natural sources, natural products have been a target of studies ranging from their total synthesis to the creation of simpler, structurally similar versions of the natural product. In recent years, total synthesis efforts have incorporated the use of various types of radical cyclization reactions in order to generate more effective syntheses in fewer steps. These radical cyclizations range from the creation of single rings by traditional methods to more complicated radical cyclizations including group-transfer radical cyclizations and radical cyclization cascades, also known as tandem radical cyclizations, resulting in multiple ring formations from a single step. Examples of each of these types of radical cyclization reactions are given below.

### 23.6.1
### Single Ring Formation – Amphidinolide E

Amphidinolide E is an 18-membered macrolide which displays cytotoxic effects against two types of murine leukemia cells (L1210 and L5178Y) *in vitro* and is derived naturally from a species of dinoflagellate. While syntheses have been attempted in the past, resulting in partial synthesis of the compound, a full synthesis had never been successfully completed until recently [108]. Kim and coworkers successfully synthesized Amphidinolide E by employing a radical cyclization of a β-alkoxy acrylate as shown in Scheme 23.58 [109]. The radical cyclization step resulted in the formation of the single oxolane ring that is found in Amphidinolide E. Following the radical cyclization step, Amphidinolide E was produced after an additional 14 steps.

### 23.6.2
### Group-Transfer Radical Cyclizations – Aphanorphine

Aphanorphine is an important natural product found naturally in a species of freshwater blue algae. It has been successfully synthesized on numerous accounts due to its structural similarity to certain analgesics including pentazocine, eptazocine, and morphine. Recently, a new synthesis of the natural product has been successfully completed through a group-transfer radical cyclization approach for the purpose of creating the 6-azabicyclo[3.2.1]octane ring system found in the natural product **246** as shown in Scheme 23.59 [110]. This group-transfer radical cyclization was preceded by

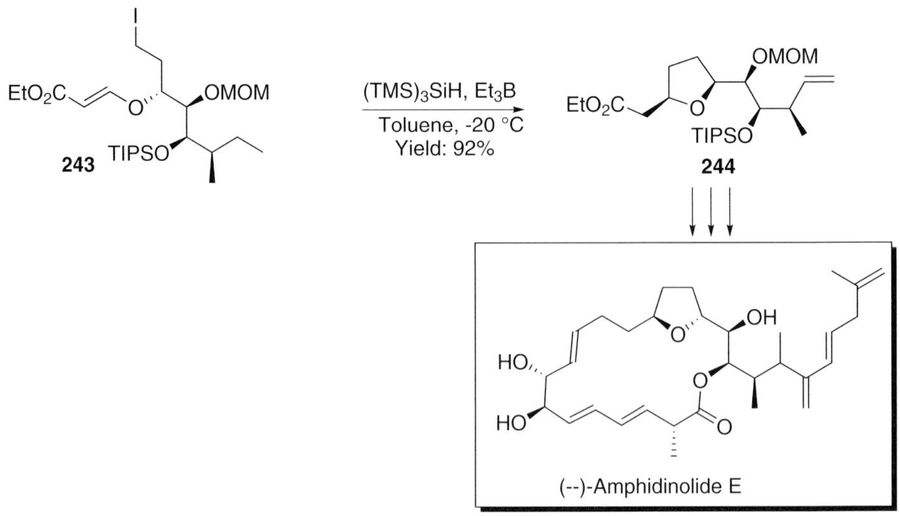

**Scheme 23.58** Single-ring formation by radical cyclization: synthesis of amphidinolide E.

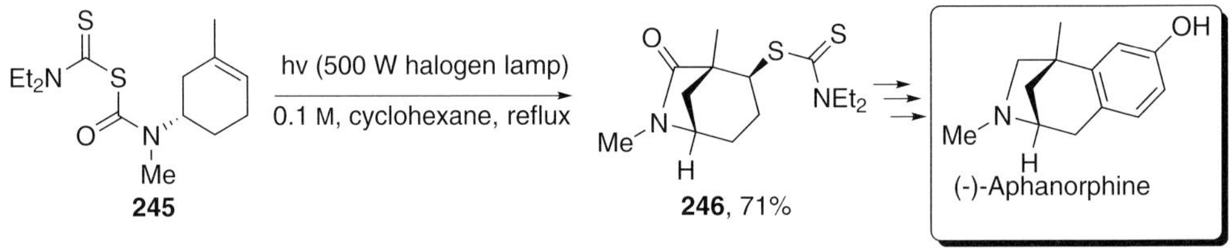

**Scheme 23.59** Group-transfer radical cyclization: synthesis of aphanorphine.

an additional 13 steps to yield a known intermediate in the aphanorphine synthesis, ultimately resulting in a successful synthesis of the natural product.

### 23.6.3
### Tandem Radical Cyclizations – Luotonin A, Azadirachtin, Aspidospermidine, and Fortucine

Luotonin A is a human DNA topoisomerase I poison that can be isolated naturally from a Chinese medicinal plant and is cytotoxic to the P388 cell line of murine leukemia by stabilizing the topoisomerase I/DNA complex. A new approach to the synthesis of Luotonin A involves a radical cyclization cascade, which is able to generate in one step both a six-membered ring and a five-membered ring [111]. Following this tandem radical cyclization step, Luotonin A was directly synthesized following a 6 h irradiation under reflux involving atom transfer, as shown in Scheme 23.60, resulting in a 43% yield.

*23.6 Applications of Radical Cyclizations in Natural Product Synthesis* | **1137**

**Scheme 23.60** Tandem radical cyclization: synthesis of luotonin A.

Azadirachtin is a highly complex natural product, isolated from the Neem tree, that exhibits potent antifeedant activity and growth inhibitory abilities against insects. Watanabe *et al.* have recently reported the synthesis of a simpler version (**250**) which has the full functionality of the natural product through the use of a tandem radical cyclization step, as shown in Scheme 23.61 [112]. This tandem radical cyclization step was the final step in generating the fully functional, structurally similar version of Azadirachtin in an impressive 90% yield as a single isomer.

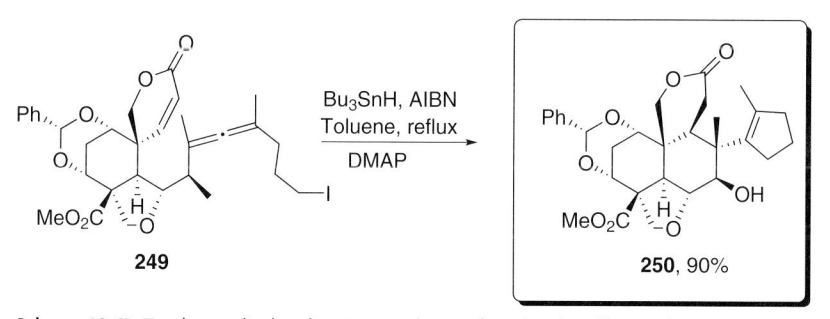

**Scheme 23.61** Tandem radical cyclization: synthesis of azadirachtin-like product.

The Aspidosperma family of alkaloids has attracted attention from synthetic chemists because of their complex structure. Zard and coworkers have developed an elegant route for the synthesis of aspidospermidine in which tandem radical cyclization is employed to prepare a key fragment (Scheme 23.62) [113]. Treatment of **251** with tributyltin hydride and 1,1′-azobis(cyclohexanecarbonitrile) (ACN) in refluxing trifluorotoluene gave the tricyclic product **252** in 53% yield, along with minor amounts of a bicyclic product. The reaction proceeds by the initial formation of an amidyl radical, which undergoes a 5-exo-trig cyclization. The resulting radical cyclizes in a 6-endo-trig mode to furnish a chlorine-stabilized radical. This radical is reduced by tin hydride and further reduction of the C—Cl bond furnishes **252**. This key intermediate was converted to aspidospermidine in four steps in good overall yields.

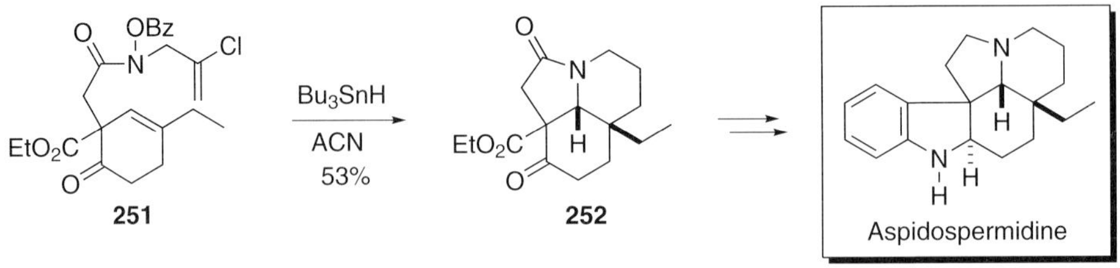

**Scheme 23.62** Tandem radical cyclizations: synthesis of aspidospermidine.

Zard and coworkers have also reported a synthesis of fortucine, a lycorine alkaloid, using a novel tandem cyclization strategy (Scheme 23.63) [114]. The key cyclization involves the treatment of **253** with dilauroyl peroxide (1.4 equiv) in refluxing 1,2-dichloroethane. The cascade begins with the formation of an N-centered amidyl radical which cyclizes on the cyclohexadiene leading to a secondary radical. The intermediate radical cyclizes onto the aromatic ring and further oxidation forms the product **254**. Three contiguous chiral centers are established during the cascade process. Compound **254** was transformed to fortucine in sex steps in modest overall yield.

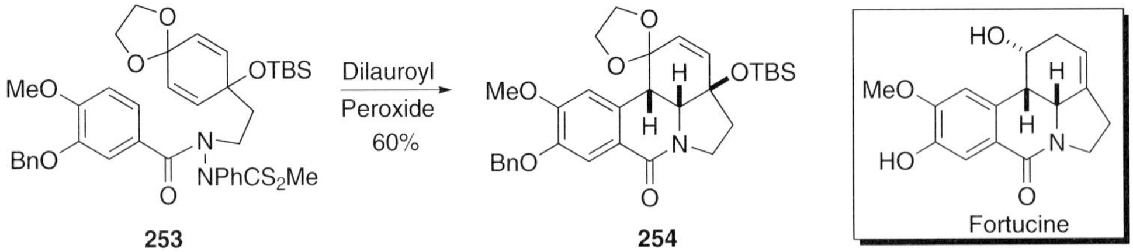

**Scheme 23.63** Tandem radical cyclizations: synthesis of fortucine.

### 23.6.4
### Manganese Triacetate-Mediated Synthesis – Mersicarpine

Kerr and coworkers have examined manganese triacetate-mediated oxidative cyclization as a strategy for the synthesis of several indole alkaloids. The synthesis of mersicarpine is shown in Scheme 23.64 [115]. Treatment of **255** with Mn(OAc)$_3$ in acetic acid provides the cyclized product **256** in 60% yield. The use of the 1,3-dicarbonyl compound was strategically important since the corresponding β-keto ester or the malonate led to difficulties is post-cyclization manipulations. Compound **256** was converted to **257** using straightforward chemistry. Further manipulations gave the target compound mersicarpine in 10 steps from **256**.

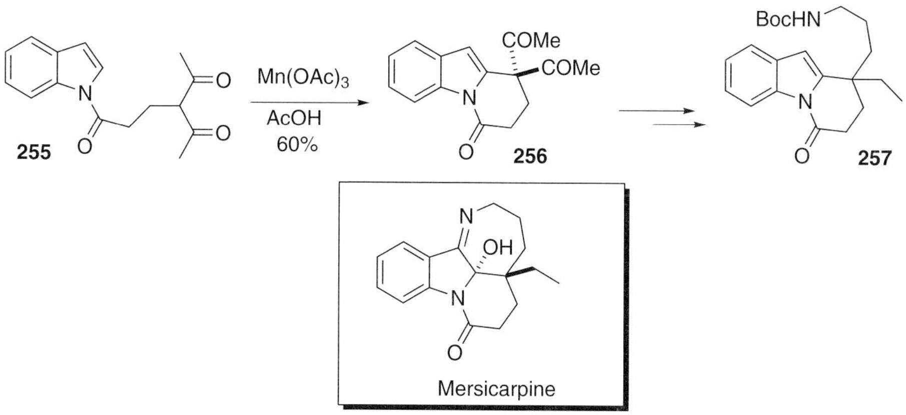

Scheme 23.64 Manganese triacetate-mediated cyclization: synthesis of mersicarpine.

## 23.6.5
### Titanium(III)-Mediated Cyclization – Smenospondiol

As illustrated in Section 23.2.5, Ti(III) mediated cyclizations are very effective for the synthesis of complex systems. Takahashi and coworkers have utilized Ti(III) cyclization as a key step in the synthesis of smenospondiol (Scheme 23.65) [116]. The reaction of epoxide **258** using Ti(III) reagent gave a mixture of the bicyclic compound **259** and monocyclic **260** in high yield and 10:1 selectivity. As discussed before, **258** undergoes epoxide ring opening followed by a 6-exo-trig and a 6-exo-dig cyclization to furnish **259**. The intermediate **259** was converted to the target smenospondiol in 12 steps.

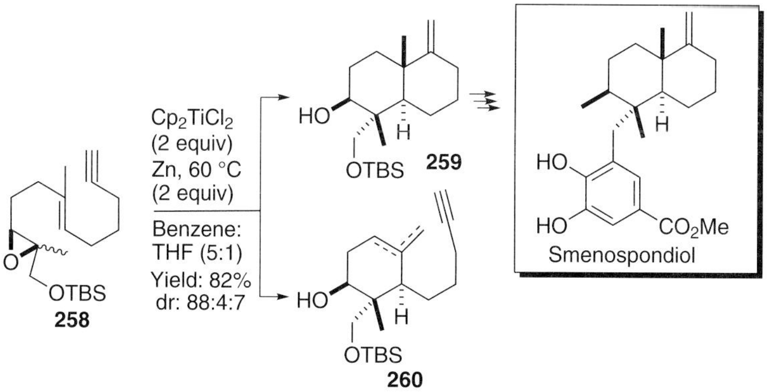

Scheme 23.65 Titanium(III)-mediated cyclizations: synthesis of smenospondiol.

## 23.6.6
### Samarium Iodide in Total Synthesis – Paeonilactone B and Platensimycin

Samarium(II) iodide-mediated radical cyclizations have played an important role in devising methods for synthesizing complex natural products and two examples are discussed here. Kilburn and coworkers have developed an elegant synthesis of paeonilactone B which features a samarium(II)-mediated radical cyclization (Scheme 23.66) [117]. The sequence involves formation of the ketyl radical from **261** that adds to the methylenecyclopropane (5-exo-trig cyclization) furnishing a [5.1.0] bicyclic intermediate (cyclopropylmethyl radical) which undergoes a rapid "endo" ring opening generating a methylenecyclohexyl radical intermediate. The cyclohexyl radical undergoes a further 5-exo-dig cyclization generating a vinyl radical which is reduced by the excess $SmI_2$ and protonated to yield the products **262** and **263** in good yield and 1:10 selectivity. The major isomer was converted to the target paeonilcatone B in 7 steps.

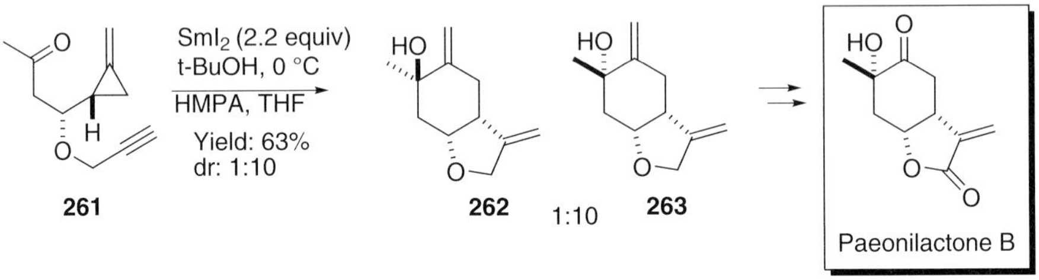

**Scheme 23.66** Samarium iodide-mediated cyclizations: synthesis of paeonilactone B.

Platensimycin has attracted a lot of attention because of its structure as well as its impressive antibacterial activity. Nicolaou and coworkers have developed several syntheses of platensimycin in racemic as well as enantiomerically pure forms (Scheme 23.67) [118]. The key reaction in the synthesis is a $SmI_2$-mediated cyclization of **264** to furnish **265** as a single diastereomer in moderate yield. The product **265** has the tricyclic core of platensimycin and was converted to the target in several steps.

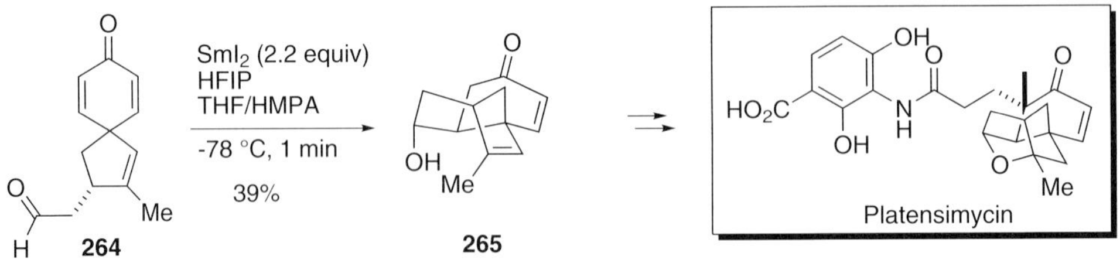

**Scheme 23.67** Samarium iodide-mediated cyclizations: synthesis of platensimycin.

## 23.7
## Experimental: Selected Procedures

### 23.7.1
### Transformation of 55 to 56, Reductive Radical Cyclization Using Tributyltin Hydride [18]

To a solution of iodoacetal **55** (213 mg, 0.49 mmol) in benzene (0.02 M, deoxygenated via nitrogen bubbling through a syringe needle for about 10 min) was added 2,2′-

azobisisobutyronitrile (AIBN, 10 mol%) and Bu$_3$SnH (0.16 mL, 0.59 mmol). The reaction mixture was stirred at reflux until all starting material had been consumed (TLC monitoring). If necessary, additional AIBN (10 mol%) was added. The solvent was replaced with ethyl acetate (0.02 M) and the mixture was stirred with KF (5 equiv) overnight. The mixture was adsorbed on silica gel, placed on a pad of clean silica gel, and eluted with ethyl acetate. Concentration and radial chromatography (5 : 1 hexane/ethyl acetate) afforded **56** (101 mg, 70%, dr 2.5 : 1).

### 23.7.2
**Transformation of 121 to 122, Manganese Triacetate-Mediated Cyclization [73]**

To a mixture of Mn(OAc)$_3$·2H$_2$O (202 mg, 0.738 mmol) and Cu(OTf)$_2$ (44.5 mg, 0.123 mmol) in boiling trifluoroethanol (8 mL) was added dropwise a solution of **121** (50.0 mg, 0.123 mmol) in trifluoroethanol (3 mL), and the mixture was heated under reflux for 9 h. Ether (5 mL) was added to the reaction mixture and the precipitated salts were removed by filtration with Celite. A saturated aqueous solution of hydroxylamine hydrochloride was added to the filtrate until the brown color of the solution faded. A 10% w/w aqueous solution of ethylenediaminetetraacetic acid disodium salt (5 mL) was added to the mixture to remove Cu(II). The organic phase was separated and the aqueous phase further extracted with AcOEt. The combined organic phase was washed successively with a saturated aqueous NaHCO$_3$ solution and brine, dried (MgSO$_4$), and concentrated. The residue was chromatographed on silica gel to give 27 mg of **122** (54% yield).

### 23.7.3
**Transformation of 186 to 187, Radical Annulation [92]**

Compound **186** (80 mg, 0.39 mmol) and diphenyl disulfide (4.0 mg, 0.02 mmol) were dissolved in 6 mL of dichloromethane and placed into a 15 mL heavy-walled glass pressure vessel. The solution was cooled to −78 °C, and 6 mL of SO$_2$ gas was condensed into the reaction mixture using a dry ice condenser. The mixture became yellow in color. The dry ice condenser was removed, and the pressure vessel was sealed and irradiated with a 450 W Hg lamp for 17 h. During the photolysis, the solution turned brown in color. The mixture was cooled to −78 °C, and the pressure vessel was slowly opened to the air and left standing in the hood at room temperature until all the SO$_2$ gas had evaporated. Volatiles were removed *in vacuo* to give 101 mg (0.36 mmol, 92% yield) of **187** as a brown oil.

### 23.7.4
**Transformation of 202 to 203, Atom-Transfer Radical Cyclization [98]**

To a solution of **202** (0.25 mmol) in dry ether (10 mL) at room temperature was added ytterbium triflate (1 equiv) in one portion under an argon atmosphere. After the mixture was stirred for 0.5 h at room temperature, it was cooled to 0 °C. After 15 min, Et$_3$B (1 M in n-hexane, 0.50 mL, 0.50 mmol) was added followed by oxygen gas

(2.5 mL) via a syringe. The reaction was monitored by TLC. On completion, the mixture was diluted with diethyl ether, and washed with water. The organic layer was dried over MgSO$_4$, filtered and concentrated. The crude residue was purified by flash column chromatography (35% EtOAc in n-hexane) to afford **203**.

### 23.7.5
**Transformation of 258 to 259 and 260, Titanium(III)-Mediated Radical Cyclization [116]**

To a stirred suspension of Cp$_2$TiCl$_2$ (73.9 mg, 0.297 mol) in dry THF (0.5 mL) and dry benzene (1.2 mL) was added activated zinc (29.2 mg, 0.446 mmol) at room temperature under argon. After being stirred for 1 h, the reaction mixture was added to a solution of epoxide **258** (50 mg, 0.149 mmol) in dry benzene (1.2 mL) at 60 °C. After being stirred for 10 min at the same temperature, the reaction mixture was diluted with ether and poured into 1 N HCl (30 mL) at 0 °C. The aqueous layer was extracted with ether (50 mL × 3). The combined organic layer was washed with brine and dried over MgSO$_4$. After removal of the solvent *in vacuo*, the residue was purified by column chromatography on silica gel to afford mixture of products. This mixture on further purification by HPLC gave pure **259**.

### 23.7.6
**Transformation of 261 to 262 and 263, Samarium Diiodide-Mediated Radical Cyclization [117]**

HMPA (30 mmol) was added to a freshly prepared solution of SmI$_2$ (3.1 mmol, 0.15 M solution) in THF to give a purple solution, which was cooled to 0 °C. Compound **261** (1.4 mmol) and *tert*-butanol (0.21 g, 2.8 mmol) in THF (30 mL) were added over 90 min and the reaction mixture were allowed to warm to room temperature. The crude mixture was washed with aqueous citric acid (2.5 g in 50 mL water) and extracted with 1 : 1 EtOAc–petrol (4 × 50 mL). The combined organic phase was washed with brine (50 mL) then water (50 mL), dried over MgSO$_4$ and concentrated *in vacuo*. The crude reaction mixture was purified by flash column chromatography, eluting with petrol, and gradually increasing the polarity to 50% Et$_2$O-petrol to give the desired bicyclic alcohols **262** and **263**.

## 23.8
### Conclusions

Radical cyclizations have played and continue to play an important role in the synthesis of molecules with complex architecture. As discussed in this chapter, there are many different ways to carry out radical cyclizations. Radical cyclizations have many attractive features such as the ability to form quaternary centers, high diastereo- and enantioselective C–C and C–X bond formations, and tandem processes. The future looks very promising for the development of new, efficient, eco-friendly, and atom economic radical cyclizations.

## Acknowledgments

Jake Zimmerman would like to thank Ohio Northern University for financial support. Mukund Sibi thanks the National Institutes of Health (GM-54656) for financial support.

## References

1. For a selected number of reviews see: (a) Curran, D.P. (1988) *Synthesis*, 417–439; (b) Curran, D.P. (1988) *Synthesis*, 489–513; (c) Jasperse, C.P., Curran, D.P. and Fevig, T.L. (1991) *Chem. Rev.*, **91**, 1237–1286; (d) Renaud, P. and Sibi, M.P. (2001) *Radicals in Organic Synthesis*, Wiley-VCH, Weinheim.

2. (a) Giese, B., Kopping, B., Gobel, T., Dickhaut, J., Thoma, G., Kulicke, K.J. and Trach, F. (1996) *Org. React.*, **48**, 301–856; (b) Bowman, W.R., Fletcher, A.J. and Potts, G.B.S. (2002) *J. Chem. Soc., Perkin Trans.*, **1**, 2747–2762. (c) Fallis, A.G. and Brinza, I.M. (1997) *Tetrahedron*, **53**, 17543–17594; (d) Yet, L. (1999) *Tetrahedron*, **55**, 9349–9403; (e) Bowman, W.R., Cloonan, M.O. and Krintel, S.L. (2001) *J. Chem. Soc., Perkin Trans.*, **1**, 2885–2902; (f) Zhang, W. (2001) *Tetrahedron*, **57**, 7237–7262; (g) Ishibashi, H., Sato, T. and Ikeda, M. (2002) *Synthesis*, 695–713; (h) Rheault, T.R. and Sibi, M.P. (2003) *Synthesis*, 803–819; (i) Lee, E. (2001) *Radicals in Organic Synthesis*, vol. 2 (eds P. Renaud and M.P. Sibi), Wiley-VCH, Weinheim, pp. 303–349; (j) Salom-Roig, X.J., Denes, F. and Renaud, P. (2004) *Synthesis*, 1903–1928.

3. Hanessian, S., Dhanoa, D.S. and Beaulieu, P.L. (1987) *Can. J. Chem.*, **65**, 1859–1866.

4. (a) Harrison, T., Myers, P.L. and Pattenden, G. (1989) *Tetrahedron*, **45**, 5247–5262; (b) Nakamura, E., Inubushi, T., Aoki, S. and Machii, D. (1991) *J. Am. Chem. Soc.*, **113**, 8980–8982; (c) Nozaki, K., Oshima, K. and Utimoto, K. (1988) *Tetrahedron Lett.*, **29**, 1041–1044; (d) Molander, G.A. and Harris, C.R. (1997) *J. Org. Chem.*, **62**, 7418–7429.

5. Yorimitsu, H., Nakamura, T., Shinokubo, H., Oshima, K., Omoto, K. and Fujimoto, H. (2000) *J. Am. Chem. Soc.*, **122**, 11041–11047.

6. Villar, F., Equey, O., Kolly-Kovac, T. and Renaud, P. (2003) *Chem. Eur. J.*, **9**, 1566–1577.

7. Lu, H. and Li, C. (2005) *Tetrahedron Lett.*, **46**, 5983–5985.

8. (a) McCarroll, A.J. and Walton, J.C. (2001) *Angew. Chem. Int. Ed.*, **40**, 2224–2248; (b) Dhimane, A.L., Fensterbank, L. and Malacria, M. (2001) *Radicals in Organic Synthesis*, vol. 2 (eds P. Renaud and M.P. Sibi) Wiley-VCH, Weinheim, pp. 350–382.

9. Patro, B. and Murphy, J.A. (2000) *Org. Lett.*, **2**, 3599–3601.

10. Toyota, M., Ilangova, A., Kashiwagi, Y. and Ihara, M. (2004) *Org. Lett.*, **6**, 3629–3632.

11. Ihara, M., Katsumata, A., Setsu, F., Tokunaga, Y. and Fukumoto, K. (1996) *J. Org. Chem.*, **61**, 677–684.

12. Lee, H.-Y., Moon, D.K. and Bahn, J.S. (2005) *Tetrahedron Lett.*, **46**, 1455–1458.

13. (a) Gomez, A.M., Company, M.D., Uriel, C., Valverde, S. and Cristobal Lopez, J. (2002) *Tetrahedron Lett.*, **43**, 4997–5000; (b) Stork, G. and Mook, R. Jr, (1986) *Tetrahedron Lett.*, **27**, 4529–4532.

14. (a) Barrero, A.F., Quílez de Moral, J.F., Sánchez, E.M. and Arteaga, J.F. (2006) *Eur. J. Org. Chem.*, 1627–1641; (b) Cuerva, J.M., Justicia, J., Oller-López, J.L., Bazdi, B. and Oltra, J.E. (2006) *Mini-Rev. Org. Chem.*, **3**, 23–35.

15 Fernández-Mateos, A., Teijón, P.H., Clemente, R.R., González, R.R. and González, F.S. (2007) *Synlett*, 2718–2722.
16 Stevens, C.B., Van Meenen, E., Masschelein, K.G.R., Eeckhout, Y., Hooghe, W., Dhondt, B., Nemykin, V.N. and Zhdankin, V.V. (2007) *Tetrahedron Lett.*, **48**, 7108–7111.
17 Kuivila, H.G. (1968) *Acc. Chem. Res.*, **1**, 299–305.
18 Friestad, G.K. and Fioroni, G.M. (2005) *Org. Lett.*, **7**, 2393–2396.
19 Jahn, U. and Rudakov, D. (2006) *Org. Lett.*, **8**, 4481–4484.
20 Clive, D.L.J., Pham, M.P. and Subedi, R. (2007) *J. Am. Chem. Soc.*, **129**, 2713–2717.
21 Clive, D.L.J. and Subedi, R. (2000) *Chem. Commun.*, 237–238.
22 (a) Chatgilialoglu, C., Ingold, K.U. and Scaiano, J.C. (1981) *J. Am. Chem. Soc.*, **103**, 7739–7742; (b) Crich, D. and Hwang, J.-T. (1998) *J. Org. Chem.*, **63**, 2765–2770; (c) Newcomb, M., Choi, S.-Y. and Horner, J.H. (1999) *J. Org. Chem.*, **64**, 1225–1231.
23 Taniguchi, T., Yonei, D., Sasaki, M., Tamura, O. and Ishibashi, H. (2008) *Tetrahedron*, **64**, 2634–2641.
24 Clark, K.B. and Griller, D. (1991) *Organometallics*, **10**, 746–750.
25 Giese, B. (1985) *Angew. Chem., Int. Ed. Engl.*, **24**, 553–565.
26 Gilman, H., Atwell, W.H., Sen, P.K. and Smith, C.L. (1965) *J. Organomet. Chem.*, **4**, 163–167.
27 Chatgilialoglu, C. (1992) *Acc. Chem. Res.*, **25**, 188–194.
28 Kanabus-Kaminska, J., Hawari, M., Griller, J.A. and Chatgilialoglu, D.C. (1987) *J. Am. Chem. Soc.*, **109**, 5267–5268.
29 Schummer, D. and Höfle, G. (1990) *Synlett*, 705–706.
30 (a) Baguley, P.A. and Walton, J.C. (1998) *Angew. Chem., Int. Ed.*, **37**, 3072–3082; (b) Gilbert, B.C. and Parsons, A.F. (2002) *J. Chem. Soc., Perkin Trans.*, **2**, 367–387; (c) Chatgilialoglu, C. (2001) *Radicals in Organic Synthesis*, vol. **1** (eds P. Renaud and M.P. Sibi), Wiley-VCH, Weinheim, pp. 28–49, ch. 1.3.
31 Gandon, L.A., Russell, A.G., Güveli, T., Brodwolf, A.E., Kariuki, B.M., Spencer, N. and Snaith, J.S. (2006) *J. Org. Chem.*, **71**, 5198–5207.
32 Pike, P., Hershberger, S. and Hershberger, J. (1985) *Tetrahedron Lett.*, **26**, 6289–6290.
33 (a) Curran, D.P., Diedericksen, U. and Palovich, M. (1997) *J. Am. Chem. Soc.*, **119**, 4797–4804; (b) Beckwith, A.L.J. and Roberts, D.H. (1986) *J. Am. Chem. Soc.*, **108**, 5893–5901; (c) Tsunoi, S., Ryu, I., Yamasaki, S., Fukushima, H., Tanaka, M., Komatsu, M. and Sonoda, N. (1996) *J. Am. Chem. Soc.*, **118**, 10670–10671; (d) Nagahara, K., Ryu, I., Komatsu, M. and Sonoda, N. (1997) *J. Am. Chem. Soc.*, **119**, 5465–5466; (e) Ryu, I. (2001) *Chem. Soc. Rev.*, **30**, 16–25.
34 Craig, P.J. and Van Elteren, J.T. (1995) *The Chemistry of Organic Germanium, Tin and Lead Compounds* (ed. S. Patai), chs. 16–17. John Wiley and Sons.
35 Bowman, W.R., Krintel, S.L. and Schilling, M.B. (2004) *Org. Biomol. Chem.*, **2**, 585–592.
36 Inoue, K., Sawada, A., Shibata, I. and Baba, A. (2001) *Tetrahedron Lett.*, **42**, 4661–4663.
37 Sawada, A., Shibata, I. and Baba, A. (2002) *J. Am. Chem. Soc.*, **124**, 906–907.
38 Takami, K., Yorimitsu, H. and Oshima, K. (2002) *Org. Lett.*, **4**, 2993–2995.
39 Hayashi, N., Shibata, I. and Baba, A. (2004) *Org. Lett.*, **6**, 4981–4983.
40 Miura, K., Tomita, M., Yamada, Y. and Hosomi, A. (2007) *J. Org. Chem.*, **72**, 787–792.
41 Ranu, B.C. and Mandal, T. (2006) *Tetrahedron Lett.*, **47**, 2859–2861.
42 (a) Yanada, R., Nishimori, N., Matsumura, A., Fujii, N. and Takemoto, Y. (2002) *Tetrahedron Lett.*, **43**, 4585–4588; (b) Yanada, R., Obika, S., Nishimori, N., Yamauchi, M. and Takemoto, Y. (2004) *Tetrahedron Lett.*, **45**, 2331–2334.
43 (a) Barton, D.H.R., Jang, D.O. and Jaszrberenyi, J.C. (1993) *J. Org. Chem.*, **58**, 6838; (b) Barton, D.H.R., Jang, D.O. and Jaszberenyi, J.C. (1992) *Tetrahedron Lett.*,

**33**, 5709–5712; (c) McCague, R., Pritchard, R.G., Stoodley, R.J. and Williamson, D.S. (1998) *Chem. Commun.*, 2691–2692; (d) Graham, S.R., Murphy, J.A. and Kennedy, A.R. (1999) *J. Chem. Soc., Perkin Trans.*, **1**, 3071–3073; (e) Tokuyama, H., Yamashita, T., Reding, M.T., Kaburagi, Y. and Fukuyama, T. (1999) *J. Am. Chem. Soc.*, **121**, 3791–3792; (f) Martin, C.G., Murphy, J.A. and Smith, C.R. (2000) *Tetrahedron Lett.*, **41**, 1833–1836; (g) Nambu, H., Anilkumar, G., Matsugi, M. and Kita, Y. (2003) *Tetrahedron*, **59**, 77–85.

44 (a) Jang, D.O. (1996) *Tetrahedron Lett.*, **37**, 5367–5368; (b) Graham, S.R., Murphy, J.A. and Coates, D. (1999) *Tetrahedron Lett.*, **40**, 2415–2416; (c) Yorimitsu, H., Shinokubo, H. and Oshima, K. (2000) *Chem. Lett.*, 104–105; (d) Kita, Y., Nambu, H., Ramesh, N.G., Anilkumar, G. and Matsugi, M. (2001) *Org. Lett.*, **3**, 1157–1160.

45 Curran, D.P. and Shen, W. (1993) *J. Am. Chem. Soc.*, **115**, 6051–6059.

46 Beaufils, F., Dénès, F. and Renaud, P. (2005) *Angew. Chem. Int. Ed.*, **44**, 5273–5275.

47 Khan, T.A., Tripoli, R., Crawford, J.J., Martin, C.G. and Murphy, J.A. (2003) *Org. Lett.*, **5**, 2971–2974.

48 (a) Crich, D. (1995) *Organosulfur Chemistry, Synthetic Aspects* (ed. P. Page), Academic, San Diego, p. 49; (b) Bertrand, M.P. and Ferreri, C. (2001) *Radicals in Organic Synthesis*, vol. **2** (eds M. Renaud and M.P. Sibi), Wiley-VCH, Weinheim, p. 485–504.

49 (a) Miyata, O., Ozawa, Y., Ninomiya, I. and Naito, T. (2000) *Tetrahedron*, **56**, 6199–6207; (b) Miyata, O., Muroya, K., Kobayashi, T., Yamanaka, R., Kajisa, S., Koide, J. and Naito, T. (2002) *Tetrahedron*, **58**, 4459–4479.

50 Beaufils, F., Denes, F., Becattini, B., Renaud, P. and Schenk, K. (2005) *Adv. Synth. Catal.*, **347**, 1587–1594.

51 Curran, D.P. and Shen, W. (1993) *J. Am. Chem. Soc.*, **115**, 6051–6059.

52 Yoshikai, K., Hayama, T., Nishimura, K., Yamada, K. and Tomioka, K. (2005) *J. Org. Chem.*, **70**, 681–683.

53 Byers, J. (2001) *Radicals in Organic Synthesis*, vol. 1 (eds P. Renaud and M.P. Sibi), Wiley-VCH, Weinheim, pp. 72–89.

54 (a) Clark, A.J. (2002) *Chem. Soc. Rev.*, **31**, 1–11; (b) Iqbal, J., Bhatia, B. and Nayyar, N.K. (1994) *Chem. Rev.*, **94**, 519–564.

55 Yang, D., Yan, Y.-L., Zheng, B.-F., Gao, Q. and Zhu, N.-Y. (2006) *Org. Lett.*, **8**, 5757–5760.

56 Edlin, C.D., Faulkner, J., Helliwell, M., Knight, C.K., Parker, J., Quayle, P. and Raftery, J. (2006) *Tetrahedron*, **62**, 3004–3015.

57 (a) Nagashima, H., Ara, K.-I., Wakamatsu, H. and Itoh, K. (1985) *J. Chem. Soc., Chem. Commun.*, 518–519; (b) Nagashima, H., Gondo, M., Masuda, S., Kondo, H., Yamaguchi, Y. and Matsubara, K. (2003) *Chem. Commun.*, 442–443.

58 Tallarico, J.A., Malnick, L.M. and Snapper, M.L. (1999) *J. Org. Chem.*, **64**, 344–345.

59 Bielawski, C.W., Louie, J. and Grubbs, R.H. (2000) *J. Am. Chem. Soc.*, **122**, 12872–12873.

60 Quayle, P., Fengas, D. and Richards, S. (2003) *Synlett*, 1797–1800.

61 Schmidt, B. and Pohler, M. (2005) *J. Organomet. Chem.*, **690**, 5552–5555.

62 Scholl, M., Ding, S., Lee, C.W. and Grubbs, R.H. (1999) *Org. Lett.*, **1**, 953–956.

63 Schmidt, B., Pohler, M. and Costisella, B. (2004) *J. Org. Chem.*, **69**, 1421–1424.

64 Nagashima, H., Seki, K., Ozaki, N., Wakamatsu, H., Itoh, K., Tomo, Y. and Tsuji, J. (1990) *J. Org. Chem.*, **55**, 985–990.

65 Wang, J. and Li, C. (2002) *J. Org. Chem.*, **67**, 1271–1276.

66 (a) Renaud, P. and Gerster, M. (1998) *Agnew. Chem., Int. Ed.*, **37**, 2562–2579; (b) Sibi, M.P., Ji, J., Sausker, J.B. and Jasperse, C.P. (1999) *J. Am. Chem. Soc.*, **121**, 7517–7526; (c) Iserloh, U., Curran, D.P. and Kanemasa, S. (1999) *Tetrahedron: Asym.*, **10**, 2417–2428; (d) Dakternieks, D., Dunn, K., Perchyonok, V.T. and

Schiesser, C.H. (1999) *Chem. Commun.*, 1665–1666; (e) Mero, C.L. and Porter, N.A. (1999) *J. Am. Chem. Soc.*, **121**, 5155–5160; (f) Porter, N.A., Feng, H. and Kavrakova, I. (1999) *Tetrahedron Lett.*, **40**, 6713–6716; (g) Sibi, M.P. and Rheault, T.R. (2000) *J. Am. Chem. Soc.*, **122**, 8873–8879; (h) Friestad, G.K. and Qin, J. (2000) *J. Am. Chem. Soc.*, **122**, 8329–8330; (i) Munakata, R., Totani, K., Takao, K.-I. and Tadano, K.-I. (2000) *Synlett*, 979–982; (j) Hayen, A., Koch, R. and Metzger, J.O. (2000) *Agnew. Chem., Int. Ed.*, **39**, 2758–2761; (k) Porter, N.A., Zhang, G. and Reed, A.D. (2000) *Tetrahedron Lett.*, **41**, 5773–5777; (l) Yang, D., Ye, X.-Y., Xu, M., Pang, K.-W. and Cheung, K.-K. (2000) *J. Am. Chem. Soc.*, **122**, 1658–1663.

67 (a) Curran, D.P., Chen, M.-H., Spletzer, E., Seong, C.M. and Chang, C.-T. (1989) *J. Am. Chem. Soc.*, **111**, 8872–8878; (b) Curran, D.P., Bosch, E., Kaplan, J. and Newcomb, M.J. (1989) *J. Org. Chem.*, **54**, 1826–1831; (c) Curran, D.P. and Chang, C.-T. (1989) *J. Org. Chem.*, **54**, 3140–3157; (d) Curran, D.P. and Tamine, J. (1991) *J. Org. Chem.*, **56**, 2746–2750.

68 (a) Curran, D.P. and Chang, C.-T. (1989) *J. Org. Chem.*, **54**, 3140–3157; (b) Milstein, D. and Stille, J.K. (1978) *J. Am. Chem. Soc.*, **100**, 3636–3638; (c) Liebner, J.E. and Jacobus, J. (1979) *J. Org. Chem.*, **44**, 449; (d) Crich, D. and Sun, S. (1996) *J. Org. Chem.*, **61**, 7200–7201; (e) Renaud, P., Lacote, E. and Quaranta, L. (1998) *Tetrahedron Lett.*, **39**, 2123–2126; (f) Edelson, B.S., Stoltz, B.M. and Corey, E.J. (1999) *Tetrahedron Lett.*, **40**, 6729–6730.

69 (a) Kharasch, M.S. and Skell, P.S. and Fisher, P. (1948) *J. Am. Chem. Soc.*, **70**, 1055–1059; (b) Zard, S.Z. (1997) *Agnew. Chem., Int. Ed. Engl.*, **36**, 672–685.

70 Ollivier, C., Bark, T. and Renaud, P. (2000) *Synthesis*, 1598–1602.

71 (a) Demir, A.S. and Emrullahoglu, M. (2007) *Curr. Org. Synth.*, **4**, 321–350;
(b) Snider, B.B. (1996) *Chem. Rev.*, **96**, 339–364.

72 Snider, B.B. and Duvall, J.R. (2004) *Org. Lett.*, **6**, 1265–1268.

73 Chikaoka, S., Toyao, A., Ogasawara, M., Tamura, O. and Ishibashi, H. (2003) *J. Org. Chem.*, **68**, 312–318.

74 Tan, X. and Chen, C. (2006) *Angew. Chem. Int. Ed.*, **45**, 4345–4348.

75 Kagan, H.B. (2003) *Tetrahedron*, **59**, 10351–10372.

76 For other selected examples of cyclizations using samarium iodide, see: (a) Inui, M., Nakazaki, A. and Kobayashi, S. (2007) *Org. Lett.*, **9**, 469–472; (b) Lam, K. and Marko, I.E. (2008) *Org. Lett.*, **10**, 2773–2776; (c) Ohno, H., Iwasaki, H., Eguchi, T. and Tanaka, T. (2004) *Chem. Commun.*, 2228–2229.

77 Aulenta, F., Berndt, M., Bruedgam, I., Hartl, H., Soergel, S. and Reissig, H.-U. (2007) *Chem. Eur. J.*, **13**, 6047–6062.

78 St. Jean, D.J., Cheng, E.P. and Bercot, E.A., (2006) *Tetrahedron Lett.*, **47**, 6225–6227.

79 Rivkin, A., Gonzalez-Lopez de Turiso, F., Nagashima, T. and Curran, D.P. (2004) *J. Org. Chem.*, **69**, 3719–3725.

80 (a) Cuerva, J.M., Justicia, J., Oller-Lopez, J.L., Bazdi, B. and Oltra, J.E. (2006) *Mini-Rev. Org. Chem.*, **3**, 23–35; (b) Nugent, W.A. and Rajanbabu, T.V. (1988) *J. Am. Chem. Soc.*, **110**, 8561–8562; (c) Barrero, A.F., Quilez del Moral, J.F., Sanchez, E.M. and Arteaga, J.F. (2006) *Eur. J. Org. Chem.*, 1627–1641; For some selected examples see: Xu, L. and Huang, X. (2008) *Tetrahedron Lett.*, **49**, 500–503;
(d) Banerjee, B. and Roy, S.C. (2005) *Synthesis*, 2913–2919; (e) Banerjee, B. and Roy, Subhas C. (2006) *Eur. J. Org. Chem.*, 489–497; (f) Justicia, J., Oller-Lopez, J.L., Campana, A.G., Oltra, J.E., Cuerva, J.M., Bunuel, E. and Cardenas, D.J. (2005) *J. Am. Chem. Soc.*, **127**, 14911–14921.

81 Roy, S.C., Rana, K.K. and Guin, C. (2002) *J. Org. Chem.*, **67**, 3242–3248.

82 Justicia, J., Campana, A.G., Bazdi, B., Robles, R., Cuerva, J.M., and Oltra, J.E.

(2008) *Adv. Synth. Catal.*, **350**, 571–576. Also see:Barrero, A.F., Cuerva, J.M., Alvarez-Manzaneda, E.J., Oltra, J.E. and Chahboun, R. (2002) *Tetrahedron Lett.*, **43**, 2793–2796.

83 Friedrich, J., Walczak, K., Dolg, M., Piestert, F., Lauterbach, T., Worgull, D. and Gansauer, A. (2008) *J. Am. Chem. Soc.*, **130**, 1788–1796.

84 For reviews see: (a) Hoffmann, N. (2007) *Pure Appl. Chem.*, **79**, 1949–1958; Kitamura, M. and Narasaka, K. (2008) *Bull. Chem. Soc. Jpn.*, **81**, 539–547. For some selected examples see: (b) Alabugin, I.V., Timokhin, V.I., Abrams, J.N., Manoharan, M., Abrams, R. and Ghiviriga, I. (2008) *J. Am. Chem. Soc.*, **130**, 10984–10995; (c) Tu, W. and Floreancig, P.E. (2007) *Org. Lett.*, **9**, 2389–2392.

85 Cho, C.H., Kim, S., Yamane, M., Miyauchi, H. and Narasaka, K. (2005) *Bull. Chem. Soc. Jpn.*, **78**, 1665–1672.

86 Tzvetkov, N.T., Arndt, T. and Mattay, J. (2007) *Tetrahedron*, **63**, 10497–10510.

87 Tsuchii, K., Ueta, Y., Kamada, N., Einaga, Y., Nomoto, A. and Ogawa, A. (2005) *Tetrahedron Lett.*, **46**, 7275–7278.

88 Tsuchii, K., Doi, M., Ogawa, I., Einaga, Y. and Ogawa, A. (2005) *Bull. Chem. Soc. Jpn.*, **78**, 1534–1548.

89 Du, W. and Curran, D.P. (2003) *Synlett*, 1299–1302.

90 Kitagawa, O., Miyaji, S., Sakuma, C. and Taguchi, T. (2004) *J. Org. Chem.*, **69**, 2607–2610.

91 Panchaud, P. and Renaud, P. (2004) *J. Org. Chem.*, **69**, 3205–3207.

92 Tsimelzon, A. and Braslau, R. (2005) *J. Org. Chem.*, **70**, 10854–10859.

93 Tsuritani, T., Shinokubo, H. and Oshima, K. (2003) *J. Org. Chem.*, **68**, 3246–3250.

94 (a) Nishida, M., Ueyama, E., Hayashi, H., Ohtake, Y., Yamaura, Y., Yanaginuma, E., Yonemitsu, O., Nishida, A. and Kawahara, N. (1994) *J. Am. Chem. Soc.*, **116**, 6455–6456; (b) Nishida, M., Hayashi, H., Yamaura, Y., Yanaginuma, E., Yonemitsu, O., Nishida, A. and Kawahara, N. (1995) *Tetrahedron Lett.*, **36**, 269–272; (c) Nishida, M., Nobuta, M., Nakaoka, K., Nishida, A. and Kawahara, N. (1995) *Tetrahedron: Asym.*, **6**, 2657–2660; (d) Kundig, E.P., Xu, L.H., Romanens, P. and Bernardinelli, G. (1996) *Synlett*, 270–272; (e) Crich, D., Suk, D.-H. and Sun, S. (2003) *Tetrahedron: Asym.*, **14**, 2861–2864; (f) Goddard, J.-P., Gomez, C., Brebion, F., Beauviere, S., Fensterbank, L. and Malacria, M. *Chem. Commun.*, **2007**, 2929–2931; (g) Sibi, M.P. and Ji, J. (1996) *J. Am. Chem. Soc.*, **118**, 3063–3064; (h) Kim, K., Okamoto, S. and Sato, F. (2001) *Org. Lett.*, **3**, 67–69.

95 Andres, C., Duque-Soladana, J.P. and Pedrosa, R. (1999) *J. Org. Chem.*, **64**, 4282–4288. For other examples see: (a) Rodriguez, V., Quintero, L. and Sartillo-Piscil, F. (2007) *Tetrahedron Lett.*, **48**, 4305–4308; (b) Kamimura, A., Omata, Y., Tanaka, K. and Shirai, M. (2003) *Tetrahedron*, **59**, 6291–6299.

96 Brebion, F., Vitale, M., Fensterbank, L. and Malacria, M. (2003) *Tetrahedron: Asym.*, **14**, 2889–2896.

97 Keum, G., Kang, S.B., Kim, Y. and Lee, E. (2004) *Org. Lett.*, **6**, 1895–1897.

98 Yang, D., Zheng, B.-F., Gu, S., Chan, P.W.H., and Zhu, N.-Y. (2003) *Tetrahedron: Asym.*, **14**, 2927–2937. Also see:Yang, D., Yan, Y.-L. Law, K.-L. and Zhu, N.-Y. (2003) *Tetrahedron*, **59**, 10465–10475.

99 Enholm, E.J., Cottone, J.S. and Allais, F. (2001) *Org. Lett.*, **3**, 145–147.

100 Barrero, A.F., Herrador, M.M., Quilez del Moral, J.F. and Valdivia, M.V. (2002) *Org. Lett.*, **4**, 1379–1382.

101 Yang, D., Gu, S., Yan, Y.L., Zhu, N.Y. and Cheung, K.K. (2001) *J. Am. Chem. Soc.*, **123**, 8612–8613.

102 Yang, D., Gu, S., Yan, Y.L., Zhao, H.W. and Zhu, N.Y. (2002) *Angew. Chem., Int. Ed. Engl.*, **41**, 3014–3017.

103 Nishida, M., Hayashi, H., Nishida, A. and Kawahara, N. (1996) *Chem. Commun.*, 579–580.

104 Hiroi, K. and Ishii, M. (2000) *Tetrahedron Lett.*, **41**, 7071–7074.

105 Curran, D.P., Liu, W. and Chen, C.H.-T. (1999) *J. Am. Chem. Soc.*, **121**, 11012–11013.

106 Petit, M., Lapierre, A.J.B. and Curran, D.P. (2005) *J. Am. Chem. Soc.*, **127**, 14994–14995.

107 Aechtner, T., Dressel, M. and Bach, T. (2004) *Angew. Chem. Int. Ed.*, **43**, 5849–5851.

108 (a) Gurjar, M.K., Mohapatra, S., Phalgune, U.D., Puranik, V.G. and Mohapatra, D.K. (2004) *Tetrahedron Lett.*, **45**, 7899–7902; (b) Heitzman, C.L., Lambert, W.T., Mertz, E., Shotwell, J.B., Tinsley, J.M., Va, P. and Roush, W.R. (2005) *Org. Lett.*, **7**, 2405–2408; (c) Marshall, J.A., Schaaf, G. and Nolting, A. (2005) *Org. Lett.*, **7**, 5331–5333.

109 Kim, C.H., An, H.J., Shin, W.K., Yu, W., Woo, S.K., Jung, S.K. and Lee, E. (2006) *Agnew. Chem., Int. Ed.*, **45**, 8019–8021.

110 Grainger, R.S. and Welsh, E.J. (2007) *Agnew. Chem., Int. Ed.*, **46**, 5377–5380.

111 Servais, A., Azzouz, M., Lopes, D., Courillon, C. and Malacria, M. (2007) *Agnew. Chem., Int. Ed.*, **46**, 576–579.

112 Watanabe, H., Mori, N., Daisuke, I., Kitahara, T. and Mori, K. (2007) *Agnew. Chem., Int. Ed.*, **46**, 1512–1516.

113 Sharp, L.A. and Zard, S.A. (2006) *Org. Lett.*, **8**, 831–834.

114 Biechy, A., Hachisu, S., Quilcet-Sire, B., Ricard, L. and Zard, S.Z. (2008) *Angew. Chem. Int. Ed.*, **47**, 1436–1438.

115 (a) Magolan, J., Carson, C.A. and Kerr, M.A. (2008) *Org. Lett.*, **10**, 1437–1440; (b) Also see: Magolan, J. and Kerr, M.A. (2006) *Org. Lett.*, **8**, 4561–4564.

116 Haruo, Y., Hasegawa, T., Tanaka, H. and Takahashi, T. (2001) *Synlett*, 1935–1937.

117 Boffey, R.J., Whittingham, W.G. and Kilburn, J.D. (2001) *J. Chem. Soc., Perkin Trans*, **1**, 487–496.

118 Nicolaou, K.C., Edmonds, D.J., Li, A. and Tria, G.S. (2007) *Angew. Chem. Int. Ed.*, **46**, 3942–3945.

# 24
# Photocyclization Reactions
*Axel G. Griesbeck*

## 24.1
### Introduction

This chapter is concerned with photochemical cyclization reactions. Carbo- and heterocyclic ring formation is a key application for photochemical reactions involving organic chromophores such as alkenes, alkynes, aromatic and heteroaromatic compounds, carbonyl and carbonyl analogs. This chapter describes the principal methods, scope, limitations, and applications of photocyclization. As most of the reactions described herein are initiated by electronically excited triplet states – and thus can be described as cyclization reactions of 1,n-biradicals – we also briefly mention intermolecular $\pi^n + \pi^m$-cycloadditions in addition to their intramolecular versions. General information on the application of photochemistry to organic synthesis can be found in recent reviews [1].

Why does electronic excitation greatly increase the chemical reactivity in multiple directions? This is not only due to the excess energy in the initially formed state but also to the electronic restructuring of the molecule. From a closed-shell structure as apparent in most organic substrates, an open-shell state is formed with one low-lying and one high-lying singly occupied orbital. This situation leads to a reactivity dichotomy: electrophiles can interact more efficiently due to the increased donor properties and nucleophiles interact more efficiently due to the increased acceptor properties of the excited state. In an extreme version, both processes can lead to full electron transfer, making excited states at the same time good oxidants and reductants, respectively. For cyclization processes, where at least two parts of the substrate have to show correlated reactivity in the ground state, excited state reactions can also combine groups with similar or even identical reactivity by a process of photochemical *umpolung*. Furthermore, different spin-isomeric excited states can exist that differ in energy, in electronic and in chemical behavior, an effect that is termed spin chemistry. Spin chemistry is an innovative new research area in photochemistry [2]. Applications that have been explicitly designed for organic synthesis are still rare – the basic phenomena, however, have been known for decades. The key feature of all processes in the context of spin chemistry is the spin conservation rule ($\Delta S = 0$). In other words, bond-forming or bond-breaking processes are strictly spin-forbidden when coupled with a change in spin multiplicity of one molecule or a set of strongly coupled molecules (a ground-state complex or an exciplex). Spin chemistry is transformed into the real world because this rule is *not* strictly observed and mechanisms exist which allow spin flips coupled with chemical reactions. The vast majority of organic molecules exist in singlet electronic ground states and there are no spin restrictions for either unimolecular processes or bimolecular reactions with other singlet molecules. Also, in the case of mono-radical addition to organic substrates in their singlet electronic ground states no spin restrictions exist and doublet states are generated which can propagate in chain reactions. A multitude of 1,$n$-biradicals with triplet multiplicity originate from triplet/singlet interactions in photochemical reactions and these triplet biradicals have been extensively studied over the last decades, both spectroscopically and theoretically. They can be formed in an intramolecular fashion thus representing the first step of a photocyclization reaction or intramolecularly in photocyclizations. These intermediates are the direct consequence of the precursor spin and their lifetimes are connected with the mode of spin inversion processes and the mechanism which leads to the formation of closed-shell products.

Photochemistry is especially valuable when applied to the synthesis of target molecules less easily available by thermal or other methods. Furthermore, selectivity has become the most important feature in synthetic organic chemistry for several decades. As far as ground state chemistry is concerned, chemo-, regio-, and stereo-selectivity are the essential factors which determine the success and the usefulness of a reaction. Spin-selectivity is now an additional possibility. Notwithstanding, it should be emphasized that spin-selectivity allows the modification of a chemical reaction solely by taking advantage of the different lifetimes and reactivities of two (or more) electronically excited states of the same molecule (spin-isomers).

Triplet states $^3A^*$ could be selectively generated by triplet–triplet energy-transfer (sensitization) and selectively deactivated by quenching. The latter process allows the study of the singlet state $^1A^*$ behavior in chemical transformations. The mechanistic scenario shown in Scheme 24.1 implies that two different products **C** and **D** arise from the spin isomers $^1A^*$ and $^3A^*$. This effect can be expressed simply in the fact that the triplet intermediate $^3X$ returns to the substrates after intersystem crossing (ISC) whereas $^1A^*$ reacts with **B** and gives the product molecule **C**. Alternatively, two

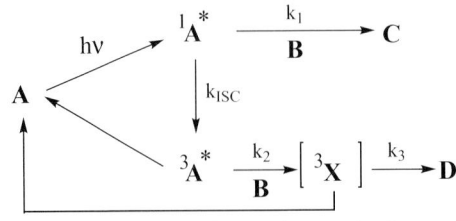

**Scheme 24.1** Kinetic scheme of intermolecular singlet and triplet photoreactions.

constitutionally different products **C** and **D** can be formed (like for example in oxa di-π-methane rearrangements) or **C** and **D** can constitute two diastereoisomers. Especially, the last case is interesting because spin-multiplicity does not only influence the reactivity of the excited molecules (due to energy and lifetime effects), but also alters the regio- and stereoselectivity of the product-forming steps. In singlet photoreactions, stereoselectivity is often controlled by the optimal geometries for radical–radical combinations, whereas in triplet photoreactions the geometries most favorable for ISC are considered to be of similar relevance. These geometries can be quite different from the former ones due to differences in spin–orbit coupling (SOC) values. These aspects play a crucial role for three important photochemical reactions for the synthesis of small-ring products (Scheme 24.2): the [2 + 2]-cycloaddition and its oxo-variant, the Paternò–Büchi reaction [3], the Norrish–Yang photocyclization [4], and the di-π-methane rearrangements [5]. 1,4-Triplet biradicals are crucial intermediates in the triplet versions of these reactions. The isomeric singlet biradicals have been detected only in the last two decades and are extremely short-lived if not highly stabilized by spin-diluting substituents [6].

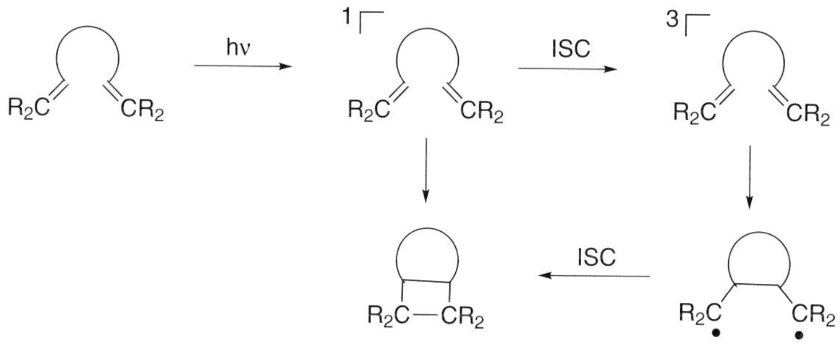

**Scheme 24.2** Biradical and concerted pathways for intramolecular photocycloaddition.

## 24.2
### Several Ways to Initiate Photocyclization

Concerted or stepwise photochemical processes can lead to the formation of cyclic products from acyclic substrates by way of photocyclization. Stepwise processes often involve biradicals (from homolytic steps like hydrogen transfer or CX bond cleavage) or radical ion pairs that originate from photoinduced electron transfer steps. In order to initiate these primary processes, the photolabile molecules have to be activated by electronic excitation. There are three major routes that can be used for achieving electronic changes:

- direct excitation in the UV and visible region,
- energy transfer sensitization using appropriate catalysts (sensitizers)
- electron transfer sensitization using PET (photoinduced electron transfer) catalysts.

## 24.2.1
### Direct Excitation and Direct or Sequential Addition

Direct excitation requires appropriate lamp and filter technology to irradiate into the relevant absorption band of the appropriate chromophore in the substrate. These experimental conditions are crucial for selective photochemistry. Prior to any photochemical experiment, the absorption spectra of substrates, reagents, and the solvents have to be compared to examine the irradiation windows [7]. The major excited singlet states in organic photochemistry are the $n\pi^*$ and $\pi\pi^*$ of alkenes, alkynes, aromatic substrates, carbonyl and carbonyl analogue groups. The generation of the initial singlet is followed by several deactivation steps that are unproductive with respect to product formation: fluorescence, intersystem-crossing, radiationless decay to the ground state, energy or electron transfer to appropriate acceptor groups either intra- or intermolecular. Fluorescence can be measured independently and the singlet lifetimes determined by time-resolution are crucial for the application of reagents in intermolecular photoreactions, for example, if reagents can be used in stoichiometric ratios or have to be used in large excess. Efficient ISC generates the excited triplets which generally have much longer lifetimes and are sensitive against energy transfer to triplet oxygen. From the reaction of excited $\pi\pi^*$ and $n\pi^*$ singlet states concerted photochemical addition processes can be expected whereas the corresponding $\pi\pi^*$ and $n\pi^*$ triplets generally react in multistep pathways due to spin restrictions.

## 24.2.2
### Direct Excitation and Group Transfer

Singlet excited $\pi\pi^*$ and $n\pi^*$ states can undergo efficient group transfer reactions. The transfer of a hydrogen atom in an intramolecular fashion is an efficient process with rate constants of $10^7$ to $10^{10}\,s^{-1}$ [8]. Intermolecular hydrogen transfer to excited singlet states can likewise compete with inter-system crossing, however, in many cases this is not synthetically useful because large excesses of hydrogen donor must be used. Nucleophile addition to $\pi\pi^*$ singlets is an alternative group transfer process and if alcohols are used in excess (as solvent) efficient ether formation is possible.

## 24.2.3
### Sensitization: Energy Transfer

Triplet excited $\pi\pi^*$ and $n\pi^*$ states can be selectively generated by energy transfer from appropriate energy donors. Prerequisites for triplet energy donors is a proper triplet energy and a high inter-system crossing quantum yield as well as low chemical reactivity of the singlet and triplet states. Benzophenone is a typical triplet sensitizer for $n\pi^*$ triplets, however, with a strong tendency for hydrogen transfer reactions. In Table 24.1 a series of the most convenient triplet sensitizers is given with singlet and triplet energies, respectively, and the intersystem-crossing quantum yields [7].

Table 24.1 Triplet sensitizers, energies and quantum yields of ISC.

| Sensitizer | $E_T$ (kJ mol$^{-1}$) | $E_S$ (kJ mol$^{-1}$) | $\Phi_{ISC}$ |
|---|---|---|---|
| Benzene | 353 | 459 | 0.25 |
| Toluene | 346 | 445 | 0.53 |
| Acetone | 332 | 372 | 0.90 |
| Acetophenone | 310 | 330 | 1.00 |
| Benzaldehyde | 301 | 323 | 1.00 |
| Triphenylamine | 291 | 362 | 0.88 |
| Benzophenone | 287 | 316 | 1.00 |
| Fluorine | 282 | 397 | 0.22 |
| Biphenyl | 274 | 418 | 0.84 |
| Phenanthrene | 260 | 346 | 0.73 |
| Styrene | 258 | 415 | 0.40 |
| Naphthalene | 253 | 385 | 0.75 |
| 2-Acetylnaphthalene | 249 | 325 | 0.84 |
| Biacetyl | 236 | 267 | 1.00 |
| Benzyl | 223 | 247 | 0.92 |
| Anthracene | 178 | 318 | 0.71 |
| Eosine | 177 | 209 | 0.33 |
| Rose Bengal | 164 | 213 | 0.61 |
| Methylene Blue | 138 | 180 | 0.52 |

### 24.2.4
### Sensitization: Electron Transfer (PET)

Electronically excited states feature, at the same time, significantly increased reduction and oxidation properties, as compared to ground states. The investigation of photoinduced electron transfer as a key step in many photochemical reactions was further stimulated by the Nobel Prize winning work of Marcus [9] and has, in the last decades, not only led to new synthetic applications [10] but to a general paradigm change in photochemistry. As has been described in detail in several reviews [11], the energetics of light-induced electron transfer can be estimated by use of a simplified version of the Rehm–Weller equation [12]. Unlike many other physical equations, this relationship immediately shows its chemical relevance: electronically excited states are concurrently much better reductants and oxidants; the actual redox behavior depends on the reaction partner. This concept can be easily explained for the carbonyl–amine system (Scheme 24.3): the high-lying lone-pair at N and the low-lying antibonding $\pi^*_{CO}$ render the amine–carbonyl combination suitable for FMO-controlled nucleophile–electrophile ground-state interaction. However, only after electronic excitation to the n $\rightarrow$ $\pi^*$ carbonyl state does an exergonic electron transfer become feasible. The same is true for less active electron donors like alkyl carboxylates, unsaturated or strained hydrocarbons, aromatic compounds and many more.

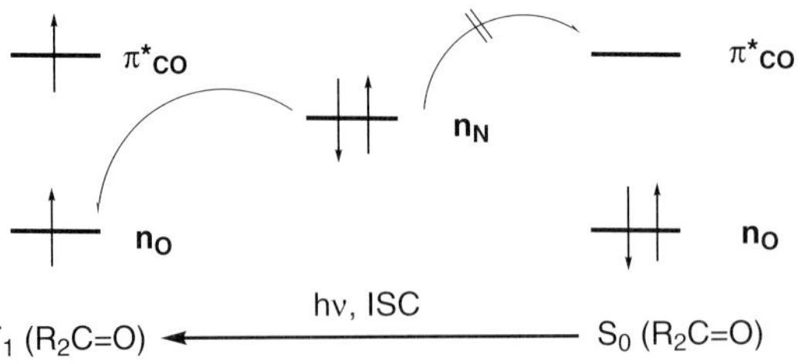

## 24 Photocyclization Reactions

$$\Delta G^0 = E^0(\text{Do}/\text{Do}^{+\cdot}) - E^0(\text{A}^{-\cdot}/\text{A}) - E_{00}$$

**Scheme 24.3** Frontier molecular orbital scheme for photoinduced electron transfer.

### 24.2.5
### Substrate Tuning

In order to perform an efficient photochemical process, that is, a photocyclization reaction, the light-absorbing substrate has to be designed carefully: (i) photoexcitation has to be limited to the substrate and neither the product(s) nor the potential long-lived reaction intermediates must absorb light in the region applied. This can be achieved by the use of monochromatic light sources or by the use of appropriate filters. Alternatively, triplet sensitizers are available that selectively transfer energy to the substrate and absorb in the red-shifted part of the UV–vis region. (ii) Non-desired electron transfer reactivity of the substrate can be suppressed by introduction of electron-withdrawing groups to reduce the donor-capability or electron-donating groups to reduce the acceptor-capability. (iii) Spin selectivity has to be investigated for intermolecular photoreactions by variation of the reagent concentration. (iv) Oxygen sensitivity is often observed for triplet photocyclization reactions where the long-lived substrate donors can transfer energy (or an electron) to triplet oxygen to generate singlet oxygen (or the superoxide radical anion).

## 24.3
## Norrish Type II Initiated Reactions: H Transfer

### 24.3.1
### Cyclization (Yang Process) Versus Cleavage

The Yang reaction is defined as the photochemical formation of cyclobutanols from acyclic or cyclic carbonyl compounds initiated by an intramolecular γ-hydrogen abstraction (Norrish Type II reaction) [13]. Acyclic carbonyl substrates often tend to cleave when electronically excited: besides the α-cleavage (Norrish Type I), β-cleavage often follows the formation of a 1,4-biradical of the 1-hydroxytetramethylene type. Thus, the Yang reaction is limited to those cases where the formation of a new carbon–carbon bond can efficiently compete with the cleavage of the β,γ-bond. Due to the experimentally established fact that the first step of this reaction sequence (i.e., the transfer of a hydrogen atom) is reversible and the

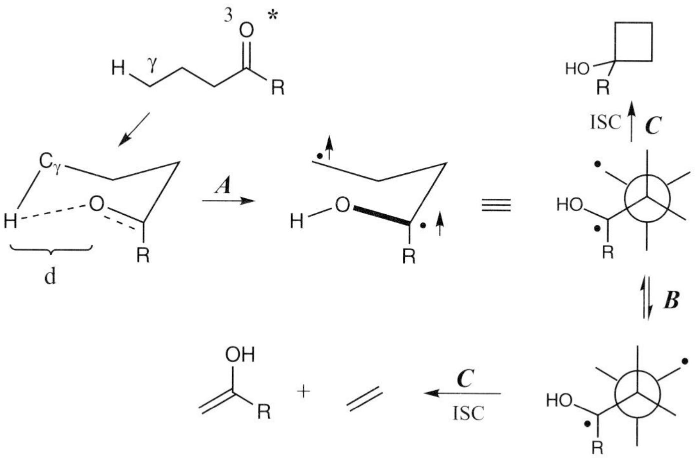

**Scheme 24.4** Mechanistic scenario for the Norrish II–Yang photocyclization.

cleavage process, however, is not, special prerequisites are necessary for high-yielding Yang reactions. The selectivity-determining factors of the Yang cyclization reaction can be described using the three selection stages A–C which are depicted in Scheme 24.4.

After electronic excitation and intersystem crossing of the carbonyl compound, a chair-like conformation initiates the Norrish Type II hydrogen transfer process (**A**). The primarily formed structure is that of a skew 1,4-triplet biradical. Biradical dynamics subsequently equilibrate syn- and anti-conformers (**B**) which have different intrinsic ISC-rates (*vide supra*) and subsequently result in the formation of cyclization and cleavage products, respectively (**C**). The geometrical prerequisites for step **A** have been extensively investigated by Scheffer and coworkers for liquid- and, especially, solid-phase photolyses [14]. Following the classical assumption of an interaction between the reactive CH-bond and the $p_y$-orbital at the carbonyl oxygen, α-substituents can adopt a pseudo-equatorial position. Following the crystal structure–solid state reactivity method, Scheffer *et al.* have determined the distance and angular requirements for photochemical γ-hydrogen transfer. The optimal C=O···Hγ distance $d$ is close to the sum of the van der Waals radii of H and O (2.72 Å), the optimal ω-angle (describing the angle by which the γ-hydrogen atom lies outside the mean plane of the carbonyl group) is $52 \pm 5°$, the optimal Δ-angle (describing the C=O···H angle) is $83 \pm 4°$, and the optimal θ-angle (describing the Cγ–H···O angle) is $115 \pm 2°$. These values have been extracted from a series of carbonyl substrates which have been analyzed in the solid state and in the fluid phase. The electronically excited carbonyl substrate has to pass through a chair-like transition state in order to achieve an efficient γ-hydrogen transfer. Norrish type I reactions always accompany this reaction and because of less restricted geometrical prerequisites often dominate. Concerning the nature of the activated CH-bond, a 1 : 30 : 200 selectivity for primary : secondary : tertiary CH is characteristic [15].

## 24.3.2
## Formation of Four-Membered Rings

The formation of cyclobutanols from the Yang reaction of electronically excited carbonyl compounds can be demonstrated nicely for photochemical amino acid transformations. In order to generate C-activated derivatives, α-amino acids are converted into N-acylated aromatic ketones. The photochemistry of these substrates **1** allows the detailed analysis of processes **B** and **C** (*vide supra*) with respect to the stereoselectivity of the Yang-cyclization [16]. In all reactions, 2-acetylamino 4-methylacetophenone, resulting from the Norrish Type II cleavage reaction, was formed to different extents (beside about 10–20% of α-cleavage products). Both the *tert*-leucine-derived substrate **1d** and the leucine-derived substrate **1b** gave solely the *cis*-diastereoisomeric cyclobutanes **2d** and **2b**. The cyclization to cleavage ratios (CCR) were also similar: 63 : 37 and 58 : 42, respectively (Scheme 24.5).

**Scheme 24.5** Amino acid-derived substrates and products of the Yang photocyclization.

The product stereochemistry requires a high 1,2-asymmetric induction which is convincingly explained by assuming a hydrogen bond formed already at the triplet biradical stage (Scheme 24.6).

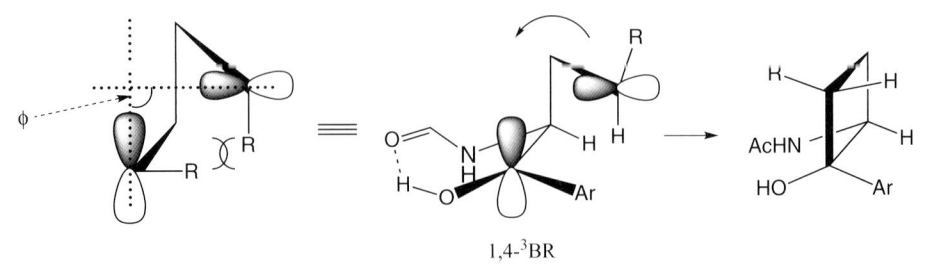

1,4-$^3$BR

**Scheme 24.6** Critical geometries for the 1,4-triplet biradical intersystem-crossing.

The lifetimes of 1-hydroxytetramethylenes reported in the literature are in the 50 ns region and thus long enough to guarantee conformational flexibility at room temperature [17]. Hydrogen bonds stabilizing Norrish II biradicals have been discussed several times based on experimental data and theoretical calculations. Calculations also resulted in the biradical $^3\mathbf{BR}$, as a global minimum. Additionally, this *syn*-biradical is in equilibrium with the anti-conformer by rotation about the C2–C3 bond. The equilibrium constant together with the hydrogen back transfer efficiency (from the syn-structure) and the orbital orientation (p,p relative to the central C–C single bond) control the cyclization/cleavage ratio. The assumption is widely accepted that, after intersystem crossing, the singlet biradicals maintain conformational memory of their triplet precursors [18], that is, the *anti*-1,4-biradical conformer gives exclusively cleavage products whereas the *syn*-1,4-biradicals can cleave and cyclize (Scheme 24.7). For $^3\mathbf{BR}$, this means that process **C** is responsible for the CCR as well as for the stereoselectivity of the formation of new stereogenic centers (*vide infra*).

The latter aspect is not relevant for substrates **1b** and **1d** where the 1,2-asymmetric induction is dictated by the hydrogen bond and no further stereogenic centers are created. A second stereogenic center is created during the photolysis of the valine-derived substrate **1c** (Scheme 24.8). *Ab initio* calculations for the *syn*-biradical $^3\mathbf{BRc'}$ resulted in a global minimum conformation with a strong hydrogen bond (C=O···HO: 1.80 Å, C(sp$_2$)–C(sp$_2$): 2.91 Å). In $^3\mathbf{BRc''}$ two gauche interactions exist, whereas in the alternative structure $^3\mathbf{BRc''}$ three gauche interactions destabilize the conformer. The latter intermediate is expected to isomerize into its anti-isomer which subsequently undergoes cleavage. Another mechanistic alternative is the assumption of a highly stereoselective hydrogen abstraction from only one of the two diastereotopic methyl groups. At this stage of our research we could not distinguish between these two alternatives. As described already in detail, spin–orbit coupling controlled geometries are controlling product formation. Calculations resulted in large SOC values for syn-geometries and lower values for anti-geometries [19]. Assuming that in the syn-conformer the hydroxy group localized at the benzylic radical is sterically more demanding, the SOC principle (Scheme 24.7) predicts the correct relative product stereochemistry for C3–C4 (cis-configuration with respect to the methyl and hydroxy groups).

Summarizing the experimental results and the proposed reaction model, a three-stage selection protocol results which enables the prediction of the correct chemo- and stereoselectivity. The hydrogen abstraction (step **A**) differentiates between diastereotopic hydrogens due to the strict geometrical prerequisites, biradical dynamics (step **B**, i.e., rotation about C1–C2 and C2–C3) control the 1,2-asymmetric induction *as well as* the cyclization/cleavage ratio. Eventually, SOC-controlled ISC (step **C**) leads directly to the formation of the final C–C bond and thus determines the stereochemistry of the conversion of the two radical centers into new stereogenic centers.

### 24.3.3
### Formation of Three-Membered Rings: Direct vs. Indirect

Cyclopropanes can also be formed by the Norrish–Yang reaction if a hydrogen abstraction from the β-position with respect to the carbonyl group is feasible, resulting in 1,3-biradicals. The geometrical parameters of the corresponding five-membered transition state, especially the O–H–C bond angle, differ substantially from the ideal parameters reported by Scheffer *et al.* [20]. Therefore, it is not

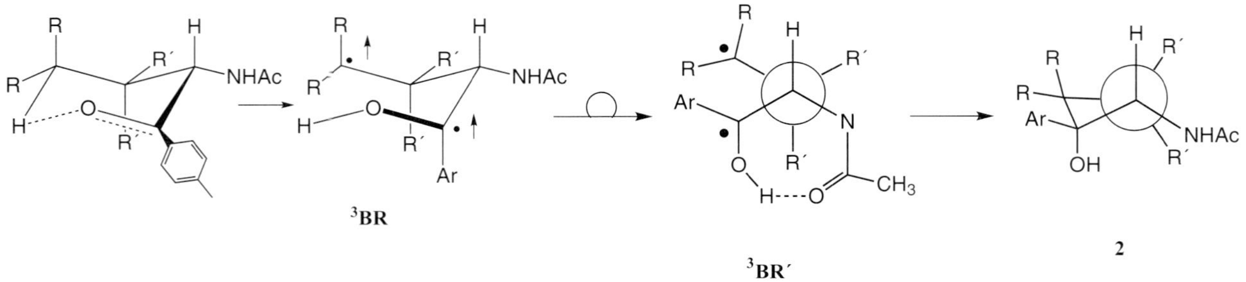

**Scheme 24.7** Norrish II–Yang photocyclization of aromatic α-amido ketones.

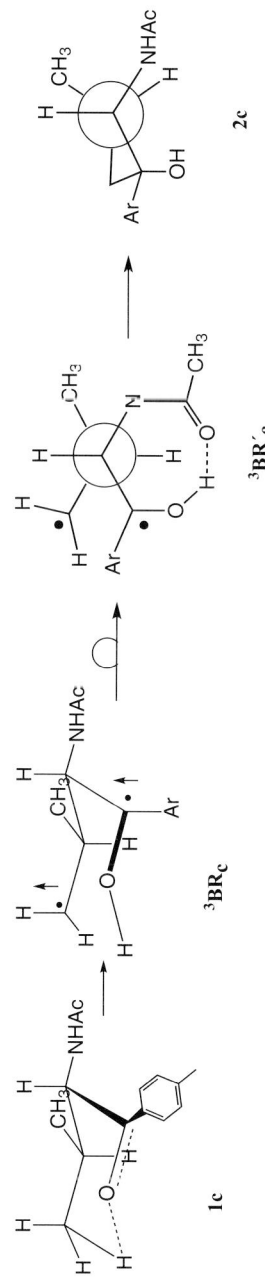

**Scheme 24.8** Norrish II–Yang photocyclization of the valine-derived α-amido ketones.

surprising that only a few syntheses of cyclopropanes via the Norrish–Yang reaction have been reported [21]. In most of these examples, the initial step appears to be a photoinduced electron transfer and a proton shift follows the electron–hole separation. These conditions limit the preparative scope because the electron-donating functional groups which are often required to initiate the PET often cause a considerable sensitivity of the cyclopropanes to oxidative ring opening. Intensively investigated compounds with this behavior are β-dialkylamino propiophenones **3** [22]. Irradiation, electron and subsequent proton transfer results in the triplet 1,3-biradicals **4**. After ISC back to the singlet state, cyclopropanes **5** are formed by cyclization; theses are sensitive to triplet oxygen and are converted into α,β-unsaturated ketones **6** (Scheme 24.9).

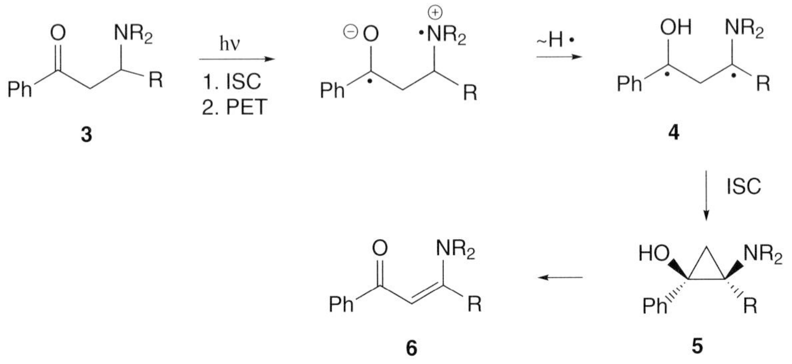

**Scheme 24.9** Photocyclization of β-amino ketones to cyclopropanols.

The cyclization to **5** proceeds with high diastereoselectivity. From enantiomerically pure aminoketones **3** only one enantiomer of the cyclopropanes is obtained [23]. This accounts for the fact that the biradicals **4** are short-lived and no rotation about the C–C-single bonds occurs, which conserves the stereogenic information of the reactant **3**.

Another way to generate a 1,3-biradical as the logical precursor to cyclopropane formation is a cascade reaction termed the spin center shift process [24]. This method is based on a typical radical behavior that is specifically designed for the 1,4-biradical formed in the classical Norrish Type II reaction (abstraction of a γ-hydrogen atom): a strong leaving group (nucleofug) X at the carbon atom adjacent to the carbonyl group (**7**) facilitates the elimination of HX at the stage of the initial 1,4-biradical, a process which can be as fast or even faster that the otherwise dominant cleavage and cyclization pathways. Starting from a triplet 1,4-biradical, the elimination of a closed-shell molecule consequently leads to a triplet 1,3-biradical which, after ISC, combines to give a cyclopropane derivative **8** (Scheme 24.10).

The spin shift approach is flexible in terms of the carbon chain and tolerates also strained substrates which lead to even higher strained products. Cyclopropyl groups in the starting materials survive the two-step reaction without ring-opening and highly strained bicyclic hydrocarbons such as bicyclo[2.1.0]pentanes **10** and spiropentanes **12** are accessible (Scheme 24.11).

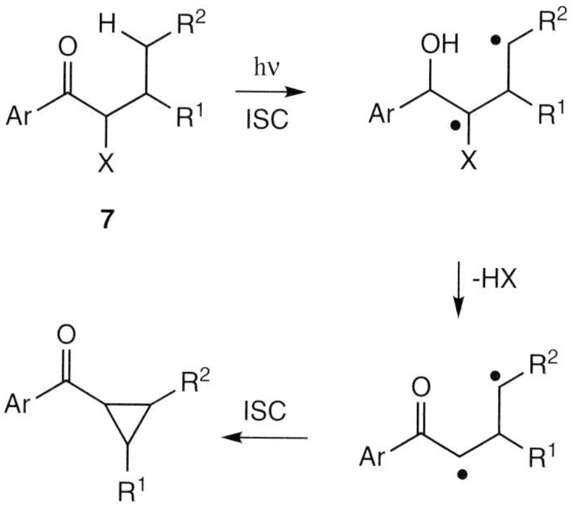

**Scheme 24.10** Mechanistic scenario for the spin-shift approach to cyclopropanes.

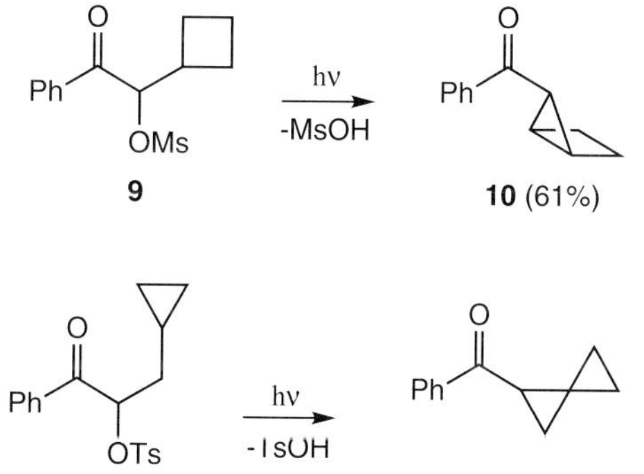

**Scheme 24.11** Applications of the spin-shift approach to strained cyclopropanes.

## 24.3.4
### Formation of Larger Ring Products

Proper substrate design with respect to constrained configuration and conformation can accomplish also the formation of products with medium and large rings via hydrogen abstraction and subsequent biradical combination. The PET process which constitutes an electronic trick to lead reactivity far away from the excited chromophore is described in Section 24.4.2. The Norrish II hydrogen transfer which is an efficient and often dominant process can be used for the formation of three-membered rings by the spin center shift process. If, on the other hand, a radical is produced that undergoes a rapid secondary reaction, that is, a cyclopropyl carbonyl ring opening, the resulting 1,7-biradical can cyclize to give an hydroxycycloheptane. In an example reported by Kraus and Wu, α-keto esters **13** were irradiated and gave cleavage and seven-membered ring products **14** and **15**, respectively

**Scheme 24.12** Norrish II reaction followed by cyclopropylcarbinyl ring-opening.

(Scheme 24.12) [25]. Phenyl substitution of the cyclopropane ring is one trick to accelerate the cyclopropyl carbonyl ring opening.

The hydrogen transfer process can also be directed to a remote position by use of constraining ring systems in the chain connecting the chromphoric carbonyl group and the potential H donor position. This has been realized for phenylglyoxylates **16**. The abstraction of γ-hydrogen was not observed for these substrates and consequently no cyclopropane ring-opened products (Scheme 24.13). Due to the

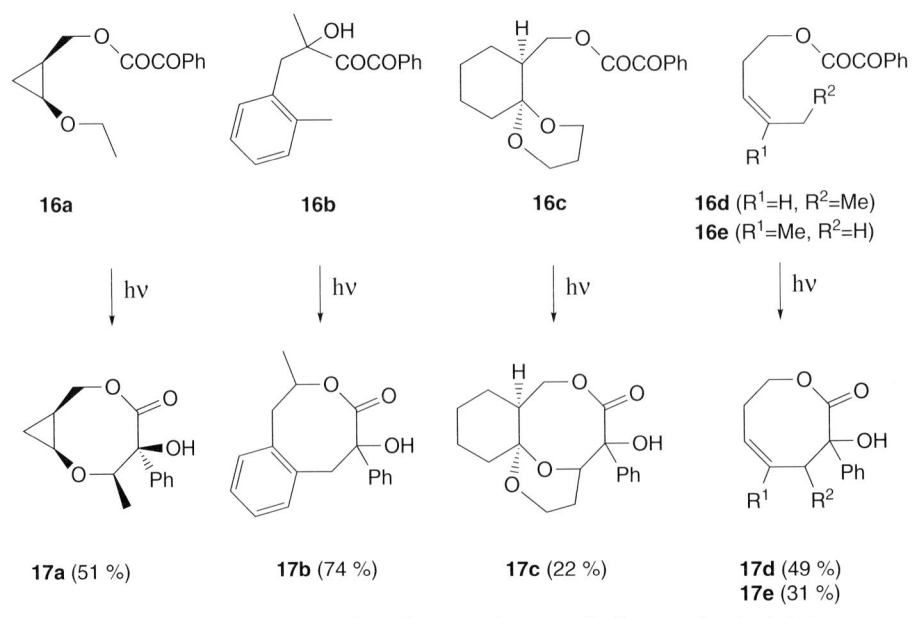

**Scheme 24.13** Eight-membered ring formation by H-transfer from restricted substrates.

geometrical restrictions from the connecting cyclopropane block, the 1,9-hydrogen transfer is favored.

Macrocyclic products are thus available by remote functionalization, a concept proposed by Breslow [26]. The concept is depicted in Scheme 24.14: 1,4-functionalized benzophenone-4-carboxylic esters **A** with a hydrogen acceptor at C1 and a hydrogen donor chain at the para-position are, under dilute reaction conditions, ideal precursors for the synthesis of macrocyclic *para*-cyclophanes **B**. The regioselectivity of the hydrogen abstraction is low and chain hydrogens between C-10 and C-19 of the carbon chain are active. This low regioselectivity was improved by introduction of a conformationally rigid steroid skeleton in place of the flexible alkyl chain [27].

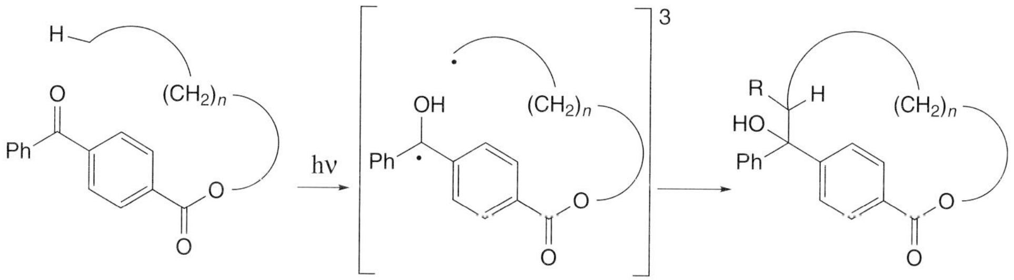

**Scheme 24.14** Principle of remote hydrogen transfer.

## 24.4
## Photoinduced Electron Transfer Cyclizations

### 24.4.1
### Formation of Three- to Six-membered Rings

The formation of three- to six-membered rings is a domain of photoinduced group transfer reactions such as hydrogen transfer. In many cases, however, improved regio- and stereoselectivity can be achieved if the site of activation is clearly defined by a photoinduced electron transfer. This concept is described herein (as a model system) in more detail for ring-forming reactions of isoindoline-1,3-diones (phthalimides), that are easily available either via solvent-free condensation of phthalic anhydrides with amines or via coupling methods under less drastic conditions and are versatile electron-acceptors in PET reactions. N-Substituted phthalimides typically absorb in the 295 nm range with extinction coefficients around $10^3$. The quantum yields for intersystem crossing $\Phi_{ISC}$ change significantly with the substitution on the imide nitrogen, for example, $\Phi_{ISC} = 0.5$ for N-isobutylphthalimide and $\Phi_{ISC} < 0.01$ for N-arylphthalimides [28]. Nevertheless, efficient population of the triplet state is possible by sensitization. With a triplet energy $E_T$ of 293–300 kJ mol$^{-1}$ and a ground state reduction potential around $E° = -1.85$ V vs. Fc/Fc$^+$, electronically excited phthalimides are potent electron acceptors [29]. The rich photochemistry of this chromophore has recently been reviewed [30]. The 1,4-biradicals formed by γ-CH transfer can undergo several subsequent reactions, among them secondary H transfer, cyclization, or fragmentation. The excited imido group is, at the same time, an efficient electron acceptor and can be reduced by numerous donor groups. The two routes can be distinguished for amino acid derivatives of phthalimides **17** and **19** from valine and aspartate (Scheme 24.15).

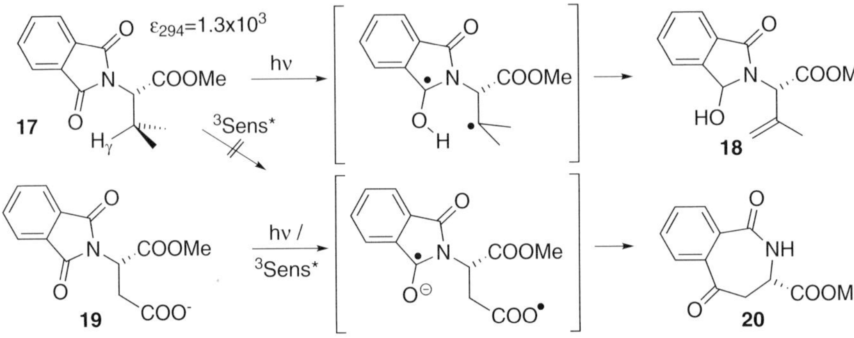

**Scheme 24.15** Photochemistry of N-phthaloyl valine and methyl aspartate.

The photophysical and photochemical properties of the N-phthaloylvaline methyl ester **17** were studied by nanosecond laser flash photolysis ($\lambda_{exc} = 248$ or 308 nm) [31]. The quantum yield of fluorescence is low ($\Phi_F = 10^{-2}$), whereas that of phosphorescence at $-196\,°C$ is large (0.5). Formation of singlet molecular oxygen ($^1\Delta_g$-$^1O_2$) was observed in several air- or oxygen-saturated solvents at room temperature with substantial quantum yields ($\Phi_\Delta = 0.47$ in acetonitrile). The triplet properties were examined at room temperature and in ethanol at low temperatures: triplet acetone, acetophenone and xanthone in acetonitrile are quenched by **17** via energy transfer; the rate constant is almost diffusion-controlled and somewhat smaller for benzophenone. No products were formed, however, when the triplet of **17** was generated by these sensitizers, only direct irradiation led to the formation of **18** with a quantum yield of 0.2. From the photophysical data, the formation of the photoisomerization product (the isodehydrovaline derivative **2**) via the observable π,π triplet state appeared unlikely and an nπ* singlet or an upper excited nπ* triplet pathway was therefore proposed. In contrast to the behavior of **17**, the aspartate derivative **19** gave the benzazepine-1,5-dione **20** in good yield, either by direct excitation or by triplet sensitization. This ring-enlargement most probably involves the formation of an intermediate hydroxyazetidine that converts to the seven-membered heterocycle. Following this first example of intramolecular photodecarboxylation, this route was developed into a powerful tool [32]. Due to the plethora of ω-amino acids from the pool of naturally occurring compounds or available by synthetic methods, structurally diverse substrates can be easily converted into imides. The phthaloyl derivatives of α-amino acids undergo efficient photodecarboxylation, resulting in the corresponding amines (for the exception of sulfur-containing substrates– *vide infra*), β-amino acids are converted to benzazepines, and γ-amino acids to benzopyrrolizidines. The glutamic acid derivative **21** resulted in the formation of a diastereomeric mixture of benzopyrrolizidinones **22** that were converted via acyliminium cation chemistry into the allylated pyrrolizidine **23** (Scheme 24.16) in diastereomerically pure form (with ee > 98%) [33].

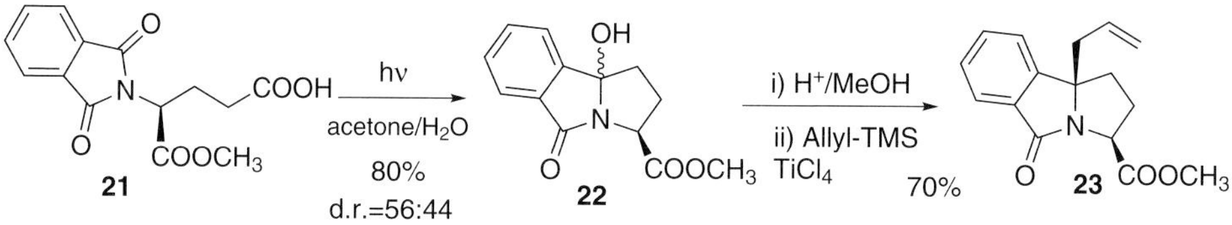

**Scheme 24.16** Photodecarboxylative cyclization of N-phthaloyl methyl glutamate.

## 24.4.2
### Formation of Larger Ring Products

A large number of PET-induced photocyclizations have been reported in the literature. A most prominent field of application is the synthesis of complex polycyclic or macrocyclic ring-systems, still a challenge in synthetic organic chemistry [34]. PET cyclizations of donor-substituted phenylglyoxylates have been intensively studied by the groups of Neckers and Hasegawa. For the thioether derivatives **24** the efficiency of the cyclization to give the macrocycles **25** depended on the linker chain length (Scheme 24.17) [35]. With longer carbon tether ($n = 2$–10) the yield dropped steadily and secondary reductive dimerizations or Norrish Type II processes became competitive. The $C_{11}$-linked substrate solely underwent dimerization and cleavage.

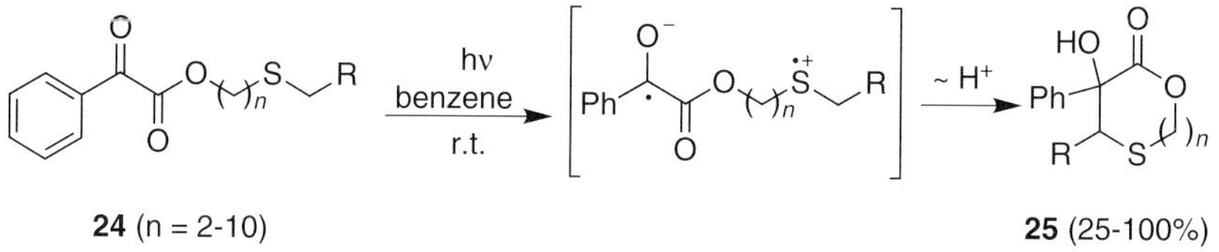

**Scheme 24.17** PET-cyclization of thioalkyl α-keto esters.

In this context the photocyclization of N-phthaloyl methionine (**26**) is worthy of mention (Scheme 24.18). When irradiated in pure acetone, this compound gave the tetracyclic lactone **27** in high yields [36]. This reaction is unusual in the sense that

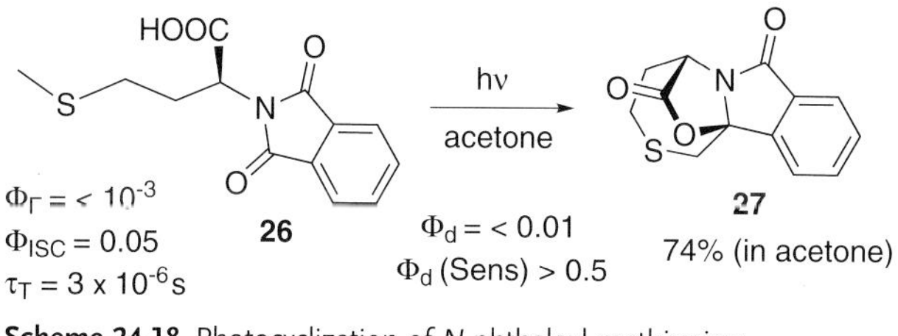

$\Phi_r = < 10^{-3}$
$\Phi_{ISC} = 0.05$
$\tau_T = 3 \times 10^{-6}$ s

$\Phi_d = < 0.01$
$\Phi_d$ (Sens) > 0.5

74% (in acetone)

**Scheme 24.18** Photocyclization of N-phthaloyl methionine.

photolysis of unprotected N-acyl amino acids normally leads to efficient α-decarboxylation. Thus, electron transfer reactions involving thioalkyl groups can efficiently compete with carboxylate activation.

From the investigation of other sulfur-substituted ω-amino acids, it became clear that the linker separating the primary electron donor ($Do^1$, i.e., the thioether group) and the terminal donor ($Do^2$, i.e., the carboxylate) is crucial. In **28**, one to four-carbon chains separates $Do^1$ and $Do^2$ (Scheme 24.19). A distance dependence of the decarboxylation efficiency can be extracted from the data for the eight different sulfur-substituted substrates. This also becomes apparent from the yields of the photocyclization products indicating that, for compounds with longer spacers between sulfur and the carboxylate anion, alternative photochemistry competes with or (in the case of the methionine derivative **26**) completely supresses decarboxylation.

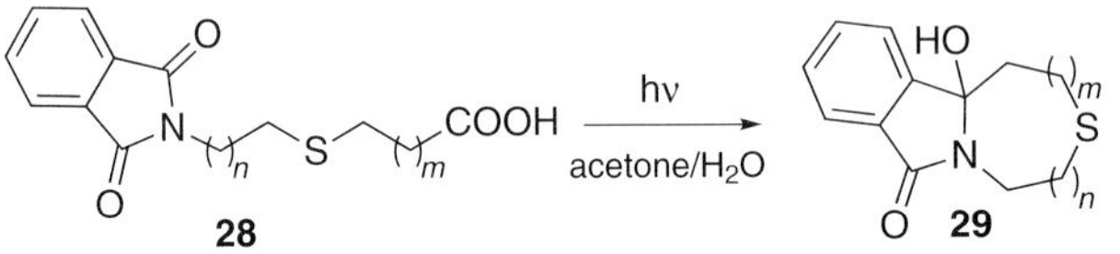

|   | n = | m = | ring size | yield(%) |
|---|---|---|---|---|
| a | 0 | 0 | 5 | 95 |
| b | 0 | 1 | 6 | 20 |
| c | 1 | 0 | 6 | 61 |
| d | 1 | 1 | 7 | 11 |
| e | 2 | 0 | 7 | 62 |
| f | 2 | 1 | 8 | 15 |
| g | 3 | 0 | 8 | 64 |

**Scheme 24.19** PET-cyclization of N-phthaloylalkylamino thioalkylcarboxylates.

The corresponding amino-substituted phenylglyoxylates **30** have been studied by Hasegawa and Yamazaki [37]. According to the results described for thioethers, photocyclization exclusively occurred at the remote α position to nitrogen (Scheme 24.20). This remarkably high regioselectivity has been rationalized by the increased acidity of the α-CH of the radical cation in combination with the close donor–acceptor geometry prior to cyclization. The photoreaction of the dibenzylated compound ($R^1 = Ph$, $R^2 = Bn$) proceeded efficiently due to the benzylic stabilization with a quantum yield for its disappearance ($\Phi_d$) of unity. The cyclization diastereoselectivity is low and a 1 : 1 isomeric mixture of **31** was obtained in a total yield of 73%.

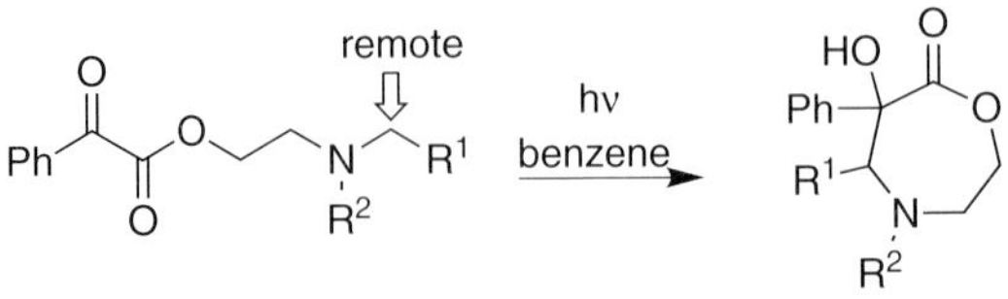

**Scheme 24.20** PET-cyclization of aminoalkyl α-keto esters.

Numerous photochemical reactions of alkenyl phenylglyoxylates have, additionally, been studied as potential PET substrates. Photocyclization is possible when the alkenyl group is situated at a proper distance and in a suitable configuration [38]. In many cases, dimerization, Norrish Type II cleavage and Paternò–Büchi reactions became dominant. Consequently, synthetic applications based on these derivatives are limited. In recent publications, Hasegawa and coworkers were able to show that aromatic β-oxoesters are also capable of macrocyclization reactions initiated via photoinduced electron transfer. Upon irradiation in benzene, the thioether linked substrates **32** gave the eight-membered thiolactones **33** in moderate yields (Scheme 24.21) [39]. Cyclization occurred exclusively at the remote position α to the heteroatom.

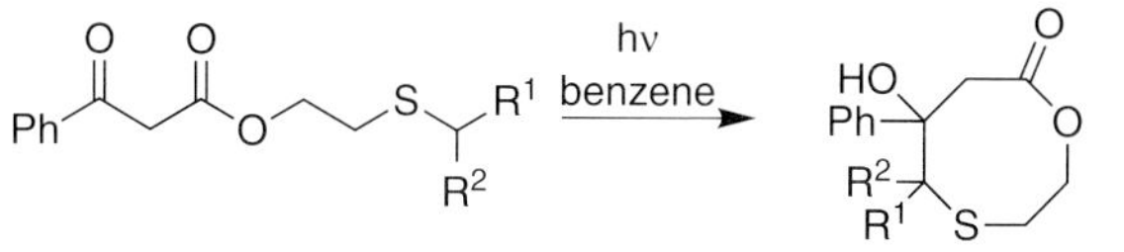

**Scheme 24.21** PET-cyclization of thioalkyl β-keto esters.

The regioisomeric azalactones **35** and lactones **36** were obtained when the corresponding amino analogues **34** were irradiated in benzene (Scheme 24.22) [40]. The photoproducts arising from hydrogen transfer from the proximate side with respect to the nitrogen donor group were formed in low yields, and their formation was totally suppressed for the twofold N-benzyl-substituted starting materials. The rigidity of the chromophore-bridge-donor system is partly compensated for most of these derivatives which allows a competing proton transfer from both sides of the nitrogen radical cation.

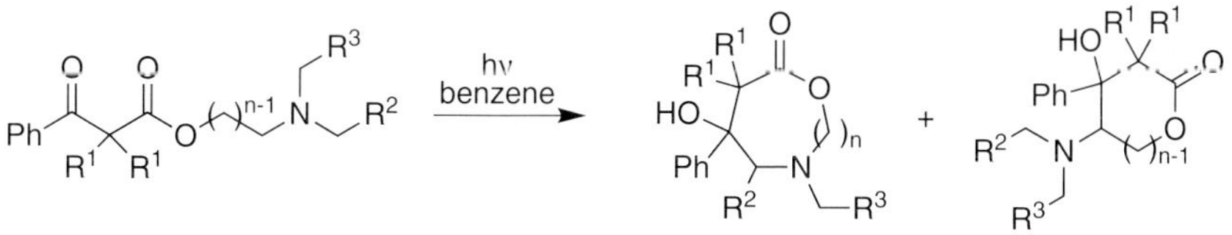

**Scheme 24.22** PET-cyclization of aminoalkyl β-keto esters.

PET reactions of aliphatic dialkylamino acetoacetates are much more sensitive towards the substitution pattern at the nitrogen atom and only derivatives carrying at least one benzyl group undergo effective cyclization [41]. Similar behavior was observed for aromatic γ-oxoesters and only substrates incorporating terminal benzylamino substituents gave the corresponding azolactones [42].

As already described for amino acid derivatives, electronically excited phthalimides are capable of efficient electron transfer with oxidation of thioether groups in the side-chain [43] with formation of the corresponding unbranched azathiacyclols as major products in moderate to good yields, thus reflecting the higher kinetic acidity of the primary vs. secondary α-CH group. Using the phthalimide/methylthioether pair, a variety of photochemical transformations resulting in medium- and macrocyclic sulfur-containing amines, lactams, lactones, and crown ether analogues, respectively, with a maximum ring-size of 38 atoms have been described [44]. Scheme 24.23 summarizes some selected examples.

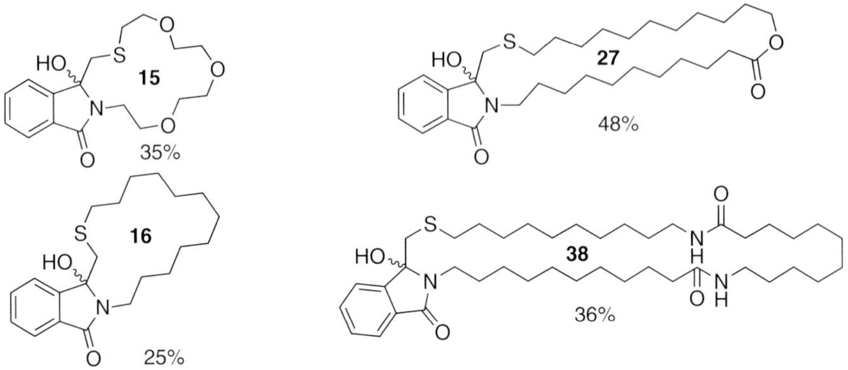

**Scheme 24.23** Macrocyclic products from the photolysis of phthalimides.

Photoinduced electron transfer reactions of aminoalkyl-substituted phthalimides are highly exergonic and proceed with high reaction rates, following the Rehm–Weller protocol. The product spectrum parallels that of the thioether case (*vide supra*) although yields were in general lower [45]. The progress of intramolecular PET reactions involving alkenyl phthalimides is essentially influenced by the solvent [46]. Upon irradiation in MeCN, $[\pi^2 + \sigma^2]$-addition to the C(O)–N bond takes place and benzazepinediones are obtained. In alcohol, the intermediary formed radical cation is trapped in an anti-Markovnikov fashion, depending on the polarity as well as the nucleophilicity of the solvent [47]. Recently, Xue *et al.* described an interesting modification of the latter process using tetrachlorophthalimides with remote hydroxyalkyl substituents (**37**) [48]. During photolysis in benzene and in the presence of alkenes, the alkene radical cation intermediate was trapped by the terminal hydroxy function, followed by an intramolecular radical–radical combination to give macrocyclic lactones, for example, **38** in 50% yield (Scheme 24.24).

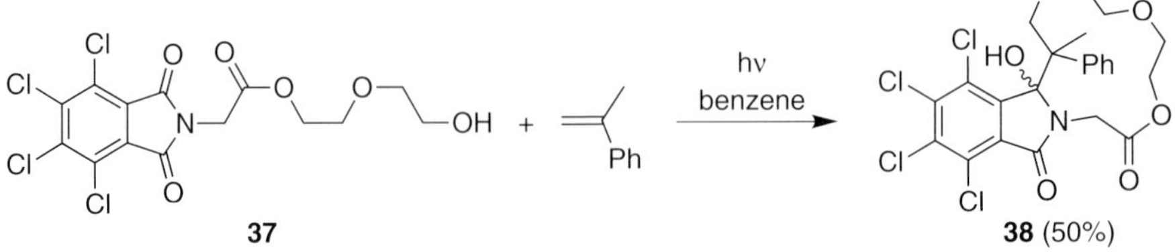

**Scheme 24.24** PET-addition and intramolecular trapping of tetrachlorophthalimides.

Griesbeck and coworkers have developed the decarboxylative photocyclization of phthalimido ω-alkylcarboxylates **39** as a versatile route to macrocyclic ring systems **40** (*vide supra*). The carboxylate serves as an electron donor and $CO_2$ is eliminated during the course of the reaction. Applying this concept, the syntheses of medium- and macrocyclic amines, polyethers, lactams, lactones, as well as cycloalkynes were accessible but the limitations concerning functional groups or maximum ring size have not been exactly explored yet (Scheme 24.25) [49, 50].

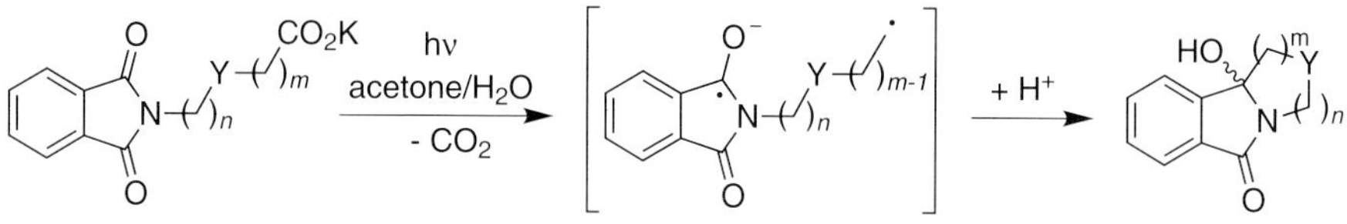

**Scheme 24.25** General route to macrocycles by photodecarboxylative cyclization.

The efficiency and selectivity of the cyclization can be improved by incorporating a suitable leaving group in the α-position to the electron donor. Trimethylsilyl-, trialkylstannyl or carboxylate functions have been utilized for this strategy [51].

In contrast to intramolecular PET where a radical ion pair is formed that undergoes the cyclization mode either by ionic recombination or, more often, by radical combination, oxidative and reductive PET are intermolecular processes, where PET-catalysis is used to produce either radical cations or radical anions of the substrate. Subsequent ring formation is followed by reformation of the neutral closed-shell states either by back electron transfer from or to the sensitizer (catalyst) radical ion. In many cases, the reactivity of the initially formed radical ion pair is reduced by use of a mediator in addition to the substrate/sensitizer system. If a twofold electron transfer is desired, sacrificial electron donors or acceptors have to be added (Scheme 24.26).

### 24.4.2.1 Oxidative PET

The efficient photoinduced electron transfer initiated cyclization of α-trimethylsilylmethyl amines has been systematically evaluated by Pandey and coworkers [52]. Using this strategy, various mono- or bicyclic amines have been constructed via the trimethylsilylmethyl amine radical cations (Scheme 24.27). The photosystem employed to generate this reactive intermediate utilized 1,4-dicyanonaphthalene (DCN) as the light harvesting electron acceptor. Extrusion of the α-trimethylsilyl group and subsequent cyclization to a tethered π-functionality gave the corresponding N-heterocyclic products **42**, **44**. In a series of publications, Pandey *et al.* described a variety of synthetic applications of this PET-mediated cyclization and, for example, the syntheses of (±)-isoretronecanol, (±)-epilupinine, and 1-N-iminosugars have been realized.

A similar radical cyclization reaction of unsaturated amino acid derivatives has been recently reported by Steckhan and coworkers [53]. The PET-catalyzed cyclization reaction proceeds under mild conditions using 9,10-dicyanoanthracene (DCA) as

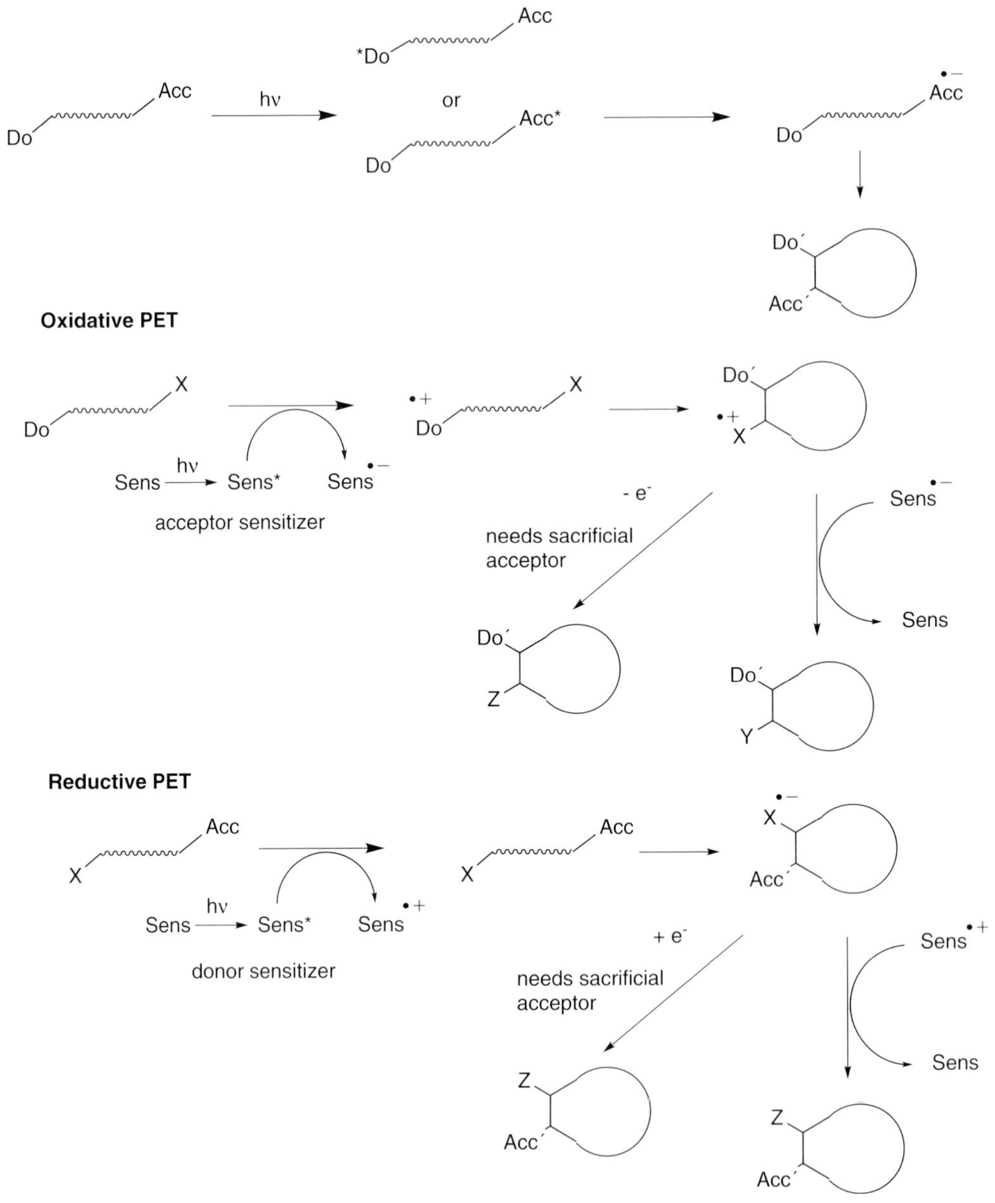

**1170** | *24 Photocyclization Reactions*

**Oxidative PET**

**Reductive PET**

**Scheme 24.26** Principles of oxidative and reductive PET-cyclization.

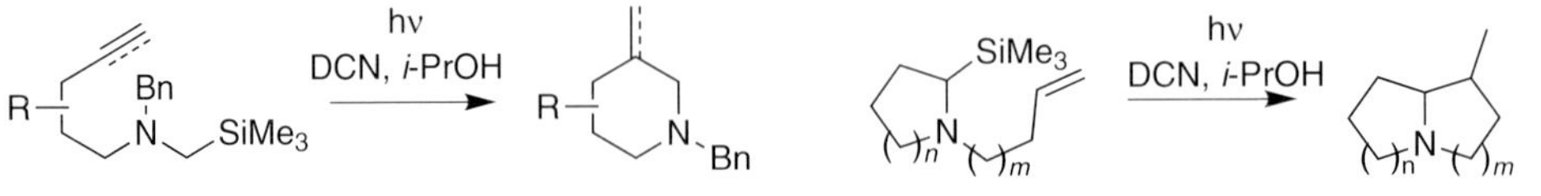

**41** → (hv, DCN, *i*-PrOH) → **42**

**43** (n, m = 1, 2) → (hv, DCN, *i*-PrOH) → **44** (85-90%)

**Scheme 24.27** PET-cyclization of trimethylsilylmethylamines.

sensitizer and biphenyl (BP) as co-sensitizer. The diastereoselectivity of the cyclization step was found to be moderate to high depending on the substitution pattern of the starting material. In almost all cases examined, the trans-diastereoisomer was predominately formed. An elegant example of the latter reaction is its application to peptide chemistry for inducing structural changes within the peptide chain. Peptide **45** readily cyclized to the 4-methylproline-containing derivative **46** in 64% yield (Scheme 24.28). The diastereoisomeric ratio was found as 2:1 in favor of the cis-isomer for this example. The cis-selectivity reflects the approach of the intermediate radical towards the double bond which is determined by the predominant conformation of the peptide chain.

PET activations of cyclic tertiary amines utilizing 1,4-dicyanonaphthalene (DCN) as electron acceptor have been studied by Pandey and coworkers [54]. The iminium cationic intermediates are generated via an electron–proton–electron (E–P–E) transfer sequence in a highly regiospecific fashion and as an example, tetrahydro-1,3-oxazines **48** were synthesized from substrates **47** with high control of regio- and stereoselectivity (Scheme 24.29).

PET cyclizations of β,γ-unsaturated oximes **49** have been established by Armesto and Horspool [55]. This 9,10-dicyanoanthracene (DCA) sensitized reaction provides an efficient route for the synthesis of dihydroisoxazole derivatives **50** in reasonable to good yields (Scheme 24.30). The key step in the mechanistic scenario is an electron transfer from the oxime group to DCA in its excited singlet state. The radical cationic intermediate thus generated undergoes subsequent exo-cyclization with the olefinic moiety. Further proton and electron transfer steps complete the reaction. A major limitation of this process is the restriction to molecules incorporating two aryl groups.

Mattay and coworkers examined the regio- and stereoselective PET-cyclization of unsaturated silyl enol ethers **51** using DCA, or alternatively DCN, as sensitizer [56]. The regiochemistry (6-endo vs. 5-exo, i.e., **52** versus **53**) of the cyclization could be controlled via the solvent applied. In the absence of a nucleophilic solvent, such as in alcohols, the cyclization of the siloxy radical cation dominates, whereas the presence of a nucleophile favors a reaction pathway to **52** via the corresponding α-keto radical (Scheme 24.31). The resulting stereoselective cis ring junction is due to a favored reactive chair-like conformer with a pseudo-axial arrangement of the substituents.

A versatile strategy for efficient intramolecular α-arylation of ketones was achieved by the reaction of silyl enol ethers with PET-generated arene radical cations. This strategy involved one-electron transfer from the excited methoxy-substituted arenes to ground-state DCN [57]. Pandey and coworkers reported the construction of five- to eight-membered benzannulated as well as benzospiroannulated compounds using this approach (Scheme 24.32). The course of the reaction can be controlled via the silyl enol ether obtained from the starting ketone **54**. The thermodynamically controlled silyl enol ether gave the branched cyclization products **55**, whereas kinetically controlled silyl enol ethers led to the unbranched substrates **56**.

Demuth *et al.* recently developed an efficient radical cationic cyclization of functionalized polyalkenes using 1,4-dicyano-tetramethylbenzene (DCTMB) as an acceptor and biphenyl (BP) as co-sensitizer [58]. The transformation is, in general,

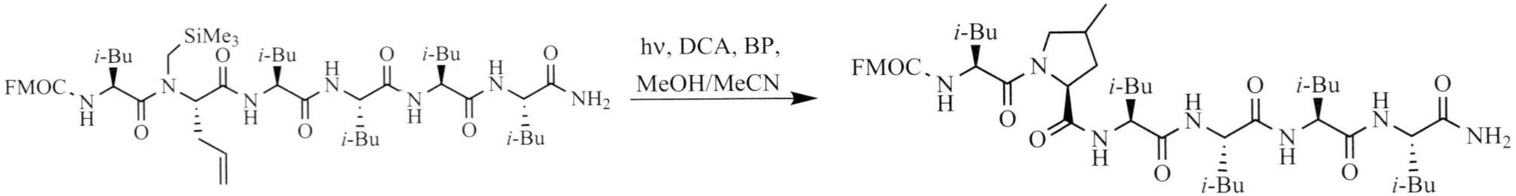


45

46 (64%)

**Scheme 24.28** PET-cyclization of an oligopeptide using the desilylation path.

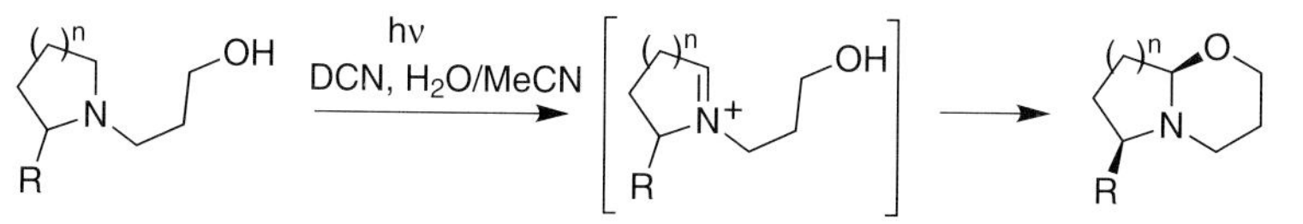


**47** (n = 1, 2; R = Alkyl) **48** (88-92%)

**Scheme 24.29** Oxidative PET-cyclization with nucleophilic alcohol trap.

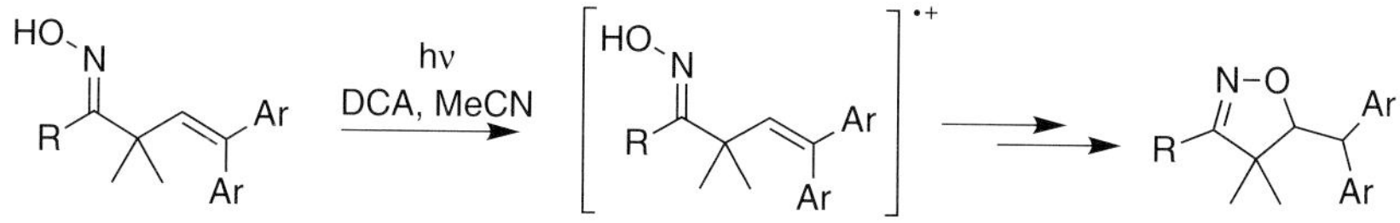


**49** **50** (35-64%)

**Scheme 24.30** Oxidative PET-cyclization with nucleophilic oxime trap.

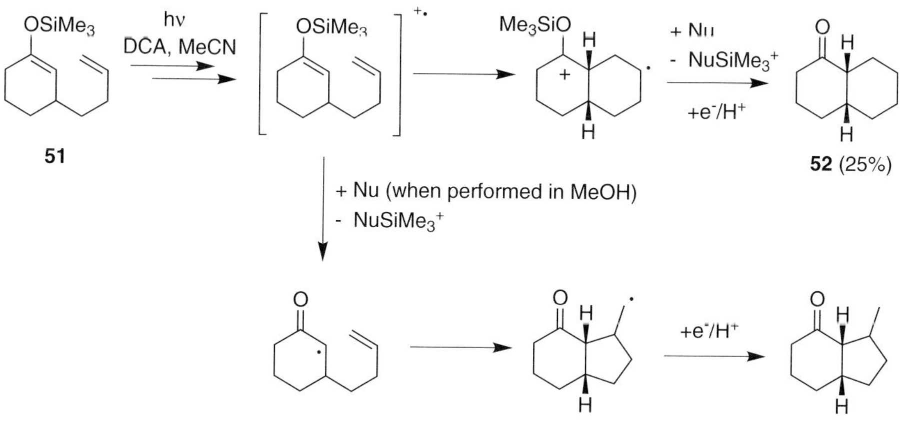


**53** (64%)

**Scheme 24.31** Oxidative PET-cyclization of silyl enolethers by with nucleophilic alkene trap.

highly stereo- and chemoselective and the substitution pattern of the polyalkene allows the construction of either five- or six-membered rings (Scheme 24.33). So far, this methodology has been applied for the synthesis of several natural products such as stypoldione, hydroxyspongianone and abietanes, respectively [58].

#### 24.4.2.2 **Reductive PET**

The synthesis of bicyclic cyclopentanols **61** via photoreductive cyclization of δ,ε-unsaturated ketones **60** has been developed by Cossy and Portella using hexamethylphosporic triamide (HMPA) or triethylamine (TEA) as electron donor [59]. The photocyclization proceeded remarkably efficiently when HMPA was used as donor and solvent (Scheme 24.34). Furthermore, only one stereoisomer was obtained carrying methyl and hydroxy groups in trans-configuration. In contrast, the yields

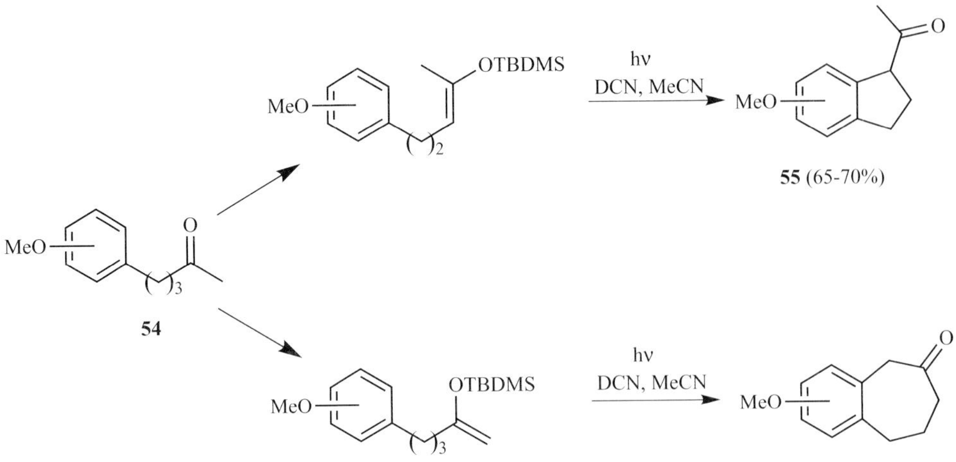

**Scheme 24.32** Oxidative PET-cyclization of silyl enolethers by nucleophilic arene trap.

for cyclization dropped significantly when TEA in a polar solvent such as acetonitrile was used. As an example, the yield for **61** ($n = 1$) decreased from 81% in HMPA to 50% in TEA/MeCN.

Mattay and coworkers used triethylamine as the electron donor in tandem cyclopropyl carbonyl ring opening/cyclization reactions of α-cyclopropylketones **62**. [60]. The initial electron transfer to the ketone moiety is followed by subsequent cyclopropylcarbinyl–homoallyl rearrangement yielding a distonic radical anion **50**. With an appropriate unsaturated side chain attached to the molecule both annelated and spirocyclic ring systems **63** are accessible in moderate yields. Scheme 24.35 shows two representative examples, the formation of a spiro[5.4]decane **65** from the homoallylic substrate **64**, and a propellane **67** synthesis from the cyclopropanated bicyclo[4.3.0]nonane **66**. This methodology is analogous to the hydrogen transfer initiated reaction, as described for substrate **13** in Scheme 24.12.

Pandey and coworkers have developed two useful photosystems for one-electron reductive chemistry and applied them to the activation of aldehyde or ketones tethered with activated olefins **68** [61] and α,β-unsaturated ketones **70**, respectively (Scheme 24.36). The central step in this complex mechanistic scenario is the PET involving a sacrificial electron donor (see Scheme 24.26). The 9,10-dicyanoanthracene radical anion (DCA$^{•-}$) thus generated undergoes a secondary thermal electron transfer to the unsaturated ketone. The resulting carbon-centered radical or radical anionic intermediate, subsequently cyclizes stereoselectively with a proximate olefin. As sacrificial electron donors, the ascorbate ion, in combination with 1,5-dimethoxynaphthalene (DMN) as primary electron donor, or triphenylphoshine in combination with DCA have been applied.

Highly selective cascade cyclizations of terpenoid polyalkenes via photoinduced electron transfer have been accomplished by Demuth and coworkers [62]. 1,4-Dicyanotetramethylbenzene (DCTMB) and biphenyl (BP) were applied as electron–acceptor

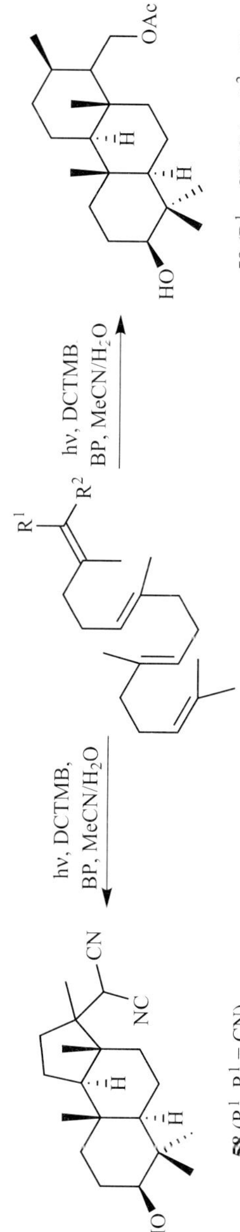


**58** ($R^1$, $R^1$ = CN)   **57**   **59** ($R^1$ = $CH_2OAc$, $R^2$ = H)

**Scheme 24.33** Oxidative biomimetic PET-polycyclization.

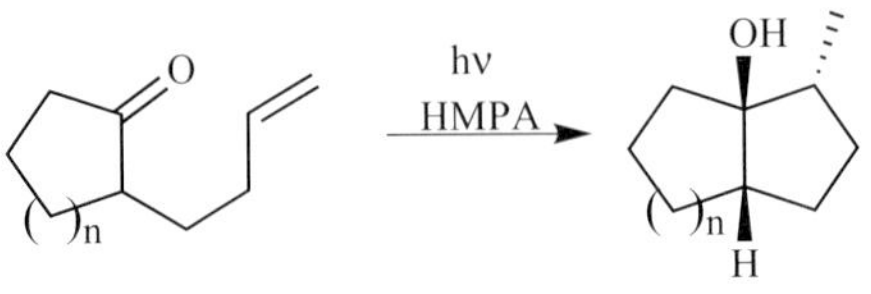
O
hν
HMPA
OH
( )n
H

**60** (n = 1, 2) **61** (76-81%)

**Scheme 24.34** Reductive PET-cyclization of cycloalkanones.

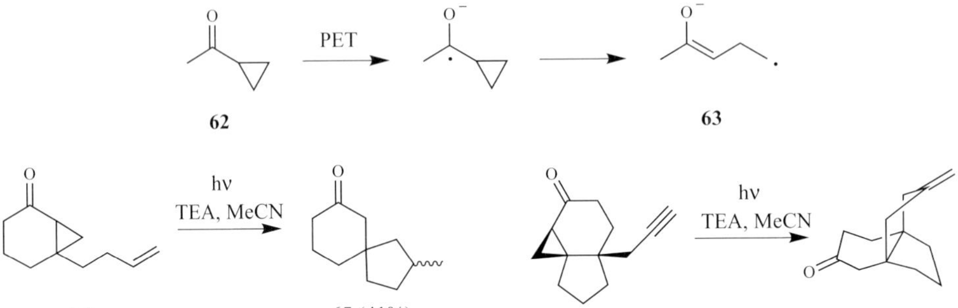
O
PET
O⁻
hν
TEA, MeCN
hν
TEA, MeCN

**62** **63**

**64** **65** (41%) **66** **67** (74%)

**Scheme 24.35** Reductive PET coupled with cyclopropylcarbinyl ring opening.

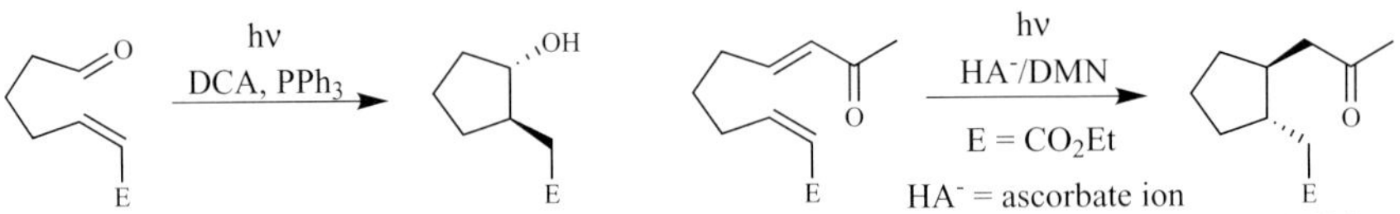
O
hν
DCA, PPh3
OH
E
hν
HA⁻/DMN
E = CO2Et
HA⁻ = ascorbate ion

**68** **69** (80%) **70** **71** (79%)

**Scheme 24.36** Reductive PET cyclization with sacrificial electron donors.

couple and a spirocyclic dioxinone/(−)-menthone pair was used as the chiral auxiliary. Although this directing group was placed in a remote location from the initiation site of the reaction, cyclization gave the enantiomerically pure cyclic terpenoids **73** through exclusive diastereofacial differentiation (Scheme 24.37). The efficient remote asym-

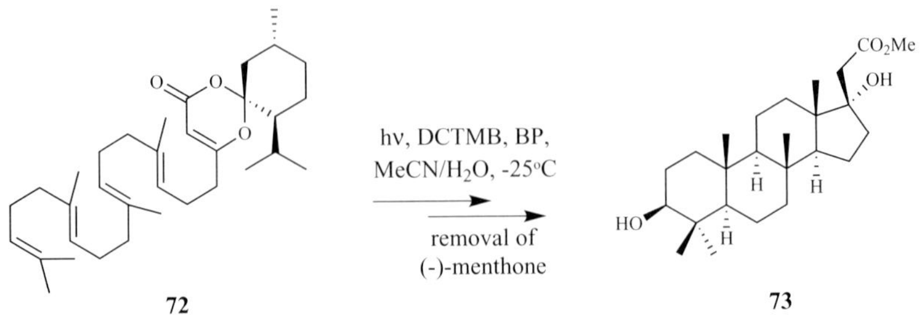
O
hν, DCTMB, BP,
MeCN/H2O, -25°C
removal of
(-)-menthone
CO2Me
OH
HO
H

**72** **73**

**Scheme 24.37** Reductive biomimetic PET-polycyclization.

metric induction is hereby most likely induced by enantioselective chiral folding of the polyalkene chain. It is noteworthy that eight stereogenic centers were created in a single step and only 2 out of 256 possible isomers were formed.

The phenomenon of memory of chirality [63] was recently extended to PET-cyclization reactions. An interesting example was the highly selective synthesis of pyrrolobenzodiazepines and photolysis of the proline derivatives **74** gave **75** with high enantiomeric excess (*ee*) values (Scheme 24.38) [64]. A reasonable explanation of this remarkably high memory effect was based on the restricted rotations about the amide C—N and the arene—N bonds in **74**. Consequently, when the *ortho*-phenylene linker was replaced by the conformationally more flexible ethylene linker, chirality memory vanished and a mixture of stereoisomeric products was obtained.

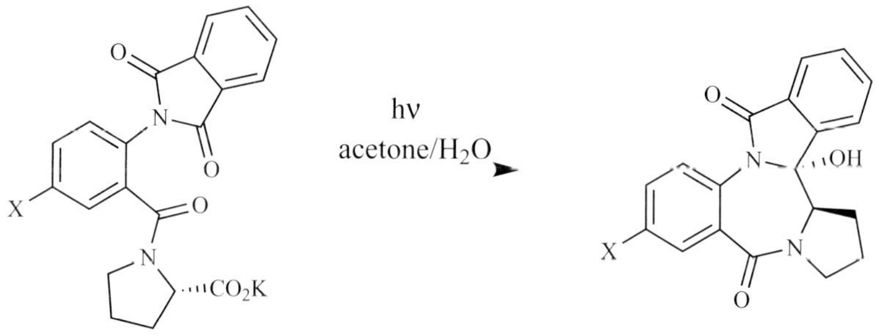

**Scheme 24.38** PET-photodecarboxylative cyclization with memory of chirality.

## 24.5
## π–π-Interactions

### 24.5.1
### $\pi^n$–$\pi^m$-Cycloadditions

Strictly speaking, $\pi^n$–$\pi^m$-cycloadditions are not defined as photocyclization processes and thus are only briefly discussed here. From an orbital symmetry point of view, suprafacial/suprafacial cycloadditions of the $\pi^2 + \pi^2$ and $\pi^4 + \pi^4$ are widely applied in organic synthesis. The $\pi^2 + \pi^2$-cycloadditions have been described in recent reviews and can be characterized by the relevant substrates as ene–ene [65], ene–enone [66] and ene–enone cycloadditions (and the carbonyl analogue) [3]. The combination of two dienes under photolysis conditions can result in and $\pi^2 + \pi^2$- (cyclobutanes), $\pi^4 + \pi^2$-(cyclohexenes), and $\pi^4 + \pi^4$- (cycloocta-1,5-dienes) products [67]. As a special variant of the ene–enone protocol, the photocycloaddition of alkenes to the β-hydroxy enone tautomers of 1,3-dicarbonyl compounds is described as the deMayo reaction. Aromatic substrates can also be used as ene components, leading to different addition modes which are characterized with

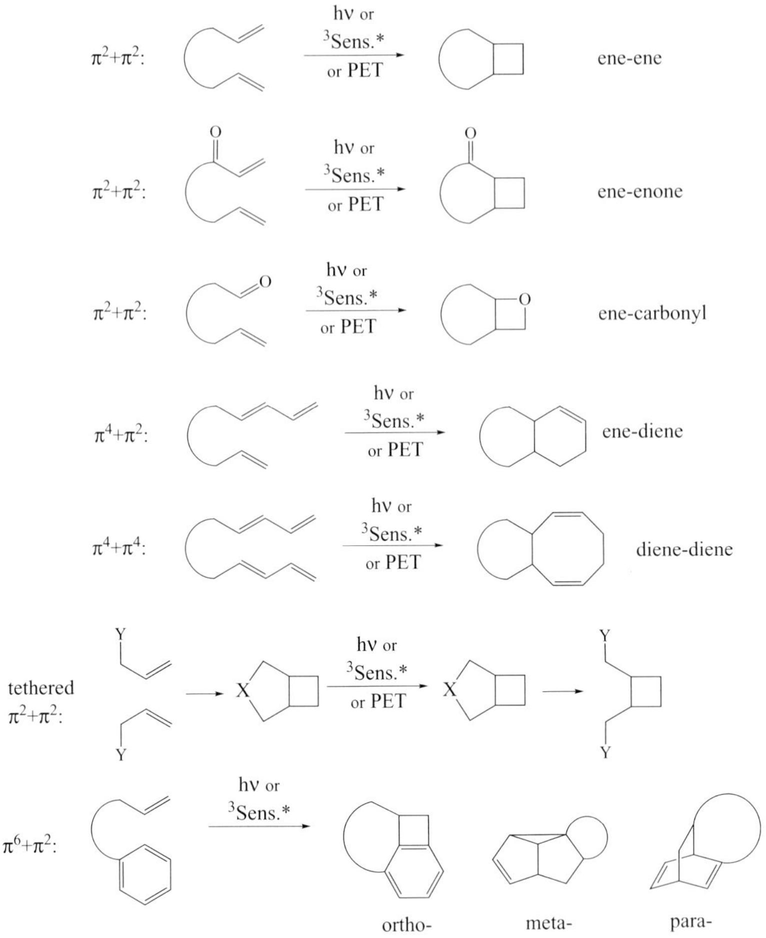

**Scheme 24.39** General scheme for $\pi^n + \pi^m$-photocycloadditions.

respect to the positions involved in the aromatic system as ortho(1,2)-, meta(1,3)-, and para(1,4) aromatic photocycloadditions (Scheme 24.39) [68]. Beside direct excitation of the π-chromophore or triplet sensitization, photoinduced electron transfer is a versatile method for activation of $\pi^n$–$\pi^m$-substrate combinations. Intermolecular reactions often suffer from the short lifetimes of the excited singlets and the tethering concept has been developed for this purpose. The spatial orientation of two allylic alcohols by silicon tethering leads to excellent yields and selectivities in photocycloadditions [69]. The tether group is most convenient if a non-photochemical cleavage follows the photocycloaddition.

Intramolecular variants of these reactions are especially valuable because they offer access to highly complex polycyclic products. An illustrative example is the synthesis of the pheromone grandisol **80** using two different substrates **76**, **77** and Cu-mediated photocycloaddition (Scheme 24.40) [70].

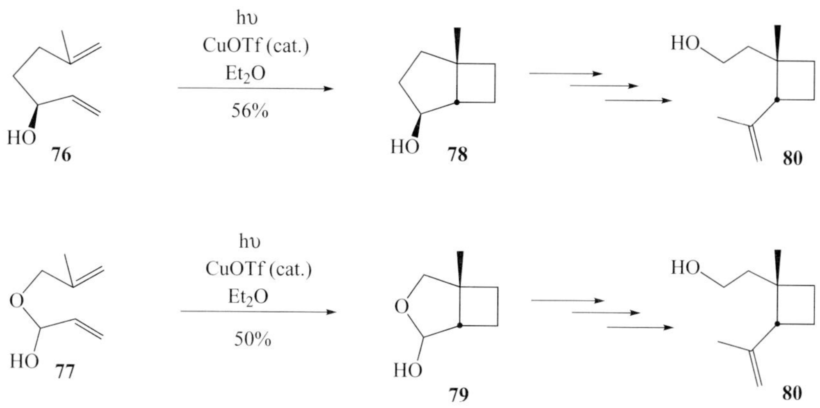

**Scheme 24.40** Cu-catalyzed intramolecular photocycloaddition to grandisol.

### 24.5.2
### $\pi^2$-$\pi^2$-Rearrangements: Di-π-Methane Rearrangement

The photoinduced interaction between two $\pi^2$ systems, as indicated in Scheme 24.39, can also lead to rearrangement products. An archetype process is the di-π-methane rearrangement where penta-1,4-dienes are photochemically converted into vinylcyclopropanes [71]. A $\pi^2 + \pi^2$-cycloaddition would result in the formation of highly strained bicyclo[2.1.0]pentanes ("housanes"). The rearrangement at the stage of the triplet biradical results in reaction products with considerably lower strain energy (Scheme 24.41). Because the rearrangement, that can be described as a 1,2-acyl shift, is also observed under direct excitation, a concerted pathway has also been considered [72]. An analogous rearrangement of a 1,5-hexadiene has also been described as a di-π-ethane rearrangement [73] and also a tri-p-methane was described recently [74].

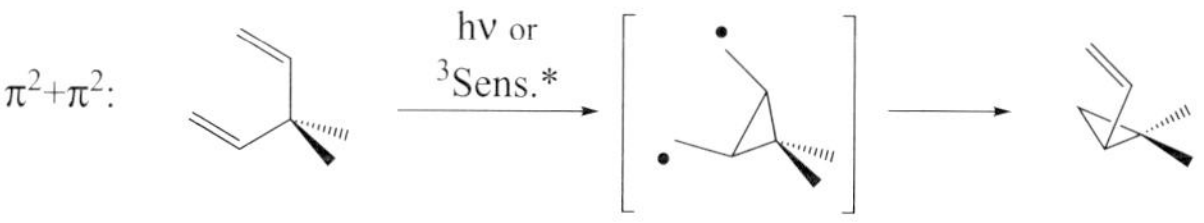

**Scheme 24.41** Mechanistic scenario of the di-π-methane rearrangement.

Applications of the di-π-methane rearrangement are numerous and in many cases conformationally restricted substrates are applied to enable rapid bond formation and to avoid deactivation of the excited states by E/Z-isomerization. Thus, the bicyclo[2.2.1]heptadienes and the bicyclo[2.2.2]-octadienes are most prominent among the starting materials for di-π-methane rearrangement [75]. The di-π-methane rearrangement and the heteroanalogue versions can be processed in the solution phase, in contrained media (zeolites, polymer films), and also in the solid state. An illustrative example is the photolysis of the dibenzobarrelene derivative **81** in the crystalline state resulting in a regioisomeric mixture (3 : 1) of dibenzodiquinanes **82a** and **82b** in high yields (Scheme 24.42) [76].

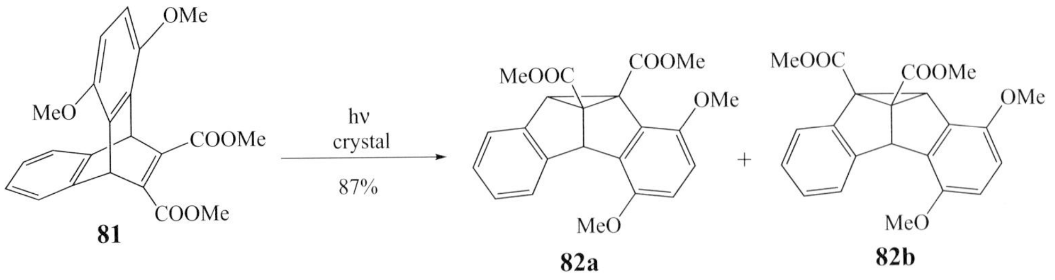

**Scheme 24.42** Barrelene di-π-methane rearrangement in the solid state.

### 24.5.3
### π²–π²-Rearrangements: Oxa-di-π- Methane Rearrangement

The 1-oxa version of the di-π-methane (ODPM) rearrangement is the most important heterovariant of the DPM rearrangement. The reaction constitutes a typical triplet carbonyl reaction and it is therefore not advisable to perform the process under direct excitation. This is merely due to the n,π* nature of the first excited singlet state which shows typical carbonyl reactivity like hydrogen abstraction or α-bond cleavage. The triplet state is of π,π* nature and prefers π reactivity, that is, addition to π-systems or electrophilic additions. Mono- and bicyclic oxa-di-π-methane substrates are the most versatile starting materials due to the favorable geometrical orientation of the C=O and the C=C double bonds as well as the restricted rotation about the C=C double bond (E/Z-isomerization) which is the favorable thermalization route of acyclic substrates. Cyclohexa-1,3-dien-5-ones 83 undergo an ODPM reaarangement upon direct excitation [77] in highly polar solvents like trifluoroethanol due to a change in order of the excited states leading to a π,π* state as the lowest excited singlet state (Scheme 24.43) which, after intersystem crossing, results in the formation of the ³π,π* state.

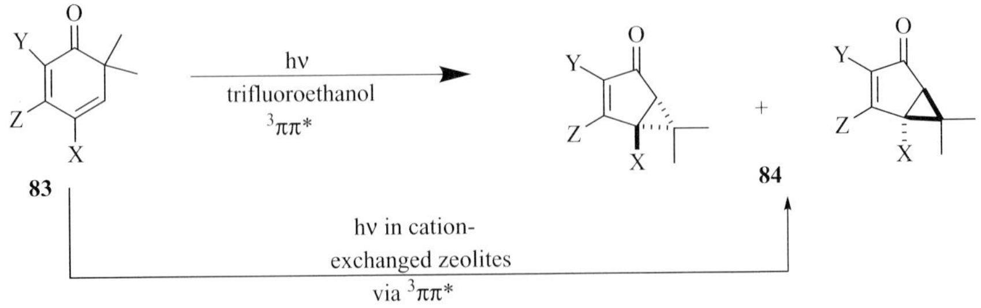

**Scheme 24.43** Oxa-di-π-methane rearrangement of cyclohexadienones.

When the ODPM rearrangement of 2,4-cyclohexadienones is performed in constrained environments such as in zeolites [78], not only is the diastereoselectivity of the cyclopropane formation influenced but also the asymmetric induction exhibited by a chiral auxiliary covalently connected to the chromophore. The photochemistry of 2,4-cyclohexadienones is known to be strongly solvent-dependent with the α-cleavage preferred in nonpolar solvents (presumably via the nπ*-state) and ODPM preferred in polar solvents (presumably via the ππ*-state). State switching can also be induced in alkali-exchanged Y zeolites [79]. The diastereoselectivity of the ODPM rearrangement of **85** leading to the diastereoisomers **86** and **87**, respectively, depends on the alkali metal used in the Y-zeolite host (Scheme 24.44).

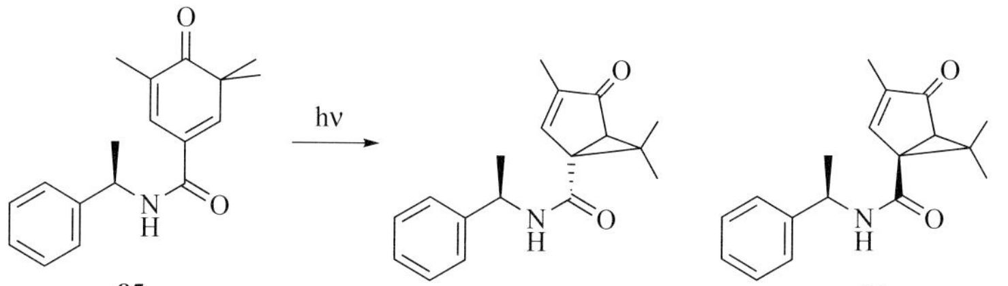

solution: d.e. = 5% (in trifluoroethanol)
LiY zeolite: d.e. = 17%
NaY zeolite: d.e. = 59%
KY zeolite: d.e. = 53%
RbY zeolite: d.e. = 39%
CsY zeolite: d.e. = 9%

**Scheme 24.44** Oxa-di-π-methane rearrangement of a chiral cyclohexadienone.

Cyclopentenylmethylketones undergo ODPM rearrangement (1,2-acyl shift) as well as 1,3-acyl shift (Scheme 24.45) reactions in the solution phase [80]. Schaffner and coworkers carried out gas-phase photolysis of the cyclopentenylmethyl ketones **88** in the presence of scavengers [81]. Their studies shed light on the mixed type of mechanism responsible for ODPM rearrangement (to give **89**) and also the 1,3-acyl shift reaction (to give **90**). These studies show that the ODPM rearrangement can be performed even in the gas phase and this enhances the general scope of the reaction.

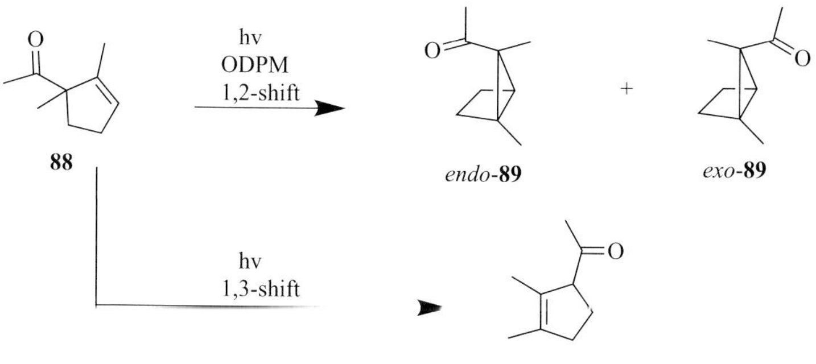

**Scheme 24.45** Competitive 1,2- versus 1,3-shift in 3-acylcyclopentenes.

Cerfontain and coworkers [82] investigated a series of acyclic β,γ-unsaturated ketones as substrates for the ODPM rearrangement. These compounds underwent efficient E → Z isomerization (free rotor effect) and only a slow ODPM rearrangement. In another study, they [83] synthesized several substrates **91** including (Scheme 24.46) β,γ- and γ,δ-dienones and studied their photochemistry.

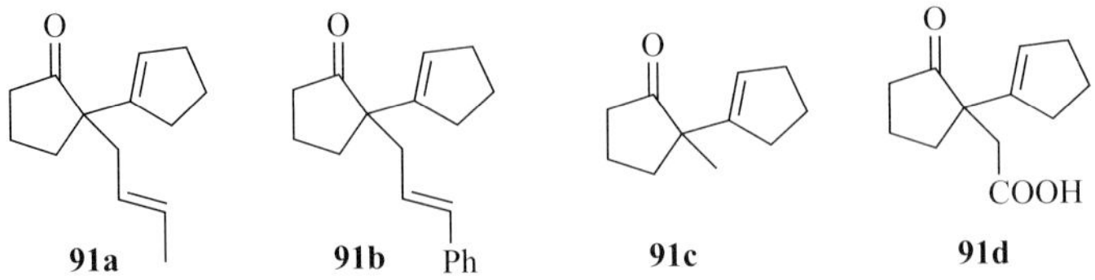

**Scheme 24.46** Oxa-di-π-methane reactive substrates with cyclic alkene unit.

Selective sensitization gave ODPM rearrangement products, whereas direct photolysis resulted in efficient E → Z isomerization and inefficient 1,3-acyl shift processes. The results also showed that γ,δ- double bonds or benzoyl groups present in these systems do not act as intramolecular triplet quenchers. Margaretha and coworkers [84] reported an illustrative example of ODPM rearrangement with a starting material **92** containing two carbonyl groups, where the double bond is situated α,β to one and β,γ to the other. Product **93** (Scheme 24.47) was derived from ODPM rearrangement with excellent quantum yield ($\Phi = 0.67$) and high selectivity under a variety of photolytic conditions.

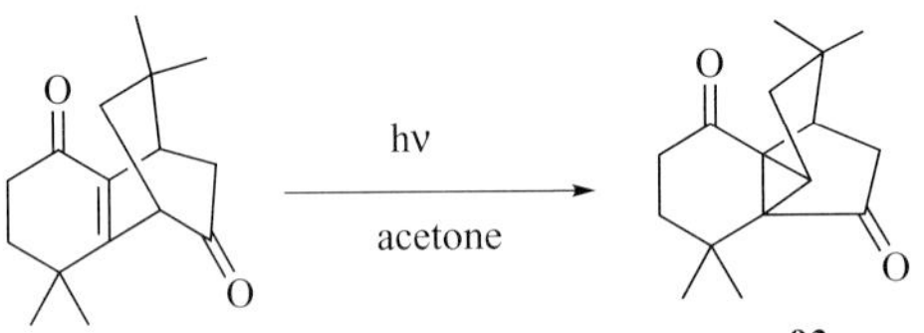

**Scheme 24.47** Oxa-di-π-methane rearrangement of a terpene-derived substrate.

Highly strained substrates can be transformed into even more highly strained isomers by ODPM rearrangement. This has been shown by Murata et al. for the synthesis of an azulene valence isomer [85]. Albeit the photochemical reaction yielded only 20–25% of the bicyclo[1.1.0]butane derivative **96**, the synthesis of the precursor cyclobutene **95** is straightforward from the bicyclo[3.3.0]octenone **94** (Scheme 24.48). Triplet sensitization of this compound reduces competitive

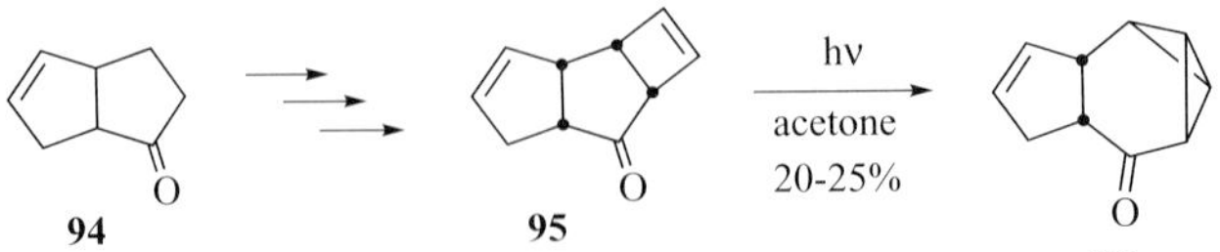

**Scheme 24.48** Oxa-di-π-methane rearrangement of a cyclobutene derivative.

pathways because the alkene excitation is excluded. Benzoannulated azulene valence isomers can also be synthesized by this approach [86].

The photolyses of conformationally fixed β,γ-unsaturated ketones can also be performed in constrained media. This was described for the ODPM rearrangement of bicyclo[2.2.1]heptenone and bicyclo[2.2.2]octenone in MY zeolites [87]. It is remarkable that direct irradiation of the guest-incorporated thallium-Y (TlY) zeolites at 254 nm results in higher yields of ODPM products than the triplet-sensitized solution photolyses. This indicates that the triplet carbonyl species are generated more efficiently in the supercages of the zeolites, presumably because of the presence of heavy atoms. The ODPM rearrangement of bicyclo[2.2.1]heptenone **99** leads to the highly strained housane derivative **100** (Scheme 24.49).

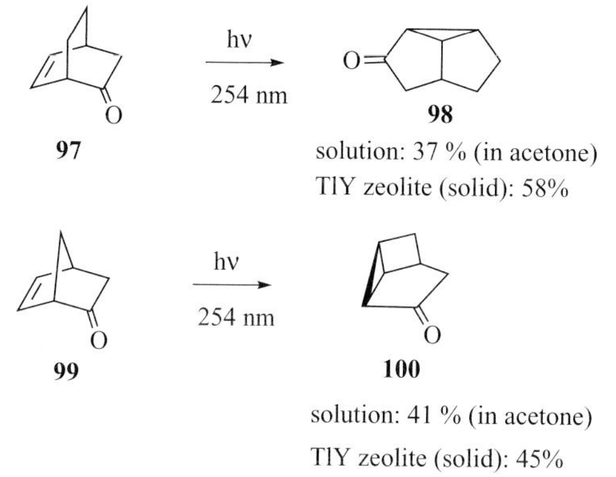

**Scheme 24.49** Oxa-di-π-methane rearrangements in zeolites.

Under normal circumstances, direct excitation of the carbonyl chromophore in bicyclo[2.2.2]octenones induces α-carbonyl C–C-bond cleavage (Norrish Type I process) and the biradical generated can undergo a variety of subsequent reactions. This dichotomy of photochemical reactions depending on the spin state of the excited substrate is nicely corroborated in a report by Liao *et al.* on the photochemistry of 3,3-dialkoxybicyclo[2.2.2]oct-5-en-2-ones **101** (Scheme 24.50) [88]. The ether-bridged substrates, when irradiated in acetone solutions, form highly functionalized tetracyclic

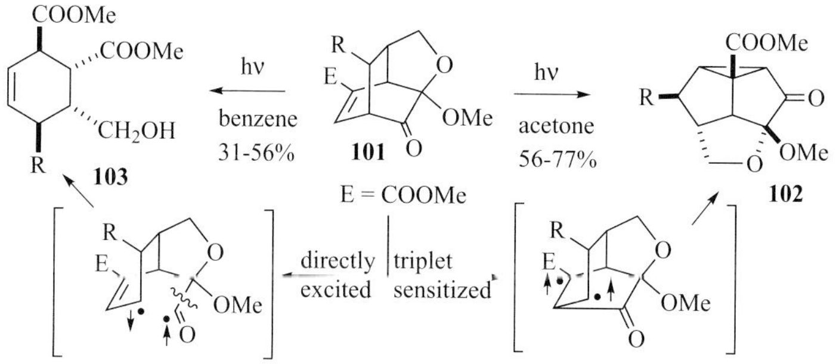

**Scheme 24.50** Photochemistry of bicyclo[2.2.2]octenones.

products **102**, whereas direct excitation (using the same wavelength as for sensitization – the sensitizer serving therefore also as an internal filter) in benzene solution resulted in decarbonylation and formation of functionalized cyclohexenes **103**.

If rapid bond formation at the stage of the Norrish I fragments competes with decarbonylation, 1,3-acyl shift products are formed as the major components from direct photolysis. In the case of the cyclobuteno-annulated bicyclo[2.2.2]octenone **104**, 1,3-acyl shift results in the cyclobutanone **105**, whereas ODPM (1,2-acyl shift) rearrangement (photolysis in acetone) results in the cyclobutenodiquinane **106** in 44% yield (Scheme 24.51) [89].

**Scheme 24.51** Singlet–triplet competition in the photochemistry of β,γ-unsaturated enones.

In the case of efficient (i.e., proceeding with high quantum yields) and rapid ODPM reactions leading to light-absorbing materials, these products can sometimes not be isolated because of further photochemical transformations. When the epoxyenone **107** is photolysed in acetone in order to generate an oxirane for carbonyl transformation, an appreciable amount of side-product **109** was isolated, originating from a photochemical rearrangement of the desired reaction product **108** (Scheme 24.52) [90].

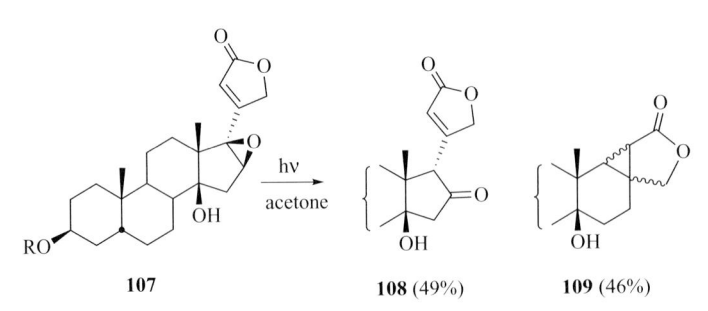

**Scheme 24.52** Oxa-di-π-methane rearrangement of an epoxyenone.

A general approach to the synthesis of linear polyquinanes with a cis–trans–cis ring junction of the central tricyclopentanoid structure from annulated bicyclo[2.2.2] octenones was reported by Singh and coworkers [91]. An extensive set of compounds were reacted in a primary thermal Diels–Alder reaction between cyclohexadienones and cyclic as well as bicyclic dienophiles such as spiroannulated cyclopentadienes, pentafulvene, indene, cycloalkenes and -alkadienes, and norbornadiene. The cycloadducts **110** were rearranged by acetone-sensitization following the di-π-methane

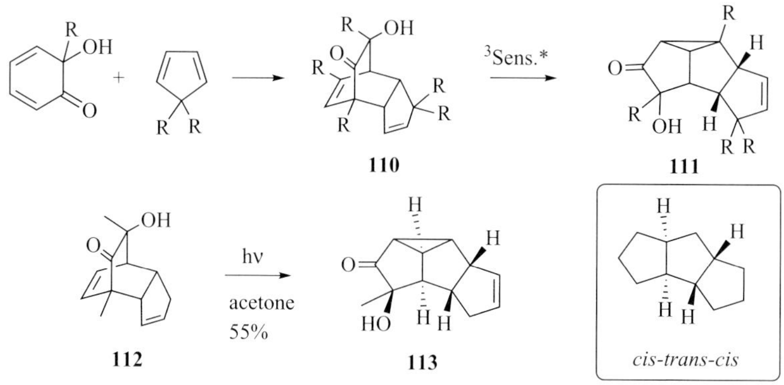

**Scheme 24.53** Oxa-di-π-methane rearrangement of cyclohexadienone-cyclopentadiene adducts.

route. Neither the intramolecular [2 + 2]-photocycloaddition (a prominent side-reaction in bridged and conformationally rigid dienes) nor the 1,3-acyl shift competed with the triquinane **111** formation. The dimethylated starting material **112** resulted in the triquinane **113** in 55% yield. The proton-catalyzed ring opening of the cyclopropane unit has been studied as a further route to accomplish the synthesis of linear polyquinanes (Scheme 24.53).

The propellane-type terpene modhephene **117** is an excellent example of a structurally highly complex target molecule which can be traced back retrosynthetically to a rather simple Diels–Alder cycloadduct. The central bridging five-membered ring must be bridgehead-connected in the substrate prior to the di-π-methane rearrangement [92]. The required bicyclo[2.2.2]octenone **115** is generated from the bicyclooctenone **114** by Grignard addition of methyl magnesium bromide and subsequent addition of a masked ketene (Scheme 24.54). Triplet-sensitized photolysis of **115** led to the formation of the propellane system **116** in moderate yield. The total synthesis of the natural product modhephene was accomplished in five further steps.

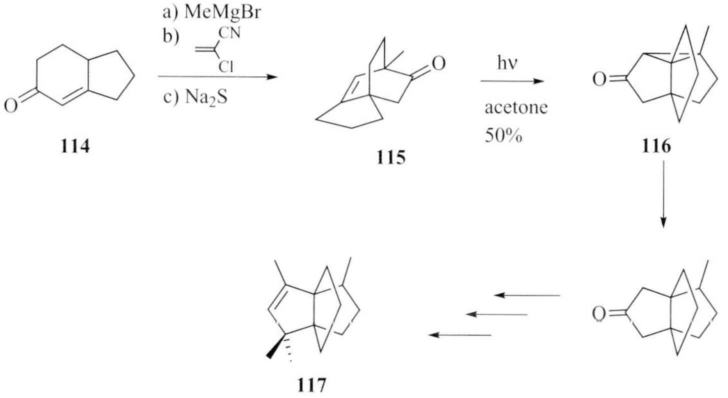

**Scheme 24.54** Oxa-di-π-methane rearrangement as key step for the synthesis of modhephene.

A chemoenzymatic route to the enantiomerically pure linear triquinane (−)-hirsutene **121** can be accomplished starting from toluene via an ODPM rearrangement as the photochemical key step [93]. Microbial dihydroxylation of toluene provided the enantiomerically pure 1,2-dihydrocatechol which was reacted with 3-cyclopentenone to the endo-product **118** in a high-pressure [4 + 2]-cycloaddition. Further modifications gave the substrate **119** for the ODPM rearrangement which was performed by triplet sensitization and led to the triquinane **120** in 80% yield (at 71% conversion). The total synthesis of the (−)-enantiomer of hirsutene was completed by reductive cyclopropane ring opening, carbonyl transposition, and methylenation (Scheme 24.55).

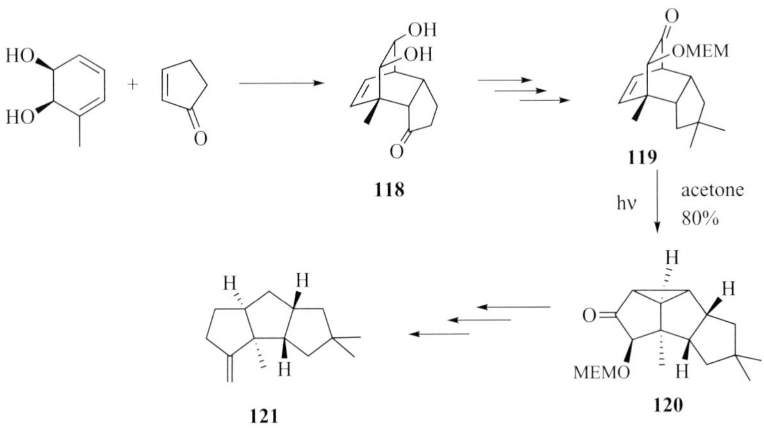

**Scheme 24.55** Oxa-di-π-methane rearrangement as key step for the synthesis of hirsutene.

Angularly annulated triquinanes can also be obtained by the ODPM route. This was shown for the terpene isocomene **125** by Uyehara *et al.* [94]. In contrast to the approaches described above, annulation of the third five-membered ring is disconnected from the ODPM rearrangement and follows a classical enolate alkylation procedure. The bridgehead-alkylated bicyclo[2.2.2]octenone **122** is converted into the tricyclic ketone **123** which, using a traditional α-alkylation procedure and subsequent functionalization/defunctionalization(cyclopropane ring-cleavage) steps led to the angularly annulated triquinane **124** (Scheme 24.56). Thus, the correct alignment of the activated $C_3$-chain and the degree of substitution at the α-carbonyl atom determines the type of annulation, that is, from the gem-dimethyl substrate **126**, the [3.3.3]-propellane ring system (as existing in modhephene, *vide supra*) becomes available.

Cyclohexadienones are not only ODPM-active substrates, but can also be easily converted into more reactive tricyclic β,γ-unsaturated ketones by intramolecular thermal [4 + 2]-cycloaddition. As versatile starting materials for this methodology, 6-alkenyl-2,4-cyclohexadien-1-ones **127** were converted into the annulated tricyclic β,γ-unsaturated ketones **128** and subsequently irradiated in the presence of

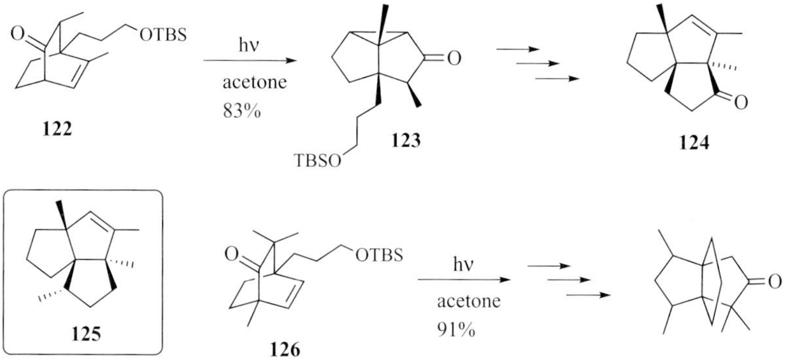

**Scheme 24.56** Oxa-di-π-methane rearrangement as key step for the synthesis of isocomene.

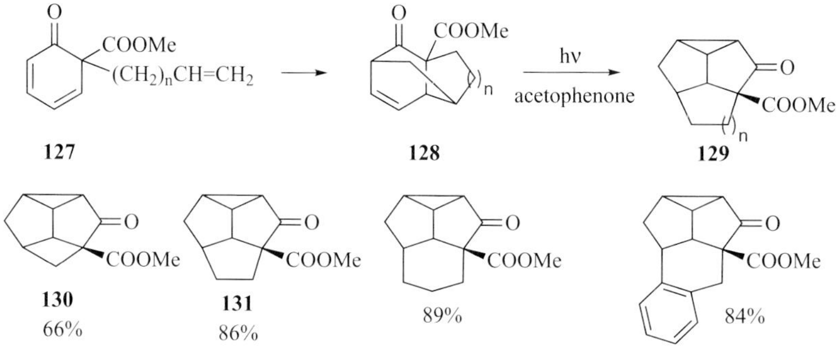

**Scheme 24.57** Oxa-di-π-methane rearrangement of a intramolecular cyclohexadien-1-one adduct.

acetophenone as triplet sensitizer to yield the tetracyclic products **129** (Scheme 24.57) [95]. Even the highly strained tetracyclic ketone **130**, combining a diquinane with annulated cyclopropane and cyclobutane ring systems was isolated in 66% yield after long irradiation times (48 h). The perhydrotriquinacene skeleton (as in **131**) is available in excellent yield as well as ring-enlarged or benzoannulated derivatives.

Bridging bicyclo[2.2.2]octenones by unsaturated hydrocarbon chains at positions C3 and C7 or C8, respectively, leads to ODPM substrates which, upon triplet sensitization, result in bridged diquinanes [96]. Depending on the relative position of the hydrocarbon linker, more or less strained products are formed, which can undergo further photochemical conversion. An interesting light-induced reaction was documented with the intermediate **133** (by ODPM from the $C_3$-bridged bicyclo [2.2.2]octenone **132**) leading to an 1,4-acyl shift with formation of a half-cage structure **134**, that is, a type of oxa-di-π-ethan rearrangement. No such reaction was observed with the corresponding epoxide **135**. Photochemical stability was also observed for the 1,3-linked diquinane **136**, resulting from the triplet-sensitized ODPM rearangement of the C3–C8-bridged bicyclo[2.2.2]octenone. Also the corresponding epoxide **137** was generated by the same process (Scheme 24.58).

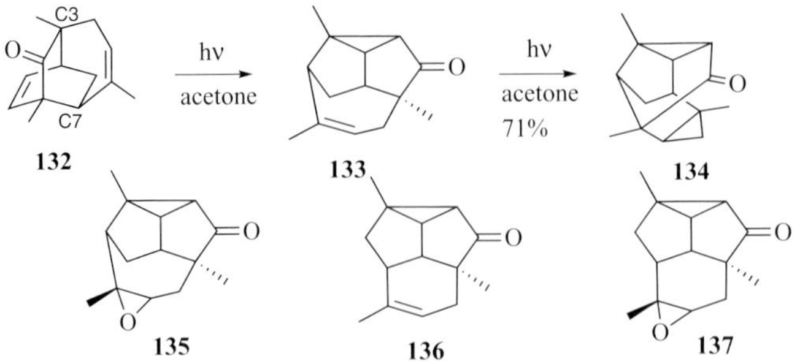

**Scheme 24.58** Complex bridged diquinanes by oxa-di-π-methane rearrangement.

The ultimate challenge for an efficient synthetic pathway to complex chiral target molecules is the generation of enantiomerically pure products, either by use of enantiomerically pure starting materials or by application of asymmetric induction using prochiral substrates. The parent bicyclo[2.2.2]octen-one **138** is available in enantiomerically pure form by a simple resolution procedure from the racemic mixture involving the acetal diastereoisomers **139** formed with tartaric acid [97]. Irradiation of the separated parent compound enantiomers **138** under conditions of triplet sensitization leads to the expected diquinanes **140** in high enantiomeric purity (Scheme 24.59) [97]. Another approach to enantiomerically pure ODPM substrates is described for bicyclo[2.2.1]heptenone [98].

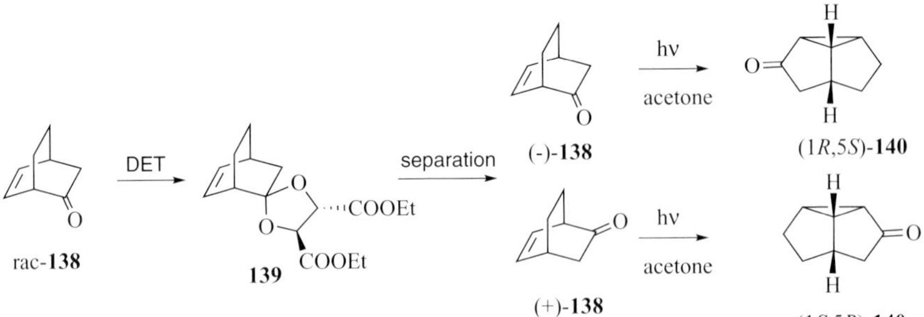

**Scheme 24.59** Oxa-di-π-methane rearrangement of the enantiopure bicyclo[2.2.2]octen-one.

This concept has been applied for the synthesis of the structurally complex and highly oxyfunctionalized triquinane (−)-coriolin (**143**) [99]. Two carbonyl groups, both in the right position for 1,2-acyl shift were present in the trimethyl-functionalized bicyclo[2.2.2]octenone **141**. With a site-selectivity of 85% the expected regioisomeric tricyclic dione **142** was formed as a mixture of epimers (Scheme 24.60). Subsequent transformations involving the annulation of the third five-membered ring as well as epoxidation and hydroxylation steps led to the

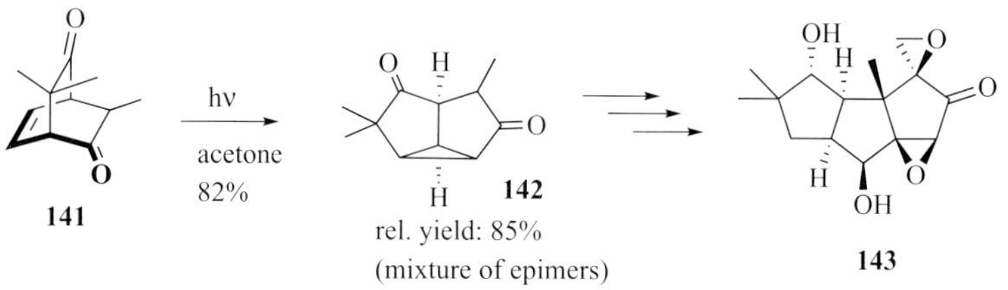

**Scheme 24.60** Oxa-di–methane rearrangement as key step for the synthesis of coriolin.

desired natural product coriolin. An impressive collection of additional target molecules has been prepared following the concept described above, for example, iridoid terpene structures [100], the aglucons of forsythide [101], loganine [102], and silphiperfolenone (vide infra) [103].

In spite of the successful applications of ODPM rearrangement for the total synthesis of complex polyoxyfunctionalized target molecules, two major limitations exist: heterocyclic three-membered rings are not easily obtained and β,γ-unsaturated aldehydes are ODPM-unreactive in the majority of cases. These limitations have been recently questioned by Armesto and coworkers. A thorough investigation of the photochemistry of β,γ-unsaturated aldehydes revealed that these species, contrary to literature opinion, are also reactive substrates for the ODPM rearrangement [104]. In the course of photoinduced electron transfer 2-nitrogen-DPM (2-ADPM) rearrangement (thus involving the radical cation of the substrate and not the electronically excited carbonyl analogue), aziridines were also isolated [105]. Crucial for the success of ODPM (as likewise ADPM) rearrangements appears to be at least one aryl substituent at the alkene part of the DPM substructure. In Scheme 24.61 the ODPM synthesis of cyclopropane carbaldehydes **145** from the β,γ-unsaturated aldehydes **144** is highlighted. Acetophenone or m-methoxyacetophenone, respectively, were used as triplet sensitizers.

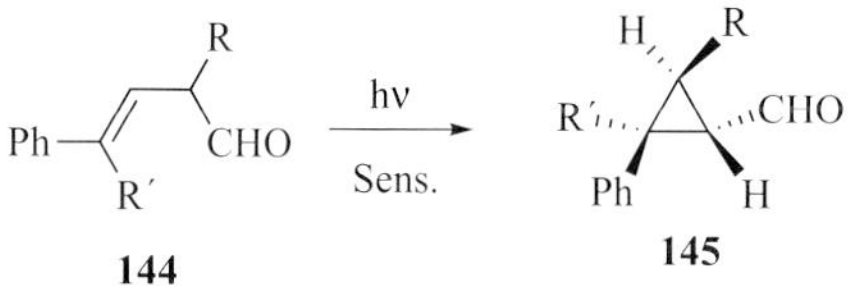

| | |
|---|---|
| R=iPr, R′=H | 21% |
| R=iPr, R′=Ph | 23% |
| R=Et, R′=H | 22% |
| R=Et, R′=Ph | 24% |

**Scheme 24.61** Oxa-di-π-methane rearrangement of β,γ-unsaturated aldehydes.

## 24.6
## $\pi^6$-Photocyclization

### 24.6.1
### Classical Examples

A multitude of $\pi^n$-photocyclization reactions exist and can be used for the generation of cyclobutenes (by $\pi^4$-photocyclization of 1,3-dienes), cyclohexadienes (by $\pi^6$-photocyclization of 1,3,5-trienes), and larger ring systems. Acyclic 1,3-dienes tend to thermalize very efficiently and this cyclobutene formation occurs with low quantum yields [106]. These systems can already be photochromic if the products bear at least one additional π-chromophor [107]. Photochromism is often a consequence of a photocyclization because the formation of one or two new bonds in the course of the reaction is accompanied by a strong blue-shift in absorption (if the chromophore is directly involved in the cyclization process). A true and useful photochromic system is of course more: substrate and product must be reversibly switchable, either with a photochemical or a thermal back reaction (P-type and T-type photochromic material). The archetype photochromic process is the diarylethene T-type switch of cis-stilbene **146** which is photochemically converted into the dihydrophenanthrene **147** [108]. The dihydroaromatic system obtained here is thermally unstable and easily oxidized to the corresponding phenanthrene **148** by mild oxidants such as oxygen or iodine (Scheme 24.62). This reaction protocol can be extended to the synthesis of higher condensed aromatic hydrocarbons such as hexahelicene **150** from the 2-styrylbenzophenanthrene **149** [109]. Thus, on the one hand the $\pi^6$-photocyclization is an excellent synthetic method for the generation of complex polycyclic structures, on the other hand the reversibility of the photochromic system is often disturbed by oxidative steps

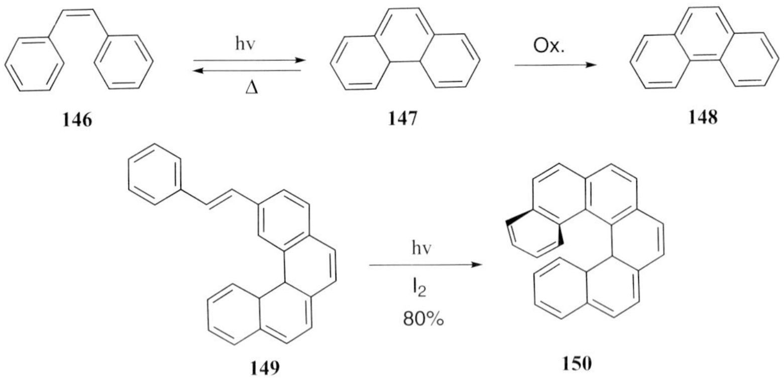

**Scheme 24.62** Photochromic diarylethenes and oxidative termination steps.

## 24.6.2
### Configurationally Restrained 1,2-Diarylethenes

More elaborate diarylethenes that avoid E/Z-photoisomerization and are suitably red-shifted as well as fatigue resistant, are the 1,2-cyclopentano-fixed 1,2-diarylethenes that have been investigated in remarkable detail in the last decade [110]. The key idea of this class of photochromic compounds is that thermalization by isomerization is prohibited by 1,2-ring junction and oxidative degradation of the cyclization stage is prohibited by alkylation of the corresponding positions. The introduction of a perfluorinated 1,2-cyclopenteno bridge (e.g., in **151**) has substantially increased the

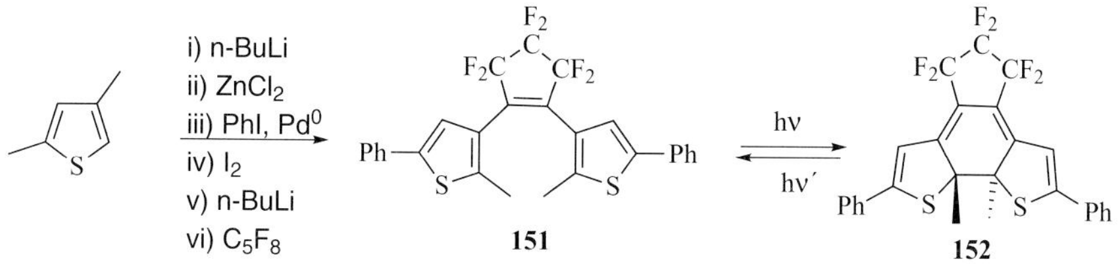

**Scheme 24.63** Photochromic diarylethenes with perfluorocyclopenteno bridge.

fatigue resistance (Scheme 24.63) [111]. The corresponding ring-closed isomer **152** is stable at room temperature and can only be ring-opened by photolysis, thus representing a true P-type photochromic material. Numerous applications have appeared recently for this photocyclization/ring-opening system in the fields of molecular switches and memory devices.

## 24.7 Experimental: Selected Procedures

### 24.7.1 Oxa-di-π-Methane-Rearrangement

#### 24.7.1.1 Synthesis of Enantiomerically Pure Tricyclo[3.3.0.0.0]octane-3-ones [97]

A solution of 10 g of the β,γ-unsaturated bicyclic ketone **138** in 1000 mL of acetone was purged with argon and irradiated in a water-cooled quartz vessel placed in a Rayonet RPR-208 photochemical reactor equipped with RUL-3000 lamps ($\lambda \sim 300$ nm). Irradiation was continued for 72 h and the reaction was monitored by thin layer chromatography. After 72 h of irradiation, the conversion was ~98% and the only detectable compound was the ODPM rearrangement product, tricyclo[3.3.0.0.0]octane-3-one **140**. The solvent was distilled off and the residue was chromatographed over silica gel using a benzene/ether mixture. The product which was eluted (8.6 g) was further distilled under vacuum (50 °C/1 mm) to get the pure ODPM rearrangement product (tricyclic ketone **140**) in 81% yield with 99.5% purity. The quantum efficiency was $\Phi = 1.0$.

#### 24.7.1.2 Synthesis of 4,4,9,9-Tetramethyltetracyclo[6.4.0.0.0]dodecane-7,12-dione [84]

A solution of 1.25 g of 8,8,9,9-tetramethyl-1,3,4,6,7,8-hexahydro-1,4-ethanonaphthalene-2,5-dione **92** (a δ-oxo-β,γ-unsaturated ketone; cf. Scheme 24.17 in 300 mL cyclohexane was degassed by bubbling nitrogen through the solution and was irradiated for 3 h in a Rayonet photochemical reactor ($\lambda \sim 300$ nm). The reaction was monitored by TLC and only one product was detectable. Evaporation of the solvent and recrystallization of the crude photolysate afforded the ODPM rearrangement product **93** in 78% yield. The quantum yield of the reaction was $\Phi = 0.67$.

Solvents like acetonitrile, benzene, or acetone and sensitizers like acetone or xanthone, respectively, can be employed for this reaction without altering the yield of the ODPM rearrangement product.

### 24.7.2
### Paternò–Büchi Photocycloaddition

#### 24.7.2.1 Photocycloaddition of Benzaldehyde to 2,4-Dimethyl-5-methoxyoxazole with Subsequent Ring Opening

2,4-Dimethyl-5-methoxyoxazole **153** (5 mmol) and benzaldehyde (5 mmol) were dissolved in 50 mL of benzene. The solution was transferred to a vacuum-jacketed quartz tube and degassed with a steady stream of $N_2$ gas (Scheme 24.64). The reaction mixture was irradiated at 10 °C in a Rayonet photoreactor (RPR-208, $\lambda = 300 \pm 10$ nm) for 24 h. The solvent was evaporated (40 °C, 20 torr) and the crude product was purified by preparative thick-layer chromatography (EA : H = 1 : 4) to give 0.59 g (72%) of the oxetane **154** as a colorless oil [112].

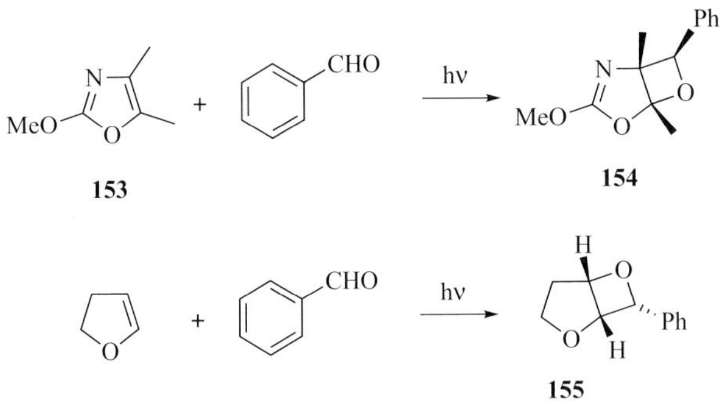

**Scheme 24.64** Paternò–Büchi reactions with oxazole and dihydrofuran.

#### 24.7.2.2 Photocycloaddition of Benzaldehyde to 2,3-Dihydrofuran [308 nmXeCl Excimer Photolysis on the 1mol Scale]

The Paternò–Büchi [2 + 2]-photocycloaddition of 2,3-dihydrofuran (70.0 g, 1 mol) and benzaldehyde (74.0 g, 0.7 mol) in ethanol was performed in an XeCl excimer preparative radiation system. The excimer reactor, which has been constructed for scaleable preparative organic photoreactions, consists of a commercially available 60 cm 3 kW XeCl excimer lamp which is operated in a vertical position and surrounded by a falling-film set-up. The latter consists of a cooled solvent reservoir (1.5 to 5 L solvent volume), a magnetic stirrer and a solvent supply tube which transports the solution to the edge of the falling film glass barrel. Two apertures are used for inert gas supply and temperature and pH-determination. After 14 h of irradiation, the solvent was evaporated and 130 g of the oxetane **155** was isolated (75%, purified by distillation, b.p. 127–129 °C/10 Torr) [113].

# References

1. (a) Griesbeck, A.G. and Mattay, J. (2005) *Synthetic Organic Photochemistry, Molecular and Supramolecular Photochemistry*, vol. 14, Marcel Dekker, New York; (b) Hoffmann, N. (2008) *Chem. Rev.*, **108**, 1052–1103; (c) Ramamurthy, V. and Schanze, K.S. (1997) *Organic Photochemistry, Molecular and Supramolecular Photochemestry*, vol. 1, Marcel Dekker, New York; (d) Ramamurthy, V. and Schanze, K.S. (1998) *Organic and Inorganic Photochemistry, Molecular and Supramolecular Photochemistry*, vol. 2, Marcel Dekker, New York; (e) Ramamurthy, V. and Schanze, K.S. (1999) *Organic Molecular Photochemistry, Molecular and Supramolecular Photochemistry*, vol. 3, Marcel Dekker, New York; (f) Ramamurthy, V. and Schanze, K.S. (2000) *Organic, Physical and Materials Photochemistry, Molecular and Supramolecular Photochemistry*, vol. 6, Marcel Dekker, New York; (g) Ramamurthy, V. and Schanze, K.S. (2002) *Photochemistry of Organic Molecules in Isotropic and Anisotropic Media, Molecular and Supramolecular Photochemistry*, vol. 9, Marcel Dekker, New York; (h) Inoue, Y. and Ramamurthy, V. (2004) *Chiral Photochemistry, Molecular and Supramolecular Photochemistry*, vol. 11, Marcel Dekker, New York; (i) Ramamurthy, V. and Schanze, K.S. (2006) *Organic Photochemistry and Photophysics, Molecular and Supramolecular Photochemistry*, vol. 14, Marcel Dekker, New York; (j) Horspool, W.M. and Lenci, F. (2004) *Handbook of Photochemistry and Photobiology*, vol. 2, CRC Press, Boca Raton; (k) Nalwa, H.S. (2003) *CRC Handbook of Organic Photochemistry and Photobiology*, vol. 2, American Scientific Publishers, Stevenson Ranch, CA; (l) Carreira, E.M. and Griesbeck, A.G. (2001) Special Issue on Organic Photochemistry, *Synthesis* Special Issue, III.
2. (a) Buchachenko, A.L. (2000) *Pure Appl. Chem.*, **72**, 2243–2258; (b) Bargon, J. (2006) *Photochem. Photobiol. Sci.*, **5**, 970–978; (c) Hayshi, H. (2002) *J. Chin. Chem. Soc.*, **49**, 137–160; (d) Wolff, H.-J., Buerssner, D. and Steiner, U. (1995) *Pure Appl. Chem.*, **67**, 167–174.
3. Griesbeck, A.G. (2005) *Synthetic Organic Photochemistry, Molecular and Supramolecular Photochemistry*, vol. 14 (eds A.G. Griesbeck and J. Mattay), Marcel Dekker, New York, pp. 89–139.
4. Wessig, P. and Mühling, O. (2005) *Synthetic Organic Photochemistry, Molecular and Supramolecular Photochemistry*, vol. 14 (eds A.G. Griesbeck and J. Mattay), Marcel Dekker, New York, pp. 41–87.
5. (a) Armesto, D., Ortiz, M.J. and Agarrabeita, A.R. (2005) *Synthetic Organic Photochemistry, Molecular and Supramolecular Photochemistry*, vol. 14 (eds A.G. Griesbeck and J. Mattay), Marcel Dekker, New York, pp. 161–187; (b) Rao, V.J. and Griesbeck, A.G. (2005) in *Synthetic Organic Photochemistry, Molecular and Supramolecular Photochemistry*, vol. 14 (eds A.G. Griesbeck and J. Mattay), Marcel Dekker, New York, pp. 189–210.
6. Abe, M., Adam, W., Borden, W.T., Hattori, M., Horvat, D.A., Nojima, M., Nozaki, K. and Wirz, J. (2004) *J. Am. Chem. Soc.*, **126**, 574–582.
7. Griesbeck, A.G. and Mattay, J. (1994) *Photochemical Key Steps in Organic Chemistry*, VCH, Weinheim.
8. Wagner, P.J. (2005) *Synthetic Organic Photochemistry, Molecular and Supramolecular Photochemistry*, vol. 14 (eds A.G. Griesbeck and J. Mattay), Marcel Dekker, New York, pp. 11–39.
9. Marcus, R.A. (1993) *Angew. Chem. Int. Ed. Engl.*, **32**, 1111–1121.

10 Griesbeck, A.G., Hoffmann, N. and Warzecha, K.-D. (2007) *Acc. Chem. Res.*, **40**, 128–140.

11 Kavarnos, G.J. (1990) *Top. Curr. Chem.*, **156**, 21–58.

12 Rehm, D. and Weller, A.H. (1969) *Ber. Bunsen-Ges. Phys. Chem.*, **73**, 834–839.

13 (a) Norrish, R.G.W. and Appleyard, M.E.S. (1934) *J. Chem. Soc.*, 874–880; (b) Yang, N.C. and Yang, D.-D.H. (1958) *J. Am. Chem. Soc.*, **80**, 2913–2914.

14 (a) Ariel, S., Ramamurthy, V., Scheffer, J.R. and Trotter, J. (1983) *J. Am. Chem. Soc.*, **105**, 6959–6960; (b) Scheffer, J.R., Garcia-Garibay, M. and Nalamasu, O. (1987) *Org. Photochem.*, **8**, 249–347.

15 Wagner, P. J. (1995) in *Handbook of Photochemistry and Photobiology*, vol. 1 (eds W.M. Horspool and P.-S. Song), CRC Press, Boca Raton, pp. 449–470 and 471–483.

16 Griesbeck, A.G. and Heckroth, H. (2002) *J. Am. Chem. Soc.*, **124**, 396–403.

17 Johnston, L.J. and Scaiano, J.C. (1989) *Chem. Rev.*, **89**, 521–547.

18 Scaiano, J.C. (1982) *Tetrahedron*, **38**, 819–824.

19 Klessinger, M. (1997) *Pure Appl. Chem.*, **69**, 773–778.

20 Ihmels, H. and Scheffer, J.R. (1999) *Tetrahedron*, **55**, 885–907.

21 (a) Yoshioka, M., Miyazoe, S. and Hasegawa, T. (1993) *J. Chem. Soc. Perkin Trans. 1*, 2781–2786; (b) Weigel, W. and Wagner, P.J. (1996) *J. Am. Chem. Soc.*, **118**, 12858–12859; (c) Weigel, W., Schiller, S. and Henning, H.-G. (1997) *Tetrahedron*, **53**, 7855–7866.

22 (a) Roth, H.J., El Raie, M.H. and Schrauth, T. (1974) *Arch. Pharm.*, **307**, 584–595; (b) Abdul-Baki, A., Rotter, F., Schrauth, T. and Roth, H.J. (1978) *Arch. Pharm.*, **311**, 341–345; (c) Roth, H.F. and George, H. (1970) *Arch. Pharm.*, **303**, 725–732.

23 Weigel, W., Schiller, S., Reck, G. and Henning, H.-G. (1993) *Tetrahedron Lett.*, **42**, 6737–6740.

24 (a) Wessig, P. and Mühling, O. (2001) *Angew. Chem. Int. Ed.*, **40**, 1064–1065; (b) Wessig, P., Glombitza, C., Müller, G. and Teubner, J. (2004) *J. Org. Chem.*, **69**, 7582–7591; (c) Mühling, O. and Wessig, P. (2006) *Photochem. Photobiol. Sci.*, **5**, 1000–1006; (d) Wessig, P. and Mühling, O. (2007) *Eur. J. Org. Chem.*, 2219–2232.

25 Kraus, G.A. and Wu, Y. (1992) *J. Am. Chem. Soc.*, **114**, 8705–8707.

26 Breslow, R. and Winnik, M.A. (1969) *J. Am. Chem. Soc.*, **91**, 3083–3084.

27 Breslow, R. and Baldwin, S.W. (1970) *J. Am. Chem. Soc.*, **92**, 732–734.

28 Warzecha, K.-D., Görner, H. and Griesbeck, A.G. (2006) *J. Phys. Chem. A*, **110**, 3356–3363.

29 Wintgens, V., Valat, P., Kossanyi, J., Biczok, L., Demeter, A. and Bérces, T. (1994) *J. Chem. Soc., Faraday Trans.*, **90**, 411–421.

30 (a) Yoon, U.C. and Mariano, P.S. (2001) *Acc. Chem. Res.*, **34**, 523–532; (b) Griesbeck, A.G., Kramer, W. and Oelgemoeller, M. (1999) *Synlett*, 1169–1178.

31 Griesbeck, A.G. and Görner, H. (1999) *J. Photochem. Photobiol. A: Chem.*, **129**, 111–119.

32 Griesbeck, A.G., Henz, A., Peters, K., Peters, E.-M. and von Schnering, H.G. (1995) *Angew. Chem. Int. Ed. Engl.*, **34**, 474–476.

33 Griesbeck, A.G., Nerowski, F. and Lex, J. (1999) *J. Org. Chem.*, **64**, 5213–5217.

34 (a) Izatt, R.M. and Christensen, J.J. (eds) (1973) *Progress in Macrocyclic Chemistry*, vol. 3, Wiley, New York, vol. 2, 1981, vol. 3, 1987; (b) Thebtaranonth, C. and Thebtaranonth, Y. (1994) *Cyclization Reactions*, CRC Press, Boca Raton; (c) Ross-Murphy, S.B. and Stepto, R.F.T. (1996) *Large Ring Molecules*, John Wiley & Sons, New York, p. 599.

35 (a) Hu, S. and Neckers, D.C. (1997) *Tetrahedron*, **53**, 2751–2766; (b) Hu, S. and Neckers, D.C. (1997) *Tetrahedron*, **53**, 7165–7180.

36 Griesbeck, A.G., Mauder, H., Müller, I., Peters, K., Peters, E.-M. and von

Schnering, H.G. (1993) *Tetrahedron Lett.*, **34**, 453–456.

37 Hasegawa, T. and Yamazaki, Y. (1998) *Tetrahedron*, **54**, 12223–12232.

38 Hu, S. and Neckers, D.C. (1997) *J. Org. Chem.*, **62**, 6820–6826.

39 Yamazaki, Y., Miyagawa, T. and Hasegawa, T. (1997) *J. Chem. Soc., Perkin Trans. 1*, 2979–2981.

40 (a) Hasegawa, T., Ogawa, T., Miyata, K., Karakizawa, A., Komiyama, M., Nishizawa, K. and Yoshioka, M. (1990) *J. Chem. Soc., Perkin Trans. 1*, 901–905; (b) Hasegawa, T., Miyata, K., Ogawa, T., Yoshihara, N. and Yoshioka, M. (1985) *Chem. Commun.*, 363–364.

41 Hasegawa, T., Yasuda, N. and Yoshioka, M. (1996) *J. Phys. Org. Chem.*, **9**, 221–226.

42 Hasegawa, T., Mukai, K., Mizukoshi, K. and Yoshioka, M. (1990) *Bull. Chem. Soc. Jpn.*, **63**, 3348–3350.

43 (a) Hatanaka, Y., Sato, Y., Nakai, H., Wada, M., Mizoguchi, T. and Kanaoka, Y. (1992) *Liebigs Ann. Chem.*, 1113–1123; (b) Sato, Y., Nakai, H., Wada, M., Mizoguchi, T., Hatanaka, Y., Migata, Y. and Kanaoka, Y. (1985) *Liebigs Ann. Chem.*, 1099–1118.

44 (a) Sato, Y., Nakai, H., Wada, M., Mizoguchi, T., Hatanaka, Y. and Kanaoka, Y. (1992) *Chem. Pharm. Bull.*, **40**, 3174–3180; (b) Wada, M., Nakai, H., Aoe, K., Kotera, K., Sato, Y., Hatanaka, Y. and Kanaoka, Y. (1983) *Tetrahedron*, **39**, 1273–1279; (c) Wada, M., Nakai, H., Sato, Y., Hatanaka, Y. and Kanaoka, Y. (1983) *Chem. Pharm. Bull.*, **31**, 429–435.

45 (a) Machida, M., Takechi, H. and Kanaoka, Y. (1982) *Chem. Pharm. Bull.*, **30**, 1579–1587; (b) Coyle, J.D., Smart, L.E., Challiner, J.F. and Haws, E.J. (1985) *J. Chem. Soc. Perkin Trans. 1*, 121–129; (c) Coyle, J.D. and Newport, G.L. (1979) *Synthesis*, 381–382.

46 (a) Maruyama, K., Ogawa, T., Kubo, Y. and Araki, T. (1985) *J. Chem. Soc. Perkin Trans. 1*, 2025–2031; (b) Machida, M., Oda, K. and Kanaoka, Y. (1985) *Tetrahedron*, **41**, 4995–5001; (c) Mazzocchi, P.H. and Fritz, G. (1986) *J. Org. Chem.*, **51**, 5362–5364;

(d) Maruyama, K. and Kubo, Y. (1981) *J. Org. Chem.*, **46**, 3612–3622.

47 Mazzocchi, P.H., Shook, D. and Liu, L. (1987) *Heterocycles*, **26**, 1165–1167.

48 Xue, J., Zhu, L., Fun, H.-K. and Xu, J.-H. (2000) *Tetrahedron Lett.*, **41**, 8553–8557.

49 (a) Griesbeck, A.G., Kramer, W. and Oelgemöller, M. (1999) *Synlett*, 1169–1178; (b) Griesbeck, A.G., Henz, A., Kramer, W., Lex, J., Nerowski, F., Oelgemöller, M., Peters, K. and Peters, E.-M. (1997) *Helv. Chim. Acta*, **80**, 912–933.

50 (a) Griesbeck, A.G., Heinrich, T., Oelgemöller, M., Molis, A. and Heidtmann, A. (2002) *Helv. Chim. Acta*, **85**, 4561–4578; (b) Yoo, D.J., Kim, E.Y., Oelgemöller, M. and Shim, S.C. (2001) *Heterocycles*, **54**, 1049–1055.

51 (a) Yoon, U.C. and Mariano, P.S. (2001) *Acc. Chem. Res.*, **34**, 523–533; (b) Yoon, U.C., Jin, Y.X., Oh, S.W., Cho, D.W., Park, K.H. and Mariano, P.S. (2002) *J. Photochem. Photobiol. A: Chem.*, **150**, 77–84; (c) Griesbeck, A.G., Oelgemöller, M., Lex, J., Haeuseler, A. and Schmittel, M. (2001) *Eur. J. Org. Chem.*, 1831–1843.

52 (a) Pandey, G., Reddy, G.D. and Chakrabarti, D. (1996) *J. Chem. Soc., Perkin Trans. 1*, 219–224; (b) Pandey, G. and Kapur, M. (2000) *Tetrahedron Lett.*, **41**, 8821–8824; (c) Pandey, G. and Kapur, M. (2001) *Synthesis*, 1263–1267; (d) Pandey, G. and Kapur, M. (2002) *Org. Lett.*, **4**, 3883–3886.

53 Jonas, M., Blechert, S. and Steckham, E. (2001) *J. Org. Chem.*, **66**, 6896–6904.

54 (a) Pandey, G. and Gadre, S.R. (2003) *ARKIVOC*, 45–54; (b) Pandey, G., Das, P. and Reddy, P.Y. (2000) *Eur. J. Org. Chem.*, 657–664.

55 (a) Horspool, W.M., Hynd, G. and Ixkes, U. (1999) *Tetrahedron Lett.*, **40**, 8295–8298; (b) Armesto, D., Ramos, A., Oritz, M.J., Horspool, W.M., Mancheño, M.J., Caballero, O. and Mayoral, E.P. (1997) *J. Chem. Soc., Perkin Trans. 1*, 1535–1541.

56 (a) Ackermann, L., Heidbreder, A., Wurche, F., Klärner, F.-G. and Mattay, J.

(1999) *J. Chem. Soc., Perkin Trans. 2*, 863–869; (b) Hintz, S., Mattay, J., van Eldik, R. and Fu, W.-F. (1998) *Eur. J. Org. Chem.*, 1583–1596; (c) Hintz, S., Fröhlich, R. and Mattay, J. (1996) *Tetrahedron Lett.*, **37**, 7349–7352.

57 (a) Pandey, G., Karthikeyan, M. and Murugan, A. (1998) *J. Org. Chem.*, **63**, 2867–2872; (b) Pandey, G., Murugan, A. and Balakrishnan, M. (2002) *Chem. Commun.*, 624–625.

58 (a) Goeller, F., Heinemann, C. and Demuth, M. (2001) *Synthesis*, 1114–1116; (b) Heinemann, C. and Demuth, M. (1999) *J. Am. Chem. Soc.*, **121**, 4894–4895; (c) Heinemann, C., Xing, X., Warzecha, K.-D., Ritterskamp, P., Görner, H. and Demuth, M. (1998) *Pure Appl. Chem.*, **70**, 2167–2176; (d) Heinemann, C. and Demuth, M. (1997) *J. Am. Chem. Soc.*, **119**, 1129–1130.

59 Belotti, D., Cossy, J., Pete, J.P. and Portella, C. (1986) *J. Org. Chem.*, **51**, 4196–4200.

60 (a) Fagnoni, M., Schmoldt, P., Kirschberg, T. and Mattay, J. (1998) *Tetrahedron*, **54**, 6427–6444; (b) Kirschberg, T. and Mattay, J. (1996) *J. Org. Chem.*, **61**, 8885–8896.

61 (a) Pandey, G., Hajra, S., Ghorai, M.K. and Kumar, K.R. (1997) *J. Am. Chem. Soc.*, **119**, 8777–8787; (b) Pandey, G., Hajra, S., Ghorai, M.K. and Kumar, K.R. (1997) *J. Org. Chem.*, **62**, 5966–5973.

62 Rosales, V., Zambrano, J. and Demuth, M. (2004) *Eur. J. Org. Chem.*, 1798–1802.

63 Fuji, K. and Kawabata, T. (1998) *Chem. Eur. J.*, **4**, 373–376.

64 (a) Griesbeck, A.G., Kramer, W. and Lex, J. (2001) *Synthesis*, 1159–1166; (b) Griesbeck, A.G., Kramer, W. and Lex, J. (2001) *Angew. Chem. Int. Ed. Engl.*, **40**, 577–579; (c) Griesbeck, A.G., Kramer, W., Bartoschek, A. and Schmickler, H. (2001) *Org. Lett.*, **3**, 537–539.

65 Fleming, S.A. (2005) *Synthetic Organic Photochemistry, Molecular and Supramolecular Photochemistry*, vol. 14 (eds A.G. Griesbeck and J. Mattay), Marcel Dekker, New York, pp. 141–160.

66 Margaretha, P. (2005) *Synthetic Organic Photochemistry, Molecular and Supramolecular Photochemistry*, vol. 14 (eds A.G. Griesbeck and J. Mattay), Marcel Dekker, New York, pp. 211–237.

67 Sieburth, S.McN. (2005) *Synthetic Organic Photochemistry, Molecular and Supramolecular Photochemistry*, vol. 14 (eds A.G. Griesbeck and J. Mattay), Marcel Dekker, New York, pp. 239–268.

68 Hoffmann, N. (2005) *Synthetic Organic Photochemistry, Molecular and Supramolecular Photochemistry*, vol. 14 (eds A.G. Griesbeck and J. Mattay), Marcel Dekker, New York, pp. 529–552.

69 (a) Ward, S.C. and Fleming, S.A. (1994) *J. Org. Chem.*, **59**, 6476–6479; (b) Bradford, C.L., Fleming, S.A. and Ward, S.C. (1995) *Tetrahedron Lett.*, **36**, 4189–4192.

70 (a) Langer, K. and Mattay, J. (1995) *J. Org. Chem.*, **60**, 7256–7266; (b) Panda, J. and Ghosh, S. (1999) *Tetrahedron Lett.*, **40**, 6693–6694.

71 Zimmerman, H.E. and Armesto, D. (1996) *Chem. Rev.*, **96**, 3065–3112.

72 Nair, V., Nandakumar, M.V., Anilkumar, G.N., Maliakal, D., Vairamani, M., Prabhakar, S. and Rath, N.P. (2000) *J. Chem. Soc., Perkin Trans. 1*, 3795–3798.

73 Leitich, J., Heise, I. and Schaffner, K. (1995) *Can. J. Chem.*, **73**, 1785–1793.

74 Zimmerman, H.E. and Cirkva, V. (2001) *J. Org. Chem.*, **66**, 1839–1851.

75 Ramaiah, D., Sajimon, M.C., Joseph, J. and George, M.V. (2005) *Chem. Soc. Rev.*, **34**, 48–57.

76 Ihmels, H., Mohrschladt, C.J., Grimme, J.W. and Quast, H. (2001) *Synthesis*, 1175–1180.

77 (a) Hart, H., Collins, P.M. and Warring, A.J. (1966) *J. Am. Chem. Soc.*, **88**, 1005–1013; (b) Hart, H. and Murray, R.K. (1970) *J. Org. Chem.*, **35**, 1535–1542.

78 Uppili, S. and Ramamurthy, V. (2002) *Org. Lett.*, **4**, 87–90.

79 Uppili, S., Takagi, S., Sunoj, R.B., Lakshminarasimhan, P., Chandrasekhar, J. and Ramamurthy, V. (2001) *Tetrahedron Lett.*, **42**, 2079–2083.

80  Mirbach, M.J., Henne, A. and Schaffner, K. (1978) *J. Am. Chem. Soc.*, **100**, 7127–7128.
81  Reimann, B., Sadler, D.E. and Schaffner, K. (1986) *J. Am. Chem. Soc.*, **108**, 5527–5530.
82  van der Weerdt, A.J.A. and Cerfontain, H. (1977) *Rec. Trav. Chim. Pays-Bas*, **96**, 247–248.
83  Koppes, J.C.M.M. and Cerfontain, H. (1985) *Rec. Trav. Chim. Pays-Bas*, **104**, 272–276.
84  Kilger, R., Körner, W. and Maragretha, P. (1984) *Helv. Chim. Acta*, **67**, 1493–1495.
85  Sugihara, Y., Sugimura, T. and Murata, I. (1981) *J. Am. Chem. Soc.*, **103**, 6738–6739.
86  Sugihara, Y., Sugimura, T. and Murata, I. (1980) *Chem. Lett.*, 1103–1106.
87  Sadeghpoor, R., Ghandi, M., Najafi, H.M. and Farzaneh, F. (1998) *Chem. Commun.*, 329–330.
88  Lee, T.-H., Rao, P.D. and Liao, C.-C. (1999) *Chem. Commun.*, 801–802.
89  Mehta, G. and Srikrishna, A. (1979) *Tetrahedron Lett.*, 3187–3190.
90  Ishii, K., Hashimoto, T., Sakamoto, M., Taira, Z. and Asakawa, Y. (1988) *Chem. Lett.*, 609–612.
91  Singh, V.K., Deota, P.T. and Bedekar, A.V. (1992) *J. Chem. Soc., Perkin Trans. 1*, 903–912.
92  (a) Mehta, G. and Subrahmanyam, D. (1985) *J. Chem. Soc., Chem. Commun.*, 768–769; (b) Mehta, G. and Subrahmanyam, D. (1991) *J. Chem. Soc., Perkin Trans. 1*, 395–401.
93  Banwell, M.G., Edwards, A.J., Harfoot, G.J. and Jolliffe, K.A. (2002) *J. Chem. Soc., Perkin Trans. 1*, 2439–2441.
94  Uyehara, T., Murayama, T., Sakai, K., Onda, K., Ueno, M. and Sato, T. (1998) *Bull. Chem. Soc. Jpn.*, **71**, 231–242.
95  Schultz, A.G., Lavieri, F.P. and Snead, T.E. (1985) *J. Org. Chem.*, **50**, 3086–3091.
96  Hayakawa, K., Schmid, H. and Fráter, G. (1977) *Helv. Chim. Acta*, **60**, 561–577.
97  Demuth, M., Raghavan, P.R., Carter, C., Nakano, K. and Schaffner, K. (1980) *Helv. Chim. Acta*, **63**, 2434–2439.
98  Corey, E.J., Guzman-Perez, A. and Loh, T.-P. (1994) *J. Am. Chem. Soc.*, **116**, 3611.
99  Demuth, M., Ritterskamp, P., Weigt, E. and Schaffner, K. (1986) *J. Am. Chem. Soc.*, **108**, 4149–4154.
100  Demuth, M. and Schaffner, K. (1982) *Angew. Chem. Int. Ed. Engl.*, **21**, 820–825.
101  Liao, C.-C. and Wei, C.-P. (1989) *Tetrahedron Lett.*, **30**, 2255–2256.
102  Demuth, M., Chandrasekhar, S. and Schaffner, K. (1984) *J. Am. Chem. Soc.*, **106**, 1092–1095.
103  Demuth, M. and Hinsken, W. (1988) *Helv. Chim. Acta*, **71**, 569–576.
104  Armesto, D., Ortiz, M.J., Agarrabeitia, A.R. and Aparicio-Lara, S. (2001) *Synthesis*, 1149–1158.
105  Armesto, D., Caballero, O. and Amador, U. (1997) *J. Am. Chem. Soc.*, **119**, 12659–12666.
106  (a) Srinivasan, R. (1968) *J. Am. Chem. Soc.*, **90**, 4498–4499; (b) Celani, P., Bernardi, F., Olivucci, M. and Robb, M.A. (1995) *J. Chem. Phys.*, **102**, 5733–5742.
107  Shrestha, S.M., Nagashima, H., Yokoyama, Y. and Yokoyama, Y. (2003) *Bull. Chem. Soc. Jpn.*, **76**, 363–367.
108  (a) Laarhoven, W.H. (1989) *Org. Photochem.*, **19**, 163–308; (b) Waldeck, D.H. (1991) *Chem. Rev.*, **91**, 415–436.
109  Laarhoven, W.H. and Prinsen, W.J.C. (1984) *Top. Curr. Chem.*, **125**, 63–130.
110  Irie, M. (2000) *Chem. Rev.*, **100**, 1685–1716.
111  Hanazawa, M., Sumiya, R., Horikawa, Y. and Irie, M. (1992) *Chem. Commun.*, 206–207.
112  Griesbeck, A.G. and Bondock, S. (2003) *Can. J. Chem.*, **81**, 555–559.
113  Griesbeck, A.G., Maptue, N., Bondock, S. and Oelgemöller, M. (2003) *Photochem. Photobiol. Sci.*, **2**, 450–451.

# 25
# Asymmetric Organocatalyzed Cyclization Reactions
*Liu-Zhu Gong, Jun Jiang, Meng-Xia Xue, and Shi-Wei Luo*

## 25.1
## Introduction

Asymmetric organocatalysis, in which an asymmetric reaction is accelerated with a catalytic amount of small organic molecules, is a rapidly growing research field of modern asymmetric synthesis [1]. Central to this has been the development of catalytic asymmetric procedures that allow efficient construction of structurally complex chiral compounds. As chiral cyclic motifs appear widely in natural products and pharmaceutically relevant substances, the organocatalytic asymmetric cyclization reaction represents one of most important synthetic tools which provides versatile approaches to access chiral cyclic compounds, in particular, those for the creation of complicated cyclic systems. This chapter is focused on the important advances made in the last five years in organocatalytic asymmetric cyclization reactions. A period of five years is quite short, but has still witnessed tremendously important advances in this field from research groups worldwide. General concepts for the invention of new asymmetric organocatalytic cyclization reactions are diverse including enamine- and iminium catalysis, Brønsted acid catalysis, Lewis base catalysis, Brønsted-acid–Lewis base bifunctional catalysis, and nucleophilic carbene catalysis [1]. Since this is not a comprehensive review, we have selected the most representative reactions and protocols that generate five-membered or larger cycles. Specifically, asymmetric transformations that are difficult to be realized using metal-based catalysts or consist of new concepts in the catalyst design will be preferentially selected with emphasis on the organocatalytic protocols, which enable complex molecules to be concisely synthesized. Other excellent organocatalytic reactions, such as asymmetric organocatalytic epoxidation and aziridination reactions [2, 3], and Staudinger reactions [4] that give rise to cyclic compounds smaller than five-membered rings will not be included in this chapter.

## 25.2
### Enamine-Catalyzed Asymmetric Cyclization Reactions

The general principal for enamine-catalyzed cyclization reaction is illustrated in Scheme 25.1 [5]. The carbonyl group of a multiply functionalized organic molecule such as **1** first condenses with a chiral amine **2** to generate an enamine intermediate **I** that subsequently reacts with the other electrophilic functional group intramolecularly leading to the formation of a cyclic compound **3** and regeneration of the amine catalyst **2**.

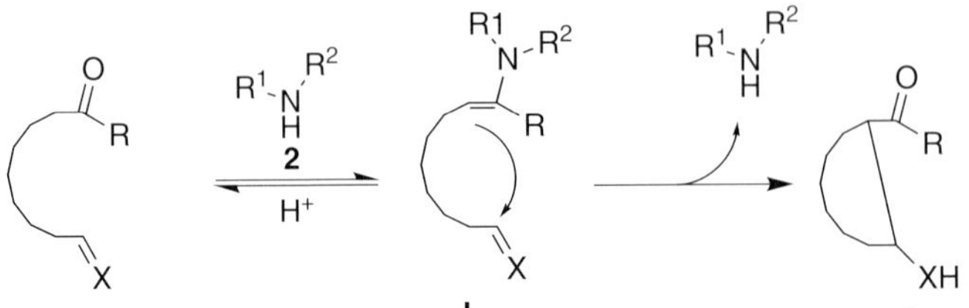

**Scheme 25.1** General strategy for the enamine-catalyzed cyclization.

### 25.2.1
#### Intramolecular Aldol Reactions

The intramolecular direct aldol reaction of a trione **4** using proline as a catalyst furnishing Wieland–Meischer ketone **6** as a sole product, which has widespread applications as an important synthon for the synthesis of steroids, was first reported over 30 years ago (Scheme 25.2) [6]. Theoretical and experimental studies on the mechanism revealed that the reaction proceeded through enamine catalysis [7]. Recent efforts have been focused on searching for new organic catalysts, leading to a variety of updated procedures for the asymmetric synthesis of Wieland–Meischer ketone derivatives with a slight enhancement in both the enantioselectivity and synthetic efficiency [8].

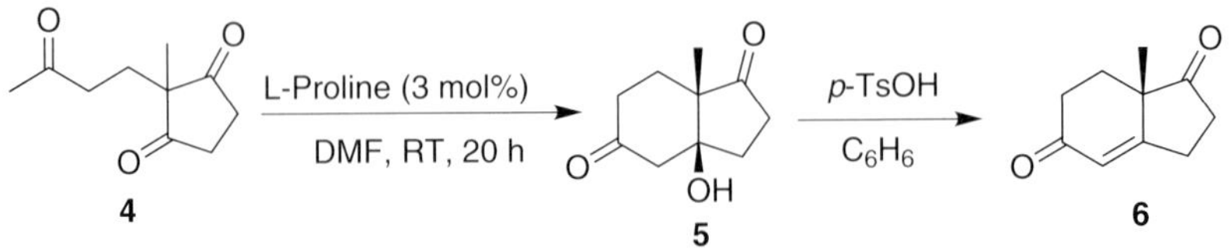

**Scheme 25.2** The synthesis of Wieland–Meischer ketone.

Direct intramolecular aldolization of aliphatic dials **7** with proline catalyst were described by List's group, providing access to 2-hydroxycyclohexanecarbaldehyde derivatives **8** with almost perfect stereoselectivities (Scheme 25.3) [9].

Iwabuchi reported that highly enantioenriched *endo*-8-hydroxybicyclo[3.3.1]nonan-2-one **10** could be obtained from an intramolecular aldol reaction of **9** under the

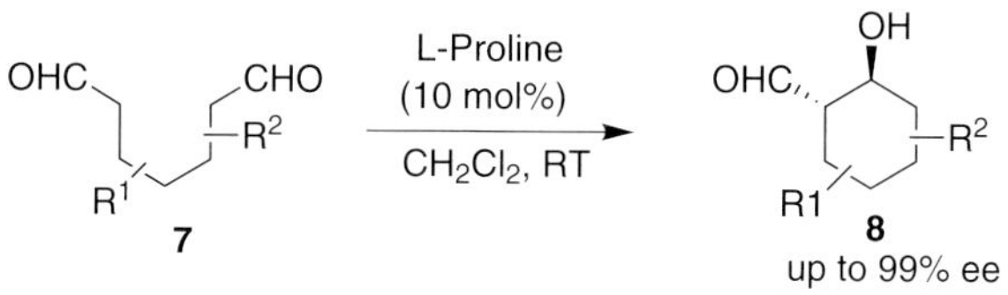

## 25.2 Enamine-Catalyzed Asymmetric Cyclization Reactions

**Scheme 25.3** Intramolecular direct aldol reaction of dials.

influence of 5 mol% of (4R, 2S)-tetrabutylammonium 4-TBDPSoxy-prolinate **11** (Scheme 25.4) [10].

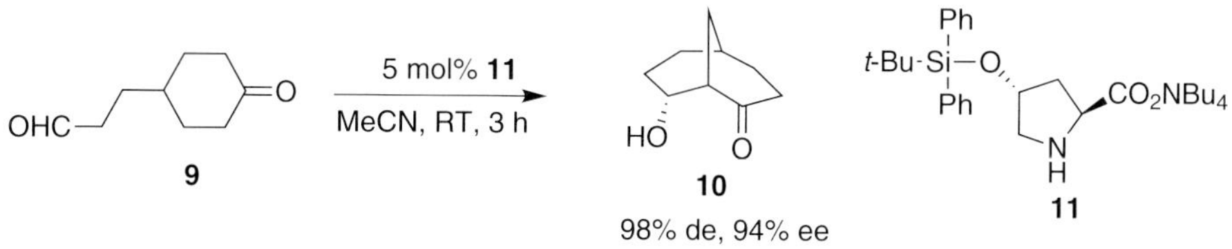

**Scheme 25.4** Synthesis of endo-8-hydroxybicyclo[3.3.1]nonan-2-one via intramolecular aldolization.

(2S,3R)-3-hydroxy-3-methylproline **12** is incorporated in natural products, polyoxypeptins A and B, thereby is of great synthetic importance. Interestingly, **12** itself serves as an efficient catalyst for an intramolecular aldol reaction of **13**, after a subsequent oxidation with NaClO$_2$ to generate the precursor **14** of (2S,3R)-3-hydroxy-3-methylproline **12** with synthetically useful levels of stereoselectivity (Scheme 25.5) [11].

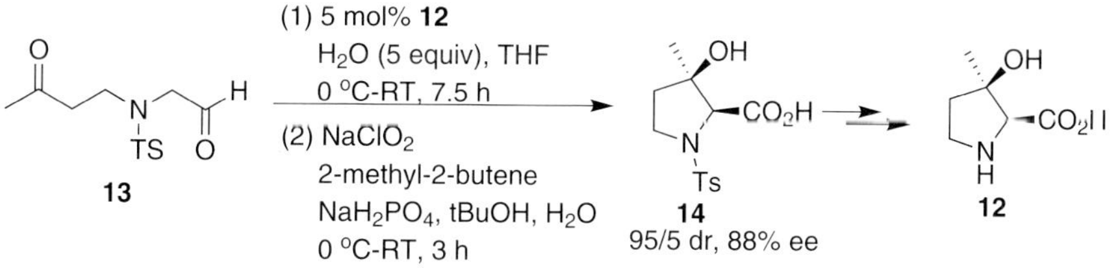

**Scheme 25.5** Synthetic approach to (2S,3R)-3-hydroxy-3-methylproline from the intramolecular aldol reaction.

A tricyclic compound **16** that constitutes the core structure of naturally occurring platensimysin **17** was accessed from a stereoselective intramolecular aldol reaction of **15**, which was conducted for 5 days in DMF in the presence of stoichiometric amounts of L-proline (Scheme 25.6) [12].

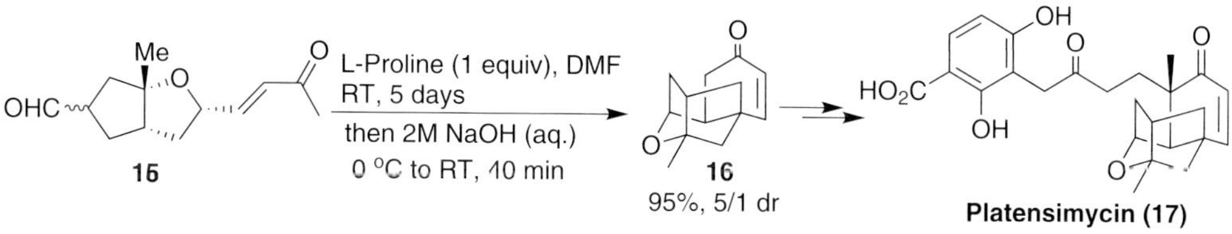

**Scheme 25.6** Construction of core structure of the platensimysin using the intramolecular aldol reaction.

## 25.2.2
### Intramolecular Michael Additions

The intramolecular Michael reaction is one of the most important synthetic methods that enable the efficient construction of cyclic systems. List and coworkers described the first organocatalytic enantioselective intramolecular reaction of formyl enones **19** with MacMillan's catalyst **18a** [13], leading to the generation of 2-(2-oxoalkyl)cyclopentanecarbaldehydes **20** with high stereochemical outcomes (Scheme 25.7) [14].

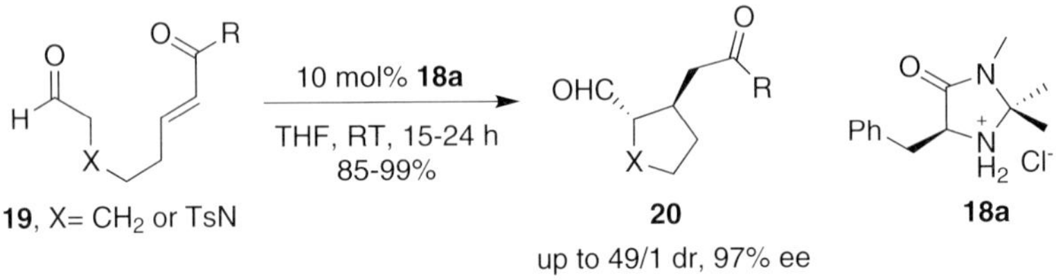

**Scheme 25.7** Intramolecular Michael reaction of formyl enones.

Hayashi disclosed that an enantioselective desymmetrization of 4-substituted-4-(3-formylpropyl)cyclohexa-2,5-dien-1-ones **21** via an intramolecular Michael addition catalyzed by 10 mol% of **23** yielded chiral bicyclic aldehydes **22** in high optical purity that may be potentially applicable to the synthesis of natural products. Interestingly, in contrast to MacMillan's catalyst **18a** that favored anti-selective intramolecular Michael addition of **19**, the organocatalyst **23** preferentially furnished a syn-selective Michael addition of formyl enones structurally similar to **19**, resulting in the formation of cis-diastereomers of 2-(2-oxoalkyl)cyclopentanecarbaldehydes **20** (Scheme 25.8) [15].

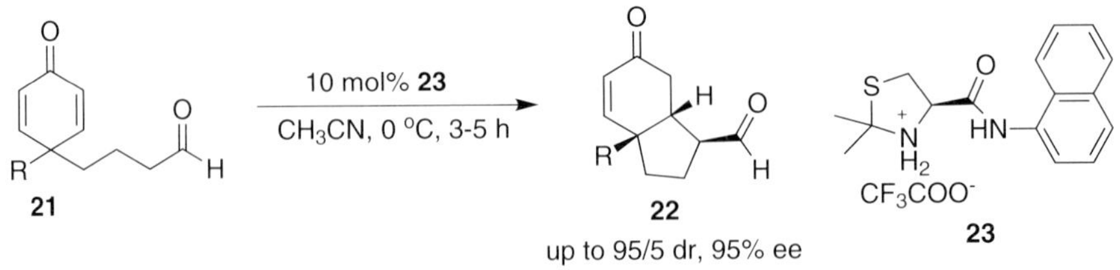

**Scheme 25.8** Enantioselective desymmetrization via intramolecular Michael addition.

The potential of intramolecular Michael additions in the total synthesis of natural products was first demonstrated by Mangion and MacMillan [16]. In the presence of 30 mol% L-proline, the formyl enal **24** underwent an intramolecular Michael addition reaction, after *in situ* acylation by exposure of the reaction mixture to acetyl anhydride, DMAP, and pyridine, to afford iridoid **25** in a high yield. Following readily available synthetic procedures, the bicyclic compound **25** was transformed into **26**, an important precursor for the accomplishment of the total synthesis of (−)-Littoralisone **27** and (−)-Brasoside **28** (Scheme 25.9) [16].

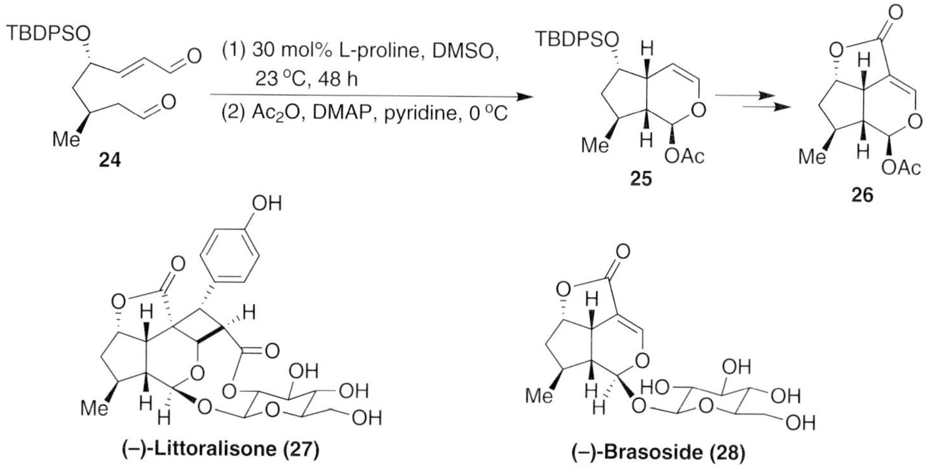

**Scheme 25.9** Total synthesis of (−)-Littoralisone **27** and (−)-Brasoside **28** from intramolecular Michael addition.

### 25.2.3
### Asymmetric Cycloaddition and Domino Reactions

Asymmetric Diels–Alder reactions via enamine catalysis were pioneered by Barbas *et al.*, who reported the first organocatalytic Diels–Alder reaction of α,β-unsaturated ketones **30** with electron-deficient olefins **29** using either L-proline or **31** as a chiral catalyst [17]. This reaction was proposed to proceed via [4 + 2] cycloaddition of dienes that were generated from the secondary amine and enones with dienophiles (Scheme 25.10) [17].

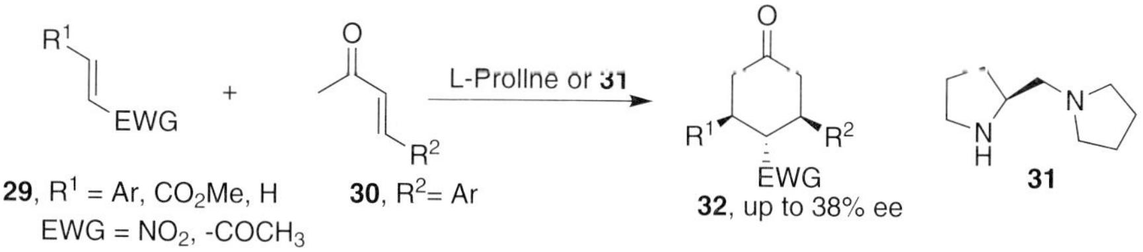

**Scheme 25.10** Direct Diels–Alder reaction of α,β-unsaturated ketones with electron-deficient olefins.

Following a similar strategy, a three-component Diels–Alder reaction of enones **33** with dienophiles formed *in situ* from Knoevenagel reactions of aldehydes **34** and Meldrum's acids **35** could be realized in the presence of amine catalysts. (*R*)-5,5-dimethylthiazolidine-4-carboxylic acid (DMTC, **37**) offered the best results in terms of yield and stereoselectivity (Scheme 25.11) [18].

An enantioselective organocatalytic formal [4 + 2] reaction between α,β-unsaturated aldehydes **39** and **40** under the promotion of prolinol derivatives **38a** and **38b** furnished optically active cyclohexa-1,3-dienyl aldehydes **41** with up to >99%ee.

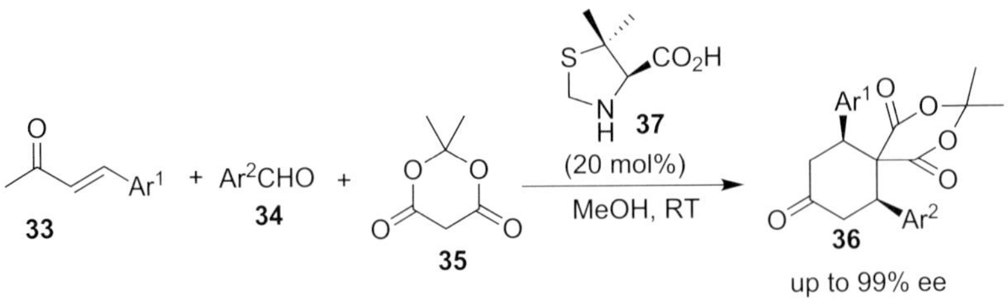

**Scheme 25.11** Three-component Diels–Alder reaction of enones with aldehydes and Meldrum's acids.

The self-[4 + 2]-cycloaddition of (*E*)-3-formylallyl acetate **42** with substoichiomeric amounts of L-proline and triethylamine at −20 °C yielded a *cis*-cyclohexa-1,3-dienyl aldehyde **43** with 95% ee. An enantioselective total synthesis of (+)-palitantin **44** was accomplished commencing with this reaction (Scheme 25.12) [19].

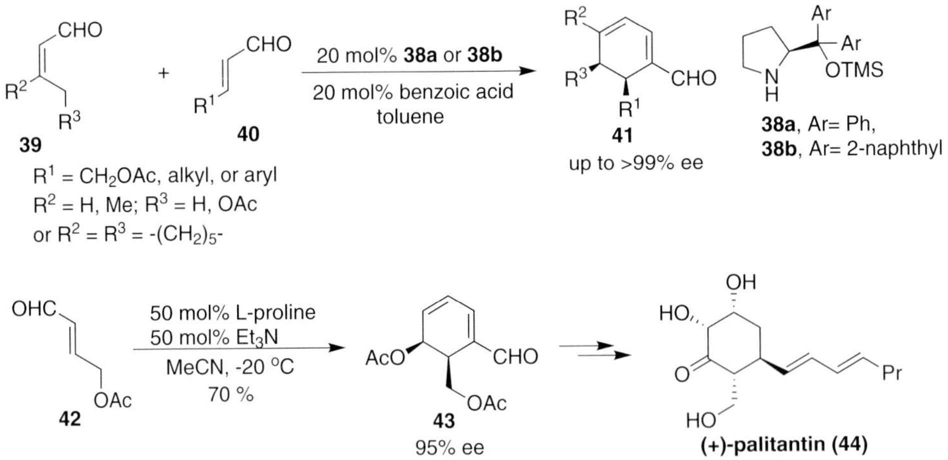

**Scheme 25.12** Formal [4 + 2]-reaction of α,β-unsaturated aldehydes.

Córdova reported that cyclic enones **45** could undergo an enantioselective Diels–Alder reaction with nitroolefins (**46**) under the combined promotion of 30 mol% each of chiral diamine **31** and 2,4-dinitrobenzenesulfonic acid **48**, to produce bicyclic compounds **47** with fairly good stereochemical outcomes (Scheme 25.13) [20].

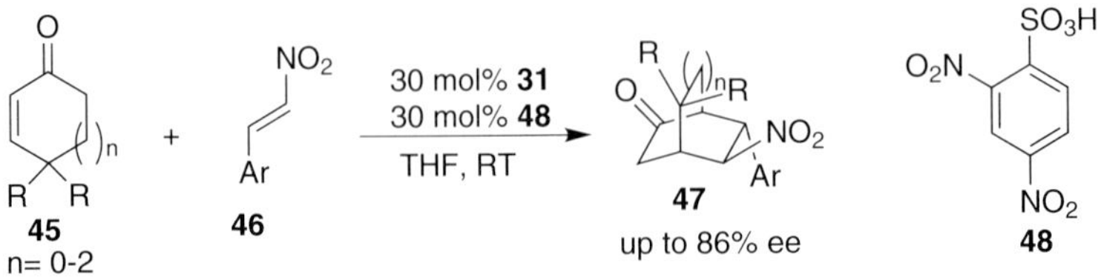

**Scheme 25.13** Direct Diels–Alder reaction with nitroolefins with cyclic enones.

An organocatalytic tandem O-nitroso aldol/Michael reaction of cyclic enones **45** with substituted nitrosobenzenes **49** using pyrrolidine-based tetrazole **51** as a catalyst produced nitroso Diels–Alder adducts **50** with almost perfect enantioselectivity (Scheme 25.14) [21].

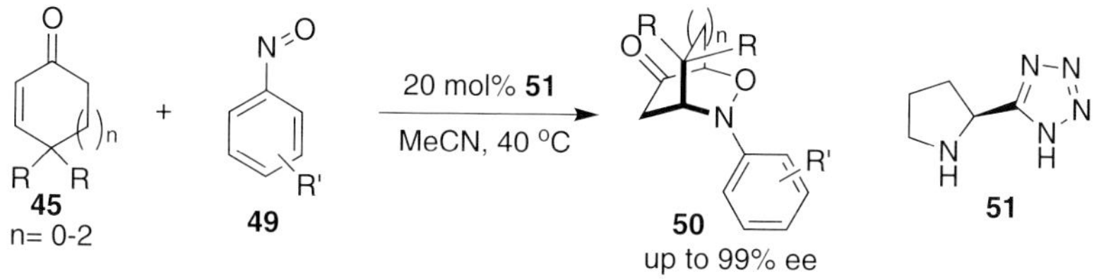

**Scheme 25.14** Tandem O-nitroso aldol/Michael reaction of cyclic enones with nitrosobenzenes.

The presence of 20 mol% of L-proline enabled a one-pot and three-component aza-Diels–Alder reaction of cyclic enones **45** with anilines **52** and aqueous formaldehyde **53** to give bicyclic compounds **54** in good yields with excellent enantiopurity (Scheme 25.15) [22].

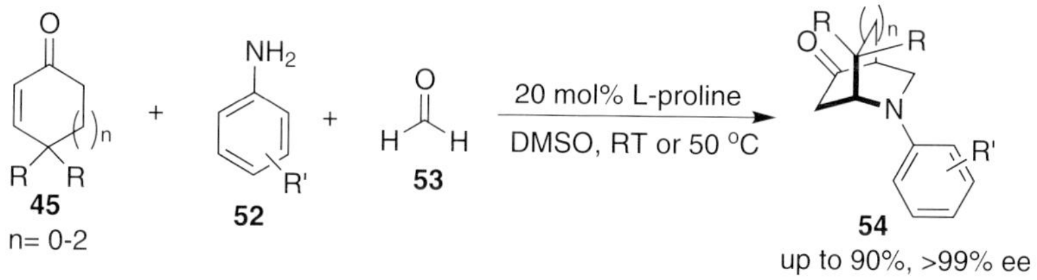

**Scheme 25.15** Direct aza-Diels-Alder reaction catalyzed by L proline.

Itoh *et al.* established an aza-Diels–Alder reaction of 9-tosyl-3,4-dihydro-b-carboline **55** with 3-ethyl-3-buten-2-one **56** catalyzed by 30 mol% of L-proline, furnishing a heterotetracyclic compound **57** with greater than 99% de and 99% ee. The conversion of the compound **57** to ent-dihydrocorynantheol **58** could be accomplished in three steps by using readily available procedures (Scheme 25.16) [23].

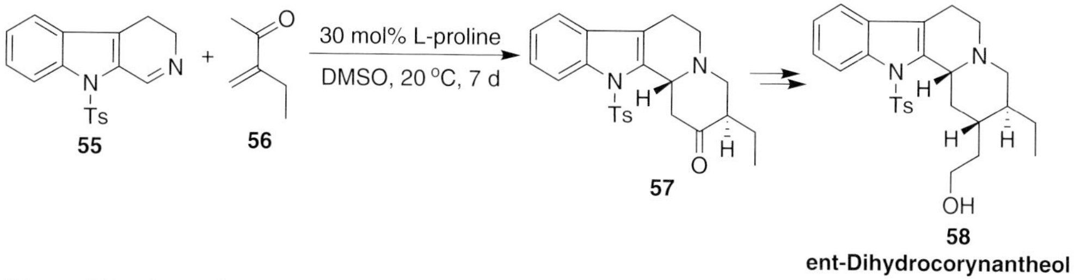

**Scheme 25.16** Aza-Diels–Alder reaction of 9-tosyl-3,4-dihydro-b-carboline with 3-ethyl-3-buten-2-one.

Jørgensen and Juhl reported an enantioselective inverse-electron-demand hetero-Diels–Alder reaction of aldehydes **59** with enones **60** catalyzed by a pyrolidine derivative **61**, after subjection to oxidation with PCC, giving chiral lactones **62** with high enantiomeric excess. The aldehydes **59** were considered to be activated by the formation of a chiral enamine with secondary amine **61** that served as a dienophile for the [4 + 2] cycloaddition with enones via an assumed intermediate **II** (Scheme 25.17) [24].

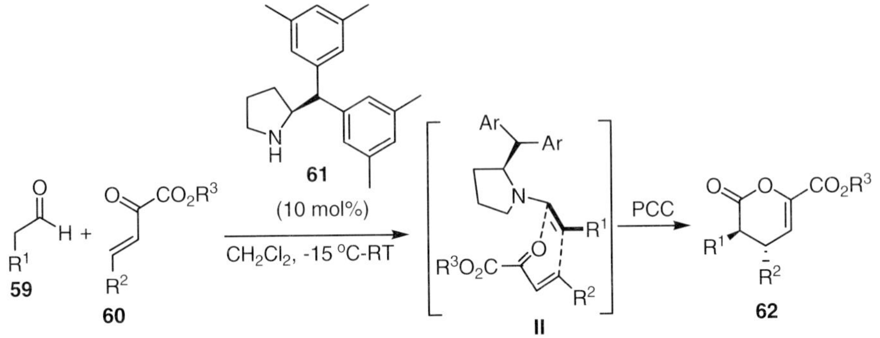

**Scheme 25.17** Enantioselective inverse-electron-demand hetero-Diels–Alder reaction.

An organocatalytic asymmetric formal [3 + 3]-cycloaddition of (*E*)-but-2-enal **40a** was reported by Hong *et al.* L-Proline served as the chiral catalyst to promote the reaction with up to 95% ee. This procedure has been exploited in the asymmetric synthesis of (−)-isopulegol and (−)-cubebaol (Scheme 25.18) [25].

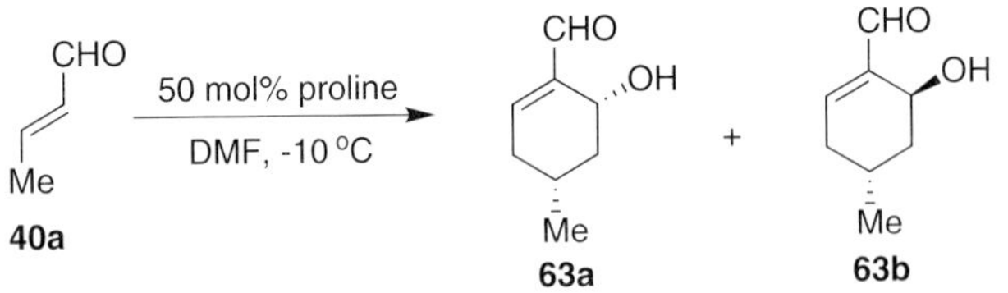

**Scheme 25.18** Asymmetric formal [3 + 3]-cycloaddition of the enal.

A tandem Michael/Henry reaction of pentan-1,5-dial **64** with nitroolefins catalyzed by 10 mol% of the prolinol derivative **38a** gave rise to tetrasubstituted cyclohexanes **65** bearing four contiguous stereogenic centers with perfect levels of stereochemical control (Scheme 25.19) [26].

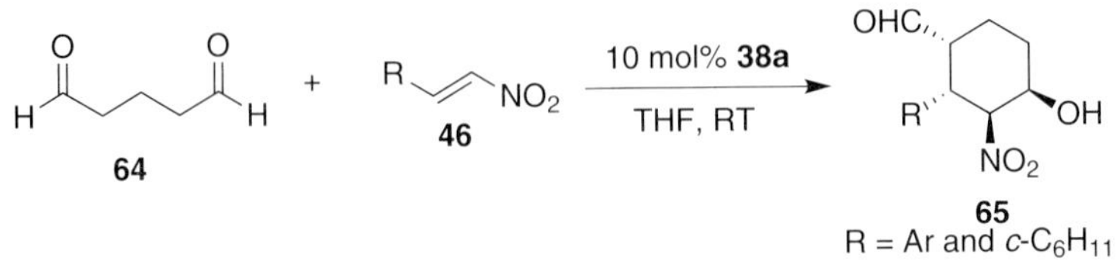

**Scheme 25.19** Cascade Michael/Henry reactions.

## 25.3
## Iminium-Catalyzed Asymmetric Cyclization Reactions

### 25.3.1
### Asymmetric Cycloadditions

In 2000, MacMillan reported the first highly enantioselective organocatalytic Diels–Alder reaction of dienes **66** with α,β-unsaturated aldehydes **40**, using a chiral imidazolidione **18a** as the catalyst [27a]. The [4 + 2]-cycloaddition reaction is accelerated by LUMO-lowering activation of α,β-unsaturated aldehydes through the reversible formation of iminium ions with chiral amine compound [13]. Similarly, α,β-unsaturated ketones **68** can also be activated by using the chiral amine catalyst **18b**, enabling the enantioseletive Diels–Alder reaction with up to 98% ee (Scheme 25.20) [27b].

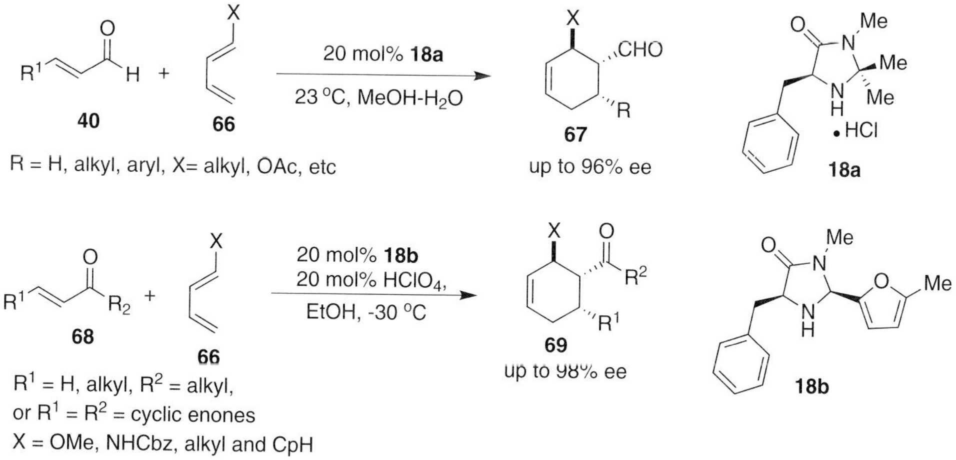

**Scheme 25.20** Asymmetric Diels–Alder reactions.

LUMO-lowering catalysis has also been applied to 1,3-dipolar cycloaddition reactions of nitrone **70** to α,β-unsaturated aldehydes **40**. The presence of 20 mol% of $HClO_4$ salt of chiral imidazolidione **18c** was capable of catalyzing the enantioselective dipolar addition that favorably yielded endo-isoxazolidines **71** with high enantioselectivities of up to 99% ee (Scheme 25.21) [28].

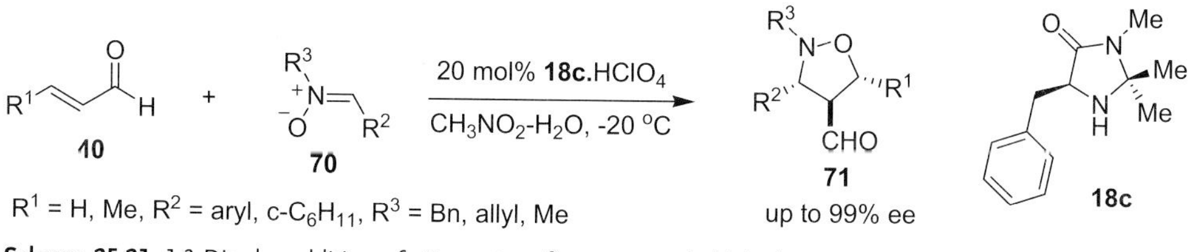

**Scheme 25.21** 1,3-Dipolar addition of nitrone to α,β-unsaturated aldehydes.

LUMO-lowering activation is a general concept widely used in the design of chiral organocatalysts for cycloaddition reactions involving unsaturated aldehydes and ketones. Some chiral amine-based organic molecules **38** and **72–75** turned out to be catalysts of choice for the Diels–Alder and 1,3-dipolar addition reactions involving nitrone, azomethine ylide- and azomethine imine dipoles, and others, which are all accelerated by lowering the LUMO of the π-system through the reversible formation of iminium ions (Scheme 25.22) [29].

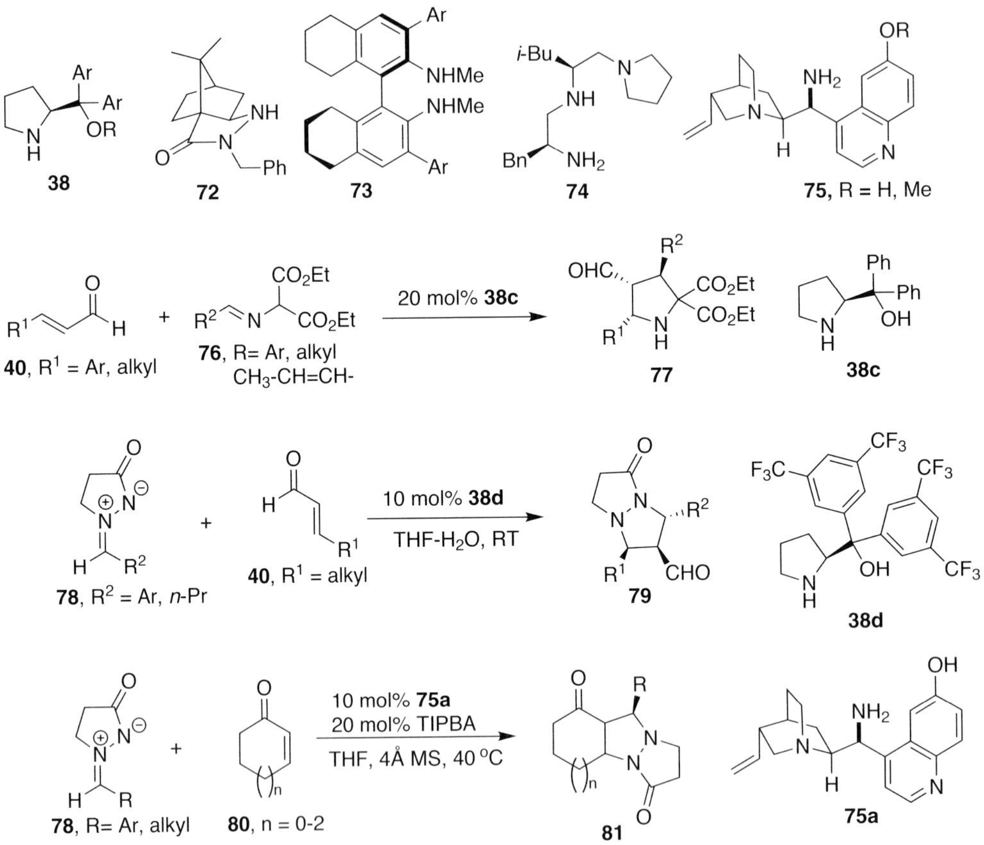

**Scheme 25.22** Organocatalysts for various 1,3-dipolar addition reactions.

The synthetic uses of organocatalytic asymmetric Diels–Alder reactions have been illustrated in the total synthesis of (+)-hapalindole Q. An asymmetric Diels–Alder reaction of an enal **82** with a diene **83** mediated with 40 mol% of MacMillan's catalyst **18a** proceeded in a 1/1 mixture of DMF and MeOH containing 5% water to give, after 1.5 days the adducts in 35% overall yield, favoring the endo-diastereomer **84a** with 85/15 dr and 93% ee. Starting with the diastereomer **84a**, a relatively concise total synthesis of (+)-Hapalindole Q (**85**) was accomplished in 1.7% overall yield with 93% ee (Scheme 25.23) [30].

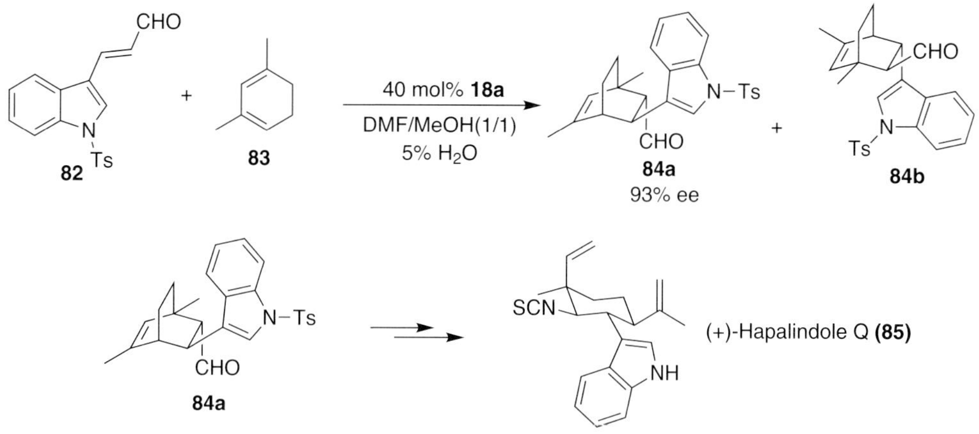

## 25.3 Iminium-Catalyzed Asymmetric Cyclization Reactions

**Scheme 25.23** Organocatalytic Diels–Alder reaction for the total synthesis of (+)-Hapalindole Q.

Organocatalytic asymmetric intramolecular Diels–Alder reactions and their application in the enantioselective synthesis of solanapyrone D have also been described by MacMillan [31]. Diels–Alder reactions of trienes **86** catalyzed by 20 mol% of imidazolidione **18a** generated the bicyclic adducts **87** in high dr and ee. The use of trienal **86a** as a substrate formed all the stereogenetic centers of a skeleton present in solanapyrone D (**89**), and thereby paved a way to concisely access this target molecule in six steps from trienal **86a** and in nine steps from commercially available chemicals (Scheme 25.24) [31].

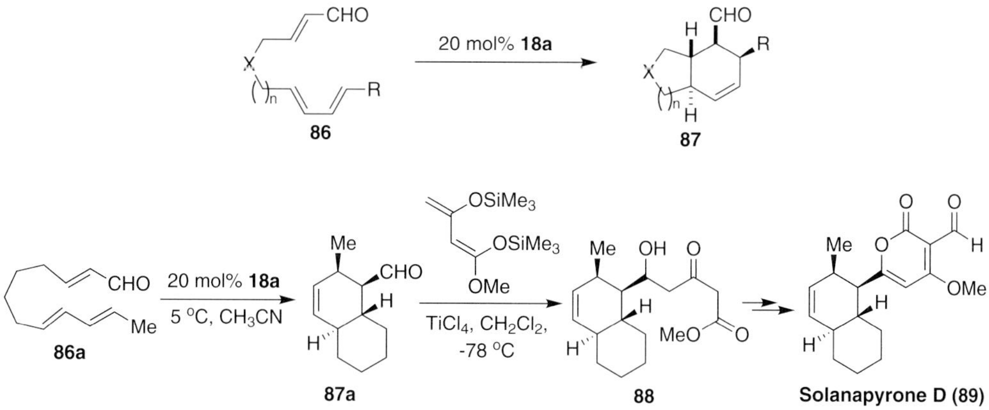

**Scheme 25.24** Total synthesis of solanapyrone D.

An elegant example of the application of iminium catalysis in asymmetric [4 + 3]-cycloaddition was reported by Harmata and coworkers. Reactions of furan derivatives **90** with silyloxypentadienals **91** in the presence of MacMillan's catalyst **18c** resulted in the formation of [4 + 3]-cycloadducts **92** in good yields with enantioselectivities ranging from 81% to 90% ee, depending on the substituents of both reaction components (Scheme 25.25) [32].

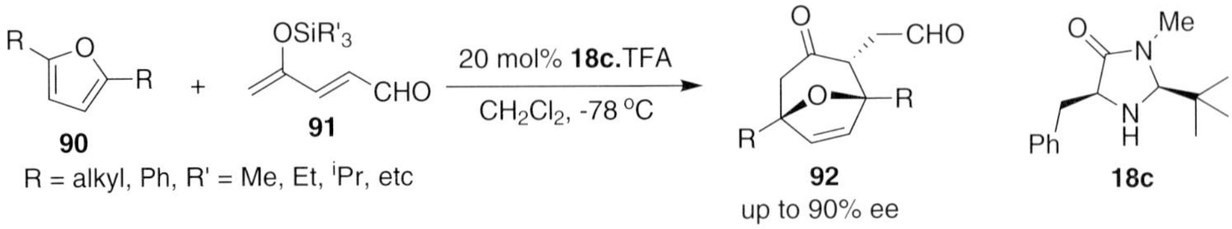

**Scheme 25.25** [4 + 3]-cycloaddition reactions.

### 25.3.2
### Asymmetric Intramolecular 1,4-Additions

Intramolecular Michael addition reactions accelerated by iminium ions have also emerged to be a powerful platform for building ring systems. An organocatalytic intramolecular aza-Michael addition reaction of carbamates **93** bearing a remote unsaturated aldehyde under the promotion of the prolinol derivative **38e** occurred smoothly to give heterocycles **94** with very high enantioselectivity. This procedure enabled concise enantioselective syntheses of alkaloids, such as (+)-sedamine **96**, (+)-allosamine **98**, and (+)-coniine **99** (Scheme 25.26) [33].

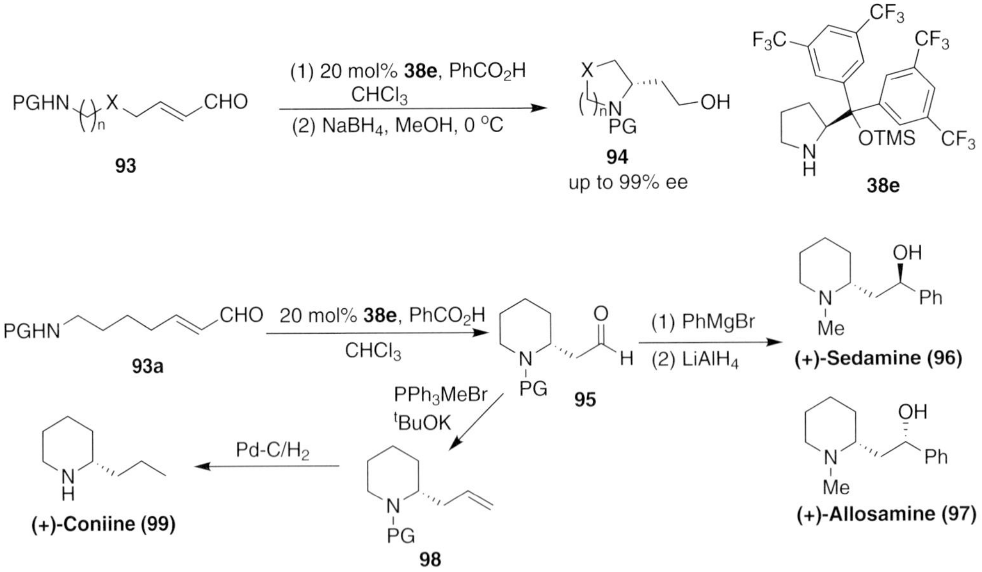

**Scheme 25.26** Intramolecular aza-Michael additions and total syntheses of alkaloids.

Xiao and coworkers recently reported an asymmetric intramolecular Friedel–Crafts reaction of indolyl α,β-unsaturated aldehydes **100** under the catalysis of MacMillan's catalyst **18c** combined with 3,5-dinitrobenzoic acid **102**, which provided a straightforward synthetic pathway to access enantioenriched tetrahydropyrano[3,4-b]indoles **101** that commonly present in natural products and biologically active compounds (Scheme 25.27) [34].

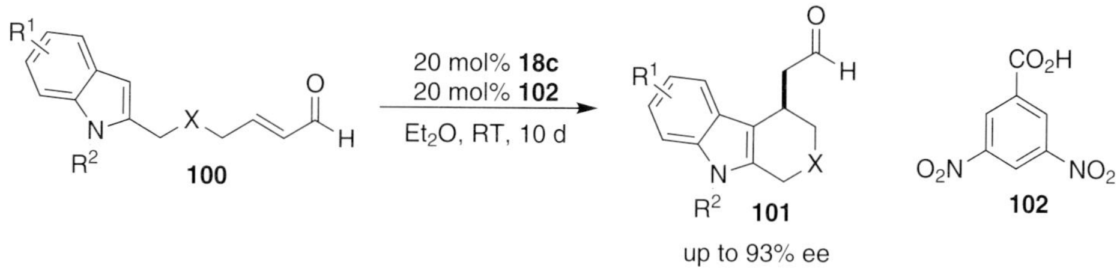

Scheme 25.27 Intramolecular Friedel–Crafts alkylation reaction.

### 25.3.3
### Asymmetric Domino Reactions

Domino reactions have been considered important transformations applied to the construction of cyclic compounds. MacMillan et al. established a domino Michael addition/cyclic amination reaction of tryptamine derivatives **103** with enals **40** on the basis of LUMO-lowering catalytic strategy by using their own catalyst **18c**, yielding pyrroloindolines **104** with high levels of stereoselectivities. Basically, this cascade reaction started with an iminium-catalyzed Friedel–Crafts reaction of tryptamine **103** with enals **40**, followed by an intramolecular amination (Scheme 25.28) [35].

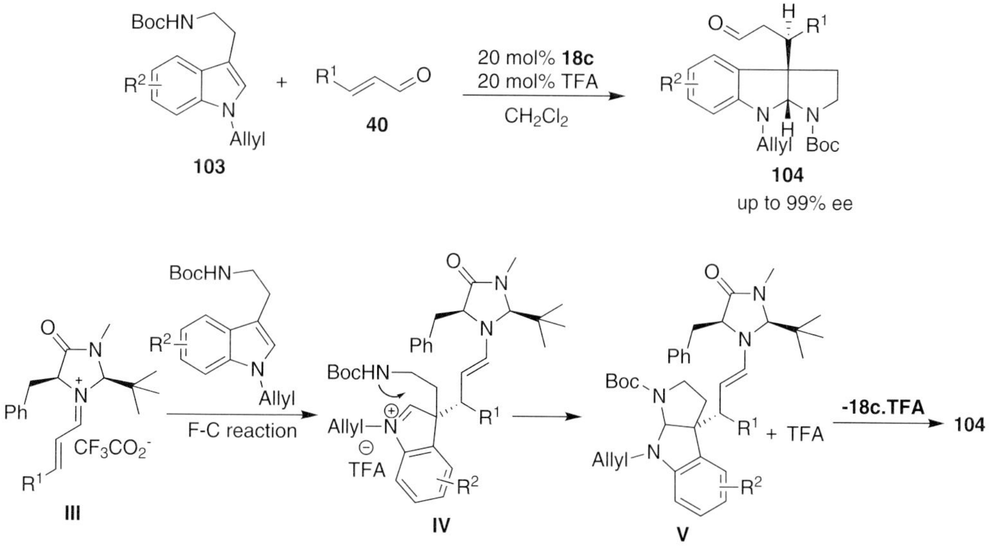

Scheme 25.28 Reaction mechanism of the iminium catalyzed cyclization reaction.

The pyrroloindolines **104** are highly synthetically useful in the total synthesis of alkaloids. Accordingly, (−)-flustramine B **107** was accomplished in six steps commencing with the cascade reaction of **103a** with acrylaldehyde in a 58% overall yield (Scheme 25.29) [35].

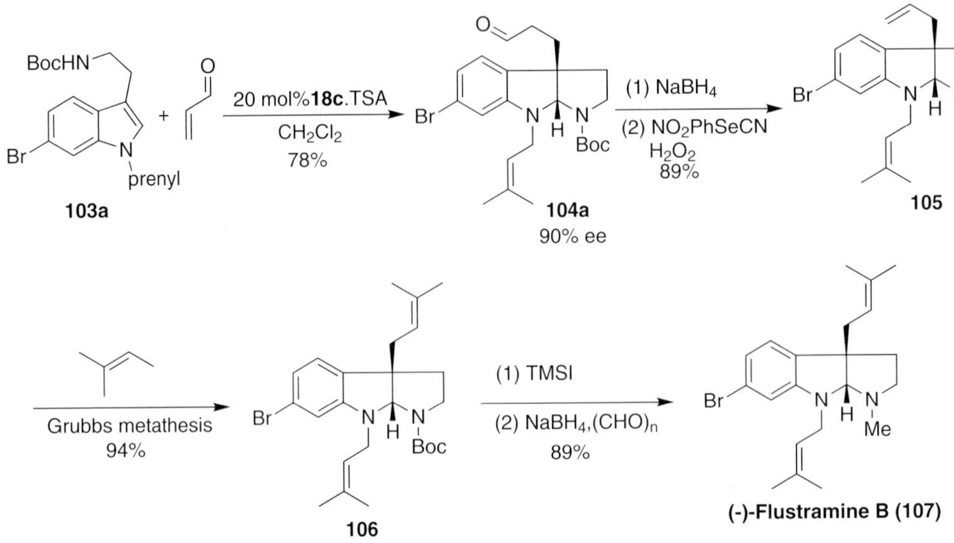

**Scheme 25.29** Total synthesis of (−)-flustramine B **107**.

The creation of carbon–carbon bonds with the control of five stereogenic centers was recently accomplished through iminium-catalyzed domino Michael/Henry reactions of enals **40** with a dinitroalkane **108** using prolinol derivative **38e** as a chiral catalyst under the assistance of catalytic amount of DABCO (Scheme 25.30) [36].

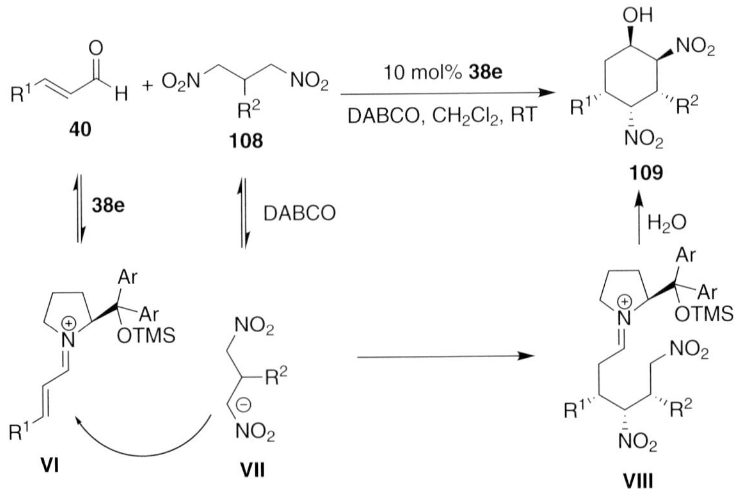

**Scheme 25.30** Domino Michael/Henry reactions.

## 25.4
## Sequential Catalysis with Chiral Amines for Domino Cyclization Reactions

As enamine and iminium ions are reversibly transformed into each other, a cascade reaction could be designed for the construction of cyclic compounds via either of sequential enamine/iminium, enamine/enamine, or iminium/enamine catalysis.

### 25.4.1
### Enamine/Iminium Catalytic Sequences

The enantioselective cyclization reactions by sequential enamine/iminium catalysis were pioneered by Enders et al., who recently reported a three-component domino reaction of aldehydes 59 and α,β-unsaturated aldehydes 40 with nitroolefins 46 catalyzed by the prolinol derivative 38a producing cyclohexene derivatives 110 with concomitant creation of four contiguous stereogenic centers with perfect stereocontrol (Scheme 25.31) [37].

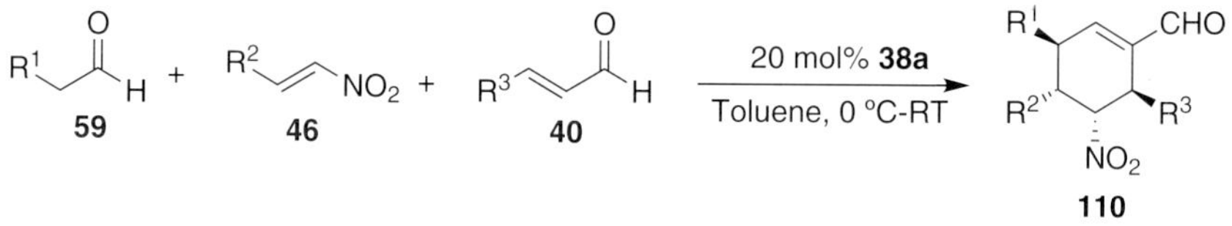

**Scheme 25.31** Three-component domino reaction catalyzed by the prolinol derivative.

A reaction mechanism proposed by Enders is shown in Scheme 25.32. The organic compound 38a initially catalyzed the Michael addition by forming an active enamine

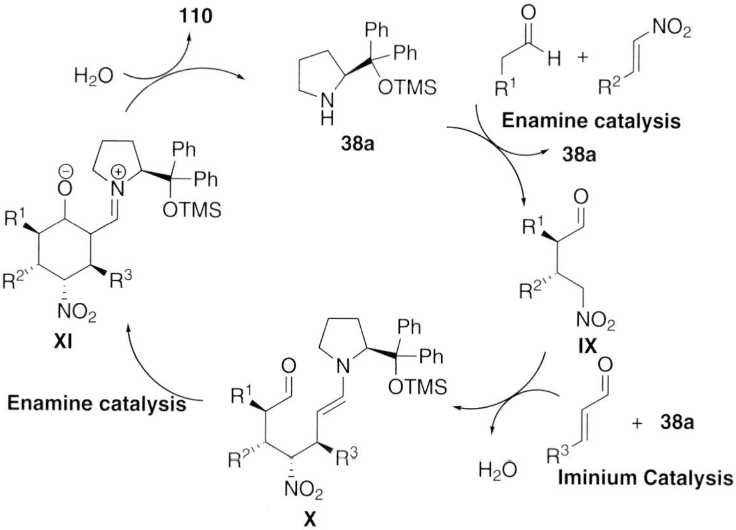

**Scheme 25.32** The mechanism of the domino reaction catalyzed by the prolinol derivative 38a.

with the aldehyde **59** furnishing a compound **IX** that reacted with unsaturated aldehydes **40** by iminium catalysis to give intermediates **X**. The intermediates **X** underwent sequentially an enamine-catalyzed intramolecular aldol reaction, which was followed by dehydration, providing products **110** [37]. This domino reaction, using reversible enamine- and iminium catalysis, indicated a general concept that could be expanded to the discovery of new cascade reactions for the construction of a cyclic skeleton.

The power of this reaction in the construction of polycyclic compounds has been amenable to the rapid synthesis of tricyclic frameworks. A mixture of aldehydes bearing a remote diene **111** and α,β-unsaturated aldehydes **40** with nitroolefins **46** first underwent a cascade cyclization reaction under the promotion of **38a**, and then an intramolecular Diels–Alder reaction by exposure to dimethylaluminum chloride at −78 °C in one-pot, giving rise to the tricyclic products **113** and **114** with perfect stereoselectivity (Scheme 25.33) [38].

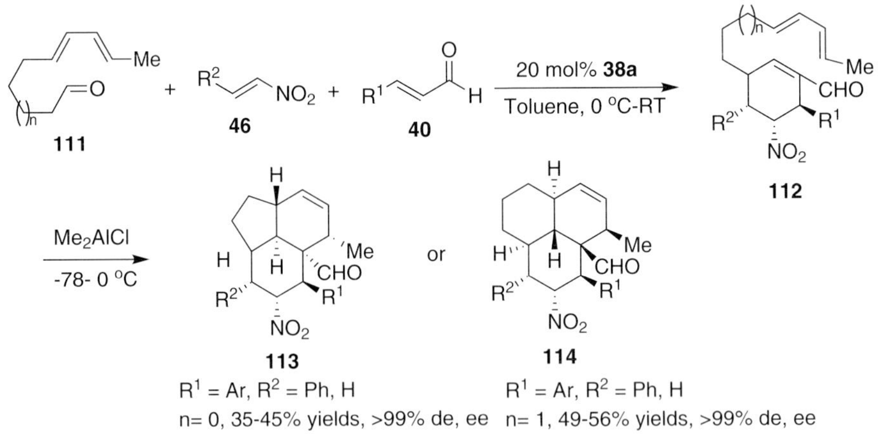

**Scheme 25.33** Synthesis of polycyclic compounds with three-component domino reactions.

## 25.4.2
### Enamine/Enamine Catalytic Sequences

Tang recently reported an enantioseletive formal [3 + 3]-annulation of cyclic ketones **115** with enones **60** in the presence of pyrrolidine derivatives. The bicyclic products **116** actually originated from the sequential enamine-catalyzed Michael addition and intramolecular aldol reactions via intermediates **XII–XIV**. N-(pyrrolidin-2-ylmethyl) trifluoromethanesulfonamide **117** exhibited superior stereocontrol ability. This procedure provides an efficient entry to [3.3.1]-bicyclic structures **116** in high optical purity (Scheme 25.34) [39].

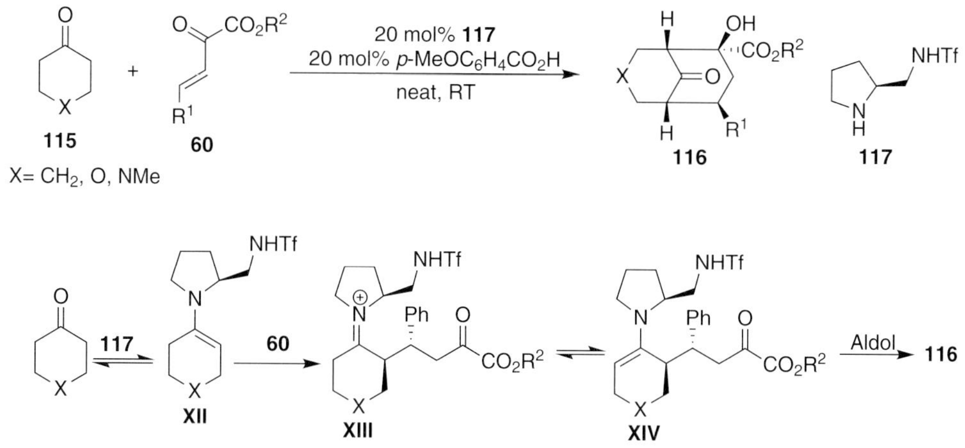

## 25.4 Sequential Catalysis with Chiral Amines for Domino Cyclization Reactions

**Scheme 25.34** Enantioseletive formal [3 + 3]-annulation of cyclic ketones with enones and the mechanism.

### 25.4.3
### Iminium/Enamine Catalytic Sequences

Jørgensen described a cyclization reaction between enones **118** and α-keto esters **119**, which proceeded through cascade iminium-catalyzed Michael addition/enamine-catalyzed intramolecular aldolization reactions, highly enantioselectively giving multiply functionalized cyclohexanones **120** under the influence of **121** (Scheme 25.35) [40].

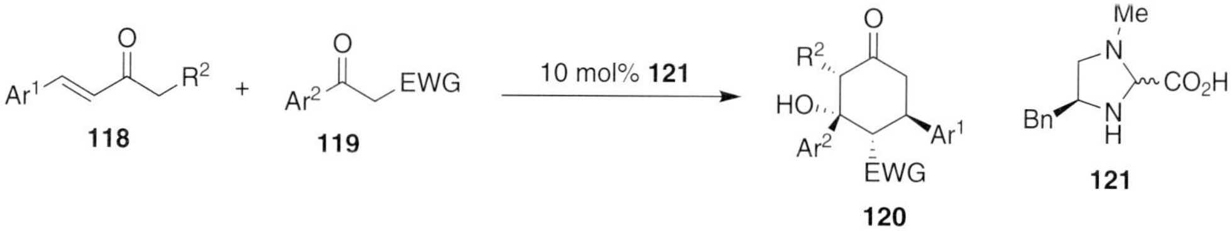

**Scheme 25.35** Cascade Michael addition/intramolecular aldolization reactions.

List's group proposed that the amine catalyst could first accelerate a reduction of enal enones **122** with Hantzsch ester **123** by forming an iminium ion **XV** to generate an enamine intermediate **XVI** that then undergoes an intramolecular Michael addition to afford cyclopentane derivatives **124**. Accordingly, the use of MacMillan's catalyst **18a** enabled this reductive Michael addition to furnish **124** with high enantioselectivities ranging from 86% to 97% ee (Scheme 25.36) [41].

A three-component domino cyclization reaction of α,β-unsaturated aldehydes **40** and **40'** with acidic methylene compounds **125** offered an efficient synthetic method to access optically pure cyclohex-1-enecarbaldehydes **126**. The iminium/iminium/

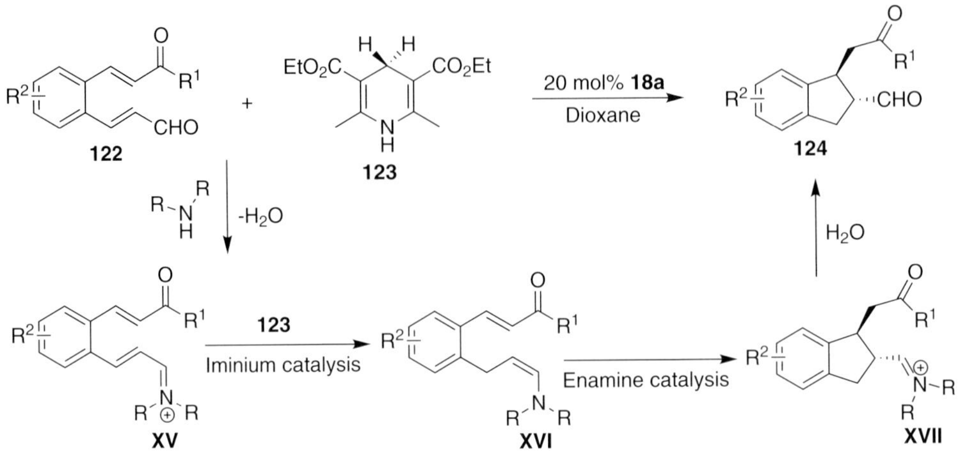

**Scheme 25.36** Asymmetric reductive Michael addition.

enamine sequential activation of unsaturated aldehydes using catalyst **38e** was considered to account for the triple cascade reaction (Scheme 25.37) [42].

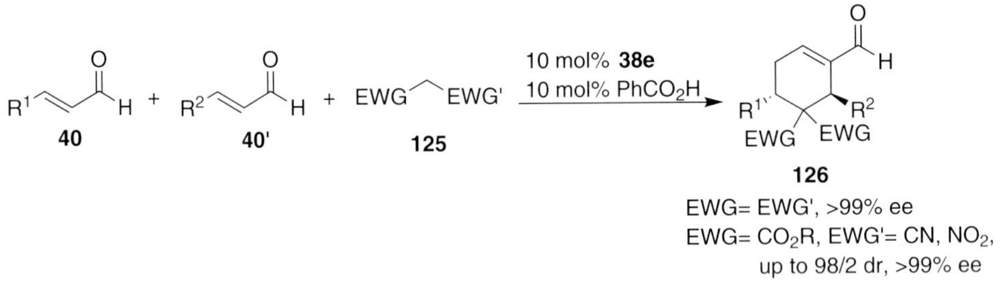

**Scheme 25.37** Three-component domino cyclization reaction of α,β-unsaturated aldehydes with acidic methylene compounds.

Highly functionalized chiral cyclopentane derivatives **128** have been readily accessed with stereoselective cascade double Michael addition reactions of **40** with **127**, which were sequentially catalyzed by iminium and enamine generated from amine catalyst **38a** and the formyl group of enals **40** (Scheme 25.38) [43].

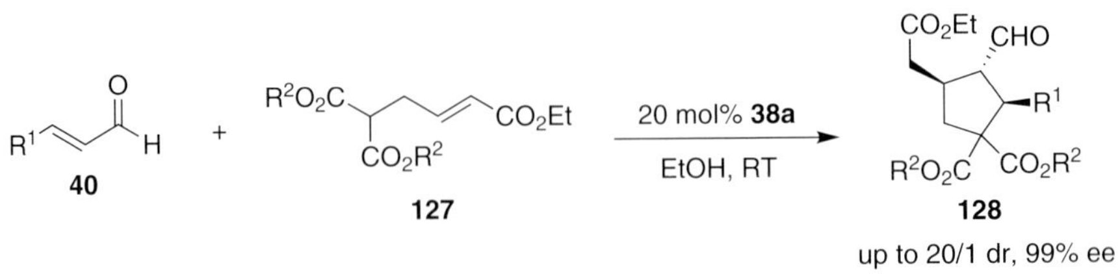

**Scheme 25.38** Synthesis of chiral cyclopentane derivatives through cascade double Michael additions.

## 25.4 Sequential Catalysis with Chiral Amines for Domino Cyclization Reactions

Michael aldol-dehydration reactions of enals **40** with **129** took place in the presence of organocatalyst **38f**, offering an efficient and highly enantioselective synthesis of highly functionalized cyclopentenes **130** (Scheme 25.39) [44].

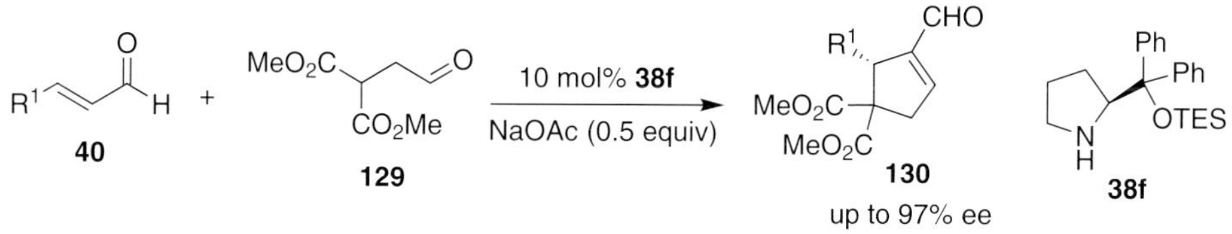

**Scheme 25.39** Enantioselective synthesis of highly functionalized cyclopentenes.

Synthetic methods for the preparation of chiral heterocyclic compounds in a single operation have been developed by using the sequential iminium/enamine activation concept. Wang et al. found a cascade sulfa-Michael aldol-dehydration reaction of **131** with enal **40** that was catalyzed by chiral amine **38e** to afford chiral thiochromenes **132** with high levels of enantioselectivities (Scheme 25.40) [45].

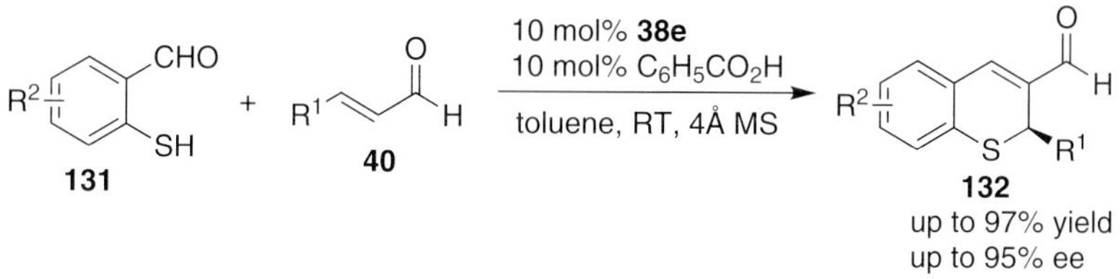

**Scheme 25.40** Cascade sulfa-Michael–aldol–dehydration reactions for the synthesis of thiochromenes.

Independently, Jørgensen and coworkers reported a domino-sulfa-Michael aldol reaction of enal **40** with 2-mercapto-1-phenylethanone **133** under the catalysis of 10 mol% of **38e**, preferentially generating tetrahydrothiophenes **134** with up to 95% ee under acidic conditions and **135** with up to 84% ee under basic conditions (Scheme 25.41) [46].

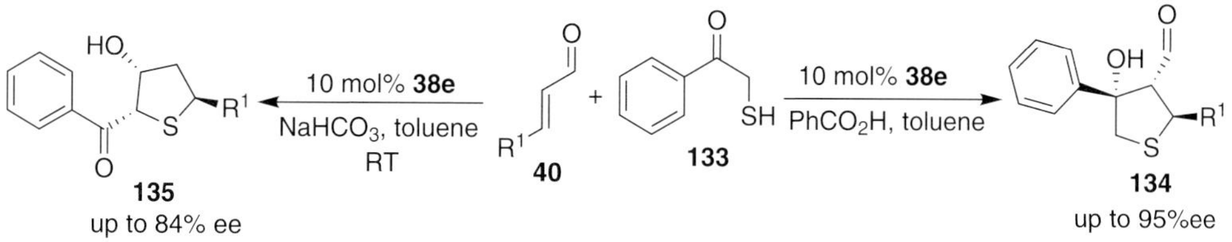

**Scheme 25.41** Cascade sulfa-Michael–aldol reactions.

As wang et al. indicated, conceptually similar oxa- and aza-Michael aldol-dehydration reactions of unsaturated aldehydes **40** with either 2-N-protected amino benzaldehydes **136** or salicylic aldehydes **137** could be realized by using **38a** or **38f** as a chiral

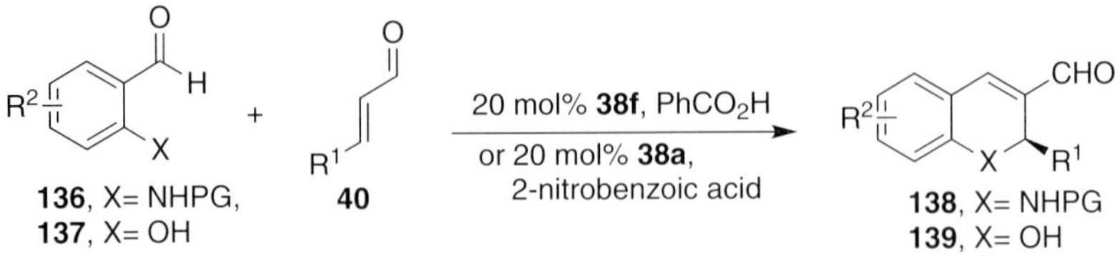

**Scheme 25.42** Oxa- and aza-Michael–aldol–dehydration reactions.

catalyst, leading to the formation of 1,2-dihydroqunolines **138** and chromenes **139**, respectively with high levels of enantioselectivities (Scheme 25.42) [47].

## 25.5
### Brønsted Acid-Catalyzed Asymmetric Cyclization Reactions

Brønsted acids are able to form hydrogen bonds with carbonyl and imine compounds through the modes shown in Schemes 25.43 and 25.44 [48]. Accordingly, the carbonyl compounds can be activated with Brønsted acids by forming hydrogen bonds with oxygen, while the imines can be activated with Brønsted acids either by forming iminium ions or hydrogen bonds, depending on the acidity of the Brønsted acids (Scheme 25.43). The Brønsted acids are also capable of enhancing the reactivity of the carbon–carbon double bond of α,β-unsaturated carbonyls, either by the formation of a hydrogen bond with carbonyl oxygen or by iminium ions (Scheme 25.44). Thus, the design of Brønsted acid-catalyzed cyclization reactions can be expected to be based on these activation modes.

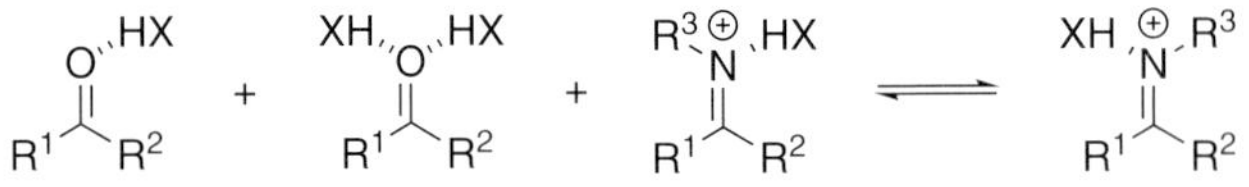

**Scheme 25.43** Activation of carbonyl and imine with Brønsted acid.

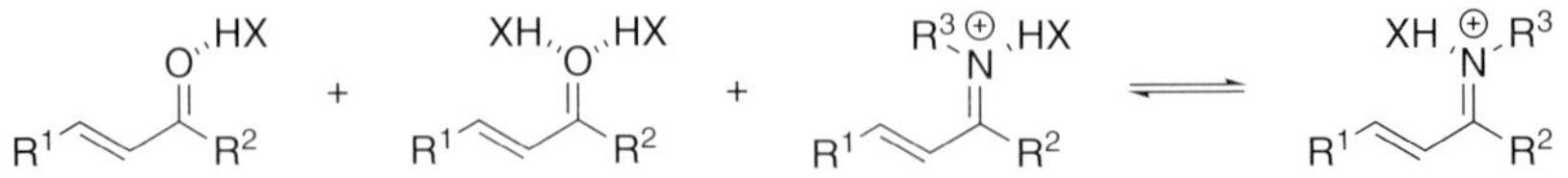

**Scheme 25.44** Activation of carbon–carbon double bond with Brønsted acid.

### 25.5.1
#### Diels–Alder and Hetero-Diels–Alder Reactions

The first example of a Brønsted acid-catalyzed asymmetric Diels–Alder reaction was reported by Göbel's group. An axially chiral amininium ion **140** was found to be capable of promoting a Diels–Alder reaction of diene **141** with diketone **142**, albeit with a modest enantioselectivity. As shown in **XVIII**, protons of **140** serve as activators

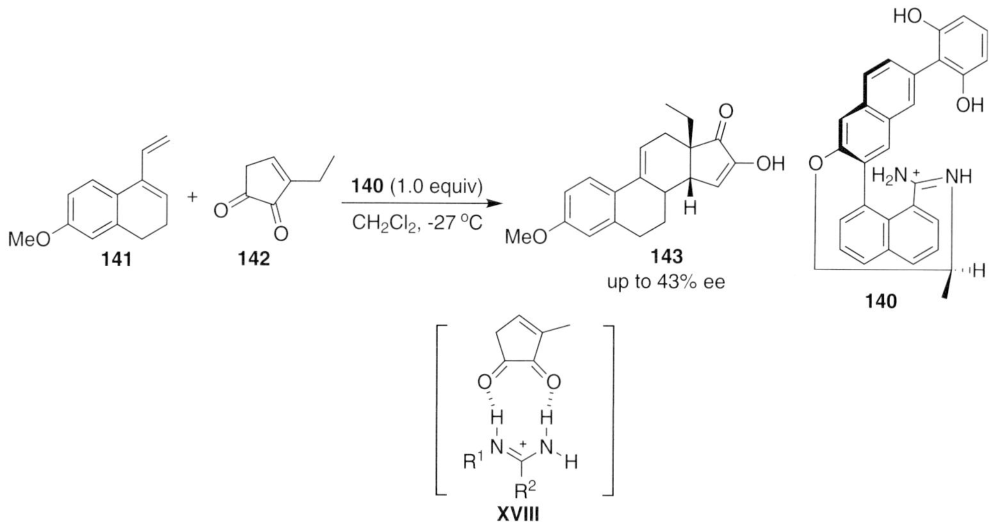

**Scheme 25.45** Brønsted acid promoted Diels–Alder reactions

of the dienophile by forming hydrogen bonds with diketone **142** to lower the LUMO (Scheme 25.45) [49].

Rawal established a highly enantioselective organocatalytic hetero-Diels–Alder reaction of aldehydes **144** with a diene **145** using chiral alcohols as catalysts. TADDOL **146** and axially chiral biaryl diols **147** have shown high catalytic activity and delivered high optical purity to 2,3-dihydropyranones **148** after work-up under standard conditions (Scheme 25.46) [50].

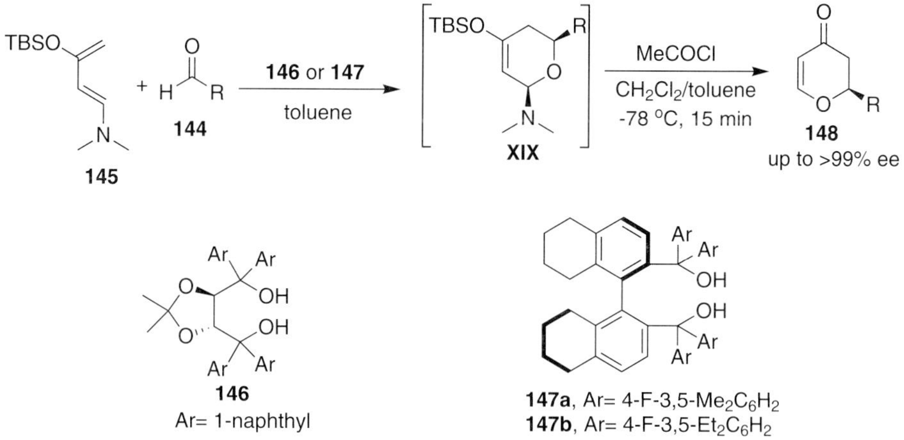

**Scheme 25.46** TADDOL catalyzed hetero-Diels–Alder reactions.

Ding and coworkers demonstrated a TADDOL catalyzed hetero-Diels–Alder reaction of Brassard's diene **149** with the aldehydes **144**, yielding lactones **150** highly enantioselectively. Significantly, a natural product, (S)-(+)-dihydrokawain

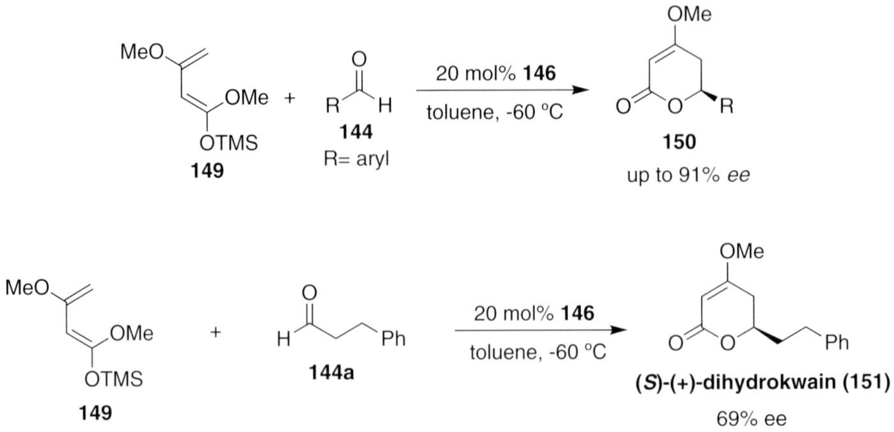

Scheme 25.47 TADDOL catalyzed hetero-Diels–Alder reaction of Brassard's diene with aldehydes.

151 could be obtained directly from the reaction of Brassard's diene with 3-phenylpropanal 144a in 50% yield with 69% ee (Scheme 25.47) [51].

Asymmetric Diels–Alder reactions of acroleins 152 with Rawal's diene 145 proceeded smoothly under the catalysis of 20 mol% of TADDOL 146, after a sequential one-pot reduction with lithium aluminum hydride and removal of TBS with hydrofluoride, to yield optically active cyclohexenones 153 with up to 92% ee (Scheme 25.48) [52].

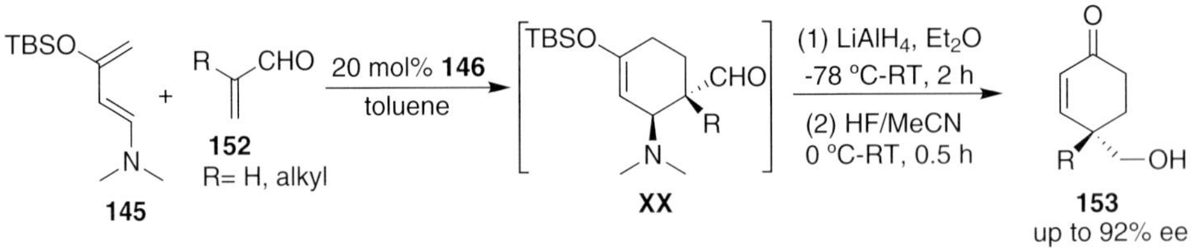

Scheme 25.48 Diels–Alder reactions of acroleins with Rawal's diene.

Investigations by Yamamoto's group into exploiting stronger Brønsted acids to catalyze Diels–Alder reactions indicated that BINOL-based chiral N-triflyl phosphoramide 154 exhibited high catalytic efficiency for the Diels–Alder reaction of silyloxydienes 155 with 3-pentenone 30a, forming cyclohexene derivatives 156 in high optical purity (Scheme 25.49) [53].

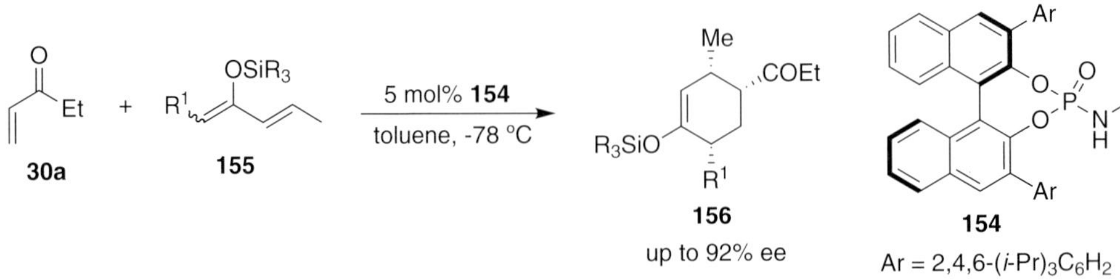

Scheme 25.49 Diels–Alder reaction of silyloxydienes with 3-pentenone.

## 25.5 Brønsted Acid-Catalyzed Asymmetric Cyclization Reactions

The same group reported that 3,3′-disubstituted BINOLs were capable of promoting the nitroso-Diels–Alder reaction of nitrobenzene derivatives **49** with dienes **158** prepared from cyclohexenone derivatives and mopholine. The presence of 30 mol% of tris-(m-xylyl)silylbinaphthol **157** furnished the reaction with up to 92% ee (Scheme 25.50) [54].

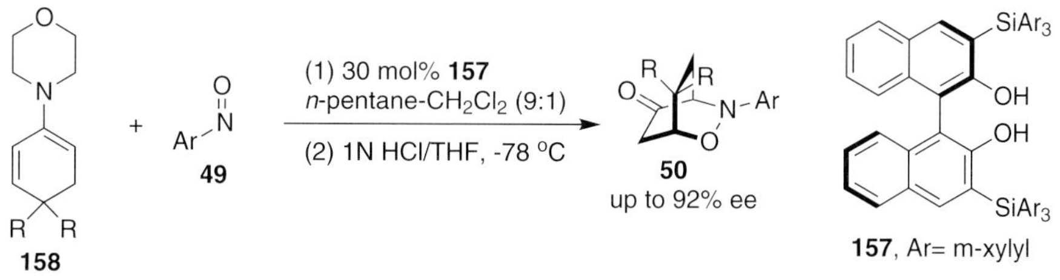

**Scheme 25.50** Nitroso-Diels–Alder reaction of nitrobenzenes with dienes.

Akiyama disclosed a Brønsted acid-catalyzed aza-Diels–Alder reaction of Brassard's diene **149** with Imines **159**. A pyridinium salt of chiral phosphoric acids **160** exhibited relatively higher catalytic activity and enantioselectivity than the corresponding free phosphoric acids because the Brassard's Diene **149** is unstable under strong acidic conditions. The presence of 3 mol% of **160** is sufficient to catalyze the cycloaddition reaction with up to 99% ee (Scheme 25.51) [55].

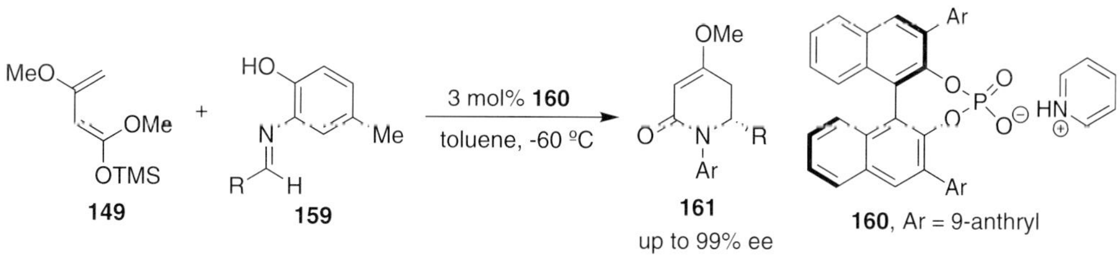

**Scheme 25.51** Aza-hetero-Diels–Alder reactions Brassard's Diene with Imines.

Asymmetric inverse electron-demand aza-Diels–Alder reaction of electron-rich alkenes **163** with aldimines **162** was realized by the same group using the phosphoric acid **165a** bearing bulky substituents, leading to the production of tetrahydroquinolines **164** with high enantioselectivity (Scheme 25.52) [56].

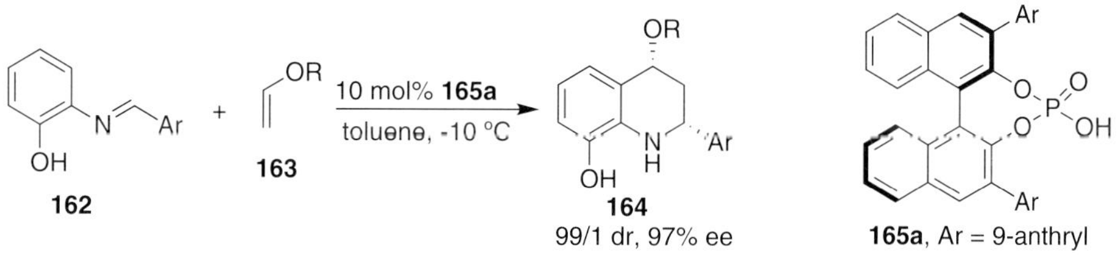

**Scheme 25.52** Inverse electron-demand aza-Diels–Alder reaction.

Gong and Rueping independently disclosed an enantioselective direct aza-Diels–Alder reaction of aldimines **167** with cyclohexenone **80a** by the use of Brønsted acid catalysis. Either an $H_8$-BINOL- or BINOL-based phosphoric acid (**165b** or **166a**) had high catalytic activity for the cycloaddition reaction with good stereoselectivity (Scheme 25.53) [57].

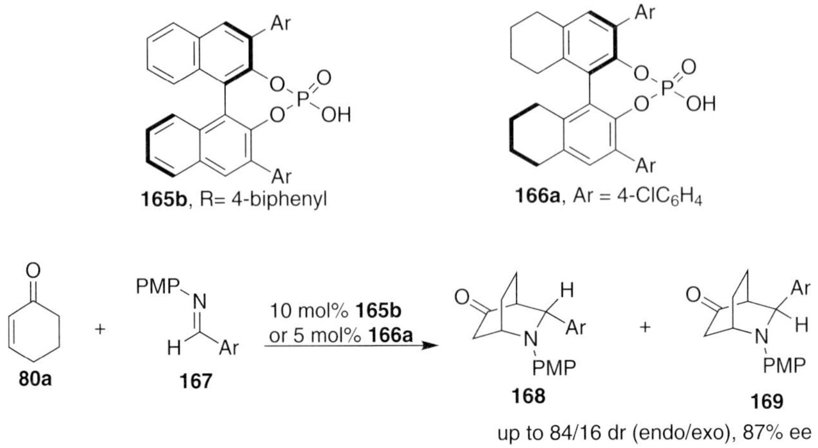
165b, R= 4-biphenyl
166a, Ar = 4-ClC6H4
80a
167
10 mol% 165b
or 5 mol% 166a
168
169
up to 84/16 dr (endo/exo), 87% ee

**Scheme 25.53** Phosphoric acid catalyzed direct hetero-Diels–Alder reaction.

### 25.5.2
### Pictet–Spengler Reactions

Jacobsen showed that asymmetric acyl-Pictet–Spengler reactions of tryptamine-derived imines **170** in the presence of stoichiometric amounts of acetyl chloride and a catalytic amount of the chiral thiourea-based organocatalyst **171a** took place smoothly to provide cyclization products **172** with high enantioselectivities (Scheme 25.54) [58].

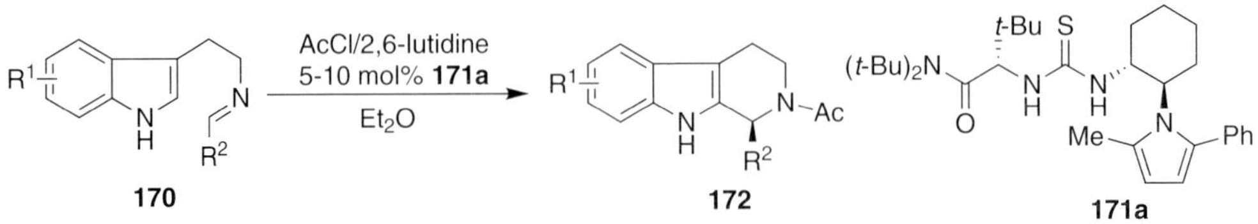
AcCl/2,6-lutidine
5-10 mol% 171a
Et2O
170
172
171a

**Scheme 25.54** Acyl Pictet–Spengler reactions of tryptamine-derived imines.

The same group demonstrated that the thiourea catalysis was also applicable to the acyl-Pictet–Spengler reaction of β-indolyl hydroxylactams **173**. The pyrrole-thiourea derivative **171b** was the optimal catalyst, delivering up to 97% ee for the cyclization products **174**. A straightforward enantioselective total synthesis of a natural product, (+)-harmicine **175** was achieved in only four steps starting with

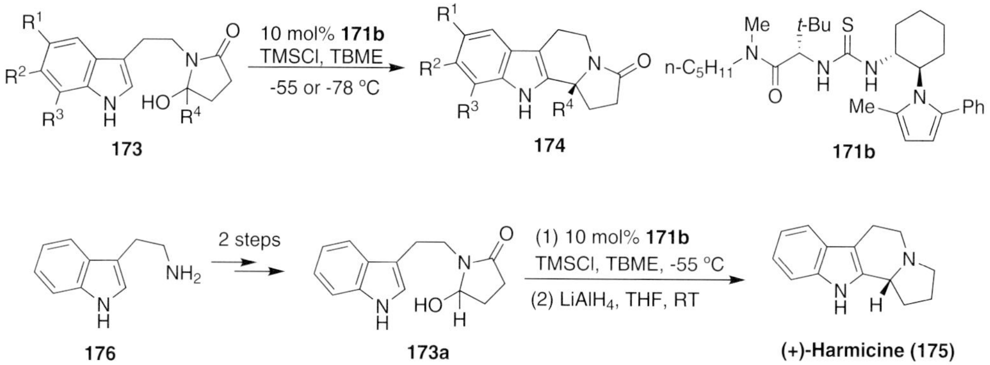

## 25.5 Brønsted Acid-Catalyzed Asymmetric Cyclization Reactions

**Scheme 25.55** Acyl Pictet–Spengler reactions of β-indolyl hydroxylactams and application to the synthesis of (+)-harmicine.

the acyl-Pictet–Spengler reaction of the commercially available tryptamine **176** in 61% overall yield and with 97% ee (Scheme 25.55) [59].

List reported that chiral phosphoric acid **165c** can serve as a good catalyst for the Pictet–Spengler reaction of tryptamine-derived imines with no requirement for stoichiometric amounts of acyl chloride for the activation of the imine. The protocol tolerates a wide range of aromatic and aliphatic aldehydes with varying enantioselectivities, depending on the structure of the substrates used (Scheme 25.56) [60].

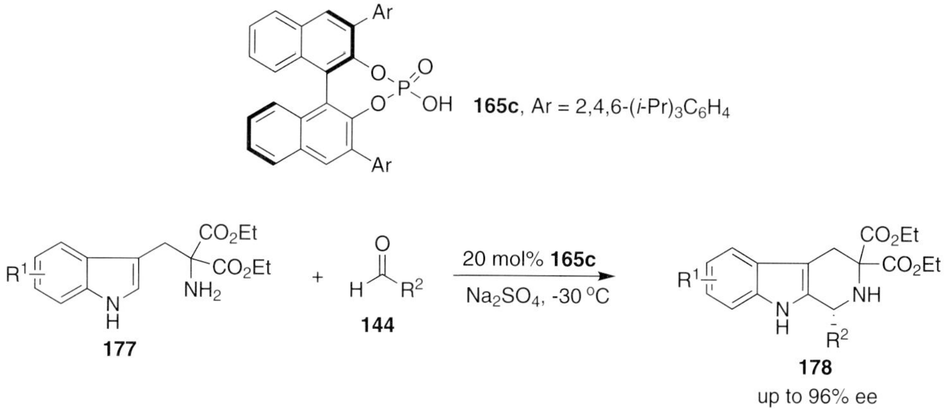

**Scheme 25.56** Phosphoric acid catalyzed Pictet–Spengler reaction of tryptamine-derived imines.

Hiemstra introduced a Pictet–Spengler reaction of sulfenyliminium ions formed from N-sulfenyltryptamine **179** and aldehydes **144** catalyzed by chiral Brønsted acids. Chiral phosphoric acid **165c** delivered the highest levels of enantioselectivity (Scheme 25.57) [61]. The stereoselectivity may arise from the asymmetric counterion-directed catalysis [62].

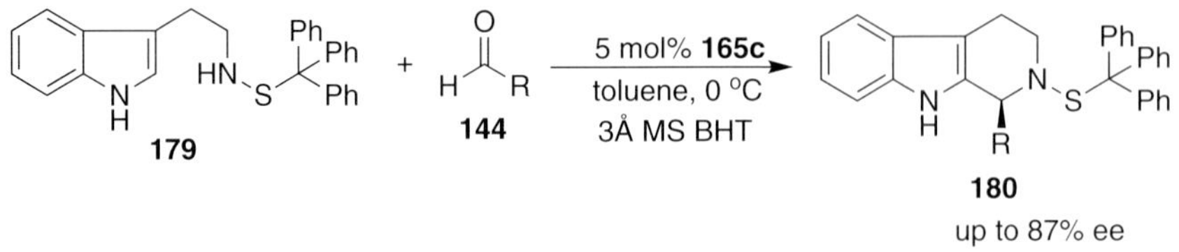

**Scheme 25.57** Pictet–Spengler reaction of sulfenyliminium ions formed from N-sulfenyltryptamine and aldehydes.

### 25.5.3
### Nazarov Reactions

Rueping reported the first organocatalytic Nazarov reaction of **181** by 2 mol% of **182**, which generated cyclopentenones **183** in good yields with excellent enantioselectivities of up to 98% ee (Scheme 25.58) [63].

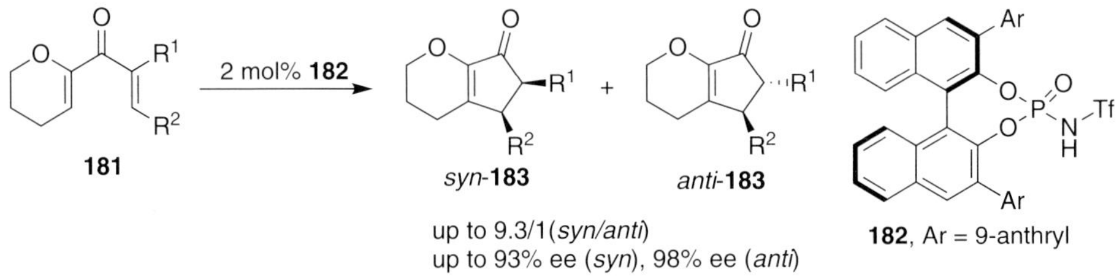

**Scheme 25.58** Asymmetric Nazarov reaction.

### 25.5.4
### Multicomponent and Domino Reactions

The Biginelli reaction represents one of the most useful multicomponent reactions that offers an efficient way to access multifunctionalized 3,4-dihydropyrimidin-2-(1H)-ones (DHPMs, **184**) and related heterocyclic compounds that show a wide scope of important pharmacological properties and make up a large family of medicinally relevant compounds. Gong found that the use of 10 mol% of chiral phosphoric acid **166b** furnished a smooth Biginelli reaction of aldehydes **144**, urea or thiourea (**185a** or **185b**), and β-keto esters **186** with high enantioselectivity (Scheme 25.59) [64].

List reported that Brønsted acids were able to catalyze a cascade aldol/reduction reaction of diketones, primary amine, Hantzsch ester, providing cyclic compounds. In the presence of 10 mol% of **165c**, 2,6-diketones **187** could be transformed into cyclohexylamines **188** by their reactions with **52a** and **123** with up to 96% ee (Scheme 25.60) [65].

Gong *et al.* reporteded that the presence of 10 mol% of H$_8$-BINOL-based phosphoric acid **166c** afforded an enantioselective three-component cyclization reaction of cinnamaldehydes **40**, aniline **52**, and 1,3-dicarbonyls **189**, allowing the straightforward synthesis of enantioenriched 4-aryl substituted 1,4-dihydropyridines **190** with high enantioselectivities of up to 98% ee (Scheme 25.61) [66].

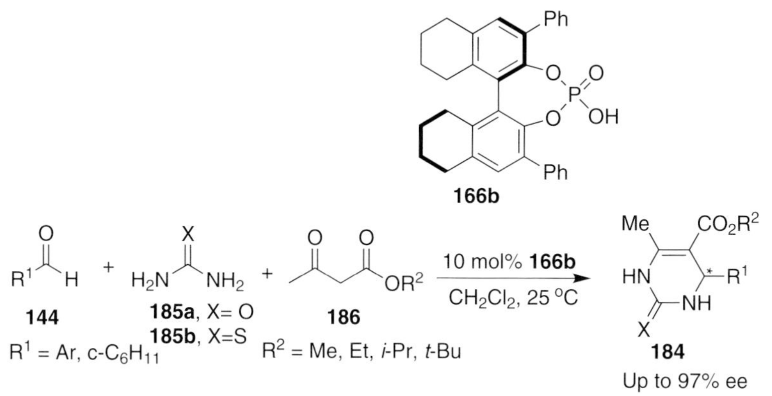

**Scheme 25.59** Asymmetric organocatalytic Biginelli reaction.

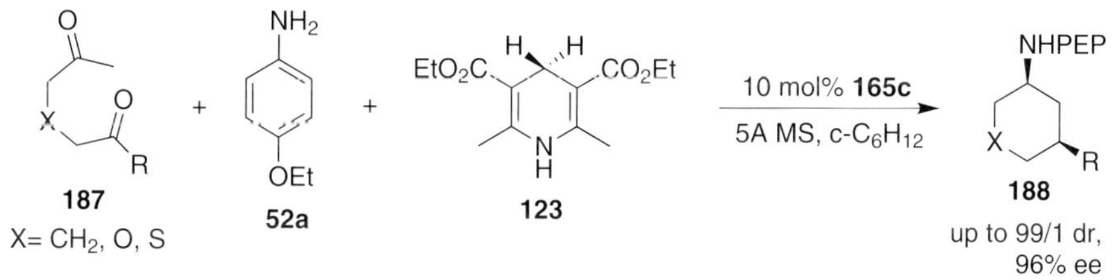

**Scheme 25.60** Asymmetric synthesis of cyclohexylamines by cascade aldol/reduction reactions.

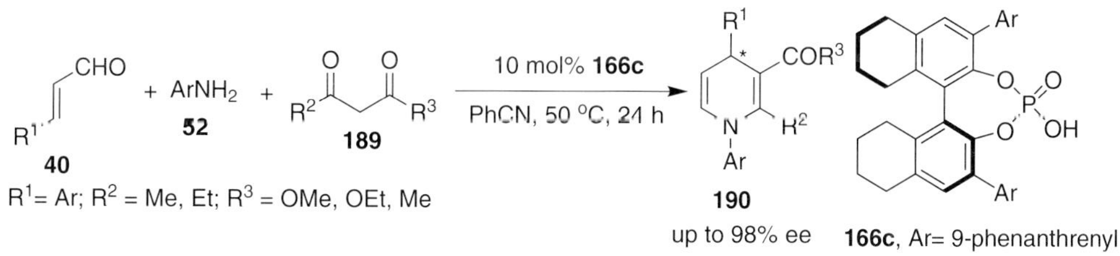

**Scheme 25.61** Catalytic asymmetric synthesis of 4-aryl substituted 1,4-dihydropyridines.

## 25.6
## Lewis Base-Catalyzed Asymmetric Cycloadditions

### 25.6.1
### Cycloaddition Reactions Catalyzed by Chiral Phosphines

Chiral phosphines have long been widely used for metal-based catalysis, while trisubstituted phosphines acting as nucleoplic catalysts have been applied to organic reactions quite recently [67]. Lu pioneered a [3 + 2]-cycloaddition of allenes **191** with enones **68** to generate multiply substituted cyclopentenes **192** in the presence of stoichiometric amounts of phosphines [68]. The enantioseletive version of the reaction was first accomplished by Zhang using chiral phosphine catalyst

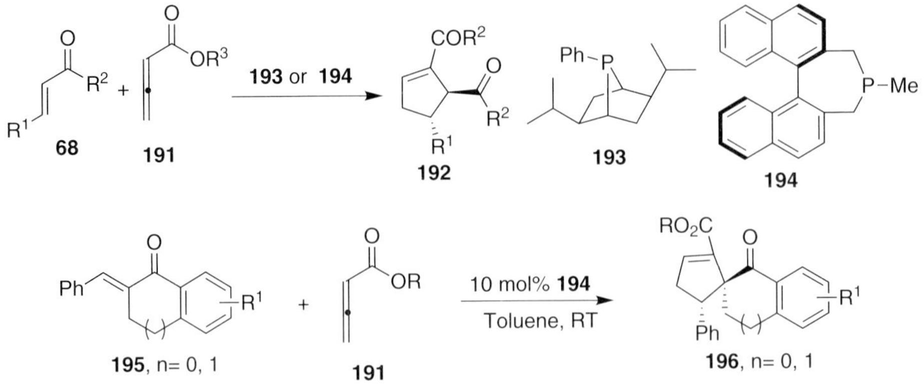

**Scheme 25.62** Asymmetric catalytic [3 + 2]-cycloaddition of allenes with enones.

**193** [69]. Fu indicated that a binaphthyl-based phosphepine **194** afforded a highly stereoselective annulation tolerating a wider scope of enones **68** in comparison with the similar asymmetric variant with **193** [70]. Notably, spirocyclic compounds **196** could be obtained with high levels of stereoselectivity using this protocol with **195** (Scheme 25.62).

Miller et al. reported that multifunctional protected amino ester-bonded phosphines such as **197** were quite promising catalysts for the highly enantioseletive annulations of enones **198** with benzyl penta-2,3-dienoate **199**, preferentially furnishing cyclopentenes **200** and spirocyclic compounds **201–203**, which are structurally different from those obtained with the binaphthyl-based phosphine **194** (Scheme 25.63) [71].

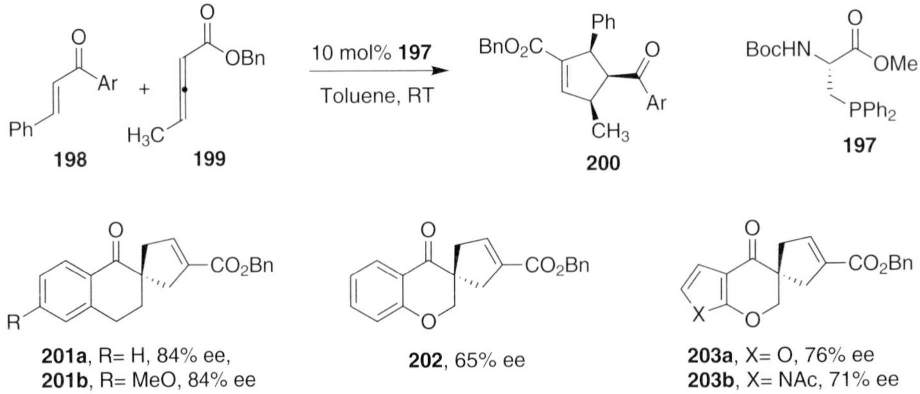

**Scheme 25.63** Enantioseletive annulations of enones with an allene catalyzed by amino ester-bonded phosphines.

Kwon reported a [4 + 2]-annulation of imines **205** with diethyl 2-vinylidenesuccinate **204** using tributylphosphine as a catalyst, producing tetrahydropyridine derivatives **206** [72]. Fu established the first enantioselective variant under the catalysis of **194** with high diastereo- and enantioselectivities (Scheme 25.64) [73].

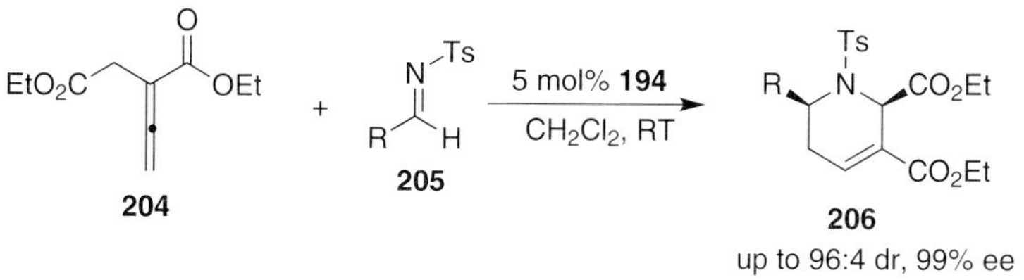

**Scheme 25.64** Asymmetric [4 + 2]-annulation of imines with diethyl 2-vinylidenesuccinate.

### 25.6.2
### Cycloaddition Reactions Catalyzed by Cinchona Alkaloids

In addition to being the base used for the deprotonation of acidic carbon nucleophiles, facilitating a large number of organic reactions beginning with the quinine-catalyzed cyanation of aldehyde in 1912 [74], Cinchona alkaloids serve as nucleophilic catalysts that have long been recognized and employed in organic reactions [4, 75]. Lectka disclosed that under the catalysis of o-quinidines **210** [4 + 2]-cycloaddition reactions of benzoylqunidine **207** with ketenes formed *in situ* from acyl chlorides **208** and Hünig's base furnished chiral benzodioxin-2-one derivatives **209** with almost complete stereochemical control in some cases (Scheme 25.65) [76].

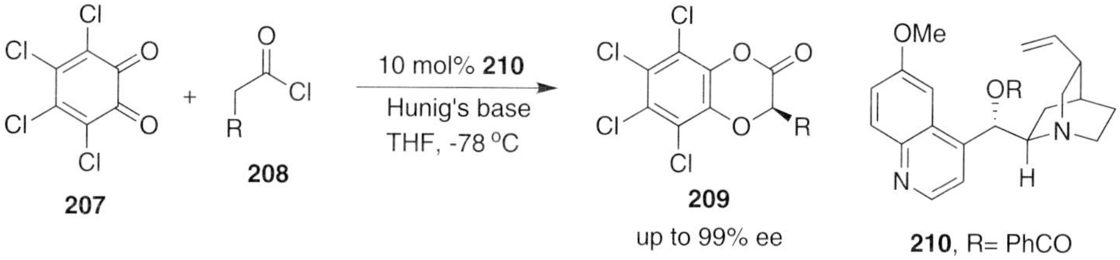

**Scheme 25.65** Asymmetric [4 + 2]-cycloaddition reactions of benzoylquinidine with ketenes.

The application of the same catalytic concept to [4 + 2]-annulations of ketenes with o-benzoquinone imides **211** was also successful in providing benzoxazinones **212** with perfect enantioselectivity (Scheme 25.66) [77].

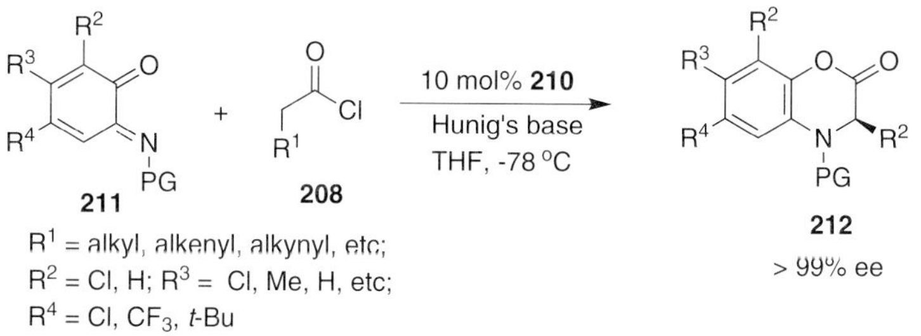

$R^1$ = alkyl, alkenyl, alkynyl, etc;
$R^2$ = Cl, H; $R^3$ = Cl, Me, H, etc;
$R^4$ = Cl, CF$_3$, t-Bu

**Scheme 25.66** Asymmetric [4 + 2]-annulations of ketenes with o-benzoquinone imides.

The [4 + 2]-annulation of ketenes with *o*-benzoquinone diimides **213** proceeded sluggishly to give mainly some by-products in the presence of benzoylquinidine **210**. Interestingly, the use of $Zn(OTf)_2$ as a co-catalyst was able to inhibit the by-products and the desired [4 + 2]-adducts **214** could be attained in high yields with complete stereochemical control (Scheme 25.67) [78].

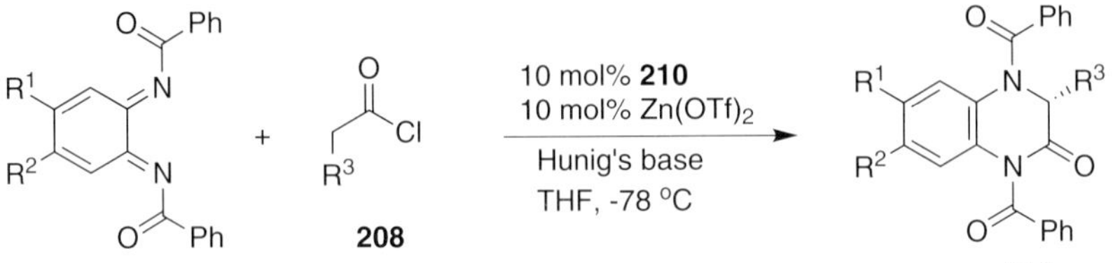


**213** **208** **214**

> 99% ee

$R^1 = R^2 = H, Cl, R^3 = alkyl$

**Scheme 25.67** Asymmetric [4 + 2]-annulations of ketenes with o-benzoquinone diimides.

Nelson established [4 + 2]-cycloadditions of ketenes with *N*-thioacyl imines **215** catalyzed cooperatively by *O*-trimethylsilylquinine **216** and $LiClO_4$, furnishing *cis*-thiazinones **217** with excellent diastereo- and enantioselectivities. Conversion of chiral thiazinones **217** to α,β-dipeptide and β-amino aldehyde derivatives was accomplished by readily available reactions (Scheme 25.68) [79].

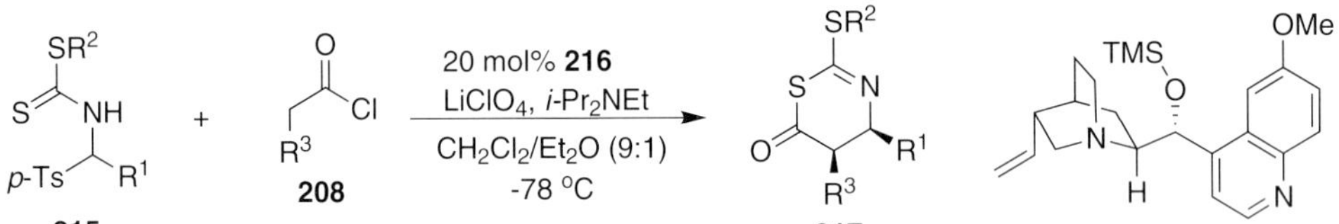


**215** **208** **217** **216**

>97/3 dr, >98% ee

**Scheme 25.68** Asymmetric [4 + 2]-annulations of ketenes with *N*-thioacyl imines.

## 25.7 Asymmetric Cyclization Reactions by Bifunctional Catalysts

During catalysis, Brønsted acid–Lewis base bifunctional catalysts are able to activate acceptors by forming hydrogen bonds and to activate reaction donors by taking advantage of their basic functionality to form a hydrogen bond with or to deprotonate the acidic hydrogen of nucleophiles, and thereby facilitate some cascade reactions that produce cyclic compounds.

Wang reported a cascade Michael-Aldol reaction between 2-mercaptobenzaldehydes **131** with α,β-unsaturated oxazolidinones **218** that proceeded under the promotion of 1 mol% of bifuncitonal catalyst **219** to form benzothiopyrans **220** with complete diastereochemical control and high enantioselectivities [80a]. Extension of

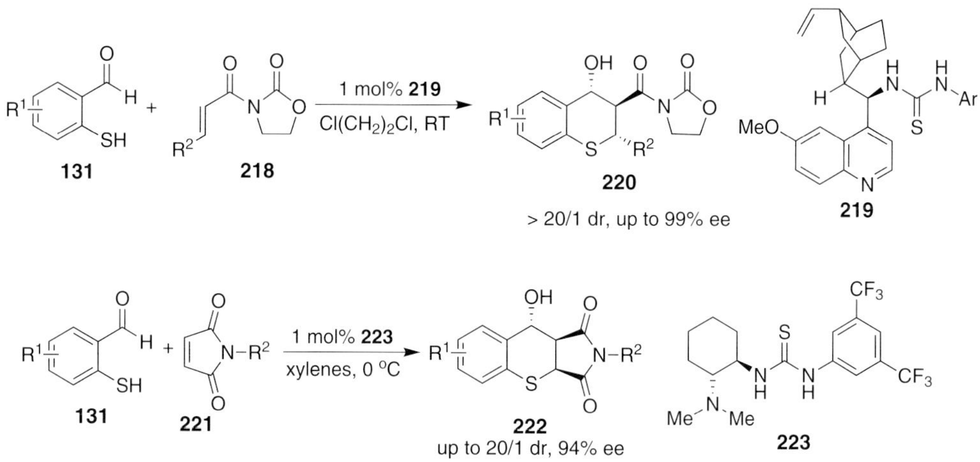

**Scheme 25.69** Cascade Michael–Aldol reactions catalyzed by binfuncitonal catalysts.

this catalysis to reactions of maleimides **221** with 2-mercaptobenzaldehydes **131** was also successful with high diastereo- and enantioselectivities using thiourea **223** as a catalyst (Scheme 25.69) [80b].

An asymmetric Diels–Alder reaction of 3-hydroxy-2-pyrones **224** with enones **225** has recently been described by Deng using cinchona alkaloids-based bifunctional catalysts. DHQD-PHN **226** was the most selective organocatalyst and afforded the Diels–Alder reaction of different enones with 4-substituted 3-hydroxy-2-pyrones to favorably furnish endo 2-oxabicyclo[2.2.2]oct-5-en-3-one derivatives **227** with high stereoselectivities (Scheme 25.70) [81].

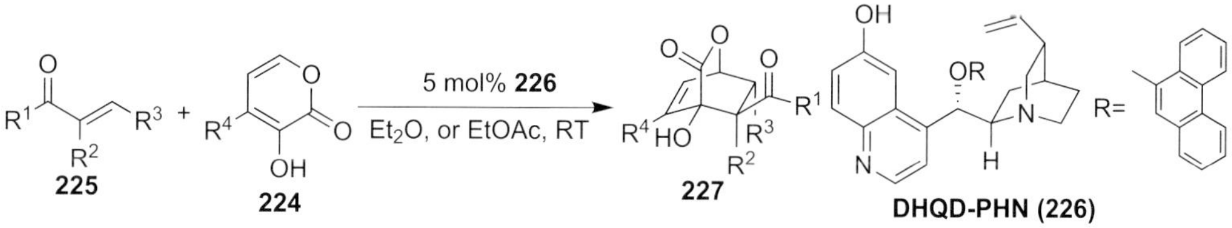

**Scheme 25.70** Asymmetric Diels–Alder reaction of 3-hydroxy-2-pyrones with enones.

## 25.8
## Asymmetric Cyclizations Catalyzed by Chiral Nucleophilic Carbenes

### 25.8.1
### Intramolecular Stetter Reactions

The pioneering studies by Breslow [82] on the thiazolium salt-catalyzed benzoin reaction and recent significant advances in nucleophilic carbene catalysis have demonstrated that nucleophilic carbenes can be used to convert aldehydes to umpolung reagents by forming Breslow- or extended Breslow-intermediates that

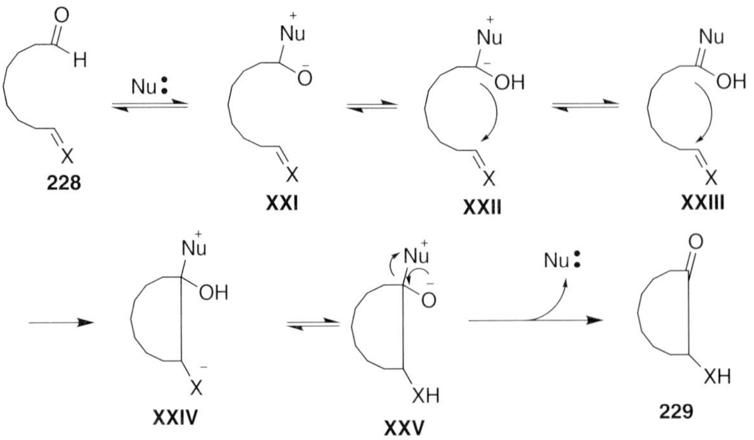

**Scheme 25.71** General concept for carbine-catalyzed cyclization reaction.

attack prochiral acceptors intramolecularly to provide, principally, access to chiral cyclic compounds (Scheme 25.71).

Rovis *et al.* have designed a variety of chiral triazolium salts that were treated with base to generate chiral carbenes capable of promoting intramolecular Stetter reactions. The chiral carbene originated from triazolium salt **230a** afforded the most enantioselective intramolecular Stetter reaction of **231** with 82%–97% ee (Scheme 25.72) [83].

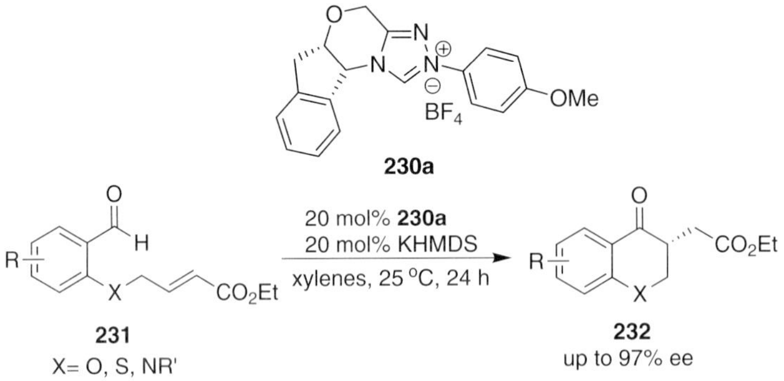

**Scheme 25.72** Intramolecular Stetter reaction.

In the presence of 2 equiv of triethylamine, the chiral triazolium salt **230b** effectively catalyzes the intramolecular Stetter reaction of aromatic vinylogous carbonates **233–235** that provided access to chiral dihydroindenones **236**, benzofuranones **237**, and benzothiophenones **238** containing a quaternary stereogenic center with excellent enantioselectivity. The chiral carbene formed by the treatment of chiral triazolium **230b** with a catalytic amount of KHMDS showed good catalytic activity and excellent enantioselectivity for aliphatic substrates **239**, leading to the generation of cyclopentanones **240** with up to 99% ee (Scheme 25.73) [84].

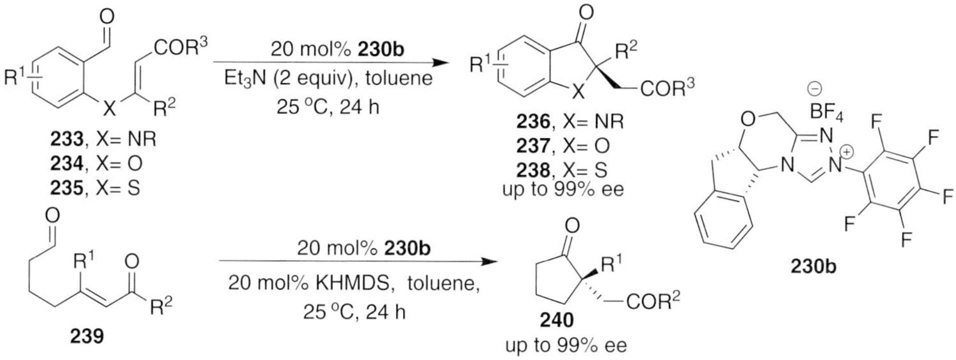

25.8 Asymmetric Cyclizations Catalyzed by Chiral Nucleophilic Carbenes | 1231

Scheme 25.73 Intramolecular Stetter reaction for the synthesis of five-membered cycles.

## 25.8.2
### Benzoin Reactions

Enders et al. reported that asymmetric intramolecular cross benzoin reactions of aldehyde ketones **241** by chiral carbenes generated by exposure of chiral triazolium **242** to tert-BuOK gave rise to α-hydroxy-α-alkyl tetralones **243** with up to 98% enantiomeric excess (Scheme 25.74) [85].

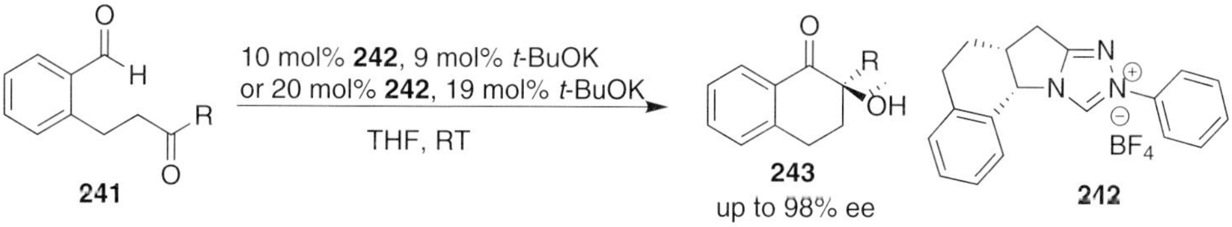

Scheme 25.74 Intramolecular cross benzoin reactions.

## 25.8.3
### Cycloaddition Reactions

Bode proposed that the extended Breslow intermediate formed from enal and N-heterocyclic carbenes **244** can isomerize into a catalyst-bonded enol **XXVI** or enolate **XXVII** that may be trapped with electrophiles and thereby lead to new carbon–carbon bond-forming reactions (Scheme 25.75) [86].

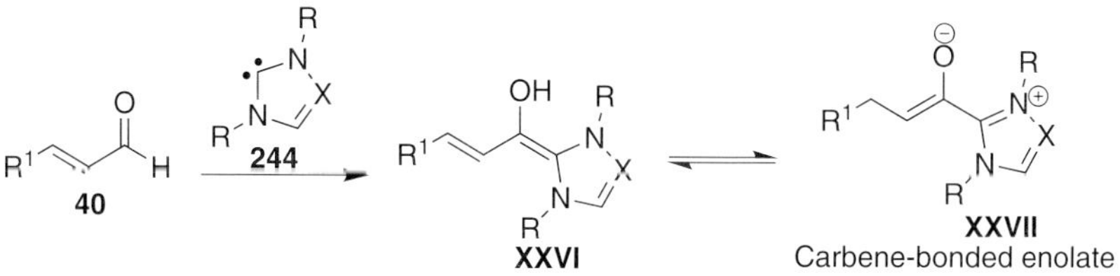

Scheme 25.75 Generation of carbene catalyst-bonded enol and enolate.

Given the reactivity of carbene-bonded enolate (**XXVII**) toward electron-deficient dienes, Bode initiated an asymmetric aza-Diels–Alder reaction of 1,3-aza-dienes **245** with *trans*-4-oxo-2-butenones **246** catalyzed by chiral N-heterocyclic carbenes. A chiral carbene generated *in situ* from the deprotonation of **230c** with DIPEA showed high stereoselectivity and tolerated a broad scope of 1,3-aza-dienes and *trans*-4-oxo-2-butenones, allowing a catalytic asymmetric synthesis of structurally diverse chiral dihydropyridinones **247** with high enantioselectivities (Scheme 25.76) [86].

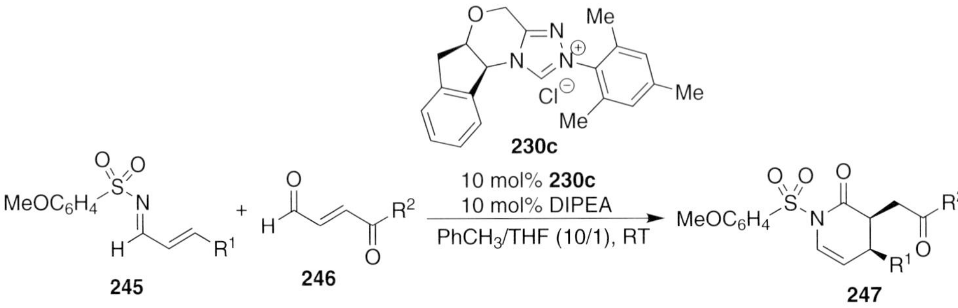

**Scheme 25.76** Aza-Diels–Alder reaction of 1,3-aza-dienes with *trans*-4-oxo-2-butenones.

Alternatively, chiral enolate dienophiles could be catalytically generated from α-chloroaldehydes **248** with NHC-catalysis. The same group disclosed a hetero-Diels–Alder reaction of racemic α-chloroaldehydes **248** with enones **68** in the presence of as little as 0.5 or 2 mol% of the carbene catalyst formed from **230c**, giving optically active 2,3-dihydropyranones **249** (Scheme 25.77) [87].

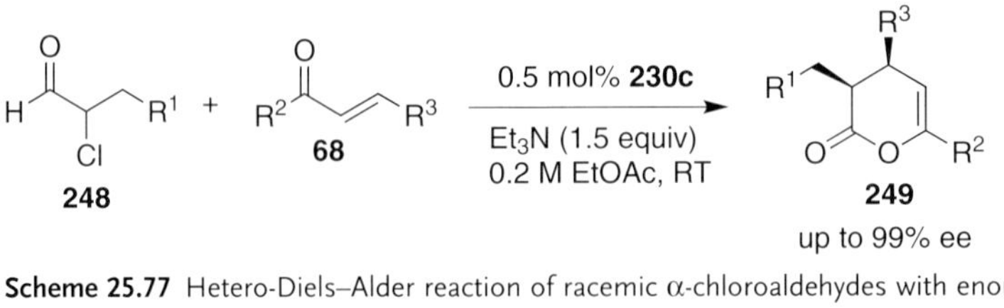

**Scheme 25.77** Hetero-Diels–Alder reaction of racemic α-chloroaldehydes with enones.

An enantioselective annulation of α,β-unsaturated aldehydes **40** with enone **68** catalyzed by **230c** was developed by Bode's group, which was revealed to proceed via a cascade reaction sequence of a carbene-catalyzed asymmetric intermolecular aldehyde–ketone cross Benzoin reaction/oxy-Cope rearrangement, furnishing cyclopentenes **250** in favor of *cis*-diastereomers with perfect enantioselectivity (Scheme 25.78) [88].

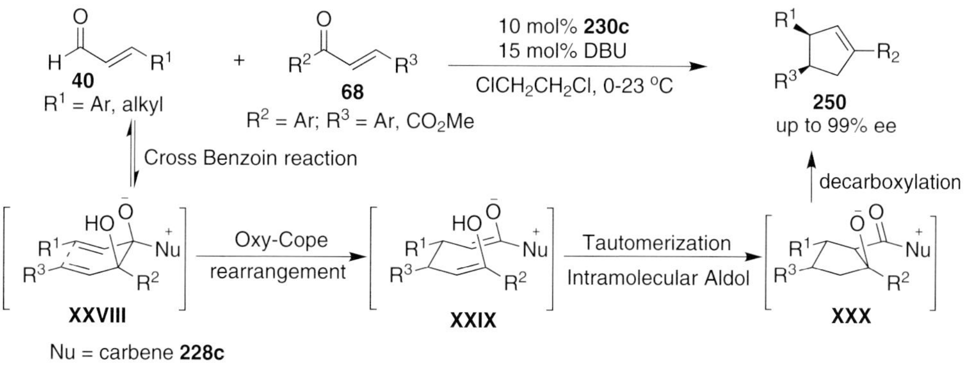

**Scheme 25.78** Carbene-catalyzed enantioselective annulation of α,β-unsaturated aldehydes.

## 25.9
## Conclusion

Since the seminal works on the use of chiral organic molecules to catalyze asymmetric transformations at the beginning of this century, asymmetric organocatalysis has rapidly grown to be a research field that is receiving increasing interest. More and more exciting advances have emerged to refute a traditional wisdom that catalytic synthesis of optically pure compounds relies predominantly on chiral metal complexes and enzymes and show that asymmetric organocatalysis is unarguably the third powerful tool. Notably, the construction of highly functionalized chiral cyclic molecules with multiple stereogenic centers, which has long remained a challenge for metal catalysis, has been at least partially realized by using diverse concepts of organocatalysis, leading to the concise and enantioselective total synthesis of some natural products. It can be predicted that new organocatalytic cyclization reactions, particularly procedures used for building core cyclic structures of natural products and biologically active substances, will be discovered in the future by taking advantage of the disciplines summarized in this chapter.

## 25.10
## Experimental: Selected Procedures

### 25.10.1
### Synthesis of Compounds 8 via Intramolecular Aldol Reaction [9]

Dicarbonyl **7** (1 mmol) was dissolved in dry dichloromethane (10 mL) and treated with L- or D-proline (12 mg, 0.1 mmol, 10%). The mixture was stirred at room temperature until the starting material had disappeared (8–16 h). Aldols **8** can be isolated after standard aqueous work-up.

## 25.10.2
### Synthesis of Compounds 19 via Intramolecular Michael Additions [14]

Ketoaldehyde **19** (123 mg, 0.8 mmol) was dissolved in dry THF (8 mL) and treated with **18a** (20.4 mg, 0.08 mmol, 10%). The resulting mixture was stirred under argon at room temperature until the starting material had disappeared (15 h). The solution was then concentrated and filtered through silica gel. The products **20** were isolated through a standard technique.

## 25.10.3
### Typical Procedure for the Preparation of 62 [24]

The aldehyde **59** (0.50 mmol) and the enone (1.00 mmol) **60** were dissolved in 0.5 mL of $CH_2Cl_2$ and cooled to $-15\,°C$. The catalyst **61** (0.05 mmol) was added, followed by the introduction of 50 mg of silica gel and the mixture was allowed to warm to room temperature with stirring overnight. The equilibrium mixture was isolated by FC (silica gel, gradient $CH_2Cl_2$ to 15% $Et_2O/CH_2Cl_2$). Oxidation of the mixture was performed in $CH_2Cl_2$ by adding 1 equiv of PCC at room temperature. After 1 h, another 1 equiv of PCC was added and after 2 h the lactone **62** was isolated in 65% yield by FC with $CH_2Cl_2$ as the eluent.

## 25.10.4
### Synthesis of 66 via Organocatalytic Diels–Alder Reaction by 18a [27a]

To a solution of catalyst **18a** in $CH_3OH/H_2O$ (95/5 v/v, 1 M) was added the α,β-unsaturated aldehyde **40**. The solution was stirred for 1–2 min before addition of the appropriate diene **66**. After the reaction was complete, the reaction mixture was diluted with $Et_2O$ and washed successively with $H_2O$ and brine. The organic layer was dried ($Na_2SO_4$), filtered, and concentrated. Hydrolysis of the product dimethyl acetal was performed by stirring the crude product mixture in $THF:H_2O:CHCl_3$ (1:1:2) for 2 h at room temperature, followed by neutralization with saturated aqueous $NaHCO_3$ and extraction with $Et_2O$. Purification of the adduct **67** was accomplished by silica gel chromatography.

## 25.10.5
### Typical Procedure for the Synthesis of 104 [35]

An amber 2-dram vial equipped with a magnetic stir bar, containing 20 mol% of catalyst **18c** and 20 mol% of TFA and tryptamine **103** was charged with methylene chloride and water, then placed in a bath at the appropriate temperature. The solution was stirred for 5 min before addition of the α,β-unsaturated aldehyde **40**. The resulting suspension was stirred at constant temperature until complete consumption of the indole was observed, as determined by TLC. To the reaction mixture was then added buffer (pH = 7.0). The aqueous layer was extracted with diethyl ether and the combined organic layers were concentrated *in vacuo*. The resulting residue was purified by silica gel chromatography to afford the title compounds **104**.

## 25.10.6
### Synthesis of 110 via Three-Component Domino Reaction [37]

To a solution of catalyst **38a** (65 mg, 0.20 mmol) and nitroolefin **46** (1.00 mmol, 1.00 equiv) in toluene (0.8 mL) were added aldehyde **59** (1.20 mmol, 1.20 equiv) and α,β-unsaturated aldehyde **40** (1.05 mmol, 1.05 equiv) at 0 °C. After 1 h the solution was allowed to reach room temperature and stirred until complete conversion of the starting materials (16–24 h, monitored by GC). The reaction mixture was directly purified by flash column chromatography on silica gel (ethyl acetate/pentane, 1: 8–1: 6) to afford the product **110** as the pure major diastereomer.

## 25.10.7
### Synthesis of 2,3-Dihydropyranones 148 Through Hetero-Diels–Alder Reaction by TADDOL [50a]

To a solution of TADDOL **146** (0.1 mmol) and the aldehyde (1.0 mmol) in toluene (0.5 mL) was added diene (**145**, 0.5 mmol) at −78 °C. After being stirred at −78 °C for 48 h, the mixture was diluted with $CH_2Cl_2$ (2.0 mL) and acetyl chloride (1.0 mmol) was added dropwise at −78 °C. The resultant mixture was stirred for an additional 15 min and then separated by chromatography on silica gel.

## 25.10.8
### Synthesis of 161 via Phosphoric Acid-Catalyzed Hetero-Diels–Alder Reaction [55]

A solution of Brassard's diene **149** (25 µL, 0.115 mmol) was added dropwise to a solution of imine **159** (0.152 mmol) and cyclic phosphoric acid pyridinium salt **160** (3.9 mg, 0.005 mmol) in mesitylene (5 mL) at −40 °C over 1 min. After 2 h, the same amount of **149** was added similarly. The addition was repeated four times every 2 h until a total of 100 µL (0.460 mmol) of the Brassard's diene was introduced. After being stirred at the temperature for 11 h, the reaction mixture was quenched with saturated aqueous $NaHCO_3$ at −40 °C. The mixture was extracted with ethyl acetate. The combined organic layers were washed with brine, dried over anhydrous $Na_2SO_4$, and concentrated to dryness. The resultant oil was treated with THF–1 N HCl (4 mL/1 mL) at 0 °C for 10 min. The solution was extracted with ethyl acetate. The combined organic layers were washed with brine, dried over anhydrous $Na_2SO_4$, and concentrated to dryness. The crude product was passed through a short column of silica gel (hexane/EtOAc = 10/1–3/1–1/1) to give an oil. The mixture and benzoic acid (19.2 mg, 0.157 mmol) were dissolved in mesitylene (2 mL) and stirred at 110 °C for 12 h. The reaction mixture was purified by column chromatography on silica gel (hexane/EtOAc = 3/1–1/1) to give the product **161**.

## 25.10.9
### Synthesis of 172 via an Asymmetric Pictet–Spengler Reaction [58]

To a vial charged with molecular sieves (250 mg, 3 Å spherical) which were flame-dried *in vacuo* and cooled to 23 °C under nitrogen, was added sequentially dichloromethane (1.25 mL) and tryptamine (40 mg, 0.25 mmol, 1.0 equiv). Aldehyde (0.275 mmol, 1.0 equiv) was added dropwise via syringe to the suspension, which was allowed to stand at 23 °C for 7 h and swirled occasionally to ensure mixing of the contents. The resulting solution was filtered by cannula transfer to a flame-dried round-bottomed flask. The desiccant was rinsed twice with dichloromethane (5 mL) and the combined rinses were combined with the filtrate by cannula transfer. The solution was concentrated *in vacuo*, yielding the imine **170** as a pale brown oil, which was dissolved in diethyl ether (5.0 mL). Catalyst **171a** (6.7 mg, 0.013 mmol, 5 mol%) was added, and the solution was cooled to −78 °C in a dry ice/acetone bath. 2,6-Lutidine (29 mL, 0.25 mmol, 1.0 equiv) and acetyl chloride (18 mL, 0.25 mmol, 1.0 equiv.) were added sequentially dropwise with a syringe. The mixture was stirred at −78 °C for 5 min, and then warmed to −30 °C and stirred for an additional 22 h. The resulting heterogeneous mixture was allowed to warm to 23 °C and concentrated *in vacuo*. The residue was purified by column chromatography on silica gel, eluting with 20% ethyl acetate in dichloromethane, yielding the product **172**.

### 25.10.10
**Synthesis of DHPMs via an Asymmetric Biginelli Reaction [64]**

After a solution of aldehyde **144** (0.2 mmol), thiourea (0.24 mmol), and **166b** (0.02 mmol) in $CH_2Cl_2$ (3 mL) was stirred at 25 °C for 2 h, a β-keto ester **186** (0.6 mmol) was added. After being stirred at 25 °C for 6 days (monitored by TLC), the reaction mixture was diluted with EtOAc (3 mL). After removal of solvent, the residue was purified by flash column chromatography on a silica gel (eluent: petroleum ether: ethyl acetate = 6 : 1–3 : 1) to yield **184**.

### 25.10.11
**Typical Procedure for the Synthesis of 192 [70]**

In a glove box, a solution of **194** (14.7 mg, 0.040 mmol) in toluene (0.5 mL) was added to a stirring solution of the enone **68** (0.400 mmol) and ethyl 2,3-butanedienoate **191** (56 μL, 0.48 mmol) in toluene (1.5 mL). The mixture was stirred at ambient temperature for 16 h, and then the product **192** was obtained directly by flash chromatography.

### 25.10.12
**Typical Procedure for the Synthesis of 212 [77]**

A solution of the quinone imide **211** (0.12 mmol) in THF (2 mL) was added to a reaction flask containing acid chloride **208** (0.12 mmol), Hünig's base (0.12 mmol), and **210** (0.012 mmol) at −78 °C. After stirring for 6 h, the reaction was concentrated *in vacuo* and the crude residue was purified by column chromatography on silica gel (EtOAc/hexanes as eluent).

## 25.10.13
### Typical Procedure for Diels–Alder Reaction Catalyzed by 226 for the Synthesis of 227 [81]

At room temperature, to a solution of 2-pyrones **224** (0.3 mmol) and catalyst **226** (0.015–0.030 mmol, 5–10 mol%) in diethyl ether (3.0 mL) was added dienophiles **225** (0.6–1.2 mmol, 2–4 equiv). The reaction mixture was kept at a certain reaction temperature for 11–72 h. The crude reaction mixture was passed through a short plug of silica gel to remove the catalyst, and the silica gel plug was washed with diethyl ether or ethyl acetate (3.0–4.0 mL). The eluent was concentrated *in vacuo*, and the residue was subjected to silica gel flash chromatography.

## 25.10.14
### Intramolecular Stetter Reaction for the Synthesis of 232 [83]

A flame-dried round-bottom flask was charged with triazolium salt **230a** (0.2 equiv) and xylenes (2 mL). To this solution was added KHMDS (0.5 M in toluene) (0.2 equiv) via syringe and the solution was stirred at ambient temperature for 15 min. Substrate **231** (1 equiv) was added and allowed to stir for 24 h at ambient temperature. The reaction was then poured into a column of silica gel and eluted with a suitable solution of ethyl acetate in hexanes (typically 1:6). Evaporation of solvent afforded analytically pure product **232**.

## 25.10.15
### Synthesis of α-Hydroxy-α-alkyl Tetralones 243 via the Benzoin Reaction [85]

Precatalyst **242** (20.6 mg, 0.055 mmol, 20 mol%) was suspended in anhydrous THF (1.7 mL) in a Schlenk tube under argon at room temperature. A solution of freshly sublimed *tert*-BuOK (5.9 mg, 0.052 mmol, 19 mol%) in anhydrous THF (0.6 mL) was added slowly, and the solution was stirred for 5 min. Aldehyde-ketone **241** (60 mg, 0.275 mmol) was dissolved in anhydrous THF (0.5 mL) and added to the carbene solution. The reaction mixture was stirred for 48 h, diluted with $CH_2Cl_2$, quenched with water, extracted two times with $CH_2Cl_2$, and dried over $MgSO_4$. The solvent was evaporated and the crude product was purified by flash chromatography on silica gel (dichloromethane/n-pentane 2:1) to yield **243**.

## 25.10.16
### Synthesis of Chiral Dihydropyridinones 247 via an Aza-Diels–Alder Reaction Catalyzed by Carbene [86]

Into an oven-dried vial (2.0 mL) was added the enal **246** (0.054 mmol, 1.1 equiv), imine **245** (0.049 mmol, 1.0 equiv), and triazolium salt **230c** (0.005 mmol, 0.10 equiv). The vial was crimped with a Teflon-lined crimp seal. To this mixture were added 1.0 mL of 10/1 toluene/THF (0.05 M) and diisopropylethylamine (0.8 µL, 0.005 mmol, 0.10 equiv). The resulting solution was stirred at room temperature for

23 h. The reaction mixture was concentrated under reduced pressure, and the residue was purified by PTLC (2 : 1 hexane/EtOAc) to afford the product **247**.

## Abbreviations

| | |
|---|---|
| DMF | *N,N*-dimethylformamide |
| THF | tetrahydrofuran |
| RT | room temperature |
| Ts | *p*-tolylsulfonyl |
| DMSO | dimethyl sulfoxide |
| Ac | acetyl |
| DMAP | 4-dimethylamino pyridine |
| EWG | electron-withdrawing group |
| TBDPS | *tert*-butyldiphenylsilyl |
| TMS | trimethylsilyl |
| Ph | phenyl |
| PCC | pyridinium chlorochromate |
| LUMO | lowest unoccupied molecular orbital |
| TIPBA | 2,4,6-triisopropylbenzenesulfonic acid |
| TFA | trifluoroacetic acid |
| TSA | *p*-tolylsulfonic acid |
| PMP | *p*-methoxyphenyl |
| TBME | *tert*-butylmethyl ether |
| KHMDS | potassium hexamethyldisilazane |
| DIPEA | di(isopropyl)ethyl amine |

## References

1. (a) List, B. (2007) *Chem. Rev.*, **107** (12), special issue for organocatalysis; (b) Berkessel, A. and Gröger, H., (2005) *Asymmetric Organocatalysis: From Biomimetic Concepts to Applications in Asymmetric Synthesis*, Wiley-VCH, Weinheim; (c) Houk, K.N. and List, B. (2004) *Acc. Chem. Res.*, **37**, (8), special issue for organocatalysis.
2. (a) Shi, Y. (2004) *Acc. Chem. Res.*, **37**, 488–496; (b) Yang, D. (2004) *Acc. Chem. Res.*, **37**, 497–505.
3. (a) McGarrigle, E.M., Myers, E.L., Illa, O., Shaw, M.A., Riches, S.L. and Aggarwal, V.K. (2007) *Chem. Rev.*, **107**, 5841–5883; (b) Li, A.-H., Dai, L.-X. and Aggarwal, V.K. (1997) *Chem. Rev.*, **97**, 2341–2372.
4. France, S., Guerin, D.J., Miller, S.J. and Lectka, T. (2003) *Chem. Rev.*, **103**, 2985–3012.
5. Mukherjee, S., Yang, J.W., Hoffmann, S. and List, B. (2007) *Chem. Rev.*, **107**, 5471–5569.
6. (a) Hajos, Z.G. and Parrish, D.R. (1974) *J. Org. Chem.*, **39**, 1615–1621; (b) Eder, U., Sauer, R. and Wiechert, R. (1971) *Angew. Chem.*, **83**, 492–493; (1971) *Angew. Chem. Int. Ed.*, **10**, 496–497.
7. (a) Hoang, L., Bahmanyar, S., Houk, K.N. and List, B. (2003) *J. Am. Chem. Soc.*, **125**, 16–17; (b) List, B., Hoang, L. and Martin,

H.J. (2004) *Proc. Natl. Acad. Sci. USA*, **101**, 5839–5842.

8  (a) Bui, T. and Barbas, C.F. III (2000) *Tetrahedron Lett.*, **41**, 6951–6954; (b) Davis, S.G., Sheppard, R.L., Smith, A.D. and Thomson, J.E. (2005) *Chem. Commun.*, 3802–3804; (c) Nagamine, T., Inomata, K., Endo, Y. and Paquette, L. (2007) *J. Org. Chem.*, **72**, 123–131; (d) Kanger, T., Kriis, K., Laars, M., Kailas, T., Muurisepp, A.-M., Pehk, T. and Lopp, M. (2007) *J. Org. Chem.*, **72**, 5168–5173; (e) Ramachary, D.B. and Kshor, M. (2007) *J. Org. Chem.*, **72**, 5056–5068; (f) Hayashi, Y., Sekizawa, H., Yamaguchi, J. and Gotoh, H. (2007) *J. Org. Chem.*, **72**, 6493–6499.

9  Pidathala, C., Hoang, L., Vignola, N. and List, B. (2003) *Angew. Chem.*, **115**, 2891–2894; (2003) *Angew. Chem. Int. Ed.*, **42**, 2785–2788.

10  Itagaki, N., Kimura, M., Sugahara, T. and Iwabuchi, Y. (2005) *Org. Lett.*, **7**, 4185–4188.

11  Yoshitomi, Y., Makino, K. and Hamada, Y. (2007) *Org. Lett.*, **9**, 2457–2460.

12  Li, P., Payette, J.N. and Yamamoto, H. (2007) *J. Am. Chem. Soc.*, **129**, 9534–9535.

13  Ledais, G. and MacMillan, D.W.C. (2006) *Aldrichimica Acta*, **39**, 79–87.

14  Fonseca, M.T.H. and List, B. (2004) *Angew. Chem.*, **116**, 4048–4050; (2004) *Angew. Chem. Int. Ed.*, **43**, 3958–3960.

15  Hayashi, Y., Gotoh, H., Tamura, T., Yamaguchi, H., Masui, R. and Shoji, M. (2005) *J. Am. Chem. Soc.*, **127**, 16028–16029.

16  Mangion, I.K. and MacMillan F D.W.C. (2005) *J. Am. Chem. Soc.*, **127**, 3696–3697.

17  (a) Thayumanavan, R., Dhevalapally, B., Sakthivel, K., Tanaka, F. and Barbas, C.F. III (2002) *Tetrahedron Lett.*, **43**, 3817–3820; (b) Ramachary, D.B., Chowdari, N.S. and Barbas, C.F. III (2002) *Tetrahedron Lett.*, **43**, 6743–6746.

18  (a) Ramachary, D.B. and Barbas, C.F. III (2004) *Chem. Eur. J.*, **10**, 5323–5331; (b) Ramachary, D.B., Anebouselvy, K., Chowdari, N.S. and Barbas, C.F. III (2004) *J. Org. Chem.*, **69**, 5838–5849.

19  Hong, B.-C., Wu, M.F., Tseng, H.-C., Huang, G.-F., Su, C.-F. and Liao, J.-H. (2007) *J. Org. Chem.*, **72**, 8459–8471.

20  Sundén, H., Rios, R., Xu, Y., Eriksson, L. and Córdova, A. (2007) *Adv. Synth. Catal.*, **349**, 2549–2555.

21  Yamamoto, Y., Momiyama, N. and Yamamoto, H. (2004) *J. Am. Chem. Soc.*, **126**, 5962–5968.

22  Sundén, H., Ibrahem, I., Eriksson, L. and Córdova, A. (2005) *Angew. Chem.*, **117**, 4955–4958; (2005) *Angew. Chem. Int. Ed.*, **44**, 4877–4880.

23  Itoh, T., Yokoya, M., Miyauchi, K., Nagata, K. and Ohsawa, A. (2006) *Org. Lett.*, **8**, 1533–1535.

24  Juhl, K. and Jørgensen, K.A. (2003) *Angew. Chem.*, **115**, 1536–1539; (2003) *Angew. Chem. Int. Ed.*, **42**, 1498–1501.

25  Hong, B.C., Wu, M.F., Tseng, H.C. and Liao, J.H. (2006) *Org. Lett.*, **8**, 2217–2220.

26  Hayashi, Y., Okano, T., Aratake, S. and Hazelard, D. (2007) *Angew. Chem.*, **119**, 5010–5013; (2007) *Angew. Chem. Int. Ed.*, **46**, 4922–4925.

27  (a) Ahrendt, K.A., Borths, C.J. and MacMillan, D.W.C. (2000) *J. Am. Chem. Soc.*, **122**, 4243–4244; (b) Northrup, A.B. and MacMillan, D.W.C. (2002) *J. Am. Chem. Soc.*, **124**, 2458–2460.

28  Jen, W.S., Wiener, J.J.M. and MacMillan, D.W.C. (2000) *J. Am. Chem. Soc.*, **122**, 9874–9875.

29  (a) Ishihara, K. and Nakano, K. (2005) *J. Am. Chem. Soc.*, **127**, 10504–10505; (b) Lemay, M. and Ogilvie, W.W. (2005) *Org. Lett.*, **7**, 4141–4144; (c) Kano, T., Tanaka, Y. and Maruoka, K. (2006) *Org. Lett.*, **8**, 2687–2689; (d) Kano, T., Tanaka, Y. and Maruoka, K. (2007) *Chem. Asian J.*, **2**, 1161–1165; (e) Gotoh, H. and Hayashi, Y. (2007) *Org. Lett.*, **9**, 2859–2862; (f) Chen, W., Yuan, X.H., Li, R., Du, W., Wu, Y., Ding, L.S. and Chen, Y.-C. (2006) *Adv. Synth. Catal.*, **348**, 1818–1822; (g) Chen, W., Du, W., Duan, Y.Z., Wu, Y., Yang, S.Y. and Chen, Y.-C. (2007) *Angew. Chem.*, **119**, 7811–7814; (2007) *Angew. Chem. Int. Ed.*, **46**, 7667–7670. (h) Vicario, J.L., Reboredo, S.,

Badía, D. and Carrillo, L. (2007) *Angew. Chem.*, **119**, 5260–5262; (2007) *Angew. Chem., Int. Ed.*, **46**, 5168–5170; (i) Ibrahem, I., Rios, R., Vesely, J. and Cordova, A. (2007) *Tetrahedron Lett.*, **48**, 6252–6257.

30 Kinsman, A.C. and Kerr, M.A. (2003) *J. Am. Chem. Soc.*, **125**, 14120–14125.

31 Wilson, R.M., Jen, W.S. and MacMillan, D.W.C. (2005) *J. Am. Chem. Soc.*, **127**, 11616–11617.

32 Harmata, M., Ghosh, S.K., Hong, X., Wacharasindhu, S. and Kirchhoefer, P. (2003) *J. Am. Chem. Soc.*, **125**, 2058–2059.

33 Fustero, S., Jimenez, D., Moscardo, J., Catalan, S. and del Pozo F C. (2007) *Org. Lett.*, **9**, 5283–5286.

34 Li, C.-F., Liu, H., Liao, J., Cao, Y.J., Liu, X.-P. and Xiao, W.-J. (2007) *Org. Lett.*, **9**, 1847–1850.

35 Austin, J.F., Kim, S.G., Sinz, C.J., Xiao, W.J. and MacMillan, D.W.C. (2004) *PNAS*, **101**, 5482–5487.

36 Reyes, E., Jiang, H., Milelli, A., Elsner, P., Hazell, R.G. and Jørgensen, K.A. (2007) *Angew. Chem.*, **119**, 9362–9365; (2007) *Angew. Chem. Int. Ed.*, **46**, 9202–9205.

37 Enders, D., Hüttl, M.R.M., Grondal, C. and Raabe, G. (2006) *Nature*, **441**, 861–863.

38 Enders, D., Hüttl, M.R.M., Runsink, J., Raabe, G. and Wendt, B. (2007) *Angew. Chem.*, **119**, 471–473; (2007) *Angew. Chem. Int. Ed.*, **46**, 467–469.

39 Cao, C.L., Sun, X.L., Kang, Y.B. and Tang, Y. (2007) *Org. Lett.*, **9**, 4151–4154.

40 (a) Halland, N., Aburel, P.S. and Jørgensen, K.A. (2004) *Angew. Chem.*, **116**, 1292–1297; (2004) *Angew. Chem. Int. Ed.*, **43**, 1272–1277; (b) Pulkkinen, J., Aburel, P.S., Halland, N. and Jørgensen, K.A. (2004) *Adv. Synth. Catal.*, **346**, 1077–1080.

41 Yang, J.W., Hechavarria Fonseca, M.T. and List, B. (2005) *J. Am. Chem. Soc.*, **127**, 15036–15037.

42 Carlone, A., Cabrera, S., Marigo, M. and Jørgensen, K.A. (2007) *Angew. Chem.*, **119**, 1119–1122; (2007) *Angew. Chem. Int. Ed.*, **46**, 1101–1104.

43 Zu, L., Li, H., Xie, H., Wang, J., Jiang, W., Tang, Y. and Wang, W. (2007) *Angew. Chem.*, **119**, 3806–3808; (2007) *Angew. Chem. Int. Ed.*, **46**, 3732–3734.

44 Wang, J., Li, H., Xie, H., Zu, L., Shen, X. and Wang, W. (2007) *Angew. Chem.*, **119**, 9208–9211; (2007) *Angew. Chem. Int. Ed.*, **46**, 9050–9053.

45 Wang, W., Li, H., Wang, J. and Zu, L. (2006) *J. Am. Chem. Soc.*, **128**, 10354–10355.

46 Brandau, S., Maerten, E., and Jørgensen, K.A., (2006) *J. Am. Chem. Soc.*, **128**, 14986–14991.

47 (a) Li, H., Wang, J., Xie, H., Zu, L., Jiang, W., Duesler, E. and Wang, W. (2007) *Org. Lett.*, **9**, (2007) 965–968; (b) Sundén, H., Ibrahem, I., Zhao, G.-L., Eriksson, L. and Córdova, A. (2007) *Chem. Eur. J.*, **13**, 574–581.

48 Pihko, P.M. (2004) *Angew. Chem. Int. Ed.*, **43**, 2062–2064.

49 Schuster, T., Bauch, M., Durner, G. and Göbel, M.W. (2000) *Org. Lett.*, **2**, 179–181.

50 (a) Huang, Y., Unni, A.K., Thadani, A.N. and Rawal, V.H. (2003) *Nature*, **424**, 146–146; (b) Unni, A.K., Takenaka, N., Yamamoto, H. and Rawal, V.H. (2005) *J. Am. Chem. Soc.*, **127**, 1336–1337.

51 Du, H., Zhao, D. and Ding, K. (2004) *Chem. Eur. J.*, **10**, 5964–5970.

52 Thadani, A.N., Stankovic, A.R. and Rawal, V.H. (2004) *PNAS*, **101**, 5846–5850.

53 Nakashima, D. and Yamamoto, H. (2006) *J. Am. Chem. Soc.*, **128**, 9626–9627.

54 Momiyama, D., Yamamoto, Y. and Yamamoto, H. (2007) *J. Am. Chem. Soc.*, **129**, 1190–1195.

55 Itoh, J., Fuchibe, K. and Akiyama, T. (2006) *Angew. Chem.*, **118**, 4914–4916; (2006) *Angew. Chem. Int. Ed.*, **45**, 4796–4798.

56 Akiyama, T., Morita, H. and Fuchibe, K. (2006) *J. Am. Chem. Soc.*, **128**, 13070–13071.

57 (a) Liu, H., Cun, L.-F., Mi, A.-Q., Jiang, Y.-Z. and Gong, L.-Z. (2006) *Org. Lett.*, **8**, 6023–6026; (b) Rueping, M. and Azap, C. (2006) *Angew. Chem.*, **118**, 7996–7999; (2006) *Angew. Chem. Int. Ed.*, **45**, 7832–7835.

58 Taylor, M.S. and Jacobsen, E.N. (2004) *J. Am. Chem. Soc.*, **126**, 10558–10559.

59 Raheem, I.T., Thiara, P.S., Peterson, E.A. and Jacobsen, E.N. (2007) *J. Am. Chem. Soc.*, **129**, 13404–13405.

60 Seayad, J., Seayad, A.M. and List, B. (2006) *J. Am. Chem. Soc.*, **128**, 1086–1087.

61 Wanner, M.J., van der Haas R.N.S. Cuba, K.R., van Maarseveen, J.H. and Hiemstra, H. (2007) *Angew. Chem.*, **119**, 7629–7631; *Angew. Chem. Int. Ed.*, **46**, 7485–7487.

62 Mayer, S. and List, B. (2006) *Angew. Chem.*, **118**, 4299–4301; (2006) *Angew. Chem. Int. Ed.*, **45**, 4193–4195.

63 Rueping, M., Ieawauman, W., Antonchick, A.P. and Nachtsheim, B.J. (2007) *Angew. Chem.*, **119**, 2143–2146; (2007) *Angew. Chem. Int. Ed.*, **46**, 2097–2100.

64 Chen, X.-H., Xu, X.-Y., Liu, H., Cun, L.-F., Gong, L.-Z. et al. (2006) *J. Am. Chem. Soc.*, **128**, 14802–14803.

65 Zhou, J. and List, B. (2007) *J. Am. Chem. Soc.*, **129**, 7498–7499.

66 Jiang, J., Yu, J., Sun, X.X., Rao, Q.Q. and Gong, L.Z. (2008) *Angew. Chem.*, **120**, 2492–2496; (2008) *Angew. Chem. Int. Ed.*, **47**, 2458–2462.

67 Methot, J.L. and Roush, W.R. (2004) *Adv. Synth. Catal.*, **346**, 1035–1050.

68 Zhang, C. and Lu, X. (1995) *J. Org. Chem.*, **60**, 2906–2908;Lu, X., Zhang, C. and Xu, Z. (2001) *Acc. Chem. Res.*, **34**, 535–544.

69 Zhu, G., Chen, Z., Jiang, Q., Xiao, D., Cao, P. and Zhang, X. (1997) *J. Am. Chem. Soc.*, **119**, 3836–3837.

70 Wilson, J.E. and Fu, G.C. (2006) *Angew. Chem.*, **118**, 1454–1457; (2006) *Angew. Chem. Int. Ed.*, **45**, 1426–1429.

71 Cower, B.J. and Miller, S.J. (2007) *J. Am.Chem. Soc.*, **129**, 10988–10989.

72 Zhu, X.-F., Lan, J. and Known, O. (2003) *J. Am. Chem. Soc.*, **125**, 4716–4717.

73 Wurz, R.P. and Fu, G.C. (2005) *J. Am. Chem. Soc.*, **127**, 12234–12235.

74 Bredig, G. and Fiske, W.S. (1912) *Biochem. Z*, 7.

75 (a) Chen, Y. McDdaid F P. and Deng, L. (2003) *Chem. Rev.*, **103**, 2965–2984; (b) Tian, S.-K., Chen, Y., Huang, J.F., Tang, L., McDdaid, P. and Deng, L. (2004) *Acc. Chem. Res.*, **37**, 621–631.

76 Bekele, T., Shah, M.H., Wolfer, J., Abraham, C.J., Weatherwax, A. and Lectka, T. (2006) *J. Am. Chem. Soc.*, **128**, 1810–1811.

77 Wolfer, J., Bekele, T., Abraham, C.J., Dogo-Isonagie F C. and Lectka, T. (2006) *Angew. Chem.*, **118**, 7558–7560; (2006) *Angew. Chem. Int. Ed.*, **45**, 7398–7400.

78 Abraham, C.J., Paull, D.H., Scerba, M.T., Grebinski, J.W. and Lectka, T. (2006) *J. Am. Chem. Soc.*, **128**, 13370–13371.

79 Xu, X., Wang, K. and Nelson, S. (2007) *J. Am. Chem. Soc.*, **129**, 11690–11691.

80 (a) Zu, L., Wang, J., Li, H., Xie, H., Jiang, W. and Wang, W. (2007) *J. Am. Chem. Soc.*, **129**, 1036–1037; (b) Zu, L.-S., Xie, H.-X., Li, H., Wang, J., Jiang, W. and Wang, W. (2007) *Adv. Synth. Catal.*, **349**, 1882–1886.

81 Wang, Y., Li, H., Wang, Y.-Q., Liu, Y., Foxman, B.M. and Deng, L. (2007) *J. Am. Chem. Soc.*, **129**, 6364–6365.

82 (a) Breslow, R. (1958) *J. Am. Chem. Soc.*, **80**, 3719; (b) Breslow, R. and Kim, R. (1994) *Tetrahedron Lett.*, **35**, 699–702; (c) Breslow, R. and Schmuck, C. (1996) *Tetrahedron Lett.*, **37**, 8241–8242.

83 Kerr, M.S., Read de Alaniz, J. and Rovis, T. (2002) *J. Am. Chem. Soc.*, **124**, 10298–10299.

84 Kerr, M.S. and Rovis, T. (2004) *J. Am. Chem. Soc.*, **126**, 8876–8877.

85 Enders, D., Niemeler, O. and Balensiefer, T. (2006) *Angew. Chem.*, **118**, 1491–1495; *Angew. Chem. Int. Ed.*, **45**, (2006) 1463–1467.

86 He, M., Struble, J.R. and Bode, J.W. (2006) *J. Am. Chem. Soc.*, **128**, 8418–8420.

87 He, M., Uc, G.J. and Bode, J.W. (2006) *J. Am. Chem. Soc.*, **128**, 15088–15089.

88 Chiang, P.C., Kaeobamrung, J. and Bode, J.W. (2007) *J. Am. Chem. Soc.*, **129**, 3520–3521.

# Index

**1**

[1 + 2] cycloaddition 754
1-(1-alkynyl)cyclopropyl ketones 780
1-benzylidene-3-oxopyrazolidin-1-ium-2-ide, preparation 160
1,2-additions, aldehydes 1029–1030
– imines 1030–1031
– intramolecular *see* intramolecular 1,2-addition reactions
– ketones 1029–1030
– RCFC 1029–1031
1,2-diarylethenes 1190–1191
1,3-cyclohexadienes, [2 + 2 + 2] cycloadditions 381–391
1,3-dipolar cycloaddition 87–168, 917–949
– α,β-unsaturated aldehydes 95–101
– α,β-unsaturated carbonyl compounds 91–95
– asymmetric 95–101, 108–109
– aziridines 123–124
– azomethine imines 126–133
– azomethine ylides 107–126
– carbonyl ylides 134–147
– Co-catalyst 140–143
– diazo compounds 154–158
– electron-deficient olefins 101–102
– electrophilic activation of alkynes 124–126, 143–147
– hydroxy-directed 148–151
– inverse electron demand 103–105
– Kinugasa reaction 105–107
– Lewis-acid catalysis 88–107, 123, 155–158
– münchnones 121–122
– nitrile imines 153–154
– nitrile oxides 152–153
– nitroalkenes 117–119
– nitrones 88–107
– organocatalysis 95–96, 119–121

– tandem carbonyl ylide formation 134–140
– transition metal catalyzed 121–122
– vinyl sulfones 117–119
– ylides 107–126, 134–147
– *see also* catalytic dipolar cycloadditions of alkynes
1,4 additions
– catalyzed 219
– intramolecular 1,4-addition reactions *see* intramolecular 1,4-addition reactions
1,5-enynes
– additions of carbon nucleophiles 656–657
– alkoxycyclizations 647–649
– hydroxycyclizations 647–649
– skeletal rearrangement 667
1,6-enynes
– additions of carbon nucleophiles 650–654
– additions of other nucleophiles 654–656
– alkoxycyclizations 644–647
– cyclopropanation 667–670
– endocyclic rearrangement 665–666
– hydroxycyclizations 644–647
– skeletal rearrangement 657–661
1,7-enynes
– alkoxycyclizations 644–647
– hydroxycyclizations 644–647
– skeletal rearrangement 661–663
1,$n$-enynes 625–686
– addition of nucleophiles 644–657
– alder-ene 626–635
– alkoxy-substituted 644
– alkoxycyclizations 644–647
– alkynes 671–672
– amination 649–650
– cyclobutenes 663–665
– cycloisomerizations 626–644
– cyclopropanations 667–670
– furans 671–672

– hydroxy-substituted  644
– hydroxycyclizations  644–647
– propargylic 1,2- and 1,3-acyl migration  635–643
– skeletal rearrangement  657–672
– transition metal catalysis  625–686

**2**
[2 + m + n] cycloaddition  754–755
[2 + 2 + 2] cyclization reactions  251–255
[2 + 2 + 2] cycloadditions *see* transition metal-mediated [2 + 2 + 2] cycloadditions
[2 + 2 + 2] transition metal  367–406
2-aminomalonate  123

**3**
[3 + 1] cycloaddition, methylenecyclopropanes  755–756
[3 + 2 + 2] cycloaddition, methylenecyclopropanes  794–797
[3 + 2] cycloaddition  756–773
– activated cyclopropanes  763–772
– alkylidenecyclopropanes  756–761
– asymmetric  131–133
– methylenecyclopropanes  756–761
– N-acylhydrazones with olefins  131–133
– vinylcyclopropanes  772–773
– vinylidenecyclopropanes  761–763
[(3 + 3) + 1] cycloaddition  798
[3 + 3] cycloaddition, activated cyclopropanes  774–777
[3 + 4] cycloaddition  777–778
3-acylcyclopropenes  749–753
3-alkoylcarbonylcyclopropenes  749
3-oxopyrazolidin-1-ium-2-ides  127–131

**4**
[4 + 1] cycloaddition  778–779
[4 + 2] cycloaddition
– 1-(1-alkynyl)cyclopropyl ketones  780
– cascade process  799–800
– Heck reaction  799–800

**5**
[5 + 1 + 2 + 1] cycloaddition  794
[5 + 1] cycloaddition  780–783
[5 + 2 + 1] cycloaddition  794
[5 + 2] cycloaddition, VCPs  784–793

**α**
α-amino acid esters, asymmetric 1,3-dipolar cycloaddition  108–109
α-halocarbanions, cyclopropanations  694–698

α,β-unsaturated aldehydes
– asymmetric 1,3-dipolar cycloaddition  95–101
– organocatalysis  95–96
α,β-unsaturated carbonyl compounds
– 1,3-dipolar cycloaddition  91–95, 109–116
– aldehydes  884–890
– ancillary coordinating group  91–95
– hydrosilanes  884–890
– Lewis acid catalysis  91–95, 157–158

**β**
β-h elimination *see* cyclometallation

**π**
π bonds, C–C bonds  1036–1041
π–π-interactions
– di-π-methane rearrangement  1179–1180
– oxa-di-π- methane rearrangement  1180–1189
– $\pi^2$–$\pi^2$-rearrangements  1179–1189
– $\pi^v$–$\pi^\mu$-cycloadditions  1177–1179
$\pi^6$-photocyclization
– classical examples  1189–1190
– configurationally restrained 1,2-diarylethenes  1190–1191

**a**
acetals
– alkynes  483–500
– cyclization reactions  483–500
acids
– Brønsted *see* Brønsted acid
– carboxylic acids *see* carboxylic acids
– Lewis *see* Lewis acid catalysis
– phosphorous acid  1111–1112
activated cyclopropanes
– [3 + 2] cycloaddition  763–772
– [3 + 3] cycloaddition  774–777
activation
– alcohols  1029–1036
– alkyl halides  1026
– C–H bond  991–1025 (*see also* C–H activation)
– carbonyl derivatives  1029–1036 (*see also* Friedel–Crafts type cyclizations)
– cyclization reactions  991–1025
acyl alkylidenecyclopropanes  744–746
acyl halides, macrolactams  1073–1074
acyl methylenecyclopropanes  744–746
acyl–oxygen bond
– Corey–Nicolaou macrolactonization  1056–1058
– Keck macrolactonization  1063–1065

- macrolactones  1056–1066
- Masamune macrolactonization  1058–1059
- Mukaiyama macrolactonization  1059–1061
- Shiina macrolactonization  1065–1066
- Yamaguchi macrolactonization  1062–1063

acylimidazoles, macrolactams  1074
acylpalladations, cyclic carbometallation  257–261
addition reactions
- 1,2-additions *see* 1,2-additions
- aldol reaction  169–177
- cycloaddition *see* cycloaddition
- imino ene reactions  182–186
- intramolecular 1,2  169–199
- intramolecular 1,4  199–217
- Michael addition  200–210
- nucleophilic addition to imino groups  186–191
- nucleophilic addition to nitriles  194–197

AIBN *see* azobisisobutyronitrile
alcohols
- activation  1029–1036
- as nucleophiles  954–958, 973–976
- cyclic coupling reactions  330–344
- direct substitution  1033–1035
- electrophilic cyclizations  954–958, 973–976
- Friedel–Crafts type cyclizations  1029–1036 (*see also* RCFC)
- RCFC  1033–1035

aldehydes
- 1,2 additions  1029–1030
- [2 + 2 + 2] cycloadditions  385–387
- α,β-unsaturated aldehydes  95–101
- α,β-unsaturated carbonyl compounds  884–890
- as nucleophiles  963–964, 976–977
- aziridinations  713–720
- electrophilic cyclizations  963–964, 976–977
- epoxidations  701–713
- hydrosilanes  884–890
- RCFC  1029–1030
- tether  884–890

alder-ene, cycloisomerizations  626–635
aldimines, asymmetric 1,3-dipolar cycloaddition  108–109
aldol reaction
- dicarbonyl compounds  169–172
- intramolecular 1,2-addition reactions  169–172
- vinylogous  173–177

aliphatic tether  483–486

alkaloids, catalysis by  1227–1228
alkanes, diazoalkanes  690–691
alkene-vinylcyclopropanes  794
alkenes
- alkene-vinylcyclopropanes  794
- as substrates  813–820
- asymmetric metathesis  552–554
- aziridinations  713–717
- cascade processes  462–465
- cyclic carbometallation  228–238
- cyclization reactions  457–465
- cyclopropanations  688–700
- epoxidations, metal-catalyzed  701–705
- epoxidations, organo-catalyzed  706–710
- Heck reaction  230–233
- hydroarylation reactions  460–462
- imino ene reactions  182–186
- intramolecular 1,2-addition reactions  177–181
- intramolecular carbometallation, catalytic  233–238
- intramolecular carbometallation, stoichiometric  228–230
- metathesis  552–554
- nitroalkenes  117–119
- RCFC  1036–1038
- ring expansion cyclizations  813–820
- silylformylation  853–858
- tethered nucleophiles  457–460
- transition metal catalysis  457–465

alkoxy-substituted, 1,$n$-enynes  644
alkoxycyclizations
- 1,5-enynes  647–649
- 1,6-enynes  644–647
- 1,7-enynes  644–647
- addition of nucleophiles  644–647

alkyl
- C–H bond activation  1005–1017
- C–O bond  1066–1070
- cyclization reactions  1005–1017
- halides, activation  1026
- macrolactonization  1066–1070

alkyl C–H bond activation  1005–1017
- C–C bond formation  1005–1009
- C–N bond formation  1010–1016
- C–O bond formation  1016–1017
- other metals  1009–1010, 1015–1016
- palladium  1008–1009, 1013–1015
- rhodium  1005–1008, 1011–1013

alkyl C–O bond
- macrolactones  1066–1070
- Mitsunobu reaction  1066–1068
- other methods  1068–1070

alkyl carbon–oxygen bond *see* alkyl C–O bond

alkyl halides 1026
alkylidenecyclopropanes 743–748
– [3 + 2] cycloaddition 756–761
– acyl 744–746
alkyne substituted carbonyl compounds, 1,2-addition 217
alkynes
– 1,3-dipolar cycloaddition 124–126, 143–147
– [2 + 2 + 2] cycloadditions 368–372
– acetals 483–500
– aliphatic tether 483–486
– aromatic tether 486–500
– as nucleophiles 971
– as substrates 827–831
– azides 919–920 (see also catalytic dipolar cycloadditions of alkynes)
– azomethine ylides 124–126
– carbonyl ylides 143–147
– carbonyls 177–181, 483–500
– catalysis 239–243
– construction of functionalized benzenes 368–372
– cyclic carbometallation 238–243
– cyclic coupling reactions 354–356
– cyclization reactions 466–504, 902–904
– cycloadditions 917–949 (see also catalytic dipolar cycloadditions of alkynes)
– domino sp–sp² 475–479
– electrophilic activation 124–126, 143–147
– electrophilic cyclizations 971
– epoxides 483–500
– furans 671–672
– heteroatom-substituted 368–372 (see also benzene derivatives)
– imines 483–500
– imino ene reactions 182–186
– intramolecular 1,2-addition reactions 177–181
– Lewis acid catalysis 157–158
– phenol synthesis 671–672
– RCFC 1039–1041
– ring-closing metathesis 599–624
– ring expansion cyclizations 827–831
– silylative carbamoylation 902–904
– silylformylation 853–858
– skeletal rearrangement 671–672
– stoichiometric methods 238–239
– tethered nucleophiles 466–475
– thioacetals 483–500
– transition metal catalysis 466–504
– transition metals mediated 197–199
alkynols, silylative lactonization 890–898
alkynylboronates 369

– [2 + 2 + 2] cycloadditions 381
alkynylcyclopropanes 738–740
alkynylhalides 370–372
alkynylphosphines 370
alkynylsilanes 368–369
allenes
– [2 + 2 + 2] cycloadditions 384
– carbonyls 512–513
– cyclic carbometallation 243–245
– cyclization reactions 504–515
– hydroarylation reactions 513–515
– tethered nucleophiles 504–511
– transition metal catalysis 504–515
allenylcyclopropanes 738–740
allylation reactions
– carbonucleophiles 271–301
– cyclic aminations 282–294
– cyclic O- or S- 294–301
– intramolecular 271–314
– iridium-catalyzed 301–304
– nickel-catalyzed 304
– palladium-catalyzed 271–301
– rhodium-catalyzed 304–305
– transition metal-catalyzed 271–314
aluminum
– catalyzed Diels–Alder reactions 2–7, 21–23
– chiral complexes 5–7
amidation
– direct see direct amidation
– macrolactams 1073–1086
amides
– as nucleophiles 967, 980–981
– electrophilic cyclizations 967, 980–981
– ynamides 370
amidines 967–968
aminations
– 1,n-enynes 649–650
– addition of nucleophiles 649–650
– cyclic allylic 282–294
amines
– as nucleophiles 965–966, 980
– asymmetric cyclizations 1213–1218
– catalysis 14–17, 39–41, 708–709, 1213–1218
– chirality 14–17, 39–41, 1213–1218
– cyclic coupling reactions 330–344
– Diels–Alder reactions 14–17
– electrophilic cyclizations 965–966, 980
– epoxides 708–709
– organocatalysis 1213–1218
– oxa-Diels–Alder reactions 39–41
– primary 16–17
– secondary 14–16, 39–41
amphidinolide E 1135

angular [N]-phenylenes 375–376
anhydrides, macrolactams 1074–1077
annulation reactions 1124–1126
– see also free radical cyclizations
annulative silylcarbonylation 890–898
anti-addition 850–853
aphanorphine 1135–1136
arene substituted carbonyl compounds, 1,2-addition 218
arenes
– cyclic carbometallation 243
– intramolecular 1,2-addition reactions 177–181
aromatic tether 486–500
aryl
– 1,2-diarylethenes 1190–1191
– C–H bond activation 994–1004
– C–N bond formation 1086
– configurationally restrained 1190–1191
– cyclization reactions 994–1004
– groups as nucleophiles 972
– groups electrophilic cyclizations 972
– hydroarylation reactions 460–462, 513–515
– macrolactams 1086
– $\pi^6$-photocyclization 1189–1191
aryl C–H bond activation 994–1004
– gold 1002–1003
– other metals 1004
– palladium 997–1002
– rhodium 994–996
aryl C–N bond
– macrolactams 1086
– see also C–N bonds
arynes, [2 + 2 + 2] cycloadditions 381–382
aspartic acid, epoxidations 710
aspidospermidine 1102, 1137–1138
asymmetric [3 + 2] cycloaddition 131–133
asymmetric 1,3-dipolar cycloaddition 95–101, 108–121
asymmetric Diels–Alder reactions 1–58
– aza- 59–86
– inverse electron demand 71–73, 80
– organocatalysis 14–21
– oxa- 37–41
– selected procedures 46–49
– using a chiral Ag complex 83
– using a chiral Zr complex 83
asymmetric intramolecular cyclization, catalysis 219
asymmetric Kinugasa reaction 105–107
asymmetric olefin metathesis 552–554
asymmetric organocatalyzed cyclizations 1199–1241

– Brønsted acid-catalyzed see Brønsted acid-catalyzed asymmetric cyclizations
– catalyzed by bifunctional catalysts 1228–1229
– catalyzed by chiral amines 1213–1218
– catalyzed by chiral nucleophilic carbenes see chirality, nucleophilic
– domino see domino cyclization reactions
– enamine-catalyzed see enamine-catalyzed
– iminium-catalyzed 1207–1212 (see also iminium catalysis)
– Lewis base-catalyzed see Lewis acid catalysis
– selected procedures 1233–1237
asymmetric ring-closing metathesis 547–554
atom-transfer cyclizations 1113–1117
– copper 1113–1115
– organotin 1116
– peroxides 1116–1117
– ruthenium 1115–1116
– see also free radical cyclizations
aza Diels–Alder reactions 59–86
– asymmetric 59–86
– catalysis 59–86
– chiral Brønsted acid catalysis 76–82
– chiral Lewis acid catalysis 59–76
– chiral metal catalyst 59–76
– chiral organocatalysis 76–82
– lanthanides 60
azadirachtin 1136–1138
azide–alkyne cycloaddition 919–920
– ruthenium-catalyzed 935–937
azides
– alkyne cycloaddition 919–920
– as nucleophiles 969
– electrophilic cyclizations 969
– in situ generated 924–925
– sulfonyl 929–930
aziridinations 713–720
– alkenes 713–717
– carbenes 718–719
– imines 717–720
– mediated 713–716
– metal-catalyzed 713–716
– metal-free 716–717
– ylides 719–720
aziridines 123–124
azobisisobutyronitrile (AIBN) 1099–1148
azomethine
– imines 126–133
– ylides 107–126
azomethine imines 126–133
– [3 + 2] cycloaddition 131–133
– 3-oxopyrazolidin-1-ium-2-ides 127–131

azomethine ylides 107–126
– aziridines 123–124

## b

bases
– intramolecular 1,4-addition reactions 200–203
– Lewis 1225–1228
– Michael addition 200–203
benzene derivatives
– [2 + 2 + 2] cycloadditions 368–381
– alkynylboronates 369
– alkynylhalides 370–372
– alkynylphosphines 370
– alkynylsilanes 368–369
– construction 368–372
– naphthalene core 372–373
– polyphenylenes *see* polyphenylenes
– ynamides 370
benzoin reactions 1231
benzynes, transient 372–373
bifunctional catalysts 1228–1229
biological profiling 932–934
biological screening 930–931
biosystems, macrolactones 1070–1071
biscyclopropylidenes 382–384
bonds
– acyl–oxygen, formation 1056–1066
– C–C $\pi$ 1036–1041
– C–C, cleavage 123–124
– C–C, formation 1005–1009
– C–H activation 991–1025
– C–N, formation 1010–1016, 1086
– C–O, formation 1016–1017, 1066–1070
– $\pi$ bonds 1036–1041
boron
– alkynylboronates 369
– catalyzed Diels–Alder reactions 2–7, 21–23, 59–76
– chiral complexes 2–5, 46–47
– organoboron 315–320
boronates, alkynylboronates 369
boron complexes, preparation 46
Brønsted acid
– asymmetric organocatalyzed cyclizations 1218–1225
– Diels–Alder reactions 17–19, 76–82
– RCFC 1039
Brønsted acid-catalyzed asymmetric cyclizations 1218–1225
– Diels–Alder 1218–1222
– domino reactions 1224–1225
– hetero-Diels–Alder 1218–1222
– multicomponent 1224–1225
– Nazarov 1224
– Pictet–Spengler 1222–1224
*tert*-butyl diazoacetate, preparation 160

## c

C–C bond cleavage 123–124
C–C bond formation 1005–1009
– other metals 1009–1010
– palladium 1008–1009
– rhodium 1005–1008
C–C bonds
– intramolecular hydrosilylation 849–853
– $\pi$ bonds 1036–1041
C–H activation 991–1025
– alkyl 1005–1017
– aryl 994–1004
– C–C bond formation 1005–1009
– C–N bond formation 1010–1016
– C–O bond formation 1016–1017
– gold 1002–1003
– other metals 1004, 1009–1010, 1015–1016
– palladium 997–1002, 1008–1009, 1013–1015
– rhodium 994–996, 1005–1008, 1011–1013
– selected procedures 1017–1018
C–N bond formation 1010–1016
– aryl 1086
– other metals 1015–1016
– palladium 1013–1015
– rhodium 1011–1013
C–O bond formation 1016–1017
– alkyl 1066–1070
C=C bonds, [2 + 2 + 2] cycloadditions 382–385
C=X bonds, [2 + 2 + 2] cycloadditions 385–391
carbamates
– as nucleophiles 962–963, 967–968
– electrophilic cyclizations 962–963, 967–968
carbamoylation 902–904
carbenes
– asymmetric organocatalyzed cyclizations 1229–1233
– aziridinations 718–719
– catalysis 1229–1233
– Fischer-carbene complexes 835–836
– nucleophilic 1229–1233
carbenoids, metalocarbenoids 698–700
carbocycles
– dienes 531–542
– enynes 562–573
– ring-closing metathesis 531–542

– ring-opening metathesis–ring-closing
   metathesis   542–547
– see also cyclization reactions
carbolithiation
– cyclic carbometallation   229–230, 239, 247
carbometallation, cyclic see cyclic
   carbometallation
carbon-centered nucleophiles
– alkynes   971
– aryl groups   972
– cyclization of compounds containing active
   methine groups   970–971
– electrophilic cyclizations   970–972,
   982–983
– halocyclizations   970–972
– selenocyclizations   982–983
carbon dioxide, [2 + 2 + 2]
   cycloadditions   385–387
carbon disulfide, [2 + 2 + 2]
   cycloadditions   387–388
carbonates, electrophilic cyclizations
   962–963
carbonucleophiles   962–963
– cyclic allylations   271–301
carbonyl compounds
– α,β-unsaturated see α,β-unsaturated carbonyl
   compounds
– activation see activation; RCFC
– addition   169–181
– aldol reaction   169–177
– alkenes   177–181 (see also alkenes)
– alkynes see alkynes
– allenes see allenes
– arenes   177–181 (see also arenes)
– C–C unsaturated bond   881–884
– carbonyl ylides see carbonyl ylides
– cyclization reactions   483–500
– epoxidations   710–713
– Friedel–Crafts type cyclizations see RCFC
– hydrosilanes   881–884
– intramolecular 1,2-addition reactions
   169–181
– tether   881–884
carbonyl ylides   134–147
– tandem 1,3-dipolar cycloaddition   134–140
carbonylation, silyl   890–898
carbopalladation, cyclic
   carbometallation   227–260
carboxylic acids
– as nucleophiles   958–962, 977–979
– derivatives   958–962, 977–979, 1026–1029
– electrophilic cyclizations   958–962,
   977–979
– RCFC   1026–1029

cascade processes
– cyclization reactions   462–465, 800–803
– cycloisomerizations   746–748
– hydrometallation-initiated
   cyclizations   890–904
– see also cascade reactions incorporating Co
cascade reactions incorporating CO   890–904
– 1,6-diynes   890–895
– 1,6-enynes   896–898
– annulative silylcarbonylation   890–898
– silylative carbamoylation of alkynes
   902–904
– silylative lactonization of alkynols   899–902
catalysis
– 1,3-dipolar cycloaddition   89–95, 134–139,
   152–153
– 1,4-addition reactions   219
– [2 + 2 + 2] cycloadditions   377–379
– alkenes   457–465
– alkyne carbometallations   239–243
– alkynes   466–504, 602–605, 917–949
– allenes   504–515
– amines   14–16, 39–41, 708–709, 1213–1218
– asymmetric   1–86, 1199–1241
– asymmetric intramolecular cyclization
   219
– asymmetric olefin metathesis   552–554
– asymmetric organocatalyzed cyclizations see
   asymmetric organocatalyzed cyclizations
– aza Diels–Alder reactions   59–86
– azides   917–949
– boron   59–76
– Brønsted acid   17–19, 76–82
– carbenes   1229–1233
– chiral amines   14–16, 1213–1218
– chiral nucleophilic carbenes   1229–1233
– chiral, lanthanides   60
– Cinchona alkaloids   1227–1228
– copper   7–9, 28–33, 105–107, 920–935
– cyclization reactions   169–226, 457–526
– cycloaddition see cycloaddition
– Diels–Alder reactions   1–86
– dirhodium(II)   134–139
– enantioselective   1044–1046
– enynes   625–686
– enzyme   1070–1071
– epoxidations   706–710
– Friedel–Crafts type cyclizations   1044–1046
– fused- and bridged-ring forming
   cyclizations   247–249
– iminium salts   707–708
– inexpensive   377–378, 602
– intramolecular 1,2-addition reactions
   191–194

- intramolecular 1,4-addition reactions 205–210, 216–217
- intramolecular alkene carbometallation 233–238
- intramolecular allylation reactions *see* allylation reactions
- iron 703–704
- ketones 706–707
- Kinugasa reaction 105–107
- lanthanides 60, 705
- Lewis acid *see* Lewis acid catalysis
- Lewis base 1225–1228
- manganese 703
- metal-based 2–13, 21–37, 59–76
- metals, other 1004, 1009–1010, 1015–1016
- microwave 378–379
- molybdenum 547–552, 603–604
- Mortreux "instant" system 602
- münchnones 121–122
- nitrile oxides 91–95, 152–153
- nitriles 191–194
- nitrones 89–95
- nucleophilic carbenes 1229–1233
- nucleophilic cyclizations 644–657
- organocatalysis *see* organocatalysis
- oxa-Diels–Alder reactions 21–41
- palladium 33–36, 271–301, 448–449, 813–831, 876, 997–1002, 1008–1009, 1013–1015
- phase-transfer 709
- phosphines 1225–1227
- platinum 33–36, 704–705
- polymer-supported 554–557
- primary amines 16–17
- PTC 709
- RCAM 602–605
- rhodium 33–36, 134–139, 304–305, 377, 377–379, 427–439, 879–898, 994–996, 1005–1008, 1011–1013
- ring-closing metathesis 547–554
- ring expansion cyclizations 813–842 (*see also* ring expansion cyclizations)
- ruthenium 378–379, 419–424, 552–554, 704–705, 832–834, 935–944
- secondary amines 14–16, 39–41
- titanium 23–26, 407–414, 701–702
- transition metals *see* transition metals catalysis
- tungsten alkylidyne catalysts 602–603
- vanadium 702–703

catalytic dipolar cycloadditions of alkynes 917–949
- azide–alkyne 919–920
- azides 917–949
- catalysts 920–924
- copper-catalyzed 920–935
- CuAAC *see* copper, catalytic dipolar cycloadditions
- *in situ* generated azides 924–925
- isoxazoles 934–935
- ligands 920–924
- mechanistic aspects 925–929, 938–944
- nitrile oxides 917–949
- ruthenium-catalyzed 935–944
- sulfonyl azides 929–930

catalytic sequences
- enamine/enamine 1214–1215
- enamine/iminium 1213–1214
- iminium/enamine 1215–1218

cationic catalysis 377

chirality
- aluminum complexes 5–7, 21–23
- amines 14–16, 1213–1218
- asymmetric 1,3-dipolar cycloaddition 87–168
- asymmetric catalysis 2–14
- asymmetric organocatalyzed cyclizations *see* asymmetric organocatalyzed cyclizations
- boron 59–76
- boron complexes 2–7, 21–23
- Brønsted acid 17–19, 76–82
- carbenes 1229–1233
- catalysis *see* catalysis
- chromium complexes 10–11, 26–28
- Co-catalyst 140–143
- cobalt complexes 10–11
- copper complexes 7–9, 28–33
- Diels–Alder as key step 41–45
- Diels–Alder reactions 2–14, 21–37, 59–86
- indium complexes 5–7, 21–23
- iron complexes 11
- lanthanides 60
- Lewis acid 2–13, 21–37, 59–76, 89–95, 140–143
- magnesium complexes 9
- metal-based 2–13, 21–37, 59–76
- molybdenum catalyst 547–552
- nucleophilic 1229–1233
- organocatalysis 76–82, 1199–1241
- phosphines 1225–1227
- primary amines 16–17
- rare earth metal complexes 12–14, 36–37
- rhodium complexes 33–36
- ruthenium complexes 11–12
- secondary amines 14–16, 39–41
- silver 83
- titanium complexes 23–26
- transition metal complexes 10–12

– zinc complexes   9, 28–33
– zirconium   83
chromium
– catalyzed Diels–Alder reactions   26–28
– chiral complexes   10–11
Cinchona alkaloids   1227–1228
circular, polycyclization   251–257
cleavage
– C–C bond   123–124
– photocyclization reactions   1154–1155
click reaction   918
Co-catalyst   140–143
Co-cyclization see transition metal-mediated [2 + 2 + 2] cycloadditions
cobalt
– [2 + 2 + 2] cycloadditions   378–379
– chiral complexes   10–11
– cyclometallation   425–427
– microwave   378–379
complexes
– chiral aluminum   5–7
– chiral boron   2–7, 21–23
– chiral chromium   10–11, 26–28
– chiral cobalt   10–11
– chiral copper   7–9, 28–33
– chiral indium   5–7
– chiral iron   11
– chiral magnesium   9
– chiral palladium   33–36
– chiral platinum   33–36
– chiral rare earth metal   12–14, 36–37
– chiral rhodium   33–36
– chiral ruthenium   11–12
– chiral titanium   10, 23–26
– chiral transition metal   10–12
– chiral zinc   9
– chiral zirconium   23–26
– Diels–Alder reactions   2–14, 21–37
compound libraries   930–931
configurationally restrained, 1,2-diarylethenes   1190–1191
conjugate additions, RCFC   1033
coordination spheres, transition metal   615
copper
– atom-transfer   1113–1115
– catalytic dipolar cycloadditions   920–935
– catalyzed Diels–Alder reactions   7–9, 28–33
– Kinugasa reaction   105–107
– radical cyclizations   1113–1115
copper-catalyzed azide–alkyne cycloaddition (CuAAC)   920–935
– biological screening   930–931
– compound libraries   930–931
– examples of application   930–936
– mechanistic aspects   925–929
– natural products   932–934
– synthesis of glycoconjugates   931–932
– with in situ generated azides   924–925
Corey–Nicolaou macrolactonization   1056–1058
coupling reactions
– cyclometallation   444
– domino sp–$sp^2$   475–479
– see also cyclic coupling reactions
CuAAC see copper-catalyzed azide–alkyne cycloaddition
cyclic allylations   271–314
cyclic allylic aminations   282–294
cyclic carbometallation   227–270
– acylpalladations   257–261
– alkenes   228–238
– alkynes   238–243
– allenes   243–245
– arenes   243
– living organometallic processes   245–257
– selected procedures   262–263
– stoichiometric intramolecular   228–230
– see also metallation
cyclic coupling reactions   315–366
– alcohols   330–344
– alkynes   354–356
– amines   330–344
– Heck reaction   344–354
– organoboron   315–320
– organosilicon   328–330
– organotin   321–328
– selected procedures   357–361
cyclic O- or S-allylation   294–301
cyclization reactions
– alkenes   457–465 (see also alkenes)
– alkynes   466–504, 599–624 (see also alkynes)
– allenes   504–515 (see also allenes)
– allylations see allylation reactions
– asymmetric   1199–1241
– aziridinations see aziridinations
– C–H activation   991–1025
– carbometallation see cyclic carbometallation
– carbon–heteroatom   169–226
– carbonyl compounds   483–500
– cascade processes see cascade processes
– catalyzed   169–226, 457–526, 991–1025, 1199–1241
– coupling see cyclic coupling reactions
– cycloaddition see cycloaddition
– cycloisomerizations see cycloisomerization
– cyclopropanations see cyclopropanations
– diastereoselective radical   1126–1129
– dienes   531–562

- domino sp–sp² 475–479
- electrophilic 951–990
- enantioselective 379–381
- enantioselective radical 1129–1135
- enynes 562–592, 625–686
- epoxidations *see* epoxidations
- free radical 1099–1148
- Friedel–Crafts 1025–1055
- halocyclizations 953–972
- hydrometallation-initiated 843–915
- lactams, macro- 1072–1086
- lactones, macro- 1055–1071
- macrolactams 1072–1086
- macrolactones 1055–1071
- metallation *see* cyclometallation
- methathesis 527–624
- migration reactions 479–482
- multicomponent 794–798
- multiple reactions 245–257
- organocatalyzed 1199–1241
- photochemical 1122–1124
- photocyclization 1149–1198
- polycyclization *see* polycyclization
- radical 1099–1148
- RCFC 1025–1055
- ring expansion 813–842
- selected procedures 905–909
- selenocyclizations 972–983
- self-feeding 843–858
- single ring formation 1100–1101, 1135
- tandem 1101–1105, 1136–1138
- transition metal catalysis 457–526, 813–842 (*see also* transition metal catalysis)
- Yamamoto 502–504
cyclizations based on C–H activation 991–1025
- *see also* C–H activation
cyclo-oligomerizations, RCAM 616–618
cycloaddition
- 1-(1-alkynyl)cyclopropyl ketones 780
- [1 + 2] 754
- 1,3-dipolar *see* 1,3-dipolar cycloaddition
- [2 + m + n] 754–755
- [2 + 2 + 2] 367–406
- [3 + 1] 755–756
- [3 + 2 + 2] 794–797
- [3 + 2] *see* [3 + 2] cycloaddition
- [(3 + 3) + 1] 798
- [3 + 3] 774–777
- [3 + 4] 777–778
- [4 + 1] 778–779
- [4 + 2] *see* [4 + 2] cycloaddition
- [5 + 1 + 2 + 1] 794
- [5 + 1] 780–783
- [5 + 2 + 1] 794

- [5 + 2] 784–793
- activated cyclopropanes 763–772, 774–777
- alkene-vinylcyclopropanes 794
- alkylidenecyclopropanes 756–761
- alkynes 791–793, 917–949
- allenes 791–793
- asymmetric 87–168, 1207–1210 (*see also* asymmetric organocatalyzed cyclizations)
- azide–alkyne 919–920
- azides 917–949
- carbenes 1231–1233
- catalytic 89–95, 917–949, 1225–1228
- chiral nucleophilic carbenes 1231–1233
- conjugate 1033
- copper-catalyzed 920–935
- cyclopropanes 754–794
- dipolar *see* catalytic dipolar cycloadditions of alkynes
- iminium-catalyzed 1207–1210
- intramolecular 1,2-addition reactions 169–199
- intramolecular 1,4-addition reactions 199–217
- Lewis acid catalysis 89–90
- methylenecyclopropanes 755–761, 794–797
- microwave 378–379
- multicomponent 794–798
- nitrile oxides 917–949
- nitrones 88–107
- nucleophiles 1231–1233
- $\pi^v$–$\pi^\mu$- 1177–1179
- ruthenium-catalyzed 935–944
- transition metals mediated 197–199, 367–406
- vinylcyclopropanes 784–793
- vinylidenecyclopropanes 761–763
cycloalkene-ynes, ring-opening metathesis–ring-closing metathesis 578–585
cyclobutenes, 1,n-enynes 663–665
cyclodimerizations, RCAM 616–618
cyclohexenes, [2 + 2 + 2] cycloadditions 391–396
cycloisomerization
- 3-acylcyclopropenes 749–753
- 3-alkoylcarbonylcyclopropenes 749
- acyl alkylidenecyclopropanes 744–746
- acyl methylenecyclopropanes 744–746
- alkylidenecyclopropanes 743–748
- alkynylcyclopropanes 738–740
- allenylcyclopropanes 738–740
- cascade process 746–748
- cyclization reactions 457–526, 733–754
- cyclopentenes 733–736

- cyclopropanes 733–749
- cyclopropenes 749–754
- cyclopropenyl imines 753–754
- enynes *see* 1,n-enynes
- five-membered heterocycles 736–738
- furans 749–753
- methylenecyclopropanes 743–748
- pyrroloheterocycles 753–754
- transition metal catalysis 626–644
- vinylcyclopropanes 733–736, 738–740
- vinylidenecyclopropanes 749
- *see also* cyclometallation
cyclometallation 407–456
- cobalt-catalyzed 425–427
- iridium-catalyzed 439–440
- nickel-catalyzed 440–448
- palladium-catalyzed 448–449
- rhodium-catalyzed 427–439
- ruthenium-catalyzed 419–424
- selected procedures 450–452
- titanium-catalyzed 407–414
- zirconium-catalyzed 414–419
- *see also* cycloisomerization
cyclopentenes 733–736
cyclopropanations 688–700
- 1,6-enynes 667–670
- α-halocarbanions 694–698
- decomposition of diazo compounds 690–694
- diazoalkanes 690–691
- diazocarbonyl compounds 691–694
- halomethylmetal compounds 688–690
- metalocarbenoids 698–700
- polycyclization 255–257
- transition metal-catalyzed 690–694
- ylides 694–698
cyclopropane-containing compounds 733–812
- *see also* cycloisomerization
cyclopropanes 733–812
- activated 763–772, 774–777
- selected procedures 804–806
- *see also* cycloaddition; cycloisomerization
cyclopropene-containing compounds 733–812
- *see also* cycloisomerization
cyclopropenyl imines 753–754
cyclotrimerization, RCAM 616–618
- *see also* transition metal-mediated [2 + 2 + 2] cycloadditions

## d

DA reaction *see* Diels–Alder reaction
Darzens reaction 712–713

DCC *see* dicyclohexylcarbodiimide
decomposition, diazo compounds 690–694
density functional theory 625–686
DFT *see* density functional theory
di-π-methane rearrangement 1179–1180
- oxa-di-π- methane rearrangement 1180–1189
diastereoselective, radical cyclizations 1126–1129
diastereoselective radical cyclizations 1126–1129
- *see also* free radical cyclizations
diazo compounds
- 1,3-dipolar cycloaddition 154–158
- α,β-unsaturated carbonyl compounds 155–157
- alkynes 157–158
- cyclopropanations 690–694
- diazoalkanes 690–691
- diazocarbonyls 691–694
- Lewis acid catalysis 155–158
diazoalkanes 690–691
dicarbonyl compounds
- aldol reaction 169–172
- intramolecular 1,2-addition reactions 169–172
dicyclohexylcarbodiimide (DCC) 1079–1080
Diels–Alder (DA) reactions
- application 41–45
- asymmetric 2–58
- asymmetric cyclizations 1218–1222
- aza *see* aza Diels–Alder reactions
- bifunctional organocatalysis via hydrogen bonding 19–21
- Brønsted acid-catalyzed 17–19, 1218–1222
- catalysis 1–86
- chiral aluminum complexes 2–7
- chiral boron complexes 2–7
- chiral copper complexes 7–9
- chiral indium complexes 2–7
- chiral magnesium complexes 7–9
- chiral primary amine catalyzed 16–17
- chiral rare earth metal complexes 12–14
- chiral secondary amine catalyzed 14–16
- chiral transition metal complexes 10–12
- chiral zinc complexes 7–9
- enantioselective 41–45
- hetero *see* hetero Diels–Alder reactions
- hydrogen bonding *see* bifunctional catalysts
- Lewis acid catalyzed, asymmetric 2–13
- organocatalysis 14–21
- oxa *see* oxa-Diels–Alder reactions
dienes
- 1,3-cyclohexadienes 381–391

- asymmetric 547–554
- carbocycles 531–542
- catalytic 547–554
- cyclization reactions 867–877
- heterocycles 531–542
- hexadienes 381–391
- hydrosilanes 867–877
- molybdenum catalyst 547–552
- natural products 557–562
- polymer-supported catalysts 554–557
- ring-closing metathesis 531–562
- ring-opening metathesis–ring-closing metathesis 542–547
- ruthenium catalyst 552–554

dienyl, ring expansion cyclizations 820–827
dienynes, ring-closing metathesis 573–578
dimer, RCAM *see* cyclodimerizations
dipolar cycloadditions 87–168
- alkynes 917–949
- azomethine imines 126–133
- azomethine ylides 107–126
- carbonyl ylides 134–146
- diazo compounds 154–157
- nitril imines 153–154
- nitril oxides 148–152
- nitrones 88–108
- selected procedures 159–160, 944–945

dipolar Lewis acid catalysis 89–90
direct amidation 1073–1086
- acyl halides 1073–1074
- acylimidazoles 1074
- anhydrides 1074–1077
- esters 1077–1086
- multi-step strategy 1077–1079
- one-pot strategy 1079–1080

direct excitation 1152
dirhodium(II)
- catalyzed asymmetric 1,3-dipolar cycloaddition 134–139
- Co-catalyst chiral Lewis acid catalysis 140–143

diynes 858–863
domino cyclization reactions
- catalyzed by Brønsted acids 1224–1225
- catalyzed by chiral amines 1213–1218
- catalyzed by enamines 1203–1206
- catalyzed by iminiums 1211–1212
- enamine/enamine catalytic sequences 1214–1215
- enamine/iminium catalytic sequences 1213–1214
- iminium/enamine catalytic sequences 1215–1218

domino sp–sp$^2$ 475–479

double bonds
- [2 + 2 + 2] cycloadditions 382–391
- C=X 385–391
- C=C 382–385
- self-feeding cyclizations 843–849

dumbbell, polycyclization 251–257

## e

electron-deficient olefins, asymmetric 1,3-dipolar cycloaddition 101–102
electron transfer *see* photoinduced electron transfer
electrophilic activation
- 1,3-dipolar cycloaddition 124–126, 143–147
- alkynes 124–126, 143–147
- azomethine ylides 124–126
- carbonyl ylides 143–147

electrophilic azomethine ylides 123–126
- C–C bond cleavage 123–124

electrophilic cyclizations 951–990
- halocyclizations 953–972 (*see also* halocyclizations)
- mechanism 952–953
- nucleophiles 951–990
- regioselectivity 952–953
- selected procedures 984
- selenocyclizations 972–983 (*see also* selenocyclizations)
- stereoselectivity 952–953

enamine-catalyzed 1200–1206
- domino reactions 1203–1206
- intramolecular aldol reactions 1200–1201
- intramolecular Michael additions 1202–1203

enamine/enamine catalytic sequences 1214–1215
enamine/iminium catalytic sequences 1213–1214
enantioselective [2 + 2 + 2] cycloadditions 379–381, 391
enantioselective 1,3-dipolar cycloaddition 88–158
enantioselective Diels–Alder reactions 1–86
enantioselective epoxidations 710
enantioselective Friedel–Crafts type cyclizations 1044–1046
enantioselective radical cyclizations 1129–1135
- *see also* free radical cyclizations

endocyclic rearrangement
- 1,6-enynes 665–666
- skeletal 665–666

energy transfer, photocyclization
  initiation 1152–1153
enyne-yne 612–614
enynes
– carbocycles 562–573
– cyclization reactions 863–867
– cycloalkene-ynes 578–585
– cycloisomerizations 626–644
– dienynes 573–578
– heterocycles 562–573
– hydrosilanes 863–867
– nucleophilic cyclizations 644–657
– ring-closing metathesis 562–592
– transition metal catalysis 625–686
– see 1,n-enynes
enzyme catalysis, macrolactams 1070–1071
epoxidations 701–713
– alkenes 701–710
– alkynes 483–500
– aspartic acid 710
  carbonyl compounds 710 713
– catalyzed by amines 708–709
– catalyzed by iminium salts 707–708
– catalyzed by ketones 706–707
– catalyzed by metals 701–705
– darzens reaction 712–713
– enantioselective 710
– iron-catalyzed 703–704
– lanthanide-catalyzed 705
– manganese-catalyzed 703
– organocatalysis 706–710
– peptides 710
– phase-transfer catalysis 709
– platinum-catalyzed 704–705
– RCFC 1035–1036
– ruthenium-catalyzed 704–705
– titanium-catalyzed 701–702
– vanadium-catalyzed 702–703
– ylides 710–712
esters
– α-amino acid esters 108–109
– macrolactams 1077–1086
– multi-step strategy 1077–1079
– one-pot strategy 1079–1080
ethers, as nucleophiles 954–958
excitation, photocyclization initiation 1152

## f

Fischer-carbene complexes, ring expansion
  cyclizations 835–836
five-membered heterocycles 736–738
fluorenes, silafluorenes 374–375
fortucine 1136–1138
four-membered rings 1156–1157, 1163–1165

free radical cyclizations 1099–1148
– annulation reactions 1124–1126
– atom-transfer 1113–1117
– definition 1099–1105
– diastereoselective 1126–1129
– enantioselective 1129–1135
– group-transfer 1135–1136
– manganese triacetate-mediated
  1117–1119
– methods 1105–1124
– natural product synthesis 1135–1140
– photochemical 1122–1124
– reductive 1105–1113
– samarium diiodide-mediated 1119–1120
– selected procedures 1140–1142
– single ring formation 1100–1101, 1135
– tandem 1101–1105, 1136–1138
– titanium (III)-mediated 1120–1122
Friedel–Crafts type cyclizations 1025–1055
– alcohols 1029–1036
– C–C π bonds 1036–1041
– carbonyl derivatives 1029–1036
– catalytic enantioselective RCFCs
  1044–1046
– classic intramolecular 1026–1029
– miscellaneous reactions 1042–1043
– RCFC reactions see ring-closing Friedel–
  Crafts
– selected procedures 1047–1049
functionalized alkenes, cyclization reactions
  see alkenes
functionalized alkynes, cyclization reactions
  see alkynes
functionalized allenes, cyclization reactions
  see allenes
furans
– acylcyclopropenes 749–753
– alkoylcarbonylcyclopropenes 749
– alkynes 671–672
– cycloisomerization 749, 749–753
– iodofurans 964–965
– skeletal rearrangement 671–672

## g

germanium, reductive cyclizations
  1109–1110
glycoconjugates 931–932
gold
– aryl C–H bond activation 1002–1003
– ring expansion cyclizations 834–835
group transfer, photocyclization
  initiation 1152
group-transfer radical cyclizations
  1135–1136

## h

halides
- activations of alkyl halides   1026
- acyl   1073–1074
- alkynylhalides   370–372
- macrolactams   1073–1074
- RCFC   1026
halocyclizations   953–972
- carbon-centered nucleophiles   970–972
- nitrogen-centered nucleophiles   965–969
- oxygen-centered nucleophiles   954–965
- selenium-centered nucleophiles   969–970
- sulfur-centered nucleophiles   969–970
halomethylmetal compounds, cyclopropanations   688–690
HDA reactions *see* hetero Diels–Alder reactions
Heck reaction
- cascade processes   799–800
- cyclic carbometallation   230–233
- cyclic coupling reactions   344–354
- double   245–257
- fused and bridged   247–249
- intramolecular   230–233
- triple   245–257
- *see also* Mizoroki–Heck coupling reaction
helicenes   373–374
hetero Diels–Alder (HDA) reactions
- asymmetric cyclizations   1218–1222
- Brønsted acid-catalyzed   1218–1222
- oxa   21–41
heteroaromatics, [2 + 2 + 2] cycloadditions   384–385
heteroatom-substituted alkynes, construction of functionalized benzenes   368–372
heterocycles *see* cyclization reactions
- dienes   531–542
- enynes   562–573
- five-membered   736–738
- pyrrolo-   753–754
- ring-closing metathesis   531–542
- ring-opening metathesis–ring-closing metathesis   542–547
heterocyclic [2 + 2 + 2] cycloadditions   381–391
heterocyclic 1,3-cyclohexadienes   381–391
heterocyclic cyclohexenes   391–396
heterocyclic cyclopropanes   736–738
heteronucleophiles
- addition   210–214
- intramolecular 1,4-addition reactions   210–214
hexadienes, 1,3-cyclohexadienes   381–391
hydroarylation reactions
- alkenes   460–462
- allenes   513–515
hydrogen bonding
- bifunctional organocatalysis of asymmetric diels–Alder reaction   19–21
- oxa-Diels–Alder reactions   37–39
hydrometallation-initiated cyclization reactions   843–916
- cascade reactions   890–904
- hydrosilanes   858–890 (*see also* hydrosilanes)
- selected procedures   905–909
- self-feeding   843–858 (*see also* self-feeding cyclizations)
- tied by outsourcing reagents   858–890
hydrosilanes
- aldehydes   884–890
- C–C unsaturated bond   881–884
- carbonyl compounds   881–884
- dienes   867–877
- diynes   858–863
- enynes   863–867
- hydrometallation-initiated cyclizations   858–890
- tether   881–884
- three C–C unsaturated bonds   877–881
hydrosilylation
- anti-addition   850–853
- intramolecular   843–849
- syn-addition   849–850
- to a C–C triple bond   849–853
- to a double bond   843–849
hydroxy-substituted, 1,$n$-enynes   644
hydroxycyclizations
- 1,5-enynes   647–649
- addition of nucleophiles   644–647

## i

imidates
- 1,3-dipolar cycloaddition   123
- 2-aminomalonate   123
- as nucleophiles   980–981
- electrophilic cyclizations   980–981
- Lewis acid catalyzed   123
imines
- 1,2-additions   1030–1031
- 1,3-dipolar cycloaddition   126–133, 147–154
- [2 + 2 + 2] cycloadditions   388–391
- addition   181–199
- aldimines   108–109
- as nucleophiles   968–969, 981–982
- aziridinations   717–720
- azomethine imines   126–133
- cyclization reactions   483–500

- cyclopropenyl imines   753–754
- electrophilic cyclizations   968–969, 981–982
- epoxidations   701–713
- iminium salts   707–708
- intramolecular 1,2-addition reactions   181–191
- nitrile imines   147–154
- RCFC   1030–1031

iminium catalysis
- asymmetric cyclization reactions   1207–1212
- asymmetric cycloadditions   1207–1210
- asymmetric domino reactions   1211–1212
- asymmetric intramolecular 1,4-additions   1210–1211

iminium/enamine catalytic sequences   1215–1218

imino compounds
- alkene or alkyne substituted   182–186
- nucleophilic addition   186–191, 218

imino ene reactions   182–186
*in situ* generated azides   924–925

indium
- chiral complexes   5–7
- reductive cyclizations   1110–1111

indium catalyzed Diels–Alder reactions   2–7
- oxa-   21–23

"instant" system, Mortreux   602

intramolecular 1,2-addition reactions   169–199
- addition of carbonyl compounds   169–181 (*see also* carbonyl compounds)
- addition of imines   181–191
- addition of nitriles   191–199
- aldol reaction   169–172

intramolecular 1,4-addition reactions   199–217, 1210–1211
- addition of carbanions   200–203
- addition of heteronucleophiles   210–214
- asymmetric   216–217, 1210–1211
- catalysis   216–217
- iminium-catalyzed   1210–1211
- Michael addition   200–210
- other acceptors   214–216

intramolecular aldol reactions   1200–1201
intramolecular alkene carbometallation   233–238

intramolecular allylation reactions   271–314
- iridium catalyzed   301–304
- nickel catalyzed   304
- palladium catalyzed   271–301
- rhodium catalyzed   304–305
- selected procedures   305–309

- *see also* allylation reactions

intramolecular carbometallation   228–238
intramolecular cyclization *see* Friedel–Crafts type cyclizations
intramolecular Heck reaction   230–233, 245–257
intramolecular hydrosilylation *see* hydrosilylation
intramolecular Michael additions   1202–1203
intramolecular silylformylation *see* silylformylation
intramolecular Stetter reactions   1229–1231

inverse electron demand
- asymmetric aza Diels–Alder reactions   71–73, 80
- nitrones   103–105

iodofurans   964–965

iridium
- catalyzed intramolecular allylation reactions   301–304
- cyclometallation   439–440

iron
- catalyst for epoxidations   703–704
- chiral complexes   11

isocyanates, [2 + 2 + 2] cycloadditions   388–391

isomerization, cycloisomerization *see* cyclometallation

isothiocyanates, [2 + 2 + 2] cycloadditions   387–388

isoureas
- as nucleophiles   967–968
- electrophilic cyclizations   967–968

isoxazoles
- synthesis, copper-catalyzed   934–935
- synthesis, ruthenium-catalyzed   937–938

# k

Keck macrolactonization   1063–1065

ketones
- 1-(1-alkynyl)cyclopropyl ketones   780
- 1,2 additions   1029–1030
- [2 + 2 + 2] cycloadditions   385–387
- as catalysts   706–707
- as nucleophiles   963–964, 976–977
- electrophilic cyclizations   963–964, 976–977
- RCFC   1029–1030

Kinugasa reaction   105–107

# l

lactames *see* macrolactams
lactones *see* macrolactones
lactonization   890–898

lanthanides
– catalyst  60
– epoxidations  705
Lewis acid catalysis
– 1,3-dipolar cycloaddition  89–101, 123–124
– α,β-unsaturated carbonyl compounds  91–95, 155–157
– alkynes  157–158
– asymmetric Diels–Alder reactions  2–13
– asymmetric organocatalyzed cyclizations  1225–1228
– asymmetric oxa-Diels–Alder reactions  21–37
– aza Diels–Alder reactions  59–76
– aziridines  123–124
– azomethines ylides  123–124
– chiral  2–13, 21–37, 59–76, 89–95
– Co-catalyst dirhodium(II)  140–143
– diazo compounds  155–158
– Diels–Alder reactions  2–13
– electrophilic azomethine ylides  123–124
– imidate of 2-aminomalonate  123
– intramolecular 1,4-addition reactions  203–205
– metal-based  2–13, 21–37, 59–76
– Michael addition  203–205
– nitrones  89–95
– oxa-Diels–Alder reactions  21–37
– RCFC  1042–1044
ligands, copper-catalyzed cycloadditions  920–924
lindlar reduction, RCAM  599–624
lithiations
– carbo  229–230, 239, 247
living organometallic processes  245–257
luotonin A  1136–1138

## m

macrocyclics, RCAM  616–618
macrolactams  1072–1086
– acyl halides  1073–1074
– acylimidazoles  1074
– anhydrides  1074–1077
– esters  1077–1086
– multi-step strategy  1077–1079
– one-pot strategy  1079–1080
– selected procedures  1088–1092
– via aryl C–N bond formation  1086
– via direct amidation  1073–1086
macrolactones  1055–1071
– Corey–Nicolaou  1056–1058
– in biosystems  1070–1071
– Keck  1063–1065
– Masamune  1058–1059

– Mitsunobu  1066–1068
– Mukaiyama  1059–1061
– selected procedures  1088–1092
– Shiina  1065–1066
– via formation of an acyl–oxygen bond  1056–1066
– via formation of an alkyl C–O bond  1066–1070
– Yamaguchi  1062–1063
magnesium, chiral complexes  9
magnesium catalyzed Diels–Alder reactions  7–9
magnesium(II) 1,3-dipolar cycloaddition, nitrile oxides  148–151
manganese
– catalyst for epoxidations  703
– radical cyclizations  1117–1119, 1138–1139
– see also free radical cyclizations
Masamune macrolactonization  1058–1059
MCP see methylenecyclopropanes
mediated [2 + 2 + 2] cycloadditions, transition metal-  367–406
mediated aziridinations  713–716
mediated free radical cyclizations see free radical cyclizations
mediated manganese triacetate cyclizations  1117–1119
mediated samarium diiodide cyclizations  1119–1120
mediated titanium (III) cyclizations  1120–1122
mersicarpine  1138–1139
metal catalysts
– aza Diels–Alder reactions  59–76
– aziridinations  713–716
– chiral  2–13, 21–37, 59–76
– Diels–Alder reactions  2–13
– epoxidations  701–705
– oxa-Diels–Alder reactions  21–37
metal-containing carbonyl ylides  143–147
metal-free aziridinations  716–717
metallation
– cyclic  227–270, 407–456
– see also cyclic carbometallation
metalocarbenoids, cyclopropanations  698–700
metals
– alkyl C–H bond activation  1009–1010, 1015–1016
– aryl C–H bond activation  1004
– C–C bond formation  1009–1010
– C–N bond formation  1015–1016
metathesis
– alkynes see ring-closing metathesis, alkynes

– dienes 531–562
– enynes 562–592
– ring-closing see ring-closing metathesis
– ring-opening–ring-closing 542–547, 578–585
methines
– active methine groups 970–971
– as nucleophiles 970–971
– azomethines 107–126
– electrophilic cyclizations 970–971
methylenecyclopropanes 743–748
– [3 + 1] cycloaddition 755–756
– [3 + 2] cycloaddition 756–761
– acyl 744–746
Michael addition
– asymmetric cyclizations 1202–1203
– catalyzed 205–210, 671
– enamine catalysis 1202–1203
– furans 671
– intramolecular 200–210, 1202–1203
– mediated by bases 200–203
– mediated by Lewis acids 203–205
Michael addition inititated ring closure (MIRC) 687
microwave cycloadditions 378–379
migration reactions, cyclization/functional group 479–482
MIRC see Michael addition inititated ring closure
Mitsunobu reaction 1066–1068
Mizoroki–Heck coupling reaction 1039–1040
– see also Heck reaction
molybdenum
– catalysis 547–552
– RCAM 603–604
– ring-closing metathesis 547–552
Mortreux-type catalyst 602
Mukaiyama macrolactonization 1059–1061
multi-step strategy, esters 1077–1079
multicomponent cyclization reactions 794–798
multiple cyclization reactions, carbometallation 245–257
münchnones, transition metal catalyzed 1,3-DC 121–122

## n

N-acylhydrazones, [3 + 2] cycloaddition 131–133
N-arylidene amino acid ester, preparation 159
N-benzylidenebenzylamine N-oxide, preparation 159

N-metalated azomethine ylides 108–119, 124–126
naphthalene core
– construction 372–373
– transient benzynes 372–373
natural product synthesis 557–562, 586–592
– amphidinolide E 1135
– aphanorphine 1135–1136
– aspidospermidine 1102, 1137–1138
– azadirachtin 1137
– fortucine 1138
– free radical cyclizations 1135–1140
– group-transfer radical cyclizations 1135–1136
– luotonin A 1136–1137
– manganese triacetate-mediated 1137
– mersicarpine 1138–1139
– single ring formation 1135
– tandem radical cyclizations 1136–1138
natural products
– biological profiling 932–934
– dienes 557–562
– enynes 586–592
– free radical cyclizations in 1135–1140
– modifying 932–934
Nazarov reactions 1224
nickel
– [2 + 2 + 2] cycloadditions 378–379
– catalyzed intramolecular allylation reactions 304
– cyclometallation 440–448
– microwave 378–379
nitrile imines, 1,3-dipolar cycloaddition 147–154
nitrile oxides 147–154
– 1,3-dipolar cycloaddition 147–154
– catalysis 152–153
– magnesium(ii) 148–151
nitriles
– 1,2-addition 218–219
– acid-catalyzed reactions 191–192
– addition 191–199
– base-catalyzed reactions 192–194
– intramolecular 1,2-addition reactions 191–199
– nucleophilic addition 194–197
– transition metals mediated 197–199
nitroalkenes, asymmetric 1,3-dipolar cycloaddition 117–119
nitrogen-centered nucleophiles
– cyclization of amides 967, 980 981
– cyclization of amidines 967–968
– cyclization of amines 965–966, 980
– cyclization of azides 969

- cyclization of carbamates  967–968
- cyclization of imidates  980–981
- cyclization of imines  968–969, 981–982
- cyclization of isoureas  967–968
- cyclization of oximes  981–982
- cyclization of ureas  967–968
- electrophilic cyclizations  965–969, 979–982
- halocyclizations  965–969
- selenocyclizations  979–982

nitrogen-containing heterocyclic compounds, optically active  59

nitrones
- 1,3-dipolar cycloaddition  88–107
- asymmetric 1,3-dipolar cycloaddition  95–101
- inverse electron demand  103–105
- Kinugasa reaction  105–107
- Lewis acid catalysis  89–95
- organocatalysis  95–96

Norrish type I reactions  1154
Norrish type II initiated reactions: H transfer  1154–1163
- cleavage  1154–1155
- formation of four-membered rings  1156–1157
- formation of larger rings  1161–1163
- formation of three-membered rings  1157–1161
- Yang process  1154–1155

nucleophiles
- addition to 1,5-enynes  656–657
- addition to 1,6-enynes  650–654
- addition to imino groups  186–191
- addition to nitriles  194–197
- alkenes  457–460
- alkynes  466–475
- allenes  504–511
- asymmetric cyclizations  1229–1233
- carbenes  1229–1233
- carbon-centered *see* carbon-centered nucleophiles
- carbonucleophiles  271–301, 650–654
- chiral  1229–1233
- cyclization reactions  457–460, 466–475, 625–686, 951–990
- enynes  644–657
- heteronucleophiles  210–214
- in electrophilic cyclizations  951–990
- nitrogen-centered *see* nitrogen-centered nucleophiles
- organocatalysis  1229–1233
- oxygen-centered *see* oxygen-centered nucleophiles
- selenium-centered *see* selenium-centered nucleophiles
- sulfur-centered *see* sulfur-centered nucleophiles
- tethered  457–460, 466–475, 504–511

# o

olefins
- [3 + 2] cycloaddition  131–133
- electron-deficient  101–102
- *see also* alkenes

oligomer, RCAM *see* cyclo-oligomerizations

one-pot strategy
- esters  1079–1080
- macrolactams  1079–1080

optically active nitrogen-containing heterocyclic compounds  59

organoboron, cyclic coupling reactions  315–320

organocatalysis
- α,β-unsaturated aldehydes  95–96
- asymmetric 1,3-dipolar cycloaddition  95–96, 119–121
- asymmetric cyclizations *see* asymmetric organocatalyzed cyclizations
- bifunctional via hydrogen bonding  19–21
- chiral  14–21, 37–41, 76–82
- Diels–Alder reactions  14–21 (*see also* aza Diels–Alder reactions; oxa-Diels–Alder reactions)
- epoxidations  706–710
- nitrones  95–96

organometallic reagent *see* cyclometallation

organosilicon, cyclic coupling reactions  328–330

organotin
- atom-transfer  1116
- cyclic coupling reactions  321–328
- radical cyclizations  1116

outsourcing reagents
- hydrometallation-initiated cyclizations  858–890
- *see also* hydrosilanes

oxa-di-π-methane rearrangement  1180–1189

oxa-Diels–Alder reactions  21–41
- asymmetric  21–41
- catalysis  21–41
- chiral aluminum complexes  21–23
- chiral boron complexes  21–23
- chiral chromium complexes  26–28
- chiral copper complexes  28–33
- chiral indium complexes  21–23
- chiral palladium complexes  33–36
- chiral platinum complexes  33–36

– chiral rare earth metal complexes 36–37
– chiral rhodium complexes 33–36
– chiral secondary amine 39–41
– chiral titanium complexes 23–26
– chiral zinc complexes 28–33
– chiral zirconium complexes 23–26
– hydrogen bonding 37–39
– Lewis acid catalyzed, asymmetric 21–37
– organocatalysis 37–41
oxidative PET 1169–1173
oxides, nitrile 147–154
oximes, as nucleophiles 981–982
oxygen-centered nucleophiles
– cyclization of alcohols 954–958, 973–976
– cyclization of aldehydes 963–964, 976–977
– cyclization of carbamates 962–963
– cyclization of carbonates 962–963
– cyclization of carboxylic acids and derivatives 958–962, 977–979
– cyclization of ethers 954–958
– cyclization of ketones 963–964, 976–977
– electrophilic cyclizations 954–965, 973–979
– halocyclizations 954–965
– iodofurans 964–965
– selenocyclizations 973–979

**p**
palladations
– acyl 257–261
– carbo 227–260
– multiple reactions 245
– silylpalladation 876
palladium
– alkyl C–H bond activation 1008–1009, 1013–1015
– aryl C–H bond activation 997–1002
– catalyzed Diels–Alder reactions 33–36
– catalyzed intramolecular allylation reactions 271–301
– cyclometallation 448–449
– ring expansion cyclizations 813–831
– silylpalladation 876
peptides, epoxidations 710
peroxides 1116–1117
PET see photoinduced electron transfer
phase-transfer catalysis 709
phenylenes 373–376
phosphines, alkynylphosphines 370
phosphorous acid, reductive cyclizations 1111–1112
photochemical cyclizations, radical cyclizations 1122–1124

photocyclization initiation
– direct excitation 1152
– electron transfer 1153–1154, 1163–1176
– energy transfer 1152–1153
– group transfer 1152
– sensitization 1152–1154
– substrate tuning 1154
photocyclization reactions 1149–1198
– $\pi^6$- see $\pi^6$-photocyclization
– H transfer 1154–1163
– initiation ways 1151–1154
– Norrish type II initiated see Norrish type II initiated reactions
– $\pi$–$\pi$-interactions see $\pi$–$\pi$-interactions
– selected procedures 1191–1192
photoinduced electron transfer (PET) 1163–1177
– formation of larger rings 1165–1177
– formation of three- to six-membered rings 1163–1165
– oxidative 1169–1173
– reductive 1173–1177
– sensitization 1153–1154
photosensitization
– electron transfer 1153–1154
– energy transfer 1152–1153
– see also photocyclization initiation
Pictet–Spengler reactions 1222–1224
platinum
– catalysis epoxidations 704–705
– catalyzed Diels–Alder reactions 33–36
$\pi^\nu$–$\pi^\mu$-cycloadditions 1177–1179
polycyclization 245–257
– [2 + 2 + 2] 251–255
– circular 251–257
– cyclopropanations 255–257
– dumbbell 251–257
– fused- and bridged-ring forming 247–249
– spiro 249–250
– stoichiometric 247
– transannular 250–251
– zipper 250–251
polymer-supported catalysts 554–557
polyphenylenes
– angular [N]-phenylenes 375–376
– construction 373–376
– helicenes 373–374
– silafluorenes 374–375
– see also benzene derivatives
propargylic 1,2- and 1,3-acyl migration 635–643
PTC see phase-transfer catalysis
pyrroloheterocycles 753–754

## r

radical cyclizations
– definition   1099–1105
– free *see* free radical cyclizations
rare earths
– chiral metal complexes   12–14
– Diels–Alder reactions   12–14
– oxa-Diels–Alder reactions   36–37
RCAM *see* ring-closing metathesis of alkynes
RCFC *see* ring-closing Friedel–Crafts
RCM *see* ring-closing metathesis
rearrangements
– endocyclic *see* endocyclic rearrangement
– $\pi^2$–$\pi^2$ *see* $\pi^2$–$\pi^2$-rearrangements
– skeletal *see* skeletal rearrangement
reducing agents
– germanium   1109–1110
– indium   1110–1111
– phosphorous acid   1111–1112
– silicon   1108–1109
– thiols   1112–1113
– tin hydride   1106–1108
reductive cyclizations   1105–1113
– germanium   1109–1110
– indium   1110–1111
– phosphorous acid   1111–1112
– silicon   1108–1109
– thiols   1112–1113
– tin hydride   1106–1108
– *see also* free radical cyclizations
reductive PET   1173–1177
regioselectivity, electrophilic cyclizations   952–953
rhodation
– silylrhodation   879, 893, 895, 898
rhodium
– [2 + 2 + 2] cycloadditions   377
– alkyl C–H bond activation   1005–1008, 1011–1013
– aryl C–H bond activation   994–996
– C–C bond formation   1005–1008
– C–N bond formation   1011–1013
– catalyzed Diels–Alder reactions   33–36
– catalyzed intramolecular allylation reactions   304–305
– cationic   377
– cyclometallation   427–439
– microwave   378–379
– silylrhodation   879–898
ring-closing Friedel–Crafts (RCFC)   1025–1055
– 1,2-addition   1029–1031
– acetals   1031–1032
– alcohols, direct substitution   1033–1035

– aldehydes   1029–1030
– alkenes   1036–1038
– alkynes   1039–1041
– Brønsted acids   1039
– carboxylic acid derivatives   1026–1029
– catalytic   1044–1046
– enantioselective   1044–1046
– epoxide derivatives   1035–1036
– imines   1030–1031
– ketones   1029–1030
– $s_n1$ type reactions   1033–1035
– via activations of alkyl halides   1026
– via conjugate additions   1033
– *see also* Friedel–Crafts type cyclizations
ring-closing metathesis   527–598
– alkynes   599–620
– asymmetric   547–554
– carbocycles   531–542, 562–573
– catalytic   547–554
– cycloalkene-ynes   578–585
– dienes   531–562
– dienynes   573–578
– enynes   562–592
– heterocycles   531–542, 562–573
– natural products   557–562, 586–592
– polymer-supported catalysts   554–557
– ring-opening metathesis–ring-closing metathesis   542–547, 578–585
– selected procedures   592, 618–620
ring-closing metathesis of alkynes (RCAM)   599–624
– alkyne–alkyne   605–612
– catalyst systems   602–605
– comparison reaction systems   604–605
– complex molecule synthesis   605–614
– cyclo-oligomerizations   616–618
– cyclodimerizations   616–618
– cyclotrimerizations   616–618
– enyne-yne   612–614
– molybdenum-based catalysts   603–604
– Mortreux "instant" system   602
– synthetic strategy   600–602
– transition metal   615
– tungsten alkylidyne catalysts   602–603
ring expansion cyclizations
– alkenyl substrates   813–820
– alkynyl substrates   827–831
– dienyl substrates   820–827
– Fischer-carbene complexes   835–836
– gold-catalyzed   834–835
– palladium-catalyzed   813–831
– ruthenium-catalyzed   832–834
– selected procedures   837–839
– transition metal-catalyzed   813–842

ring-opening metathesis–ring-closing
  metathesis
– dienes   542–547
– enynes   578–585
rings
– five-membered   736–738, 1163–1165
– four-membered   1156–1157, 1163–1165
– larger   1161–1163, 1165–1177
– six-membered   1163–1165
– three-membered   687–732, 1157–1161, 1163–1165
ruthenium
– [2 + 2 + 2] cycloadditions   378–379
– asymmetric olefin metathesis   552–554
– atom-transfer   1115–1116
– azide-alkyne   935–937
– catalytic dipolar cycloadditions   935–944
– chiral complexes   11–12
– cyclometallation   419–424
– epoxidations   704–705
– isoxazoles   937–938
– mechanistic studies   938–944
– microwave   378–379
– radical cyclizations   1115–1116
– ring expansion cyclizations   832–834

### s

samarium diiodide-mediated
  cyclizations   1119–1120
– see also free radical cyclizations
selenium
selenium-centered nucleophiles   969–970, 982
selenocyclizations   972–983
– carbon-centered nucleophiles   982–983
– nitrogen-centered nucleophiles   979–982
– oxygen-centered nucleophiles   973–979
– selenium-centered nucleophiles   982
– sulfur-centered nucleophiles   982
self-feeding cyclizations   843–858
– alkenes   853–858
– alkynes   853–858
– C–C triple bond   849–853
– double bond   843–849
– intramolecular hydrosilylation   843–849
– silylformylation   853–858
sensitization see photosensitization
sequences see catalytic sequences
Shiina macrolactonization   1065–1066
silafluorenes   374–375
silanes, alkynylsilanes   368–369
silicon
– organosilicon   328–330
– reductive cyclizations   1108–1109
silylative carbamoylation, alkynes   902–904

silylative lactonization, alkynols   899–902
silylcarbonylation   890–898
silylformylation   853–858
silylpalladation   876
silylrhodation   879, 893, 895, 898
single ring formation   1100–1101
– natural product synthesis   1135
– see also free radical cyclizations
skeletal rearrangement
– 1,5-enynes   667
– 1,6-enynes   657–661, 665–666
– 1,7-enynes   661–663
– 1,$n$-enynes   657–672
– alkynes   671–672
– cyclobutenes   663–665
– endocyclic   665–666
– furans   671–672
– single- and double-cleavage   657–661
$s_N$1 type reactions   1033–1035
solid support, microwave   379
spiro-mode polycyclization   249–250
stereochemistry
– intramolecular 1,2-addition reactions   169–199
– intramolecular 1,4-addition reactions   199–217
stereoselectivity, electrophilic
  cyclizations   952–953
Stetter reactions   1229–1231
Stille coupling   318–345
stoichiometric intramolecular
  carbometallation
– alkenes   228–230
– alkynes   238–239
stoichiometric polycyclization   247
substrate tuning, photocyclization
  initiation   1154
substrates
– alkenyl   813–820
– alkynyl   827–831
– dienyl   820–827
– for ring expansion cyclizations   813–831
sulfonyl azides   929–930
sulfur-centered nucleophiles   969–970, 982
Suzuki coupling, cyclic coupling
  reactions   315–357
syn-addition   849–850

### t

tandem carbonyl ylide formation-1,3-DC
– asymmetric   134–139
– three-component   139–140
tandem cyclizations   1101–1105
– natural product synthesis   1136–1138

– *see also* free radical cyclizations
termination, intramolecular catalytic alkene carbometallation  233–238
tether
– aliphatic  483–486
– aromatic  486–500
– hydrosilanes  881–884
tethered nucleophiles
– alkenes  457–460
– alkynes  466–475
– allenes  504–511
– cyclization reactions  457–460, 466–475
thioacetals
– alkynes  483–500
– cyclization reactions  483–500
thiols, reductive cyclizations  1112–1113
three-membered rings  687–732
tin
– organotin  321–328, 1116
– reductive cyclizations  1106–1108, 1116
– tin hydride  1106–1108
titanium
– catalysis epoxidations  701–702
– catalyzed Diels–Alder reactions  23–26
– chiral complexes  10, 23–26
– cyclometallation  407–414
– (III)-mediated cyclizations  1120–1122
– *see also* free radical cyclizations
total synthesis, Diels–Alder as key step  41–45
transannular polycyclization  250–251
transient benzynes, construction of naphthalene core  372–373
transition metal catalysis, enynes  625–686
transition metal-mediated [2 + 2 + 2] cycloadditions  367–406
– 1,3-cyclohexadienes  381–391
– benzene derivatives  368–381
– catalytic systems, new  377–379
– cyclohexenes  391–396
– enantioselective  379–381, 391
– heterocyclic  381–391
– selected procedures  397–400
– unsaturated  381–391
transition metals
– 1,3-dipolar cycloaddition  121–122
– alkenes  457–465
– alkynes  466–504, 615
– allenes  504–515
– asymmetric catalysis  10–12
– chiral complexes  10–12
– coordination spheres  615
– cyclization reactions  457–526, 843–916 (*see also* cyclization reactions)
– cycloaddition  197–199

– cycloisomerizations  626–644
– cyclometallation  407–456 (*see also* cyclometallation)
– cyclopropanations  690–694
– Diels–Alder reactions  10–12
– enynes  625–686
– hydrometallation-initiated  843–916
– intramolecular 1,2-addition reactions  197–199
– intramolecular allylation reactions  271–314 (*see also* allylation reactions)
– münchnones  121–122
– nucleophilic cyclizations  644–657
– ring-closing metathesis  527–598, 615
– ring expansion cyclizations  813–842
transition metal catalysis  121–122
– alkenes  457–465
– alkynes  466–504
– allenes  504–515
– allylation reactions  271–314
– cyclization reactions  457–526, 813–842
– cycloisomerization  626–644
– cyclopropanations  690–694
triacetate-mediated cyclizations  1117–1119
trimer, RCAM *see* cyclotrimerization
triple bonds
– [2 + 2 + 2] cycloadditions  381–382
– self-feeding cyclizations  849–853
tungsten alkylidyne catalysts  602–603

**u**

unsaturated [2 + 2 + 2] cycloadditions  381–391
unsaturated carbonyl compounds  881–884
– *see also* α,β-unsaturated carbonyl compounds
ureas, electrophilic cyclizations  967–968

**v**

vanadium, catalyst for epoxidations  702–703
vinylcyclopropanes (VCPs)
– [3 + 2] cycloaddition  772–773
– [5 + 2] cycloaddition  784–793
– alkene-  794
– cycloisomerization  733–736, 738–740
– hetero-analogues  736–738
vinylidenecyclopropanes  749
– [3 + 2] cycloaddition  761–763
vinylogous aldol reaction  173–177
vinyl sulfones, asymmetric 1,3-dipolar cycloaddition  117–119

**y**

Yamaguchi macrolactonization  1062–1063
Yamamoto cyclization reactions  502–504

Yang process, photocyclization
  reactions 1154–1155
ylides
– 1,3-dipolar cycloaddition 107–126
– 2-aminomalonate 123
– α,β-unsaturated carbonyl compounds
  109–116
– aldimines 108–109
– aziridinations 719–720
– azomethine 107–126
– carbonyl 134–147
– cyclopropanations 694–698
– electrophilic 123–126
– electrophilic activation 143–147
– epoxidations 710–712
– Lewis acid catalyzed 123–124
– N-metalated 108–119
– nitroalkenes 117–119
– organocatalysis 119–121
– tandem 1,3-dipolar cycloaddition 134–140
– vinyl sulfones 117–119
ynamides 370

**z**

zinc catalyzed Diels–Alder reactions 7–9,
  28–33
zipper mode polycyclizations 250–251
zirconium
– catalyzed Diels–Alder reactions 23–26
– cyclometallation 414–419